Stanley O. Kochman

Single Variable Calculus
Concepts, Applications and Theory

Fourth Edition

Cover Art: courtesy of Darryl Johnson.

Pearson Learning Solutions, 501 Boylston Street, Suite 900, Boston, MA 02116
A Pearson Education Company
www.pearsoned.com

Printed in Canada

1 2 3 4 5 6 7 8 9 10 V0CR 16 15 14 13 12 11

0002000102707755553

MHB

ISBN 10: 1-256-29759-3
ISBN 13: 978-1-256-29759-8

Contents

Preface

Calculus studies functions. Differential calculus studies their rates of change, called derivatives, while integral calculus studies areas under their graphs. Mathematicians view calculus as a beautiful, coherent theory which forms the basis of the branch of mathematics called analysis. Differential calculus leads to the study of differential equations. Generalizing calculus to the complex numbers is called analytic function theory. Combining calculus with linear algebra produces functional analysis. In contrast, engineers, scientists and social scientists study changing phenomena in the real world. They have found that the quantitative analysis of their problems using calculus often leads to fundamental new insights.

There are three aspects of calculus: *Concepts, Applications* and *Theory*. Each course and each student has unique motivations and expectations for studying calculus. A liberal arts student may be content with an intuitive understanding of the basic concepts. A science or engineering student may, in addition, want to study applications with little interest in the theory. On the other hand, a mathematics student may want to supplement her intuitive understanding of calculus with a study of the underlying theory. She may have little interest in the applications. In this text, each chapter is divided into three parts which develop these aspects of the chapter's subject matter. This design allows the text to be used conveniently by liberal arts, science, engineering and mathematics students. Each course can select its own custom blend of applications and theory to enhance the intuitive understanding. An individual student may supplement the course by reading additional sections of applications or theory which are not studied by the entire class. However, all readers are expected to read the entire *Concepts* portion of each chapter. After reading a section of *Concepts*, a student may read a section of *Applications* or *Theory* before proceeding to the next section of *Concepts*. Instructions for creating such a blend are presented in the introduction to each chapter.

The *Concepts* portion of each chapter develops an understanding of each concept through an intuitive presentation followed by methods of analyzing concrete examples. The emphasis is on intuitive understanding. When rigor helps in developing and clarifying ideas and facts, then it is included. However, in many cases there is an intuitive explanation which has logical gaps of a technical nature. This has led mathematicians to develop a logically complete approach which may obscure the underlying intuitive ideas. In this case, we present the intuitive ideas. However, we make it clear which of our statements are logically sound and which are only intuitively valid. The emphasis is on ideas, examples and problem solving. However, we use the standard mathematical style of definitions, theorems and corollaries. Moreover, when we give a logically complete argument we call it a proof. This method of exposition encourages the reader to organize her thoughts in a systematic manner and to distinguish logical truths from intuition. In addition, it prepares the reader for the style that she will encounter in more advanced mathematics courses. Thus the reader will not merely learn new facts; she will also reach a higher level of mathematical maturity. Every reader is expected to study all the sections in the *Concepts* portion of each chapter.

The *Applications* part of each chapter presents concrete applications of the mathematical ideas and theorems which were studied in the *Concepts* portion of that chapter. In some of these sections, we present interpretations of these concepts in specific scientific or economic contexts. The presentation is accessible to everyone – no prior knowledge of physics or economics is assumed. The object of these sections is to show how abstract mathematics is applied to relevant situations in the real world. However, we do not attempt to give a thorough exposition of the area of application. Other sections of *Applications* are devoted to mathematical subjects which are not required for understanding the basics of calculus. These sections may be important to readers who are learning calculus for use in other subjects. For example, elasticity is important in economics, and hyperbolic functions are important in engineering. However, the reader can achieve a thorough understanding of calculus without studying either of these subjects. Each section of *Applications* begins by listing the prerequisite sections. These prerequisites are usually sections from the *Concepts* portion of the chapter but may also include earlier sections of *Applications*. All sections of the *Applications* portion of each chapter are optional. However, all readers are encouraged to read those applications which interest them. The basic constructions of calculus are motivated by applications. Conversely, calculus leads to fundamental new insights into these applications. Therefore, a calculus course is not complete without the study of some applications.

The *Theory* portion of each chapter is a mathematically complete treatment of topics relevant to the chapter. Some sections present alternate expositions of material presented intuitively in the *Concepts* part of the chapter. With this intuitive understanding in hand, the reader is prepared for a logically complete, less intuitive development of the same topic. Other sections of *Theory* introduce additional concepts relevant to the mathematics of the chapter. Throughout the *Theory* sections, abstract concepts and arguments are clarified with concrete examples. The sections of *Theory* are intended for readers with an interest in calculus as mathematics. All of these sections are optional and will likely be omitted by many readers. Each section begins by listing prerequisite sections from the *Concepts* portion of the chapter and previous *Theory* sections.

The best method for learning mathematics is to *do it!* Listening to lectures and reading books can only transmit vague ideas of the subject matter. A clear understanding comes through problem solving. Routine exercises are at first solved by referring to relevant material in the text and lecture notes. This results in a clear understanding of the ideas being illustrated. Thereafter, the "routine exercises" do indeed become routine. The reader is then ready to solve more difficult problems which require some ingenuity. Some readers will enjoy the challenge of problems which require substantial thought and experimentation before a solution is discovered. The nature of the exercises vary in the three parts of each chapter. Exercises in *Concepts* sections are mostly computational and reinforce understanding by applying concepts to examples. The exercises in *Applications* sections contain mathematical and word problems which stress the application of calculus to other subjects. Problems in *Theory* sections train the reader to apply theory to examples and to use the theorems of the section, in a straightforward manner, to prove new results. Each section has a large number of exercises which are divided into *Basic Exercises* and *Challenging Problems*. The *Basic Exercises* are mostly routine problems which all readers should be able to solve. However, some exercises are more complicated or require a little ingenuity. The *Challenging Problems* make the text more interesting for the inquisitive reader. Some of these problems are difficult but readily solvable. Other problems are very difficult and require new ideas for their solutions. The reader should be confident that she has mastered the material if she can solve the *Basic Exercises*. Readers should not expect to be able to solve all the *Challenging Problems*.

Here is a brief overview of the development of the major concepts of calculus. Calculus of one variable studies real valued functions defined on an interval. Historically, a function was viewed as a curve, or as the path of a point moving in the plane. Differential calculus studies the rate of change of a function. The operation of computing this rate of change is called differentiation. Differentiation can be interpreted as finding the slope of the tangent lines to a curve or as computing the velocity of a moving point. Integral calculus studies the problem of computing the area of a plane figure. The operation of computing the area bounded by a curve is called integration. Nontrivial studies of tangent lines and areas were made in Greece as early as the sixth century BCE.

The unified subject of calculus, however, is usually associated with the discovery of the Fundamental Theorem of Calculus by Isaac Newton and Gottfried Wilhelm Leibniz in the second half of the seventeenth century. This theorem states that differentiation and integration are inverse operations. Both Newton and Leibniz developed methods for differentiating any function described by an equation. By viewing the problem of integration as the inverse of differentiation, they computed areas in a systematic manner. In addition, they used infinite series to make calculus computations routine for even the most complicated functions. Using the methods of Newton and Leibniz, most of the methods and applications of calculus we study today were completed in the eighteenth century by Leonhard Euler and the Bernoulli brothers.

The first modern calculus text was published by Euler in 1748. At the turn of the nineteenth century, the study of Fourier series led to the realization that the logical foundation of calculus was not properly understood. For example, abstract functions and the definition of a limit had not yet been formulated. This was done by Bernard Bolzano in the beginning of the nineteenth century. In 1821, the first mathematically rigorous textbook on calculus was published by Augustin–Louis Cauchy. Even this presentation, however, lacked a logical construction of the real numbers. Two such constructions were made by Georg Cantor and Richard Dedekind in 1872. Thus after 2500 years, the geometric computations of the Greeks were fully developed into the rich, complete and logically rigorous subject of calculus.

Special Features

Motivating Examples

The concepts and methods of calculus involve many new ideas. To help the reader understand them, many subsections are introduced by a *Motivating Example*. The exposition that follows indicates how the material presented is motivated by that example. Other examples in the subsection illustrate material after it is presented.

Basic Exercises

Each section concludes with a large number of *Basic Exercises*. These problems include routine analogues of examples from the section, straightforward problems and problems which require some ingenuity. These problems reinforce the concepts of the section and allow the reader to identify which topics she understands and which topics require further study and clarification. Every reader should do a selection of these problems.

Challenging Problems

Each section includes several challenging problems for the more ambitious reader. Some of these problems can be solved from a thorough understanding of the section. Other problems, however, are very difficult and require much thought as well as new ideas for their solutions. Working on these problems gives the reader a taste of the nature of mathematical research.

Review Exercises

At the end of each chapter there are *Review Exercises*. They refer to the material of the *Concepts* portion of the chapter and other specified sections. Each set begins with a set of challenging True/False questions which determine whether the reader has a thorough understanding of the concepts of the chapter. Solution of the remaining exercises may require combining ideas from different sections of the chapter.

History of Calculus

The historical development of the concepts of calculus is very different from the development of the subject in modern texts. It is impossible to describe the details of this history meaningfully without a prior knowledge of the subject. To give the reader a rudimentary understanding of the history of calculus, we have included *Historical Remarks* throughout this text.

Student Manual

The Student Manual divides each section into the topics covered. After a brief review of each topic, it presents the strategy for solving typical problems. Then it lists the examples of these problems that the author presents in class. The enclosed CD contains detailed solutions of these problems in PowerPoint. They appear on the screen with color graphics and are explained by an audio accompaniment. A copy of each solution in pdf format is included for printing two slides per page.

Solutions Manual

Valery Mishkin has produced solutions manuals for this text. These manuals contain detailed solutions of all the Basic Exercises for all *Concepts* sections and many of the *Applications* and *Theory* sections. Students may use these solutions to verify the correctness of their work or for help on those exercises they can not solve themselves.

The Five Chapters

We give a brief description of the content of each chapter.

Chapter 1: Functions and Limits

The *Concepts* portion of this chapter begins with a study of functions. Trigonometric functions are defined and their inverses constructed. Limits are defined and their properties derived. In Chapters 2 and 3, limits are used to determine the slopes of tangent lines and to compute areas. We conclude with a study of continuous functions. The *Applications* section gives examples of functions in various fields of study. The *Theory* portion begins with a study of sets and functions. The language and notation of set theory is used in many of the theory sections of this text. Then the topology of the real line is developed, culminating in proofs of the Intermediate Value Theorem and the Maximum Value Theorem.

Chapter 2: Differentiation

The derivative is defined as the slope of the tangent line to the graph of a function in the *Concepts* part of this chapter. The derivatives of the basic functions introduced in Chapter 1 are computed. Various "rules" are established which compute the derivative of any function constructed from these basic functions. The Mean Value Theorem is studied and applied to identify several properties of the graph of a function including its minimum and maximum values. The sign of the second derivative is interpreted to determine the shape of the graph. These properties, as well as intercepts and asymptotes, are used to sketch graphs. Our ability to compute derivatives is used to evaluate limits by L'Hôpital's Rule. In the *Applications* portion, the derivative of a function is interpreted as a measure of its rate of change. Illustrations include examples from physics and economics. The derivative is applied to estimate the change in the value of a function and to approximate its roots. The derivative is used to find and identify maxima and minima in practical problems. The *Theory* section contains proofs of several results related to the Mean Value Theorem which were presented intuitively in the Concepts portion of the chapter.

Chapter 3: Integration

The *Concepts* portion formulates a precise definition of area, called the definite integral. The Fundamental Theorem of Calculus is proved. It shows how to compute integrals from our knowledge of finding derivatives. These methods are extended to compute simple volumes. The *Applications* component interprets integrals in various contexts of physics and geometry. In addition, methods of approximating the value of an integral are developed. The *Theory* segment proves the existence of definite integrals of continuous functions. Integral invariants are introduced to identify quantities, other than area, as definite integrals.

Chapter 4: Logarithms and Exponentials

The logarithm and exponential functions are studied from the calculus viewpoint in the *Concepts* portion of this chapter. Methods of computing integrals are presented where the logarithm and inverse trigonometric functions play improtant roles. Improper integrals are introduced to compute areas of nonbounded regions. The *Applications* sections show how a principle describing the behavior of a situation can often be formulated as a differential equation. Solving this equation gives a precise description of the situation. Methods are presented to solve many types of first and second order differential equations. The solutions often involve exponential and trigonometric functions. Applications are given to a variety of contexts. The *Theory* portion uses the topological methods of Chapter 1 to study fixed points. This viewpoint is applied to prove an existence theorem for solutions of differential equations.

Chapter 5: Infinite Series

The *Concepts* portion studies power series as generalizations of polynomials. Power series make the solution of a calculus problem involving complicated functions as easy as the solution of the problem for polynomials. In particular, we find power series representations of integrals. The use of an infinite series to approximate its sum is also emphasized. These methods are generalized to find power series solutions of differential equations in the *Applications* section. The *Theory* portion studies numerical infinite series as well as power series of a complex variable.

Multivariable Calculus

This text provides a solid background in the calculus of real valued functions of one variable. The next step in studying calculus is to extend these ideas to real valued functions of several variables: multivariate functions of one variable and multivariate functions of several variables. For example, the real valued function $g(t) = (x(t), y(t), z(t))$ of one variable describes a curve. On the other hand, the graph of the real valued function $z = f(z, y)$ of two variables as well as the multivariate function $g(s, t) = (x(s, t), y(s, t), z(s, t))$ of two variables describe surfaces. For those readers who enjoy learning calculus from this text, the text *Multivariable Calculus: Concepts, Applications and Theory* presents a comprehensive exposition of the methods of calculus which are used to study these functions. That text uses the same approach as this one and has the same special features.

The multivariable text is organized into four chapters, numbered six to nine. Chapter 6 studies curves using methods from calculus and algebra. The *Applications* sections contain a detailed exposition of Kepler's laws while the *Theory* sections contain a comprehensive treatment of curvature. Chapter 7 studies differential calculus of functions of several variables including the identification of local extrema. The *Applications* portion includes applications to physics as well as a section on Lagrange multipliers with applications to economics. The *Theory* sections study topological concepts, the inverse and implicit function theorems and holomorphic functions of a complex variable. Chapter 8 is devoted to multiple integrals. Their definition is motivated by the computation of volume while their value is computed by identifying them with iterated integrals. The *Applications* sections are devoted to methods of approximating the values of multiple integrals as well as applications to physics. Chapter 9 studies line and surface integrals. Green's Theorem identifies the line integral along the boundary of a region in the plane as a double integral over that region. This result is generalized to Stokes' Theorem for two dimensional regions in three dimensions and to Gauss' Theorem for three dimensional regions. The *Applications* sections present applications to physics. The *Theory* section proves Cauchy's integral theorem as an application of Green's Theorem to holomorphic functions. The Fundamental Theorem of Algebra is deduced as a corollary. In addition, the equivalence of holomorphic and analytic functions is established.

Acknowledgments

The author thanks all the students and instructors who have provided corrections and constructive criticism of the previous editions of this text. The feedback from Eli Brettler, Robert Burns, Kunquan Lan and Valery Mishkin have been especially useful. In particular, I am very grateful to Valery Mishkin for producing the *Student Manuals*. In addition I am very appreciative of my wife's efforts in editing the original manuscript.

Revised Editions

The Third Edition has several significant improvements over the original text. These include two of the special features listed above.[1] The Fourth Edition corrects errors that were discovered in the Third Edition.

- Errors in the original text have been corrected.
- The exposition of many topics has been improved.
- New figures and examples have been added.
- Motivating examples have been added to introduce new concepts.
- Additional exercises have been added to most sections.
- Review exercises have been added at the end of each chapter.
- An index has been created.
- The pages now have wide margins that contain most of the figures.
- Each section is divided into subsections. Each subsection focuses on a specific aspect of the topic of the section.
- In each section, definitions, theorems, figures, examples, etc. are labelled in one sequence. In the original text, each of these items was labelled by its own sequence. Now only displayed equations are labelled by a separate sequence.
- Two sections of the original text have been deleted: the section on applications of trigonometry in Chapter 1 and the section on applications of integration to economics in Chapter 3.

The sections of the original text listed below have been divided into two sections.

Chapter 1

⋆ The section on limits is divided into a section which defines limits and a section which computes their values.

Chapter 3

⋆ The section that defines the definite integral is divided into a section devoted to sigma notation and Riemann sums followed by a section that defines the definite integral and establishes its properties.

⋆ The section on methods of integration is divided into a section on indefinite integrals and simple methods of integration as well as a section on change of variables in integrals.

⋆ The abstract concept of integral invariants has been removed from the section on computing volume to a new section in the *Theory* portion of the chapter.

Chapter 4

⋆ The section on integration of trigonometric functions is divided into a section devoted to powers of trigonometric functions and a section on trigonometric substitutions.

⋆ The section on integrating rational functions is divided into a section on integrating rational functions of x and a section on integrating other types of rational functions.

Chapter 5

⋆ The section on sequences is divided into a section on sequence properties and a section on convergence of sequences.

⋆ The section on Taylor series is divided into a section on Taylor series at $x = 0$ and a section on general Taylor series at $x = c$.

---

[1] Some of the discussion below is only meaningful to instructors who have used the original text.

Chapter 1

Functions and Limits

1.1 Introduction

Calculus studies real valued functions of a real variable. This type of function is visualized as its graph in the plane which is usually a curve. If the graph of a function can be described as the path of a point moving in the plane, then we call the function continuous. Calculus gives methods for computing many properties of curves, such as the slopes of its tangent lines, the area under the curve and the length of the curve. The idea underlying all of these computations is the same. In a systematic way, we approximate the computation we require with simpler computations which we can perform. Taking the limit of these simpler computations, we obtain the value of the required computation. In this chapter, we study functions, continuity and limits. These concepts will be used in the exposition of calculus in the subsequent chapters.

The *Concepts* portion of this chapter exposits basic information about functions, trigonometry, limits and continuity. Section 2 introduces several ideas from set theory. In particular, we describe basic subsets of the real numbers $\Re$, including the various types of intervals which are used in calculus. A function is defined with emphasis on the importance of specifying its domain. A function is more than a formula! Our statements about functions are illustrated through geometric interpretations of their graphs. Section 3 is devoted to algebraic operations on functions and composite functions. A one–to–one function is introduced to describe those functions which have an inverse. We will see that the domain of a function, as well as its formula, determine whether a given function has an inverse. Section 4, applies the concepts of the preceding two sections to trigonometric functions. We define the six trigonometric functions and list their properties. Then we restrict their domains to define inverse trigonometric functions. The trigonometric functions and their inverses are important examples which will be used to illustrate calculus constructions. Section 5 develops an intuitive understanding of limits as well as an ability to use the definition of a limit to justify the values of simple limits. The emphasis in these justifications is the use of pictures to motivate algebraic formulas. The basic properties of limits are derived in Section 6. These properties are used to easily derive other limits from the limits established in Section 5. In particular, we derive the trigonometric limits which are used to compute the derivatives of trigonometric functions in the next chapter. Section 7 is devoted to continuous functions and their properties. In particular, we study two fundamental properties of continuous functions given by the Intermediate Value and the Maximum Value Theorems.

The *Applications* section of this chapter, Section 8, previews some of the contexts and functions which are used in later chapters to illustrate calculus applications.

We begin the *Theory* portion of this chapter in Section 9 with the study of constructions involving sets and functions. Set theory is the language of mathematics. It allows us to make our statements concise, precise and clear. The next three sections study special types of subsets of the real line $\Re$ and give an introduction to the branch of mathematics called *topology*. The goal of these sections is to prove the Intermediate Value and Maximum Value Theorems. Based on the description of $\Re$ as the set of decimals, we prove that the real numbers are complete in Section 10. We also introduce the topological concepts of open sets, closed sets, sequences, greatest lower bounds and least upper bounds. Section 11 studies connected subsets of $\Re$ and uses connectedness to prove the Intermediate Value Theorem. Section 12 studies compact subsets of $\Re$ and uses compactness to prove the Maximum Value Theorem.

The flow chart in Figure 1.1.1 indicates how the sections of the *Applications* and *Theory* portions of this chapter depend on preceding sections.

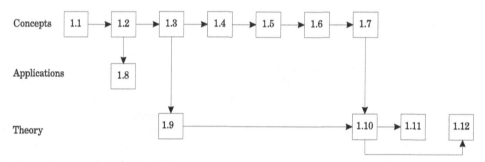

Figure 1.1.1 The Sections of Chapter 1

1.2   Functions and their Graphs

This section is devoted to the study of sets, functions and their geometric interpretations. In the first subsection we define the term *set* as the mathematical name for a collection of objects. In calculus, the sets we study are collections of numbers. Therefore, the examples in this section will be mostly of this type. In particular, we introduce intervals of numbers and depict them on the number line. We define the term *subset* to denote part of a given set and the term *union* to denote the combination of two given sets. In the second subsection, we introduce the symbol ∞ (infinity) to describe unbounded intervals. We explain that ∞ is not a number, because only some arithmetic operations make sense for ∞. In the third subsection, we study *functions* which describe the relationship between two variables: an independent variable x and a dependent variable y. We emphasize that a function consists of two pieces of data: a rule describing how the value of y depends on the value of x as well as the domain which specifies the values of the dependent variable x under consideration. We depict a function by its *graph*. In particular, we visualize various features of a function by the corresponding features of its graph.

Sets

Mathematics is not just computing numbers and solving problems. Equally important is the use of language and notation to pose a problem in a clear and precise manner. This same language and notation is used to communicate the solution of the problem to others. In particular, when we use numbers, we need to specify which numbers we are talking about. The following example indicates the importance of this specification.

Motivating Example 1.2.1 Tickets are being sold for a basketball game. The seats are arranged in three rows with 50 seats per row.

A seats are sold in row 1 for s dollars each,
B seats are sold in row 2 for t dollars each,
C seats are sold in row 3 for u dollars each.

A statistical analysis is planned to determine how the number of seats sold A, B, C and the total amount of money collected P depend on the pricing of the seats s, t, u.
(a) Describe the allowable values of each of these variables.
(b) Determine a formula for P, and explain how the description of the variables in part (a) clarifies this formula.

Solution (a) A, B, C represent seats sold and therefore must be whole numbers between 0 and 50. The variables s, t, u, P represent money and are decimals greater than or equal to zero with two digits to the right of the decimal point.
(b) The sale of A seats from row 1, at s dollars each, generates sA dollars of revenue. The sale of B seats from row 2, at t dollars each, generates tB dollars of revenue. The sale of C seats from row 3, at u dollars each, generates uC dollars of revenue. Thus the total revenue P from these sales is given by:

$$P = sA + tB + uC . \tag{1.2.1}$$

Someone who sees formula (1.2.1) might think that the variables can be any numbers. However in the context where we are applying this formula, it makes no sense for A to have a value of $\frac{7}{3}$, -5 or 73. It also makes no sense for t to have a value of 2.473 or $\sqrt{3}$. Therefore the specification of the allowable values of the variables in (a) explains how formula (1.2.1) is to be used. □

We introduce the mathematical term *set* to denote the objects under consideration. We also give notation to denote that an object is or is not included in a given set.

Definition 1.2.2 (a) *A set S is a collection of objects.*
(b) *$x \in S$ means that the object x is a member of the set S.*
(c) *$y \notin S$ means that the object y is not a member of the set S.*

Although we are usually interested in sets of numbers when studying calculus, we can define sets in many other contexts. Note in Example 5 how a finite set is described by listing its elements in braces.

Examples 1.2.3 (1) In Example 1.2.1, let N denote the set of whole numbers between 0 and 50. Then $A \in N$, $B \in N$ and $C \in N$ while $\frac{7}{3} \notin N$, $-5 \notin N$ and $73 \notin N$. Let D denote the set of decimals greater than or equal to zero with two digits to the right of the decimal point. Then $s \in D$, $t \in D$, $u \in D$ and $P \in D$. However, $2.473 \notin D$ and $\sqrt{3} \notin D$.

(2) Let S denote the set of all students in the classroom. If *Joe* is one of these students but *Jane* is not, write *Joe* $\in S$ and *Jane* $\notin S$.

(3) Let B denote the set of all books in the library. If *War and Peace* is in the library but *The Tin Drum* is not, write *War and Peace* $\in B$ and *The Tin Drum* $\notin B$.

(4) $\Re$ denotes the set of all real numbers. Then $5 \in \Re$, $-7.21 \in \Re$, $\sqrt{3} \in \Re$. However, *Joe* $\notin \Re$, *War and Peace* $\notin \Re$ and *blue* $\notin \Re$.

(5)   The set U whose elements are 1 and 2 can be depicted as $U = \{1, 2\}$.   □

Consider the set of real numbers $\Re$. The easiest way to denote a real numbers is as a decimal. These include whole numbers such as -5 and 9, finite decimals such as 4.328 and -5.04 as well as infinite decimals such as $\sqrt{2}$, π and $4.8\overline{372}$. The bar over the digits 372 indicates that they are repeated:

$$4.8\overline{372} = 4.8372372372372...$$

Consider the decimals $2.\overline{9} = 2.99999...$ and 3. Since the difference between them is zero, they are equal: $3 = 2.\overline{9}$. This motivates the following definition.

Definition 1.2.4 *The set of real numbers $\Re$ consists of all decimals which do not end in an infinite sequence of nines.*

Sometimes we start with a large set S and then decide that we only want to use the part of S called T. We introduce the term *subset* to denote this construction.

Definition 1.2.5 *The set T is called a subset of the set S if every element of T is an element of S. We write $T \subset S$.*

We illustrate the use of the subset construction in the sets of Examples 1.2.3.

Examples 1.2.6 **(1)** The set S of all students in the classroom has the subsets:

$$
\begin{aligned}
M &= \textit{all males in the classroom,} \\
F &= \textit{all females in the classroom,} \\
E &= \textit{all seniors in the classroom.}
\end{aligned}
$$

Write $M \subset S$, $F \subset S$ and $E \subset S$. Of course, S has many other subsets as well.

(2)   The set B of all books in the library has many subsets which include:

$$
\begin{aligned}
E &= \textit{all books on economics in the library,} \\
P &= \textit{all paperback books in the library} \\
R &= \textit{all red books in the library.}
\end{aligned}
$$

Write $E \subset B$, $P \subset B$ and $R \subset B$.

(3)   Show that the empty set $\emptyset$ which has no elements is a subset of every set S.

Solution We must show that every element of $\emptyset$ is an element of S. This is an easy job because there are no elements of $\emptyset$ which must be shown to be elements of S! Therefore $\emptyset \subset S$.

(4)   List all the subsets of $U = \{1, 2\}$.

Solution Since U has two elements, a subset of U has at most two elements.

The empty set $\emptyset$ is the only subset of U with no elements.
U has two subsets with one element: $\{1\}$ and $\{2\}$.
The only subset of U with two elements is U itself.

To summarize, U has four subsets: $\emptyset$, $\{1\}$, $\{2\}$ and U.   □

There are four important types of numbers which we use. Each of them is a set of numbers, and we use the subset notation to specify these four subsets of $\Re$.

$$N \;=\; the\ set\ of\ natural\ numbers \qquad\qquad (1.2.2)$$

consists of all non-negative whole numbers such as 0, 1, 2, 173 and $5,298,468$. However, $-3 \notin N$, $\frac{2}{5} \notin N$ and $\sqrt{5} \notin N$.

$$Z \;=\; the\ set\ of\ integers \qquad\qquad (1.2.3)$$

consists of all whole numbers such as -432, -1, 0, 5 and 971. However, $\frac{9}{4} \notin Z$ and $\sqrt{7} \notin Z$. Note that $N \subset Z \subset \Re$.

$$Q \;=\; the\ set\ of\ rational\ numbers \qquad\qquad (1.2.4)$$

consists of all fractions such as $\frac{2}{7}$, $\frac{53}{1} = 53$, $-\frac{94}{21} = -4\frac{10}{21}$ and $\frac{0}{7} = 0$. However, $\pi \notin Q$ and $\sqrt{2} \notin Q$. Observe that $N \subset Z \subset Q \subset \Re$.

$$I \;=\; the\ set\ of\ irrational\ numbers \qquad\qquad (1.2.5)$$

consists of all real numbers which are not rational such as $\sqrt{2}$ and π. However, $8 \notin I$ and $-\frac{2}{3} \notin I$.

We often use the set of all numbers which lie between a and b where $a \leq b$. This type of subset of $\Re$ is called an *interval*. There are four kinds of intervals depending on which of the endpoints a, b are included.

$$
\begin{aligned}
[a,b] \;&=\; the\ set\ of\ real\ numbers\ x\ with\ a \leq x \leq b, \\
(a,b) \;&=\; the\ set\ of\ real\ numbers\ x\ with\ a < x < b, \\
[a,b) \;&=\; the\ set\ of\ real\ numbers\ x\ with\ a \leq x < b\ and \\
(a,b] \;&=\; the\ set\ of\ real\ numbers\ x\ with\ a < x \leq b.
\end{aligned}
$$

We call a and b the *endpoints* of these intervals. For example, the interval $(-3,5]$ has endpoints -3, 5. $[a,b]$ is called a *closed interval* and contains both of its endpoints. (a,b) is called an *open interval* and contains neither of its endpoints. Each of $[a,b)$, $(a,b]$ is called a *half–open, half–closed interval* and contains one of its two endpoints.

Figure 1.2.7 The Real Number Line

We depict the set of real numbers $\Re$ as the *number line* in Figure 1.2.7. A positive number A has position A units to the right of 0 while a negative number $-B$ has position B units to the left of 0. A subset S of $\Re$ is depicted by drawing the part of the number line consisting of those numbers in S. We illustrate this procedure by depicting several intervals. In these pictures, a solid circle indicates that an endpoint is included while a hollow circle indicates that an endpoint is excluded. The *length* of an interval with endpoints a and b is the geometric length $b - a$ of the interval viewed as a subset of the real number line.

Examples 1.2.8 Depict each interval on the real number line.

(1) $[-7,-6]$

> **Solution** This interval of length $-6 - (-7) = 1$ is shown in Figure 1.2.9. The endpoints -7 and -6 are depicted by solid dots to show that they are included in this interval.

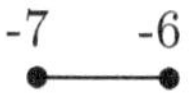

-7   -6

Fig 1.2.9 $[-7,6]$

Fig 1.2.10 $(-4, -2)$

Fig 1.2.11 $[-1, 0)$

Fig 1.2.12 $(2, 5]$

(2) $(-4, -2)$

> **Solution** This interval of length $-2 - (-4) = 2$ is shown in Figure 1.2.10. The endpoints -4 and -2 are depicted by hollow dots to show that they are not included in this interval.

(3) $[-1, 0)$

> **Solution** This interval of length $0 - (-1) = 1$ is shown in Figure 1.2.11. The left endpoint -1 is depicted by a solid dot because $-1 \in [-1, 0)$ while the right endpoint 0 is depicted by a hollow dot because $0 \notin [-1, 0)$.

(4) $(2, 5]$

> **Solution** This interval of length $5 - 2 = 3$ is shown in Figure 1.2.12. The left endpoint 2 is depicted by a hollow dot because $2 \notin (2, 5]$ while the right endpoint 5 is depicted by a solid dot because $5 \in (2, 5]$.   □

Sometimes we want to combine two sets S and T into one set called their *union*. This union will contain all the elements of S and all the elements of T including those objects which are in both S and T.

Definition 1.2.13 *Let S and T be two sets. Define the union U of S and T by letting $x \in U$ if $x \in S$ or $x \in T$. We write $U = S \cup T$.*

We illustrate the construction $S \cup T$ when S and T are sets of numbers.

Examples 1.2.14 (1) $\{-3, 4, 7\} \cup \{1, 4, 9, 12\} = \{-3, 1, 4, 7, 9, 12\}$.

(2) $Q \cup I = \Re$ because every number is either rational or irrational.

(3) $N \cup Z = Z$ because every natural number is also an integer. In fact, $S \cup T = T$ whenever $S \subset T$.

(4) $[1, 4) \cup (6, 8)$ consists of two disjoint intervals.

(5) $(-3, 8) \cup (2, 7] = (-3, 8)$ because the second interval is a subset of the first one.

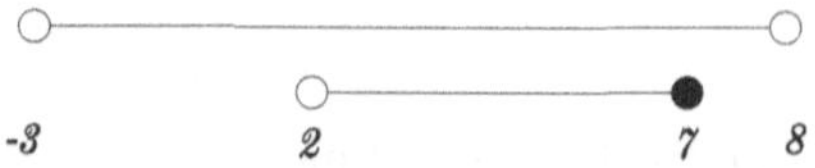

(6) $(-7, 4] \cup (2, 11) = (-7, 11)$ identifies the union of these two overlapping intervals as a single interval.   □

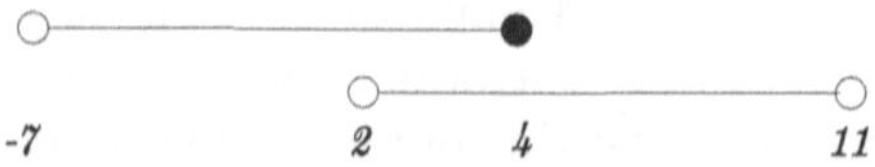

The Symbol ∞

In depicting intervals of infinite length and in evaluating certain limits, it is useful to have the symbol ∞ (infinity) to denote an object to the right of the number line which is larger than every number in $\Re$. We also introduce the symbol $-\infty$ (minus infinity) to denote an object to the left of the number line which is smaller than every number in $\Re$. See Figure 1.2.15. Thus for every $x \in \Re$: $-\infty < x < \infty$.

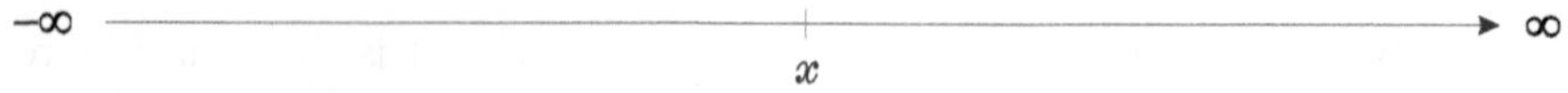

$$-\infty \longleftarrow \underset{x}{\rule{0pt}{0pt}\quad\quad\quad|\quad\quad\quad} \longrightarrow \infty$$

Figure 1.2.15 The Symbols $-\infty$ and ∞

We use the symbols $-\infty$ and ∞ to define four types of intervals of infinite length:

$$
\begin{aligned}
[a, \infty) &= \text{the set of } x \in \Re \text{ with } x \ge a; \\
(a, \infty) &= \text{the set of } x \in \Re \text{ with } x > a; \\
(-\infty, b] &= \text{the set of } x \in \Re \text{ with } x \le b; \\
(-\infty, b) &= \text{the set of } x \in \Re \text{ with } x < b.
\end{aligned}
$$

Examples 1.2.16 Depict each set as a subset of the number line.

(1) $(-\infty, 2]$

Solution This set consists of all numbers less than or equal to 2 as depicted in Figure 1.2.17. There is a solid dot at 2 because 2 is included in this interval.

Fig 1.2.17 $(-\infty, 2]$

(2) $(-\infty, 2)$

Solution This set consists of all numbers less than 2 as depicted in Figure 1.2.18. There is a hollow dot at 2 because 2 is not included in this interval.

Fig 1.2.18 $(-\infty, 2)$

(3) $[5, \infty)$

Solution This set consists of all numbers greater than or equal to 5 as depicted in Figure 1.2.19. There is a solid dot at 5 because 5 is included in this interval.

Figure 1.2.19 $[5, \infty)$

(4) $(5, \infty)$

Solution This set consists of all numbers greater than 5. See Figure 1.2.20. There is a hollow dot at 5 because 5 is not included in this interval.

Figure 1.2.20 $(5, \infty)$

(5) Let S be the set of numbers x that satisfy $\frac{x+1}{x-5} \ge 0$.

Solution This quotient is positive when either the numerator and denominator are both positive or both negative. The former case occurs when $x > 5$ while the latter case occurs when $x < -1$. This quotient is zero when the numerator is zero, i.e. when $x = -1$. Hence

$$
S = (-\infty, -1) \cup (5, \infty) \cup \{-1\} = (-\infty, -1] \cup (5, \infty) .
$$

This disjoint union of intervals is depicted on the number line as:

The symbols ∞ and $-\infty$ are not in $\Re$. Some arithmetic operations are defined for these symbols while others are not. Think of ∞ as a very large positive number and $-\infty$ as a very small negative number. (Note that a small negative number means $-1,000,000$ not $-.00001$.) Let $n \in \Re$. The following sums are defined:

$$
n + \infty = \infty, \quad n + -\infty = -\infty, \quad \infty + \infty = \infty, \quad -\infty + -\infty = -\infty.
$$

The justification for letting $n + \infty$ equal ∞ is: n plus a very large positive number is a very large positive number. For example, if $n = 18$ then $18 + 3,000,000$ is a very large positive number. The justification for letting $-\infty + -\infty$ equal $-\infty$ is: the sum of two very small negative numbers is a very small negative number. For example, $-5,000,000 + -7,000,000,000$ is a very small negative number. The definitions of the other two sums above are justified similarly.

However, $\infty + -\infty$ is not defined because a very large positive number plus a very small negative number could be any kind of number. For example:

$$
\begin{aligned}
10^{100} + -10^{100} &= 0, \\
10^{200} + -10^{100} &= 10^{100}(10^{100} - 1) \approx 10^{200}, \\
10^{100} + -10^{200} &= -10^{100}(10^{100} - 1) \approx -10^{200}.
\end{aligned}
$$

In particular, if we were to define $\infty + -\infty = 0$, then the associative property of addition would fail:

$$(\infty + \infty) + -\infty = \infty + -\infty = 0 \quad \text{while} \quad \infty + (\infty + -\infty) = \infty + 0 = \infty.$$

Thus we conclude that $\infty + -\infty$ is not defined.

Let $n \in \Re$ with $n > 0$. The following products are defined:

$$n \times \infty = -n \times -\infty = \infty, \qquad n \times -\infty = -n \times \infty = -\infty,$$
$$\infty \times \infty = -\infty \times -\infty = \infty, \qquad \infty \times -\infty = -\infty \times \infty = -\infty.$$

The justification for letting $-\infty \times \infty$ equal $-\infty$ is: a very small negative number times a very large positive number is a very small negative number. For example, $-25,000,000 \times 43,000,000,000$ is a very small negative number. The definitions of the other seven products above are justified similarly.

Since a very large number times zero equals zero, you might want to define $\infty \times 0 = 0$. However, in that case we would have the equation

$$0 = \infty \times 0 = \infty \times [1 + (-1)] = \infty \times 1 + \infty \times (-1) = \infty + -\infty.$$

As we observed above, this equation is not compatible with the associative property of addition. Thus $\infty \times 0$ can not be defined. Similarly, $-\infty \times 0$ can not be defined.

Let $n \in \Re$ with $n > 0$. The following quotients are defined:

$$\frac{\infty}{n} = \frac{-\infty}{-n} = \infty, \qquad \frac{-\infty}{n} = \frac{\infty}{-n} = -\infty.$$

The justification for letting $\frac{\infty}{n}$ equal ∞ is: a very large positive number divided by a positive number is a very large positive number. For example, if $n = 46$ then $\frac{58,000,000}{46}$ is a very large positive number.

You might be tempted to define $\frac{\infty}{\infty} = 1$. However, this would imply

$$\infty \times \left(\frac{\infty}{\infty}\right) = \infty \times 1 = \infty \quad \text{while} \quad \infty \times \left(\frac{\infty}{\infty}\right) = \frac{\infty \times \infty}{\infty} = \frac{\infty}{\infty} = 1.$$

Thus $\frac{\infty}{\infty}$ is not defined. Similarly, $\frac{-\infty}{-\infty}$, $\frac{\infty}{-\infty}$ and $\frac{-\infty}{\infty}$ are also not defined.

You might also be tempted to define $\frac{n}{0} = \infty$ for $n > 0$. However, this would imply

$$-1 \times \left(\frac{n}{0}\right) = -1 \times \infty = -\infty \quad \text{while} \quad -1 \times \left(\frac{n}{0}\right) = \frac{n}{-1 \times 0} = \frac{n}{0} = \infty.$$

Thus $\frac{n}{0}$ is not defined for $n > 0$. Similarly, $\frac{\infty}{0}$, $\frac{-\infty}{0}$ and $\frac{-n}{0}$ are not defined.

We summarize our discussion. The following definition lists all arithmetic operations involving ∞ and $-\infty$ which are defined. The note which follows it, lists the arithmetic operations which can not be defined.

Definition 1.2.21 *Let $k, m, n \in \Re$ with $m > 0$ and $n < 0$. Define*

$$
\begin{array}{llll}
k + \infty = \infty & k + -\infty = -\infty & \infty + \infty = \infty & -\infty + -\infty = -\infty \\[4pt]
\infty \times \infty = \infty & -\infty \times \infty = -\infty & -\infty \times -\infty = \infty & \frac{k}{\pm\infty} = 0 \\[4pt]
m \times \infty = \infty & n \times \infty = -\infty & m \times -\infty = -\infty & n \times -\infty = \infty \\[4pt]
\frac{\infty}{m} = \infty & \frac{\infty}{n} = -\infty & \frac{-\infty}{m} = -\infty & \frac{-\infty}{n} = \infty.
\end{array}
$$

Note The following arithmetic operations *can not* be defined:

$$\infty + -\infty, \qquad \pm\infty \times 0, \qquad \frac{\pm\infty}{\pm\infty}, \qquad \frac{\pm\infty}{0}, \qquad \frac{k}{0}.$$

It is important to realize the limitations of the symbols ∞ and $-\infty$: they relate to numbers with respect to ordering but one must be very careful not to perform "illegal" arithmetic operations with these symbols.

Functions

In this subsection we study functions which are two variables related by an equation. The following example shows that the equation alone is not sufficient to analyze the situation. It is also important to specify the values of the variables we are using.

Motivating Example 1.2.22 The radius r and the area A of a circle are related by the equation

$$A = \pi r^2. \qquad (1.2.6)$$

This formula shows how the value of the radius of a circle determines the value of its area. For example, if $r = 3$ then $A = 9\pi$. Observe how equation (1.2.6) is used differently in each of the following contexts.

(a) If we think of equation (1.2.6) as giving the area of a circle then we want the radius r to be a positive number.
(b) Equation (1.2.6) is a mathematical equation which makes sense for every number r, including negative numbers.
(c) If we want to study circles which can be cut from a square sheet of metal with side 20 cm long, we would only use equation (1.2.6) for $0 \le r \le 10$.

Thus formula (1.2.6) does not give a complete description of the relationship between the quantities r and A. We also need to know which values of r are to be used. □

The term *function* denotes a rule which shows how the value of a dependent variable y depends on the value of an independent variable x as well as a context which specifies the values of x under consideration. The three contexts of the preceding example will be described by three different functions which are all based on the same formula. Note that we want a function to have a well-defined rule which assigns exactly one value of y to each value of x under consideration.

Definition 1.2.23 (a) *A function $f : D \to C$, or f for short, is a rule which associates a unique element $f(x)$ of the set C to each element x of the set D.*
(b) *We call D the domain of the function f, and we call C the codomain of f. When C is a subset of the real numbers $\Re$, we call f a real–valued function.*
(c) *If x represents an object in the domain of f, we call x an independent variable. If y represents an object in the codomain of f, we call y a dependent variable.*

When $f : D \to C$ is a function and $y = f(x)$, then x is called an independent variable because we are free to choose any object of the domain D as the value of x. However, the function f determines the value of y from the value of x, i.e. the value of y depends on the value of x. That is why y is called a dependent variable. The codomain C is used to inform us of the type of object that f assigns to each element of its domain. The specification of the codomain $C = \Re$ may seem superfluous for most of our applications which deal with real valued functions. However, the codomain is essential for more general functions where C tells us the nature of the object $f(x)$.

Examples 1.2.24 (1)   Consider the three functions of Example 1.2.22 which are based upon the formula $A = \pi r^2$. In all cases r is the independent variable and A is the dependent variable.

   (a)   The first function f_1 considers r to be the radius of a circle. Thus f_1 assigns each positive number r to the number πr^2, and we write $f_1(r) = \pi r^2$. The domain of f_1 is the set of positive numbers $(0, \infty)$, i.e. $f_1 : (0, \infty) \to \Re$.

   (b)   The second function f_2 allows r to be any real number. That is, f_2 assigns each number r to the number πr^2, and we write $f_2(r) = \pi r^2$. The domain of f_2 is $\Re$, i.e. $f_2 : \Re \longrightarrow \Re$.

   (c)   The third function f_3 is only defined on radii r of circles which can be cut from a square with side 20 cm, i.e $0 \leq r \leq 10$. Then f_3 assigns each such r to the number πr^2. The domain of f_3 is $[0, 10]$, i.e. $f_3 : [0, 10] \longrightarrow \Re$.

(2) (a)   Let g be the real valued function with domain $[0, \infty)$ which assigns each non–negative number x to its positive square root. We write $g(x) = \sqrt{x}$.

   (b)   If we changed the function g to the rule g_1 with domain $\Re$ given by $g_1(x) = \sqrt{x}$ then g_1 is not a function because g_1 is not defined on all elements of its domain. That is, g_1 is not defined on negative numbers.

(3)   Let h be the function with domain $\Re$ which leaves positive numbers in tact but removes the minus sign from in front of negative numbers. We denote $h(x)$ as $|x|$. For example, $h(12) = |12| = 12$ and $h(-8) = |-8| = 8$. A more formal definition of h is given by:

$$h(x) \;=\; |x| \;=\; \left\{ \begin{array}{ll} x & if \ x \geq 0 \\ -x & if \ x < 0 \end{array} \right\}. \tag{1.2.7}$$

We call h the *absolute value function*. An alternate description of h is given by:

$$h(x) \;=\; |x| \;=\; \sqrt{x^2}. \tag{1.2.8}$$

(4) (a)   Let k be the rule with domain $[-2, 2]$ which assigns each number x, with $-2 \leq x \leq 2$, to the number y such that $x^2 + y^2 = 4$. This rule is not a function because if we solve for y, we obtain

$$y = \pm\sqrt{4 - x^2}.$$

Thus, all numbers in the subset $(-2, 2)$ of the domain of k are assigned to two numbers by this rule while our definition of a function requires that each number in the domain be assigned to a *unique* number.

   (b)   With this insight, we can modify the definition of k to create a function. Let k_1 be the function with domain $[-2, 2]$ defined by

$$k_1(x) = \sqrt{4 - x^2}.$$

   (c)   Alternatively, we can define a function k_2 with domain $[-2, 2]$ by

$$k_2(x) = -\sqrt{4 - x^2}.$$

(5)   Consider an object which moves in the plane P for thirty seconds. Construct a function m with domain the time interval $[0, 30]$ and codomain the plane P. The function $m : [0, 30] \to P$ is defined by letting $m(t)$ denote the position of this object at time t.

(6) Consider the process of assigning grades to each student in the class. Let S denotes the set of these students, and let $G = \{A, B, C, D, E, F\}$ denote the set of possible grades. The process of assigning grades can be described as a function $q : S \rightarrow G$ with domain S and codomain G. That is, if $x \in S$ is a student then $q(x) \in G$ is the grade that student x receives.

(7) Although the rule of a function is specified by a formula for most calculus examples, the rule of a function can also be given by a table. In particular, statistics studies functions defined by tables of data produced by an experiment. For example, the table below records the amount of sewage in parts per million (ppm) found in the water at Clearwater Beach on each of the first ten days of August. This table is a function $f : D \rightarrow N$ with domain the set of integers $D = \{1, 2, 3, 4, 5, 6, 7, 8, 9, 10\}$ and codomain the set of natural numbers N. That is, $f(n)$ is the sewage in ppm at Clearwater Beach on the nth day of August. For example, we see from the table that $f(3) = 12$. This means there were 12 ppm of sewage in the water at Clearwater Beach on August 3rd.

Day	1	2	3	4	5	6	7	8	9	10
Sewage (ppm)	9	11	12	11	13	14	17	16	15	13

$\square$

The codomain C of a function f gives little information about the particular function. For example, $C = \Re$ would be a suitable codomain for every real valued function. However, the set of all values $f(x)$ of the function is an important feature of the specific function under consideration. We call this set the *range* of f.

Definition 1.2.25 *The range of the function $f : D \rightarrow C$ with domain D is the subset $f(D)$ of the codomain C which consists of all $f(x)$ for $x \in D$.*

Observe the distinction between the codomain C and the range $f(D)$ of the function $f : D \rightarrow C$. The codomain specifies the type of object that f assigns to each object of its domain. That is, C is the set of potential values of f. On the other hand the range of f is the set of actual values of f. In particular, not every element of the codomain is necessarily in the range. Also note that although each element of the domain of a function must be assigned to a unique element of its range, more than one element of the domain can be assigned to the same element of the range. For example, the constant function $f : \Re \rightarrow \Re$ with $f(x) = 3$ has codomain $\Re$ indicating that f is a real valued function. However, the range of f consists only of the number 3, i.e. every element of the domain of f is assigned to 3. We describe the ranges of the functions of Examples 1.2.24.

Examples 1.2.26 (1) (a) Let $f_1(r) = \pi r^2$ with domain $(0, \infty)$. Observe that:

    (i) $\pi r^2 > 0$ for all positive numbers r;

    (ii) any positive number A can be written as πr^2 where $r = \sqrt{A/\pi}$, i.e. $f_1(\sqrt{A/\pi}) = A$.

    It follows that Range $f_1 = (0, \infty)$. Geometrically, we are saying that every positive number A is the area of a circle.

  (b) Let $f_2(r) = \pi r^2$ with domain $\Re$. Observe that:

    (i) $\pi r^2 \geq 0$ for all numbers r;

    (ii) any non–negative number y can be written as πr^2 where r is either $\sqrt{y/\pi}$ or $-\sqrt{y/\pi}$, i.e. $f_2(\sqrt{y/\pi}) = f_2(-\sqrt{y/\pi}) = A$.

    It follows that Range $f_2 = [0, \infty)$.

(c) Let $f_3(r) = \pi r^2$ with domain $[0, 10]$. Observe that:

   (i) if $0 \leq r \leq 10$ then $0 \leq \pi r^2 \leq 100\pi$;

   (ii) if $0 \leq A \leq 100\pi$, then $0 \leq \sqrt{A/\pi} \leq 10$ and $f_3(\sqrt{A/\pi}) = A$.

   It follows that Range $f_3 = [0, 100\pi]$. This statement can be interpreted as saying that we can cut a circle of any area between 0 and 100π cm^2 from a square sheet of metal whose side is 20 cm long.

(2) The square root function $g(x) = \sqrt{x}$ with domain $[0, \infty)$ has range $[0, \infty)$ because:

   (i) the square root of a number means its positive square root;

   (ii) all non–negative numbers y are square roots, i.e. $y = g(y^2)$.

(3) The absolute value function $h(x) = |x|$ with domain $\Re$ has range $[0, \infty)$ because:

   (i) the absolute value of all numbers are non–negative;

   (ii) if $y \geq 0$ then $h(y) = h(-y) = y$.

(4) (a) The range of the function $k_1(x) = \sqrt{4 - x^2}$ with domain $[-2, 2]$ is determined as follows.

   (i) If x is between -2 and 2, then x^2 is between 0 and 4. Hence $4 - x^2$ is also between 0 and 4. Therefore, $k_1(x) = \sqrt{4 - x^2}$ is between 0 and 2.

   (ii) Given any y between 0 and 2, we can solve the equation $y = \sqrt{4 - x^2}$ for x to obtain $x = \pm\sqrt{4 - y^2}$. Then $k_1(\sqrt{4 - y^2}) = k_1(-\sqrt{4 - y^2}) = y$.

   Thus, the range of k_1 is $[0, 2]$.

(b) The range of the function $k_2(x) = -\sqrt{4 - x^2}$ with domain $[-2, 2]$ is determined as follows.

   (i) If x is between -2 and 2, then x^2 is between 0 and 4. Hence $4 - x^2$ is also between 0 and 4. Therefore $k_2(x) = -\sqrt{4 - x^2}$ is between -2 and 0.

   (ii) Given any y between -2 and 0, we can solve the equation $y = -\sqrt{4 - x^2}$ for x to obtain $x = \pm\sqrt{4 - y^2}$. Then $k_2(\sqrt{4 - y^2}) = k_2(-\sqrt{4 - y^2}) = y$.

   Thus, the range of k_2 is $[-2, 0]$.

(5) If $m(t)$ denotes the position of an object in the plane P at time t for $0 \leq t \leq 30$, then the range of m is the path of this object during the time interval $[0, 30]$.

(6) If $q(x)$ is the grade assigned to student x, then the range of q is a subset of the codomain $G = \{A, B, C, D, E, F\}$. It is unreasonable to expect that the range of q is the set $\{A\}$, i.e. every student will receive an A. However, we can realistically hope that the range of q is $\{A, B, C\}$.

(7) The function q defined by the table in Example 1.2.24 (7) assigns each number of the top row to the number beneath it in the bottom row. Hence the range R of q is the set of numbers in the bottom row: $R = \{9, 11, 12, 13, 14, 15, 16, 17\}$. □

Since real valued functions are the primary functions studied in this text, we limit our considerations from now on to this important case. When dealing with these functions, we are frequently given only a formula $f(x)$ with no specification of the domain. The implicit understanding is that the domain is the largest set on which this formula defines a function. Be aware of two phenomena which place restrictions on the domain of f: we can not divide by zero and the even root of a negative number is not defined. Therefore:

- when the formula for $f(x)$ is a fraction, we must exclude numbers from the domain which make the denominator zero;

- when the formula for $f(x)$ involves an even root of an expression, we must exclude numbers from the domain which make this expression negative.

Examples 1.2.27 Find the domain of each of the following functions.

(1) $f(x) = 3x^2 - 2x + 1$

Solution Since the formula for $f(x)$ makes sense for every number x, the domain of f is entire set of real numbers $\Re$.

(2) $g(x) = \frac{5x}{x^2-4}$

Solution The domain of g consists of all numbers x of $\Re$ except those for which the denominator $x^2 - 4 = 0$, i.e. we must exclude $+2$ and -2 from the domain of g. Using interval notation:

$$\text{Domain } g \ = \ (-\infty, -2) \cup (-2, 2) \cup (2, \infty).$$

(3) $h(x) = \sqrt{4x - 8}$

Solution The domain of h equals all numbers x for which make the expression $4x - 8$ under the square root sign greater than or equal to zero. That is, we require $4x \geq 8$ or $x \geq 2$. Hence Domain $h = [2, \infty)$.

(4) $j(x) = \left(\frac{x}{x^3-1}\right)^{3/4}$.

Solution In this example we have both a fraction and an even root. Note

$$x^3 - 1 = (x-1)(x^2 + x + 1) \quad and \quad x^2 + x + 1 = \left(x + \frac{1}{2}\right)^2 + \frac{3}{4} > 0$$

for all x. Therefore, we must exclude 1 from the domain of j where the denominator of the fraction is zero. In addition, for the fraction to be greater than or equal to zero, both the numerator and denominator must have the same sign. Thus, either $x \geq 0$, $x > 1$ or $x \leq 0$, $x < 1$, i.e. $x > 1$ or $x \leq 0$. Using interval notation:

$$\text{Domain } j \ = \ (1, \infty) \cup (-\infty, 0] \ . \qquad \square$$

Graph of a Function

We associate a graph to a real valued function which is usually a curve. The graph turns the equation $y = f(x)$ which defines the function into a picture that we can visualize. Moreover, algebraic properties of the equation $y = f(x)$ correspond to geometric properties of the graph. In particular, we show how the domain and range of the function can be determined from its graph. Recall how a point of the coordinate plane is described by a pair of numbers. The *coordinate plane* is vertical and contains the real number line in a horizontal position at "ground level". We call this number line the x–axis. Draw a vertical coordinate line, called the y–axis, which points upwards and passes through the point 0 on the x–axis. Let P denote a point in this plane. Assign an x–coordinate s to P as the number on the x–axis directly above or below P. Assign a y–coordinate t to P as the altitude of P. The y–coordinate of P is positive if P lies above the x–axis (ground level), and the y–coordinate of P is negative if P lies below the x–axis (ground level). We call (s, t) the *coordinates* of P and write

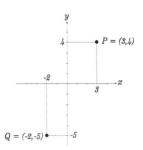

Figure 1.2.28
Graphing Points

Figure 1.2.29
$f_1(x) = \pi x^2$

$P = (s, t)$. The point $(0, 0)$, where the x–axis and y–axis intersect, is called the *origin*. For example, the points $P = (3, 4)$ and $Q = (-2, -5)$ are plotted in Figure 1.2.28.

To graph the function f, we place a dot at the point $(x, f(x))$, above or below x, for each x in the domain of f. The set of these dots is called the *graph* of f. For nice functions, this graph is a curve which lies above and below the domain of f.

Definition 1.2.31 *Let f be a function with domain D. The graph of f consists of the set of points $(x, f(x))$ in the coordinate plane for $x \in D$.*

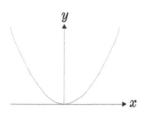

Figure 1.2.30
$f_2(x) = \pi x^2$

We construct the graphs of the real valued functions we studied in Examples 1.2.24.

Examples 1.2.32 Sketch the graph of each function.

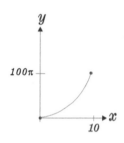

Figure 1.2.33
$f_3(x) = \pi x^2$

(1) **(a)** The function $f_1 : (0, \infty) \to \Re$ defined by $f_1(x) = x^2$ has graph the right half of the parabola $y = x^2$ where $x > 0$. See Figure 1.2.29.
(b) The function $f_2 : \Re \to \Re$ defined by $f_2(x) = x^2$ has graph the entire parabola $y = x^2$. See Figure 1.2.30.
(c) The function $f_3 : [0, 10] \to \Re$ defined by $f_3(x) = x^2$ has graph the piece of the parabola $y = x^2$ which lies above the interval $[0, 10]$ on the x–axis. See Figure 1.2.33.

(2) Consider the function $g : [0, \infty) \longrightarrow \Re$ defined by $g(x) = \sqrt{x}$. Note that if $y = \sqrt{x}$, then $y^2 = x$ with $y \geq 0$. Hence the graph of g is the top half of the parabola $x = y^2$. See Figure 1.2.34.

Figure 1.2.34
$g(x) = \sqrt{x}$

(3) The absolute value function $h : \Re \to \Re$ is described by $h(x) = x$ for $x \geq 0$ and $h(x) = -x$ for $x < 0$. Hence the V-shaped graph of h consists of the half–line $y = x$ for $x \geq 0$ and the half-line $y = -x$ for $x < 0$. See Figure 1.2.35.

(4) **(a)** The function $k_1 : [-2, 2] \to \Re$ defined by $y = k_1(x) = \sqrt{4 - x^2}$ can be described as $y^2 = 4 - x^2$ for $y \geq 0$. Hence the graph of k_1 is the upper half of the circle $x^2 + y^2 = 4$ depicted in Figure 1.2.37.
(b) The function $k_2 : [-2, 2] \to \Re$ defined by $y = k_2(x) = -\sqrt{4 - x^2}$ can be described as $y^2 = 4 - x^2$ for $y \leq 0$. Hence the graph of k_2 is the lower half of the circle $x^2 + y^2 = 4$ depicted in Figure 1.2.38. □

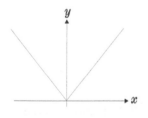

Figure 1.2.35
$h(x) = |x|$

We can use the graph G of a function f to determine the domain and range of f. This is often a more practical way of determining the range of f than by analyzing the formula $f(x)$. Recall that the graph G consists of all $(x, f(x))$ for x in the domain D of f. Therefore, the domain D of f consists of all x for which $(x, y) = (x, f(x)) \in G$, That is, D is the vertical projection of the graph G onto the x–axis. Analogously, the range R of f consists of all y for which $(x, y) = (x, f(x)) \in G$. Equivalently, R is the horizontal projection of the graph G onto the y-axis.

Figure 1.2.37
$k_1(x) = \sqrt{4 - x^2}$

Examples 1.2.36 Use the graph of each function to determine its domain and range.

(1) **(a)** Consider the graph of the function f_1 in Figure 1.2.29. Since this graph lies above the right half of the x–axis, Domain $f_1 = (0, \infty)$. Since this graph lies to the right of the upper half of the y–axis, Range $f_1 = (0, \infty)$.
(b) Consider the graph of the function f_2 in Figure 1.2.30. Since this graph lies above the entire x–axis, Domain $f_2 = \Re$. Since this graph lies to the left and right of the upper half of the y–axis, Range $f_2 = [0, \infty)$.
(c) Consider the graph of the function f_3 in Figure 1.2.33. Since this graph lies

above the interval $[0, 10]$ on the x–axis, that interval is the domain of f_3. Since this graph lies to the right of the interval $[0, 100\pi]$ on the y–axis, this interval is the range of f_3.

(2) Consider the function whose graph is depicted in Figures 1.2.39 and 1.2.41. The domain of f is the vertical projection of this graph onto the x–axis, as indicated by the vertical arrows in Figure 1.2.39. This is the interval $[-9, 8]$. The range of f is the horizontal projection of this graph onto the y–axis, as indicated by the horizontal arrows in Figure 1.2.41. This is the interval $[-3, 5]$. □

Given a curve C in the plane, we can easily determine whether it is the graph of a function f. The domain D of f would have to be the projection of C onto the x–axis. For each $s \in D$, $f(s)$ would be the y–coordinate t of the point $(s, t) \in C$ which lies on the vertical line through the point s on the x–axis. Recall that a function has to assign *exactly one number* to *each point* of its domain. Therefore this procedure for finding $f(s)$ defines a function f with domain D if *every vertical line* through D intersects the curve C in *exactly one point*.

Examples 1.2.40 (1) Consider the curve in Figure 1.2.42. The vertical projection of this curve on the x–axis is the interval $[1, 3]$. This curve defines a function with domain $[1, 3]$ because every dotted vertical line through the interval $[1, 3]$ on the x–axis intersects the curve in exactly one point.

(2) Consider the curve in Figure 1.2.44. Its vertical projection on the x–axis is the closed interval $[2, 5]$. This curve does not define a function because the dotted vertical lines through the center section of the curve intersect the curve in more than one point. □

Let f, g be functions whose domain D has the property that $x \in D$ implies $-x \in D$. If $f(-x) = f(x)$, we call f an *even function*. If $g(-x) = -g(x)$, we call g an *odd function*. For example, even powers of x, such as $f(x) = x^8$, are even functions while odd powers of x, such as $g(x) = x^7$, are odd functions.

Examples 1.2.43 Determine whether each function is even or odd.

(1) $f(x) = -8x^5 + 9x^3 - 4x$

Solution Since $f(-x) = -8(-x)^5 + 9(-x)^3 - 4(-x) = 8x^5 - 9x^3 + 4x = -f(x)$, f is an odd function.

(2) $g(x) = \frac{9x^6 - 7x^4}{5x^2 + 6}$

Solution Since $g(-x) = \frac{9(-x)^6 - 7(-x)^4}{5(-x)^2 + 6} = \frac{9x^6 - 7x^4}{5x^2 + 6} = g(x)$, g is an even function.

(3) $h(x) = 6x^3 + 1$

Solution Note $h(-x) = 6(-x)^3 + 1 = -6x^3 + 1$ is neither $h(x)$ nor $-h(x)$. Hence h is neither an even function nor an odd function. □

The graph of an even function f is symmetric with respect to the y–axis, i.e. if $(a, f(a))$ is a point on the graph, then the point $(-a, f(-a)) = (-a, f(a))$ is also on the graph. See the left diagram in Figure 1.2.45. The graph of an odd function g is symmetric with respect to the origin, i.e. if $(a, g(a))$ is a point on the graph then the point $(-a, g(-a)) = (-a, -g(a))$ is also on the graph. See the right diagram in Figure 1.2.45.

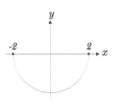

Figure 1.2.38 (4b)
$k_2(x) = -\sqrt{4 - x^2}$

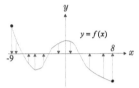

Figure 1.2.39
Domain f

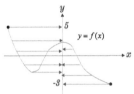

Figure 1.2.41
Range f

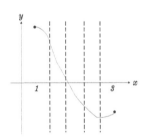

Figure 1.2.42
A Curve which Defines
a Function

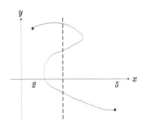

Figure 1.2.44
A Curve which Does
Not Define a Function

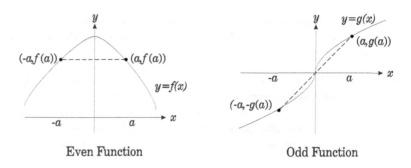

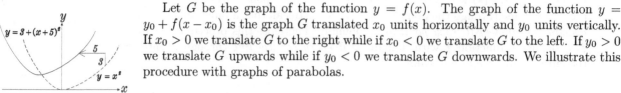

Even Function   Odd Function

Figure 1.2.45 Symmetry in Graphs

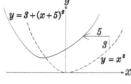

Figure 1.2.46
Translation of $y = x^2$

Let G be the graph of the function $y = f(x)$. The graph of the function $y = y_0 + f(x - x_0)$ is the graph G translated x_0 units horizontally and y_0 units vertically. If $x_0 > 0$ we translate G to the right while if $x_0 < 0$ we translate G to the left. If $y_0 > 0$ we translate G upwards while if $y_0 < 0$ we translate G downwards. We illustrate this procedure with graphs of parabolas.

Examples 1.2.47 (1) Find the equation of the parabola P obtained by translating the parabola $y = x^2$ five units to the left and three units upwards.

Solution Since we are translating five units to the left, we let $x_0 = -5$. Since we are translating three units upwards, we let $y_0 = +3$. See Figure 1.2.46. Hence the equation of P is $y = 3 + (x + 5)^2$.

(2) Sketch the graph G of the parabola $y = x^2 - 8x + 7$.

Solution We complete the square:

$$y = x^2 - 8x + 7 = (x - 4)^2 - 9 \,.$$

Hence G is the graph of the parabola $y = x^2$ translated four units to the right and nine units downwards. See Figure 1.2.48.   □

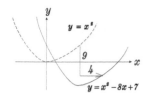

Figure 1.2.48
$y = x^2 - 8x + 7$

Historical Remarks

Curves have been studied since the times of the Greeks ca. 500–200 BCE. They constructed curves either as a locus (set) of points, an intersection of two surfaces or as the path of a point moving under given constraints. For example, a parabola is the locus of points equidistant from a fixed point and a fixed line. They defined a conic section as the curve obtained as the intersection of a plane and a cone. Archimedes studied a spiral which he defined as the curve traced out by a point which moves at a constant speed along a half–line as the half–line revolves around its endpoint at a constant speed. These viewpoints provided the Greeks with only a limited number of examples to study.

In 1637, René Descartes and Pierre de Fermat independently introduced the coordinate plane. Descartes used this point of view to associate an equation with a given curve. He then translated a geometric problem about the curve into a corresponding algebraic problem about its equation. He would solve the algebraic problem, thereby obtaining a solution of the original geometric problem. For example, the intersection of a line and a parabola can be determined by algebraically finding the simultaneous solutions of their equations. Fermat started with an equation and then constructed its graph. He used the geometric properties of this graph to deduce algebraic properties of the original equation. For example, the solution of two linear equations in two unknowns can be determined by finding the intersection point of the two lines in the plane which represent the two equations. This type of mathematics is called *analytic geometry*.

The subsequent development of calculus studied the properties of curves using the equations, or infinite series, which represent them. The concept of a function was introduced by Leonhard Euler in his calculus textbook of 1748, where he viewed the dependent variable as an "analytical expression" in the independent variable. In his text of 1755 he gave the modern definition of a function which does not require an equation to express the relationship between the dependent and independent variables. From then on calculus was formulated as solving problems regarding the variables of functions rather than solving problems about curves. Solving problems about curves became an application of calculus. This new point of view had immediate practical implications. In 1747, Jean d'Alembert published a paper on the motion of a plucked string. He insisted that the methods of calculus which he used in his solution required that the initial position of the string be described by a single equation. Euler argued that calculus in general, and d'Alembert's solution in particular, applied to more general functions. For example, consider a string of unit length that is plucked a distance of one unit at its midpoint. Its shape, as depicted in Figure 1.2.49, is the graph of the function

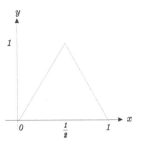

Fig 1.2.49
Plucked String

$$f(x) = \left\{ \begin{array}{ll} 2x & \text{if } 0 \leq x \leq \frac{1}{2} \\ 2 - 2x & \text{if } \frac{1}{2} \leq x \leq 1 \end{array} \right\}$$

with domain $[0, 1]$. D'Alembert considered this example unacceptable, while Euler correctly insisted that the methods of calculus applied.

Summary

At this stage the reader should understand the concepts of a set and a subset. In particular, she should be familiar with the set of real numbers $\Re$ and the subsets of natural numbers, integers, rational numbers and irrational numbers. In addition, the notation for the various types of intervals of $\Re$ should be understood. When using the symbols ∞ and $-\infty$ she should be aware of their limitations.

The reader should also understand the concepts of a function, its domain and its range. In particular, she should be aware that a formula does not determine a function until a suitable domain is specified. The construction of the graph of a function should be understood. In addition, the reader should be able to determine when a given curve is the graph of a function. In that case, she should be able to find the domain and range of that function from its graph.

In determining the domain of a function constructed from polynomials, such as $f(x) = \sqrt{x^2 - x - 6}$, it may be necessary to factor a polynomial. Factoring polynomials will also be used in the computations of other sections. Here are the factorizations some common polynomials.

$$\begin{array}{rcll} x^2 - a^2 & = & (x - a)(x + a) & (1.2.9) \\ x^2 + (a + b)x + ab & = & (x + a)(x + b) & (1.2.10) \\ x^3 + a^3 & = & (x + a)(x^2 - ax + a^2) & (1.2.11) \\ x^3 - a^3 & = & (x - a)(x^2 + ax + a^2) & (1.2.12) \\ x^3 + 3ax^2 + 3a^2x + a^3 & = & (x + a)^3 & (1.2.13) \end{array}$$

Basic Exercises

1. For each set below:
      (i) List three elements of the set.
      (ii) List three objects which are not in the set.
      (iii) List two subsets of the set.
(a) Let C denote the set of objects in your closet.
(b) Let W denote the set of objects in your wallet.
(c) Let U denote the set of countries which are members of the United Nations.
(d) Let R denote the seven colors of the rainbow.

2. Let Q be the set of all quadrilaterals in the coordinate plane.
    Let S be the set of all squares in the coordinate plane.
    Let R be the set of all rhombuses in the coordinate plane.
    Let P be the set of all parallelograms in the coordinate plane.
    Let E be the set of all rectangles in the coordinate plane.
List all subset relations among these five sets.

3. Write each number as an element of $\Re$, i.e. as a decimal which does not end in an infinite sequence of nines.
(a) $-\frac{3}{2}$ **(b)** $\frac{2}{3}$ **(c)** $\frac{71}{1000}$ **(d)** $\frac{3}{7}$ **(e)** $-\frac{2}{11}$ **(f)** $5.872\overline{9}$ **(g)** $9.9\overline{9}$ **(h)** $-4799.\overline{9}$

4. Determine whether each number is a natural number, an integer, a rational number or an irrational number. Note that some of these numbers may have more than one such characterization.
(a) 3 **(b)** -6 **(c)** 0 **(d)** $\frac{2}{9}$ **(e)** $-\frac{3}{11}$
(f) $\sqrt{5}$ **(g)** $\sqrt{81}$ **(h)** -5.821 **(i)** $3.7999...$ **(j)** $\frac{\pi}{3.14}$

5. Place a dot on the number line to indicate the position of each number.
(a) 5 **(b)** -3 **(c)** $\frac{5}{2}$ **(d)** $-\frac{2}{3}$ **(e)** -3.24 **(f)** $4.\overline{16}$ **(g)** $\sqrt{2}$ **(h)** π

6. For each interval below:
      (i) find the endpoints of the interval;
      (ii) find the length of the interval;
      (iii) sketch the interval as a subset of the number line.
(a) $(2,9)$ **(b)** $[-3,2]$ **(c)** $(-8,-1]$ **(d)** $[-4,7)$
(e) $(1,\infty)$ **(f)** $(-\infty,6]$ **(g)** $[-5,\infty)$ **(h)** $(-\infty,-2)$

7. Determine which of these intervals contains the number 3.
(a) $[1,5)$ **(b)** $[-3,0]$ **(c)** $[3,7]$ **(d)** $[-2,3)$ **(e)** $(3,\infty)$ **(f)** $(-\infty,3]$
Sketch each of these intervals as a subset of the real number line.

8. Write each subset of $\Re$ as a specific interval.
(a) A is the set of all negative numbers.
(b) B is the set of all numbers whose square is less than four.
(c) C is the set of areas of all circles whose radius is three or less.
(d) D is the set of areas of all squares whose side is larger than 5 and at most 7.
(e) E is the set of negative numbers whose square is larger than 16.

9. Simplify each union of intervals whenever possible.
(a) $[-3,4)\cup(-2,6)$ **(b)** $[-3,5]\cup(-2,1]$ **(c)** $(-4,1]\cup(1,9]$
(d) $(-6,-4)\cup[-2,3]$ **(e)** $[2,\infty)\cup(-1,3]$ **(f)** $(-\infty,5)\cup[2,6]$
(g) $(-\infty,4)\cup[4,\infty)$ **(h)** $(-\infty,-6]\cup(-5,\infty)$ **(i)** $[-2,4)\cup(-3,5]$

10. Write the solution of each inequality as a union of intervals. Then depict the solution as a subset of the number line.
(a) $x+7\geq 0$ **(b)** $x^2-3x-28<0$ **(c)** $\frac{x+3}{x-2}\leq 0$ **(d)** $\frac{x}{x^2-x-6}>0$

11. Show that $S\subset T$ if and only if $S\cup T=T$.   **12.** Show that $\emptyset\cup S=S$.

13. Determine whether each statement is true or false. Justify your answers.

(a) $3 < \infty$ (b) $5 + \infty = \infty$ (c) $-\infty < -2$ (d) $3 \times \infty = \infty$

(e) $\frac{\infty}{\infty} = 1$ (f) $0 \times \infty = 0$ (g) $\frac{-7}{\infty} = 0$ (h) $\frac{3}{0} = \infty$

(i) $(-\infty)^2 = \infty$ (j) $3 \times \infty - 2 \times \infty = \infty$ (k) $\frac{\infty}{\infty} = \frac{-\infty}{-\infty}$ (l) $\frac{\infty}{-5} = \frac{-\infty}{10}$

14. Determine whether each expression makes sense. If it does, determine its value.

(a) $2 \times \infty + 7 \times \infty - 25$ (b) $9 \times \infty - 4 \times \infty + 18$ (c) $2 \times \infty \times \infty$

(d) $(-\infty) \times (-\infty) \times \infty$ (e) $(2 \times \infty - 3 \times \infty + 6 \times \infty) \times \infty$ (f) $\frac{6 \times \infty - 6 \times \infty}{\infty}$

(g) $\frac{5 \times \infty + 8 \times \infty}{10}$ (h) $(-2 \times \infty - 3 \times \infty - 6 \times \infty) \times \infty$ (i) $\frac{5 - \infty}{5 - \infty}$

15. Let $n \in \Re$ with $n > 0$. Give an intuitive justification for each definition.

(a) $n + -\infty = -\infty$ (b) $\infty + \infty = \infty$ (c) $n \times -\infty = -\infty$

(d) $-\infty \times -\infty = \infty$ (e) $\frac{\infty}{-n} = -\infty$ (f) $\frac{-\infty}{-n} = \infty$

16. Determine whether each rule defines a function. If it is a function, state its domain and a possible codomain.

(a) $f(x) = \sqrt[3]{x}$ for all numbers x.

(b) $g(x) = \sqrt[4]{x}$ for all numbers x.

(c) For each negative number x, $h(x)$ equals all solutions y of the equation $3x + 2y = 1$.

(d) For each positive number x, $j(x)$ equals all solutions y of the equation $x^3 + y^3 = 1$.

(e) For each number x, $k(x)$ equals all solutions y of the equation $x^4 + y^4 = 1$.

(f) For each number x strictly between -2 and 4, define $m(x) = x^4 + x^2 + 1$.

(g) For each integer x in the interval $(-1, 5]$ define $n(x) = \frac{1}{1 + \sqrt{x}}$.

(h) For each number x, $p(x)$ equals all solutions y of the equation $x = |y|$.

17. A car enters the New York State Thruway at mile marker 200 and drives west for three hours at a constant speed of 60 miles per hour. (The mile markers increase in this direction.) Define a function which gives the position of the car t hours after it begins this trip.

18. A penny thrown from the roof of a 240 foot high building falls with height $-16t^2 - 32t + 240$ feet after t seconds. Define a function which gives the height of this penny as it falls.

19. A farmer is constructing a rectangular chicken pen with perimeter 40 meters by placing a fence on the boundary of the pen. He has three choices of fence material:

     (a) prefabricated sections that are 1 meter wide;

     (b) boards that are 10 cm wide;

     (c) plastic fencing that can be cut to any width.

For each type of fence, define an appropriate function of the length of the pen which gives its area. Be sure to specify the domain of each function.

20. Find the range of each function. K is the set of all nonzero numbers.

(a) $f : [1, 2] \to \Re$ by $f(x) = x^3$.   (b) $g : [-1, 3] \to \Re$ by $g(x) = 10 - x$.

(c) $h : [-2, 5] \to \Re$ by $h(x) = x^2$.   (d) $j : [3, 11] \to \Re$ by $j(x) = \sqrt{2x + 3}$.

(e) $k : \Re \to \Re$ by $k(x) = \frac{1}{x^2 + 1}$.   (f) $m : K \to \Re$ by $m(x) = \frac{x}{|x|}$

21. Let $f(x) = \frac{1}{x}$ with domain D. Find the range of the function f in each case.

(a) $D = [1, 5)$ (b) $D = (0, 2]$ (c) $D = (3, \infty)$ (d) $D = (0, \infty)$

22. Determine the domain and range of each of the following functions.

(a) f assigns each number to four times itself.

(b) g assigns the side of each square to its perimeter.

(c) Consider squares that can be cut from a 10 cm $\times$ 30 cm sheet of paper. The function h assigns the length of the side of each of these squares to its perimeter.

23. Find the largest domain for which each formula defines a function.

(a) $f(x) = 3x^5 - 7x^3 + 8$ (b) $g(x) = \frac{x}{x + 5}$ (c) $h(x) = \sqrt{8 - x}$

(d) $j(x) = \sqrt[3]{x^3 + 1}$ **(e)** $k(x) = \frac{x^2-1}{x^2-x-6}$ **(f)** $m(x) = \sqrt{x^2 + 4x - 12}$

(g) $n(x) = \sqrt{x^2 - x + 1}$ **(h)** $p(x) = \sqrt{\frac{x+2}{x-3}}$ **(i)** $q(x) = \frac{x^2-13x+36}{x^4-13x^2+36}$

(j) $r(x) = \left(\frac{(x+1)^2}{x^2-2x-8}\right)^{5/6}$ **(k)** $s(x) = \sqrt{\frac{x^5-32}{x^4-81}}$ **(l)** $t(x) = \sqrt[4]{\frac{x+7}{x^3-4x}}$

24. Sketch each point in the plane.
  (a) $(2,4)$ **(b)** $(3,-1)$ **(c)** $(0,5)$ **(d)** $(-3,0)$ **(e)** $(-4,1)$ **(f)** $(-2,-3)$

25. Sketch the graph of each function.
  (a) f has domain $[1,4]$ and $f(x) = 2x + 3$.
  (b) g has domain $(-3,0)$ and $g(x) = x^2 + 1$.
  (c) h has domain $[-2,5)$ and $h(x) = |x|$.
  (d) j has domain $[1,3]$ and $j(x) = |x - 1| + 2$.
  (e) k has domain $(-2,4]$ and $k(x) = |x + 1| - 1$.
  (f) m has domain $[-4,4]$ and $m(x) = \sqrt{4 - x}$.

26. Determine whether each curve is the graph of a function. If it is, determine its domain and range.

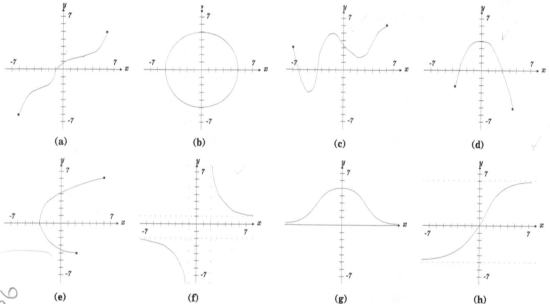

(a)   (b)   (c)   (d)

(e)   (f)   (g)   (h)

27. Determine whether each function is even, odd or neither.
  a) $f(x) = 3x^7 - 5x^3 + 8x$ **b)** $g(x) = 9x^6 - 7x^4 + 2$ **c)** $h(x) = \frac{x^3+1}{x^4+1}$
  d) $i(x) = \frac{6x^4-5x^2+2}{3x^5+8x}$ **e)** $j(x) = \left(x^7 + 4x^3 - 6x\right)^{4/5}$ **f)** $k(x) = \sqrt[3]{x^5 - x}$
  g) $m(x) = \left(x^4 + 2x - 5\right)^{2/3}$ **h)** $n(x) = \left(x^8 - 3x^6 - 9\right)^{7/9}$ **i)** $p(x) = x^5 + \frac{x^2+1}{x^3+x}$

28. Determine if each graph represents an even function, an odd function or neither.

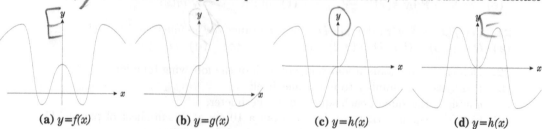

(a) $y=f(x)$ **(b)** $y=g(x)$ **(c)** $y=h(x)$ **(d)** $y=h(x)$

29. Find the formula for the given translation of each function.
  (a) $f(x) = x^2$ translated 3 units up and 6 units right.
  (b) $g(x) = x^3 + 5x + 7$ translated 2 units down and 9 units left.

(c) $h(x) = \frac{x}{5x^2+9}$ translated 8 units up and 4 units left.

(d) $k(x) = \sqrt{x^5 - 2x^2 + 3}$ translated 7 units down and 5 units right.

30. Identify each function as a translation of a parabola, and sketch its graph.
(a) $y = x^2 + 6x - 1$ (b) $y = 4x - x^2$ (c) $y = x^2 - 8x + 18$ (d) $y = 12 - x^2 - 2x$

Challenging Problems

1. (a) How many subsets are there of a set with two elements?

(b) How many subsets are there of a set with three elements?

(c) Find a formula for the number of subsets of a set with n elements.

2. How do we know that $\sqrt{2}$ is not a rational number?

3. The decimal $n = 0.888...$ is a rational number. Write n as a quotient $\frac{p}{q}$ of natural numbers p and q.

4. Consider the operation on an interval $[a, b]$ of dividing the interval into three equal pieces and removing the middle piece. The remaining two pieces are $[a, a + (b - a)/3]$ and $[a + 2(b - a)/3, b]$.

Start with the interval $[0, 1]$ and perform this operation. Perform this operation again on each of the two resulting intervals. Perform this operation again on each of the four resulting intervals. Continue this procedure indefinitely. Let C denote the set of numbers which are never removed. Of course, C contains many rational numbers such as 0, 1, $\frac{1}{3}$ and $\frac{7}{9}$. Describe all the numbers in C.

5. For each number x, let $G(x)$ denote the largest integer which is less than or equal to x. For example $G(3) = 3$, $G(5\frac{2}{9}) = 5$, $G(\sqrt{2}) = 1$ and $G(\pi) = 3$. Determine the graph of the function G.

6. Let F denote the function with domain $[0, 1]$ defined by $F(x) = 1$ if x is a rational number while $F(x) = 2$ if x is an irrational number. Sketch how the graph of F looks to the human eye. Your drawing does not look like the graph of a function. Why is your drawing the graph of a function?

7. Find the largest domain for which the formula $f(x) = \sqrt{\frac{4-x^2}{x-1}}$ defines a function. What is the range of f?

8. Sketch the graph of the function f with domain $\Re$ defined by $f(x) = |1 - x^2|$. What is the range of f?

9. Sketch a curve, with no vertical line segments, such that there is no way to cut the curve into a finite number of pieces with each piece the graph of a function.

1.3 Operations on Functions

A complicated function is often constructed from simple functions using various operations. We can analyze a given complicated function by combining the analyses of its simple components. This viewpoint will be applied in Section 6 to compute limits, in Chapter 2 to compute derivatives and in Chapter 3 to compute integrals. We begin by introducing the algebraic operations of addition, multiplication and division of functions. Then we introduce the idea of a composite function where a function is applied to the number produced by a second one. In the final subsection we form the inverse of a function $y = f(x)$ to try to make x a function of y.

Algebraic Operations

A simple operation on a function f with domain D is to create a new function f_0 which uses the same formula as f but is only defined on some of the elements of D. We call f_0 a *restriction* of f because we restrict the values on which f_0 is defined. This construction will be used in defining quotients, composites and inverses of functions.

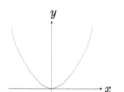

Figure 1.3.1
$y = f(x)$

Definition 1.3.2 *Let* $f : D \to C$ *be a function. If* D_0 *is a subset of* D, *then the restriction of* f *to* D_0 *is the function* $f_0 : D_0 \to C$ *defined by* $f_0(x) = f(x)$ *for* $x \in D_0$. *We write* $f_0 = f \mid D_0$.

Observe how the graphs of $f : D \to C$ and $f_0 = f \mid D_0$ are related. The graph of f lies above and below its domain D on the x–axis. The graph of f_0 is the part of the graph of f which lies above and below the subset D_0 of D.

Figure 1.3.3
$y = f_0(x)$

Examples 1.3.4 (1) The graph of $f(x) = x^2$ with domain $D = \Re$ in Figure 1.3.1 is the entire parabola $y = x^2$ which lies above all the points of the x–axis. The graph of $f_0 = f \mid [0, \infty)$ in Figure 1.3.3 is the right half of this parabola which lies above the right half D_0 of the x–axis.

(2) Consider the graph of $h(x) = |x|$ in Figure 1.3.5 with domain $\Re$. The restriction $h_+ = h \mid [0, \infty)$ is given by $h_+(x) = x$. Its graph in Figure 1.3.6 is the right half of the vee shaped graph of $h(x) = |x|$ which is the right half of the line $y = x$.

Similarly, the restriction of h_- of h to $(-\infty, 0]$ is given by $h_-(x) = -x$. Its graph is the left half of the vee shaped graph of $h(x) = |x|$ which is the left half of the line $y = -x$ depicted in Figure 1.3.8.   □

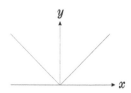

Figure 1.3.5
$h(x) = |x|$

Every operation that is defined for two numbers is also defined for two real valued functions f and g with the same domain. For example, we can add and multiply numbers. Hence we can add and multiply f and g as $f(x) + g(x)$ and $f(x)g(x)$. We can also divide the numbers $\frac{A}{B}$ as long as B is nonzero. Thus we can divide f by g as $\frac{f(x)}{g(x)}$ when $g(x)$ is never zero, i.e. 0 is not in the range of g.

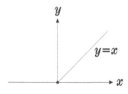

Figure 1.3.6
$h_+(x) = x$

Definition 1.3.7 *Let* f *and* g *be real valued functions with the same domain* D.
(a) *Define the real valued function* $f + g$ *with domain* D *by*

$$(f + g)(x) = f(x) + g(x) .$$

(b) *Define the real valued function* fg *with domain* D *by*

$$(fg)(x) = f(x)g(x) .$$

(c) *Assume that* $0 \notin$ *Range* g. *Define the real valued function* $\frac{f}{g}$ *with domain* D *by*

$$\left(\frac{f}{g}\right)(x) = \frac{f(x)}{g(x)} .$$

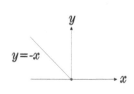

Figure 1.3.8
$h_-(x) = -x$

Let f and g be two real valued functions with domain D. Consider the problem of defining $\frac{f}{g}$ when zero is in the Range of g. Then $\frac{f(x)}{g(x)}$ is only defined for those x where $g(x)$ is nonzero. Define D_0 as the subset of D consisting of those x where $g(x) \neq 0$. Let $f_0 = f \mid D_0$ and $g_0 = g \mid D_0$. Then $g_0(x) = g(x)$ is never zero for $x \in D_0$. Hence the quotient $\frac{f_0}{g_0}$ is defined.

Examples 1.3.9 (1) Let $f(x) = x^2$ and $g(x) = x^2 - 1$. Find $f + g$, fg, $3f - 7g$, $\frac{f}{g}$.

Solution $f(x)$ and $g(x)$ are defined for every number x, so f and g have the same domain $\Re$. Then for $x \in \Re$:

$$\begin{aligned}
(f+g)(x) &= f(x) + g(x) = x^2 + (x^2 - 1) = 2x^2 - 1, \\
(fg)(x) &= f(x)g(x) = x^2(x^2 - 1) = x^4 - x^2 .
\end{aligned}$$

To define $3f - 7g$ think of 3 and 7 as constant functions with domain $\Re$. Then for $x \in \Re$:

$$(3f - 7g)(x) = (3f)(x) + (-7g)(x) = 3f(x) - 7g(x) = 3x^2 - 7(x^2 - 1) = 7 - 4x^2 .$$

Observe that $g(x) = x^2 - 1 = (x + 1)(x - 1) = 0$ when $x = -1$ or $x = +1$. Hence $\frac{f}{g}$ is not defined. Let D_0 be $\Re$ with the numbers -1 and $+1$ removed, i.e. $D_0 = (-\infty, -1) \cup (-1, 1) \cup (1, \infty)$. Let f_0 denote the restriction of f to D_0, and let g_0 denote the restriction of g to D_0. Then $\frac{f_0}{g_0}$ is defined by:

$$\left(\frac{f_0}{g_0} \right)(x) = \frac{f_0(x)}{g_0(x)} = \frac{f(x)}{g(x)} = \frac{x^2}{x^2 - 1} \quad \text{for } x \in D_0.$$

(2) Let $F(x) = \sqrt{x - 1}$ and $G(x) = \sqrt{3 - x}$. Find $F + G$, FG and $\frac{F}{G}$.

Solution Observe that the domain of F is the interval $D_1 = [1, \infty)$ where the expression $x - 1$ under the square root sign is greater than or equal to zero. The domain of G is the interval $D_2 = (-\infty, 3]$ where the expression $3 - x$ under the square root sign is greater than or equal to zero. The functions $F + G$, FG and $\frac{F}{G}$ are not defined because F and G have different domains. Let $D = [1, 3]$ where both F and G are defined. Let $f = F \mid D$, and let $g = G \mid D$. Then $f + g$ and fg are defined with domain $[1, 3]$ by:

$$\begin{aligned}
(f+g)(x) &= f(x) + g(x) = \sqrt{x - 1} + \sqrt{3 - x}, \\
(fg)(x) &= f(x)g(x) = \sqrt{(x - 1)(3 - x)} = \sqrt{4x - x^2 - 3} .
\end{aligned}$$

Since $g(3) = 0$, $\frac{f}{g}$ is not defined on D. However, if we let $D_0 = [1, 3)$, $f_0 = f \mid D_0$ and $g_0 = g \mid D_0$, then $\frac{f_0}{g_0}$ is defined with domain D_0 by:

$$\left(\frac{f_0}{g_0} \right)(x) = \frac{f_0(x)}{g_0(x)} = \frac{F(x)}{G(x)} = \sqrt{\frac{x - 1}{3 - x}} \quad \text{for } x \in D_0. \qquad \square$$

The operations of adding, multiplying and dividing functions have been defined using the usual operations of adding, multiplying and dividing numbers. Hence these operations on functions have the standard properties of arithmetic of numbers:

$$\begin{array}{llll}
f + g &=& g + f, & \qquad fg = gf & \text{(commutative properties)}; \\
f + (g + h) &=& (f + g) + h, & \qquad f(gh) = (fg)h & \text{(associative properties)}; \\
f(g + h) &=& fg + fh & & \text{(distributive property)}.
\end{array}$$

Composite Functions

A powerful method for using the functions f and g to create a new function is to start with a number x in the domain of g and form $g(x)$. Then apply the function f to the number $g(x)$ to obtain $f(g(x))$. This function $f \circ g$ is called the *composite function* of f and g. Figure 1.3.10 gives a pictorial representation of how the composite function $f \circ g$ is defined in two steps. First, we use the function g to take x to $g(x)$. Then we use the function f to take $g(x)$ to $f(g(x))$. Note that the composite function $f \circ g$ is only defined when f is defined on every $g(x)$, i.e. when the domain of f contains the range of g. Equivalently, we require that Range g be a subset of Domain f. Thus $f(g(x))$ is defined whenever $g(x)$ is defined, i.e. the domain of the composite function $f \circ g$ is the domain of g. Also observe that the values $f(g(x))$ of this composite function are also values of f, i.e. the range of $f \circ g$ is a subset of the range of f. Hence we can take the codomain of $f \circ g$ to be the codomain of f.

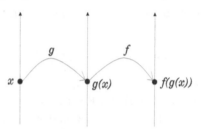

Figure 1.3.10 The Composite Function $f \circ g$

Definition 1.3.11 *Let f and g be two functions such that Range g is a subset of Domain f. Define the composite function $f \circ g$ with*

$$\text{Domain } f \circ g = \text{Domain } g \quad and \quad \text{Codomain } f \circ g = \text{Codomain } f$$
$$by \quad (f \circ g)(x) \quad = \quad f(g(x)).$$

Write $y = g(x)$ and $z = f(y)$. To obtain the formula for $(f \circ g)(x) = f(g(x))$, substitute the expression $g(x)$ for the variable y in the formula $f(y)$. This substitution is defined by our assumption that the range of g is a subset of the domain of f.

Examples 1.3.12 (1) Consider the functions $f(u) = \frac{1}{u^2+1}$ with domain $[0, \infty)$ and $g(x) = x^4$ with domain $[-1, 1]$.

(a) Show that $f \circ g$ is defined, and find a formula for $(f \circ g)(x)$.
(b) Show that $g \circ f$ is defined, and find a formula for $(g \circ f)(u)$.

Solution   (a) The range of g is $[0, 1]$ which is a subset of the domain $[0, \infty)$ of f. Hence $f \circ g$ is defined. The domain of $f \circ g$ is the domain of g which is $[-1, 1]$. Then

$$(f \circ g)(x) = f(g(x)) = f(x^4) = \frac{1}{(x^4)^2 + 1} = \frac{1}{x^8 + 1}.$$

(b) The range of f is $(0, 1]$ which is a subset of the domain $[-1, 1]$ of g. Hence $g \circ f$ is defined. The domain of $g \circ f$ is the domain of f which is $[0, \infty)$. Then

$$(g \circ f)(u) = g(f(u)) = g\left(\frac{1}{u^2+1}\right) = \left(\frac{1}{u^2+1}\right)^4 = \frac{1}{(u^2+1)^4}.$$

(2) Consider the two functions $F(s) = 5s^2 - 1$ with domain $\Re$ and $G(t) = \sqrt{4t-1}$ with domain $\left[\frac{1}{4}, \infty\right)$.

(a) Show that $F \circ G$ is defined, and find a formula for $(F \circ G)(t)$.
(b) Show that $G \circ F$ is not defined.

Solution   (a) The range of G is a subset of $\Re$ which is the domain of F. Hence $F \circ G$ is defined. The domain of $F \circ G$ is the domain of G which is $\left[\frac{1}{4}, \infty\right)$. Then

$$(F \circ G)(t) = F(G(t)) = F(\sqrt{4t-1}) = 5\sqrt{4t-1}^2 - 1 = 20t - 6.$$

(b) The range of F equals $[-1, \infty)$ which is not a subset of the domain $\left[\frac{1}{4}, \infty\right)$ of G. Therefore, $G \circ F$ is not defined. $\qquad\qquad \square$

The preceding example raises the question of how to define $f \circ g$ when the range of g is not contained in the domain of f. The problem occurs for those x in the domain D of g for which $f(g(x))$ is not defined. Hence if we eliminate those x from the domain of g, then we will be able to define this composite function. That is, let g_0 be the restriction of g to the set D_0 of those x in D where $f(g(x))$ is defined. Then $f \circ g_0$ is defined with domain D_0.

Examples 1.3.13 (1) Let $f(x) = \frac{1}{x}$ and $g(x) = x^2 - 4$. Define the composite of f with a restriction of g.

Solution Note that f is defined for every nonzero number, i.e. the domain of f is $(-\infty, 0) \cup (0, \infty)$. The function g is defined for every number and the values of $g(x)$ are all numbers greater than or equal to -4. That is, Domain $g = \Re$ and Range $g = [-4, \infty)$. Since the range g contains zero which is not in the domain of f, the composite function $f \circ g$ is not defined. To define such a composite we form a restriction of g to eliminate those numbers where

$$0 = g(x) = x^2 - 4 = (x + 2)(x - 2),$$

i.e. we remove -2 and $+2$ from the domain $\Re$ of g. Thus let

$$D_0 = (-\infty, -2) \cup (-2, 2) \cup (2, \infty),$$

and let $g_0 = g \mid D_0$. Then $g_0(x)$ is never zero for $x \in D_0$ and $f \circ g_0$ is defined by

$$(f \circ g_0)(x) = f(g_0(x)) = f(g(x)) = f(x^2 - 4) = \frac{1}{x^2 - 4} \quad \text{for } x \in D_0.$$

(2) In Example 1.3.12 (2) we studied the functions $F(s) = 5s^2 - 1$ with domain $\Re$ and $G(t) = \sqrt{4t - 1}$ with domain $\left[\frac{1}{4}, \infty\right)$. Define the composite of G with a restriction of F.

Solution We showed in Example 1.3.12 (2) that $G \circ F$ is not defined because Range $F = [-1, \infty)$ is not contained in Domain $G = \left[\frac{1}{4}, \infty\right)$. The domain of the polynomial F is $\Re$. Let D_0 be the subset of those $s \in \Re$ where $F(s)$ is in the domain of G, i.e. where $F(s) = 5s^2 - 1 \geq \frac{1}{4}$. That is, we require

$$0 \geq \frac{5}{4} - 5s^2 = 5\left(\frac{1}{4} - s^2\right) = 5\left(\frac{1}{2} + s\right)\left(\frac{1}{2} - s\right)$$

which occurs when $s \leq -\frac{1}{2}$ or $s \geq \frac{1}{2}$. Thus let $D_0 = \left(-\infty, -\frac{1}{2}\right] \cup \left[\frac{1}{2}, \infty\right)$ and $F_0 = F \mid D_0$. Then $G \circ F_0$ is defined with domain D_0, and for $s \in D_0$:

$$(G \circ F_0)(s) = G(F_0(s)) = G(5s^2 - 1) = \sqrt{4(5s^2 - 1) - 1} = \sqrt{20s^2 - 5}. \quad \square$$

Inverse Functions

When $y = f(x)$ is a function with domain D and range R, f takes each $x \in D$ to $y \in R$. We want to define a function f^{-1} which reverses this process: f^{-1} takes each $y \in R$ to $x \in D$ as in the left diagram of Figure 1.3.26. f^{-1} is called the *inverse* of f. Unfortunately, the following example shows that the f^{-1} defined by this process is not always a function.

Motivating Example 1.3.14 A car is traveling on Highway 401 between Windsor and Toronto for three hours. Let $y = f(t)$ denote the position of the car on the highway at time t hours. Then $f : [0, 3] \to \Re$ is a function. Say we know that the closest the car is to Windsor on this trip is 50 km and the farthest it is from Windsor on this trip is 200 km. Then the range of f is $[50, 200]$. Define f^{-1} which takes each point $y \in [50, 200]$ on the highway to the time $t \in [0, 3]$ where $f(t) = y$. That is, every point on the highway traveled by the car is assigned by f^{-1} to the time when the car is at that position. Recall that a function must assign exactly one value to each element of its domain. Let's examine whether this criterion is met in each of the following scenarios.

Say the car starts by heading east and passes through the 100 km marker at 1 hr. Then it turns around, heading west, and passes through the 100 km marker again at 2 hrs. In this case $f(1) = 100$ and $f(2) = 100$. Hence $f^{-1}(100) = 1$ and $f^{-1}(100) = 2$, so f^{-1} is not a function. See the left diagram in Figure 1.3.15.

On the other hand, if the car never turns around during its trip, then it never passes through the same point on the highway twice. In this case $t = f^{-1}(y)$ will be the unique time t when the car is at km marker y on the highway. See the right diagram in Figure 1.3.15.

We summarize our conclusions. f^{-1} is a function if for each point y on the highway, the only way that the car is at position y at times t_1 and t_2 is when t_1 equals t_2. That is, f^{-1} is a function if $f(t_1) = f(t_2)$ implies that $t_1 = t_2$. □

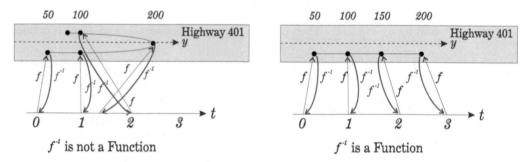

Figure 1.3.15 Two Trips on Highway 401

We give the name *one–to–one* to the condition we discovered in the preceding example that determines whether f^{-1} is a function.

Definition 1.3.16 *A function $f : D \to C$ is called one–to–one if $x_1 \in D$ and $x_2 \in D$ with $f(x_1) = f(x_2)$ implies $x_1 = x_2$.*

We give a formal proof to validate the observation made in the example above: a function f is one–to–one if and only if f^{-1} is a function.

Proposition 1.3.17 *Let $f : D \to C$ be a function with range R. Define $f^{-1}(y) = x$ when $y \in R$ and $y = f(x)$. Then f^{-1} is a function with domain R and range D if and only if f is one–to–one. In this case, we call f^{-1} the inverse function of f.*

Proof Assume that f^{-1} is a function. Let $x_1 \in D$ and $x_2 \in D$ with $y = f(x_1) = f(x_2)$. Then $f^{-1}(y) = x_1$ and $f^{-1}(y) = x_2$. Since the function f^{-1} must assign a unique value to y, it follows that $x_1 = x_2$.

Assume that f is one–to–one. f^{-1} is a function with domain R if f^{-1} assigns each $y \in R$ to a unique number. If y is in the range R of f, then $y = f(x)$ for some $x \in D$, i.e. $f^{-1}(y) = x$. Say $y \in R$ with $f^{-1}(y) = x_1$ and $f^{-1}(y) = x_2$. Then $f(x_1) = y$ and $f(x_2) = y$. Since f is one–to–one, $x_1 = x_2$. Therefore, f^{-1} is a function.

When f^{-1} is a function, its domain consists of all y in C of the form $f(x)$, i.e. the domain of f^{-1} is the range R of f. The range of f^{-1} is the set of all $x = f^{-1}(y)$ in D. However, each $x \in D$ equals $f^{-1}(f(x))$. Hence the range of f^{-1} is D. $\qquad\square$

Proposition 1.3.17 makes two statements. First, if the function f is not one–to–one, then f^{-1} is not a function. Second, if f is one–to–one, then f^{-1} is a function with

$$\text{Domain } f^{-1} \ = \ \text{Range } f \quad \text{and} \quad \text{Range } f^{-1} \ = \ \text{Domain } f. \qquad (1.3.1)$$

We illustrate this proposition with the following examples.

Examples 1.3.18 Determine whether each function has an inverse function. If it does, determine its domain, range and formula.

(1) Let $f : [0,4] \to \Re$ by $y = f(x) = x + 2$.

  Solution

  - We check whether f is one–to–one. If $f(x_1) = f(x_2)$, then $x_1 + 2 = x_2 + 2$ and $x_1 = x_2$. Hence f is one–to–one, and f^{-1} is a function by Prop. 1.3.17.

  - The domain of f^{-1} is the range of f which is $[2,6]$.

  - The range of f^{-1} is the domain of f which is $[0,4]$.

  - By definition, if $y = f(x) = x + 2$, then $f^{-1}(y) = x = y - 2$. In words, the operation of subtracting 2 undoes the operation of adding 2. Note that we can use any name for the variable that gives the formula for the function f^{-1}. That is f^{-1} is the function which takes a number in its domain and subtracts two from that number. When we write $f^{-1}(y) = y - 2$ we are calling the initial number y. However, we usually prefer to call the initial number (the independent variable) x and write $f^{-1}(x) = x - 2$.

  - Note the symbol f^{-1} does not have the same meaning as the reciprocal $\frac{1}{f}$. In this example $\left(\frac{1}{f}\right)(x) = \frac{1}{x+2}$ while $f^{-1}(x) = x - 2$.

(2) Let $g : \Re \to \Re$ by $y = g(x) = |x|$.

  Solution We check whether g is one–to–one. If $g(x_1) = g(x_2)$, then $|x_1| = |x_2|$ and $x_1 = \pm x_2$. For example, $g(-3) = |-3| = 3 = |3| = g(3)$. Thus g is not one–to–one, and by Proposition 1.3.17 g^{-1} is not a function.

(3) Let $y = h(x) = \frac{2x-1}{x-1}$ with domain all numbers other than 1.

  Solution

  - We determine whether h is one–to–one. If $h(x_1) = h(x_2)$, then

$$\begin{array}{rcll}
\frac{2x_1-1}{x_1-1} & = & \frac{2x_2-1}{x_2-1} & \\
(2x_1 - 1)(x_2 - 1) & = & (2x_2 - 1)(x_1 - 1) & [\text{cross} - \text{multiplying}] \\
2x_1x_2 - 2x_1 - x_2 + 1 & = & 2x_1x_2 - 2x_2 - x_1 + 1 & [\text{multiplying out}] \\
-2x_1 - x_2 & = & -2x_2 - x_1 & [\text{cancelling } 2x_1x_2 + 1] \\
x_2 & = & x_1. & [\text{adding } 2x_1 + 2x_2]
\end{array}$$

  Thus h is one–to–one, and h^{-1} is a function by Proposition 1.3.17.

  - The domain of h^{-1} is the range of h. Observe that

$$h(x) = 2 + \frac{1}{x - 1} \ .$$

  The values of $\frac{1}{x-1}$ are all nonzero numbers. Hence the values of $h(x)$ are all numbers other than 2, i.e. Domain h^{-1} = Range h = $(-\infty, 2) \cup (2, \infty)$.

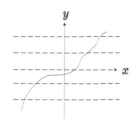

Figure 1.3.19
$y = F(x)$

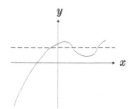

Figure 1.3.20
$y = G(x)$

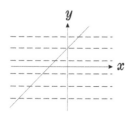

Figure 1.3.21
$y = f(x) = x + 2$

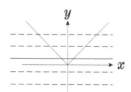

Figure 1.3.23
$y = g(x) = |x|$

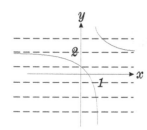

Figure 1.3.25
$y = h(x) = \frac{2x-1}{x-1}$

- Range h^{-1} = Domain h = $(-\infty, 1) \cup (1, \infty)$.

- By definition, $h^{-1}(y) = x$ if $y = h(x) = \frac{2x-1}{x-1}$. We solve this equation for x:

$$
\begin{aligned}
y &= \tfrac{2x-1}{x-1} \\
yx - y &= 2x - 1 && [\text{cross} - \text{multiplying}] \\
x(y - 2) &= y - 1 && [\text{rearranging the equation}] \\
h^{-1}(y) = x &= \tfrac{y-1}{y-2} && [\text{dividing by } y - 2]
\end{aligned}
$$

We usually prefer to use the letter x for the independent variable. In this notation, $h^{-1}(x) = \frac{x-1}{x-2}$.

(4) Let $k : \Re \to \Re$ by $y = k(x) = x^2$.

Solution We check whether k is one–to–one. If $k(x_1) = k(x_2)$, then $x_1^2 = x_2^2$ and $x_1 = \pm x_2$. For example, $k(-5) = (-5)^2 = 25 = 5^2 = k(5)$. Thus k is not one–to–one. By Proposition 1.3.17, k^{-1} is not a function. □

We can determine whether f is one–to–one from the graph of f. If there are two numbers x_1, x_2 in the domain of f such that $f(x_1) = f(x_2) = c$ then the points (x_1, c) and (x_2, c) lie on the graph of f. That is, the horizontal line $y = c$ intersects the graph of f in at least two points. Note that if c is not in the range of f then the line $y = c$ will not intersect the graph of f. Thus, we have established the following result.

Proposition 1.3.22 *A function f is one–to–one if and only if every horizontal line which intersects the graph of f crosses the graph of f at exactly one point.*

We apply the criterion of Proposition 1.3.22 to the graphs of the functions in Figures 1.3.19 and 1.3.20. F is one–to–one because every horizontal line intersects its graph in exactly one point. On the other hand, G not one–to–one because the dotted horizontal line intersects its graph in three points. We apply this criterion to the functions of Examples 1.3.18.

Examples 1.3.24 Use the graph of each function to determine if it is one–to–one.

(1) Let $f : [0, 4] \to \Re$ by $y = f(x) = x + 2$.

Solution In Figure 1.3.21, every horizontal line intersects the graph of $y = f(x) = x + 2$ in exactly one point. By Proposition 1.3.22, f is one–to–one.

(2) Let $g : \Re \to \Re$ by $y = g(x) = |x|$.

Solution In Figure 1.3.23, every line above the x–axis intersects the graph of $y = g(x) = |x|$ in two points. By Proposition 1.3.22, g is not one–to–one.

(3) Let h be the function

$$y = h(x) = \frac{2x-1}{x-1}.$$

Solution Note $y = h(x) = 2 + \frac{1}{x-1}$. Hence the graph of h in Figure 1.3.25 is the hyperbola $y = \frac{1}{x}$ translated one unit to the right and two units upwards. The line $y = 2$ does not intersect this graph while every other horizontal line intersects this graph in exactly one point. By Proposition 1.3.22, h is one–to–one.

(4) Let $k : \Re \to \Re$ by $y = k(x) = x^2$.

Solution In Figure 1.3.27, every line above the x–axis intersects the graph of $y = k(x) = x^2$ in two points. By Proposition 1.3.22, k is not one–to–one.   □

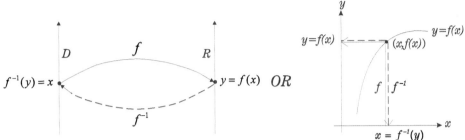

Figure 1.3.26 Inverse Function

Let f be a one–to–one function. The relationship between f and f^{-1} is:

$$y = f(x) \text{ if and only if } x = f^{-1}(y). \qquad (1.3.2)$$

In other words, f takes x to y if and only if f^{-1} takes y to x as depicted in the left diagram of Figure 1.3.26. Alternatively, we illustrate this relationship on the graph of f in the right diagram of Figure 1.3.26. Let D be the domain of f, and let R be its range. It follows from (1.3.2) or from either of the diagrams in Figure 1.3.26 that:

$$
\begin{array}{llll}
(f^{-1} \circ f)(x) & = & f^{-1}(f(x)) = f^{-1}(y) = x & \text{for all } x \in D, \\
(f \circ f^{-1})(y) & = & f(f^{-1}(y)) = f(x) = y & \text{for all } y \in R.
\end{array}
\qquad (1.3.3)
$$

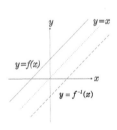

Figure 1.3.27
$y = k(x) = x^2$

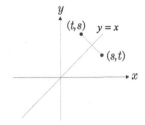

Reflection in the Line $y=x$

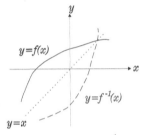

The Graph of f^{-1}

Figure 1.3.28 Constructing the Graph of f^{-1}

If f is a one–to–one function, we can derive the graph of f^{-1} from the graph of f. If $f(s) = t$, then the point (s,t) lies on the graph of f. Since $f^{-1}(t) = s$, the point (t,s) lies on the graph of f^{-1}. Thus, we obtain the graph of f^{-1} by interchanging the coordinates of each point on the graph of f. The operation of sending (s,t) to (t,s) is described geometrically in the left diagram of Figure 1.3.28 as reflection in the line $y = x$. Therefore:

the graph of f^{-1} is the reflection of the graph of f in the line $y = x$.

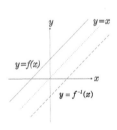

Figure 1.3.29
$y = f(x) = x + 2$

We apply this principle in the right diagram of Figure 1.3.28 to obtain the dotted graph of f^{-1} from the graph of f. We construct the graphs of the inverse functions of $f(x) = x + 2$ and $h(x) = \frac{2x-1}{x-1}$ by this procedure in Figures 1.3.29 and 1.3.30.

If a function f is not one–to–one, we can shrink the domain of f to define a restriction f_0 of f which is one–to–one. Then f_0 will have an inverse. There may be many ways to shrink the domain of f and select one–to–one restrictions f_0. However, consideration of the graph of f may suggest a natural domain of a one–to–one restriction. We illustrate this procedure with two examples.

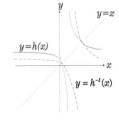

Figure 1.3.30
$y = h(x) = \frac{2x-1}{x-1}$

Examples 1.3.31 Show that each of the following functions is not one–to–one. Select two one–to–one restrictions of each function and describe their inverse functions.

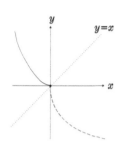

Figure 1.3.32
$f_1(x) = x^2$

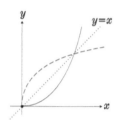

Figure 1.3.33
$f_2(x) = x^2$

$-x - 3 = x + 3$

$y = -x - 3$

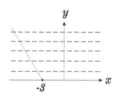

Figure 1.3.34
$y = g(x) = |x + 3|$

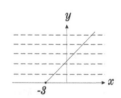

Figure 1.3.35
$y = g_1(x)$

(another graph)

Figure 1.3.36
$y = g_2(x)$

(1)   Let $f(x) = x^2$ with domain the set of all real numbers $\Re$.

> **Solution** As observed above, f is not one–to–one. For example, $f(-3) = 9 = f(3)$. To select a one–to–one restriction of f, consider the graph of f in Figure 1.3.27. The fact that f is not one–to–one is depicted in this graph by the phenomenon that every horizontal line $y = c$, for $c > 0$, intersects the graph in two points: one to the left of the y–axis and one to the right of the y–axis. Thus, there are two natural restrictions of f.
>
> Define $D_1 = (-\infty, 0]$. The graph of $f_1 = f \mid D_1$ is the left half of the graph of f depicted as the solid curve in Figure 1.3.32. Since no horizontal line intersects the graph of f_1 in more than one point, f_1 is one–to–one, and f_1^{-1} is defined. The domain of f_1^{-1} is the range of f_1 which is $[0, \infty)$. Then $x = f_1^{-1}(y)$ if $y = f_1(x) = f(x) = x^2$ with $x \le 0$, i.e. $f_1^{-1}(y) = x = -\sqrt{y}$ or
>
> $$f_1^{-1}(x) = -\sqrt{x}\,.$$
>
> The graph of f_1^{-1} is the reflection of the graph of f_1 about the line $y = x$. It is shown as the dotted curve in Figure 1.3.32.
>
> Alternatively define $D_2 = [0, \infty)$. The graph of $f_2 = f \mid D_2$ is the right half of the graph of f depicted as the solid curve in Figure 1.3.33. Since no horizontal line intersects the graph of f_2 in more than one point, f_2 is one–to–one and f_2^{-1} is defined. The domain of f_2^{-1} is the range of f_2 which is $[0, \infty)$. Then $x = f_2^{-1}(y)$ if $y = f_2(x) = f(x) = x^2$ with $x \ge 0$, i.e. $f_2^{-1}(y) = x = +\sqrt{y}$ or
>
> $$f_2^{-1}(x) = +\sqrt{x}.$$
>
> The graph of f_2^{-1} is the reflection of the graph of f_2 about the line $y = x$. It is shown as the dotted curve in Figure 1.3.33.

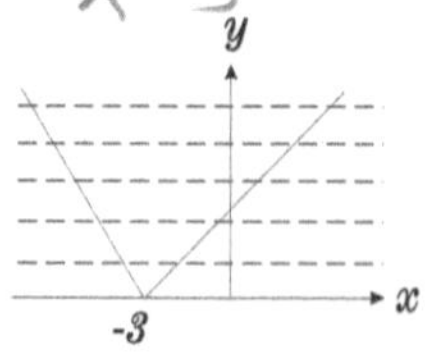

(2)   Let $g(x) = |x + 3|$ with domain the set of all real numbers $\Re$.

> **Solution** Since $|2| = |-2| = 2$, we have $g(-1) = g(-5) = 2$ and g is not one–to–one. The graph of g in Figure 1.3.34 is the graph of the absolute value function shifted three units to the left. We see that g is not one–to–one because horizontal line above the x–axis intersects the graph of g in two points.
>
> Let $D_1 = (-\infty, -3]$. The graph of $g_1 = g \mid D_1$ is the left half of the graph of g. See Figure 1.3.35. Note that g_1 is one–to–one because every horizontal line intersects the graph of g_1 in at most one point. Thus g_1^{-1} is defined. The domain of g_1^{-1} is the range of g_1 which is $[0, \infty)$. For $x \le -3$, we have $x = g_1^{-1}(y)$ if $y = g_1(x) = g(x) = |x + 3| = -(x + 3)$, i.e. $g_1^{-1}(y) = x = 3 - y$ or
>
> $$g_1^{-1}(x) = 3 - x\,.$$
>
> Alternatively, let $D_2 = [-3, \infty)$. The graph of $g_2 = g \mid D_2$ is the right half of the graph of g. See Figure 1.3.36. Observe that g_2 is one–to–one because every horizontal line intersects the graph of g_2 in at most one point. Thus g_2^{-1} is defined. The domain of g_2^{-1} is the range of g_2 which is $[0, \infty)$. For $x \ge -3$, $x = g_2^{-1}(y)$ if $y = g_2(x) = g(x) = |x + 3| = x + 3$, i.e. $g_2^{-1}(y) = x = y - 3$ or
>
> $$g_2^{-1}(x) = x - 3\,. \qquad \square$$

We will use this procedure of shrinking the domain to define inverse trigonometric functions in the next section.

Summary

The reader should be able to add, multiply and divide real valued functions. If f/g is not defined, she should know how to replace g by a restriction g_0 so that f/g_0 is defined. She should also understand the meaning of the composite function $f \circ g$. If $f \circ g$ is not defined, she should know how to replace g by a restriction g_1 so that $f \circ g_1$ is defined.

The reader should be able to test a function f, using algebra or geometry, to see if it is one–to–one, thereby determining whether the inverse function f^{-1} exists. If f^{-1} does exist, she should know how to obtain its formula from the formula of f. In addition, she should be able to construct the graph of f^{-1} from the graph of f. If f is not one–to–one, then the reader should know how to select a restriction f_0 of f which is one–to–one and has an inverse.

Basic Exercises

1. Let f be the function with domain $\Re$ defined by $f(x) = 1 - x^2$. Graph the restriction of f to each of the following domains.
   (a) $[0, \infty)$ (b) $(-\infty, 0]$ (c) $[-1, 3]$ (d) $(2, 5]$.

2. Let g be the function with domain $\Re$ defined by $g(x) = |4 - x|$. Graph the restriction of g to each of the following domains.
   (a) $(0, \infty)$ (b) $(-\infty, 0]$ (c) $(4, \infty]$ (d) $(-\infty, 4]$.

3. Let h be the function with domain $\Re$ defined by $h(x) = \left\{ \begin{array}{ll} 2x + 1 & \text{for } x \geq 1 \\ 1 - 2x & \text{for } x < 1 \end{array} \right\}$.
   Graph the restriction of h to each of the following domains.
   (a) $(0, \infty)$ (b) $(-\infty, 0]$ (c) $[1, \infty]$ (d) $(-\infty, 1]$.

4. Let k be the function with domain $[0, 2]$ with $k(x)$ defined as the largest integer which is less than or equal to x.
   (a) Graph the function k.   (b) Graph the restriction of k to $[0, 2)$.
   (c) Graph the restriction of k to $[0, 1)$.   (d) Graph the restriction of k to $[0, 1]$.

5. A ball is thrown upwards from the edge of the roof of a 192 foot tall building and falls to the ground. The height of the ball from the ground $s(t)$ at time t seconds is given by $s(t) = -16t^2 + 64t + 192$ feet.
   (a) Find the domain of s.
   (b) Describe the function, in terms of s, which gives the height of the ball at time t seconds while the ball is rising.
   (c) Describe the function, in terms of s, which gives the height of the ball at time t seconds while the ball is falling.

6. A bar of iron at temperature 1000°C is placed in a lake at temperature 20°C. The temperature $T(t)$ of this bar after t minutes is given by: $T(t) = 2 + 980 \times 10^{-.02t}$ °C.
   (a) Find the domain of T.
   (b) Describe the function which gives the temperature of the bar during the first hour in terms of T.
   (c) Describe the function which gives the temperature of the bar during the first day in terms of T.
   (d) Describe the function which gives the temperature of the bar until it cools to 100°C in terms of T.

7. For each pair of functions f, g below, find a formula and domain for each of the following functions.
   (i) $f + g$ (ii) fg (iii) f/g (iv) $3f - 2g$ (v) g^3 (vi) $\frac{f+g}{fg}$

(a) Let f, g have domain $[1, \infty)$ with $f(x) = x^2$ and $g(x) = 5x + 3$.

(b) Let f, g have domain $\Re$ with $f(x) = \sqrt{x^2 + 1}$ and $g(x) = x^2 + 1$.

(c) Let f, g have domain $[2, \infty)$ with $f(x) = \frac{3x+2}{x^2-1}$ and $g(x) = 2 - 2x$.

8. For each pair of functions f, g, define restrictions f_0 of f and g_0 of g for which f_0/g_0 is defined. Then determine the formula for f_0/g_0.

(a) $f(x) = x^2 - x - 6$ and $g(x) = x^2 + 3x - 10$ both with domain $\Re$.

(b) $f(x) = x^3 - 1$ and $g(x) = x^3 + 1$ both with domain $\Re$.

(c) $f(x) = \sqrt{x^2 - 4}$ and $g(x) = \sqrt{x^2 - 8x + 16}$ both with domain $[2, \infty)$.

(d) $f(x) = x$ and $g(x) = |x|$ both with domain $\Re$.

(e) $f(x) = 1 - x^2$ and $g(x) = \sqrt{1 - x^2}$ both with domain $[0, 1]$.

9. When a constant force of F units acts on an object which is moving along the coordinate line for a distance of s units, the work W done by this force on this object is given by $W = Fs$. Consider the situation where an object starts at the origin and moves right for three units under a constant force of six units, then moves right for five units under a constant force of two units and then moves right for four units under a constant force of three units.

(a) Describe the force $F(s)$ on this object as a function of its position s.

(b) Let $W(s)$ be the work done by this force as a function of the object's position s. Write $W(s)$ as a product of $F(s)$ and another function.

(c) Consider the situation where a second force G also acts on this object with a constant value of four units for the first half of its journey and a constant value of six units for the second half of its journey. Describe $G(s)$ as a function of the position s of the object. Then describe the work $V(s)$ done by these forces in the form $V(s) = p(s)F(s) + q(s)G(s)$.

10. Let C be the set of students in a class. Let F_1, F_2, F_3 be real valued functions which assigns each student an integer grade between 0 and 10 on each of three quizzes. Use F_1, F_2 and F_3 to describe the function G which assigns each student his/her average grade on these quizzes.

11. Let f, g and h be real valued functions with the same domain. Verify each of these properties of function arithmetic.

  (a) $f + g = g + f$ (b) $fg = gf$ (c) $f + (g + h) = (f + g) + h$
  (d) $f(gh) = (fg)h$ (e) $f(g + h) = fg + fh$

12. Let f be a real valued function. For k a natural number, define f^k as the product of f with itself k times.

  (a) If $f(x) = x^2$, find $f^5(x)$.   (b) If $g(x) = 2x - 3$, find $g^3(x)$.
  (c) If $h(x) = x + \frac{1}{x}$, find $h^4(x)$.   (d) If $k(x) = \frac{\sqrt{x}+1}{\sqrt{x}-1}$, find $k^2(x)$

13. Use the notation of the previous exercise. Let f, g be real valued functions with the same domain, and let h, k be natural numbers. Verify each of these statements.

  (a) $(f^h)^k = f^{hk}$ (b) $(f^h)(f^k) = f^{h+k}$
  (c) $(fg)^k = f^k g^k$ (d) $(f + g)^2 = f^2 + 2fg + g^2$

14. For each pair of functions f, g, explain why $f \circ g$ is defined. Then find a formula for $(f \circ g)(x)$.

  (a) $f(x) = 3x + 4$ and $g(x) = 2 - 5x$.   (b) $f(x) = x^2 + 1$ and $g(x) = x^2 - 1$.
  (c) $f(x) = \frac{1}{x^2 + x + 1}$ and $g(x) = x^2 - 4$.   (d) $f(x) = \sqrt{x^2 + 1}$ and $g(x) = 5 - x^2$.

  (e) $f(x) = \sqrt{\frac{2-x}{x+3}}$ and $g(x) = \frac{1}{x^2+1}$.   (f) $f(x) = \sqrt{1 + x}$ and $g(x) = |x|$.

15. For each pair of functions f, g, determine whether $f \circ g$ is defined. If it is defined, find a formula for $f \circ g$. If not, define a restriction g_0 of g such that $f \circ g_0$ is defined, and find a formula for $f \circ g_0$.

(a) f and g have domain $\Re$ with $f(x) = x^2 + 3x - 4$ and $g(x) = 2x - 1$.

(b) f and g have domain $[0, \infty)$ with $f(x) = \sqrt{x}$ and $g(x) = 3 - 2x$.
(c) f has domain $\Re$ with $f(x) = \frac{x}{x^2+1}$, and g has domain $[1, \infty)$ with $g(x) = \sqrt{x - 1}$.
(d) $f(x) = \frac{x+2}{x^2-16}$ has domain $[5, \infty)$, and $g(x) = \sqrt{x + 1}$ has domain $[-1, \infty)$.
(e) $f(x) = \left(x^2 - x - 1\right)^{2/3}$ has domain $[-1, \infty)$, and $g(x) = x^{3/4}$ has domain $[0, \infty)$.

16. Let $f(s)$ be the function which adds three units to the length s of the side of a square. Let $g(s)$ be the function which doubles the length s of the side of a square. Let $A(s)$ be the function which gives the area of a square of side s. Express each of the following functions as a composite function using the functions f, g and A:
(a) the area $B(s)$ of a square whose side has length three plus twice the length s of a given square;
(b) the area $C(s)$ of a square whose side has length six plus twice the length s of a given square;
(c) the area $D(s)$ of a square whose side is eight times as long as the length s of a given square.

17. Assume all the composite functions below are defined. Decide whether each statement is true or false. If it is true, verify it. If it is false give an example where it fails.
    (a) $f \circ (g_1 + g_2) = f \circ g_1 + f \circ g_2$ **(b)** $(f_1 + f_2) \circ g = f_1 \circ g + f_2 \circ g$
    (c) $f \circ (g \circ h) = (f \circ g) \circ h$ **(d)** $f \circ (g_1 g_2) = (f \circ g_1)(f \circ g_2)$
    (e) $(f_1 f_2) \circ g = (f_1 \circ g)(f_2 \circ g)$ **(f)** $f \circ \frac{g}{h} = \frac{f \circ g}{h}$

18. Determine a domain D_0 where $f \circ f$ is defined, and find the formula for $(f \circ f)(x)$.
    (a) $f(x) = x^2 + 1$ **(b)** $f(x) = \sqrt{x}$ **(c)** $f(x) = \frac{1}{x}$ **(d)** $f(x) = \frac{x-1}{x+1}$

19. Let h and k be natural numbers. Define $f^{(k)}$ as the composite of f with itself k times. Assume that all the composite functions below are defined. Decide whether each statement is true or false. If it is true, verify it. If it is false give an example where it fails.
    (a) $(f^{(h)})^{(k)} = f^{(hk)}$ **(b)** $f^{(h)} \circ f^{(k)} = f^{(h+k)}$
    (c) $(f \circ g)^{(k)} = f^{(k)} \circ g^{(k)}$ **(d)** $(f + g)^{(2)} = f^{(2)} + 2f \circ g + g^{(2)}$

20. Let $f(x)$ be the function that adds half a percent of interest to the balance x of savings accounts with 6% interest compounded monthly. Use the notation of the preceding exercise to describe the balance of the savings account
    (a) in three months;   **(b)** after one year.

21. Determine whether each function is one–to–one.
    (a) $f(x) = x^{1/3}$ **(b)** $g(x) = (x - 1)^{1/4}$ **(c)** $h(x) = (x + 4)^4$
    (d) $k(x) = \frac{2x+5}{3x-1}$ **(e)** $m(x) = \frac{x^2}{x^2+1}$ **(f)** $n(x) = \sqrt{1 - x^2}$

22. **(i)** Give conditions on the domain of each function which makes it one–to–one.
(ii) Under these conditions, describe the inverse function.
    (a) $H(t)$ is the altitude of a plane t seconds after takeoff.
    (b) $L(t)$ is the length of a side of an ice cube at time t.
    (c) $V(t)$ is the number of cubic meters of grain in a silo at time t.
    (d) $S(t)$ is the speed of a car at time t.
    (e) $A(x)$ is the area of a circle of radius x.
    (f) $W(x)$ is the volume of cube of side x.
    (g) $P(x)$ is the number of people sick with the flu on day x.
    (h) $Q(n)$ is the probability that when a coin is tossed n times, heads appears
         upwards on every toss.

23. You may assume that each function is one–to–one. Find the domain and range of its inverse function.

(a) $f(x) = 5x - 3$ (b) $g(x) = 4 - x^2$ with domain $[0, \infty)$
(c) $h(x) = \frac{5x+3}{x+4}$ (d) $k(x) = \sqrt{x^4 + x^2 + 1}$ with domain $[0, \infty)$
(e) $m(x) = \sqrt{x^3 + 1}$ (f) $n(x) = |3x + 7|$ with domain $[-1, \infty)$

24. Find the formula for the inverse function of each function of Exercise 23.

25. For each function below:
  (i) show that the inverse function exists;
  (ii) find the domain of the inverse function;
  (iii) find a formula for the inverse function.
(a) Let f have domain $\Re$ with $f(x) = 7x + 2$.
(b) Let g have domain $\Re$ with $g(x) = x^3 + 1$.
(c) Let h have domain $[0, \infty)$ with $h(x) = \sqrt[3]{x^2 + 2x + 1}$.
(d) Let j have domain $[0, \infty)$ with $j(x) = \frac{1}{x^2+1}$.
(e) Let k have domain all numbers other than $1/3$ with $k(x) = \frac{2x+5}{3x-1}$.

26. Use the graph of each function to determine whether it is one–to–one. If the function is one–to–one, sketch the graph of its inverse function.

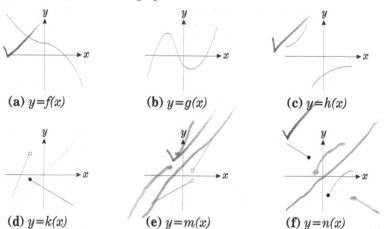

(a) $y=f(x)$ (b) $y=g(x)$ (c) $y=h(x)$

(d) $y=k(x)$ (e) $y=m(x)$ (f) $y=n(x)$

27. Consider the following curve.

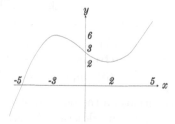

For each interval below, this curve defines a function with that interval as its domain. Determine which of these functions are one–to–one. For each one–to–one function, sketch the graph of its inverse function.
    (a) $\Re$ (b) $[-5, -3]$ (c) $[-3, 5]$ (d) $[-3, 2]$

28. If S is a set, define the *identity function* $1_S : S \longrightarrow S$ of S by $1_S(x) = x$ for $x \in S$. Let $f : D \to C$ be a function, and let $g : D \to C$ be a one–to–one function. Verify each statement.
    (a) $1_S \circ 1_S = 1_S$ (b) $f \circ 1_D = f$ (c) $1_C \circ f = f$ (d) $g^{-1} \circ g = 1_D$
    (e) $g \circ g^{-1} = 1_C$ (f) $g \circ g^{-1} \circ g = g$ (g) $g^{-1} \circ g \circ g^{-1} = g^{-1}$

29. Let $f : D \to C$ and $g : C \to E$ be one–to–one functions.
    (a) Show $g \circ f$ is one–to–one.   (b) Show $f^{-1} \circ g^{-1}$ is defined.
    (c) Show $(g \circ f)^{-1} = f^{-1} \circ g^{-1}$.

30. For each function below:

(i) show the function is not one–to–one;

(ii) find a subset of its domain D such that the function restricted to D is one–to–one;

(iii) find the domain of the inverse of this restriction of the original function;

(iv) find a formula for the inverse of this restriction of the original function.

    (a) Let f have domain $\Re$ with $f(x) = |x|$.

    (b) Let g have domain $\Re$ with $g(x) = x^2 - 5$.

    (c) Let h have domain $\Re$ with $h(x) = x^2 + 2x + 3$.

    (d) Let j have domain $[-3/2, 3/2]$ with $j(x) = \sqrt{9 - 4x^2}$.

    (e) Let k have domain $\Re$ with $k(x) = (x + 6)^{2/3}$.

[handwritten: $x = \sqrt{9 - 4y^2}$]

[handwritten: $x = (y + 6)^{2/3}$]

[handwritten: $\sqrt{x^3} - 6 = y$]

Challenging Problems

1. The arithmetic of functions does not have all of the familiar properties of the arithmetic of real numbers. For example, find two specific functions f and g with domain $\Re$ such that $fg = 0$ although neither $f = 0$ nor $g = 0$.

2. Usually we can find a restriction g_0 of g so that f/g_0 is defined. Find an example of a function g, whose graph contains no line segment of the x–axis, such that the f/g_0 is not defined for every restriction g_0 of g to an interval of positive length.

3. Let f, g_1, g_2 be functions such that $f \circ g_1$ and $f \circ g_2$ are defined with $f \circ g_1 = f \circ g_2$. Find a condition on the function f which implies that $g_1 = g_2$.

4. Let f be a function with domain D.

(a) We say f is increasing if whenever $x_1, x_2 \in D$ with $x_1 < x_2$, then $f(x_1) < f(x_2)$. Show that an increasing function has an inverse which is also increasing.

(b) Similarly, we say f is decreasing if whenever $x_1, x_2 \in D$ with $x_1 < x_2$, then $f(x_1) > f(x_2)$. Show that a decreasing function has an inverse which is also decreasing.

5. Find an example of a nonzero function f such that $f \circ f$ is defined and $f \circ f = 0$.

1.4 Trigonometric Functions

We recall the definition and basic properties of the six trigonometric functions. Our presentation of this material is brief, as we assume that the reader has learned this subject previously. A feature which may be new is the use of the unit of radians to measure angles. We define trigonometric functions of all angles, even those which are not acute. In the second subsection we summarize the properties that these functions satisfy. In the third subsection we construct the graphs of the trigonometric functions as well as the graphs of more complicated functions which they define. In the fourth subsection we restrict the domain of the trigonometric functions to define their inverses. We determine the properties and graphs of these inverse trigonometric functions. Our treatments of trigonometric and inverse trigonometric functions illustrate the constructions of Section 3. Both the trigonometric and inverse trigonometric functions play fundamental roles in calculus.

Definitions of the Trigonometric Functions

We begin with a discussion on measuring angles. There are two units which are used: degrees and radians. Think of a rod of unit length with one end hinged to the origin of the coordinate plane. A unit for measuring angles is determined by stating how many angular units there are in one revolution of this unit rod: 360 degrees, written

360°, or 2π radians (rdns). We use the convention that an angle determined by a counterclockwise revolution of the unit rod is positive while an angle determined by a clockwise revolution of the unit rod is negative.

Examples 1.4.1 Determine the size of each angle in degrees and in radians.

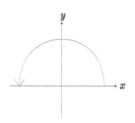

Figure 1.4.2
The Angle α

(1) Let α be half a counterclockwise revolution of the unit rod shown in Figure 1.4.2.

Solution Since one counterclockwise revolution of the unit rod is 360° or 2π radians, half of this revolution is $\frac{1}{2}(360°) = 180°$ or $\frac{1}{2}(2\pi$ radians$) = \pi$ radians.

(2) Let β be one third a clockwise revolution of the unit rod depicted in Figure 1.4.3.

Solution Since one clockwise revolution of the unit rod is $-360°$ or -2π rdns, one third of this revolution is $\frac{1}{3}(-360°) = -120°$ or $\frac{1}{3}(-2\pi$ rdns$) = -\frac{2\pi}{3}$ rdns.

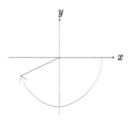

Figure 1.4.3
The Angle β

(3) Let γ be 4 counterclockwise revolutions of the unit rod depicted in Figure 1.4.4.

Solution Since one counterclockwise revolution of the unit rod is 360° or 2π rdns, four of these revolutions is $4(360°) = 1440°$ or $4(2\pi$ rdns$) = 8\pi$ rdns. □

Since π *radians* equals 180°, each degree is $\frac{\pi}{180}$ radians and each radian is $\frac{180}{\pi}$ degrees. Hence radians and degrees are related by the formulas:

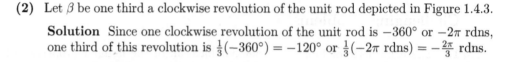

$$radians = \left(\frac{\pi}{180}\right) \times degrees \qquad (1.4.1)$$

$$degrees = \left(\frac{180}{\pi}\right) \times radians. \qquad (1.4.2)$$

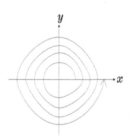

Figure 1.4.4
The Angle γ

Examples 1.4.5 (1) Write the angle $\theta = 45°$ in radians.

Solution By (1.4.1): $\theta = \frac{\pi}{180}(45)$ *radians* $= \frac{\pi}{4}$ radians.

(2) Write the angle $\phi = -\frac{\pi}{10}$ radians in degrees.

Solution By (1.4.2): $\phi = \frac{180}{\pi}\left(-\frac{\pi}{10}\right)° = -18°$.

(3) Write the angle $\psi = 1$ radian in degrees.

Solution By (1.4.2): $\psi = \frac{180}{\pi}(1)° = \frac{180}{\pi}° \approx 57.3°$. □

Consider the sector of a circle in Figure 1.4.6 of radius r with angle θ radians. Since the whole circle is a sector with angle 2π radians, the sector in Figure 1.4.6 is $\frac{\theta}{2\pi}$ of a whole circle. Therefore, the length s of the arc of this sector is $\frac{\theta}{2\pi}$ times the circumference $2\pi r$ of the entire circle. That is,

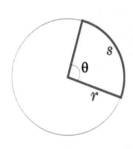

Figure 1.4.6
Sector of θ Radians

$$s = \frac{\theta}{2\pi}(2\pi r) = \theta r \quad (\theta \text{ measured in radians}). \qquad (1.4.3)$$

Examples 1.4.7 (1) Find the length s of the arc of a circle of radius 10 and angle $\frac{\pi}{6}$ radians.

Solution By (1.4.3), $s = \frac{\pi}{6}(10) = \frac{5\pi}{3}$.

(2) Find the length S of the arc of a circle of radius 12 and angle 135°.

Solution $135° = 135\frac{\pi}{180}$ rdns $= \frac{3}{4}\pi$ rdns. By (1.4.3): $S = \left(\frac{3}{4}\pi\right)12 = 9\pi$. □

If the angle ϕ is measured in degrees, then ϕ equals $\frac{\pi}{180}\phi$ radians, and $s = \left(\frac{\pi}{180}\phi\right)r = \frac{\pi}{180}\phi r$. This formula is more complicated than formula (1.4.3). Since the equation for arclength is the basis for all formulas in calculus involving trigonometric functions, we obtain simpler formulas by using radians to measure angles and formula (1.4.3) to compute arclength. This explains why we use the unit of radians to measure angles throughout this text.

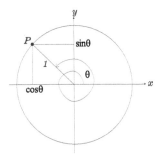

Figure 1.4.8 Definition of $\sin\theta$ and $\cos\theta$

Let θ denote an angle determined by revolving the unit rod around the origin in the coordinate plane. Observe that the point P at the end of this rod is one unit from the origin and thus lies on the unit circle $x^2 + y^2 = 1$ in Figure 1.4.8. The coordinates of the point P define the sine and cosine of the angle θ. Then $\sin\theta$ and $\cos\theta$ are used to define the other four trigonometric functions of θ.

Definition 1.4.9 *Let P be the point on the unit circle $x^2 + y^2 = 1$ obtained by revolving a unit rod by the angle θ. Then $\cos\theta$, $\sin\theta$ are defined as the coordinates of P:*

$$P = (\cos\theta, \sin\theta) \tag{1.4.4}$$

If $\sin\theta \neq 0$, define the cotangent and cosecant functions of θ by:

$$\cot\theta \;=\; \frac{\cos\theta}{\sin\theta}, \qquad \csc\theta \;=\; \frac{1}{\sin\theta} . \tag{1.4.5}$$

If $\cos\theta \neq 0$, define the tangent and secant functions of θ by:

$$\tan\theta \;=\; \frac{\sin\theta}{\cos\theta}, \qquad \sec\theta \;=\; \frac{1}{\cos\theta} . \tag{1.4.6}$$

The six functions of this definition are called the *trigonometric functions*. Since the sine and cosine are defined for all angles,

$$\text{Domain } \sin\theta \;=\; \text{Domain } \cos\theta \;=\; \Re.$$

Note that $\sin\theta = 0$ when P lies on the x–axis. This occurs when θ is of the form $N\pi$ where N is an integer. Hence

$$\text{Domain } \cot\theta \;=\; \text{Domain } \csc\theta \text{ consists of all numbers in } \Re \text{ not of the form } N\pi.$$

Similarly, $\cos\theta = 0$ when P lies on the y–axis. This occurs when θ is of the form $\frac{\pi}{2} + N\pi$ where N is an integer. Hence

$$\text{Domain } \tan\theta \;=\; \text{Domain } \sec\theta \text{ consists of all numbers in } \Re \text{ not of the form } \frac{\pi}{2} + N\pi.$$

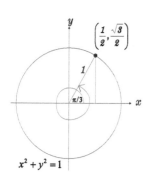

Figure 1.4.11 $\theta = \frac{7\pi}{3}$

Figure 1.4.12 $\theta = \frac{7\pi}{6}$

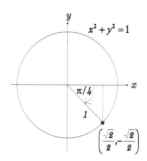

Fig 1.4.13 $\theta = -\frac{\pi}{4}$

Examples 1.4.10 (1) Find the six trigonometric functions of the angle $\frac{7\pi}{3}$.

Solution Let $\theta = \frac{7\pi}{3}$, as depicted in Figure 1.4.11. From the 30–60–90 triangle in the first quadrant, we see that $P = (1/2, \sqrt{3}/2)$. Therefore,

$$\sin\frac{7\pi}{3} = \frac{\sqrt{3}}{2} \quad \text{and} \quad \cos\frac{7\pi}{3} = \frac{1}{2}.$$

Then

$$\tan\frac{7\pi}{3} = \frac{\sin 7\pi/3}{\cos 7\pi/3} = \frac{\sqrt{3}/2}{1/2} = \sqrt{3}, \quad \sec\frac{7\pi}{3} = \frac{1}{\cos 7\pi/3} = \frac{1}{1/2} = 2,$$

$$\cot\frac{7\pi}{3} = \frac{\cos 7\pi/3}{\sin 7\pi/3} = \frac{1/2}{\sqrt{3}/2} = \frac{1}{\sqrt{3}}, \quad \csc\frac{7\pi}{3} = \frac{1}{\sin 7\pi/3} = \frac{1}{\sqrt{3}/2} = \frac{2}{\sqrt{3}}.$$

(2) Find the six trigonometric functions of the angle $\frac{7\pi}{6}$.

Solution Let $\theta = \frac{7\pi}{6}$, as depicted in Figure 1.4.12. From the 30–60–90 triangle in the third quadrant, we see that $P = (-\sqrt{3}/2, -1/2)$. Therefore,

$$\sin\frac{7\pi}{6} = -\frac{1}{2} \quad \text{and} \quad \cos\frac{7\pi}{6} = -\frac{\sqrt{3}}{2}.$$

Then

$$\tan\frac{7\pi}{6} = \frac{\sin 7\pi/6}{\cos 7\pi/6} = \frac{-1/2}{-\sqrt{3}/2} = \frac{1}{\sqrt{3}}, \quad \sec\frac{7\pi}{6} = \frac{1}{\cos 7\pi/6} = \frac{1}{-\sqrt{3}/2} = -\frac{2}{\sqrt{3}},$$

$$\cot\frac{7\pi}{6} = \frac{\cos 7\pi/6}{\sin 7\pi/6} = \frac{-\sqrt{3}/2}{-1/2} = \sqrt{3}, \quad \csc\frac{7\pi}{6} = \frac{1}{\sin 7\pi/6} = \frac{1}{-1/2} = -2.$$

(3) Find the six trigonometric functions of the angle $-\frac{\pi}{4}$.

Solution Let $\theta = -\frac{\pi}{4}$ as depicted in Figure 1.4.13. From, the isosceles right triangle in the fourth quadrant, we see that $P = (\sqrt{2}/2, -\sqrt{2}/2)$. Therefore,

$$\sin-\frac{\pi}{4} = -\frac{\sqrt{2}}{2} \quad \text{and} \quad \cos-\frac{\pi}{4} = \frac{\sqrt{2}}{2}.$$

Then

$$\tan-\frac{\pi}{4} = \frac{\sin -\pi/4}{\cos -\pi/4} = \frac{-\sqrt{2}/2}{\sqrt{2}/2} = -1, \quad \sec-\frac{\pi}{4} = \frac{1}{\cos -\pi/4} = \frac{1}{\sqrt{2}/2} = \sqrt{2},$$

$$\cot-\frac{\pi}{4} = \frac{\cos -\pi/4}{\sin -\pi/4} = \frac{\sqrt{2}/2}{-\sqrt{2}/2} = -1, \quad \csc-\frac{\pi}{4} = \frac{1}{\sin -\pi/4} = \frac{1}{-\sqrt{2}/2} = -\sqrt{2}.$$

(4) Let f and g be two functions with domain $\Re$ defined by

$$f(x) = \sin x \quad \text{and} \quad g(x) = x^2 + 1.$$

Find $3f - 5g$, fg, $\frac{f}{g}, \frac{g}{f}$, $f \circ g$ and $g \circ f$.

Solution f, g have the same domain $\Re$. Hence the functions below are defined.

$$(3f - 5g)(x) = 3f(x) - 5g(x) = 3\sin x - 5x^2 - 5,$$

$$(fg)(x) = f(x)g(x) = (x^2 + 1)\sin x,$$

$$\left(\frac{f}{g}\right)(x) = \frac{f(x)}{g(x)} = \frac{\sin x}{x^2 + 1},$$

$$(f \circ g)(x) = f(g(x)) = \sin g(x) = \sin(x^2 + 1),$$

$$(g \circ f)(x) = g(f(x)) = g(\sin x) = \sin^2 x + 1.$$

To define $(g/f)(x)$ we replace f, g by their restrictions f_0, g_0 to the set $\Re$ with all numbers $N\pi$ excluded, for $N \in Z$. Then $f_0(x) = \sin x$ is never zero and

$$\left(\frac{g_0}{f_0}\right)(x) = \frac{g_0(x)}{f_0(x)} = \frac{g(x)}{f(x)} = \frac{x^2 + 1}{\sin x} \quad \text{for } x \neq N\pi. \qquad \square$$

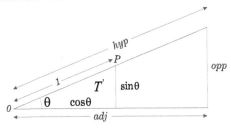

Figure 1.4.14 Triangle T

When θ is an acute angle, we can compute the trigonometric functions of θ from a right triangle T with θ as one of its angles. Let hyp denote the length of the hypotenuse of T, let adj denote the length of the side of T which is adjacent to the angle θ and let opp denote the length of the side of T which is opposite the angle θ. The case where $hyp > 1$ is illustrated in Figure 1.4.14. The triangle T is similar to the triangle T' in the first quadrant which is used to define $\sin \theta$ and $\cos \theta$. Since the endpoint P of the unit rod OP lies on the unit circle with center the origin, the length $|OP|$ of the hypotenuse of T' is one. By the definition of sine and cosine, the length of the opposite side of T' equals the y–coordinate of P which is $\sin \theta$ while the length of the adjacent side of T' equals the x–coordinate of P which is $\cos \theta$. The ratios of corresponding sides of the similar triangles T' and T are equal:

$$\sin \theta \;=\; \frac{\sin \theta}{1} \;=\; \frac{opp}{hyp}, \qquad \cos \theta \;=\; \frac{\cos \theta}{1} \;=\; \frac{adj}{hyp} \;. \qquad (1.4.7)$$

By the definitions of the other trigonometric functions:

$$
\begin{aligned}
\tan \theta &= \frac{\sin \theta}{\cos \theta} = \frac{opp/hyp}{adj/hyp} = \frac{opp}{adj}, \\
\cot \theta &= \frac{\cos \theta}{\sin \theta} = \frac{adj/hyp}{opp/hyp} = \frac{adj}{opp}, \\
\sec \theta &= \frac{1}{\cos \theta} = \frac{1}{adj/hyp} = \frac{hyp}{adj}, \\
\csc \theta &= \frac{1}{\sin \theta} = \frac{1}{opp/hyp} = \frac{hyp}{opp} \;.
\end{aligned}
\qquad (1.4.8)
$$

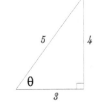

Figure 1.4.15
Example 1

Examples 1.4.16 (1) Find the trigonometric functions of the angle θ in Fig. 1.4.15.

Solution Note that $hyp = 5$, $opp = 4$ and $adj = 3$. Therefore,

$$\sin \theta = \frac{4}{5}, \quad \cos \theta = \frac{3}{5}, \quad \tan \theta = \frac{4}{3}, \quad \cot \theta = \frac{3}{4}, \quad \sec \theta = \frac{5}{3}, \quad \csc \theta = \frac{5}{4}.$$

(2) Find the trigonometric functions of the angle $\frac{\pi}{4}$.

Solution Consider the isosceles right triangle whose legs have length one depicted in Figure 1.4.17. Its two equal acute angles equal $\frac{\pi}{4}$. We have $adj = opp = 1$, and $hyp = \sqrt{2}$ by the Pythagorean Theorem. Thus

$$\sin \frac{\pi}{4} = \frac{1}{\sqrt{2}}, \quad \cos \frac{\pi}{4} = \frac{1}{\sqrt{2}}, \quad \tan \frac{\pi}{4} = 1, \quad \cot \frac{\pi}{4} = 1, \quad \sec \frac{\pi}{4} = \sqrt{2}, \quad \csc \frac{\pi}{4} = \sqrt{2}.$$

Figure 1.4.17
Example 2

(3) Find the trigonometric functions of the angles $\frac{\pi}{6}$ and $\frac{\pi}{3}$.

Solution Consider the 30–60–90 right triangle depicted in Figure 1.4.18. Its side opposite the angle $\frac{\pi}{6}$ has length one, and its side opposite the angle $\frac{\pi}{3}$ has length $\sqrt{3}$. By the Pythagorean Theorem, $hyp = 2$. For the angle $\frac{\pi}{6}$: we have $adj = \sqrt{3}$ and $opp = 1$. Thus

$$\sin \frac{\pi}{6} = \frac{1}{2}, \quad \cos \frac{\pi}{6} = \frac{\sqrt{3}}{2}, \quad \tan \frac{\pi}{6} = \frac{1}{\sqrt{3}}, \quad \cot \frac{\pi}{6} = \sqrt{3}, \quad \sec \frac{\pi}{6} = \frac{2}{\sqrt{3}}, \quad \csc \frac{\pi}{6} = 2.$$

For the angle $\frac{\pi}{3}$: we have $adj = 1$ and $opp = \sqrt{3}$. Thus

$$\sin \frac{\pi}{3} = \frac{\sqrt{3}}{2}, \quad \cos \frac{\pi}{3} = \frac{1}{2}, \quad \tan \frac{\pi}{3} = \sqrt{3}, \quad \cot \frac{\pi}{3} = \frac{1}{\sqrt{3}}, \quad \sec \frac{\pi}{3} = 2, \quad \csc \frac{\pi}{3} = \frac{2}{\sqrt{3}}. \square$$

Figure 1.4.18
Example 3

Trigonometric Identities

A *trigonometric identity* is an equation involving trigonometric functions which is true for each angle in the domains of all the functions. These identities are useful in applications of trigonometry. In particular, we will use them extensively to develop methods of integration. The more complicated identities are stated without proof.

When we take the unit rod which describes the angle θ and revolve it N revolutions clockwise or counterclockwise, the rod returns to its original position. That is, the point P in Figure 1.4.8 which represents the angle θ also represents the angles $\theta + 2N\pi$ for all integers N. Hence the trigonometric functions of θ and $\theta + 2N\pi$ are equal:

$$\sin(\theta + 2N\pi) = \sin\theta \quad \cos(\theta + 2N\pi) = \cos\theta \quad \tan(\theta + 2N\pi) = \tan\theta$$
$$\cot(\theta + 2N\pi) = \cot\theta \quad \sec(\theta + 2N\pi) = \sec\theta \quad \csc(\theta + 2N\pi) = \csc\theta \tag{1.4.9}$$

We say that the six trigonometric functions are *periodic* with period 2π.

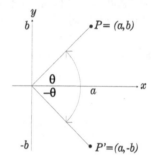

Figure 1.4.19 Trigonometric Functions of $-\theta$

Let $P = (a, b)$ represent the angle θ on the unit circle. Figure 1.4.19 shows that the angle $-\theta$ is represented by $P' = (a, -b)$. By the definitions of $\sin(-\theta)$ and $\cos(-\theta)$:

$$\sin(-\theta) = -b = -\sin\theta \quad \text{and} \quad \cos(-\theta) = a = \cos\theta . \tag{1.4.10}$$

It follows from the definitions of the other trigonometric functions that

$$\begin{aligned}
\tan(-\theta) &= \frac{\sin(-\theta)}{\cos(-\theta)} = \frac{-\sin\theta}{\cos\theta} = -\tan\theta, \\
\cot(-\theta) &= \frac{\cos(-\theta)}{\sin(-\theta)} = \frac{\cos\theta}{-\sin\theta} = -\cot\theta, \\
\sec(-\theta) &= \frac{1}{\cos(-\theta)} = \frac{1}{\cos\theta} = \sec\theta, \\
\csc(-\theta) &= \frac{1}{\sin(-\theta)} = \frac{1}{-\sin\theta} = -\csc\theta .
\end{aligned} \tag{1.4.11}$$

Thus $\cos x$, $\sec x$ are even functions while $\sin x$, $\tan x$, $\cot x$, $\csc x$ are odd functions.

By definition, $\sec\theta$ and $\csc\theta$ are the reciprocals of $\cos\theta$ and $\sin\theta$. Note the definition of $\cot x$ is the reciprocal of the definition of $\tan x$. Thus

$$\cot\theta = \frac{1}{\tan\theta}, \quad \sec\theta = \frac{1}{\cos\theta}, \quad \csc\theta = \frac{1}{\sin\theta} . \tag{1.4.12}$$

Since $P = (\cos\theta, \sin\theta)$ lies on the unit circle $x^2 + y^2 = 1$,

$$\cos^2\theta + \sin^2\theta = 1. \tag{1.4.13}$$

Divide this equation by $\cos^2\theta$ or by $\sin^2\theta$ to obtain:

$$1 + \tan^2\theta = \sec^2\theta, \qquad \cot^2\theta + 1 = \csc^2\theta . \tag{1.4.14}$$

The preceding three formulas are called the *Pythagorean identities*. Two deep formulas give the sine and cosine of the sum of two angles:

$$\sin(\theta + \phi) = \sin\theta\cos\phi + \cos\theta\sin\phi, \tag{1.4.15}$$

$$\cos(\theta + \phi) = \cos\theta\cos\phi - \sin\theta\sin\phi. \tag{1.4.16}$$

We assume the reader has seen the verification of these formulas previously. When $\theta = \phi$ we obtain the double angle formulas:

$$\sin 2\theta = 2\sin\theta\cos\theta, \tag{1.4.17}$$

$$\cos 2\theta = \cos^2\theta - \sin^2\theta = 2\cos^2\theta - 1 = 1 - 2\sin^2\theta. \tag{1.4.18}$$

The alternate forms for the preceding identity are obtained by applying the Pythagorean identity (1.4.13). By (1.4.10),

$$\sin(\theta - \phi) = \sin\theta\cos(-\phi) + \cos\theta\sin(-\phi) = \sin\theta\cos\phi - \cos\theta\sin\phi, \tag{1.4.19}$$

$$\cos(\theta - \phi) = \cos\theta\cos(-\phi) - \sin\theta\sin(-\phi) = \cos\theta\cos\phi + \sin\theta\sin\phi. \tag{1.4.20}$$

Add (1.4.19) to (1.4.15) and divide by two:

$$\sin\theta\cos\phi = \frac{1}{2}\sin(\theta + \phi) + \frac{1}{2}\sin(\theta - \phi). \tag{1.4.21}$$

Similarly, add or subtract (1.4.16) and (1.4.20) and divide by two:

$$\cos\theta\cos\phi = \frac{1}{2}\cos(\theta + \phi) + \frac{1}{2}\cos(\theta - \phi), \tag{1.4.22}$$

$$\sin\theta\sin\phi = \frac{1}{2}\cos(\theta - \phi) - \frac{1}{2}\cos(\theta + \phi). \tag{1.4.23}$$

Sometimes we require an equation which relate the sides and angles of the general triangle depicted in Figure 1.4.20. The most common of these formulas are:

$$\frac{\sin\alpha}{a} = \frac{\sin\beta}{b} = \frac{\sin\gamma}{c} \quad \text{(Law of Sines)}; \tag{1.4.24}$$

$$c^2 = a^2 + b^2 - 2ab\cos\gamma \quad \text{(Law of Cosines)}. \tag{1.4.25}$$

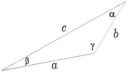

Figure 1.4.20
General Triangle

In a right triangle $\cos\gamma = \cos\frac{\pi}{2} = 0$, and the law of cosines becomes the Pythagorean Theorem.

Examples 1.4.21 (1) Show that $\cos\theta = \sin\left(\theta + \frac{\pi}{2}\right)$.

Solution Expand $\sin\left(\theta + \frac{\pi}{2}\right)$ by (1.4.15):

$$\sin\left(\theta + \frac{\pi}{2}\right) = \sin\theta\cos\frac{\pi}{2} + \cos\theta\sin\frac{\pi}{2} = (\sin\theta)(0) + (\cos\theta)(1) = \cos\theta.$$

(2) Show that $\tan\theta$ is periodic with period π.

Solution We must show $\tan(\theta + \pi) = \tan\theta$. By the definition of the tangent:

$$\tan(\theta + \pi) = \frac{\sin(\theta+\pi)}{\cos(\theta+\pi)} = \frac{\sin\theta\cos\pi + \cos\theta\sin\pi}{\cos\theta\cos\pi - \sin\theta\sin\pi} \quad \text{[by (1.4.15) and (1.4.16)]}$$

$$= \frac{(\sin\theta)(-1) + (\cos\theta)(0)}{(\cos\theta)(-1) - (\sin\theta)(0)} = \frac{-\sin\theta}{-\cos\theta} = \tan\theta.$$

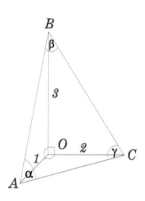

Figure 1.4.22
The Tetrahedron T

(3) The tetrahedron T of Figure 1.4.22 has three faces which are right triangles with legs OA of length 1, OB of length 3 and OC of length 2. Find the cosines of the angles α, β, γ of the triangular face ABC.

Solution By the Pythagorean Theorem, the length of AB is $\sqrt{1^2 + 3^2} = \sqrt{10}$, the length of AC is $\sqrt{1^2 + 2^2} = \sqrt{5}$ and the length of BC is $\sqrt{2^2 + 3^2} = \sqrt{13}$. Apply the law of cosines to triangle ABC three times:

$$|BC|^2 = |AB|^2 + |AC|^2 - 2|AB|\,|AC|\cos\alpha$$
$$13 = 10 + 5 - 2\sqrt{10}\sqrt{5}\cos\alpha \quad and \quad \cos\alpha = \frac{1}{\sqrt{50}}$$
$$|AC|^2 = |AB|^2 + |BC|^2 - 2|AB|\,|BC|\cos\beta$$
$$5 = 10 + 13 - 2\sqrt{10}\sqrt{13}\cos\beta \quad and \quad \cos\beta = \frac{9}{\sqrt{130}}$$
$$|AB|^2 = |AC|^2 + |BC|^2 - 2|AB|\,|BC|\cos\gamma$$
$$10 = 5 + 13 - 2\sqrt{5}\sqrt{13}\cos\gamma \quad and \quad \cos\gamma = \frac{4}{\sqrt{65}}$$

Graphs of Trigonometric Functions

We begin by recalling the graphs of the six trigonometric functions. Then we use the graph of $y = \sin x$ to construct related functions with interesting graphs.

Since each of the six trigonometric functions is periodic with period 2π, we only need to plot the piece of each graph on the interval $[0, 2\pi]$. The remainder of each graph is obtained by translating this piece horizontally to the left and right by multiples of 2π units. The resulting graphs are sketched in Figure 1.4.25. Note that the graphs of $y = \sin x$ and $y = \cos x$ are waves which oscillate between the lines $y = -1$ and $y = +1$, i.e.

$$-1 \leq \sin x \leq +1, \qquad -1 \leq \cos x \leq +1 . \tag{1.4.26}$$

Figure 1.4.23
$f(x) = 4\sin x$

These inequalities are consequences of the Pythagorean identity: $\sin^2 x + \cos^2 x = 1$. Observe that the graphs of $y = \cot x = \frac{\cos x}{\sin x}$ and $y = \csc x = \frac{1}{\sin x}$ approach the vertical line $x = N\pi$ at each multiple of π where their denominator $\sin x$ vanishes. We call these lines *vertical asymptotes* of the graphs. Similarly, the graphs of $y = \tan x = \frac{\sin x}{\cos x}$ and $y = \sec x = \frac{1}{\cos x}$ have vertical asymptotes $x = N\pi + \frac{\pi}{2}$ for each integer N where their denominator $\cos x$ vanishes. The functions $y = \cos x$ and $y = \sec x$ are even. Hence their graphs are symmetric with respect to the y-axis. The functions $y = \sin x$, $y = \tan x$, $y = \cot x$ and $y = \csc x$ are odd. Hence their graphs are symmetric with respect to the origin.

The graphs of functions $f(x) = A\sin Bx$ are derived from the graph of the sine function. The symbols A and B represent numbers or functions of x. First, consider these graphs when $B = 1$. Think of the graph of the sine function as waves of height one. The factor A in $f(x) = A\sin x$ changes the height of these waves to A. For example, the graph of $f(x) = 4\sin x$ depicted in Figure 1.4.23 has waves of height 4. If $A = A(x)$ is a function of x, begin by graphing the functions $y = A(x)$ and $y = -A(x)$ as dotted curves. Then graph the waves of the sine function between these two dotted curves. For example, the graph of

$$F(x) = x\sin x$$

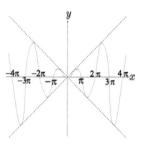

Figure 1.4.24
$F(x) = x\sin x$

is constructed in Figure 1.4.24 by drawing the waves of the graph of $y = \sin x$ between the two dotted lines $y = x$ and $y = -x$ which determine the heights of the waves.

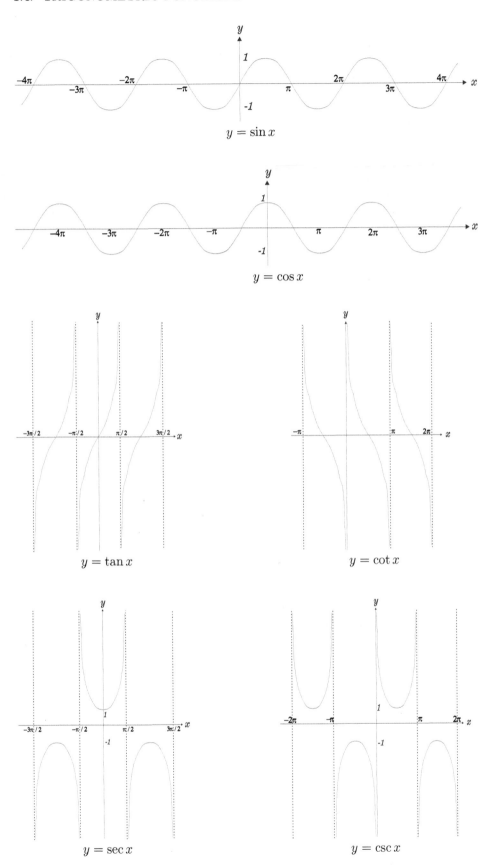

Figure 1.4.25 Graphs of the Trigonometric Functions

Now consider the graph of $f(x) = A \sin Bx$ when $A = 1$ The role of B in the graph of $f(x) = \sin Bx$ is to alter the frequency of the waves. For example, in the graph of $g(x) = \sin 3x$ in Figure 1.4.26 the waves occur three times as fast as in the graph of the usual sine function. In Figure 1.4.27, we construct the graph of

$$G(x) = \sin \frac{1}{x}$$

with domain all numbers $\Re$ except 0. Let $x > 0$. As x approaches 0, its reciprocal $\theta = 1/x$ gets larger and larger. Therefore, the graph of $y = G(x)$, as x approaches 0, has all of the waves of the graph of $y = \sin\theta$, $\theta > 0$. These waves become narrower and narrower as x approaches 0. When $x > 1/\pi$, the graph of G has the one wave of $\sin\theta$ for $0 < \theta < \pi$. Since $G(-x) = -G(x)$, the function G is odd, and the same phenomena occur when x is negative.

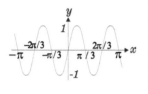

Figure 1.4.26
$g(x) = \sin 3x$

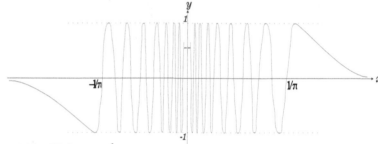

Figure 1.4.27 $G(x) = \sin\frac{1}{x}$

We combine these ideas to graph $y = f(x) = A \sin Bx$ when neither A nor B equals one. For example, the graph of $y = 4\sin 3x$ has waves of height four as in the graph of $y = 4\sin x$. These waves occur three times faster than in the graph of the usual sine function as in the graph of $y = \sin 3x$. See Figure 1.4.28. We conclude by using the graphs of $F(x) = x\sin x$ and $G(x) = \sin(1/x)$ to graph the function

$$H(x) = x \sin \frac{1}{x}$$

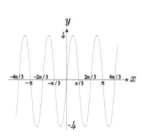

Figure 1.4.28
$y = 4\sin 3x$

in Figure 1.4.29. H has domain all nonzero numbers. We graph $y = H(x)$ for $x > 0$ by drawing the waves of $y = \sin(1/x)$ between the dotted lines $y = x$ and $y = -x$ which determine their heights. There are an infinite number of these waves near the origin as in the graph of $y = \sin(1/x)$. The behavior of the graph of $y = H(x)$, for x large, will be determined in Section 6. Since $H(-x) = H(x)$, the function H is even. Its graph for $x < 0$ is obtained by reflecting its graph for $x > 0$ about the y–axis.

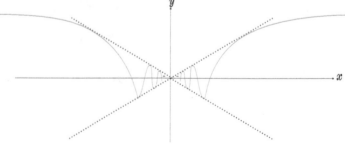

Figure 1.4.29 $H(x) = x\sin\frac{1}{x}$

Inverse Trigonometric Functions

We define inverse functions of the trigonometric functions by restricting their domains. These inverse trigonometric functions are important in calculus. They even arise in the solutions of some problems which seem to have no relation to trigonometry. The following example gives a noncalculus application.

Motivating Example 1.4.30 A plane is approaching an airport from the west. The air traffic controller makes radar readings each minute from two sources: one at the airport and another from a point on the ground 10 km west of the airport. Each reading gives the distance of the plane from the instrument. The air traffic controller wants to use these readings to determine the altitude of the plane. Suggest an efficient procedure for making this computation.

Solution Consider this situation as depicted in Figure 1.4.31. A is the airport, P is the plane, R is the location of the second radar site and Q is the point on the ground directly below the plane. The altitude h of the plane is $|PQ|$. Here is a strategy for solving this problem.

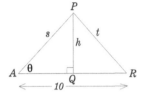

Step I Use the triangle APR to compute a trigonometric function of the angle of elevation θ of the plane from the airport.

Step II Deduce the angle of elevation θ.

Figure 1.4.31 Plane Approaching Airport

Step III Use the right triangle APQ to compute the altitude of the plane.

To implement Step I, apply the law of cosines to $\triangle APR$ and solve for $\cos\theta$. Let $s = |AP|$ as measured by the radar at the airport, and let $t = |RP|$ as measured at the other radar site R. All of s, t, h are measured in km. By the law of cosines:

$$
\begin{aligned}
t^2 &= 10^2 + s^2 - 2(10)(s)\cos\theta \\
\cos\theta &= \frac{100 + s^2 - t^2}{20s} .
\end{aligned}
\tag{1.4.27}
$$

To implement Step II, use the graph of $y = \cos\theta$ to find θ from the value of y computed in Step I.

To implement Step III, use the value of $\sin\theta$ given by the right triangle APQ:

$$
\sin\theta = \frac{h}{s} \quad \text{and} \quad h = s\sin\theta
\tag{1.4.28}
$$

We test this procedure with the radar readings: $s = 10$ km and $t = 10$ km. Step I is to compute $\cos\theta$ by (1.4.27):

$$
\cos\theta = \frac{100 + 10^2 - 10^2}{20(10)} = \frac{100}{200} = \frac{1}{2} .
$$

Step II is to use the graph of $y = \cos\theta$ to find the angle $\theta > 0$ whose cosine is $\frac{1}{2}$. There are many such angles: $\frac{\pi}{3}$, $\frac{5\pi}{3}$, $\frac{7\pi}{3}$, $\frac{11\pi}{3}$, etc. How do we know which angle to choose? Since θ is an angle of a triangle it must be between 0 and π radians. Hence the appropriate choice of θ is $\theta = \frac{5\pi}{3}$. Step III is to compute h from (1.4.28):

$$
h = (10)\sin\frac{\pi}{3} = 10\frac{\sqrt{3}}{2} = 5\sqrt{3} \text{ km.}
$$

The determination of θ from $\cos\theta$ in the second step was awkward. To make this step routine, define a function $g(y) = \theta$ which assigns every number y between -1 and $+1$ to the unique angle θ between 0 and π radians with $\cos\theta = y$. Now we have a clear statement of Step II: compute $\theta = g(y)$. □

Let's look more closely at the function g we defined in the preceding example:

$$g(y) = \theta \quad \text{if} \quad 0 \le \theta \le \pi \quad \text{and} \quad \cos \theta = y.$$

The part of the statement " $g(y) = \theta$ if $\cos \theta = y$" says that g is the inverse function of the cosine function. The condition "$0 \le \theta \le \pi$" says that we are not looking at the entire cosine function with domain $\Re$, but rather at its restriction to the interval $[0, \pi]$. As we observed, we can not use the entire cosine function because there are many angles θ with $\cos \theta$ equal to any $y \in [-1, 1]$, i.e. the entire cosine function is not one–to–one. Recall the "restriction of domain" construction of the last subsection of Section 3. We used it to define the inverse of a function which is not one–to–one. The definition of g is an application of this construction to the function $y = \cos \theta$.

All six of the trigonometric functions are not one–to–one because they are periodic: they have the same value at θ and at $\theta + 2\pi$. Therefore we use the "restriction of domain" construction to define an inverse function of each one. First we choose a set of numbers D_0 on which the particular trigonometric function is one–to–one. Then we construct the inverse function of this restriction.

Let's start with the function $f(\theta) = \cos \theta$ which we studied above. We decided to restrict f to the interval $D_0 = [0, \pi]$ where the cosine function is one–to–one. We verify this from the graph of this restriction depicted in the left diagram of Figure 1.4.32: no horizontal line intersects this graph in more than one point. If f_0 is the restriction of the cosine function to D_0, then its inverse function, denoted $g(y)$ above, is called the arccos function:

$$\text{arccos } y = \theta \quad \text{if} \quad \theta \in [0, \pi] \quad \text{and} \quad y = \cos \theta.$$

That is, arccos y means the angle θ whose cosine is y. The domain of the arccos function is the range of the cosine function which is $[-1, 1]$. The range of the arccos function is the domain $D_0 = [0, \pi]$ of the restriction of the cosine function we are using. The graph of $y = \arccos x$ is the reflection about the line $y = x$ of the graph of $y = \cos x$ with domain $[0, \pi]$ shown in the right diagram of Figure 1.4.32.

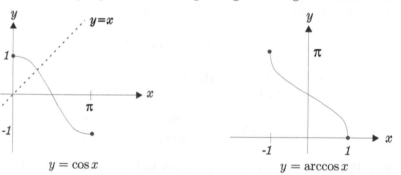

$$y = \cos x \qquad\qquad\qquad y = \arccos x$$

Figure 1.4.32 Defining an Inverse of the Cosine Function

Examples 1.4.33 (1) Find arccos $\frac{\sqrt{2}}{2}$.

Solution By definition, $\theta = \arccos \frac{\sqrt{2}}{2}$ if $\cos \theta = \frac{\sqrt{2}}{2}$ and $0 \le \theta \le \pi$. We know from an isosceles right triangle that $\cos \frac{\pi}{4} = \frac{\sqrt{2}}{2}$. Hence arccos $\frac{\sqrt{2}}{2} = \frac{\pi}{4}$.

(2) Find arccos $-\frac{\sqrt{3}}{2}$.

Solution By definition, $\theta = \arccos -\frac{\sqrt{3}}{2}$ if $\cos \theta = -\frac{\sqrt{3}}{2}$ and $0 \le \theta \le \pi$. From a 30–60–90 right triangle we see that θ is an angle in the second quadrant which makes an angle of $\frac{\pi}{6}$ with the negative x–axis, i.e. arccos $-\frac{\sqrt{3}}{2} = \frac{5\pi}{6}$.

(3) Find $\cos\left(\arccos\frac{3}{17}\right)$.

Solution We are asked to find the cosine of the angle whose cosine is $\frac{3}{17}$. The question itself gives the answer: $\cos\left(\arccos\frac{3}{17}\right) = \frac{3}{17}$.

(4) Find $\arccos\left(\cos-\frac{\pi}{2}\right)$.

Solution $\arccos\cos-\frac{\pi}{2} = \arccos 0 = \frac{\pi}{2}$ because $0 \leq \frac{\pi}{2} \leq \pi$.

Warning: $\arccos\left(\cos-\frac{\pi}{2}\right) = -\frac{\pi}{2}$ is **wrong** because the angle $-\frac{\pi}{2}$ is not in the range $[0, \pi]$ of the arccos function. □

Now consider the function $y = \sin\theta$ with domain $\Re$. The restriction of this function to $D_1 = \left[-\frac{\pi}{2}, \frac{\pi}{2}\right]$ is one–to–one. We verify this from the graph of this restriction depicted in the left diagram of Figure 1.4.34: no horizontal line intersects this graph in more than one point. The inverse of this restriction is called the arcsin function:

$$\arcsin y = \theta \quad \text{if} \quad \theta \in \left[-\frac{\pi}{2}, \frac{\pi}{2}\right] \quad \text{and} \quad y = \sin\theta.$$

Note $\arcsin y$ means the angle θ whose sine is y. The domain of the arcsin function is the range of the sine function which is $[-1, 1]$. The range of the arcsin function is the domain $D_1 = \left[-\frac{\pi}{2}, \frac{\pi}{2}\right]$ of the restriction of the sine function we are using. The graph of $y = \arcsin x$ is the reflection about the line $y = x$ of the graph of $y = \sin x$ with domain $\left[-\frac{\pi}{2}, \frac{\pi}{2}\right]$ shown in the right diagram of Figure 1.4.34.

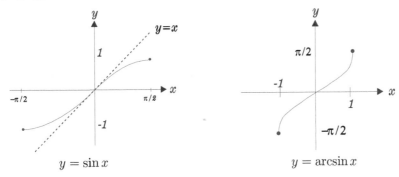

$$y = \sin x \qquad\qquad y = \arcsin x$$

Figure 1.4.34 Defining an Inverse of the Sine Function

Examples 1.4.35 (1) Find $\arcsin\frac{1}{2}$.

Solution By definition, $\theta = \arcsin\frac{1}{2}$ if $\sin\theta = \frac{1}{2}$ and $-\frac{\pi}{2} \leq \theta \leq \frac{\pi}{2}$. We know from a 30–60–90 right triangle that $\sin\frac{\pi}{6} = \frac{1}{2}$. Hence $\arcsin\frac{1}{2} = \frac{\pi}{6}$.

(2) Find $\arcsin-1$.

Solution By definition, $\theta = \arcsin-1$ if $\sin\theta = -1$ and $-\frac{\pi}{2} \leq \theta \leq \frac{\pi}{2}$. We know that $\sin-\frac{\pi}{2} = -1$. Hence $\arcsin-1 = -\frac{\pi}{2}$.

(3) Find $\sin\left(\arcsin\frac{8}{9}\right)$.

Solution We are asked to find the sine of the angle whose sine is $\frac{8}{9}$. The question itself gives the answer: $\sin\left(\arcsin\frac{8}{9}\right) = \frac{8}{9}$.

(4) Find $\arcsin\left(\sin\frac{6\pi}{7}\right)$.

Solution By definition, $\theta = \arcsin\sin\frac{6\pi}{7}$ if $\sin\theta = \sin\frac{6\pi}{7}$ and $-\frac{\pi}{2} \leq \theta \leq \frac{\pi}{2}$. Note that $\frac{6\pi}{7}$ is not between $-\frac{\pi}{2}$ and $\frac{\pi}{2}$. However, $\sin\frac{6\pi}{7} = \sin\left(\pi - \frac{6\pi}{7}\right) = \sin\frac{\pi}{7}$. Hence $\theta = \arcsin\left(\sin\frac{6\pi}{7}\right) = \frac{\pi}{7}$ because $-\frac{\pi}{2} \leq \frac{\pi}{7} \leq \frac{\pi}{2}$.

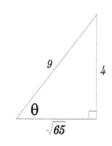

Figure 1.4.36
$\sin\theta = \frac{4}{9}$

(5) Find $\cos\left(\arcsin\frac{4}{9}\right)$.

>**Solution** Let $\theta = \arcsin\frac{4}{9}$. Then $-\frac{\pi}{2} \le \theta \le \frac{\pi}{2}$ and $\sin\theta = \frac{4}{9}$. Since $\sin\theta > 0$, we have $0 \le \theta \le \frac{\pi}{2}$. Consider the right triangle with angle θ in Figure 1.4.32 whose opposite side has length 4 and whose hypotenuse has length 9. By the Pythagorean Theorem, the adjacent side has length $\sqrt{9^2 - 4^2} = \sqrt{65}$. Hence $\cos\left(\arcsin\frac{4}{9}\right) = \cos\theta = \frac{\sqrt{65}}{9}$. □

Next, consider the function $y = \tan\theta$ with domain all numbers which are not odd multiples of $\frac{\pi}{2}$. The restriction of the tangent function to $D_2 = \left(-\frac{\pi}{2}, \frac{\pi}{2}\right)$ is one-to-one. We verify this from the graph of this restriction as depicted in the left diagram of Figure 1.4.37: each horizontal line intersects this graph in exactly one point. The inverse of this restriction is called the arctan function:

$$\arctan y = \theta \quad \text{if} \quad \theta \in \left(-\frac{\pi}{2}, \frac{\pi}{2}\right) \quad \text{and} \quad y = \tan\theta.$$

Note $\arctan y$ means the angle θ whose tangent is y. The domain of the arctan function is the range of the tangent function which is $\Re$. The range of the arctan function is the domain $D_2 = \left(-\frac{\pi}{2}, \frac{\pi}{2}\right)$ of the restriction of the tangent function we are using. The graph of $y = \arctan x$ is the reflection about the line $y = x$ of the graph of $y = \tan x$ with domain $\left(-\frac{\pi}{2}, \frac{\pi}{2}\right)$ shown in the right diagram of Figure 1.4.37. The vertical asymptotes of the graph of $y = \tan x$ at $x = \pm\frac{\pi}{2}$, when reflected about the line $y = x$, become *horizontal asymptotes* $y = \pm\frac{\pi}{2}$ of the graph of $y = \arctan x$. That is, the graph of $y = \arctan x$ approaches the line $y = -\frac{\pi}{2}$ on the left and the line $y = \frac{\pi}{2}$ on the right. This graph has the interesting property that it has different horizontal asymptotes on the left and the right.

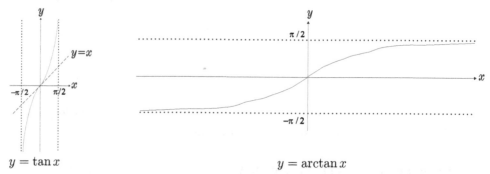

$y = \tan x$ $y = \arctan x$

Figure 1.4.37 Defining an Inverse of the Tangent Function

Examples 1.4.38 (1) Find $\arctan\sqrt{3}$.

>**Solution** By definition, $\theta = \arctan\sqrt{3}$ if $\tan\theta = \sqrt{3}$ and $-\frac{\pi}{2} < \theta < \frac{\pi}{2}$. From a 30–6–90 right triangle, we see that $\tan\frac{\pi}{3} = \sqrt{3}$. Hence $\arctan\sqrt{3} = \frac{\pi}{3}$.

(2) Find $\arctan -1$.

>**Solution** By definition, $\theta = \arctan -1$ if $\tan\theta = -1$ and $-\frac{\pi}{2} < \theta < \frac{\pi}{2}$. We know from an isosceles right triangle that $\tan -\frac{\pi}{4} = -1$. Hence $\arctan -1 = -\frac{\pi}{4}$.

(3) Find $\tan(\arctan 3)$.

>**Solution** We are asked to find the tangent of the angle whose tangent is 3. The question itself gives the answer: $\tan(\arctan 3) = 3$.

(4) Find $\arctan\left(\tan\frac{6\pi}{7}\right)$.

>**Solution** By definition, $\theta = \arctan\left(\tan\frac{6\pi}{7}\right)$ if $\tan\theta = \tan\frac{6\pi}{7}$ and $-\frac{\pi}{2} < \theta < \frac{\pi}{2}$. Note that $\frac{6\pi}{7}$ is not between $-\frac{\pi}{2}$ and $\frac{\pi}{2}$. Since the tangent function is periodic

with period π, we have $\tan \frac{6\pi}{7} = \tan\left(\frac{6\pi}{7} - \pi\right) = \tan -\frac{\pi}{7}$. Hence $\arctan\left(\tan\frac{6\pi}{7}\right) = \arctan\left(\tan -\frac{\pi}{7}\right) = -\frac{\pi}{7}$ because $-\frac{\pi}{2} < -\frac{\pi}{7} < \frac{\pi}{2}$.

(5) Find $\sin\left(\arctan \frac{x+1}{x-1}\right)$.

Solution   Let $\theta = \arctan \frac{x+1}{x-1}$. Then $-\frac{\pi}{2} < \theta < \frac{\pi}{2}$ and $\tan\theta = \frac{x+1}{x-1}$. Consider the right triangle with angle θ in Figure 1.4.39 having opposite side of length $x + 1$ and adjacent side of length $x - 1$. By the Pythagorean Theorem, the hypotenuse has length $\sqrt{(x+1)^2 + (x-1)^2} = \sqrt{2x^2 + 2}$. Hence $\sin\left(\arctan \frac{x+1}{x-1}\right) = \sin\theta = \frac{x+1}{\sqrt{2x^2+2}}$.   □

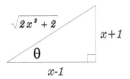

Figure 1.4.39
$\tan\theta = \frac{x+1}{x-1}$

The analysis of the function $y = \cot\theta$ is analogous to the analysis of $y = \tan\theta$. The restriction of the cotangent function to $D_3 = (0, \pi)$, whose graph is sketched in the left diagram of Figure 1.4.40, is one–to–one. The inverse of this restriction is called the arccot function:

$$\operatorname{arccot} y = \theta \quad \text{if} \quad \theta \in (0, \pi) \quad \text{and} \quad y = \cot\theta.$$

Note $\operatorname{arccot} y$ means the angle θ whose cotangent is y. The domain of the arccot function is the range of the cotangent function which is $\Re$. The range of the arccot function is the domain $D_3 = (0, \pi)$ of the restriction of the cotangent function we are using. The graph of $y = \operatorname{arccot} x$ is the reflection about the line $y = x$ of the graph of $y = \cot x$ with domain $(0, \pi)$ shown in the right diagram of Figure 1.4.40. This graph has the line $y = \pi$ as a horizontal asymptote on the left and the x–axis as a horizontal asymptote on the right.

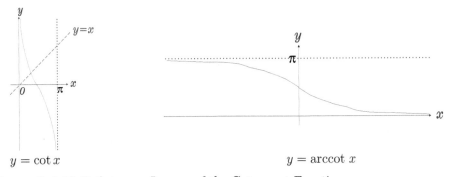

$y = \cot x$ $y = \operatorname{arccot} x$

Figure 1.4.40 Defining an Inverse of the Cotangent Function

Now consider the function $y = \sec\theta$ with domain all numbers which are not odd multiples of $\frac{\pi}{2}$. The restriction of the secant function to $D_4 = \left[0, \frac{\pi}{2}\right) \cup \left(\frac{\pi}{2}, \pi\right]$ is one–to–one. We verify this from the graph of this restriction depicted in the left diagram of Figure 1.4.41. The inverse of this restriction is called the arcsec function:

$$\operatorname{arcsec} y = \theta \quad \text{if} \quad \theta \in \left[0, \frac{\pi}{2}\right) \cup \left(\frac{\pi}{2}, \pi\right] \quad \text{and} \quad y = \sec\theta.$$

Note $\operatorname{arcsec} y$ means the angle θ whose secant is y. The domain of the arcsec function is the range of the secant function which is $(-\infty, -1] \cup [1, \infty)$. The range of the arcsec function is the domain D_4 of the restriction of the secant function we are using. The graph of $y = \operatorname{arcsec} x$ is the reflection about the line $y = x$ of the graph of $y = \sec x$ with domain D_4 shown in the right diagram of Figure 1.4.41. Note that the vertical asymptote of the graph of $y = \sec x$ at $x = \frac{\pi}{2}$, when reflected about the line $y = x$, becomes the horizontal asymptote $y = \frac{\pi}{2}$ of the graph of $y = \operatorname{arcsec} x$.

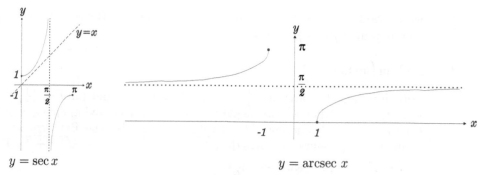

$y = \sec x$ $y = \text{arcsec } x$

Figure 1.4.41 Defining an Inverse of the Secant Function

The last of the trigonometric functions is $y = \csc \theta$ with domain all numbers which are not multiples of π. The restriction of this function to $D_5 = \left[-\frac{\pi}{2}, 0\right) \cup \left(0, \frac{\pi}{2}\right]$ is one–to–one. We verify this from the graph of this restriction depicted in the left diagram of Figure 1.4.42. The inverse of this restriction is called the arccsc function:

$$\text{arccsc } y = \theta \quad \text{if} \quad \theta \in \left[-\tfrac{\pi}{2}, 0\right) \cup \left(0, \tfrac{\pi}{2}\right] \quad \text{and} \quad y = \csc \theta.$$

Note arccsc y means the angle θ whose cosecant is y. The domain of the arccsc function is the range of the cosecant function which is $(-\infty, -1] \cup [1, \infty)$. The range of the arccsc function is the domain D_5 of the restriction of the cosecant function we are using. The graph of $y = \text{arccsc } x$ is the reflection about the line $y = x$ of the graph of $y = \csc x$ with domain D_5 shown in the right diagram of Figure 1.4.42. Note that the y–axis is a vertical asymptote of the graph of $y = \csc x$. When reflected about the line $y = x$, the x–axis becomes the horizontal asymptote of the graph of $y = \text{arccsc } x$.

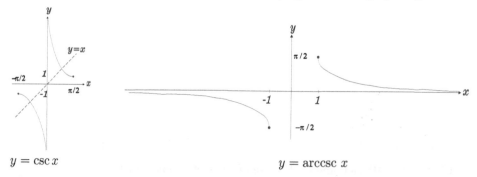

$y = \csc x$ $y = \text{arccsc } x$

Figure 1.4.42 Defining an Inverse of the Cosecant Function

Examples 1.4.43 (1) Find arcsec $\sqrt{2}$.

Solution By definition, $\theta = \text{arcsec } \sqrt{2}$ if $\sec \theta = \sqrt{2}$ and $0 < \theta < \pi$, $\theta \neq \frac{\pi}{2}$. From an isosceles right triangle, we see that $\sec \frac{\pi}{4} = \sqrt{2}$. Hence arcsec $\sqrt{2} = \frac{\pi}{4}$.

(2) Find arccot $-\sqrt{3}$.

Solution By definition, $\theta = \text{arccot } -\sqrt{3}$ if $\cot \theta = -\sqrt{3}$ and $0 < \theta < \pi$. We know from a 30–60–90 right triangle that $\cot \frac{5\pi}{6} = -\sqrt{3}$. Hence arccot $-\sqrt{3} = \frac{5\pi}{6}$.

(3) Find csc (arccsc 7).

Solution We are asked to find the cosecant of the angle whose cosecant is 7. The question itself gives the answer: csc (arccsc 7) = 7.

(4) Find arcsec $\sec\left(-\frac{3\pi}{8}\right)$.

Solution By definition, $\theta = \mathrm{arcsec}\left(\sec -\frac{3\pi}{8}\right)$ if $\sec\theta = \sec -\frac{3\pi}{8}$ and $0 < \theta < \pi$, $\theta \neq \frac{\pi}{2}$. Since $\sec -\frac{3\pi}{8} = \sec\frac{3\pi}{8}$, we have $\mathrm{arcsec}\left(\sec -\frac{3\pi}{8}\right) = \mathrm{arcsec}\left(\sec\frac{3\pi}{8}\right) = \frac{3\pi}{8}$ because $0 < \frac{3\pi}{8} < \pi$.

(5) Compute $\sin\left(\mathrm{arccot}\,\frac{6}{5}\right)$.

Solution Let $\theta = \mathrm{arccot}\,\frac{6}{5}$. Then $0 < \theta < \pi$ and $\cot\theta = \frac{6}{5}$. Since $\cot\theta > 0$, we have $0 < \theta < \frac{\pi}{2}$. Draw a right triangle with angle θ in Figure 1.4.44 with adjacent side of length 6 and opposite side of length 5. By the Pythagorean Theorem, the hypotenuse has length $\sqrt{6^2 + 5^2} = \sqrt{61}$. Hence $\sin\left(\mathrm{arccot}\,\frac{6}{5}\right) = \sin\theta = \frac{5}{\sqrt{61}}$.

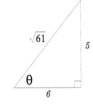

Figure 1.4.44
$\cot\theta = \frac{6}{5}$

(6) Find $\sec\left[\mathrm{arccsc}\,(3x^2 + 2)\right]$.

Solution Let $\theta = \mathrm{arccsc}\,(3x^2 + 2)$. Then $0 < \theta < \frac{\pi}{2}$ and $\csc\theta = 3x^2 + 2$. Consider the right triangle with angle θ in Figure 1.4.45 having hypotenuse of length $3x^2 + 2$ and opposite side of length one. By the Pythagorean Theorem, the adjacent side has length $\sqrt{(3x^2 + 2)^2 - 1^2} = \sqrt{9x^4 + 12x^2 + 3}$. Hence

$$\sec\left[\mathrm{arccsc}\,(3x^2 + 2)\right] = \sec\theta = \frac{3x^2 + 2}{\sqrt{9x^4 + 12x^2 + 3}}\,. \qquad \square$$

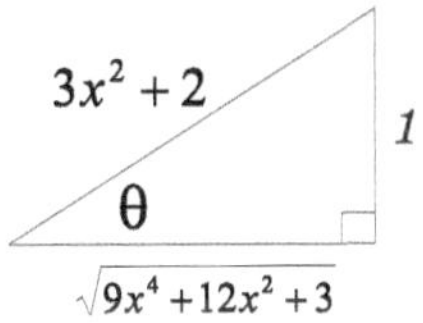

Figure 1.4.45
$\csc\theta = 3x^2 + 2$

In most contexts, the inverse function of $f(x)$ is denoted as $f^{-1}(x)$. Here too, some people use this notation for the inverse trigonometric functions:

$$\arcsin x = \sin^{-1}x, \qquad \arccos x = \cos^{-1}x, \qquad \arctan x = \tan^{-1}x,$$
$$\mathrm{arccot}\,x = \cot^{-1}x, \qquad \mathrm{arcsec}\,x = \sec^{-1}x, \qquad \mathrm{arccsc}\,x = \csc^{-1}x.$$

Be careful: $\sin^{-1}x$ denotes the inverse function of $\sin x$, not the reciprocal $\frac{1}{\sin x}$ of $\sin x$. The domains and ranges of the inverse trigonometric functions are summarized in the following table. This information was derived in the discussion above.

$$
\begin{aligned}
\arcsin x &= \sin^{-1}x \\
\text{domain} &= [-1, 1] \\
\text{range} &= [-\pi/2, \pi/2]
\end{aligned}
\qquad
\begin{aligned}
\arccos x &= \cos^{-1}x \\
\text{domain} &= [-1, 1] \\
\text{range} &= [0, \pi]
\end{aligned}
$$

$$
\begin{aligned}
\arctan x &= \tan^{-1}x \\
\text{domain} &= \Re \\
\text{range} &= (-\pi/2, \pi/2)
\end{aligned}
\qquad
\begin{aligned}
\mathrm{arccot}\,x &= \cot^{-1}x \\
\text{domain} &= \Re \\
\text{range} &= (0, \pi)
\end{aligned}
$$

$$
\begin{aligned}
\mathrm{arcsec}\,x &= \sec^{-1}x \\
\text{domain} &= (-\infty, -1] \cup [1, \infty) \\
\text{range} &= [0, \pi/2) \cup (\pi/2, \pi]
\end{aligned}
\qquad
\begin{aligned}
\mathrm{arccsc}\,x &= \csc^{-1}x \\
\text{domain} &= (-\infty, -1] \cup [1, \infty) \\
\text{range} &= [-\pi/2, 0) \cup (0, \pi/2]
\end{aligned}
$$

Historical Remarks

Degrees were defined by the Babylonians in the fourth century BCE to introduce a coordinate system in the sky. In the second century BCE, astronomical problems led the Greek mathematician Hipparchus of Bithynia to study the chord function

$$\mathrm{crd}(\theta) = 2R\sin\frac{\theta}{2},$$

the length of a chord of angle θ in a circle of radius R depicted in Figure 1.4.46. In the second century CE, Claudius Ptolemy wrote the *Almagest*, a fundamental work of astronomy, which includes a table of the chord function. Ptolemy also derives many properties of the chord function, including formulas equivalent to the usual one for $\sin(\theta - \phi)$ and the law of cosines. He also developed spherical trigonometry to solve spherical triangles. He applied these methods in his astronomical calculations.

In the 3rd century CE, the Chinese mathematician Liu Hui used similar triangles to compute distances. These computations can be interpreted as describing the unknown opposite side of a large right triangle as the known length of the adjacent side times the tangent of the angle θ. He used a small triangle, similar to the large original one, to compute the value of $\tan\theta$. This led to the construction of tangent tables in China in the 8th century. In the 4th century CE, the Indian book *Surya-Siddhanta* appeared. It contains tables of the half–chord of the double of an angle θ in a circle of radius one. This is precisely $\sin\theta$. Trigonometric knowledge came to the Islamic world from India and China through military conquests. In the 9th century, Ahmad ibn Abdallah al-Marwazi Habas al-Hasib introduced all six of the trigonometric functions. Tables of these functions were constructed and used for both astronomical computations as well as calculations of distances between earthly objects.

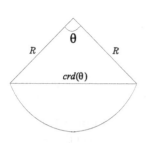

Figure 1.4.46
The Chord Function

Trigonometry in medieval Europe was learned from translations of Ptolemy's Almagest and Islamic texts. Its use, however, was limited to astronomical calculations. Progress was made in the Renaissance, including the discovery of the law of sines by Johannes Müller in 1463. The terms *tangent*, *secant*, *sine complement* (cosine), *tangent complement* (cotangent) and *secant complement* (cosecant) were introduced by Thomas Finck in 1583. Trigonometric tables were improved to the extent that a 15 place table begun by Georg Joachim Rheticus was completed fifty years later by Otho and Bartholomew Pitiscus in 1613. The need to do arithmetic with these fifteen digit numbers in astronomy and navigation was one of the motivations of John Napier when he introduced logarithms in 1614. Logarithms convert multiplication and division into addition and subtraction.

Inverse trigonometric functions were first studied as infinite series. The series for the arctangent was discovered by the Indian mathematician Kerala Gargya Nilakantha in 1500 using geometric methods. This series, as well as the one for the arcsin, were rediscovered independently by Isaac Newton in 1669, James Gregory in 1670 and Gottfried Wilhelm Leibniz in 1676 in their studies of calculus.

Summary

The reader should be able to measure angles in radians. Definitions, computations and graphs of the six trigonometric functions should be understood, and the basic trigonometric identities should be memorized. In addition, she should be able to construct the graphs of more complicated trigonometric functions of the form $f(x) = A\sin Bx$.

The reader should understand how to restrict each of the trigonometric functions to construct its inverse. She should know the domains, ranges and graphs of the inverse trigonometric functions and be able to compute an inverse trigonometric function of a number in its domain.

Basic Exercises

1. Write each angle in degrees and in radians.
    (a) α is two clockwise revolutions of the unit rod.
    (b) β is a quarter counterclockwise revolution of the unit rod.
    (c) γ is three fifths of a clockwise revolution of the unit rod.
    (d) δ is five counterclockwise revolutions of the unit rod.
    (e) ϵ is one clockwise revolution of the unit rod.

2. Write each angle in radians.
(a) 180° (b) 30° (c) −18° (d) 360° (e) 10° (f) −240° (g) 40° (h) 225°

3. Write each angle in degrees.
(a) $\pi/4$ radians (b) 3π radians (c) $-2\pi/3$ radians (d) π radians
(e) 3 radians (f) $5\pi/6$ radians (g) $\pi/10$ radians (h) $-\pi/6$ radians

4. In each case, find the arclength of the sector of the circle of radius R and angle θ.
(a) $R = 4$, $\theta = 2\pi/3$ radians (b) $R = 3$, $\theta = 60°$
(c) $R = 10$, $\theta = 11\pi/6$ radians (d) $R = 6$, $\theta = 210°$

5. Find the domain of each function.
(a) $f(x) = 5\sin^3 x - 9\cos^4 x$ (b) $g(x) = \frac{9\sin x}{7\tan^2 x + 3\cot^2 x + 1}$
(c) $h(x) = \frac{\cot x}{2\sec x + 1}$ (d) $k(x) = \sqrt{\csc x}$

6. Calculate those trigonometric functions which are defined for each angle.
(a) $\pi/3$ radians (b) π radians (c) $5\pi/4$ radians (d) $-\pi/6$ radians
(e) $2\pi/3$ radians (f) $15\pi/2$ radians (g) $-17\pi/6$ radians (h) $21\pi/4$ radians

7. Calculate the six trigonometric functions of the angles θ, ϕ, ψ in the triangles below.

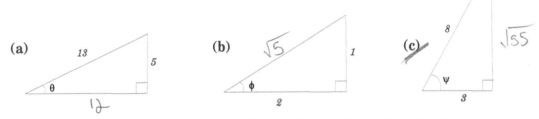

(a) 13 5 θ 12

(b) $\sqrt{5}$ 1 ϕ 2

(c) 8 $\sqrt{55}$ ψ 3

8. A six foot tall man casts an eight foot shadow. What is the angle of elevation θ of the sun? (θ is the angle between the ground and a line from the ground to the sun.)

9. A tourist is standing 50 meters from a tall building. The angle of elevation α of the building is 80°. How high is the building? (α is the angle between the ground and the line from the tourist to the top of the building.)

10. A pilot flies over a landmark which he knows is ten miles from the airport. The angle of depression β of the airport is 10°. (β is the angle between the line from the airplane to the airport and the horizontal plane through the airplane.)
(a) What is the altitude of the airplane?
(b) What is the distance between the airplane and the airport?

11. A hiker standing at the top of a 100 meter high cliff sees a bear crossing a river on the plain below. The angle of depression γ of the bear is 30°. (γ is the angle between the line from the hiker to the bear and the horizontal plane through the hiker.)
(a) How far is the bear from the bottom of the cliff?
(b) What is the distance between the bear and the hiker?

12. For each pair of functions f, g, find formulas for the functions $f + g$, fg, f/g, $f \circ g$ and $g \circ f$. Determine a suitable domain for each of these functions.
(a) $f(x) = \tan x$, $g(x) = \sqrt{x}$ (b) $f(x) = x^2 - 1$, $g(x) = \sec x$
(c) $f(x) = \cos x$, $g(x) = 1/x$ (d) $f(x) = \frac{x}{x^2+1}$, $g(x) = \cot x$

13. Determine whether each function is an even function, an odd function or neither.
(a) $m(x) = \sin x \cos x$ (b) $n(x) = \tan x \sec x + 1$
(c) $p(x) = \frac{11x^7 + 3x^3}{x + \cot x}$ (d) $q(x) = \frac{\sin x + \tan x}{\sec x + \cos x}$

14. Show that $\sin\theta = \cos(\theta - \pi/2)$.

15. Show that $\cot\theta$ is periodic with period π.

16. **(a)** Find an identity for $\tan(\theta + \phi)$ in terms of $\tan \theta$ and $\tan \phi$.
(b) Use this identity to derive a formula for $\tan 2\theta$.

17. **(a)** Find a formula for $\sin 3\theta$ in terms of $\sin \theta$.
(b) Find a formula for $\cos 3\theta$ in terms of $\cos \theta$.

18. Find a formula for each expression in terms of $\cos 2\theta$ and $\cos 4\theta$ which contains no products of trigonometric functions.
    (a) $\sin^2 \theta$ **(b)** $\cos^2 \theta$ **(c)** $\sin^4 \theta$ **(d)** $\cos^4 \theta$

19. Find all angles and the lengths of all sides in each triangle below.

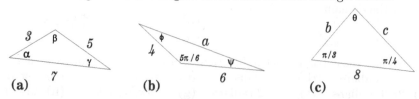

    (a)   **(b)**   **(c)**

20. Find the tangents of the angles between the diagonal AB of a cube and the three diagonals of its square faces which begin at A.

21. Sketch the graphs of each function.
    (a) $f(x) = 3 \sin x$ **(b)** $g(x) = \sin 4x$ **(c)** $h(x) = 3 \sin 4x$
    (d) $k(x) = 5 \sin x$ **(e)** $m(x) = \sin 2x$ **(f)** $n(x) = 5 \sin 2x$

22. Sketch the graph of each function.
    (a) $f(x) = x^2 \sin x$ **(b)** $g(x) = \sin \frac{1}{x^2}$ **(c)** $h(x) = x^2 \sin \frac{1}{x^2}$
    (d) $k(x) = \frac{\sin x}{1+x}$ **(e)** $m(x) = \frac{\sin x}{1+x^2}$ **(f)** $n(x) = \sin \frac{\pi}{1+x^2}$

23. Sketch the graph of each function.
    (a) $f(x) = 3 \cos 2x$ **(b)** $g(x) = 2 \cot 4x$ **(c)** $h(x) = 5 \csc 3x$
    (d) $j(x) = x^2 \cos x$ **(e)** $k(x) = \cos \frac{1}{x}$ **(f)** $m(x) = x^2 \cos \frac{1}{x}$

24. Find each angle.
(a) $\arccos \frac{1}{2}$ (b) $\arccos 1$ (c) $\arccos -\frac{\sqrt{2}}{2}$ (d) $\arccos -1$ (e) $\arccos \frac{\sqrt{3}}{2}$ (f) $\arccos 0$

25. Find each angle.
(a) $\arcsin \frac{\sqrt{2}}{2}$ (b) $\arcsin -\frac{1}{2}$ (c) $\arcsin 1$ (d) $\arcsin -\frac{\sqrt{3}}{2}$ (e) $\arcsin -\frac{1}{\sqrt{2}}$ (f) $\arcsin 0$

26. Find each angle.
(a) $\arctan \frac{1}{\sqrt{3}}$ (b) $\arctan 1$ (c) $\arctan -\sqrt{3}$ (d) $\arctan -\frac{\sqrt{3}}{3}$ (e) $\arctan 0$

27. Find each angle.
(a) $\operatorname{arccot} \frac{1}{\sqrt{3}}$ (b) $\operatorname{arcsec} 1$ (c) $\operatorname{arccsc} \sqrt{2}$ (d) $\operatorname{arcsec} -2$ (e) $\operatorname{arccot} -1$ (f) $\operatorname{arccsc} -\frac{2}{\sqrt{3}}$

28. Evaluate each expression.
    (a) $\tan(\arcsin 1/\sqrt{2})$ **(b)** $\cos(\arccos 1/10)$ **(c)** $\sec(\arctan 7)$
    (d) $\cot(\arcsin -5/13)$ **(e)** $\sin(\operatorname{arccsc} 3)$ **(f)** $\operatorname{arccot}(\cot -\pi/4)$
    (g) $\arccos(\cos -\pi/10)$ **(h)** $\arcsin(\sin 6\pi/7)$ **(i)** $\operatorname{arcsec}(\tan \pi/4)$

29. Find a formula for each function which does not involve trigonometric functions.
    (a) $f(x) = \sec\left(\arctan \frac{1}{x}\right)$ **(b)** $g(x) = \cot\left(\arccos \frac{x}{x+1}\right)$
    (c) $h(x) = \sin\left(\arctan \sqrt{x}\right)$ **(d)** $k(x) = \cos\left[\arcsin(2x - 1)\right]$
    (e) $m(x) = \tan\left[\operatorname{arcsec}(x^2 + 1)\right]$ **(f)** $n(x) = \csc\left[\operatorname{arccot}(\sqrt{x} - 1)\right]$

30. A man two meters tall is walking away from a six meter high street lamp. He measures the distance from his head H to the shadow of his head S on the ground as x meters. Let θ denote the angle between HS and the ground.
    (a) Find $\sin \theta$.   **(b)** Find the angle θ as a function of x.
    (c) Use the answer to (b) to find the length of his shadow L as a function of x.

31. A child holds a ball of string whose loose end is tied to a balloon which rises vertically from a spot on the ground ten meters from the child. The child knows how many meters x of string are between himself and the balloon. Let ϕ denote the angle between the string and the ground.
(a) Find $\sec \phi$.   **(b)** Find the angle ϕ as a function of x.
(c) Use the answer to (b) to find the altitude h of the balloon as a function of x.

32. The bottom of a crane is at the point P on the ground. The crane raises its fifteen meter long arm PQ. Let x denote the distance of the point on the ground under Q from P. Let ψ denote the angle between the arm of the crane and the ground.
(a) Find $\cos \psi$.   **(b)** Find the angle ψ as a function of x.
(c) Use the answer to (b) to find the height y of Q from the ground as a function of x.

33. Verify that the chord function of Hipparchus is given by $\operatorname{crd}(\theta) = 2R \sin \frac{\theta}{2}$.

Challenging Problems

1. Find a formula for the area of a sector of a circle of radius R and angle θ where θ is measured in radians.

2. Find formulas for $\sin \frac{\theta}{2}$ and $\cos \frac{\theta}{2}$ in terms of $\sin \theta$ and $\cos \theta$.

3. Find all angles θ which satisfy the equation $2 \sin^2 \theta - \sin \theta - 1 = 0$.

4. (a) Let A, B, C be the angles of a triangle which is not a right triangle. Show

$$\tan A + \tan B + \tan C = \tan A \, \tan B \, \tan C.$$

(b) Show that $\arctan 1 + \arctan 2 + \arctan 3 = \pi$.

5. For those who are familiar with complex numbers, define $E(\theta) = \cos \theta + i \sin \theta$.
(a) Show that $E(\theta + \phi) = E(\theta)E(\phi)$.   **(b)** Show that $E(\theta - \phi) = E(\theta)/E(\phi)$.
(c) Show that $E(n\theta) = E(\theta)^n$ for all integers n.

1.5 Defining Limits

We explain the concept of the limit of a function and show how to use intuition to compute the values of various types of limits. Then we give precise geometric definitions. These definitions are applied to justify our intuitive computations. The first subsection studies numerical limits. The second subsection study limits at infinity while the third subsection studies limits whose values are infinite. The section concludes with an addendum which gives algebraic interpretations of our geometric definitions.

Basic Limits

The following example illustrates how a limit is used to understand the behavior of a function near a point where its value is undefined. This motivates the intuitive definition of a limit which we use to evaluate several limits. Then we present a proper geometric definition of a limit and apply it to justify the values of several simple limits.

Motivating Example 1.5.1 Consider the functions

$$g(x) = \frac{x + 1}{\sqrt{x} - 1} \quad \text{and} \quad f(x) = \frac{x - 1}{\sqrt{x} - 1}.$$

The domain of these functions consists of all $x \geq 0$ where $\sqrt{x}$ is defined and excludes $x = 1$ where their denominators are zero. That is, the domain of f and g is $D =$

$[0, 1) \cup (1, \infty)$. As x approaches 1, the numerator of $g(x)$ approaches two and the denominator of $g(x)$ approaches zero. Therefore the values of the fraction $g(x)$ become infinite. However, the behavior of $f(x)$, as x approaches one, is unclear because both the numerator and denominator of $f(x)$ approach zero. To analyze this situation, we use a calculator to compute the values of $f(x)$ for several numbers x close to one.

x	0	2	.5	1.5	.9	1.1	.99	1.01	.999	1.001
$f(x)$	1	2.4142	1.7071	2.2247	1.9487	2.0488	1.9950	2.0050	1.9995	2.0005

This table indicates that as x approaches one, the values of $f(x)$ approach two. We say the limit of $f(x)$, as x approaches one, equals two even though $f(2)$ is not defined.

We simplify the formula for $f(x)$ to clarify the situation and justify our observations. Write the numerator as the difference of two squares and factor it:

$$f(x) = \frac{x-1}{\sqrt{x}-1} = \frac{\sqrt{x}^2 - 1^2}{\sqrt{x}-1} = \frac{(\sqrt{x}-1)(\sqrt{x}+1)}{\sqrt{x}-1} = \sqrt{x}+1.$$

Let $h(x) = \sqrt{x} + 1$ with domain $[0, \infty)$. The preceding computation can be phrased as saying that $f(x) = h(x)$ for every non–negative number x other than one. In addition, $h(1) = 2$ while $f(1)$ is not defined. As x approaches one, the values of $f(x) = h(x) = \sqrt{x} + 1$ approach $h(1) = 2$. Consider the graph of $y = f(x)$ in Figure 1.5.2 which consists of the top half of the parabola $y = h(x) = \sqrt{x} + 1$ with the point $(1, 2)$ removed. When x_1 on the x–axis approaches 1 from the left, the point $f(x_1)$ on the y–axis rises to height 2. When x_2 on the x–axis approaches 1 from the right, the point $f(x_2)$ on the y–axis falls to height 2.   □

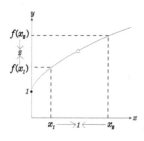

Figure 1.5.2
$y = f(x)$

We introduce notation to describe the situation of the preceding example: when x approaches c, the values of $f(x)$ approach L. First, we introduce notation to describe how c relates to the domain of f. Let c be a point of an interval I. Assume that the domain of f includes all points of the interval I, except possibly the point c. That is, $f(c)$ may not be defined. Assume that if $x \in I$, with $x \neq c$, then $f(x)$ is close to L. In fact, assume that as this number x gets closer and closer to c, the values of $f(x)$ get closer and closer to L. If c is an interior point of I write:

$$\lim_{x \to c} f(x) = L.$$

When c is the left endpoint of I, we are checking that for $x > c$ the values of $f(x)$ get closer and closer to L as x gets closer and closer to c. In this case write:

$$\lim_{x \to c^+} f(x) = L.$$

Similarly, when c is the right endpoint of I, we are checking that for $x < c$ the values of $f(x)$ get closer and closer to L as x gets closer and closer to c. In this case write:

$$\lim_{x \to c^-} f(x) = L.$$

$\lim_{x \to c} f(x) = L$ is called a two–sided limit because x approaches c from both the left and the right. The other two limits are called one–sided limits. In particular, $\lim_{x \to c^+} f(x) = L$ is called a right–hand limit because x approaches c from the right while $\lim_{x \to c^-} f(x) = L$ is called a left–hand limit because x approaches c from the left. Note that in evaluating these limits we do not care whether $f(c) = L$ or even if $f(c)$ is defined. We only want to know whether the values of $f(x)$, for $x \neq c$, approach L as x approaches c, but does not equal c.

The preceding definition of a limit is intuitive. Before making this definition precise, we consider several examples.

Examples 1.5.3 (1) Let $f(x) = 3x + 1$ with domain $\Re$. Let $c = 2$ and $I = \Re$. As x gets closer and closer to 2, $f(x) = 3x + 1$ gets closer and closer to $3 \cdot 2 + 1 = 7$. Hence $\lim_{x \to 2}(3x + 1) = 7$. This behavior is typical of functions defined by formulas.

(2) Let $g(x) = \frac{x}{|x|}$ for $x \neq 0$ and $g(0) = 1$ with domain $\Re$. Observe that

$$g(x) = \left\{ \begin{array}{ll} 1 & \text{if } x \geq 0 \\ -1 & \text{if } x < 0 \end{array} \right\}.$$ The graph of g is depicted in Figure 1.5.4.

(a) Let $I_1 = (0, \infty)$. Since $|x| = x$ for $x > 0$, $g(x) = 1$ for $x \in I_1$. Therefore, $\lim_{x \to 0^+} g(x) = 1$.

(b) Let $I_2 = (-\infty, 0)$. Since $|x| = -x$ for $x < 0$, $g(x) = -1$ for $x \in I_2$. Therefore, $\lim_{x \to 0^-} g(x) = -1$ even though $g(0) = +1$.

(c) Let $I_3 = \Re$. Since there are positive numbers x arbitrarily close to 0 with $g(x) = 1$ and negative numbers x' arbitrarily close to 0 with $g(x') = -1$, neither 1 nor -1 can be the limit of g as x approaches 0. Therefore, $\lim_{x \to 0} g(x)$ does not exist even though $g(0) = 1$.

(d) If we redefine $g(0)$ in any way whatsoever then the statements in (a), (b), (c) remain valid since limits, as x approaches 0, ignore the value of $g(0)$.

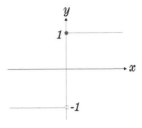

Figure 1.5.4
$g(x) = \frac{x}{|x|}$

The behavior of $g(x)$ is typical of a function whose graph consists of two curves with a gap between them.

(3) (a) Let k be the function with domain $\Re$ defined by

$$k(x) = \left\{ \begin{array}{ll} 1 & \text{if } x \text{ is a rational number} \\ 0 & \text{if } x \text{ is an irrational number} \end{array} \right\}.$$

The graph of k in Figure 1.5.5 looks like two horizontal dotted lines: $y = 0$ and $y = 1$. Note that the dots on these lines are so close together that the spaces between them are not visible under any magnification. Let c be any number. There are rational numbers r arbitrarily close to c with $k(r) = 1$, and there are irrational numbers i arbitrarily close to c with $k(i) = 0$. Thus, $\lim_{x \to c} k(x)$ does not exist even though $k(c)$ is defined. This function is as bad as possible when it comes to evaluating limits.

Figure 1.5.5
$y = k(x)$

(b) Modify the definition of k to define the function k_1 with domain $\Re$ as follows:

$$k_1(x) = \left\{ \begin{array}{ll} x & \text{if } x \text{ is a rational number} \\ 0 & \text{if } x \text{ is an irrational number} \end{array} \right\}.$$

The graph of k_1 in Figure 1.5.6 looks like two dotted lines: $y = 0$ and $y = x$. Again, the dots on these lines are so close together that the spaces between them are not visible under any magnification. Observe that $\lim_{x \to 0} k_1(x) = 0$ because both the dotted lines $y = 0$ and $y = x$ approach the origin as x approaches 0.

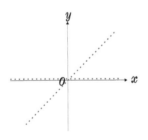

Figure 1.5.6
$y = k_1(x)$

(4) (a) Let G be the function with domain all numbers of $\Re$, except 0, defined by $G(x) = \sin \frac{1}{x}$. We constructed the graph of G in Figure 1.4.27 of the preceding section. Let $I = \Re$, and let $c = 0$. As x approaches 0, the graph of G moves in waves between -1 and $+1$. Therefore, $\lim_{x \to 0} G(x)$ does not exist.

(b) Let H be the function with domain all numbers of $\Re$, except 0, defined by $H(x) = x \sin \frac{1}{x}$. We constructed the graph of H in Figure 1.4.29 of the preceding section. As x approaches 0, the graph of H moves in waves between the lines $y = x$ and $y = -x$. Since both of these lines approach the origin as x approaches 0, the graph of H approaches the origin as well. Therefore, $\lim_{x \to 0} H(x) = 0$. $\square$

Our treatment of limits has been intuitive. To make our statements precise, we must define the terms "x is near c" and "$f(x)$ is near L". We do this in terms of the graph of f. Requiring $f(x)$ to be near L is equivalent to requiring that the point $(x, f(x))$ on the graph of f be near the horizontal line $y = L$. That is, $(x, f(x))$ should lie inside a narrow horizontal strip containing the line $y = L$. In particular, each positive number ϵ defines a horizontal strip $S_\epsilon(L)$ as follows:

$$S_\epsilon(L) \text{ equals all } (x, y) \text{ with } L - \epsilon < y < L + \epsilon \,.$$

The strip $S_\epsilon(L)$ is depicted as the shaded region in Figure 1.5.7. The lines bounding this region are dotted to indicate they are not included in $S_\epsilon(L)$. Let $(x, f(x)) \in S_\epsilon(L)$. When the value of ϵ decreases, the strip $S_\epsilon(L)$ narrows, and $f(x)$ is closer to L.

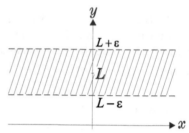

Figure 1.5.7 The Horizontal Strip $S_\epsilon(L)$

We use a positive number δ to specify numbers that are near c. Let $I_\delta(c)$ denote those numbers of the interval I, other than c, which are within δ units of c:

$$I_\delta(c) = \text{ set of } x \in I \text{ such that } x \in (c - \delta, c) \cup (c, c + \delta) \,.$$

When δ is small compared to the length of I, the three possibilities for $I_\delta(c)$ are depicted in Figure 1.5.8. They correspond to the cases where c is an interior point of I, c is the left endpoint of I or c is the right endpoint of I. Note that the smaller the value of δ, the smaller the interval $I_\delta(c)$, and the closer the points of $I_\delta(c)$ are to c. We now have the notation to make a precise definition of a limit.

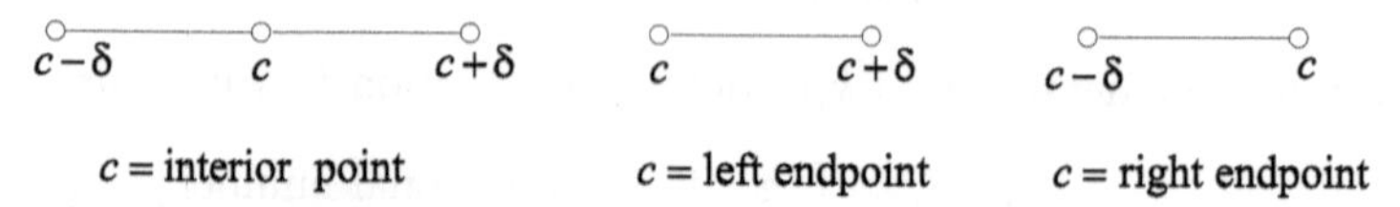

$$
\begin{array}{lll}
\underset{c-\delta \quad\quad c \quad\quad c+\delta}{\circ\!-\!-\!-\!-\!-\!\circ\!-\!-\!-\!-\!-\!\circ} & \underset{c \quad\quad c+\delta}{\circ\!-\!-\!-\!-\!-\!\circ} & \underset{c-\delta \quad\quad c}{\circ\!-\!-\!-\!-\!-\!\circ}
\end{array}
$$

$\quad\quad c = \text{interior point} \quad\quad\quad c = \text{left endpoint} \quad\quad c = \text{right endpoint}$

Figure 1.5.8 The Sets $I_\delta(c)$

Definition 1.5.9 *Let I be an interval of finite length with either $c \in I$ or c an endpoint of I. Let f be a function whose domain contains all numbers of I other than c. The limit of $f(x)$, as x approaches c, equals L if for every $\epsilon > 0$, there is a corresponding $\delta > 0$ such that the graph of f restricted to $I_\delta(c)$ is contained in the strip $S_\epsilon(L)$.*

Note c may, or may not, be in the domain of f.

The horizontal strip $S_\epsilon(L)$ is a thin target consisting of points (x, y) with y near L. To establish that $\lim\limits_{x \to c} f(x) = L$ we must show that for any such target, it is possible to choose a small piece of the graph of f near the point $(c, f(c))$ which lies inside the target. The graph of f restricted to $I_\delta(c)$ designates this small piece of the graph of f. When such an $I_\delta(c)$ can be chosen, $f(x)$ is near L for $x \in I_\delta(c)$. Figure 1.5.10 illustrates this definition when c is an interior point of I, c is the left endpoint of I or c is the right endpoint of I.

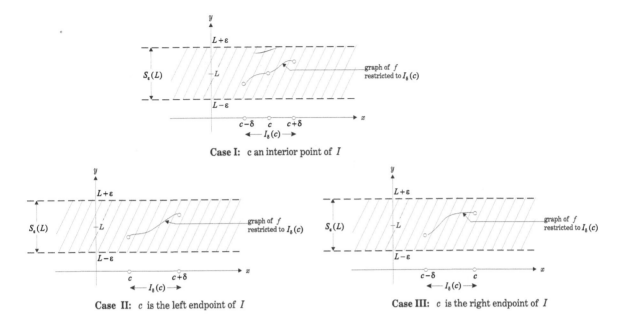

Figure 1.5.10 The Graph of f Restricted to $I_\delta(c)$ is Contained in $S_\epsilon(L)$

Observe that the choice of $I_\delta(c)$ depends on the given target $S_\epsilon(L)$: as $S_\epsilon(L)$ becomes narrower, $I_\delta(c)$ becomes smaller. In addition, given $S_\epsilon(L)$, there is never a unique choice for $I_\delta(c)$. If the graph of f restricted to $I_\delta(c)$ is contained in the strip $S_\epsilon(L)$ and $0 < \delta' < \delta$, then $I_{\delta'}(c) \subset I_\delta(c)$ and the graph of f restricted to $I_{\delta'}(c)$ is also contained in the strip $S_\epsilon(L)$.

We illustrate how to apply Definition 1.5.9 to verify the values of specific limits. The first two easy examples will be used in the next section to compute more complicated limits.

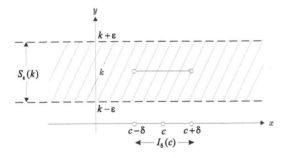

Figure 1.5.11 The Graph of $f(x) = k$ restricted to $I_\delta(c)$

Examples 1.5.12 (1) Let c be any number. Find $\lim_{x \to c} k$.

Solution Note that $f(x) = k$ is the constant function with domain $\Re$ whose graph is the horizontal line $y = k$. For any $\epsilon > 0$, the entire graph of f is contained in the strip $S_\epsilon(k)$ as shown in Figure 1.5.11. Thus any choice of δ will satisfy Definition 1.5.9 and show that

$$\lim_{x \to c} k = k. \qquad (1.5.1)$$

(2) Let c be any number. Find $\lim_{x \to c} x$.

Solution $f(x) = x$ has domain $\Re$. For any $\epsilon > 0$, the point $(x, f(x)) = (x, x)$ lies in $S_\epsilon(c)$ if $x \in I_\epsilon(c)$. See Figure 1.5.13. Thus the choice $\delta = \epsilon$ satisfies Definition 1.5.9, and

$$\lim_{x \to c} x = c. \qquad (1.5.2)$$

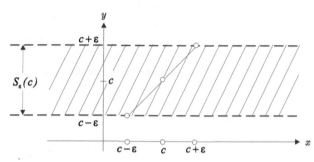

Figure 1.5.13 $f(x) = x$ restricted to $I_\epsilon(c)$

(3)   Verify that $\lim_{x \to 2}(3x + 1) = 7$, the limit of Example 1.5.3 (1).

Solution Given $\epsilon > 0$ we must specify how to choose a corresponding $\delta > 0$ such that if $x \in I_\delta(2) = (2 - \delta, 2) \cup (2, 2 + \delta)$, then $(x, 3x + 1) \in S_\epsilon(7)$, i.e. $3x + 1 \in (7 - \epsilon, 7 + \epsilon)$. Thus, we must find δ such that if

$$2 - \delta \; < \; x \; < \; 2 + \delta \text{ and } x \neq 2,$$

then

$$7 - \epsilon \; < \; 3x + 1 \; < \; 7 + \epsilon.$$

The latter inequality is equivalent to:

$$6 - \epsilon < \quad 3x \quad < 6 + \epsilon \quad \text{or} \qquad \text{[subtracting } 1\text{]}$$

$$2 - \epsilon/3 < \quad x \quad < 2 + \epsilon/3. \qquad \text{[dividing by } 3\text{]}$$

Thus, take $\delta = \epsilon/3$: if $x \in I_{\epsilon/3}(2) \subset (2 - \epsilon/3, 2 + \epsilon/3)$ then $3x + 1 \in (7 - \epsilon, 7 + \epsilon)$, i.e: $(x, 3x + 1) \in S_\epsilon(7)$. See Fig. 1.5.14. Since the formula $\delta = \epsilon/3$ produces a δ for every ϵ so that Def. 1.5.9 is satisfied, we have verified that $\lim_{x \to 2}(3x + 1) = 7$. $\square$

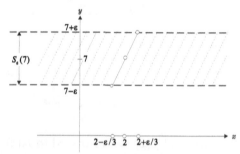

Figure 1.5.14 $f(x) = 3x + 1$ restricted to $I_{\epsilon/3}(2)$

Limits as x Approaches Infinity

We generalize the definition of a limit to the cases where x approaches infinity or x approaches minus infinity. First, we discuss the intuitive meaning of these limits and then give precise definitions.

The phrase "x approaches ∞" means that x gets larger and larger. Assume that $f(x)$ is defined when x is large, say $x > b$. We say the limit of $f(x)$, as x approaches ∞, equals L when the values of $f(x)$ get closer and closer to L as x gets larger and larger. Since the graph of f consists of the points $(x, f(x))$, this graph approaches the horizontal line $y = L$ as x gets large. The graph of f is said to have the line $y = L$ as a *horizontal asymptote on the right*. See the right diagram in Figure 1.5.15. We write:

$$\lim_{x \to \infty} f(x) = L.$$

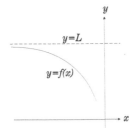

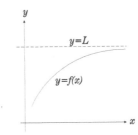

Horizontal Asymptote on the Left   Horizontal Asymptote on the Right

Figure 1.5.15 Horizontal Asymptotes

The phrase "x approaches $-\infty$" means that x gets smaller and smaller. (Just as "large numbers" refer to numbers on the right end of the number line, so "small numbers" refer to numbers on the left end of the number line. For example, $-1,000,000$ is a small number while -0.000001 is a large negative number.) Assume that $f(x)$ is defined when x is small, say $x < a$. We say the limit of $f(x)$, as x approaches $-\infty$, equals L when the values of $f(x)$ get closer and closer to L as x gets smaller and smaller. Since the graph of f consists of the points $(x, f(x))$, this graph approaches the horizontal line $y = L$ as x gets small. The graph of f is said to have the line $y = L$ as a *horizontal asymptote on the left*. See the left diagram in Figure 1.5.15. We write:

$$\lim_{x \to -\infty} f(x) = L.$$

Before giving precise definitions of these limits, we consider two examples.

Examples 1.5.17 (1) Let n be a positive integer. Let h be the function with domain all nonzero numbers defined by $h(x) = \frac{1}{x^n}$. If $n = 2s$ is even, the graph of h is given in Fig. 1.5.16. This graph has the x–axis as a horizontal asymptote on both the left and the right. If $n = 2t + 1$ is odd, the graph of h is given in Fig. 1.5.18. This graph also has the x–axis as a horizontal asymptote on both the left and the right. Thus in all cases:

$$\lim_{x \to \infty} \frac{1}{x^n} = 0 \quad \text{and} \quad \lim_{x \to -\infty} \frac{1}{x^n} = 0.$$

(2) Consider the function $y = k(x) = \arctan x$ whose graph is depicted in Fig. 1.4.37. This graph has the line $y = -\frac{\pi}{2}$ as a horizontal asymptote on the left and the line $y = \frac{\pi}{2}$ as a horizontal asymptote on the right. Hence

$$\lim_{x \to -\infty} \arctan x = -\frac{\pi}{2} \quad \text{and} \quad \lim_{x \to \infty} \arctan x = \frac{\pi}{2}.$$

In particular, $y = k(x) = \arctan x$ has different horizontal asymptotes on the left and the right. □

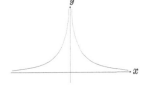

Figure 1.5.16
$y = \frac{1}{x^{2s}}$

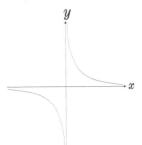

Figure 1.5.18
$y = \frac{1}{x^{2t+1}}$

To make a precise definition of a limit as x approaches ∞, we need to specify numbers which are near ∞. The larger the number, the closer it is to ∞. Thus define

$$I_\delta(\infty) = \text{set of } x \in I \text{ such that } x > \delta \tag{1.5.3}$$

to denote numbers near ∞. The larger the value of δ, the smaller the interval $I_\delta(\infty)$, and the closer the points of $I_\delta(\infty)$ are to ∞. Similarly, to make a precise definition of a limit as x approaches $-\infty$ we need to specify numbers which are near $-\infty$. The smaller the number, the closer it is to $-\infty$. Thus define

$$I_\delta(-\infty) \;=\; \text{set of } x \in I \text{ such that } x < -\delta \qquad (1.5.4)$$

to denote numbers near $-\infty$. The larger the value of δ, the smaller the interval $I_\delta(-\infty)$, and the closer the points of $I_\delta(-\infty)$ are to $-\infty$. These intervals are depicted in Figure 1.5.19.

$$\underset{\delta}{\circ\!\!-\!\!\!-\!\!\!-\!\!\!\longrightarrow} \qquad\qquad\qquad \underset{-\delta}{-\!\!\!-\!\!\!-\!\!\!-\!\!\!\circ}$$

$$I_\delta(\infty) \text{ when } (\delta,\infty) \subset I \qquad\qquad I_\delta(-\infty) \text{ when } (-\infty,-\delta) \subset I$$

Figure 1.5.19 The Intervals $I_\delta(c)$ when $c = \pm\infty$

Definition 1.5.20 (a) *Let f be a function whose domain contains an interval (a,∞). We say $\lim\limits_{x\to\infty} f(x) = L$ if for every $\epsilon > 0$, there is a corresponding $\delta > 0$ such that the graph of f restricted to the interval $I_\delta(\infty)$ is contained in the strip $S_\epsilon(L)$.*
(b) *Let f be a function whose domain contains an interval $(-\infty,b)$. We say $\lim\limits_{x\to-\infty} f(x) = L$ if for every $\epsilon > 0$, there is a corresponding $\delta > 0$ such that the graph of f restricted to the interval $I_\delta(-\infty)$ is contained in the strip $S_\epsilon(L)$.*

The horizontal strip $S_\epsilon(L)$ is a thin target consisting of points (x,y) with y near L. To establish that $\lim\limits_{x\to\infty} f(x) = L$ we must show that for any such target, it is possible to choose a piece of the graph of $y = f(x)$, for $x > \delta$, which lies inside this target. The graph of f restricted to $I_\delta(\infty)$ designates this piece of the graph of f. When such an $I_\delta(\infty)$ can be chosen, $f(x)$ is within ϵ units of L for $x \in I_\delta(\infty)$. The right diagram in Figure 1.5.21 illustrates this situation. Observe that the choice of $I_\delta(\infty)$ depends on the given target $S_\epsilon(L)$: as $S_\epsilon(L)$ becomes narrower, $I_\delta(\infty)$ becomes smaller. In addition, given $S_\epsilon(L)$, there is never a unique choice for $I_\delta(\infty)$. If the graph of f restricted to $I_\delta(\infty)$ is contained in the strip $S_\epsilon(L)$ and $\delta' > \delta$, then $I_{\delta'}(\infty) \subset I_\delta(\infty)$ and the graph of f restricted to $I_{\delta'}(\infty)$ is also contained in the strip $S_\epsilon(L)$.

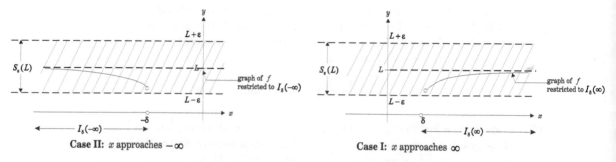

Case II: x approaches $-\infty$ Case I: x approaches ∞

Figure 1.5.21 Graph of f Restricted to $I_\delta(c)$ is Contained in $S_\epsilon(L)$ when $c = \pm\infty$

Similarly, to establish $\lim\limits_{x\to-\infty} f(x) = L$ we must show that for any target $S_\epsilon(L)$, it is possible to choose a piece of the graph of $y = f(x)$, for $x < -\delta$, which lies inside this target. The graph of f restricted to $I_\delta(-\infty)$ designates this piece of the graph of f. When such an $I_\delta(-\infty)$ can be chosen, $f(x)$ is within ϵ units of L for $x \in I_\delta(-\infty)$. The left diagram in Figure 1.5.21 illustrates this situation. Observe that the choice of $I_\delta(-\infty)$ depends on the given target $S_\epsilon(L)$: as $S_\epsilon(L)$ becomes narrower, $I_\delta(-\infty)$ becomes smaller. In addition, given $S_\epsilon(L)$, there is never a unique choice for $I_\delta(-\infty)$. If the graph of f restricted to $I_\delta(-\infty)$ is contained in the strip $S_\epsilon(L)$ and $\delta' > \delta$, then

$I_{\delta'}(-\infty) \subset I_\delta(-\infty)$ and the graph of f restricted to $I_{\delta'}(-\infty)$ is also contained in the strip $S_\epsilon(L)$.

We illustrate how to apply Definition 1.5.20 to establish limits as x approaches either $+\infty$ or $-\infty$.

[handwritten: $x > \delta$ $\frac{1}{x} < e$ $\frac{1}{e} < x$]

Examples 1.5.22 (1) Find $\displaystyle\lim_{x \to \infty} \frac{1}{x}$.

Solution As x gets larger and larger, its reciprocal $f(x) = \frac{1}{x}$ gets smaller and smaller. Thus, we show that $\displaystyle\lim_{x \to \infty} \frac{1}{x} = 0$. To establish this limit by Def. 1.5.20 we must show that given $\epsilon > 0$, we can find a corresponding $\delta > 0$ such that if $x \in I_\delta(\infty)$ then $(x, \frac{1}{x}) \in S_\epsilon(0)$. That is, we must find $\delta > 0$ such that if $x > \delta$ then $\frac{1}{x} < \epsilon$. Clearly $\delta = \frac{1}{\epsilon}$ satisfies this requirement. See Figure 1.5.23.

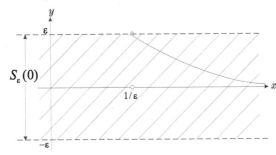

Figure 1.5.23 $f(x) = \frac{1}{x}$ Restricted to $I_{1/\epsilon}(\infty)$

(2) Find $\displaystyle\lim_{x \to -\infty} \frac{1}{x^2}$.

Solution As x approaches $-\infty$, $g(x) = \frac{1}{x^2}$ gets closer and closer to zero. Thus, we show $\displaystyle\lim_{x \to \infty} \frac{1}{x^2} = 0$. Given $\epsilon > 0$ we must find a corresponding $\delta > 0$ such that if $x \in I_\delta(-\infty)$ then $(x, \frac{1}{x^2}) \in S_\epsilon(0)$. That is, we must find $\delta > 0$ such that if $x < -\delta$ then $\frac{1}{x^2} < \epsilon$. The latter inequality is equivalent to $\frac{1}{\epsilon} < x^2$ or $x < -\frac{1}{\sqrt{\epsilon}}$. Hence $\delta = \frac{1}{\sqrt{\epsilon}}$ meets the requirements of Definition 1.5.20. See Figure 1.5.24. $\square$

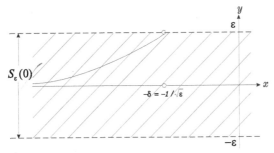

Figure 1.5.24 $g(x) = \frac{1}{x^2}$ Restricted to $I_\delta(-\infty)$

Infinite Limits

The final generalization of the definition of a limit is to infinite limits. We begin with an intuitive discussion and then give a precise definition.

The function $f(x)$ has limit ∞ at $x = c$ if the values of $f(x)$ get larger and larger as x gets closer and closer to c. Note that this statement makes sense in three cases: c is a number, $c = \infty$ or $c = -\infty$. When c is a number, the graph of f rises steeply to approach the line $x = c$ as x approaches c. We say this graph has the line $x = c$ as a *vertical asymptote*. See the left diagram in Figure 1.5.25. When $c = \infty$, we rephrase this definition: $f(x)$ has limit ∞ at $x = \infty$ if the values of $f(x)$ get larger and larger as x gets larger and larger. In this case, the graph of f rises without bound as x moves towards the right end of the x–axis. See the middle diagram in Figure 1.5.25. We also rephrase this definition when $c = -\infty$: $f(x)$ has limit ∞ at $x = -\infty$ if the values of $f(x)$ get larger and larger as x gets smaller and smaller. In this case, the graph of f rises without bound as x moves towards the left end of the x–axis. See the right diagram in Figure 1.5.25. In all three cases we write:

$$\lim_{x \to c} f(x) = \infty .$$

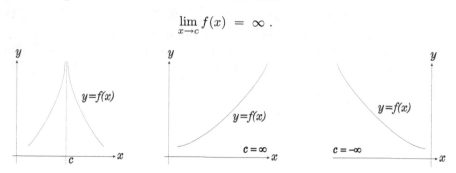

Figure 1.5.25 Infinite Limits with Value ∞

Similarly, the function $f(x)$ has limit $-\infty$ at $x = c$ if the values of $f(x)$ get smaller and smaller as x gets closer and closer to c. This statement makes sense in three cases: c is a number, $c = \infty$ or $c = -\infty$. When c is a number, the graph of f falls steeply to approach the line $x = c$ as x approaches c. Here too we say that this graph has the line $x = c$ as a *vertical asymptote*. See the left diagram in Figure 1.5.26. When $c = \infty$, we rephrase this definition: $f(x)$ has limit $-\infty$ at $x = \infty$ if the values of $f(x)$ get smaller and smaller as x gets larger and larger. In this case, the graph of f falls without bound as x moves towards the right end of the x–axis. See the middle diagram in Figure 1.5.26. We also rephrase this definition when $c = -\infty$: $f(x)$ has limit $-\infty$ at $x = -\infty$ if the values of $f(x)$ get smaller and smaller as x gets smaller and smaller. In this case, the graph of f falls without bound as x moves towards the left end of the x–axis. See the right diagram in Fig. 1.5.26. In all three cases we write:

$$\lim_{x \to c} f(x) = -\infty .$$

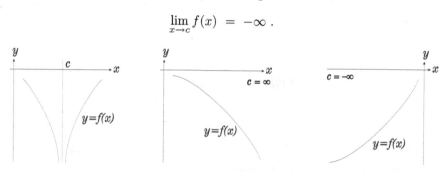

Figure 1.5.26 Infinite Limits with Value $-\infty$

Before giving a precise definition of infinite limits, we consider two examples.

Examples 1.5.27 (1) Let n be a positive integer. Let $h(x) = \frac{1}{x^n}$ be the function with domain all nonzero numbers. If $n = 2s$ is even, the graph of h is depicted in Figure 1.5.16. This graph has the y–axis as a vertical asymptote. Hence

$$\lim_{x \to 0} \frac{1}{x^{2s}} = \infty.$$

If $n = 2t + 1$ is odd, the graph of h is depicted in Figure 1.5.18. As x approaches zero from the left the graph of f falls to its vertical asymptote, the y–axis. However, when x approaches zero from the right the graph of f rises to its vertical asymptote, the y–axis. Therefore $\lim\limits_{x \to 0} \dfrac{1}{x^{2t+1}}$ does not exist. However, both one–sided limits exist:

$$\lim_{x \to 0^-} \frac{1}{x^{2t+1}} = -\infty \quad \text{and} \quad \lim_{x \to 0^+} \frac{1}{x^{2t+1}} = \infty.$$

(2) Consider the graph of $f(x) = \tan x$ given in Figure 1.4.25. As x approaches $\frac{\pi}{2}$ from the left, the graph of f rises to the vertical asymptote $x = \frac{\pi}{2}$. However, as x approaches $\frac{\pi}{2}$ from the right, the graph of x falls to the the vertical asymptote $x = \frac{\pi}{2}$. Hence $\lim\limits_{x \to \frac{\pi}{2}} \tan x$ does not exist while the two one–sided limits exist:

$$\lim_{x \to \frac{\pi}{2}^-} \tan x = \infty \quad \text{and} \quad \lim_{x \to \frac{\pi}{2}^+} \tan x = -\infty. \qquad \square$$

To make a precise definition of an infinite limit we must define the term "$f(x)$ is near ∞". We do this in terms of the graph of f. Requiring $f(x)$ to be near ∞ is equivalent to requiring that the point $(x, f(x))$ on the graph of f be above a high horizontal line $y = \epsilon$. That is, $(x, f(x))$ should lie inside the half–plane $y > \epsilon$. Denote this half–plane by:

$$S_\epsilon(\infty) \;=\; \text{the set of all } (x, y) \text{ with } y > \epsilon.$$

$S_\epsilon(\infty)$ is depicted as the shaded region in the right diagram of Figure 1.5.28. Note the larger the value of ϵ, the narrower the half–plane $S_\epsilon(\infty)$, and the closer the y–coordinates of the points of $S_\epsilon(\infty)$ are to ∞.

Similarly, requiring $f(x)$ to be near $-\infty$ is equivalent to requiring that the point $(x, f(x))$ on the graph of f be below a low horizontal line $y = -\epsilon$. That is, $(x, f(x))$ should lie inside the half–plane $y < -\epsilon$. Denote this half–plane by:

$$S_\epsilon(-\infty) \;=\; \text{the set of all } (x, y) \text{ with } y < -\epsilon.$$

$S_\epsilon(-\infty)$ is depicted as the shaded region in the left diagram of Figure 1.5.28. Note the larger the value of ϵ, the narrower the half–plane $S_\epsilon(-\infty)$, and the closer the y–cordinates of the points of $S_\epsilon(-\infty)$ are to $-\infty$.

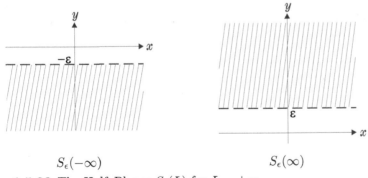

$$S_\epsilon(-\infty) \qquad\qquad S_\epsilon(\infty)$$

Figure 1.5.28 The Half–Planes $S_\epsilon(L)$ for $L = \pm\infty$

The following definition uses this notation to give precise formulations of the statements: "$f(x)$ is near ∞ when x is near c" and "$f(x)$ is near $-\infty$ when x is near c."

Definition 1.5.29 *Let I be an open interval of finite or infinite length with c either an interior point of I or an endpoint of I. Let f be a function whose domain contains* *all numbers of the interval I other than c.*

(a) *The limit of $f(x)$, as x approaches c, equals ∞ if for every $\epsilon > 0$, there is a corresponding $\delta > 0$ such that the graph of f restricted to $I_\delta(c)$ is contained in the half–plane $S_\epsilon(\infty)$.*

(b) *The limit of $f(x)$, as x approaches c, equals $-\infty$ if for every $\epsilon > 0$, there is a corresponding $\delta > 0$ such that the graph of f restricted to $I_\delta(c)$ is contained in the half–plane $S_\epsilon(-\infty)$.*

Notes (1) Each part of this definition applies to the three cases: c is a number, $c = \infty$ and $c = -\infty$.

(2) When c is a number, c may, or may not, be in the domain of f. In this case, this definition applies to two sided limits (when c is an interior point of I), to left–hand limits (when c is the right endpoint of I) and to right–hand limits (when c is the left endpoint of I).

The half–plane $S_\epsilon(\infty)$ is a thin target consisting of points (x, y) with y near ∞. To establish that $\lim_{x \to c} f(x) = \infty$ we must show that for any such target, it is possible to choose a piece of the graph of f near $x = c$ which lies inside the target. The graph of f restricted to $I_\delta(c)$ designates this piece of the graph of f. When such an $I_\delta(c)$ can be chosen, $f(x)$ is near ∞ for $x \in I_\delta(c)$. Figure 1.5.30 illustrates this definition.

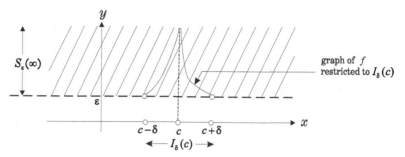

Figure 1.5.30 Graph of f Restricted to $I_\delta(c)$ is Contained in $S_\epsilon(\infty)$

Observe that the choice of $I_\delta(c)$ depends on the given target $S_\epsilon(\infty)$: as $S_\epsilon(\infty)$ becomes narrower, $I_\delta(c)$ becomes smaller. In addition, given $S_\epsilon(\infty)$, there is never a unique choice of $I_\delta(c)$. If the graph of f restricted to $I_\delta(c)$ is contained in the half–plane $S_\epsilon(\infty)$ and $I_{\delta'}(c) \subset I_\delta(c)$ then the graph of $f \mid I_{\delta'}(c)$ is also contained in $S_\epsilon(\infty)$. Similar remarks appy to the definition of $\lim_{x \to c} f(x) = -\infty$. See Figure 1.5.31.

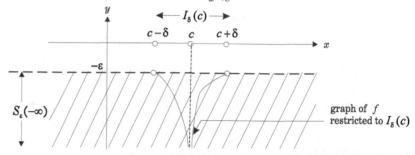

Figure 1.5.31 Graph of f Restricted to $I_\delta(c)$ is Contained in $S_\epsilon(-\infty)$

We use Definition 1.5.29 to justify the values of two infinite limits.

Examples 1.5.32 (1) Find $\lim_{x \to 0} \dfrac{1}{x^2}$.

Solution As x approaches zero, $f(x) = \frac{1}{x^2}$ gets larger and larger. Thus, we show that $\lim_{x \to 0} \frac{1}{x^2} = \infty$. Given $\epsilon > 0$ we must find a corresponding $\delta > 0$ such that if $x \in I_\delta(0)$ then $(x, \frac{1}{x^2}) \in S_\epsilon(0)$. That is, we must find $\delta > 0$ such that if $-\delta < x < \delta$ then $\frac{1}{x^2} > \epsilon$. This is equivalent to $x^2 < \frac{1}{\epsilon}$ or $-\frac{1}{\sqrt{\epsilon}} < x < \frac{1}{\sqrt{\epsilon}}$. Thus choosing $\delta = \frac{1}{\sqrt{\epsilon}}$ satisfies the requirements of Definition 1.5.29. See Fig. 1.5.33.

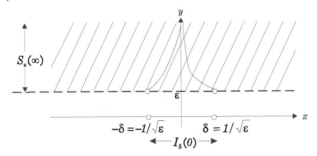

Figure 1.5.33 $f(x) = \frac{1}{x^2}$ Restricted to $I_\delta(0)$

(2) Find $\lim\limits_{x \to -\infty} x^3$.

Solution As x gets smaller and smaller, $g(x) = x^3$ also gets smaller and smaller. Thus, we show that $\lim\limits_{x \to -\infty} x^3 = -\infty$. To apply Definition 1.5.29 we must show that given $\epsilon > 0$ there is a corresponding $\delta > 0$ such that if $x \in I_\delta(-\infty)$ then $(x, x^3) \in S_\epsilon(-\infty)$. That is, we must find $\delta > 0$ such that if $x < -\delta$ then $x^3 < -\epsilon$. Clearly $\delta = \sqrt[3]{\epsilon}$ satisfies this requirement. See Figure 1.5.34. □

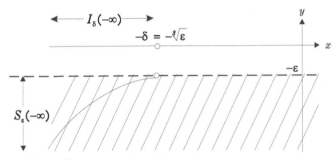

Figure 1.5.34 $g(x) = x^3$ Restricted to $I_\delta(-\infty)$

Addendum

Our definitions of limits are geometric interpretations of the usual algebraic formulations. We intentionally avoided these formulations as they are more difficult to understand. However, our discussion would not be complete without including them. The key observation in translating our geometric definitions into algebraic ones is: the distance between the points A, B on the number line is $|B - A|$. Thus the condition

"the point $(x, f(x))$, on the graph of f, is in the horizontal strip $S_\epsilon(L)$"

can be reformulated algebraically in one of three ways depending on whether L is a number, $L = -\infty$ or $L = +\infty$:

- $[L \in \Re\ ]$ the distance between $f(x)$ and L is less than ϵ, i.e. $|f(x) - L| < \epsilon$;

- $[L = -\infty\ ]$ $(x, f(x))$ lies below the line $y = -\epsilon$, i.e. $f(x) < -\epsilon$;

- $[L = \infty\ ]$ $(x, f(x))$ lies above the line $y = \epsilon$, i.e. $f(x) > \epsilon$.

Recall that the domain of f contains all numbers of the interval I, except possibly c. Let δ be a small positive number. Then the condition

$$x \in I_\delta(c)$$

can be reformulated algebraically in one of three ways depending on whether c is a number, $c = -\infty$ or $c = +\infty$:

- $[c \in \Re\ ]$ $x \neq c$ and the distance between x and c is less than δ, i.e.

  (a) $0 < |x - c| < \delta$ if c is an interior point of I;

  (b) $c < x < c + \delta$ if c is the left endpoint of I;

  (c) $c - \delta < x < c$ if c is the right endpoint of I;

- $[c = -\infty\ ]$ x lies to the left of $-\delta$, i.e $x < -\delta$;

- $[c = \infty\ ]$ x lies to the right of δ, i.e $x > \delta$.

We summarize our geometric definitions of limits as one statement which covers all the cases. Then we restate the definition using the above algebraic reformulations of the statements $(x, f(x)) \in S_\epsilon(L)$ and $x \in I_\delta(c)$,

Definition 1.5.35 *Let I be an interval with either $c \in I$ or c an endpoint of I. Let f be a function whose domain contains all numbers of the interval I other than c. The limit of $f(x)$, as x approaches c, equals L if for every $\epsilon > 0$ there is a corresponding $\delta > 0$ such that the graph of f restricted to $I_\delta(c)$ lies in the horizontal strip $S_\epsilon(L)$. Equivalently, for every $\epsilon > 0$, there is a corresponding $\delta > 0$ such that the appropriate condition holds:*

$$
\textit{if}\
\begin{cases}
0 < |x - c| < \delta & \textit{when } c \in \Re \textit{ or } c \\
 & \textit{is in interior } I \\
c < x < c + \delta & \textit{when } c \in \Re \textit{ or } c \textit{ is} \\
 & \textit{left endpoint of } I \\
c - \delta < x < c & \textit{when } c \in \Re \textit{ or } c \textit{ is} \\
 & \textit{right endpoint of } I \\
x < -\delta & \textit{when } c = -\infty \\
x > \delta & \textit{when } c = \infty
\end{cases}
\ \textit{then}\
\begin{cases}
|f(x) - L| < \epsilon & \textit{when } L \in \Re \\
f(x) < -\epsilon & \textit{when } L = -\infty \\
f(x) > \epsilon & \textit{when } L = \infty
\end{cases}
$$

There are fifteen cases of this definition since for each of the five possibilities for c, there are three possibilities for L. Thus, for a specified c and L the appropriate one of these fifteen cases must be selected. For example, if $c = \infty$ and $L = -23$ then we select the last case on the left ($c = \infty$) and the first case on the right ($L \in \Re$). It should be understood that this definition is equivalent to our earlier geometric ones. Thus either form of these definitions can be used to construct rigorous arguments.

Historical Remarks

The calculus that Newton and Leibniz developed in the late seventeenth century was based on manipulating "fluxions" and "infinitesmals". Thus, calculus was a subject without a logical foundation. Since these methods consistently produced correct solutions to important problems in mathematics and science, this situation was ignored by Euler, the Bernoulli brothers and most other eighteenth century mathematicians. The Anglican bishop George Berkeley assessed this situation in 1734 as follows.

> *He who can digest a second or third fluxion, a second or third difference,*
> *need not, methinks, be squeamish about any point in divinity.*

In 1754 Jean d'Alembert proposed the point of view that the derivative be defined as a limit. This suggestion was ignored by other mathematicians for the remainder of eighteenth century! It was not until Cauchy's three textbooks of the 1820s that limits were used systematically to define integrals and derivatives and to verify their properties. This provided the first logically sound development of calculus. From then on, all mathematicians have taken Cauchy's point of view.

Summary

The reader should understand the intuitive meaning of limits and should be able to evaluate limits from the graph or formula of a function. The precise definition of a limit should be understood so that it can be applied directly to verify simple limits.

Basic Exercises

1. In each case use a calculator to compute $f(x)$ for the indicated values of x near c. Then guess the value of the limit of $f(x)$ as x approaches c.

(a) Let $f(x) = \frac{x^2-x-2}{x-2}$ with $c = 2$. Compute $f(1.9)$, $f(1.99)$, $f(2.1)$ and $f(2.01)$.

(b) Let $f(x) = \frac{\sqrt{x+5}-4}{x-11}$ with $c = 11$. Compute $f(10)$, $f(10.9)$, $f(10.99)$, $f(12)$, $f(11.1)$ and $f(11.01)$.

(c) Let $f(x) = \frac{\sin x}{x}$ with $c = 0$. Compute $f(1)$, $f(0.1)$, $f(-0.5)$ and $f(-0.02)$.

(d) Let $f(x) = \frac{1-\cos x}{x}$ with $c = 0$. Compute $f(0.1)$, $f(0.01)$, $f(-0.3)$ and $f(-0.02)$.

(e) Let $f(x) = \frac{2^x-1}{x}$ with $c = 0$. Compute $f(0.1)$, $f(0.01)$, $f(.001)$, $f(-0.1)$, $f(-0.01)$ and $f(-0.001)$.

2. **(a)** Determine whether $\lim\limits_{x \to c} f(x)$ exists for each function depicted below. If the limit exists, find its value.

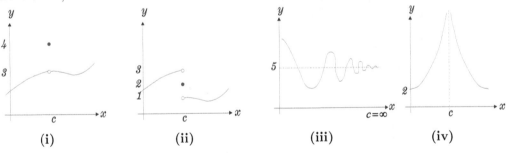

(i)   (ii)   (iii)   (iv)

(b) Determine whether $\lim\limits_{x \to c^+} f(x)$ exists for each function depicted below. If the limit exists, find its value.

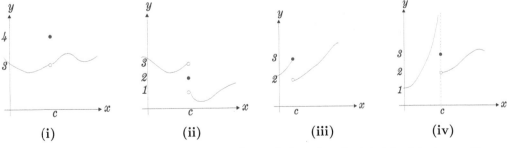

(i)   (ii)   (iii)   (iv)

(c) Determine whether $\lim\limits_{x \to c^-} f(x)$ exists for each function depicted in (b) above. If the limit exists, find its value.

3. Use intuitive methods to find the value of each limit.

(a) $\lim\limits_{x\to 3}(3x^2-12)$ (b) $\lim\limits_{x\to 2}\dfrac{x}{x^2-1}$ (c) $\lim\limits_{x\to -1}\cos(\pi x)$

(d) $\lim\limits_{x\to 7^-}\sqrt{7-x}$ (e) $\lim\limits_{x\to -5^+}\dfrac{2}{1+\sqrt{5+x}}$ (f) $\lim\limits_{x\to 0}\dfrac{\cos x}{x}$

4. Use intuitive methods to determine whether the limits $\lim\limits_{x\to c^-}f(x)$, $\lim\limits_{x\to c^+}f(x)$ and $\lim\limits_{x\to c}f(x)$ exist. Find the values of those limits which exist.

(a) $c=3$ and $f(x)=\begin{cases}2x+4 & \text{if } x\ge 3\\ 4-2x & \text{if } x<3\end{cases}$

(b) $c=-3$ and $f(x)=\begin{cases}x^2 & \text{if } x\ge -3\\ 3-2x & \text{if } x<-3\end{cases}$

(c) $c=0$, $f(x)=\begin{cases}x+1 & \text{if } x>0\\ 2 & \text{if } x=0\\ 2x+3 & \text{if } x<0\end{cases}$ (d) $c=2$, $f(x)=\begin{cases}x^2-1 & \text{if } x>2\\ 4 & \text{if } x=2\\ 7-x^2 & \text{if } x<2\end{cases}$

(e) $c=-1$ and $f(x)=\dfrac{|x+1|}{x+1}$ (f) $c=4$ and $f(x)=\dfrac{|2x-8|}{4-x}$

5. Use intuitive methods to determine the values of c for which $\lim\limits_{x\to c}f(x)$ exists.

(a) $f(x)=\begin{cases}3x-2 & \text{if } x \text{ is rational}\\ 5+3x & \text{if } x \text{ is irrational}\end{cases}$ (b) $f(x)=\begin{cases}4x-6 & \text{if } x \text{ is rational}\\ 6-4x & \text{if } x \text{ is irrational}\end{cases}$

(c) $f(x)=\begin{cases}x^2+1 & \text{if } x \text{ is rational}\\ \cos x & \text{if } x \text{ is irrational}\end{cases}$ (d) $f(x)=\begin{cases}\sin x & \text{if } x \text{ is rational}\\ \cos x & \text{if } x \text{ is irrational}\end{cases}$

6. Use intuitive methods to determine whether the limits $\lim\limits_{x\to 0^-}f(x)$, $\lim\limits_{x\to 0^+}f(x)$ and $\lim\limits_{x\to 0}f(x)$ exist. Find the value of those limits which exist.

(a) $f(x)=\cos\frac{1}{x}$ (b) $f(x)=x\cos\frac{1}{x}$ (c) $f(x)=\sin\frac{1}{x^2}$ (d) $f(x)=x^2\sin\frac{1}{x^2}$
(e) $f(x)=\arctan\frac{1}{x}$

7. Sketch each subset of the real number line.
(a) $I_2(5)$ with $I=[0,10]$ (b) $I_1(-6)$ with $I=\Re$ (c) $I_2(5)$ with $I=[5,9)$
(d) $I_3(-4)$ with $I=[-8,-4)$ (e) $I_4(2)$ with $I=[0,\infty)$ (f) $I_3(-1)$ with $I=(-3,1]$
(g) $I_7(-\infty)$ with $I=(-\infty,0]$ (h) $I_6(\infty)$ with $I=\Re$ (i) $I_3(2)$ with $I=(1,6)$

8. Sketch each region in the plane.
(a) $S_2(9)$ (b) $S_1(-4)$ (c) $S_8(\infty)$ (d) $S_5(-\infty)$

9. For each limit and value of ϵ, find a value of δ which meets the requirements of Definition 1.5.9.
(a) $\lim\limits_{x\to 7}x=7$ and $\epsilon=0.4$ (b) $\lim\limits_{x\to -2}3x=-6$ and $\epsilon=0.6$
(c) $\lim\limits_{x\to 3}(5x-4)=11$ and $\epsilon=0.3$ (d) $\lim\limits_{x\to 0}x^2=0$ and $\epsilon=\frac{1}{4}$
(e) $\lim\limits_{x\to 9}\sqrt{x}=3$ and $\epsilon=0.2$ (f) $\lim\limits_{x\to 1}\dfrac{1}{x}=1$ and $\epsilon=0.1$

10. Use Definition 1.5.9 to verify each limit.
(a) $\lim\limits_{x\to 2}(5x-1)=9$ (b) $\lim\limits_{x\to -3}(4x+7)=-5$ (c) $\lim\limits_{x\to 6}(3x-8)=10$
(d) $\lim\limits_{x\to -1}(5-4x)=9$ (e) $\lim\limits_{x\to 4}(-3x-7)=-19$ (f) $\lim\limits_{x\to -5}(-2x-9)=1$

11. Determine whether each number is "large", "small" or neither.
(a) .000003 (b) $5,000,000$ (c) $-.000006$ (d) $-9,000,000,000$ (e) 10^{-30} (f) -10^{50}

12. Use intuitive methods to determine whether the limits $\lim\limits_{x\to \infty}f(x)$ and $\lim\limits_{x\to -\infty}f(x)$ exist. Find the values of those limits which exist.

(a) $f(x)=\frac{1}{x^2+1}$ (b) $f(x)=\frac{1}{\sqrt[3]{x}}$ (c) $f(x)=\frac{|x|}{x+1}$ (d) $f(x)=\sin x$
(e) $f(x)=\cos x$ (f) $f(x)=\sin\frac{1}{x}$ (g) $f(x)=\cos\frac{1}{x}$ (h) $f(x)=\operatorname{arccot} x$

(i) $f(x) = \operatorname{arcsec} x$ (j) $f(x) = \operatorname{arccsc} x$ (k) $f(x) = \frac{\sin x}{x}$ (l) $f(x) = \frac{\tan x}{x}$

13. For each limit and value of ϵ find a value of δ which meets the requirements of Definition 1.5.20.

(a) $\lim\limits_{x \to \infty} \frac{3}{x} = 0$ and $\epsilon = 0.2$ (b) $\lim\limits_{x \to -\infty} \frac{4}{x^2} = 0$ and $\epsilon = \frac{1}{9}$

(c) $\lim\limits_{x \to \infty} \frac{16}{\sqrt{x}} = 0$ and $\epsilon = 0.1$ (d) $\lim\limits_{x \to -\infty} \frac{1}{x + 10} = 0$ and $\epsilon = 0.1$

14. Use Definition 1.5.20 to verify each limit.

(a) $\lim\limits_{x \to \infty} \frac{1}{x^2} = 0$ (b) $\lim\limits_{x \to -\infty} \frac{1}{x} = 0$ (c) $\lim\limits_{x \to \infty} \frac{1}{x^3} = 0$ (d) $\lim\limits_{x \to -\infty} \frac{1}{x^4} = 0$

15. Use intuitive methods to determine whether the limits $\lim\limits_{x \to c^-} f(x)$, $\lim\limits_{x \to c^+} f(x)$ and $\lim\limits_{x \to c} f(x)$ exist. Find the values of those limits which exist.

(a) $f(x) = \frac{1}{x^4}$ and $c = 0$ (b) $f(x) = \frac{1}{x^2 - 1}$ and $c = -1$

(c) $f(x) = \frac{x+5}{x^3 - x}$ and $c = 1$ (d) $f(x) = \cot x$ and $c = \pi$

(e) $f(x) = \sec x$ and $c = \frac{\pi}{2}$ (f) $f(x) = \csc x$ and $c = 0$

16. Use Definition 1.5.29 to verify each limit.

(a) $\lim\limits_{x \to 0} \frac{5}{x^4} = \infty$ (b) $\lim\limits_{x \to 0^-} \frac{3}{x} = -\infty$ (c) $\lim\limits_{x \to 0^+} \frac{4}{x^3} = \infty$

(d) $\lim\limits_{x \to \infty} x^6 = \infty$ (e) $\lim\limits_{x \to -\infty} \sqrt[3]{x} = -\infty$ (f) $\lim\limits_{x \to \infty} (x^4 + x^2 + 1) = \infty$

17. Sketch the variation of Figures 1.5.10, 1.5.21 or 1.5.30 which illustrates the definition of $\lim\limits_{x \to c} f(x) = L$ in each case:

(a) $c \in \Re$ is the left endpoint of I and $L = \infty$;

(b) $c \in \Re$ is the left endpoint of I and $L = -\infty$;

(c) $c \in \Re$ is the right endpoint of I and $L = \infty$;

(d) $c \in \Re$ is the right endpoint of I and $L = -\infty$;

(e) $c = \infty$ and $L = \infty$;   (f) $c = \infty$ and $L = -\infty$;

(g) $c = -\infty$ and $L = \infty$;   (h) $c = -\infty$ and $L = -\infty$.

18. (a) Sketch the variation of Figure 1.5.11 which illustrates $\lim\limits_{x \to \infty} k = k$.

(b) Sketch the variation of Figure 1.5.13 which illustrates $\lim\limits_{x \to \infty} x = \infty$.

19. State the algebraic variation of the definition of each limit.

(a) $\lim\limits_{x \to 2} (x^2 + x + 1) = 7$ (b) $\lim\limits_{x \to -3} \frac{1}{x^2 + 6x + 9} = \infty$ (c) $\lim\limits_{x \to -\infty} \frac{x}{x^2 + 5} = 0$

(d) $\lim\limits_{x \to 1^+} \frac{1}{1 - x^2} = -\infty$ (e) $\lim\limits_{x \to 4^-} \frac{|x - 4|}{x - 4} = -1$ (f) $\lim\limits_{x \to \infty} \frac{x^2}{1 - x} = -\infty$

Challenging Problems

Use the definition of the limit to verify each limit.

1. $\lim\limits_{x \to c} \sqrt{x} = \sqrt{c}$ 2. $\lim\limits_{x \to c} x^2 = c^2$ 3. $\lim\limits_{x \to c} \frac{1}{x} = \frac{1}{c}$ for $c \neq 0$

1.6   Computing Limits

In the preceding section we learned how to compute straightforward limits intuitively. We also learned how to define limits and apply those definitions to justify intuitive computations of simple limits. This section is devoted to develop methods to justify the intuitive computations of complicated limits. In addition, we show how to evaluate limits whose values can not be found using intuition alone. The approach we use is to establish eight fundamental limit properties in the first subsection. In the second subsection, we show how to apply these limit properties to evaluate specific limits. The last subsection is devoted to computing trigonometric limits. In particular, two fundamental trigonometric limits are derived. They are used in Chapter 2 to compute the derivatives of trigonometric functions.

Limit Properties

We could apply the definition of a limit to verify complicated limits. However, such verifications would be difficult and unnecessary. The simpler procedure, which we follow, is to establish eight basic properties of limits from the definition of a limit. Then we apply these properties to determine complicated limits from the simple limits established in Section 5. We state each limit property and discuss its meaning. The proofs of these properties, from the definition of a limit, are given in the appendices of this section and of Section 7. We give one example of each property and postpone systematic applications of these properties to the next two subsections.

Property 1   *When a limit exists, it is unique. That is,*

$$\text{if } \lim_{x \to c} f(x) = L \text{ and } \lim_{x \to c} f(x) = L', \text{ then } L = L'.$$

Notes **(1)** Each of c, L is either a number or one of the symbols $\pm\infty$.
(2) This property is also true for one–sided limits.

This property of limits is reasonable. If the values of $f(x)$ are closer and closer to L as x gets closer and closer to c, then these values of $f(x)$ can not be getting close to any other number. This property can be used to show that a limit does not exist.

For example, let $f(x) = \left\{ \begin{array}{ll} 1 & \text{if } x \text{ is rational} \\ 0 & \text{if } x \text{ is irrational} \end{array} \right\}$ with c any number. There are rational numbers arbitrarily close to c. Hence if $\lim_{x \to c} f(x)$ exists, it must have value one. On the other hand, there are irrational numbers arbitrarily close to c. Hence if $\lim_{x \to c} f(x)$ exists, it must have value zero. By Property 1, $\lim_{x \to c} f(x)$ can not exist.

Property 2   *The existence of a two sided limit is equivalent to the equality of the two one–sided limits. That is, if $c \in \Re$, then*

$$\lim_{x \to c^-} f(x) = \lim_{x \to c^+} f(x) = L \text{ if and only if } \lim_{x \to c} f(x) = L.$$

Note L is either a number or one of the symbols $\pm\infty$.

This property of limits is also reasonable. The existence of $\lim_{x \to c} f(x) = L$ means that when x is near c, the value of $f(x)$ is near L. There are two types of such numbers x: those x which are less than c and those x which are greater than c. Saying that $f(x)$ is near L when $x < c$ is near c is saying that the left–hand limit $\lim_{x \to c^-} f(x)$ has value L. Saying that $f(x)$ is near L when $x > c$ is near c is saying that the right–hand limit $\lim_{x \to c^+} f(x)$ has value L.

For example, let $f(x) = |x|$ with $c = 0$. Then $|x| = x$ for $x \geq 0$. Therefore, $\lim_{x \to 0^+} |x| = \lim_{x \to 0^+} x = 0$. Now $|x| = -x$ for $x \leq 0$. Thus, $\lim_{x \to 0^-} |x| = \lim_{x \to 0^-} -x = 0$. Hence $\lim_{x \to 0} |x|$ exists and has value 0.

Property 3   *The limit of a sum is the sum of the limits:*

$$\text{if } A+B \text{ is defined, } \lim_{x \to c} f(x) = A \text{ and } \lim_{x \to c} g(x) = B \text{ then } \lim_{x \to c} [f(x) + g(x)] = A+B.$$

Notes **(1)** Each of A, B, c is either a number or one of the symbols $\pm\infty$.
(2) Note that $A + B$ is defined as long as $A + B$ is neither $\infty + -\infty$ nor $-\infty + \infty$.
(3) This property is also true for one–sided limits.

This property is common sense. If the values of $f(x)$ approach A as x approaches c and the values of $g(x)$ approach B as x approaches c, then the values of $f(x) + g(x)$ approach $A + B$ as x approaches c.

For example, we know that $\lim_{x \to 4} x = 4$ and $\lim_{x \to 4} 7 = 7$. Hence
$$\lim_{x \to 4} (x + 7) = \lim_{x \to 4} x + \lim_{x \to 4} 7 = 4 + 7 = 11.$$

Property 4 *The limit of a product is the product of the limits:*

$$\text{if } AB \text{ is defined, } \lim_{x \to c} f(x) = A \text{ and } \lim_{x \to c} g(x) = B \text{ then } \lim_{x \to c} [f(x)g(x)] = AB.$$

Notes **(1)** Each of A, B, c is either a number or one of the symbols $\pm\infty$.
(2) Note that AB is defined as long as AB is neither $(\pm\infty)(0)$ nor $(0)(\pm\infty)$.
(3) This property is also true for one–sided limits.

This property is also common sense. If the values of $f(x)$ approach A as x approaches c and the values of $g(x)$ approach B as x approaches c, then the values of $f(x)g(x)$ approach AB as x approaches c.

For example, we know that $\lim_{x \to c} x = c$. Hence $\lim_{x \to c} x^2 = \left(\lim_{x \to c} x\right)\left(\lim_{x \to c} x\right) = (c)(c) = c^2$.

Property 5 *The limit of a quotient is the quotient of the limits:*

$$\text{if } \frac{A}{B} \text{ is defined, } \lim_{x \to c} f(x) = A \text{ and } \lim_{x \to c} g(x) = B \text{ then } \lim_{x \to c} \frac{f(x)}{g(x)} = \frac{A}{B}.$$

Notes **(1)** Each of A, B, c is either a number or one of the symbols $\pm\infty$.
(2) Note that $\frac{A}{B}$ is defined as long as $B \neq 0$ and $\frac{A}{B} \neq \frac{\pm\infty}{\pm\infty}$.
(3) This property is also true for one–sided limits.

Property 5 is also reasonable. If the values of $f(x)$ approach A as x approaches c and the values of $g(x)$ approach B as x approaches c, then the values of $\frac{f(x)}{g(x)}$ approach $\frac{A}{B}$.

For example, we know that $\lim_{x \to 15} (6x) = (6)(15) = 90$ by Property 4 and

$\lim_{x \to 15} (x - 5) = 15 - 5 = 10$ by Property 3. Hence $\lim_{x \to 15} \frac{6x}{x - 5} = \dfrac{\lim_{x \to 15} (6x)}{\lim_{x \to 15} (x - 5)} = \frac{90}{10} = 9.$

Property 6 *The limit of a composite function $h \circ g$ equals h applied to the limit of g:*
if $B, L \in \Re$, $\lim_{x \to c} g(x) = B$, $\lim_{x \to B} h(x) = L$ and $h(B) = L$ then

$$\lim_{x \to c} h(g(x)) = h\left(\lim_{x \to c} g(x)\right) = L.$$

Notes **(1)** c is either a number or one of the symbols $\pm\infty$.
(2) This property is also true if the limit of $g(x)$ is a one–sided limit.

This property agrees with our intuition. Let x approach c. Since $\lim_{x \to c} g(x) = B$, the values of $g(x)$ are close to B. Even though $x \neq c$, it may happen that $g(x) = B$. In this case, $h(g(x)) = h(B) = L$. If $g(x) \neq B$, then $h(g(x))$ is close to L because $\lim_{x \to B} h(x) = L$. In both cases, $h(g(x))$ is close to L, and $\lim_{x \to c} h(g(x)) = L$.

For example, we showed above that $\lim_{x \to 15} \frac{6x}{x - 5} = 9$ by Property 5, and $\lim_{x \to 9} x^2 = 81$

by Property 4. Hence $\lim_{x \to 15} \left(\frac{6x}{x - 5}\right)^2 = \left(\lim_{x \to 15} \frac{6x}{x - 5}\right)^2 = 9^2 = 81.$

Property 7 *Say the graph of a function g lies between the graphs of the functions f and h which have the same limit L as x approaches c. Then g also has limit L as x approaches c. That is, if there is an $I_\delta(c)$ such that*

$$f(x) \le g(x) \le h(x) \quad for \quad x \in I_\delta(c) \quad and \quad \lim_{x \to c} f(x) = \lim_{x \to c} h(x) = L \quad then \quad \lim_{x \to c} g(x) = L.$$

Notes (1) Each of c, L is either a number or one of the symbols $\pm\infty$.
(2) This property is also true for one–sided limits.

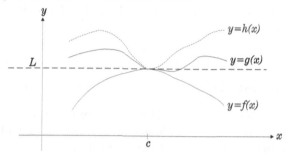

Figure 1.6.1 Property 7

Property 7 is called the *Pinching Theorem* because the graph of g is pinched between the graphs of f and h as in Figure 1.6.1. Since the graphs of f and h approach the line $y = L$ as x approaches c, the graph of g must also approach this line. Alternatively, when x is near c, we know that $g(x)$ is between $f(x)$ and $h(x)$. Since $f(x)$ and $h(x)$ are both near L, $g(x)$ must also be near L.

For example, the function $g(x) = \left\{ \begin{array}{ll} x^2 & \text{if } x \text{ is rational} \\ 0 & \text{if } x \text{ is irrational} \end{array} \right\}$ satisfies $0 \le g(x) \le x^2$ for all x. Since $\lim_{x \to 0} x^2 = \lim_{x \to 0} 0 = 0$, it follows from the Pinching Theorem that $\lim_{x \to 0} g(x) = 0$.

Property 8 *Let f be a one–to–one function with domain the interval I. Assume that $\lim_{x \to c} f(x)$ exists and equals $f(c)$ for $c \in I$. If $L \in f(I)$, then $\lim_{x \to L} f^{-1}(x) = f^{-1}(L)$.*

Note If L is an interior point of $f(I)$, then $\lim_{x \to L} f^{-1}(x) = f^{-1}(L)$. On the other hand, if L is the left endpoint of $f(I)$ then $\lim_{x \to L^+} f^{-1}(x) = f^{-1}(L)$ while if L is the right endpoint of $f(I)$ then $\lim_{x \to L^-} f^{-1}(x) = f^{-1}(L)$.

This property allows us to compute the limit of an inverse functions if we know the limit of the original function. It is useful because even the inverse functions of simple functions may be complicated. For example, consider the one–to–one function $f(x) = x^2$ with domain $I = [0, \infty)$. By Property 4: $\lim_{x \to c} x^2 = c^2$ for $c > 0$ and $\lim_{x \to 0^+} x^2 = 0$. Note $f^{-1}(x) = \sqrt{x}$ with domain $[0, \infty)$. Thus Property 8 applies with $L = c^2$ and $c = \sqrt{L}$:

$$\lim_{x \to L} \sqrt{x} = \sqrt{L} \quad for \quad L > 0 \quad and \quad \lim_{x \to 0^+} \sqrt{x} = 0 . \qquad (1.6.1)$$

Applying Limit Properties

We begin by using limit properties to evaluate limits of polynomials and rational functions. Then we find limits of functions involving roots.

In Examples 1.5.12 (1), (2) we used the definition of a limit to justify that:

$$\lim_{x \to c} k = k, \quad and \quad \lim_{x \to c} x = c \qquad (1.6.2)$$

for $c \in \Re$. Every polynomial is a sum of products of constant functions $f(x) = k$ and the function $g(x) = x$. Therefore we deduce the values of limits of polynomials from the two limits of (1.6.2) by using Property 3 (the limit of a sum is the sum of the limits) and Property 4 (the limit of a product is the product of the limits).

Examples 1.6.2 (1) Find $\lim\limits_{x \to -5} x^3$.

Solution By Property 4:

$$\lim_{x \to -5} x^3 = \left(\lim_{x \to -5} x \right) \left(\lim_{x \to -5} x \right) \left(\lim_{x \to -5} x \right) = (-5)(-5)(-5) = -125 \ .$$

(2) Find $\lim\limits_{x \to 3} (2x^3 - 5x + 4)$.

Solution By Properties 3 and 4:

$$\lim_{x \to 3} (2x^3 - 5x + 4) = \lim_{x \to 3} (2x^3) + \lim_{x \to 3} (-5x) + \lim_{x \to 3} 4$$
$$= \left(\lim_{x \to 3} 2 \right) \left(\lim_{x \to 3} x \right) \left(\lim_{x \to 3} x \right) \left(\lim_{x \to 3} x \right) + \left(\lim_{x \to 3} -5 \right) \left(\lim_{x \to 3} x \right) + \lim_{x \to 3} 4$$
$$= (2)(3)(3)(3) + (-5)(3) + 4 = 43 \qquad \square$$

A *rational function* is the name for a quotient of polynomials $R(x) = \frac{p(x)}{q(x)}$. Since we can compute limits of the polynomials $p(x)$ and $q(x)$ as above, we can find the limit of $R(x)$, as x approaches c, by Property 5 when $q(c) \neq 0$ (the limit of a quotient is the quotient of the limits). When $p(c) = q(c) = 0$, direct substitution of $x = c$ into a rational function $\frac{p(x)}{q(x)}$ produces $\frac{0}{0}$. In this case, Property 5 does not apply and our intuition does not help in evaluating $\lim\limits_{x \to c} \dfrac{p(x)}{q(x)}$. Note $x = c$ is a root of both the numerator $p(x)$ and the denominator $q(x)$. Hence $x - c$ is a common factor of these polynomials. Therefore we use the following procedure to compute this limit. First, simplify $\frac{p(x)}{q(x)}$ by factoring $p(x)$ and $q(x)$. Then cancel all common factors of $x - c$ in the numerator and denominator. The resulting reduced rational function and $\frac{p(x)}{q(x)}$ are equal at all x other than c. By Property 5, substitution of $x = c$ into this reduced rational function gives the value of the limit of $\frac{p(x)}{q(x)}$ as x approaches c.

Examples 1.6.3 (1) Evaluate $\lim\limits_{x \to -2} \dfrac{3x^2 - 7}{5x^2 + 2x - 4}$.

Solution By Properties 5 and 3, 4:

$$\lim_{x \to -2} \frac{3x^2 - 7}{5x^2 + 2x - 4} = \frac{\lim\limits_{x \to -2} (3x^2 - 7)}{\lim\limits_{x \to -2} (5x^2 + 2x - 4)} = \frac{3(-2)^2 - 7}{5(-2)^2 + 2(-2) - 4} = \frac{5}{12}.$$

(2) Evaluate $\lim\limits_{x \to 3} \dfrac{x^2 - x - 6}{x^2 + 2x - 15}$.

Solution Direct substitution of $x = 3$ into this rational function produces $\frac{9-3-6}{9+6-15} = \frac{0}{0}$. Hence we factor the numerator and denominator, cancel the common factor $x - 3$ and apply Property 5:

$$\lim_{x \to 3} \frac{x^2 - x - 6}{x^2 + 2x - 15} = \lim_{x \to 3} \frac{(x-3)(x+2)}{(x-3)(x+5)} = \lim_{x \to 3} \frac{x+2}{x+5} = \frac{\lim\limits_{x \to 3} (x+2)}{\lim\limits_{x \to 3} (x+5)} = \frac{5}{8}.$$

(3) Evaluate $\displaystyle\lim_{x\to-1}\frac{x^4+4x^3+6x^2+4x+1}{x^4-2x^2+1}$.

Solution Direct substitution of $x=-1$ into this rational function produces $\frac{1-4+6-4+1}{1-2+1}=\frac{0}{0}$. We factor the numerator and denominator, cancel the common factors of $x+1$ and apply Property 5:

$$\lim_{x\to-1}\frac{x^4+4x^3+6x^2+4x+1}{x^4-2x^2+1}\;=\;\lim_{x\to-1}\frac{(x+1)^4}{(x^2-1)^2}\;=\;\lim_{x\to-1}\frac{(x+1)^4}{(x+1)^2(x-1)^2}$$

$$=\;\lim_{x\to-1}\frac{(x+1)^2}{(x-1)^2}\;=\;\frac{\displaystyle\lim_{x\to-1}(x+1)^2}{\displaystyle\lim_{x\to-1}(x-1)^2}\;=\;\frac{0}{4}=0.\qquad\square$$

To compute the limit of a rational function as x approaches ∞ or $-\infty$, divide the numerator and denominator by x^n where n is the degree of the denominator. Then apply Property 5 to evaluate the limit of this fraction.

Examples 1.6.4 (1) Find $\displaystyle\lim_{x\to\infty}\frac{3x^4-5x^2+7}{5x^4-2x^3+9}$.

Solution The denominator is a degree 4 polynomial, so divide numerator and denominator by x^4 and apply Property 5:

$$\lim_{x\to\infty}\frac{3x^4-5x^2+7}{5x^4-2x^3+9}\;=\;\lim_{x\to\infty}\frac{3-5/x^2+7/x^4}{5-2/x+9/x^4}\;=\;\frac{\displaystyle\lim_{x\to\infty}(3-5/x^2+7/x^4)}{\displaystyle\lim_{x\to\infty}(5-2/x+9/x^4)}$$

$$=\;\frac{3-0+0}{5-0+0}=\frac{3}{5}$$

by Properties 3, 4 and Example 1.5.22 (1).

(2) Find $\displaystyle\lim_{x\to-\infty}\frac{100x^2+6x-5}{x^3+x^2-8}$.

Solution The denominator has degree 3. Therefore, divide numerator and denominator by x^3:

$$\lim_{x\to-\infty}\frac{100x^2+6x-5}{x^3+x^2-8}\;=\;\lim_{x\to-\infty}\frac{100/x+6/x^2-5/x^3}{1+1/x-8/x^3}=\frac{0+0-0}{1+0-0}=\frac{0}{1}=0\;.$$

(3) Find $\displaystyle\lim_{x\to\infty}\frac{x^5-9x^3+2}{8x^4-7x^3+1}$.

Solution The denominator has degree 4. Hence we divide numerator and denominator by x^4:

$$\lim_{x\to\infty}\frac{x^5-9x^3+2}{8x^4-7x^3+1}=\lim_{x\to\infty}\frac{x-9/x+2/x^4}{8-7/x+1/x^4}=\frac{\infty-0+0}{8-0+0}=\infty.\qquad\square$$

The next group of examples involve the square root function $h(x)=\sqrt{x}$. Recall that we used Property 8 to establish:

$$\lim_{x\to c}\sqrt{x}\;=\;\sqrt{c}\;=\;h(c)\ \text{ for } c>0\ \text{ and }\ \lim_{x\to 0^+}\sqrt{x}\;=\;0\;=\;h(0)\;.$$

In particular, h satisfies the hypotheses of Property 6 (the limit of h of $g(x)$ is h of the limit of $g(x)$). Some limits of fractions involving square roots, as x approaches c, produce $\frac{0}{0}$ when substituting $x=c$. Therefore Property 5 does not apply, and our intuition does not help to evaluate this limit. If $A - B$ appears in this fraction where A and/or B involve square roots, we multiply by $\frac{A+B}{A+B}$. Then we use the formula

$$(A - B)(A + B) = A^2 - B^2 \tag{1.6.3}$$

to rewrite this limit. The numerator and denominator of the resulting fraction have a common factor. After cancellation of this common factor, the resulting limit can be evaluated by applying Property 5. This trick is called "rationalization".

Examples 1.6.5 (1) Evaluate $\lim\limits_{x \to 4} \left(6\sqrt{x + 5} - 2\sqrt{6x + 1}\right)$.

Solution By Properties 2 and 3:

$$\lim_{x \to 4} \left(6\sqrt{x + 5} - 2\sqrt{6x + 1}\right) = \left(\lim_{x \to 4} 6\right)\left(\lim_{x \to 4} \sqrt{x + 5}\right) + \left(\lim_{x \to 4} -2\right)\left(\lim_{x \to 4} \sqrt{6x + 1}\right)$$
$$= 6 \lim_{x \to 4} \sqrt{x + 5} - 2 \lim_{x \to 4} \sqrt{6x + 1} .$$

We apply Property 6 and then Properties 3, 4 to evaluate the remaining limits:

$$\lim_{x \to 4} \left(6\sqrt{x + 5} - 2\sqrt{6x + 1}\right) = 6\sqrt{\lim_{x \to 4} (x + 5)} - 2\sqrt{\lim_{x \to 4} (6x + 1)}$$
$$= 6\sqrt{4 + 5} - 2\sqrt{6(4) + 1} = (6)(3) - (2)(5) = 8.$$

(2) Evaluate $\lim\limits_{x \to 3} \dfrac{\sqrt{6x - 14} - \sqrt{x + 1}}{x - 3}$.

Solution Substitution of $x = 3$ in this fraction produces $\frac{\sqrt{4} - \sqrt{4}}{3 - 3} = \frac{0}{0}$. Therefore we use the rationalization method: multiply this fraction by $\frac{\sqrt{6x - 14} + \sqrt{x + 1}}{\sqrt{6x - 14} + \sqrt{x + 1}}$ and cancel the common factor of the numerator and denominator.

$$\lim_{x \to 3} \frac{\sqrt{6x - 14} - \sqrt{x + 1}}{x - 3} = \lim_{x \to 3} \frac{\sqrt{6x - 14} - \sqrt{x + 1}}{x - 3} \cdot \frac{\sqrt{6x - 14} + \sqrt{x + 1}}{\sqrt{6x - 14} + \sqrt{x + 1}}$$
$$= \lim_{x \to 3} \frac{(6x - 14) - (x + 1)}{(x - 3)(\sqrt{6x - 14} + \sqrt{x + 1})} = \lim_{x \to 3} \frac{5(x - 3)}{(x - 3)(\sqrt{6x - 14} + \sqrt{x + 1})}$$
$$= \lim_{x \to 3} \frac{5}{\sqrt{6x - 14} + \sqrt{x + 1}} = \frac{5}{\sqrt{4} + \sqrt{4}} = \frac{5}{4}.$$

(3) Evaluate $\lim\limits_{x \to 2} \dfrac{x^2 - x - 2}{\sqrt{3x + 10} - 2x}$.

Solution Substitution of $x = 2$ in this fraction produces $\frac{4 - 2 - 2}{\sqrt{16} - 4} = \frac{0}{0}$. Hence we use the rationalization method: multiply this fraction by $\frac{\sqrt{3x + 10} + 2x}{\sqrt{3x + 10} + 2x}$ and cancel the common factor of the numerator and denominator.

$$\lim_{x \to 2} \frac{x^2 - x - 2}{\sqrt{3x + 10} - 2x} = \lim_{x \to 2} \frac{x^2 - x - 2}{\sqrt{3x + 10} - 2x} \cdot \frac{\sqrt{3x + 10} + 2x}{\sqrt{3x + 10} + 2x}$$
$$= \lim_{x \to 2} \frac{(x^2 - x - 2)(\sqrt{3x + 10} + 2x)}{(3x + 10) - 4x^2} = \lim_{x \to 2} \frac{(x - 2)(x + 1)(\sqrt{3x + 10} + 2x)}{-(4x + 5)(x - 2)}$$
$$= \lim_{x \to 2} -\frac{(x + 1)(\sqrt{3x + 10} + 2x)}{4x + 5} = -\frac{3(\sqrt{16} + 4)}{13} = -\frac{24}{13}. \qquad \square$$

Now we apply Property 8 to evaluate limits of functions involving roots and fractional exponents.

Examples 1.6.6 (1) (a) Show $\lim\limits_{x \to L} {}^{2n+1}\!\sqrt{x} = {}^{2n+1}\!\sqrt{L}$ for $n > 0$ an integer and $L \in \Re$.

(b) Let $n > 0$ be an even integer. Show $\lim\limits_{x \to L} {}^{2n}\!\sqrt{x} = {}^{2n}\!\sqrt{L}$ for $L > 0$ and $\lim\limits_{x \to 0^+} {}^{2n}\!\sqrt{x} = 0$.

Solution (a) The function $f(x) = x^{2n+1}$ with domain $\Re$ is one–to–one. By Property 4, $\lim\limits_{x \to c} x^{2n+1} = c^{2n+1} = f(c)$ for $c \in \Re$. Note $f^{-1}(x) = {}^{2n+1}\!\sqrt{x}$ with domain $\Re$. Hence Property 8 applies, and for $L \in \Re$:

$$\lim_{x \to L} {}^{2n+1}\!\sqrt{x} \;=\; \lim_{x \to L} f^{-1}(x) \;=\; f^{-1}(L) \;=\; {}^{2n+1}\!\sqrt{L} \ .$$

(b) An analogous argument applies using the one–to–one function $g(x) = x^{2n}$ with domain $[0, \infty)$.

(2) Evaluate $\lim\limits_{x \to 64} \dfrac{x^{5/6} + x^{2/3}}{x^{1/6} + x^{3/2}}$.

Solution When p, q are positive integers, the fractional power $x^{p/q}$ is the product of p copies of $\sqrt[q]{x}$. Thus we can use the preceding example to evaluate this limit. By Properties 5, 3 and 4:

$$
\lim_{x \to 64} \frac{x^{5/6} + x^{2/3}}{x^{1/6} + x^{3/2}} = \frac{\lim\limits_{x \to 64}(x^{5/6} + x^{2/3})}{\lim\limits_{x \to 64}(x^{1/6} + x^{3/2})} = \frac{\lim\limits_{x \to 64} \sqrt[6]{x^5} + \lim\limits_{x \to 64} \sqrt[3]{x^2}}{\lim\limits_{x \to 64} \sqrt[6]{x} + \lim\limits_{x \to 64} \sqrt{x^3}}
$$

$$
= \frac{(\lim\limits_{x \to 64} \sqrt[6]{x})^5 + (\lim\limits_{x \to 64} \sqrt[3]{x})^2}{\lim\limits_{x \to 64} \sqrt[6]{x} + (\lim\limits_{x \to 64} \sqrt{x})^3}
$$

$$
= \frac{(2)^5 + (4)^2}{2 + (8)^3} = \frac{48}{514} = \frac{24}{257} \qquad \text{[by Example 1]} \qquad \square
$$

We conclude this subsection by establishing the limit which we found by intuitive methods in Example 1.5.3 (3b).

Example 1.6.7 Find $\lim\limits_{x \to 0} f(x)$ where $f(x) = \left\{ \begin{array}{ll} x & \text{if } x \text{ is rational} \\ 0 & \text{if } x \text{ is irrational} \end{array} \right\}$.

Solution Recall that the graph of f, depicted in Figure 1.5.6, consists of the dotted lines $y = 0$ and $y = x$. Hence $0 \le f(x) \le x$ for $x \ge 0$. We know that $\lim\limits_{x \to 0^+} 0 = \lim\limits_{x \to 0^+} x = 0$. By the Pinching Theorem:

$$\lim_{x \to 0^+} f(x) = 0 \ .$$

On the other hand, $x \le f(x) \le 0$ for $x \le 0$. Since $\lim\limits_{x \to 0^-} 0 = \lim\limits_{x \to 0^-} x = 0$, it follows from the Pinching Theorem that:

$$\lim_{x \to 0^-} f(x) = 0 \ .$$

Since both one–sided limits have value zero, $\lim\limits_{x \to 0} f(x) = 0$ by Property 2. $\qquad \square$

Trigonometric Limits

We begin by using limit properties to justify simple trigonometric limits. Then we establish two fundamental limits and apply them to evaluate several nontrivial trigonometric limits.

Examples 1.6.8 (1) Show $\lim\limits_{\theta \to 0} \sin\theta = 0$.

Solution Consider the sector of the unit circle with angle 2θ in Figure 1.6.9. Since the hypotenuse of each right triangle is a radius of the circle, it has length one. Therefore, the opposite side of each right triangle has length $\sin\theta$. By (1.4.3), the length of the arc of this sector is 2θ. Since the line segment PQ is the shortest path between the points P and Q, the length of the line segment PQ is less than the length of the arc PQ, i.e.

$$0 \leq 2\sin\theta \leq 2\theta \quad \text{and} \quad 0 \leq \sin\theta \leq \theta.$$

Since $\lim\limits_{\theta \to 0^+} 0 = \lim\limits_{\theta \to 0^+} \theta = 0$, it follows from the Pinching Theorem that

$$\lim\limits_{\theta \to 0^+} \sin\theta = 0.$$

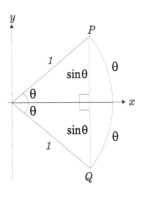

Figure 1.6.9
Establishing $\lim\limits_{\theta \to 0} \sin\theta$

Let $\phi = -\theta$. Since the sine function is odd,

$$\lim\limits_{\theta \to 0^-} \sin\theta = \lim\limits_{-\theta \to 0^+} -\sin(-\theta) = -\lim\limits_{\phi \to 0^+} \sin\phi = 0.$$

The left and right hand limits both equal 0. Hence $\lim\limits_{\theta \to 0} \sin\theta = 0$ by Property 2.

(2) Show that $\lim\limits_{\theta \to 0} \cos\theta = 1$.

Solution If $-\pi/2 \leq \theta \leq \pi/2$, then $0 \leq \cos\theta \leq 1$. It follows that

$$\cos^2\theta \leq \cos\theta \leq 1 \tag{1.6.4}$$

for $\theta \in [-\pi/2, \pi/2]$. Note

$$\lim\limits_{\theta \to 0} \cos^2\theta = \lim\limits_{\theta \to 0}(1 - \sin^2\theta) = 1 - 0^2 = 1 \quad \text{and} \quad \lim\limits_{\theta \to 0} 1 = 1.$$

Apply the Pinching Theorem to (1.6.4) to conclude that $\lim\limits_{\theta \to 0} \cos\theta = 1$.

(3) Evaluate $\lim\limits_{x \to 0} x\sin\dfrac{1}{x}$.

Solution Note $-1 \leq \sin\frac{1}{x} \leq +1$ for all positive numbers x. Multiply by x:

$$-x \leq x\sin\frac{1}{x} \leq x$$

for $x \geq 0$. Observe that $\lim\limits_{x \to 0^+} -x = \lim\limits_{x \to 0^+} x = 0$. By the Pinching Theorem:

$$\lim\limits_{x \to 0^+} x\sin\frac{1}{x} = 0.$$

Let $\theta = -x$. Since the sine function is odd,

$$\lim\limits_{x \to 0^-} x\sin\frac{1}{x} = \lim\limits_{-x \to 0^+} (-x)\sin\frac{1}{-x} = \lim\limits_{\theta \to 0^+} \theta\sin\frac{1}{\theta} = 0.$$

The left and right hand limits both equal 0. Hence $\lim\limits_{x \to 0} x\sin\dfrac{1}{x} = 0$ by Property 2.

(4) Find $\lim\limits_{x \to \infty} \cos\left(\dfrac{x}{x^2+1}\right)$.

Solution Divide numerator and denominator by x^2:

$$\lim\limits_{x \to \infty} \frac{x}{x^2+1} = \lim\limits_{x \to \infty} \frac{1/x}{1 + 1/x^2} = \frac{0}{1+0} = 0.$$

Since $\lim\limits_{\theta\to 0}\cos\theta = 1 = \cos 0$, we can apply Property 6 to evaluate our limit:

$$\lim_{x\to\infty}\cos\left(\frac{x}{x^2+1}\right) = \cos\left(\lim_{x\to\infty}\frac{x}{x^2+1}\right) = \cos 0 = 1.$$

(5)   Find $\lim\limits_{\theta\to 0}\dfrac{\tan^2\theta}{1-\sec\theta}$.

Solution   Substitution of $\theta = 0$ in this fraction produces $\frac{0}{1-1} = \frac{0}{0}$. Hence we can not apply Property 5, and our intuition does not help in evaluating this limit. Instead, rewrite the numerator using the Pythagorean identity $\tan^2\theta = \sec^2\theta - 1$. Then the numerator factors, and we cancel the common factor from the numerator and denominator.

$$\lim_{\theta\to 0}\frac{\tan^2\theta}{1-\sec\theta} = \lim_{\theta\to 0}\frac{\sec^2\theta-1}{1-\sec\theta} = \lim_{\theta\to 0}\frac{(\sec\theta+1)(\sec\theta-1)}{1-\sec\theta}$$
$$= -\lim_{\theta\to 0}(\sec\theta+1) = -(1+1) = -2 \qquad\qquad \square$$

We establish two limits which are useful in evaluating nontrivial trigonometric limits.

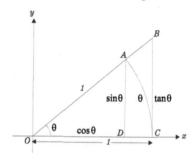

Figure 1.6.10 Establishing $\lim\limits_{\theta\to 0}\dfrac{\sin\theta}{\theta}$

Theorem 1.6.11   **(a)** $\lim\limits_{\theta\to 0}\dfrac{\sin\theta}{\theta} = 1$ **(b)** $\lim\limits_{\theta\to 0}\dfrac{1-\cos\theta}{\theta} = 0$

Proof   **(a)** Consider the sector of the unit circle in Figure 1.6.10 with $0 < \theta < \pi/2$. Let T_1 denote the small triangle OAD, and let T_2 denote the large triangle OBC. Since the hypotenuse of T_1 is a radius of the circle, it has length one. Then the opposite side of T_1 has length $\sin\theta$, and the adjacent side has length $\cos\theta$. Thus, T_1 has area $\frac{1}{2}\sin\theta\cos\theta$. Since the adjacent side of T_2 is a radius of the circle, it has length one. Therefore, the opposite side of T_2 has length $\tan\theta$, and T_2 has area $\frac{1}{2}(1)\tan\theta$. The sector $S = 0AC$ of the circle is $\frac{\theta}{2\pi}$ of the entire circle. Therefore, S has area $\pi(1)^2 \cdot \frac{\theta}{2\pi} = \frac{1}{2}\theta$. Since T_1 is contained in S and S is contained in T_2, we have:

$$\begin{array}{ccccc}
\text{area } T_1 & \leq & \text{area } S & \leq & \text{area } T_2 \\
\frac{1}{2}\sin\theta\cos\theta & \leq & \frac{1}{2}\theta & \leq & \frac{1}{2}\tan\theta = \frac{1}{2}\frac{\sin\theta}{\cos\theta} \\
\cos\theta & \leq & \frac{\theta}{\sin\theta} & \leq & \frac{1}{\cos\theta} \qquad \text{[dividing by } \frac{1}{2}\sin\theta\text{]}\\
\frac{1}{\cos\theta} & \geq & \frac{\sin\theta}{\theta} & \geq & \cos\theta \qquad \text{[taking reciprocals]}
\end{array}$$

Since $\lim\limits_{\theta\to 0^+}\dfrac{1}{\cos\theta} = \lim\limits_{\theta\to 0^+}\cos\theta = 1$, it follows from the Pinching Theorem that

$$\lim_{\theta\to 0^+}\frac{\sin\theta}{\theta} = 1.$$

Let $\phi = -\theta$. Since the sine function is odd,

$$\lim_{\theta \to 0^-} \frac{\sin \theta}{\theta} = \lim_{-\theta \to 0^+} \frac{\sin(-\theta)}{-\theta} \lim_{\phi \to 0^+} \frac{\sin \phi}{\phi} = 1.$$

The left and right hand limits both equal one. Hence $\lim\limits_{\theta \to 0} \dfrac{\sin \theta}{\theta} = 1$ by Property 2.

(b) We use the Pythagorean identity $1 - \cos^2 \theta = \sin^2 \theta$ to establish this limit from the limit determined in (a). First, multiply this limit by $\frac{1+\cos \theta}{1+\cos \theta}$:

$$\lim_{\theta \to 0} \frac{1 - \cos \theta}{\theta} = \lim_{\theta \to 0} \frac{1 - \cos \theta}{\theta} \cdot \frac{1 + \cos \theta}{1 + \cos \theta} = \lim_{\theta \to 0} \frac{1 - \cos^2 \theta}{\theta(1 + \cos \theta)}$$

$$= \lim_{\theta \to 0} \frac{\sin^2 \theta}{\theta(1 + \cos \theta)} = \lim_{\theta \to 0} \frac{\sin \theta}{\theta} \cdot \lim_{\theta \to 0} \frac{\sin \theta}{1 + \cos \theta} = 1 \cdot \frac{0}{1 + 1} = 0. \quad \square$$

Since all six trigonometric functions can be written in terms of sine and cosine, the two limits of Theorem 1.6.11 can be used to evaluate other trigonometric limits.

Examples 1.6.12 (1) Find $\lim\limits_{\theta \to 0} \dfrac{\tan 3\theta}{\theta}$.

Solution First, rewrite this limit in terms of $\sin 3\theta$ and $\cos 3\theta$:

$$\lim_{\theta \to 0} \frac{\tan 3\theta}{\theta} = \lim_{\theta \to 0} \frac{\sin 3\theta}{\theta \cos 3\theta}$$

by the definition of $\tan 3\theta$. We want to apply Theorem 1.6.11(a) to the angle 3θ. However, that limit requires the denominator of the fraction to equal the angle 3θ. Thus we multiply and divide by 3 and then apply Property 4:

$$\lim_{\theta \to 0} \frac{\tan 3\theta}{\theta} = \lim_{\theta \to 0} \left(\frac{\sin 3\theta}{3\theta} \cdot \frac{3}{\cos 3\theta} \right) = \lim_{\theta \to 0} \frac{\sin 3\theta}{3\theta} \cdot \lim_{\theta \to 0} \frac{3}{\cos 3\theta} = 1 \cdot \frac{3}{1} = 3$$

(2) Find $\lim\limits_{\theta \to 0} (\csc \theta - \cot \theta)$.

Solution We write this differences of trigonometric functions as a quotient to which the limits of Theorem 1.6.11 apply. By the definitions of $\csc \theta$ and $\cot \theta$:

$$\lim_{\theta \to 0} (\csc \theta - \cot \theta) = \lim_{\theta \to 0} \left(\frac{1}{\sin \theta} - \frac{\cos \theta}{\sin \theta} \right) = \lim_{\theta \to 0} \frac{1 - \cos \theta}{\sin \theta}$$

$$= \lim_{\theta \to 0} \frac{1 - \cos \theta}{\theta} \cdot \frac{\theta}{\sin \theta} \qquad \text{[multiplying and dividing by } \theta \text{]}$$

$$= \lim_{\theta \to 0} \frac{1 - \cos \theta}{\theta} \cdot \lim_{\theta \to 0} \frac{\theta}{\sin \theta} = 0 \cdot 1 = 0.$$

(3) Find $\lim\limits_{\theta \to \frac{\pi}{2}} (2\theta - \pi) \sec \theta$.

Solution We know nothing about limits of trigonometric functions when the angle approaches a nonzero number. Therefore, change variables from θ to $\phi = \theta - \frac{\pi}{2}$ because ϕ approaches 0 when θ approaches $\frac{\pi}{2}$. Since $\theta = \phi + \frac{\pi}{2}$:

$$\lim_{\theta \to \frac{\pi}{2}} (2\theta - \pi) \sec \theta = \lim_{\phi \to 0} [(2\phi + \pi) - \pi] \sec \left(\phi + \frac{\pi}{2} \right) = \lim_{\phi \to 0} 2\phi \sec \left(\phi + \frac{\pi}{2} \right)$$

$$= \lim_{\phi \to 0} \frac{2\phi}{\cos \left(\phi + \frac{\pi}{2} \right)} \qquad \text{[by definition of the secant]}$$

$$= \lim_{\phi \to 0} \frac{2\phi}{\cos \phi \cos \frac{\pi}{2} - \sin \phi \sin \frac{\pi}{2}} \qquad \text{[by identity (1.4.16)]}$$

$$= \lim_{\phi \to 0} \frac{2\phi}{(\cos \phi)(0) - (\sin \phi)(1)} = \lim_{\phi \to 0} \frac{2\phi}{-\sin \phi}$$

$$= -2 \lim_{\phi \to 0} \frac{\phi}{\sin \phi} = (-2)(1) = -2 \text{ [by Theorem (1.6.11)(a)]}$$

(4) Find $\lim_{\theta \to 0} \dfrac{\theta}{1 - \cos \theta}$.

Solution This fraction is the reciprocal of the fraction $\frac{1-\cos\theta}{\theta}$. Therefore, our intuition says that this limit should be the reciprocal of the limit $\lim_{\theta \to 0} \dfrac{1 - \cos \theta}{\theta} = 0$ of Theorem 1.6.11(b). Since $\frac{1}{0}$ is not defined, we suspect that $\lim_{\theta \to 0} \dfrac{\theta}{1 - \cos \theta}$ does not exist. To justify this observation, note that the two one–sided limits have different values. The denominator $1 - \cos \theta$ of this fraction is always positive for $-\pi < \theta < \pi$ and $\theta \neq 0$. Hence the right–hand limit is $+\infty$ because the numerator θ approaches zero through positive numbers while the left–hand limit is $-\infty$ because the numerator θ approaches zero through negative numbers. By Property 2, $\lim_{\theta \to 0} \dfrac{\theta}{1 - \cos \theta}$ does not exist.   □

Summary

The reader should know the statements of the eight limit properties and be able to apply them to examples. In particular, she should be able to compute limits of rational functions and limits of quotients involving differences of square roots. In addition, she should be able to compute limits of trigonometric functions using trigonometric identities and the two fundamental trigonometric limits of Theorem 1.6.11.

Basic Exercises

1. Use Property 1 or 2 to show that each limit does not exist.

(a) $\lim_{x \to 3} f(x)$ where $f(x) = x$ for x rational and $f(x) = 0$ for x irrational.

(b) $\lim_{x \to 2} g(x)$ where $g(x) = x^2$ for x rational and $g(x) = \frac{1}{x^2}$ for x irrational.

(c) $\lim_{x \to \pi} h(x)$ where $h(x) = \sin x$ for x rational and $h(x) = \cos x$ for x irrational.

(d) $\lim_{x \to 0} \dfrac{|x|}{x}$ **(e)** $\lim_{x \to -2} \dfrac{|x^2 - 4|}{x + 2}$ **(f)** $\lim_{x \to 0} \arctan \dfrac{1}{x}$ **(g)** $\lim_{x \to 0} \cot x$ **(h)** $\lim_{x \to \pi} \sec x$

2. Use Property 2 to determine if each limit exists. If the limit exists, find its value.

(a) $\lim_{x \to 4} |x - 4|$ **(b)** $\lim_{x \to -5} |x^2 + 2x - 15|$ **(c)** $\lim_{x \to 0} |\sin x|$ **(d)** $\lim_{x \to 0} \arctan \dfrac{1}{x}$

(e) $\lim_{x \to -3} f(x)$ where $f(x) = x + 3$ for $x \geq -3$ and $f(x) = 3 - x$ for $x < -3$.

(f) $\lim_{x \to 2} g(x)$ where $g(x) = x - 2$ for $x \geq 2$ and $g(x) = 2 - x$ for $x < 2$.

(g) $\lim_{x \to 0} h(x)$ where $h(x) = \sin x$ for $x \geq 0$ and $h(x) = \tan x$ for $x < 0$.

(h) $\lim_{x \to 0} k(x)$ where $k(x) = \cos x$ for $x \geq 0$ and $k(x) = \sin x$ for $x < 0$.

3. Explain how Property 6 determines the value of each limit.

(a) $\lim_{x \to 2} \sqrt{x^2 + 3x + 6}$ **(b)** $\lim_{x \to 0} \sin \dfrac{x}{x^2 + 1}$ **(c)** $\lim_{x \to 2} \cos \dfrac{x^2 - x - 2}{x^2 + x + 2}$

(d) $\lim_{x \to -2} \left| \dfrac{x^3 + 2x^2}{x + 1} \right|$ **(e)** $\lim_{x \to 0} \sqrt{\sin x^2}$ **(f)** $\lim_{x \to \pi} |\cos x|$

(g) $\lim_{x \to -4} \sqrt{|x^3|}$ **(h)** $\lim_{x \to 0} \sqrt{\sin |x|}$ **(i)** $\lim_{x \to 0} \cos \sqrt{\tan \sqrt[3]{x}}$

4. Use the Pinching Theorem to determine the value of each limit.

(a) $\lim_{x \to 0} x^2 \cos \dfrac{1}{x}$ (b) $\lim_{x \to 0} x \cos \dfrac{1}{x}$ (c) $\lim_{x \to 0} x^5 \sin \dfrac{1}{x^3}$ (d) $\lim_{x \to 0} x^4 \sin \dfrac{1}{x^7}$

(e) $\lim_{x \to 0} f(x)$ where $f(x) = 0$ for x rational and $f(x) = x^6$ for x irrational

(f) $\lim_{x \to 1} g(x)$ where $g(x) = x^3$ for x rational and $g(x) = x^4$ for x irrational

(g) $\lim_{x \to 0} h(x)$ where $h(x) = x$ for x rational and $h(x) = \sin x$ for x irrational

Evaluate these limits. At each step, state the limit property you are using.

5. $\lim_{x \to -4} x^3$ **6.** $\lim_{x \to 3} (5x^2 - 17)$ **7.** $\lim_{x \to 2} (x^2 + 1)^5$

8. $\lim_{x \to 2} (x^4 - 3x^2 + 8)$ **9.** $\lim_{x \to -2} \dfrac{x^2 - x - 2}{x^2 + x - 1}$ **10.** $\lim_{x \to 2} \dfrac{x - 2}{x^2 - x - 2}$

11. $\lim_{x \to -1} \dfrac{x + 1}{x^3 + 1}$ **12.** $\lim_{x \to -5} \dfrac{x^2 - 3x - 10}{x^2 - 9x + 20}$ **13.** $\lim_{x \to 4} \dfrac{x^3 - 12x^2 + 48x - 64}{x^3 - 64}$

14. $\lim_{x \to \infty} \dfrac{5x^4 - 2x + 3}{3x^2 + 7}$ **15.** $\lim_{x \to -2} \dfrac{x + 2}{x^2 + x + 2}$ **16.** $\lim_{x \to \infty} \dfrac{3x^2 - 2x + 5}{4x^2 + 1}$

17. $\lim_{x \to \infty} \dfrac{2x^2 - 3x + 1}{5x^3 - 2x + 4}$ **18.** $\lim_{x \to 1} \dfrac{x^4 - 4x^3 + 6x^2 - 4x + 1}{x^3 - 3x^2 + 3x - 1}$

19. $\lim_{x \to -\infty} \dfrac{x^4 - 3x + 7}{2x^4 + 4x^2 - 3}$ **20.** $\lim_{x \to -\infty} \dfrac{x^3 + x^2 + x + 1}{36x^2 - 9x + 81}$

21. $\lim_{x \to -\infty} \dfrac{80x^3 - 19x^2 + 32x - 29}{x^4 + 1}$ **22.** $\lim_{x \to \infty} \dfrac{24x^5 - 3x^2 + 9}{8x^5 - 16x^4 + 5x^2 - 7}$

23. $\lim_{x \to -3} \sqrt{x^2 + 4x + 7}$ **24.** $\lim_{x \to 0^+} \dfrac{1}{\sqrt{x}}$ **25.** $\lim_{x \to 3} \sqrt{\dfrac{x^2 - x - 6}{8x - x^2 - 15}}$

26. $\lim_{x \to 1} \dfrac{x - 1}{\sqrt{x} - 1}$ **27.** $\lim_{x \to 5} \dfrac{\sqrt{4x + 5} - x}{x^2 - 4x - 5}$ **28.** $\lim_{x \to -4} \dfrac{x^2 - 2x - 24}{\sqrt{4 - 3x} - 4}$

29. $\lim_{x \to 2} \dfrac{x^2 - 4}{\sqrt{7x + 2} - \sqrt{10x - 4}}$ **30.** $\lim_{x \to 2} \dfrac{\sqrt{3x^2 - 3} - \sqrt{x^2 + 5}}{x - 2}$

31. $\lim_{x \to -3} \dfrac{x^2 + 2x - 3}{\sqrt{3 - 2x} - \sqrt{x + 12}}$ **32.** $\lim_{x \to 1} \dfrac{\sqrt{3x + 6} - \sqrt{5x + 4}}{x^2 + 2x - 3}$

33. $\lim_{x \to 6} \dfrac{\sqrt{3x + 7} - 5}{7 - \sqrt{8x + 1}}$ **34.** $\lim_{x \to -2} \dfrac{\sqrt{1 - 4x} - \sqrt{x + 11}}{\sqrt{2 - x} - \sqrt{x + 2}}$

35. $\lim_{x \to 3} \dfrac{\sqrt{3x + 7} - \sqrt{5x + 1}}{\sqrt{x + 6} - \sqrt{2x + 3}}$ **36.** $\lim_{x \to 2} \dfrac{\sqrt{3x - 2} - \sqrt{x + 2}}{\sqrt{4x + 1} - \sqrt{6x - 3}}$

37. $\lim_{x \to 1} \dfrac{\sqrt[4]{x} - 1}{x - 1}$ **38.** $\lim_{x \to 1} \sin\left(\dfrac{x^2 - 2x + 1}{x^2 + 2x - 3}\right)$

39. $\lim_{x \to 2} \sqrt[5]{x^3 + 3x^2 + 4x + 4}$ **40.** $\lim_{x \to 729} \dfrac{x^{2/3} - 4x^{1/6} + 1}{x^{5/6} + \sqrt{x} + 5}$

41. $\lim_{x \to 16} \left(\dfrac{5\sqrt{x} + 7}{\sqrt{x} + 4}\right)^{2/3}$ **42.** $\lim_{\theta \to 0} (2\cos^3 \theta - 3\cos^2 \theta + 4)$

43. $\lim_{x \to -\infty} \sin\left(\dfrac{x^2 + x + 1}{x^3 - 8}\right)$ **44.** $\lim_{x \to 0} x^2 \sin \sqrt{x^4 + x^2 + 1}$

45. $\lim_{x \to -\infty} \dfrac{\cos x}{x}$ **46.** $\lim_{x \to \infty} \dfrac{\sin x + \cos x}{x^2}$ **47.** $\lim_{x \to 1} \sqrt{\cos \dfrac{x^2 - 1}{x^2 + 1}}$ **48.** $\lim_{x \to 0} x \cos \dfrac{1}{x}$

49. $\lim_{\theta \to 0} \tan \theta$ **50.** $\lim_{\theta \to 0} \dfrac{\theta}{\sin 5\theta}$ **51.** $\lim_{\theta \to 1} \sin(\theta^2 - 1)$ **52.** $\lim_{\theta \to 0} \dfrac{\tan 4\theta}{\tan 7\theta}$

53. $\lim_{\theta \to 0} \dfrac{1 - \cos^2 \theta}{\theta^2}$ **54.** $\lim_{\theta \to 0} \sec \theta$ **55.** $\lim_{\theta \to \frac{\pi}{2}} \dfrac{\cot^2 \theta}{1 - \csc \theta}$ **56.** $\lim_{\theta \to \frac{\pi}{4}} \dfrac{\cos 2\theta}{\sin \theta - \cos \theta}$

57. $\lim\limits_{\theta \to 0} \theta \cot 2\theta$ **58.** $\lim\limits_{\theta \to \pi} \dfrac{\sin \theta}{\pi - \theta}$ **59.** $\lim\limits_{\theta \to 0} \dfrac{\sec \theta - 1}{\sin \theta}$ **60.** $\lim\limits_{\theta \to \pi/2} (\pi - 2\theta) \tan \theta$

61. $\lim\limits_{\theta \to 0} \theta \csc 3\theta$ **62.** $\lim\limits_{\theta \to 0} (\tan^3 \theta - \sec^3 \theta)$ **63.** $\lim\limits_{\theta \to \pi/2} (\sec \theta - \tan \theta)$

64. $\lim\limits_{\theta \to \pi} \dfrac{\tan \theta}{\cos \theta + 1}$ **65.** $\lim\limits_{\theta \to 0} \dfrac{1 - \cos \theta}{\theta^2}$ **66.** $\lim\limits_{\theta \to 0} \theta \csc \theta$

67. $\lim\limits_{\theta \to 0} \dfrac{\sec \theta - 1}{\sin \theta \sec \theta}$ **68.** $\lim\limits_{\theta \to \pi} (\csc \theta - \cot \theta)$ **69.** $\lim\limits_{\theta \to \pi} (\pi - \theta) \csc \theta$

70. $\lim\limits_{\theta \to 0} \dfrac{\sin \theta}{\theta + \tan \theta}$ **71.** $\lim\limits_{\theta \to \pi} \dfrac{1 + \csc \theta}{\cot^2 \theta}$ **72.** $\lim\limits_{\theta \to 0} \dfrac{\sec \theta}{\theta}$

Appendix: Verification of the Limit Properties

We use the definition of a limit to verify the first seven properties of limits which we stated and used in this section. Property 8 is proved in the appendix to Section 7. Except in Property 2, c is either a number in $\Re$ or one of the symbols $\pm\infty$.

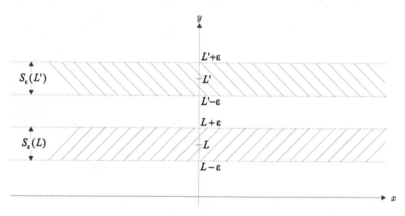

Figure 1.6.13 Property 1 with $L, L' \in \Re$

Property 1 *If* $\lim\limits_{x \to c} f(x) = L$ *and* $\lim\limits_{x \to c} f(x) = L'$ *then* $L = L'$.

Proof Name L' and L so that $L' \geq L$. Say it is possible to have $L' > L$. We show how to find positive numbers ϵ and ϵ' so that the horizontal strips $S_\epsilon(L)$ and $S_{\epsilon'}(L')$ are disjoint. There are three cases.

- Consider the case $L, L' \in \Re$. Let $\epsilon = (L' - L)/4$. Then the horizontal strips $S_\epsilon(L)$ and $S_\epsilon(L')$ are disjoint. See Figure 1.6.13.

- Next, consider the case $L = -\infty$. Select any positive number ϵ'. If $L' \in \Re$. select $\epsilon > 0$ such that $-\epsilon < L' - \epsilon'$. If $L' = \infty$, select $\epsilon > 0$ arbitrarily. These horizontal strips $S_\epsilon(L)$ and $S_{\epsilon'}(-\infty)$ are disjoint.

- Consider the case $L' = \infty$ and $L \in \Re$. Select any positive number ϵ. Then select $\epsilon' > 0$ with $\epsilon' > L + \epsilon$. These horizontal strips $S_\epsilon(L)$ and $S_{\epsilon'}(\infty)$ are disjoint.

In all cases, find $\delta, \delta' > 0$ such that the graph of f restricted to $I_\delta(c)$ lies in $S_\epsilon(L)$ and the graph of f restricted to $I_{\delta'}(c)$ lies in $S_\epsilon(L')$. Note that one of $I_\delta(c), I_{\delta'}(c)$ is a subset of the other. If x is in both $I_\delta(c)$ and $I_{\delta'}(c)$, then $f(x)$ is in both $S_\epsilon(L)$ and $S_{\epsilon'}(L')$. This is impossible because these horizontal strips are disjoint. Thus it is not possible to have $L' > L$. Hence L' must equal L. □

Property 2 *If* $c \in \Re$, *then* $\lim\limits_{x \to c^-} f(x) = \lim\limits_{x \to c^+} f(x) = L$ *if and only if* $\lim\limits_{x \to c} f(x) = L$.

Proof Assume $\lim_{x \to c} f(x) = L$. If a horizontal strip $S_\epsilon(L)$ is given, there is $\delta > 0$ such that the graph of f restricted to $I_\delta(c)$ lies in $S_\epsilon(L)$. Therefore, the graph of f restricted to both $(c - \delta, c)$ and $(c, c + \delta)$ lies in $S_\epsilon(L)$. Hence $\lim_{x \to c^+} f(x) = \lim_{x \to c^-} f(x) = L$.

Assume $\lim_{x \to c^+} f(x) = \lim_{x \to c^-} f(x) = L$. If a horizontal strip $S_\epsilon(L)$ is given, there are $\delta_1, \delta_2 > 0$ such that the graph of f restricted to each of the intervals $(c - \delta_1, c)$ and $(c, c + \delta_2)$ lies in $S_\epsilon(L)$. Let δ be the smaller of δ_1 and δ_2. Observe that

$$I_\delta(c) = (c - \delta, c) \cup (c, c + \delta) \subset (c - \delta_1, c) \cup (c, c + \delta_2).$$

Therefore, the graph of f restricted to $I_\delta(c)$ lies in $S_\epsilon(L)$, and $\lim_{x \to c} f(x) = L$. $\quad\square$

The next three properties deal with the arithmetic of limits. From the geometric point of view, observe that we can do "vertical arithmetic" on our horizontal strips as follows. If $(x, y) \in S_\epsilon(L)$ and $(x, y') \in S_{\epsilon'}(L')$ then

$$
\begin{aligned}
(x, y) + (x, y') &= (x, y + y'), \\
(x, y) \cdot (x, y') &= (x, y \cdot y') \quad \text{and} \\
(x, y) \div (x, y') &= (x, y \div y') \text{ if } y' \neq 0.
\end{aligned}
$$

Thus, $S_\epsilon(L) + S_{\epsilon'}(L')$ and $S_\epsilon(L) \cdot S_{\epsilon'}(L')$ are defined. If $S_{\epsilon'}(L')$ does not intersect the x–axis and $(x, y') \in S_{\epsilon'}(L')$ then $y' \neq 0$. Therefore, $S_\epsilon(L) \div S_{\epsilon'}(L')$ is also defined in this case. We use this vertical arithmetic in the proofs of the next three properties.

Property 3 *If $A + B$ is defined,* $\lim_{x \to c} f(x) = A$ *and* $\lim_{x \to c} g(x) = B$, *then*

$$\lim_{x \to c} [f(x) + g(x)] = A + B.$$

Proof Consider the case where $A, B \in \Re$. Say we are given a horizontal strip $S_\epsilon(A + B)$. Consider the horizontal strips $S_{\epsilon/2}(A)$ and $S_{\epsilon/2}(B)$. The sum of these two strips is contained in the strip $S_\epsilon(A + B)$. Find $\delta_1, \delta_2 > 0$ such that the graph of f restricted to $I_{\delta_1}(c)$ lies inside $S_{\epsilon/2}(A)$, and the graph of g restricted to $I_{\delta_2}(c)$ lies inside $S_{\epsilon/2}(B)$. Choose δ to be either δ_1 or δ_2 so that $I_\delta(c)$ is the smaller of $I_{\delta_1}(c)$ and $I_{\delta_2}(c)$. Then the graph of $f + g$ restricted to $I_\delta(c)$, being the sum of two graphs inside the strips $S_{\epsilon/2}(A)$ and $S_{\epsilon/2}(B)$, must lie inside the strip $S_\epsilon(A + B)$. By definition, $\lim_{x \to c} [f(x) + g(x)] = A + B$.

The proof when one or both of A, B is $\pm\infty$ is given as Exercise 2. $\quad\square$

Property 4 *If AB is defined,* $\lim_{x \to c} f(x) = A$ *and* $\lim_{x \to c} g(x) = B$, *then*

$$\lim_{x \to c} [f(x)g(x)] = AB.$$

Proof Consider the case where $A, B > 0$ are numbers. Say we are given a horizontal strip $S_\epsilon(AB)$. We want to find $\epsilon' > 0$ such the product of the horizontal strips $S_{\epsilon'}(A)$ and $S_{\epsilon'}(B)$ lies inside $S_\epsilon(AB)$. If $(x, f(x)) \in S_{\epsilon'}(A)$ and $(x, g(x)) \in S_{\epsilon'}(B)$, then

$$A - \epsilon' < f(x) < A + \epsilon' \quad \text{and} \quad B - \epsilon' < g(x) < B + \epsilon'.$$

If ϵ' is less than 1, A, B then multiply these two inequalities:

$$
\begin{aligned}
(A - \epsilon')(B - \epsilon') &< f(x)g(x) < (A + \epsilon')(B + \epsilon'), \\
AB - \epsilon'(A + B - \epsilon') &< f(x)g(x) < AB + \epsilon'(A + B + \epsilon'), \\
AB - \epsilon'(A + B + 1) &< f(x)g(x) < AB + \epsilon'(A + B + 1).
\end{aligned}
$$

Note the replacement above of $-\epsilon'(A + B - \epsilon')$ by $-\epsilon'(A + B + 1)$ above replaces the positive number ϵ'^2 by the negative number $-\epsilon'$. If we choose $\epsilon' < \epsilon/(A + B + 1)$, then

$$AB - \epsilon < f(x)g(x) < AB + \epsilon.$$

To summarize: if we choose $\epsilon' > 0$ to be less than 1, A, B, $\epsilon/(A + B + 1)$ then the product of the strips $S_{\epsilon'}(A)$ and $S_{\epsilon'}(B)$ lies inside the strip $S_\epsilon(AB)$.

Find $\delta_1, \delta_2 > 0$ such that the graph of f restricted to $I_{\delta_1}(c)$ lies inside the strip $S_{\epsilon'}(A)$, and the graph of g restricted to $I_{\delta_2}(c)$ lies inside the strip $S_{\epsilon'}(B)$. Choose δ to be either δ_1 or δ_2 so that $I_\delta(c)$ is the smaller of $I_{\delta_1}(c)$ and $I_{\delta_2}(c)$. Then the graph of fg restricted to $I_\delta(c)$, being the product of two graphs inside the strips $S_{\epsilon'}(A)$ and $S_{\epsilon'}(B)$, must be inside the strip $S_\epsilon(AB)$. By definition, $\lim_{x \to c}[f(x)g(x)] = AB$.

The verification when A and/or B is a number less than or equal to zero or $\pm\infty$ is given as Exercise 3.   □

Property 5 *If $\frac{A}{B}$ is defined,* $\lim_{x \to c} f(x) = A$ *and* $\lim_{x \to c} g(x) = B$, *then*

$$\lim_{x \to c} \frac{f(x)}{g(x)} = \frac{A}{B}.$$

Proof Consider the case where $A, B \in \Re$ with $B \neq 0$. We verify $\lim_{x \to c}[1/g(x)] = 1/B$. It then follows from Property 4 that $\lim_{x \to c}[f(x)/g(x)] = A/B$. Assume $B > 0$. Say we are given a horizontal strip $S_\epsilon(1/B)$. We show below how to select $\epsilon' > 0$ such that the reciprocal of the strip $S_{\epsilon'}(B)$ is a subset of the strip $S_\epsilon(1/B)$. Then choose $\delta > 0$ so that the graph of g restricted to $I_\delta(c)$ lies inside the strip $S_{\epsilon'}(B)$. It follows that the graph of $1/g$ restricted to $I_\delta(c)$ lies inside the strip $S_\epsilon(1/B)$. By definition, $\lim_{x \to c}[1/g(x)] = 1/B$. It remains to show how to choose ϵ'. Assume $\epsilon' < B$. If $(x, g(x)) \in S_{\epsilon'}(B)$ then

$$B - \epsilon' \; < \; g(x) \; < \; B + \epsilon' \quad \text{and} \quad \frac{1}{B + \epsilon'} \; < \; \frac{1}{g(x)} \; < \; \frac{1}{B - \epsilon'}.$$

We will show how to select ϵ' so that

$$\frac{1}{B} - \epsilon < \frac{1}{B + \epsilon'} \quad \text{and} \quad \frac{1}{B - \epsilon'} < \frac{1}{B} + \epsilon. \qquad (1.6.5)$$

It follows that
$$\frac{1}{B} - \epsilon \; < \; \frac{1}{g(x)} \; < \; \frac{1}{B} + \epsilon$$

and $(x, 1/g(x)) \in S_\epsilon(1/B)$ as required. To determine how to select ϵ', multiply the first inequality of (1.6.5) by $B(B + \epsilon')$ and the second one by $B(B - \epsilon')$:

$$B + \epsilon' - \epsilon B(B + \epsilon') < B, \qquad B < B - \epsilon' + \epsilon B(B - \epsilon');$$
$$\epsilon'(1 - \epsilon B) < \epsilon B^2, \qquad \epsilon'(1 + \epsilon B) < \epsilon B^2;$$
$$\epsilon' < \frac{\epsilon B^2}{1 - \epsilon B}, \qquad \epsilon' < \frac{\epsilon B^2}{1 + \epsilon B}. \qquad (1.6.6)$$

[We assume that $\epsilon < 1/B$. If not, replace ϵ by the smaller number $\epsilon_0 = 1/(2B)$ because $S_{\epsilon_0}(1/B) \subset S_\epsilon(1/B)$.] Since $0 < 1 - \epsilon B < 1 + \epsilon B$, the second inequality in (1.6.6) implies the first one. Therefore, let

$$\epsilon' = \frac{\epsilon B^2}{2(1 + \epsilon B)}.$$

Then both of the inequalities in (1.6.6) hold, and the inequalities in (1.6.5) are satisfied. Therefore, the reciprocal of the strip $S_{\epsilon'}(B)$ lies in the strip $S_\epsilon(1/B)$ as required. This completes the proof of the case when B is positive.

If B is negative then $-B > 0$. Therefore, $\displaystyle\lim_{x \to c} \frac{1}{g(x)} = -\lim_{x \to c} \frac{1}{-g(x)} = -\frac{1}{-B} = \frac{1}{B}$.

The cases where one of A, B is $\pm\infty$ are given as Exercise 4. $\square$

Property 6 *If B, $L \in \Re$, $\displaystyle\lim_{x \to c} g(x) = B$, $\displaystyle\lim_{x \to B} h(x) = L$ and $h(B) = L$, then*

$$\lim_{x \to c} h(g(x)) = h\left(\lim_{x \to c} g(x)\right) = L.$$

Proof Say we are given a horizontal strip $S_\epsilon(L)$. Find $\delta' > 0$ such that the graph of h restricted to $(B - \delta', B + \delta')$ lies in the strip $S_\epsilon(L)$. In particular, $(B, h(B))$ lies in this strip by the assumption $h(B) = L$. Thus,

$$h(B - \delta', B + \delta') \subset (L - \epsilon, L + \epsilon).$$

Now find $\delta > 0$ such that the graph of g restricted to $I_\delta(c)$ lies in the strip $S_{\delta'}(B)$, i.e.

$$g(I_\delta(c)) \subset (B - \delta', B + \delta').$$

Combine the preceding two inclusions:

$$h(g(I_\delta(c))) \subset h(B - \delta', B + \delta') \subset (L - \epsilon, L + \epsilon).$$

Hence the graph of $h \circ g$ restricted to $I_\delta(c)$ lies in $S_\epsilon(L)$. By definition, $\displaystyle\lim_{x \to c} h(g(x)) = L$. $\square$

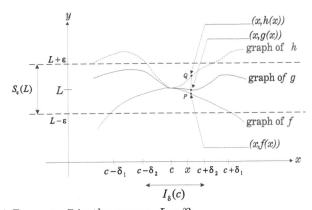

Figure 1.6.14 Property 7 in the case c, $L \in \Re$

Property 7 *Assume there is an interval $I_\delta(c)$ such that*

$$f(x) \leq g(x) \leq h(x) \quad for \quad x \in I_\delta(c).$$

If $\displaystyle\lim_{x \to c} f(x) = \lim_{x \to c} h(x) = L$, then $\displaystyle\lim_{x \to c} g(x) = L$.

Proof Say we are given a horizontal strip $S_\epsilon(L)$. Find δ_1, $\delta_2 > 0$ such that the graph of f restricted to $I_{\delta_1}(c)$ and the graph of g restricted to $I_{\delta_2}(c)$ both lie in $S_\epsilon(L)$. Choose δ to be either δ_1 or δ_2 so that $I_\delta(c)$ is the smaller of $I_{\delta_1}(c)$ and $I_{\delta_2}(c)$. If $x \in I_\delta(c)$, then $S_\epsilon(L)$ contains the points $P = (x, f(x))$ and $Q = (x, h(x))$. The case where c and L are numbers is depicted in Figure 1.6.14. Therefore $S_\epsilon(L)$ must contain all points, such as $(x, g(x))$, which lie on the vertical line segment PQ. Therefore the graph of g restricted to $I_\delta(c)$ is contained in $S_\epsilon(L)$, and $\displaystyle\lim_{x \to c} g(x) = L$. $\square$

Exercises

1. Sketch the disjoint strips $S_\epsilon(L)$ and $S_{\epsilon'}(L')$ used to prove Property 1 when
(a) $L \in \Re$ and $L' = \infty$; **(b)** $L = -\infty$ and $L' \in \Re$; **(c)** $L = -\infty$ and $L' = \infty$.

2. Show that Property 3 is valid when
(a) $A \in \Re$, $B = \infty$; **(b)** $A \in \Re$, $B = -\infty$; **(c)** $A = B = \infty$; **(d)** $A = B = -\infty$.

3. Show that Property 4 is valid when
(a) A, $B \in \Re$ with $A \leq 0$ or $B \leq 0$; **(b)** $A = \pm\infty$ and $B \in \Re$ with $B \neq 0$;
(c) A and B are both $\pm\infty$.

4. Show that Property 5 is valid when
(a) $A = \pm\infty$ and $B \in \Re$ with $B \neq 0$; **(b)** $A \in \Re$ and $B = \pm\infty$.

5. Sketch the analogue of Figure 1.6.14 illustrating Property 7 when
(a) $c = \infty$, $L = \infty$; **(b)** $c = \infty$, $L = -\infty$; **(c)** $c = \infty$, $L \in \Re$; **(d)** $c \in \Re$, $L = \infty$.

1.7   Continuous Functions

A continuous function is one whose graph can be drawn without removing the pencil from the paper. In the first subsection, a precise definition of a continuous function is formulated using limits. In the second subsection, we distinguish between two types of behavior that cause a function to fail to be continuous. These concepts are illustrated with a variety of examples, including some where it is not obvious if the given function is continuous. In the third subsection, we show that five of the limit properties of Section 6 determine analogous properties for continuous functions. These properties are used to construct complicated continuous functions from simple ones. The fourth subsection studies the Intermediate Value Theorem with applications to making approximations. The concluding subsection is devoted to the Maximum Value Theorem. The proofs of these two theorems, using topological properties of the real line, are given in Sections 11 and 12 of the Theory portion of this chapter. The appendix to this section applies the Intermediate and Maximum Value Theorems to establish Limit Property 8 which computes limits of inverse functions.

Defining Continuity

A continuous function at $x = c$ should behave nicely for values of x near c. The following example illustrates how to detect this good behavior using $\lim_{x \to c} f(x)$.

Motivating Example 1.7.1 Consider the functions $f(x) = \left\{ \begin{array}{ll} \frac{x^2-4}{x-2} & \text{if } x \neq 2 \\ 4 & \text{if } x = 2 \end{array} \right\}$,

$g(x) = \left\{ \begin{array}{ll} \frac{x^2-4}{x-2} & \text{if } x \neq 2 \\ -4 & \text{if } x = 2 \end{array} \right\}$ and $h(x) = \left\{ \begin{array}{ll} \frac{|x^2-4|}{x-2} & \text{if } x \neq 2 \\ 4 & \text{if } x = 2 \end{array} \right\}$.

We compute the limit of each of these functions as x approaches 2:

$$\lim_{x \to 2} f(x) = \lim_{x \to 2} \frac{x^2-4}{x-2} = \lim_{x \to 2} \frac{(x-2)(x+2)}{x-2} = \lim_{x \to 2}(x+2) = 2+2 = 4 = f(2);$$

$$\lim_{x \to 2} g(x) = \lim_{x \to 2} \frac{x^2-4}{x-2} = \lim_{x \to 2} \frac{(x-2)(x+2)}{x-2} = \lim_{x \to 2}(x+2) = 2+2 = 4 \neq g(2);$$

$$\lim_{x \to 2} h(x) = \lim_{x \to 2} \frac{|x^2-4|}{x-2} = \lim_{x \to 2} \frac{|x-2|(x+2)}{x-2} = \left[\lim_{x \to 2}(x+2)\right] \left[\lim_{x \to 2} \frac{|x-2|}{x-2}\right]$$

$$= 4 \lim_{x \to 2} \frac{|x-2|}{x-2}.$$

Observe that $\frac{|x-2|}{x-2}$ equals -1 for $x < 2$ and equals $+1$ for $x > 2$. Hence $\lim_{x \to 2^-} h(x) = -4$, $\lim_{x \to 2^+} h(x) = 4$ and $\lim_{x \to 2} h(x)$ does not exist.

What causes the different behavior of these functions that have similar definitions? That is, why is $\lim_{x\to 2} f(x)$ just $f(2)$, while $\lim_{x\to 2} g(x)$ is not $g(2)$ and $\lim_{x\to 2} h(x)$ is not $h(2)$? The answer can be seen from the graphs of these functions in Figures 1.7.2, 1.7.3 and 1.7.5. Each of the graphs of f and g approaches the point $(2,4)$ from the left and the right as x approaches 2. However, $(2,4) = (2, f(2))$ while $(2,4) \neq (2, g(2))$. On the other hand, the graph of h approaches different numbers from the left and the right. We introduce the word *continuous* to describe the behavior of the function f at $x = 2$, i.e. the limit of $f(x)$, as x approaches 2, is just $f(2)$. Functions like g, h whose limits at $x = 2$ are not $g(2)$, $h(2)$ are called *discontinuous* at $x = 2$. □

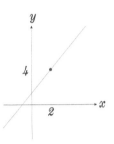

Fig 1.7.2 $y = f(x)$

Let f be a function whose domain is an interval I. To formulate a precise definition that f is continuous, we specify our requirements at each number c of its domain. We want f to be nice near $x = c$, i.e. we want the limit of f, as x approaches c, to exist and equal a *number* L. In addition, we want $f(c)$ to be defined in a reasonable manner: that is, we require that $f(c) = L$. We call f *continuous at* $x = c$ if f satisfies these conditions. If f is continuous at each point of its domain I, then we call f a continuous function. Since I may contain its left or right endpoints, there are three cases of this definition.

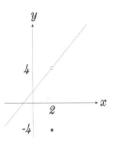

Fig 1.7.3 $y = g(x)$

Definition 1.7.4 *Let f be a function with domain an interval I of positive length.*
  (a) *f is continuous at $c \in I$ if the appropriate condition is satisfied.*
    (•) *If c is an interior point of I, then $\lim_{x\to c} f(x)$ exists and equals $f(c)$.*
    (•) *If c is the left endpoint of I, then $\lim_{x\to c^+} f(x)$ exists and equals $f(c)$.*
    (•) *If c is the right endpoint of I, then $\lim_{x\to c^-} f(c)$ exists and equals $f(c)$.*
  (b) *If f is not continuous at $c \in I$, we say f is discontinuous at $x = c$.*
  (c) *f is a continuous function if f is continuous at every $c \in I$.*
  (d) *Let g be a function whose domain is a union of disjoint intervals of positive length. g is called continuous if the restriction of g to each of these intervals is continuous.*

Note By our definition, f is not discontinuous at a point which is not in its domain.

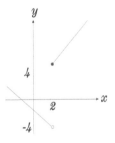

Fig 1.7.5 $y = h(x)$

The functions f, g, h of Example 1.7.1 all have domain $I = \Re$. All three are continuous at $x = c$ when $c \neq 2$:

$$\lim_{x\to c} f(x) = \lim_{x\to c}(x + 2) = c + 2 = f(c) \qquad \text{for } c \neq 2;$$
$$\lim_{x\to c} g(x) = \lim_{x\to c}(x + 2) = c + 2 = g(c) \qquad \text{for } c \neq 2;$$
$$\lim_{x\to c} h(x) = \lim_{x\to c}(x + 2) = c + 2 = h(c) \qquad \text{for } c > 2;$$
$$\lim_{x\to c} h(x) = \lim_{x\to c} -(x + 2) = -(c + 2) = h(c) \qquad \text{for } c < 2.$$

We showed above that f is continuous at $x = 2$ while g and h are discontinuous at $x = 2$. Thus f is a continuous function while g and h are not. In general, we compute the limit of a function at $x = c$ to check whether the function is continuous there.

Examples 1.7.6 Determine the points where each function is continuous.

(1) Let $f(x) = 2x^3 - 3x + 7$ with domain $I = \Re$.
    Solution For each $c \in \Re$, $\lim_{x\to c}(2x^3 - 3x + 7) = 2c^3 - 3c + 7 = f(c)$. Therefore, f is a continuous function.

(2) Let $g(x) = \frac{1}{x}$ with domain $I = (0, \infty)$.

Solution The graph of g is shown in Figure 1.7.7. Since $\lim\limits_{x \to c} \dfrac{1}{x} = \dfrac{1}{c} = g(c)$ for all $c > 0$, g is a continuous function. The problem that g has at $x = 0$ is not relevant to our considerations: although 0 is the left endpoint of the domain $I = (0, \infty)$ of g, the point 0 is not in the domain of g.

Fig 1.7.7 $g(x) = \frac{1}{x}$

(3) (a) Let k be the function with domain $\Re$ defined by $k(x) = \left\{ \begin{array}{ll} \sin\frac{1}{x} & \text{if } x \neq 0 \\ 0 & \text{if } x = 0 \end{array} \right\}$.

Solution This function was studied in Example 1.5.3 (4a), and its graph is given in Figure 1.7.8. If $c \neq 0$, then $\lim\limits_{x \to c} k(x) = \sin\dfrac{1}{c} = k(c)$, and k is continuous at $x = c$. Since $\lim\limits_{x \to 0} k(x)$ does not exist, k is discontinuous at $x = 0$. Thus k is not a continuous function.

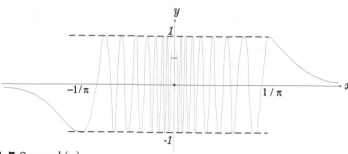

Figure 1.7.8 $y = k(x)$

(b) Let K be the function with domain $\Re$ defined by $K(x) = \left\{ \begin{array}{ll} x\sin\frac{1}{x} & \text{if } x \neq 0 \\ 0 & \text{if } x = 0 \end{array} \right\}$.

Solution This function was studied in Example 1.5.3 (4b), and its graph is given in Figure 1.7.9. Clearly K is continuous at $x = c$ for all $c \neq 0$. In Example 1.5.3(4b), we showed that $\lim\limits_{x \to 0} K(x) = 0 = K(0)$. Therefore, K is also continuous at $x = 0$, and K is a continuous function.

We verify that the graph of K has the horizontal asymptote $y = 1$ on both the left and the right. Let $\theta = \frac{1}{x}$. Note that θ approaches zero through positive numbers when x approaches ∞ while θ approaches zero through negative numbers when x approaches $-\infty$. Therefore,

$$\lim_{x \to \infty} K(x) = \lim_{x \to \infty} x\sin\frac{1}{x} = \lim_{x \to \infty} \frac{\sin\frac{1}{x}}{\frac{1}{x}} = \lim_{\theta \to 0^+} \frac{\sin\theta}{\theta} = 1,$$

$$\lim_{x \to -\infty} K(x) = \lim_{x \to -\infty} x\sin\frac{1}{x} = \lim_{x \to -\infty} \frac{\sin\frac{1}{x}}{\frac{1}{x}} = \lim_{\theta \to 0^-} \frac{\sin\theta}{\theta} = 1 .$$

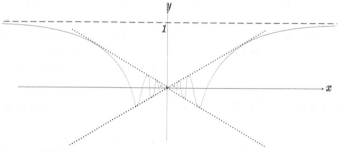

Figure 1.7.9 $y = K(x)$

(4) Let $m(x) = \left\{ \begin{array}{ll} 1 & \text{if } x \text{ is a rational number} \\ 0 & \text{if } x \text{ is an irrational number} \end{array} \right\}$ with domain $\Re$.

Solution This function was studied in Example 1.5.3 (3a), and its graph is given in Figure 1.7.10. Since $\lim_{x \to c} m(x)$ does not exist for all $c \in \Re$, m is discontinuous at every $x = c$. □

Discontinuous Functions

Fig 1.7.10 $y = m(x)$

A discontinuous function at $x = c$ behaves badly there. The following example distinguishes between bad behavior at $x = c$ itself and bad behavior near $x = c$.

Motivating Example 1.7.11 Return to the functions g and h of Example 1.7.1. We observed that both functions are discontinuous at $x = 2$. We describe how this discontinuity is seen in their graphs which are depicted in Figures 1.7.3 and 1.7.5.

The graph of g has the odd feature that although its graph approaches the point $(2, 4)$, as x approaches 2, there is a hole in the graph at $(2, 4)$. The dot which might have been there appears at $(2, -4)$. We say that g has a *removable discontinuity* at $x = 2$ because we can make the function g continuous by moving the dot at $(2, -4)$ on its graph to the location $(2, 4)$, i.e. by changing the definition of $g(2)$ from -4 to 4. This change in the definition of $g(2)$ removes the problem that g has at $x = 2$.

On the other hand, the graph of h approaches the point $(2, -4)$ when x approaches 2 from the left while the graph of h approaches the point $(2, 4)$ when x approaches 2 from the right. Notice how the graph of h jumps from the point $(2, -4)$ to the point $(2, 4)$ at $x = 2$. We say that h has an *essential discontinuity* at $x = 2$. No change in the definition of $h(2)$ will make this function continuous at $x = 2$. □

The formal definitions of removable and essential discontinuities generalize the situations described above on the graphs of g and h.

Definition 1.7.12 *Assume the function f, with domain the interval I, is discontinuous at $x = c \in I$.*
(a) *If $\lim_{x \to c} f(x)$ exists and equals the number L, but $L \neq f(c)$, we say f has a removable discontinuity at $x = c$.*
(b) *If $\lim_{x \to c} f(x)$ is not a number, we say f has an essential discontinuity at $x = c$.*

Notes (1) If f has a removable discontinuity at $x = c$, we can modify the definition of $f(c)$ to define a new function F which is continuous at $x = c$. Specifically, let

$$F(x) = \left\{ \begin{array}{ll} f(x) & \text{if } x \neq c \\ L & \text{if } x = c \end{array} \right\}.$$

(2) There are two ways that f can have an essential discontinuity at $x = c$. Either $\lim_{x \to c} f(x)$ does not exist or $\lim_{x \to c} f(x) = \pm\infty$.

The following examples emphasize that a function with a removable discontinuity at $x = c$ is nice near $x = c$ but has the misfortune that $f(c)$ is defined in a strange manner. On the other hand, a function with an essential discontinuity at $x = c$ has an intrinsic problem there. No minor modification of the definition of this function will define a new function which is continuous at $x = c$.

Examples 1.7.14 (1) Let $f(x) = \left\{ \begin{array}{ll} x^2 & \text{if } x \neq 1 \\ 2 & \text{if } x = 1 \end{array} \right\}$ with domain $\Re$. The graph of f is depicted in Figure 1.7.13. Then $\lim_{x \to 1} x^2 = 1$ while $f(1) = 2$. Thus, f has

Fig 1.7.13 $y = f(x)$

a removable discontinuity at $x = 1$. Define $F(x) = \left\{ \begin{array}{ll} f(x) & \text{if } x \neq 1 \\ 1 & \text{if } x = 1 \end{array} \right\}$ with domain $\Re$. Then F is continuous at $x = 1$. Note $F(x) = x^2$ for all $x \in \Re$.

(2) (a) Define $g(x) = \left\{ \begin{array}{ll} \frac{1}{x} & \text{if } x \neq 0 \\ 0 & \text{if } x = 0 \end{array} \right\}$ with domain $\Re$. The graph of g is given in Figure 1.7.15. Since $\lim\limits_{x \to 0} g(x)$ DNE, g has an essential discontinuity at $x = 0$.

(b) Define $G(x) = \frac{1}{x}$ with domain all $x \in \Re$ for $x \neq 0$. The graph of G is depicted in Figure 1.7.16. The domain of G is $(-\infty, 0) \cup (0, \infty)$. Note G is a continuous function because G is continuous at each point of its domain. We do *not* say G is discontinuous at $x = 0$ because zero is not in the domain of G.

(3) Consider the function h of Example 1.7.1. It has an essential discontinuity at $x = 2$ where the graph of h jumps from height -4 at the left of $x = 2$ to height $+4$ at the right of $x = 2$. The one–sided limits of h exist but have different values. This type of essential discontinuity is called a *jump discontinuity*.

(4) (a) The function k of Example 1.7.6 (3a) has an essential discontinuity at $x = 0$.

(b) The grpah of the function $k_1(x) = \left\{ \begin{array}{ll} x \sin \frac{1}{x} & \text{if } x \neq 0 \\ 1 & \text{if } x = 0 \end{array} \right\}$ with domain $\Re$ is sketched in Figure 1.7.17. Since $\lim\limits_{x \to 0} k_1(x) = 0$ and $k_1(0) = 1$, this function has a removable discontinuity at $x = 0$. Define $K(x) = \left\{ \begin{array}{ll} k_1(x) & \text{if } x \neq 0 \\ 0 & \text{if } x = 0 \end{array} \right\}$. The function K is continuous at $x = 0$.   □

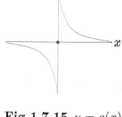

Fig 1.7.15 $y = g(x)$

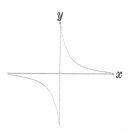

Fig 1.7.16 $y = G(x)$

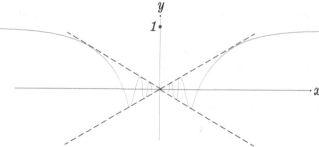

Figure 1.7.17 $y = k_1(x)$

Properties of Continuous Functions

Five of the limit properties of Section 6 translate into properties of continuous functions. Limit properties were used in Section 6 to evaluate complicated limits from simple ones. So too, the properties of continuous functions are used to construct complicated continuous functions from simple ones. We prove Properties 1 and 5. The proofs of the other properties are left for the exercises.

Property 1 *The sum of continuous functions is continuous: if f and g are continuous on the interval I, then $f + g$ is also continuous on I.*

Proof For each number c in I: $\lim\limits_{x \to c} f(x) = f(c)$ and $\lim\limits_{x \to c} g(x) = g(c)$. Hence

$$\lim_{x \to c} [f(x) + g(x)] = \lim_{x \to c} f(x) + \lim_{x \to c} g(x) = f(c) + g(c) = (f + g)(c) .$$

Thus $f + g$ is continuous at $x = c$.   □

For example, we know $f(x) = x$ and $g(x) = 3$ are continuous on $\Re$. Hence their sum $(f + g)(x) = x + 3$ is also continuous on $\Re$.

Property 2 *The product of continuous functions is continuous: if f and g are continuous on the interval I, then fg is also continuous on I.*

For example, we know $f(x) = x$ and $g(x) = x$ are continuous on $\Re$. Hence their product $(fg)(x) = x^2$ is also continuous on $\Re$.

Property 3 *The quotient of continuous functions is continuous. If f and g are continuous on the interval I and $g(x) \neq 0$ for $x \in I$, then $\frac{f}{g}$ is also continuous on I.*

For example, we know $f(x) = x + 7$ and $g(x) = x^2 + 3$ are continuous on $\Re$ by Properties 1 and 2. Moreover, $g(x) \geq 3$ and is never zero. Hence the quotient $\left(\frac{f}{g}\right)(x) = \frac{x+7}{x^2+3}$ is also continuous on $\Re$.

Property 4 *The composite of continuous functions is continuous. Let f and g be continuous functions such that the domain of f contains the range of g. Then $f \circ g$ is continuous with domain equal to the domain of g.*

Note The assumption that the domain of f contains the range of g is necessary for the composite function $f \circ g$ to be defined.

For example, we know $f(x) = x^2$ and $g(x) = x + 3$ are continuous on $\Re$ by Properties 1 and 2. Hence their composite $(f \circ g)(x) = f(g(x)) = f(x+3) = (x+3)^2$ is also continuous on $\Re$.

Property 5 *The inverse of a continuous one–to–one function is continuous. Let f be a continuous one–to–one function with domain the interval I and range R. Then f^{-1} is a continuous function with domain R.*

Proof By the definition of continuity, $\lim_{x \to c} f(x)$ exists and equals $f(c)$ for every $c \in I$. Let $L \in R$. By Property 8 of limits, $\lim_{x \to L} f^{-1}(x)$ exists and equals $f^{-1}(L)$. That is, f^{-1} is continuous at each $L \in R$, and f^{-1} is continuous. $\square$

For example, $f(x) = x^2$ with domain $[0, \infty)$ is a continuous one–to–one function with range $R = [0, \infty)$. Hence $f^{-1}(x) = \sqrt{x}$ is continuous with domain $[0, \infty)$.

The following examples use these five properties to verify that large classes of common functions are continuous.

Examples 1.7.18 (1) Every constant function $f(x) = k$ with domain $\Re$ is continuous because by Example 1.5.12 (1): $\lim_{x \to c} f(x) = k = f(c)$ for all $c \in \Re$.

(2) $g(x) = x$ with domain $\Re$ is continuous because by Example 1.5.12 (2): $\lim_{x \to c} g(x) = c = g(c)$ for all $c \in \Re$.

(3) Every polynomial function, with domain $\Re$, can be constructed from the above functions f and g by using sums and products. Hence all polynomial functions are continuous by Properties 1 and 2.

(4) Recall that a rational function $r(x) = \frac{p(x)}{q(x)}$ is a quotient of two polynomials $p(x)$ and $q(x)$. By Property 3, this rational function is continuous on any interval which contains no roots of the polynomial q.

(5) The function $h(x) = x^n$ with domain $[0, \infty)$ for n an even integer or domain $\Re$ for n an odd integer is continuous and one–to–one. By Property 5, the inverse function $h^{-1}(x) = \sqrt[n]{x}$ is also continuous. By Properties 2 and 3, the functions

$k(x) = x^{p/q}$ are continuous for every rational number $\frac{p}{q}$. Therefore Property 4 implies that the following functions are continuous:

$$\sqrt[5]{2x^3 + x^2 + 4}, \qquad \left(\frac{3x+5}{x^2+x+1}\right)^{9/7}, \qquad \sqrt{5x^{2/5} - 7x^{7/3} + 4x^{4/9} - 1}.$$

(6)   Show that all six trigonometric functions are continuous.

Solution Let $\theta \in \Re$ with $y = x - \theta$. Note y approaches zero when x approaches θ. Then $x = y + \theta$ and by Examples 1.6.8 (1), (2):

$$
\begin{aligned}
\lim_{x \to \theta} \sin x &= \lim_{y \to 0} \sin(y + \theta) = \lim_{y \to 0} (\sin y \cos \theta + \cos y \sin \theta) \\
&= (\cos \theta) \lim_{y \to 0} \sin y + (\sin \theta) \lim_{y \to 0} \cos y = (\cos \theta)(0) + (\sin \theta)(1) = \sin \theta.
\end{aligned}
$$

Hence $\sin x$ is continuous for $x \in \Re$. Note $\cos x$ is also continuous on $\Re$ because

$$\lim_{x \to \theta} \cos x = \lim_{x \to \theta} \sin(x + \pi/2) = \lim_{x + \pi/2 \to \theta + \pi/2} \sin(x + \pi/2) = \sin(\theta + \pi/2) = \cos \theta.$$

By Property 3, the quotients of continuous functions $\tan x = \frac{\sin x}{\cos x}$, $\cot x = \frac{\cos x}{\sin x}$, $\sec x = \frac{1}{\cos x}$ and $\csc x = \frac{1}{\sin x}$ are continuous on their domains.

(7)   Show that all six inverse trigonometric functions are continuous.

Solution Each of these functions is the inverse of a one–to–one function which is the restriction of a trigonometric function. By Example 6, these restrictions are continuous. The domains of the restrictions of $\sin x$, $\cos x$, $\tan x$ and $\cot x$ are intervals. By Property 5, their inverse functions $\arcsin x$, $\arccos x$, $\arctan x$ and $\operatorname{arccot} x$ are continuous.

The function arcsec has a technical problem: it is the inverse of the restriction of the secant function with domain $\left[0, \frac{\pi}{2}\right) \cup \left(\frac{\pi}{2}, \pi\right]$ which is not an interval. Hence we can not apply Property 5 which requires the original function to have domain an interval. Observe that the domain of the arcsecant function is $(-\infty, -1] \cup [1, \infty)$. Moreover, arcsec x restricted to $(-\infty, -1]$ is the inverse of the secant function restricted to $\left(\frac{\pi}{2}, \pi\right]$ while arcsec x restricted to $[1, \infty)$ is the inverse of the secant function restricted to $\left[0, \frac{\pi}{2}\right)$. Each of these restrictions of the arcsecant function is continuous by Property 5, and therefore the entire arcsecant function is continuous. By an analogous argument, $\operatorname{arccsc} x$ is also continuous. We leave the details as an exercise. $\square$

Intermediate Value Theorem (IVT)

Consider two points $P = (a, f(a))$ and $Q = (b, f(b))$ on the graph of a continuous function f with domain an interval. Intuitively, the Intermediate Value Theorem (or IVT for short) says that a curve drawn from P to Q, without lifting the pencil from the paper, must cross every horizontal line $y = k$ which lies between P and Q. These are the numbers k with $f(a) < k < f(b)$. In other words, all numbers k between $A = f(a)$ and $B = f(b)$ must be in the range of f. For example, the graph of f in Figure 1.7.20 crosses the line $y = k$ three times.

Theorem 1.7.19 (Intermediate Value Theorem) *Let f be a continuous function whose domain is an interval. Assume the range of f contains the numbers A and B with $A < B$. Then the range of f contains the entire interval $[A, B]$.*

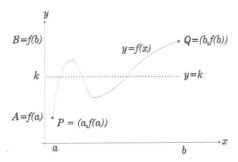

Figure 1.7.20 Intermediate Value Theorem

A proof of the Intermediate Value Theorem is given in Section 11 using the topological concept of connectedness. The following corollary is an alternate formulation.

Corollary 1.7.21 *Let f be a continuous function whose domain is an interval I. Assume that a and b are numbers in I with $f(a) < f(b)$. If $f(a) < k < f(b)$, then there is a number c in I such that $f(c) = k$.*

Proof Apply the Intermediate Value Theorem to the function f whose domain is the closed interval J with endpoints a and b. Let $A = f(a)$ and $B = f(b)$. Since $k \in [A, B]$, the Intermediate Value Theorem says that k is in the range of f, i.e. there is $c \in J$ with $f(c) = k$. Since J is a subset of I, the number c is in I. □

The first example below shows that when $A = f(a) < B = f(b)$, we have $a < b$ in some examples and $a > b$ in others. The second example illustrates the necessity of the hypotheses of the Intermediate Value Theorem. Its conclusion may fail if the function f is not continuous or if the domain of f is not an interval.

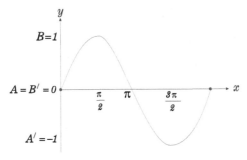

Figure 1.7.22 The Function $f(x) = \sin x$

Examples 1.7.23 (1) Let $f(x) = \sin x$ with $a = 0 < b = \frac{\pi}{2}$. Then

$$A = f(a) = f(0) = 0 < B = f(b) = f\left(\frac{\pi}{2}\right) = 1.$$

By the IVT, every number between 0 and 1 is the sine of an angle between 0 rdns and $\frac{\pi}{2}$ rdns. Alternatively, let $b' = \pi < a' = \frac{3\pi}{2}$. Then

$$A' = f(a') = f\left(\frac{3\pi}{2}\right) = -1 < B' = f(b') = f(\pi) = 0.$$

In this case, the IVT says that every number between -1 and 0 is the sine of an angle between π rdns and $\frac{3\pi}{2}$ rdns. See Figure 1.7.22.

(2) **(a)** Let $g(x) = \left\{ \begin{array}{ll} \frac{1}{x} & \text{if } x \neq 0 \\ 1 & \text{if } x = 0 \end{array} \right\}$ with domain $[-1, 1]$. Then $A = g(-1) = -1 < 0 < B = g(1) = 1$. However, there is no number c between -1 and

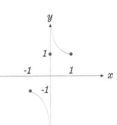

Fig 1.7.24 $y = g(x)$

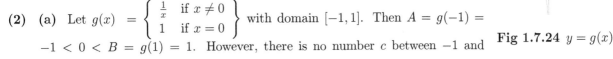

1 with $g(c) = 0$. The nonexistence of c can be seen from the graph of g in Figure 1.7.24: the graph of g passes from below the x–axis to above the x–axis without crossing the x–axis. This does not contradict the IVT because the function g is not continuous at $x = 0$.

(b) Let $G(x) = \frac{1}{x}$ with domain all nonzero numbers. Again $A = G(-1) = -1 < 0 < B = G(1) = 1$, but there is no number c between -1 and 1 with $G(c) = 0$. This does not contradict the IVT because the domain of G is $(-\infty, 0) \cup (0, \infty)$ which is not an interval. See Figure 1.7.25. $\square$

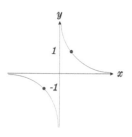

Fig 1.7.25 $y = G(x)$

Let f be a continuous one–to–one function with Domain f an interval and $k \in$ Range f. Examples 1.7.27 use the Intermediate Value Theorem to approximate a solution of

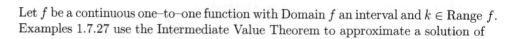
$$f(x) = k.$$

We want to approximate a number r which satisfies $f(r) = k$. First, find a_0 and b_0 with $f(a_0) < k < f(b_0)$. By the IVT, r is between a_0 and b_0. This estimates r with error less than $b_0 - a_0$. Now compute $f\left(\frac{a_0+b_0}{2}\right)$.

- If $f\left(\frac{a_0+b_0}{2}\right) < k < f(b_0)$, then r is between $\frac{a_0+b_0}{2}$ and b_0 by the IVT. In this case, let $a_1 = \frac{a_0+b_0}{2}$ and $b_1 = b_0$.

- If $f\left(\frac{a_0+b_0}{2}\right) > k > f(a_0)$, then r is between a_0 and $\frac{a_0+b_0}{2}$ by the IVT. In this case, let $a_1 = a_0$ and $b_1 = \frac{a_0+b_0}{2}$.

In both cases, r is between a_1 and b_1. Thus we have estimated r with error less than $b_1 - a_1 = \frac{b_0-a_0}{2}$. Next, compute $f\left(\frac{a_1+b_1}{2}\right)$ and argue as above: r is either between a_1 and $\frac{a_1+b_1}{2}$ or r is between $\frac{a_1+b_1}{2}$ and b_1. This estimates r with error less than $\frac{b_1-a_1}{2} = \frac{b_0-a_0}{4}$. Iterate this process n times to estimate r with error less than $\frac{b_0-a_0}{2^n}$. This procedure to estimate r to any required accuracy is called the *bisection method*.

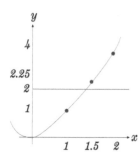

Examples 1.7.27 (1) Apply the bisection method once to estimate $\sqrt{2}$.

Solution Consider the continuous function $f(x) = x^2$ with domain $\Re$. We want to estimate the solution $r = \sqrt{2}$ of $f(x) = 2$. Since

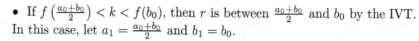
$$f(1) = 1 < 2 < f(2) = 4,$$

it follows from the IVT that there is a number r between 1 and 2 whose square equals two, i.e. $\sqrt{2}$ exists! To apply the bisection method, note that $1 = f(1) < 2 < 2.25 = f(1.5)$. Hence $1 < \sqrt{2} < 1.5$. See Figure 1.7.26.

Fig 1.7.26 $f(x) = x^2$

(2) Iterate the bisection method twice to estimate a root r of $h(x) = x^4 + x - 1$.

Solution A root of this polynomial is a solution of $h(x) = x^4 + x - 1 = 0$. Note $h(0) = -1 < 0 < 1 = h(1)$. By the IVT, h has a root r between 0 and 1. Observe $\frac{1}{2}(0 + 1) = 0.5$ and $h(0.5) = -0.4375 < 0 < 1 = h(1)$. By the IVT,

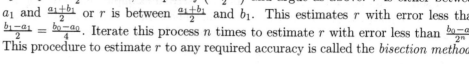
$$0.5 < r < 1.$$

Now $\frac{1}{2}(0.5 + 1) = 0.75$ and $h(0.75) = 0.066 > 0 > -0.4375 = h(0.5)$. Hence

$$0.5 < r < 0.75$$

by the IVT. See Figure 1.7.28.

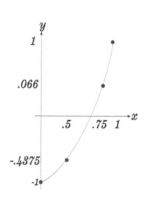

Figure 1.7.28
$h(x) = x^4 + x - 1$

(3) Iterate the bisection method three times to approximate $\sqrt[3]{2}$.

Solution Note $\sqrt[3]{2}$ is a root of the polynomial $P(x) = x^3 - 2$. Since $P(1) = -1 < 0 < P(2) = 6$, it follows from the IVT that
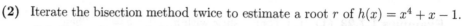

$$1 < \sqrt[3]{2} < 2.$$

Note $\frac{1}{2}(1+2) = 1.5$ and $P(1) = -1 < 0 < P(1.5) = 1.375$. By the IVT,

$$1 < \sqrt[3]{2} < 1.5.$$

$\frac{1}{2}(1+1.5) = 1.25$ and $P(1.25) = -0.047 < 0 < P(1.5) = 1.375$. By the IVT,

$$1.25 < \sqrt[3]{2} < 1.5.$$

$\frac{1}{2}(1.25+1.5) = 1.375$ and $P(1.25) = -0.47 < 0 < P(1.375) = 0.60$. By the IVT,

$$1.25 < \sqrt[3]{2} < 1.375. \qquad \square$$

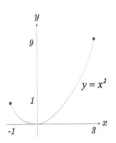

Fig 1.7.29 $f(x) = x^2$

In Section 5.2 we will prove the following stronger form of the Intermediate Value Theorem. The key point is that intervals are the only subsets J of $\Re$ which have the property: if $A, B \in J$ with $A < B$, then $[A, B] \subset J$. Since the range of a continuous function has this property, it must be an interval.

Corollary 1.7.31 *Let f be a continuous function whose domain is an interval. Then the range of f is also an interval.*

The following examples apply this corollary to continuous functions whose domains are a variety of intervals. We compute the range of each function by projecting its graph onto the y–axis.

Examples 1.7.33 Find the range of each continuous function.

(1) Let $f(x) = x^2$ with domain $[-1, 3]$.

Solution The range of f is the projection onto the y–axis of its graph depicted in Figure 1.7.29. The lowest point on this curve is at the origin and the highest point is the right endpoint $(3, 9)$. Hence the range of f is the interval $[0, 9]$.

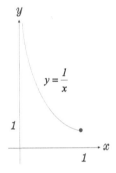

Fig 1.7.30 $g(x) = \frac{1}{x}$

(2) Let $g(x) = \frac{1}{x}$ with domain $(0, 1]$.

Solution Note the graph of g is part of the right branch of the hyperbola $y = \frac{1}{x}$. The range of g is the projection onto the y–axis of its graph in Figure 1.7.30. This graph has no highest point, and its lowest point is its right endpoint $(1, 1)$. Hence the range of g is the interval $[1, \infty)$.

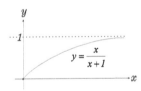

Figure 1.7.32 $h(x) = \frac{x}{x+1}$

(3) Let $h(x) = \frac{x}{x+1}$ with domain $[0, \infty)$.

Solution Note the graph of h, in Figure 1.7.32, starts at the origin and rises to its horizontal asymptote $y = 1$. The range of h is the projection of this graph onto the y–axis. Hence the range of h is the interval $[0, 1)$.

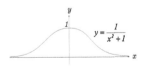

Figure 1.7.34 $k(x) = \frac{1}{x^2+1}$

(4) Let $k(x) = \frac{1}{x^2+1}$ with domain $\Re$.

Solution Note the graph of k in Figure 1.7.34 rises from its left horizontal asymptote, the x–axis, to the point $(0, 1)$. Then it falls towards its right horizontal asymptote, the x–axis. The range of k is the projection of this graph onto the y–axis. Hence the range of k is the interval $(0, 1]$. $\qquad \square$

Maximum Value Theorem

The Maximum Value Theorem says the graph of a continuous function, whose domain is a closed interval $[a, b]$, has a highest point and a lowest point. This makes sense: if you draw a continuous curve from a starting point to an endpoint then there must be a highest point $(M, f(M))$ and lowest point $(m, f(m))$ on that curve. $f(m)$ is called the *minimum value* of f and $f(M)$ is called the *maximum value* of f. See Fig. 1.7.36.

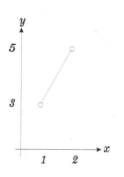

Figure 1.7.35
$f(x) = 2x + 1$

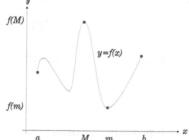

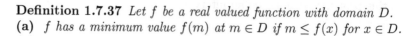

Figure 1.7.36 Maximum Value Theorem

Definition 1.7.37 *Let f be a real valued function with domain D.*
(a) *f has a minimum value $f(m)$ at $m \in D$ if $m \leq f(x)$ for $x \in D$.*
(b) *f has a maximum value $f(M)$ at $M \in D$ if $f(x) \leq M$ for $x \in D$.*

The following examples show that if the domain of f is not a closed interval, then f may not have a minimum or maximum value. In particular, this situation arises when the domain of f is either an interval which does not contain both its endpoints or an interval of infinite length.

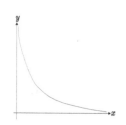

Fig 1.7.38 $y = g(x)$

Examples 1.7.40 (1) The graph of the continuous function $f(x) = 2x + 1$ with domain $(1, 2)$ is depicted in Figure 1.7.35. This function has neither a minimum nor a maximum value. In particular, 3 is not the minimum value of f because $x = 1$ is not in the domain of f. Similarly, 5 is not the maximum value of f because $x = 2$ is not in the domain of f.

(2) (a) The continuous function $g(x) = \frac{1}{x}$ with domain $(0, \infty)$ has no maximum value and no minimum value. See Figure 1.7.38. In particular, 0 is not the minimum value of g because there is no number x with $g(x) = \frac{1}{x} = 0$.

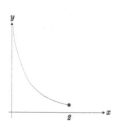

Fig 1.7.39 $y = h(x)$

(b) From the graph of the continuous function $h(x) = \frac{1}{x}$ with domain $(0, 2]$ in Fig. 1.7.39, we see h has minimum value $\frac{1}{2}$ at $x = 2$ but no maximum value.

(c) From the graph of the continuous function $k(x) = \frac{1}{x}$ with domain $(0, 2)$ in Fig. 1.7.41, we see k has no maximum value and no minimum value. In particular, $\frac{1}{2}$ is not the minimum value of k because 2 is not in Domain k.   □

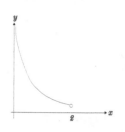

Fig 1.7.41 $y = k(x)$

We give a formal statement of the Maximum Value Theorem.

Theorem 1.7.42 (Maximum Value Theorem) *Let f be a continuous function whose domain is the closed interval $[a, b]$. Then there are $m, M \in [a, b]$ such that*

$$f(m) \leq f(x) \leq f(M)   \quad \text{for all } x \in [a, b].$$

This theorem is called the Maximum Value Theorem even though it speaks about minimum values as well as maximum values of continuous functions. A proof of the Maximum Value Theorem is given in Section 12 using the topological concept of compactness. In Chapter 2, we use derivatives to identify the minimum and maximum values of functions. The following examples illustrate the Maximum Value Theorem.

Examples 1.7.43 (1) Let $f(x) = 1 - x^2$ with domain $[-1, 2]$. The graph of f is given in Figure 1.7.44. Clearly $(0, 1)$ is the highest point on this graph, and $(2, -3)$ is the lowest point on this graph. Therefore, the maximum value 1 of $f(x)$ occurs with $M = 0$, and the minimum value -3 of $f(x)$ occurs with $m = 2$.

(2) Let $g(x) = \sin x$ with domain $[0, 4\pi]$. The graph of g is given in Figure 1.7.46. The points $(\pi/2, 1)$ and $(5\pi/2, 1)$ are the highest points on this graph, and the maximum value of $g(x)$ is 1. The lowest points on the graph are $(3\pi/2, -1)$ and $(7\pi/2, -1)$, and the minimum value of $g(x)$ is -1. This example illustrates that either the minimum or maximum value of a continuous function can be achieved at more than one point. In this case $M_1 = \frac{\pi}{2}$ and $M_2 = \frac{5\pi}{2}$ with $f(M_1) = f(M_2) = 1$ while $m_1 = \frac{3\pi}{2}$, $m_2 = \frac{7\pi}{2}$ with $f(m_1) = f(m_2) = -1$. □

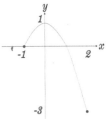

Figure 1.7.44
$f(x) = 1 - x^2$

We can apply both the Intermediate Value and Maximum Value Theorems to a continuous function whose domain is a closed interval to describe its range.

Corollary 1.7.45 *Let f be a continuous function with domain the closed interval $[a, b]$. Then there are $m, M \in [a, b]$ such that Range f is the closed interval $[f(m), f(M)]$.*

Proof By the Maximum Value Theorem, f has a minimum value $f(m)$ and a maximum value $f(M)$, i.e. $f(m) \leq f(x) \leq f(M)$ for $x \in [a, b]$. Hence Range f is contained in the closed interval $[f(m), f(M)]$. By the IVT, the closed interval $[f(m), f(M)]$ is contained in Range f. Thus Range f equals $[f(m), f(M)]$. □

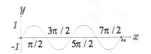

Figure 1.7.46
$g(x) = \sin x$

Note in Examples 1.7.43 that Range f is the closed interval $[f(m), f(M)] = [-3, 1]$ and Range g is the closed interval $[f(m_1), f(M_1)] = [-1, 1]$.

Historical Remarks

As discussed in the Historical Remarks of Section 2, until the middle of the eighteenth century mathematicians worked with continuous functions defined by one equation. More general functions were introduced by Leonhard Euler in 1748 to include examples, such as a plucked string, which are defined by different equations on different parts of their domains. Euler called such functions, as well as curves sketched from no equation, "discontinuous"! Functions which are not continuous by our definition were called "discontiguous" in that era. (Such curves consist of pieces whose ends do not meet, i.e. the ends are not contiguous.)

The first attempt to define a continuous function was made by Louis Arbogast in 1791. He suggested that a function be called continuous if it satisfies the Intermediate Value Theorem. He also noticed that the solutions of partial differential equations are often discontinuous functions. In 1822, Joseph Fourier published a book which developed the subject of Fourier series to solve partial differential equations. He observed that the functions describing the boundary conditions may have jump discontinuities. Moreover, even if these functions are continuous, the solution to the partial differential equation often has jump discontinuities. The first example of a function which is not continuous anywhere was given by Peter Lejeune Dirichlet in 1829. We analyzed his example in Example 1.7.6 (4).

The standard definition of a continuous function was first formulated by Bernard Bolzano in a private pamphlet in the early nineteenth century. He used this definition to prove the Intermediate Value Theorem. In his textbook of 1821, Augustin–Louis Cauchy gave this definition and proved the basic properties of continuous functions. It is not clear whether Cauchy was aware of the earlier work of Bolzano.

Summary

The reader should understand what it means for a function to be continuous at a point as well as the definition of a continuous function. She should be able to identify continuous functions and characterize points of discontinuity as being removable or essential. She should know the properties of continuous functions and be able to use them to establish the continuity of complicated functions from simpler ones. The statements of the Intermediate Value Theorem (IVT) and the Maximum Value Theorem should be understood. The reader should know how to approximate roots using the bisection method and be able to apply the IVT to determine the range of a function.

Basic Exercises

1. For each function whose graph is depicted below, determine all numbers in the function's domain where it is continuous. For each point of discontinuity, determine whether the discontinuity is removable or essential.

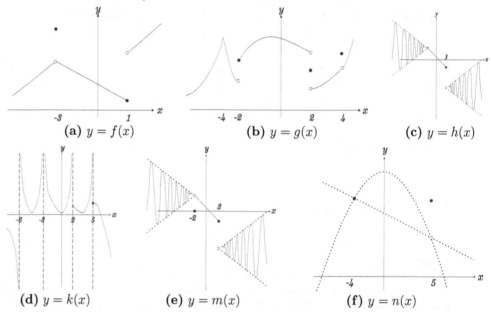

(a) $y = f(x)$ (b) $y = g(x)$ (c) $y = h(x)$

(d) $y = k(x)$ (e) $y = m(x)$ (f) $y = n(x)$

For each function in Exercises 2 to 20 determine where the function is continuous. For each point of discontinuity, determine whether the discontinuity is removable or essential.

2. $a(x) = x^3 - x^2 + 6$ with domain $\Re$. **3.** $b(x) = \frac{x^2-1}{x^2+1}$ with domain $\Re$.

4. $c(x) = \left\{ \begin{array}{ll} \frac{x^2}{x^4-1} & \text{if } x \neq \pm 1 \\ 0 & \text{if } x = \pm 1 \end{array} \right\}$ with domain $\Re$.

5. $d(x) = \left\{ \begin{array}{ll} \frac{2x^2-3x-2}{x^2-5x+6} & \text{if } x \neq 2, 3 \\ 1 & \text{if } x = 2, 3 \end{array} \right\}$ with domain $\Re$.

6. e has domain $\Re$ with $e(x) = \frac{x^2-1}{x^4-1}$ if $x \neq \pm 1$, $e(-1) = -\frac{1}{2}$ and $e(1) = \frac{1}{2}$.

7. $f(x) = \left\{ \begin{array}{ll} \frac{x^4+2x^2-24}{x^2-4} & \text{if } x \neq \pm 2 \\ 8 & \text{if } x = \pm 2 \end{array} \right\}$ with domain $\Re$.

8. g has domain $\Re$ with $g(x) = \frac{x^2-x-6}{x^2+2x-15}$ if $x \neq -5, 3$; $g(-5) = 0$ and $g(3) = 1$.

9. $h(x) = \sin \sqrt{x}$ with domain $[0, \infty)$. **10.** $i(x) = \left\{ \begin{array}{ll} \frac{|x|}{x} & \text{if } x \neq 0 \\ 0 & \text{if } x = 0 \end{array} \right\}$ with domain $\Re$.

11. j has domain $\Re$ with $j(x) = \frac{|x+3|}{x+3}$ if $x \neq -3$ and $j(-3) = 1$.

12. $k(x) = \begin{cases} \frac{|x^2-x-20|}{x^2-x-20} & \text{if } x \neq -4, \ 5 \\ 1 & \text{if } x = -4, \ 5 \end{cases}$ with domain $\Re$.

13. m has domain $\Re$ with $m(x) = \frac{|x^2-8x+16|}{x^2-8x+16}$ if $x \neq \pm 4$, $m(4) = 1$ and $m(-4) = -1$.

14. $n(x) = \sqrt{x^3 + 7x + 8}$ with domain $[0, \infty)$.   **15.** $p(x) = \cos \frac{1}{x}$ with domain $(0, \infty)$.

16. q has domain $\Re$ with $q(x) = x^2 \cos \frac{1}{x}$ if $x \neq 0$ and $q(0) = 1$.

17. r has domain $\Re$ with $r(x) = \arctan \frac{1}{x^2}$ if $x \neq 0$ and $r(0) = 0$.

18. s has domain $\Re$ with $s(x) = \arctan \frac{1}{x}$ if $x \neq 0$ and $s(0) = 0$.

19. $t(x) = \begin{cases} x & \text{if } x \text{ is a rational number} \\ 0 & \text{if } x \text{ is an irrational number} \end{cases}$ with domain $\Re$.

20. $u(x) = \begin{cases} x^2 & \text{if } x \text{ is a rational number} \\ 3x + 28 & \text{if } x \text{ is an irrational number} \end{cases}$ with domain $\Re$.

21. Each of these functions has a removable discontinuity at $x = 1$. Change the value of the function at $x = 1$ to define a new function which is continuous at $x = 1$.

(a) $f(x) = \frac{x^2-1}{x^3-x^2+x-1}$ for $x \neq 1$ and $f(1) = 2$.

(b) $g(x) = \frac{x^4-4x^3+6x^2-4x+1}{x^3-3x^2+3x-1}$ for $x \neq 1$ and $g(1) = 1$.

(c) $h(x) = \arctan \frac{1}{x^2-2x+1}$ for $x \neq 1$ and $h(1) = 0$.

(d) $k(x) = x^2$ for $x < 1$, $k(1) = 0$ and $k(x) = x$ for $x > 1$.

(e) $m(x) = \frac{|x^2-2x+1|}{x^2-2x+1}$ for $x \neq 1$ and $m(1) = -1$.

22. Show that $f(x) = |x|$ is a continuous function with domain $\Re$.

23. Show that the function $y = \operatorname{arccsc} x$ is continuous.

Use the properties of continuous functions to determine that each function in Exercises 24 to 33 is continuous. State the properties you use.

24. $f(x) = (2x^5 - 3x + 1)^4$ **25.** $f(x) = 7 \sin x - 9 \cos x$ **26.** $f(x) = x \sin x$

27. $g(x) = \cos(4x^2 - 3)$ **28.** $f(x) = \frac{x}{2+\sin x}$ **29.** $f(x) = |\cos x|$

30. $k(x) = \frac{\sqrt{x^2+1}}{x^4+1}$ **31.** $m(x) = \arcsin \frac{x^2}{x^2+1}$ **32.** $n(x) = (\arctan x)^{2/3}$

33. h has domain $\Re$ with $h(x) = \frac{x \sin(1/x)}{x^2+1}$ if $x \neq 0$ and $h(0) = 0$.

34. Verify Property 2 of continuous functions: the product of two continuous functions is continuous.

35. Verify Property 3 of continuous functions: the quotient of two continuous functions is continuous if zero is not in the range of the denominator.

36. Verify Property 4 of continuous functions: the composite of two continuous functions is continuous.

37. Each of these functions is continuous and has domain $\Re$. According to the IVT, which numbers must be in the range of each function?

(a) $f(-2) = 5$ and $f(6) = 11$.   **(b)** $g(3) = -4$ and g is an odd function.

(c) $h(-3) = 4$ and $h(2) = -1$.   **(d)** The maximum value of the odd function k is 7.

38. Use the IVT to show each equation has a solution in the interval $(0, 1)$.

(a) $x^4 + x^2 - 1 = 0$ **(b)** $2x + 3\sqrt{x} = 4$ **(c)** $\frac{5x+7}{2x+1} = 6$ **(d)** $\frac{\cos \pi x}{x} = 2$

39. Show that the conclusion of the IVT is false for each function below. Why does each of these functions not contradict the IVT?

(a) $f(x) = \frac{1}{x^2-1}$ with $A = f(0)$ and $B = f(2)$.

(b) $g(x) = \frac{|x-4|}{x-4}$ for $x \neq 4$, $g(4) = 0$ with $A = g(1)$ and $B = g(9)$.

(c) $h(x) = \sec x$ with $A = h(0)$ and $B = h(\pi)$.

(d) $k(x) = \left\{ \begin{array}{ll} 1 + x^2 & \text{for } x \text{ irrational} \\ -x^2 & \text{for } x \text{ rational} \end{array} \right\}$ with $A = k(1)$ and $B = k(\sqrt{2})$.

(e) $m(x) = \frac{1-\cos x}{x}$ for $x \neq 0$, $m(0) = 1$ with $A = m(0)$ and $B = m\left(\frac{\pi}{2}\right)$.

(f) $n(x) = \arctan \frac{1}{x}$ with $A = n(-1)$ and $B = n(1)$.

40. Use the bisection method to estimate each irrational number to the nearest tenth.

(a) $\sqrt{3}$ (b) $\sqrt{8}$ (c) $\sqrt{11}$ (d) $\sqrt[3]{2}$ (e) $\sqrt[3]{5}$ (f) $\sqrt[4]{21}$

41. Each of these polynomials has a root between 0 and 1. Use the bisection method to estimate this root to the nearest tenth.

(a) $x^3 + x - 1$ (b) $x^4 + 2x^2 - 1$ (c) $5x^3 - x - 2$ (d) $x^4 - 3x^2 + 1$
(e) $x^3 - 2x^2 - 3x + 1$

42. Sketch the graph of each function. Then apply the IVT to determine its range.

(a) $f(x) = 4 - x^2$ with domain $[-2, 3]$ (b) $g(x) = \frac{1}{x}$ with domain $(1, \infty)$

(c) $h(x) = \tan x$ with domain $\left[0, \frac{\pi}{2}\right)$ (d) $k(x) = \sqrt{1 - x^2}$ with domain $\left[-1, -\frac{\sqrt{3}}{2}\right]$

(e) $m(x) = \sin x$ with domain $[-3\pi, 4\pi]$ (f) $n(x) = \sqrt{x^2 + 9}$ with domain $[-1, 4]$

43. Let f be a continuous function whose domain is a union of n intervals. The range of f is the union of two nonoverlapping closed intervals. Find all possible values of n.

44. Find all examples of continuous functions with domain $\Re$ that have a finite range.

45. Let f be a continuous function with domain the closed interval $[a, b]$. Assume that the range of f is contained in $[a, b]$. Show that f has a fixed point k, i.e. show that there is $k \in [a, b]$ with $f(k) = k$.

46. Find the minimum and maximum values of each function. Specify all values m in the domain where the minimum value occurs and all values M in the domain where the maximum value occurs.

(a) $f(x) = x^2 - 2x + 5$ with domain $[-3, 2]$.   (b) $g(x) = |x + 5|$ with domain $[-14, 4]$.
(c) $h(x) = 3 \cos 2x$ with domain $\left[-\frac{\pi}{4}, \frac{\pi}{3}\right]$.   (d) $k(x) = \sqrt{25 - x^2}$ with domain $[-4, 3]$.
(e) $p(x) = 3 - x^2 + 4x$ with domain $[-1, 5]$.   (f) $q(x) = \frac{1}{x^2+1}$ with domain $[-3, 6]$.

47. Each function below has no maximum value. Explain why each of these functions does not contradict the Maximum Value Theorem.

(a) $f(x) = x^2$ with domain $\Re$.   (b) $g(x) = 5x + 3$ with domain $(-1, 5)$.
(c) h has domain $[-1, 3]$ with $h(x) = \frac{1}{x^2}$ for $x \neq 0$ and $h(0) = 0$.
(d) $k(x) = \frac{1}{x^2-1}$ with domain $(1, 4]$.   (e) $m(x) = \tan x$ with domain $[0, \pi/2)$.

(f) n has domain $[-1, 1]$ with $n(x) = \left\{ \begin{array}{ll} 2x + 3 & \text{if } -1 \leq x < 0 \\ 0 & \text{if } x = 0 \\ 3x - 2 & \text{if } 0 < x \leq 1 \end{array} \right\}$.

Challenging Problems

1. Let f be a continuous function with domain $[a, b)$ such that $\lim_{x \to b^-} f(x) = \infty$. Which numbers must be in the range of f? Justify your answer.

2. Give an example of a continuous function f with domain $[0, 1]$ and maximum value V such that the graph of f contains no horizontal line segment and there are an infinite number of $x \in [0, 1]$ with $f(x) = V$.

Appendix: Limits of Inverse Functions

We use the Intermediate Value and the Maximum Value Theorems to establish a lemma which determines the range of a one–to–one continuous function with domain a closed interval $[a, b]$. It says that the values of $f(x)$ either increase from $f(a)$ to $f(b)$ or decrease from $f(a)$ to $f(b)$. We apply this lemma to prove the eighth limit property of Section 6 which computes the limit of an inverse function.

Lemma 1.7.47 *Let f be a one–to–one continuous function with domain the closed interval $[a, b]$. Then $f[a, b]$ is the closed interval with endpoints $f(a)$ and $f(b)$.*

Proof By Corollary 1.7.45, there are $m, M \in [a, b]$ such that the range of f is the interval $[f(m), f(M)]$. If $f(a) < f(b)$, we show $a = m$ and $b = M$. If $a < m$, then $f(m) < f(a) < f(b)$. Apply the IVT to the interval $[m, b]$: there is $c \in [m, b]$ with $f(c) = f(a)$. This contradicts the hypothesis that f is one–to–one. Hence $m = a$. That is, the graph of f can not go down from height $f(a)$ to height $f(m)$ and then rise to a higher height $f(b)$ without crossing height $f(a)$ again. See the left diagram in Figure 1.7.48. Similarly, if $M < b$ then $f(a) < f(b) < f(M)$. Apply the IVT to the interval $[a, M]$: there is $e \in [a, M]$ with $f(e) = f(b)$. This contradicts the hypothesis that f is one–to–one. Thus $M = b$. That is, the graph of f can not rise from height $f(a)$ to height $f(M)$ and then fall to height $f(b)$ without having crossed height $f(b)$ on the interval $[a, M]$. See the right diagram in Figure 1.7.48. Therefore $f[a, b] = [f(m), f(M)] = [f(a), f(b)]$. If $f(a) > f(b)$, an analogous argument shows that $f[a, b] = [f(b), f(a)]$. We leave the proof as an exercise. $\square$

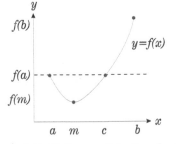
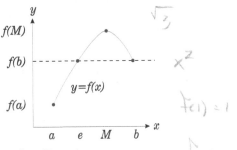
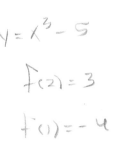

Figure 1.7.48 Impossible Graphs for a One-to-One Function

We establish the limit of f^{-1} from the definition of a limit by applying Lemma 1.7.47 to find an interval on which the graph of $y = f^{-1}(x)$ lies in a given horizontal strip. Note how the statement of Property 8 simplifies using the terminology of continuity.

Property 8 *Let f be a continuous one–to–one function with domain the interval I. Then the function f^{-1} is continuous on $f(I)$, i.e. $\lim\limits_{x \to L} f^{-1}(x) = f^{-1}(L)$ for $L \in f(I)$.*

Proof Consider the case where c is an interior point of the interval I with $L = f(c)$. We use Definition 1.5.9 to show that $\lim\limits_{x \to L} f^{-1}(x) = c = f^{-1}(L)$. Given $\epsilon > 0$, we find $\delta > 0$ such that the graph of f^{-1} restricted to $(L - \delta, L + \delta)$ is contained in the horizontal strip $S_\epsilon(c)$, i.e.

$$f^{-1}(L - \delta, L + \delta) \subset (c - \epsilon, c + \epsilon).$$

Replace ϵ by a smaller positive number, if necessary, so that $[c - \epsilon, c + \epsilon]$ is contained in I. Assume that $f(c - \epsilon) < f(c + \epsilon)$. The case $f(c - \epsilon) > f(c + \epsilon)$ is given as an exercise. Apply Lemma 1.7.47 with $a = c - \epsilon$ and $b = c + \epsilon$:

$$f[c - \epsilon, c + \epsilon] = [f(c - \epsilon), f(c + \epsilon)], \quad \text{i.e.} \quad f^{-1}[f(c - \epsilon), f(c + \epsilon)] = [c - \epsilon, c + \epsilon].$$

Since f^{-1} is one–to–one, $f(c-\epsilon)$ and $f(c+\epsilon)$ are the only numbers which f^{-1} sends to $c-\epsilon$ and $c+\epsilon$. Thus

$$f(c-\epsilon, c+\epsilon) = (f(c-\epsilon), f(c+\epsilon)), \quad \text{i.e.} \quad f^{-1}(f(c-\epsilon), f(c+\epsilon)) = (c-\epsilon, c+\epsilon).$$

Since $L = f(c) \in f(c-\epsilon, c+\epsilon)$, select an interval $(L-\delta, L+\delta)$, with $\delta > 0$, contained in $(f(c-\epsilon), f(c+\epsilon))$. (For example, take δ to be the smaller of the positive numbers $L - f(c-\epsilon)$ and $f(c+\epsilon) - L$.) See the left diagram in Figure 1.7.49. Then

$$f^{-1}(L-\delta, L+\delta) \subset f^{-1}(f(c-\epsilon), f(c+\epsilon)) = (c-\epsilon, c+\epsilon), \quad \text{as required.}$$

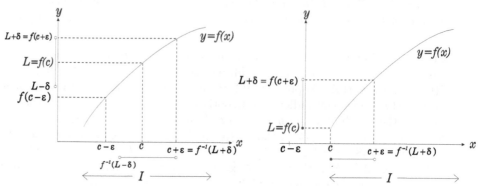

Figure 1.7.49 Selecting δ

Consider the case where c is the left endpoint of the interval I. We use Def. 1.5.9 to establish the limit of $f^{-1}(x)$ as x approaches L. Given $\epsilon > 0$, replace ϵ by a smaller positive number, if necessary, so that $[c, c+\epsilon]$ is contained in I. Assume $L = f(c) < f(c+\epsilon)$. The proof of the case $L > f(c+\epsilon)$ is given as an exercise. We find $\delta > 0$ such that

$$f^{-1}[L, L+\delta) \subset (c-\epsilon, c+\epsilon)$$

to satisfy the requirements of Def. 1.5.9 and establish $\lim\limits_{x \to L^+} f^{-1}(x) = c = f^{-1}(L)$. Apply Lemma 1.7.47 with $a = c$ and $b = c+\epsilon$: $f[c, c+\epsilon]$ is the interval $[f(c), f(c+\epsilon)] = [L, f(c+\epsilon)]$. Hence

$$f^{-1}[L, f(c+\epsilon)] = [c, c+\epsilon].$$

Since f^{-1} is one–to–one, $f(c+\epsilon)$ is the only number which f^{-1} sends to $c+\epsilon$. Thus

$$f^{-1}[L, f(c+\epsilon)) = [c, c+\epsilon).$$

Take $\delta = f(c+\epsilon) - L$. See the right diagram in Fig. 1.7.49. Then $f(c+\epsilon) = L+\delta$ and

$$f^{-1}[L, L+\delta) = f^{-1}[L, f(c+\epsilon)) = [c, c+\epsilon) \subset (c-\epsilon, c+\epsilon), \quad \text{as required.}$$

Verification of Property 8 in the case where c is the right endpoint of the interval I is given as an exercise. $\qquad\qquad\qquad \square$

Exercises

1. Prove the following case of Lemma 1.7.47: if f is a continuous one–to–one function with domain $[a, b]$ such that $f(a) > f(b)$, then $f[a, b] = [f(b), f(a)]$.

2. Prove Property 8 in the case where c is an interior point of the domain I of f and $\epsilon > 0$ is given with $f(c-\epsilon) > f(c+\epsilon)$.

3. Prove the case $\lim\limits_{x \to L^+} f^{-1}(x) = c$ of Property 8 when c is the left endpoint of the domain I of f and $\epsilon > 0$ is given with $L = f(c) > f(c+\epsilon)$.

4. Prove Property 8 in the case where c is the right endpoint of the domain I of f.

Applications

1.8   Functions in the Sciences and Social Sciences

Prerequisite: Section 1.2

This section presents several examples of functions that arise in the sciences and social sciences which can be analyzed using calculus. We describe in general terms how calculus is used to study these functions. These functions will be used as examples in applications sections of later chapters. This section, however, is meant merely as an introduction and preview . It is not a prerequisite for any future sections.

Mechanics

The subject of mechanics studies the motion of an object which is acted upon by a force. There are four important functions associated with an object that moves in a straight line. The domain of these functions is the time interval I for which we observe the object. If we study the motion of a car from 1:00 p.m. to 7:00 p.m. then $I = [1, 7]$. If we study the motion of the car from now onwards then $I = [0, \infty)$ where now is designated as time zero. These four functions are:

$$
\begin{aligned}
F(t) &= \text{ the force acting on the object at time } t, \\
s(t) &= \text{ the position of the object at time } t, \\
v(t) &= \text{ the velocity of the object at time } t, \\
a(t) &= \text{ the acceleration of the object at time } t.
\end{aligned}
$$

The position, velocity and acceleration functions are related by the calculus concepts of derivative and integral. As we shall see, given one of these functions, we can compute the other two.

Often in mechanics we know a physical principle which describes the force acting on an object. If the mass of the object is a constant m, then Newton's Second Law of Motion says that:

$$ F(t) = m\, a(t). $$

Since we know the force, we also know the acceleration of the object. We can then use calculus to determine its velocity and position functions. For example, consider a mass m attached to one end of a spring whose other end is attached to a wall as in Figure 1.8.1. The system is initially at rest at an *equilibrium position*. We denote this point as zero on the line of motion of the object. The mass is then moved and released. The physical principle in this situation is that the force the spring exerts on the object is proportional to the distance between the object and its equilibrium position.

$$ F(t) = -k^2\, s(t). $$

The minus sign indicates that the force is in the opposite direction of the displacement. Then

$$ a(t) = -\frac{k^2}{m}\, s(t) = -K^2 s(t) $$

where $K = k/\sqrt{m}$. Calculus can be used to show that

$$ s(t) = \frac{v(0)}{K} \sin Kt + s(0) \cos Kt. $$

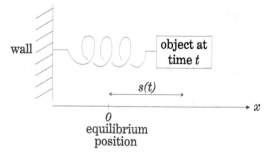

Figure 1.8.1 A Spring

Fluid Statics

Consider an object submerged in a liquid of depth d such as a lake. Let h denote the depth of the object beneath the surface of the fluid. The pressure P on the object is a function of h with domain $[0, d]$. A principle of fluid statics states that

$$P(h) = \delta h$$

where δ is the weight of the liquid per unit volume. Since the pressure on the object is defined as the force on the object per unit area, the force on a thin submerged horizontal plate will be δ times the area of the plate times its depth. If the object is not in a horizontal position or if the object has a shape other than a thin plate, then calculus can be used to compute the force which acts on the entire object. For example, this method can be used to compute the force exerted by water on a dam. In addition, calculus can be used to verify Archimedes Principle: if an object of volume V is underwater in a stable position, then the weight of the object equals the weight $V\delta$ of the liquid displaced by the object. Submarines use this principle to fill their ballast tanks with the appropriate amount of water to stabilize their position at the desired depth.

Biology

Consider bacteria in a petrie dish filled with agar. Experimentation shows that bacteria reproduce at a rate proportional to the amount present. Let $y(t)$ be the function with domain $[0, \infty)$ defined as the number of bacteria present at time t. Using calculus, we will see that

$$y(t) = y(0)e^{kt}$$

where e is a special irrational number which is approximately 2.7.

Chemistry

Consider a chemical C which is decomposing through a chemical reaction. The concentration c of C is a function of time t with domain $[0, \infty)$. When call this chemical reaction a *first order reaction* if the rate of change of c is proportional to the value of c. Let k denote this constant of proportionality. Since C is decomposing, k will be negative. Using integral calculus, we can show that

$$\ln \frac{c(t)}{c(0)} = kt$$

where ln denotes the logarithm to the base e. Chemists identify reactions as being first order by plotting values of $\ln \frac{c(t)}{c(0)}$ at various times. If these points lie on a straight line, then the reaction is first order. Solving the above equation for $c(t)$ yields

$$c(t) = c(0)e^{kt},$$

which says that the concentration of C is changing exponentially. An example of this situation occurs when sucrose decomposes into glucose and fructose.

When the rate of change of c is proportional to c^2 we call the chemical reaction a *second order reaction*. Let k denote the constant of proportionality. Applying integral calculus to this situation shows that

$$\frac{c(t) - c(0)}{c(0)c(t)} = kt.$$

Thus, a reaction can be identified as being second order by plotting $\frac{c(t)-c(0)}{c(0)c(t)}$ at various times t. If these points lie on a straight line, then the reaction is second order. In this case, we can solve for $c(t)$ to obtain:

$$c(t) = \frac{c(0)}{1 - kc(0)t}.$$

Thus, the graph of the concentration of C is a hyperbola. An example of a second order reaction is the decomposition of nitrosyl chloride into nitric oxide and chlorine.

Economics

Consider a factory which manufactures one type of object. Let q denote the number of these objects manufactured per week. There are three important functions of q:

$$
\begin{aligned}
R(q) &= \text{ \textit{the revenue received for manufacturing } q \textit{ objects};} \\
C(q) &= \text{ \textit{the cost of manufacturing } q \textit{ objects};} \\
P(q) &= \text{ \textit{the profit earned by manufacturing } q \textit{ objects};}
\end{aligned}
$$

Assume that $R(q)$ and $C(q)$ are the restrictions to the set of integers of functions with domain $[0, \infty)$. By definition, the profit is the revenue received minus the cost:

$$P(q) = R(q) - C(q).$$

An important problem in economics is to determine the value of q which will maximize the profit. (Note that when q becomes too large, the cost will increase substantially. For example, workers will be paid double wages for working overtime.) Calculus can be used to find the largest and least values of a function. When calculus is applied to this problem, one sees easily that the maximum profit is achieved at that value of q where the manufacturing one additional object results in the same revenue as cost. This is a simple practical criterion that the factory can use to determine the optimum level of production.

Probability

Consider an experiment whose result is a real number x. Let F be the function with domain $\Re$ defined by letting $F(x)$ be the probability that the outcome of the experiment is less than or equal to x. F is called the *cumulative probability function*.

The rate of change of the cumulative probability function F is called the *probability density function f*. In calculus terminology f is the derivative of F. The Fundamental Theorem of Calculus applied to this situation says that the probability P_a^b that the result x of the experiment will lie in the interval $[a, b]$ equals the area under the probability density function curve $y = f(x)$ bounded by the lines $x = a$, $x = b$ and the x–axis. For example, consider a commuter train which is scheduled to arrive at 9:00 a.m. The experiment consists of observing the actual time of arrival of this train each day. If the probability density function for this experiment is given by Figure 1.8.2, then the shaded area equals the probability $P_{8:55}^{9:05}$ that the train will arrive between 8:55 and 9:05.

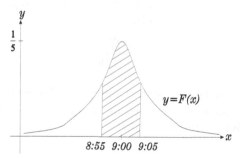

Figure 1.8.2 Area Equal to $P_{8:55}^{9:05}$

Theory

1.9 Sets and Functions

Prerequisite: Section 1.3

Set theory is the language of mathematics. It gives a precise way to specify the collection of objects being studied. In the first subsection we introduce three fundamental operations on sets: union, intersection and complement. These operations allow us to manipulate collections of objects to create new collections from given ones. Their nice properties make them especially useful. We illustrate these constructions with subsets of the real line and the plane. In the second subsection we determine how functions and inverse functions interact with these operations.

Set Operations

The first fundamental operation is the amalgamation of two given sets into a set called their union which we defined in Definition 1.2.13. We rephrase this definition to add a feature that specifies the context in which we are working.

Definition 1.9.1 *Let A and B be two subsets of the set U. The union of A and B, denoted $A \cup B$, consists of all elements of U which are either in A or in B.*

The set U is called the *universal set*. In the examples in this chapter, U will usually be either the real numbers $\Re$ or the plane $\Re^2$. Note that by our definition $A \cup B$ includes elements which are in both A and B. In Examples 1.2.14 we illustrated the union of two subsets of $\Re$. We now present examples of unions of subsets of the plane.

Examples 1.9.3 $U = \Re^2$ in the following examples.

(1) Let A be the line $y = x + 2$, and let B be the parabola $y = x^2$. Then $A \cup B$ is the set of points that are either on this line or on this parabola as depicted in Figure 1.9.2.

(2) Let C be the set of points in the disc of radius one with center the origin. Let D be the set of points inside the square with vertices $(0,0)$, $(0,2)$, $(2,2)$ and $(2,0)$. Then $C \cup D$ is the entire shaded area in Figure 1.9.4. □

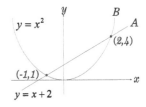

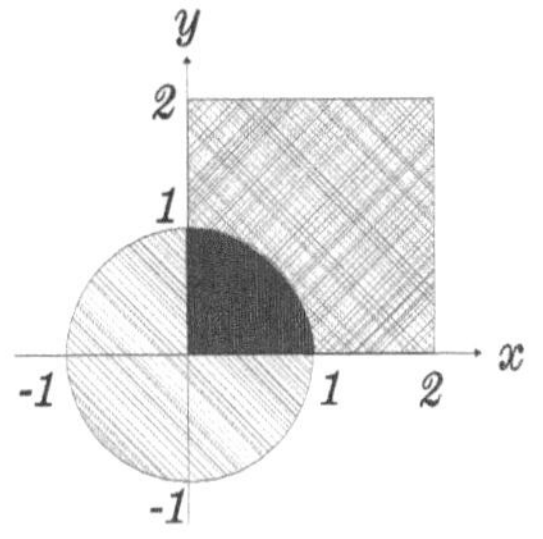

Figure 1.9.2
$A \cup B$

The second fundamental operation is the extraction of the common elements of two given sets.

Definition 1.9.5 *Let A and B be two subsets of the set U. The intersection of A and B, denoted $A \cap B$, consists of all elements of U which are in both A and B.*

Note If there are no elements which are in both A and B then we write $A \cap B = \emptyset$. The set $\emptyset$ with no elements is called the *empty set*.

Figure 1.9.4
$C \cup D$

We consider examples of the intersection of subsets of the line and the plane.

Examples 1.9.6 $U = \Re$ in the first two examples.

(1) Let $A = [-1, 2]$ and $B = (3, 5)$ be the two intervals depicted below. Since these two intervals do not overlap, $A \cap B = \emptyset$.

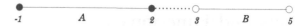

(2) Let $C = (1, 5]$ and $D = (4, 9)$ be the intervals depicted below. These interval overlap in the interval $C \cap D = (4, 5]$.

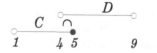

$U = \Re^2$ in the next two examples.

(3) Let A be the line $y = x + 2$, and let B be the parabola $y = x^2$ depicted in Figure 1.9.2. Then $A \cap B$ consists of the two points $(-1, 1)$ and $(2, 4)$ where this line and parabola intersect.

(4) Let C and D be the disc and square of of Example 1.9.3 (4) depicted in Figure 1.9.4. They overlap in the heavily shaded quarter circle $C \cap D$ in the first quadrant.
$\square$

Before we state the properties of union and intersection, we need a method for proving that two sets A and B are equal. The idea is that A and B must be equal if we show that every element of A is an element of B and every element of B is an element of A. This procedure for demonstrating that $A = B$ is called the *Extension Principle*.

Extension Principle *Let A and B be subsets of the set U. Then $A = B$ if and only if $A \subset B$ and $B \subset A$.*

We now state the properties of union and intersection which we refer to as "laws". These properties are proved by using the Extension Principle.

Proposition 1.9.7 *Let A, B and C be subsets of the set U. Then*

(1) $A \cup B = B \cup A$ *and* $A \cap B = B \cap A$ *(commutative laws);*

(2) $(A \cup B) \cup C = A \cup (B \cup C)$ *and* $(A \cap B) \cap C = A \cap (B \cap C)$ *(associative laws);*

(3) $A \cup \emptyset = A$ *and* $A \cap U = A$ *(identity laws);*

(4) $A \cup U = U$ *and* $A \cap \emptyset = \emptyset$;

(5) $A \cup A = A$ *and* $A \cap A = A$ *(idempotent laws);*

(6) $A \cap (B \cup C) = (A \cap B) \cup (A \cap C)$ *and* $A \cup (B \cap C) = (A \cup B) \cap (A \cup C)$
(distributive laws).

Proof We verify the first distributive law. The straightforward verification of the other properties by the Extension Principle is left for the exercises.

To show that $A \cap (B \cup C) \subset (A \cap B) \cup (A \cap C)$, we show that every element $x \in A \cap (B \cup C)$ must be an element of $(A \cap B) \cup (A \cap C)$. Now x is an element of both A and $B \cup C$. Hence x is an element of either B or C. If $x \in B$ then $x \in A \cap B$, while if $x \in C$ then $x \in A \cap C$. Thus, x is either in $A \cap B$ or in $A \cap C$, i.e. $x \in (A \cap B) \cup (A \cap C)$ as required.

To show that $(A \cap B) \cup (A \cap C) \subset A \cap (B \cup C)$, we show that every element $y \in (A \cap B) \cup (A \cap C)$ must be an element of $A \cap (B \cup C)$. Now y is either an element of $A \cap B$ or an element of $A \cap C$. In both cases, $y \in A$. In the first case $y \in B$ while in the second case $y \in C$. Thus, $y \in B \cup C$. Since $y \in A$ and $y \in B \cup C$, it follow that $y \in A \cap (B \cup C)$ as required. $\qquad\square$

It follows from the associative laws that we can define the union and intersection of more than two sets without using parentheses to indicate the order in which the operations are to be performed. Thus,

$$A_1 \cup \cdots \cup A_n \quad and \quad A_1 \cap \cdots \cap A_n$$

are defined. In fact we can define infinite unions and intersections as well.

Definition 1.9.8 *Assume that we are given a subset A_n of U for each positive integer n. Define the union of the A_n as the subset of U consisting of all elements which are in at least one A_n. We denote this union by*

$$\bigcup_{n=1}^{\infty} A_n = A_1 \cup \cdots \cup A_n \cup \cdots.$$

Similarly, define the intersection of the A_n as the subset of U consisting of those elements of U which are in all the A_n. We denote this intersection by

$$\bigcap_{n=1}^{\infty} A_n = A_1 \cap \cdots \cap A_n \cap \cdots.$$

We illustrate infinite unions and intersections with subsets of the line and subsets of the plane.

Examples 1.9.9 Let $U = \Re$ in the first two examples.

(1) Let A_n be the set with one element, the number n. Then $\bigcup_{n=1}^{\infty} A_n$ is the set of positive integers.

(2) Let $A_n = (-1/n, 1/n)$. Then $\bigcap_{n=1}^{\infty} A_n$ is the set whose only element is the number 0.

Let $U = \Re^2$ in the next two examples.

(3) Let A_n be the set of points inside the square with vertices $(0,0)$, $(n,0)$, (n,n) and $(0,n)$. Then $\bigcup_{n=1}^{\infty} A_n$ is the entire first quadrant.

(4) Let A_n be the horizontal strip depicted in Figure 1.9.10 which consists of all points (x,y) such that $y \in \left(L - \frac{1}{n}, L + \frac{1}{n}\right)$. Then $\bigcap_{n=1}^{\infty} A_n$ is the line $y = L$. $\qquad\square$

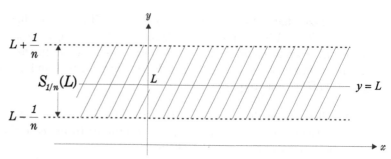

Figure 1.9.10 The Strip A_n

The third fundamental operation is the construction the set of elements which are not in a given set.

Definition 1.9.11 *Let A be a subset of the set U. Define the complement A^C of A as the set of all elements of U which are not elements of A.*

Note The definition of A^C depends not only on A but also on U.

We illustrate the complement operation with subsets of the line and the plane.

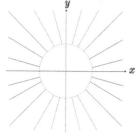

Examples 1.9.13 (1) If $U = \Re$ then $(3, \infty)^C = (-\infty, 3]$.

(2) If $U = \Re$ then $[0, 1]^C = -(\infty, 0) \cup (1, \infty)$, while if $U = \Re^2$ then $[0, 1]^C$ equals the entire plane with a unit long slit.

(3) Let $U = \Re^2$ with A the set of points inside or on the unit circle with center at the origin. Then A^C is the shaded region in Figure 1.9.12 consisting of all points outside this circle.   □

Figure 1.9.12 A^C

The complement operation has the following properties.

Proposition 1.9.14 *Let A and B be two subsets of the set U. Then*

(1) $A \cup A^C = U$ *and* $A \cap A^C = \emptyset$;

(2) $(A \cup B)^C = A^C \cap B^C$ *and* $(A \cap B)^C = A^C \cup B^C$ *(DeMorgan laws)*;

(3) $(A^C)^C = A$ *(reflexive law)*;

(4) $U^C = \emptyset$ *and* $\emptyset^C = U$.

Proof We prove the first DeMorgan law and leave the proofs of the remaining properties for the exercises. We use the Extension Principle to demonstrate that the two sets $(A \cup B)^C$ and $A^C \cap B^C$ are equal.

To show that $(A \cup B)^C \subset A^C \cap B^C$, we show that every element x of $(A \cup B)^C$ is an element of $A^C \cap B^C$. Since x is not an element of $A \cup B$, x is neither an element of A nor an element of B. Hence $x \in A^C$ and $x \in B^C$, i.e. $x \in A^C \cap B^C$ as required.

To show that $A^C \cap B^C \subset (A \cup B)^C$, we show that every element y of $A^C \cap B^C$ is an element of $(A \cup B)^C$. Now y is an element of both A^C and B^C. Hence y is neither an element of A nor an element of B. Thus, y is not an element of $A \cup B$, i.e. $y \in (A \cup B)^C$ as required.   □

Functions and Set Operations

We describe what happens when a function f or the relation f^{-1} is applied to one of the fundamental operations of sets. Let A be a subset of the domain of f. We use the notation $f(A)$ to denote the set of all $f(x)$ for x an element of A. Equivalently, $f(A)$ is the range of the restriction of f to A. We begin by applying f to a union, intersection or complement of sets.

Proposition 1.9.15 *Let f be a function with domain U and range U'. Let A, B be subsets of U. Then*

(1) $f(A \cup B) = f(A) \cup f(B)$;

(2) $f(A \cap B) \subset f(A) \cap f(B)$;

(3) $f(A)^C \subset f(A^C)$.

Proof **(1)** An element of $f(A \cup B)$ has the form $f(x)$ where $x \in A \cup B$. Such an x is either in A or in B. Hence $f(x)$ is either in $f(A)$ or in $f(B)$, i.e $f(x) \in f(A) \cup f(B)$. Therefore, $f(A \cup B) \subset f(A) \cup f(B)$.
An element of $f(A) \cup f(B)$ is of the form $f(y)$ where y is an element of either A or B. Thus $y \in A \cup B$, and $f(y) \in f(A \cup B)$. Therefore $f(A) \cup f(B) \subset f(A \cup B)$. By the Extension Principle, $f(A \cup B) = f(A) \cup f(B)$.
(2) An element of $f(A \cap B)$ is of the form $f(x)$ where $x \in A \cap B$. Hence x is an element of both A and B, and $f(x)$ is an element of both $f(A)$ and $f(B)$. That is, $f(x) \in f(A) \cap f(B)$. Therefore, $f(A \cap B) \subset f(A) \cap f(B)$.
(3) An element y of $f(A)^C$ is an element of U' which is not of the form $f(a)$ for any $a \in A$. Since $f(U) = U'$, it is possible to write y as $f(x)$. By our first observation x is not an element of A, i.e. $x \in A^C$. Hence $y = f(x) \in f(A^C)$, and $f(A)^C \subset f(A^C)$. $\square$

In Properties 2 and 3 of Proposition 1.9.15 we have equality in some examples but only inclusion in others. See Exercises 11 and 12. If U' is larger than the range of f, then Property 3 will be false as in Example 3(b) below.

Examples 1.9.16 Let f be the function with domain $\Re$ defined by $f(x) = x^2$. Let $A = [-3, 2]$ and $B = [1, 5]$. Then

$$f(A) = [0, 9] \quad and \quad f(B) = [1, 25].$$

(1) $A \cup B = [-3, 5]$ and $f(A) \cup f(B) = [0, 25]$. Thus,

$$f(A \cup B) = f([-3, 5]) = [0, 25] = f(A) \cup f(B)$$

as stated in Property 1.

(2) $A \cap B = [1, 2]$ and $f(A \cap B) = [1, 4]$. Thus,

$$f(A \cap B) = [1, 4] \subset [1, 9] = [0, 9] \cap [1, 25] = f(A) \cap f(B)$$

as stated in Property 2. Note that in this example $f(A \cap B)$ and $f(A) \cap f(B)$ are not equal.

(3a) If $U = \Re$ and $U' = [0, \infty)$, then $f(U) = U'$. We have $A^C = (-\infty, -3) \cup (2, \infty)$ and

$$f(A^C) = (4, \infty) \subset (9, \infty) = [0, 9]^C = f(A)^C$$

as stated in Property 3. Note that $A^C = [-3, 2]^C$ is taken in $U = \Re$ while $f(A)^C = [0, 9]^C$ is taken in $U' = [0, \infty)$. In this case $f(A)^C \subset f(A^C)$, but the two sets are not equal.

(3b) If $U = U' = \Re$ then the range $f(U)$ of f equals $[0, \infty)$ which does not equal U'. $f(A^C) = (4, \infty)$ as above. However,

$$f(A)^C = [0, 9]^C = (-\infty, 0) \cup (9, \infty).$$

$f(A)^C$ is not a subset of $f(A^C)$ in this case. In this example, all complements are taken in $U = U' = \Re$. Note that Property 3 does not apply because we chose U' to be larger than the range of f. $\qquad\square$

Recall that the definition $f : D \to C$ of a function includes its domain D and its codomain C. The codomain is a set that contains all values of f. If A is a subset of C, we define $f^{-1}(A)$ as the set of all elements of D which f sends into A.

Definition 1.9.17 *Let f be a function with A a subset of the codomain of f. Define $f^{-1}(A)$ as the set of all elements x of the domain of f such that $f(x) \in A$.*

Observe that the standard notation we use is a bit confusing. The notation $f^{-1}(A)$ does not mean the inverse function f^{-1} applied to A. Even if f is not one–to–one and f^{-1} is not defined, the set $f^{-1}(A)$ is still defined. However, if the function f^{-1} is defined, then Property 6 below says that both of these interpretations of $f^{-1}(A)$ are the same. Also note that the codomain C may be larger than the range of f. In that case A may contain elements which are not of the form $f(x)$. Nevertheless, $f^{-1}(A)$ is still defined as in the second example below.

Examples 1.9.18 Let $f : \Re \to \Re$ by $f(x) = x^2$.

(1) The set $f^{-1}([1, 4))$ consists of all numbers whose square is between 1 and 4, i.e. $f^{-1}([1, 4)) = (-2, -1] \cup [1, 2)$.

(2) The set $f^{-1}([-10, 9])$ consists of those numbers whose square is between -10 and 9, i.e. $f^{-1}([-10, 9]) = [0, 3]$. $\qquad\square$

The operation f^{-1} has the following properties.

Proposition 1.9.19 *Let $f : U \to U'$ be a function with A and B subsets of U'. Then*

(1) $f^{-1}(A \cup B) = f^{-1}(A) \cup f^{-1}(B);$

(2) $f^{-1}(A \cap B) = f^{-1}(A) \cap f^{-1}(B);$

(3) $f^{-1}(A^C) = f^{-1}(A)^C;$

(4) $f(f^{-1}(A)) \subset A;$

(5) *if S is a subset of U then $S \subset f^{-1}(f(S))$.*

(6) *If f is one–to–one, then the inverse function f^{-1} is defined. If A is a subset of the range of f, then $f^{-1}(A)$ equals the function f^{-1} applied to the set A.*

Proof We prove (3) through (6), leaving the proofs of (1) and (2) for the exercises. **(3)** If $x \in f^{-1}(A^C)$, then $f(x) \in A^C$, i.e. $f(x)$ is not an element of A. Thus, x is not an element of $f^{-1}(A)$. That is, $x \in f^{-1}(A)^C$. Therefore,

$$f^{-1}(A^C) \subset f^{-1}(A)^C.$$

If $y \in f^{-1}(A)^C$, then y is not an element of $f^{-1}(A)$. Thus, $f(y)$ is not an element of A, i.e $f(y) \in A^C$. Therefore, $y \in f^{-1}(A^C)$, and

$$f^{-1}(A)^C \subset f^{-1}(A^C).$$

By the Extension Principle, $f^{-1}(A^C) = f^{-1}(A)^C$.

(4) A typical element of $f(f^{-1}(A))$ is $f(x)$ with $x \in f^{-1}(A)$. Thus $f(x) \in A$, i.e. every element of $f(f^{-1}(A))$ is an element of A.

(5) If $x \in S$, then $f(x) \in f(S)$. Hence $x \in f^{-1}(f(S))$. Thus, $S \subset f(f^{-1}(S))$.

(6) $f^{-1}(A)$ is the set of all $x \in U$ such that $y = f(x) \in A$, i.e. $f^{-1}(A)$ is the set of all $x = f^{-1}(y)$ for $y \in A$. This is the definition of the function f^{-1} applied to the set A. $\qquad\square$

The inclusions in Properties 4 and 5 are equalities in some examples. Note that in Property 6 we need that A is a subset of the range of f, because otherwise we can not apply the function f^{-1} to A.

Examples 1.9.20 (1) Let f be the function with domain $\Re$ defined by $f(x) = x^2$. Note that f is not one–to–one and the function f^{-1} is not defined. Let $A = (1, 16)$ and $B = (-4, 9]$. Then

$$f^{-1}(A) = (-4, -1) \cup (1, 4), \quad and \quad f^{-1}(B) = [-3, 3] \ .$$

Note that $A \cup B = (-4, 16)$, and

$$f^{-1}(A \cup B) = (-4, 4) = f^{-1}(A) \cup f^{-1}(B)$$

as stated in Property 1. Observe that $A \cap B = (1, 9]$, and

$$f^{-1}(A \cap B) = [-3, -1) \cup (1, 3] = f^{-1}(A) \cap f^{-1}(B)$$

as stated in Property 2. Note that $A^C = (-\infty, 1] \cup [16, \infty)$, and

$$f^{-1}(A^C) = (-\infty, -4] \cup [-1, 1] \cup [4, \infty) = f^{-1}(A)^C$$

as stated in Property 3. Observe that Property 4 is valid with an equality for A but only with an inclusion for B:

$$f(f^{-1}(A)) = (1, 16) = A \quad while \quad f(f^{-1}(B)) = [0, 9] \subset B = (-4, 9] \ .$$

Property 5 is valid with an equality for $(-3, 3)$, but is only valid with an inclusion for $[0, 3)$:

$$f^{-1}(f(-3, 3)) = f^{-1}([0, 9)) = (-3, 3) \quad and$$
$$f^{-1}(f[0, 3)) = f^{-1}([0, 9)) = (-3, 3) \supset [0, 3] \ .$$

(2) Let g be the function with domain $[0, \infty)$ defined by $g(x) = x^2$. Then g is one–to–one with range $[0, \infty)$. Thus the inverse function g^{-1} is defined with domain $[0, \infty)$ and $g^{-1}(x) = \sqrt{x}$. We verify Property 6 for $A = (1, 16)$: the function g^{-1} applied to A equals $(1, 4)$ which is the same as $g^{-1}(A)$. $\qquad\square$

Summary

The reader should understand the concepts of set union, intersection and complement. The basic properties of these operations should be known. Given a function f, she should understand what happens when we apply f or f^{-1} to a union, an intersection or a complement. In addition, she should be able to apply the Extension Principle to prove that two sets are equal.

Basic Exercises

1. Let $U = \Re$. For each pair of sets A and B below, find $A \cup B$, $A \cap B$, A^C and B^C.
  (a) $A = (-1, 7]$ and $B = [2, 11)$.   **(b)** $A = (-\infty, 8)$ and $B = [-3, \infty)$.
  (c) $A = (-1, 5)$ and $B = [5, 9]$.   **(d)** $A = (-\infty, -3) \cup [3, \infty)$ and $B = (-5, 8]$.
  (e) $A = [-4, 3] \cup (7, 11)$ and $B = (0, 5) \cup [10, 15)$.

2. Let $U = \Re^2$. For each pair of sets A and B, sketch $A \cup B$, $A \cap B$, A^C and B^C.
(a) A is the set of points above the x–axis, and B is the set of points inside the unit circle with center at the origin.
(b) A is the set of points inside the triangle with vertices $(0, 1)$, $(-1, -1)$ and $(1, -1)$. B is the set of points inside the unit circle with center at the origin.
(c) A is the set of points on the line $y = x$, and B is the set of points inside the square with vertices $(1, 1)$, $(1, -1)$, $(-1, -1)$, $(-1, 1)$.
(d) A is the set of point on the parabola $y = x^2 - 1$, and B is the set of points on the parabola $y = 1 - x^2$.
(e) A is the set of points below the graph of the function $y = 1 - |x|$, and B is the set of points above the graph of the function $y = |x| - 1$.

3. Verify the six properties of union and intersection listed in Proposition 1.9.7 for each of the following examples.
(a) $U = \Re$, $A = (-7, 5]$, $B = (-2, 8]$ and $C = [0, 11)$.
(b) $U = \Re$, $A = Q$ the set of rational numbers, $B = I$ the set of irrational numbers and $C = Z$ the set of integers.
(c) $U = \Re^2$. Let A be the set of all points to the left of the y–axis, let B be the set of all points inside the unit circle with center at the origin and let C be the set of all points below the x–axis.
(d) $U = \Re^2$, A is the first quadrant, B is the disc with radius one center the origin and C is the upper half plane.
(e) U is the set of polynomials, A is the set of polynomials of degree at most three, B is the set of even polynomials and C is the set of odd polynomials.

4. Use the Extension Principle to prove the properties of union and intersection stated in Proposition 1.9.7:
  (a) $A \cup B = B \cup A$ **(b)** $A \cap B = B \cap A$
  (c) $(A \cup B) \cup C = A \cup (B \cup C)$ **(d)** $(A \cap B) \cap C = A \cap (B \cap C)$
  (e) $A \cup \phi = A$ **(f)** $A \cap U = A$ **(g)** $A \cup U = U$ **(h)** $A \cap \phi = \phi$
  (i) $A \cup A = A$ **(j)** $A \cap A = A$ **(k)** $A \cup (B \cap C) = (A \cup B) \cap (A \cup C)$

5. Describe each of the following sets.
  (a) $\bigcup_{n=1}^{\infty} \left(\frac{1}{n}, n \right)$ **(b)** $\bigcap_{n=1}^{\infty} \left(-\frac{1}{n}, 1 + \frac{1}{n} \right)$
  (c) $\bigcup_{n=1}^{\infty} A_n$ where A_n is the set of numbers x such that nx is an integer.
  (d) $\bigcap_{n=1}^{\infty} B_n$ where B_n is the disc of radius $1 + \frac{1}{n}$ and center the origin.
  (e) $\bigcup_{n=1}^{\infty} C_n$ where C_n is the annulus with center the orgin, outer radius one and inner radius $\frac{1}{n}$.
  (f) $\bigcap_{n=1}^{\infty} D_n$ where D_n is the set of points in the plane which lie on or below the parabola $y = \frac{x^2}{n}$.

6. Verify the four properties of complementation listed in Proposition 1.9.14 for each of the following examples.
(a) $U = \Re$, $A = (-\infty, 5]$ and $B = [1, 11)$.
(b) $U = \Re$, $A = Q$, the set of rational numbers and $B = I$, the set of irrational numbers.
(c) $U = \Re^2$. Let A be the set of points between the lines $y = -4$ and $y = 3$. Let B be the set of points between the lines $x = -2$ and $x = 1$.

(d) $U = \Re^2$, let A be the first quadrant and let B be the second quadrant where each quadrant contains its boundary points.

(e) Let U be the set of polynomials of degree at most two. Let A be the set of polynomials with root $x = 1$, and let B be the set of polynomials with root $x = 2$.

7. Use the Extension Principle to prove the properties of the complement operation stated in Proposition 1.9.14.

    (a) $A \cup A^C = U$ **(b)** $A \cap A^C = \phi$ **(c)** $(A \cap B)^C = A^C \cup B^C$
    (d) $(A^C)^C = A$ **(e)** $U^C = \phi$ **(f)** $\phi^C = U$

8. Assume that $A \subset B$ are two subsets of U. Use the Extension Principle to prove each of the following statements:

    (a) $A \cup B = B$;   **(b)** $A \cap B = A$;   **(c)** $B^C \subset A^C$.

9. Verify the three properties of Proposition 1.9.15 for each of the following examples.

    (a) $U = U' = \Re$, $A = (-1, 5]$, $B = (3, 7)$ and $f(x) = 2x + 1$.
    (b) $U = U' = [0, \infty)$, $A = [0, 9]$, $B = (4, 16)$ and $f(x) = \sqrt{x}$.
    (c) $U = \Re$, $U' = (-\infty, 10]$, $A = (-5, 3]$, $B = (-2, 4]$ and $f(x) = 10 - x^2$.
    (d) U is the xy–plane, $U' = \Re$, A is the second quadrant, B is the third quadrant and $f(x, y) = xy$. Each quadrant contains its axes.
    (e) U is the xy–plane, $U' = \Re$, A is the x–axis, B is the y–axis and $f(x, y) = \frac{x+y}{1+y^2}$.

10. In each case, find $f^{-1}(A)$.

    (a) $f(x) = 2x + 3$ with domain $\Re$ and $A = [5, 11)$.
    (b) $f(x) = 7 - 3x$ with domain $[0, \infty)$ and $A = (4, 16]$.
    (c) $f(x) = 14 - x^2$ with domain $\Re$ and $A = [-11, 5]$.
    (d) $f(x) = 14 - x^2$ with domain $[4, \infty)$ and $A = [-11, 5]$.
    (e) $f(x) = \frac{1}{x}$ with domain all nonzero numbers and $A = (-3, 9)$.

11. Verify the first five properties of the operation f^{-1} listed in Proposition 1.9.19 for each of the following examples.

    (a) $U = \Re$, $U' = \Re$, $A = [0, 10)$, $B = (-15, 8]$, $S = [0, 3)$ and $f(x) = 1 - x^4$.
    (b) $U = (0, \infty)$, $U' = \Re$, $A = [1, 9)$, $B = (0, 7]$, $S = (1, 5]$ and $f(x) = 1/x$.
    (c) $U = [0, 3\pi]$, $U' = \Re$, $A = [0, \infty)$, $B = (-\infty, 0]$, $S = \left(\frac{\pi}{2}, \pi\right]$ and $f(x) = \cos x$.
    (d) U is the xy–plane, $U' = \Re$, $A = [0, \infty)$, $B = (-\infty, 2]$, S is the circle $x^2 + y^2 \le 4$ and $f(x, y) = y$.
    (e) U is the xy–plane, $U' = \Re$, $A = [0, 2]$, $B = (-\infty, 1]$, S is the right half of the x–axis, i.e. all points $(x, 0)$ with $x \ge 0$, and $f(x, y) = x + y$.

12. Use the Extension Principle to prove the following properties of f^{-1} stated in Proposition 1.9.19.

    (a) $f^{-1}(A \cup B) = f^{-1}(A) \cup f^{-1}(B)$ **(b)** $f^{-1}(A \cap B) = f^{-1}(A) \cap f^{-1}(B)$

13. Let $f(x) = x^2$ have domain $\Re$ with $A = [0, 2]$ and $B = [1, 3]$. Show that for this example, equality holds in Property 2 of Proposition 1.9.15: $f(A \cap B) = f(A) \cap f(B)$.

14. Let $f(x) = x^2$ have domain and range equal to $[0, \infty)$ with $A = [0, 2]$. Show that for this example, equality holds in Property 3 of Proposition 1.9.15: $f(A)^C = f(A^C)$.

Challenging Problems

1. Determine identities to rewrite each of the expressions below. Prove your identity using the Extension Principle. The set A and each of the B_n, $n \ge 1$, are subsets of the set U. f is a function with domain U, and the C_n are subsets of codomain f.

$$\textbf{(a) } A \cap \left(\bigcup_{n=1}^{\infty} B_n \right) \qquad \textbf{(b) } A \cup \left(\bigcap_{n=1}^{\infty} B_n \right) \qquad \textbf{(c) } \left(\bigcup_{n=1}^{\infty} B_n \right)^C \qquad \textbf{(d) } \left(\bigcap_{n=1}^{\infty} B_n \right)^C$$

$$\textbf{(e) } f\left(\bigcup_{n=1}^{\infty} B_n \right) \qquad \textbf{(f) } f\left(\bigcap_{n=1}^{\infty} B_n \right) \qquad \textbf{(g) } f^{-1}\left(\bigcup_{n=1}^{\infty} C_n \right) \qquad \textbf{(h) } f^{-1}\left(\bigcap_{n=1}^{\infty} C_n \right)$$

2. Let S be a set. Define the power set $P(S)$ of S as the set of all subsets of S (including the empty set $\emptyset$ and S).
(a) List the elements of the power set $P(S)$ when $S = \{1, 2\}$.
(b) If S is a finite set with n elements, how many elements are in the power set $P(S)$?
(c) If $S \subset T$, show that $P(S) \subset P(T)$.
Let S_n be a subset of U for each $n \geq 1$.

(d) Show that $P\left(\bigcap_{n=1}^{\infty} S_n \right) = \bigcap_{n=1}^{\infty} P(S_n).$ **(e)** Show that $\bigcup_{n=1}^{\infty} P(S_n) \subset P\left(\bigcup_{n=1}^{\infty} S_n \right).$

(f) Give an example where the latter two sets are not equal.

3. If x and y are elements of a set S, then the precise definition of the ordered pair (x, y) is: $(x, y) = \{\{x\}, \{x, y\}\}$.
Show that $(x_1, y_1) = (x_2, y_2)$ if and only if $x_1 = x_2$ and $y_1 = y_2$.

4. If X and Y are sets then define $X \times Y$ as the set of all ordered pairs (x, y) where $x \in X$ and $y \in Y$. Let A, B be subsets of X with D a subset of Y.
(a) Show that $(A \cup B) \times D = (A \times D) \cup (B \times D)$.
(b) Show that $(A \cap B) \times D = (A \times D) \cap (B \times D)$.
(c) Find an identity for $(A \times D)^C$.

1.10   Topology of the Real Line

Prerequisites: Sections 1.7 and 1.9

Geometry is the study of rigid objects, constructed say from metal. Topology is the study of these same objects viewed as flexible objects, constructed say from rubber. Thus, in geometry a circle is very different from a square, and a large circle is very different from a small one. In topology, however, all circles and squares are the same because the same piece of rubber can be stretched into all these shapes. In the next three sections we introduce several fundamental concepts from topology as they apply to the real line. In this section, we study open sets, closed sets, continuous functions, bounded sets and completeness. We prove the real numbers are complete. Then we reformulate this result in terms of Cauchy sequences.

Open and Closed Sets

The neighborhood of a point P is defined as the set of points within a certain distance of P. An open set is defined to generalize open intervals. Then a closed set is defined as the complement of an open set. Closed sets include closed intervals as well as more complicated sets.

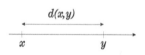

Fig 1.10.1 Distance Between x and y

We begin by introducing notation for the *distance* between two numbers on the real number line. This distance is the larger of the two numbers minus the smaller one. Thus the distance between x and y is either $x - y$ or $y - x = -(x - y)$. See Figure 1.10.1. We use the absolute value to denote this non–negative number.

Definition 1.10.2 *For $x, y \in \Re$, the distance $d(x, y)$ between x and y is given by:*

$$d(x, y) = |x - y|.$$

Distance has the following three properties.

Proposition 1.10.3 *Let $x, y, z \in \Re$. Then*

(a) $d(x, y) = d(y, x)$;

(b) $d(x, y) \geq 0$ *while* $d(x, y) = 0$ *if and only if* $x = y$;

(c) $d(x, z) \leq d(x, y) + d(y, z)$ (*Triangle Inequality*).

Proof (a) $d(x, y) = |x - y| = |-(x - y)| = |y - x| = d(y, x)$.
(b) $d(x, y) = |x - y| \geq 0$. Also $d(x, y) = |x - y| = 0$ if and only if $x - y = 0$, i.e. $x = y$.
(c) For $a, b \in \Re$,

$$|a + b|^2 = a^2 + 2ab + b^2 \leq a^2 + |2ab| + b^2 = |a|^2 + 2|a||b| + |b|^2 = (|a| + |b|)^2.$$

The inequality above is just the observation $x \leq |x|$, applied to $x = 2ab$. Take the positive square root of the inequality above:

$$|a + b| \quad \leq \quad |a| + |b| \text{ for } a, b \in \Re. \qquad (1.10.1)$$

Thus $d(x, z) = |x - z| = |(x - y) + (y - z)| \quad \leq \quad |x - y| + |y - z| = d(x, y) + d(y, z)$.□

The triangle inequality gets its name from the generalization of this formula to three points in the plane. In that context the points x, y, z determine the triangle T in Figure 1.10.4. The triangle inequality says that the length of the side xz of T is less than or equal to the sum of the lengths of other two sides xy and yz of T. In other words, the line segment xz is the shortest path between the points x and z. The triangle inequality is used to show that x, z are close to each other by showing they are both close to y.

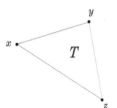

Figure 1.10.4
Triangle Inequality

The set of points within a distance r of x is called the *neighborhood* of x of radius r. When r is small, this neighborhood consists of points which are close to x. We introduce notation for these neighborhoods.

Definition 1.10.5 *Let $x \in \Re$, and let r be a positive number. The neighborhood $N(x; r)$ of x with radius r is the open interval $(x - r, x + r)$.*

We reformulate the definition of a neighborhood to a form which can be used to define neighborhoods in higher dimensions.

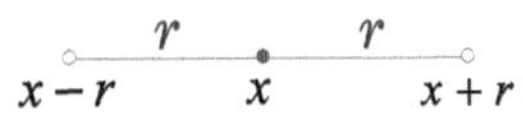

Fig 1.10.6 $N(x; r)$

Proposition 1.10.7 *The neighborhood $N(x; r)$ consists of the $y \in \Re$ with $d(y, x) < r$.*

Proof Since $d(x + r, x) = d(x - r, x) = r$, the interval $(x - r, x + r) = N(x; r)$ consists precisely of those points y with $d(y, x) < r$. See Figure 1.10.6.   □

A prototype *open set* in the real line $\Re$ is an open interval (a, b). We define a subset of $\Re$ to be open if it contains a neighborhood of each of its points.

Definition 1.10.8 *A subset U of $\Re$ is called open if for every $x \in U$ there is a neighborhood $N(x; r)$ of x which is a subset of U.*

Note we can choose different values of r for different choices of x. In particular, we have to choose r to be small when x is near an endpoint of U.

Examples 1.10.9 (1) $\Re$ itself is an open set because $N(x; 1) \subset \Re$ for $x \in \Re$.

(2)   An open interval (a, b) is an open set. If $x \in (a, b)$, let r be the smaller of the numbers $d(x, a)$ and $d(x, b)$. Then $N(x; r)$ is a subset of (a, b).

(3)   The empty set $\emptyset$ is open because there are no points $x \in \emptyset$ for which we have to find a neighborhood $N(x; r)$ contained in $\emptyset$.   □

Open sets have the following properties.

Proposition 1.10.10 (a)   *If U, V are open sets, then $U \cap V$ is also open.*

(b)   *If U, V are open sets, then $U \cup V$ is also open.*

(c)   *If $U_1, \ldots, U_n, \ldots$ are open sets, then $\displaystyle\bigcup_{n=1}^{\infty} U_n$ is also open.*

Proof (a)   Let $x \in U \cap V$. Then x is an element of both U and V. Find a neighborhood $N(x; r_1)$ which is a subset of U and a neighborhood $N(x; r_2)$ which is a subset of V. Let r be the smaller of the numbers r_1 and r_2. Then

$$N(x; r) \subset N(x; r_1) \subset U \quad \text{and} \quad N(x; r) \subset N(x; r_2) \subset V.$$

Hence $N(x; r)$ is a subset of $U \cap V$, and $U \cap V$ is an open set.

(b)   Let $x \in U \cup V$. Then x is either an element of U or an element of V. Find a neighborhood $N(x; r)$ that is a subset of U in the first case or a subset of V in the second case. Then $N(x; r)$ is a subset of $U \cup V$. Hence $U \cup V$ is an open set.

(c)   Let $x \in \displaystyle\bigcup_{n=1}^{\infty} U_n$. Then $x \in U_k$ for some k. Find a neighborhood $N(x; r)$ that is a

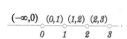

$(-\infty, 0)\ \ (0,1)\ \ (1,2)\ \ (2,3)$
$\quad\quad 0 \quad\ 1 \quad\ 2 \quad\ 3$

Figure 1.10.11
Complement of N

subset of U_k. Then $N(x; r)$ is a subset of $\displaystyle\bigcup_{n=1}^{\infty} U_n$. Hence $\displaystyle\bigcup_{n=1}^{\infty} U_n$ is an open set.   □

Examples 1.10.12 (1) $(-3, 4) \cup (7, 23)$ is an open set because it is the union of two open sets.

(2)   The complement of the natural numbers N is an open set because it is the union of the open sets $(-\infty, 0)$ and $(n - 1, n)$ for $n \geq 1$. See Figure 1.10.11.

(3)   The interval $[8, 17)$ is not an open set. Every neighborhood $N(8; r)$ contains numbers less than 8 and is therefore not a subset of $[8, 17)$.

(4)   The set $\{4\}$ whose only element is 4 is not an open set because every neighborhood $N(4; r)$ contains numbers other than 4.

$$U_1$$
$$U_2$$
$$U_3$$
$$\vdots$$

$-1 \quad\quad 0 \quad\quad 1$

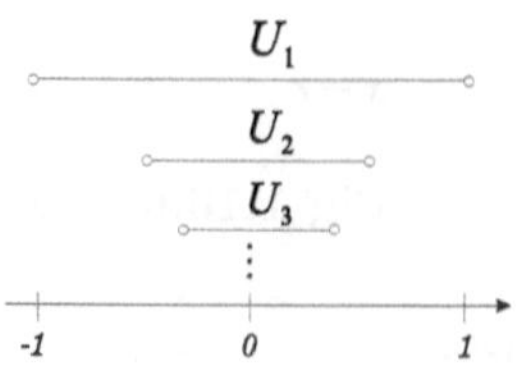

Figure 1.10.13
Infinite Intersection
of Open Sets

(5)   An infinite intersection of open sets may not be open. For example, if $U_n = (-1/n, 1/n)$, then $\displaystyle\bigcap_{n=1}^{\infty} U_n = \{0\}$ which is not an open set. See Figure 1.10.13.   □

You have seen all the examples of open sets! We leave it as an exercise to show that every open set is the union of disjoint open intervals. We now turn our attention to *closed sets*. You might guess that closed intervals $[a, b]$ are the prototype closed sets. However, more complicated sets are also called closed.

Definition 1.10.14 *A subset F of $\Re$ is called closed if its complement F^C is open.*

As we shall see, by this definition the properties of open and closed sets correspond under the complement operation.

Examples 1.10.15 (1) $\Re$ is a closed set because $\Re^C = \emptyset$ is open.

(2) $\emptyset$ is a closed set because $\emptyset^C = \Re$ is open.

(3) A closed interval $[a, b]$ is a closed set because $[a, b]^C = (-\infty, a) \cup (b, \infty)$ is open.

(4) The set $\{x\}$ is closed because its complement is the open set $(-\infty, x) \cup (x, \infty)$.

(5) The natural numbers N is closed because N^C is the union of the open intervals $(-\infty, 0)$ and $(n, n+1)$ for $n = 0, 1, 2, \ldots$ which is open by Example 1.10.12 (2).

(6) $[a, b)$ is not a closed set because $[a, b)^C = (-\infty, a) \cup [b, \infty)$ is not open. $\qquad\square$

Closed sets have the following properties.

Proposition 1.10.16 (a) *If F, G are closed sets, then $F \cup G$ is also closed.*

(b) *If F, G are closed sets, then $F \cap G$ is also closed.*

(c) *If $F_1, \ldots, F_n, \ldots$ are closed sets, then $\bigcap_{n=1}^{\infty} F_n$ is also closed.*

Proof These properties are obtained by taking the complements of the properties of open sets in Proposition 1.10.10.

(a) F^C and G^C are open sets. By Proposition 1.10.10(a), $F^C \cap G^C$ is also open. Therefore its complement is a closed set:

$$\left(F^C \cap G^C\right)^C = (F^C)^C \cup (G^C)^C = F \cup G.$$

(b) F^C and G^C are open sets. By Proposition 1.10.10(b), $F^C \cup G^C$ is also open. Therefore its complement is a closed set:

$$\left(F^C \cup G^C\right)^C = (F^C)^C \cap (G^C)^C = F \cap G.$$

(c) Each F_n^C is an open set. By Proposition 1.10.10(c), $\bigcup_{n=1}^{\infty} F_n^C$ is also open. Hence its complement is a closed set:

$$\left(\bigcup_{n=1}^{\infty} F_n^C\right)^C = \bigcap_{n=1}^{\infty} (F_n^C)^C = \bigcap_{n=1}^{\infty} F_n. \qquad\square$$

Proposition 1.10.16(c) has interesting consequences. As illustrated in the third example below, it can be used to construct complicated closed sets.

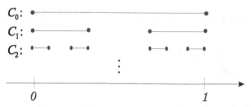

Figure 1.10.17 Cantor Set

Examples 1.10.18 (1) $[-5, -2] \cup [0, 7]$ is closed, because it is the union of two closed sets.

(2) An infinite union of closed sets may not be closed. For example, $\bigcup_{n=1}^{\infty} [1/n, 1] = (0, 1]$ is not closed since its complement $(0, 1]^C = (-\infty, 0] \cup (1, \infty)$ is not open.

(3) The Cantor set C is a complicated closed set defined as $C = \bigcap_{n=0}^{\infty} C_n$ with C_n a closed set given as the union of 2^n closed intervals. By Prop. 1.10.16(c), C is a closed set. The C_n, for $n \geq 0$, are defined recursively. That is, let $C_0 = [0, 1]$. If C_k has been defined, define C_{k+1} by dividing each closed interval of C_k into thirds and removing the open interval given by the middle third. For example,

$$C_1 = \left[0, \frac{1}{3}\right] \cup \left[\frac{2}{3}, 1\right],$$

$$C_2 = \left[0, \frac{1}{9}\right] \cup \left[\frac{2}{9}, \frac{1}{3}\right] \cup \left[\frac{2}{3}, \frac{7}{9}\right] \cup \left[\frac{8}{9}, 1\right].$$

See Figure 1.10.17. If we write the numbers between 0 and 1 as base three decimals, then C is constructed from $[0, 1]$ by removing all decimals which contain the digit 1. Thus, C consists of all base three decimals between zero and one which contain only the digits 0 and 2. □

A *sequence* is an infinite list of numbers. We say a sequence has *limit L* if the corresponding points on the number line get closer and closer to L. See Figure 1.10.19.

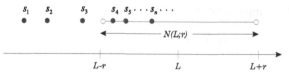

Figure 1.10.19 A Convergent Sequence

Definition 1.10.20 *The sequence $s = \{s_n\}$ is the infinite list of numbers*

$$s_1, s_2, \ldots, s_n, \ldots$$

We say this sequence converges to the number L if given any neighborhood $N(L; r)$, all but a finite number of elements of the sequence are in $N(L; r)$. L is called the limit of the sequence. A sequence which does not converge is said to diverge.

Examples 1.10.21 (1) The sequence $s_n = 1/n$ converges and has limit 0. Given any neighborhood $N(0; r)$, choose a positive integer $k > \frac{1}{r}$. Then $\frac{1}{k} < r$, and all numbers of the sequence, except $1, \frac{1}{2}, \ldots, \frac{1}{k-1}$ are in $N(0; r)$. Thus, this sequence has limit 0.

(2) The sequence $s_n = \frac{n-1}{n}$ converges and has limit 1. Observe that $s_n = 1 - \frac{1}{n}$. Say we are given a neighborhood $N(1; r)$. Choose a positive integer $k > \frac{1}{r}$. Then $\frac{1}{k} < r$, and all numbers of the sequence except $0, \frac{1}{2}, \ldots, \frac{k-2}{k-1}$ are in $N(1; r)$. Hence this sequence has limit 1.

(3) The sequence $s_n = n$ diverges. Note ∞ is not the limit of this sequence, because the limit of a sequence must be a number, and ∞ is not a number.

(4) The sequence $s_n = (-1)^n$ diverges. In particular, 1 is not the limit of this sequence because every neighborhood $N(1; r)$, with $r < 2$, does not contain the odd terms of this sequence. Similarly, -1 is not the limit of this sequence because every neighborhood $N(-1; r)$, with $r < 2$, does not contain the even terms of this sequence. □

We use sequences to give an alternate description of closed sets.

Proposition 1.10.22 *A subset F of $\Re$ is closed if and only if whenever $\{s_n\}$ is a convergent sequence of points of F with limit L, L is also an element of F.*

Proof Let $\{s_n\}$ be a sequence of points of a closed set F with limit L. If L were not in F, then L would be an element of the open set F^C. There would be a neighborhood $N(L; r)$ of L contained in F^C. No elements of the sequence $\{s_n\}$ are in this neighborhood. Since this is impossible, L must be an element of F.

Assume F has the property that whenever $\{s_n\}$ is a sequence of points of F which converges to L, it follows that $L \in F$. We show F^C is an open set. If F^C were not open, there would be $x \in F^C$ such that every neighborhood $N(x; r)$ contains an element of F. For each n, let $s_n \in F$ be an element of $N(x; 1/n)$. The sequence $\{s_n\}$ consists of elements of F and converges to x. Therefore $x \in F$ which is false. Hence F^C must be open, and F is closed. □

Examples 1.10.23 (1) $[0, 1)$ is not a closed set because the sequence $s_n = \frac{n-1}{n}$ is a sequence of points of $[0, 1)$ which converges to 1 and $1 \notin [0, 1)$.

(2) Let $x \in [0, 1]$ be a base three decimal whose digits are all 0s and 2s. We give a rigorous argument that x is an element of the Cantor set C. Let s_n be the truncation of the decimal x consisting of its first n digits. Then s_n is the endpoint of one of the 2^n closed intervals whose union is C_n. Hence $s_n \in C$, and $\{s_n\}$ is a sequence of points of the closed set C with limit x. By Proposition 1.10.22, $x \in C$. □

Continuous Functions

We reformulate the definition of continuity given in Section 7 in terms of closed sets and in terms of sequences. These descriptions allow us to generalize the definition of continuity from functions with domain a closed interval to functions whose domain is a closed set. This viewpoint will be used in the proofs of the Intermediate Value and Maximum Value Theorems.

Recall from Section 7 that a function f, with domain a closed interval D, is continuous if

$$\lim_{x \to c} f(x) = f(c) \qquad (1.10.2)$$

for all $c \in D$. By Definition 1.5.35 this means that for every $\epsilon > 0$, there is a corresponding $\delta > 0$ such that if $x \in D$ with $|x - c| < \delta$, then $|f(x) - f(c)| < \epsilon$, i.e.

$$\text{if } x \in D \text{ and } d(x, c) < \delta, \text{ then } d(f(x), f(c)) < \epsilon. \qquad (1.10.3)$$

In other words, for all $c \in D$ and $\epsilon > 0$ there is a corresponding $\delta > 0$ such that:

$$\text{if } x \in N(c; \delta) \cap D, \text{ then } f(x) \in N(f(c); \epsilon). \qquad (1.10.4)$$

This condition makes sense for any closed set D, not just for closed intervals. The following proposition lists two other conditions which are equivalent to this one. We use these conditions to define a continuous function whose domain is any closed set.

Proposition 1.10.24 *Let f be a real valued function with domain a closed subset D of $\Re$. Then the following conditions are equivalent.*

(a) *If $c \in D$ and $N(f(c); \epsilon)$ is a neighborhood of $f(c)$, then there is a neighborhood $N(c; \delta)$ of c such that*

$$f(N(c; \delta) \cap D) \subset N(f(c); \epsilon).$$

(b) *If $\{s_n\}$ is a convergent sequence of points of D with limit c, then the sequence $\{f(s_n)\}$ is also convergent with limit $f(c)$.*

(c) *If G is a closed subset of $\Re$, then $f^{-1}(G)$ is also closed.*

Proof (a) implies (b). Since D is a closed set, $c \in D$. By (a), given any neighborhood $N(f(c); \epsilon)$, there is a neighborhood $N(c; \delta)$ such that $f(N(c; \delta) \cap D) \subset N(f(c); \epsilon)$. All but a finite number of points of the sequence $\{s_n\}$ are in $N(c; \delta)$. Hence all but a finite number of points of the sequence $\{f(s_n)\}$ are in $N(f(c); \epsilon)$. Therefore the sequence $\{f(s_n)\}$ has limit $f(c)$.

(b) implies (c). Let G be a closed subset of $\Re$. We use Proposition 1.10.22 to show $f^{-1}(G)$ is closed. Let $\{s_n\}$ be a sequence of elements of $f^{-1}(G)$ with limit c. By (b), $\{f(s_n)\}$ is a sequence of elements of G with limit $f(c)$. Since G is closed, $f(c) \in G$. Hence $c \in f^{-1}(G)$, and $f^{-1}(G)$ is closed.

(c) implies (a). Let $c \in D$, and let $\epsilon > 0$. By (c), $A = f^{-1}(N(f(c); \epsilon)^C)$ is a closed set. Since c is an element of the open set A^C, there is a neighborhood $N(c; \delta)$ which is contained in A^C. Then $f(N(c; \delta) \cap D) \subset N(f(c); \epsilon)$. □

We use this proposition to extend the definition of a continuous function to functions whose domain is a closed set.

Definition 1.10.25 *Let f be a real valued function with domain a closed set D. If f satisfies any of the three equivalent conditions of the preceding proposition, we call f a continuous function.*

We could prove an analogue of Proposition 1.10.24 for functions whose domain is an open set. Since we will not need this result for our applications, we relegate it to the exercises. In fact, by using the concepts of relative open and relative closed sets, Proposition 1.10.24 generalizes to functions whose domain is any subset of $\Re$. We give this generalization as a Challenging Problem. In the following examples we use the equivalence of condition (1.10.2) and Proposition 1.10.24 (b) to evaluate limits of complicated sequences. That is, the limit of a function applied to a sequence with limit c equals the limit of the function at $x = c$.

Examples 1.10.26 (1) Let $s_n = \cos\left(\frac{1}{n^2+1}\right)$. The sequence $\left\{\frac{1}{n^2+1}\right\}$ has limit zero and $f(x) = \cos x$ is a continuous function. Hence the sequence $s_n = f\left(\frac{1}{n^2+1}\right)$ converges to $f(0) = 1$.

(2) Let $t_n = n \sin \frac{3}{n}$. The function $g(x) = \left\{ \begin{array}{ll} \frac{\sin 3x}{x} & \text{if } x \neq 0 \\ 3 & \text{if } x = 0 \end{array} \right\}$ with domain $\Re$ is continuous. In particular,

$$\lim_{x \to 0} g(x) = 3 \lim_{x \to 0} \frac{\sin 3x}{3x} = (3)(1) = 3 = g(0) .$$

The sequence $\left\{\frac{1}{n}\right\}$ has limit zero. Therefore the sequence $t_n = g\left(\frac{1}{n}\right)$ converges to $g(0) = 3$. □

Completeness

In this section we show that the set of real numbers, defined as the set of all decimals, is complete. That is, there are no gaps in the real number line where numbers are missing. We give two equivalent formulations of a complete set: in terms of least upper bounds and in terms of Cauchy sequences. We identify the complete subsets of $\Re$ as the closed subsets.

In contrast to the set of real numbers, the set of rational numbers is not complete because there are gaps where irrational numbers, such as π and $\sqrt{2}$, are missing.

Motivating Example 1.10.27 We want to explain that $\sqrt{2}$ is missing from the set of rational numbers Q but is present in the set of real numbers $\Re$. Consider the set S of positive rational numbers whose square is less than two. There are many rational numbers b, such as 3 or 5, which are larger than every number in S. We call each choice of b an *upper bound* of S. Note that there is no choice of a *rational number* B which is the smallest of all these upper bounds b. That is, the non–existence of a smallest upper bound of S indicates the presence of a hole in the rational numbers at $\sqrt{2}$. However, if we work with the real numbers $\Re$, then there is a *real number* $B = \sqrt{2}$ which is the smallest of all these upper bounds b. See Figure 1.10.28. □

Figure 1.10.28
Q is not Complete

Using the preceding example as a model, we define a set of numbers to be complete if each of its subsets with an upper bound has a smallest one while each of its subsets with a lower bound has a largest one.

Definition 1.10.29 *Let S be a subset of $\Re$.*

(a) *S is bounded above if there is a number b such that $x \leq b$ for $x \in S$. We call b an upper bound of S.*

(b) *S is bounded below if there is a number w such that $w \leq x$ for $x \in S$. We call w a lower bound of S.*

(c) *S is bounded if it is both bounded above and bounded below.*

(d) *B is a least upper bound of S if B is an upper bound of S and $B \leq b$ for every upper bound b of S. We write $B = \text{lub}(S)$.*

(e) *W is a greatest lower bound of S if W is a lower bound of S, and $w \leq W$ for every lower bound w of S. We write $W = \text{glb}(S)$.*

(f)   *A subset T of $\Re$ is complete if:*
   (1) every subset S of T which is bounded above has a least upper bound which is an element of T;
   (2) every subset S of T which is bounded below has a greatest lower bound which is an element of T.

We interpret these definitions in terms of the real number line. An upper bound of S is a number b which lies to the right of all the numbers in S. The least upper bound of S is the smallest number B which lies to the right of all the numbers in S. On the other hand, a lower bound of S is a number w which lies to the left of all the numbers in S. The greatest lower bound of S is the largest number W which lies to the left of all the numbers in S. This interpretation for $S = [a, c]$ is illustrated in Figure 1.10.30.

Figure 1.10.30 Least Upper Bound and Greatest Lower Bound of $S = [a, c]$

Observe that if S is bounded above, there are an infinite number of upper bounds of S. If b is one upper bound, then every number greater than b is also an upper bound. Similarly, if S is bounded below, S has an infinite number of lower bounds.

Figure 1.10.32
Upper and Lower
Bounds for $(1, 5)$

Examples 1.10.31 (1)   Let $S = (1, 5)$. S is bounded and has many upper bounds such as 6, 8 and 9. The number 5 is the least upper bound of S. S has many lower bounds such as 0, -3 and -4. The number 1 is the greatest lower bound of S. Thus, S is bounded. See Figure 1.10.32. Note that S is not complete because the set S itself is bounded above but does not contain its least upper bound 5. (Alternatively, S is bounded below but does not contain its greatest lower bound 1.)

(2)   Let $T = (-\infty, 7]$. T is bounded above and has least upper bound 7. However, T is not bounded below.

(3)   Let $U = (8, \infty)$. U is bounded below and has greatest lower bound 8. However, U is not bounded above. Observe that U is not complete because U itself is bounded below but does not contain its greatest lower bound 8.

(4)   The sets $\Re$, Q and Z are neither bounded above nor bounded below. We show below that the set of all real numbers $\Re$ is complete. In the motivating example above, we showed that the set of rational numbers Q is not complete. Observe that the set of integers Z is complete because:
   (1) a subset of Z which is bounded above has a largest element which is its least upper bound;
   (2) a subset of Z which is bounded below has a smallest element which is its greatest lower bound.   □

Before proceeding with our study of completeness, we digress to show that least upper bounds and greatest lower bounds are unique when they exist.

Proposition 1.10.33 *Let S be a subset of the real numbers $\Re$.*

(a)   *If S has a least upper bound, then it is unique.*

(b)   *If S has a greatest lower bound, then it is unique.*

Proof (a) Let B_1 and B_2 be two least upper bounds of S. Since B_1 is a least upper bound of S and B_2 is an upper bound of S, we have $B_1 \leq B_2$. Conversely, B_2 is a least upper bound of S and B_1 is an upper bound of S, so $B_2 \leq B_1$. Thus $B_1 = B_2$.
(b) We leave the analogous proof of this case for the exercises. □

View a real number $x \in \Re$ as a decimal. We show $\Re$ is complete.

Theorem 1.10.34 (Completeness Theorem) *The set of real numbers $\Re$ is complete.*

Proof Let S be a nonempty subset of the set of real numbers $\Re$ which is bounded above. We determine the decimal expansion of the least upper bound B of S,

$$B = b_0.b_1b_2b_3 \cdots,$$

one digit at a time. Select b_0 to be an integer and the b_n, $n \geq 1$, as integers between 0 and 9. Let $b_0 + 1$ be the smallest integer which is an upper bound of S. Assume $b_0 \geq 0$. (If $b_0 < 0$, a similar argument applies.) There must be an element of S in $(b_0, b_0 + 1]$. Divide the interval $I_0 = (b_0, b_0 + 1]$ into ten subintervals:

$$(b_0.0, b_0.1], \ (b_0.1, b_0.2], \ (b_0.2, b_0.3], \ (b_0.3, b_0.4], \ (b_0.4, b_0.5],$$

$$(b_0.5, b_0.6], \ (b_0.6, b_0.7], \ (b_0.7, b_0.8], \ (b_0.8, b_0.9], \ (b_0.9, b_0 + 1].$$

Let b_1 be the largest integer between 0 and 9 such that

$$I_1 = (b_0.b_1, b_0.(b_1 + 1)]$$

contains an element of S. [When $b_1 = 9$, the symbol $b_0.(b_1 + 1)$ denotes $(b_0 + 1).0$.] Similarly, divide the interval I_1 into ten subintervals to choose the next subinterval I_2 and to determine the next digit b_2 in the decimal expansion of B. Iterate this procedure to determine intervals

$$I_n = (u_n, u_n + 10^{-n}]$$

and the digits b_n in the decimal expansion of B. Observe $B \in I_n$ for each n.

We show that B is an upper bound of S. If not, there would be $x \in S$ with $x > B$. Let n be the smallest positive integer with $10^{-n} < x - B$. Then $B \in I_n = (u_n, u_n + 10^{-n}]$, while $x > B + 10^{-n} > u_n + 10^{-n}$. This is impossible because u_n was chosen to be the largest number for which the subinterval I_n contains an element of S. Thus B must be an upper bound of S.

Now we show that B is the least upper bound of S. Let b be an upper bound of S. We show $B \leq b$. For each n, there is an element x_n of S with $x_n \in I_n$. Since B is also in I_n, the distance between x_n and B is less than 10^{-n}. Therefore,

$$b \geq x_n > u_n \geq B - 10^{-n}.$$

Since this inequality is true for each positive integer n, it follows that $b \geq B$.

The same reasoning applies to show that a set of real numbers T which is bounded below has a greatest lower bound. However, we use a trick to derive this result from the preceding one. Let S be the set of all numbers $-x$ for $x \in T$. If w is a lower bound of T then $-w$ is an upper bound of S. By the above argument, S has a least upper bound B. Then $W = -B$ is the greatest lower bound of T. □

We use the Completeness Theorem to show that all closed sets are complete.

Corollary 1.10.35 *Every closed subset F of $\Re$ is complete.*

Proof By the Completeness Theorem, every subset S of F which is bounded above has a least upper bound $B \in \Re$. If B were not in F, there would be a neighborhood $N(B; r)$ of B contained in the open set F^C. Then $B - r/2$ would be an upper bound of F, and $B - r/2 < B$. This contradicts the fact that B is the least upper bound of F. Therefore $B \in F$. Similarly every subset T of F which is bounded below has a greatest lower bound which is an element of F. We leave the proof as an exercise. $\square$

Now we have many examples of complete sets: closed intervals $[a, b]$, the set of integers Z and even the Canter set C.

There is an alternate method to describe the completeness of a set S. If there were a number L missing from S, then we could construct a convergent sequence of points of S whose limit L is not in S. Thus a set S is complete if every sequence of numbers in S which should converge does in fact converge to a number in S.

Motivating Example 1.10.36 We use a sequence to show that the set of rational numbers Q is not complete because it is missing $\sqrt{2}$. Let $\sqrt{2} = 1.b_1 b_2 \cdots$ be its decimal expansion. Let $x_n = 1.b_1 \cdots b_n$ be the approximation of $\sqrt{2}$ which uses only the first n of these digits. Then $\{x_n\}$ is a convergent sequence of rational numbers. However, this sequence converges to $\sqrt{2}$ which is an irrational number and not an element of Q. (The fact $\sqrt{2} \in \Re$ illustrates that $\Re$ is complete.) $\square$

To make this viewpoint useful, we need a criterion to determine when a sequence converges. Moreover, this criterion should not involve the limit of the sequence. We define the *Cauchy sequence* criterion and show that it specifies which sequences converge.

Definition 1.10.37 *A sequence $\{x_n\}$ of numbers is called a Cauchy sequence if for every $\epsilon > 0$ there is a positive integer N such that if $m, n \geq N$ then $d(x_m, x_n) < \epsilon$.*

Recall that a sequence $\{x_n\}$ converges to a number L if every neighborhood $N(L; \epsilon)$ contains all but a finite number of the x_n. Thus a convergent sequence clusters about its limit L, while the numbers of a Cauchy sequence cluster around each other. The following theorem says that these two types of behavior are exactly the same.

Theorem 1.10.38 *A sequence $\{x_n\}$ converges if and only if it is a Cauchy sequence.*

Proof Assume $\{x_n\}$ is a convergent sequence with limit L. Find a positive integer N such that if $n \geq N$, then $d(x_n, L) < \frac{\epsilon}{2}$. Let $m, n \geq N$. By the triangle inequality, $\{x_n\}$ is a Cauchy sequence:

$$d(x_m, x_n) \leq d(x_m, L) + d(L, x_n) < \frac{\epsilon}{2} + \frac{\epsilon}{2} = \epsilon.$$

Assume $\{x_n\}$ is a Cauchy sequence. We use the completeness property of the real numbers to find the limit of this sequence. For each $n \geq 1$ we recursively define a positive integer $k(n)$ and two numbers a_n, b_n which have these three properties:

(1) $k(n) \leq k(n+1)$ and $a_n \leq a_{n+1} \leq b_{n+1} \leq b_n$;

(2) $b_n - a_n \leq \frac{2}{n}$;

(3) if $p \geq k(n)$ then $x_p \in [a_n, b_n]$.

Apply the Cauchy criterion with $\epsilon = 1$ to find $k(1)$ such that if $p, q \geq k(1)$, then $d(x_p, x_q) < 1$. In particular, $d(x_p, x_{k(1)}) < 1$. Let $a_1 = x_{k(1)} - 1$ and $b_1 = x_{k(1)} + 1$. If $k(n-1)$, a_{n-1}, b_{n-1} have been defined, apply the Cauchy criterion with $\epsilon = \frac{1}{n}$ to find $k(n) \geq k(n-1)$ such that if $p, q \geq k(n)$, then $d(x_p, x_q) < \frac{1}{n}$. In particular, $d(x_p, x_{k(n)}) < \frac{1}{n}$. Define a_n to be the larger of the two numbers a_{n-1} and $x_{k(n)} - \frac{1}{n}$. Define b_n to be the smaller of the two numbers b_{n-1} and $x_{k(n)} + \frac{1}{n}$. Clearly these $k(n)$, a_n, b_n satisfy the above three requirements.

By Property 1, the set A of all a_n, for $n \geq 1$, is bounded above by any b_p, and A has a least upper bound U by the Completeness Theorem. Also, by Property 1 the set B of all b_n, for $n \geq 1$, is bounded below by any a_q, and B has a greatest lower bound L by the Completeness Theorem. Since $L, U \in [a_n, b_n]$ for $n \geq 1$, it follows from Property 2 that $L = U$. See Figure 1.10.39. We show the sequence $\{x_n\}$ converges to L. Let $\epsilon > 0$. Select $n > \frac{2}{\epsilon}$. If $p \geq k(n)$ then $x_p, L \in [a_n, b_n]$ by Property 3. By Property 2: $d(x_p, L) \leq b_n - a_n \leq \frac{2}{n} < \epsilon$. Thus the sequence $\{x_n\}$ converges to L. $\square$

$$U = L$$

$$a_1 \; a_2 \; \cdots \; a_n \; a_{n+1} \; \cdots \quad \cdots \; b_{n+1} \; b_n \cdots \; b_2 \; b_1$$

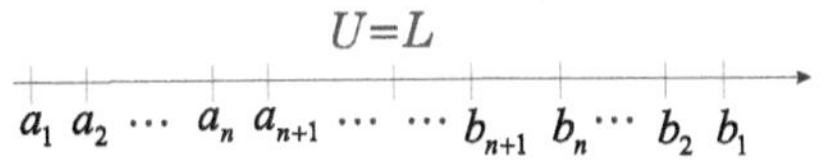

Figure 1.10.39 Locating the Limit of a Cauchy Sequence

The following corollary ties together our results on completeness. It says that the approach of using Cauchy sequences to test for completeness is equivalent to the definition using least upper bounds and greatest lower bounds. In addition, we identify the complete subsets of $\Re$ as the closed sets.

Corollary 1.10.40 *Let S be a subset of $\Re$. The following statements are equivalent:*
  (a) *S is complete.*
  (b) *Every Cauchy sequence of points of S converges to a point of S.*
  (c) *S is a closed set.*

Proof **(a) implies (b)** If $\{x_n\}$ is a Cauchy sequence of points of S, then this sequence has a limit L. Let S_- be the set of those x_n which are less than L, and let S_+ be the set of those x_n which are greater than or equal to L. At least one of S_-, S_+ is infinite. If S_- is infinite, then L is its least upper bound while if S_+ is infinite, then L is its greatest lower bound. Since S is complete, $L \in S$. Hence the Cauchy sequence $\{x_n\}$ converges to an element of S.
(b) implies (c) Let $\{x_n\}$ be a sequence of points of S which converges to L. By Theorem 1.10.38, this sequence is a Cauchy sequence. By condition (b), $L \in S$. Hence S is a closed set.
(c) implies (a) This is Corollary 1.10.35. $\square$

The usual way to specify an irrational number x is to give its decimal expansion. Alternatively we can specify x by giving a Cauchy sequence of rational numbers with limit x. For example, let $x = b_0.b_1b_2 \cdots$ be the decimal expansion of x. Define $x_n = b_0.b_1 \cdots b_n$ as the approximation of x using the first n digits in its decimal expansion. Then $\{x_n\}$ is an example of a Cauchy sequence with limit x.

Summary

The meaning and examples of neighborhoods, open sets and closed sets should be understood. The reader should know the basic properties of open and closed sets and be able to apply them to examples. She should understand what it means for a sequence to converge or diverge. In particular, she should be able to determine whether a specific sequence converges or diverges. She should understand the three formulations of continuity for functions whose domain is a closed set. The meaning of upper bounds, lower bounds, least upper bounds and greatest lower bounds should be known. The completeness property of $\Re$ given by the Completeness Theorem as well as its reformulation in terms of Cauchy sequences should be understood.

Basic Exercises

1. Sketch each neighborhood on the real number line.
(a) $N(0; 2)$ **(b)** $N(5; 1/2)$ **(c)** $N(-3; 1)$

2. Evaluate each of these distances.
(a) $d(2, 7)$ **(b)** $d(-2, -7)$ **(c)** $d(-4, 3)$

3. Verify the three properties of distance listed in Proposition 1.10.3 in the case $x = -5$, $y = 4$, $z = -1$.

4. Show $d(rx, ry) = |r| d(x, y)$ for any three numbers r, x, $y \in \Re$.

5. Determine whether each set is open and whether each set is closed.
(a) $A = (-3, 5)$;   **(b)** $B = [2, 6]$;   **(c)** $C = (-1, 4]$;
(d) $D = (-\infty, 5]$;   **(e)** $E = (-3, \infty)$;   **(f)** $F = (-3, 5) \cup [2, 6]$;
(g) $G = (-\infty, 2) \cup [-1, \infty)$;   **(h)** $H = (-\infty, -5] \cup [0, \infty)$; **(i)** $I = \bigcup_{n=1}^{\infty} (-n, 1 - n)$;
(j) $J = \bigcup_{n=1}^{\infty} [-n, 1 - n]$;   **(k)** $K = \bigcup_{n=1}^{\infty} \left[\frac{1}{n}, \frac{n}{n+1} \right]$;   **(l)** $L = \bigcap_{n=1}^{\infty} \left[-\frac{1}{n}, \frac{n+1}{n} \right]$;
(m) $M = \bigcap_{n=1}^{\infty} \left(-\frac{1}{n}, \frac{n+1}{n} \right)$;   **(n)** $N = \bigcap_{n=1}^{\infty} \left(-\frac{n}{n+1}, \frac{n+1}{n} \right)$;   **(o)** $V = \{3\}$
(p) P is the union of the set of integers and the set of numbers $\frac{n}{n+1}$ for $n \geq 1$;
(q) Q is the set of rational numbers;
(r) Z is the set of integers;
(s) S is the set of points in the Cantor set which are less than $\frac{1}{2}$.
(t) T is the set of points in the Cantor set which are greater than or equal to $\frac{1}{2}$.
(u) U is the set of points in the Cantor set which are less than $\frac{1}{3}$.

6. Determine whether each sequence converges or diverges. If it converges, find its limit.
(a) $a_n = \frac{n}{n^2 + 1}$ **(b)** $b_n = \frac{2n^3 - 5}{5n^3 + 1}$ **(c)** $c_n = \frac{3n^4 + 5}{7n^2 + 1}$ **(d)** $d_n = \sqrt{\frac{n^2 + 1}{4n^2 + 2}}$
(e) $e_n = \sec \left(\frac{\pi n}{n+1} \right)$ **(f)** $f_n = n^2 \tan \left(\frac{1}{5n^2 + 1} \right)$ **(g)** $g_n = (-1)^n$ **(h)** $h_n = (-1)^n \frac{1}{n}$
(i) $i_n = \frac{\sin n}{n}$ **(j)** $j_n = n \sin \frac{1}{n}$ **(k)** $k_n = \arctan n$ **(l)** $s_n = \sqrt{n}$

7. For each non–closed set in Exercise 5 above, find a convergent sequence of elements in that set whose limit is not in the set.

8. Verify the following analogue of Proposition 1.10.24.
(a) Let f be a function whose domain D is an open subset of $\Re$. Then the following conditions are equivalent.
     (i) If $c \in D$ and $N(f(c); \epsilon)$ is a neighborhood of $f(c)$, then there is a neighborhood $N(c; \delta)$ of c such that $f(N(c; \delta)) \subset N(f(c); \epsilon)$.
     (ii) If $\{s_n\}$ is a convergent sequence of points of D with limit $L \in D$, then the sequence $\{f(s_n)\}$ is also convergent with limit $f(L)$.
     (iii) If U is an open subset of $\Re$, then $f^{-1}(U)$ is an open subset of D.
(b) Show that if D is an open interval, then these three conditions are equivalent to the definition of a continuous function in Section 1.7: $\lim_{x \to c} f(x) = f(c)$ for all $c \in D$.

9. Show that the limit of a convergent sequence $\{s_n\}$ is unique.

10. Determine whether each set is bounded below and whether it is bounded above. In (a)–(f), if the set is bounded below determine its greatest lower bound while if the set is bounded above determine its least upper bound.

(a) $A = (-3, 5]$ **(b)** $B = (-5, -1] \cup [6, 11)$
(c) C equals the Cantor set.   **(d)** $D = (-\infty, 2] \cup (7, \infty)$
(e) E is the set of numbers $\frac{n}{n+1}$ for n a positive integer.
(f) F is the set of numbers $(-1)^n \frac{n}{n+1}$ for n a positive integer.
(g) G is the set of numbers $\frac{2n^3+6}{3n^2+7}$ for n a positive integer.
(h) H is a convergent sequence of numbers.
(i) I is the set of roots of the polynomial $x^7 - 5x^4 + 2x^3 - 6x + 8$.

11. Prove Proposition 1.10.33(b): if W_1 and W_2 are both greatest lower bounds of a set of real numbers S, then $W_1 = W_2$.

12. Complete the proof of Completeness Theorem 1.10.34 by proving the case $b_0 < 0$.

13. Complete the proof of Corollary 1.10.35 by showing that if F is a closed subset of $\Re$ which is bounded below, then there is an element $W \in F$ which is a greatest lower bound of F.

14. Determine whether each sequence is a Cauchy sequence. Justify your answers.
(a) $a_n = \frac{1}{n}$ **(b)** $b_n = \frac{n}{n+1}$ **(c)** $c_n = \frac{n^2}{6n+1}$ **(d)** $d_n = n \sin \frac{1}{n}$ **(e)** $e_n = \cot \frac{\pi}{n}$

15. Determine whether each set is complete.
(a) Z **(b)** $B = (-\infty, 4]$ **(c)** $C = (-4, 7]$ **(d)** $D = [-8, -2) \cup (-5, 3]$
(e) $E = \bigcup_{n=1}^{\infty} \left[0, \frac{n}{n+1}\right]$ **(f)** $F = \bigcap_{n=1}^{\infty} \left(-\frac{1}{n}, \frac{n+1}{n}\right)$
(g) G is the set of 0 and all $\frac{1}{n}$ for $n \in Z$.
(h) H is the set of 0 and all x–intercepts of the graph of $y = \sin \frac{1}{x}$

Challenging Problems

1. Show every open subset of $\Re$ is a union of neighborhoods.

2. Show the only subsets of $\Re$ which are both open and closed are $\Re$ and $\emptyset$.

3. Show every nonempty open subset U of $\Re$ can be written as a union of disjoint open intervals: $U = (a_1, b_1) \cup \cdots \cup (a_n, b_n) \cup \cdots$.

4. Let S be a subset of $\Re$. We say a subset U of S is *relatively open in S* if there is an open subset U' of $\Re$ such that $U = U' \cap S$. Similarly, we say a subset G of S is *relatively closed in S* if there is a closed subset G' of $\Re$ such that $G = G' \cap S$.
(a) Give an example of a subset S of $\Re$ and a subset U of S, such that U is relatively open in S, but U is not open in $\Re$. Give an example of another subset G of this set S such that G is relatively closed in S, but G is not closed in $\Re$.
(b) Show if S is an open subset of $\Re$, then a subset U of S is relatively open in S if and only if U is open in $\Re$.
(c) Show if S is a closed subset of $\Re$, then a subset G of S is relatively closed in S if and only if G is closed in $\Re$.
(d) Show if U, V are relatively open subsets of S, then $U \cup V$ and $U \cap V$ are also relatively open subsets of S.
(e) Show if F, G are relatively closed subsets of S, then $F \cup G$ and $F \cap G$ are also relatively closed subsets of S.

5. Let f be a function whose domain contains $[1, \infty)$. Define the sequence $s_n = f(n)$. Show that s is a convergent sequence if $\lim_{x \to \infty} f(x)$ exists and equals a number L.

6. Prove the following generalization of Proposition 1.10.24 which uses the terminology of Exercise 4. Let f be a function whose domain D is any subset of $\Re$. Then the following conditions are equivalent.
(a) If $c \in D$ and $N(f(c); \epsilon)$ is a neighborhood of $f(c)$, then there is a neighborhood $N(c; \delta)$ of c such that $f(N(c; \delta) \cap D) \subset N(f(c); \epsilon)$.

(b) If $\{s_n\}$ is a convergent sequence of elements of D with limit $L \in D$, then the sequence $\{f(s_n)\}$ is also convergent with limit $f(L)$.

(c) If G is a closed subset of $\Re$, then $f^{-1}(G)$ is a relatively closed subset of D.

(d) If U is an open subset of $\Re$, then $f^{-1}(U)$ is a relatively open subset of D.

7. Let $\{x_n\}$ be a Cauchy sequence with limit L. Let N be a positive integer such that $m, n \geq N$ implies $d(x_m, x_n) < \epsilon$. Prove if $n \geq N$, then $d(x_n, L) \leq \epsilon$.

1.11   Connectedness and the IVT

Prerequisite: Section 1.10

As the word suggests, a *connected* subset of $\Re$ consists of one piece. After defining this concept, we show that the image of a closed connected set under a continuous function is also connected. This fact is used to prove the Intermediate Value Theorem.

A set that is not connected is called *disconnected*. We define a disconnected subset C of $\Re$ by specifying the meaning of the phrase: "C consists of more than one piece." We do this by defining the phrase: "C can be broken up into two pieces." Of course, breaking up a set such as $[0,1]$ into $[0, 1/2]$ and $(1/2, 1]$ is not what we have in mind! We want our two pieces to be separated. However, two disjoint closed sets are separated, and we use this idea to make the general definition. That is, we call C disconnected if we can break up C into two pieces, C_1 and C_2, which are contained in two disjoint closed sets F and G as in Figure 1.11.1.

Figure 1.11.1 Disconnected Sets

There is, however, one problem with this proposed definition. We want the set $C = (-1, 0) \cup (0, 1)$ to be disconnected. However, we can not separate the two pieces $(-1, 0)$ and $(0, 1)$ by including them in two disjoint closed sets F and G. Thus, we do not insist that the closed sets $F = [-1, 0]$ and $G = [0, 1]$ be disjoint. Rather, we require that $C \cap F = (-1, 0)$ and $C \cap G = (0, 1)$ be disjoint.

Definition 1.11.2 *A subset C of $\Re$ is disconnected if there are two closed subsets F and G of $\Re$ such that:*

(a) $C \subset F \cup G$;   (b) $C \cap F$, $C \cap G$ are nonempty;   (c) $C \cap F$, $C \cap G$ are disjoint. C is connected if C is not disconnected.

This definition meets our criteria. We have defined a set C to be disconnected if it can be broken up into the two nonempty, disjoint, separated pieces $C \cap F$ and $C \cap G$.

Examples 1.11.3 (1) Show $A = (0, 1] \cup (2, 3)$ is disconnected.

> **Solution** Choose the closed sets $F = [0, 1]$ and $G = [2, 3]$. Then $A \subset F \cup G$ and $F \cap A = (0, 1]$, $G \cap A = (2, 3)$ are nonempty disjoint sets.

(2) As noted above, $(-1, 0) \cup (0, 1)$ is disconnected.

(3) Show the set of rational numbers Q is disconnected.

> **Solution** Choose the closed sets $F = \left(-\infty, \sqrt{2}\,\right]$ and $G = \left[\sqrt{2}, \infty\right)$. Then $Q \cap F$ is the set of rational numbers less than $\sqrt{2}$, and $Q \cap G$ is the set of rational numbers greater than $\sqrt{2}$. These sets are nonempty and disjoint.

(4) Show the Cantor set C is disconnected.

> **Solution** Since $\frac{1}{2}$ is not in C, choose the closed sets $F = [0, 1/2]$ and $G = [1/2, 1]$. Then $F \cap C$, $G \cap C$ are nonempty and disjoint.

(5) Show the set $B = \{r\}$ is connected.

> **Solution** A set with two disjoint nonempty subsets has at least two elements. Since B has only one element, it is not disconnected, i.e. B is connected. $\square$

We can produce many examples of disconnected sets. However, Example 5 above is our only example of a connected set. We use the Completeness Theorem to show that all intervals are connected.

Proposition 1.11.4 *Every interval I is connected.*

Proof Assume I is disconnected. We deduce a contradiction. Let F, G be closed sets such that $I \subset F \cup G$, and $I \cap F$, $I \cap G$ are nonempty and disjoint. Choose $a \in I \cap F$ and $b \in I \cap G$. Interchange the names of F and G, if necessary, so that $a < b$.

$$\text{Let } S \text{ be the set of } x \in I \cap F \text{ such that } x < b.$$

Note $a \in S$, and S is bounded above by b. By the Completeness Theorem, S has a least upper bound c. Since $a \le c \le b$ and a, b are in the interval I, it follows that $c \in I$. Either $c \in F$ or $c \in G$. If $c \in F$, then c is in the open set G^C. Find a neighborhood $N(c; r)$ which is a subset of G^C. Then every number of $N(c; r)$, including $c + \frac{r}{2}$, is in S. This contradicts the fact that c is an upper bound of S. Therefore $c \in G$. Since c is in the open set F^C, find a neighborhood $N(c; t)$ which is a subset of F^C. None of the numbers in $N(c; t)$ are in S. Hence there are smaller upper bounds of S than c, such as $c - \frac{t}{2}$. This is impossible because c is the least upper bound of S. Thus our assumption that I is disconnected is wrong, i.e. I is connected. $\square$

In fact, intervals are the only connected subsets of $\Re$. We leave the proof as a Challenging Problem. We show next that the continuous image of a closed connected set is connected.

Proposition 1.11.5 *Let f be a continuous real valued function whose domain D is a closed connected subset of $\Re$. Then the range $f(D)$ of f is also connected.*

Proof Assume $f(D)$ is disconnected. Let F and G be closed sets such that $f(D) \subset F \cup G$, and $f(D) \cap F$, $f(D) \cap G$ are nonempty and disjoint. Since f is continuous with a closed domain, the sets $f^{-1}(F)$ and $f^{-1}(G)$ are closed. Since $f(D) \subset F \cup G$,

$$D = f^{-1}(f(D)) = f^{-1}(F \cup G) = f^{-1}(F) \cup f^{-1}(G).$$

If $d_1 \in D$ with $f(d_1) \in f(D) \cap F$, then $d_1 \in f^{-1}(F)$. If $d_2 \in D$ with $f(d_2) \in f(D) \cap G$, then $d_2 \in f^{-1}(G)$. Hence $f^{-1}(F)$ and $f^{-1}(G)$ are nonempty. Also,

$$f^{-1}(F) \cap f^{-1}(G) = f^{-1}(F \cap G) = f^{-1}[(F \cap f(D)) \cap (G \cap f(D))] = f^{-1}(\emptyset) = \emptyset.$$

Thus $f^{-1}(F)$ and $f^{-1}(G)$ are disjoint. We have shown that the connected set D is disconnected! This contradiction implies that our assumption "$f(D)$ is disconnected" is false. Therefore $f(D)$ is connected. $\square$

This proposition is a special case of the more general fact: the continuous image of a connected set is connected. Since we do not need this result here, we give it as a

Challenging Problem. The Intermediate Value Theorem is an immediate consequence of Proposition 1.11.5.

Theorem 1.7.19 (Intermediate Value Theorem) *Let f be a continuous function whose domain is an interval. Assume the range of f contains the numbers A and B with $A < B$. Then the range of f contains the entire interval $[A, B]$.*

Proof Let $A = f(a)$ and $B = f(b)$. Interchange the definitions of a and b, if necessary, so that $a < b$. Assume $c \in [A, B]$ is not in $f[a, b]$. Choose the closed sets. $F = (-\infty, c]$ and $G = [c, \infty)$. Since $F \cup G = \Re$, $f[a, b] \subset F \cup G$. Note $A \in f[a, b] \cap F$ and $B \in f[a, b] \cap G$. Hence both $f[a, b] \cap F$ and $f[a, b] \cap G$ are nonempty. $f[a, b] \cap F$ and $f[a, b] \cap G$ are disjoint because $c \notin f[a, b]$. Thus $f[a, b]$ is disconnected which contradicts Proposition 1.11.5. Therefore every $c \in [A, B]$ is in $f[a, b]$, a subset of the range of f, i.e. $[A, B] \subset$ Range f. $\square$

Summary

The meaning of a connected set and a disconnected set should be understood. The reader should be able to demonstrate why a given subset of $\Re$ is either connected or disconnected. She should know why intervals are connected and why the continuous image of a connected closed set is connected. The proof of the Intermediate Value Theorem should be understood.

Basic Exercises

1. Determine whether each set is connected or disconnected:
   (a) $[5, 11)$;   **(b)** $(-\infty, 8)$;   **(c)** $[4, \infty)$;
   (d) $(-7, 2] \cup [5, 9)$;   **(e)** $[-1, -2] \cup [-3, 5]$; **(f)** $(-4, 2) \cup [2, 6)$;
   (g) $(1, 4) \cup (4, 7)$;   **(h)** the set of integers Z; **(i)** the set I of irrational numbers;
   (j) the complement Z^C of the set of integers;
   (k) the set consisting of 0 and all $\frac{1}{n}$ for n a positive integer.

2. Describe all finite connected subsets of $\Re$.

3. Show $A \subset \Re$ is disconnected if and only if there are open sets U, V in $\Re$, such that:
   (i) $A \subset U \cup V$ **(ii)** $U \cap A$, $V \cap A$ are nonempty;   **(iii)** $U \cap A$, $V \cap A$ are disjoint.

4. Let A, B be connected subsets of $\Re$ with $A \cap B \neq \emptyset$. Show $A \cup B$ is also connected.

5. Show if A, B are connected subsets of $\Re$, then $A \cap B$ is also connected.

6. Give an example of a continuous function with disconnected domain whose range is connected.

Challenging Problems

1. Show every connected subset of $\Re$ is an interval.

2. **(a)** Show the image of an interval under a continuous function is connected.
   (b) Let f be a continuous real valued function with domain an interval. Show Range f is also an interval.

3. Use the terminology of Challenging Problem 4 of Section 10 to show these four statements are equivalent for a subset A of $\Re$.
   (a) A is disconnected.
   (b) $A = U \cup V$ where U, V are nonempty disjoint relatively open subsets of A.
   (c) $A = F \cup G$ where F, G are nonempty disjoint relatively closed subsets of A.
   (d) There is a nonempty subset B of A, $B \neq A$, which is both relatively open and relatively closed in A.

4. Show A is a disconnected subset of $\Re$ if and only if there is a continuous real–valued function f with domain A and range $\{0, 1\}$.

5. Show if f is a continuous real valued function whose domain is a connected subset of $\Re$, then the range of f is also connected.

6.(a) Let A be a subset of $\Re$. Show there is a smallest closed set $\overline{A}$ which contains A. The set $\overline{A}$ is called the *closure* of A.
(b) If A is a connected subset of $\Re$ and $A \subset B \subset \overline{A}$, show B is also connected.

1.12 Compactness and the Maximum Value Theorem

Prerequisite: Section 1.10

A compact set is defined in terms of sequences. We prove the Bolzano–Weierstrass Theorem which says that compactness is equivalent to being closed and bounded. Then we show the continuous image of a compact set is compact. The Maximum Value Theorem is an easy consequence of these two results.

A *compact set* is a subset C of $\Re$ with the feature that every sequence in C has an infinite number of terms which cluster about a number of L of C. See Figure 1.12.1. We call these terms a subsequence, and this subsequence has L as its limit.

Figure 1.12.1 A Compact Set C

Definition 1.12.2 *Let* $s = \{s_n\}$ *be a sequence. A subsequence of s is a sequence which consists of terms of s:*

$$s_{k_1}, s_{k_2}, \ldots, s_{k_n}, \ldots \quad \text{with } k_1 < k_2 < \cdots < k_n < \cdots.$$

Example 1.12.3 Let $s = \{s_n\} = \left\{\frac{1}{n}\right\}$ be the sequence $1, \frac{1}{2}, \frac{1}{3}, \frac{1}{4}, \ldots, \frac{1}{n}, \ldots$
Then s has the subsequence $\{s_{2n}\} = \left\{\frac{1}{2n}\right\} \frac{1}{2}, \frac{1}{4}, \frac{1}{6}, \frac{1}{8}, \ldots, \frac{1}{2n}, \ldots$
s also has the subsequence $\{s_{4n+1}\} = \left\{\frac{1}{4n+1}\right\} \frac{1}{5}, \frac{1}{9}, \frac{1}{13}, \frac{1}{17}, \ldots, \frac{1}{4n+1}, \ldots$ $\square$

We now have the terminology to define a compact set.

Definition 1.12.4 *A subset C of $\Re$ is called compact if every sequence $\{s_n\}$ of elements of C has a convergent subsequence with limit an element of C.*

We apply this definition to show that certain sets are not compact. However, it is not reasonable to show directly from this definition that a nontrivial set is compact.

Examples 1.12.5 (1) Show a finite set of numbers $C = \{c_1, \ldots, c_t\}$ is compact.

> **Solution** A sequence $s = \{s_n\}$ in C is an infinite list. Thus there must be some element of C, say c_r, which is listed an infinite number of times. The elements of s which equal c_r define a subsequence with limit c_r. Therefore every sequence in C has a convergent subsequence, and C is compact.

(2) Show the interval $[1, 4)$ is not compact.

> **Solution** Note every subsequence of the sequence $\{4 - 1/n\}$ converges to 4 and $4 \notin [1, 4)$. Hence $[1, 4)$ is not compact.

(3)   Show the interval $[-5, \infty)$ is not compact.

Solution Note the sequence $\{n\}$ has no convergent subsequence. Hence $[-5, \infty)$ is not compact. $\square$

To find examples of infinite compact sets we need an alternate description of compact sets which is easier to apply to examples. A practical description of compact sets is given by the Bolzano–Weierstrass Theorem.

Theorem 1.12.6 (Bolzano–Weierstrass Theorem)   *A subset C of the real numbers $\Re$ is compact if and only if C is a bounded closed set.*

Proof Assume C is a compact set. Let s be a sequence of elements of C with limit L. Since C is compact, s has a subsequence which converges to an element of C. The terms of this subsequence are also terms of the sequence s. Hence this subsequence has limit L, and $L \in C$. Thus C is a closed set. If C does not have an upper bound, then for each positive integer n, there is a number $s_n \in C$ with $s_n > n$. Clearly the sequence $\{s_n\}$ does not have a convergent subsequence. Thus, C must have an upper bound. Similarly C must have a lower bound, and C is bounded.

Assume C is a bounded closed set. Let $s = \{s_n\}$ be a sequence of elements of C. Since C is bounded we can find an interval $(a, b]$ with a, b integers and $C \subset (a, b]$. Divide this interval into subintervals of unit length to find an interval $I_0 = (b_0, b_0 + 1]$, with b_0 an integer, which contains an infinite number of elements of the sequence s. Assume $b_0 \geq 0$. (If $b_0 < 0$, a similar argument applies. We leave the proof of this case as an exercise.) Proceed as in the proof of the Completeness Theorem: divide I_0 into ten subintervals of length $\frac{1}{10}$ and select one $I_1 = \left(u_1, u_1 + \frac{1}{10}\right]$ which contains an infinite number of elements of s. Write $u_1 = b_0.b_1$. Iterate this process to obtain intervals $I_n = (u_n, u_n + 10^{-n}]$ which contain an infinite number of elements of s. Write $u_n = b_0.b_1 \cdots b_n$. Since I_n contains an infinite number of elements of s, define a subsequence $t = \{t_n\}$ of s by selecting $t_n = s_{k_n} \in I_n$ with $k_n > k_{n-1}$. Since

$$I_0 \supset I_1 \supset I_2 \supset \cdots \supset I_{n-1} \supset I_n,$$

we have $s_{k_n} \in I_q$ for $q \leq n$. Let L be the number with decimal expansion

$$L = b_0.b_1 b_2 \cdots b_n \cdots.$$

We show the sequence t converges to L. Let $N(L; r)$ be a neighborhood of L. Select a positive integer n with $10^{-n} < r$. Since the distance between L and any point of I_n is at most 10^{-n}, we have

$$t_k \in I_k \subset I_n \subset N(L; 10^{-n}) \subset N(L; r) \text{ for } k \geq n.$$

Hence the sequence t converges to L. Since each $t_n = s_{k_n} \in C$ and C is a closed set, it follows that $L \in C$. Thus C is compact. $\square$

We use the Bolzano–Weierstrass Theorem to find examples of infinite compact sets.

Examples 1.12.7 (1)   Every closed interval $[a, b]$ is compact, because it is closed and bounded.

(2)   Let F be the set consisting of 0 and all the numbers $1/n$ for n a positive integer. Clearly F is bounded. The complement of F is the open set

$$(-\infty, 0) \cup (1, \infty) \cup \bigcup_{n=1}^{\infty} \left(\frac{1}{n+1}, \frac{1}{n}\right).$$

Thus F is a closed set. Therefore F is compact.

(3) The Cantor set C is a subset of $[0,1]$. Hence C is bounded. We showed in Example 1.10.18 (3) that C is a closed set. Thus C is compact. □

We show that the continuous image of a compact set is compact.

Proposition 1.12.8 *Let f be a continuous function whose domain C is a compact subset of $\Re$. Then the range $f(C)$ of f is also compact.*

Proof Let $s = \{s_n\}$ be a sequence of elements of $f(C)$. We show s has a subsequence which converges to an element of $f(C)$. Write $s_n = f(t_n)$ with $t_n \in C$. The sequence $t = \{t_n\}$ in the compact set C must have a subsequence $\{t_{k_n}\}$ with limit $L \in C$. By the Bolzano–Weierstrass Theorem, C is a closed set. Since f is a continuous function, the sequence $\{f(t_{k_n})\} = \{s_{k_n}\}$ converges to $f(L) \in f(C)$ by Proposition 1.10.24. Since this sequence is a subsequence of s, we have shown that $f(C)$ is compact. □

Example 1.12.9 Let B be the set of $\sin x$ for x an element of any one of the three sets A of Examples 1.12.7. Since A is compact and the sine function is continuous, B is compact by Proposition 1.12.8. □

We are now ready to prove the Maximum Value Theorem. We prove a more general version than the theorem stated in Section 7. That is, we allow the domain of our function to be any compact set – it need not be a closed interval.

Theorem 1.12.10 (Maximum Value Theorem) *Let f be a continuous function with domain a compact set C. Then there are numbers m, M in C such that for all $x \in C$:*

$$f(m) \le f(x) \le f(M). \tag{1.12.1}$$

Proof By Proposition 1.12.8, $f(C)$ is compact. Therefore $f(C)$ is closed and bounded. Let B be the least upper bound of $f(C)$, and let W be the greatest lower bound of $f(C)$. Since $f(C)$ is closed, it is complete by Corollary 1.10.35. Hence B, W are elements of $f(C)$. Write $B = f(M)$ and $W = f(m)$. Equation (1.12.1) is the statement that $f(m)$ is a lower bound of $f(C)$ and $f(M)$ is an upper bound of $f(C)$. □

Historical Remarks

The discipline of *topology* developed in the twentieth century from several diverse concepts which were exposited in the nineteenth century, often with little regard for rigor. The branch of topology called *point set topology* incorporates the concepts we studied in Sections 10–12. It arose from the "proofs" of the Intermediate Value Theorem by Bernard Bolzano and Augustin–Louis Cauchy in the early nineteenth century and from the independent rigorous constructions of the real numbers by Georg Cantor, Richard Dedekind, Eduard Heine and Charles Meray in 1872.

Bolzano's "proof" of the Intermediate Value Theorem introduced several new concepts. (The major problem with his work is the lack of a rigorous construction of the real numbers. Once this construction was made in 1872, it was straightforward to complete Bolzano's proof.) Bolzano proved that a set of real numbers with an upper bound has a least upper bound, Theorem 1.10.34. This was generalized by Karl Weierstrass in the 1860s who proved that a set of real numbers in which every sequence has a convergent subsequence is the same as a set which is closed and bounded, Theorem 1.12.6. This theorem is now known as the Bolzano-Weierstrass Theorem. Their proofs are similar to the ones we presented. In particular, they introduced the modern definition of continuity, the completeness property of the real numbers and the concept of a compact set. The work of Heine in the 1870s and of Emile Borel in 1894 led to an alternative description of compactness in terms of open covers given in Problem 1.

The construction of the real numbers from the rational numbers is based on the concept of completeness. Our formulation in Theorem 1.10.34 follows Dedekind's approach. We can partition the rational numbers into two sets A and B such that every element of A is less than every element of B, yet A has no largest element and B has no smallest element. (For example, let A be the set of rational numbers which are negative or have square less than two. Let B be the set of positive rational numbers which have square greater than two.) However, it is not possible to partition the set of real numbers $\Re$ in this way. This is Dedekind's definition that $\Re$ *is complete*. Cantor's study of the properties of $\Re$ include the first attempt to define a connected set.

In 1906, Grace and William Young published *The Theory of Sets of Points*. They abstracted many of the nineteenth century topological ideas such as closed sets and connected sets. In addition, they generalized these concepts as well as the Bolzano–Weierstrass and Heine–Borel Theorems to the plane. This was followed by the work of Maurice Fréchet who generalized these concepts to higher dimensions and to sets of functions. The use of open sets in topology and their role in defining a general topological space was initiated by Felix Hausdorff in his 1914 text, *Foundations of Set Theory*. For example, he defined a function to be continuous if the inverse image of an open set is always open. He used this formulation to prove the continuous image of a connected set is connected, Theorem 1.11.5, and the continuous image of a compact set is compact, Theorem 1.12.8. He noted that the former result implies the Intermediate Value Theorem, and that the latter result implies the Maximum Value Theorem. Our exposition, in Sections 11 and 12, follows this approach.

Summary

The definition of a compact set should be understood. The reader should be able to apply it to show that a set is not compact. She should know the proof of the Bolzano–Weierstrass Theorem and be able to apply it to determine whether a given set is compact. The proof of the Maximum Value Theorem should also be understood.

Basic Exercises

1. Define two different subsequences of each sequence.

    (a) $\{n^3\}$ **(b)** $\left\{\frac{1}{\sqrt{n^2+1}}\right\}$ **(c)** $\left\{\sin\frac{\pi}{n}\right\}$

2. Give an example of a divergent sequence which has a convergent subsequence.

3. Let s be a sequence with limit L. Show every subsequence of s also has limit L.

4. **(a)** Use the definition of compactness to show each of these sets is not compact.
    (i) Z **(ii)** $A = (-3, 5)$ **(iii)** $B = (2, 9]$ **(iv)** $C = (-\infty, 0]$ **(v)** Q
    (vi) D is the set of all rational numbers x with $0 \leq x \leq 1$.
    (vii) E is the set of all reciprocals of positive integers.
    (viii) F is the set of all x–intercepts of the graph of $y = \sin x$.
    (ix) G is the set of all x–intercepts of the graph of $y = \sin\frac{1}{x}$.
    (x) H is the set of all numbers $n \sin\frac{1}{n}$ for n a positive integer.
  (b) Use the Bolzano–Weierstrass Theorem to show the above sets are not compact.

5. Determine whether each of these subsets of $\Re$ is compact:
(a) $[-3, 2]$; **(b)** $(4, 7)$; **(c)** $[-3, \infty)$; **(d)** $[-1, 1] \cup [3, 8]$; **(e)** $[2, 7) \cup (7, 9]$;
(f) $[-8, -2) \cup (-4, 3]$;   **(g)** Z^C;   **(h)** the complement of the Cantor set;
(i) $\bigcap_{n=1}^{\infty}\left(-\frac{1}{n}, \frac{n+1}{n}\right)$;

(j) the set consisting of 1 as well as all the numbers $\frac{n}{n+1}$ for n a positive integer.

6. Show each of these subsets of $\Re$ is compact:

(a) A is the set of all numbers $2x^3 - 5x + 7$ where x is either 0 or of the form $\frac{1}{n^k}$ for n, k positive integers.

(b) B is the set of all numbers $\cos\frac{2\pi}{n}$ for n a positive integer.

(c) D is the set of square roots of all the numbers in the Cantor set.

(d) E is the range of the function $f(x) = \frac{2x^3 - 5x + 7}{x^6 + 5x^4 + 9x^2 + 8}$ with domain $[-1, 1]$.

(e) F is the range of the function $g(x) = 7\cos^3 x - 9\sin^5 x + 17$ with domain $[0, \pi]$.

7. Let F be a closed subset of a compact set C. Use the definition of compactness to show F is compact.

8. Let f be a continuous function whose domain is a compact set D. Show f is a closed function, i.e. if G is a closed subset of D then $f(G)$ is closed.

9. Complete the proof of the Bolzano–Weierstrass Theorem for the case $b_0 < 0$.

10. Let $s = \{s_n\}$ be a sequence of real numbers. A number P is called an *accumulation point* of s if every neighborhood $N(P; r)$ contains an infinite number of elements of s.

(a) Show P is an accumulation point of the sequence s if and only if s has a subsequence which converges to P.

(b) Show a subset C of $\Re$ is compact if and only if every sequence in C has an accumulation point in C.

Challenging Problems

1. Let $A \subset \Re$. We call the set of open sets $\mathbf{U} = \{U_1, \ldots, U_n, \ldots\}$ an *open cover* of A if

$$A \subset U_1 \cup \cdots \cup U_n \cup \cdots$$

Prove the Heine–Borel Theorem: A is compact if and only if every open cover $\mathbf{U}$ of A has a finite subcover, i.e. a subset $\{U_{k_1}, \ldots, U_{k_t}\}$ of $\mathbf{U}$ such that $A \subset U_{k_1} \cup \cdots \cup U_{k_t}$.

2. (a) Show if $\mathbf{U}$ is an open cover of a compact set C, there is a positive number r such that each neighborhood $N(x; r)$, with $x \in C$, is contained in an element U of $\mathbf{U}$.

(b) Let f be a continuous real–valued function with domain a compact set C. Given $\epsilon > 0$, show there is a number $\delta > 0$ such that

$$f(N(x; \delta)) \subset N(f(x); \epsilon) \text{ for all } x \in C. \tag{1.12.2}$$

(c) Find an example of a continuous function f whose domain is not compact and a number $\epsilon > 0$ such that there is no number $\delta > 0$ which satisfies (1.12.2).

3. Let B be a bounded subset of $\Re$.

(a) Use the notation of Section 11, Problem 6 to show the closure $\overline{B}$ of B is compact.

(b) Show if f is a real–valued continuous function with domain B, there is a unique continuous function F with domain $\overline{B}$ such that F restricted to B equals f.

1.13   Review Exercises for Chapter 1

Decide if each of these 50 statements is True or False. Justify your answers.

1. $-\frac{3}{5} \in N \cup Q$ **2.** $I \cup Q = \Re$ **3.** $[-3, -2) \cup [-2, 3] = (-3, 3)$

4. $(-5, 2) \cup (-2, 5) = (-5, 5)$

5. If $f(x) = \sqrt{x^2 - 4x + 5}$, we take the domain of f to be $\Re$.

6. If $g(x) = \frac{x^2 - 1}{x^4 - 1}$, we take the domain of g to be $\Re$.

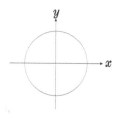

7. $f(x) = \sqrt{1 - x^2}$ has the graph depicted in the margin.

8. If $P(x)$ is a polynomial of degree three, then the range of P is $\Re$.

9. If f is a one–to–one function, then f^{-1} is also one–to–one.

10. $(\arctan x)^3 + y^3 = 1$ defines a function with domain $\Re$.

11. The range of $f(x) = \frac{x^2}{1 + x^2}$ is $\Re$.   **12.** The domain of $g(x) = \sec \frac{x^2}{x^2 + 1}$ is $\Re$.

13. Let $f(x) = \tan x$ and $g(x) = \cos x$, both with the standard domain of $\tan x$. Give $\sin x$ its standard domain. Then $(fg)(x) = \sin x$.

14. If f and g have domain $\Re$, then $f \circ g = g \circ f$.

15. If $f(x) = \frac{1}{x^2}$, then $(f \circ f)(x) = x^4$ with domain $\Re$.

16. If f is a one–to–one function, then $f^{-1}(x) = \frac{1}{f(x)}$.

17. If $g(x) = \sin x$, then $g^{-1}(x) = \csc x$.

18. If $h(x) = \cos x$, then $h^{-1}(x)$ exists and is called $\arccos x$.

19. If $f(x) = -x$ with domain $(-1, 1]$, then $f^{-1}(x) = -x$ with domain $[-1, 1)$.

20. $f \circ g$ is one–to–one if and only if f and g are both one–to–one.

21. A bicycle with wheels of radius 40 cm travels 10 full rotations of its wheels. This bicycle has traveled $10(40)(360) = 144,000$ cm.

22. A pie of circumference 20π in. is cut into ten equal slices. The area of each slice is 10π in^2.

23. $\cot \theta$ is defined when the sides of the angle θ, in standard position, are not both on the x–axis.

24. The function $\tan \circ \sin$ is defined with domain $\Re$.

25. $f(\theta) = \frac{1}{\sin \theta - \cos \theta}$ has domain all $\theta \in \Re$ not of the form $\frac{\pi}{4} + 2n\pi$ with n an integer.

26. $\arctan 1 = \frac{\arcsin 1}{\arccos 1}$ **27.** $\sin(\arcsin x + \arccos x) = 1$ for $-1 \leq x \leq 1$.

28. The function $f(x) = \cos \frac{1}{x}$ has the following graph.

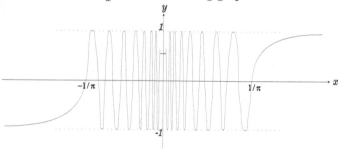

29. The function $g(x) = x \cos \frac{1}{x}$ has the following graph.

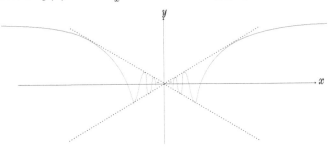

30. If $\lim_{x \to 3^-} f(x) = -4$ and $\lim_{x \to 3^+} f(x) = +4$, then $\lim_{x \to 3} |f(x)|$ exists.

31. $\lim_{x \to \infty} f(x) = \lim_{x \to 0} f\left(\frac{1}{x}\right)$.

32. If $\lim_{x \to \infty} [f(x) + g(x)]$ and $\lim_{x \to \infty} f(x)$ exist, then $\lim_{x \to \infty} g(x)$ exists.

33. If $g(x) = x^2$, then $\lim_{x \to -\infty} g(x) = \infty$ because $g(I_\epsilon(-\infty)) \subset S_\epsilon(\infty)$ for $\epsilon \geq 1$.

34. Let $f(x) = 4x - 7$. We know $\lim_{x \to 3} f(x) = 5$ because $f(I_{\epsilon/4}(3)) \subset S_\epsilon(5)$ for $\epsilon > 0$.

35. $\lim_{x \to 2} \dfrac{x^2 - 4}{x^3 - 8}$ does not exist.   **36.** $\lim_{\theta \to \pi} \cot \theta$ does not exist.

37. $\lim_{\theta \to \frac{\pi}{2}^+} \dfrac{|\tan \theta|}{\tan \theta}$ does not exist.   **38.** If $f(x) = x^2$ and $g(x) = \frac{|x|}{x}$, then $\lim_{x \to 0} (f \circ g)(x) = 1$.

The domain of the function f in Exercises 39, 40 is $(-\infty, c) \cup (c, \infty)$ with $c \in \Re$.
39. If $\lim_{x \to c} f(x) = +\infty$, then f is not one–to–one.
40. If f is continuous and $\lim_{x \to c} f(x) = +\infty$, then f is not one–to–one.

41. Let $f(x) = \begin{cases} x & \text{for } x \text{ rational} \\ x^2 & \text{for } x \text{ irrational} \end{cases}$. Then $\lim_{x \to 3} [f(x) \sin(\pi x)] = 0$.

42. $f(x) = \frac{1}{x^2 - 1}$ is not a continuous function.

43. $g(x) = \begin{cases} \frac{\sin x}{x} & \text{for } x \neq 0 \\ 1 & \text{for } x = 0 \end{cases}$ is a continuous function.

44. $h(x) = \begin{cases} \frac{1 + \cos x}{x - \pi} & \text{for } x \neq \pi \\ 1 & \text{for } x = \pi \end{cases}$ has an essential discontinuity at $x = \pi$.

45. $p(x) = \begin{cases} x^2 - 4 & \text{for } x \text{ rational} \\ 4 - x^2 & \text{for } x \text{ irrational} \end{cases}$ has no removable discontinuities.

46. If f is a continuous function, then it is impossible for the range of f to be $(-\infty, -1) \cup (1, \infty)$.

47. Its impossible for a continuous function to have domain $[0, 2]$ and range $(-1, 1)$.

48. It is impossible for a continuous function to have domain $(0, 2)$ and range $[-1, 1]$.

49. If the range of a continuous function is an interval, then its domain is an interval.

50. If the range of a continuous one–to–one function is an interval, then its domain is also an interval.

Solve each of the following problems.

51. Let $f(x) = \frac{2x-1}{5x+3}$. Find $\lim_{x \to 1} f^{-1}(x)$.

52. Let $f(x) = x^2$ and let $g(x) = \arctan \frac{1}{x}$ for $x \neq 0$ with $g(0) = 0$.
(a) Does $g(x)$ have a removable or essential discontinuity at $x = 0$?
(b) Does $(f \circ g)(x)$ have a removable or essential discontinuity at $x = 0$?

53. Define $f : [0, \infty) \to [0, \infty)$ by $f(x) = \frac{1}{x^2+1}$.
(a) Is f^{-1} a function?   **(b)** Find the domain and range of f^{-1}.

54. Find the range of the function $h(x) = \frac{\sin^2 x}{x}$ with domain $\left[\frac{\pi}{2}, \infty\right)$.

55. **(a)** Show $f(x) = \arcsin x$ is a one–to–one function.
(b) Find the formula and domain for $f^{-1}(x)$.

56. Let $g(x) = x^2 - 4$ for x rational, and let $g(x) = \frac{1}{x^2-4}$ for x irrational.
(a) Is g a function with domain $\Re$?   **(b)** Determine where g is continuous.

57. Let $h(x) = x$ if x is an integer, let $h(x) = x^2$ if x is a rational number, and let $h(x) = x^3$ if x is an irrational number. Is h a function with domain $\Re$?

58. Find an angle θ with $\sin \theta = -\frac{1}{2}$ and $\sec \theta = \frac{2}{\sqrt{3}}$.

59. Simplify the expression $\sin \left(\arctan \frac{x}{x^2+1} + \operatorname{arccot} \frac{x}{x^2+1} \right)$.

60. Find a formula for the composite function $\tan \circ \arcsin$.

61. Use the definition of the limit to show: **(a)** $\lim\limits_{x \to 1} \dfrac{1}{x} = 1$; **(b)** $\lim\limits_{x \to -\infty} x^4 = \infty$.

62. Let $f(x) = \sqrt{3 - x}$ and $g(x) = \sqrt{3 + x}$.
Write (Domain f) $\cup$ (Domain g) in interval notation. Simplify your answer.

63. Find $\lim\limits_{x \to 4} \sqrt{\dfrac{x^2 - x - 12}{x^2 + x - 20}}$.   **64.** Find $\lim\limits_{\theta \to 0} \dfrac{\sqrt{3 + \cos \theta} - \sqrt{4 - \sin \theta}}{\theta}$.

65. Find $\lim\limits_{\theta \to \pi^+} (\csc \theta + \cot \theta)$.   **66.** Find $\lim\limits_{x \to \infty} \left(\sqrt{x^2 + 4} - \sqrt{x^2 + 1} \right)$.

67. Show the function $f(x) = \begin{cases} \frac{x^2-9}{x-3} & \text{for } x < 3 \\ 6 & \text{for } x = 3 \\ \frac{24x-72}{x^2-2x-3} & \text{for } x > 3 \end{cases}$ is continuous at $x = 3$.

68. Let $g(x) = \frac{|x \sin x|}{x \sin x}$ for $x \neq 0$ with $g(0) = 0$. Determine whether g is continuous, has a removable discontinuity or has an essential discontinuity at $x = 0$.

69. Estimate each root of $p(x) = x^3 + x^2 - 4x + 1$ to the nearest $\frac{1}{4}$.

70. What does the Maximum Value Theorem say about the function $f(x) = \frac{|2x+1|}{2x+1}$ on the interval $[-1, 1]$?

71. Let $f(x) = \sqrt{x}$, $g(x) = \arcsin x$ and $h(x) = \arccos x$.
(a) Is the function $f \circ g$ defined?   **(b)** Is the function $f \circ h$ defined?

72. Assume that fg and $\frac{f}{g}$ are defined with $(fg)(x) = x^3 + 2x^2 + x$ and $\left(\frac{f}{g}\right)(x) = x$. Find $f(x)$ and $g(x)$. What must be true about the domain of g?

73. f is continuous, one–to–one, Domain $f = \Re$ and $f(3) = 2$. Find $\lim\limits_{x \to 2} \dfrac{1}{f^{-1}(x)}$.

74. Find the domain and range of $f(x) = \frac{1}{\arctan x}$.

75. What is the conclusion of the Maximum Value Theorem for the function $f(x) = \frac{1}{3x^8 + 2x^4 + 7x^2 + 5}$ with domain $[0, 1]$?

Chapter 2

Differentiation

2.1 Introduction

This chapter establishes the basic results of differential calculus. The core of this subject consists of fundamental methods for computing the slopes of tangent lines to the graph of $y = f(x)$. These slopes are the values of the derivative of f. Mathematical applications of these computations include curve sketching, finding the maximum and minimum values of functions and computing limits. Applications to the sciences and social sciences arise from interpreting the derivative as the rate at which the function is changing. Thus differential calculus is a method for studying quantities which change.

The first part of the *Concepts* portion of this chapter defines derivatives and derives methods for computing them. Section 2 defines the slope of a tangent line T to the graph of f as the limit of slopes of secant lines which approximate T. The methods of evaluating limits, developed in Section 1.6, are used to compute slopes of tangent lines for simple functions. Section 3 establishes formulas for the derivatives of sums, products and quotients of functions whose derivatives are known. In Section 4, the chain rule is used to compute the derivative of a composite function. The formulas of Sections 3 and 4 allow us to find the derivatives of complicated functions from the derivatives of the simple functions computed in Section 2. In this way we compute the derivatives of polynomials, rational functions and functions constructed from them using fractional powers. In Section 5, we compute the derivatives of the trigonometric and inverse trigonometric functions.

The second part of the *Concepts* portion of this chapter is devoted to the Mean Value Theorem and its mathematical applications. The Mean Value Theorem itself is presented in Section 6. The First Derivative Test is derived to locate and identify maxima and minima. Section 7 presents a systematic method to assemble information about a function f to sketch its graph. This information includes the asymptotes of f, which are defined as limits. In addition, the Mean Value Theorem interprets the first derivative of f to determine where f is increasing, where f is decreasing, and where f has local maxima and minima. The second derivative is also interpreted to determine the concavity and inflection points of f. Section 8 is devoted to L'Hôpital's Rule, a method which uses our ability to compute derivates to evaluate nontrivial limits.

Section 9 of the *Applications* portion of this chapter lays the groundwork for applying derivatives to various contexts by interpreting the derivative of f as its rate of change. We illustrate this viewpoint by showing that the derivative describes the relationship between certain fundamental concepts in physics and economics. Section 10 uses the chain rule to solve related rates problems. Two applications of the derivative to make approximations are presented in Section 11. First, differentials are used to approximate the change in the value of a function. Then we present the Newton–Raphson method to approximate the roots of a function. In Section 12, we compute absolute minima and absolute maxima to solve practical problems.

The *Theory* portion of this chapter consists of Section 13 which gives rigorous proofs of three theorems. Two of them, the Mean Value Theorem and L'Hôpital's Rule, were presented intuitively in Sections 6 and 8. We also prove Cauchy's Mean Value Theorem, a generalization of the Mean Value Theorem to curves. This theorem is needed to prove L'Hôpital's Rule.

The flow chart in Figure 2.1.1 indicates the dependence of the sections of the *Applications* and *Theory* portions of this chapter on preceding sections.

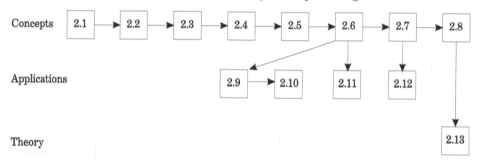

Figure 2.1.1 The Sections of Chapter 2

2.2   Tangent Lines and Derivatives

The derivative of the function f at the point x_0 of its domain is the slope of the tangent line to the graph of f at the point $P = (x_0, f(x_0))$. We begin by giving a precise definition of this tangent line. Our definition gives the value of its slope as a limit, and we compute the equation of the tangent line by evaluating this limit. In the second subsection, we study the derivative as a function and show that a derivative can be computed by evaluating the limit that defines it. We show that if the derivative of f exists, then f is continuous. In the last subsection, we define one–sided derivatives. We apply them to study half–tangent lines at endpoints of graphs and to identify corners, called *cusps*, on the graph of f.

Tangent and Normal Lines

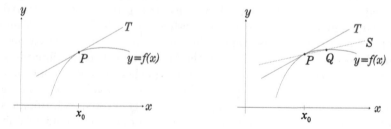

Figure 2.2.1 The Tangent Line T and a Secant Line S

Consider the problem of finding the equation of the tangent line T to the graph of the function $y = f(x)$ at the point $P = (x_0, f(x_0))$ as depicted in the left diagram of

Figure 2.2.1. If we knew the slope m of T, then the equation of T would be given by
the point–slope formula:
$$y = f(x_0) + m(x - x_0) . \tag{2.2.1}$$
To compute the slope of a line we need to know two points on the line, and we only
know one point P on the line T. The key idea is that we can approximate the slope m
of T by choosing a point Q on the graph of f near P. The *secant line* S through the
points P and Q approximates T, and the slope of S approximates the slope m of T.
See the right diagram in Figure 2.2.1.

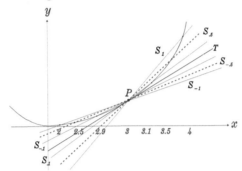

Figure 2.2.2 Six Secant Lines to $f(x) = x^2$ at $x = 3$

Motivating Example 2.2.3 Find the equation of the tangent line T, at the point
$P = (3, 9)$, to the graph of the function $f(x) = x^2$ with domain $\Re$.

Solution The graph of f is the parabola $y = x^2$. If $Q = (x, y)$ is a point on this
parabola, then the slope of the secant line PQ is:
$$\text{slope } PQ = \frac{y - 9}{x - 3} .$$

We make several choices of the point Q to compare the slopes of the secant lines they
determine. To avoid confusion we let Q_h designate the point on the parabola $y = x^2$
whose x–coordinate is h units from $x = 3$, i.e.
$$Q_h = (3 + h, f(3 + h)) = (3 + h, (3 + h)^2).$$

Note that Q_h is to the right of $x = 3$ for $h > 0$ while Q_h is to the left of $x = 3$ for
$h < 0$. We designate the corresponding secant line PQ_h by S_h. In the following table
we select three points $Q_1, Q_{.5}, Q_{.1}$ on this parabola to the right of P and three points
$Q_{-1}, Q_{-.5}, Q_{-.1}$ on this parabola to the left of P. For these six Q_h, we compute the
slopes of the corresponding secant lines PQ_h.

h	1	0.5	0.1	-1	-0.5	-0.1
$3 + h$	4	3.5	3.1	2	2.5	2.9
$y = (3 + h)^2$	16	12.25	9.61	4	6.25	8.41
Q_h	$Q_1 =$ $(4, 16)$	$Q_{.5} =$ $(3.5, 12.25)$	$Q_{.1} =$ $(3.1, 9.61)$	$Q_{-1} =$ $(2, 4)$	$Q_{-.5} =$ $(2.5, 6.25)$	$Q_{-.1} =$ $(2.9, 8.41)$
slope S_h	7	6.5	6.1	5	5.5	5.9

For example, if $h = +.1$, then $3 + h = 3.1$, $y = (3 + h)^2 = (3.1)^2 = 9.61$ and
$Q_{.1} = (x, y) = (3.1, 9.61)$. Hence the slope of the secant line $S_{.1} = PQ_{.1}$ is:
$$\frac{9.61 - 9}{3.1 - 3} = \frac{0.61}{.1} = 6.1$$

These six secant lines depicted in Figure 2.2.2 indicate that the closer the point Q_h is
to P, the better the secant line S_h approximates T. Then the table above suggests:

as $3 + h$ is chosen closer and closer to 3, slope PQ_h gets closer and closer to 6.

Thus it seems that the slope m of T is 6. Since $P = (3,9)$ is a point on T, the equation of T is given by the point–slope formula:

$$y \;=\; 9 + 6(x - 3) \;=\; 6x - 9 \;.$$

To complete this derivation of the equation of T we must justify the observation we made above that as $3 + h$ gets closer and closer to 3, the slope of the secant line PQ_h gets closer and closer to 6. Note that saying $3 + h$ gets closer and closer to 3 is equivalent to saying that h gets closer and closer to zero. Thus we must show that

$$\lim_{h \to 0} (\text{slope } PQ_h) \;=\; 6.$$

Since $P = (3,9)$ and $Q_h = (3 + h, (3 + h)^2)$,

$$\text{slope } PQ_h \;=\; \frac{(3+h)^2 - 9}{(3+h) - 3} \;=\; \frac{(h^2 + 6h + 9) - 9}{h} \;=\; \frac{h^2 + 6h}{h} \;=\; \frac{h(h+6)}{h} \;=\; h + 6 \;.$$

Hence $\qquad\qquad$ $\text{slope } T \;=\; \displaystyle\lim_{h \to 0} (\text{slope } PQ_h) \;=\; \lim_{h \to 0} (h+6) \;=\; 6.$ $\qquad$ □

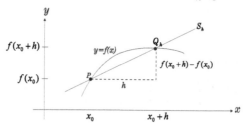

Figure 2.2.4 Computing Slope S_h

The method used to compute the slope of the tangent line of the parabola in Example 2.2.3 generalizes to compute the slope of the tangent line to any function $y = f(x)$ at $x = x_0$. The point on the graph of f at $x = x_0$ is $P = (x_0, f(x_0))$. Let $x_0 + h$ be a point on the x–axis near x_0, i.e h is a small number. The point on the graph of f at $x = x_0 + h$ is $Q_h = (x_0 + h, f(x_0 + h))$. The slope of the secant line $S_h = PQ_h$ through P and Q_h, as depicted in Figure 2.2.4, is given by:

$$\text{slope } S_h \;=\; \frac{f(x_0 + h) - f(x_0)}{(x_0 + h) - x_0} \;=\; \frac{f(x_0 + h) - f(x_0)}{h}.$$

Note the closer Q_h is to P, the better the slope of S_h approximates the slope m of T. The number $|h|$ determines how close Q_h is to P: the smaller $|h|$ is, the closer Q_h is to P. Thus, as h approaches zero, S_h approaches T. See Figure 2.2.5. Therefore,

$$m \;=\; \text{slope } T \;=\; \lim_{Q_h \to P} (\text{slope } S_h) \;=\; \lim_{h \to 0} (\text{slope } S_h) \;=\; \lim_{h \to 0} \frac{f(x_0 + h) - f(x_0)}{h}.$$

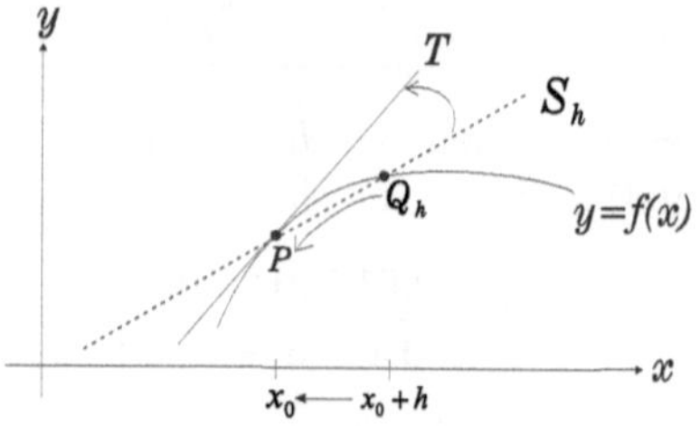

Figure 2.2.5 S_h approaches T as Q_h approaches P

The preceding analysis assumes we know that the tangent line T to the graph of f exists at $x = x_0$. Then we used our intuition to describe the slope m of T as the limit above. However, we do not have a definition of a tangent line to a curve and certainly can not assume that a tangent line exists at every point of the graph of every function. Therefore, we use the above limit for m to *define* the tangent line T of f at $x = x_0$. When the limit exists, we say that T exists and has slope m.

Definition 2.2.6 *Let f be a real valued function whose domain includes the open interval I with $x_0 \in I$. The tangent line T to the graph of f at the point $P = (x_0, f(x_0))$ exists if the limit*

$$m = \lim_{h \to 0} \frac{f(x_0 + h) - f(x_0)}{h} \qquad (2.2.2)$$

exists with m a number. In this case, T is the line with equation

$$y = f(x_0) + m(x - x_0). \qquad (2.2.3)$$

We say f is differentiable at $x = x_0$. The number m is called the derivative of f at x_0.

Observe that we require the domain of f to contain an open interval which contains x_0 so that the two–sided limit (2.2.2) is defined. In the following examples we use limit (2.2.2) to establish the existence of several tangent lines and compute their slopes.

Examples 2.2.7 (1) Find the equation of the tangent line T to the graph of the function $f(x) = x^3$ at the point $P = (2, 8)$.

Solution By Definition 2.2.6, the slope m of the tangent line T is given by:

$$m = \lim_{h \to 0} \frac{f(2 + h) - f(2)}{h} = \lim_{h \to 0} \frac{(2 + h)^3 - 8}{h} = \lim_{h \to 0} \frac{(h^3 + 6h^2 + 12h + 8) - 8}{h}$$

$$= \lim_{h \to 0} \frac{h^3 + 6h^2 + 12h}{h} = \lim_{h \to 0} \frac{h(h^2 + 6h + 12)}{h} = \lim_{h \to 0} (h^2 + 6h + 12) = 12 \ .$$

Thus T exists, and by (2.2.3) the equation of T is:

$$y = 8 + 12(x - 2) \quad \text{or} \quad y = 12x - 16 \ .$$

(2) Let $g(x) = \frac{1}{x}$ with domain $(0, \infty)$. Find the equation of of the tangent line T to the graph of g at the point $P = \left(2, \frac{1}{2}\right)$.

Solution By Definition 2.2.6, the slope m of the tangent line T is given by:

$$m = \lim_{h \to 0} \frac{g(2 + h) - g(2)}{h} = \lim_{h \to 0} \frac{\frac{1}{2+h} - \frac{1}{2}}{h} = \lim_{h \to 0} \frac{\frac{2 - (2+h)}{2(2+h)}}{h} = \lim_{h \to 0} \frac{-h}{2h(2 + h)}$$

$$= \lim_{h \to 0} \frac{-1}{2(2 + h)} = -\frac{1}{4}.$$

Thus T exists, and by (2.2.3) the equation of T is:

$$y = \frac{1}{2} - \frac{1}{4}(x - 2) \quad \text{or} \quad y = 1 - \frac{x}{4}.$$

(3) Let $k(x) = \sqrt{x}$ with domain $[0, \infty)$. Find the equation of the tangent line T to the graph of k at the point $P = (16, 4)$.

Solution By Definition 2.2.6, the slope m of the tangent line T is given by:

$$m = \lim_{h \to 0} \frac{k(16 + h) - k(16)}{h} = \lim_{h \to 0} \frac{\sqrt{16 + h} - 4}{h}.$$

To evaluate this limit we "rationalize the numerator":

$$m \;=\; \lim_{h \to 0} \frac{\sqrt{16+h}-4}{h} \cdot \frac{\sqrt{16+h}+4}{\sqrt{16+h}+4} \;=\; \lim_{h \to 0} \frac{\sqrt{16+h}^{\,2}-4^2}{h(\sqrt{16+h}+4)}$$

$$\;=\; \lim_{h \to 0} \frac{(16+h)-16}{h(\sqrt{16+h}+4)} \;=\; \lim_{h \to 0} \frac{h}{h(\sqrt{16+h}+4)} \;=\; \lim_{h \to 0} \frac{1}{\sqrt{16+h}+4} \;=\; \frac{1}{8}.$$

Thus T exists, and by (2.2.3) the equation of T is:

$$y = 4 + \frac{1}{8}(x-16) \quad \text{or} \quad y = \frac{x}{8} + 2.$$

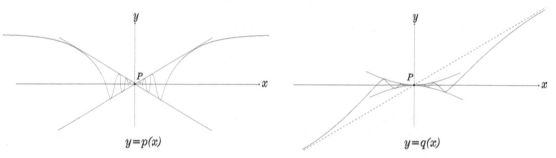

Figure 2.2.8 Example 4

(4) Define two functions p and q with domain $\Re$ by:

$$p(x) = \left\{ \begin{array}{ll} x \sin \frac{1}{x} & \text{if } x \neq 0 \\ 0 & \text{if } x = 0 \end{array} \right\} \quad \text{and} \quad q(x) = \left\{ \begin{array}{ll} x^2 \sin \frac{1}{x} & \text{if } x \neq 0 \\ 0 & \text{if } x = 0 \end{array} \right\}.$$

Determine whether the graphs of p and q have tangent lines at the origin.

Solution The graphs of these functions are sketched in Figure 2.2.8. The graph of p has a tangent line at the origin if the following limit exists:

$$\lim_{h \to 0} \frac{p(0+h)-p(0)}{h} = \lim_{h \to 0} \frac{h \sin \frac{1}{h} - 0}{h} = \lim_{h \to 0} \sin \frac{1}{h}.$$

In Example 1.5.3(4a) we showed that this limit does not exist. Therefore, p does not have a tangent line at the origin. On the other hand, the graph of q has a tangent line at the origin if the following limit exists:

$$m = \lim_{h \to 0} \frac{q(0+h)-q(0)}{h} = \lim_{h \to 0} \frac{h^2 \sin \frac{1}{h} - 0}{h} = \lim_{h \to 0} h \sin \frac{1}{h}.$$

Recall this limit exists and equals zero by Example 1.5.3(4b). Therefore, the tangent line T to the graph of q at the origin exists and has slope 0, i.e. T is the x–axis. □

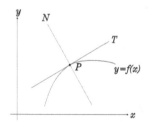

Figure 2.2.9
Normal Line N

The line N in Figure 2.2.9 which passes through the point P on the graph of f and is perpendicular to the tangent line T is called the *normal line* to the graph of f at P. The slope of N is the negative reciprocal of the slope m of T:

$$\text{slope } N = -\frac{1}{m}. \tag{2.2.4}$$

Examples 2.2.10 Find the equation of the normal line N for each of Examples 2.2.7.

Solutions **(1)** For $f(x) = x^3$, the slope of the tangent line T at the point $P = (2, 8)$ is 12. Hence the slope of N is $-\frac{1}{12}$. By the point–slope formula, N has equation:

$$y = 8 - \frac{1}{12}(x - 2) \quad \text{or} \quad y = \frac{49}{6} - \frac{x}{12}.$$

(2) For $g(x) = \frac{1}{x}$, the slope of the tangent line T at the point $P = \left(2, \frac{1}{2}\right)$ is $-\frac{1}{4}$. Hence the slope of N is $-\frac{1}{-1/4} = 4$. By the point–slope formula, N has equation:

$$y = \frac{1}{2} + 4(x - 2) \quad \text{or} \quad y = 4x - \frac{15}{2}.$$

(3) For $k(x) = \sqrt{x}$, the slope of the tangent line T at the point $P = (16, 4)$ is $\frac{1}{8}$. Hence the slope of N is $-\frac{1}{1/8} = -8$. By the point–slope formula, N has equation:

$$y = 4 - 8(x - 16) \quad \text{or} \quad y = 132 - 8x.$$

(4) The tangent line to $q(x) = x^2 \sin\frac{1}{x}$ at the origin is the x-axis. Therefore, the normal line at the origin is the y–axis. $\qquad\square$

The Derivative

We have limited our computations to calculate the slope of one specific tangent line T to the graph of f. However, if the graph of f is a smooth curve, there are tangent lines T_x at each point $P = (x, f(x))$ on this curve. The slope of T_x is called the derivative of f at x. That is, the derivative is the function with the same domain D as f which assigns the slope of the tangent line T_x to each $x \in D$.

Definition 2.2.11 *Let f be a real valued function whose domain D is a union of open intervals. If the tangent line T_x exists at every point $(x, f(x))$ on the graph of f, we say that f is a differentiable function. The derivative of f is the real valued function f' with domain D defined by*

$$f'(x) = \text{slope } T_x = \lim_{h \to 0} \frac{f(x + h) - f(x)}{h}. \tag{2.2.5}$$

Notes **(1)** We require the domain D of f to be a union of open intervals so that the two–sided limit (2.2.5) is defined for each $x \in D$.
(2) By Definition 2.2.6, the values $f'(x)$ of the derivative are numbers and not one of the symobls $-\infty$ or ∞.
(3) In derivative notation, the equation of the tangent line T_c is:

$$y = f(c) + f'(c)(x - c). \tag{2.2.6}$$

There are several common notations for the derivative of f:

$$f'(x) = f^{(1)}(x) = Df(x) = \frac{dy}{dx} = y' = y^{(1)} = \overset{\bullet}{y}.$$

The notation $\frac{dy}{dx}$ for the derivative was introduced by Leibniz who considered the derivative to be the quotient of two "infinitesmal quantities" dy and dx. Newton introduced the notation $\overset{\bullet}{y}$ for the derivative. This was convenient for his applications to mechanics, where no more than two derivatives are needed. The computational methods we used to calculate the derivative of f at a specific point $P = (x_0, f(x_0))$ can also be used to compute the derivative of f at an arbitrary point $(x, f(x))$.

Examples 2.2.12 (1) Find the derivative of the function $f(x) = x$ with domain $\Re$.

Solution By Definition 2.2.11,

$$f'(x) = \lim_{h \to 0} \frac{f(x+h) - f(x)}{h} = \lim_{h \to 0} \frac{(x+h) - x}{h} = \lim_{h \to 0} \frac{h}{h} = \lim_{h \to 0} 1 = 1.$$

(2) Find the derivative of the function $g(x) = x^2$ with domain $\Re$.

Solution By Definition 2.2.11,

$$\begin{aligned}
g'(x) &= \lim_{h \to 0} \frac{g(x+h) - g(x)}{h} = \lim_{h \to 0} \frac{(x+h)^2 - x^2}{h} = \lim_{h \to 0} \frac{(x^2 + 2xh + h^2) - x^2}{h} \\
&= \lim_{h \to 0} \frac{2xh + h^2}{h} = \lim_{h \to 0} \frac{h(2x + h)}{h} = \lim_{h \to 0} (2x + h) = 2x.
\end{aligned}$$

(3) Find the derivative of the function $i(x) = x^3$ with domain $\Re$.

Solution By Definition 2.2.11,

$$\begin{aligned}
i'(x) &= \lim_{h \to 0} \frac{i(x+h) - i(x)}{h} = \lim_{h \to 0} \frac{(x+h)^3 - x^3}{h} \\
&= \lim_{h \to 0} \frac{(x^3 + 3x^2 h + 3xh^2 + h^3) - x^3}{h} = \lim_{h \to 0} \frac{3x^2 h + 3xh^2 + h^3}{h} \\
&= \lim_{h \to 0} \frac{h(3x^2 + 3xh + h^2)}{h} = \lim_{h \to 0} (3x^2 + 3xh + h^2) = 3x^2.
\end{aligned}$$

(4) Find the derivative of the function $j(x) = \frac{1}{x}$ with domain all nonzero numbers.

Solution Use Definition 2.2.11 and subtract the fractions in the numerator:

$$\begin{aligned}
j'(x) &= \lim_{h \to 0} \frac{j(x+h) - j(x)}{h} = \lim_{h \to 0} \frac{\frac{1}{x+h} - \frac{1}{x}}{h} = \lim_{h \to 0} \frac{x - (x+h)}{hx(x+h)} \\
&= \lim_{h \to 0} \frac{-h}{hx(x+h)} = \lim_{h \to 0} \frac{-1}{x(x+h)} = -\frac{1}{x^2}.
\end{aligned}$$

(5) Find the derivative of $k(x) = \frac{x}{3x-4}$ with domain $(-\infty, 4/3) \cup (4/3, \infty)$.

Solution Use Definition 2.2.11 and subtract the fractions in the numerator:

$$\begin{aligned}
k'(x) &= \lim_{h \to 0} \frac{k(x+h) - k(x)}{h} = \lim_{h \to 0} \frac{\frac{x+h}{3(x+h)-4} - \frac{x}{3x-4}}{h} \\
&= \lim_{h \to 0} \frac{(x+h)(3x-4) - x(3x+3h-4)}{h(3x-4)(3x+3h-4)} \\
&= \lim_{h \to 0} \frac{(3x^2 + 3xh - 4x - 4h) - (3x^2 + 3xh - 4x)}{h(3x-4)(3x+3h-4)} \\
&= \lim_{h \to 0} \frac{-4h}{h(3x-4)(3x+3h-4)} \\
&= \lim_{h \to 0} \frac{-4}{(3x-4)(3x+3h-4)} = -\frac{4}{(3x-4)^2}.
\end{aligned}$$

(6) Find the derivative of the function $m(x) = \sqrt{x}$ with domain $(0, \infty)$.

Solution Use Definition 2.2.11 and rationalize the numerator:

$$m'(x) = \lim_{h \to 0} \frac{m(x+h) - m(x)}{h} = \lim_{h \to 0} \frac{\sqrt{x+h} - \sqrt{x}}{h}$$

$$
\begin{aligned}
&= \lim_{h \to 0} \frac{\sqrt{x+h} - \sqrt{x}}{h} \cdot \frac{\sqrt{x+h} + \sqrt{x}}{\sqrt{x+h} + \sqrt{x}} \\
&= \lim_{h \to 0} \frac{\sqrt{x+h}^2 - \sqrt{x}^2}{h(\sqrt{x+h} + \sqrt{x})} = \lim_{h \to 0} \frac{(x+h) - x}{h(\sqrt{x+h} + \sqrt{x})} \\
&= \lim_{h \to 0} \frac{h}{h(\sqrt{x+h} + \sqrt{x})} = \lim_{h \to 0} \frac{1}{\sqrt{x+h} + \sqrt{x}} = \frac{1}{2\sqrt{x}}.
\end{aligned}
$$

(7) Find the derivative of the function $n(x) = \sqrt{2x+5}$ with domain $\left(-\frac{5}{2}, \infty\right)$.

Solution Use Definition 2.2.11 and rationalize the numerator:

$$
\begin{aligned}
n'(x) &= \lim_{h \to 0} \frac{m(x+h) - m(x)}{h} = \lim_{h \to 0} \frac{\sqrt{2(x+h)+5} - \sqrt{2x+5}}{h} \\
&= \lim_{h \to 0} \frac{\sqrt{2x+2h+5} - \sqrt{2x+5}}{h} \cdot \frac{\sqrt{2x+2h+5} + \sqrt{2x+5}}{\sqrt{2x+2h+5} + \sqrt{2x+5}} \\
&= \lim_{h \to 0} \frac{\sqrt{2x+2h+5}^2 - \sqrt{2x+5}^2}{h(\sqrt{2x+2h+5} + \sqrt{2x+5})} = \lim_{h \to 0} \frac{(2x+2h+5) - (2x+5)}{h(\sqrt{2x+2h+5} + \sqrt{2x+5})} \\
&= \lim_{h \to 0} \frac{2h}{h(\sqrt{2x+2h+5} + \sqrt{2x+5})} = \lim_{h \to 0} \frac{2}{\sqrt{2x+2h+5} + \sqrt{2x+5}} \\
&= \frac{2}{2\sqrt{2x+5}} = \frac{1}{\sqrt{2x+5}}. \qquad \square
\end{aligned}
$$

Intuitively, it is clear that if the tangent line to the graph of a function f exists at $x = c$, then the graph of f is smooth near $(c, f(c))$. In particular, we show that f must be continuous at $x = c$.

Proposition 2.2.13 *Let f be a real valued function whose domain contains an open interval I. If $c \in I$ and f is differentiable at $x = c$, then f is continuous at $x = c$.*

Proof Observe that

$$
\begin{aligned}
\lim_{h \to 0} f(c+h) - f(c) &= \lim_{h \to 0} [f(c+h) - f(c)] = \lim_{h \to 0} h \cdot \lim_{h \to 0} \frac{f(c+h) - f(c)}{h} \\
&= 0 \cdot f'(c) = 0.
\end{aligned}
$$

Hence $\qquad \lim_{h \to 0} f(c+h) = f(c)$.

Let $x = c + h$. Note "h approaches 0" is equivalent to "x approaches c". Thus the preceding limit says $\lim_{x \to c} f(x) = f(c)$. Hence f is continuous at $x = c$. $\qquad \square$

Proposition 2.2.13 applies to all points of the domain of a differentiable function.

Corollary 2.2.14 *A differentiable function with domain an open interval is continuous.*

It follows from the Proposition 2.2.13 that if f is not continuous at $x = c$, then the derivative $f'(c)$ does not exist. Therefore, our examples of discontinuous functions in Section 1.7 give us examples of functions whose derivatives do not exist.

Examples 2.2.15 (1) Consider the function $f(x) = \left\{ \begin{array}{ll} \frac{x}{|x|} & \text{if } x \neq 0 \\ 1 & \text{if } x = 0 \end{array} \right\}$. Note that $f(x) = 1$ for $x \geq 0$ and $f(x) = -1$ for $x < 0$. Hence f has a jump discontinuity at $x = 0$ as depicted in Figure 1.5.4. Therefore, $f'(0)$ does not exist.

(2) The function $h(x) = \left\{ \begin{array}{ll} \sin \frac{1}{x} & \text{if } x \neq 0 \\ 0 & \text{if } x = 0 \end{array} \right\}$ of Example 1.7.6 (3a) has an essential discontinuity at $x = 0$. Therefore, $h'(0)$ does not exist.

(3) The function $g(x) = \left\{ \begin{array}{ll} 1 & \text{if } x \text{ is a rational number} \\ 0 & \text{if } x \text{ is an irrational number} \end{array} \right\}$ of Example 1.7.6 (4) has an essential discontinuity and g' does not exist at every number x. □

The following examples show that the converse of Proposition 2.2.13 is false. That is, a function which is continuous at $x = c$ may not be differentiable at $x = c$. For example, it may have a corner where a tangent line does not exist.

Examples 2.2.16 (1) Show the continuous function $g(x) = |x|$ is not differentiable at $x = 0$.

Solution By Definition 2.2.11:

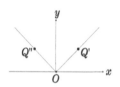

Figure 2.2.17
$y = g(x) = |x|$

$$g'(0) = \lim_{h \to 0} \frac{g(h) - g(0)}{h} = \lim_{h \to 0} \frac{|h| - 0}{h} = \lim_{h \to 0} \frac{|h|}{h} \,.$$

This two sided limit does not exist because the left–hand limit is -1 while the right–hand limit is $+1$. The graph of g in Fig. 2.2.17 has a corner at the origin.

(2) In Example 2.2.7 (4) we showed the continuous function
$f(x) = \left\{ \begin{array}{ll} x \sin \frac{1}{x} & \text{for } x \neq 0 \\ 0 & \text{for } x = 0 \end{array} \right\}$ is not differentiable at $x = 0$. □

One–Sided Derivatives

If the domain of a function f is an interval, then its derivative is not defined at the endpoints of that interval. The derivative of f also does not exist at a corner on the graph of f. Although the tangent lines to the graph of f do not exist at these points, there may be half–tangent lines there.

Motivating Examples 2.2.18 (1) Consider the function $f(x) = \sqrt{x}$ with domain $[0, \infty)$. By Definition 2.2.11:

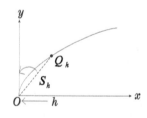

Figure 2.2.19
$y = f(x) = \sqrt{x}$

$$f'(x) = \lim_{h \to 0} \frac{f(x + h) - f(x)}{h} = \lim_{h \to 0} \frac{\sqrt{x + h} - \sqrt{x}}{h} \,.$$

In Example 2.2.12 (6) we evaluated this limit as $\frac{1}{2\sqrt{x}}$ for $x > 0$. However,

$$f'(0) = \lim_{h \to 0} \frac{\sqrt{0 + h} - \sqrt{0}}{h} = \lim_{h \to 0} \frac{\sqrt{h}}{h} = \lim_{h \to 0} \frac{1}{\sqrt{h}}$$

is not defined because $\sqrt{h}$ is not defined for $h < 0$. However, the right–hand limit is defined:

$$\lim_{h \to 0^+} \frac{1}{\sqrt{h}} = +\infty \,.$$

In other words, the limit of the slopes of the secant lines $S_h = OQ_h$ on the graph of f, for $h > 0$, is the upper half of the y–axis. See Figure 2.2.19.

(2) Consider the function $g(x) = |x|$ with domain $\Re$. In Example 2.2.16 (1) we showed the two–sided limit for $g'(0)$ does not exist, and hence the tangent line to the graph of g does not exist at $x = 0$. See Figure 2.2.17. In particular, we showed that the left–hand limit has value -1 while the right–hand limit has value $+1$. This means that the limit of the secant lines OQ'', to the left of the origin, is the upper half of the line $y = -x$ while the limit of the secant lines OQ', to the right of the origin, is the upper half of the line $y = x$. □

The first example above is typical of the behavior of the graph of a function at an endpoint of its domain: there may be a half–tangent line there. The second example is typical of the behavior of the graph of a function at a corner: the left and right half–tangent line have different slopes. These examples motivate the need to define half–tangent lines. Their slopes are called one–sided derivatives.

In general, the right derivative of f at $x = c$ is the limit of the slopes of secant lines joining the point $P = (c, f(c))$ with points $Q' = (c + h, f(c + h))$, $h > 0$, to the right of P on the graph of f. The limit of these secant lines S' is called the *right half–tangent line* T' at P. The left derivative of f at $x = c$ is the limit of the slopes of secant lines joining the point $P = (c, f(c))$ with points $Q'' = (c - h, f(c - h))$, $h > 0$, to the left of P on the graph of f. The limit of these secant lines S'' is called the *left half–tangent line* T'' at P. See Figure 2.2.20.

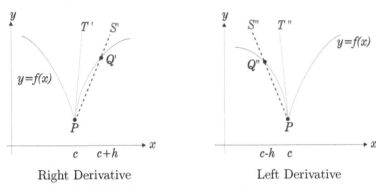

Right Derivative   **Left Derivative**

Figure 2.2.20 One–Sided Derivatives

Definition 2.2.21 *Let c be a number in the domain of the real valued function f.*
(a) *If the domain of f contains an interval $(c - r, c]$ with $r > 0$, then the left derivative of f at $x = c$ is*

$$f'_-(c) = \lim_{h \to 0^-} \frac{f(c + h) - f(c)}{h}.$$

When this limit exists and is finite, the left half–tangent line of the graph of f at $x = c$ has domain $(-\infty, c]$ and equation

$$y = f(c) + f'_-(c)(x - c) .$$

When $f'_-(c) = \pm\infty$, then the left half–tangent line of the graph of f at $x = c$ is vertical.
(b) *If the domain of f contains an interval $[c, c + r)$ with $r > 0$, then the right derivative of f at $x = c$ is*

$$f'_+(c) = \lim_{h \to 0^+} \frac{f(c + h) - f(c)}{h}.$$

When this limit exists and is finite, the right half–tangent line of the graph of f at $x = c$ has domain $[c, \infty)$ and equation

$$y = f(c) + f'_+(c)(x - c) .$$

When $f'_+(c) = \pm\infty$, then the right half–tangent line of the graph of f at $x = c$ is vertical.

(c) *f is differentiable on the interval I if $f'(c)$ exists at each interior point $c \in I$ and the appropriate one–sided derivative of f exists and is finite at each endpoint of I.*

Notes: **(1)** The definition of the derivative as a limit in Definition 2.2.11 requires $f'(c)$ to be a number. When the value of this limit is either $+\infty$ or $-\infty$, we say the derivative $f'(c)$ does not exist. This is necessary to make $f'(x)$ a real valued function. However, the preceding definition allows the one–sided derivatives, $f'_+(c)$ and $f'_-(c)$, to have values $+\infty$ or $-\infty$. This may be convenient, as in Example 2.2.18 (1) above.
(2) We interpret Limit Property 2 for the limit which defines the derivative at $x = c$.
   (a) If the two sided limit $f'(c)$ exists, then the left derivative $f'_-(c)$ and the right derivative $f'_+(c)$ exist with $f'(c) = f'_-(c) = f'_+(c)$.
   (b) If the left derivative $f'_-(c)$ and the right derivative $f'_+(c)$ exist and equal the *same number*, then the two–sided limit $f'(c)$ exists with $f'(c) = f'_-(c) = f'_+(c)$.

    The next result allows us to compute a one–sided derivative of f at $x = c$ from the knowledge of $f'(x)$ for x near c. We use this proposition to find the value of a one–sided derivative without having to evaluate the limit in the preceding definition. The proof of this proposition is given in Section 6 using the Mean Value Theorem.

Proposition 2.2.22 *Let $f(x)$ be a continuous function with domain the closed interval $[a, b]$, $a < b$. Assume $f'(x)$ is continuous on the open interval (a, b).*
(a) *If $\lim\limits_{x \to a^+} f'(x)$ exists, then $f'_+(a)$ exists and*

$$f'_+(a) = \lim_{x \to a^+} f'(x).$$

(b) *If $\lim\limits_{x \to b^-} f'(x)$ exists, then $f'_-(b)$ exists and*

$$f'_-(b) = \lim_{x \to b^-} f'(x).$$

    In the following examples, we use Proposition 2.2.22 to compute the required one–sided derivatives.

Examples 2.2.23 (1) Let $f(x) = x^2$ with domain $[1, 5]$. Find the equations of the left half–tangent line at $x = 1$ and the right half–tangent line at $x = 5$.

Solution For $1 < x < 5$, we know $f'(x) = 2x$. However, neither $f'(1)$ nor $f'(5)$ exist because 1 and 5 are not interior points of the domain of f. Nevertheless, the right derivative at $x = 1$ exists by Proposition 2.2.22(a):

$$f'_+(1) = \lim_{x \to 1^+} f'(x) = \lim_{x \to 1^+} 2x = 2.$$

Hence the right half–tangent line T' at $x = 1$ has equation

$$y = 1 + 2(x - 1) = 2x - 1$$

with domain $[1, \infty)$. The left derivative at $x = 5$ exists by Proposition 2.2.22(b):

$$f'_-(5) = \lim_{h \to 5^-} f'(x) = \lim_{h \to 5^-} 2x = 10.$$

Hence the left half–tangent line T'' at $x = 5$ has equation

$$y = 25 + 10(x - 5) = 10x - 25$$

with domain $(-\infty, 5]$. See Figure 2.2.24.

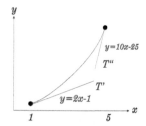

Figure 2.2.24
$y = f(x) = x^2$
with Domain $[1, 5]$

(2) **(a)** Is the function $h(x) = \sqrt[3]{x}$ with domain $[-1, 1]$ differentiable ?

  (b) Is the function $k(x) = \sqrt[3]{x}$ with domain $[0, 1]$ differentiable ?

  Solution **(a)** We shall see in Section 4 that $h'(x)$ exists for $x \neq 0$ with

$$h'(x) = \frac{1}{3x^{2/3}} \ .$$

However, $h'(0)$ does not exist because a derivative at an interior point of the domain of h must have a number as its value, not $+\infty$. Hence h is *not* differentiable. See Figure 2.2.28.

  (b) As in (a), $k'(x)$ exists with $k'(x) = \frac{1}{3x^{2/3}}$ for $x \in (0, 1)$. By Proposition 2.2.22 the one–sided derivatives at the endpoints of the domain $[0, 1]$ of k both exist:

$$k'_-(1) \ = \ \lim_{x \to 1^-} \frac{1}{x^{2/3}} = 1 \quad \text{and} \quad k'_+(0) \ = \ \lim_{x \to 0^+} \frac{1}{x^{2/3}} = +\infty \ .$$

Since a one-sided derivative may have infinite value, k *is* differentiable. □

Corollary 2.2.14 says that a differentiable function on an open interval is continuous. This result extends to closed intervals and half–open, half–closed intervals.

Corollary 2.2.25 *A differentiable function f with domain an interval is continuous.*

Proof Let $a < b$ be the endpoints of the domain I of f. By Proposition 2.2.13, f is continuous at each interior point of I. If $a \in I$, then $f'_+(a)$ is finite and

$$\lim_{h \to 0^+} f(a+h) - f(a) = \lim_{h \to 0^+} h \cdot \lim_{h \to 0^+} \frac{f(a+h) - f(a)}{h} = 0 \cdot f'_+(a) = 0.$$

Hence $\lim_{h \to 0^+} f(a+h) = f(a)$, i.e. $\lim_{x \to a^+} f(x) = f(a)$. Thus f is continuous at $x = a$. Similarly if $b \in I$, f is continuous at $x = b$. We leave the proof as an exercise. □

A *cusp* is the mathematical term used to describe a (curved) corner on the graph of a continuous function f. At a cusp the two one–sided derivatives of f have different values. In particular, the graph of f has a *vertical cusp*, where one of the one–sided derivatives has value $+\infty$ while the other one has value $-\infty$. A related phenomenon is a point $(c, f(c))$ where the graph of f has a *vertical tangent*, i.e. the slope of the tangent line is infinite. These phenomena are illustrated in Figure 2.2.26.

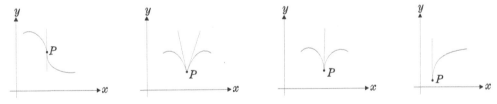

Vertical Tangent   Cusp   Vertical Cusp   One–Sided Vertical Tangent

Figure 2.2.26 Cusps and Vertical Tangents

Definition 2.2.27 *Let f be a continuous real valued function whose domain includes the open interval I with $c \in I$.*

(a) *A vertical tangent occurs at $x = c$ when $f'_+(c) = f'_-(c)$ and these one sided derivatives both have value $+\infty$ or both have value $-\infty$.*

(b) *A cusp occurs at $x = c$ when $f'_+(c)$ and $f'_-(c)$ exist and have different values.*

(c) *A vertical cusp occurs at $x = c$ when either*

$$f'_-(c) = -\infty, \quad f'_+(c) = +\infty \quad \text{or} \quad f'_-(c) = +\infty, \quad f'_+(c) = -\infty.$$

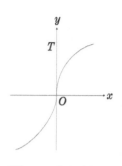

Figure 2.2.28
$f(x) = \sqrt[3]{x}$

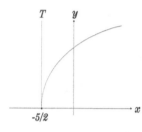

Figure 2.2.30
$n(x) = \sqrt{2x + 5}$

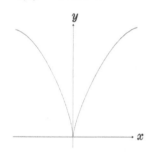

Figure 2.2.31
$p(x) = x^{2/3}$

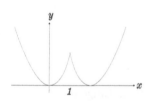

Figure 2.2.32
$y = g(x)$

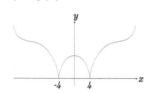

Figure 2.2.33
$h(x) = (x^2 - 16)^{2/3}$

(d) *Consider the case where the domain of f is an interval D. A one–sided vertical tangent occurs at an endpoint $x = c$ of D when $c \in D$ and the one–sided derivative of f at $x = c$ equals either $+\infty$ or $-\infty$.*

These phenomena are illustrated in the examples below. In particular, we will see that vertical tangents and vertical cusps occur for functions with fractional exponents which lie between zero and one.

Examples 2.2.29 (1) Let $f(x) = \sqrt[3]{x} = x^{1/3}$ with domain $\Re$. In Example 2.2.23 (2) we showed the left and right derivatives of f at $x = 0$ have value $+\infty$. Hence the graph of f has the y–axis as a vertical tangent at the origin. See Fig. 2.2.28.

(2) Let $n(x) = \sqrt{2x + 5}$ with domain $[-5/2, \infty)$. We saw in Example 2.2.12 (7) that

$$n'(x) = \frac{1}{\sqrt{2x + 5}} \quad \text{for } x > -\frac{5}{2}.$$

By Proposition 2.2.22(a) $n'_+(-5/2) = \lim_{x \to -5/2^+} n'(x) = +\infty$. Therefore n has a one–sided vertical tangent T at $x = -\frac{5}{2}$. See Figure 2.2.30.

(3) Let $p(x) = x^{2/3}$ with domain $\Re$. We will see in Section 4 that this continuous function has derivative:

$$p'(x) = \frac{2}{3x^{1/3}} \quad \text{for } x \neq 0.$$

By Prop. 2.2.22, $p'_+(0) = \lim_{x \to 0^+} p'(x) = +\infty$ while $p'_-(0) = \lim_{x \to 0^-} p'(x) = -\infty$. Hence $p(x)$ has a vertical cusp at $x = 0$. Graph p is depicted in Figure 2.2.31.

(4) Let $g(x)$ be the continuous function with domain $\Re$ defined by $g(x) = \left\{ \begin{array}{ll} (x - 2)^2 & \text{if } x \geq 1 \\ x^2 & \text{if } x < 1 \end{array} \right\}$. Then $g'(x) = \left\{ \begin{array}{ll} 2x - 4 & \text{if } x > 1 \\ 2x & \text{if } x < 1 \end{array} \right\}$, and by Proposition 2.2.22(a):

$$g'_-(1) = \lim_{x \to 1^-} g'(x) = 2 \lim_{x \to 1^-} 2x = 2,$$
$$g'_+(1) = \lim_{x \to 1^+} g'(x) = \lim_{x \to 1^+} (2x - 4) = -2.$$

Therefore, g has a cusp at $x = 1$. See Figure 2.2.32.

(5) Let $h(x) = (x^2 - 16)^{2/3}$ with domain $\Re$. We will see in Section 4 that

$$h'(x) = \frac{4x}{3(x^2 - 16)^{1/3}}.$$

Let c equal either -4 or $+4$. By Proposition 2.2.22 $h'_-(c) = \lim_{x \to c^-} h'(x) = -\infty$, and $h'_+(c) = \lim_{x \to c^+} h'(x) = +\infty$. Therefore, h has vertical cusps at $x = -4$ and at $x = +4$. The graph of h is depicted in Figure 2.2.33.

(6) Consider the function
$$m(x) = |x^2 - 4x - 21|$$

with domain $\Re$. To find the derivative of m, rewrite the formula for $m(x)$ to eliminate the absolute value. Note $x^2 - 4x - 21 = (x - 7)(x + 3)$. Therefore,

$$m(x) = \left\{ \begin{array}{ll} x^2 - 4x - 21 & \text{if } x \leq -3 \text{ or } x \geq 7 \\ -(x^2 - 4x - 21) & \text{if } -3 < x < 7 \end{array} \right\}.$$

Hence
$$m'(x) = \left\{ \begin{array}{ll} 2x - 4 & \text{if } x < -3 \text{ or } x > 7 \\ -2x + 4 & \text{if } -3 < x < 7 \end{array} \right\}.$$

By Proposition 2.2.22:

$$\begin{aligned}
m'_-(-3) &= \lim_{x \to -3^-} m'(x) = \lim_{x \to -3^-} (2x - 4) = -10, \quad \text{while} \\
m'_+(-3) &= \lim_{x \to -3^+} m'(x) = \lim_{x \to -3^+} (-2x + 4) = +10; \\
m'_-(7) &= \lim_{x \to 7^-} m'(x) = \lim_{x \to 7^-} (-2x + 4) = -10, \quad \text{while} \\
m'_+(7) &= \lim_{x \to 7^+} m'(x) = \lim_{x \to 7^+} (2x - 4) = +10.
\end{aligned}$$

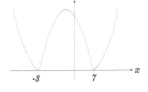

Figure 2.2.34
$m(x) = |x^2 - 4x - 21|$

Thus $m'(-3)$ and $m'(7)$ do not exist, and m has cusps at these two points. The graph of m is depicted in Figure 2.2.34.

(7) Let $j(x) = \left\{ \begin{array}{ll} \frac{|x|}{x}(x+1) & \text{if } x \neq 0 \\ 0 & \text{if } x = 0 \end{array} \right\}$ with domain $\Re$. Note $\frac{|x|}{x}$ equals -1 for $x < 0$ and equals $+1$ for $x > 0$. Hence $j(x) = \left\{ \begin{array}{ll} -(x+1) & \text{if } x < 0 \\ 0 & \text{if } x = 0 \\ x+1 & \text{if } x > 0 \end{array} \right\}$, and

$j'(x) = \left\{ \begin{array}{ll} -1 & \text{if } x < 0 \\ +1 & \text{if } x > 0 \end{array} \right\}$. By Proposition 2.2.22, $j'_-(0) = -1$ and $j'_+(0) = +1$. Hence $j'(0)$ does not exist. However, j does not have a cusp at $x = 0$ because the function j is not continuous there. See Figure 2.2.35. □

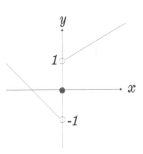

Figure 2.2.35
$j(x) = \frac{|x|}{x}(x+1)$

Historical Remarks

The construction of tangent lines and the study of their properties was begun by the Greeks as one aspect of their work on the geometry of conic sections. In particular, Apollonius's eight books of *Conics*, written about 200 BCE, contain many results about tangent lines. In addition, Archimedes's book *On Spirals* contains the first tangent line construction for a curve other than a conic section.

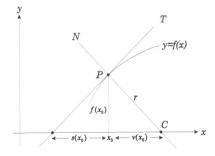

Figure 2.2.36 Subtangent $s(x_0)$ and Subnormal $v(x_0)$

The systematic study of tangent lines, however, began in the seventeenth century. The first results were stated in terms of sub-tangents and sub-normals of a curve. Let T be the tangent line, and let N be the normal line to the graph of $y = f(x)$ at $P = (x_0, f(x_0))$. The subtangent $s(x_0)$ is the distance between x_0 and the intersection of the tangent line T with the x–axis. The subnormal $v(x_0)$ is the distance between x_0 and the intersection C of the normal line N with the x–axis. See Figure 2.2.36. In the 1620s, Pierre de Fermat computed the sub–tangent $s(x_0)$ of a polynomial $f(x)$.

His method was to compute the quotient

$$\frac{hf(x_0)}{f(x_0 + h) - f(x_0)},$$

cancel the common factor h from the numerator and denominator, and then set h equal to zero. This computation is equivalent to computing the derivative since

$$f'(x_0) = \frac{f(x_0)}{s(x_0)}.$$

Fermat never justified this procedure by explaining that h is a small number which approaches zero. Simultaneously, René Descartes discovered that the subnormal $v(x_0)$ could be computed from the point where the circle with center C and radius r intersects the graph at only one point. This method was reduced to an algorithm for polynomials by Johann Hudde in 1659. At the same time, René de Sluse produced an algorithm for computing the subtangent of a function defined implicitly by a polynomial equation in two variables. In all of this work, the tangent line was only defined intuitively.

The idea of computing the slope of the tangent line T from the slope of secant lines S which approximate T was introduced by Isaac Barrow in his lectures at Cambridge in the 1660s. He applied this method to functions defined implicitly by a polynomial in two variables: $p(x, y) = 0$. If (x_0, y_0) and $(x_0 + h, y_0 + k)$ are on this curve, they define a secant line S with slope $\frac{k}{h}$. See Figure 2.2.37. Barrow computed the difference

$$p(x_0 + h, y_0 + k) - p(x_0, y_0) = 0 - 0 = 0. \tag{2.2.7}$$

Every summand of this difference is divisible by either h or k. Taking h and k to be small, he would ignore any term with a factor of h^2, k^2 or hk because those terms are much smaller than those divisible by only h or k. The remaining terms of equation (2.2.7) give $a(x_0, y_0)h + b(x_0, y_0)k \approx 0$. Then

$$slope\ S = \frac{k}{h} \approx -\frac{a(x_0, y_0)}{b(x_0, y_0)} = m.$$

This number m is the slope of the tangent line T. Implicit in Barrow's approach is the definition of the tangent line as the limit of secant lines. Isaac Newton may have attended these lectures of Barrow. If so, this would have started Newton in his development of the methods of calculus.

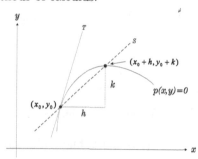

Figure 2.2.37 Barrow's Characteristic Triangle

Summary

The reader should understand the meaning of the derivative of a function f at $x = c$ both as the slope of a tangent line and as the value of a limit. She should be able to compute the derivatives of simple functions by evaluating this limit. The use of derivatives to find the equations of tangent and normal lines to curves should be known. The computation of one sided derivatives and the identification of vertical tangents, cusps and vertical cusps should be understood. In addition, she should be able to identify points where the derivative does not exist.

Basic Exercises

If you happen to know formulas for finding derivatives, you may not use them in solving the exercises of this section.

1. For each function f, find the slope of the secant line through the points P, Q. Round off your arithmetic to two decimal places.

(a) $f(x) = 3x^2 - 1$ with $P = (2, 11)$ and $Q = (2.1, 12.23)$.

(b) $f(x) = x^3$ with $P = (3, 27)$ and $Q = (2.8, 21.95)$.

(c) $f(x) = \frac{2x-10}{5x+3}$ with $P = (1, -1)$ and $Q = (1.1, -.92)$.

(d) $f(x) = \sqrt{7x^2 + 8}$ with $P = (2, 6)$ and $Q = (1.9, 5.77)$.

(e) $f(x) = \frac{1}{\sqrt{5x+26}}$ with $P = (-2, 1/4)$ and $Q = (-2.2, .26)$.

2. Find the point Q_h on the graph of each function.

(a) $f(x) = x^3$ with $x_0 = 4$ and $h = 1$.

(b) $g(x) = x^2 + 3x - 5$ with $x_0 = 5$ and $h = 0.5$.

(c) $k(x) = \frac{x}{x+2}$ with $x_0 = -1$ and $h = -0.3$.

(d) $m(x) = \sqrt{2x + 5}$ with $x_0 = 2$ and $h = -0.4$.

(e) $n(x) = \sqrt{\frac{7x-3}{x-1}}$ with $x_0 = 3$ and $h = 0.2$.

(f) $p(x) = \frac{\sqrt{x}}{1+\sqrt{x}}$ with $x_0 = 1$ and $h = -0.19$.

3. Approximate the slopes of the secant lines PQ_h of the graph of each function where P is the point on the graph at $x = c$.

(a) $f(x) = x^2$ with $c = 2$ and $h = 1$, $h = 0.5$, $h = -0.2$, $h = -0.1$.

(b) $g(x) = x^4 + x^3 - 40$ with $c = -3$ and $h = 0.4$, $h = 0.1$, $h = -1$, $h = -0.3$.

(c) $p(x) = \frac{1}{x}$ with $c = 5$ and $h = 0.3$, $h = 0.2$, $h = -0.5$, $h = -0.1$.

(d) $q(x) = \sqrt{x}$ with $c = 9$ and $h = 0.8$, $h = 0.4$, $h = -0.6$, $h = -0.2$.

(e) $r(x) = \frac{x}{x+1}$ with $c = -2$ and $h = 0.7$, $h = 0.5$, $h = -0.4$, $h = -0.1$.

4. For each function of Exercise 3, use (2.2.6) to find the derivative at $x = c$.

5. For each function of Exercise 1, use (2.2.6) to find the equation of the tangent line T to the graph at P.

6. For each function of Exercise 1, find the equation of the normal line N to the graph at P.

7. Let T, N be the tangent and normal lines to the graph of $y = f(x)$ at $x = c$.

(a) If T has slope 6, find the slope of N.

(b) If N has slope -3, find the slope of T.

(c) If T has equation $2x + 5y = 10$, find the slope of N.

(d) If N has equation $3x - 7y = 8$, find the slope of T.

(e) If T has equation $y = 5x - 3$ and $c = 2$, find the equation of N.

(f) If N has equation $x = 4y + 9$ and $c = 5$, find the equation of T.

8. Find the derivative of each polynomial function:

(a) $f(x) = 7x - 5$;   (b) $g(x) = 5x^2 - 2$;   (c) $h(x) = 6 + 3x - x^2$;

(d) $j(x) = 4x^3 - x + 1$;   (e) $k(x) = 2x^3 + 3x^2 - 7$;   (f) $m(x) = x^4 + 6x - 4$.

9. Find the derivative of each rational function:

(a) $f(x) = \frac{1}{x}$;   (b) $g(x) = \frac{4x}{7x-2}$;   (c) $h(x) = \frac{5-8x}{3x+4}$;

(d) $j(x) = \frac{x^2}{2x+1}$;   (e) $k(x) = \frac{2x^2-3}{5x^2+6}$;   (f) $m(x) = \frac{x^3}{x^2+1}$.

10. Find the derivative of each function:

(a) $f(x) = \sqrt{x}$;   (b) $g(x) = \sqrt{3x - 4}$;   (c) $h(x) = \sqrt{x^2 - 8}$;

(d) $j(x) = \frac{1}{\sqrt{2x+1}}$;   (e) $k(x) = \sqrt{\frac{4x+5}{3x-2}}$;   (f) $m(x) = \sqrt{\frac{x}{x^2-9}}$.

11. Show that if $f'(c)$ exists, then $f'(c) = \lim\limits_{x \to c} \dfrac{f(x) - f(c)}{x - c}$.

12. Is the following statement true or false? Justify your answer.

If f is a continuous function, then f is differentiable.

13. For each function, determine all points where its derivative exists. At each point where the derivative exists, find its value. At all other points, explain why the derivative does not exist.

(a) f has domain $\Re$ and is defined by $f(x) = |3x - 2|$.
(b) g has domain $\Re$ and is defined by $g(x) = |7 + 5x|$.
(c) h has domain $\Re$ and is defined by $h(x) = |x^2 - 1|$.
(d) i has domain $(-\infty, 4/7]$ and is defined by $i(x) = \sqrt{4 - 7x}$.

(e) j has domain $\Re$ and is defined by $j(x) = \begin{cases} 8x + 3 & \text{if } x \geq 1 \\ 4x^2 + 7 & \text{if } x < 1 \end{cases}$.

(f) k has domain $\Re$ and is defined by $k(x) = \begin{cases} 5x - 2 & \text{if } x \geq 4 \\ 34 - 4x & \text{if } x < 4 \end{cases}$.

(g) m has domain $\Re$ and is defined by $m(x) = \begin{cases} 3x + 1 & \text{if } x \geq 2 \\ 3x - 8 & \text{if } x < 2 \end{cases}$.

(h) n has domain $[-3/2, \infty)$ and is defined by $n(x) = \sqrt{3 + 2x}$.

(i) p has domain $\Re$ and is defined by $p(x) = \begin{cases} x^2 - 2x + 1 & \text{if } x \leq 1 \\ -x^2 + 2x - 1 & \text{if } x > 1 \end{cases}$.

(j) q has domain $\Re$ and is defined by $q(x) = \begin{cases} x^3 - 5x^2 + 1 & \text{if } x \leq 2 \\ x^4 - 40x + 50 & \text{if } x > 2 \end{cases}$.

(k) r has domain $\Re$ and is defined by $r(x) = \begin{cases} x^2 & \text{if } x \text{ is rational} \\ 0 & \text{if } x \text{ is irrational} \end{cases}$.

(l) s has domain $\Re$ and is defined by $s(x) = \begin{cases} x^3 + x^2 + 5 & \text{if } x \text{ is rational} \\ 5 + 2x & \text{if } x \text{ is irrational} \end{cases}$.

14. Assume $f'_-(c)$ and $f'_+(c)$ exist. Show that f is continuous at $x = c$.

15. Give a specific example of a continuous function $y = f(x)$ with domain an interval and a point c in the interior of its domain where $f'_-(c)$ and $f'_+(c)$ both do not exist.

16. Evaluate each one–sided derivative.

(a) Find $f'_+(1)$ where $f(x) = x^3 - 10$ with domain $[1, 8]$.
(b) Find $g'_-(9)$ where $g(x) = (4x - 5)^2$ with domain $[0, 9]$.
(c) Find $h'_+(6)$ where $h(x) = \sqrt{x - 6}$.
(d) Find $i'_-(3)$ where $i(x) = \sqrt{3 - x}$.
(e) Find $j'_+(-2)$ and $j'_-(-2)$ where $j(x) = |x + 2|$.
(f) Find $k'_+(1/5)$ and $k'_-(1/5)$ where $k(x) = |5x - 1|$.
(g) Find $m'_+(2)$ and $m'_-(2)$ where $m(x) = |x^2 - 8x + 12|$.
(h) Find $n'_+(-3)$ and $n'_-(5)$ where $n(x) = |15 + 2x - x^2|$.

17. Find the half–tangent lines, with their domains, that are defined at $x = c$.
(a) $f(x) = 1 - 4x^2$ with domain $[-1, 2]$ and $c = 2$.   (b) $g(x) = \sqrt{9 - x}$ and $c = 9$.
(c) $h(x) = \frac{x}{3x+5}$ with domain $[1, 7]$ and $c = 1$.   (d) $k(x) = |x^2 - 5x - 14|$ and $c = 7$.
(e) $p(x) = -\sqrt{4 + x}$ and $c = -4$.   (f) $q(x) = |x^2 + 3x - 10|$ and $c = -5$.

18. Determine whether each function is differentiable.
(a) $f(x) = x^2 - 3x + 5$ with domain $[-1, 7]$. (b) $g(x) = \sqrt{4 + x}$ with domain $[-4, \infty)$.

(c) $h(x) = x^{2/3}$ with domain $[-1, 1]$.   **(d)** $k(x) = x^{2/3}$ with domain $[0, 8]$.
(e) $m(x) = |x + 2|$ with domain $[-5, 5]$. **(f)** $n(x) = |x^2 - x - 20|$ with domain $[-3, 3]$.
(g) $p(x) = \sqrt{1 - x^2}$ with domain $[-1, 1]$. **(h)** $q(x) = \sqrt{2 + x - x^2}$ with domain $[-1, 2]$.

19. Find where the derivative of each function does not exist. Identify each point as a vertical tangent, one–sided vertical tangent, cusp, vertical cusp, or none of these.

    (a) $f(x) = |3x + 7|$ **(b)** $g(x) = |4 - 5x|$ **(c)** $h(x) = |x^2 + x - 30|$
    (d) $j(x) = |x^2 - 2x - 15|$ **(e)** $k(x) = \sqrt[3]{x - 6}$ **(f)** $m(x) = \sqrt{x + 7}$ for $x \geq -7$

    (g) $n(x) = \sqrt{8 - x}$ for $x \leq 8$ **(h)** $p(x) = \begin{cases} \sqrt{x - 3} & \text{if } x \geq 3 \\ \sqrt{3 - x} & \text{if } x < 3 \end{cases}$

    (i) $q(x) = \begin{cases} 2x^2 - 3x + 1 & \text{if } x \geq 1 \\ 1 + 2x - 3x^2 & \text{if } x < 1 \end{cases}$ **(j)** $r(x) = \begin{cases} x^2 - 3x + 1 & \text{if } x \geq 2 \\ 5x - x^2 - 7 & \text{if } x < 2 \end{cases}$

20. Let f be a continuous function with domain the interval $[a, b]$. Let $c \in (a, b)$. Is each statement true or false? Justify your answers.
(a) If f has a vertical tangent at $x = c$, then f is not differentiable.
(b) If f has a cusp at $x = c$, then f is not differentiable.
(c) If f has a vertical cusp at $x = c$, then f is not differentiable.
(d) If f has a right half–tangent line at $x = a$ and a left half–tangent line at $x = b$, then f is not differentiable.
(e) If f has a vertical right half–tangent line at $x = a$ and a vertical left half–tangent line at $x = b$, then f is not differentiable.

21. For each graph, determine the points of the function's domain where the two sided derivative does not exist and where each one–sided derivative does not exist.

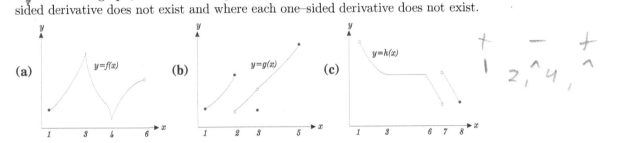

22. Let f be a differentiable function on an interval I which contains its right endpoint b. Complete the proof of Corollary 2.2.25 by showing $f(x)$ is continuous at $x = b$.

Challenging Problems

1. Find the derivative of each function with domain $\Re$:
    (a) $f(x) = \sqrt[3]{x}$,   **(b)** $g(x) = x^{2/3}$.

2. Find an example of a function f with domain $[a, b]$ such that $f'_+(a)$, $f'(c)$, for $c \in (a, b)$, and $f'_-(b)$ exist but f is not continuous at $x = a$ and at $x = b$.

2.3 Computing Derivatives

In the preceding section we computed the derivatives of simple functions by evaluating the limits which define their derivatives. However, this procedure is prohibitively difficult for complicated functions. In this section, we derive several formulas for computing derivatives. Using these formulas, we can easily compute the derivatives of functions which are constructed from simple ones using sums, products and quotients. In particular, we can compute the derivative of any polynomial or rational function. In the last subsection we define and compute higher derivatives.

Derivatives of Sums

We show the derivative is a linear function. In particular, the derivative of a sum is the sum of the derivatives.

Proposition 2.3.1 *Let f, g be differentiable functions with the same domain, and let a, b be two numbers. Then $af + bg$ is a differentiable function with*

$$D(af + bg) = aDf + bDg.$$

Proof We use the definition of the derivative to compute $D(af + bg)$. We break up this limit into the sum of two limits, $D(af) + D(bg)$, using the fact that the limit of a sum is the sum of the limits.

$$
\begin{aligned}
D(af + bg)(x) &= \lim_{h \to 0} \frac{(af + bg)(x + h) - (af + bg)(x)}{h} \\
&= \lim_{h \to 0} \frac{[af(x + h) + bg(x + h)] - [af(x) + bg(x)]}{h} \\
&= \lim_{h \to 0} \left[\frac{af(x + h) - af(x)}{h} + \frac{bg(x + h) - bg(x)}{h} \right] \\
&= \lim_{h \to 0} \frac{af(x + h) - af(x)}{h} + \lim_{h \to 0} \frac{bg(x + h) - bg(x)}{h} \\
&= a \lim_{h \to 0} \frac{f(x + h) - f(x)}{h} + b \lim_{h \to 0} \frac{g(x + h) - g(x)}{h} \\
&= aDf(x) + bDg(x).
\end{aligned}
$$

Prop. 2.3.1 computes new derivatives from the derivatives computed in Section 2.

Examples 2.3.2 (1) Find $D(7x^2 - 5x)$.

    Solution By Examples 2.2.12 (1), (2): $D(x) = 1$ and $D(x^2) = 2x$. Therefore,

$$D(7x^2 - 5x) = 7D(x^2) - 5D(x) = 7(2x) - 5(1) = 14x - 5.$$

(2) Find $D(\frac{4}{x} - 6\sqrt{x})$.

    Solution By Examples 2.2.12 (4), (6): $D\left(\frac{1}{x}\right) = -\frac{1}{x^2}$ and $D(\sqrt{x}) = \frac{1}{2\sqrt{x}}$. Hence

$$D\left(\frac{4}{x} - 6\sqrt{x}\right) = 4D\left(\frac{1}{x}\right) - 6D(\sqrt{x}) = -\frac{4}{x^2} - \frac{3}{\sqrt{x}}.$$

Derivatives of Products

We find a formula for the derivative of a product of functions. Thus formula is called the "product rule", the "Leibniz rule", or the "derivation property". The proof writes the limit which defines the derivative of a product as the sum of two limits. Each of these limits is related to the derivative of one of the factors.

Proposition 2.3.3 *Let f, g be differentiable functions with the same domain. Then fg is a differentiable function with*

$$D(fg) = fDg + gDf.$$

Proof Use the definition of the derivative to compute $D(fg)$:

$$D(fg)(x) = \lim_{h \to 0} \frac{(fg)(x+h) - (fg)(x)}{h} = \lim_{h \to 0} \frac{f(x+h)g(x+h) - f(x)g(x)}{h}.$$

To evaluate this limit we use a trick: add and subtract $f(x)g(x+h)$ to the numerator of this fraction. This writes the limit as the sum of two limits whose values we know.

$$
\begin{aligned}
D(fg)(x) &= \lim_{h \to 0} \frac{[f(x+h)g(x+h) - f(x)g(x+h)] + [f(x)g(x+h) - f(x)g(x)]}{h} \\
&= \lim_{h \to 0} \frac{f(x+h)g(x+h) - f(x)g(x+h)}{h} + \lim_{h \to 0} \frac{f(x)g(x+h) - f(x)g(x)}{h} \\
&= \lim_{h \to 0} \frac{g(x+h)\,[f(x+h) - f(x)]}{h} + \lim_{h \to 0} \frac{f(x)\,[g(x+h) - g(x)]}{h} \\
&= \lim_{h \to 0} g(x+h) \cdot \lim_{h \to 0} \frac{f(x+h) - f(x)}{h} + f(x) \cdot \lim_{h \to 0} \frac{g(x+h) - g(x)}{h} \\
&= g(x)Df(x) + f(x)Dg(x). \qquad \Box
\end{aligned}
$$

We use the product rule to compute the derivative of x^n for any integer $n > 0$.

Corollary 2.3.4 *If n is a non–negative integer, then*

$$D(x^n) = nx^{n-1}. \qquad (2.3.1)$$

Proof We already know that this formula is true for $n = 0$, $n = 1$, $n = 2$ and $n = 3$:

$$D(1) = 0, \quad D(x) = 1 = 1x^0, \quad D(x^2) = 2x, \quad D(x^3) = 3x^2.$$

If this formula is not always true, let n be the smallest positive integer for which formula (2.3.1) fails. Thus, $n \geq 4$. By the choice of n, formula (2.3.1) is true for $D(x^{n-1})$. By the product rule:

$$
\begin{aligned}
D(x^n) &= D(x \cdot x^{n-1}) = xD(x^{n-1}) + x^{n-1}D(x) = x(n-1)x^{n-2} + x^{n-1}1 \\
&= (n-1)x^{n-1} + x^{n-1} = nx^{n-1}.
\end{aligned}
$$

Thus formula (2.3.1) is true for $D(x^n)$. This is a contradiction! Therefore formula (2.3.1) must be true for all positive integers n. $\qquad \Box$

We can compute the derivatives of polynomials as well as other products.

Examples 2.3.5 (1) Find $D(6x^4 - 3x^2 + 5)$.

Solution We use the linear property of the derivative and then apply (2.3.1):

$$D(6x^4 - 3x^2 + 5) = 6D(x^4) - 3D(x^2) + 5D(1) = 6(4x^3) - 3(2x) + 5(0) = 24x^3 - 6x.$$

(2) Find $D(2x^7 - 5x^3 + 16x)$.

Solution Again, use the linear property of the derivative and then apply (2.3.1):

$$
\begin{aligned}
D(2x^7 - 5x^3 + 16x) &= 2D(x^7) - 5D(x^3) + 16D(x) = 2(7x^6) - 5(3x^2) + 16(1) \\
&= 14x^6 - 15x^2 + 16.
\end{aligned}
$$

(3) Find $D(x^{9/2})$.

Solution By the product rule:

$$
\begin{aligned}
D(x^{9/2}) &= D(x^4 \sqrt{x}) = x^4 D(\sqrt{x}) + \sqrt{x} D(x^4) = x^4 \frac{1}{2\sqrt{x}} + \sqrt{x}(4x^3) \\
&= \frac{1}{2}x^{7/2} + 4x^{7/2} = \frac{9}{2}x^{7/2}.
\end{aligned}
$$

(4) Find $D\left(\frac{1}{\sqrt{x}}\right)$.

Solution By the product rule:

$$
\begin{aligned}
D\left(\frac{1}{\sqrt{x}}\right) &= D\left(\frac{1}{x}\sqrt{x}\right) = \frac{1}{x}D(\sqrt{x}) + \sqrt{x}D\left(\frac{1}{x}\right) \\
&= \frac{1}{x}\frac{1}{2\sqrt{x}} + \sqrt{x}\left(-\frac{1}{x^2}\right) = \frac{1}{2x\sqrt{x}} - \frac{1}{x\sqrt{x}} = -\frac{1}{2x\sqrt{x}}. \qquad \square
\end{aligned}
$$

Derivatives of Quotients

Next, we derive the formula for the derivative of a quotient. This formula is called the "quotient rule." The proof applies the product rule by viewing a quotient as the product of its numerator and the reciprocal of its denominator.

Proposition 2.3.6 *Let f, g be differentiable functions with the same domain I. Assume $g(x) \neq 0$ for all $x \in I$. Then f/g is a differentiable function with*

$$
D\left(\frac{f}{g}\right) = \frac{gDf - fDg}{g^2}.
$$

Proof We begin by deriving the formula for $D(1/g)$ from the definition of the derivative. Then we use the product rule to deduce the formula for $D\left(\frac{f}{g}\right) = D\left(\frac{1}{g}\cdot f\right)$.

$$
\begin{aligned}
D\left(\frac{1}{g}\right)(x) &= \lim_{h\to 0}\frac{\frac{1}{g}(x+h) - \frac{1}{g}(x)}{h} = \lim_{h\to 0}\frac{\frac{1}{g(x+h)} - \frac{1}{g(x)}}{h} = \lim_{h\to 0}\frac{g(x) - g(x+h)}{hg(x+h)g(x)} \\
&= -\lim_{h\to 0}\frac{g(x+h) - g(x)}{h}\cdot \lim_{h\to 0}\frac{1}{g(x+h)g(x)} = -Dg(x)\cdot\frac{1}{g(x)^2} \\
&= -\frac{Dg(x)}{g(x)^2}.
\end{aligned}
$$

By the product rule,

$$
D\left(\frac{f}{g}\right) = D\left(\frac{1}{g}\cdot f\right) = \frac{1}{g}Df + fD\left(\frac{1}{g}\right) = \frac{Df}{g} + f\left(-\frac{Dg}{g^2}\right) = \frac{gDf - fDg}{g^2}. \qquad \square
$$

We use the quotient rule to compute the derivative of x^n for n a negative integer.

Corollary 2.3.7 *If n is any integer then*

$$
D(x^n) = nx^{n-1} \tag{2.3.2}
$$

Proof We know this formula is true when n is non–negative. Let n be negative. Then $-n$ is positive, so we know that $D(x^{-n}) = -nx^{-n-1}$. By the quotient rule:

$$
D(x^n) = D\left(\frac{1}{x^{-n}}\right) = \frac{x^{-n}D(1) - 1D(x^{-n})}{(x^{-n})^2} = \frac{0 - (-nx^{-n-1})}{x^{-2n}} = nx^{n-1}. \qquad \square
$$

We can compute the derivatives of rational functions as well as other quotients.

Examples 2.3.8 (1) Find $D\left(\frac{2x^6-3}{5x^4+7}\right)$.

Solution By the quotient rule:

$$D\left(\frac{2x^6 - 3}{5x^4 + 7}\right) = \frac{(5x^4 + 7)D(2x^6 - 3) - (2x^6 - 3)D(5x^4 + 7)}{(5x^4 + 7)^2}$$

$$= \frac{(5x^4 + 7)(12x^5) - (2x^6 - 3)(20x^3)}{(5x^4 + 7)^2}$$

$$= \frac{20x^9 + 84x^5 + 60x^3}{(5x^4 + 7)^2}$$

(2) Find $D\left(\frac{\sqrt{x}+1}{4x^3+2}\right)$.

Solution By the quotient rule:

$$D\left(\frac{\sqrt{x}+1}{4x^3+2}\right) = \frac{(4x^3 + 2)D(\sqrt{x}+1) - (\sqrt{x}+1)D(4x^3+2)}{(4x^3+2)^2}$$

$$= \frac{(4x^3 + 2)\frac{1}{2\sqrt{x}} - (\sqrt{x}+1)(12x^2)}{(4x^3+2)^2} \cdot \frac{\sqrt{x}}{\sqrt{x}} = \frac{1 - 10x^3 - 12x^2\sqrt{x}}{\sqrt{x}(4x^3+2)^2}$$

(3) Find $D\left(\frac{4}{x^2} - \frac{7}{x^5}\right)$.

Solution Use the linear property of the derivative and apply (2.3.2):

$$D\left(\frac{4}{x^2} - \frac{7}{x^5}\right) = 4D(x^{-2}) - 7D(x^{-5}) = 4(-2x^{-3}) - 7(-5x^{-6})$$

$$= -8x^{-3} + 35x^{-6} . \qquad \square$$

Higher Derivatives

We will see that there are applications where it is useful to know the derivative of the derivative. In fact, we can iterate the derivative operation as many times as desired. If we iterate the derivative operation n times on a function f, the resulting function is called the nth*-derivative* of f. It is denoted

$$D^n f(x) = f^{(n)}(x) = y^{(n)} = \frac{d^n y}{dx^n}.$$

Additional notations for the second derivative include

$$D^2 f(x) = f''(x) = y'' = \ddot{y} .$$

These iterated derivatives are called *higher derivatives*.

Examples 2.3.9 (1) Find all the higher derivatives of $f(x) = 2x^3 - 5x^2 + 4x + 6$.

Solution By (2.3.1):

$$Df(x) = 6x^2 - 10x + 4, \quad D^2 f(x) = 12x - 10, \quad D^3 f(x) = 12,$$
$$D^n f(x) = 0 \text{ for } n \geq 4.$$

(2) Find the second and third derivatives of $g(x) = x^{-2}$.

Solution By (2.3.2):

$$Dg(x) = -2x^{-3}, \quad D^2 g(x) = 6x^{-4}, \quad D^3 g(x) = -24x^{-5}.$$

Note that there is no positive integer n for which $D^n g = 0$.

(3) Find the second derivative of $h(x) = \dfrac{x+1}{x-1}$.

   Solution Apply the quotient rule:

$$Dh(x) = \frac{(x-1)(1)-(x+1)(1)}{(x-1)^2} = -\frac{2}{(x-1)^2} = -\frac{2}{x^2-2x+1},$$

$$D^2 h(x) = -\frac{(x^2-2x+1)(0)-2(2x-2)}{(x^2-2x+1)^2} = \frac{4(x-1)}{(x-1)^4} = \frac{4}{(x-1)^3}. \qquad \square$$

Summary

The reader should know how to compute the derivatives of the sum, product and quotients of functions with known derivatives. In addition, she should be able to compute higher derivatives of these functions.

Basic Exercises

1. Find the derivative of each of these sums.
   (a) $f(x) = 7x^4 - 9x^2 + 6$ **(b)** $g(x) = 5x^2 - 8\sqrt{x}$ **(c)** $h(x) = \frac{10}{x} - x^3 + 1$
   (d) $k(x) = 5x^6 - 4x^3 + 2$ **(e)** $m(x) = 6\sqrt{x} - \frac{2}{x^3}$ **(f)** $p(x) = 3x^5 + 4\sqrt{x} + \frac{5}{x^2}$

2. Find the derivative of each of these products.
   (a) $f(x) = (x^3 + x - 1)(x^2 - x + 1)$ **(b)** $g(x) = \sqrt{x}(3x^5 + 5x^3 + 7)$
   (c) $h(x) = \frac{1}{x}\left(2x^6 - 9x^2 + 4\right)$ **(d)** $k(x) = (\sqrt{x} + x - 4)(\sqrt{x} - 3x + 7)$
   (e) $m(x) = \left(\frac{5}{x} - 8x^3 + 6\right)\left(\frac{9}{x} + 4x^2 - 3\right)$ **(f)** $p(x) = \left(6\sqrt{x} - \frac{4}{x}\right)\left(\frac{2}{x^2} + \frac{6}{x^3}\right)$

3. Find the derivative of each of these quotients.
   (a) $f(x) = \frac{5x+1}{3x-2}$ **(b)** $g(x) = \frac{4x^5-3x+1}{x+8\sqrt{x}+1}$ **(c)** $h(x) = \frac{x}{x^2+1}$
   (d) $k(x) = \frac{1}{x^4+x^2+1}$ **(e)** $m(x) = \frac{6\sqrt{x}}{x^3+4}$ **(f)** $p(x) = \frac{4\sqrt{x}+3}{5-2\sqrt{x}}$

4. Find the derivative of each function.
   (a) $f(x) = (2x^2 + 3x + 4)(x^3 - 5x + 2)(x^4 - 3x^2 + 6)$
   (b) $g(x) = (7x^4 - 2x^{-3} - 5\sqrt{x} - 1)^2$ **(c)** $h(x) = (5x^2 - 6\sqrt{x} + 7)^3$
   (d) $k(x) = \frac{(x^2+1)(6\sqrt{x}+3)}{8\sqrt{x}+9}$ **(e)** $m(x) = \frac{7x^5+1}{\sqrt{x}(3x^2-2)}$
   (f) $p(x) = \frac{(x^2+\sqrt{x}+1)(6x-\sqrt{x}+5)}{(5x^3-1)(4\sqrt{x}-2)}$ **(g)** $q(x) = \frac{x^{-7/2}+1}{x^{5/2}+1}$

5. Find a formula for the derivative of each function.
   (a) $f(x)^2$ **(b)** $f(x)^3$ **(c)** $f(x)g(x)h(x)$ **(d)** $\frac{1}{f(x)^2}$
   (e) $\frac{1}{f(x)^3}$ **(f)** $\frac{f(x)g(x)}{h(x)}$ **(g)** $\frac{f(x)}{g(x)h(x)}$ **(h)** $\frac{f(x)g(x)}{h(x)k(x)}$

6. Use the product rule, as in Example 2.3.5 (3), to differentiate each function.
(a) $f(x) = x^{5/2}$ **(b)** $g(x) = x^{7/2}$ **(c)** $h(x) = x^{-9/2}$ **(d)** $k(x) = x^{-3/2}$

7. Each function in Exercise 6 can be written as a quotient with numerator $\sqrt{x}$. Use the quotient rule to find the derivative of each function.

8. (a) Use the product rule to show that for every integer n: $D\left(x^{\frac{n}{2}}\right) = \frac{n}{2}x^{\frac{n}{2}-1}$.
(b) Use the quotient rule to verify this formula.

9. Use Exercise 8 to find the derivative of each function.
   (a) $f(x) = 4x^{-7/2} - 6x^{5/2} + 3$ **(b)** $g(x) = (2x^{11/2} - 4x^{-3/2})(8x^{-9/2} + 12x^{7/2})$
   (c) $h(x) = \frac{10x^{5/2}+7}{4x^{-13/2}+5x^2-1}$ **(d)** $k(x) = (x^3 - 6x^{15/2} + 9)(4x^{-17/2} + 7x^3 + 8)$
   (e) $m(x) = \frac{x^2+6x^{-3/2}}{x^{7/2}-x^3+1}$ **(f)** $p(x) = (2x^{7/2} - 3x^{-9/2})^2$

10.   Find the equation of the tangent line to the graph of each function at the indicated point:

(a) $f(x) = 2x^3 - 5x + 4$ at $x = 1$; (b) $g(x) = 4x^{-1} - 64x^{-4} + 3$ at $x = 2$;
(c) $h(x) = x^{5/2}$ at $x = 4$; (d) $j(x) = \frac{3x^3 - 8}{2x^2 - 4}$ at $x = 2$;
(e) $k(x) = \frac{1}{4\sqrt{x}-5}$ at $x = 9$; (f) $m(x) = (x^{-9/2} + 1)(x^{5/2} + 2)$ at $x = 1$.

11. Find the equation of the normal line of each function at the indicated point:
(a) $f(x) = 2x^4 - 48x + 32$ at $x = 2$; (b) $g(x) = \frac{4}{x^3} + \frac{3}{x^2} - \frac{6}{x}$ at $x = 1$;
(c) $h(x) = x^{-3/2}$ at $x = 9$; (d) $j(x) = \frac{4x^2 - 5}{x^3 + 7}$ at $x = -2$;
(e) $k(x) = \frac{3\sqrt{x}+5}{2\sqrt{x}-2}$ at $x = 16$; (f) $m(x) = \frac{4x^{7/2}+2}{2x^{-3/2}+1}$ at $x = 1$.

12. Find the equation of the right half–tangent line of each function at $x = 0$:
(a) $f(x) = x^{5/2}$; (b) $g(x) = (4x^{3/2} - 8x^{5/2} + 1)(x^2 + 5x + 2)$;
(c) $h(x) = \frac{2x^{5/2}+x+1}{4\sqrt{x}+3x+2}$; (d) $k(x) = \frac{6x^{7/2}+3}{2x^{3/2}+2x+1}$.

13. Find the second derivative of each function.
(a) $f(x) = 2x^7 - 3x^5 + 9x - 2$ (b) $g(x) = 6x^3 + 2x^{-11} + 3x^{-7} + 4$
(c) $h(x) = \sqrt{x}(5x^4 + 8)$ (d) $i(x) = \frac{1}{x^3 + 2}$
(e) $j(x) = \frac{3x}{7x^2 - 5}$ (f) $k(x) = \frac{\sqrt{x}}{\sqrt{x}+1}$

14. Find all the higher derivatives of each function.
(a) $f(x) = 3x^2 - 8$ (b) $g(x) = 4x^3 - 9x + 7$ (c) $h(x) = 2x^4 - 8x^2 + 4x - 9$
(d) $i(x) = x^{-1}$ (e) $j(x) = \sqrt{x}$ (f) $k(x) = \frac{x}{x+1}$
(g) $m(x) = \frac{x-1}{x+1}$ (h) $p(x) = \frac{1}{2x+3}$ (i) $q(x) = x\sqrt{x}$

Challenging Problems

1. Let $F = f_1 \cdots f_n$ be the product of n functions with the same domain. Find a formula for $(DF)/F$.

2. Let f, g be two functions with domain an open interval I such that $f''(x)$ and $g''(x)$ exist for $x \in I$. Assume $g(x) \neq 0$ for all $x \in I$. Find formulas for the second derivatives of these functions: (a) fg; (b) $\frac{f}{g}$.

2.4 The Chain Rule

The *chain rule* is a formula for computing the derivative of a composite function. Our derivation of the chain rule does not apply to all cases, and a general proof is deferred to the appendix of this section. The chain rule, together with the differentiation formulas of the preceding section, allow us to compute the derivatives of even the most complicated functions. In the last subsection, we apply the chain rule to find the derivative of a function which is defined implicitly.

Composites of Two Functions

We want a formula for the derivative of $f \circ g$ in terms of the derivatives of f and g. The general procedure is illustrated by the computations of the next example. In that example, we will be working with various combinations of variables. Hence we rephrase the definition of the derivative $p'(x)$. We rename h as Δx, the change in x, because h denotes the amount that x has changed between the points x and $x + h$ on the x–axis. We rename $p(x + h) - p(x)$ as Δy, the change in y, because $p(x + h) - p(x)$ denotes the amount that y has changed between the points $(x, p(x))$ and $(x + h, p(x + h))$ on the graph of $y = p(x)$. See Figure 2.4.1. Using this notation:

$$p'(x) = \lim_{h \to 0} \frac{p(x + h) - p(x)}{h} = \lim_{\Delta x \to 0} \frac{\Delta y}{\Delta x}. \qquad (2.4.1)$$

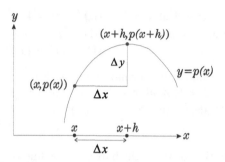

Figure 2.4.1 Delta Notation

Motivating Example 2.4.2 Find the derivative of the function $y = k(x) = \sqrt{7x^3 - 5}$.

Solution We write k as the composite function $f \circ g$ where

$$u = g(x) = 7x^3 - 5, \quad y = f(u) = \sqrt{u} \quad \text{and} \quad y = k(x) = f(g(x)) = \sqrt{7x^3 - 5}\,.$$

Using the notation of (2.4.1):

$$k'(x) = \lim_{\Delta x \to 0} \frac{\Delta y}{\Delta x}, \qquad g'(x) = \lim_{\Delta x \to 0} \frac{\Delta u}{\Delta x}, \qquad f'(u) = \lim_{\Delta u \to 0} \frac{\Delta y}{\Delta u}\,.$$

When Δx is small,

$$\frac{\Delta y}{\Delta x} \approx k'(x) \quad \text{and} \quad \frac{\Delta u}{\Delta x} \approx g'(x) = D(7x^3 - 5) = 21x^2\,. \qquad (2.4.2)$$

When Δx is small, Δu is also small because $u = g(x) = 7x^3 - 5$ is a continuous function. Hence

$$\frac{\Delta y}{\Delta u} \approx f'(u) = D(\sqrt{u}) = \frac{1}{2\sqrt{u}} = \frac{1}{2\sqrt{7x^3 - 5}}\,. \qquad (2.4.3)$$

Combine equations (2.4.2) and (2.4.3):

$$k'(x) \approx \frac{\Delta y}{\Delta x} = \frac{\Delta y}{\Delta u} \cdot \frac{\Delta u}{\Delta x} \approx \frac{1}{2\sqrt{7x^3 - 5}} \cdot 21x^2 = \frac{21x^2}{2\sqrt{7x^3 - 5}}\,.$$

We can make the error in this approximation smaller than any given number by choosing Δx to be sufficiently small. Hence this approximation must be an equality:

$$k'(x) = \frac{21x^2}{2\sqrt{7x^3 - 5}}\,. \qquad \square$$

The procedure we used to find the derivative of the specific composite function $k = f \circ g$ in the preceding example applies to any composite function. Assume that f and g are differentiable functions such that the range of g is contained in the domain of f. Then the composite function $k = f \circ g$ is defined. Recall that the function k sends the number x to the number $y = f(g(x))$ in two steps. First, x is sent to the number $u = g(x)$. Then the number u is sent to $y = f(u)$ which is $f(g(x))$. Using the notation of (2.4.1):

$$k'(x) = \lim_{\Delta x \to 0} \frac{\Delta y}{\Delta x}, \qquad g'(x) = \lim_{\Delta x \to 0} \frac{\Delta u}{\Delta x}, \qquad f'(u) = \lim_{\Delta u \to 0} \frac{\Delta y}{\Delta u}\,.$$

When Δx is small,

$$\frac{\Delta y}{\Delta x} \approx k'(x) \quad \text{and} \quad \frac{\Delta u}{\Delta x} \approx g'(x)\,. \qquad (2.4.4)$$

In other words, the slope $\frac{\Delta y}{\Delta x}$ of the secant line of $y = k(x)$ is approximately the slope $k'(x)$ of the tangent line as in the left diagram of Figure 2.4.3. Also, the slope $\frac{\Delta u}{\Delta x}$ of the secant line of $u = g(x)$ is approximately the slope $g'(x)$ of the tangent line as in the center diagram of Figure 2.4.3. Since g is a differentiable function, g is continuous by Theorem 2.2.13. Hence when Δx is small, Δu is also small. Thus

$$\frac{\Delta y}{\Delta u} \approx f'(u) . \tag{2.4.5}$$

That is, the slope $\frac{\Delta y}{\Delta u}$ of the secant line of $y = f(u)$ is approximately the slope $f'(u)$ of the tangent line as in the right diagram of Figure 2.4.3. Combine equations (2.4.4) and (2.4.5):

$$k'(x) \approx \frac{\Delta y}{\Delta x} = \frac{\Delta y}{\Delta u} \cdot \frac{\Delta u}{\Delta x} \approx f'(u) \cdot g'(x) = f'(g(x)) \cdot g'(x) .$$

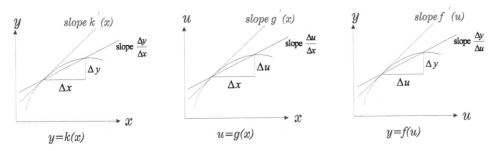

Figure 2.4.3 Tangent and Secant Lines

We can make the error in this approximation smaller than any given number by choosing Δx to be sufficiently small. Hence this approximation must be an equality:

$$k'(x) = f'(g(x)) \cdot g'(x) .$$

This formula for the derivative of the composite function $f \circ g$ is called the *chain rule*. We summarize our conclusions in the following theorem.

Theorem 2.4.4 (Chain Rule) *Assume f and g are differentiable functions such that the range of g is contained in the domain of f. Then the composite function $f \circ g$ is also differentiable with*

$$(f \circ g)'(x) = f'(g(x)) \cdot g'(x) . \tag{2.4.6}$$

The formulation of the chain rule in Leibniz notation reflects our derivation of the formula and is convenient for applications to specific functions:

$$\frac{dy}{dx} = \frac{dy}{du} \cdot \frac{du}{dx} . \tag{2.4.7}$$

This notation is only suggestive. *Derivatives are not fractions*, and we can not cancel du in the product of the two derivatives on the right.

Note that our derivation of the chain rule was made under the assumption that the change Δu in $u = g(x)$ is nonzero. This assumption is often true, as in the case when g is one–to–one. However, in some examples this assumption is false. For example, when $g(x) = c$ is a constant function, Δu is always zero! We defer the proof of the general case of the chain rule to the appendix of this section.

A common application of the chain rule is to the case $y = g(x)^n$ where n is an integer. Then $y = u^n$ with $u = g(x)$, and

$$
\begin{aligned}
\frac{d}{dx}[g(x)^n] &= \frac{dy}{dx} = \frac{dy}{du} \cdot \frac{du}{dx} = \frac{d}{du}(u^n) \cdot \frac{d}{dx}[g(x)] = nu^{n-1}g'(x) \\
&= ng(x)^{n-1}g'(x).
\end{aligned}
\tag{2.4.8}
$$

Examples 2.4.5 (1) Find the derivative of $h(x) = (3x^2 + 7)^{11}$.

Solution Write $y = h(x) = u^{11}$ with $u = 3x^2 + 7$. By the chain rule (2.4.8):

$$
h'(x) = \frac{dy}{dx} = \frac{dy}{du} \cdot \frac{du}{dx} = 11u^{10}(6x) = 11(3x^2 + 7)^{10}(6x) = 66x(3x^2 + 7)^{10}.
$$

(2) Find the derivative of $h(x) = \dfrac{1}{(5x^7 - 9)^{17}}$.

Solution Write $y = h(x) = u^{-17}$ with $u = 5x^7 - 9$. By the chain rule (2.4.8):

$$
h'(x) = \frac{dy}{dx} = \frac{dy}{du} \cdot \frac{du}{dx} = -17u^{-18}(35x^6) = -17(5x^7-9)^{-18} \cdot 35x^6 = -\frac{595x^6}{(5x^7 - 9)^{18}}.
$$

(3) Find the derivative of $h(x) = \sqrt{\dfrac{2x^4 - 7}{5x^3 + 6}}$.

Solution Write $y = h(x) = \sqrt{u}$ where $u = \frac{2x^4-7}{5x^3+6}$. By the quotient rule:

$$
\begin{aligned}
\frac{du}{dx} &= \frac{(5x^3 + 6)D(2x^4 - 7) - (2x^4 - 7)D(5x^3 + 6)}{(5x^3 + 6)^2} \\
&= \frac{(5x^3 + 6)(8x^3) - (2x^4 - 7)(15x^2)}{(5x^3 + 6)^2} = \frac{10x^6 + 48x^3 + 105x^2}{(5x^3 + 6)^2}.
\end{aligned}
$$

By the chain rule (2.4.7):

$$
\begin{aligned}
h'(x) &= \frac{dy}{dx} = \frac{dy}{du} \cdot \frac{du}{dx} = \frac{1}{2\sqrt{u}} \cdot \frac{du}{dx} \\
&= \frac{1}{2\sqrt{\frac{2x^4-7}{5x^3+6}}} \cdot \frac{10x^6 + 48x^3 + 105x^2}{(5x^3 + 6)^2} = \frac{10x^6 + 48x^3 + 105x^2}{2(5x^3 + 6)^{3/2}\sqrt{2x^4 - 7}}.
\end{aligned}
$$

(4) Find the derivative of $h(x) = 9x^{7/2} - 3x^{5/2} + 8x^{-3/2}$.

Solution Write $y = h(x) = 9u^7 - 3u^5 + 8u^{-3}$ with $u = \sqrt{x} = x^{1/2}$. By the chain rule (2.4.7):

$$
\begin{aligned}
h'(x) &= \frac{dy}{dx} = \frac{dy}{du} \cdot \frac{du}{dx} = \left(63u^6 - 15u^4 - 24u^{-4}\right)\frac{1}{2\sqrt{x}} \\
&= \left(63x^3 - 15x^2 - 24x^{-2}\right)\frac{1}{2}x^{-1/2} = \frac{63}{2}x^{5/2} - \frac{15}{2}x^{3/2} - 12x^{-5/2}.
\end{aligned}
$$

(5) Find the derivative of $h(x) = \dfrac{3(2x^5 - 7)^2 - 1}{4(2x^5 - 7)^6 + 3}$.

Solution Write $y = h(u) = \frac{3u^2-1}{4u^6+3}$ with $u = 2x^5 - 7$. By the chain rule (2.4.7):

$$
\begin{aligned}
h'(x) &= \frac{dy}{dx} = \frac{dy}{du} \cdot \frac{du}{dx} \\
&= \frac{6u(4u^6 + 3) - 24u^5(3u^2 - 1)}{(4u^6 + 3)^2} \, 10x^4 \qquad\qquad \text{[by the quotient rule]} \\
&= \frac{6\left[2x^5 - 7\right]\left[4(2x^5 - 7)^6 + 3\right] - 24\left[2x^5 - 7\right]^5\left[3(2x^5 - 7)^2 - 1\right]}{\left[4(2x^5 - 7)^6 + 3\right]^2} \, 10x^4. \ \square
\end{aligned}
$$

Composites of Several Functions

The chain rule can be iterated to compute the derivative of the composite of three or more functions. This is done by inserting parentheses so that in each step we take the derivative of the composite of two functions. For example,

$$
\begin{aligned}
D(f \circ g \circ h)(x) &= D(f \circ (g \circ h))(x) = Df((g \circ h)(x)) \cdot D(g \circ h)(x) \\
&= Df(g(h(x))) \cdot Dg(h(x)) \cdot Dh(x) \ . \tag{2.4.9}
\end{aligned}
$$

Write $y = f(g(h(x))) = f(u)$ where $u = g(h(x)) = g(w)$ with $w = h(x)$. The preceding formula can be restated in Leibniz notation:

$$
\frac{dy}{dx} = \frac{dy}{du} \cdot \frac{du}{dw} \cdot \frac{dw}{dx}. \tag{2.4.10}
$$

Examples 2.4.6 (1) Find the derivative of

$$
k(x) = 7(3x^2 - 8)^{9/2} - 2(3x^2 - 8)^{3/2} + 6(3x^2 - 8)^{-5/2}.
$$

Solution Write $y = k(x) = 7u^9 - 2u^3 + 6u^{-5}$ where $u = (3x^2 - 8)^{1/2}$. By the chain rule:

$$
\begin{aligned}
k'(x) &= \frac{dy}{dx} = \frac{dy}{du} \cdot \frac{du}{dx} = \left[63u^8 - 6u^2 - 30u^{-6}\right]\frac{du}{dx} \\
&= \left\{ 63\left[(3x^2 - 8)^{1/2}\right]^8 - 6\left[(3x^2 - 8)^{1/2}\right]^2 - 30\left[(3x^2 - 8)^{1/2}\right]^{-6} \right\}\frac{du}{dx} \\
&= \left[63(3x^2 - 8)^4 - 6(3x^2 - 8) - 30(3x^2 - 8)^{-3}\right]\frac{du}{dx}. \tag{2.4.11}
\end{aligned}
$$

Write $u = \sqrt{w}$ where $w = 3x^2 - 8$. Apply the chain rule again:

$$
\frac{du}{dx} = \frac{du}{dw} \cdot \frac{dw}{dx} = \frac{1}{2\sqrt{w}}(6x) = \frac{3x}{\sqrt{3x^2 - 8}}.
$$

Substitute the value of $\frac{du}{dx}$ into (2.4.11):

$$
\begin{aligned}
k'(x) &= \left[63(3x^2 - 8)^4 - 6(3x^2 - 8) - 30(3x^2 - 8)^{-3}\right]\frac{3x}{\sqrt{3x^2 - 8}} \\
&= 3x\left[63(3x^2 - 8)^{7/2} - 6(3x^2 - 8)^{1/2} - 30(3x^2 - 8)^{-7/2}\right].
\end{aligned}
$$

(2) Find the derivative of $p(x) = \left(\dfrac{6\sqrt{x} - 1}{8\sqrt{x} + 3}\right)^5$.

Solution Write $y = p(x) = u^5$ where $u = \frac{6\sqrt{x}-1}{8\sqrt{x}+3}$. By the chain rule:

$$
p'(x) = \frac{dy}{dx} = \frac{dy}{du} \cdot \frac{du}{dx} = 5u^4 \frac{du}{dx} = 5\left(\frac{6\sqrt{x} - 1}{8\sqrt{x} + 3}\right)^4 \frac{du}{dx}. \tag{2.4.12}
$$

Write $u = \frac{6w-1}{8w+3}$ where $w = \sqrt{x}$. Apply the chain rule again:

$$
\begin{aligned}
\frac{du}{dx} &= \frac{du}{dw} \cdot \frac{dw}{dx} = \frac{6(8w + 3) - 8(6w - 1)}{(8w + 3)^2}\frac{1}{2\sqrt{x}} \qquad \text{[by the quotient rule]} \\
&= \frac{13}{(8\sqrt{x} + 3)^2\sqrt{x}}.
\end{aligned}
$$

Substitute this value of $\frac{du}{dx}$ into (2.4.12):

$$
p'(x) = 5\left(\frac{6\sqrt{x} - 1}{8\sqrt{x} + 3}\right)^4 \frac{13}{(8\sqrt{x} + 3)^2\sqrt{x}} = \frac{65(6\sqrt{x} - 1)^4}{\sqrt{x}(8\sqrt{x} + 3)^6}. \qquad \square
$$

Implicit Differentiation

An important application of the chain rule is the determination of the derivative of a function which is defined implicitly. That is, given an equation in x and y, we use the chain rule to find a formula for $\frac{dy}{dx}$ without solving the given equation for y.

Motivating Example 2.4.7 Find the slopes of the tangent lines T_1, T_2 at the points $P_1 = \left(\frac{3}{5}, \frac{4}{5}\right)$, $P_2 = \left(\frac{12}{13}, -\frac{5}{13}\right)$ on the circle of radius one with center the origin.

Solution This circle has equation

$$x^2 + y^2 = 1 .$$ (2.4.13)

We could solve this equation for y:

$$y = g_1(x) = +\sqrt{1 - x^2} = \left(1 - x^2\right)^{1/2} \text{ or } y = g_2(x) = -\sqrt{1 - x^2} = -\left(1 - x^2\right)^{1/2} .$$

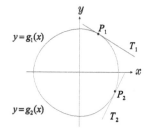

Figure 2.4.8
Tangent Lines of the Circle $x^2 + y^2 = 1$

We use the function g_1 to study the point P_1 on the upper semi–circle and the function g_2 to study the point P_2 on the lower semi–circle. See Figure 2.4.8. Now we could use the chain rule to compute the slopes of T_1 and T_2 as the derivatives $g_1'\left(\frac{3}{5}\right)$ and $g_2'\left(\frac{12}{13}\right)$. However, there is an alternate more elegant solution which computes both slopes simultaneously. Take the derivative with respect to x of equation (2.4.13):

$$\frac{d}{dx}\left(x^2 + y^2\right) = \frac{d}{dx}(1)$$

$$2x + \frac{d}{dx}\left(y^2\right) = 0$$ (2.4.14)

Think of $y = g_1(x)$ and $y = g_2(x)$ as expressions in x. Although we know the formulas for $g_1(x)$ and $g_2(x)$, that information is not required. Let $u = y^2$. By the chain rule:

$$\frac{d}{dx}\left(y^2\right) = \frac{du}{dx} = \frac{du}{dy} \cdot \frac{dy}{dx} = \frac{d}{dy}(y^2) \cdot \frac{dy}{dx} = 2y\frac{dy}{dx}$$

Then equation (2.4.14) becomes:

$$2x + 2y\frac{dy}{dx} = 0.$$

This is a linear equation in $\frac{dy}{dx}$ which we solve for $\frac{dy}{dx}$:

$$\frac{dy}{dx} = -\frac{x}{y}$$ (2.4.15)

Thus, the slope of the tangent line T_1 at $(x, y) = \left(\frac{3}{5}, \frac{4}{5}\right)$ is $-\frac{x}{y} = -\frac{3/5}{4/5} = -\frac{3}{4}$ while the slope of the tangent line T_2 at $(x, y) = \left(\frac{12}{13}, -\frac{5}{13}\right)$ is $-\frac{x}{y} = -\frac{12/13}{-5/13} = \frac{12}{5}$. □

The preceding method of computing $\frac{dy}{dx}$ is called *implicit differentiation*. It begins with an equation relating x and y. The set of points (x, y) which satisfy this equation determine a curve in the plane. This curve defines a function $y = g(x)$ with a suitable domain. Taking the derivative of the original equation with respect to x results in a linear equation which can be solved for $\frac{dy}{dx} = g'(x)$. The knowledge of an explicit formula for the function $y = g(x)$ is not needed. In complicated examples it may be impossible to solve for y to obtain an explicit formula for $g(x)$. Note that the formula obtained for $\frac{dy}{dx}$ by implicit differentiation will involve both x's and y's.

Examples 2.4.9 (1) Find $\frac{dy}{dx}$ for the function $y = g(x)$ which satisfies the equation

$$2x^3 - 5x^2y^4 + 4y^5 = 7.$$

Solution Take the derivative with respect to x of the given equation:

$$\frac{d}{dx}\left(2x^3 - 5x^2y^4 + 4y^5\right) = \frac{d}{dx}(7).$$

Use the product rule to differentiate the second summand on the left:

$$6x^2 - 5(2x)(y^4) - 5x^2\frac{d}{dx}(y^4) + 4\frac{d}{dx}(y^5) = 0. \qquad (2.4.16)$$

Let $u = y^4$ and $v = y^5$. By the chain rule:

$$\frac{d}{dx}\left(y^4\right) = \frac{du}{dx} = \frac{du}{dy}\frac{dy}{dx} = \frac{d}{dy}\left(y^4\right)\cdot\frac{dy}{dx} = 4y^3\frac{dy}{dx}$$

$$\frac{d}{dx}\left(y^5\right) = \frac{dv}{dx} = \frac{dv}{dy}\frac{dy}{dx} = \frac{d}{dy}\left(y^5\right)\cdot\frac{dy}{dx} = 5y^4\frac{dy}{dx}.$$

Substitute the values of these derivatives into equation (2.4.16):

$$6x^2 - 10xy^4 - 5x^2\left(4y^3\frac{dy}{dx}\right) + 4\left(5y^4\frac{dy}{dx}\right) = 0.$$

Solve this linear equation for $\frac{dy}{dx}$:

$$\left(-20x^2y^3 + 20y^4\right)\frac{dy}{dx} = 10xy^4 - 6x^2,$$

$$\frac{dy}{dx} = \frac{5xy^4 - 3x^2}{10y^4 - 10x^2y^3}.$$

(2) Find $\frac{dy}{dx}$ for the function $y = h(x)$ which satisfies the equation

$$x^2y = \frac{3}{xy^4 + 7}.$$

Solution Take the derivative with respect to x of the given equation:

$$\frac{d}{dx}\left(x^2y\right) = \frac{d}{dx}\left(\frac{3}{xy^4 + 7}\right) = 3\frac{d}{dx}\left[(xy^4 + 7)^{-1}\right].$$

Use the product rule to differentiate the left side of this equation and the chain rule to differentiate the right side of this equation.

$$2x\cdot y + x^2\cdot\frac{dy}{dx} = 3(-1)(xy^4 + 7)^{-2}\frac{d}{dx}(xy^4 + 7)$$

$$= \frac{-3}{(xy^4 + 7)^2}\left[(1)y^4 + x\frac{d}{dx}(y^4)\right] \qquad \text{[by the product rule]}$$

$$= \frac{-3y^4 - 3x(4y^3\frac{dy}{dx})}{(xy^4 + 7)^2} \qquad \text{[by the chain rule]}$$

Solve this linear equation for $\frac{dy}{dx}$ by bringing all summands with a factor $\frac{dy}{dx}$ to the left side of the equation and bringing all other summands to the right side:

$$\left[x^2 + \frac{12xy^3}{(xy^4 + 7)^2}\right]\frac{dy}{dx} = -2xy - \frac{3y^4}{(xy^4 + 7)^2}$$

$$\frac{dy}{dx} = \frac{-2xy - \frac{3y^4}{(xy^4+7)^2}}{x^2 + \frac{12xy^3}{(xy^4+7)^2}} = -\frac{2xy(xy^4 + 7)^2 + 3y^4}{x^2(xy^4 + 7)^2 + 12xy^3}$$

multiplying numerator and denominator by $(xy^4 + 7)^2$.

(3) Find the derivative of the function $y = k(x)$ which satisfies the equation

$$xy^2 + 1 = \sqrt{x^2 + y}.$$

Solution Take the derivative of this equation with respect to x, using the product rule on the left side and the chain rule on the right side:

$$\frac{d}{dx}(xy^2 + 1) = \frac{d}{dx}(x^2 + y)^{1/2}$$

$$1 \cdot y^2 + x \cdot 2y\frac{dy}{dx} = \frac{1}{2}(x^2 + y)^{-1/2}\left(2x + \frac{dy}{dx}\right)$$

Solve this linear equation for $\frac{dy}{dx}$ by bringing all summands with a factor $\frac{dy}{dx}$ to the left side of the equation and bringing all other summands to the right side:

$$\left[2xy - \frac{1}{2}(x^2 + y)^{-1/2}\right]\frac{dy}{dx} = x(x^2 + y)^{-1/2} - y^2$$

$$\frac{dy}{dx} = \frac{x(x^2 + y)^{-1/2} - y^2}{2xy - \frac{1}{2}(x^2 + y)^{-1/2}} = \frac{2x - 2y^2\sqrt{x^2 + y}}{4xy\sqrt{x^2 + y} - 1}$$

multiplying numerator and denominator by $2\sqrt{x^2 + y}$.

(4) Find the second derivative of the circle of radius one and center the origin.

Solution Recall that each semi–circle is given by a function $y = g(x)$ which satisfies the equation

$$x^2 + y^2 = 1. \tag{2.4.17}$$

We used implicit differentiation to establish the derivative in (2.4.15) as:

$$\frac{dy}{dx} = -\frac{x}{y}. \tag{2.4.18}$$

Take the derivative of this equation with respect to x and evaluate the right side of the resulting equation by the quotient rule:

$$\frac{d^2y}{dx^2} = -\frac{d}{dx}\left(\frac{x}{y}\right) = -\frac{1 \cdot y - x \cdot \frac{dy}{dx}}{y^2}.$$

Now use equation (2.4.18) to substitute $-\frac{x}{y}$ for $\frac{dy}{dx}$:

$$\frac{d^2y}{dx^2} = -\frac{y - x\left(-\frac{x}{y}\right)}{y^2} = -\frac{y + \frac{x^2}{y}}{y^2}$$

$$= -\frac{y^2 + x^2}{y^3} \qquad \text{[multiplying numerator and denominator by } y]$$

$$= -\frac{1}{y^3} \qquad \text{[using (2.4.17)]}$$

(5) Find the second derivative of the function $y = m(x)$ which satisfies the equation

$$xy - 1 = x + y.$$

Solution Find $\frac{dy}{dx}$ by taking the derivative with respect to x of the given equation. By the product rule:

$$\frac{d}{dx}(xy - 1) = \frac{d}{dx}(x + y)$$

$$1 \cdot y + x \cdot \frac{dy}{dx} = 1 + \frac{dy}{dx}$$

$$(x-1)\frac{dy}{dx} = 1 - y$$

$$\frac{dy}{dx} = \frac{1-y}{x-1}. \qquad (2.4.19)$$

To find $\frac{d^2y}{dx^2}$ take the derivative of this equation. By the quotient rule:

$$\frac{d^2y}{dx^2} = \frac{d}{dx}\left(\frac{1-y}{x-1}\right) = \frac{(x-1)\frac{d}{dx}(1-y) - (1-y)\frac{d}{dx}(x-1)}{(x-1)^2}$$

$$= \frac{(x-1)(-\frac{dy}{dx}) - (1-y)(1)}{(x-1)^2}.$$

Substitute the value of $\frac{dy}{dx}$ given by (2.4.19) into the preceding equation:

$$\frac{d^2y}{dx^2} = \frac{-(x-1)\left(\frac{1-y}{x-1}\right) - (1-y)}{(x-1)^2} = \frac{-2(1-y)}{(x-1)^2} = \frac{2y-2}{(x-1)^2}. \qquad \square$$

We use implicit differentiation, to compute the derivative of rational powers of x.

Proposition 2.4.10 *Let n be any rational number. Then*

$$D(x^n) = nx^{n-1}.$$

Proof. We already know that the above formula for $D(x^n)$ holds when n is an integer. If n is a rational number, we can write $n = \frac{p}{q}$ where p, q are integers with $q \neq 0$. Then $y = x^n = x^{p/q}$ and

$$y^q = x^p.$$

Take the derivative with respect to x of this equation. Use the chain rule to evaluate the derivative on the left side of the equation:

$$\frac{d}{dx}(y^q) = \frac{d}{dx}(x^p)$$

$$qy^{q-1}\frac{dy}{dx} = px^{p-1} = px^{nq-1}$$

$$\frac{dy}{dx} = \frac{px^{nq-1}}{qy^{q-1}} = \frac{p}{q}\frac{x^{nq-1}}{x^{n(q-1)}} = nx^{nq-1-n(q-1)} = nx^{n-1}. \qquad \square$$

We apply this proposition in the following examples.

Examples 2.4.11 (1) Find the derivative of $h(x) = 4x^{2/3} - 2x^{8/7} + 9x^{-4/5}$.

Solution By the preceding proposition:

$$h'(x) = \frac{8}{3}x^{-1/3} - \frac{16}{7}x^{1/7} - \frac{36}{5}x^{-9/5}.$$

(2) Find the derivative of the function $y = j(x)$ with domain $[0, \infty)$ defined by

$$j(x) = \sqrt[6]{2x^3 + 7x + 9}.$$

Solution Observe $y = j(x) = u^{1/6}$ where $u = 2x^3 + 7x + 9$. By the chain rule:

$$j'(x) = \frac{dy}{dx} = \frac{dy}{du} \cdot \frac{du}{dx} = \frac{1}{6}u^{-5/6}\frac{du}{dx} = \frac{1}{6}(2x^3 + 7x + 9)^{-5/6}(6x^2 + 7).$$

(3) Find the derivative of the function $y = k(x)$ which satisfies the equation

$$x^{2/3} + y^{2/3} = (xy)^{1/4} + 1.$$

Solution To find $\frac{dy}{dx}$, take the derivative with respect to x of the given equation. By the product and chain rules:

$$\frac{d}{dx}\left(x^{2/3} + y^{2/3}\right) = \frac{d}{dx}\left[(xy)^{1/4} + 1\right] = \frac{d}{dx}\left(x^{1/4}y^{1/4} + 1\right)$$

$$\frac{2}{3}x^{-1/3} + \frac{2}{3}y^{-1/3}\frac{dy}{dx} = \frac{1}{4}x^{-3/4}y^{1/4} + x^{1/4}\left(\frac{1}{4}y^{-3/4}\frac{dy}{dx}\right)$$

$$\left(\frac{2}{3}y^{-1/3} - \frac{1}{4}x^{1/4}y^{-3/4}\right)\frac{dy}{dx} = \frac{1}{4}x^{-3/4}y^{1/4} - \frac{2}{3}x^{-1/3}$$

$$\frac{dy}{dx} = \frac{\frac{1}{4}x^{-3/4}y^{1/4} - \frac{2}{3}x^{-1/3}}{\frac{2}{3}y^{-1/3} - \frac{1}{4}x^{1/4}y^{-3/4}} \qquad \square$$

Historical Remarks

The chain rule and the formulas of Section 2.3 were discovered independently by Isaac Newton in the 1660s and by Gottfried Wilhelm Leibniz in the 1670s. We briefly present their approaches and illustrate them with the quotient and chain rules.

Newton's approach to derivatives considers all variables z to depend on time t. He calls the rate of change of z, its derivative with respect to t, the *fluxion* of z which he denotes as $\dot{z}$. He computes the derivative of a function $y = f(x)$ as

$$f'(x) = \frac{\dot{y}}{\dot{x}}.$$

(From our point of view, this follows from the chain rule: $\frac{dy}{dt} = \frac{dy}{dx}\frac{dx}{dt}$.) He computes the derivatives of polynomials and then uses the methods of Sections 3 and 4 to compute the derivatives of functions constructed explicitly and implicitly from polynomials. Newton proves the quotient rule in the following manner. Let $u = g(x)$ and $v = h(x)$ be polynomials with $y = f(x) = \frac{g(x)}{h(x)} = \frac{u}{v}$ a rational function. Since $u = yv$ is a polynomial, he computes $\dot{u} = \dot{y}\,v + y\,\dot{v}$. Then

$$f'(x) = \frac{\dot{y}}{\dot{x}} = \frac{\frac{\dot{u}-y\dot{v}}{v}}{\dot{x}} = \frac{\frac{\dot{u}}{\dot{x}} - y\frac{\dot{v}}{\dot{x}}}{v} = \frac{g'(x) - \frac{g(x)}{h(x)}h'(x)}{h(x)} = \frac{g'(x)h(x) - g(x)h'(x)}{h(x)^2}.$$

Newton did not have a general formulation of the chain rule. Nevertheless, he could compute the derivative of composite functions. For example, to compute the derivative of $y = f(x) = \sqrt{x^2 + 3x - 5}$, Newton would write $y^2 = x^2 + 3x - 5$. Since this equation involves only polynomials, he would compute

$$2y\,\dot{y} = 2x\,\dot{x} + 3\,\dot{x}, \qquad \frac{2y\,\dot{y}}{\dot{x}} = 2x + 3,$$

$$f'(x) = \frac{\dot{y}}{\dot{x}} = \frac{2x+3}{2y} = \frac{2x+3}{2\sqrt{x^2 + 3x - 5}}.$$

Leibniz's approach to the derivative of $y = f(x)$ begins with the approximation

$$f'(x) \approx \frac{\Delta y}{\Delta x}.$$

However, he does not take the limit of this fraction as Δx goes to zero. Instead, he asserts that since this approximation becomes increasingly better as Δx gets smaller, the exact value of $f'(x)$ is obtained by taking Δx and Δy to be the *infinitesmaly small* quantities dx and dy. Then $f'(x)$ is the *quotient* of these infinitesmals:

$$f'(x) = \frac{dy}{dx}.$$

Let y be a function of u, and let u be a function of x. The chain rule then becomes a trivial application of the arithmetic of fractions:

$$\frac{dy}{dx} = \frac{dy}{du}\frac{du}{dx}.$$

The other rules for differentiation are also easy consequences of this method. For example, Leibniz proves the quotient rule as follows. Let $y = \frac{u}{v}$ then

$$\frac{dy}{dx} = \frac{(y+dy)-y}{dx} = \frac{\frac{u+du}{v+dv}-\frac{u}{v}}{dx} = \frac{v(u+du)-u(v+dv)}{v(v+dv)dx} = \frac{vdu-udv}{(v^2+vdv)dx}$$

$$= \frac{v\frac{du}{dx}-u\frac{dv}{dx}}{v^2+vdv} = \frac{v\frac{du}{dx}-u\frac{dv}{dx}}{v^2}.$$

Leibniz ignores vdv in the sum $v^2 + vdv$, as vdv is *infinitely small* compared to v^2.

Newton's approach to methods of differentiation suffers from its ad hoc nature and its limitation to functions constructed from polynomials. On the other hand, Leibniz's approach is easy, general and gives an intuitive understanding of the derivative formulas. For these reasons Leibniz's approach was preferred, outside of England, and led to many advances and applications of calculus in the eighteenth century. However, Leibniz's approach is based on the use of infinitesmals which is totally lacking in mathematical rigor. Our rigorous derivation of these formulas using limits was first given in textbooks written by Augustin–Louis Cauchy in the 1820s.[1]

Summary

The reader should be able to use the chain rule to calculate the derivative of composites of two or more functions. She should also know how to compute the first and second derivatives of functions which are defined implicitly. In addition, she should be able to differentiate functions which involve fractional exponents.

Basic Exercises

1. **(a)** Find Δx and Δy when $y = x^2 + 1$, $x = 3$ and $h = \frac{1}{2}$.
(b) Find Δx and Δu when $u = \frac{1}{x}$, $x = 4$ and $h = -1$.
(c) Find Δy when $y = \frac{u}{u+1}$, $u = 2$ and $\Delta u = -\frac{1}{3}$.
(d) Find Δz when $z = \sqrt{4u+1}$, $u = 2$ and $\Delta u = 4$.
(e) Find Δx when $y = 3x - 2$, $x = 4$ and $\Delta y = \frac{1}{5}$.

2. **(a)** Find $\frac{\Delta y}{\Delta x}$ when $y = 3 - x^2$, $x = 3$ and $h = \frac{1}{3}$.
(b) Find $\frac{\Delta u}{\Delta x}$ when $u = \frac{1}{1+x^2}$, $x = 1$ and $h = 1$.
(c) Find $\frac{\Delta y}{\Delta u}$ when $y = u^3$, $u = 2$ and $\Delta u = -\frac{2}{3}$.
(d) Find $\frac{\Delta w}{\Delta u}$ when $w = \frac{1}{2u+1}$, $u = 4$ and $\Delta u = -\frac{1}{2}$.
(e) Find $\frac{\Delta z}{\Delta w}$ when $z = \sqrt{3w+1}$, $w = 5$ and $\Delta w = 3$.

---

[1]Cauchy gave the derivation of the chain rule, presented at the beginning of this section. He did not realize the gap in his proof and the need for the proof given in the appendix to this section.

3. Find the derivative of each function.

(a) $y = (5x^7 - 2x^5 + 7)^{18}$

(b) $y = \sqrt{3x^8 + 7x^4 + 3}$

(c) $y = \frac{1}{\sqrt{9x^6 + 5x^2 + 4}}$

(d) $y = \left(\frac{3x^5 - 2}{7x^2 + 9}\right)^{4/9}$

(e) $y = 3(2x^2 - 8)^{-4/3}$

(f) $y = \left(4\sqrt{3x^2 + 1} - 5\right)\left(7\sqrt{3x^2 + 1} + 8\right)$

(g) $y = \frac{1}{(5x^4 - 2)^8}$

(h) $y = (5\sqrt{x} - 1)^{9/7}(8\sqrt{x} + 6)^{7/9}$

(i) $y = \frac{\sqrt{x} + \sqrt[8]{x}}{\sqrt[4]{x} + \sqrt[3]{x}}$

(j) $y = \frac{(4x^3 - 12)^{5/4}}{(7x^6 + 8)^{2/7}}$

(k) $y = \sqrt{1 + (x^3 + 2)^9}$

(l) $y = (5 - \sqrt{2x^5 + 7})^8$

(m) $y = \left(\frac{\sqrt[3]{x^3 + 1} + 5}{4\sqrt[3]{x^3 + 1} + 2}\right)^{2/5}$

(n) $y = \sqrt[4]{\frac{5\sqrt{x} + 7}{2\sqrt{x} + 3}}$

(o) $y = \sqrt{1 + \sqrt{2 + \sqrt{3 + x}}}$

(p) $y = (7 + (8 - (x^4 + 6)^7)^5)^9$

(q) $y = \left(\frac{(1 + x^8)^{7/3}}{2 + (1 + x^8)^{7/3}}\right)^6$

(r) $y = \left[5 - \sqrt{4 + (x^2 + 5)^8}\right]^7$

(s) $y = \sqrt{\sqrt[3]{x^2 + 1} + \sqrt[4]{x^2 + 1}}$

(t) $y = \sqrt{\frac{1 - 12(x^7 + 5)^{3/4}}{2 + 18(x^7 + 5)^{2/3}}}$

4. Find the tangent line of each function at the indicated point:

(a) $y = (17 - x^2)^{10}$ at $x = 4$;

(b) $y = \sqrt{x^2 - x + 7}$ at $x = 2$;

(c) $y = \frac{\sqrt{x}}{\sqrt[3]{x} + \sqrt[8]{x}}$ at $x = 64$;

(d) $y = (9 - (1 + (10 - x^2)^7)^3)^4$ at $x = -3$;

(e) $y = (4 - 2\sqrt{x} + 9x^{3/2} - x^{5/2})^3$ at $x = 9$.

5. Find the normal line of each function at the indicated point:

(a) $y = \sqrt{\frac{x}{x+3}}$ at $x = 1$;

(b) $y = (11 + x^2)^{2/3}$ at $x = -4$;

(c) $y = \sqrt{8 + (9 - x^3)^4}$ at $x = 2$;

(d) $y = (6 - \sqrt{x^2 + 7})(15 - 2\sqrt{x^2 + 7})$ at $x = 3$;

(e) $y = \frac{\sqrt{x}}{4 - \sqrt{x}}$ at $x = 36$.

6. Find the points at which the derivative of each function does not exist. Identify each of these points as a vertical tangent, cusp or vertical cusp.

(a) $f(x) = (x - 5)^{3/4}$

(b) $g(x) = (x + 4)^{2/7}$

(c) $h(x) = (x^2 - 6x + 9)^{1/5}$

(d) $j(x) = (x^2 - 6x + 8)^{1/3}$

(e) $k(x) = \sqrt{4x^2 - 4x + 1}$

(f) $m(x) = \sqrt[3]{4x^2 - 4x + 1}$

7. Assume that f is a differentiable function. In each case, find the derivative $\frac{dy}{dx}$ in terms of x, y, f and f'.

(a) $y = f(x)^3$

(b) $y = f(x^4)$

(c) $y = f(\sqrt{x})^5$

(d) $y = \sqrt{f(x)}$

(e) $y = f(f(x))$

(f) $y = \frac{1}{1 + f(3x^2 + 2)}$

(g) $y = \frac{\sqrt{f(x)}}{f(x) + f(\sqrt{x})}$

(h) $y = \sqrt{f(\sqrt{x})}$

8. State the chain rule for the composite of four functions.

9. Let f, g be differentiable functions with domain $\Re$. We are given that

$$\begin{array}{llll}
f(1) = 3 & f'(1) = -2 & g(1) = 2 & g'(1) = 5 \\
f(2) = 2 & f'(2) = 3 & g(2) = 1 & g'(2) = 4 \\
f(3) = 4 & f'(3) = -7 & g(3) = 4 & g'(3) = -1 \\
f(4) = 1 & f'(4) = 6 & g(4) = 3 & g'(4) = 8
\end{array}$$

Find (a) $D(f \circ g)(1)$; (b) $D(f \circ g)(4)$; (c) $D(g \circ f)(3)$; (d) $D(g \circ f)(2)$.

10. Let f, g, h be differentiable functions with domain $\Re$. We are given that

$$\begin{array}{llllll}
f(1) = 2 & f'(1) = 3 & g(1) = 4 & g'(1) = 7 & h(1) = 3 & h'(1) = -2 \\
f(2) = 3 & f'(2) = -2 & g(2) = 3 & g'(2) = 3 & h(2) = 4 & h'(2) = -1 \\
f(3) = 4 & f'(3) = -5 & g(3) = 2 & g'(3) = -3 & h(3) = 1 & h'(3) = 3 \\
f(4) = 1 & f'(4) = 1 & g(4) = 1 & g'(4) = -4 & h(4) = 2 & h'(4) = 6
\end{array}$$

Find (a) $D(f \circ g \circ h)(4)$; (b) $D(g \circ f \circ h)(1)$; (c) $D(h \circ g \circ f)(3)$; (d) $D(h \circ f \circ g)(2)$.

11. Find the derivative of each implicitly defined function.

(a) $x^4 + y^4 = 3$

(b) $5x^3y^6 - 7x^2y^4 = 8$

(c) $6xy^4 - 7x^5y = 2$

(d) $\sqrt{3x^2 + 4y^2} = 5x^2y^2 + 1$

(e) $(4y^5 - 7)(6x^5 + 2) = 7xy - 1$

(f) $\frac{5x^3 - 6}{8y + 9} = 7x^2y - 2$

(g) $(5x^6 - 2x^4y^7)^{5/3} = 8x^2y - 6$

(h) $\sqrt[6]{5y - 3} = 3xy^2 + 1$

(i) $x^{8/5} + y^{2/5} = x^{6/5}y^{4/5} + 1$

12. Find the equation of the tangent line to each implicitly defined function at the indicated point:

(a) $5x^{2/3} + 4y^{2/3} = 21$ at $(1, 8)$;

(b) $4x^3y^2 - 2x^2y^3 = 40$ at $(2, -1)$;

(c) $\sqrt{2xy^4 + 3} = 5xy - 12$ at $(3, 1)$;

(d) $(x^4y^2 - 6)(3x^2y + 1) = 8xy^2 + 2$ at $(-1, 4)$;

(e) $x^2y - 9 = \frac{2y^3 + 6}{3x^3 - 4}$ at $(2, 3)$.

13. Find the equation of the normal line to each implicitly defined function at the indicated point.

(a) $x^{1/3} + y^{1/3} = 2$ at $(125, -27)$;

(b) $\sqrt{x} + \sqrt{y} = xy - 5$ at $(9, 1)$;

(c) $3 + 9\sqrt{x^2 + y^2} = xy^2$ at $(3, -4)$;

(d) $\frac{4xy}{x^2 + y^2} = x + y - 2$ at $(-1, 1)$;

(e) $x^{3/4}y^{1/4} + 11 = x - y$ at $(81, 16)$.

14. Find the second derivative of each implicitly defined function.

(a) $x^3 + y^3 = 1$

(b) $x^2 + y^2 = xy + 3$

(c) $\frac{x+1}{y+1} = y^2 + 1$

(d) $\sqrt{y + 2} = y^3 + 5x$

(e) $x^{3/4}y^{3/4} + 1 = x^3 + 12$

(f) $\sqrt{x} + \sqrt{y} = x + 1$

15. Assume f is a differentiable function. In each case, find the derivative $\frac{dy}{dx}$ in terms of x, y, f and f'.

(a) $f(xy) = x^3 + y^3 + 2$

(b) $f(x + y) = x^3y^2 + 7$

(c) $f(y)^2 = x^2y - 3$

(d) $f(x^2 + y^2) = xy + 1$

(e) $f(\sqrt{x + y})^3 = 8x^5 - 3$

(f) $f(x^2) + f(y^2) = f(xy) + 1$

Challenging Problems

1. Let $f^{[n]}(x)$ denote the composite of f with itself n times. Find a formula for the derivative of $f^{[n]}(x)$.

2. Find the first, second and third derivatives for each of the following implicitly defined functions. (a) $x^3 + y^3 = 1$ (b) $xy + 7 = 4x^6 + 3y$

Appendix: Proof of the Chain Rule

We introduce a fundamental method for analyzing derivatives and apply it to prove the chain rule. Recall that the derivative, the slope of the tangent line T, is defined as the limit of the slope of the secant line S as S approaches T. We introduce a *difference function* to represent the difference between the derivative and the slope of the secant line S. This difference function has limit zero as S approaches T.

Theorem 2.4.12 (Chain Rule) *Assume f and g are differentiable functions such that the range of g is contained in the domain of f. Then the composite function $f \circ g$ is also differentiable with*

$$(f \circ g)'(x) = f'(g(x)) \cdot g'(x).$$

Proof Since $f'(u)$ is defined as the limit of $\frac{f(u+k) - f(u)}{k}$ as k approaches zero, the difference function

$$F(u, k) \quad = \quad f'(u) - \frac{f(u + k) - f(u)}{k} \qquad \text{(2.4.20)}$$

has the property : $\lim_{k \to 0} F(u, k) \quad = \quad 0.$

Similarly, since $g'(x)$ exists, the difference function

$$G(x, h) \;=\; g'(x) - \frac{g(x+h) - g(x)}{h} \qquad (2.4.21)$$

has the property : $\lim_{h \to 0} G(x, h) \;=\; 0.$

Rewrite equations (2.4.20) and (2.4.21) as:

$$f(u + k) - f(u) \;=\; f'(u)k - F(u, k)k \qquad \text{and} \qquad (2.4.22)$$
$$g(x + h) - g(x) \;=\; g'(x)h - G(x, h)h. \qquad (2.4.23)$$

We now turn our attention to the composite function $f \circ g$. Since $f'(g(x))$ exists, consider the special case of (2.4.22) where $u = g(x)$ and $k = g(x + h) - g(x)$:

$$f(g(x) + k) - f(g(x)) \;=\; f'(g(x))k - F(g(x), k)k,$$
$$f(g(x+h)) - f(g(x)) \;=\; f'(g(x)) \left[g(x+h) - g(x)\right] - F(g(x), k)\left[g(x+h) - g(x)\right].$$

Use (2.4.23) to substitute $g'(x)h - G(x, h)h$ for the first $g(x + h) - g(x)$:

$$f(g(x+h)) - f(g(x)) = f'(g(x)) \left[g'(x)h - G(x, h)h\right] - F(g(x), k)\left[g(x+h) - g(x)\right].$$

Divide this equation by h:

$$\frac{f(g(x+h)) - f(g(x))}{h} = f'(g(x)) \left[g'(x) - G(x, h)\right] - F(g(x), k)\left[\frac{g(x+h) - g(x)}{h}\right].$$

Now take the limit of this equation as h approaches 0:

$$\lim_{h \to 0} \frac{f(g(x+h)) - f(g(x))}{h} \;=\; \lim_{h \to 0} f'(g(x))g'(x) - \lim_{h \to 0} \left[f'(g(x))G(x, h)\right]$$
$$- \lim_{h \to 0} F(g(x), k) \left[\frac{g(x+h) - g(x)}{h}\right]$$
$$(f \circ g)'(x) \;=\; f'(g(x))g'(x) - f'(g(x)) \lim_{h \to 0} G(x, h)$$
$$- \lim_{h \to 0} F(g(x), k) \lim_{h \to 0} \frac{g(x+h) - g(x)}{h}.$$

Since g is differentiable, it is continuous by Prop. 2.2.13. Hence $k = g(x + h) - g(x)$ approaches zero as h approaches zero. Thus, the previous equation becomes:

$$(f \circ g)'(x) \;=\; f'(g(x))g'(x) - f'(g(x)) \cdot 0 - \lim_{k \to 0} F(g(x), k) \cdot g'(x)$$
$$=\; f'(g(x))g'(x) - 0 \cdot g'(x) = f'(g(x))g'(x). \qquad \square$$

2.5   Derivatives of Trigonometric Functions

In the first subsection, we compute the derivatives of the six trigonometric functions. In the second subsection implicit differentiation is used to determine the derivative of an inverse function in terms of the derivative of the original function. This method is applied in the third subsection to find the derivatives of $\arcsin x$, $\arctan x$ and $\operatorname{arcsec} x$.

The Six Trigonometric Functions

We find the derivatives of the six trigonometric functions. Since these formulas are very important, we present their derivation in full detail before illustrating their use with several examples. The derivative of the sine function is computed directly from the definition of the derivative. The resulting limit is evaluated from the trigonometric limits of Section 1.6. The derivative of the sine function determines the derivatives of the other five trigonometric functions by the methods of the preceding two sections.

Theorem 2.5.1
(a) $D(\sin x) = \cos x$ (b) $D(\cos x) = -\sin x$
(c) $D(\tan x) = \sec^2 x$ (d) $D(\cot x) = -\csc^2 x$
(e) $D(\sec x) = \tan x \sec x$ (f) $D(\csc x) = -\cot x \csc x$

Proof (a) We compute the derivative of $f(x) = \sin x$ from the definition of the derivative. We use the following two limits which we determined in Theorem 1.6.11:

$$\lim_{h \to 0} \frac{\sin h}{h} = 1 \quad \text{and} \quad \lim_{h \to 0} \frac{\cos h - 1}{h} = 0.$$

By the definition of the derivative:

$$
\begin{aligned}
f'(x) &= \lim_{h \to 0} \frac{\sin(x+h) - \sin x}{h} = \lim_{h \to 0} \frac{\sin x \cos h + \cos x \sin h - \sin x}{h} && \text{[by (1.4.15)]}\\
&= \lim_{h \to 0} \frac{\sin x(\cos h - 1)}{h} + \lim_{h \to 0} \frac{\cos x \sin h}{h} = \sin x \lim_{h \to 0} \frac{\cos h - 1}{h} + \cos x \lim_{h \to 0} \frac{\sin h}{h}\\
&= (\sin x)(0) + (\cos x)(1) = \cos x.
\end{aligned}
$$

(b) We use implicit differentiation to compute the derivative of the function $y = \cos x$. Take the derivative with respect to x of the Pythagorean identity (1.4.13):

$$
\begin{aligned}
1 &= \cos^2 x + \sin^2 x = y^2 + \sin^2 x \quad \text{sub } y = \cos x\\
\frac{d}{dx}(1) &= \frac{d}{dx}\left(y^2 + \sin^2 x\right) \quad s\\
0 &= 2y\frac{dy}{dx} + 2\sin x D(\sin x) = 2y\frac{dy}{dx} + 2\sin x \cos x && \text{[by (a)]}
\end{aligned}
$$

Solve this linear equation for $\frac{dy}{dx}$:

$$\frac{dy}{dx} = -\frac{\sin x \cos x}{y} = -\frac{\sin x \cos x}{\cos x} = -\sin x.$$

(c) Take the derivative of the definition (1.4.6) of the tangent function:

$$
\begin{aligned}
D(\tan x) &= D\left(\frac{\sin x}{\cos x}\right) = \frac{D(\sin x)\cos x - \sin x D(\cos x)}{\cos^2 x} && \text{[by the quotient rule]}\\
&= \frac{(\cos x)(\cos x) - (\sin x)(-\sin x)}{\cos^2 x} && \text{[by (a), (b)]}\\
&= \frac{\cos^2 x + \sin^2 x}{\cos^2 x} = \frac{1}{\cos^2 x} && \text{[by Pythagorean identity (1.4.13)]}\\
&= \sec^2 x && \text{[by definition of sec (1.4.6)]}
\end{aligned}
$$

(d) Similarly, we differentiate the definition (1.4.5) of the cotangent function:

$$
\begin{aligned}
D(\cot x) &= D\left(\frac{\cos x}{\sin x}\right) = \frac{D(\cos x)\sin x - \cos x D(\sin x)}{\sin^2 x} && \text{[by the quotient rule]}\\
&= \frac{(-\sin x)(\sin x) - (\cos x)(\cos x)}{\sin^2 x} && \text{[by (a), (b)]}\\
&= -\frac{\sin^2 x + \cos^2 x}{\sin^2 x} = -\frac{1}{\sin^2 x} && \text{[by Pythagorean identity (1.4.13)]}\\
&= -\csc^2 x && \text{[by definition of csc (1.4.5)]}
\end{aligned}
$$

(e), (f) Recall the definitions (1.4.6) and (1.4.5) of the secant and cosecant functions:

$$\sec x = \frac{1}{\cos x} = (\cos x)^{-1} \quad \text{and} \quad \csc x = \frac{1}{\sin x} = (\sin x)^{-1}.$$

Apply the chain rule to compute the derivatives of these two functions:

$$
\begin{aligned}
D(\sec x) &= -(\cos x)^{-2}D(\cos x) & D(\csc x) &= -(\sin x)^{-2}D(\sin x) \\
&= -(\cos x)^{-2}(-\sin x) & &= -(\sin x)^{-2}\cos x \\
&= \frac{\sin x}{\cos x}\cdot\frac{1}{\cos x} & &= -\frac{\cos x}{\sin x}\cdot\frac{1}{\sin x} \\
&= \tan x \sec x & &= -\cot x \csc x \qquad \square
\end{aligned}
$$

The following examples illustrate how the methods of the preceding two sections are used to compute derivatives which involve trigonometric functions.

Examples 2.5.2 (1) Find the derivative of $h(x) = \cos(5x^4 - 7)$.

Solution Write $y = h(x) = \cos u$ where $u = 5x^4 - 7$. By the chain rule:

$$
\begin{aligned}
h'(x) &= \frac{dy}{dx} = \frac{dy}{du}\cdot\frac{du}{dx} = \frac{d}{du}(\cos u)\frac{d}{dx}\left(5x^4 - 7\right) \\
&= (-\sin u)\left(20x^3\right) = -20x^3\sin(5x^4 - 7).
\end{aligned}
$$

(2) Find the derivative of $j(x) = \sin(3x^2 + 1)\tan(3x^2 + 1)$.

Solution By the product rule:

$$
\begin{aligned}
j'(x) &= D\left[\sin(3x^2 + 1)\right]\tan(3x^2 + 1) + \sin(3x^2 + 1)D\left[\tan(3x^2 + 1)\right] \\
&= \cos(3x^2 + 1)D(3x^2 + 1)\tan(3x^2 + 1) \\
&\quad + \sin(3x^2 + 1)\sec^2(3x^2 + 1)D(3x^2 + 1) \qquad \text{[by the chain rule]} \\
&= 6x\cos(3x^2 + 1)\tan(3x^2 + 1) + 6x\sin(3x^2 + 1)\sec^2(3x^2 + 1) \\
&= 6x\sin(3x^2 + 1) + 6x\tan(3x^2 + 1)\sec(3x^2 + 1)
\end{aligned}
$$

(3) Find the derivative of $k(x) = \cot^3(5x^3 - 7x + 4)$.

Solution Write $y = k(x) = u^3$ where $u = \cot w$ and $w = 5x^3 - 7x + 4$. By the chain rule:

$$
\begin{aligned}
k'(x) &= \frac{dy}{dx} = \frac{dy}{du}\cdot\frac{du}{dw}\cdot\frac{dw}{dx} = \frac{d}{du}\left(u^3\right)\frac{d}{dw}\left(\cot w\right)\frac{d}{dx}\left(5x^3 - 7x + 4\right) \\
&= \left(3u^2\right)\left(-\csc^2 w\right)\left(15x^2 - 7\right) \\
&= -3\left(15x^2 - 7\right)\cot^2\left(5x^3 - 7x + 4\right)\csc^2\left(5x^3 - 7x + 4\right). \qquad \square
\end{aligned}
$$

Derivatives of Inverse Functions

Our next project is to compute the derivatives of the inverse trigonometric functions. We lay the groundwork in this subsection by considering the general problem of computing the derivative of an inverse function.

Motivating Example 2.5.3 Find the derivative of $y = \sqrt[4]{x}$.

Solution Consider $y = \sqrt[4]{x}$ as the inverse function $y = f^{-1}(x)$ of the function $x = f(y) = y^4$. We find $\frac{dy}{dx}$ by implicit differentiation:

$$
\begin{aligned}
\frac{d}{dx}(x) &= \frac{d}{dx}(y^4) \\
1 &= 4y^3\frac{dy}{dx} \\
\frac{dy}{dx} &= \frac{1}{4y^3}
\end{aligned}
$$

To describe the value of this derivative in terms of x, substitute $y = \sqrt[4]{x}$:

$$\frac{dy}{dx} = \frac{1}{4(\sqrt[4]{x})^3} .$$

This answer agrees with the formula $D(x^{1/4}) = \frac{1}{4}x^{-3/4}$ of Proposition 2.4.10. $\quad\square$

The analysis of the preceding example applies to any function $y = f^{-1}(x)$ which is the inverse function of $x = f(y)$. Assume that both f and f^{-1} are differentiable functions. We find $\frac{dy}{dx}$ by implicit differentiation:

$$\frac{d}{dx}(x) = \frac{d}{dx}[f(y)]$$

$$1 = Df(y)\frac{dy}{dx}$$

$$\frac{dy}{dx} = \frac{1}{Df(y)}$$

To write this derivative in terms of x, substitute $y = f^{-1}(x)$:

$$D(f^{-1})(x) = \frac{1}{Df(f^{-1}(x))} .$$

In the following theorem we give an alternate derivation of the preceding formula which also demonstrates that f^{-1} is differentiable when f has nonzero derivative.

Theorem 2.5.4 *Let f be a differentiable one–to–one function whose domain and range are open intervals. If the derivative of f is nonzero, then the inverse function f^{-1} is differentiable and*

$$D(f^{-1})(x) = \frac{1}{Df(f^{-1}(x))}. \tag{2.5.1}$$

Proof Let x_0 be in domain f^{-1}. Define $x = x_0 + h$. Note that x approaches x_0 when h approaches zero. Then $h = x - x_0$, and we rewrite the definition of $Df^{-1}(x_0)$ as:

$$Df^{-1}(x_0) = \lim_{h\to 0}\frac{f^{-1}(x_0 + h) - f^{-1}(x_0)}{h} = \lim_{x\to x_0}\frac{f^{-1}(x) - f^{-1}(x_0)}{x - x_0} \tag{2.5.2}$$

Let $y = f^{-1}(x)$ and $y_0 = f^{-1}(x_0)$. Then $x = f(y)$ and $x_0 = f(y_0)$. Since f is differentiable, f is continuous by Proposition 2.2.13. By Property 5 of continuous functions, f^{-1} is continuous. Therefore, $f^{-1}(x) = y$ approaches $f^{-1}(x_0) = y_0$ when x approaches x_0. Hence equation (2.5.2) becomes:

$$Df^{-1}(x_0) = \lim_{y\to y_0}\frac{y - y_0}{f(y) - f(y_0)} = \frac{1}{\lim\limits_{y\to y_0}\frac{f(y)-f(y_0)}{y-y_0}} = \frac{1}{Df(y_0)} = \frac{1}{Df(f^{-1}(x_0))}. \quad\square$$

Note Formula (2.5.1) is motivated by Leibniz notation where the derivative of $y = f(x)$ is denoted $\frac{dy}{dx}$ while the derivative of $x = f^{-1}(y)$ is denoted $\frac{dx}{dy}$. If we could treat Leibniz notation as fractions, we would have:

$$Df^{-1} = \frac{dx}{dy} = \frac{1}{\frac{dy}{dx}} = \frac{1}{Df} .$$

We can justify formula (2.5.1) for the derivative of $f^{-1}(x)$ from the graph of $y = f^{-1}(x)$. Recall from Section 1.3 that (s, t) lies on the graph of $y = f^{-1}(x)$ if and only if (t, s) lies on the graph of $y = f(x)$. Equivalently, the graph of $y = f^{-1}(x)$ is the reflection of the graph of $y = f(x)$ about the line $y = x$. Hence the reflection about the line $y = x$ of the tangent line $y = mx + b$ to the graph of $y = f(x)$ at the point (t, s) gives the tangent line $x = my + b$ (or $y = \frac{1}{m}x - \frac{b}{m}$) to the graph of $y = f^{-1}(x)$ at the point (s, t). See Figure 2.5.5. By definition, the slope m of the tangent line $y = mx + b$ of f is the derivative $Df(t) = Df(f^{-1}(s))$ while the slope $\frac{1}{m}$ of the tangent line $y = \frac{1}{m}x - \frac{b}{m}$ of f^{-1} is the derivative $Df^{-1}(s)$. That is,

$$Df^{-1}(s) = \frac{1}{m} = \frac{1}{Df(f^{-1}(s))}.$$

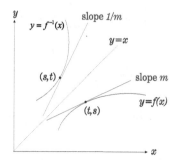

Figure 2.5.5 Tangent Lines of the Graphs of f and f^{-1}

Before applying this formula to inverse trigonometric functions in the next subsection, we illustrate its use in two simpler examples.

Examples 2.5.6 (1) Let $f(x) = x^4 + x^2 + 1$ with domain $(0, \infty)$. Note that $f(2) = 21$. Find $Df^{-1}(21)$.

Solution To apply Theorem 2.5.4, we must show that f is one–to–one. Assume that $0 < A \leq B$ with $f(A) = f(B)$. Then

$$A^4 + A^2 + 1 = B^4 + B^2 + 1$$
$$A^2 - B^2 = B^4 - A^4 = -(A^2 - B^2)(A^2 + B^2)$$

Either $A^2 = B^2$ or we can divide this equation by $A^2 - B^2$ to obtain $1 = -(A^2 + B^2)$ which is false. Hence $A^2 = B^2$ and $A = B$. Thus f is one–to–one, and Theorem 2.5.4 applies. By formula (2.5.1):

$$Df^{-1}(21) = \frac{1}{Df(f^{-1}(21))}.$$

Since $f(2) = 21$, it follows that $f^{-1}(21) = 2$. Hence $Df^{-1}(21) = \frac{1}{Df(2)}$. Observe that $Df(x) = 4x^3 + 2x$, and $Df(2) = 36$. Therefore $Df^{-1}(21) = \frac{1}{36}$.

(2) Use Theorem 2.5.4 to find $D(\sqrt{x})$.

Solution Let $x = f(y) = y^2$ with domain $(0, \infty)$. Then f is a one–to–one function with range $(0, \infty)$. Hence f^{-1} is defined with domain $(0, \infty)$ and is given by $y = f^{-1}(x) = \sqrt{x}$. By formula (2.5.1):

$$Df^{-1}(x) = \frac{1}{Df(f^{-1}(x))} = \frac{1}{Df(y)} = \frac{1}{2y} = \frac{1}{2\sqrt{x}}.$$

This answer agrees with our computation of $D(\sqrt{x})$ in Example 2.2.12 (6).   □

Derivatives of Inverse Trigonometric Functions

We apply Theorem 2.5.4 to compute the derivatives of the inverse trigonometric functions $\arcsin x$, $\arctan x$, $\text{arcsec } x$ from the known derivatives of $\sin x$, $\tan x$, $\sec x$.

Proposition 2.5.7

(a) $D(\arcsin x) = \frac{1}{\sqrt{1-x^2}}$ **(b)** $D(\arctan x) = \frac{1}{1+x^2}$ **(c)** $D(\text{arcsec } x) = \frac{1}{|x|\sqrt{x^2-1}}$

Proof (a) Recall that $f(y) = \sin y$, with domain $\left[-\frac{\pi}{2}, +\frac{\pi}{2}\right]$, is one–to–one with range $[-1, +1]$. Its inverse function $y = f^{-1}(x) = \arcsin x$ has domain $[-1, +1]$. Note that $Df(y) = \cos y$ is nonzero for $y \in (-\pi/2, +\pi/2)$. By Theorem 2.5.4, $f^{-1}(x) = \arcsin x$ is differentiable for $x \in (-1, 1)$ with

$$D(\arcsin x) = Df^{-1}(x) = \frac{1}{Df(f^{-1}(x))} = \frac{1}{\cos f^{-1}(x)} = \frac{1}{\cos(\arcsin x)} = \frac{1}{\cos y}.$$

The value of $\cos y$ is computed from the right triangle in Fig. 2.5.8 as $\sqrt{1-x^2}$. Hence

$$D(\arcsin x) = \frac{1}{\cos y} = \frac{1}{\sqrt{1-x^2}}.$$

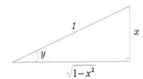

Figure 2.5.8
$y = \arcsin x$

(b) Recall that $g(y) = \tan y$ with domain $\left(-\frac{\pi}{2}, +\frac{\pi}{2}\right)$, is one–to–one with range $\mathfrak{R}$. Its inverse function $y = g^{-1}(x) = \arctan x$ has domain $\mathfrak{R}$. Note $Dg(y) = \sec^2 y \neq 0$ for $y \in \left(-\frac{\pi}{2}, +\frac{\pi}{2}\right)$. By Theorem 2.5.4, $g^{-1}(x) = \arctan x$ is differentiable for $x \in \mathfrak{R}$ with

$$D(\arctan x) = Dg^{-1}(x) = \frac{1}{Dg(g^{-1}(x))} = \frac{1}{\sec^2 g^{-1}(x)} = \frac{1}{\sec^2(\arctan x)} = \frac{1}{\sec^2 y}.$$

Using the right triangle in Figure 2.5.9, $\sec y = \frac{\sqrt{1+x^2}}{1} = \sqrt{1+x^2}$. Hence

$$D(\arctan x) = \frac{1}{\sec^2 y} = \frac{1}{1+x^2}.$$

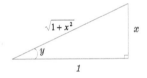

Figure 2.5.9
$y = \arctan x,$

(c) Recall that $h(y) = \sec y$, with domain $[0, \pi/2) \cup (\pi/2, \pi]$, is one–to–one with range $(-\infty, -1] \cup [1, \infty)$. Its inverse function $y = h^{-1}(x) = \text{arcsec } x$ has domain $(-\infty, -1] \cup [1, \infty)$. Note that $Dh(y) = \tan y \sec y$ is nonzero for $y \in (0, \pi/2) \cup (\pi/2, \pi)$. By Theorem 2.5.4, $h^{-1}(x) = \text{arcsec } x$ is differentiable for $x \in (-\infty, -1) \cup (1, \infty)$ with

Figure 2.5.10
$y = \text{arcsec } x$

$$
\begin{aligned}
D(\text{arcsec } x) &= Dh^{-1}(x) = \frac{1}{Dh(h^{-1}(x))} = \frac{1}{\sec h^{-1}(x) \tan h^{-1}(x)} \\
&= \frac{1}{\sec(\text{arcsec } x) \tan(\text{arcsec } x)} = \frac{1}{x \tan(\text{arcsec } x)} = \frac{1}{x \tan y}.
\end{aligned}
$$

Note that $\tan^2 y = \sec^2 y - 1 = x^2 - 1$. For x positive, $\tan y = \frac{\sqrt{x^2-1}}{1} = +\sqrt{x^2-1}$ by the right triangle in Figure 2.5.10. When $x = \sec y$ is negative, $y = \text{arcsec } x \in (\pi/2, \pi]$. Therefore, $\tan y$ is negative: $\tan y = -\sqrt{x^2 - 1}$. In both cases, $x \tan y$ is positive. Thus

$$x \tan y = |x|\sqrt{x^2-1}.$$

Hence
$$D(\text{arcsec } x) = \frac{1}{x \tan y} = \frac{1}{|x|\sqrt{x^2-1}}. \qquad \square$$

We use the methods of the preceding two sections to find derivatives of functions which involve inverse trigonometric functions.

Examples 2.5.11 (1) Find the derivative of $f(x) = (1 + x^2) \arctan x$.

Solution By the product rule:

$$
\begin{aligned}
f'(x) &= D(1 + x^2) \arctan x + (1 + x^2) D(\arctan x) \\
&= 2x \arctan x + (1 + x^2) \frac{1}{1 + x^2} = 2x \arctan x + 1.
\end{aligned}
$$

(2) Find the derivative of $h(x) = \arcsin(1 - x^2)$.

Solution Write $y = h(x) = \arcsin u$ where $u = 1 - x^2$. By the chain rule:

$$
\begin{aligned}
h'(x) &= \frac{dy}{dx} = \frac{dy}{du} \cdot \frac{du}{dx} = \frac{d}{du}(\arcsin u) \frac{d}{dx}(1 - x^2) = \frac{1}{\sqrt{1 - u^2}}(-2x) \\
&= -\frac{2x}{\sqrt{1 - (1 - x^2)^2}} = -\frac{2x}{\sqrt{2x^2 - x^4}}.
\end{aligned}
$$

(3) Find the derivative of $k(x) = \sqrt{\arctan(4x + 1)}$.

Solution Write $y = k(x) = \sqrt{u}$ where $u = \arctan w$ and $w = 4x + 1$. By the chain rule:

$$
\begin{aligned}
k'(x) &= \frac{dy}{dx} = \frac{dy}{du} \cdot \frac{du}{dw} \cdot \frac{dw}{dx} = \frac{d}{du}\left(\sqrt{u}\right) \frac{d}{dw}(\arctan w) \frac{d}{dx}(4x + 1) \\
&= \left(\frac{1}{2\sqrt{u}}\right)\left(\frac{1}{1 + w^2}\right)(4) = \frac{2}{\sqrt{\arctan(4x + 1)}} \cdot \frac{1}{1 + (4x + 1)^2} \\
&= \frac{1}{(8x^2 + 4x + 1)\sqrt{\arctan(4x + 1)}}. \qquad \qquad \square
\end{aligned}
$$

Summary

The reader should memorize the nine derivatives computed in this section, which are listed in Theorem 2.5.1 and Proposition 2.5.7. She should be able to use these formulas in conjunction with the methods of the preceding two sections to compute more complicated derivatives. In addition, she should understand the method used to determine the derivatives of inverse functions as we will be using this method for important computations in Chapter 4.

Basic Exercises

1. Find the derivative of each function.
   (a) $y = \tan(5x + 3)$ **(b)** $y = \sec^2 x$ **(c)** $y = \cos^3(6x^3 - 7)$
   (d) $y = \csc^4(8x^4 - 5)$ **(e)** $y = \sin(x^2 + 1) \sec(x^4 - 1)$
   (f) $y = \cot^4(5x - 2) \csc^3(5x - 2)$ **(g)** $y = \frac{\sin(3x^3 + 4)}{\tan(6x^2 + 3)}$
   (h) $y = \frac{\cot(2x^6 - 1)}{3x^5 + 1}$ **(i)** $y = \tan\sqrt{4x^2 - 3}$
   (j) $y = \sqrt{\sin(6x^7 + 5)}$ **(k)** $y = \sqrt{\cot\sqrt{x^2 + 1}}$

2. Find the derivative $\frac{dy}{dx}$ for each implicitly defined function.
   (a) $\sin(x^2 + y^2) = xy + 1$ **(b)** $\tan^2(x + y) = x^2y^2 - 3$ **(c)** $\csc(xy) = \sqrt{xy^2 + x^2y}$
   (d) $\cos\sqrt{xy} = x^2y^2 + 1$ **(e)** $1 + \cot\frac{x}{y} = \sec\frac{y}{x}$ **(f)** $1 + \sin(xy) = \tan(x + y)$

3. Find the equation of the tangent line to the graph of each function at the given point.
   (a) $y = \sin x$ at $x = \frac{\pi}{6}$ **(b)** $y = \tan(3x)$ at $x = \frac{\pi}{4}$
   (c) $y = \sec^3(2x)$ at $x = \frac{2\pi}{3}$ **(d)** $y = \sqrt{\csc(\pi x^2)}$ at $x = \frac{1}{2}$
   (e) $x = \frac{\sin y}{2 + \cos y}$ at $(0, \pi)$ **(f)** $\sin\pi(x + y) = xy - 2$ at $(2, 1)$

4. Find the equation of the normal line to the graph of each function at the given point.
(a) $y = \cos x$ at $x = \frac{\pi}{4}$ **(b)** $y = \csc x \cot x$ at $x = \frac{\pi}{3}$ **(c)** $y = \sqrt{\tan x}$ at $x = \frac{\pi}{4}$
(d) $\cos \sqrt{x+y} = xy + 2$ at $(-1, 1)$ **(e)** $\sec^2(xy) = x^2 + y^2$ at $(0, 1)$

5. **(a)** Find all the higher derivatives of the function $f(x) = \sin x$.
(b) Find all the higher derivatives of the function $g(x) = \cos x$.

6. Show that each of the following functions is one–to–one. Then find the indicated derivative of its inverse function.

(a) $f(x) = x^6 - 60$ with domain $(0, \infty)$. Find $D(f^{-1})(4)$.   *Prac more*
(b) $g(x) = \cos x$ with domain $(0, \pi)$. Find $D(g^{-1})(0)$.
(c) $h(x) = \frac{x}{x^2+2}$ with domain $(0, 1)$. Find $D(h^{-1})\left(\frac{1}{3}\right)$.
(d) $k(x) = x^6 + x^2 + 1$ with domain $(0, \infty)$. Find $D(k^{-1})(69)$.
(e) $m(x) = \sqrt{\arctan x}$ with domain $\left(0, \frac{\pi}{2}\right)$. Find $D(m^{-1})(1)$.

7. **(a)** Find the derivative of $y = \sqrt[3]{x}$, for $x \in \mathfrak{R}$, as the inverse function of $x = y^3$.
(b) Find the derivative of $y = \sqrt{x^2+1}$ as the inverse function of $x = \sqrt{y^2-1}$.
(c) Find the derivative of $y = \frac{x}{x+1}$ as the inverse function of $x = \frac{y}{1-y}$.
(d) Find the derivative of $y = \sqrt{\arcsin x}$ as the inverse function of $x = \sin y^2$.
(e) Find the derivative of $y = \arctan \sqrt{x}$ as the inverse function of $x = \tan^2 y$.

8. Let f be a differentiable one–to–one function.
(a) If $y = 7x + 2$ is the tangent line to the graph of $y = f(x)$ at the point $(1, 9)$, find the equation of the tangent line to the graph of $y = f^{-1}(x)$ at the point $(9, 1)$.
(b) If $y = 5x - 8$ is the tangent line to the graph of $y = f^{-1}(x)$ at the point $(3, 7)$,   *Prac*
find the equation of the tangent line to the graph of $y = f(x)$ at the point $(7, 3)$.
(c) If $y = 7 - 3x$ is the normal line to the graph of $y = f(x)$ at the point $(2, 1)$, find   *more*
the equation of the tangent line to the graph of $y = f^{-1}(x)$ at the point $(1, 2)$.
(d) If $y = 4x - 6$ is the normal line to the graph of $y = f(x)$ at the point $(5, 14)$, find
the equation of the normal line to the graph of $y = f^{-1}(x)$ at the point $(14, 5)$.
(e) If $y = 10 - 3x$ is the normal line to the graph of $y = f^{-1}(x)$ at the point $(3, 1)$,
find the equation of the tangent line to the graph of $y = f(x)$ at the point $(1, 3)$.

9. Let f, g be differentiable one–to–one functions. Find formulas for each derivative
in terms of f, g, f' and g'.   **(a)** $D(f^{-1}g^{-1})(x)$ **(b)** $D\left(\frac{f^{-1}}{g^{-1}}\right)(x)$

(c) $D(f^{-1} \circ g^{-1})(x)$ **(d)** $D(ff^{-1})(x)$ **(e)** $D\left(\frac{f^{-1}}{fg^{-1}}\right)(x)$

10. Find the derivative of each function.
(a) $y = \arctan(5x^2 - 1)$ **(b)** $y = \sqrt[5]{\arcsin x}$ **(c)** $y = [\operatorname{arcsec}(6x - 5)]^4$
(d) $y = \arcsin(3x + 2)\arctan(3x + 2)$ **(e)** $y = \frac{\operatorname{arcsec}(4x^2+7)}{4x^2+7}$
(f) $y = \frac{\arctan x}{\arcsin x}$ **(g)** $y = (\operatorname{arcsec} x)(\arcsin 2x)$
(h) $y = \arctan(\arctan x)$ **(i)** $y = \arcsin(\tan x)$
(j) $y = \operatorname{arcsec} \sqrt{x}$ **(k)** $y = \arctan \sqrt{\arcsin x}$

11. Find the derivative $\frac{dy}{dx}$ for each implicitly defined function.
(a) $x \arcsin y + y \arcsin x = 1$ **(b)** $\operatorname{arcsec}(xy) = xy^2 + 3$
(c) $\arctan(x + y) = \arcsin(xy) - 1$ **(d)** $\arcsin(2x - 3y) = x^4y + 5$
(e) $\arctan(x^8y^6 + 2) = (x + y)^{2/3}$ **(f)** $\arctan(xy) = \arcsin(x^2 + y^2)$

12. Find the equation of the tangent line to the graph of each function at the given point.
(a) $y = x \arcsin x$ at $x = \frac{1}{2}$ **(b)** $y = (\operatorname{arcsec} x)^2$ at $x = \sqrt{2}$
(c) $y = \frac{\arctan 4x}{\tan \pi x}$ at $x = \frac{1}{4}$ **(d)** $y = \arctan(x^2 - 1)$ at $x = 1$
(e) $y = \arcsin(3x + 7)$ at $x = -\frac{13}{6}$ **(f)** $y = \operatorname{arcsec}(\tan \pi x)$ at $x = -\frac{1}{4}$

13. Find the equation of the normal line to the graph of each function at the given po

(a) $y = \frac{x}{\arctan x}$ at $x = \sqrt{3}$ (b) $y = \arcsin(x^2)$ at $x = \frac{\sqrt{2}}{2}$

(c) $y = (\arcsin x)(\arctan 2x)$ at $x = \frac{1}{2}$ (d) $y = \frac{1}{\text{arcsec } x}$ at $x = \sqrt{2}$

(e) $y = (\tan \pi x)(\arctan 4x)$ at $x = \frac{1}{4}$ (f) $y = \frac{\text{arcsin } x}{\arctan 2x}$ at $x = \frac{1}{2}$

Challenging Problem

1. Compute the derivatives of: **(a)** $y = \arccos x$; **(b)** $y = \text{arccot } x$; **(c)** $y = \text{arccsc } x$.

2.6 Mean Value Theorem

Many important applications of the derivative, presented in Sections 7 to 13, are based on the Mean Value Theorem. This result says that there is a tangent line parallel to every secant of the graph of a differentiable function. In the first subsection, we prepare for studying the Mean Value Theorem by showing that the tangent line at a local maximum or at a local minimum must be horizontal. The second subsection is devoted to understanding the Mean Value Theorem itself. We deduce several important consequences of the Mean Value Theorem in the third subsection. In particular, we show that the sign of the derivative determines whether a function is increasing or decreasing. In the last subsection, we use this information, in the form of the First Derivative Test, to find and identify local minima and local maxima.

Local Extrema

We show that the first derivative allows us to identify the "hill tops" and "valley bottoms" of the graph of a continuous function as points where the first derivative either does not exist or has value zero. This result is used in Section 12 to find the minimum and maximum values of a function. It is also used to derive the Mean Value Theorem in the next subsection. First, we introduce the mathematical terminology *local maximum* for a hill top and *local minimum* for a valley bottom.

Definition 2.6.1 *Let f be a continuous function.*
(a) *f has a local maximum at $x = M$ if there is an open interval I, contained in the domain of f, such that $M \in I$ and $f(x) \leq f(M)$ for $x \in I$.*
(b) *f has a local minimum at $x = m$ if there is an open interval I, contained in the domain of f, such that $m \in I$ and $f(x) \geq f(m)$ for $x \in I$.*
(c) *f has a local extremum at $x = c$ if it is either a local minimum or a local maximum.*

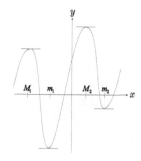

Figure 2.6.2 Local Extrema

Note the function in Figure 2.6.2 has four local extrema: local maxima at $x = M_1$, $x = M_2$ and local minima at $x = m_1$, $x = m_2$.

Motivating Example 2.6.3 Find the derivative of $f(x) = 1 - x^2$ at its local maximum $x = 0$.

Solution The graph of f is the parabola $y = 1 - x^2$, sketched in Figure 2.6.4, which has a local maximum at $x = 0$. The derivative of f at $x = 0$ is the slope $f'(0)$ of its tangent line T at $(0,1)$. The slope of T is approximated by the slope of the secant line joining the points $(0,1)$ and $(h, 1 - h^2)$ on the graph of f. The slope K_h of this secant line is given by:

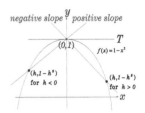

$$K_h = \frac{(1 - h^2) - 1}{h - 0} = \frac{-h^2}{h} = -h .$$

Thus K_h is positive when $h < 0$, i.e. when the point $(h, 1 - h^2)$ is to the left of $(0,1)$. On the other hand, K_h is negative when $h > 0$, i.e. when the point $(h, 1 - h^2)$ is to the right of $(0,1)$. See Figure 2.6.4. There is only one number $f'(0)$ which is approximated by both negative and positive numbers: $f'(0) = 0$.[2] $\quad\square$

Figure 2.6.4
Local Maximum
at $x = 0$

The theorem below generalizes the analysis of this example to a local extremum at $x = c$ of any function f where $f'(c)$ exists. For example, the function in Figure 2.6.2 has horizontal tangent lines at its four local extrema. On the other hand, $f'(c)$ may not exist at a local extremum. For example, the function of Figure 2.6.5 has cusps at its two local extrema where the derivative does not exist.

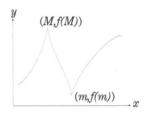

Theorem 2.6.6 *Let f be a continuous function with domain (a, b) which has a local extremum at $x = c$. If f is differentiable at $x = c$, then $f'(c) = 0$.*

Figure 2.6.5
Local Extrema at Cusps

Proof Assume f is differentiable and has a local maximum at $x = c$. Let I be an open interval containing c such that $f(x) \leq f(c)$ for $x \in I$. The analysis Example 2.6.3 applies to the tangent line T of the graph of $y = f(x)$ at $x = c$. Assume $c + h \in I$. The left secant line in Figure 2.6.7 joining the points $(c + h, f(c + h))$ and $(c, f(c))$, for $h < 0$, goes from the lower point $(c + h, f(c + h))$ on the left to the higher point $(c, f(c))$ on the right and must have positive slope. The slope of this secant line

$$\frac{f(c + h) - f(c)}{h} \geq 0 \quad \text{for } h < 0$$

because the numerator and denominator are both negative. Hence the left derivative of f at $x = c$ must be greater than or equal to zero.

$$f'_-(c) = \lim_{h \to 0^-} \frac{f(c + h) - f(c)}{h} \geq 0.$$

On the other hand, the right secant line in Figure 2.6.7 joining the points $(c, f(c))$ and $(c + h, f(c + h))$, for $h > 0$, goes from the higher point $(c, f(c))$ on the left to the lower point $(c + h, f(c + h))$ on the right and must have negative slope. The slope of this secant line

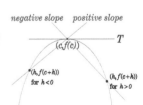

$$\frac{f(c + h) - f(c)}{h} \leq 0 \quad \text{for } h > 0$$

because the numerator is negative while the denominator is positive. Hence the right derivative of f at $x = c$ must be less than or equal to zero:

$$f'_+(c) = \lim_{h \to 0^+} \frac{f(c + h) - f(c)}{h} \leq 0.$$

Figure 2.6.7
Local Maximum at $x = c$

Since $f'(c)$ exists, its value is the same as the values of the two one–sided derivatives:

$$0 \geq f'_+(c) = f'(c) = f'_-(c) \geq 0, \quad \text{and} \quad f'(c) = 0.$$

---

[2]Of course, this conclusion could be obtained quickly by computing $f'(x) = -2x$ with $f'(0) = 0$.

A similar analysis could be used to show the tangent line of the graph of $y = f(x)$ is horizontal at a local minimum. However, we deduce this result from the preceding case. Assume f is differentiable and has a local minimum at $x = k$. Then $y = g(x) = -f(x)$ is differentiable, and g has a local maximum at $x = k$. By the preceding case, $g'(k) = 0$. Hence $f'(k) = -g'(k) = 0$.   □

Recall the Maximum Value Theorem says that a continuous function f, with domain a closed interval, has a minimum value and a maximum value, i.e. a lowest point and a highest point on its graph. The first two examples below compare the values of f' at the minimum and maximum values of f with the values of f' at the relative minima and relative maxima of f. In particular, if f has a minimum or maximum value at an endpoint $x = a$ or $x = b$, then the one sided derivative at that point can have any value. Note that by Definition 2.6.1 we do not call an endpoint of the domain of f a local extremum.

Figure 2.6.8
$f(x) = x$

Examples 2.6.9 (1) Consider the graph of the function $f(x) = x$ with domain $[1, 2]$ depicted in Figure 2.6.8. The function f has its minimum value at $x = 1$ and its maximum value at $x = 2$. However, f has no local extrema. Note $f'(c) = 1$ at every point, and there is no point $x = c$ where $f'(c) = 0$.

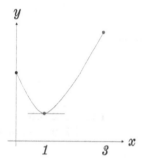

Figure 2.6.10
$g(x) = x^2 - 2x + 3$

(2) Consider the graph of the function $g(x) = x^2 - 2x + 3$ with domain $[0, 3]$. Since $g(x) = (x - 1)^2 + 2$, the graph of g is a piece of a parabola. See Figure 2.6.10. The minimum value of g occurs at $x = 1$, and the maximum value of g occurs at $x = 3$. Since 1 is an interior point of $[0, 3]$, it is a local minimum and the tangent line at $x = 1$ is horizontal. Since $g'(x) = 2x - 2$, the tangent line at $x = 3$ has slope $g'(3) = 4$ and is not horizontal. This does not contradict Theorem 2.6.6 because 3 is an endpoint of the interval $[0, 3]$ and is not a local maximum.

(3) Consider the graph of $h(x) = \cos x$ with domain $[-\pi/2, 3\pi/2]$ depicted in Figure 2.6.11. h has two local extrema: a local maximum at the point $x = 0$ and a local minimum at the point $x = \pi$. Since $h'(x) = -\sin x$, the tangent lines to the graph of h are horizontal at both of these points.   □

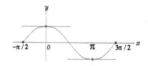

Figure 2.6.11
$h(x) = \cos x$

Mean Value Theorem

The Mean Value Theorem says that a differentiable function has a tangent line parallel to each of its secant lines. This theorem is a consequence of the special case, called Rolle's Theorem, where the secant line is horizontal. We deduce Rolle's Theorem from Theorem 2.6.6 and show how the Mean Value Theorem follows. Then we prove Proposition 2.2.22 which we used in Section 2 to compute one–sided derivatives.

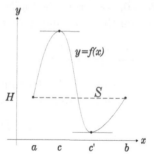

Figure 2.6.12 Rolle's Theorem

Rolle's Theorem considers a differentiable function defined on a closed interval $[a, b]$ such that $f(a) = f(b) = H$. The secant line S joining the two endpoints (a, H)

and (b, H) of the graph of f is horizontal. Rolle's Theorem says that f must have a horizontal tangent line at some point $c \in (a, b)$. See Figure 2.6.12. This is reasonable. If the curve rises to the right of (a, H) to a height greater than H, then it must eventually come back down to height H at (b, H). At the point where the curve turns to go down it will have a horizontal tangent. Similarly, if the curve falls to the right of (a, H), then it must have a horizontal tangent at the point where it turns to go back up to (b, H). The formal proof, however, uses Theorem 2.6.6 and does not rely on geometric intuition.

Theorem 2.6.13 (Rolle's Theorem) *Let f be a continuous function with domain the closed interval $[a, b]$, $a < b$. Assume f is differentiable on the open interval (a, b). If $f(a) = f(b)$, then there is at least one number $c \in (a, b)$ such that $f'(c) = 0$.*

Proof By the Maximum Value Theorem, f has a minimum value at $x = m$ and a maximum value at $x = M$. If either m or M is an interior point of $[a, b]$, then the derivative of f is zero there by Theorem 2.6.6. Now consider the case where both m and M are endpoints of $[a, b]$. The assumption $f(a) = f(b)$ can be rewritten as $f(m) = f(M)$. Hence f is a constant function, and $f'(c) = 0$ for all $c \in (a, b)$. $\square$

Observe that when Rolle's Theorem applies, there may be many points on the graph of f which have a horizontal tangent. For example, in Figure 2.6.12 there are two points, $x = c$ and $x = c'$, where the tangent line to the graph of f is horizontal.

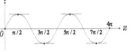

Examples 2.6.15 In each example, determine whether Rolle's Theorem applies. If it applies, find all numbers c where the derivative of the function is zero.

Figure 2.6.14
$f(x) = \sin x$

(1) Let $f(x) = \sin x$ with domain $[0, 4\pi]$.

Solution Note f is continuous, and f is differentiable on $(0, 4\pi)$. Since $f(0) = f(4\pi) = 0$, Rolle's Theorem applies. The derivative $f'(x) = \cos x$ is zero at $x = \pi/2$, $x = 3\pi/2$, $x = 5\pi/2$ and $x = 7\pi/2$. See Figure 2.6.14.

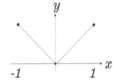

(2) Let $g(x) = x^3 - x^2 + 3$ with domain $[0, 1]$.

Solution Observe that g is continuous, and g is differentiable on $(0, 1)$. Since $g(0) = g(1) = 3$, Rolle's Theorem applies. Although we do not know the graph of g, there must be at least one number c between 0 and 1 with $g'(c) = 0$. To find these values of c we set $g'(x) = 3x^2 - 2x$ equal to zero:

Figure 2.6.16
$h(x) = |x|$

$$0 = 3x^2 - 2x = x(3x - 2).$$

Thus, $x = 0$ or $x = \frac{2}{3}$. Since c must be in $(0, 1)$, we see that there is a unique value for c: $c = \frac{2}{3}$.

(3) Let $h(x) = |x|$ with domain $[-1, 1]$.

Solution Although h is continuous and $h(-1) = h(1) = 1$, Rolle's Theorem does not apply to h because h is not differentiable at $x = 0$. In fact, there is no point on the graph of h, in Figure 2.6.16, with a horizontal tangent line. However, there is a cusp on the graph of h at $x = 0$ where the curve has a local minimum without having a horizontal tangent. $\square$

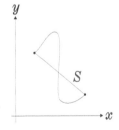

Figure 2.6.17
$y = g(x)$

Let f be a function with domain $[a, b]$ which satisfies the hypotheses of Rolle's Theorem. The conclusion of Rolle's Theorem says that there is a point $c \in (a, b)$ where the tangent line T to the graph of f is horizontal. Equivalently, there is a point $c \in (a, b)$ where the tangent line T is parallel to the horizontal secant line S joining

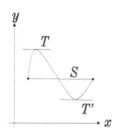

Figure 2.6.18
g Rotated

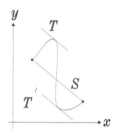

Figure 2.6.19
$y = g(x)$

the endpoints $(a, f(a))$ and $(b, f(b))$ of the graph of f. Now consider any continuous function g with domain $[a, b]$ which is differentiable on the open interval (a, b). Let S be the secant line obtained by joining the endpoints $(a, g(a))$ and $(b, g(b))$ of the graph of g as in Figure 2.6.17. Rotate the graph of g so that the secant line S becomes horizontal as in Figure 2.6.18. By Rolle's Theorem, this rotated graph has a tangent line T, which is horizontal. Now rotate the graph back to its original position as in Figure 2.6.19. T will still be tangent to the graph and will still be parallel to the secant line S. This geometric fact is called the Mean Value Theorem. However, the Mean Value Theorem is usually phrased without using geometry. Recall that two parallel lines, such as S and T, have the same slope. The slope of T is $f'(c)$, the slope of S is $\frac{f(b)-f(a)}{b-a}$, and these two numbers are equal.

Theorem 2.6.20 (Mean Value Theorem) *Let f be a continuous function with domain the closed interval $[a, b]$, $a < b$. Assume f is differentiable on the open interval (a, b). Then there is at least one number $c \in (a, b)$ such that*

$$f'(c) = \frac{f(b) - f(a)}{b - a}. \qquad \qquad \square$$

There is a problem with our derivation of the Mean Value Theorem. When we rotate the graph of f to make the secant line S horizontal, the resulting curve may not be a function. Even if it is a function, it may have vertical tangent lines, i.e. points where the derivative does not exist. Both of these problems occur when rotating the graph of f in Figure 2.6.21. We defer a rigorous proof of the Mean Value Theorem to Section 13. Observe that when $f(a) = f(b)$ the Mean Value Theorem becomes Rolle's Theorem, i.e. the Mean Value Theorem produces a number $c \in (a, b)$ where $f'(c) = 0$.

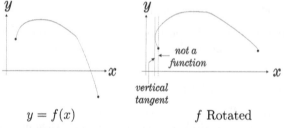

$y = f(x)$ f Rotated

Figure 2.6.21 Rotated Graph with Problems

Examples 2.6.22 For each function, find all numbers c which satisfy the conclusion of the Mean Value Theorem.

(1) Let $f(x) = x^3$ with domain $[-3, 3]$.

Solution The function $f(x) = x^3$ is continuous on $[-3, 3]$ and differentiable on $(-3, 3)$. Hence the Mean Value Theorem applies to f and says: there is at least one number $c \in (-3, 3)$ such that

$$f'(c) = 3c^2 = \frac{f(3) - f(-3)}{3 - (-3)} = \frac{27 - (-27)}{6} = \frac{54}{6} = 9.$$

Then $c^2 = 3$, and $c = \pm\sqrt{3}$ give two points in the open interval $(-3, 3)$ where the tangent line is parallel to the secant line joining $(-3, -27)$ and $(3, 27)$.

(2) Let $g(x) = \arctan x$ with domain $[0, 1]$.

Solution The function $g(x) = \arctan x$ is continuous on $[0, 1]$ and differentiable on $(0, 1)$. Hence the Mean Value Theorem applies to h and says: there is at least one number $c \in (0, 1)$ such that

$$g'(c) = \frac{1}{1 + c^2} = \frac{g(1) - g(0)}{1 - 0} = \frac{\frac{\pi}{4} - 0}{1} = \frac{\pi}{4}.$$

Cross-multiplying: $\pi(1 + c^2) = 4$ and $c^2 = \frac{4}{\pi} - 1 = \frac{4 - \pi}{\pi}$. Hence $c = \sqrt{\frac{4 - \pi}{\pi}}$ is the only number in the open interval $(0, 1)$ where the tangent line is parallel to the secant line joining $(0, 0)$ and $\left(1, \frac{\pi}{4}\right)$. Note $c = -\sqrt{\frac{4 - \pi}{\pi}}$ is not in $(0, 1)$.   □

Recall that in Section 2 we used Proposition 2.2.22 to compute a one sided derivative of f at $x = c$ as the limit of $f'(x)$ as x approaches c. This result is an easy consequence of the Mean Value Theorem.

Proposition 2.2.22 Let $f(x)$ be a continuous function with domain the closed interval $[a, b]$, $a < b$. Assume $f'(x)$ is continuous on the open interval (a, b).
(a) If $\lim\limits_{x \to a^+} f'(x)$ exists, then $f'_+(a)$ exists and $f'_+(a) = \lim\limits_{x \to a^+} f'(x)$.
(b) If $\lim\limits_{x \to b^-} f'(x)$ exists, then $f'_-(b)$ exists and $f'_-(b) = \lim\limits_{x \to b^-} f'(x)$.

Proof (a) By definition of the one sided derivative $f'_+(a)$: $f'_+(a) = \lim\limits_{h \to 0^+} \dfrac{f(a + h) - f(a)}{h}$. By the Mean Value Theorem there is $c_h \in (a, a + h)$ such that $f'(c_h) = \frac{f(a+h)-f(a)}{(a+h)-a} = \frac{f(a+h)-f(a)}{h}$, i.e. $f(a + h) - f(a) = hf'(c_h)$. Thus

$$f'_+(a) = \lim_{h \to 0^+} \frac{hf'(c_h)}{h} = \lim_{h \to 0^+} f'(c_h) = \lim_{x \to a^+} f'(x)$$

because c_h approaches a from the right as h approaches zero from the right.

(b) By definition of the one sided derivative $f'_-(b)$: $f'_-(b) = \lim\limits_{h \to 0^-} \dfrac{f(b + h) - f(b)}{h}$. By the Mean Value Theorem there is $e_h \in (b + h, b)$ such that $f'(e_h) = \frac{f(b)-f(b+h)}{b-(b+h)} = \frac{f(b)-f(b+h)}{-h}$, i.e. $f(b + h) - f(b) = hf'(e_h)$. Thus

$$f'_-(b) = \lim_{h \to 0^-} \frac{hf'(e_h)}{h} = \lim_{h \to 0^-} f'(e_h) = \lim_{x \to b^-} f'(x)$$

because e_h approaches b from the left as h approaches zero from the left.   □

Applications

We present two applications of the Mean Value Theorem. A fundamental calculus problem is to find functions with a given derivative. If we find one such function, the Mean Value Theorem determines all of them. We also use the Mean Value Theorem to find where a differentiable function is increasing and where it is decreasing.

We begin with a special case of of the first application. We know that the derivative of the zero function is zero. We determine all functions that have derivative zero.

Corollary 2.6.23 *Let f be a differentiable function with domain an open interval such that $Df = 0$. Then f is a constant function.*

Proof Let a and b be two points in the domain of f with $a < b$. By the Mean Value Theorem, there is a number $c \in (a, b)$ such that

$$0 = Df(c) = \frac{f(b) - f(a)}{b - a}.$$

Hence the numerator $f(b) - f(a)$ equals zero, and $f(b) = f(a)$. Thus f is a constant function.   □

We use this corollary to determine all functions that have the same derivative as f.

Corollary 2.6.24 *Let f, g be differentiable functions with domain the same open interval (a, b) such that $Df = Dg$. Then there is a number k such that*

$$g(x) = f(x) + k \quad \text{for all } x \in (a, b).$$

Proof Since $D(g - f) = Dg - Df = 0$, Corollary 2.6.23 says that $g(x) - f(x)$ is a constant function. Write $g(x) - f(x) = k$.   □

Consider the problem of finding *all* functions whose derivative is a given function $f(x)$ when we have found one solution $s(x)$ with $s'(x) = f(x)$. By Corollary 2.6.24 the functions $s(x) + k$, for $k \in \Re$, give *all* solutions of the problem. We say that $y = s(x) + k$ is the *general solution* of the *differential equation* $y' = f(x)$. When we are given one value of the solution, say $s(x_0) = y_0$, we can solve for the value of the constant k, say $k = k_0$. We say $y = s(x) + k_0$ is the solution of the *initial value problem* $y' = f(x)$, $y(x_0) = y_0$.

Examples 2.6.25 (1) (a) Solve the differential equation $y' = 8x^3$.

Solution Since $D(x^4) = 4x^3$, it follows that $s(x) = 2x^4$ is one solution. Hence the general solution of the differential equation $y' = 8x^3$ is $y = 2x^4 + k$ for $k \in \Re$.

(b) Solve the initial value problem $y' = 8x^3$, $y(0) = 5$.

Solution By (a), $y = 2x^4 + k$ gives all functions with $y' = 8x^3$. Then

$$5 = y(0) = 2(0^4) + k = k.$$

Thus the solution of this initial value problem is $y = 2x^4 + 5$.

(2) (a) Solve the differential equation $y' = \sin x$.

Solution Since $D(\cos x) = -\sin x$, $s(x) = -\cos x$ is one solution. Therefore the general solution of this differential equation is $y = -\cos x + k$ for $k \in \Re$.

(b) Solve the initial value problem $y' = \sin x$, $y(\pi) = 4$.

Solution By (a), $y = -\cos x + k$ gives all functions with $y' = \sin x$. Then

$$4 = y(\pi) = -\cos \pi + k = -(-1) + k = 1 + k, \quad \text{and } k = 3.$$

Therefore, the solution of this initial value problem is $y = 3 - \cos x$.   □

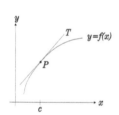

Figure 2.6.26
$f'(c) > 0$

We turn our attention to the problem of determining where a function is increasing and where it is decreasing. We begin with a precise definition of these terms.

Definition 2.6.28 *Let f be a function whose domain contains an interval I.*
(a) *f is increasing on I if $f(x_1) < f(x_2)$ for $x_1 < x_2$ any two numbers in I.*
(b) *f is decreasing on I if $f(x_1) > f(x_2)$ for $x_1 < x_2$ any two numbers in I.*

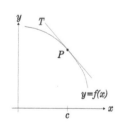

Figure 2.6.27
$f'(c) < 0$

For example, the function in Figure 2.6.26 is increasing, while the function in Figure 2.6.27 is decreasing. An increasing function is also called *monotone increasing* or *strictly monotone increasing* Similarly, a decreasing function is also called *monotone decreasing* or *strictly monotone decreasing*. The following consequence of the Mean Value Theorem says that the sign of the first derivative indicates where a function is increasing and where it is decreasing.

Corollary 2.6.29 *Let f be a differentiable function whose domain contains the open interval (a, b).*

(a) *If $f'(x) > 0$ for $x \in (a, b)$, then f is increasing on (a, b).*

(b) *If $f'(x) < 0$ for $x \in (a, b)$, then f is decreasing on (a, b).*

Proof Let $x_1 < x_2$ be two numbers in (a, b). Apply the Mean Value Theorem to f on the interval $[x_1, x_2]$: there is $c \in (x_1, x_2)$ such that

$$\frac{f(x_2) - f(x_1)}{x_2 - x_1} = f'(c). \tag{2.6.1}$$

(a) Since $f'(c) > 0$ and the denominator of the fraction in (2.6.1) is a positive number, the numerator must be positive too. Thus $f(x_2) > f(x_1)$, and f is increasing.

(b) Since $f'(c) < 0$ and the denominator of the fraction in (2.6.1) is a positive number, the numerator must be negative. Thus $f(x_2) < f(x_1)$, and f is decreasing. □

A function may be increasing on parts of its domain and decreasing on other parts. In the following examples, we plot the sign of the first derivative on a number line to determine where the function is increasing and where it is decreasing.

Examples 2.6.30 Find where each function is increasing and where it is decreasing.

(1) Let $f(x) = x^3$ with domain $\Re$.

Solution f has derivative $f'(x) = 3x^2$. Since $f'(x) > 0$ on $(-\infty, 0)$ and on $(0, \infty)$, f is increasing on each of these intervals. Moreover, if $x_1 < 0 < x_2$, then

$$f(x_1) < f(0) = 0 < f(x_2).$$

Therefore f is an increasing function on the entire real line $\Re$.

(2) Let $g(x) = x^3 - 12x + 1$ with domain $\Re$.

Solution g has derivative

$$g'(x) = 3x^2 - 12 = 3(x^2 - 4) = 3(x - 2)(x + 2).$$

Thus
$$g'(x) > 0 \quad \text{if} \quad x < -2,$$
$$g'(x) < 0 \quad \text{if} \quad -2 < x < 2,$$
$$g'(x) > 0 \quad \text{if} \quad 2 < x.$$

Therefore g is increasing on the interval $(-\infty, -2)$, g is decreasing on the interval $(-2, 2)$, and g is increasing on the interval $(2, \infty)$. We summarize our conclusions on the number line in Figure 2.6.31.

Figure 2.6.31 $g(x) = x^3 - 12x + 1$

(3) Let $h(x) = 3x^5 - 25x^3 + 60x - 7$ with domain $\Re$.

Solution h has derivative

$$h'(x) = 15x^4 - 75x^2 + 60 = 15(x^4 - 5x^2 + 4) = 5(x^2 - 1)(x^2 - 4)$$
$$= 5(x - 1)(x + 1)(x - 2)(x + 2).$$

Observe that $h'(x)$ is positive on the intervals $(-\infty, -2)$, $(-1, 1)$ and $(2, \infty)$. Hence h is increasing on these three intervals. On the other hand, $h'(x)$ is negative on the intervals $(-2, -1)$ and $(1, 2)$. Therefore h is decreasing on these two intervals. We summarize these conclusions in Figure 2.6.32.   □

Figure 2.6.32 $h(x) = 3x^5 - 25x^3 + 60x - 7$

First Derivative Test

Consider the problem of finding and identifying the local extrema of a continuous function f. Theorem 2.6.6 locates the local extrema: they occur at points where $f'(x)$ is zero or does not exist. We introduce the First Derivative Test to identify each of them as a local minimum or a local maximum. We introduce the term *critical point* to denote a point where a local extremum may occur.

Figure 2.6.33
$g(x) = \sin x$

Definition 2.6.34 *Let f be a continuous function with c an interior point of its domain. $x = c$ is a critical point of f if either $f'(c) = 0$ or $f'(c)$ does not exist.*

Although every local extremum is a critical point by Theorem 2.6.6, not every critical point is a local extremum. (See Examples 5 and 6 below.) In the following examples we use the graph of a function to identify each of its critical points as a local minimum, a local maximum or neither.

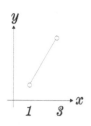

Figure 2.6.35
$h(x) = 2x + 1$

Examples 2.6.36 Find and identify the critical points of each function.

(1) $g(x) = \sin x$ with domain $(0, 4\pi)$.

Solution Since $g'(x) = \cos x$, the function g has four critical points at $\frac{\pi}{2}$, $\frac{3\pi}{2}$, $\frac{5\pi}{2}$ and $\frac{7\pi}{2}$ where $g'(x)$ is zero. From the graph of g in Figure 2.6.33, we see that g has local maxima at $x = \pi/2$ and $x = 5\pi/2$ as well as local minima at $x = 3\pi/2$ and $x = 7\pi/2$.

(2) $h(x) = 2x + 1$ with domain $(1, 3)$.

Solution Since $h'(x) = 2$, the function h has no critical points. Hence h has no local extrema. See Figure 2.6.35.

Figure 2.6.37
$k(x) = |x|$

(3) $k(x) = |x|$ with domain $\mathfrak{R}$.

Solution $x = 0$ is a critical point of k because $k'(0)$ does not exist. For $x < 0$, $k(x) = -x$ with $k'(x) = -1$ while for $x > 0$, $k(x) = x$ with $k'(x) = +1$. Thus $x = 0$ is the only critical point of k. We see from the graph of k in Figure 2.6.37 that $x = 0$ is a local minimum of k.

Figure 2.6.38
$m(x) = x^{2/3}$

(4) $m(x) = x^{2/3}$ with domain $\mathfrak{R}$.

Solution For $x \neq 0$,

$$m'(x) = \frac{2}{3x^{1/3}}$$

which is never zero. Moreover, $m'(0)$ does not exist. Hence $x = 0$ is the only critical point of m. We see from the graph of m in Figure 2.6.38 that $x = 0$ is a local minimum.

(5) $s(x) = x^3$ with domain $\Re$.

    Solution $s'(x) = 3x^2$. Hence $x = 0$, where the derivative of s is zero, is the only critical point of s. From the graph of s in Figure 2.6.39, we see that $x = 0$ is not a local extremum.

(6) $t(x) = \sqrt[3]{x} = x^{1/3}$ with domain $\Re$.

    Solution For $x \neq 0$,

$$t'(x) = \frac{1}{3}x^{-2/3} = \frac{1}{3x^{2/3}}.$$

    which is never 0. Since $t'(0)$ does not exist, $x = 0$ is the only critical point of t. However, the graph of t in Fig. 2.6.40 shows $x = 0$ is not a local extremum. $\square$

Figure 2.6.39
$s(x) = x^3$

Figure 2.6.40
$t(x) = x^{1/3}$

We want to identify the critical points of complicated functions whose graphs we do not know. The following scenario motivates the method we use. Imagine that we are taking a scenic walk along the path from left to right depicted in Figure 2.6.41. First, we approach hill A. We walk uphill, rest for a minute at $R1$, then continue uphill until we reach the top A. Then we walk downhill, leaving the hill behind. We stop for a snack at $R2$, then continue downhill until we arrive at the bottom of valley B. We walk uphill out of the valley, stop for a brief rest at $R3$, and continue walking uphill to the top of hill C. We walk downhill from C, stop to rest again at $R4$ and continue walking downhill. We rephrase these observations of our scenic walk in mathematical terms. Let the path of our walk be the graph of the function f. The tops of hills A, C are local maxima of f and the bottom of valley B is a local minimum of f.

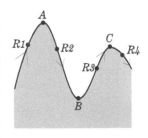

Figure 2.6.41
A Scenic Walk

(a) Before we reach the tops of hills A, C we are walking uphill. Hence $f(x)$ is increasing there. After we leave the tops of these hills we are walking downhill. Hence $f(x)$ is decreasing there.

(b) Before we reach the bottom of valley B we are walking downhill. Hence $f(x)$ is decreasing there. After we leave the bottom of this valley we are walking uphill. Hence $f(x)$ is increasing there.

(c1) Before and after the rest stops $R1$, $R3$ we are walking uphill. Hence $f(x)$ is increasing there.

(c2) Before and after the rest stops $R2$, $R4$ we are walking downhill. Hence $f(x)$ is decreasing there.

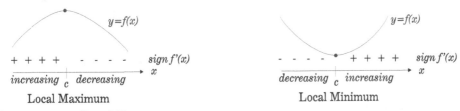

Figure 2.6.42 Local Extrema

The observations from our scenic walk apply to a critical point $x = c$ of any continuous function f. There are four possibilities.

(a) If f is increasing to the left of $x = c$ and decreasing to the right of $x = c$, as in the left diagram of Figure 2.6.42, then $x = c$ must be a local maximum.

(b) If f is decreasing to the left of $x = c$ and increasing to the right of $x = c$, as in the right diagram of Figure 2.6.42, then $x = c$ must be a local minimum.

(c1) If f is increasing both to the left of $x = c$ and to the right of $x = c$, as in the left diagram of Figure 2.6.43, then $x = c$ is not a local extremum.

(c2) If f is decreasing both to the left of $x = c$ and to the right of $x = c$, as in the right diagram of Figure 2.6.43, then $x = c$ is not a local extremum.

This method of identifying the critical points of f is called the *First Derivative Test*. We summarize our conclusions in the following theorem. We use the criterion $f'(x) > 0$ to identify that f is increasing and criterion $f'(x) < 0$ to identify that f is decreasing.

Figure 2.6.43 No Local Extrema

Theorem 2.6.44 (First Derivative Test) *The continuous function f has a critical point at $x = c$.*

(a) *Assume there is $r > 0$ such that*

$$f'(x) > 0 \quad for \ x \in (c - r, c) \quad and \quad f'(x) < 0 \quad for \ x \in (c, c + r).$$

Then f has a local maximum at $x = c$.

(b) *Assume there is $r > 0$ such that*

$$f'(x) < 0 \quad for \ x \in (c - r, c) \quad and \quad f'(x) > 0 \quad for \ x \in (c, c + r).$$

Then f has a local minimum at $x = c$.

(c) *Assume there is $r > 0$ such that either*

$$f'(x) > 0 \quad for \ x \in (c - r, c) \cup (c, c + r) \quad or$$
$$f'(x) < 0 \quad for \ x \in (c - r, c) \cup (c, c + r).$$

Then f does not have a local extremum at $x = c$.   □

We illustrate the procedure for finding and identifying critical points in the following examples. We compute the first derivative. Critical points occur wherever the first derivative does not exist or has value zero. We plot the sign of the first derivative on a number line and apply the First Derivative Test to identify each critical point.

Examples 2.6.45 Find and identify the critical points of each function.

(1) $f(x) = x^3 - 3x^2 - 72x - 64$ with domain $\mathfrak{R}$.

Solution We factor $f'(x)$ to determine where it vanishes.

$$f'(x) = 3x^2 - 6x - 72 = 3(x^2 - 2x - 24) = 3(x + 4)(x - 6).$$

Thus $x = -4$ and $x = 6$ are the two critical points of f. Observe that

$$f'(x) > 0 \ \text{ if } \ x < -4 \text{ or } x > 6 \ \text{ while } \ f'(x) < 0 \text{ if } \ -4 < x < 6.$$

Figure 2.6.46 $f(x) = x^3 - 3x^2 - 72x - 64$

Since f is increasing for $x < -4$ and decreasing for $-4 < x < 6$, it follows from the First Derivative Test that $x = -4$ is a local maximum. Since f is decreasing for $-4 < x < 6$ and increasing for $x > 6$, it follows from the First Derivative Test that $x = 6$ is a local minimum. See Figure 2.6.46.

(2) $g(x) = \dfrac{1}{x^2 - x}$ has domain all numbers other than 0 and 1.

Solution Write $g(x) = (x^2 - x)^{-1}$. By the chain rule:

$$g'(x) = -(x^2 - x)^{-2}(2x - 1) = \frac{1 - 2x}{(x^2 - x)^2}.$$

The only critical point of g occurs at $x = \frac{1}{2}$, where the numerator vanishes. In particular, $x = 0$ and $x = 1$ are not critical points because they are not in the domain of g. Observe that the denominator of $g'(x)$ is always positive. Hence $g'(x)$ has the same sign as its numerator:

$$g'(x) > 0 \text{ if } x < \frac{1}{2} \text{ and } g'(x) < 0 \text{ if } x > \frac{1}{2}.$$

See Figure 2.6.47. Since g is increasing for $x < \frac{1}{2}$ and decreasing for $x > \frac{1}{2}$, it follows form the First Derivative Test that $x = \frac{1}{2}$ is a local maximum.

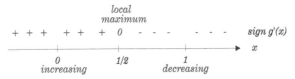

Figure 2.6.47 $g(x) = \frac{1}{x^2 - x}$

(3) $h(x) = \cos^{2/3} x$ with domain $[-\pi, \pi]$.

Solution By the chain rule:

$$h'(x) = \frac{2}{3}\left(\cos^{-1/3} x\right) D(\cos x) = \frac{2}{3}\left(\cos^{-1/3} x\right)(-\sin x) = -\frac{2\sin x}{3\cos^{1/3} x}.$$

Thus $x = 0$ is a critical point, because the numerator of $h'(x)$ vanishes there and $h'(0) = 0$. Also, $x = \pm\frac{\pi}{2}$ are critical points because the denominator of $h'(x)$ vanishes there and $h'\left(-\frac{\pi}{2}\right)$, $h'\left(\frac{\pi}{2}\right)$ do not exist. Observe that the numerator $2\sin x$ of $h'(x)$ is positive if $0 < x < \pi$ and is negative if $-\pi < x < 0$. The denominator $3\cos^{1/3} x$ of $h'(x)$ is positive if $-\frac{\pi}{2} < x < \frac{\pi}{2}$, while it is negative if $-\pi < x < -\frac{\pi}{2}$ or $\frac{\pi}{2} < x < \pi$. Using this information, we depict the sign of $h'(x)$ on the number line in Figure 2.6.48. We identify the three critical points by the First Derivative Test. Since h is decreasing for $-\pi < x < -\frac{\pi}{2}$ and increasing for $-\frac{\pi}{2} < x < 0$, it follows that $x = -\frac{\pi}{2}$ is a local minimum. Since h is increasing for $-\frac{\pi}{2} < x < 0$ and decreasing for $0 < x < \frac{\pi}{2}$, we see that $x = 0$ is a local maximum. Since h is decreasing for $0 < x < \frac{\pi}{2}$ and increasing for $\frac{\pi}{2} < x < \pi$, it follows that $x = \frac{\pi}{2}$ is a local minimum.

	local minimum		local maximum		local minimum	
$- - - - -$	DNE	$+ + +$	0	$- - -$	DNE	$+ + + +$	sign $h'(x)$

$-\pi$ *decreasing* $-\frac{\pi}{2}$ *increasing* 0 *decreasing* $\frac{\pi}{2}$ *increasing* π

Figure 2.6.48 $h(x) = \cos^{2/3} x$

(4) $k(x) = |x^2 - 2x - 15|$ with domain $\Re$.

   Solution Since $k(x) = |(x+3)(x-5)|$,

$$k(x) = \left\{ \begin{array}{ll} x^2 - 2x - 15 & \text{if } x \leq -3 \text{ or } x \geq 5 \\ -(x^2 - 2x - 15) & \text{if } -3 < x < 5 \end{array} \right\},$$

$$k'(x) = \left\{ \begin{array}{ll} 2x - 2 & \text{if } x < -3 \text{ or } x > 5 \\ -2x + 2 & \text{if } -3 < x < 5 \end{array} \right\}$$

and $k'(-3)$, $k'(5)$ do not exist. Observe that $2x - 2$ does not vanish for any number less than -3 or greater than 5. On the other hand, $-2x + 2$ vanishes for $x = 1$ with $-3 < 1 < 5$. Thus we have three critical points: $x = 1$, where $k'(x)$ vanishes, as well as $x = -3$, $x = 5$, where $k'(x)$ does not exist. Note

$$k'(x) > 0 \text{ if } -3 < x < 1 \text{ or } x > 5 \quad \text{while} \quad k'(x) < 0 \text{ if } x < -3 \text{ or } 1 < x < 5.$$

See Figure 2.6.49. By the First Derivative Test, $x = -3$ and $x = 5$ are local minima while $x = 1$ is a local maximum. □

Figure 2.6.49 $k(x) = |x^2 - 2x - 15|$

Historical Remarks

The Mean Value Theorem was formulated for power series by Joseph Louis Lagrange in 1797. In his 1823 text, Louis Augustin Cauchy proves the Mean Value Thoerem when $f'(x)$ is continuous on the interval $[a, b]$. Let m be the minimum value of $f'(x)$, and let M be its maximum value on this interval. Cauchy shows that

$$m \leq \frac{f(b) - f(a)}{b - a} \leq M.$$

By the Intermediate Value Theorem, there must be a number $c \in [a, b]$ where $f'(c) = \frac{f(b)-f(a)}{b-a}$. Cauchy gives the applications of this section as corollaries of the Mean Value Theorem. In 1868, Ossian Bonnet published a proof of the Mean Value Theorem which does not require $f'(x)$ to be continuous.

   In his 1829 text Cauchy generalizes the Mean Value Theorem to "Cauchy's Mean Value Theorem" which we present in Section 13.

Summary

The reader should know the statements of Theorem 2.6.6, Rolle's Theorem, the Mean Value Theorem and its three corollaries. She should be able to apply each of these results to examples. In addition, she should be able to find the critical points of a function and use the First Derivative Test to identify them.

Basic Exercises

1. Argue as in Example 2.6.3 to find the value of the derivative of $f(x) = x^2$ at its local minimum $x = 0$.

2. Does the argument of Example 2.6.3 establish the value of the derivative of $f(x) = x^3$ at $x = 0$?

3. Argue as in Example 2.6.3 to show that if f is a differentiable function which has a local minimum at $x = c$, then $f'(c) = 0$.

4. Graph each function to find its local extrema:
(a) $f(x) = x^2 + 6x$ with domain $[-4, 5]$;   (b) $g(x) = 5 - x^3$ with domain $[-4, 1]$;
(c) $h(x) = \cos x$ with domain $[0, 3\pi]$;   (d) $j(x) = |x - 4|$ with domain $[0, 6]$;
(e) $k(x) = 9 - x^{2/3}$ with domain $[-8, 1]$;   (f) $m(x) = x^{1/3}$ with domain $[-2, 3]$.

5. For each function of Example 4, find the derivative at its local extrema.

6. Graph each function to find its minimum and maximum values. Determine the value of the derivative at each of these two points. Explain why the derivative is zero, nonzero or does not exist in each case.
(a) $f(x) = x^2$ with domain $[-1, 3]$.   (b) $g(x) = \sin x$ with domain $[0, 2\pi]$.
(c) $h(x) = x^3$ with domain $[-1, 2]$.   (d) $j(x) = \sec x$ with domain $[-\pi/4, \pi/3]$.
(e) $k(x) = (x - 2)^2$ with domain $[-2, 4]$.   (f) $m(x) = |2 - x|$ with domain $[1, 5]$.

7. Determine whether Rolle's Theorem applies to each function. If Rolle's Theorem applies, determine all points in the appropriate open interval where the derivative vanishes. If it does not apply, explain why.
(a) $f(x) = 6 - x^2$ with domain $[-2, 2]$.   (b) $g(x) = \tan x$ with domain $[0, \pi]$.
(c) $h(x) = x^3 - 2x^2 + 5$ with domain $[0, 2]$.   (d) $j(x) = \frac{1}{|x^2+1|}$ with domain $[-3, 3]$.
(e) $k(x) = x^{2/3}$ with domain $[-1, 1]$.   (f) $m(x) = |5 - x|$ with domain $[1, 9]$.

8. Let f and g be continuous functions having domain $[a, b]$ with $a < b$. Assume f and g are differentiable on (a, b). If $f(a) = g(a)$ and $f(b) = g(b)$ show there is at least one number $c \in (a, b)$ such that $f'(c) = g'(c)$.

9. For each function, find all numbers c in the appropriate open interval which satisfy the conclusion of the Mean Value Theorem.
(a) $f(x) = x^2$ for $x \in [1, 5]$.   (b) $g(x) = x^3$ for $x \in [-1, 4]$.
(c) $h(x) = \cos x$ for $x \in [-\pi, 5\pi]$.   (d) $j(x) = x^3 - 3x^2 - 5x + 4$ for $x \in [1, 2]$.
(e) $k(x) = \arctan x$ for $x \in [0, 1]$.   (f) $m(x) = \sqrt{3 - x}$ for $x \in [-1, 3]$.

10. Find *all* functions with each of the indicated derivatives.
(a) $f'(x) = 3x^4 - 5x^2 + 6$ (b) $g'(x) = \frac{4}{x^3}$ (c) $h'(x) = 12\sqrt{x}$
(d) $j'(x) = 5\cos 6x$ (e) $k'(x) = \frac{4}{\sqrt{1-x^2}}$ (f) $m'(x) = \frac{\sec^2\sqrt{x}}{\sqrt{x}}$

11. In each case, find the unique function which satisfies the given conditions.
(a) $f'(x) = 4x^3 - 6x + 4$ and $f(0) = 5$.   (b) $g'(x) = 6x^{2/3} - 9x^{3/2}$ and $g(1) = 2$.
(c) $h'(x) = \frac{8}{1+x^2}$ and $h(1) = \pi$.   (d) $j'(x) = 2\sec^2 x$ and $i(0) = 3$.
(e) $k'(x) = \frac{4x}{\sqrt{1-x^2}}$ and $j(0) = 7$.   (f) $m'(x) = \frac{8}{x^2+1}$ and $m(1) = 0$.

12. Let $s(x)$ be one solution of the differential equation $y'' = f(x)$.
(a) Show the general solution for y' is $y' = s'(x) + A$ for $A \in \Re$.
(b) Show the general solution for y is $y = s(x) + Ax + B$ for $A, B \in \Re$.

13. Use Exercise 12 to find the general solution of each differential equation.
(a) $f''(x) = -32$ (b) $g''(x) = 24x^2 - 6$ (c) $h''(x) = \sin x$
(d) $j''(x) = \sqrt{x}$ (e) $k''(x) = \frac{1}{x^4}$ (f) $m''(x) = \frac{x}{(1+x^2)^2}$

14. Use Exercise 12 to find each function which meets the given conditions.
(a) $f''(x) = -32$ with $f'(0) = 3$, $f(0) = -2$.
(b) $g''(x) = 90x^4$ with $g'(0) = -1$, $g(0) = 5$.
(c) $h''(x) = \frac{80}{x^3}$ with $h'(1) = 4$, $h(1) = 7$.
(d) $j''(x) = 112\sqrt[3]{x}$ with $j'(0) = -6$, $j(0) = -4$.
(e) $k''(x) = \cos x$ with $k'(0) = 1$, $k(0) = -1$.
(f) $m''(x) = \frac{4x}{(1-x^2)^{3/2}}$ with $m'(0) = 10$, $m(0) = 9$.

15. Determine the intervals on which each function is increasing and the intervals on which it is decreasing. Summarize your findings on a number line.

(a) $f(x) = 3x^2 - 24x + 5$ for $x \in \Re$.   (b) $g(x) = x^3 - 6x^2 - 36x + 1$ for $x \in \Re$.

(c) $h(x) = x^3 - 3x^2 + 3x + 8$ for $x \in \Re$.   (d) $i(x) = 3x^5 - 50x^3 + 135x + 7$ for $x \in$

(e) $j(x) = \sqrt{x^2 + 1}$ for $x \in \Re$.   (f) $k(x) = \frac{x+1}{x^2+3}$ for $x \in \Re$.

(g) $m(x) = \cos(3x + 1)$ for $x \in [-1, 1]$.   (h) $n(x) = \sin(x^2 - 1)$ for $x \in [-1, 2]$.

(i) $p(x) = \arctan(x^2 - 4x + 3)$ for $x \in \Re$.   (j) $q(x) = |1 - x^2|$ for $x \in \Re$.

16. Find the critical points of each function. Use the graph of the function to identify each critical point.

(a) $f(x) = x^2 + 8x$ (b) $g(x) = \cos x$ with domain $(-\pi, 5\pi)$ (c) $h(x) = 5 - 3x$

(d) $j(x) = |6 - x|$ (e) $k(x) = x^{3/5}$ (f) $m(x) = x^{4/7}$

17. Find the critical points of each function. Use the First Derivative Test to identify each of them as a local maximum, a local minimum or neither.

(a) $f(x) = x^2 - 8x + 3$ (b) $g(x) = 2x^3 + 15x^2 + 36x + 17$

(c) $h(x) = 3x^5 - 25x^3 + 60x + 9$ (d) $j(x) = \frac{x}{4x^2+1}$

(e) $k(x) = \frac{3x^2+2}{x^2-1}$ (f) $m(x) = \frac{x^3}{2x^3-1}$

(g) $p(x) = (x + 3)^{3/7}$ (h) $n(x) = \sin(x^2)$ with domain $(0, 3)$

(i) $q(x) = (2x - 1)^{4/9}$ (j) $r(x) = (x^2 - 5x + 4)^{4/5}$

(k) $s(x) = |x^2 - 6x - 16|$ (l) $t(x) = |x^4 - 5x^2 + 4|$

18. Show $2 \arcsin \sqrt{x} = \arcsin(2x - 1) + \frac{\pi}{2}$ for $0 \le x \le 1$.

Challenging Problems

1. Let f be a differentiable function whose domain is an open interval I. Assume there are n points $x_1, \ldots, x_n \in I$ such that $f(x_1) = \cdots = f(x_n) = 0$.

(a) Show that there are at least $n - 1$ points in I where the derivative of f is zero.

(b) Assume that the kth derivative of f exists on the interval I. At how many points of I must the kth derivative of f equals zero?

2. Show that for $x \ge 0$: $\frac{1}{1+x} \ge 1 - x^2$.

Hint: Consider the function $f(x) = \frac{1}{1+x} - (1 - x)$.

3. Let A, B, C, D be four constants.

(a) Use Rolle's Theorem to show that a cubic polynomial $f(x) = Ax^3 + Bx^2 + Cx + D$ has at most three distinct real roots.

(b) What are the maximum number of distinct real roots of a cubic polynomial of the form $g(x) = x^3 + C^2x + D$? Must this polynomial have that many distinct real roots?

2.7   Graphing Functions

To sketch the graph of a given function, we compile information about its graph. We begin in the first subsection with a study of the intercepts of a graph. The second subsection introduces three types of asymptotes and methods for locating them. The third subsection studies concavity and inflection points which describe the relationship of a graph to its tangent lines. In the fourth subsection, we assemble information about the function f to sketch its graph. In addition to intercepts, asymptotes and concavity we also use information supplied by the first derivative exposited in Sections 2 and 6. The sign of $f'(x)$ determines where f is increasing and where f is decreasing. We locate the critical points and use the First Derivative Test to identify the local maxima and minima. We also examine the points where $f'(x)$ does not exist to identify vertical tangents and cusps.

Intercepts

The intercepts of a function are the points where its graph crosses the coordinate axes. After defining these terms, we show how to find the intercepts of a given function.

Definition 2.7.1 *Let G be the graph of the function f.*
(a) *The values of x where G crosses the x–axis are called the x–intercepts of f.*
(b) *If G crosses the y–axis at the point $(0, y_0)$, then y_0 is called the y–intercept of f.*

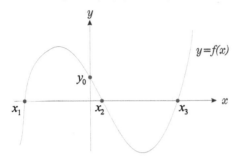

Figure 2.7.2 Intercepts of f

In Figure 2.7.2, the x-intercepts of f are x_1, x_2, x_3 while the y–intercept of f is y_0. We find the x–intercepts of f by solving the equation $f(x) = 0$. If f is a polynomial, the x–intercepts of f are the roots of f. If 0 is in the domain of f, the y–intercept of f is merely the value of $f(0)$. If 0 is not in the domain of f, f has no y–intercept.

Examples 2.7.3 Determine the intercepts of each function.

(1) Let $f(x) = x^3 - x^2 - 17x - 15$.

Solution The integer roots of $f(x)$ are factors of the constant term -15, i.e. they are either ± 1, ± 3, ± 5 or ± 15. Since $f(1) = -32 \neq 0$, 1 is not a root of $f(x)$. However, $f(-1) = 0$, and -1 is a root of $f(x)$. Hence $f(x)$ is divisible by $x + 1$. Divide $f(x)$ by $x + 1$:

$$f(x) = (x + 1)(x^2 - 2x - 15) = (x + 1)(x - 5)(x + 3).$$

Thus, $x = -1$, $x = 5$, $x = -3$ are the x–intercept of f. The y–intercept of f is $f(0) = -15$.

(2) Let $g(x) = \frac{x^2 - 9}{x^2 - 25}$.

Solution Since a fraction is zero precisely when its numerator vanishes, the x–intercepts of g are the solutions to the equation $x^2 - 9 = 0$. Thus, the x–intercepts of g are $x = -3$ and $x = +3$. The y–intercept of g is $g(0) = \frac{9}{25}$.

(3) Let $k(x) = \frac{1}{x^2 + x}$.

Solution Since the numerator of this fraction never vanishes, k has no x–intercepts. Since the denominator of $k(x)$ vanishes at $x = 0$, the number 0 is not in the domain of k. Hence k has no y–intercept. □

Asymptotes

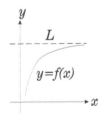

Figure 2.7.4
Horizontal Asymptote

An asymptote of a function is a line L which the graph of the function approaches as a limit. There are three types of asymptotes: horizontal (when L is horizontal), as in Figure 2.7.4, vertical (when L is vertical), as in Figure 2.7.6, and oblique (when L is neither horizontal nor vertical), as in Figure 2.7.7. We begin by defining a horizontal asymptote. In Figure 2.7.4 the right end of the graph approaches its horizontal asymptote L. In other examples, the left end of the graph may approach its horizontal asymptote. For example, see Figure 2.7.9.

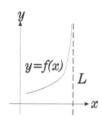

Figure 2.7.6
Vertical Asymptote

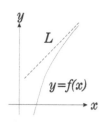

Figure 2.7.7
Oblique Asymptote

Definition 2.7.5 (a) *The function f has the line $y = c$ as a horizontal asymptote on the left if*

$$\lim_{x \to -\infty} f(x) = c.$$

(b) *The function f has the line $y = c$ as a horizontal asymptote on the right if*

$$\lim_{x \to \infty} f(x) = c.$$

Consider a rational function f. That is, $f(x)$ is a quotient of two polynomials:

$$f(x) = \frac{ax^m + \ lower\ powers\ of\ x}{bx^n + \ lower\ powers\ of\ x}.$$

(1) If the degree m of the numerator is less than the degree n of the denominator, then f has the x–axis, $y = 0$, as a horizontal asymptote on both the left and the right.

(2) If the degrees m of the numerator and n of the denominator are equal, then f has the line $y = \frac{a}{b}$ as a horizontal asymptote on both the left and the right.

(3) If the degree m of the numerator is greater than the degree n of the denominator, then f has no horizontal asymptote.

The trick for verifying these rules for horizontal asymptotes is to divide both the numerator and denominator by x^n, the highest power of x which appears in the denominator of $f(x)$. We use this trick in the first three examples below. Alternatively, these horizontal asymptotes can be determined by using the three rules above.

Examples 2.7.8 (1) Find the horizontal asymptotes of $f(x) = \frac{3x^4 - 2x^3 - 17}{6x^5 + 8x^2 + 3}$.

Solution Divide numerator and denominator by x^5, the largest power of x in the denominator of $f(x)$:

$$f(x) = \frac{3/x - 2/x^2 - 17/x^5}{6 + 8/x^3 + 3/x^5}.$$

Thus $\displaystyle \lim_{x \to -\infty} f(x) = \frac{-0 - 0 + 0}{6 - 0 - 0} = 0$ and $\displaystyle \lim_{x \to \infty} f(x) = \frac{0 - 0 - 0}{6 + 0 + 0} = 0.$

Hence the x–axis is a horizontal asymptote of f on both the left and right.

(2) Find the horizontal asymptotes of the function $g(x) = \frac{24x^3 - x^2 + 7}{8x^3 + 5x - 3}$.

Solution Divide numerator and denominator by x^3, the largest power of x in the denominator of $g(x)$:

$$g(x) = \frac{24 - 1/x + 7/x^3}{8 + 5/x^2 - 3/x^3}.$$

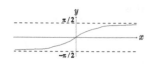

Figure 2.7.9
$p(x) = \arctan x$

Then $\displaystyle \lim_{x \to -\infty} g(x) = \frac{24 + 0 - 0}{8 + 0 + 0} = 3$ and $\displaystyle \lim_{x \to \infty} g(x) = \frac{24 - 0 + 0}{8 + 0 - 0} = 3.$

Hence the line $y = 3$ is a horizontal asymptote of g on both the left and right.

(3) Find the horizontal asymptotes of the function $m(x) = \frac{2x^3 + x^2 + 1}{x^2 + 1}$.

Solution Divide numerator and denominator by x^2, the largest power of x in the denominator of $m(x)$:

$$m(x) = \frac{2x + 1 + 1/x^2}{1 + 1/x^2}.$$

Hence $\displaystyle\lim_{x \to -\infty} m(x) = \frac{-\infty + 1 + 0}{1 + 0} = -\infty$ and $\displaystyle\lim_{x \to \infty} m(x) = \frac{\infty + 1 + 0}{1 + 0} = \infty$.

Thus, m has no horizontal asymptotes.

(4) Find the horizontal asymptotes of the function $p(x) = \arctan x$.

Solution Consider the graph of the function $p(x) = \arctan x$ in Figure 2.7.9. We see from this graph that

$$\lim_{x \to -\infty} p(x) = -\frac{\pi}{2} \quad \text{and} \quad \lim_{x \to \infty} p(x) = \frac{\pi}{2}.$$

Therefore, p has horizontal asymptote $y = -\frac{\pi}{2}$ on the left and the horizontal asymptote $y = \frac{\pi}{2}$ on the right. $\qquad\Box$

Next we define a vertical asymptote L of a function. Observe that the graph can approach L from the left, as in Figure 2.7.6, from the right, as in Figure 2.7.10, or from both the left and the right. In this case, there are four possibilities. The graph may approach the top of L on both sides of L, as in Figure 2.7.11. The graph may approach the bottom of L on both sides of L, as in Figure 2.7.12. Alternatively, the graph may approach the top of L on one side of L and approach the bottom of L on the other side of L, as in Figures 2.7.13 and 2.7.15.

Definition 2.7.14 *A continuous function f has the line $x = k$ as a vertical asymptote if at least one of the following limits occurs:*

$$\lim_{x \to k^-} f(x) = \infty, \quad \lim_{x \to k^-} f(x) = -\infty, \quad \lim_{x \to k^+} f(x) = \infty \quad or \quad \lim_{x \to k^+} f(x) = -\infty.$$

Note The number k may not be in the domain of f.

A rational function f often has vertical asymptotes. Write the equation of f in reduced form, i.e. cancel all common factors of the numerator and denominator. Then vertical asymptotes occur where the denominator of f vanishes.

Examples 2.7.16 (1) Find the vertical asymptotes of the function $f(x) = \frac{x^2 - 8}{x^2 - 16}$.

Solution The denominator $x^2 - 16 = (x - 4)(x + 4)$ of f vanishes when $x = -4$ or $x = +4$. Therefore, the lines $x = -4$ and $x = +4$ are vertical asymptotes of f.

(2) Find the vertical asymptotes of the function $g(x) = \frac{x^2 - 1}{x^4 + 1}$.

Solution Since the denominator $x^4 + 1$ of g has no real roots, g has no vertical asymptotes.

(3) Find the vertical asymptotes of the function $h(x) = \frac{x^3 - 1}{x^4 - 1}$.

Solution The denominator $x^4 - 1$ of h vanishes at $x = -1$ and $x = +1$. However, $x = 1$ is also a root of the numerator $x^3 - 1$. Thus, the fraction in the definition of h is not in reduced form. In fact,

$$h(x) = \frac{(x - 1)(x^2 + x + 1)}{(x - 1)(x^3 + x^2 + x + 1)} = \frac{x^2 + x + 1}{x^3 + x^2 + x + 1}.$$

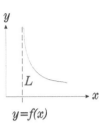

$y = f(x)$

Figure 2.7.10
$y = f(x)$

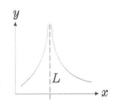

Figure 2.7.11
$y = g(x)$

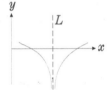

Figure 2.7.12
$y = h(x)$

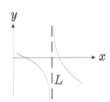

Figure 2.7.13
$y = k(x)$

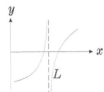

Figure 2.7.15
$y = m(x)$

In this reduced representation, the denominator $x^3 + x^2 + x + 1 = (x+1)(x^2+1)$ of h only vanishes at $x = -1$. Therefore, the line $x = -1$ is a vertical asymptote. Note $x = +1$ is not in Domain h, but the line $x = +1$ is *not* a vertical asymptote.

(4) Find the vertical asymptotes of the function $t(x) = \tan x$.

Solution Recall graph $t(x) = \tan x$ depicted in Figure 2.7.17. This graph indicates that t has a vertical asymptote at $x = (2k+1)\pi/2$ for each integer k. $\square$

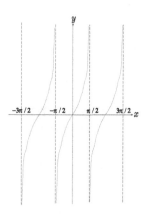

Figure 2.7.17
$t(x) = \tan x$

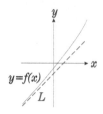

Figure 2.7.18
Oblique Asymptote

The third type of asymptote is an oblique asymptote L. The right end of the graph may approach the line L as in Figure 2.7.7. Alternatively, the left end of the graph may approach the line L as in Figure 2.7.18.

Definition 2.7.19 (a) *The function f has the line $y = mx + b$ as an oblique asymptote on the left if*
$$\lim_{x \to -\infty} [f(x) - (mx + b)] = 0.$$

(b) *The function f has the line $y = mx + b$ as an oblique asymptote on the right if*
$$\lim_{x \to \infty} [f(x) - (mx + b)] = 0.$$

Oblique asymptotes of rational functions occur when the numerator has degree one larger than the degree of the denominator. In this case we determine the equation of the oblique asymptote by dividing the denominator into the numerator.

Examples 2.7.20 (1) Find the oblique asymptotes of the function $f(x) = \frac{5x^4 + 3x^3 + 7x - 8}{x^3}$.

Solution Divide the numerator by the denominator x^3:

$$f(x) \quad = \quad 5x + 3 + \frac{7x - 8}{x^3}.$$

Then
$$\lim_{x \to -\infty} [f(x) - (5x+3)] \quad = \quad \lim_{x \to -\infty} \frac{7x - 8}{x^3} = 0$$

and
$$\lim_{x \to \infty} [f(x) - (5x+3)] \quad = \quad \lim_{x \to \infty} \frac{7x - 8}{x^3} = 0.$$

Thus, the line $y = 5x + 3$ is an oblique asymptote of f on both the left and right.

(2) Find the oblique asymptotes of the function $g(x) = \frac{2x^3 - x^2 + 4}{x^2 + 1}$.

Solution Divide the denominator $x^2 + 1$ of $g(x)$ into its numerator $2x^3 - x^2 + 4$:

$$g(x) \quad = \quad 2x - 1 + \frac{-2x + 5}{x^2 + 1}.$$

Then
$$\lim_{x \to -\infty} [g(x) - (2x-1)] \quad = \quad \lim_{x \to -\infty} \frac{-2x + 5}{x^2 + 1} = 0$$

and
$$\lim_{x \to \infty} [g(x) - (2x-1)] \quad = \quad \lim_{x \to \infty} \frac{-2x + 5}{x^2 + 1} = 0.$$

Thus, $y = 2x - 1$ is an oblique asymptote of g on both the left and right. $\square$

Concavity and Inflection Points

The sign of the first derivative of f determines where the graph of f is increasing and where it is decreasing. Analogously, the sign of the second derivative of f determines where the graph of f increases from its tangent line and where it decreases from its tangent line. The relationship of a graph to its tangent lines is called concavity. The graph is *concave up* where it lies above its tangent line as at P_1 in the left graph of Figure 2.7.21. The graph is *concave down* where it lies below its tangent line as at P_2 in the second graph of Figure 2.7.21. A point on the graph where the concavity changes is called an *inflection point*. For example, in the right graph of Figure 2.7.21 there are three inflection points. At P_3 the concavity changes from up to down while at P_4 the concavity changes from down to up. Note that the graph of h lies above the left half–tangent line T' at P and lies below the right half–tangent line T'' at P. Since the concavity at P changes from up to down, P is also an inflection point of h. Concavity helps determine the shape of a graph. For example, if f is an increasing function which is concave up, then its graph has the shape of the left graph in Figure 2.7.21. On the other hand, if g is an increasing function which is concave down, then its graph has the shape of the middle graph in Figure 2.7.21.

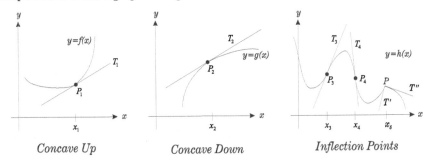

Concave Up   Concave Down   Inflection Points

Figure 2.7.21 Graphs and Tangent Lines

Definition 2.7.22 *Let f be a continuous function with domain D.*

(a) *Assume $f'(c)$ exists, $c \in D$, with T the tangent line to the graph of f at $x = c$.*
  (i) *f is concave up at $x = c$ if there is an open interval J contained in D such that $c \in J$ and the graph of f restricted to J lies above T.*
  (ii) *f is concave down at $x = c$ if there is an open interval J contained in D such that $c \in J$ and the graph of f restricted to J lies below T.*

(b) *Assume $f'_-(c)$ and $f'_+(c)$ exist. f has an inflection point at $x = c$ if there is $(a, b) \subset D$, with $c \in (a, b)$, such that the graphs of f restricted to $(a, c]$ and f restricted to $[c, b)$ lie on opposite sides of the half–tangent lines at $x = c$.*

Note Assume $f'(c)$ exists. Then $x = c$ is an inflection point of f if there is an open interval $J = (a, b)$ contained in D, with $c \in J$, such that the graph of f restricted to $(a, c]$ and the graph of f restricted to $[c, b)$ lie on opposite sides of T.

The following theorem shows how the sign of the second derivative determines where a function is concave up, concave down and has inflection points.

Theorem 2.7.23 *Let f be a continuous function on an open interval I with $c \in I$.*

(a) *If $f''(x)$ exists for $x \in I$ and is continuous at $x = c$ with $f''(c) > 0$, then f is concave up at $x = c$.*

(b) *If $f''(x)$ exists for $x \in I$ and is continuous at $x = c$ with $f''(c) < 0$, then f is concave down at $x = c$.*

(c) *Assume $f''(x)$ exists for $x \in I$, $x \neq c$, and the following limits exist:*

$$f'_-(c) = \lim_{x \to c^-} f'(x), \quad f'_+(c) = \lim_{x \to c^+} f'(x), \quad f''_-(c) = \lim_{x \to c^-} f''(x), \quad f''_+(c) = \lim_{x \to c^+} f''(x).$$

If $f''_-(c)$ and $f''_+(c)$ are nonzero with opposite signs, then $x = c$ is an inflection point.

Proof **(a)** Since $f''(c) > 0$ and $f''(x)$ is continuous at $x = c$, there is an open interval J, with $c \in J$, such that $f''(x) > 0$ for $x \in J$. Let $T(x) = f(c) + f'(c)(x - c)$ denote the tangent line T to the graph of f at $x = c$. Define $g(x)$ as the vertical distance from the point $(x, T(x))$ on the tangent line T to the point $(x, f(x))$ on graph f:

$$g(x) \; = \; f(x) - T(x) \; = \; f(x) - [f(c) + f'(c)(x - c)] \; .$$

See Figure 2.7.24. We show $g(x) \geq 0$, for $x \in J$, which means the graph of f lies above the tangent line T over the interval J on the x–axis. Observe that

$$
\begin{aligned}
g'(x) &= f'(x) - f'(c) & (2.7.1)\\
D\,[g'(x)] &= g''(x) = f''(x) > 0 \ \text{ for } x \in J.
\end{aligned}
$$

Hence $g'(x)$ is an increasing function on the interval J. By (2.7.1), $g'(c) = f'(c) - f'(c) = 0$. Hence $g'(x)$ is negative for $x \in J$ with $x < c$ while $g'(x)$ is positive for $x \in J$ with $x > c$. See Figure 2.7.25. By the First Derivative Test, $g(x)$ has a local minimum at $x = c$. That is, $g(x) \geq g(c) = 0$ for $x \in J$, and the graph of f lies above T over the interval J.

(b) By (a), the function $-f(x)$ is concave up. Hence $f(x)$ is concave down.

(c) Consider the case where $f''_-(c) = g''_-(c) > 0$ and $f''_+(c) = g''_+(c) < 0$. By the argument of (a), $g(x) \geq g(c) = 0$ for $x \leq c$, and the graph of f to the left of $x = c$ lies above the left half–tangent line at $x = c$. By the argument of (b), the graph of f to the right of $x = c$ lies below the right half–tangent line at $x = c$. Hence $x = c$ is an inflection point. A similar argument applies when $f''_-(c) < 0$ and $f''_+(c) > 0$. $\square$

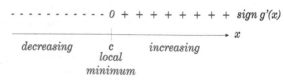

Figure 2.7.25 g' has a local minimum at $x = c$

To summarize, an inflection point occurs where the concavity of f changes. Equivalently, an inflection point occurs where the sign of $f''(x)$ changes. Observe that if $x = c$ is an inflection point of f where $f''(c)$ exists and $f''(x)$ is continuous at $x = c$, then $f''(c) = 0$. However, just as it is *not correct* to say that a local extremum is a value of x where $f'(x) = 0$, so too is it *not correct* to say that an inflection point is a value of x where $f''(x) = 0$. If $f''(c) = 0$, then f does not necessarily have an inflection point at $x = c$ because the sign of f'' may be the same on both sides of $x = c$ as in Example 2 below. Moreover, f'' need not exist at an inflection point as in Examples 3 and 6 below.

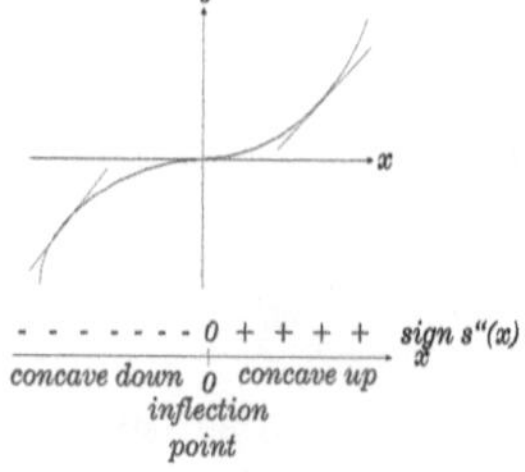

Examples 2.7.27 Determine where each function is concave up and concave down. Then find all the inflection points.

(1) Let $s(x) = x^3$.

Solution We have $s'(x) = 3x^2$, and $s''(x) = 6x$. Thus, $s''(x)$ is positive if $x > 0$ and $s''(x)$ is negative if $x < 0$. Therefore, s is concave up if $x > 0$, and s is concave down if $x < 0$. Since the concavity of s changes at $x = 0$, s has an inflection point at $x = 0$. See Figure 2.7.26.

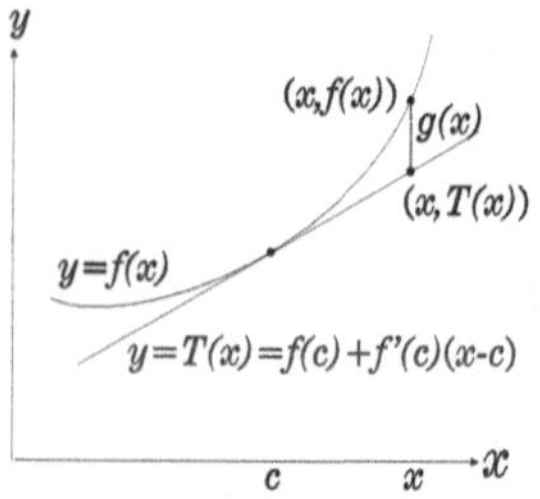

Figure 2.7.24
Definition of $g(x)$

Figure 2.7.26
$s(x) = x^3$

(2) Let $t(x) = x^4$.

Solution Note $t'(x) = 4x^3$, and $t''(x) = 12x^2$. Thus for all nonzero numbers x, $t''(x)$ is positive, and t is concave up. Therefore $x = 0$ is not an inflection point although $t''(0) = 0$. In fact, t is concave up at $x = 0$. See Figure 2.7.28.

(3) Let $u(x) = \sqrt[3]{x}$.

Solution Note $u(x) = x^{1/3}$, $u'(x) = \frac{1}{3}x^{-2/3}$ and

$$u''(x) = -\frac{2}{9}x^{-5/3} = -\frac{2}{9x^{5/3}}.$$

Hence $u''(x) > 0$ for $x < 0$, $u''(x) < 0$ for $x > 0$, and $u''(0)$ does not exist. Thus u is concave up for $x < 0$ and concave down for $x > 0$. Since the concavity of u changes from up to down at $x = 0$, it is an inflection point. See Figure 2.7.29.

(4) Let $f(x) = 2x^4 - 12x^2 + 5x - 7$.

Solution We have $f'(x) = 8x^3 - 24x + 5$ and

$$f''(x) = 24x^2 - 24 = 24(x^2 - 1) = 24(x + 1)(x - 1).$$

Thus $f''(x)$ is positive if $x < -1$ or $x > 1$, and f is concave up there. Also $f''(x)$ is negative if $-1 < x < 1$, and f is concave down on $(-1, 1)$. Since the concavity of f changes from up to down at $x = -1$, f has an inflection point there. Since the concavity of f changes from down to up at $x = 1$, f has an inflection point there. See Figure 2.7.30.

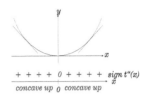

Figure 2.7.28
$t(x) = x^4$

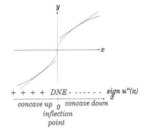

Figure 2.7.29
$u(x) = \sqrt[3]{x}$

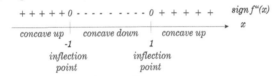

Figure 2.7.30 $f(x) = 2x^4 - 12x^2 + 5x - 7$

(5) Let $g(x) = \frac{x}{x^2 - 9}$.

Solution By the quotient rule: $g'(x) = \frac{1(x^2 - 9) - x(2x)}{(x^2 - 9)^2} = -\frac{x^2 + 9}{(x^2 - 9)^2}$. Apply the quotient rule again:

$$g''(x) = -\frac{2x(x^2 - 9)^2 - (x^2 + 9)2(x^2 - 9)(2x)}{(x^2 - 9)^4}$$

$$= -\frac{2x(x^2 - 9) - (x^2 + 9)(4x)}{(x^2 - 9)^3} \quad \text{[dividing numerator and denominator by } x^2 - 9]$$

$$= \frac{2x^3 + 54x}{(x^2 - 9)^3} = \frac{2x(x^2 + 27)}{(x - 3)^3(x + 3)^3}.$$

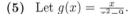

Figure 2.7.31 $g(x) = \frac{x}{x^2 - 9}$

Note $x^2 + 27 \geq 27 > 0$ for all x. Hence $g''(x) > 0$ and g is concave up for $-3 < x < 0$ or $x > 3$. On the other hand, $g''(x) < 0$ and g is concave down for $x < -3$ or $0 < x < 3$. Hence $x = 0$ is an inflection point because the concavity changes from up to down there. Although the concavity of g changes at $x = -3$ and at $x = 3$, these are not inflection points, because they are not in the domain of g. In fact, g has vertical asymptotes at $x = \pm 3$. See Figure 2.7.31.

(6) Let $h(x) = |x^2 - 4|$.

Solution Note $h(x) = |(x-2)(x+2)| = \left\{ \begin{array}{ll} x^2 - 4 & \text{if } x \le -2 \text{ or } x \ge 2 \\ -(x^2 - 4) & \text{if } -2 < x < 2 \end{array} \right\}$.

Hence

$$h'(x) = \left\{ \begin{array}{ll} 2x & \text{if } x < -2 \text{ or } x > 2 \\ -2x & \text{if } -2 < x < 2 \end{array} \right\}, \quad h''(x) = \left\{ \begin{array}{ll} 2 & \text{if } x < -2 \text{ if } x > 2 \\ -2 & \text{if } -2 < x < 4 \end{array} \right\}$$

and $h''(-2)$, $h''(2)$ do not exist. Thus, h is concave up for $x < -2$ or $x > 2$, while h is concave down for $-2 < x < 2$. Therefore, h has an inflection point at $x = -2$ where the concavity changes from up to down and at $x = 2$ where the concavity changes from down to up. See Figure 2.7.32. □

Figure 2.7.32 $h(x) = |x^2 - 4|$

Constructing Graphs

The graph of a continuous function f is constructed by combining a variety of clues which describe its important features. For each function, we perform the following five step procedure to assemble these clues.

I. Intercepts Determine the x–intercepts and the y–intercept.

II. Asymptotes Determine the vertical asymptotes as well as the horizontal or oblique asymptotes.

III. First Derivative Compute the first derivative. Plot its sign on a number line to determine where the function is increasing and where it is decreasing. Locate the critical points, and identify them by the First Derivative Test.

IV. Vertical Tangents and Cusps Examine the one-sided derivatives at all points in the domain where the derivative does not exist. Identify these points as cusps, vertical cusps or vertical tangents.

V. Concavity and Inflection Points Compute the second derivative. Plot its sign on a number line to determine where the function is concave up and where it is concave down. Identify the inflection points as those points in the domain of f where the sign of the second derivative changes.

We use all this information about a function to sketch its graph. This procedure is illustrated in the following examples.

Examples 2.7.33 Sketch the graph of each function.

(1) Let $f(x) = x^3 + 9x^2 + 24x - 34$ with domain $\mathfrak{R}$.

Solution We perform the 5 steps above to determine features of the graph of f.

   I. To find the x–intercepts we factor the polynomial f. Since the coefficients of f have sum zero, 1 is a root of f and $f(x)$ is divisible by $x - 1$: $f(x) = (x-1)(x^2 + 10x + 34)$. Since $10^2 - 4(1)(34) = -36 < 0$, $x^2 + 10x + 34$ has no real roots by the quadratic formula. Thus $x = 1$ is the only x–intercept of f. The y–intercept occurs at $y = f(0) = -34$.

II. f has no asymptotes.

III. $f'(x) = 3x^2 + 18x + 24 = 3(x^2 + 6x + 8) = 3(x+4)(x+2)$. The sign of $f'(x)$ is plotted in Figure 2.7.34. We see that $f'(x) > 0$ for $x < -4$ or $x > -2$ and f is increasing there. Since $f'(x) < 0$ for $-4 < x < -2$, f is decreasing there. f has critical points at $x = -4$ and at $x = -2$ where $f'(x)$ is zero. By the First Derivative Test, f has a local maximum at $x = -4$ and a local minimum at $x = -2$. Observe that $f(-4) = -50$ and $f(-2) = -54$.

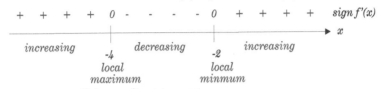

Figure 2.7.34 $f'(x) = 3x^2 + 18x + 24$

IV. f has no vertical tangents or cusps.

V. $f''(x) = 6x + 18 = 6(x + 3)$. The sign of $f''(x)$ is plotted in Figure 2.7.35. Since $f''(x) < 0$ for $x < -3$, f is concave down there. Since $f''(x) > 0$ for $x > -3$, f is concave up there. Hence f has an inflection point at $x = -3$ where the concavity changes from down to up. Observe that $f(-3) = -52$.

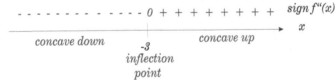

Figure 2.7.35 $f''(x) = 6x + 18x$

We conclude that the graph of f rises concave down from the left to a local maximum at $(-4, -50)$. Then it falls, passing through the inflection point at $(-3, -52)$ where it becomes concave up. It reaches a local minimum at $(-2, -54)$ and rises to cross the y–axis at $y = -34$ and the x–axis at $x = 1$. This graph is depicted in Figure 2.7.36.

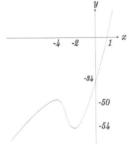

x–intercept: $x = 1$
y–intercept: $y = -34$
no asymptotes
increasing: $x < -4$ or $x > -2$
decreasing: $-4 < x < -2$
local maximum: $(-4, -50)$
local minimum: $(-2, -54)$
no cusps or vertical tangents
concave down: $x < -3$
concave up: $x > -3$
inflection point: $(-3, -52)$

Figure 2.7.36 $f(x) = x^3 + 9x^2 + 24x - 34$

(2) Let $g(x) = \frac{x^2 - 1}{x^2 - 4}$.

Solution The domain of g is $\Re$ excluding $x = \pm 2$ where the denominator of $g(x)$ vanishes. We perform the five steps above to determine features of the graph of g.

I. The graph of g has x–intercepts where its numerator $x^2 - 1 = (x+1)(x-1)$ vanishes, i.e. at $x = -1$ and at $x = +1$. The y–intercept is $g(0) = \frac{1}{4}$.

II. We find the horizontal asymptotes of g:

$$\lim_{x \to -\infty} \frac{x^2 - 1}{x^2 - 4} = \lim_{x \to -\infty} \frac{1 - 1/x^2}{1 - 4/x^2} = 1 \quad \text{and} \quad \lim_{x \to \infty} \frac{x^2 - 1}{x^2 - 4} = \lim_{x \to \infty} \frac{1 - 1/x^2}{1 - 4/x^2} = 1.$$

Hence g has the horizontal asymptote $y = 1$ on both the left and the right. g has vertical asymptotes where its denominator $x^2 - 4 = (x + 2)(x - 2)$ vanishes: at $x = -2$ and at $x = 2$.

III. By the quotient rule:

$$g'(x) \;=\; \frac{2x(x^2 - 4) - 2x(x^2 - 1)}{(x^2 - 4)^2} \;=\; \frac{-6x}{(x^2 - 4)^2}$$

The sign of $g'(x)$, depicted in Figure 2.7.37, is the sign of its numerator $-6x$ since its denominator is always positive. Thus g is increasing for $x < 0$ and decreasing for $x > 0$. By the First Derivative Test, $x = 0$ is a local maximum. Note that $x = \pm 2$ are not critical points because they are not in the domain of g. In fact, the graph of g has vertical asymptotes there.

Figure 2.7.37 $g'(x) = \frac{-6x}{(x^2-4)^2}$

IV. g has no vertical tangents or cusps.

V. By the quotient rule:

$$
\begin{aligned}
g''(x) \;&=\; \frac{-6(x^2 - 4)^2 + (6x)2(x^2 - 4)(2x)}{(x^2 - 4)^4} \\[2mm]
&=\; \frac{-6(x^2 - 4) + 24x^2}{(x^2 - 4)^3} \quad \text{[dividing numerator, denominator by } x^2 - 4] \\[2mm]
&=\; \frac{18x^2 + 24}{(x + 2)^3(x - 2)^3} \;.
\end{aligned}
$$

The numerator of this fraction is always positive. Hence the sign of $g''(x)$ is the same as the sign of its denominator as depicted in Figure 2.7.38. Thus g is concave up for $x < -2$ or $x > 2$ while g is concave down for $-2 < x < 2$. Since $x = \pm 2$ are not in the domain of g, g has no inflection points.

Figure 2.7.38 $g''(x) = \frac{18x^2+24}{(x^2-4)^3}$

We conclude that the graph of g begins at the left above the horizontal asymptote $y = 1$ and rises concave up to the left of the vertical asymptote $x = -2$. Then it rises concave down from the right of this vertical asymptote, crossing the x–axis at $(-1, 0)$, to a local maximum at $\left(0, \frac{1}{4}\right)$. It falls from there, crossing the x–axis at $(1, 0)$ to the left of the vertical asymptote $x = 2$. It falls concave up from the right of this vertical asymptote to approach the horizontal asymptote $y = 1$ from above. This graph is depicted in Figure 2.7.39. Observe that g is an even function, i.e. $g(-x) = g(x)$. Hence the graph of g is symmetric with respect to the y–axis.

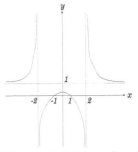

x–intercepts: $x = \pm 1$
y–intercept: $y = 1/4$
horizontal asymptote: $y = 1$ on left and right
vertical asymptotes: $x = \pm 2$
increasing: $x < 0$
decreasing: $x > 0$
local maximum: $(0, 1/4)$
no cusps or vertical tangents
concave up: $x < -2$ or $x > 2$
concave down: $-2 < x < 2$
no inflection points

Figure 2.7.39 $g(x) = \frac{x^2-1}{x^2-4}$

(3) Let $h(x) = (x^2 - 1)^{2/3}$ with domain $\Re$.

Solution We perform the five steps above to assemble information about the graph of h.

I. The graph of $h(x) = (x+1)^{2/3}(x-1)^{2/3}$ has x–intercepts at $x = -1$ and at $x = +1$. h has the y–intercept $y = h(0) = 1$.

II. h has no asymptotes.

III. By the chain rule:

$$h'(x) = \frac{2}{3}(x^2 - 1)^{-1/3}(2x) = \frac{4x}{3(x+1)^{1/3}(x-1)^{1/3}} \ .$$

The sign of $h'(x)$ is depicted in Figure 2.7.40. Hence h is increasing for $-1 < x < 0$ or $x > 1$ while h is decreasing for $x < -1$ or $0 < x < 1$. There is a critical point at $x = 0$ where the first derivative vanishes. It is a local maximum by the First Derivative Test. There are also critical points at $x = \pm 1$ where the first derivative does not exist. They are local minima by the First Derivative Test.

$$
\begin{array}{ccccccccc}
- - - & DNE & + + + & 0 & - - - & DNE & + + + & sign\ h'(x) \\
\end{array}
$$

$$
\begin{array}{cccccc}
decreasing & -1 & increasing & 0 & decreasing & 1 & increasing \\
 & local & & local & & local & \\
 & minimum & & maximum & & minimum & \\
\end{array}
$$

Figure 2.7.40 $h'(x) = \frac{4x}{3(x^2-1)^{1/3}}$

IV. Observe that

$$h'_-(-1) = \lim_{x \to -1^-} h'(x) = \lim_{x \to -1^-} \frac{4x}{3(x^2-1)^{1/3}} = \frac{-4}{+0} = -\infty,$$

$$h'_+(-1) = \lim_{x \to -1^+} h'(x) = \lim_{x \to -1^+} \frac{4x}{3(x^2-1)^{1/3}} = \frac{-4}{-0} = +\infty,$$

$$h'_-(1) = \lim_{x \to 1^-} h'(x) = \lim_{x \to 1^-} \frac{4x}{3(x^2-1)^{1/3}} = \frac{+4}{-0} = -\infty,$$

$$h'_+(1) = \lim_{x \to 1^+} h'(x) = \lim_{x \to 1^+} \frac{4x}{3(x^2-1)^{1/3}} = \frac{+4}{+0} = +\infty.$$

Hence h has vertical cusps at $x = -1$ and at $x = 1$.

V. By the quotient rule:

$$
\begin{aligned}
h''(x) &= \frac{4(x^2-1)^{1/3} - 4x\frac{1}{3}(x^2-1)^{-2/3}(2x)}{3(x^2-1)^{2/3}} \\
&= \frac{12(x^2-1) - 4x(2x)}{9(x^2-1)^{4/3}} \quad \text{[multiplying numerator and denominator by } 3(x^2-1)^{2/3}] \\
&= \frac{4x^2 - 12}{9(x^2-1)^{4/3}} = \frac{4(x+\sqrt{3})(x-\sqrt{3})}{9(x+1)^{4/3}(x-1)^{4/3}}
\end{aligned}
$$

The denominator of this fraction is always greater than or equal to zero. Hence the sign of $h''(x)$ is the same as the sign of its numerator as depicted in Figure 2.7.41. Hence h is concave up for $x < -\sqrt{3}$ or $x > \sqrt{3}$ while h is concave down for $-\sqrt{3} < x < -1$, $-1 < x < 1$ or $1 < x < \sqrt{3}$. The concavity changes from up to down at $x = -\sqrt{3}$ while the concavity changes from down to up at $x = \sqrt{3}$. Hence h has inflection points at $x = \pm\sqrt{3}$. On the other hand, $x = -1$ and $x = 1$ are not inflection points because the concavity does not change there. Note $h(-\sqrt{3}) = h(\sqrt{3}) = \sqrt[3]{4}$.

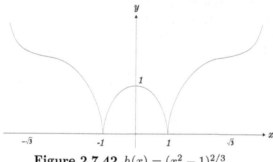

Figure 2.7.41 $h''(x) = \frac{4(x^2-3)}{9(x^2-1)^{4/3}}$

Thus the graph of h starts at the left and decreases concave up to the inflection point at $(-\sqrt{3}, \sqrt[3]{4})$. Then it becomes concave down and continues decreasing to the local minimum at the vertical cusp $(-1, 0)$. From there it rises concave down to the local maximum at $(0, 1)$. It then falls to the local minimum at $(1, 0)$ where the graph has a second vertical cusp. From there it continues concave down and rises to the inflection point at $(\sqrt{3}, \sqrt[3]{4})$. Then it becomes concave up and continues to rise. This graph is depicted in Figure 2.7.42. Since $h(-x) = h(x)$ the function h is even, and its graph is symmetric with respect to the y–axis.

x–intercepts: $x = \pm 1$
y–intercept: $y = 1$
asymptotes: none
increasing: $-1 < x < 0$ or $x > 1$
decreasing: $x < -1$ or $0 < x < 1$
local maximum: $(0, 1)$
local minima: $(-1, 0)$ and $(1, 0)$
vertical cusps: $x = -1$ and $x = 1$
concave up: $x < -\sqrt{3}$ or $x > \sqrt{3}$
concave down: $-\sqrt{3} < x < \sqrt{3}$
inflection points: $(-\sqrt{3}, \sqrt[3]{4})$
and $(\sqrt{3}, \sqrt[3]{4})$

Figure 2.7.42 $h(x) = (x^2 - 1)^{2/3}$

(4) Let $k(x) = \arctan \frac{x^2}{x^2-1}$.

Solution The domain of k consists of all numbers, other than ± 1, where the denominator $x^2 - 1$ vanishes. Observe that $k(-x) = k(x)$, and k is an even function. We perform the 5 steps above to determine features of the graph of k.

I. Since 0 is the only angle with tangent 0, the x–intercepts of k occur when $\frac{x^2}{x^2-1} = 0$, i.e. at $x = 0$. The y–intercept of k is $y = k(0) = \arctan 0 = 0$.

II. Observe that

$$\lim_{x \to -\infty} k(x) = \lim_{x \to -\infty} \arctan \frac{x^2}{x^2 - 1} = \arctan\left(\lim_{x \to -\infty} \frac{x^2}{x^2 - 1} \right) = \arctan 1 = \frac{\pi}{4}.$$

Since k is an even function, $\lim_{x \to \infty} k(x) = \frac{\pi}{4}$ too. Hence k has the horizontal asymptote $y = \frac{\pi}{4}$ on both the left and the right. Note that $\frac{x^2}{x^2-1}$ has vertical asymptotes at $x = \pm 1$. However,

$$\lim_{x \to -1^-} k(x) = \lim_{x \to -1^-} \arctan \frac{x^2}{x^2 - 1} = \arctan\left(\lim_{x \to -1^-} \frac{x^2}{x^2 - 1} \right)$$

$$= \arctan \infty = \frac{\pi}{2},$$

$$\lim_{x \to -1^+} k(x) = \lim_{x \to -1^+} \arctan \frac{x^2}{x^2 - 1} = \arctan \left(\lim_{x \to -1^+} \frac{x^2}{x^2 - 1} \right)$$

$$= \arctan -\infty = -\frac{\pi}{2},$$

Since k is an even function,

$$\lim_{x \to 1^-} k(x) = \lim_{x \to -1^+} k(x) = -\frac{\pi}{2} \quad \text{and} \quad \lim_{x \to 1^+} k(x) = \lim_{x \to -1^-} k(x) = \frac{\pi}{2}.$$

Thus k has jump discontinuities at $x = \pm 1$ and no vertical asymptotes.

III. By the chain rule:

$$k'(x) = \frac{1}{1 + \left(\frac{x^2}{x^2 - 1} \right)^2} \frac{d}{dx} \left(\frac{x^2}{x^2 - 1} \right)$$

$$= \frac{1}{1 + \frac{x^4}{(x^2 - 1)^2}} \frac{2x(x^2 - 1) - x^2(2x)}{(x^2 - 1)^2} \quad \text{[by the quotient rule]}$$

$$= \frac{-2x}{(x^2 - 1)^2 + x^4}$$

The denominator of $k'(x)$ is always positive. Hence $k'(x)$ has the same sign as its numerator. See Figure 2.7.43. Thus k is increasing for $x < 0$ and decreasing for $x > 0$. k has a critical point at $x = 0$ where its derivative vanishes. By the First Derivative Test, it is a local maximum.

Figure 2.7.43 $k'(x) = \frac{-2x}{(x^2-1)^2 + x^4}$

IV. k has no vertical tangents or cusps.

V. By the quotient rule:

$$k''(x) = \frac{-2\left[(x^2 - 1)^2 + x^4\right] + 2x\left[2(x^2 - 1)(2x) + 4x^3\right]}{[(x^2 - 1)^2 + x^4]^2}$$

$$= \frac{-2(2x^4 - 2x^2 + 1) + 2x(8x^3 - 4x)}{[(x^2 - 1)^2 + x^4]^2} = \frac{2(6x^4 - 2x^2 - 1)}{[(x^2 - 1)^2 + x^4]^2}$$

By the quadratic formula, the numerator of $k''(x)$ has roots $x^2 = \frac{1 \pm \sqrt{7}}{6} = \frac{1 + \sqrt{7}}{6}$ and $x = \pm\sqrt{\frac{1+\sqrt{7}}{6}}$. Hence

$$k''(x) = \frac{2}{[(x^2 - 1)^2 + x^4]^2} \left(x - \sqrt{\frac{1 + \sqrt{7}}{6}} \right) \left(x + \sqrt{\frac{1 + \sqrt{7}}{6}} \right).$$

$\frac{2}{[(x^2-1)^2+x^4]^2}$ is always positive, and the sign of $k''(x)$ is given in Figure 2.7.44. Hence k is concave up for $x < -\sqrt{\frac{1+\sqrt{7}}{6}}$ or $x > \sqrt{\frac{1+\sqrt{7}}{6}}$ while k is concave down for $-\sqrt{\frac{1+\sqrt{7}}{6}} < x < \sqrt{\frac{1+\sqrt{7}}{6}}$. Since the concavity changes at $x = \pm\sqrt{\frac{1+\sqrt{7}}{6}}$, these are two inflection points of k. Note that $\sqrt{\frac{1+\sqrt{7}}{6}} \approx 0.78$.

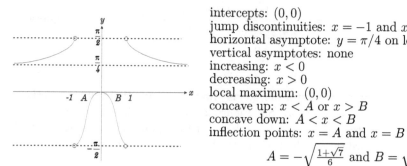

Figure 2.7.44 $k''(x) = \frac{2}{[(x^2-1)^2+x^4]^2}\left(x - \sqrt{\frac{1+\sqrt{7}}{6}}\right)\left(x + \sqrt{\frac{1+\sqrt{7}}{6}}\right)$

We conclude that the graph of k rises concave up from its horizontal asymptote $y = \frac{\pi}{4}$ on the left towards the point $\left(-1, \frac{\pi}{2}\right)$. Then it jumps down to the point $\left(-1, -\frac{\pi}{2}\right)$ and rises concave up towards its local maximum at the origin, changing its concavity to down at $x = -\sqrt{\frac{1+\sqrt{7}}{6}}$. It decreases from the origin towards the point $\left(1, -\frac{\pi}{2}\right)$, changing its concavity to up at $x = \sqrt{\frac{1+\sqrt{7}}{6}}$. Then it jumps up to the point $\left(1, \frac{\pi}{2}\right)$ and decreases concave up to its horizontal asymptote $y = \frac{\pi}{4}$ on the right. This graph is depicted in Figure 2.7.45.

intercepts: $(0,0)$
jump discontinuities: $x = -1$ and $x = 1$
horizontal asymptote: $y = \pi/4$ on left and right
vertical asymptotes: none
increasing: $x < 0$
decreasing: $x > 0$
local maximum: $(0,0)$
concave up: $x < A$ or $x > B$
concave down: $A < x < B$
inflection points: $x = A$ and $x = B$

$$A = -\sqrt{\frac{1+\sqrt{7}}{6}} \text{ and } B = \sqrt{\frac{1+\sqrt{7}}{6}}$$

Figure 2.7.45 $k(x) = \arctan\frac{x^2}{x^2-1}$

(5) Let $m(x) = |x^2 + 3x - 28|$ with domain $\Re$.

Solution We perform the 5 steps above to determine features of the graph of m.

I. Since $m(x) = |(x + 7)(x - 4)|$, x–intercepts occur at $x = -7$ and at $x = 4$. The y–intercept is $m(0) = 28$.

II. m has no asymptotes.

III. Observe that $m(x) = \left\{ \begin{array}{ll} x^2 + 3x - 28 & \text{for } x \le -7 \text{ or } x \ge 4 \\ -(x^2 + 3x - 28) & \text{for } -7 < x < 4 \end{array} \right\}$. Hence

$$m'(x) = \left\{ \begin{array}{ll} 2x + 3 & \text{for } x < -7 \text{ or } x > 4 \\ -2x - 3 & \text{for } -7 < x < 4 \end{array} \right\}$$

and $m'(-7)$, $m'(4)$ do not exist. The sign of $m'(x)$ is depicted in Fig. 2.7.46. Hence m is increasing for $-7 < x < -\frac{3}{2}$ or $x > 4$ while m is decreasing for $x < -7$ or $-\frac{3}{2} < x < 4$. m has a critical point at $x = -\frac{3}{2}$ where $m'(x)$ vanishes. It is a local maximum by the First Derivative Test. m also has critical points at $x = -7$ and $x = 4$ where $m'(x)$ does not exist. These are local minima by the First Derivative Test. Note that $m\left(-\frac{3}{2}\right) = 30\frac{1}{4}$.

```
          - - -    DNE    + + +     0     - - -    DNE    + + +    sign m'(x)
     ─────────────────┼──────────────┼──────────────┼──────────────────► x
       decreasing    -7   increasing  -3/2  decreasing  4  increasing
                     local               local              local
                   minimum             maximum            minimum
```

Figure 2.7.46 $m'(x)$

IV. Observe that

$$m'_-(-7) = \lim_{x \to -7^-} m'(x) = \lim_{x \to -7^-} (2x + 3) = -11,$$

$$m'_+(-7) = \lim_{x \to -7^+} m'(x) = \lim_{x \to -7^+} (-2x - 3) = +11,$$

$$m'_-(4) = \lim_{x \to 4^-} m'(x) = \lim_{x \to 4^-} (-2x - 3) = -11,$$

$$m'_+(4) = \lim_{x \to 4^+} m'(x) = \lim_{x \to 4^+} (2x + 3) = +11.$$

Hence m has cusps at $x = -7$ and at $x = 4$.

V. Note that

$$m''(x) = \left\{ \begin{array}{ll} 2 & \text{for } x < -7 \text{ or } x > 4 \\ -2 & \text{for } -7 < x < 4 \end{array} \right\}$$

and $m''(-7)$, $m''(4)$ do not exist. The sign of $m''(x)$ is depicted in Figure 2.7.47. Hence m is concave up for $x < -7$ or $x > 4$ while m is concave down for $-7 < x < 4$. Therefore m has inflection points at $x = -7$ where the concavity changes from up to down and at $x = 4$ where the concavity changes from down to up.

$$+\ +\ +\ +\quad DNE \quad -\ -\ -\ -\ -\ -\quad DNE \quad +\ +\ +\ +\qquad sign\ m''(x)$$

concave up   *concave down*   *concave up* $\longrightarrow x$

-7   4

inflection   *inflection*

point   *point*

Figure 2.7.47 $m''(x)$

We conclude that the graph of m decreases concave up at the left to a cusp at $x = -7$. For $-7 < x < 4$, it is concave down, increasing to a local maximum at $x = -\frac{3}{2}$ and then decreasing to a cusp at $x = 4$. It increases concave up for $x > 4$. This graph is depicted in Figure 2.7.48. □

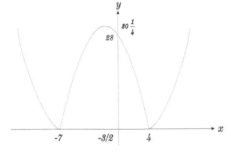

x–intercepts: $x = -7$ and $x = 4$
y–intercept: $y = 28$
asymptotes: none
increasing: $-7 < x < -3/2$ or $x > 4$
decreasing: $x < -7$ or $-3/2 < x < 4$
local maximum: $\left(-1\frac{1}{2}, 30\frac{1}{4}\right)$
local minima: $(-7, 0)$ and $(4, 0)$
cusps: $(-7, 0)$ and $(4, 0)$
concave up: $x < -7$ or $x > 4$
concave down: $-7 < x < 4$
inflection points: $x = -7$ and $x = 4$

Figure 2.7.48 $m(x) = \left| x^2 + 3x - 28 \right|$

Summary

The reader should know how to compute the x–intercepts and y–intercept of a given function. She should be able to determine whether a function has horizontal, vertical or oblique asymptotes and find the equations of those asymptotes. She should use the sign of the first derivative to determine where the function is increasing and decreasing. She should apply the First Derivative Test to locate and identify the local extrema. Cusps and vertical asymptotes are identified at points where the first derivative does not exist by comparing the values of the one–sided derivatives. She should use the sign of second derivative to determine where the function is concave up, concave down and has inflection points. She should be able to assemble all this information to sketch the graph of the function. Of course, this assembly process requires understanding exactly what each of these pieces of information is saying about the graph.

Basic Exercises

1. Find the x–intercepts and y–intercept for each function.

(a) $f(x) = x^3 - 2x^2 - 8x$ **(b)** $g(x) = \frac{x^2+3x+2}{x^2+5}$ **(c)** $h(x) = \frac{x^2+x+1}{x^2-1}$

(d) $j(x) = \frac{x^3+1}{5x^2+3x}$ **(e)** $k(x) = \sqrt{3x^2+6}$ **(f)** $m(x) = \arcsin(x^2 - 1)$

2. How can we say that a function has at most one y–intercept when the circle $x^2 + y^2 = 1$ has two y–intercepts at $y = -1$ and at $y = +1$?

3. **(a)** Give an example of a degree three polynomial with exactly one x–intercept.
(b) Give an example of a degree three polynomial with exactly two x–intercepts.
(c) Give an example of a degree three polynomial with exactly three x–intercepts.
(d) Why are there no degree three polynomials without x–intercepts?

4. Find the horizontal asymptotes for each function.

(a) $f(x) = \frac{3x^4-2x+7}{x^4+5x^2-2}$ **(b)** $g(x) = \frac{2x^3-3x^2+1}{5x^6-4x^4+3}$ **(c)** $h(x) = \frac{8x^6-3x^3+9}{4x^4-5x^2+3}$

(d) $j(x) = \frac{5}{2x^3+x+1}$ **(e)** $k(x) = \sqrt{\frac{9x^4+7x+8}{3x^4-5x^2+4}}$ **(f)** $m(x) = \cos\left(\frac{x}{x^3+1}\right)$

(g) $n(x) = \arctan x^2 + 1$ **(h)** $p(x) = \cot\left[\arcsin\left(\frac{3x^2+2}{5x^2+6}\right)\right]$ **(i)** $q(x) = \arctan\frac{x}{x+1}$

5. Prove the rules for horizontal asymptotes of a rational function $f(x) = \frac{p(x)}{q(x)}$.
(a) If the degree of $p(x)$ is less than the degree of $q(x)$, then f has the x–axis as a horizontal asymptote on both the left and the right.
(b) Let $p(x)$ and $q(x)$ have the same degree n. Write $p(x) = ax^n +$ *lower powers of x* and write $q(x) = bx^n +$ *lower powers of x*. Then f has the line $y = \frac{a}{b}$ as a horizontal asymptote on both the left and the right.
(c) If degree $p(x)$ is greater than degree $q(x)$, then f has no horizontal asymptotes.

6. Find the vertical asymptotes for each function.

(a) $f(x) = \frac{x^2-1}{x^2-x-12}$ **(b)** $g(x) = \frac{3x+4}{x^3-6x^2+11x-6}$ **(c)** $h(x) = \frac{x^3+1}{x^4-1}$

(d) $j(x) = \cot \pi x$ **(e)** $k(x) = \sec(2x + 1)$ **(f)** $m(x) = \frac{1}{\arctan x}$

7. Find the oblique asymptotes for each function.

(a) $f(x) = \frac{5x^2-2x+3}{x}$ **(b)** $g(x) = \frac{6x^4-x^3+7}{2x^3+1}$ **(c)** $h(x) = \frac{2x^5-6x^2+9}{x^3-4}$

(d) $j(x) = \frac{4x^3+5x-1}{2x^4+3}$ **(e)** $k(x) = \frac{2x^3+4x+8}{x^2-x+2}$ **(f)** $m(x) = \frac{2x^6-3x^2+5}{x^5+1}$

8. Find all the asymptotes for each function.

(a) $f(x) = \frac{x}{x^2-9}$ **(b)** $g(x) = \frac{2x^3+x-7}{x^3-x^2-6x}$ **(c)** $h(x) = \frac{8x^4-x^2+6}{x^3-1}$

(d) $j(x) = \frac{3x^2-x+1}{x^2-3x-10}$ **(e)** $k(x) = |\arctan x|$ **(f)** $m(x) = \sin\left(\frac{5x+3}{x^2-4}\right)$

(g) $n(x) = \sec\left(\frac{\pi}{x^2+1}\right)$ **(h)** $p(x) = \arctan\left(\frac{x^3}{x^2-1}\right)$ **(i)** $q(x) = \frac{\sin x}{x}$

(j) $r(x) = \tan\frac{\pi x^2}{3x^2+1}$ **(k)** $s(x) = \cot\frac{\pi x^2}{4x^2+1}$ **(l)** $t(x) = \arcsin\frac{\arctan x}{\pi}$

9. Find where each function is concave up, concave down and has inflection points.

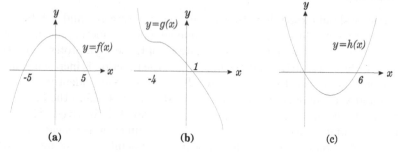

       (a)   **(b)**   **(c)**

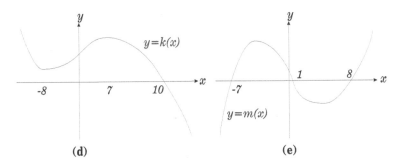

(d)   **(e)**

10. Solve each implicit equation for y to find all the asymptotes.
(a) $xy + x + y = 1$ **(b)** $\tan(xy) = 1$ **(c)** $x^2y + xy + y = 4$ **(d)** $x^4y - y + x^5 = 1$
(e) $\sin(xy)\cos\frac{y}{x} - \cos(xy)\sin\frac{y}{x} = 1$ **(f)** $4\arctan(xy + x + y + 1) = \pi$

11. Determine where each function is concave up and where it is concave down. Then identify all inflection points.
  (a) $f(x) = x^3 - 2x^2 + 6x - 4$ **(b)** $g(x) = x^4 - 6x^3 + 12x^2 - 5x + 7$
  (c) $h(x) = x^6 - 5x^4 + 15x^2 - 3$ **(d)** $j(x) = \frac{1}{4x^2-9}$
  (e) $k(x) = \frac{x}{3x^2+1}$ **(f)** $m(x) = \frac{2x^2+1}{x^2-1}$
  (g) $n(x) = \cos x$ with domain $[0, 2\pi]$ **(h)** $p(x) = \arctan x$
  (i) $q(x) = |x^2 - 4x - 12|$ **(j)** $r(x) = \begin{cases} 1 - x^2 & \text{if } x \leq 1 \\ x^3 - 2x^2 + 4x - 3 & \text{if } x > 1 \end{cases}$.

12. Give an example of a function f and a point $x = c$ of its domain where the graph of f is concave up even though $f''(c)$ is not positive.

13. Show that f is concave up at $x = c$ if and only if $-f$ is concave down at $x = c$.

14. Give a direct proof of Theorem 2.7.23(b) analogous to the proof of Theorem 2.7.23(a).

15. Prove Theorem 2.7.23(c) in the case $f''_-(c) < 0$ and $f''_+(c) > 0$.

Sketch the graph of each function.

16. $y = x^3 + 5x$ **17.** $y = 9x - x^3$ **18.** $y = 4x^2 - x^4$
19. $y = x^3 + 3x^2 - 24x + 20$ **20.** $y = 3x^5 - 10x^3 + 15x - 8$ **21.** $y = \frac{1}{x^2+9}$
22. $y = \frac{x}{x^2+4}$ **23.** $y = \frac{x^2}{x^2+1}$ **24.** $y = \frac{1}{3-x^3}$
25. $y = \frac{x}{5-x^2}$ **26.** $y = \frac{x^2}{2-x^2}$ **27.** $y = \frac{x^3}{1-x^2}$
28. $y = \frac{1}{x(3-x^2)}$ **28.** $y = \frac{x}{x^3+1}$ **30.** $y = \frac{x^2}{x^3+8}$
31. $y = \frac{x^3}{x^3+6}$ **32.** $y = \frac{x}{1-1000x^3}$ **33.** $y = \frac{1}{x-x^4}$
34. $y = \frac{x}{x^4+16}$ **35.** $y = \frac{x^2}{x^4+1}$ **36.** $y = x\sqrt{1-x^2}$
37. $y = \frac{\sqrt{4-x^2}}{x}$ **38.** $y = \frac{x^2}{\sqrt{x^2-9}}$ **39.** $y = \frac{1}{\sqrt{5+x^2}}$
40. $y = \sin^2 x$ **41.** $y = \cos x \sin^2 x$ **42.** $y = |x^2 + 6x - 27|$
43. $y = |x^4 - 13x^2 + 36|$ **44.** $y = (9 - x^2)^{3/5}$ **45.** $y = (25 - x^2)^{4/7}$
46. $y = \left(\frac{x+5}{x+3}\right)^{4/5}$ **47.** $y = \left(\frac{x^2}{x^2-4}\right)^{3/7}$ **48.** $y = \arctan\frac{x}{4-x^2}$
49. $y = \arctan\frac{x^2-9}{x^2-16}$ **50.** $y = \arcsin\frac{x}{x^2+1}$ **51.** $y = \arcsin\frac{x^2}{x^2+4}$

Challenging Problems

1. A function $f(x)$ is said to have the function $A(x)$ as an asymptote if either
$$\lim_{x \to -\infty} [f(x) - A(x)] = 0 \quad \text{or} \quad \lim_{x \to \infty} [f(x) - A(x)] = 0.$$
Find polynomial functions which are asymptotes of each function.
  (a) $f(x) = \frac{2x^5 - 3x^3 - 4x + 5}{x^2}$ **(b)** $g(x) = \frac{5x^4 - 3x^2 + 8}{x^2 + 1}$ **(c)** $h(x) = \frac{2x^6 - 5x^4 + 8x - 3}{x^3 - 5}$
  (d) $j(x) = \frac{2x^5 + 4x^3 + 2}{x^2 - 2x + 3}$ **(e)** $k(x) = x^2 \arctan x$

2. Sketch the graph of each implicit function.

(a) $y^2 - x^3 - x^2 = 0$ (b) $y^2 - 4x^5 - 8x^4 = 0$ (c) $x^2y^2 - 2x = 3$

(d) $y^2 - 5x^3 + 3x^4 = 0$ (e) $x^3y^2 - 8x - 7 = 0$ (f) $y^2 - 4x^4 + 4x^6 = 0$

(g) $y^2 - x^2 + x^4 = 0$ (h) $3y^2 - 10xy^2 - 10x^2 - 10x^3 = 0$

2.8   L'Hôpital's Rule

There are two types of limits. Straightforward limits, as x approaches c, can be evaluated by substituting $x = c$. However, there are also nontrivial limits, called *indeterminate forms*. When we substitute $x = c$ into this type of limit we obtain an arithmetic expression, such as $\frac{0}{0}$ or $\frac{\infty}{\infty}$ which is not defined. In particular, every derivative is defined as a limit which is an indeterminate form:

$$f'(c) = \lim_{h \to 0} \frac{f(c+h) - f(c)}{h} = \frac{f(c+0) - f(c)}{0} = \frac{0}{0}.$$

That is, every derivative we compute gives us the value of a limit which is an indeterminate form. In this section, we study L'Hôpital's Rule which allows us to rewrite a given limit, which is an indeterminate form, as a limit of derivatives we can compute. In the first subsection, we introduce the simplest form of this rule which applies to the indeterminate form $\frac{0}{0}$. In the second subsection, we show that this rule also applies to the indeterminate form $\frac{\infty}{\infty}$. The third subsection extends L'Hôpital's Rule to limits as x approaches infinity.

L'Hôpital's Rule for the Indeterminate Form $\frac{0}{0}$

As we observed, every derivative is an example of a limit which is the indeterminate form $\frac{0}{0}$. The following example illustrates how a limit, which is the indeterminate form $\frac{0}{0}$, can be rewritten as a limit of derivatives. Then we use our ability to compute derivatives to evaluate the given limit. This procedure is called L'Hôpital's Rule which we state and apply to several examples.

Motivating Example 2.8.1 Evaluate $\displaystyle \lim_{x \to 1} \frac{\sin \pi x}{x + \cos \pi x}$.

Solution Observe that if we substitute $x = 1$ into $\frac{\sin \pi x}{x + \cos \pi x}$ we obtain $\frac{0}{1 + (-1)} = \frac{0}{0}$ which is not defined. We want to evaluate the given limit by writing it in terms of limits that define derivatives. These types of limits have a denominator h which approaches zero. If x approaches 1, then $h = x - 1$ approaches zero. Thus we substitute $x = h + 1$ into our limit to change variables from x to h. Then we divide numerator and denominator by h:

$$\lim_{x \to 1} \frac{\sin \pi x}{x + \cos \pi x} = \lim_{h \to 0} \frac{\sin \pi (h+1)}{(h+1) + \cos \pi (h+1)} = \lim_{h \to 0} \frac{\frac{\sin \pi (h+1)}{h}}{\frac{(h+1) + \cos \pi (h+1)}{h}} = \lim_{h \to 0} \frac{\frac{f(h+1)}{h}}{\frac{g(h+1)}{h}}$$

where $f(x) = \sin \pi x$ and $g(x) = x + \cos \pi x$. Since $f(1) = \sin \pi = 0$ and $g(1) = 1 + \cos \pi = 0$, write this limit as the quotient of limits which define $f'(1)$ and $g'(1)$:

$$\lim_{x \to 1} \frac{\sin \pi x}{x + \cos \pi x} = \lim_{h \to 0} \frac{\frac{f(h+1) - f(1)}{h}}{\frac{g(h+1) - g(1)}{h}} = \frac{\displaystyle \lim_{h \to 0} \frac{f(h+1) - f(1)}{h}}{\displaystyle \lim_{h \to 0} \frac{g(h+1) - g(1)}{h}} = \frac{f'(1)}{g'(1)}.$$

By the chain rule: $f'(x) = \pi \cos \pi x$ and $g'(x) = 1 - \pi \sin \pi x$. Hence $f'(1) = \pi \cos \pi = -\pi$ and $g'(1) = 1 - \pi \sin \pi = 1$. Thus

$$\lim_{x \to 1} \frac{\sin \pi x}{x + \cos \pi x} = \frac{-\pi}{1} = -\pi. \qquad \square$$

The preceding method of evaluating a limit generalizes to $\lim_{x \to c} \dfrac{f(x)}{g(x)}$ where f, g are differentiable fucntions with $f(c) = g(c) = 0$ and $g'(c) \neq 0$. In this case:

$$\lim_{x \to c} \frac{f(x)}{g(x)} = \frac{f'(c)}{g'(c)}.$$

This formula is called *L'Hôpital's Rule*. In fact, it is true in greater generality. We do not need to assume $\frac{f'(c)}{g'(c)}$ is defined. It suffices to assume $\lim_{x \to c} \dfrac{f'(x)}{g'(x)}$ exists. Then L'Hôpital's Rule says:

$$\lim_{x \to c} \frac{f(x)}{g(x)} = \lim_{x \to c} \frac{f'(x)}{g'(x)}.$$

In particular, L'Hôpital's Rule can even be used when $f'(c) = g'(c) = 0$. We state L'Hôpital's Rule but defer the proof to Section 13.

Theorem 2.8.2 [L'Hôpital's Rule for $\frac{0}{0}$] *Let f, g be differentiable functions whose domains include the interval (a, b) with $a, b \in \Re$ and $c \in (a, b)$. Assume*

$$f(c) = g(c) = 0.$$

If $\lim_{x \to c} \dfrac{f'(x)}{g'(x)}$ exists, then $\lim_{x \to c} \dfrac{f(x)}{g(x)}$ exists and

$$\lim_{x \to c} \frac{f(x)}{g(x)} = \lim_{x \to c} \frac{f'(x)}{g'(x)}. \qquad \square$$

We give several examples which apply L'Hôpital's Rule. In Example 2, we emphasize that L'Hôpital's Rule only applies to limits where substitution of $x = c$ produces the indeterminate value $\frac{0}{0}$. If the rule is applied to other limits, then it usually produces the wrong answer. Also, note that the derivatives of the numerator and denominator in L'Hôpital's Rule are computed separately. L'Hôpital's Rule *does not* tell us to compute the derivative of the entire fraction by the quotient rule. Examples 4 and 5 show how L'Hôpital's Rule can be used to evaluate limits when direct substitution produces the indeterminate form $0 \cdot \infty$ or $\infty - \infty$.

Examples 2.8.3 (1) Evaluate $\lim_{x \to 0} \dfrac{\sin 5x}{\sin 3x}$.

Solution The functions $\sin 5x$ and $\sin 3x$ are differentiable. Substitution of $x = 0$ in this limit produces $\frac{\sin 0}{\sin 0} = \frac{0}{0}$ and L'Hôpital's Rule applies:

$$\lim_{x \to 0} \frac{\sin 5x}{\sin 3x} = \lim_{x \to 0} \frac{D(\sin 5x)}{D(\sin 3x)} = \lim_{x \to 0} \frac{5 \cos 5x}{3 \cos 3x} = \frac{5 \cos 0}{3 \cos 0} = \frac{5}{3}.$$

(2) Evaluate $\lim_{x \to 0} \dfrac{\cos 5x}{\cos 3x}$.

Solution Substitution of $x = 0$ in this limit produces the value $\frac{\cos 0}{\cos 0} = \frac{1}{1} = 1$ of this limit by Limit Property 5. In particular, L'Hôpital's Rule does not apply to this limit. Consider the limit of the quotient of derivatives of L'Hôpital's Rule:

$$\lim_{x \to 0} \frac{D(\cos 5x)}{D(\cos 3x)} = \lim_{x \to 0} \frac{-5 \sin 5x}{-3 \sin 3x} = \frac{5}{3} \lim_{x \to 0} \frac{\sin 5x}{\sin 3x} = \left(\frac{5}{3}\right)\left(\frac{5}{3}\right) = \frac{25}{9}$$

by Example 1. Thus the use of L'Hôpital's Rule on this example, which does not satisfy the hypotheses of the rule, produces the wrong answer.

(3) Evaluate $\displaystyle\lim_{x \to 1} \frac{x^3 - 4x^2 + 5x - 2}{x^{16} - 2x^8 + 1}$.

Solution The functions $x^3 - 4x^2 + 5x - 2$ and $x^{16} - 2x^8 + 1$ are differentiable. Substitution of $x = 1$ in this limit produces $\frac{0}{0}$ and L'Hôpital's Rule applies:

$$\lim_{x \to 1} \frac{x^3 - 4x^2 + 5x - 2}{x^{16} - 2x^8 + 1} = \lim_{x \to 1} \frac{D(x^3 - 4x^2 + 5x - 2)}{D(x^{16} - 2x^8 + 1)} = \lim_{x \to 1} \frac{3x^2 - 8x + 5}{16x^{15} - 16x^7}.$$

Substitution of $x = 1$ produces $\frac{0}{0}$. Hence we apply L'Hôpital's Rule again:

$$\lim_{x \to 1} \frac{x^3 - 4x^2 + 5x - 2}{x^{16} - 2x^8 + 1} = \lim_{x \to 1} \frac{D(3x^2 - 8x + 5)}{D(16x^{15} - 16x^7)} = \lim_{x \to 1} \frac{6x - 8}{240x^{14} - 112x^6} = \frac{-2}{128} = -$$

(4) Evaluate $\displaystyle\lim_{x \to \frac{1}{2}} (2x - 1)\tan(\pi x)$.

Solution Substitution of $x = \frac{1}{2}$ into this limit produces the indeterminate form $0 \cdot \infty$. Rewrite this limit as the limit of a quotient where L'Hôpital's Rule applies:

$$\lim_{x \to \frac{1}{2}} (2x - 1)\tan(\pi x) = \lim_{x \to \frac{1}{2}} \frac{2x - 1}{1/\tan(\pi x)} = \lim_{x \to \frac{1}{2}} \frac{2x - 1}{\cot(\pi x)}.$$

Both the functions $2x - 1$ and $\cot(\pi x)$ are differentiable on the interval $(0, 1)$. Substituting $x = \frac{1}{2}$ into $\frac{2x-1}{\cot(\pi x)}$ produces $\frac{0}{0}$. Hence L'Hôpital's Rule applies:

$$\lim_{x \to \frac{1}{2}} (2x - 1)\tan(\pi x) = \lim_{x \to \frac{1}{2}} \frac{D[2x - 1]}{D[\cot(\pi x)]} = \lim_{x \to \frac{1}{2}} \frac{2}{-\pi \csc^2(\pi x)} = -\frac{2}{\pi(1)^2} = -\frac{2}{\pi}.$$

(5) Evaluate $\displaystyle\lim_{x \to 0} \left(\frac{1}{x} - \cot x\right)$.

Solution Substitution of $x = 0$ into $\frac{1}{x} - \cot x$ produces the indeterminate form $\infty - \infty$. We write $\frac{1}{x} - \cot x$ as a fraction to which L'Hôpital's Rule applies:

$$\lim_{x \to 0} \left(\frac{1}{x} - \cot x\right) = \lim_{x \to 0} \left(\frac{1}{x} - \frac{\cos x}{\sin x}\right) = \lim_{x \to 0} \frac{\sin x - x \cos x}{x \sin x}.$$

The numerator and denominator of the preceding fraction are differentiable and have value zero at $x = 0$. Thus L'Hôpital's Rule applies:

$$\lim_{x \to 0} \left(\frac{1}{x} - \cot x\right) = \lim_{x \to 0} \frac{D(\sin x - x \cos x)}{D(x \sin x)} = \lim_{x \to 0} \frac{x \sin x}{\sin x + x \cos x}.$$

Substitution of $x = 0$ in the latter fraction produces $\frac{0}{0}$. Hence we apply L'Hôpital's Rule a second time:

$$\lim_{x \to 0} \left(\frac{1}{x} - \cot x\right) = \lim_{x \to 0} \frac{D(x \sin x)}{D(\sin x + x \cos x)} = \lim_{x \to 0} \frac{\sin x + x \cos x}{2 \cos x - x \sin x} = \frac{0 + 0}{2 - 0} = 0. \ \square$$

L'Hôpital's Rule for the Indeterminate Form $\frac{\infty}{\infty}$

We show L'Hôpital's Rule applies to limits which are the indeterminate form $\frac{\infty}{\infty}$. The following example indicates how to apply L'Hôpital's Rule of Theorem 2.8.2 to this type of limit.

Motivating Example 2.8.4 Evaluate $\lim\limits_{x \to 1} \dfrac{\cot \pi x}{1 + \csc \pi x}$.

Solution Substitution of $x = 1$ into this limit produces the indeterminate form $\frac{\infty}{\infty}$. Since $\frac{1}{\infty} = 0$, we rewrite the given fraction as the quotient of the reciprocals of its denominator and numerator:

$$\lim_{x \to 1} \frac{\cot \pi x}{1 + \csc \pi x} = \lim_{x \to 1} \frac{\frac{1}{1 + \csc \pi x}}{\frac{1}{\cot \pi x}} . \tag{2.8.1}$$

Substitution of $x = 1$ into the latter fraction produces $\frac{0}{0}$ and L'Hôpital's Rule applies. For greater clarity, write $f(x) = \cot \pi x$ and $g(x) = 1 + \csc \pi x$. Then

$$
\begin{aligned}
\lim_{x \to 1} \frac{\cot \pi x}{1 + \csc \pi x} &= \lim_{x \to 1} \frac{f(x)}{g(x)} = \lim_{x \to 1} \frac{\frac{1}{g(x)}}{\frac{1}{f(x)}} && [\text{by } (2.8.1)] \\[2mm]
&= \lim_{x \to 1} \frac{D\left[\frac{1}{g(x)}\right]}{D\left[\frac{1}{f(x)}\right]} && [\text{by L'Hôpital's Rule}] \\[2mm]
&= \lim_{x \to 1} \frac{D\left[g(x)^{-1}\right]}{D\left[f(x)^{-1}\right]} = \lim_{x \to 1} \frac{-\frac{g'(x)}{g(x)^2}}{-\frac{f'(x)}{f(x)^2}} && [\text{by the chain rule}] \\[2mm]
&= \lim_{x \to 1} \left(\frac{g'(x)}{f'(x)} \cdot \frac{f(x)^2}{g(x)^2}\right) .
\end{aligned}
$$

If we assume the given limit exists and has nonzero value L, then the preceding equation determines the value of L. That equation says:

$$L = \lim_{x \to 1} \frac{g'(x)}{f'(x)} \cdot L^2 .$$

Multiply this equation by $\dfrac{1}{L} \lim\limits_{x \to 1} \dfrac{f'(x)}{g'(x)}$:

$$\lim_{x \to 1} \frac{f'(x)}{g'(x)} = L = \lim_{x \to 1} \frac{f(x)}{g(x)} . \tag{2.8.2}$$

By the chain rule: $f'(x) = -\pi \csc^2 \pi x$ and $g'(x) = -\pi \csc \pi x \cot \pi x$. Hence the preceding equation becomes:

$$
\begin{aligned}
\lim_{x \to 1} \frac{\cot \pi x}{1 + \csc \pi x} &= L = \lim_{x \to 1} \frac{f'(x)}{g'(x)} = \lim_{x \to 1} \frac{-\pi \csc^2 \pi x}{-\pi \csc \pi x \cot \pi x} = \lim_{x \to 1} \frac{\csc \pi x}{\cot \pi x} \\[2mm]
&= \lim_{x \to 1} \frac{\csc \pi x}{\cot \pi x} \cdot \frac{\sin \pi x}{\sin \pi x} = \lim_{x \to 1} \frac{1}{\cos \pi x} = \frac{1}{-1} = -1 . \qquad \square
\end{aligned}
$$

Equation (2.8.2) is true for any limit $\lim\limits_{x \to c} \dfrac{f(x)}{g(x)}$ which is the indeterminate form $\frac{\infty}{\infty}$ and does not require the assumption that this limit exists and is nonzero. We state this result but defer its proof to Section 13.

Theorem 2.8.5 [L'Hôpital's Rule for $\frac{\infty}{\infty}$]   *Let f, g be differentiable functions whose domains include $(a,c) \cup (c,b)$. Assume each of*

$$\lim_{x \to c^-} f(x), \quad \lim_{x \to c^+} f(x), \quad \lim_{x \to c^-} g(x), \quad \lim_{x \to c^+} g(x)$$

exists and has value $-\infty$ or $+\infty$. If $\lim\limits_{x \to c} \dfrac{f'(x)}{g'(x)}$ exists, then $\lim\limits_{x \to c} \dfrac{f(x)}{g(x)}$ exists and

$$\lim_{x \to c} \frac{f(x)}{g(x)} = \lim_{x \to c} \frac{f'(x)}{g'(x)}.$$ $\square$

Note The hypothesis of this theorem about the one–sided limits of $f(x)$ and $g(x)$ is often described by saying "$\lim\limits_{x \to c} \dfrac{f(x)}{g(x)} = \dfrac{\infty}{\infty}$" even in examples where this two–sided limit is $-\dfrac{\infty}{\infty}$ or does not exist, i.e. $\lim\limits_{x \to c^-} f(x) = \infty$, $\lim\limits_{x \to c^+} f(x) = -\infty$ and $\lim\limits_{x \to c} g(x) = \infty$.

We show how to apply this form of L'Hôpital's Rule.

Examples 2.8.6 (1)   Evaluate $\lim\limits_{x \to 0} \dfrac{1 + x^{-2/3}}{x + x^{-4/5}}$.

Solution The limits, as x approaches 0, of the numerator and denominator of this fraction both equal $+\infty$. Hence L'Hôpital's Rule applies to this limit:

$$\lim_{x \to 0} \frac{1 + x^{-2/3}}{x + x^{-4/5}} = \lim_{x \to 0} \frac{D\left(1 + x^{-2/3}\right)}{D\left(x + x^{-4/5}\right)} = \lim_{x \to 0} \frac{-\frac{2}{3}x^{-5/3}}{1 - \frac{4}{5}x^{-9/5}} = -\frac{2}{3} \lim_{x \to 0} \frac{x^{-5/3}}{1 - \frac{4}{5}x^{-9/5}}.$$

This limit is also the indeterminate form $\frac{\infty}{\infty}$. That is:

$$\lim_{x \to 0^-} x^{-5/3} = -\infty, \qquad \lim_{x \to 0^+} x^{-5/3} = +\infty,$$

$$\lim_{x \to 0^-} \left(1 - \frac{4}{5}x^{-9/5}\right) = +\infty, \qquad \lim_{x \to 0^+} \left(1 - \frac{4}{5}x^{-9/5}\right) = -\infty.$$

Hence we apply L'Hôpital's Rule again:

$$\lim_{x \to 0} \frac{1 + x^{-2/3}}{x + x^{-4/5}} = -\frac{2}{3} \lim_{x \to 0} \frac{D\left(x^{-5/3}\right)}{D\left(1 - \frac{4}{5}x^{-9/5}\right)} = -\frac{2}{3} \lim_{x \to 0} \frac{-\frac{5}{3}x^{-8/3}}{\left(-\frac{4}{5}\right)\left(-\frac{9}{5}\right)x^{-14/5}}$$

$$= \frac{10/9}{36/25} \lim_{x \to 0} \frac{x^{-8/3}}{x^{-14/5}} = \frac{125}{162} \lim_{x \to 0} x^{14/5 - 8/3} = \frac{125}{162} \lim_{x \to 0} x^{2/15} = 0.$$

(2) Evaluate $\lim\limits_{x \to 0} x(1 - \cot x)$.

Solution This limit is the indeterminate form $(0)(\infty)$. We evaluate it by writing it as a fraction where L'Hôpital's Rule applies:

$$\lim_{x \to 0} x(1 - \cot x) = \lim_{x \to 0} \frac{1 - \cot x}{x^{-1}}.$$

The latter limit is the indeterminate form $\frac{\infty}{\infty}$. That is:

$$\lim_{x \to 0^-} (1 - \cot x) = +\infty, \ \lim_{x \to 0^+} (1 - \cot x) = -\infty, \ \lim_{x \to 0^-} x^{-1} = -\infty, \ \lim_{x \to 0^+} x^{-1} = +\infty.$$

Hence L'Hôpital's Rule applies. Since $\csc x = \frac{1}{\sin x}$,

$$\lim_{x \to 0} x(1 - \cot x) = \lim_{x \to 0} \frac{D(1 - \cot x)}{D\left(x^{-1}\right)} = \lim_{x \to 0} \frac{\csc^2 x}{-x^{-2}} = -\lim_{x \to 0} \frac{x^2}{\sin^2 x}$$

$$= -\left(\lim_{x \to 0} \frac{x}{\sin x}\right)^2 = -(1)^2 = -1 \qquad \text{by Theorem 1.6.11(a)}.$$

Alternatively, we can apply L'Hôpital's Rule twice to evaluate $\lim\limits_{x \to 0} \dfrac{x^2}{\sin^2 x}$. $\square$

L'Hôpital's Rule when x Approaches $\pm\infty$

The forms of L'Hôpital's Rule which we have used to evaluate limits, as x approaches c, of the indeterminate forms $\frac{0}{0}$ and $\frac{\infty}{\infty}$ require that c be a number. L'Hôpital's can also be used to evaluate these indeterminate forms when $c = +\infty$ and when $c = -\infty$.

Theorem 2.8.7 [L'Hôpital's Rule for $c = \pm\infty$] **(a)** *Let f, g be differentiable functions whose domains include the interval (a, ∞). Assume*

$$\lim_{x \to \infty} f(x) = \lim_{x \to \infty} g(x) = L \text{ where } L \text{ is either } 0, \ -\infty \text{ or } +\infty.$$

If $\displaystyle\lim_{x \to \infty} \frac{f'(x)}{g'(x)}$ exists, then $\displaystyle\lim_{x \to \infty} \frac{f(x)}{g(x)}$ exists and $\displaystyle\lim_{x \to \infty} \frac{f(x)}{g(x)} = \lim_{x \to \infty} \frac{f'(x)}{g'(x)}$.
(b) *Let f, g be differentiable functions whose domains include the interval $(-\infty, b)$. Assume $\displaystyle\lim_{x \to -\infty} f(x) = \lim_{x \to -\infty} g(x) = L$ where L is either 0, $-\infty$ or $+\infty$.*

If $\displaystyle\lim_{x \to -\infty} \frac{f'(x)}{g'(x)}$ exists, then $\displaystyle\lim_{x \to -\infty} \frac{f(x)}{g(x)}$ exists and $\displaystyle\lim_{x \to -\infty} \frac{f(x)}{g(x)} = \lim_{x \to -\infty} \frac{f'(x)}{g'(x)}$.

Proof (a) To apply the L'Hôpital's Rules of the preceding subsections, change variables from x to $u = \frac{1}{x}$ because u approaches zero from the right as x approaches infinity.

$$\lim_{x \to \infty} \frac{f(x)}{g(x)} = \lim_{u \to 0^+} \frac{f\left(\frac{1}{u}\right)}{g\left(\frac{1}{u}\right)} = \lim_{u \to 0^+} \frac{f\left(u^{-1}\right)}{g\left(u^{-1}\right)}. \tag{2.8.3}$$

By hypothesis,

$$\lim_{u \to 0^+} f\left(\frac{1}{u}\right) = \lim_{x \to \infty} f(x) = L \text{ and } \lim_{u \to 0^+} g\left(\frac{1}{u}\right) = \lim_{x \to \infty} g(x) = L$$

where L is either 0, $-\infty$ or $+\infty$. Hence L'Hôpital's Rule of Theorem 2.8.2 or 2.8.5 applies to the right limit in (2.8.3).[3] Use the chain rule to evaluate the derivatives of the numerator and denominator:

$$\lim_{x \to \infty} \frac{f(x)}{g(x)} = \lim_{u \to 0^+} \frac{\frac{d}{du}\left[f\left(u^{-1}\right)\right]}{\frac{d}{du}\left[g\left(u^{-1}\right)\right]} = \lim_{u \to 0^+} \frac{f'\left(u^{-1}\right)\frac{d}{du}\left(u^{-1}\right)}{g'\left(u^{-1}\right)\frac{d}{du}\left(u^{-1}\right)}$$

$$= \lim_{u \to 0^+} \frac{f'\left(u^{-1}\right)\left(-u^{-2}\right)}{g'\left(u^{-1}\right)\left(-u^{-2}\right)} = \lim_{u \to 0^+} \frac{f'\left(u^{-1}\right)}{g'\left(u^{-1}\right)} = \lim_{x \to \infty} \frac{f'(x)}{g'(x)}.$$

(b) Change variables from x to $v = -x$ because v approaches $+\infty$ when x approaches $-\infty$. Apply L'Hôpital's Rule of (a) to evaluate the limit in v:

$$\lim_{x \to -\infty} \frac{f(x)}{g(x)} = \lim_{v \to +\infty} \frac{f(-v)}{g(-v)}. \tag{2.8.4}$$

By hypothesis,

$$\lim_{v \to +\infty} f(-v) = \lim_{x \to -\infty} f(x) = L, \text{ and } \lim_{v \to +\infty} g(-v) = \lim_{x \to -\infty} g(x) = L.$$

Hence L'Hôpital's Rule of (a) applies to the right limit of (2.8.4). By the chain rule:

$$\lim_{x \to -\infty} \frac{f(x)}{g(x)} = \lim_{v \to +\infty} \frac{\frac{d}{dv}f(-v)}{\frac{d}{dv}g(-v)} = \lim_{v \to +\infty} \frac{-f'(-v)}{-g'(-v)} = \lim_{x \to -\infty} \frac{f'(x)}{g'(x)}. \qquad \square$$

We apply this form of L'Hôpital's Rule to several examples.

---

[3] We shall show in Section 13 that Theorem 2.8.2 is also true for one–sided limits.

Examples 2.8.8 (1) Evaluate $\displaystyle\lim_{x\to\infty}\frac{\sqrt{3x+1}}{x^2-x+4}$.

Solution The differentiable functions $\sqrt{3x+1}$ and x^2-x+4 both have limit ∞, as x approaches ∞. Thus L'Hôpital's Rule applies:

$$\lim_{x\to\infty}\frac{\sqrt{3x+1}}{x^2-x+4}=\lim_{x\to\infty}\frac{D(\sqrt{3x+1})}{D(x^2-x+4)}=\lim_{x\to\infty}\frac{\frac{3}{2\sqrt{3x+1}}}{2x-1}$$

$$=\lim_{x\to\infty}\frac{3}{(4x-2)\sqrt{3x+1}}=\frac{3}{\infty}=0.$$

(2) Evaluate $\displaystyle\lim_{x\to\infty}\frac{\arctan 2x}{\arctan 3x}$.

Solution Recall $\arctan x$ has horizontal asymptote $y=\frac{\pi}{2}$ on the right. Hence

$$\lim_{x\to\infty}\frac{\arctan 2x}{\arctan 3x}=\frac{\lim\limits_{x\to\infty}\arctan 2x}{\lim\limits_{x\to\infty}\arctan 3x}=\frac{\pi/2}{\pi/2}=1.$$

In particular L'Hôpital's Rule does not apply. The limit of the quotient of the derivatives of L'Hôpital's Rule is:

$$\lim_{x\to\infty}\frac{D(\arctan 2x)}{D(\arctan 3x)}=\lim_{x\to\infty}\frac{2/\left[1+(2x)^2\right]}{3/\left[1+(3x)^2\right]}=\lim_{x\to\infty}\frac{2(1+9x^2)}{3(1+4x^2)}=\left(\frac{2}{3}\right)\left(\frac{9}{4}\right)=\frac{3}{2}.$$

This is the wrong answer because we are not entitled to use L'Hôpital's Rule to evaluate this limit.

(3) Evaluate $\displaystyle\lim_{x\to\infty}x\sin\frac{1}{x}$.

Solution As x approaches ∞, $\sin\frac{1}{x}$ approaches $\sin 0=0$. Thus our limit is the indeterminate form $\infty\cdot 0$. To evaluate it by L'Hôpital's Rule, rewrite this product as a quotient:

$$\lim_{x\to\infty}x\sin\frac{1}{x}=\lim_{x\to\infty}\frac{\sin(x^{-1})}{x^{-1}}.$$

The differentiable functions $\sin(x^{-1})$ and x^{-1} both have limit zero as x approaches infinity. Hence L'Hôpital's Rule applies to the right limit. By the chain rule:

$$\lim_{x\to\infty}x\sin\frac{1}{x}=\lim_{x\to\infty}\frac{D\left[\sin(x^{-1})\right]}{D\left[x^{-1}\right]}=\lim_{x\to\infty}\frac{-x^{-2}\cos(x^{-1})}{-x^{-2}}$$

$$=\lim_{x\to\infty}\cos(x^{-1})=\cos 0=1.$$

(4) Evaluate $\displaystyle\lim_{x\to\infty}\sqrt{\frac{x^2-1}{x^2+1}}$.

Solution This example demonstrates that sometimes L'Hôpital's Rule does not succeed in evaluating a limit, and we have to resort to another method. Substitution into the limit

$$\lim_{x\to\infty}\sqrt{\frac{x^2-1}{x^2+1}}=\lim_{x\to\infty}\frac{\sqrt{x^2-1}}{\sqrt{x^2+1}}$$

produces the indeterminate value $\frac{\infty}{\infty}$. Thus we can apply L'Hôpital's Rule. By the chain rule:

$$\lim_{x\to\infty}\sqrt{\frac{x^2-1}{x^2+1}}=\lim_{x\to\infty}\frac{D(\sqrt{x^2-1})}{D(\sqrt{x^2+1})}=\lim_{x\to\infty}\frac{\frac{2x}{2\sqrt{x^2-1}}}{\frac{2x}{2\sqrt{x^2+1}}}=\lim_{x\to\infty}\frac{\sqrt{x^2+1}}{\sqrt{x^2-1}}$$

Substitution into the latter limit also produces the indeterminate form $\frac{\infty}{\infty}$, so we apply L'Hôpital's Rule a second time:

$$\lim_{x\to\infty} \sqrt{\frac{x^2-1}{x^2+1}} = \lim_{x\to\infty} \frac{D(\sqrt{x^2+1})}{D(\sqrt{x^2-1})} = \lim_{x\to\infty} \frac{\frac{2x}{2\sqrt{x^2+1}}}{\frac{2x}{2\sqrt{x^2-1}}} = \lim_{x\to\infty} \sqrt{\frac{x^2-1}{x^2+1}}$$

which is the original limit. Thus L'Hôpital's Rule has led us in a circle and does not evaluate our limit. Note the limit we obtained after applying L'Hôpital's Rule once is the reciprocal of the original limit. *If we knew that our limit existed and had value L*, then we would have $L = 1/L$, $L^2 = 1$ and $L = 1$. (Our limit is a limit of positive numbers and can not have value -1.) The easiest way to evaluate our limit, without assuming it exists, is to divide numerator and denominator of the rational function in the original limit by x^2:

$$\lim_{x\to\infty} \sqrt{\frac{x^2-1}{x^2+1}} = \lim_{x\to\infty} \sqrt{\frac{1-1/x^2}{1+1/x^2}} = \sqrt{\lim_{x\to\infty} \frac{1-1/x^2}{1+1/x^2}} = \sqrt{\frac{1-0}{1+0}} = 1. \qquad \square$$

Historical Remarks

L'Hôpital's Rule was discovered by John Bernoulli. He was hired by the Marquis de l'Hôpital to teach him mathematics and to send him the results of his research. Under the terms of this agreement the Marquis was entitled to use Bernoulli's work in any manner. In 1696, the Marquis published the first calculus textbook which contained the result we call L'Hôpital's Rule for the indeterminate form $\frac{0}{0}$. Since limits were unknown at that time, the rule was stated and proved in Leibniz notation: when $f(c) = g(c) = 0$,

$$\frac{f(c+dx)}{g(c+dx)} = \frac{f(c+dx) - f(c)}{g(c+dx) - g(c)} = \frac{f'(c)dx}{g'(c)dx} = \frac{f'(c)}{g'(c)}.$$

Summary

The reader should know the statements of all the cases of L'Hôpital's Rule. In particular, she should understand that L'Hôpital's Rule only applies to limits with an indeterminate value, and she should be able to use L'Hôpital's Rule to evaluate them.

Basic Exercises

Evaluate each of the following 32 limits.

1. $\displaystyle\lim_{x\to2} \frac{x^3 - 4x^2 + 2x + 4}{x^5 - 9x - 14}$

2. $\displaystyle\lim_{x\to-3} \frac{x^4 - 2x^3 + 7x - 11}{x^4 - 5x^2 + 4x - 9}$

3. $\displaystyle\lim_{x\to-1} \frac{x^3 + x^2 - x - 1}{x^4 + 2x^3 + 3x^2 + 4x + 2}$

4. $\displaystyle\lim_{x\to1} \frac{x^4 - 2x^3 + 2x - 1}{x^6 - 3x^5 + 3x^4 - 3x^2 + 3x - 1}$

5. $\displaystyle\lim_{x\to0} \frac{\sin^2 6x}{\sin^2 7x}$

6. $\displaystyle\lim_{x\to0} \frac{\cos^8 x}{\cos^2 3x}$

7. $\displaystyle\lim_{x\to0} \frac{\tan^3 5x}{x^3 - 4x^2 + 7x - 9}$

8. $\displaystyle\lim_{x\to\pi/2} \frac{\tan 3x}{\tan 7x}$

9. $\displaystyle\lim_{x\to0} \frac{\csc 5x}{\cot 9x}$

10. $\displaystyle\lim_{x\to\pi} \frac{\sec 3x}{\cot 6x}$

11. $\displaystyle\lim_{x\to0} \frac{\arctan 7x}{\arcsin 8x}$

12. $\displaystyle\lim_{x\to0} x \cot 2x$

13. $\displaystyle\lim_{x\to0} \sin 3x \csc 5x$

14. $\displaystyle\lim_{x\to5} (x^2 - 25) \cot \pi x$

15. $\displaystyle\lim_{x\to\pi/2} \cos 4x \tan 3x$

16. $\displaystyle\lim_{x\to1} \left(\frac{x}{x^2-1} - \frac{x^2}{x^4 - 4x^3 + 6x^2 - 4x + 1} \right)$

17. $\lim\limits_{x \to 2} \left(\dfrac{1}{x^2 - 4} - \dfrac{x}{(x-2)^3} \right)$ **18.** $\lim\limits_{x \to \pi/2} \left(5 \tan 5x - 3 \tan 3x \right)$

19. $\lim\limits_{x \to 0} \left(\cot 4x - \dfrac{1}{x^2} \right)$ **20.** $\lim\limits_{x \to \infty} \sqrt{x} \sin(1/\sqrt{x})$

21. $\lim\limits_{x \to -\infty} \left(\sqrt{5 - 2x} - \sqrt{7 - 2x} \right)$ **22.** $\lim\limits_{x \to \infty} \sqrt{\dfrac{3x^2 + 1}{5x^2 + 2}}$

23. $\lim\limits_{x \to \infty} \dfrac{\sin x}{x}$ **24.** $\lim\limits_{x \to \infty} \dfrac{\arctan x}{\sqrt{1 + x^5}}$

25. $\lim\limits_{x \to 0} \dfrac{x - \sin x}{x \cos x - x}$ **26.** $\lim\limits_{x \to 0} \dfrac{\arctan x - x}{\arcsin x - x}$

27. $\lim\limits_{x \to \infty} x(2 \arctan x - \pi)$ **28.** $\lim\limits_{x \to \infty} x \cot \dfrac{\pi}{x}$

29. $\lim\limits_{x \to \infty} x \tan \dfrac{\pi}{x}$ **30.** $\lim\limits_{x \to 0} \dfrac{x \sin x}{1 - \cos x}$

31. $\lim\limits_{x \to 0} \dfrac{2 + x^2 - 2\sqrt{1 + x^2}}{1 - \cos(x^2)}$ **32.** $\lim\limits_{x \to 0} \dfrac{1 - \cos\left(x^{10}\right)}{x^{20}}$

33.   Show that if f and g are differentiable at $x = c$ with $f(c) = g(c) = 0$ and $g'(c) \neq 0$, then $\lim\limits_{x \to c} \dfrac{f(x)}{g(x)}$ exists and $\lim\limits_{x \to c} \dfrac{f(x)}{g(x)} = \dfrac{f'(c)}{g'(c)}$.

34. Let f, g be differentiable functions whose domains include $(a, c) \cup (c, b)$. Assume $\lim\limits_{x \to c} f(x) = \lim\limits_{x \to c} g(x) = \infty$. Use Theorem 2.8.2 to show that if $\lim\limits_{x \to c} \dfrac{f(x)}{g(x)}$ and $\lim\limits_{x \to c} \dfrac{f'(x)}{g'(x)}$ exist, then $\lim\limits_{x \to c} \dfrac{f(x)}{g(x)} = \lim\limits_{x \to c} \dfrac{f'(x)}{g'(x)}$.

35. Let f, g be differentiable functions whose domains include the interval (a, ∞). Assume $\lim\limits_{x \to \infty} f(x) = \infty$ and $\lim\limits_{x \to \infty} g(x) = -\infty$. Use the theorems of this section to show that if $\lim\limits_{x \to \infty} \dfrac{f'(x)}{g'(x)}$ exists, then $\lim\limits_{x \to \infty} \dfrac{f(x)}{g(x)}$ exists and $\lim\limits_{x \to \infty} \dfrac{f(x)}{g(x)} = \lim\limits_{x \to \infty} \dfrac{f'(x)}{g'(x)}$.

Challenging Problems

1.   Give an example of a limit $\lim\limits_{x \to c} \dfrac{f(x)}{g(x)}$ which can be evaluated by the L'Hôpital's Rule of Theorem 2.8.2 but not by the L'Hôpital's Rule of Exercise 33.

2. Let $f(x) = \left\{ \begin{array}{ll} x^2 & \text{if } x \text{ is rational} \\ x^3 & \text{if } x \text{ is irrational} \end{array} \right\}$ and $g(x) = \left\{ \begin{array}{ll} x & \text{if } x \text{ is rational} \\ \sin x & \text{if } x \text{ is irrational} \end{array} \right\}$.

Show $\lim\limits_{x \to 0} \dfrac{f(x)}{g(x)}$ can be evaluated by the L'Hôpital's Rule of Exercise 33 but not by the L'Hôpital's Rule of Theorem 2.8.2.

Applications

2.9 The Derivative as a Rate of Change

Prerequisite: Section 2.6

Science, engineering and economics study quantities which change over time. In the first subsection, we identify the rate of change of a quantity as its derivative with respect to time. In the last two subsections, we illustrate this viewpoint with two applications. First, we consider an object which is moving in a straight line. We interpret the derivative of its position as its velocity and the second derivative as its acceleration. In particular, we study the motion of an object moving vertically under gravity. The last subsection gives an application to economics. We use the derivative to maximize profit when the cost and revenue functions are known.

Average and Instantaneous Rates of Change

We define the (instantaneous) rate of change of a quantity y as the limit of its average rate of change. Then we identify the instantaneous rate of change as a derivative. The following example illustrates that the amount of change of y is distinct from its rate of change.

Motivating Example 2.9.1 A tree grows 15 meters in 25 years while a flower grows 4 centimeters in 2 weeks. The height of the tree increases by 0.6 meters per year or 60 centimeters per year. On the other hand, the height of the flower increases by 104 centimeters per year. Even though the change in height of the tree is much greater than the change in height of the flower, the rate of growth of the flower is greater than the rate of growth of the tree. □

To compute the average change of $y = f(x)$ for $a \le x \le b$ we divide the change Δy of y by the change Δx of x. That is, we divide $\Delta y = f(b) - f(a)$ by $\Delta x = b - a$.

Definition 2.9.2 *Let* $y = f(x)$ *be a function whose domain contains the interval* $[a, b]$. *The average rate of change of* f *on* $[a, b]$ *is defined as*

$$\frac{\Delta y}{\Delta x} = \frac{f(b) - f(a)}{b - a}.$$

The average rate of change tells us how much y changes per unit change in x as x increases from $x = a$ to $x = b$.

Examples 2.9.3 (1) Every 30 minutes a biologist estimates the number of bacteria in a petri dish. The following table records these estimates over a 6 hour period.

Time	Bacteria	Time	Bacteria
$9:00$	$1,000$	$12:30$	$7,200$
$9:30$	$1,400$	$1:00$	$7,100$
$10:00$	$2,100$	$1:30$	$6,000$
$10:30$	$3,300$	$2:00$	$4,000$
$11:00$	$5,400$	$2:30$	$3,800$
$11:30$	$6,500$	$3:00$	$3,700$
$12:00$	$7,000$		

Compute the average rate of bacteria growth over each half–hour interval.

Solution During the interval from 9:00 to 9:30, the average rate of growth is:

$$\frac{1,400 - 1,000}{9:30 - 9:00} = \frac{400}{30} = 13\frac{1}{3} \text{ bacteria/minute} = 800 \text{ bacteria/hour.}$$

The analogous computations for each of the twelve half–hour intervals from 9 : 00 to 3 : 00 are given in the following table.

Time Interval	Average Rate of Change of Bacteria	Time Interval	Average Rate of Change of Bacteria
$[9:00, 9:30]$	$+800$ bacteria/hr	$[12:00, 12:30]$	$+400$ bacteria/hr
$[9:30, 10:00]$	$+1,400$ bacteria/hr	$[12:30, 1:00]$	-200 bacteria/hr
$[10:00, 10:30]$	$+2,400$ bacteria/hr	$[1:00, 1:30]$	$-2,200$ bacteria/hr
$[10:30, 11:00]$	$+4,200$ bacteria/hr	$[1:30, 2:00]$	$-4,000$ bacteria/hr
$[11:00, 11:30]$	$+2,200$ bacteria/hr	$[2:00, 2:30]$	-400 bacteria/hr
$[11:30, 12:00]$	$+1,000$ bacteria/hr	$[2:30, 3:00]$	-200 bacteria/hr

(2) An arrow is shot in the air. Until it hits the ground 8 seconds later, its height at t seconds is observed to be $s(t) = 39.2t - 4.9t^2$ meters. Compute the average velocity of the arrow during each second of its flight.

Solution The term velocity is used instead of speed to indicate that the sign of the velocity gives the direction of the arrow. When the velocity is positive the arrow is moving upwards, while the velocity is negative when the arrow is moving downwards. The absolute value of the velocity is the speed of the arrow. For example, the average velocity of the arrow during the first second is

$$\frac{s(1) - s(0)}{1 - 0} = \frac{34.3 - 0}{1} = +34.3 \text{ meters/second.}$$

The computations of the average velocity of the arrow for each of the eight seconds of its flight are given in the following table.

Time Interval	Average Velocity	Time Interval	Average Velocity
$[0, 1]$	$+34.3$ meters/second	$[4, 5]$	-4.9 meters/second
$[1, 2]$	$+24.5$ meters/second	$[5, 6]$	-14.7 meters/second
$[2, 3]$	$+14.7$ meters/second	$[6, 7]$	-24.5 meters/second
$[3, 4]$	$+4.9$ meters/second	$[7, 8]$	-34.3 meters/second

In the first example above, where $y = f(x)$ is measured for a sequence of values of x, the average rate of change is all we can determine from the observed data. However in the second example above, we know the values of $y = s(t)$ for all values of t. Hence we can deduce more precise information on the rate at which y is changing. In this case the rate of change of y at t is approximated by the average rate of change of y on the interval between t and $t + h$. The smaller the value of h, the better the approximation. We define the *instantaneous rate of change* of y with respect to t as the limit of the average rate of change of y between t and $t + h$, as h approaches zero.

Definition 2.9.4 *Let $y = f(x)$ be a function with domain an interval. The instantaneous rate of change of y with respect to x is*

$$\lim_{\Delta x \to 0} \frac{\Delta y}{\Delta x} = \lim_{h \to 0} \frac{f(x + h) - f(x)}{h} = f'(x) = \frac{dy}{dx}.$$

The *rate of change of y* with respect to x refers to the instantaneous rate of change of y with respect to x. Its units are the units of y divided by the units of x.

Examples 2.9.5 (1) Find the rate of change of the area A of a circular oil spill with respect to its radius r.

Solution The area of the oil spill is $A = \pi r^2$. Its rate of change is:

$$\frac{dA}{dr} = 2\pi r.$$

(2) If p is the probability of arriving at work on time, then the probability of arriving at work on time on exactly three days of a five day work week is $W = 10p^3(1-p)^2$. Find the instantaneous rate of change of W with respect to p.

Solution By the product rule:

$$\frac{dW}{dp} = 30p^2(1-p)^2 - 20p^3(1-p).$$

(3) The gross national product G is observed to vary over time t, measured in years, with a periodic pattern given by:

$$G = 2.3 \times 10^{12} + 8.7 \times 10^{10} \sin(.712t) \text{ dollars.}$$

Find the rate of change of the gross national product with respect to time.

Solution By the chain rule:

$$\frac{dG}{dt} = 0 + (8.7)(.712) \times 10^{10} \cos(.712t) = 6.2 \times 10^{10} \cos(.712t) \text{ dollars per year.} \square$$

Motion Along a Straight Line

Consider an object moving in a straight line. Place coordinates on this line, and let $s(t)$ denote the position of the object at time t. The average rate of change in position of the object over a time interval $\frac{\Delta s}{\Delta t}$ is its change in position divided by the change in time, i.e. the *average velocity* of the object. Thus the instantaneous rate of change in position with respect to time is the (*instantaneous*) *velocity* $v(t)$ of the object:

$$v(t) = \frac{ds}{dt}. \tag{2.9.1}$$

The absolute value of the velocity is the *speed* of the object, while the sign of the velocity denotes the direction of motion. If the velocity is positive, the object is moving in the direction of the coordinate line. If the velocity is negative, the object is moving in the opposite direction. When the velocity is zero, the object stops.

The average rate of change of velocity of the object over a time interval is its change in velocity divided by the change in time $\frac{\Delta v}{\Delta t}$, i.e. the *average acceleration* of the object. Thus the instantaneous rate of change in velocity with respect to time is the (*instantaneous*) *acceleration* $a(t)$ of the object:

$$a(t) = \frac{dv}{dt} = \frac{d^2s}{dt^2}. \tag{2.9.2}$$

The sign of the acceleration denotes the direction in which the velocity is changing. If the velocity and acceleration have the same sign, the speed of the object is increasing. In this case, there is a force acting on the object, pushing it in the direction it is moving. If the velocity and acceleration have opposite signs, the speed of the object is decreasing. In this case, there is a force acting on the object, pulling it in the opposite direction of its motion. The following table summarizes this relationship. To simplify the terminology in this table we assume that the coordinate line, along which the object is moving, points from left to right.

sign $v(t)$	+	+	−	−
sign $a(t)$	+	−	+	−
	moving right speed increasing	moving right speed decreasing	moving left speed decreasing	moving left speed increasing

In each example below, we are given the position $s(t)$ of a moving object at time t. We find the velocity $v(t)$ and acceleration $a(t)$ of the object. By sketching the signs of $v(t)$ and $a(t)$ on a time axis, we are able to determine the direction of motion and how the speed of the object is changing for each value of t. We depict our conclusions with an annotated sketch of the motion of the object above its line of motion. Note the first two examples make sense even when the time t is negative. The time now is $t = 0$. Time in the future is positive while time in the past is negative.

Figure 2.9.7
$v(t) = 3t^2 - 6t - 24$
and $a(t) = 6t - 6$

Examples 2.9.6 (1) Describe the motion of an object which moves in a straight line with position $s(t) = t^3 - 3t^2 - 24t + 20$ for $t \in \Re$.

Solution The velocity and acceleration of this object are:

$$
\begin{aligned}
v(t) &= s'(t) = 3t^2 - 6t - 24 = 3(t-4)(t+2) \text{ and} \\
a(t) &= v'(t) = 6t - 6 = 6(t-1).
\end{aligned}
$$

The signs of the velocity and acceleration are shown in Figure 2.9.7. When $t < -2$ the object is moving right and slowing down. It stops at $t = -2$ and moves to the left with increasing speed for $-2 < t < 1$. For $1 < t < 4$, the object continues moving left but slows down. The object stops at $t = 4$ and then moves to the right with increasing speed for $t > 4$. These conclusions are depicted in Figure 2.9.8.

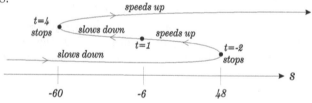

Figure 2.9.8 $s(t) = t^3 - 3t^2 - 24t + 20$

(2) Describe the motion of an object which moves along a straight line with position $s(t) = 3t^5 - 20t^3 + 30$ for $t \in \Re$.

Solution The velocity and acceleration of this object are:

$$
\begin{aligned}
v(t) &= s'(t) = 15t^4 - 60t^2 = 15t^2(t^2 - 4) = 15t^2(t-2)(t+2) \text{ and} \\
a(t) &= v'(t) = 60t^3 - 120t = 60t(t^2 - 2) = 60t(t-\sqrt{2})(t+\sqrt{2}).
\end{aligned}
$$

The signs of the velocity and acceleration are shown in Figure 2.9.9.

Figure 2.9.9 $v(t) = 15t^4 - 60t^2$ and $a(t) = 60t^3 - 120t$

When $t < -2$ the object is moving right and slowing down. It stops at $t = -2$ and moves left with increasing speed for $-2 < t < -\sqrt{2}$. For $-\sqrt{2} < t < 0$ the object continues to move left but slows down. At $t = 0$ the object stops but continues to move left for $0 < t < \sqrt{2}$ with increasing speed. For $\sqrt{2} < t < 2$ the object moves left and slows down. It stops at $t = 2$ and moves to the right with increasing speed for $t > 2$. The motion of this object is depicted in Figure 2.9.10.

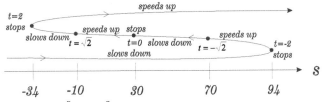

Figure 2.9.10 $s(t) = 3t^5 - 20t^3 + 30$

(3) One end of a vertical spring is fastened to the floor. The loose end of the spring is pulled upwards and released. The displacement $s(t)$ of the loose end of the spring from its equilibrium position at time t is given by $s(t) = 5\cos \pi t$ for $t \geq 0$. Describe the motion of the loose end of the spring.

Solution The velocity and acceleration of the loose end of the spring are:

$$
\begin{aligned}
v(t) &= s'(t) = -5\pi \sin \pi t \ \text{ and} \\
a(t) &= v'(t) = -5\pi^2 \cos \pi t.
\end{aligned}
$$

Thus, $v(t) < 0$ when $2n < t < 2n + 1$ for integers $n \geq 0$, while $v(t) > 0$ when $2n - 1 < t < 2n$ for integers $n > 0$. The acceleration $a(t) > 0$ when $2n + \frac{1}{2} < t < 2n + \frac{3}{2}$ for integers $n \geq 0$. Also, $a(t) < 0$ when $0 \leq t < \frac{1}{2}$ or $2n - \frac{1}{2} < t < 2n + \frac{1}{2}$ for integers $n \geq 1$. These conclusions are shown in Figure 2.9.11. Since the functions $\sin \pi t$ and $\cos \pi t$ are periodic with period 2, the loose end of the spring moves in a periodic pattern with period 2. For each integer $n \geq 0$ the loose end of the spring starts from rest and moves downwards with increasing speed for $2n < t < 2n + \frac{1}{2}$. For $2n + \frac{1}{2} < t < 2n + 1$, the loose end of the spring continues to move downwards but slows down. The spring stops at $t = 2n + 1$, and its loose end moves upwards with increasing speed for $2n + 1 < t < 2n + \frac{3}{2}$. For $2n + \frac{3}{2} < t < 2n + 2$ the loose end of the spring continues moving upwards but slows down. At $t = 2n + 2$ the spring stops. Then it repeats this pattern of motion. □

```
0 - - - - - - - - - 0 + + + + + + 0    sign v(t)

    - - - - - 0 + + + + +  0 - - - - - -    sign a(t)
                                            → t
    +         +         +       +       +
   2n      2n+1/2    2n+1    2n+3/2   2n+2
```

Figure 2.9.11 $v(t) = -5\pi \sin \pi t$ and $a(t) = -5\pi^2 \cos \pi t$

Consider an object moving on a vertical coordinate line which points upwards. Ignoring air resistance, gravity is the only force acting on this object. Experimentation shows this object moves with a constant downwards acceleration denoted g with value

$$
a(t) = -g \approx -9.8 \text{ meters/sec}^2 \approx -32 \text{ feet/sec}^2. \qquad (2.9.3)
$$

Since $v'(t) = a(t) = -g$ and $D(-gt) = -g$, it follows from Corollary 2.6.24 that $v(t) = -gt + c$. Substituting $t = 0$, we see that $c = v(0)$, the initial velocity of the object. Thus

$$
v(t) = -gt + v(0). \qquad (2.9.4)
$$

Since $s'(t) = v(t) = -gt + v(0)$ and $D\left[-\frac{1}{2}gt^2 + v(0)t\right] = -gt + v(0)$, it follows from Corollary 2.6.24 that $s(t) = -\frac{1}{2}t^2 + v(0)t + k$. Substituting $t = 0$, we see that $k = s(0)$, the initial position of the object. Thus

$$
s(t) = -\frac{g}{2}t^2 + v(0)t + s(0). \qquad (2.9.5)
$$

We apply these formulas in the examples below.

Examples 2.9.12 (1) A soccer ball is kicked upwards at 96 feet per second.
    (a) How high does the ball travel?
    (b) How long does it take for the ball to return to the ground?

Solution Place a coordinate axis pointing upwards with the origin on the ground. Then $s(0) = 0$ and $v(0) = +96$. By (2.9.4) and (2.9.5):

$$v(t) = -32t + 96, \quad \text{and} \quad s(t) = -16t^2 + 96t.$$

(a) When the ball reaches its highest point, it stops, i.e. $v(t) = 0$:

$$0 = v(t) = -32t + 96, \quad \text{and} \quad t = \frac{96}{32} = 3 \text{ seconds.}$$

Therefore, the maximum height of the ball is $s(3) = -16(3)^2 + 96(3) = 144$ feet.
(b) The ball returns to the ground when $s(t) = 0$: ,

$$0 = s(t) = -16t^2 + 96t = 16t(-t + 6), \quad \text{and} \quad t = 6 \text{ seconds.}$$

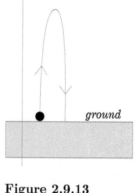

Figure 2.9.13
Soccer Ball

(2) A rock is thrown upwards at 30 meters/sec from the edge of a 100 meter high cliff.
    (a) How long does it take the rock to reach the bottom of the cliff?
    (b) Find the speed of the rock when hits the ground at the bottom of the cliff.

Solution Let the coordinate axis point upwards with origin at the bottom of the cliff. See Figure 2.9.14. Then $v(0) = +30$ and $s(0) = 100$. By (2.9.4) and (2.9.5):

$$v(t) = -9.8t + 30 \quad \text{and} \quad s(t) = -4.9t^2 + 30t + 100.$$

(a) At the bottom of the cliff, $s(t) = 0$. Thus,

$$0 = s(t) = -4.9t^2 + 30t + 100.$$

The positive root of this polynomial is when the ball hits the ground. By the quadratic formula:

$$t = \frac{-30 - \sqrt{30^2 - 4(-4.9)(100)}}{2(-4.9)} = \frac{30 + \sqrt{2860}}{9.8} \approx 8.5 \text{ seconds.}$$

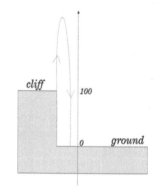

Figure 2.9.14 Rock
Thrown from a Cliff

(b) When the ball hits the ground at 8.5 seconds its velocity is

$$v(8.5) = -9.8(8.5) + 30 \approx -53 \text{ meters/second.}$$

Hence the speed of the rock is $|-53| = 53$ meters/second. □

Rates of Change in Economics

Consider a product which is being manufactured. The amount of money spent to produce q units of this product is called the *cost* $C(q)$. The amount of money received when these q units are sold is called the *revenue* $R(q)$. The *profit* $P(q)$ is the amount of money earned:

$$P(q) = R(q) - C(q). \tag{2.9.6}$$

There are two varieties of manufactured products. The first type comes in distinct units. Examples include refrigerators and cars. The second type of product is available in quantities of any size. They include flour and ribbon. When the product comes in distinct units we can calculate the changes in cost C, revenue R and profit P when one additional unit is produced. These average rates of change are called the *marginal* cost MC, *marginal revenue* MR and *marginal profit* MP: If N units were originally produced, then

$$MC(N) = C(N+1) - C(N), \qquad MR(N) = R(N+1) - R(N)$$
$$\text{and } MP(N) = P(N+1) - P(N). \qquad (2.9.7)$$

When the product is available in all quantities, we can compute the instantaneous rates of change of cost, revenue and profit. These derivatives are also called the *marginal cost* MC, *marginal revenue* MR and *marginal profit* MP:

$$MC(q) = \frac{dC}{dq}, \quad MR(q) = \frac{dR}{dq} \quad \text{and} \quad MP(q) = \frac{dP}{dq}. \qquad (2.9.8)$$

Since the derivative of a difference is the difference of the derivatives, by (2.9.6) the marginal profit equals the marginal revenue minus the marginal cost:

$$MP(q) = MR(q) - MC(q). \qquad (2.9.9)$$

The negative of the profit is called the *loss* L, i.e. $L = -P$. Taking derivatives, the *marginal loss* $ML = \frac{dL}{dq}$ is the negative of the marginal profit:

$$ML(q) = -MP(q).$$

The goal in business is to maximize profit or, equivalently, minimize loss. Consider a product which can be produced in all quantities. If the profit function $P(q)$, for $q \geq 0$, is differentiable and has a maximum value, then this maximum value occurs either at the endpoint $q = 0$ or at a point where $MP(q) = \frac{dP}{dq} = 0$. When the maximum profit occurs at $q = 0$, the business should find a different product to manufacture.

Consider the case where the product is produced in distinct quantities. The graph of the profit $P(n)$ consists of dots, one for each positive integer n. The maximum value of P occurs at the highest of these dots. To locate this dot, without constructing the graph of P, we may be able to define a differentiable function $p(x)$, with domain $[0, \infty)$, whose graph passes through each of the dots of the graph of P. That is, $P(n) = p(n)$ for each positive integer n. If p has maximum value at a positive integer m, then P will also have its maximum value there. If p has maximum value at a number in the open interval $(m, m+1)$, with m a positive integer, then $P(n)$ probably has its maximum value at either $n = m$ or $n = m+1$. We illustrate these procedures in the examples below.

Examples 2.9.15 (1) The cost $C(x)$, in thousands of dollars, of manufacturing x meters of sheet metal and the revenue $R(x)$, in thousands of dollars, from selling x meters of sheet metal are given by:

$$C(x) = \frac{5x^2 + 1,000,000}{x + 500} \quad \text{and} \quad R(x) = \frac{5,000x}{x + 500}.$$

Find the profit $P(x)$, the marginal cost, the marginal revenue and the marginal profit. Then find the value of x which maximizes the profit.

Solution By (2.9.6), the profit is given by:

$$P(x) = R(x) - C(x) = \frac{5,000x}{x + 500} - \frac{5x^2 + 1,000,000}{x + 500} = \frac{5,000x - 5x^2 - 1,000,000}{x + 500}.$$

Differentiating by the quotient rule:

$$MC(x) = \frac{dC}{dx} = \frac{10x(x + 500) - (5x^2 + 1,000,000)(1)}{(x + 500)^2}$$

$$= \frac{5x^2 + 5,000x - 1,000,000}{(x+500)^2}$$

$$MR(x) = \frac{dR}{dx} = \frac{5,000(x+500) - 5,000x(1)}{(x+500)^2} = \frac{2,500,000}{(x+500)^2}$$

$$MP(x) = \frac{dP}{dx} = \frac{(5,000 - 10x)(x+500) - (5,000x - 5x^2 - 1,000,000)(1)}{(x+500)^2}$$

$$= \frac{-5x^2 - 5,000x + 3,500,000}{(x+500)^2}.$$

The maximal profit occurs when $MP(x) = 0$. This fraction is zero when its numerator is zero:

$$0 = -5x^2 - 5,000x + 3,500,000$$
$$0 = x^2 + 1,000x - 700,000$$

By the quadratic formula, there is a unique positive solution x of this equation:

$$x = \frac{-1,000 + \sqrt{1,000^2 - 4(1)(-700,000)}}{2}$$

$$= \frac{-1,000 + \sqrt{3,800,000}}{2} = -500 + 100\sqrt{95} \approx 475 \text{ meters.}$$

Note the numerator of $P'(x) = MP(x)$ is positive for $x < 475$ and is negative for $x > 475$. Since the denominator of $P'(x)$ is always positive, $P'(x) > 0$ for $x < 475$, and $P'(x) < 0$ for $x > 475$. By the First Derivative Test, $P(x)$ has a local maximum at $x = 475$. Since $x = 475$ is the only local extremum of $P(x)$, it is a global maximum. Thus producing approximately 475 meters of sheet metal achieves the maximum profit.

(2)   The revenue, $R(n)$ dollars, received from manufacturing n stoves and the cost, $C(n)$ dollars, of manufacturing n stoves are given by:

$$R(n) = 600n - 3n^2 \quad \text{and} \quad C(n) = 3,000 + 200n.$$

Find the profit, marginal revenue, marginal cost and marginal profit. Then find the number of stoves to manufacture which maximizes the profit.

Solution   By (2.9.6), the profit is given by:

$$P(n) = R(n) - C(n) = (600n - 3n^2) - (3,000 + 200n) = 400n - 3n^2 - 3,000.$$

Then

$$MR(n) = R(n+1) - R(n) = \left[600(n+1) - 3(n+1)^2\right] - \left[600n - 3n^2\right]$$
$$= 597 - 6n$$
$$MC(n) = C(n+1) - C(n) = [3,000 + 200(n+1)] - [3,000 + 200n] = 200$$
$$MP(n) = MR(n) - MC(n) = (597 - 6n) - 200 = 397 - 6n.$$

To find the value of n which maximizes the profit, replace $P(n)$ by the differentiable function

$$p(x) = 400x - 3x^2 - 3,000$$

with domain $[0, \infty)$. The local extrema of $p(x)$ occur where $p'(x) = 0$. Now

$$p'(x) = 400 - 6x.$$

Thus, $p'(x) = 0$ when $x = 66\frac{2}{3}$. Since $p'(x) > 0$ for $x < 66\frac{2}{3}$ and $p'(x) < 0$ for $x > 66\frac{2}{3}$, p has its maximum value at $x = 66\frac{2}{3}$ by the First Derivative Test.

Therefore, $P(n)$ has its maximum value at either $n = 66$ or $n = 67$. Since $P(66) = 10,332$ and $P(67) = 10,333$, we see that the profit is largest when 67 stoves are manufactured. □

Historical Remarks

Newton's approach to mathematics was motivated by his interest in mechanics. He considered all variables as denoting positions which depend on time t. Newton's only notation for a derivative was $\dot{z}$, which denotes the derivative of z with respect to t. He called $\dot{z}$ the *fluxion of z*, because it gives the velocity of an object Q with position z at time t. His only need for higher derivatives was the second derivative $\ddot{z}$, which gives the acceleration of Q. Although Newton studied the tangent line T to the curve $f(x, y) = 0$, he denoted the slope of T as $\frac{\dot{y}}{\dot{x}}$ which emphasizes its physical, rather than its geometrical, significance. Newton's fundamental contributions to physics were based upon studying physical phenomena using the methods of calculus. Every quantity which changes with time is a candidate for study using calculus. It is interesting that many of Newton's applications of calculus, including his mathematical expositions of mechanics and optics, were made concurrently with his discovery of calculus in the years 1664 to 1676. This phenomenon is a prime example of the symbiotic relationship between good mathematics and significant applications.

Summary

The reader should know how to compute average rates of change and instantaneous rates of change of y with respect to x from given information which defines y as a function of x. Given the position function $s(t)$ of an object moving in a straight line, she should be able to find its velocity and acceleration. By comparing their signs, she should describe the motion of the object. In particular, she should be able to analyze the motion of an object which is moving vertically under gravity. In economics, she should realize that the word "marginal" refers to the derivative, and she should be able to determine where profit is maximal by setting the marginal profit equal to zero.

Basic Exercises

1. A nervous parent measures her child's temperature every 10 minutes for two hours.

Time	Temperature	Time	Temperature
3 : 00	100°F	4 : 10	101.2°F
3 : 10	100.1°F	4 : 20	100.9°F
3 : 20	100.3°F	4 : 30	100.5°F
3 : 30	100.6°F	4 : 40	100.4°F
3 : 40	101.1°F	4 : 50	100.6°F
3 : 50	101.3°F	5 : 00	100.5°F
4 : 00	101.4°F		

(a) Find the average rate of change in temperature over each ten minute interval.
(b) Find the average rate of change in temperature over each twenty minute interval.

2. A pilot records the altitude of his plane every 15 seconds after takeoff.
(a) Find the average rate of change in altitude over each 15 second interval.
(b) Find the average rate of change in altitude over each 30 second interval starting at takeoff.
(c) Find the average rate of change in altitude over each 45 second interval starting at takeoff.

Time	Altitude	Time	Altitude
0 seconds	0 meters	120 seconds	2,890 meters
15 seconds	40 meters	135 seconds	3,500 meters
30 seconds	200 meters	150 seconds	4,030 meters
45 seconds	450 meters	165 seconds	4,500 meters
60 seconds	750 meters	180 seconds	4,920 meters
75 seconds	1,180 meters	195 seconds	5,300 meters
90 seconds	1,720 meters	210 seconds	5,650 meters
105 seconds	2,350 meters	225 seconds	5,970 meters

3. A census of Bay City is taken every ten years.

Year	Population	Year	Population
1910	4,568	1960	5,306
1920	4,839	1970	7,122
1930	4,217	1980	6,940
1940	4,104	1990	6,018
1950	4,823	2000	5,856

(a) Find the average rate of change of population over each 10 year interval.
(b) Find the average rate of change of population over each 20 year interval.

4. A baseball player records his batting average at the end of each week.

Week	Average	Week	Average	Week	Average	Week	Average
1	.169	8	.271	15	.276	22	.268
2	.235	9	.275	16	.278	23	.274
3	.251	10	.272	17	.280	24	.278
4	.262	11	.277	18	.277	25	.281
5	.270	12	.279	19	.270	26	.283
6	.265	13	.275	20	.265	27	.279
7	.268	14	.273	21	.263	28	.280

(a) Find the average rate of change of batting average over each 2 week interval.
(b) Find the average rate of change of batting average over each 4 week interval.

5. The cost of manufacturing N pencils is $\sqrt{5N^2 + 75,200}$ cents. Ten thousand pencils are manufactured.
(a) Find the average rate of change of cost of manufacturing each thousand pencils.
(b) Find the average rate of change of cost of manufacturing each two thousand pencils.
(c) Find the marginal cost at each thousand pencils.

6. **(a)** Find the rate of change of the area A of a square with respect to the length of its side s.
(b) Find the rate of change of the perimeter P of a square with respect to the length of its side s.

7. **(a)** Find the rate of change of the area A of an equilateral triangle with respect to the length of its side s.
(b) Find the rate of change of the perimeter P of an equilateral triangle with respect to the length of its side s.

8. **(a)** Find the rate of change of the volume V of a cube with respect to the length of its side s.
(b) Find the rate of change of the surface area A of a cube with respect to the length of its side s.

9. **(a)** Find the rate of change of the volume V of a sphere with respect to the length of its radius r.
(b) Find the rate of change of the surface area A of a sphere with respect to the length of its radius r.

10. **(a)** Consider a cylinder of constant height h whose radius r varies. Find the rate of change of the volume V of this cylinder with respect to the length of its radius.
(b) Find the rate of change of the surface area A of the cylinder in (a) with respect to the length of its radius r.
(c) Consider a cylinder of constant radius r whose height h varies. Find the rate of change of the volume V of this cylinder with respect to the length of its height.
(d) Find the rate of change of the surface area A of the cylinder in (c) with respect to the length of its height h.

11. **(a)** Consider a cone of constant height h whose radius r varies. Find the rate of change of the volume V of this cone with respect to the length of its radius.
(b) Consider a cone of constant radius r whose height h varies. Find the rate of change of the volume V of this cone with respect to the length of its height.

12. If a constant force acts on an object, then the *work* W equals the amount of the force F times the distance s the object moves. The instantaneous rate of change of work with respect to time is called *power*. Show that in this case the power P at time t equals the value of the constant force times the velocity v of the object at time t.

13. Chemical C is decomposing in a second order reaction. The concentration $c(t)$ of C at time t is given by:

$$c(t) = \frac{500}{1 - 400t} \text{ moles per liter.}$$

Show that the rate at which C is decomposing is proportional to c^2.

In each of the following ten exercises we are given the position $s(t)$ at time t of an object which is moving along the x–axis. Find the velocity and acceleration of the object at time t. Then sketch its motion. Indicate the direction of motion, where the object is slowing down and where it is speeding up.

14. $s(t) = t^2 - 6t + 4$ for $t \in \mathcal{R}$.
16. $s(t) = t^3 + 6t^2 + 9t + 7$ for $t \in \mathcal{R}$.
18. $s(t) = 3t^5 - 5t^3 + 1$ for $t \in \mathcal{R}$.
20. $s(t) = 2t^6 - 3t^4 + 5$ for $t \in \mathcal{R}$.
22. $s(t) = 3\sin \pi t$ for $t \geq 0$.

15. $s(t) = t^3 - 6t^2 - 36t + 100$ for $t \in \mathcal{R}$.
17. $s(t) = t^4 - 18t^2 + 80$ for $t \in \mathcal{R}$.
19. $s(t) = 3t^5 - 40t^3 + 240t - 80$ for $t \in \mathcal{R}$.
21. $s(t) = \frac{t}{t^2+1}$ for $t \in \mathcal{R}$.
23. $s(t) = 4\sec \pi t$ for $-1/2 < t < 1/2$.

24. Find the equations for the position $s(t)$, the velocity $v(t)$ and the acceleration $a(t)$ for a falling object in each of the following cases.
(a) $s(0) = 4$ meters and $v(0) = -7$ meters/second.
(b) $s(1) = 8$ feet and $v(0) = 5$ feet/second.
(c) $s(4) = -5$ meters and $v(1) = 10$ meters/second.
(d) $s(0) = 2$ feet and $s(3) = 16$ feet.
(e) $s(2) = -1$ meters and $s(4) = 6$ meters.   **(f)** $s(4) = s(10) = 8$ feet.

25. A window washer's watch falls off his wrist when he is cleaning a window on the 30th floor of a building. (Assume that the watch was at the center of the 30th floor when it fell.) If each floor is 4 meters high, how long does it take the watch to hit the ground? How fast is it moving at that time?

26. A hunter shoots a bullet straight up at the head of a bird which is flying horizontally 100 meters over his head at a constant speed. If the initial velocity of the bullet is 500 meters/second and the bird is 30 cm long, how fast must the bird fly so that it is not hit by the bullet?

27. A rock is dropped into a pit. If the rock hits the bottom of the pit in 4 seconds, how deep is the pit?

28. A lady leans out a window 30 meters above the ground and tosses a blueberry muffin to her friend on the roof 5 meters above. Although the muffin comes to a stop right in front of the friend, she fails to catch it, and the muffin falls to the ground. How fast was the muffin thrown upwards? How fast is the muffin moving when it hits the ground?

29. A baseball player hits a ball four feet above the ground. The ball goes straight up in the air at 60 miles per hour. The catcher does not know where the ball is. How long does the catcher have to find the ball before it hits the ground? If the catcher does not catch the ball, how fast is the ball moving when it hits the ground?

30. A hawk attacks a dove sitting on an egg. In the ensuing fight the egg falls to the ground. If the egg hits the ground at 96 feet per second, how high is the dove's nest?

31. A soccer ball is kicked directly upwards from the ground. If the ball rises 48 feet in the first second, how high will the ball go? What was the ball's initial velocity?

32. An object is moving along the y–axis under gravity with position $s(t)$ and velocity $v(t)$ at time t. Assume the initial position of the object is at the origin. Show that

(a) $s(t) = \frac{v(0)+v(t)}{2}t$;   **(b)** $v(t)^2 = v(0)^2 - 2gs(t)$.

In each of the next five exercises we are given the cost function $C(q)$ and the revenue function $R(q)$ of a product.
   (i) Find the profit function $P(q)$.
   (ii) Find the marginal cost, marginal revenue and marginal profit.
   (iii) Find the value of q which gives the maximum profit.

33.   The cost and revenue in dollars of manufacturing q tons of sugar per day are $C(q) = 800q$ and $R(q) = q^2 - 1,000q$ for $q \in [0, 1000]$.

34. The cost and revenue in dollars of producing q barrels of oil per day are $C(q) = \frac{q^3}{1,000} + 85q$ and $R(q) = 6q^2 + 100q$ for $q \in [0, \infty)$.

35. The cost and revenue in dollars of producing q litres of cola per day are $C(q) = \frac{q}{10} + 5,000$ and $R(q) = \frac{q}{4} - \frac{q^2}{100,000}$ for $q \in [0, 250000]$.

36.   The cost and revenue in dollars of producing q gallons of vanilla per day are $C(q) = 2q + 40$ and $R(q) = \frac{4q}{1+q^2/70,000}$ for $q \in [0, \infty)$.

37. The cost and revenue in dollars for producing q tons of salt per week are $C(q) = 300q + 1,000$ and $R(q) = 8,000,000 \arctan\left(\frac{q}{10,000}\right) + 100q$ for $q \in [0, \infty)$.

In each of the next five exercises we are given the cost function $C(n)$ and the revenue function $R(n)$ of n units of a product where n is a positive integer.
   (i) Find the profit function $P(n)$.
   (ii) Find the marginal cost, marginal revenue and marginal profit.
   (iii) Find the value of n which gives the maximum profit.

38. The cost and revenue in thousands of dollars for manufacturing n cars per month are $C(n) = 15n + 1,000$ and $R(n) = 20n - n^2/10,000$ for $0 \le n \le 200,000$.

39. The cost and revenue in dollars for manufacturing n coats per week are $C(n) = 50n + 800$ and $R(n) = 90n - 3n^2/1,400$ for $0 \le n \le 42,000$.

40. The cost and revenue in dollars for manufacturing n wedding rings per day are

$C(n) = 63n^2 + 200n$ and $R(n) = n^3 + 755n$ for $0 \le n \le 50$.

41. The cost and revenue in dollars for manufacturing n dozen plates per day are $C(n) = 90\sqrt{n} + 4,000$ and $R(n) = 230n^{3/8}$ for $n \ge 0$.

42. The cost and revenue in dollars for manufacturing n radios per week are $C(n) = 2,000 + 700n$ and $R(n) = \frac{1500n}{1+n/10,000}$ for $n \ge 0$.

43. A product can be manufactured in n units for $n \in (0, \infty)$. Show maximal profit occurs where the graphs of marginal cost and marginal revenue intersect.

Challenging Problems

1. Define the *elasticity* of the differentiable function $y = f(t)$ by:

$$El_t f = \lim_{\Delta t \to 0} \frac{\Delta y / y}{\Delta t / t} = \frac{t}{f(t)} \cdot f'(t) \ .$$

Elasticity measures the relative change of y divided by the relative change in t. Verify the following properties of elasticity.

 (a) $El_t(cf) = El_t f$ (b) $El_t(f+g) = \frac{fEl_t f + gEl_t g}{f+g}$

 (c) $El_t(fg) = El_t f + El_t g$ (d) $El_t(f/g) = El_t f - El_t g$

 (e) If $u = g(t)$ then $El_t(f \circ g) = El_u f \cdot El_t g$.

 (f) If $f(t) = t^n$, then $El_t f = n$ for n a positive integer.

 (g) If $f(t) = t^n$, then $El_t f = n$ for n a negative integer.

 (h) If $f(t) = c$ is a constant function, then $El_t f = 0$.

2. $P(n)$ is the profit of manufacturing n units of a product where n is a positive integer. Give an example of a function $P(n)$ and a differentiable function $p(x)$, having domain $(0, \infty)$, with the following properties.

 (i) $P(n) = p(n)$ for n a positive integer.

 (ii) $p(x)$ has its maximum value at $x = 9.5$.

 (iii) $P(n)$ does not have its maximum value at $n = 9$ nor at $n = 10$.

2.10 Chain Rule Problems

Prerequisite: Section 2.9

In this section we consider two variable which are related by an equation, and the values of these variables vary with time. We know the rate at which one of these variables is changing, and we are asked to find the rate of change of the other variable. These problems are often called *related rates problems*.

Motivating Example 2.10.2 An oil tanker scrapes an iceberg and leaks oil to form a circular shape on the surface of the water. If the radius of this oil spill increases at 2 meters per minute, how fast is its area increasing when its radius is 30 meters?

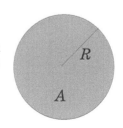

Figure 2.10.1
Oil Spill

Solution We solve this problem in seven steps.

(1) Consider the picture of the oil spill in Figure 2.10.1.

(2) Let R denote the radius of the oil spill measured in meters, and let A denote its area measured in square meters. We also need a variable for time, which does not appear in the diagram. So, let t denote time measured in minutes.

(3)   The rates at which A and R change are their derivatives with respect to time: $\frac{dA}{dt}$ and $\frac{dR}{dt}$. We are given that $\frac{dR}{dt} = +2$ is a constant function.

(4)   We are asked to find the value of $\frac{dA}{dt}$ when $R = 30$.

(5)   We know the area of a circle is given by the formula $A = \pi R^2$.

(6)   Think of A and R as functions of time t, and use implicit differentiation to take the derivative of the equation in (5) with respect to t. We do not have formulas for A and R as functions of t. Nevertheless, we apply the chain rule and use the information from (3):

$$\frac{dA}{dt} = \frac{dA}{dR}\frac{dR}{dt} = (2\pi R)\frac{dR}{dt} = (2\pi R)(2) = 4\pi R \ .$$

(7)   Now we solve the problem. When $R = 30$, the value of $\frac{dA}{dt}$ is given in (6) as $4\pi(30) = 120\pi$ square meters per minute. $\square$

Related rates problems can be quite complicated. The seven steps in the solution of the example above illustrate the general procedure. This procedure organizes the solution into seven small tasks with precise goals. As you read the description of this procedure below, compare each step with the manner it was implemented in the example above.

(1)   Draw a diagram to depict the situation described in the problem. Label all lengths which vary by a variable, and label all lengths which are constant by their value.

(2)   Introduce variables for all other quantities mentioned in the problem, such as time, areas and volumes, which are not in the diagram. Be sure to specify the units of each variable. In particular, all of these units should be compatible.

(3)   Reword *all* the given information in terms of the variables you have defined.

(4)   State the question asked in the problem in terms of these variables and their derivatives. Usually you are asked to find the derivative with respect to time of one of the variables at a given time.

(5)   Use the diagram of Step 1 as an aid to find equations which relate the variables in the problem.

(6)   Take the derivatives of these equations with respect to time using implicit differentiation and the chain rule.

(7)   Evaluate the required rate of change at the specified time.

There is a common point of confusion in solving these problems. Many of the quantities vary with time. The statement of the problem may give information about some of these quantities at a particular time $t = t_0$ and ask for the value of the rate of change of another quantity at $t = t_0$. Quantities which vary with time are not constant. Therefore variables should be introduced to represent their values. Only *after* finding all required derivatives should values be substituted for these variables. In Example 2.10.2, we only substituted $R = 30$ in Step 7. We can not substitute $R = 30$ into the formula $A = \pi R^2$ in Step 5 before taking the derivative of this equation in Step 6.

Examples 2.10.3 (1) An ice cube is melting with the length of its side decreasing at 0.2 centimeters per minute. Find the rate at which the volume of the ice cube is decreasing when the side of the ice cube is 1.5 centimeters long.

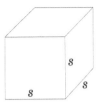

Solution We draw a picture of the ice cube in Figure 2.10.4 and label its side by the variable s which is measured in centimeters. The problem speaks of the volume of the ice cube, so we introduce the variable V to represent its volume in cubic centimeters. Let t denote time measured in minutes. We are given the rate of change of the side of the ice cube:

Figure 2.10.4
Ice Cube

$$\frac{ds}{dt} = -0.2$$

Note that this derivative is negative because s is decreasing. The problem asks:

$$\text{find } \frac{dV}{dt} \text{ when } s = 1.5 \text{ cm.}$$

The variables s and V are related by the equation for the volume of a cube:

$$V = s^3.$$

Use the chain rule to take the derivative of this equation with respect to t:

$$\frac{dV}{dt} = 3s^2\frac{ds}{dt} = 3s^2(-0.2) = -0.6s^2.$$

Thus when $s = 1.5$, the value of $\frac{dV}{dt}$ is $-0.6(1.5)^2 = -1.35 \text{ cm}^3/\text{min}$.

(2) Sand is falling into a conical pile at the rate of 200 cubic inches per minute. The radius of the circular base of this pile is increasing at 0.2 inches per minute. Find the rate of change of the height of this pile when its height is 20 inches and the diameter of its base is 30 inches.

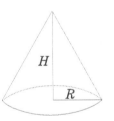

Solution We draw a picture of the pile of sand in Figure 2.10.5 with the radius of the base labelled as R and the height labelled as H. Both R, H are measured in inches. The rate at which sand falls into the pile is the rate of change of the volume of the pile. Introduce the variable V to denote the volume of the pile in cubic inches. Let t denote time measured in minutes. We are given that

Figure 2.10.5
Pile of Sand

$$\frac{dR}{dt} = +0.2 \quad \text{and} \quad \frac{dV}{dt} = +200 \qquad (2.10.1)$$

We are told : $\quad$ find $\dfrac{dH}{dt}$ when $H = 20$ and $2R = 30$, i.e. $R = 15$.

Recall that the volume of a cone is given by:

$$V = \frac{\pi}{3}R^2H.$$

Take the derivative of this equation with respect to t using the product and chain rules:

$$\frac{dV}{dt} = \frac{2\pi}{3}R\frac{dR}{dt}H + \frac{\pi}{3}R^2\frac{dH}{dt}.$$

Substitute the values of $\frac{dV}{dt}$ and $\frac{dR}{dt}$ from (2.10.1):

$$200 = \frac{0.4\pi}{3}RH + \frac{\pi}{3}R^2\frac{dH}{dt}.$$

When $H = 20$ and $R = 15$:

$$200 = \frac{0.4\pi}{3}(15)(20) + \frac{\pi}{3}(15)^2\frac{dH}{dt}$$

$$200 = 40\pi + 75\pi\frac{dH}{dt}$$

$$\frac{dH}{dt} = \frac{200 - 40\pi}{75\pi} \text{ in/min.}$$

Thus, H is increasing at approximately 0.3 in/min when $H = 20$ and $R = 15$.

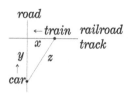

Figure 2.10.6
A Car and a Train

(3) A straight north–south road and a straight east–west railroad track cross at an intersection. A car is traveling north on this road at a constant speed of 80 km/hour while a train is traveling west on this track at 100 km/hour. Find the speed at which the car and train are approaching each other when the car is 750 meters south of the intersection and the train is 1 km east of the intersection.

Solution We draw a picture of the road and the track in Figure 2.10.6 with the car south of the intersection and the train east of the intersection. Draw the line joining the car and the train, and introduce variables x, y, z measured in kilometers. Let t denote time measured in hours. We are given that

$$\frac{dx}{dt} = -100 \text{ and } \frac{dy}{dt} = -80. \tag{2.10.2}$$

Note these derivatives are negative because x and y are decreasing. We are told:

$$\text{find } \frac{dz}{dt} \text{ when } x = 1 \text{ and } y = 0.75.$$

By the Pythagorean Theorem, the variables x, y, z are related by the equation:

$$z^2 = x^2 + y^2. \tag{2.10.3}$$

Differentiate this equation with respect to t by the chain rule and use (2.10.2):

$$2z\frac{dz}{dt} = 2x\frac{dx}{dt} + 2y\frac{dy}{dt} = 2x(-100) + 2y(-80) = -200x - 160y$$

When $x = 1$ and $y = 0.75$, equation (2.10.3) tells us that $z = 1.25$. Hence

$$2(1.25)\frac{dz}{dt} = -200(1) - 160(0.75) = -320, \text{ and } \frac{dz}{dt} = -128.$$

Thus at the specified time the car and train approach each other at 128 km/hr.

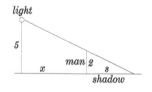

Figure 2.10.7
Shadow of a Man

(4) A man, two meters tall, walks away from a five meter high street lamp at 4 km/hr. Find the rate that the shadow of the man is increasing.

Solution In Figure 2.10.7 we draw a diagram of the ground, the man and the lamp. Draw a line from the light at the top of the lamp through the top of the man to the ground to determine the man's shadow. Label the length of the shadow s and the horizontal distance x between the man and the lamp. Both s and x are measured in meters. Measure time t in seconds. We are given that

$$\frac{dx}{dt} = 4 \text{ km/hr} = \frac{4000}{3600} \text{ m/sec} = \frac{10}{9} \text{ m/sec.} \tag{2.10.4}$$

We must find $\frac{ds}{dt}$. The small and large triangles in our diagram are similar. Hence

$$\frac{s}{s+x} = \frac{2}{5}.$$

Cross $-$ multiplying : $\quad 5s = 2s + 2x \quad$ and $\quad 3s = 2x.$

Differentiate this equation with respect to t and use (2.10.4):

$$3\frac{ds}{dt} = 2\frac{dx}{dt} = 2\left(\frac{10}{9}\right) = \frac{20}{9}.$$

Thus the length of the man's shadow is increasing at $\frac{ds}{dt} = \frac{20}{27}$ m/sec.

(5) A police car stops 20 feet opposite the front door of a house. The light on the police car rotates clockwise at 45 revolutions per minute. Find the speed of the illuminated point on the wall of the house when it is moving away from the front door and passes a point P which is 5 feet from the door.

Solution In Figure 2.10.8 we draw a diagram which includes the police car, the wall of the house and the beam of light. The distance x between P and the door D as well as the hypotenuse h are measured in feet. To interpret the given information about the rate of rotation of the light, introduce the angle θ in the triangle of Figure 2.10.8, measured in radians. Let t denote time measured in minutes. We are given that

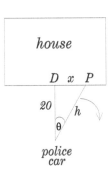

Figure 2.10.8
Light on a Police Car

$$\frac{d\theta}{dt} = 45 \text{ rev/min} = 45(2\pi)\text{rad/min} = 90\pi \text{ rad/min} \qquad (2.10.5)$$

We are told : find $\dfrac{dx}{dt}$ when $x = 5$.

From the triangle in Figure 2.10.8, we see that

$$\tan\theta = \frac{x}{20}.$$

Use the chain rule to differentiate this equation with respect to t:

$$\begin{aligned}
\sec^2\theta\frac{d\theta}{dt} &= \frac{1}{20}\frac{dx}{dt} \\
(\sec^2\theta)(90\pi) &= \frac{1}{20}\frac{dx}{dt} \qquad \text{[by (2.10.5)]} \\
\frac{dx}{dt} &= 1800\pi\sec^2\theta
\end{aligned}$$

When $x = 5$, the length of the hypotenuse h in Figure 2.10.8 is $5\sqrt{17}$ by the Pythagorean Theorem. Then $\sec\theta = \frac{5\sqrt{17}}{20} = \frac{\sqrt{17}}{4}$, and

$$\frac{dx}{dt} = 1800\pi\left(\frac{\sqrt{17}}{4}\right)^2 = \frac{3825\,\pi}{2} \text{ ft/min.}$$

(6) A fisherman is standing in a boat which is moving north at 9.9 km/hr. A salmon, 15 meters northwest of the boat, bites his hook and swims northwest at 4.5 km/hour. If the fisherman holds his rod at a height of 2 meters, how fast is he letting out his line 10 seconds after the salmon is hooked?

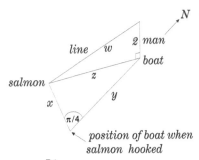

Figure 2.10.9 Salmon on a Line

Solution We draw a three–dimensional diagram in Figure 2.10.9 including the man and the fishing line as well as the paths of the boat and the salmon. The distances x, y, z, w in the diagram are measured in meters. The time t is measured in seconds from when the salmon is hooked. We are given that

$$x(0) = 15, \qquad y(0) = 0, \qquad\qquad (2.10$$

$$\frac{dx}{dt} = 4.5 \text{ km/hr} = \frac{4500}{3600} \text{ m/sec} = 1.25 \text{ m/sec}, \qquad (2.10$$

$$\frac{dy}{dt} = 9.9 \text{ km/hr} = \frac{9900}{3600} \text{ m/sec} = 2.75 \text{ m/sec}. \qquad (2.10$$

We are told :   find $\dfrac{dw}{dt}$ when $t = 10$.

Apply the Law of Cosines to the triangle at water level in Figure 2.10.9:

$$z^2 = x^2 + y^2 - 2xy \cos \frac{\pi}{4} . \qquad (2.10.9)$$

We solve the differential equations (2.10.7) and (2.10.8) using Corollary 2.6.24:

$$x = 1.25t + A \text{ and } y = 2.75t + B.$$

By (2.10.6), $A = 15$ and $B = 0$. Thus $x = 1.25t + 15$ and $y = 2.75t$. Substitute these values of x, y into equation (2.10.9):

$$z^2 = (1.25t + 15)^2 + (2.75t)^2 - 2(1.25t + 15)(2.75t)\left(\frac{\sqrt{2}}{2}\right)$$

$$z^2 = (1.25t + 15)^2 + 7.5625t^2 - \sqrt{2}(3.4375t^2 + 41.25t) \qquad (2.10.10)$$

Apply the Pythagorean Theorem to the vertical right triangle in Figure 2.10.9:

$$w^2 = z^2 + 2^2$$

By (2.10.10):

$$w^2 = (1.25t + 15)^2 + 7.5625t^2 - \sqrt{2}(3.4375t^2 + 41.25t) + 4 \qquad (2.10.11)$$

Use implicit differentiation to differentiate this equation with respect to t:

$$2w\frac{dw}{dt} = 2.5(1.25t + 15) + 15.125t - \sqrt{2}(6.875t + 41.25). \qquad (2.10.12)$$

We want to substitute $t = 10$ into this equation. However, we first need to find the value of w from (2.10.11) when $t = 10$:

$$w^2 = (12.5 + 15)^2 + 756.25 - \sqrt{2}(343.75 + 412.5) + 4 \approx 447$$

and $w \approx 21.14$. Now substitute $t = 10$ into equation (2.10.12):

$$2(21.14)\frac{dw}{dt} \approx 2.5(12.5 + 15) + 151.25 - \sqrt{2}(68.75 + 41.25) \approx 64.44 \quad \text{and}$$

$$\frac{dw}{dt} \approx 1.5 \text{ m/sec}.$$

Thus, if the salmon can keep swimming at 4.5 km/hr for 10 seconds, the line will probably break, and the salmon will escape!   □

Summary

The reader should be able to solve a related rates problem. With the aid of a diagram, she should introduce variables to interpret the given information and the required information in terms of these variables and their derivatives. Then she should differentiate the appropriate equation using implicit differentiation and the chain rule. By substituting the given values into the resulting equation, she should compute the required rate of change.

Basic Exercises

Use a calculator to approximate the solutions of these exercises.

1. In each case below an object is moving along a curve $y = f(x)$ in the coordinate plane. Distances are measured in meters, and time is measured in seconds.
(a) An object is moving along the curve $y = x^2$ with $\frac{dx}{dt} = 4$ m/sec. Find the rate of change of the distance between the object and the origin when the object is at $(3, 9)$.
(b) An object is moving along the curve $y = \sqrt{5x^2 + 4}$ with $\frac{dy}{dt} = 7t^3$ m/sec. When $t = 3$ seconds, the object is at the point $(1, 3)$. Find the rate of change of the distance between the object and the point $(-2, 1)$ at that time.
(c) An object is moving around the circle $x^2 + y^2 = 169$ with $\frac{dx}{dt} = \sin 4t$ m/sec. When $t = \frac{\pi}{24}$ seconds, the object is at the point $(12, 5)$. Find the rate of change of the distance between the object and the point $(2, 1)$ at that time.
(d) An object is moving around the ellipse $\frac{x^2}{8} + \frac{y^2}{32} = 1$ with $\frac{dy}{dt} = t^4 - 3$ m/sec. When $t = 2$ seconds, the object is at the point $(2, -4)$. Find the rate of change of the distance between the object and the origin at that time.

2. A pipe is leaking water which is making a circular puddle on the floor. The puddle has constant depth 0.6 cm and a radius which increases at 2 cm per minute. Find the rate of the leak when the puddle has a radius of 1.5 meters.

3. A machine blows up a spherical balloon by pumping air into the balloon at 20 cubic centimeters per second. Find the rate of increase of the radius of the balloon when the balloon has a diameter of 30 centimeters.

4. City planners have decided to develop their city in the shape of a square which increases at three square miles per year. Find the rate of increase of the dimensions of the city when its area is 16 square miles.

5. A spider is constructing a circular web at 1.5 square cm per minute. Find the rate of increase of the radius of the web when the radius is 5 cm.

6. A grandmother is crocheting a rectangular Afghan for her grandchild. The length of the Afghan is increasing at 6 cm/hour, and its width is increasing at 5 cm/hour. Find the rate of change of the area of the Afghan when it measures 18 cm by 15 cm.

7. A free standing rectangular yeast cake is rising while it bakes in such a way that its length is increasing at 0.4 cm/minute, its width is increasing at 0.2 cm/minute and its height is increasing at 0.1 cm/minute. Find the rate of increase of the volume of the cake when its dimensions are 12 cm by 8 cm by 5 cm.

8. A flexible cylindrical container of gas is expanding on a hot day. Its radius is increasing at 5 cm/hr, and its length is increasing at 3 cm/hr. Find the rate of increase of the volume of the tank when its radius is 2 meters and its length is 4 meters.

9. Flour is flowing out of a conical funnel at 0.7 cubic centimeters per second. The height of the funnel is twice its radius. Find the rate at which the depth of flour in the funnel is decreasing when its depth is two centimeters.

10. A hemispherical sink with radius r inches is filling with water at 2 cubic inches per second. When the water is h inches deep, the volume of water is given by:

$$V = \pi r h^2 - \frac{1}{3}\pi h^3.$$

If $r = 8$ inches, how fast is the water rising when the sink becomes full?

11. In a baseball game, there is a runner on third base when the batter hits a ground ball which is fielded by the first baseman on the first base bag. The runner runs towards homeplate at 20 miles per hr. The first baseman throws the ball to the catcher at homeplate at 95 miles per hr. How fast is the ball approaching the runner when the runner is 10 feet from home plate and the ball is 25 feet from home plate?

12. One plane is flying south from Toronto at 900 kilometers per hour while a second plane at the same altitude is flying south–west from Toronto at 800 kilometers per hour. How fast are the two planes separating when the first plane is 400 kilometers from Toronto and the second plane is 700 kilometers from Toronto?

13. A boat traveling west at 20 kilometers per hour passes under a car on a 50 meter high bridge. The car is traveling north at 80 kilometers per hour. How fast is the distance between the car and the boat increasing when they are one kilometer apart?

14. A plane flies over a car at an altitude of 2,000 meters. The plane is flying north–east at 600 kilometers per hour, while the car is driving north at 100 kilometers per hour. How fast are the car and plane separating 20 minutes later?

15. A boy holds a spool of string which is attached to a kite. The kite is moving horizontally at 3 km/hour away from the boy at a constant altitude of 80 meters. How fast does the boy have to let out the string when the kite is 150 meters away? How fast is the angle of elevation of the kite decreasing at that time?

16. One night, a toddler is dropping eggs from a window of a building which is 20 feet above the ground. A neighbor, hearing the commotion, looks through her window which is 15 feet from the toddler's window, on the same wall of the building and at the same height. She shines a flashlight from her window on one of the falling eggs. How fast is the shadow of this egg moving on the ground when the egg is half–way to the ground?

17. A 4 m long ladder is leaning against a building. The bottom of the ladder is dragged along the ground, away from the building, at 3 m per min. How fast is the top of ladder moving down the side of the building when it is 1 m above the ground?

18. A 3 meter long trough of depth 60 cm is being filled with water at 20 liters per minute. In each of the following cases determine how fast the depth of the water is rising when the depth of water is 30 cm. (A liter is 1000 cubic cm.)
(a) The vertical cross–section of the trough is a square.
(b) The vertical cross–section of the trough is an equilateral triangle.
(c) The vertical cross–section of the trough is a semi–circle.

19. A car is going around an oval race track. Place the x and y axes so that the origin is located at the center of this oval. The coordinates of the car are given by:

$$x = 100\cos t \quad \text{and} \quad y = 180\sin t$$

where x, y are measured in miles and t is measured in hours. How fast is the distance from the car to the origin changing when the car is at the point $(50\sqrt{3},\ 90)$?

20. A clock has a minute hand of length 6 cm and an hour hand of length 4 cm. How fast is the end of the minute hand approaching the end of hour hand at 5:10?

21. A merry–go–round of radius 8 meters turns clockwise at 2 revolutions per minute. A man is standing on the ground at the north edge of the merry–go–round. His daughter is riding a horse on the merry–go–round 5 meters from its center. How fast is the girl approaching her father when she is west of the center of the merry–go–round?

22. A cylindrical lump of clay is being rolled out so that its radius R is decreasing at $\frac{1}{R}$ inches per minute. How fast is the length of this cylinder increasing when its radius is 2 inches and its length is 5 inches?

23. A man is in a bus which is traveling east at 60 kilometers per hour. A bird flying north–east at an altitude of 180 meters and a speed of 10 kilometers per hour passes over the bus. As the man watches this bird, how fast is its angle of elevation decreasing 15 seconds after the bird passes over the bus?

24. A lighthouse is located 3 km north of the nearest point P on a straight east–west shoreline. The light in the lighthouse rotates clockwise at 4 revolutions per minute. Find the speed of the illuminated point Q on the shoreline when Q is 1 km east of P.

2.11   Tangent Line Approximations

Prerequisite: Section 2.6

Let $y = T(x)$ be the tangent line to the graph of $y = f(x)$ at $x = x_0$. Then $T(x)$ approximates $f(x)$ for x near x_0 because the graphs of $f(x)$ and $T(x)$ have the same value and direction at $x = x_0$, i.e. $T(x_0) = f(x_0)$ and $T'(x_0) = f'(x_0)$. We use this observation to make two applications. In the first subsection, we use differentials to approximate $f(x)$, for x near x_0, by $T(x)$ as in Figure 2.11.1. In the second subsection, we use the Newton–Raphson method to approximate a root r of $y = f(x)$ near x_0 by the root s of $y = T(x)$ as in Figure 2.11.2. The formulas for bounds on the errors of these approximations are derived in the appendix of this section.

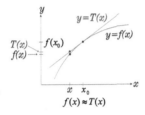

Figure 2.11.1
Approximating $f(x)$

Differentials

Differentials approximate the change in the curve $y = f(x)$ by the change in the tangent line $y = T(x)$ at $x = x_0$. This estimates the value of $f(x)$ by the value of $T(x)$. A formula is also given to bound the error of this approximation. This subsection concludes with an estimate of the percent of error in the computed value of $y = f(x)$ when we know the percent of error in the value of x we are using.

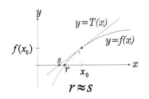

Figure 2.11.2
Approximating
the Root r

We begin by introducing notation for changes in f, denoted by the symbol Δ, and changes in the tangent line T, denoted by the symbol d.

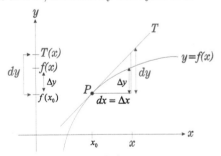

Figure 2.11.3 The Differentials dx and dy

Definition 2.11.4 *Let f be a continuous function whose domain contains the open interval I. Assume f is differentiable at $x_0 \in I$ with T the tangent line to the graph of f at the point $P = (x_0, f(x_0))$. For $x \in I$, define the differentials dx and dy by:*

$$\Delta x = dx = x - x_0, \quad \Delta y = f(x) - f(x_0), \quad dy = T(x) - T(x_0).$$

The interpretations of Δx, dx, Δy, dy in terms of the graphs of $y = f(x)$ and $y = T(x)$ are depicted in Figure 2.11.3. Next, we calculate the value of dy.

Proposition 2.11.5 *Let f be a continuous function whose domain contains the open interval I. Assume f is differentiable at $x_0 \in I$. Then*

$$dy = f'(x_0) \cdot dx. \qquad (2.11.1)$$

Proof By (2.2.6), the tangent line T to the graph of $y = f(x)$ at $x = x_0$ has equation:

$$T(x) = f(x_0) + f'(x_0)(x - x_0) = T(x_0) + f'(x_0)(x - x_0).$$

Hence $\qquad\qquad dy = T(x) - T(x_0) = f'(x_0) \cdot (x - x_0) = f'(x_0) \cdot dx.$ $\qquad$ □

Before using $T(x)$ to approximate $f(x)$, we illustrate the notation we have introduced.

Examples 2.11.6 For each function, find $T(x)$, dx, dy and Δy at x_0 for the given change Δx of x.

(1) Let $f(x) = 3x^2 - 5$ with $x_0 = 2$ and $\Delta x = +1$.

Solution Observe that $f'(x) = 6x$. Then

$$\begin{aligned}
T(x) &= f(x_0) + f'(x_0)(x - x_0) = f(2) + f'(2)(x - 2) \\
&= 7 + 12(x - 2) = 12x - 17, \\
dx &= \Delta x = +1, \\
\Delta y &= f(x_0 + \Delta x) - f(x_0) = f(3) - f(2) = 22 - 7 = 15, \\
dy &= f'(2) \cdot dx = 12 \cdot 1 = 12.
\end{aligned}$$

(2) Let $f(x) = 4x^{5/3} - 6x^{2/3} + 1$ with $x_0 = 8$ and $\Delta x = -1$.

Solution Observe that $f'(x) = \frac{20}{3}x^{2/3} - 4x^{-1/3}$. Then

$$\begin{aligned}
T(x) &= f(x_0) + f'(x_0)(x - x_0) = 105 + \frac{74}{3}(x - 8) = \frac{74}{3}x - \frac{277}{3}, \\
dx &= \Delta x = -1, \\
\Delta y &= f(x_0 + \Delta x) - f(x_0) = f(7) - f(8) \\
&= (4 \cdot 7^{5/3} - 6 \cdot 7^{2/3} + 1) - (128 - 24 + 1) \\
&= 4 \cdot 7^{5/3} - 6 \cdot 7^{2/3} - 104 \approx -23.5, \\
dy &= f'(8) \cdot dx = \frac{74}{3} \cdot (-1) = -24\frac{2}{3}.
\end{aligned}$$

(3) Let $f(x) = \tan x$ with $x_0 = \frac{\pi}{4}$ and $\Delta x = \frac{\pi}{32}$.

Solution Observe that $f'(x) = \sec^2 x$. Then

$$\begin{aligned}
T(x) &= f(x_0) + f'(x_0)(x - x_0) = 1 + 2\left(x - \frac{\pi}{4}\right) = 2x - \frac{\pi - 2}{2}, \\
dx &= \Delta x = \frac{\pi}{32}, \\
\Delta y &= f(x_0 + \Delta x) - f(x_0) = f\left(\frac{9\pi}{32}\right) - f\left(\frac{\pi}{4}\right) = \tan \frac{9\pi}{32} - 1 \approx 0.22, \\
dy &= f'\left(\frac{\pi}{4}\right) \cdot dx = 2 \cdot \frac{\pi}{32} = \frac{\pi}{16} \approx 0.20
\end{aligned}$$

$\qquad$ □

Recall the Leibniz notation $\frac{dy}{dx}$ for the derivative of $y = f(x)$ is not defined as a quotient of two quantities dy and dx. The notation $\frac{dy}{dx}$ is just one symbol which represents the derivative. Now, however, we have defined symbols dy and dx. Since dy is defined as $f'(x) \cdot dx$, the quotient of these two symbols is:

$$dy \div dx = [f'(x) \cdot dx] \div dx = f'(x) = \frac{dy}{dx}.$$

Thus, our definition of the symbols dy and dx allows us to view the derivative of f as the quotient dy divided by dx. This interprets the Leibniz notation for the derivative as a quotient. However, this quotient can not be used as the definition of the derivative. That would be circular reasoning because the symbol dy is defined in terms of the derivative. The formulas for computing derivatives translate into formulas for differentials. We summarize these formulas in the following proposition whose proof we relegate to the exercises. When there is more than one function under consideration, the symbol dy for the differential of the function $y = f(x)$ is ambiguous. Therefore, we denote dy as df so that it is clear that the differential is referring to f.

Proposition 2.11.7 *Let f, g be differentiable functions with domain the open interval I. Then*

(a) $d(f + g) = df + dg$; 　　　　(b) $d(cf) = cdf$, 　*for $c \in \Re$;*

(c) $d(f \cdot g) = g \cdot df + f \cdot dg$; 　　(d) $d\left(\frac{f}{g}\right) = \frac{g \cdot df - f \cdot dg}{g^2}$;

(e) $d(f \circ g) = f'(g(x)) \cdot dg$. 　　　　　　　　　　　　　　$\square$

Our main interest in differentials is to make approximations. Say we want to approximate the value of $f(x)$. Choose the closest point x_0 to x for which we know the values of $f(x_0)$ and $f'(x_0)$. Let T be the tangent line to the graph of $y = f(x)$ at the point $(x_0, f(x_0))$. Then the value of $T(x)$ approximates $f(x)$:

$$f(x) = f(x_0) + \Delta y \approx T(x) = f(x_0) + dy.$$

The closer x is to x_0, the better dy approximates Δy and the better $T(x)$ approximates $f(x)$. See Figure 2.11.3.

Examples 2.11.8 (1) Approximate $\sqrt{37}$.

Solution Let $f(x) = \sqrt{x}$. Since we know that $f(36) = \sqrt{36} = 6$, let

$$x_0 = 36 \quad \text{and} \quad \Delta x = dx = +1.$$

Then $f'(x) = \frac{1}{2\sqrt{x}}$ and $f'(x_0) = f'(36) = \frac{1}{12}$. Hence

$$dy = f'(36) \cdot dx = \frac{1}{12}(1) = \frac{1}{12}.$$

Therefore, $\sqrt{37} = f(37) = f(36) + \Delta y \approx f(36) + dy = 6 + \frac{1}{12} \approx 6.0833$

(2) Approximate $\sin 29°$.

Solution We know $30° = \frac{\pi}{6}$ radians, and $\sin \frac{\pi}{6} = \frac{1}{2}$. Thus, let $g(x) = \sin x$, $x_0 = \frac{\pi}{6}$ and $\Delta x = dx = -1° = -\frac{\pi}{180}$ radians.[4] Then $g'(x) = \cos x$ and

$$dy = g'\left(\frac{\pi}{6}\right) dx = \left(\cos\frac{\pi}{6}\right)\left(-\frac{\pi}{180}\right) = \left(\frac{\sqrt{3}}{2}\right)\left(-\frac{\pi}{180}\right) = -\frac{\pi\sqrt{3}}{360}.$$

Hence 　$\sin 29° = g(29°) = g\left(\frac{\pi}{6}\right) + \Delta y \approx g\left(\frac{\pi}{6}\right) + dy = \frac{1}{2} - \frac{\pi\sqrt{3}}{360} \approx 0.484885.$

---

[4]We must measure angles in radians, not degrees, for our calculus formulas to be valid.

(3)   The inside of a wooden box has the shape of a cube whose side is one meter long. The walls of the box are one centimeter thick. Estimate how much gold foil will be needed to cover the outside of this box.

Solution   Let S denote the surface area of a cube which has six faces, each a square with side of length x centimeters. Then $S = 6x^2$ and $S'(x) = 12x$. Thus

$$dS = S'(x) \cdot dx = 12x \, dx.$$

When $x = 1$ m $= 100$ cm and $dx = +1$, we have $dS = 12(100)(1) = 1,200$. Thus

$$S(101) = S(100) + \Delta S \approx S(100) + dS = 6(100^2) + 1,200 = 61,200.$$

Therefore approximately $61,200$ cm^2 $= 6.12$ m^2 of gold foil will be needed to cover the outside of this box.   □

When making an approximation, such as $f(x) \approx T(x)$, it is important to have a criterion to determine the accuracy of the approximation. Since $f(x) = f(x_0) + \Delta y$ and $T(x) = f(x_0) + dy$, the exact error of our approximation is $|f(x) - T(x)| = |\Delta y - dy|$. The following theorem gives an upper bound B of the error, i.e. a number B with

$$|f(x) - T(x)| \leq B.$$

The proof of this theorem is presented in the appendix of this section.

Theorem 2.11.9 *Consider a function f whose domain contains the open interval I such that f'' exists and is continuous on I. Let $x_0 \in I$ with T the tangent line to the graph of f at the point $(x_0, f(x_0))$. For $x \in I$, choose M greater than or equal to the maximum value of $|f''|$ on the closed interval with endpoints x_0 and x. Then*

$$|f(x) - T(x)| = |\Delta y - dy| \leq \frac{M(x - x_0)^2}{2}. \qquad (2.11.2)$$

We apply criterion (2.11.2) to determine the accuracy of the approximations we made in Examples 2.11.8 above.

Examples 2.11.10 (1)   In Example 2.11.8 (1),

$$f(x) = \sqrt{x}, \quad f'(x) = \frac{x^{-1/2}}{2} \quad \text{and} \quad f''(x) = -\frac{x^{-3/2}}{4}.$$

We let $x_0 = 36$, $x = 37$, $dx = 1$ and calculated $\sqrt{37} = f(37) \approx 6.0833$. For $x > 0$, the function $\frac{1}{4}x^{-3/2}$ has negative derivative $-\frac{3}{8x^{5/2}}$. Hence $\frac{1}{4}x^{-3/2}$ is a decreasing function. Thus on the interval $[x_0, x] = [36, 37]$:

$$|f''(x)| = \frac{1}{4x^{3/2}} \leq \frac{1}{4(36^{3/2})} = \frac{1}{864} < .0012.$$

Thus take the upper bound M of the second derivative to be .0012. By Theorem 2.11.9, the error in our approximation 6.0833 for $\sqrt{37}$ has error less than $.0012(37-36)^2/2 = .0006$. Thus, $6.0827 \leq \sqrt{37} \leq 6.0839$. In fact, $\sqrt{37} = 6.0828$ to four decimal places.

(2)   In Example 2.11.8 (2),

$$g(x) = \sin x, \quad g'(x) = \cos x \quad \text{and} \quad g''(x) = -\sin x.$$

We let $x_0 = 30° = \frac{\pi}{6}$ radians, $x = 29° = \frac{29\pi}{180}$ radians, $dx = -1° = -\frac{\pi}{180}$ radians and calculated $\sin 29° = g(29°) \approx 0.484885$. Note $|g''(x)| = |\sin x| \le 1$ for all x, so we take the upper bound M of the second derivative to be 1. A more careful analysis, however, shows that on the interval $[29°, 30°] = \left[\frac{29\pi}{180}, \frac{\pi}{6}\right]$ we can take $M = \sin \frac{\pi}{6} = 0.5$, because $\sin x$ is an increasing function on this interval. By Theorem 2.11.9, the error in our approximation 0.484885 for $\sin 29°$ is less than or equal to

$$\frac{(.5)\left(\frac{29\pi}{180} - \frac{\pi}{6}\right)^2}{2} = \frac{\left(\frac{\pi}{180}\right)^2}{4} = \frac{\pi^2}{129,600} < 0.000076.$$

Thus $.484809 < \sin 29° < .484961$. Note $\sin 29° = .4848096$ to 7 decimal places.

(3) In Example 2.11.8 (3),

$$S(x) = 6x^2, \quad S'(x) = 12x \quad \text{and} \quad S''(x) = 12.$$

We let $x_0 = 100$, $x = 101$, $dx = 1$ and calculated $S(101) \approx 61,200$. Take the upper bound M of $S''(x)$ to be 12. By Theorem 2.11.9 the error in approximating the area $S(101)$ of the outside of the box as $61,200$ cm^2 is less than or equal to

$$\frac{(12)(101 - 100)^2}{2} = 6 \text{ cm}^2.$$

Thus $61,194 \le S(101) \le 61,206$. In fact $S(101) = 61,206$ cm^2. $\square$

Let $y = f(x)$ be a differentiable function. Say we have an estimate of the value of x with error at most Δx. The percent of error $E(x)$ of this estimate is at most:

$$E(x) = 100\frac{\Delta x}{x} = 100\frac{dx}{x}. \tag{2.11.3}$$

We compute the estimate of $y = f(x)$ from this estimate of x. The percent of error $E(y)$ of this estimate of y is at most $E(y) = 100\frac{\Delta y}{y}$. By approximating Δy by the differential dy we obtain an approximation of $E(y)$:

$$E(y) = 100\frac{\Delta y}{y} \approx 100\frac{dy}{y} = 100\frac{f'(x)dx}{y} = \frac{xf'(x)}{y} \cdot 100\frac{dx}{x} = \frac{xf'(x)}{f(x)}E(x). \tag{2.11.4}$$

In performing an experiment we measure a quantity x and know the percent of error $E(x)$ of this measurement. Then we perform a computation to determine the value of the variable y. Formula (2.11.4) gives the percent of error $E(y)$ in the value of y we have computed. We illustrate this procedure.

Examples 2.11.11 (1) Approximate the percent of error in computing the volume of a cube from the measurement of its side made with an accuracy of 2%.

Solution The volume V of a cube with side s is given by $V = s^3$. Then $\frac{dV}{ds} = 3s^2$. By (2.11.4):

$$E(V) \approx \frac{s \cdot 3s^2}{s^3}E(s) = 3E(s).$$

Thus, a 2% error in the measurement of s results in an error of 6% in the computed value of V.

(2)   Greasy Motors finds that the profit in dollars of selling n cars is given by

$$P = \frac{1000n^2}{n + 500}.$$

They predict that between $4,800$ and $5,200$ cars will be sold. Therefore they use the value $n = 5,000$ in the above formula to compute their profit. Find the percent of error in this computation of P.

Solution   By the quotient rule:

$$\frac{dP}{dn} = \frac{2,000n(n + 500) - (1,000n^2)(1)}{(n + 500)^2} = \frac{1,000n^2 + 1,000,000n}{(n + 500)^2}$$

By (2.11.4):

$$E(P) \approx \left(n \cdot \frac{1,000n^2 + 1,000,000n}{(n + 500)^2} \div \frac{1,000n^2}{n + 500} \right) E(n) = \frac{n + 1,000}{n + 500} E(n).$$

Take $n = 5,000$. Since the number of cars sold varies between $4,800 = 5,000 - 200$ and $5,200 = 5,000 + 200$, take $dn = 200$. By (2.11.3), $E(n) = 100\frac{200}{5000} = 4\%$. By the equation above:

$$E(P) \approx \frac{5,000 + 1,000}{5,000 + 500}(4\%) = \frac{12}{11}(4\%) \approx 4.4\%.$$

Newton–Raphson Method

We exposit the Newton–Raphson Method, a recursive procedure for computing a sequence of increasingly better approximations of a root of $y = f(x)$. We also present a formula to estimate the errors of these approximations.

Consider a differentiable function f whose graph crosses the x–axis at $x = r$, i.e. $f(r) = 0$. We call r a *root* of f. That is, a root is the same as an x–intercept. Suppose we can not compute the exact value of r, but we know an approximation x_0 of r. Let T_{x_0} denote the tangent line to the graph of f at the point $P = (x_0, f(x_0))$. Since r is near x_0, the root r of f, where the graph of f crosses the x–axis, is approximated by the root x_1 of T_{x_0}, where the tangent line T_{x_0} crosses the x–axis. See the left diagram in Figure 2.11.12. Recall from (2.2.6) that T_{x_0} has equation:

$$y = T_{x_0}(x) = f(x_0) + f'(x_0)(x - x_0).$$

Assume T_{x_0} is not horizontal, i.e. $f'(x_0) \neq 0$. Then T_{x_0} has a root x_1 which is a solution of $T_{x_0}(x_1) = 0$:

$$
\begin{aligned}
T_{x_0}(x_1) &= f(x_0) + f'(x_0)(x_1 - x_0) = 0 \\
f'(x_0)(x_1 - x_0) &= -f(x_0) \\
x_1 - x_0 &= -\frac{f(x_0)}{f'(x_0)} \\
x_1 &= x_0 - \frac{f(x_0)}{f'(x_0)}
\end{aligned}
$$

Usually x_1 is a better approximation of r than x_0. Then the root $x_2 = x_1 - \frac{f(x_1)}{f'(x_1)}$ of T_{x_1} is an even better approximation of r. See the right diagram in Figure 2.11.12. Iteration of this procedure produces a sequence of increasingly better approximations $x_0, x_1, x_2, \ldots, x_n, \ldots$ of r.

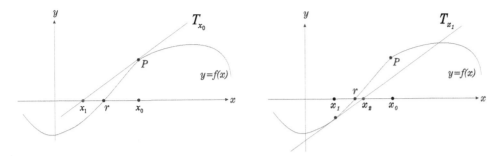

Figure 2.11.12 Newton–Raphson Method for Approximating r

Definition 2.11.13 *Let f be a differentiable function with domain the open interval I. Assume f has a root at $r \in I$. If $x_0 \in I$ is an approximation of r, then the Newton–Raphson approximations of r are defined recursively for $n \geq 0$ by:*

$$x_{n+1} = x_n - \frac{f(x_n)}{f'(x_n)} . \tag{2.11.5}$$

The use of equation (2.11.5) to produce estimates of the root r of f is called the *Newton–Raphson method*. We illustrate this method with several examples.

Examples 2.11.14 (1) Start with the approximation $x_0 = 7$ for $\sqrt{48}$, and apply the Newton–Raphson method twice to obtain better approximations of $\sqrt{48}$.

Solution Let $f(x) = x^2 - 48$. Then $f(\sqrt{48}) = 0$. Observe that $f'(x) = 2x$. The Newton–Raphson method produces a better approximation x_1 of $\sqrt{48}$:

$$x_1 = x_0 - \frac{f(x_0)}{f'(x_0)} = 7 - \frac{1}{14} = 6\frac{13}{14} \approx 6.928571429.$$

Apply the Newton–Raphson method again to obtain an even better approximation x_2 of $\sqrt{48}$:

$$x_2 = x_1 - \frac{f(x_1)}{f'(x_1)} \approx 6.928571429 - \frac{.00510205}{13.857142858} \approx 6.928203243.$$

(2) Iterate the Newton–Raphson method twice to estimate the real root of the polynomial $f(x) = x^3 - x^2 + 1$.

Solution Since $f(-1) = -1 < 0$ and $f(0) = 1 > 0$, the polynomial f has a root r between -1 and 0 by the Intermediate Value Theorem. Since $f'(x) = 3x^2 - 2x$ and $f'(0) = 0$, the tangent line T_0 is horizontal with no roots, and we cannot take $x_0 = 0$. Thus we take $x_0 = -1$. The Newton–Raphson method gives a better approximation x_1 of r:

$$x_1 = x_0 - \frac{f(x_0)}{f'(x_0)} = -1 - \frac{-1}{5} = -0.8$$

Apply the Newton–Raphson method a second time to obtain an even better approximation x_2 of r:

$$x_2 = x_1 - \frac{f(x_1)}{f'(x_1)} = -0.8 - \frac{-0.152}{3.52} \approx -0.7568$$

(3) We can use the Newton–Raphson method to approximate a *fixed point p* of a function f, i.e. a number p such that $f(p) = p$. We do this by defining

$h(x) = f(x) - x$. Then p is a fixed point of f if and only if p is a root of h. For example, iterate the Newton–Raphson method twice to approximate a fixed point of $f(x) = x^3 - x - 5$.

Solution Since $f(2) = 1$, we use $x_0 = 2$ as an estimate of a fixed point of f as well as an estimate of a root of

$$h(x) = f(x) - x = (x^3 - x - 5) - x = x^3 - 2x - 5.$$

Apply the Newton–Raphson method. Since $h'(x) = 3x^2 - 2$,

$$x_1 = x_0 - \frac{h(x_0)}{h'(x_0)} = 2 - \frac{-1}{10} = 2.1 \text{ and}$$

$$x_2 = x_1 - \frac{h(x_1)}{h'(x_1)} = 2.1 - \frac{0.061}{11.23} = 2.09457.\qquad\square$$

To evaluate the accuracy of the approximation x_n of r we estimate the error

$$E_n = x_n - r$$

of this approximation. The estimate of the error E_n given in the following proposition can be used to decide how many iterations of the Newton–Raphson method are required to compute an x_n which approximates r to the desired degree of accuracy. The proof of this theorem is presented in the appendix of this section.

Theorem 2.11.15 *Let f be a function, with domain the open interval I, whose third derivative exists and is continuous. Assume f has a root $r \in I$ with $f'(r) \neq 0$ and $f''(r) \neq 0$. Define*

$$A = \frac{f''(r)}{f'(r)}.$$

Let $x_0 \in I$ be an approximation of r. For $n \geq 1$, let x_n denote the approximation of r obtained from x_0 by iterating the Newton–Raphson method n times. Then the error $E_n = x_n - r$ of this approximation satisfies:

$$E_n \approx \frac{2}{A}\left(\frac{AE_0}{2}\right)^{2^n}.$$

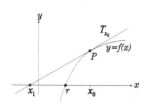

Figure 2.11.16
Bad Choice of x_0

The Newton–Raphson method does not always work. If x_0 is a bad approximation of r, as in Figure 2.11.16, do not expect that x_1 will be a better approximation of r. A more subtle problem occurs when $f'(x_0)$ is close to zero, as in Figure 2.11.17. The tangent line T_{x_0} at x_0 is almost horizontal, and x_1 is far away from x_0. In this case, even though x_0 is a good approximation of r, x_1 is a terrible approximation of r. However, Theorem 2.11.15 tells us that if $\left|\frac{AE_0}{2}\right| < 1$, then the x_n are increasingly better approximations of r.

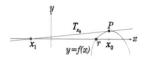

Figure 2.11.17 Near Horizontal Tangent

Corollary 2.11.18 *In the notation of Theorem 2.11.15, assume $\frac{AE_0}{2} < 1$. Then*

$$|E_{n+1}| < |E_n| \quad and \quad \lim_{n\to\infty} E_n = 0.\qquad\square$$

We illustrate the use of Theorem 2.11.15 to determine the accuracy of Newton–Raphson approximations. Since we do not know the value of r, we do not know the value of the number $A = \frac{f''(r)}{f'(r)}$ which is needed to apply Theorem 2.11.15. Nevertheless, we can analyze A as a function of r to obtain an upper bound of A.

Examples 2.11.19 Use the criterion of Theorem 2.11.15 to approximate the error in the Newton–Raphson approximations that we made in Examples 2.11.14 above.

(1) In Example 2.11.14 (1) we computed the two approximations $x_1 = 6.928571429$ and $x_2 = 6.928203243$ of $r = \sqrt{48}$ using:

$$f(x) \;=\; x^2 - 48, \; x_0 \;=\; 7 \; \text{with} \; f'(x) = 2x, \; f''(x) \;=\; 2.$$
$$\text{Hence} \qquad A \;=\; \frac{f''(r)}{f'(r)} = \frac{f''(\sqrt{48})}{f'(\sqrt{48})} = \frac{2}{2\sqrt{48}} = \frac{1}{4\sqrt{3}} < \frac{1}{4(1.5)} < \frac{1}{6}.$$

Note $6.9^2 = 47.61 < 48 < 49 = 7^2$. By the Intermediate Value Theorem, $6.9 < r < 7$. Therefore,

$$E_0 = 7 - \sqrt{48} < 7 - 6.9 = \frac{1}{10}, \quad \text{and} \quad \left| \frac{AE_0}{2} \right| < \frac{\frac{1}{6}\frac{1}{10}}{2} = \frac{1}{120} < 1.$$

By Cor. 2.11.18 our approximations x_n of r become better. By Thm. 2.11.15,

$$|E_1| \approx \left| \frac{2}{A} \left(\frac{AE_0}{2} \right)^2 \right| = \frac{|A| E_0^2}{2} < \frac{\frac{1}{6}\frac{1}{10^2}}{2} = \frac{1}{1200} \approx 0.00083.$$

Hence $x_1 = 6.92857$ approximates $\sqrt{48}$ within 0.00083. By Theorem 2.11.15:

$$|E_2| \;\approx\; \left| \frac{2}{A} \left(\frac{AE_0}{2} \right)^4 \right| = \left| \frac{2}{A} \left(\frac{AE_0}{2} \right)^2 \right| \left(\frac{AE_0}{2} \right)^2$$

$$\approx \; |E_1| \left(\frac{AE_0}{2} \right)^2 < 0.00083 \left(\frac{\frac{1}{6}\frac{1}{10}}{2} \right)^2 \approx 5.8 \times 10^{-8}.$$

Thus $x_2 = 6.928203243$ approximates $\sqrt{48}$ within 5.8×10^{-8}.

(2) In Example 2.11.14 (2) we computed the two approximations $x_1 = -0.8$ and $x_2 = -0.7568$ of the real root of the cubic polynomial

$$f(x) = x^3 - x^2 + 1 \; \text{using} \; x_0 \;=\; -1.$$

Note $f\left(-\frac{3}{4}\right) = \frac{1}{64} > 0$ and $f(-1) = -1 < 0$. By the Intermediate Value Theorem, $r \in \left(-1, -\frac{3}{4}\right)$. Hence

$$|E_0| = |x_0 - r| < \frac{1}{4}.$$

$f'(x) = 3x^2 - 2x$ and $f''(x) = 6x - 2$. We need to estimate $A = g(r)$ where

$$g(x) = \frac{f''(x)}{f'(x)} = \frac{6x - 2}{3x^2 - 2x}.$$

By the quotient rule:

$$g'(x) = \frac{(6)(3x^2 - 2x) - (6x - 2)(6x - 2)}{(3x^2 - 2x)^2} = \frac{-18x^2 + 12x - 4}{(3x^2 - 2x)^2} < 0$$

for $x \in (-1, 0)$. Hence g is a decreasing function on this interval, and

$$-\frac{8}{5} \;=\; g(-1) > A = g(r) > g\left(-\frac{3}{4}\right) = -\frac{104}{51}.$$

Therefore $\qquad \left| \dfrac{AE_0}{2} \right| \;<\; \dfrac{\frac{104}{51}\frac{1}{4}}{2} = \dfrac{13}{51} \approx 0.255 < 1.$

By Cor. 2.11.18 our approximations x_n of r become better. By Thm. 2.11.15:

$$|E_1| \approx \left| \frac{2}{A} \left(\frac{AE_0}{2} \right)^2 \right| = \frac{|A|E_0^2}{2} < \frac{\frac{104}{51} \frac{1}{4^2}}{2} < .064.$$

Thus $x_1 = -0.800$ is within 0.064 of the root r. By Theorem 2.11.15,

$$|E_2| \approx \left| \frac{2}{A} \left(\frac{AE_0}{2} \right)^4 \right| = \left| \frac{2}{A} \left(\frac{AE_0}{2} \right)^2 \right| \left(\frac{AE_0}{2} \right)^2$$

$$\approx |E_1| \left(\frac{AE_0}{2} \right)^2 \approx .064 \left(\frac{\frac{104}{51} \frac{1}{4}}{2} \right)^2 < .0041.$$

Therefore $x_2 = -0.7568$ is within 0.0041 of the root r of the polynomial $f(x)$.

(3) In Example 2.11.14 (3) we estimated a fixed point p of the polynomial $f(x) = x^2 - x - 5$. Recall we applied the Newton–Raphson method to

$$h(x) = f(x) - x = x^3 - 2x - 5 \quad \text{with} \quad x_0 = 2 .$$

The polynomial h has the fixed point p of f as a root. We obtained the estimates $x_1 = 2.1$ and $x_2 = 2.09457$ of p. Since $h(2) = -1 < 0$ and $h(2.2) = 1.248 > 0$, the root p of h must be between 2 and 2.2 by the Intermediate Value Theorem. Thus, $E_0 < 0.2$. Observe that $h'(x) = 3x^2 - 2$ and $h''(x) = 6x$. Hence

$$|A| = \left| \frac{h''(r)}{h'(r)} \right| = \frac{6r}{3r^2 - 2} < \frac{6(2.2)}{3(2)^2 - 2} = \frac{13.2}{10} = 1.32.$$

Therefore $\quad \left| \dfrac{AE_0}{2} \right| < \dfrac{(1.32)(0.2)}{2} = .132 < 1.$

By Cor. 2.11.18, the approximations x_n of r become better. By Thm. 2.11.15,

$$|E_1| \approx \frac{2}{|A|} \left(\frac{AE_0}{2} \right)^2 = \frac{|A|E_0^2}{2} < \frac{1.32(0.2)^2}{2} = .0264.$$

Thus $x_1 = 2.10$ is within 0.03 of the fixed point p of f. By Theorem 2.11.15,

$$|E_2| \approx \left| \frac{2}{A} \left(\frac{A_0}{2} \right)^4 \right| = \left| \frac{2}{A} \left(\frac{AE_0}{2} \right)^2 \right| \left(\frac{AE_0}{2} \right)^2$$

$$\approx |E_1| \left(\frac{AE_0}{2} \right)^2 < .0264 \left(\frac{(1.32)(0.2)}{2} \right)^2 < .00046.$$

Thus $x_2 = 2.09457$ is within 0.00046 of the the fixed point p of f.   □

Historical Remarks

The concept of differentials was introduced by Gottfried Wilhelm Leibniz as the cornerstone of his study of calculus. As explained in the Historical Remarks of Section 4, this viewpoint makes it easy to understand and find formulas for computing derivatives. However, Leibniz's mystical definition of a differential as an *infinitesmal* is illogical and is not mathematically acceptable. Interestingly, Figure 2.11.3 is fundamental to much of Leibniz's work. We defined the derivative as a limit whose value is the definition of the slope of the tangent line T. Therefore, we can interpret the triangle in Figure 2.11.3 as saying that the slope of T equals dy divided by dx. Leibniz, however, uses Figure 2.11.3 to define the slope of T. He shrinks the triangle to be infinitesmaly small where "$dy = \Delta y$". Then he defines the slope of T as dy divided by dx. For example, Leibniz computes the derivative of $y = x^2$ as follows.

$$\frac{dy}{dx} = \frac{\Delta y}{dx} = \frac{(x + dx)^2 - x^2}{dx} = \frac{x^2 + 2x\,dx + (dx)^2 - x^2}{dx} = 2x + dx = 2x.$$

Leibniz ignores the dx in the sum $2x + dx$ because it is insignificantly small compared to $2x$. Leibniz also introduced higher differentials $d^n x$, which he uses to define and compute higher derivatives.

A mathematically rigorous exposition of Leibniz's approach was developed by Abraham Robinson in 1960. Robinson extends the set of real numbers $\Re$ to a larger set $\Re^*$, with arithmetic operations, called the set of *hyperreal numbers*. Hyperreal numbers are generated by real numbers, *infinitesmal numbers* and *infinite numbers*. If $x^* \in \Re^*$ and $x \in \Re$ with $x^* - x$ infinitesmal, then x is called the *standard part of* x^*. He writes $x = \mathrm{st}(x^*)$. The derivative of a function f is defined by:

$$f'(c) = \mathrm{st}\left(\frac{f'(c + dx) - f(c)}{dx} \right)$$

when the quantity on the right has the same value for every nonzero infinitesmal dx. Now Leibniz style arguments, such as the one computing $D(x^2)$ above, become rigorous. This approach to calculus is called *nonstandard analysis.*.

In 1685 Isaac Newton discovered an algorithm for obtaining a sequence of increasingly better approximations of the root r of a polynomial $p(x) = a_n x^n + \cdots + a_1 x + a_0$. His method was entirely algebraic. If x_0 is an approximation of r, let $E_0 = r - x_0$ denote the error of this approximation. Then

$$
\begin{aligned}
0 &= p(r) = p(E_0 + x_0) \\
&= a_n(E_0 + x_0)^n + a_{n-1}(E_0 + x_0)^{n-1} + \cdots + a_1(E_0 + x_0) + a_0 \quad (2.11.6) \\
&= p(x_0) + na_n E_0 x_0^{n-1} + (n-1)a_{n-1} E_0 x_0^{n-2} + \cdots + a_1 E_0 \ + \text{ terms divisible by } E_0^2 \\
&= p(x_0) + E_0 p'(x_0) \ + \text{ terms divisible by } E_0^2 \ = p_1(E_0). \quad (2.11.7)
\end{aligned}
$$

If $|E_0| < 1$ is small, then E_0^2 is much smaller than $|E_0|$, and

$$
\begin{aligned}
0 &\approx p(x_0) + E_0 p'(x_0), \\
E_0 &\approx e_0 = -\frac{p(x_0)}{p'(x_0)} \quad \text{and} \\
r &= x_0 + E_0 \approx x_0 + e_0 = x_0 - \frac{p(x_0)}{p'(x_0)} = x_1.
\end{aligned}
$$

x_1 is a better approximation of r than x_0. Newton finds the next approximation x_2 of r as follows. e_0 approximates the root E_0 of the polynomial p_1 defined in (2.11.7). The above procedure produces an approximation e_1 of $E_1 = E_0 - e_0$, and $x_2 = x_0 + e_0 + e_1$. Then an approximation e_2 of $E_2 = E_1 - e_1$ is found which defines $x_3 = x_0 + e_0 + e_1 + e_2$, etc. In 1690 Joseph Raphson published a variation of this method to produce these approximations x_n. He uses the original equation (2.11.6), with x_0 replaced by x_1, to find the next approximation x_2 of r. Iteration of this procedure, produces the recursive formula (2.11.5), for the sequence of approximations x_n of r, which we call the Newton–Raphson method.

Summary

The reader should understand the meaning of the symbols Δx, Δy, dx, dy and be able to compute their values for a specific function. She should be able to use differentials to make approximations and should know how to determine their accuracy. She should understand how to apply the Newton–Raphson method to produce a sequence of approximations of a root r of a function. She should be able to verify that these approximations of r become increasingly better and should know how to determine their accuracy.

Basic Exercises

You may use a calculator for addition, subtraction, multiplication, division and the value of π in all exercises. You may also use a calculator to approximate roots, trigonometric functions and inverse trigonometric functions in Exercises 1, 5, 24, 26.

1. For each function, find dx, dy, Δy for the given values of x_0 and Δx.
(a) $f(x) = 2x^3 - 7$ with $x_0 = 3$ and $\Delta x = +1$.
(b) $g(x) = x^{5/4} - 3x^{3/4} + 5$ with $x_0 = 16$ and $\Delta x = -2$.
(c) $h(x) = \frac{x}{x^2+1}$ with $x_0 = 5$ and $\Delta x = -1$.
(d) $j(x) = \sin x$ with $x_0 = \frac{\pi}{6}$ and $\Delta x = \frac{\pi}{24}$.
(e) $k(x) = \arctan x$ with $x_0 = 1$ and $\Delta x = 2$.
(f) $m(x) = \cos^2 x$ with $x_0 = \frac{\pi}{4}$ and $\Delta x = \frac{\pi}{16}$.

2.　Assume that f and g are differentiable functions with $f(3) = 2$, $g(6) = 7$, $df = (6x^2 + 6)dx$ and $dg = (6x^2 - 6)dx$.
(a) Find $d(f + g)$.　　(b) Find $d(7f)$.　　(c) Find f and g.
(d) Find $d(fg)$.　　(e) Find $d(f/g)$.　　(f) Find $d(f \circ g)$.

3. Prove Proposition 2.11.7.

4. Use differentials to approximate each expression. Determine the accuracy of each estimate.
(a) $\sqrt{26}$　　(b) $\sqrt{48}$　　(c) $\sqrt[3]{128}$　　(d) $\sqrt[4]{80}$　　(e) $\sqrt[5]{33}$
(f) $65^{3/2}$　　(g) $213^{2/3}$　　(h) $245^{-3/4}$　　(i) $30^{-7/5}$　　(j) $62^{5/6}$

5. Use differentials to approximate each expression. Determine the accuracy of each estimate.
(a) $\sin 44°$ (b) $\cos 31°$ (c) $\tan 46°$ (d) $\sec 59°$ (e) $\arctan 1.1$ (f) $\arcsin 0.4$

6. Use differentials to approximate each expression. Use $\sqrt{3} \approx 1.732$.
(a) $\sec^2 .23\pi$　　　　　　　(b) $\sqrt{\tan .28\pi}$　　(c) $(1.02)^{100}$　　(d) $\frac{\sqrt{35}}{\sqrt{35}+18}$
(e) $\sqrt{7 + \sqrt{3 + \sqrt{1.03}}}$　(f) $\tan^{50} 1.249\pi$　(g) $\arctan \sqrt{2.96}$　(h) $\frac{3-\cot^2 .69\pi}{3+\tan^2 .69\pi}$

7. The area of a circular puddle increases from 100π cm^2 to 104π cm^2. Estimate the increase in the radius of the puddle.

8. Assume a rope is tied around the Earth's equator. If the rope is stretched by one foot and still maintains its circular shape, estimate the distance between the rope and the earth. How accurate is your estimate?

9. Estimate the volume of tin required to construct a 3 meter long cylindrical duct of diameter 20 cm whose walls have thickness $\frac{1}{8}$ cm. How accurate is your estimate?

10.　It requires one litre of paint to cover a square sheet of metal with side five meters long. On a hot summer day each side of this metal sheet expands by a quarter centimeter. Estimate how much more paint is needed to paint the metal sheet on that day. How accurate is your estimate?

11.　A cone is filled with ice cream so that it has the shape of a cone of radius 2.5 centimeters and height twelve centimeters surmounted by a hemisphere. If the radius of the cone is increased by 0.3 centimeters, estimate how much ice cream is required to fill the new cone so that it also has the shape of a cone surmounted by a hemisphere.

How accurate is your estimate?

12. A wire 30 centimeters long is bent into each of the following shapes:
  (a) a square; **(b)** a circle; **(c)** an equilateral triangle.
In each case, estimate the decrease in area of the shape if the length of the wire were only 29 centimeters.

13. The revenue R in dollars obtained by selling x gallons of milk is given by

$$R = \frac{100x}{50 + x/1000}.$$

If a company sells 2000 gallons of milk, estimate the additional revenue that results from selling an additional 10 gallons.

14. A balloon is constantly 100 feet from an observer on the ground. If the altitude of the balloon increases from 60 to 62 feet, find the increase of the angle of inclination of the balloon (the angle between the ground and the line of sight of the observer). How accurate is your estimate?

15. A violinist plucks the center of a 30 cm string moving it 3 cm perpendicular to the violin. Estimate the increase in the angle of the string at her finger if she moves the string an additional 0.2 cm from the violin. How accurate is your estimate?

16. For each function, use differentials to estimate the percent change $E(y)$ of y in terms of the percent change $E(x)$ of x.
(a) $y = x^5$ **(b)** $y = 3x^6 - 4x^3 + 8$ **(c)** $y = \frac{5x+7}{7x-2}$ **(d)** $y = \sqrt{3x^4 + 9}$ **(e)** $y = \sin(\pi x)$

17. Find the percent of error in computing the volume of a sphere when the radius is measured with error less than 3%.

18.[5] A penny is dropped from the top of a building. It reaches the ground in 5.7 seconds, measured with error less than 0.2 seconds. This measurement is used to compute the height h of the building. Find the percent of error in this value of h.

19. The gravitational force between two objects which are r miles apart is proportional to $\frac{1}{r^2}$. Find the percent of error in calculating this force in terms of the percent of error in measuring r.

20. The Valley Vitners Vine Company finds that the revenue R in dollars obtained by selling n bottles of wine is given by

$$R = \frac{500n}{n + 1000}.$$

Estimates are that between 1900 and 2100 bottles will be sold. If the value $n = 2000$ is used to calculate the revenue, find the percent of error in the resulting value of R.

21. A protractor is used to measure an acute angle of a right triangle as $30°$ with error less than 4%. If the hypotenuse is known to have length 40 cm, find the percent of error in the calculated area of this triangle.

22. A fruit store is packaging bags of 30 grapes each. Each grape has diameter between 0.95 and 1.05 centimeters. The store advertises the volume of grapes in each bag by using a value of 1 cm for the diameter of each grape to calculate the volume. Find the percent of error in the advertised value of the volume of this bag.

23. The pressure P of air in a balloon is inversely proportional to its volume V. Find the relationship between $E(P)$ and $E(V)$.

24. Each function below has a root r which is approximated by the given number x_0. In each case:

---

[5] Knowledge of Section 2.9 is required for this exercise.

(i)   derive the Newton–Raphson formula for x_{n+1} as a function of x_n;
(ii) find x_1 and x_2;
(iii) determine the accuracy of the estimates x_1 and x_2 of r.

(a) $f(x) = x^2 + x - 5$ with $x_0 = 2$.   (b) $g(x) = 2x^3 - x + 2$ with $x_0 = -1$.
(c) $h(x) = x^3 - 3x - 1$ with $x_0 = 2$.   (d) $j(x) = x^4 - 8x^2 + 11$ with $x_0 = 3$.
(e) $k(x) = x^{5/3} + x^{4/3} - 49$ with $x_0 = 8$.   (f) $m(x) = x^{7/4} - 4x^{5/4} - 1$ with $x_0 = 16$.
(g) $n(x) = 4x - 3\tan x$ with $x_0 = \pi/4$.   (h) $p(x) = 3x - 4\arctan x$ with $x_0 = 1$.

25. Iterate the Newton–Raphson method to obtain an estimate x_2 of each number. Determine the accuracy of your estimates.

(a) $\sqrt{24}$ (b) $\sqrt{38}$ (c) $\sqrt[3]{62}$ (d) $\sqrt[3]{-26}$ (e) $\sqrt[4]{17}$ (f) $\sqrt[5]{33}$

26. Each function below has a fixed point which is estimated by the given number x_0. Iterate the Newton–Raphson method to obtain a better approximation x_2 of this fixed point. Determine the accuracy of your approximations.

(a) $f(x) = x^2 - 4x + 5$ with $x_0 = 4$.   (b) $g(x) = x^3 - 4x - 3$ with $x_0 = -1$.
(c) $h(x) = x^{6/5} - 30$ with $x_0 = 32$.   (d) $j(x) = x^{5/3} - 2x^{4/3} + 7$ with $x_0 = 8$.
(e) $k(x) = \sec x - 1$ with $x_0 = \pi/3$.   (f) $m(x) = \cos x$ with $x_0 = 3/4$.

Challenging Problems

1. Let r be the real root of the cubic polynomial $x^3 - 2x - 5$. Start with the approximation 2 for r. Apply Newton's algorithm for approximating roots, as given in the Historical Remarks, three times to obtain better approximations of r.

Appendix: Error Bounds

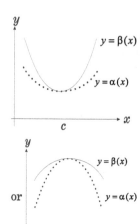

In this appendix we prove Theorems 2.11.9 and 2.11.15 which give bounds on the errors that occur when differentials and the Newton–Raphson method are used to make approximations. These results are consequences of two generalizations of the Mean Value Theorem. The Mean Value Theorem, which says $f(x) = f(c) + f'(h)(x - c)$, can be interpreted as saying that $y = f(x)$ is approximated by the constant function $y = f(c)$ with error $f'(h)(x - c)$. We show $f(x)$ is approximated by the linear function $y = f(c) + f'(c)(x - c)$ and apply this result to prove Thm. 2.11.9. Then we show $f(x)$ is approximated by the quadratic polynomial $y = f(c) + f'(c)(x - c) + \frac{f''(c)}{2}(x - c)^2$ and apply this result to prove Theorem 2.11.15.

Figure 2.11.20
Lemma 2.11.22(a)

These two generalizations of the Mean Value Theorem are consequences of the next lemma. The first part of this lemma considers the graphs of two functions α and β which intersect at $x = c$ where they both have horizontal tangent lines. Assume that α is steeper[6] than β to the left of c and β is steeper than α to the right of c. The lemma concludes that β lies above α. See the two diagrams in Figure 2.11.20. In the second part of this lemma we consider two curves α and β which intersect at $x = c$ with β steeper than α everywhere. The lemma concludes that β lies below α to the left of c and β lies above α to the right of c. See the two diagrams in Figure 2.11.21.

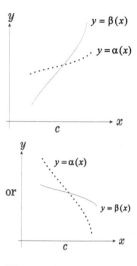

Lemma 2.11.22 *Let α, β be continuous functions with domain $I = [a, b]$ such that α, β are differentiable on (a, b). Let $c \in (a, b)$ with $\alpha(c) = \beta(c)$.*

(a) *Assume $\beta'(x) \le \alpha'(x)$ when $a < x \le c$ and $\alpha'(x) \le \beta'(x)$ when $c \le x < b$. Then $\alpha(x) \le \beta(x)$ for $x \in I$.*

(b) *Assume $\alpha'(x) \le \beta'(x)$ for $x \in (a, b)$. Then*

$$\beta(x) \le \alpha(x) \text{ when } a \le x \le c \text{ and } \alpha(x) \le \beta(x) \text{ when } c \le x \le b.$$

Figure 2.11.21
Lemma 2.11.22(b)

Proof Let J be the closed interval with endpoints x, c with $x \in I$ considered to be

[6]We use the term α is *steeper* than β to mean α has a larger derivative than β.

constant. Define
$$f(t) = \beta(t) - \alpha(t)$$

for $t \in J$. Note that $f(c) = \beta(c) - \alpha(c) = 0$. Apply the Mean Value Theorem to the function f on the interval J: there is $h \in J$ such that

$$
\begin{aligned}
f(x) &= f(c) + f'(h)(x - c) = f'(h)(x - c), \text{ i.e.} \\
\beta(x) - \alpha(x) &= [\beta'(h) - \alpha'(h)](x - c). \quad (2.11.8)
\end{aligned}
$$

(a) If $x \le c$, then $\beta'(h) - \alpha'(h)$ and $x - c$ are both less than or equal to zero. Hence $\beta(x) - \alpha(x) \ge 0$. If $x \ge c$, then $\beta'(h) - \alpha'(h)$ and $x - c$ are both greater than or equal to zero. Hence $\beta(x) - \alpha(x) \ge 0$. Thus $\beta(x) - \alpha(x) \ge 0$ for all $x \in I$.
(b) When $x \le c$, equation (2.11.8) says $\beta(x) - \alpha(x)$ is the product of $\beta'(h) - \alpha'(h)$, whch is greater than or equal to zero, and $x - c$, which is less than or equal to zero. Thus $\beta(x) - \alpha(x) \le 0$ for $x \le c$. When $x \ge c$ equation (2.11.8) says $\beta(x) - \alpha(x)$ is the product of $\beta'(h) - \alpha'(h)$ and $x - c$ which are both greater than or equal to zero. Hence $\beta(x) - \alpha(x) \ge 0$ for $x \ge c$. Thus $\beta(x) \ge \alpha(x)$ for all $x \in I$. $\qquad\square$

We use the first part of the preceding lemma to prove a refinement of the Mean Value Theorem which gives the error in approximating $y = f(x)$ by its tangent line $y = f(c) + f'(c)(x - c)$ at $x = c$.

Proposition 2.11.23 *Let f be a function, with domain the open interval I, whose second derivative exists and is continuous. If $c, x \in I$, then there is at least one number h between x and c such that:*

$$f(x) = f(c) + f'(c)(x - c) + \frac{f''(h)}{2}(x - c)^2. \quad (2.11.9)$$

Proof When $x = c$, let $h = c$. If $x \ne c$, let J be the closed interval with endpoints x and c. Apply the Maximum Value Theorem to the continuous function f'' on the interval J: f'' has a minimum value m_x and a maximum value M_x. That is:

$$m_x \le f''(t) \le M_x \quad (2.11.10)$$

for $t \in J$. Apply Lemma 2.11.22(a) to the three functions with domain J:

$$\alpha(t) = \frac{m_x}{2}(t - c)^2, \quad \beta(t) = f(t) - f'(c)(t - c) - f(c), \quad \gamma(t) = \frac{M_x}{2}(t - c)^2.$$

Note $\alpha(c) = \beta(c) = \gamma(c) = 0$. Consider x to be constant and t to be a variable. Then

$$\alpha'(t) = m_x(t - c), \quad \beta'(t) = f'(t) - f'(c), \quad \gamma'(t) = M_x(t - c). \quad (2.11.11)$$

Now consider $t \in J$ to be constant, and apply the Mean Value Theorem to the function f' on the interval with endpoints t and c: there is at least one number k_t between c and t such that

$$\beta'(t) = f'(t) - f'(c) = f''(k_t)(t - c). \quad (2.11.12)$$

By (2.11.10), (2.11.11), (2.11.12): if $t \in J$ with $t \ge c$, then $\alpha'(t) \le \beta'(t) \le \gamma'(t)$ while if $t \in J$ with $t < c$, then $t - c < 0$ and $\alpha'(t) \ge \beta'(t) \ge \gamma'(t)$. By Lemma 2.11.22(a): $\alpha(t) \le \beta(t) \le \gamma(t)$ for all $t \in J$. In particular, for $t = x$ these inequalities are:

$$m_x \frac{(x - c)^2}{2} \le f(x) - f'(c)(x - c) - f(c) \le M_x \frac{(x - c)^2}{2}$$

$$m_x \le \frac{f(x) - f'(c)(x - c) - f(c)}{(x - c)^2/2} \le M_x \qquad \text{[dividing by } (x - c)^2/2].$$

Apply the Intermediate Value Theorem to the continuous function f'' on the interval J: there is $h \in J$ with

$$f''(h) \;=\; \frac{f(x) - f'(c)(x-c) - f(c)}{(x-c)^2/2}$$

$$f''(h)\frac{(x-c)^2}{2} \;=\; f(x) - f'(c)(x-c) - f(c) \qquad \text{[multiplying by } (x-c)^2/2]. \;\square$$

We apply the preceding proposition to prove Theorem 2.11.9 and establish a bound on the error in approximating a function by its tangent line.

Theorem 2.11.9 *Consider a function f whose domain contains the open interval I such that f'' exists and is continuous on I. Let $x_0 \in I$ with T the tangent line to the graph of f at the point $(x_0, f(x_0))$. For $x \in I$, choose M greater than or equal to the maximum value of f'' on the closed interval with endpoints x_0 and x. Then*

$$|f(x) - T(x)| = |\Delta y - dy| \le \frac{M(x-x_0)^2}{2}. \qquad (2.11.13)$$

Proof Apply Proposition 2.11.23 to f with $c = x_0$: there is a number h between x_0 and x such that

$$f(x) \;=\; f(x_0) + f'(x_0)(x-x_0) + \frac{f''(h)}{2}(x-x_0)^2$$

$$\;=\; T(x) + \frac{f''(h)}{2}(x-x_0)^2, \qquad \text{[by (2.2.6)]}$$

$$\text{and } |f(x) - T(x)| \;=\; \frac{|f''(h)|}{2}(x-x_0)^2 \le \frac{M}{2}(x-x_0)^2. \qquad \square$$

We use Lemma 2.11.22(b) to further generalize the Mean Value Theorem to give the error in approximating $y = f(x)$ by the parabola $y = f(c) + f'(c)(x-c) + \frac{f''(c)}{2}(x-c)^2$.

Proposition 2.11.24 *Let f be a function with domain the open interval I whose third derivative exists and is continuous. If $x, c \in I$, then there is at least one number h between x and c such that:*

$$f(x) = f(c) + f'(c)(x-c) + \frac{f''(c)}{2}(x-c)^2 + \frac{f'''(h)}{6}(x-c)^3.$$

Proof The proof of this proposition is analogous to the proof of Proposition 2.11.23. When $x = c$, take $h = c$. If $x \ne c$, let J be the closed interval with endpoints x and c. Apply the Maximum Value Theorem to the continuous function f''' on the interval J: f''' has a minimum value m_x and a maximum value M_x. That is:

$$m_x \le f'''(t) \le M_x \qquad (2.11.14)$$

for $t \in J$. Apply Lemma 2.11.22(b) to the three functions with domain J:

$$\alpha(t) = \frac{m_x}{6}(t-c)^3, \quad \beta(t) = f(t) - f'(c)(t-c) - f''(c)\frac{(t-c)^2}{2} - f(c), \quad \gamma(t) = \frac{M_x}{6}(t-c)^3.$$

Note $\alpha(c) = \beta(c) = \gamma(c) = 0$. Consider x to be constant and t to be a variable. Then

$$\alpha'(t) = \frac{m_x}{2}(t-c)^2, \quad \beta'(t) = f'(t) - f'(c) - f''(c)(t-c), \quad \gamma'(t) = \frac{M_x}{2}(t-c)^2. \quad (2.11.15)$$

Now consider $t \in J$ to be constant, and apply Proposition 2.11.23 to the function f' on the interval with endpoints t and c: there is at least one number k_t between c and t such that

$$\beta'(t) \;=\; f'(t) - f'(c) - f''(c)(t - c) \;=\; \frac{f'''(k_t)}{2}(t - c)^2. \qquad (2.11.16)$$

By (2.11.14), (2.11.15), (2.11.16) we have $\alpha'(t) \le \beta'(t) \le \gamma'(t)$ for $t \in J$. If $x < c$, then by Lemma 2.11.22(b): $\alpha(x) \ge \beta(x) \ge \gamma(x)$, i.e.

$$m_x \frac{(x - c)^3}{6} \;\ge\; f(x) - f'(c)(x - c) - f''(c)\frac{(x - c)^2}{2} - f(c) \;\ge\; M_x \frac{(x - c)^3}{6}$$

$$m_x \;\le\; \frac{f(x) - f'(c)(x - c) - f''(c)\frac{(x-c)^2}{2} - f(c)}{(x - c)^3/6} \;\le\; M_x \qquad (2.11.17)$$

dividing by the negative number $\frac{(x-c)^3}{6}$. On the other hand, if $x > c$, then by Lemma 2.11.22(b): $\alpha(x) \le \beta(x) \le \gamma(x)$, i.e.

$$m_x \frac{(x - c)^3}{6} \;\le\; f(x) - f'(c)(x - c) - f''(c)\frac{(x - c)^2}{2} - f(c) \;\le\; M_x \frac{(x - c)^3}{6}$$

$$m_x \;\le\; \frac{f(x) - f'(c)(x - c) - f''(c)\frac{(x-c)^2}{2} - f(c)}{(x - c)^3/6} \;\le\; M_x \qquad (2.11.18)$$

dividing by the positive number $\frac{(x-c)^3}{6}$. By (2.11.17), the inequalities (2.11.18) are valid for all $x \in I$. Apply the Intermediate Value Theorem to the continuous function f''' on the interval J: there is $h \in J$ with

$$f'''(h) \;=\; \frac{f(x) - f'(c)(x - c)f''(c)\frac{(x-c)^2}{2} - f(c)}{(x - c)^3/6}$$

$$f'''(h)\frac{(x - c)^3}{6} \;=\; f(x) - f'(c)(x - c) - f''(c)\frac{(x - c)^2}{2} - f(c) \qquad \square$$

Let x_0 be an approximation of a root r of a differentiable function $f(x)$. The Newton–Raphson method approximates r by a sequence of increasingly better approximations $x_1, x_2, \ldots, x_n, \ldots$ defined recursively by

$$x_n = x_{n-1} - \frac{f(x_{n-1})}{f'(x_{n-1})} \qquad (2.11.19)$$

for $n \ge 1$. We apply the preceding proposition to prove Theorem 2.11.15 and estimate the size of the error $E_n = x_n - r$ of the approximation x_n of r.

Theorem 2.11.15 *Let f be a function, with domain the open interval I, whose third derivative exists and is continuous. Assume that f has a root $r \in I$ with $f'(r) \ne 0$ and $f''(r) \ne 0$. Define*

$$A = \frac{f''(r)}{f'(r)}.$$

Let $x_0 \in I$ be an approximation of r. For $n \ge 1$, let x_n denote the approximation of r obtained from x_0 by iterating the Newton–Raphson method n times. Then the error $E_n = x_n - r$ of this approximation satisfies:

$$E_n \approx \frac{2}{A}\left(\frac{AE_0}{2}\right)^{2^n}.$$

Proof Subtract r from both sides of equation (2.11.19) which defines x_n:

$$E_n = x_n - r = x_{n-1} - r - \frac{f(x_{n-1})}{f'(x_{n-1})} = E_{n-1} - \frac{f(x_{n-1})}{f'(x_{n-1})}.$$

Apply Proposition 2.11.24 to $f(x_{n-1})$ and Proposition 2.11.23 to $f'(x_{n-1})$. In both cases, take $c = r$. Then there are two numbers h and k between x_{n-1} and r such that

$$E_n = E_{n-1} - \frac{f(r) + f'(r)(x_{n-1} - r) + \frac{f''(r)}{2}(x_{n-1} - r)^2 + \frac{f'''(k)}{6}(x_{n-1} - r)^3}{f'(r) + f''(r)(x_{n-1} - r) + \frac{f'''(h)}{2}(x_{n-1} - r)^2}$$

$$= E_{n-1} - \frac{f'(r)E_{n-1} + \frac{f''(r)}{2}E_{n-1}^2 + \frac{f'''(k)}{6}E_{n-1}^3}{f'(r) + f''(r)E_{n-1} + \frac{f'''(h)}{2}E_{n-1}^2} \qquad [\text{because } f(r) = 0]$$

$$= \frac{f'(r)E_{n-1} + f''(r)E_{n-1}^2 + \frac{f'''(h)}{2}E_{n-1}^3}{f'(r) + f''(r)E_{n-1} + \frac{f'''(h)}{2}E_{n-1}^2} - \frac{f'(r)E_{n-1} + \frac{f''(r)}{2}E_{n-1}^2 + \frac{f'''(k)}{6}E_{n-}^3}{f'(r) + f''(r)E_{n-1} + \frac{f'''(h)}{2}E_{n-1}^2}$$

$$= \frac{\frac{f''(r)}{2}E_{n-1}^2 + \frac{f'''(h)}{2}E_{n-1}^3 - \frac{f'''(k)}{6}E_{n-1}^3}{f'(r) + f''(r)E_{n-1} + \frac{f'''(h)}{2}E_{n-1}^2}.$$

We assume that E_{n-1} is a small number. Then the terms with a factor of E_{n-1}^3 in the numerator are much smaller than the term with a factor of E_{n-1}^2 and can be ignored to make an approximation. Similarly, the terms with a factor of E_{n-1} in the denominator are much smaller than the term with no factor of E_{n-1} and can be ignored to make an approximation. Hence the preceding equation produces the estimate:

$$E_n \approx \frac{\frac{f''(r)}{2}E_{n-1}^2}{f'(r)} = \frac{A}{2}E_{n-1}^2.$$

Iterate this formula n times:

$$E_n \approx \frac{A^{2^n - 1}}{2^{2^n - 1}} E_0^{2^n} = \frac{2}{A}\left(\frac{AE_0}{2}\right)^{2^n}. \qquad \square$$

Exercises

1. Find all numbers h, k between x and c such that

$$f(x) = f(c) + f'(c)(x - c) + \frac{f''(h)}{2}(x - c)^2 \quad \text{and}$$

$$f(x) = f(c) + f'(c)(x - c) + \frac{f''(c)}{2}(x - c)^2 + \frac{f'''(k)}{6}(x - c)^3$$

for each function with the given values of x and c.
(a) $f(x) = x^3$ with $x = 3$, $c = 1$.   (b) $f(x) = x^4$ with $x = 0$, $c = -2$.
(c) $f(x) = 1/x$ with $x = 2$, $c = 1$.   (d) $f(x) = \sin x$ with $x = 0$, $c = \pi/6$.
(e) $f(x) = \cos x$ with $x = 0$, $c = \pi/4$.   (f) $f(x) = \sqrt{6 + x}$ with $x = -2$, $c = 3$.

2. Prove the following generalization of Propositions 2.11.23 and 2.11.24. Let f be a function with domain the open interval I whose $(n + 1)^{st}$ derivative exists and is continuous. If $c, x \in I$, then there is at least one number k between x and c such that:

$$f(x) = f(c) + \frac{f^{(1)}(c)}{1!}(x - c) + \frac{f^{(2)}(c)}{2!}(x - c)^2 + \cdots + \frac{f^{(n)}(c)}{n!}(x - c)^n + \frac{f^{(n+1)}(k)}{(n+1)!}(x - c)^{n+1}$$

where $n! = n(n - 1)(n - 2) \cdots (3)(2)(1)$.

3. Use the preceding exercise to write each polynomial as a polynomial in $x - 2$:
(a) $f(x) = 3x^2 - 4x + 7$;   (b) $g(x) = 5x^3 + 6x^2 + 9x - 8$;
(c) $h(x) = 2x^4 - 5x^3 + 4x^2 - 7x - 1$;   (d) $k(x) = x^5$.

2.12 Maxima and Minima

Prerequisite: Section 2.7

An important application of the derivative is to locate and identify the maximum and minimum values of a continuous function whose domain is an interval. These may occur at hills or valleys of the graph which are the local extrema we studied in Section 6. By Theorem 2.6.6, local extrema occur at critical points, i.e. points where the first derivative is either zero or does not exist. We used the First Derivative Test to identify each critical point as a local minimum (a valley), a local maximum (a hill) or neither. We begin this section with the Second Derivative Test, an alternate method to identify those critical points where the first derivative vanishes. In the second subsection, we study the problem of finding the minimum and maximum values of a function on its entire domain. These methods are applied to functions defined by equations as well as word problems.

Second Derivative Test

Let $x = c$ be a critical point of the continuous function f with domain an interval I. That is, $f'(c) = 0$ or $f'(c)$ does not exist. The First Derivative Test requires us to determine the sign of $f'(x)$ to the left and right of $x = c$. It concludes:

- if sign $f'(x)$ changes from $+$ to $-$ at $x = c$, then f has a local minimum at $x = c$;
- if sign $f'(x)$ changes from $+$ to $-$ at $x = c$, then f has a local maximum at $x = c$;
- if sign $f'(x)$ does not change at $x = c$, then f does not have a local extremum there.

However, it may be tedious to determine the sign of $f'(x)$ in order to use the First Derivative Test. In the case where $f'(c) = 0$ and $f''(c) \neq 0$, it may be easier to identify the critical point at $x = c$ by the Second Derivative Test.

Theorem 2.12.1 (Second Derivative Test) *Let f be a differentiable function with domain the open interval (a, b). Assume $c \in (a, b)$ such that $f'(c) = 0$ and $f''(c)$ exists.*

(a) *If $f''(c) > 0$, then f has a local minimum at $x = c$.*

(b) *If $f''(c) < 0$, then f has a local maximum at $x = c$.*

(c) *If $f''(x) = 0$, there is no conclusion.*

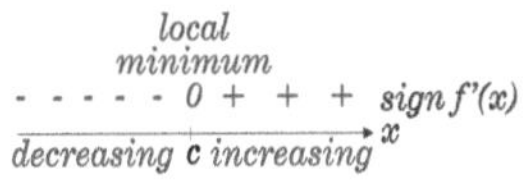

local
minimum
- - - - - 0 + + + sign $f'(x)$
                         x
decreasing c *increasing*

Figure 2.12.2
$f''(c) > 0$

Proof (a) $f''(c)$ is the derivative of $f'(x)$ at $x = c$. Hence $f''(c)$ is approximated by

$$\frac{f'(x) - f'(c)}{x - c} = \frac{f'(x)}{x - c} \tag{2.12.1}$$

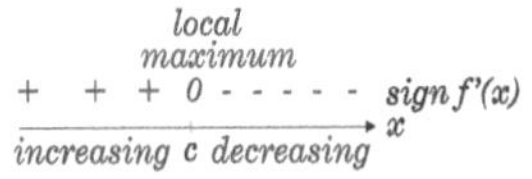

local
maximum
+ + + 0 - - - - - sign $f'(x)$
                         x
increasing c *decreasing*

Figure 2.12.3
$f''(c) < 0$

for x near c, $x \neq c$.
(a) Since $f''(c) > 0$, the quotient in (2.12.1) is positive for x near c, $x \neq c$. Hence for x near c, $f'(x) < 0$ when $x < c$ while $f'(x) > 0$ when $x > c$. See Figure 2.12.2. By the First Derivative Test, f has a local minimum at $x = c$.
(b) Since $f''(c) < 0$, the quotient in (2.12.1) is negative for x near c, $x \neq c$. Hence for x near c, $f'(x) > 0$ when $x < c$ while $f'(x) < 0$ when $x > c$. See Figure 2.12.3. By the First Derivative Test, f has a local maximum at $x = c$. □

Note the Second Derivative Test does not apply when $f'(c)$ does not exist or when $f''(c) = 0$. However, the First Derivative Test does apply to these cases. Therefore, the First Derivative Test is the preferred method in most applications.

Examples 2.12.4 Locate and identify the critical points of each function.

(1) Let $f(x) = x^3 + 9x^2 + 24x - 34$ with domain $\Re$.

Solution Observe that

$$f'(x) = 3x^2 + 18x + 24 = 3(x^2 + 6x + 8) = 3(x + 4)(x + 2).$$

Thus, f has critical points at $x = -4$ and $x = -2$ where $f'(x)$ vanishes. Note

$$f''(x) = 6x + 18.$$

We apply the Second Derivative Test. Since $f''(-4) = -6 < 0$, f has a local maximum at $x = -4$. Since $f''(-2) = 6 > 0$, f has a local minimum at $x = -2$.

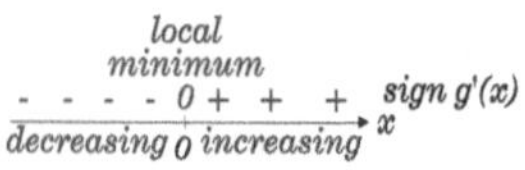

(2) Let $g(x) = x^n$ with domain $\Re$ where $n \geq 2$ is an integer.

Solution Since $g'(x) = nx^{n-1}$, g has a critical point at $x = 0$ where the first derivative vanishes. Since $g''(x) = n(n - 1)x^{n-2}$, we have $g''(0) = 0$ if $n \geq 3$. The Second Derivative Test *gives no information* about these critical points. To identify them, we use the First Derivative Test. If n is even, then $g'(x) < 0$ for $x < 0$ and $g'(x) > 0$ for $x > 0$. Thus, g has a local minimum at $x = 0$. See Figure 2.12.5. If n is odd, then $g'(x) > 0$ for both $x < 0$ and $x > 0$. Therefore, g has neither a local maximum nor a local minimum at $x = 0$. See Figure 2.12.6.

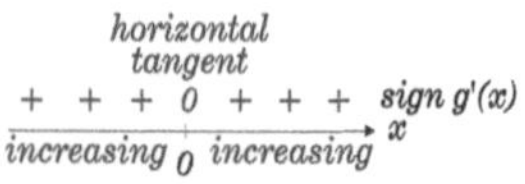

(3) Let $h(x) = x^{2/3}$ with domain $\Re$.

Solution Since

$$h'(x) = \frac{2}{3}x^{-1/3} = \frac{2}{3x^{1/3}},$$

h has a critical point at $x = 0$ where $h'(0)$ does not exist. Therefore, the Second Derivative Test *does not apply* to identify this critical point. We use the First Derivative Test. Since $h'(x) < 0$ for $x < 0$ and $h'(x) > 0$ for $x > 0$, h has a local minimum at $x = 0$. See Figure 2.12.7. □

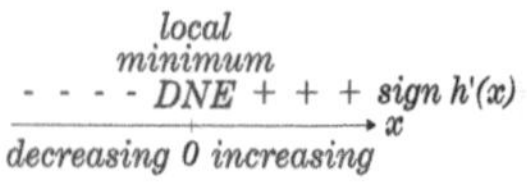

Figure 2.12.5
$g(x) = x^n$ for n Even

Figure 2.12.6
$g(x) = x^n$ for n Odd

Figure 2.12.7
$h(x) = x^{2/3}$

Global Maxima and Minima

The goal of this subsection is to find the minimum and maximum values of a function whose domain is an interval. After establishing a strategy, we apply it to functions defined by graphs, by equations and by word problems.

Consider the simplest case: a continuous function f with domain the closed interval $[a, b]$. The Maximum Value Theorem says that f has a minimum value at $x = m$ (a lowest point on its graph) and a maximum value at $x = M$ (a highest point on its graph). m or M may be an interior point of the interval $[a, b]$, i.e an element of (a, b). In this case m or M is a local extremum, hence a critical point. Alternatively, m or M may be one of the endpoints a, b of $[a, b]$. Sometimes we say f has a *global minimum* at $x = m$ and a *global maximum* at $x = M$ to distinguish these points from the local minima and local maxima.

Now consider the problem of finding the global minimum and maximum values of a continuous function f with domain an interval I which may not be closed. As the following example shows, we must consider the behavior of f near those endpoints of I which are not contained in I.

Motivating Example 2.12.8 Find the global minimum and global maximum values of $f(x) = x^3 - 6x^2 + 9x + 2$ with domain the open interval $(-1, 5)$.

Solution Begin by finding the local extrema of f. Note that
$$f'(x) = 3x^2 - 12x + 9 = 3(x^2 - 4x + 3) = 3(x - 1)(x - 3) .$$
Hence f has local extrema at $x = 1$ and at $x = 3$ where the derivative vanishes. The sign of $f'(x)$ is plotted at the left in Figure 2.12.9. By the First Derivative Test, f has a local maximum at $x = 1$ and a local minimum at $x = 3$. Using the information from the sign of $f'(x)$, we sketch the graph of f at the right in Figure 2.12.9. Note $x = 1$ is not a global maximum of f because $f(x) > f(1) = 6$ for x near 5. Also $x = 3$ is not a global minimum of f because $f(x) < f(3) = 2$ for x near -1. Observe $x = 5$ can not be the global maximum of f nor can $x = -1$ be the global minimum of f because the endpoints -1 and 5 of the interval $(-1, 5)$ are not in the domain of f. Therefore, f has neither a global minimum nor a global maximum even though f has a local minimum and a local maximum. □

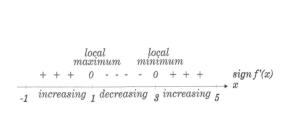

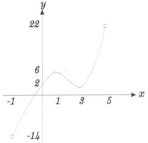

Figure 2.12.9 $f(x) = x^3 - 6x^2 + 9x + 2$

Using the preceding example as a guide, we describe how to find the global minimum and global maximum of a continuous function f whose domain is an interval I with endpoints $a < b$. We compare the following numbers: [7]

(1) $f(c)$ for c a critical point of f;   (2) $f(a)$ if $a \in I$;   (3) $f(b)$ if $b \in I$;

(4) $\lim_{x \to a^+} f(x)$ if $a \notin I$;   (5) $\lim_{x \to b^-} f(x)$ if $b \notin I$.

If the smallest number occurs in (1), (2) or (3), we have located the global minimum of f. However, if the smallest number occurs in (4) or (5), f has no global minimum. Similarly, if the largest number occurs in (1), (2) or (3), we have located the global maximum of f. However, if the largest number occurs in (4) or (5), f has no global maximum. The following examples use graphs to illustrate various phenomena that occur when we search for global minimum and maximum values of a function. In particular, a function may have the same minimum or maximum value at different points.

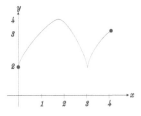

Figure 2.12.10
$y = f(x)$

Examples 2.12.11 Determine the global minimum and global maximum values of each function.

(1) Let f be the function with domain $I = [0, 4]$ whose graph is given in Figure 2.12.10.

    Solution f has a local maximum at $x = 2$ where $f'(2) = 0$. It has a local minimum at $x = 3$ where the derivative does not exist and the graph has a cusp. The value of f is 2 at both the left endpoint $x = 0$ of $[0, 4]$ and at the cusp at $x = 3$. Hence f has a global minimum at these two points. Since the value of f at the right endpoint of $[0, 4]$ is $f(4) = 3$ which is less than $f(2) = 4$, f has a global maximum at $x = 2$.

(2) Let g be the function with domain $(1, 4)$ whose graph is given in Figure 2.12.12.

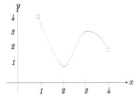

Figure 2.12.12
$y = g(x)$

[7] In this paragraph the word *number* includes the symbols $-\infty$ and $+\infty$.

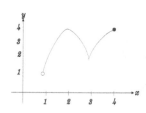

Figure 2.12.13
$y = h(x)$

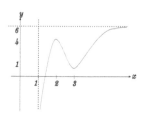

Figure 2.12.14
$y = k(x)$

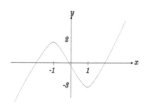

Figure 2.12.15
$y = m(x)$

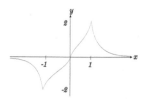

Figure 2.12.16
$y = p(x)$

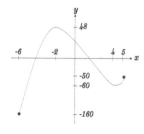

Figure 2.12.17 $g(x)$
$= x^3 - 3x^2 - 24x + 20$

Solution g has a local minimum at $x = 2$ and a local maximum at $x = 3$ where the derivative vanishes. Since $\lim_{x \to 1^+} g(x) = 4 > g(3) = 3$, g has no global maximum. Since $\lim_{x \to 4^-} g(x) = 2 > g(2) = 1$, g has a global minimum at $x = 2$.

(3) Let h be the function with domain $(1, 4]$ whose graph is given in Figure 2.12.13.

Solution Observe that h has a local maximum at $x = 2$ where the derivative vanishes. At $x = 3$, the derivative does not exist, the graph has a cusp and h has a local minimum. Since $h(2) = 4$ is the same value as $h(4) = 4$ at the right endpoint of $(1, 4]$, there is a global maximum at each of these points. Since $\lim_{x \to 1^+} h(x) = 1 < h(3) = 2$, h has no global minimum.

(4) Let k be the function with domain $(1, \infty)$ whose graph is given in Figure 2.12.14.

Solution Note k has a local maximum at $x = 2$ and a local minimum at $x = 3$ where the derivative vanishes. Observe that k decreases towards its vertical asymptote at the left endpoint $x = 1$ of $(1, \infty)$: $\lim_{x \to 1^+} k(x) = -\infty$. Hence k has no global minimum. Observe that as x approaches the right endpoint ∞ of $(1, \infty)$, the graph of k approaches its horizontal asymptote $y = 6$ from below: $\lim_{x \to \infty} k(x) = 6$. Since $k(2) = 4 < 6$, k has no global maximum.

(5) Let m be the function with domain $\Re$ whose graph is given in Figure 2.12.15.

Solution Note m has a local maximum at $x = -1$ and a local minimum at $x = 1$ where the derivative vanishes. The limit as x approaches the left endpoint $-\infty$ of $\Re = (-\infty, \infty)$ is $\lim_{x \to -\infty} m(x) = -\infty$. Hence m has no global minimum. The limit as x approaches the right endpoint ∞ of $\Re = (-\infty, \infty)$ is $\lim_{x \to \infty} m(x) = \infty$. Hence m has no global maximum.

(6) Let p be the function with domain $\Re$ whose graph is given in Figure 2.12.16.

Solution Note p has a local minimum at $x = -1$ and a local maximum at $x = 1$ where its derivative does not exist and the graph has cusps. As x approaches either the left or right endpoints of $\Re = (-\infty, \infty)$, the graph of p approaches the x–axis as a horizontal asymptote: $\lim_{x \to -\infty} p(x) = \lim_{x \to \infty} p(x) = 0$. Hence p has a global minimum at $x = -1$ where $p(-1) = -2$ and a global maximum at $x = 1$ where $p(1) = +2$.   □

In the next group of examples, we determine the global minimum and maximum values of a function f defined by an equation or a word problem. The domain of f will be an interval I with endpoints a and b. By Theorem 2.6.6, if f has a local extremum at an interior point $c \in (a, b)$, then c is a critical point of f. We compute $f'(x)$ to find the critical points. We use the First Derivative Test or the Second Derivative Test to determine whether $x = c$ is a local maximum, a local minimum or neither. Then we follow the procedure used to analyze the preceding examples. That is, we compute the values of f at the local extrema and the value(s) of f at the endpoint(s) of I which are in I. Then we compare these numbers with the limit(s) of f as x approaches the endpoint(s) of I which are not in I.

Examples 2.12.18 (1) Find the global minimum and maximum values of $g(x) = x^3 - 3x^2 - 24x + 20$ with domain $I = [-6, 5]$.

Solution Observe that

$$g'(x) = 3x^2 - 6x - 24 = 3(x^2 - 2x - 8) = 3(x - 4)(x + 2)$$

Thus, g has critical points at $x = -2$ and $x = 4$ where $g'(x)$ vanishes. We use the Second Derivative Test to identify the critical points.

$$g''(x) = 6x - 6$$

Since $g''(-2) = -18 < 0$, we conclude that $x = -2$ is a local maximum. Since $g''(4) = 18 > 0$, we see that $x = 4$ is a local minimum. Observe that both endpoints $-6, 5$ of I are contained in I. Therefore, we compare the values

$$g(-6) = -160, \quad g(-2) = 48, \quad g(4) = -60, \quad g(5) = -50.$$

Since $g(-2)$ is the largest of these numbers, g has its maximum value at $x = -2$. Since $g(-6)$ is the smallest of these numbers, g has its minimum value at $x = -6$. From this information, we sketch the graph of g in Figure 2.12.17. Note we did not use this graph to find the global maximum and minimum values of g.

(2) Find the global minimum and maximum values of $h(x) = \frac{2-x}{x^2-1}$ with domain $I = (1, 10]$.

Solution By the quotient rule:

$$h'(x) = \frac{(-1)(x^2 - 1) - (2 - x)(2x)}{(x^2 - 1)^2} = \frac{x^2 - 4x + 1}{(x^2 - 1)^2}.$$

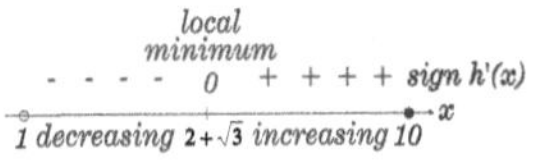

By the quadratic formula, the numerator $x^2 - 4x + 1$ has roots $x = \frac{4 \pm \sqrt{12}}{2} = 2 \pm \sqrt{3}$. Only the root $x = 2 + \sqrt{3}$ is in I. We use the First Derivative Test in Figure 2.12.19 to identify this critical point. Since $h'(x) < 0$ for $x < 2 + \sqrt{3}$ and $h'(x) > 0$ for $x > 2 + \sqrt{3}$, h has a local minimum at $x = 2 + \sqrt{3}$. The endpoint 10 of I is contained in I, while the endpoint 1 of I is not contained in I. Since the denominator of h vanishes at $x = 1$, h has a vertical asymptote there and $\lim_{x \to 1^+} h(x) = \infty$. Thus, h has no global maximum. To determine the minimum value of h compare

$$h(2 + \sqrt{3}) = -\frac{\sqrt{3}}{6 + 4\sqrt{3}} \approx -.13 \quad \text{and} \quad h(10) = -\frac{8}{99} \approx -.08$$

Since the smaller of these two numbers is $h(2 + \sqrt{3})$, h has a global minimum at $x = 2 + \sqrt{3}$. From this information, we sketch the graph of h in Figure 2.12.20. Note we did not use this graph in our analysis.

(3) Find the global minimum and maximum values of $k(x) = \frac{x^{2/3}}{x^{2/3}+1}$ with domain $I = \Re$.

Solution By the quotient rule:

$$k'(x) = \frac{\frac{2}{3}x^{-1/3}(x^{2/3} + 1) - x^{2/3}(\frac{2}{3}x^{-1/3})}{(x^{2/3} + 1)^2} = \frac{\frac{2}{3}x^{-1/3}}{(x^{2/3} + 1)^2} = \frac{2}{3x^{1/3}(x^{2/3} + 1)^2}.$$

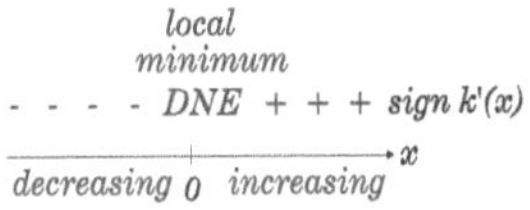

The only critical point of k occurs at $x = 0$, where $k'(x)$ does not exist. We use the First Derivative Test to identify this critical point in Figure 2.12.21. Since $k'(x) < 0$ for $x < 0$ and $k'(x) > 0$ for $x > 0$, k has a local minimum at $x = 0$. Neither of the endpoints $-\infty, \infty$ of I are in I. To evaluate the limit of k as x approaches $-\infty$ or ∞, divide the numerator and denominator of $k(x)$ by $x^{2/3}$:

$$k(x) = \frac{1}{1 + \frac{1}{x^{2/3}}} \quad \text{and} \quad \lim_{x \to -\infty} k(x) = \lim_{x \to \infty} k(x) = \frac{1}{1 + 0} = 1.$$

Since $k(0) = 0$ is less than the value 1 of these limits, k has a global minimum at $x = 0$, and k has no global maximum. Since $k'_-(0) = -\infty$ and $k'_+(0) = +\infty$,

local minimum

– – – – 0 + + + sign $h'(x)$

1 *decreasing* $2+\sqrt{3}$ *increasing* 10

Figure 2.12.19
$h'(x) = \frac{x^2-4x+1}{(x^2-1)^2}$

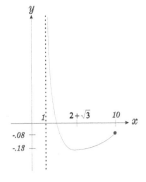

Figure 2.12.20
$h(x) = \frac{2-x}{x^2-1}$

local minimum

– – – – DNE + + + sign $k'(x)$

decreasing 0 *increasing*

Figure 2.12.21
$k'(x) = \frac{2}{3x^{1/3}(x^{2/3}+1)^2}$

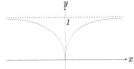

Figure 2.12.22
$k(x) = \frac{x^{2/3}}{x^{2/3}+1}$

k has a vertical cusp at the origin. The limits above indicate that k has the line $y = 1$ as a horizontal asymptote on both the left and the right. We sketch the graph of k in Figure 2.12.22. However, we did not use this graph to find the global minimum and maximum values of k.

(4) A box with no top is to be constructed from a sheet of cardboard which measures 3 meters by 8 meters. Four congruent squares are cut from the corners, and the remaining cardboard is folded into a box. Find the dimensions of the box of largest volume which can be constructed in this manner.

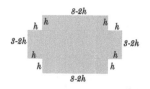

Figure 2.12.23
Cardboard

Solution Let h denote the length of the side of each of the squares which are cut from the corners of the cardboard sheet. The remaining cardboard will have the dimensions given in Figure 2.12.23. Note h will be the height of the box, $8 - 2h$ will be its length and $3 - 2h$ will be its width. For all three of these dimensions to be non–negative, we require $0 \leq h \leq \frac{3}{2}$. The volume V of this box is the product of its height, width and length:

$$V \;=\; h(3 - 2h)(8 - 2h) = 4h^3 - 22h^2 + 24h \quad \text{with domain } \left[0, \frac{3}{2}\right].$$

Then $\dfrac{dV}{dh} \;=\; 12h^2 - 44h + 24 = 4(3h^2 - 11h + 6) = 4(3h - 2)(h - 3).$

Figure 2.12.24
$V'(h) = 12h^2 - 44h + 24$

Thus, V has critical points at $h = \frac{2}{3}$ and at $h = 3$. However, only $h = \frac{2}{3}$ is in the domain of V. We use the First Derivative Test in Figure 2.12.24 to identify the behavior of V at this critical point. Since $V'(h) > 0$ for $0 < h < \frac{2}{3}$ and $V'(h) < 0$ for $\frac{2}{3} < h < \frac{3}{2}$, V has a local maximum at $h = \frac{2}{3}$. Moreover, the values of V at the endpoints $h = 0$ and $h = \frac{3}{2}$ of the domain of V are zero. Therefore, V has a global maximum at $h = \frac{2}{3}$. This box of maximum volume has height $h = \frac{2}{3}$ meters, width $3 - 2h = \frac{5}{3}$ meters and length $8 - 2h = \frac{20}{3}$ meters.

(5) A silo with volume $1,000$ cubic meters is to be constructed with cylindrical walls and a hemispheric roof. See Figure 2.12.25. The walls cost \$90 per square meter to construct, and the roof costs \$120 per square meter to construct. Find the dimensions of the most economical silo.

Figure 2.12.25 Silo

Solution Introduce the following variables to discuss this problem:

    let R denote the radius of the cylinder and the hemisphere in meters;

    let H denote the height of the cylinder in meters;

    let C denote the cost of building the silo in dollars.

The volume of the cylinder is $\pi R^2 H$, and the volume of a whole sphere of radius R is $\frac{4}{3}\pi R^3$. The volume $1,000$ of the silo is the volume of the cylinder plus half the volume of the sphere:

$$1,000 \;=\; \pi R^2 H + \frac{2}{3}\pi R^3.$$

Solve for H: $\qquad H \;=\; \dfrac{1,000 - \frac{2}{3}\pi R^3}{\pi R^2} = \dfrac{1,000}{\pi R^2} - \dfrac{2R}{3}.$ $\qquad$ (2.12.2

If we cut the cylinder along its height and fold it flat, we obtain a rectangle of dimensions H and $2\pi R$. Thus, the cylinder has surface area $2\pi RH$, and the cost to build the walls is $90(2\pi RH) = 180\pi RH$ dollars. The surface area of an entire sphere is $4\pi R^2$. Thus, the cost to build the roof is $120(2\pi R^2) = 240\pi R^2$ dollars. Therefore, the total cost to build the silo is

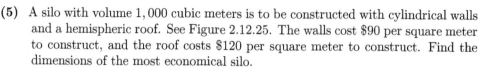

$$C = 180\pi RH + 240\pi R^2 = 180\pi R\left(\dfrac{1,000}{\pi R^2} - \dfrac{2R}{3}\right) + 240\pi R^2 = \dfrac{180,000}{R} + 120\pi R^2$$

where we used the value of H given by (2.12.2). Then

$$\frac{dC}{dR} = -\frac{180,000}{R^2} + 240\pi R.$$

C has a critical point where $\frac{dC}{dR} = 0$, i.e. where

$$240\pi R = \frac{180,000}{R^2}, \qquad R^3 = \frac{180,000}{240\pi} = \frac{750}{\pi}$$

$$R = \sqrt[3]{\frac{750}{\pi}} = 5\sqrt[3]{\frac{6}{\pi}} \approx 6.2 \text{ meters}.$$

We identify this critical point by the First Derivative Test in Figure 2.12.26. Since $C'(R) < 0$ for $0 < R < 5\sqrt[3]{\frac{6}{\pi}}$ and $C'(r) > 0$ for $5\sqrt[3]{\frac{6}{\pi}} < R$, C has a local minimum at $R = 5\sqrt[3]{\frac{6}{\pi}}$. Since C has larger value to the left and to the right of $R = 5\sqrt[3]{\frac{6}{\pi}}$, it follows that $R = 5\sqrt[3]{\frac{6}{\pi}}$ is a global minimum. By (2.12.2) the corresponding value of H at this global minimum is

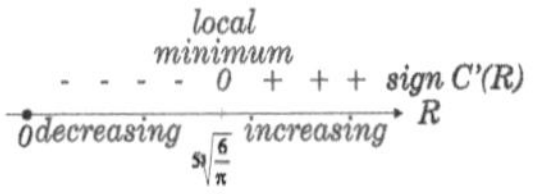

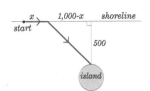

Figure 2.12.26
$\frac{dC}{dR} = 240\pi R - \frac{180,000}{R^2}$

$$H = \frac{1,000 - \frac{2}{3}\pi\left(5\sqrt[3]{6/\pi}\right)^3}{\pi\left(5\sqrt[3]{6/\pi}\right)^2} = \frac{500}{25\pi\sqrt[3]{36/\pi^2}} = \frac{20}{\sqrt[3]{36\pi}} \approx 4.1 \text{ meters}.$$

Thus the silo of least cost has radius $R = 6.2$ meters and height $H = 4.1$ meters.

(6) An island lies 500 meters south of a straight east–west shoreline. A racer starts on the shore 1,000 meters west of the island, runs eastwards along the shore at 7 km/hr and then swims in a straight line to the island at 3 km/hr. How far should the racer run to reach the island in the shortest possible time?

Figure 2.12.27
Path of the Racer

Solution Let x denote the distance in meters that the racer runs along the shore. Since the racer runs at 7000 meters/hr, the running will take $\frac{x}{7000}$ hours. Apply the Pythagorean Theorem to the right triangle in Figure 2.12.27 to see that the racer swims $\sqrt{500^2 + (1,000 - x)^2}$ meters $= \sqrt{500^2 + (1,000 - x)^2}/1000$ kilometers. This will take $\sqrt{500^2 + (1,000 - x)^2}/3000$ hours. Let T denote the time in hours that it takes for the racer to reach the island from the the the starting point. Then T has domain $[0, 1000]$ with

$$T = \frac{x}{7000} + \frac{\sqrt{500^2 + (1,000 - x)^2}}{3000}.$$

By the chain rule :
$$\frac{dT}{dx} = \frac{1}{7000} + \frac{2(1,000 - x)(-1)}{(3000)(2)\sqrt{500^2 + (1,000 - x)^2}}.$$

Critical points occur when $0 = \frac{dT}{dx}$:

$$\frac{1}{7000} = \frac{1,000 - x}{3000\sqrt{500^2 + (1,000 - x)^2}}$$

$$3\sqrt{500^2 + (1,000 - x)^2} = 7(1,000 - x) \qquad \text{[cross − multiplying]}$$

$$9\left[500^2 + (1,000 - x)^2\right] = 49(1,000 - x)^2 \qquad \text{[squaring]}$$

$$9(500^2) = 40(1,000 - x)^2$$

$$\frac{3}{\sqrt{40}}(500) = 1,000 - x \qquad \text{[taking the square − root]}$$

$$x = 1,000 - \frac{750}{\sqrt{10}} \approx 763 \text{ meters}.$$

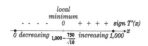

Figure 2.12.28
First Derivative Test
Applied to T

We use the First Derivative Test to identify this critical point in Figure 2.12.28. $T'(x) < 0$ for $0 < x < 1,000 - \frac{750}{\sqrt{10}}$ and $T'(x) > 0$ for $1,000 - \frac{750}{\sqrt{10}} < x < 1000$. Hence $x = 1,000 - \frac{750}{\sqrt{10}}$ meters is a local minimum. Since T has larger values to the left and right of $x = 1,000 - \frac{750}{\sqrt{10}}$, it follows that $x = 1,000 - \frac{750}{\sqrt{10}}$ is a global minimum. Thus the quickest way to reach the island is to run along the shore for 763 meters and then swim directly to the island.   □

Historical Remarks

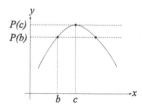

Figure 2.12.29
Fermat's Method

In the late 1620s Pierre de Fermat devised an algorithm for finding a local extremum c of a polynomial p. He observes that for b near c the horizontal line $y = p(b)$ intersects the graph of $y = p(x)$ at two points. As b approaches c, these two points approach one another and coincide when $b = c$. That is, the polynomial $p(x) - p(c)$ has a double root at $x = c$, i.e. $p(x) - p(c)$ is divisible by $(x - c)^2$. See Figure 2.12.29. To find the unknown local extremum c, Fermat divides $p(x) - p(c)$ by $x - c$, simplifies the quotient and replaces x by c producing $p'(c)$ in our notation. Then he sets this polynomial in c equal to zero and solves for c, i.e. Fermat solves the equation $p'(c) = 0$.

Both Newton and Leibniz recognized that local extrema of differentiable functions occur at points where the tangent line is horizontal. They used their methods for computing derivatives to determine the local extrema of specific functions. The Second Derivative Test was discovered by Newton's student Colin Maclaurin in 1742 using infinite series methods.

Summary

The reader should be able to apply the Second Derivative Test to identify a critical point, where the first derivative vanishes and the second derivative is nonzero, as a local maximum or a local minimum. In other cases, she should use the First Derivative Test to identify the critical point. She should know how to find the global maximum and minimum values of a function with domain an interval. In particular, she should compare the values of the function at its local extrema with the value or limit of the function at each endpoint of its domain. She should be able to apply this method to solve maxima–minima problems.

Basic Exercises

Identify the critical points of each function by the Second Derivative Test.

1. $f(x) = 2x^3 - 3x^2 - 12x + 7$ **2.** $g(x) = x^5 - 5x + 3$ **3.** $h(x) = \frac{1}{x^2 - 1}$

4. $j(x) = \frac{x^2}{x^2 + 1}$ **5.** $k(x) = \frac{1}{2 + \sin x}$ **6.** $m(x) = \arctan(x^2)$

7. State the information given by the Second Derivative Test at the critical point x_0.
  (a) $f(x) = \tan^2 x$, $x_0 = 0$ **(b)** $g(x) = (x + 7)^{2/5}$, $x_0 = -7$
  (c) $h(x) = x^3 + 6x^2 - 15x + 7$, $x_0 = -5$ **(d)** $j(x) = \cos \pi(x^2 - 1)$, $x_0 = 0$
  (e) $k(x) = x^4 - 8x^3 + 24x^2 - 32x + 16$, $x_0 = 2$ **(f)** $m(x) = \arcsin(x^3)$, $x_0 = 0$

In Exercises 8 to 19, find the global minimum and global maximum of each function.

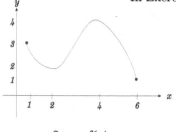

8. $y = f(x)$

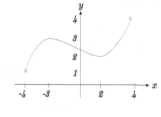

9. $y = g(x)$

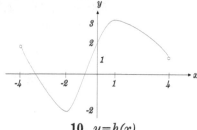

10. $y = h(x)$

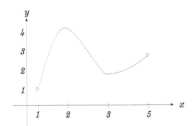

11. $y=i(x)$

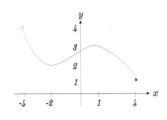

12. $y=j(x)$

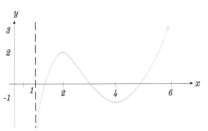

13. $y=k(x)$

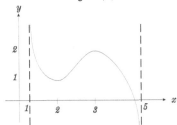

14. $y=m(x)$

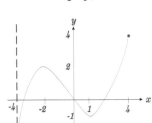

15. $y=n(x)$

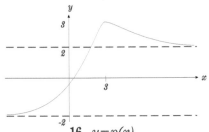

16. $y=p(x)$

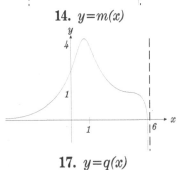

17. $y=q(x)$

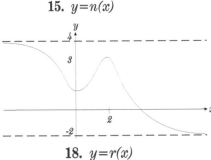

18. $y=r(x)$

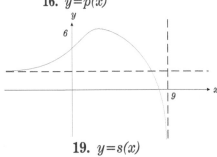

19. $y=s(x)$

Find the global minimum and global maximum of each function.

20. $f(x) = x^3 - 6x^2 - 15x + 8$ with domain $[-2, 6]$.

21. $g(x) = x^2$ with domain $(-1, 3]$.

22. $h(x) = \frac{x-3}{x^2-4}$ with domain $(2, \infty)$.

23. $j(x) = x^3 - 3x^2 + 9x - 27$ with domain $(-1, 4)$.

24. $k(x) = \frac{x}{x^2+1}$ with domain $\Re$.

25. $m(x) = x^3 + 7x^2 - 5x - 10$ with domain $(-6, 1)$.

26. $n(x) = \frac{x^{4/5}}{x^{4/5}+2}$ with domain $\Re$.

27. $p(x) = \frac{1}{1+\tan^2 x}$ with domain $(-\pi/2, \pi/2)$.

28. $q(x) = \arctan(x^2 + 1)$ with domain $\Re$.

29. $r(x) = |x^2 - 1|$ with domain $[-4, 6]$.

30. Find the numbers s and t with the largest product in each case.
   (a) The sum of s and t is 60.
   (b) The sum of the squares of s and t is 20.
   (c) The sum of the square roots of s and t is 30.
   (d) The sum of the absolute values of s and t is 10.
   (e) The sum of the reciprocals of s and t is 5.

31. In each case, find the shortest fence which meets the given conditions.
(a) The fence is a rectangle which encloses an area of 100 square meters.
(b) The fence forms three sides of a rectangle of area 150 square feet with the fourth side of the rectangle the side of a barn.
(c) The fence consists of two sides of an isosceles triangle with the third side a stone wall. The isosceles triangle encloses an area of 60 square meters.

32. In each case find the dimensions of the least expensive container.
(a) The container is a rectangular box with square bottom of volume 3 cubic meters. It is constructed from three types of materials. The material used for the bottom costs three times as much per square meter as the material used for the sides, while the material used for the top costs half as much per square meter as the material used for the sides.
(b) The container is a cylinder of volume 50 cubic cm. Its sides are constructed from thin tin which costs ten cents per square cm, while its top and bottom are constructed from thicker tin, which costs fifteen cents per square cm.
(c) The container is a paper cone which is to be filled with 10 cubic inches of soda pop. The cone is made from recycled paper which costs one cent per square foot.

33. A foot long wire is cut into two pieces. One piece will be bent into a circle, and the other piece will be bent into a square. How should the wire be cut so that the sum of the areas of the circle and square is largest? How should the wire be cut so that the sum of the areas of the circle and square is smallest?

34. In each case, find the point on the graph of f which is closest to the point P.
(a) $f(x) = 3x + 1$ and $P = (2, 10)$.   **(b)** $f(x) = x^2$ and $P = (0, 4)$.
(c) $f(x) = \sqrt{2x - 4}$ and $P = (2, 6)$.   **(d)** $f(x) = \frac{2}{x}$ and $P = (4, 5)$.

35. Find the rectangle of largest area which lies in the first quadrant, has one corner at the origin and opposite corner on the line $5x + 3y = 1$.

36. Find the right triangles in the first quadrant of largest and smallest areas that have their right angles at the origin with hypotenuses passing through the point $(4, 7)$.

37. **(a)** Find the volume of the cylinder of greatest volume which can be packed inside a spherical bag of radius R.
(b) Find the volume of the cone of greatest volume which can be packed inside a spherical bag of radius R.
(c) Find the volume of the conical bag of least volume which contains a sphere of radius R.
(d) Find the dimensions of the largest cylinder which can be packed inside a conical bag whose radius is 8 cm and whose height is 16 cm. (Assume the bases of both the cylinder and cone are horizontal.)

38. A window is constructed in the shape of a rectangle surmounted by a semi–circle. If the perimeter of the window is 10 meters, find the dimensions of the window with the largest area?

39. A 2 m high fence is parallel to the wall of a tall building. If the distance between the fence and the building is 3 m, what is the length of the shortest ladder which, when placed on the ground, passes over the fence and rests against the building?

40. A 30 inch wide strip of metal is folded lengthwise into thirds to form a feeding trough with a trapezoid cross–section. (The angles of both sides of this trough with the bottom of the trough are equal.) What is the depth of this type of trough that has the greatest cross–sectional area?

41. A kayak race starts at a point P on the shore of a circular lake. The goal is to reach the diametrically opposite point Q on the lakeshore. The rules allow carrying the kayak part of the way around the lake to a point M and then paddling to the finishing point Q. A racer knows that she can run with the kayak twice as fast as she can paddle it. How should she select M to complete the race in the shortest time?

Challenging Problems

1. An art critic is positioning himself to view a rectangular mural which is 4 m. high and 5 m. long. If the critic's eyes are 2 m. from the floor and the bottom edge of the mural is 3 m. from the floor, how far from the mural should the critic stand so that he has the greatest angle of vision? (The angle of vision is the angle between the lines joining the critic's eyes with the bottom and top edges of the mural.)

2. A pole is carried around a 90° corner in a hall. The hall on one side of the corner is 4 m wide, and the hall at the other side of the corner is 5 m wide.
(a) What is the length of the longest pole that can be carried around this corner in a horizontal position?
(b) If the halls are 3 m high, what is the longest pole that can be carried around this corner in any position?

3. One corner C of a strip of paper 10 cm wide is folded so that, after being folded, the corner C lies on the side of the strip opposite C. How should the paper be folded so that the length of the fold is least?

4. Experiments show that light travels between two points in such a way that it arrives in the least possible time.
(a) A beam of light travels from a point P to a point Q by reflection in a straight mirror. Show that the point M where the light reflects on the mirror satisfies the property that the angle of incidence θ equals the angle of reflection ϕ as shown in Figure 2.12.30.
(b) A beam of light passes from a bulb in the ceiling of an indoor swimming pool to a swimmer under the water surface of the pool. The speed of light in air is v_1 km/sec, while the speed of light in water is v_2 km/sec. Show that the path of light, as indicated in Figure 2.12.31, satisfies the condition:

$$\frac{\sin\theta_1}{\sin\theta_2} = \frac{v_1}{v_2}.$$

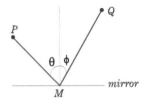

Figure 2.12.30
Reflection of Light

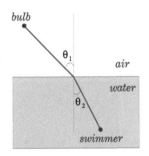

Figure 2.12.31
Refraction of Light

Theory

2.13   Mean Value Theorems and L'Hôpital's Rule

Prerequisite: Section 2.8

There are two fundamental theorems we have been using based upon intuitive justifications: the Mean Value Theorem and L'Hôpital's Rule. The Mean Value Theorem was used to solve differential equations, justify the First and Second Derivative Tests and sketch graphs. L'Hôpital's Rule was used to evaluate limits which are indeterminate forms. In the first subsection, we give a rigorous proof of the Mean Value Theorem. In the second subsection we define curves and find their tangent lines. The third subsection exposits Cauchy's Mean Value Theorem, the generalization of the Mean Value Theorem from graphs to curves. In the last subsection we use Cauchy's Mean Value Theorem to prove L'Hôpital's Rule.

Mean Value Theorem

Recall that Rolle's Theorem says that if f is continuous on $[a, b]$ and differentiable on (a, b) with $f(a) = f(b)$, then there is at least one number $a < c < b$ such that $f'(c) = 0$. In other words, Rolle's Theorem says that if the secant line joining the endpoints of the graph of f is horizontal, then the graph of f must have a horizontal tangent line. The Mean Value Theorem generalizes this result: if f is continuous on $[a, b]$ and differentiable on (a, b), then the graph of f has a tangent line that is parallel to the secant line joining the endpoints of its graph. In Section 6, we gave an intuitive justification that Rolle's Theorem implies the Mean Value Theorem. We now give a rigorous proof of this implication.

Theorem 2.13.1 [Mean Value Theorem] *Let f be a continuous function with domain the closed interval $[a, b]$. Assume f is differentiable on the open interval (a, b). Then there is at least one number $c \in (a, b)$ such that*

$$f'(c) = \frac{f(b) - f(a)}{b - a}.$$

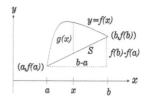

Figure 2.13.2
  Definition of $g(x)$

Proof Let S be the secant line which joins the endpoints $(a, f(a))$ and $(b, f(b))$ of the graph of f. Define $g(x)$ as the vertical distance between the point $(x, f(x))$ on the graph of f and the secant line S. See Figure 2.13.2. We find a formula for $g(x)$. Since S passes through the points $(a, f(a))$ and $(b, f(b))$, S has slope $\frac{f(b)-f(a)}{b-a}$ and equation

$$y = \left(\frac{f(b) - f(a)}{b - a} \right)(x - a) + f(a).$$

Then $g(x)$ is the difference of the y–coordinate $f(x)$ of the point $(x, f(x))$ on the graph of f and the y–coordinate $\left(\frac{f(b)-f(a)}{b-a} \right)(x - a) + f(a)$ of the secant line S:

$$g(x) = f(x) - \left[\left(\frac{f(b) - f(a)}{b - a} \right)(x - a) + f(a) \right] \quad \text{and}$$

$$g'(x) = f'(x) - \frac{f(b) - f(a)}{b - a}.$$

Apply Rolle's Theorem to g on the interval $[a, b]$. Since f is continuous on $[a, b]$ and differentiable on (a, b), it follows that g is also continuous on $[a, b]$ and differentiable on (a, b). Since the secant line and the graph of f intersect at $(a, f(a))$ and $(b, f(b))$, it follows from the definition of g that $g(a) = g(b) = 0$. Thus Rolle's Theorem applies to g, and there is at least one number $c \in (a, b)$ such that

$$0 = g'(c) = f'(c) - \frac{f(b) - f(a)}{b - a} \quad \text{and} \quad f'(c) = \frac{f(b) - f(a)}{b - a}. \qquad \square$$

Curves

The graph of a function g with domain $[a, b]$ is the set of points $(x, g(x))$ for $x \in [a, b]$. We obtain more general curves when we allow the first coordinate to be a function $f(x)$ rather than x. After presenting this definition and illustrating it, we show how to find the tangent lines of these curves.

Definition 2.13.5 *Let $f(t)$, $g(t)$ be continuous functions with domain the interval I. The set of points $(f(t), g(t))$, for $t \in I$, is called a curve. If I is the closed interval $[a, b]$, then $(f(a), g(a))$ and $(f(b), g(b))$ are called the endpoints of the curve.*

Example 1 below shows that the graph of a function is a curve. In addition, curves include circles and spirals which intersect vertical lines at more than one point and are not functions. A curve may also have a loop where it crosses itself.

Examples 2.13.6 (1)   Describe the graph of the function $y = g(x)$ with domain $[a, b]$ as a curve.

   Solution Let $f(t) = t$. Graph g is the set of $(t, g(t)) = (f(t), g(t))$, for $t \in [a, b]$.

(2)   Describe the circle C_R, with radius R and center the origin, as a curve.

   Solution Recall Definition 1.4.9 of $\cos \theta$ and $\sin \theta$: $(\cos \theta, \sin \theta)$ are the coordinates of the intersection of the line L_θ of angle θ with the circle C_1. The similar triangles in Figure 2.13.3 show that L_θ intersects the circle C_R at the point $(R \cos \theta, R \sin \theta)$. Let $f(\theta) = R \cos \theta$ and $g(\theta) = R \sin \theta$. Then the circle C_R is the curve $(f(\theta), g(\theta)) = (R \cos \theta, R \sin \theta)$ for $\theta \in [0, 2\pi]$.

(3)   Sketch the curve S defined by $(\theta \cos \theta, \theta \sin \theta)$ for $\theta \geq 0$.

   Solution By Example 2, $(\theta \cos \theta, \theta \sin \theta)$ is a point on the circle C_θ. Thus we can think of S as a "circle" whose radius θ increases from zero. That is, S is a counterclockwise circular spiral which spirals outwards from the origin. See Figure 2.13.4.

(4)   Sketch the curve C defined by $(t^3 - 3t, t^2)$ for $t \in \Re$.

   Solution Let $x = t^3 - 3t$ and $y = t^2$. For $t \geq 0$, the set of values of $y = t^2$ is the interval $[0, \infty)$. Hence $x = (t^2)^{3/2} - 3(t^2)^{1/2} = y^{3/2} - 3y^{1/2} = h(y)$ is a function of y with domain $[0, \infty)$. Then

$$h'(y) = \frac{3}{2}y^{1/2} - \frac{3}{2}y^{-1/2} = \frac{3(y - 1)}{2y^{1/2}}.$$

The sign of $h'(y)$ is given in Figure 2.13.7. Note the y–intercepts of h occur when $0 = x = t^3 - 3t = t(t^2 - 3)$, i.e. at $t = 0$ and $t = \sqrt{3}$. Hence the y–intercepts of h are $y = 0^2 = 0$ and $y = (\sqrt{3})^2 = 3$. The graph of h is sketched in Figure 2.13.8. For $t \leq 0$, $(t^3 - 3t, t^2) = (-[(-t)^3 - 3(-t)], (-t)^2)$ with $-t \geq 0$. Let $u = -t$. Hence the points of C with $t \leq 0$ are $(-h(u), u)$ for $u \geq 0$ i.e. we have the graph of h reflected about the y–axis. The entire curve C is sketched in Fig. 2.13.9. $\square$

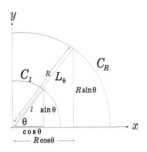

Figure 2.13.3
The Circle C_R

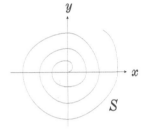

Figure 2.13.4
The Spiral S

Figure 2.13.7
$h'(y) = \frac{3(y-1)}{2y^{1/2}}$

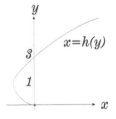

Figure 2.13.8
$x = h(y)$

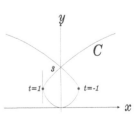

Figure 2.13.9
The Curve C

The definition of the tangent line T to a graph as the limit of secant lines also applies to define the tangent line to a curve. In particular, the slope of the tangent line is the limit of the slopes of these secant lines. See Figure 2.13.11.

Definition 2.13.10 *Let C be the curve $(f(t), g(t))$ with domain the interval I and t_0 an interior point of I. The tangent line T to the curve C at $t = t_0$ is the line through the point $(f(t_0), g(t_0))$ with slope*

$$m = \lim_{t \to t_0} \frac{g(t) - g(t_0)}{f(t) - f(t_0)}$$

when this limit is a number. When this limit is either $+\infty$ or $-\infty$, define T as the vertical line through the point $(f(t_0), g(t_0))$.

Notes **(1)** If m is a number, then T has equation $y = g(t_0) + m\,[x - f(t_0)]$.
(2) When T is vertical, it has equation $x = f(t_0)$.

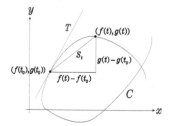

Figure 2.13.11 Tangent Line to a Curve

We evaluate the limit of the slopes of the secant lines S_t in terms of the derivatives f' and g'.

Proposition 2.13.12 *Let f, g be differentiable functions with domain the interval I such that f' is continuous at the point t_0 in the interior of I. Let C be the curve $(f(t), g(t))$ for $t \in I$. Assume $f'(t_0)$ and $g'(t_0)$ are not both zero. Then the tangent line T to the curve C at $t = t_0$ exists. If $f'(t_0) \neq 0$, then*

$$m = \text{slope } T = \frac{g'(t_0)}{f'(t_0)}.$$

If $f'(t_0) = 0$ and $g'(t_0) \neq 0$, the tangent line T is vertical.

Proof Assume first that $f'(t_0) \neq 0$. Let $t = t_0 + h$, $h \neq 0$, with S_t the secant line through the points $(f(t_0), g(t_0))$ and $(f(t), g(t)) = (f(t_0 + h), g(t_0 + h))$. By Definition 2.13.10, the tangent line T is defined as the limit of the secant lines S_t as t approaches t_0, i.e. as h approaches zero. By Figure 2.13.11,

$$\text{Slope } T = \lim_{t \to t_0} \text{Slope } S_t = \lim_{t \to t_0} \frac{g(t) - g(t_0)}{f(t) - f(t_0)} = \lim_{h \to 0} \frac{g(t_0 + h) - g(t_0)}{f(t_0 + h) - f(t_0)}.$$

Since $f'(t_0) \neq 0$ and $f'(t)$ is continuous at $x = t_0$, f is either increasing or decreasing on an open interval J containing t_0. Hence the denominator $f(t_0 + h) - f(t_0)$ of the fraction above is nonzero when h is small enough so that $t_0 + h \in J$. Divide the numerator and denominator of this fraction by h:

$$\text{Slope } T = \lim_{h \to 0} \frac{\frac{g(t_0 + h) - g(t_0)}{h}}{\frac{f(t_0 + h) - f(t_0)}{h}} = \frac{\lim\limits_{h \to 0} \dfrac{g(t_0 + h) - g(t_0)}{h}}{\lim\limits_{h \to 0} \dfrac{f(t_0 + h) - f(t_0)}{h}} = \frac{g'(t_0)}{f'(t_0)}.$$

Assume now that $f'(t_0) = 0$ and $g'(t_0) \neq 0$. Consider the curve C' defined as $(g(t), f(t))$ for $t \in I$. By the preceding case of this theorem, C' has a horizontal tangent line at the point $(g(t_0), f(t_0))$. Observe that C is the reflection of the curve C' about the line $y = x$. Therefore, C has a vertical tangent line at the point $(f(t_0), g(t_0))$ with equation $x = f(t_0)$. $\qquad\qquad\square$

The case of a point $(f(t_0), g(t_0))$ on the curve C where $f'(t_0) = g'(t_0) = 0$ is not covered by the preceding theorem. As Examples 3, 4 below show, the tangent line to C may or may not exist at such points.

Examples 2.13.13 (1) A is the spiral $(\theta \cos \theta, \theta \sin \theta)$ for $\theta \geq 0$ of Example 2.13.6 (3). Find the tangent line to the curve A at $\theta = \frac{\pi}{3}$.

Solution Let $f(\theta) = \theta \cos \theta$ and $g(\theta) = \theta \sin \theta$. By the product rule: $f'(\theta) = \cos \theta - \theta \sin \theta$ and $g'(\theta) = \sin \theta + \theta \cos \theta$. Hence T has equation:

$$
\begin{aligned}
y &= g\left(\frac{\pi}{3}\right) + \frac{g'\left(\frac{\pi}{3}\right)}{f'\left(\frac{\pi}{3}\right)}\left[x - f\left(\frac{\pi}{3}\right)\right] = \frac{\pi}{3}\frac{\sqrt{3}}{2} + \frac{\frac{\sqrt{3}}{2} + \frac{\pi}{3}\frac{1}{2}}{\frac{1}{2} - \frac{\pi}{3}\frac{\sqrt{3}}{2}}\left(x - \frac{\pi}{3}\frac{1}{2}\right) \\
&= \frac{\pi\sqrt{3}}{6} + \frac{3\sqrt{3} + \pi}{3 - \pi\sqrt{3}}\left(x - \frac{\pi}{6}\right)
\end{aligned}
$$

(2) Let B be the curve of Example 2.13.6 (4) defined by $(t^3 - 3t, t^2)$ for $t \in \Re$. Find the tangent line T to the curve B at $t = 1$.

Solution Let $f(t) = t^3 - 3t$ and $g(t) = t^2$. Then $f'(t) = 3t^2 - 3$ and $g'(t) = 2t$. Thus $f'(1) = 0$, $g'(1) = 2$, and T is vertical. T passes through the point $(f(1), g(1)) = (-2, 1)$. Hence T has equation $x = -2$. See Figure 2.13.9.

(3) Let C be the curve defined by (t^3, t^6) for $t \in \Re$. Find the tangent line to the curve C at $t = 0$.

Solution Note that $t^6 = (t^3)^2$. Hence C is the parabola which is the graph of $h(x) = x^2$. Note that $D(t^3) = 3t^2$ and $D(t^6) = 6t^5$ are both zero for $t = 0$. Although Proposition 2.13.12 does not apply, we know this parabola has the x–axis as its tangent line at the origin. See Figure 2.13.14.

Figure 2.13.14
The Curve C

(4) Let D be the curve consisting of (t^3, t^2) for $t \in \Re$. Show that the tangent line T to D does not exist at $t = 0$.

Solution Let $f(t) = t^3$ and $g(t) = t^2$ for $x \in \Re$. Observe that $f'(t) = 3t^2$ and $g'(t) = 2t$. Hence $f'(0) = g'(0) = 0$, and Proposition 2.13.12 does not apply. Since $g(t) = t^2 = (t^3)^{2/3} = f(t)^{2/3}$, the curve D is the same as the graph of the function $p(x) = x^{2/3}$ for $x \in \Re$. Thus, D has a cusp at the origin. In particular, $\lim_{t \to 0^-} \frac{g(t) - g(0)}{f(t) - f(0)} = \lim_{t \to 0^-} \frac{1}{t} = -\infty$ while $\lim_{t \to 0^+} \frac{g(t) - g(0)}{f(t) - f(0)} = \lim_{t \to 0^+} \frac{1}{t} = +\infty$. Hence $\lim_{t \to 0} \frac{g(t) - g(0)}{f(t) - f(0)}$ does not exist, and T does not exist. See Fig. 2.13.15. $\square$

Figure 2.13.15
The Curve D

We define a differentiable curve to exclude curves, such as those of Examples 3, 4 above, where Proposition 2.13.12 does not apply.

Definition 2.13.16 *Let C be the curve $(f(t), g(t))$ for $t \in [a, b]$. C is differentiable if $f'(t)$, $g'(t)$ exist for $t \in (a, b)$, and there is no $t \in (a, b)$ where $f'(t) = g'(t) = 0$.*

The curves of Examples 1, 2 above are differentiable while the curves of Examples 3, 4 are not.

Cauchy's Mean Value Theorem

The Mean Value Theorem (MVT) says that the graph G of a function with domain $[a, b]$ has a tangent line parallel to the secant line S joining the endpoints of G. See the left diagram in Figure 2.13.17. The generalization of the Mean Value Theorem to a curve C with domain $[a, b]$ is called Cauchy's Mean Value Theorem. It says that C has a tangent line which is parallel to the secant line S joining the endpoints of C. See the right diagram in Figure 2.13.17.

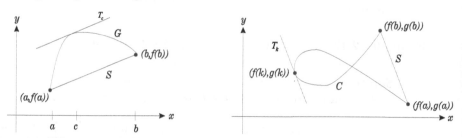

Figure 2.13.17 MVT for Graphs and Cauchy's MVT for Curves

Consider the curve C defined as $(f(t), g(t))$ for $t \in [a, b]$. We showed in Prop. 2.13.12 that if $f'(k) \neq 0$, then the slope of the tangent line T_k at $t = k$ is given by:

$$\text{slope } T_k = \frac{g'(k)}{f'(k)}. \tag{2.13.1}$$

Let S be the secant line joining the points $(f(a), g(a))$ and $(f(b), g(b))$. From Fig. 2.13.11

$$\text{slope } S = \frac{g(b) - g(a)}{f(b) - f(a)}.$$

The conclusion of Cauchy's Mean Value Theorem says that there is at least one number $k \in (a, b)$ where T_k is parallel to S. Since two lines are parallel if and only if they have the same slope, we restate the conclusion as slope T_k = slope S, i.e.

$$\frac{g'(k)}{f'(k)} = \frac{g(b) - g(a)}{f(b) - f(a)} \tag{2.13.2}$$

Observe when $f(x) = x$, the curve C is the graph of $y = g(x)$ and $f'(x) = 1$. In this case equation (2.13.2) becomes:

$$\frac{g'(k)}{1} = \frac{g(b) - g(a)}{b - a}.$$

In this special case, Cauchy's Mean Value Theorem is the Mean Value Theorem. We state Cauchy's Mean Value Theorem and use Rolle's Theorem to prove it. Instead of stating the conclusion as equation (2.13.2), we cross-multiply so that we do not have to worry that the denominators of the fractions in (2.13.2) may be zero.

Theorem 2.13.18 [Cauchy Mean Value Theorem] *Let $(f(t), g(t))$, for $t \in [a, b]$, be a differentiable curve. Then there is at least one number $k \in (a, b)$ such that*

$$f'(k) [g(b) - g(a)] = g'(k) [f(b) - f(a)]. \tag{2.13.3}$$

Proof As in the proof of the Mean Value Theorem, we construct a function h to which Rolle's Theorem applies. The function h is defined in such a way that the vanishing of its derivative at $x = k$ is equivalent to (2.13.3). Define

$$\begin{aligned} h(x) &= [f(b) - f(a)] [g(x) - g(a)] - [g(b) - g(a)] [f(x) - f(a)] \quad \text{with} \\ h'(x) &= [f(b) - f(a)] g'(x) - [g(b) - g(a)] f'(x). \end{aligned}$$

Since f and g are continuous on $[a, b]$, h is continuous on $[a, b]$. Since f and g are differentiable on (a, b), h is differentiable on (a, b). Observe that

$$
\begin{aligned}
h(a) &= [f(b) - f(a)] \cdot 0 - [g(b) - g(a)] \cdot 0 = 0 \quad \text{and} \\
h(b) &= [f(b) - f(a)] [g(b) - g(a)] - [g(b) - g(a)] [f(b) - f(a)] = 0.
\end{aligned}
$$

Thus $h(a) = h(b)$, and Rolle's Theorem applies to h. Hence there is at least one number $k \in (a, b)$ with $h'(k) = 0$:

$$
\begin{aligned}
0 = h'(k) &= [f(b) - f(a)] g'(k) - [g(b) - g(a)] f'(k) \\
[g(b) - g(a)] f'(k) &= [f(b) - f(a)] g'(k).
\end{aligned}
$$
□

We deduce the geometric interpretation of Cauchy's Mean Value Theorem which motivated its statement.

Corollary 2.13.19 *Let $(f(t), g(t))$, for $t \in [a, b]$, be a differentiable curve C with distinct endpoints. Then there is at least one number $k \in (a, b)$ where the tangent line T_k is parallel to the secant line S joining the endpoints $(f(a), g(a))$ and $(f(b), g(b))$ of C.*

Proof By Cauchy's Mean Value Theorem, there is at least one $k \in (a, b)$ where

$$
f'(k) [g(b) - g(a)] = g'(k) [f(b) - f(a)]. \tag{2.13.4}
$$

There are two cases. First consider the case where $f(a) \neq f(b)$. If $f'(k)$ were zero, then $g'(k)$ would also be zero. Since C is a differentiable curve this can not happen. Hence $f'(k) \neq 0$, and we divide equation (2.13.4) by the nonzero number $f'(k) [f(b) - f(a)]$:

$$
\frac{g(b) - g(a)}{f(b) - f(a)} = \frac{g'(k)}{f'(k)}.
$$

This is equation (2.13.2) which says that the tangent line T_k at $t = k$ is parallel to the secant line S.

Now consider the case $f(a) = f(b)$. Since the endpoints of the curve C are distinct, $g(a) \neq g(b)$. The secant line S joins two points with same x–coordinate and is therefore vertical. By (2.13.4), $f'(k) [g(b) - g(a)] = 0$ and $f'(k) = 0$. Since this curve is differentiable, $g'(k) \neq 0$. By Proposition 2.13.12, the tangent line T_k is vertical and is thus parallel to S.
□

The following examples illustrate Cauchy's MVT and its geometric interpretation.

Examples 2.13.20 In each example, find the numbers k guaranteed by Cauchy's Mean Value Theorem. Sketch the graph, its secant line S and the tangent lines T_k.

(1) Let C be the curve $(t^3 - 3t, t^2)$ for $t \in [-1, 3]$.

  Solution For this curve, $f(t) = t^3 - 3t$, $g(t) = t^2$ with $f'(t) = 3t^2 - 3$ and $g'(t) = 2t$. Cauchy's Mean Value Theorem says that there is at least one number $k \in (-1, 3)$ such that:

$$
\frac{g'(k)}{f'(k)} = \frac{2k}{3k^2 - 3} = \frac{g(3) - g(-1)}{f(3) - f(-1)} = \frac{9 - 1}{18 - 2} = \frac{8}{16} = \frac{1}{2}.
$$

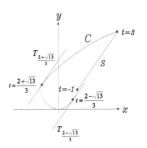

Figure 2.13.21
The Curve C

Hence $4k = 3k^2 - 3$ or $3k^2 - 4k - 3 = 0$. By the quadratic formula: $k = \frac{4 \pm \sqrt{52}}{6} = \frac{2 \pm \sqrt{13}}{3}$. The curve C, with its parallel secant and tangent lines, is sketched in Figure 2.13.21.

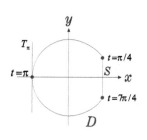

Figure 2.13.22
The Curve D

(2) Let D be the curve $(2\cos\theta, 2\sin\theta)$ for $\theta \in \left[\frac{\pi}{4}, \frac{7\pi}{4}\right]$.

Solution D is three fourths of the circle of radius two and center the origin. The secant line S joining its endpoints $(\sqrt{2}, \sqrt{2})$ and $(\sqrt{2}, -\sqrt{2})$ is vertical. The only vertical tangent line is T_π at the point $(-2, 0)$ on D. See Figure 2.13.22. $\square$

L'Hôpital's Rule

In Section 8, we used L'Hôpital's Rule to evaluate these limits of indeterminate forms:

(1) $\displaystyle\lim_{x\to c} \frac{f(x)}{g(x)}$ with $c \in \Re$ and $\frac{f(c)}{g(c)} = \frac{0}{0}$;   **(2)** $\displaystyle\lim_{x\to c} \frac{f(x)}{g(x)}$ with $c \in \Re$ and $\frac{f(c)}{g(c)} = \frac{\infty}{\infty}$;

(3) $\displaystyle\lim_{x\to -\infty} \frac{f(x)}{g(x)}$ with $\frac{f(c)}{g(c)}$ either $\frac{0}{0}$ or $\frac{\infty}{\infty}$;   **(4)** $\displaystyle\lim_{x\to \infty} \frac{f(x)}{g(x)}$ with $\frac{f(c)}{g(c)}$ either $\frac{0}{0}$ or $\frac{\infty}{\infty}$.

We showed that the first two cases of L'Hôpital's Rule imply the third and fourth cases, but deferred the proofs of the first two cases to this section. In this subsection, we use Cauchy's Mean Value Theorem to prove these two cases. We begin with L'Hôpital's Rule for the indeterminate form $\frac{0}{0}$.

Theorem 2.8.2 [L'Hôpital's Rule for $\frac{0}{0}$] *Let f, g be differentiable functions whose domain includes the interval (a, b) with $a, b \in \Re$ and $c \in (a, b)$. Assume $f(c) = g(c) = 0$. If $\displaystyle\lim_{x\to c} \frac{f'(x)}{g'(x)}$ exists, then $\displaystyle\lim_{x\to c} \frac{f(x)}{g(x)}$ exists and $\displaystyle\lim_{x\to c} \frac{f(x)}{g(x)} = \lim_{x\to c} \frac{f'(x)}{g'(x)}$.*

Proof Since $\displaystyle\lim_{x\to c} \frac{f'(x)}{g'(x)}$ exists, there must be an open interval (A, B) containing c where $\frac{f'(x)}{g'(x)}$ is defined, i.e. $g'(x) \neq 0$ for $x \in (A, B)$, $x \neq c$. Apply the Cauchy Mean Value Theorem to the curve $(g(x), f(x))$ on the interval $[A, c]$. There is a number $k \in (A, c)$ such that:

$$\frac{f'(k)}{g'(k)} = \frac{f(c) - f(A)}{g(c) - g(A)} = \frac{0 - f(A)}{0 - g(A)} = \frac{f(A)}{g(A)}.$$

Take the limit of this equation as A approaches c from the left:

$$\lim_{A\to c^-} \frac{f'(k)}{g'(k)} = \lim_{A\to c^-} \frac{f(A)}{g(A)}.$$

Since $A < k < c$, it follows that k approaches c when A approaches c. Thus, the limit on the left equals $\displaystyle\lim_{k\to c^-} \frac{f'(k)}{g'(k)}$ which we know exists. Hence the limit on the right exists. If we replace the names of the variables k and A by x, we obtain:

$$\lim_{x\to c} \frac{f'(x)}{g'(x)} = \lim_{x\to c^-} \frac{f'(x)}{g'(x)} = \lim_{x\to c^-} \frac{f(x)}{g(x)}. \tag{2.13.5}$$

Similarly, apply the Cauchy Mean Value Theorem to the curve $(g(x), f(x))$ on the interval $[c, B]$ to deduce that

$$\lim_{x\to c} \frac{f'(x)}{g'(x)} = \lim_{x\to c^+} \frac{f'(x)}{g'(x)} = \lim_{x\to c^+} \frac{f(x)}{g(x)}. \tag{2.13.6}$$

By (2.13.5) and (2.13.6), the two–sided limit $\displaystyle\lim_{x\to c} \frac{f(x)}{g(x)}$ exists and equals $\displaystyle\lim_{x\to c} \frac{f'(x)}{g'(x)}$. $\square$

We also use Cauchy's Mean Value Theorem to prove L'Hôpital's Rule for the indeterminate form $\frac{\infty}{\infty}$.

Theorem 2.8.5 (L'Hôpital's Rule for $\frac{\infty}{\infty}$) *Let f, g be differentiable functions whose domain includes $(a, c) \cup (c, b)$. Assume each of*

$$\lim_{x \to c^-} f(x), \quad \lim_{x \to c^+} f(x), \quad \lim_{x \to c^-} g(x), \quad \lim_{x \to c^+} g(x)$$

exists and has value $-\infty$ or $+\infty$. If $\lim_{x \to c} \frac{f'(x)}{g'(x)}$ exists, then $\lim_{x \to c} \frac{f(x)}{g(x)}$ exists and

$$\lim_{x \to c} \frac{f(x)}{g(x)} = \lim_{x \to c} \frac{f'(x)}{g'(x)}.$$

Proof Let $L = \lim_{x \to c} \frac{f'(x)}{g'(x)}$. Assume $\lim_{x \to c^-} \frac{f(x)}{g(x)}$ does not equal L. We deduce a contradiction thereby showing that this assumption is false. By the definition of the limit there is a horizontal strip $S_\epsilon(L)$ such that for every $x < c$, there is $x < e_x < c$ with $\frac{f(e_x)}{g(e_x)} \notin S_\epsilon(L)$. Let $A < c$ with $c - A < \frac{1}{2}$. Since $\lim_{x \to c^-} g(x) = \pm\infty$, we can choose $U(A)$ sufficiently close to c so that $A < U(A) < c$ and if $b \in (U(A), c)$, then

$$\left| \frac{f(A)}{g(b)} \right| < c - A \quad \text{and} \quad \left| \frac{g(A)}{g(b)} \right| < c - A.$$

In particular, $B_A = e_{U(A)} \in (U(A), c)$. Hence

$$0 \le \lim_{A \to c^-} \left| \frac{g(A)}{g(B_A)} \right| \le \lim_{A \to c^-} (c - A) = 0 \quad \text{and} \quad \lim_{A \to c^-} \frac{g(A)}{g(B_A)} = 0. \qquad (2.13.7)$$

Since $\left| \frac{g(A)}{g(B_A)} \right| < c - A < \frac{1}{2}$, we have $1 - \frac{g(A)}{g(B_A)} \ge 1 - \left| \frac{g(A)}{g(B_A)} \right| > 1 - \frac{1}{2} = \frac{1}{2}$. Hence

$$0 \le \lim_{A \to c^-} \left| \frac{\frac{f(A)}{g(B_A)}}{1 - \frac{g(A)}{g(B_A)}} \right| \le \lim_{A \to c^-} \frac{c - A}{1/2} = 0 \quad \text{and} \quad \lim_{A \to c^-} \frac{\frac{f(A)}{g(B_A)}}{1 - \frac{g(A)}{g(B_A)}} = 0. \qquad (2.13.8)$$

Since $\lim_{x \to c} \frac{f'(x)}{g'(x)}$ exists, there is an open interval (A_0, c) where $\frac{f'(x)}{g'(x)}$ is defined, i.e. $g'(x) \ne 0$. For $A_0 < A < c$, apply Cauchy's Mean Value Theorem to the curve $(g(x), f(x))$ on the interval $[A, B_A]$: there is $k_A \in (A, B_A)$ such that

$$\frac{f'(k_A)}{g'(k_A)} = \frac{f(B_A) - f(A)}{g(B_A) - g(A)} = \frac{\frac{f(B_A)}{g(B_A)} - \frac{f(A)}{g(B_A)}}{\frac{g(B_A)}{g(B_A)} - \frac{g(A)}{g(B_A)}} = \frac{\frac{f(B_A)}{g(B_A)} - \frac{f(A)}{g(B_A)}}{1 - \frac{g(A)}{g(B_A)}}. \qquad (2.13.9)$$

Since $A < k_A < c$, it follows that k_A approaches c from the left when A approaches c from the left. Take the limit of (2.13.9) as A approaches c from the left:

$$L = \lim_{x \to c} \frac{f'(x)}{g'(x)} = \lim_{x \to c^-} \frac{f'(x)}{g'(x)} = \lim_{A \to c^-} \frac{f'(k_A)}{g'(k_A)}$$

$$= \lim_{A \to c^-} \frac{\frac{f(B_A)}{g(B_A)}}{1 - \frac{g(A)}{g(B_A)}} - \lim_{A \to c^-} \frac{\frac{f(A)}{g(B_A)}}{1 - \frac{g(A)}{g(B_A)}} = \lim_{A \to c^-} \frac{\frac{f(B_A)}{g(B_A)}}{1 - \frac{g(A)}{g(B_A)}} - 0$$

by limit (2.13.8). By limit (2.13.7):

$$L = \lim_{A \to c^-} \frac{\frac{f(B_A)}{g(B_A)}}{1 - \frac{g(A)}{g(B_A)}} \cdot \lim_{A \to c^-} \left(1 - \frac{g(A)}{g(B_A)} \right) = \lim_{A \to c^-} \frac{f(B_A)}{g(B_A)}. \qquad (2.13.10)$$

However, none of the $\frac{f(B_A)}{g(B_A)}$ lie in the horizontal strip $S_\epsilon(L)$. Hence $\lim\limits_{A\to c^-}\dfrac{f(B_A)}{g(B_A)}$ can

not equal L, and we have a contradiction. Thus our initial assumption that $\lim\limits_{x\to c^-}\dfrac{f(x)}{g(x)}$

does not equal L must be false, i.e.

$$\lim_{x\to c^-}\frac{f(x)}{g(x)} = L = \lim_{x\to c}\frac{f'(x)}{g'(x)}.$$

Similarly, $\lim\limits_{x\to c^+}\dfrac{f(x)}{g(x)} = \lim\limits_{x\to c}\dfrac{f'(x)}{g'(x)}$. Thus $\lim\limits_{x\to c}\dfrac{f(x)}{g(x)}$ exists, and equals $\lim\limits_{x\to c}\dfrac{f'(x)}{g'(x)}$. $\square$

Summary

The reader should know the definition of a curve and be able to find its tangent lines. She should understand the geometric interpretations of the Mean Value Theorem and Cauchy's Mean Value Theorem. The use of Rolle's Theorem to prove these theorems should be known. She should also understand how Cauchy's Mean Value Theorem is used to prove L'Hôpital's Rule.

Basic Exercises

1. Describe each of the following as a curve.
(a) The graph of $y = x^2$ for $-1 \le x \le 4$.
(b) The graph of $y = \cos\theta$ for $0 \le \theta \le \pi$.
(c) The line segment joining the points $(-5, 3)$ and $(6, 4)$.
(d) The left half of the circle of radius 3 and center $(2, 0)$.
(e) The circle of radius 5 and center $(-3, 4)$.
(f) The points of the circle of radius 2 and center the origin which are not in the first quadrant.

2. Sketch each of these curves.
(a) A is the curve (t^2, t^3) for $t \in \Re$.
(b) B is the curve $(\sec\pi t, \cos\pi t)$ for $t \in \left(-\frac{1}{2}, \frac{1}{2}\right)$.
(c) C is the curve $(\sin t, \cos^2 t)$ for $t \ge 0$.
(d) D is the curve $(t^4, t^5 - 5t)$ for $t \in \Re$.
(e) E is the curve $(\arcsin t, \arccos t)$ for $0 \le t \le 1$.
(f) F is the curve $(t - \sin t, 1 - \cos t)$ for $t \ge 0$.

3. Plot points to sketch the folium of Descartes $\left(\frac{3t}{1+t^3}, \frac{3t^2}{1+t^3}\right)$ for $t \in \Re$, $t \ne -1$.

4. Find the equation of the tangent line to each curve at $t = t_0$.
(a) A is the curve $(t^5 - 20, t^3 + 3)$ and $t_0 = 2$.
(b) B is the curve $(\frac{1+t}{t^2+1}, \frac{1-t}{t^2+t})$ and $t_0 = 3$.
(c) C is the curve $(\sqrt{169 - t^2}, \sqrt{t^2 - 16})$ and $t_0 = 5$.
(d) D is the curve $(3\sin t, 3\cos t)$ and $t_0 = \frac{\pi}{2}$.
(e) E is the curve $(4\sec t, 4\tan t)$ and $t_0 = \frac{\pi}{4}$.
(f) F is the curve $(\arctan t, \arcsin t)$ and $t_0 = 0$.

5. Each pair of functions f, g below has domain a specified closed interval $[a, b]$. Define the curve C as the set of points $(f(t), g(t))$ for $t \in [a, b]$. In each case below:
   (i) find the slope of the tangent line T_t to this curve C at the point $(f(t), g(t))$;
   (ii) find all points on the curve C where the tangent line T_t is parallel to the secant line S joining the two endpoints $(f(a), g(a))$ and $(f(b), g(b))$ of the curve C.
(a) $f(t) = t^3 - 1$ and $g(t) = t^2 + 1$ with domain $[0, 3]$.
(b) $f(t) = 3t^2 - t - 5$ and $g(t) = 5t^2 - 7t + 2$ with domain $[-2, -1]$.

(c) $f(t) = \sqrt{3t^2 + 1}$ and $g(t) = \sqrt{5t^2 + 4}$ with domain $[0, 1]$.
(d) $f(t) = \sin \pi t$ and $g(t) = \cos \pi t$ with domain $\left[-\frac{1}{4}, \frac{1}{2}\right]$.
(e) $f(t) = \arcsin \frac{t}{2}$ and $g(t) = \operatorname{arcsec} t$ with domain $[1, 2]$.

6. For each pair of functions f, g with domain $[a, b]$, find the numbers $k \in (a, b)$ guaranteed by Cauchy's Mean Value Theorem where $f'(k)\,[g(b) - g(a)] = g'(k)\,[f(b) - f(a)]$.
(a) $f(t) = t^2 - 3t + 1$ and $g(t) = t^2 + 5t - 2$ with domain $[0, 3]$.
(b) $f(t) = t^3 - t^2 + t + 1$ and $g(t) = t^2 + 2t + 1$ with domain $[1, 2]$.
(c) $f(t) = \sqrt{1 + t}$ and $g(t) = \sqrt{4 - t}$ with domain $[0, 3]$.
(d) $f(t) = \sin t$ and $g(t) = \cos t$ with domain $[0, 3\pi]$.
(e) $f(t) = \arcsin t$ and $g(t) = \arctan t$ with domain $[0, 1]$.

7. Let f be a continuous function with domain $[a, b]$ which is differentiable on (a, b). Assume $f(a) = f(b)$. Show that there is a fixed number $k \in (a, b)$ with the following property: *if g is any continuous function with domain $[a, b]$ which has a nonzero derivative for every $x \in (a, b)$, then $\frac{f'(k)}{g'(k)} = \frac{f(b) - f(a)}{g(b) - g(a)}$*. In particular, the number k does not depend on g.

8. What is the error in the following attempt to prove Cauchy's Mean Value Theorem? *Apply the Mean Value Theorem to each of f and g on the interval $[a, b]$: there is at least one $k \in (a, b)$ such that $f(b) - f(a) = f'(k)(b - a)$ and $g(b) - g(a) = g'(k)(b - a)$. Divide the second equation by the first one: $\frac{g(b) - g(a)}{f(b) - f(a)} = \frac{g'(k)(b-a)}{f'(k)(b-a)} = \frac{g'(k)}{f'(k)}$.*

9. Complete the proof of Theorem 2.8.2 by showing $\displaystyle\lim_{x \to c^+} \frac{f(x)}{g(x)} = \lim_{x \to c} \frac{f'(x)}{g'(x)}$.

10. Complete the proof of Theorem 2.8.5 by showing $\displaystyle\lim_{x \to c^+} \frac{f(x)}{g(x)} = \lim_{x \to c} \frac{f'(x)}{g'(x)}$.

Challenging Problems

1. Let f and g be functions with continuous derivatives. Assume $f \circ g$ is defined with domain the open interval I.
(a) For $x, x + h \in I$, show there are b between x and $x + h$ and c between $g(x)$ and $g(x + h)$ such that $f(g(x + h)) - f(g(x)) = f'(c)g'(b)h$.
(b) Use (a) to show that $f \circ g$ is differentiable [8] with $(f \circ g)'(x) = f'(g(x))g'(x)$.

2. Prove the following generalization of L'Hôpital's Rule. Let $n \geq 2$, and assume the nth derivatives of f and g exist. Also assume $f(c) = g(c) = 0$ and $f^{(k)}(c) = g^{(k)}(c) = 0$ for $1 \leq k \leq n - 1$. If $\displaystyle\lim_{x \to c} \frac{f^{(n)}(x)}{g^{(n)}(x)}$ exists, then $\displaystyle\lim_{x \to c} \frac{f(x)}{g(x)}$ exists and

$$\lim_{x \to c} \frac{f(x)}{g(x)} = \lim_{x \to c} \frac{f^{(n)}(x)}{g^{(n)}(x)}.$$

---

[8] This easy proof of the chain rule requires that f' and g' be continuous. The more complicated proof given in the appendix to Section 4 only requires that f' and g' exist.

2.14   Review Exercises for Chapter 2

Decide if each of these 40 statements is True or False. Justify your answers.

1. A differentiable function with domain $\Re$ does not have a horizontal normal line.

2. A differentiable function, with domain an interval, may have a vertical half–tangent.

3. Continuous functions are differentiable.

4. A function can not have both a vertical tangent and a cusp.

5. A differentiable function with domain $\Re$ has no cusps.

6. f and g are differentiable functions with domain $\Re$. If $g'(2) = 0$, then $\frac{f}{g}$ has a vertical tangent at $x = 2$.

7. If $f(c) = g(c) \neq 0$ and $f'(c) = g'(c)$, then $\left(\frac{f}{g}\right)'(c) = 0$.

8. If $f(x)g(x)$ and $g(x)$ are differentiable functions with domain $\Re$, then $f(x)$ is differentiable.

9. Assume that $f \circ g$ is defined.
(a) If f has a cusp and g is differentiable, then $f \circ g$ has a cusp.
(b) If f is differentiable and g has a cusp, then $f \circ g$ has a cusp.

10. Let $g(x) = \frac{f(x^2)}{x}$ where f is a differentiable function. Then $g'(x) = 2f'(x^2) - \frac{g(x)}{x}$.

11. If f is an even differentiable function, then f' is an odd function.

12. If $x^2 + y^2 = 1$, then $\frac{dy}{dx} = -\frac{x}{y}$. Similarly, if $xy = 1$, then $\frac{dy}{dx} = -\frac{x}{y}$.

13. $f(x) = \tan^2 \sqrt{1 + x^2}$ and $g(x) = \sec^2 \sqrt{1 + x^2}$ have the same derivative.

14. The derivative of $\sin(x^2) \cos(x^2)$ is $2x \cos(2x^2)$.

15. The derivative of $f^{-1}(x)$ is $-f(x)^{-2} f'(x)$.

16. The derivative of $\arcsin(\sin \sqrt{1 - x^2})$ is $-\frac{x}{\sqrt{1-x^2}}$.

17. Let $f(x) = \operatorname{arcsec} x$ be the restriction of the arcsec function to the interval $[1, \infty)$. Then $f'(x) = \frac{1}{x\sqrt{x^2-1}}$.

18. The function $f(x) = (x + 2)^2 \arcsin x$ has no horizontal tangent line.

19. $f(x) = \frac{\arcsin x}{\arctan x}$ has the y–axis as a vertical asymptote.

20. If f is a continuous function with a local extremum at $x = c$, then $f'(c) = 0$.

21. Let f be a differentiable function with domain an open interval. If $f'(x)$ is continuous and never zero, then f is one–to–one.

22. f and g are differentiable functions with domain $\Re$. If $f' = g'$ and $f(1) = g(1)$, then $f = g$.

23. The function $f(x) = x^{10} + x^6 + x^4 + 1$ does not have a local extremum at $x = 0$.

24. The graph of an even function with domain $\Re$ has a horizontal tangent line.

25. f has a critical point at $x = c$, and $f'(c)$ does not exist. The First Derivative Test can not identify this critical point as a local minimum or a local maximum.

26. The graph of a function with a horizontal asymptote L approaches L on both the left and the right.

27. Every rational function $\frac{P(x)}{Q(x)}$ with deg $Q(x) \geq 1$ has a vertical asymptote.

28. A rational function $\frac{P(x)}{Q(x)}$ with deg $P(x) >$ deg $Q(x)$ has no horizontal asymptote.

29. If $y = f(x)$ is a continuous function with domain $\Re$ and $f''(c) = 0$, then f has an inflection point at $x = c$.

30. An increasing function can not be concave down.

31. By L'Hôpital's Rule, $\lim\limits_{x \to 1} \dfrac{x^2 - x + 1}{x^2 - 1} = \dfrac{1}{2}$.

32. By L'Hôptial's Rule, $f(x) = \dfrac{\arctan \frac{1}{x^2}}{\arctan \frac{1}{x}}$ has the y–axis as a vertical asymptote.

33. By L'Hôptial's Rule, $g(x) = \dfrac{\arcsin \frac{1}{x^2}}{\arcsin \frac{1}{x}}$ has the x–axis as a horizontal asymptote.

34. Let f, g, h be continuously differentiable functions with $f(0) = g(0) = h(0) = 0$, $f'(0) = 2$, $g'(0) = -1$ and $h'(0) = 5$. Then $\lim\limits_{x \to 0} \dfrac{f(x)g(x)}{h(x)} = 0$.

The following 3 exercises are for those who have studied Section 2.9.

35. If the average rate of change of f on the interval $[0, x]$ is $4x$, then the instantaneous rate of change of f at $x = 0$ equals four.

36. An object is moving along a straight line. If $v(2) < 0$ and $a(2) < 0$, then the object is slowing down at $t = 2$.

37. Let $y(x)$ denote the temperature in degrees centigrade of a falling meteorite at altitude x ft above the ground. At time zero, the meteorite is $100,000$ ft above the ground and falling at $5,000$ ft per second. The instantaneous rate of change of the meteorite's temperature at time 10 seconds is $-5,320y'(48,400)$ degrees per second.

The following 3 exercises are for those who have studied Section 2.12.

38. Every differentiable function with domain an interval has an absolute maximum value and an absolute minimum value.

39. A continuous function f with domain a closed interval, which is not differentiable, must have an absolute minimum value.

40. The Second Derivative Test will always identify a critical point of a polynomial.

Solve each of the following problems.

41. Use the definition of the derivative to find the derivative of $f(x) = \sqrt[4]{x}$.

42. Find the vertical tangents, cusps and vertical cusps of $f(x) = |\sqrt[3]{x^2 - 1}|$.

43. f g, h are differentiable functions with domain $\Re$. Assume $f(5) = 3$, $g(5) = -4$, $h(5) = 2$, $f'(5) = -6$, $g'(5) = 7$, $h'(5) = -1$. Let $k(x) = \frac{f(x)g(x)}{h(x)}$. Find $k'(5)$.

44. Find the equation of the normal line of the graph of $y = \frac{x^2}{\sqrt{x^2 - 9}}$ at $x = 5$.

45. $f(x) = x^2$ with domain $(-\infty, 0)$ is a one–to–one function. Find $D(f^{-1})(x)$.

46. f is a differentiable one–to–one function with domain $\Re$. Let $y = (f \circ f)^{-1}(x)$. Find $\frac{dy}{dx}$.

47. Find the tangent line to the curve $xy^2 + x^2y + xy = 8$ at the point $(2, 1)$.

48. Find the tangent line to the graph of $y = \frac{x}{(x^2 + 1)^{100}}$ at $x = 1$.

Differentiate these 3 functions.
49. $y = \sin(\sin(\sin x))$ **50.** $y = \sqrt{1 + \sqrt{1 + x^4}}$ **51.** $y = \frac{(\tan x)(\arctan x)}{1 + x^2}$

52. Let $f(x) = \left\{ \begin{array}{ll} x^2 \sin \frac{1}{x} & \text{for } x \neq 0 \\ 0 & \text{for } x = 0 \end{array} \right\}$. Identify the critical point $x = 0$ as a local minimum, a local maximum or neither. Justify your answer.

53. Let $P(x)$ be a degree four polynomial with four distinct real roots. Show that $P'(x)$ has three distinct real roots.

54. Apply the Mean Value Theorem to $f(x) = \frac{\arcsin x}{1+\arctan x}$ on the interval $[0, 1]$.

55. Find and identify the local extrema of $f(x) = \tan^8 \pi x$ with domain $\left(-\frac{1}{2}, \frac{1}{2}\right)$.

56. Solve the initial value problem $y' = \csc^3 x + \cot^2 x \csc x$ with $y\left(\frac{\pi}{4}\right) = 1$.

57. Find all the asymptotes of $f(x) = \frac{x^3+1}{x^2-2x-8}$.

58. Let $f(x) = \left\{ \begin{array}{ll} 1 - x^2 & \text{if } x \leq 0 \\ 1 + x^2 & \text{if } 0 \leq x \leq 1 \\ 3 - \sqrt{x} & \text{if } 1 \leq x \end{array} \right\}$ with domain $\mathfrak{R}$.

(a) Find the cusps of f.
(b) Find where f is concave up and where f is concave down.
(c) Locate all the inflection points of f.

59. Sketch the graph of $f(x) = \left(\frac{x}{x^2-1}\right)^{2/3}$.

60. Sketch the graph of $g(x) = \frac{\sin x}{1+\sin x}$ with domain $[-\pi, \pi]$, $x \neq \frac{-\pi}{2}$.

Evaluate each limit.

61. $\displaystyle\lim_{x \to 0} \frac{\csc x}{x}$ **62.** $\displaystyle\lim_{x \to 0} \sqrt{\frac{\sin(x^2)}{\arcsin(x^2)}}$ **63.** $\displaystyle\lim_{x \to \infty} \frac{\sqrt{x^4 + x^3 + x^2 + x + 1}}{4x^2 + 5x + 3}$

The following 3 exercises are for those who have studied Section 2.9.

64. Let $V = f(P)$ denote the volume of a balloon when the air pressure is P. If $f(8) = 12$ and $f'(8) = 5$, find the instantaneous rate of change of the air pressure with respect to the volume when $V = 12$.

65. Describe the motion of the object with $a(t) = 24t^2$ meters/sec^2, $v(0) = -6$ meters/sec and $s(0) = 5$ meters.

The following 2 exercises are for those who have studied Section 2.10.

66. A rock is dropped from the top of a 128 foot high cliff. A hiker stands at the bottom of the cliff, 48 feet from the spot where the rock lands. Find the rate of change of the angle of inclination of the rock (the angle between the ground and the line joining the hiker with the rock) when the rock is halfway to the ground.

67. A balloon floating horizontally towards the west at 384 meters/hour passes 64 meters over a child walking south at 96 meters/hour. How fast are the balloon and the child separating 30 minutes later?

The following 3 exercises are for those who have studied Section 2.12.

68. Find the absolute minimum and absolute maximum values of the function $f(x) = \arctan(1 + x^2)$ with domain $\mathfrak{R}$.

69. Sketch the graph of a continuous function f which has an absolute maximum value at an inflection point $x = c$.

70. A boy is running towards the right on the x–axis at 8 feet per second. At time $t = 0$ he is at $x = -16$ when a ball at the origin is kicked upwards along the y–axis at 64 feet per second. Find the shortest distance between the boy and the ball.

Chapter 3

Integration

3.1   Introduction

This chapter presents the foundations of integral calculus. This aspect of calculus originates with the problem of computing area. The area of a region bounded by the graph of a function f, the x–axis and two vertical lines is called the definite integral of f. We apply the Fundamental Theorem of Calculus to compute these areas. This theorem says that differentiation and integration are inverse operations. Thus, we use our expertise in finding derivatives to evaluate integrals and thereby compute areas.

The *Concepts* portion of this chapter begins in Section 2 with sigma notation, a concise method for writing sums. A region is approximated by a union of rectangles. Then sigma notation is used to estimate the area of the region as the sum of the areas of these rectangles. In Section 3, area is defined as the limit of the approximations of Section 2. This definition is applied to establish several elementary properties of area. Although areas of very simple regions can be computed from this definition, it is not practical for finding more complicated areas. In Section 4, the Fundamental Theorem of Calculus is proved and applied to compute a wide variety of areas. This method requires the determination of a function with a given derivative, called its indefinite integral. In general, evaluating an indefinite integral is not routine. In Section 5, elementary methods for solving this problem are given. Changing variables, to simplify an indefinite integral, is presented in Section 6. A more thorough study of evaluating integrals is a central topic of Chapter 4. In Section 7, several methods for computing volumes as definite integrals are studied.

The *Applications* portion of this chapter begins in Section 8 with the study of numerical methods for approximating areas. These methods are useful because there are many areas whose exact value cannot be determined. In Section 9, several invariants studied in physics are identified as definite integrals. In Section 10, definite integrals are used to compute moments and centers of mass of rods of variable density as well as plates of constant density.

Section 11, of the *Theory* portion of this chapter, is devoted to proving the existence of the definite integral of a continuous function. The definite integral of a bounded function is defined, and its value is computed for a function with a finite number of jump discontinuities. In Section 12, methods are developed to identify invariants, other than area, as definite integrals. These methods are applied to justify the formulas used to compute volumes, identify physics invariants and compute the centers of mass of rods in Sections 7, 9, 10.

The flow chart in Figure 3.1.1 indicates the dependence of the sections of the *Applications* and *Theory* portions of this chapter on preceding sections.

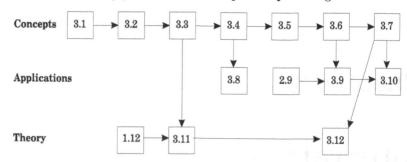

Figure 3.1.1 The Sections of Chapter 3

3.2   Riemann Sums

We begin our study of areas by developing a procedure to estimate them. We know the area of a rectangle is the product of its length and width. In the second subsection, we approximate a more complicated region S by a union of narrow rectangular strips. The sum of the areas of these rectangles is called a *Riemann sum* and gives an estimate of the area of S. The third subsection studies lower and upper Riemann sums. We use them to bound the errors of our estimates of the area of S. Riemann sums are written in *sigma notation*, a scheme for writing long sums concisely which we introduce in the first subsection.

Sigma Notation

Sigma notation describes a sum by specifying the pattern of the summands. This is much more efficient than listing all the summands. We interpret several properties of arithmetic to produce formulas for manipulating sigma notation.

Motivating Example 3.2.1 Describe the sum S of the squares of the first hundred positive integers.

Solution First, consider a simpler problem. If we were only asked to describe the sum of the squares of the first ten positive integers, we would write:

$$1^2 + 2^2 + 3^2 + 4^2 + 5^2 + 6^2 + 7^2 + 8^2 + 9^2 + 10^2.$$

However, it is tedious to write this entire sum, and it is not practical to write all 100 summands of S. Nevertheless, we can write the first few summands and last few summands with three dots between them to indicate that the intermediate summands are omitted:

$$S = 1^2 + 2^2 + 3^2 + \cdots + 98^2 + 99^2 + 100^2.$$

The pattern of the specified summands indicate the values of the omitted ones. This notation, however, is cumbersome. Moreover, it may be confusing in a complicated sum where the pattern of the summands is not obvious. We give an alternate description of S. Begin with the definition of S as the sum of the squares of the integers from 1 to 100. Choose a letter, say k, to designate one of these integers. Then S can be described as the sum of k^2 for k an integer between 1 and 100. Let's use the upper case Greek letter sigma Σ to denote sum. Then $S = \Sigma\, k^2$ for $1 \le k \le 100$. We make this notation more concise by putting the smallest value $k = 1$ of k at the bottom of the Σ and the largest value 100 of k on top of the Σ. That is, $S = \sum_{k=1}^{100} k^2$.   □

The precise definition of sigma notation adds summands $f(k)$ whose pattern is described by the function f. These summands are added over a specified set of consecutive integers k.

Definition 3.2.2 *Let p and q be integers with $p \leq q$. Let f be a function whose domain includes the integers between p and q. The sum of the values of f on these integers is denoted:*

$$\sum_{k=p}^{q} f(k) = f(p) + f(p+1) + f(p+2) + \cdots + f(q-2) + f(q-1) + f(q).$$

Observe that the variable k does not appear when the sigma notation is written as a sum. Thus $\sum_{k=p}^{q} f(k) = \sum_{n=p}^{q} f(n) = \sum_{i=p}^{q} f(i)$. For this reason, the variable used in sigma notation is called a *dummy variable*.

Examples 3.2.3 (1) Evaluate $\sum_{k=1}^{5} k^3$.

Solution Compute k^3 for $k = 1, 2, 3, 4, 5$ and add these numbers:

$$\sum_{k=1}^{5} k^3 = 1^3 + 2^3 + 3^3 + 4^3 + 5^3 = 1 + 8 + 27 + 64 + 125 = 225.$$

(2) Evaluate $\sum_{n=-1}^{3} (3n + 2)$.

Solution Compute $3n + 2$ for $n = -1, 0, 1, 2, 3$ and add these numbers:

$$\sum_{n=-1}^{3} (3n + 2) = [3(-1) + 2] + [3(0) + 2] + [3(1) + 2] + [3(2) + 2] + [3(3) + 2]$$
$$= -1 + 2 + 5 + 8 + 11 = 25.$$

(3) Factor the polynomial $P(x) = x^{2n+2} - 1$ for $n \geq 1$.

Solution Observe $P(-1) = 0$ and $P(1) = 0$. Hence $-1, +1$ are roots of $P(x)$, and $P(x)$ is divisible by $(x+1)(x-1) = x^2 - 1$. Divide $x^2 - 1$ into $x^{2n+2} - 1$:

$$P(x) = (x^2 - 1)(x^{2n} + x^{2n-2} + \cdots + x^4 + x^2 + 1) = (x+1)(x-1)\sum_{k=0}^{n} x^{2k}.$$

Note $Q(x) = \sum_{k=0}^{n} x^{2k} \geq 1$ for all $x \in \Re$. Hence $Q(x)$ has no real roots and no linear factors. However, $Q(x)$ has quadratic factors when $n \geq 1$.

(4) Yeast is placed in a vat of grape juice. Each hour the amount of CO_2 (carbon dioxide) produced doubles. If 3 cc of CO_2 are produced in the first hour, write an expression to denote the total amount T of CO_2 produced on the first day.

Solution Note the amount of CO_2 produced is 3 cc in the first hour $2 \cdot 3$ cc in the second hour, $2 \cdot (2 \cdot 3) = 2^2 \cdot 3$ cc in the third hour, $2 \cdot (2^2 \cdot 3) = 2^3 \cdot 3$ cc in the fourth hour, etc. Thus

$$T = 3 + 2 \cdot 3 + 2^2 \cdot 3 + 2^3 \cdot 3 + \cdots + 2^{23} \cdot 3 = \sum_{k=0}^{23} (2^k \cdot 3) \qquad \square$$

We present five properties of sigma notation. The first one generalizes the commutative property of addition: $A + B = B + A$. The second one generallizes the distributive property of arithmetic: $CA + CB = C(A + B)$. The third one is a restatement of the definition of multiplication: the sum of A with itself n times is nA. The fourth and fifth properties will be used in the last subsection when we approximate the area bounded by a parabola.

Proposition 3.2.4 *Let $0 < n$ and $p < q$ be integers. Let $c \in \Re$, and let f, g be functions whose domains include the integers between p and q. Then*

(a) $\sum_{k=p}^{q} [f(k) + g(k)] = \sum_{k=p}^{q} f(k) + \sum_{k=p}^{q} g(k)$ *(commutative property)*;

(b) $\sum_{k=p}^{q} [cf(k)] = c \sum_{k=p}^{q} f(k)$ *(distributive property)*;

(c) $\sum_{k=p}^{q} c = c(q - p + 1)$ *(multiplicative property)*;

(d) $\sum_{k=1}^{n} k = \frac{n(n+1)}{2}$; **(e)** $\sum_{k=1}^{n} k^2 = \frac{n(n+1)(2n+1)}{6}$.

Proof (a) This formula follows from the commutative property of addition:

$$[f(p) + g(p)] \quad + \quad [f(p+1) + g(p+1)] + \cdots + [f(q-1) + g(q-1)] + [f(q) + g(q)]$$
$$= \quad [f(p) + f(p+1) + \cdots + f(q-1) + f(q)]$$
$$+ [g(p) + g(p+1) + \cdots + g(q-1) + g(q)].$$

(b) This formula follows from the distributive property of arithmetic:

$$cf(p) + cf(p+1) + \cdots + cf(q-1) + cf(q) = c[f(p) + f(p+1) + \cdots + f(q-1) + f(q)].$$

(c) This property merely says: $(q - p + 1)c$ is the sum of c with itself $q - p + 1$ times.
(d) Let $S = \sum_{k=1}^{n} k = 1 + 2 + \cdots + (n-1) + n$. Write these summands backwards under the original sum and add:

S_n	$=$	1	$+$	2	$+$	3	$+ \cdots +$	$(n-1)$	$+$	n
S_n	$=$	n	$+$	$(n-1)$	$+$	$(n-2)$	$+ \cdots +$	2	$+$	1
$2S_n$	$=$	$(n+1)$	$+$	$(n+1)$	$+$	$(n+1)$	$+ \cdots +$	$(n+1)$	$+$	$(n+1)$

Since there are n summands on the right, $2S_n = n(n+1)$ and $S_n = n(n+1)/2$.
(e) This formula is proved in the appendix to this section. $\square$

We apply these properties in the following examples.

Examples 3.2.5 (1) Evaluate $\sum_{k=1}^{40} (4k^2 - 3k + 5)$.

Solution We use all of the above properties to evaluate this sum:

$$\sum_{k=1}^{40} (4k^2 - 3k + 5) \quad = \quad \sum_{k=1}^{40} 4k^2 + \sum_{k=1}^{40} -3k + \sum_{k=1}^{40} 5 \text{ [commutative property]}$$

$$= \quad 4 \sum_{k=1}^{40} k^2 - 3 \sum_{k=1}^{40} k + \sum_{k=1}^{40} 5 \quad \text{[distributive property]}$$

$$= \quad 4 \frac{40(41)(81)}{6} - 3 \frac{40(41)}{2} + 5(40 - 1 + 1) = 86,300$$

by properties (e), (d) and (c).

(2) Write $5 \sum_{n=0}^{8} 2 - 9 \sum_{k=0}^{8} x^k + 4 \sum_{t=0}^{8} x^{2t}$ as a single sum.

Solution We change all the dummy variables to n to combine these sums.

$$5 \sum_{n=0}^{8} 2 \quad - \quad 9 \sum_{k=0}^{8} x^k + 4 \sum_{t=0}^{8} x^{2t} = 5 \sum_{n=0}^{8} 2 - 9 \sum_{n=0}^{8} x^n + 4 \sum_{n=0}^{8} x^{2n}$$

$$= \quad \sum_{n=0}^{8} 10 + \sum_{n=0}^{8} -9x^n + \sum_{n=0}^{8} 4x^{2n} \quad \text{[distributive property]}$$

$$= \quad \sum_{n=0}^{8} \left(10 - 9x^n + 4x^{2n}\right). \quad \text{[commutative property]} \quad \square$$

General Riemann Sums

We estimate the area of a region S by dividing S into narrow strips and approximating each strip by a rectangle. The sum of the areas of these rectangles gives an estimate of the area of S. This estimate of the area of S is called a Riemann sum.

Motivating Example 3.2.6 Estimate the area A of the region S bounded by the graph of $y = f(x) = x^2$, the x–axis and the line $x = 8$ depicted in Figure 3.2.7.

Solution Divide S into four strips S_1, S_2, S_3, S_4 as in Figure 3.2.8. Approximate these strips by the rectangles R_1, R, R_3, R_4 of Figure 3.2.10:

R_1 has one side the interval $[0, 2]$ and the other side has length $f(1) = 1$;
R_2 has one side the interval $[2, 4]$ and the other side has length $f(3) = 9$;
R_3 has one side the interval $[4, 6]$ and the other side has length $f(5) = 25$;
R_4 has one side the interval $[6, 8]$ and the other side has length $f(7) = 49$. Then

$$A \approx \text{Area } R_1 + \text{Area } R_2 + \text{Area } R_3 + \text{Area } R_4$$
$$A \approx (1)(2) + (9)(2) + (25)(2) + (49)(2) = 168.$$ □

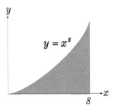

Figure 3.2.7
The Region S

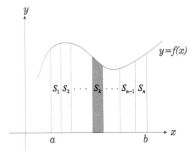

Figure 3.2.9 The Regions S_k

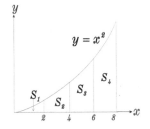

Figure 3.2.8
Subdividing Region S

In general, the region S whose area A we estimate is bounded by the graph of $y = f(x)$ on top, by the x–axis on the bottom and by the vertical lines $x = a$ on the left and $x = b$ on the right. See Figure 3.2.11. We show in Section 4 how other regions can be decomposed into a union of regions of this type. To approximate the value of A, divide S into narrow vertical strips $S_1, \ldots, S_n$ as in Fig. 3.2.9. We approximate the area of each narrow strip S_k by the area of a narrow rectangle R_k as in Fig. 3.2.14.

We digress to introduce notation to indicate the points on the x–axis where we construct vertical lines to subdivide the region R into the strips $S_1, \ldots, S_n$. There are $n + 1$ of these lines, starting with $x = a$ on the left and ending with $x = b$ on the right. A selection of $n + 1$ points from the interval $[a, b]$ on the x–axis determines the positions of these lines. A selection of these points, as in Figure 3.2.12, is called a *partition* of the interval $[a, b]$. In Example 3.2.6 we used the partition $\{0, 2, 4, 6, 8\}$ of the interval $[0, 8]$ to construct the vertical lines $x = 0$, $x = 2$, $x = 4$, $x = 6$, $x = 8$ which determine the strips S_1, S_2, S_3, S_4.

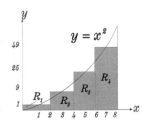

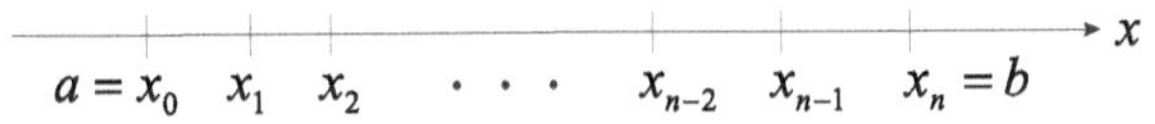

Figure 3.2.10
Rectangles
R_1, R_2, R_3, R_4

$$a = x_0 \quad x_1 \quad x_2 \quad \cdots \quad x_{n-2} \quad x_{n-1} \quad x_n = b$$

Figure 3.2.12 A Partition of the Interval $[a, b]$

Definition 3.2.13 *A partition $P = \{x_0, x_1, \ldots, x_{n-1}, x_n\}$ of the interval $[a, b]$ consists of an increasing sequence of numbers:*

$$a = x_0 \leq x_1 \leq \cdots \leq x_{n-1} \leq x_n = b.$$

The mesh of the partition P is the largest of the numbers $x_k - x_{k-1}$ for $1 \leq k \leq n$.

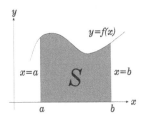

Figure 3.2.11
A General Region S

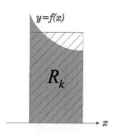

Figure 3.2.14
Approximating S_k by
the Rectangle R_k

Note For $1 \leq k \leq n$, the interval $[x_{k-1}, x_k]$, of length $x_k - x_{k-1}$, is the base of the narrow strip S_k and is one side of the rectangle R_k.

Example 3.2.15 The interval $[0, 1]$ can be partitioned into 4 subintervals in many ways. For example,

$$P_1 = \left\{0, \tfrac{1}{4}, \tfrac{1}{2}, \tfrac{3}{4}, 1\right\} \text{ has } x_0 = 0, \ x_1 = \tfrac{1}{4}, \ x_2 = \tfrac{1}{2}, \ x_3 = \tfrac{3}{4}, \ x_4 = 1.$$
$$P_2 = \left\{0, \tfrac{1}{5}, \tfrac{1}{3}, \tfrac{3}{5}, 1\right\} \text{ has } x_0 = 0, \ x_1 = \tfrac{1}{5}, \ x_2 = \tfrac{1}{3}, \ x_3 = \tfrac{3}{5}, \ x_4 = 1.$$
$$P_3 = \left\{0, \tfrac{1}{6}, \tfrac{2}{5}, \tfrac{5}{6}, 1\right\} \text{ has } x_0 = 0, \ x_1 = \tfrac{1}{6}, \ x_2 = \tfrac{2}{5}, \ x_3 = \tfrac{5}{6}, \ x_4 = 1.$$

See Figure 3.2.16. Mesh P_1 is $\tfrac{1}{4}$, mesh P_2 is $\tfrac{2}{5}$ and mesh P_3 is $\tfrac{13}{30}$. $\quad\square$

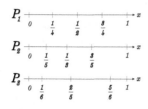

Figure 3.2.16 Three
Partitions of $[0, 1]$

The partition P_n divides the interval $[a, b]$ into n subintervals of equal length. Since the interval $[a, b]$ has length $b - a$, each subinterval of P_n has length $\tfrac{b-a}{n}$. Hence mesh $P_n = \tfrac{b-a}{n}$. In particular, $P_n = \{x_0, \dots, x_n\}$ with

$$x_k = a + k\left(\frac{b-a}{n}\right) \tag{3.2.1}$$

for $0 \leq k \leq n$. P_n is called the *regular partition* of the interval $[a, b]$ into n subintervals. In Example 3.2.6 we used the regular partition P_4 of the interval $[0, 8]$.

Example 3.2.17 Find the regular partition P_5 of the interval $[3, 10]$.

Solution Each of the five subintervals of $[3, 10]$ has length $\tfrac{10-3}{5} = 1.4$. Hence
$$x_0 = 3 + 0(1.4) = 3, \quad x_1 = 3 + 1(1.4) = 4.4, \quad x_2 = 3 + 2(1.4) = 5.8,$$
$$x_3 = 3 + 3(1.4) = 7.2, \quad x_4 = 3 + 4(1.4) = 8.6, \quad x_5 = 3 + 5(1.4) = 10.$$
Thus $P_5 = \{3, \ 4.4, \ 5.8, \ 7.2, \ 8.6, \ 10\}$. See Figure 3.2.18. $\quad\square$

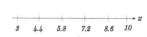

Figure 3.2.18
Regular Partition P_5
of $[3, 10]$

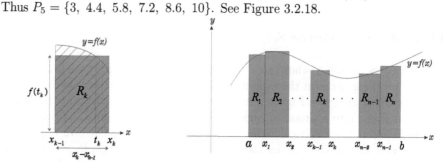

Figure 3.2.19 Approximating A by the Riemann Sum $R(P, T, f)$

Return to the problem of estimating the area A of the region S depicted in Fig. 3.2.11 bounded by the graph of $y = f(x)$ on top, the x–axis on the bottom, the line $x = a$ on the left and the line $x = b$ on the right. Selection of a partition $P = \{x_0, \dots, x_n\}$ of $[a, b]$ determines the positions and widths of the narrow rectangles $R_1, \dots, R_n$ we use to approximate the region S in the right diagram of Fig. 3.2.19. Let $1 \leq k \leq n$. The rectangle R_k, in the left diagram of Fig. 3.2.19, has its base on the interval $[x_{k-1}, x_k]$ of the x–axis. We let its length be the height $f(t_k)$ of the graph of f at a point t_k chosen from $[x_{k-1}, x_k]$. Thus rectangle R_k has width $x_k - x_{k-1}$, length $f(t_k)$ and

$$\text{Area } R_k = f(t_k)(x_k - x_{k-1}).$$

The sum of the areas of rectangles $R_1, \dots, R_n$ approximates the area A of region S:

$$A \approx \text{ Area } R_1 + \cdots + \text{Area } R_n = f(t_1)(x_1 - x_0) + \cdots + f(t_n)(x_n - x_{n-1}).$$

This approximation of A is called a *Riemann sum* and is written more concisely in sigma notation.

Definition 3.2.20 *Let f be a function with domain $[a, b]$, and let $P = \{x_0, \ldots, x_n\}$ be a partition of the interval $[a, b]$. Let $T = \{t_1, \ldots, t_n\}$ with $t_k \in [x_{k-1}, x_k]$ for $1 \le k \le n$. The Riemann sum determined by P, T and f is defined as*

$$R(P, T, f) = \sum_{k=1}^{n} f(t_k)(x_k - x_{k-1}).$$

Observe that the function f and the interval $[a, b]$ are given. They determine the region S of area A. We must select a partition P and a set of points T, from the subintervals of $[a, b]$ determined by P, to define a Riemann sum $R(P, T, f)$ which approximates A. In the following examples, however, P and T are given in the statement of the problem. In the next subsection, a strategy is presented for selecting P and T.

Examples 3.2.21 (1) Let A be the area of the region S bounded by the parabola $f(x) = x^2$, the x–axis and the line $x = 1$. Approximate A using the partition $P = \{0, \frac{1}{4}, \frac{1}{2}, \frac{3}{4}, 1\}$. Let T be the set of midpoints of the subintervals of $[0, 1]$ determined by P.

Solution Note that $x_0 = 0$, $x_1 = \frac{1}{4}$, $x_2 = \frac{1}{2}$, $x_3 = \frac{3}{4}$ and $x_4 = 1$. Then

$$t_1 = \frac{0 + \frac{1}{4}}{2} = \frac{1}{8}, \quad t_2 = \frac{\frac{1}{4} + \frac{1}{2}}{2} = \frac{3}{8}, \quad t_3 = \frac{\frac{1}{2} + \frac{3}{4}}{2} = \frac{5}{8}, \quad t_4 = \frac{\frac{3}{4} + 1}{2} = \frac{7}{8}.$$

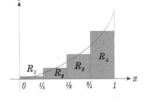

Thus $T = \{\frac{1}{8}, \frac{3}{8}, \frac{5}{8}, \frac{7}{8}\}$. The region S and the rectangles R_1, R_2, R_3, R_4 are depicted in Figure 3.2.22. Thus

Figure 3.2.22
$f(x) = x^2$

$$
\begin{aligned}
A &\approx R(P, T, f) = f(t_1)(x_1 - x_0) + f(t_2)(x_2 - x_1) \\
&\qquad\qquad\qquad\quad + f(t_3)(x_3 - x_2) + f(t_4)(x_4 - x_3) \\
&\approx \left(\frac{1}{8}\right)^2 \cdot \frac{1}{4} + \left(\frac{3}{8}\right)^2 \cdot \frac{1}{4} + \left(\frac{5}{8}\right)^2 \cdot \frac{1}{4} + \left(\frac{7}{8}\right)^2 \cdot \frac{1}{4} = \frac{21}{64}.
\end{aligned}
$$

(2) Let A' be the area of the region S' bounded by the graph of $g(x) = \cos x$, for $0 \le x \le \frac{\pi}{2}$, the x–axis and the y–axis. Approximate A' using the partition $P' = \{0, \frac{\pi}{6}, \frac{\pi}{4}, \frac{\pi}{3}, \frac{\pi}{2}\}$. Let T' be the set of midpoints of the subintervals of $[0, \frac{\pi}{2}]$ determined by P'.

Solution Note $x_0 = 0$, $x_1 = \frac{\pi}{6}$, $x_2 = \frac{\pi}{4}$, $x_3 = \frac{\pi}{3}$, $x_4 = \frac{\pi}{2}$. Hence

$$t_1 = \frac{0 + \frac{\pi}{6}}{2} = \frac{\pi}{12}, \quad t_2 = \frac{\frac{\pi}{6} + \frac{\pi}{4}}{2} = \frac{5\pi}{24}, \quad t_3 = \frac{\frac{\pi}{4} + \frac{\pi}{3}}{2} = \frac{7\pi}{24}, \quad t_4 = \frac{\frac{\pi}{3} + \frac{\pi}{2}}{2} = \frac{5\pi}{12}.$$

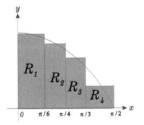

Thus $T' = \{\frac{\pi}{12}, \frac{5\pi}{24}, \frac{7\pi}{24}, \frac{5\pi}{12}\}$. The region S' and the rectangles R_1, R_2, R_3, R_4 are depicted in Figure 3.2.23. Then

Figure 3.2.23
$g(x) = \cos x$

$$
\begin{aligned}
A' &\approx R(P', T', g) = g(t_1)(x_1 - x_0) + g(t_2)(x_2 - x_1) \\
&\qquad\qquad\qquad\quad + g(t_3)(x_3 - x_2) + g(t_4)(x_4 - x_3) \\
&\approx \left(\cos \frac{\pi}{12}\right)\left(\frac{\pi}{6} - 0\right) + \left(\cos \frac{5\pi}{24}\right)\left(\frac{\pi}{4} - \frac{\pi}{6}\right) + \left(\cos \frac{7\pi}{24}\right)\left(\frac{\pi}{3} - \frac{\pi}{4}\right) \\
&\qquad\qquad\qquad\quad + \left(\cos \frac{5\pi}{12}\right)\left(\frac{\pi}{2} - \frac{\pi}{3}\right) \\
&\approx (.966)\frac{\pi}{6} + (.793)\frac{\pi}{12} + (.609)\frac{\pi}{12} + (.259)\frac{\pi}{6} \approx 1.01 \qquad\qquad \square
\end{aligned}
$$

When we compute a Riemann sum, the rectangle R_k has dimensions $f(t_k)$ and $x_k - x_{k-1} \ge 0$. When the graph of f is above the x–axis, $f(t_k) \ge 0$ and $f(t_k)(x_k - x_{k-1})$

is the area of R_k. However, when the graph of f is below the x–axis, $f(t_k) \leq 0$ and $f(t_k)(x_k - x_{k-1})$ is the negative of the area of R_k. That is, a Riemann sum of a region above the x–axis is positive while a Riemann sum of a region under the x–axis is negative. On the other hand, the area A of the region S always refers to its geometric area which is a positive number.

Example 3.2.24 Let A be the area bounded by the graph of $y = f(x) = x^2 - x$ and the x-axis for $0 \leq x \leq 1$. Use the partition $P = \left\{0, \frac{1}{3}, \frac{1}{2}, \frac{2}{3}, 1\right\}$ with $T = \left\{\frac{1}{4}, \frac{1}{3}, \frac{1}{2}, \frac{3}{4}\right\}$ to estimate A.

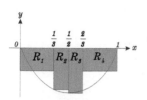

Figure 3.2.25
$f(x) = x^2 - x$

Solution Since $f\left(\frac{1}{4}\right)$, $f\left(\frac{1}{3}\right)$, $f\left(\frac{1}{2}\right)$, $f\left(\frac{3}{4}\right)$ are all negative, the rectangles R_1, R_2, R_3, R_4 lie below the x–axis, and the Riemann sum computes their areas as negative. See Figure 3.2.25. In particular,

$$
\begin{aligned}
R(P,T,f) &= f\left(\frac{1}{4}\right)\left(\frac{1}{3} - 0\right) + f\left(\frac{1}{3}\right)\left(\frac{1}{2} - \frac{1}{3}\right) + f\left(\frac{1}{2}\right)\left(\frac{2}{3} - \frac{1}{2}\right) \\
&\qquad + f\left(\frac{3}{4}\right)\left(1 - \frac{2}{3}\right) \\
&= \left(-\frac{3}{16}\right)\left(\frac{1}{3}\right) + \left(-\frac{2}{9}\right)\left(\frac{1}{6}\right) + \left(-\frac{1}{4}\right)\left(\frac{1}{6}\right) + \left(-\frac{3}{16}\right)\left(\frac{1}{3}\right) \\
&= -\frac{1}{16} - \frac{1}{27} - \frac{1}{24} - \frac{1}{16} = -\frac{11}{54}.
\end{aligned}
$$

Hence $A \approx +\frac{11}{54}$.   □

Lower and Upper Riemann Sums

Let S be the region of Figure 3.2.11 bounded by the graph of $y = f(x)$ on the top, the x–axis on the bottom, the line $x = a$ on the left and the line $x = b$ on the right. In the preceding subsection, we estimated the area A of S by Riemann sums $R(P,T,f)$ where P is a partition of $[a,b]$ and T is a selection of points from the subintervals of $[a,b]$ determined by P. Although there are many choices of T for each partition P, there are two special choices: T^* and T^{**}. The Riemann sums they define are called lower and upper Riemann sums. Their average approximates A, and we bound the error of this approximation.

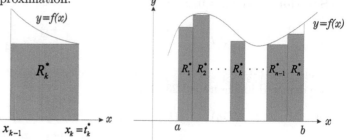

Figure 3.2.26 Lower Riemann Sum

Consider the partition $P = \{x_0, \ldots, x_n\}$ of $[a,b]$. To define a Riemann sum to estimate A we must select $t_k \in [x_{k-1}, x_k]$ for $1 \leq k \leq n$. Each of these selections determines a rectangle R_k of width $x_k - x_{k-1}$ and length $f(t_k)$ that approximates the strip S_k as in Figure 3.2.19. When f is continuous on $[x_{k-1}, x_k]$, it has a minimum value at $x = t_k^*$ and a maximum value at $x = t_k^{**}$ by the Maximum Value Theorem. Selecting $\{t_1^*, \ldots, t_n^*\}$ as T defines the *lower Riemann sum* while selecting $\{t_1^{**}, \ldots, t_n^{**}\}$ as T defines the *upper Riemann sum*. See the diagrams in Figures 3.2.26 and 3.2.27.

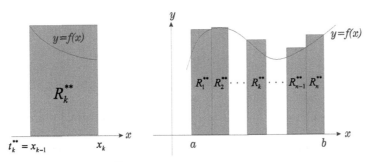

Figure 3.2.27 Upper Riemann Sum

Definition 3.2.28 *Let f be a continuous function having domain $[a, b]$ with $P = \{x_0, \ldots, x_n\}$ a partition of the interval $[a, b]$.*

(a) *Let $T^* = \{t_1^*, \ldots, t_n^*\}$ where the minimum value of $y = f(x)$ on the interval $[x_{k-1}, x_k]$ occurs at $x = t_k^*$ for $1 \le k \le n$. The lower Riemann sum of f for the partition P is the Riemann sum*

$$L(P, f) = R(P, T^*, f) = \sum_{k=1}^{n} f(t_k^*)(x_k - x_{k-1}).$$

(b) *Let $T^{**} = \{t_1^{**}, \ldots, t_n^{**}\}$ where the maximum value of $y = f(x)$ on the interval $[x_{k-1}, x_k]$ occurs at $x = t_k^{**}$ for $1 \le k \le n$. The upper Riemann sum of f for the partition P is the Riemann sum*

$$U(P, f) = R(P, T^{**}, f) = \sum_{k=1}^{n} f(t_k^{**})(x_k - x_{k-1}).$$

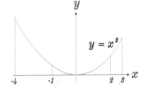

Figure 3.2.29
$f(x) = x^2$ on $[-4, 3]$

Example 3.2.30 Find $L(P, f)$ and $U(P, f)$ when $f(x) = x^2$ and $P = \{-4, -1, 2, 3\}$.

Solution Consider the graph of f on the interval $[-4, 3]$ depicted in Figure 3.2.29. Since f is decreasing on $[-4, -1]$, $t_1^* = -1$ and $t_1^{**} = -4$. On $[-1, 2]$ the smallest value of f occurs at $t_2^* = 0$ while the largest value of f occurs at $t_2^{**} = 2$. Since f is increasing on $[2, 3]$, $t_3^* = 2$ and $t_3^{**} = 3$. Hence

$$
\begin{aligned}
L(P, f) &= f(t_1^*)(x_1 - x_0) + f(t_2^*)(x_2 - x_1) + f(t_3^*)(x_3 - x_2) \\
&= f(-1)\left[-1 - (-4)\right] + f(0)\left[2 - (-1)\right] + f(2)\left[3 - 2\right] \\
&= (1)(3) + (0)(3) + (4)(1) \ = \ 7 \\
U(P, f) &= f(t_1^{**})(x_1 - x_0) + f(t_2^{**})(x_2 - x_1) + f(t_3^{**})(x_3 - x_2) \\
&= f(-4)\left[-1 - (-4)\right] + f(2)\left[2 - (-1)\right] + f(3)\left[3 - 2\right] \\
&= (16)(3) + (4)(3) + (9)(1) \ = \ 69 \qquad \square
\end{aligned}
$$

Consider the case where S lies entirely above the x–axis. The rectangle R_k^*, used to define the lower Riemann sum, has length $f(t_k^*)$ and is contained in the strip S_k. Hence the union of these rectangles is contained in S. See Fig. 3.2.26. The rectangle R_k^{**}, used to define the upper Riemann sum, has length $f(t_k^{**})$ and contains the strip S_k. Hence the union of these rectangles contains S. See Fig. 3.2.27. Thus

$$R_1^* \cup \cdots \cup R_n^* \ \subset S \subset \ R_1^{**} \cup \cdots \cup R_n^{**}$$

$$\sum_{k=1}^{n} \text{Area } R_k^* \ \le \ \text{Area } S \ \le \ \sum_{k=1}^{n} \text{Area } R_k^{**}$$

$$\sum_{k=1}^{n} f(t_k^*)(x_k - x_{k-1}) \ \le A \le \ \sum_{k=1}^{n} f(t_k^{**})(x_k - x_{k-1})$$

$$L(P, f) \ \le A \le \ U(P, f). \qquad (3.2.2)$$

Figure 3.2.31
$[L(P, f), U(P, f)]$

We use inequalities (3.2.2) to bound the error when A is estimated by the average of $L(P, f)$ and $U(P, f)$.

Proposition 3.2.32 *Let f be a continuous function with $f(x) \geq 0$ for $x \in [a, b]$. Let A be the area of the region S bounded by the graph of $y = f(x)$, the x–axis and the lines $x = a$, $x = b$. Let P be a partition of the interval $[a, b]$. Then*

$$A \approx \frac{1}{2}\left[L(P, f) + U(P, f)\right] \quad with \ error \ less \ than \quad \frac{1}{2}\left[U(P, f) - L(P, f)\right]. \quad (3.2.3)$$

Proof Since $f(x) \geq 0$ for $x \in [a, b]$, the region S lies entirely above the x–axis. Consider the interval $I = [L(P, f), U(P, f)]$ of width $w = U(P, f) - L(P, f)$. The midpoint of I is $m = \frac{1}{2}[L(P, f) + U(P, f)]$. Every number in I is within $\frac{w}{2}$ units of m. See Fig. 3.2.31. In particular, $A \in I$ by (3.2.2). Hence A is within $\frac{w}{2} = \frac{1}{2}[U(P, f) - L(P, f)]$ units of $m = \frac{1}{2}[L(P, f) + U(P, f)]$. $\square$

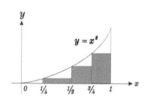

Figure 3.2.33
$L(P, f)$

We apply Proposition 3.2.32 to estimate the areas studied in Examples 3.2.21.

Examples 3.2.34 (1) Let A be the area of the region S bounded by the parabola $y = f(x) = x^2$, the x–axis and $x = 1$. Use Proposition 3.2.32 to estimate A for the partition $P = \left\{0, \frac{1}{4}, \frac{1}{2}, \frac{3}{4}, 1\right\}$.

Solution Since f is an increasing function on $[0, 1]$, $t_k^* = x_{k-1}$, the left endpoint of the interval $[x_{k-1}, x_k]$ and $t_k^{**} = x_k$, the right endpoint of the interval $[x_{k-1}, x_k]$, for $1 \leq k \leq 4$. See Figures 3.2.33 and 3.2.35. Then

$$L(P, f) = f(t_1^*)(x_1 - x_0) + f(t_2^*)(x_2 - x_1) + f(t_3^*)(x_3 - x_2) + f(t_4^*)(x_4 - x_3)$$

$$= (0)^2\left(\frac{1}{4} - 0\right) + \left(\frac{1}{4}\right)^2\left(\frac{1}{2} - \frac{1}{4}\right) + \left(\frac{1}{2}\right)^2\left(\frac{3}{4} - \frac{1}{2}\right) + \left(\frac{3}{4}\right)^2\left(1 - \frac{3}{4}\right)$$

$$= 0 + \frac{1}{64} + \frac{1}{16} + \frac{9}{64} = \frac{7}{32}$$

$$U(P, f) = f(t_1^{**})(x_1 - x_0) + f(t_2^{**})(x_2 - x_1) + f(t_3^{**})(x_3 - x_2) + f(t_4^{**})(x_4 - x_3)$$

$$= \left(\frac{1}{4}\right)^2\left(\frac{1}{4} - 0\right) + \left(\frac{1}{2}\right)^2\left(\frac{1}{2} - \frac{1}{4}\right) + \left(\frac{3}{4}\right)^2\left(\frac{3}{4} - \frac{1}{2}\right) + (1)^2\left(1 - \frac{3}{4}\right)$$

$$= \frac{1}{64} + \frac{1}{16} + \frac{9}{64} + \frac{1}{4} = \frac{15}{32}$$

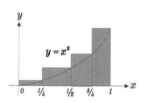

Figure 3.2.35
$U(P, f)$

By Proposition 3.2.32: $A \approx \frac{1}{2}[L(P, f) + U(P, f)] = \frac{1}{2}\left[\frac{7}{32} + \frac{15}{32}\right] = \frac{11}{32} \approx 0.344$ with error less than $\frac{1}{2}\left[\frac{15}{32} - \frac{7}{32}\right] = \frac{1}{8} = .125$.

(2) Let A' be the area of the region S' bounded by the graph of $g(x) = \cos x$, for $0 \leq x \leq \frac{\pi}{2}$, the x–axis and the y–axis. Use Proposition 3.2.32 to estimate A' for the partition $P' = \left\{0, \frac{\pi}{6}, \frac{\pi}{4}, \frac{\pi}{3}, \frac{\pi}{2}\right\}$.

Solution Since g is a decreasing function, $t_k^* = x_k^*$, the right endpoint of the interval $[x_{k-1}, x_k]$, and $t_k^{**} = x_{k-1}$, the left endpoint of the interval $[x_{k-1}, x_k]$, for $1 \leq k \leq 4$. See Figures 3.2.36 and 3.2.37. Then

$$L(P', g) = g(t_1^*)(x_1 - x_0) + g(t_2^*)(x_2 - x_1) + g(t_3^*)(x_3 - x_2) + g(t_4^*)(x_4 - x_3)$$

$$= \left(\cos\frac{\pi}{6}\right)\left(\frac{\pi}{6} - 0\right) + \left(\cos\frac{\pi}{4}\right)\left(\frac{\pi}{4} - \frac{\pi}{6}\right) + \left(\cos\frac{\pi}{3}\right)\left(\frac{\pi}{3} - \frac{\pi}{4}\right)$$

$$\qquad\qquad + \left(\cos\frac{\pi}{2}\right)\left(\frac{\pi}{2} - \frac{\pi}{3}\right)$$

$$= \frac{\pi\sqrt{3}}{12} + \frac{\pi\sqrt{2}}{24} + \frac{\pi}{24} + 0 \approx .769$$

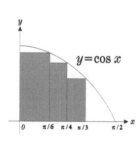

Figure 3.2.36
$L(P', g)$

$$U(P', g) = g(t_1^{**})(x_1 - x_0) + g(t_2^{**})(x_2 - x_1) + g(t_3^{**})(x_3 - x_2) + g(t_4^{**})(x_4 - x_3)$$

$$= (\cos 0)\left(\frac{\pi}{6} - 0\right) + \left(\cos \frac{\pi}{6}\right)\left(\frac{\pi}{4} - \frac{\pi}{6}\right) + \left(\cos \frac{\pi}{4}\right)\left(\frac{\pi}{3} - \frac{\pi}{4}\right)$$
$$+ \left(\cos \frac{\pi}{3}\right)\left(\frac{\pi}{2} - \frac{\pi}{3}\right)$$

$$= \frac{\pi}{6} + \frac{\pi\sqrt{3}}{24} + \frac{\pi\sqrt{2}}{24} + \frac{\pi}{12} \approx 1.197$$

By Proposition 3.2.32: $A \approx \frac{1}{2}[L(P', g) + U(P', g)] \approx \frac{1}{2}(.769 + 1.197) = .983$ with error less than $\frac{1}{2}(1.197 - .769) = .214$. □

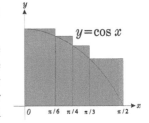

Figure 3.2.37
$U(P', g)$

Let S be the region of Figure 3.2.11 bounded by the graph of $y = f(x)$, the x–axis and the lines $x = a$, $x = b$. Assume S lies entirely above the x–axis. Suppose we want to estimate the area A of S with error less than a given number E. For simple functions, we can compute the lower and upper Riemann sums of the regular partition P_n of the interval $[a, b]$ as functions of n. By Proposition 3.2.32, the required approximation is:

$$A \approx \frac{1}{2}[L(P_n, f) + U(P_n, f)] \quad \text{when} \quad \frac{1}{2}[U(P_n, f) - L(P_n, f)] < E. \qquad (3.2.4)$$

When applying this procedure, use sigma notation to simplify the notation.

Example 3.2.38 Let S be the region in Figure 3.2.39 bounded by the parabola $f(x) = x^2$, the x–axis and the line $x = 6$. Estimate A with error less than $.001$.

Solution We use (3.2.4) to estimate this area. Let P_n be the regular partition of $[0, 6]$ into n subintervals of equal width $\frac{6}{n}$. Then $x_k = k\frac{6}{n} = \frac{6k}{n}$ for $0 \le k \le n$. Since $f(x) = x^2$ is an increasing function, t_k^* is the left endpoint $\frac{6(k-1)}{n}$ of the interval $[x_{k-1}, x_k] = \left[\frac{6(k-1)}{n}, \frac{6k}{n}\right]$, and t_k^{**} is the right endpoint $\frac{6k}{n}$. Then

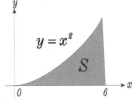

Figure 3.2.39
The Region S

$$U(P_n, f) = \sum_{k=1}^{n} f(t_k^{**})(x_k - x_{k-1}) = \sum_{k=1}^{n} \left(\frac{6k}{n}\right)^2 \frac{6}{n}$$

$$= \sum_{k=1}^{n} \frac{216 k^2}{n^3} = \frac{216}{n^3} \sum_{k=1}^{n} k^2 \qquad \text{[distributive property]}$$

$$= \frac{216}{n^3} \frac{n(n+1)(2n+1)}{6} \qquad \text{[Proposition 3.2.4(e)]}$$

$$= \frac{36(n+1)(2n+1)}{n^2} = \frac{36(2n^2 + 3n + 1)}{n^2} = 72 + \frac{108}{n} + \frac{36}{n^2}$$

$$L(P_n, f) = \sum_{k=1}^{n} f(t_k^*)(x_k - x_{k-1}) = \sum_{k=1}^{n} \left[\frac{6(k-1)}{n}\right]^2 \frac{6}{n}$$

$$= \sum_{k=1}^{n} \frac{216(k-1)^2}{n^3} = \frac{216}{n^3} \sum_{k=1}^{n}(k^2 - 2k + 1) \qquad \text{[distributive property]}$$

$$= \frac{216}{n^3} \sum_{k=1}^{n} k^2 - \frac{432}{n^3} \sum_{k=1}^{n} k + \frac{216}{n^3} \sum_{k=1}^{n} 1 \qquad \text{[commutative property]}$$

$$= \frac{216}{n^3} \frac{n(n+1)(2n+1)}{6} - \frac{432}{n^3} \frac{n(n+1)}{2} + \frac{216}{n^3}(n)(1) \quad \text{[Prop.3.2.4(e), (d), (c)]}$$

$$= \frac{36(n+1)(2n+1) - 216(n+1) + 216}{n^2} = \frac{72n^2 - 108n + 36}{n^2}$$

$$= 72 - \frac{108}{n} + \frac{36}{n^2}$$

The error in the estimate of A using (3.2.4) is:

$$\frac{1}{2}\left[U(P_n, f) - L(P_n, f)\right] = \frac{1}{2}\left[\left(72 + \frac{108}{n} + \frac{36}{n^2}\right) - \left(72 - \frac{108}{n} + \frac{36}{n^2}\right)\right] = \frac{108}{n}.$$

We want this error to be less than $.001 = \frac{1}{1,000}$, i.e. $n > 108,000$. Take $n = 108,001$:

$$\begin{aligned}
A &\approx \frac{1}{2}\left[L(P_{108,001}, f) + U(P_{108,001}, f)\right] \\
&= \frac{1}{2}\left[\left(72 + \frac{108}{108,001} + \frac{36}{108,001^2}\right) + \left(72 - \frac{108}{108,001} + \frac{36}{108,001^2}\right)\right] \\
&= 72 + \frac{36}{108,001^2} \approx 72.000
\end{aligned}$$

with error less than $.001$ by (3.2.4).   □

Let S be the region of Figure 3.2.11 bounded by the graph of $y = f(x)$, the x–axis and the lines $x = a$, $x = b$. Let P be a given partition of the interval $[a, b]$. Each choice of points T from the subintervals of $[a, b]$ determined by P defines a Riemann sum $R(P, T, f)$. The following proposition says that the lower Riemann sum is the smallest of these Riemann sums, and the upper Riemann sum is the largest one.

Proposition 3.2.41 *Let f be a continuous function with domain $[a, b]$. Let P be a partition of the interval $[a, b]$. For any choice of T:*

$$L(P, f) \leq R(P, T, f) \leq U(P, f). \tag{3.2.5}$$

Proof The partition P divides S into n strips $S_1, \ldots, S_n$ as in Figure 3.2.19. Consider the three rectangles in Figure 3.2.40 which approximate the strip S_k. They all have the interval $[x_{k-1}, x_k]$ on the x–axis as one side. Their lengths satisfy:

$$f(t_k^*) \leq f(t_k) \leq f(t_k^{**})$$

because $f(t_k^*)$ is the minimum value of f on $[x_{k-1}, x_k]$ while $f(t_k^{**})$ is the maximum value of f on $[x_{k-1}, x_k]$. Multiply the preceding inequalities by $x_k - x_{k-1} \geq 0$:

$$f(t_k^*)(x_k - x_{k-1}) \leq f(t_k)(x_k - x_{k-1}) \leq f(t_k^{**})(x_k - x_{k-1})$$

as indicated by the areas of the three rectangles in Figure 3.2.40. Sum these inequalities:

$$\sum_{k=1}^n f(t_k^*)(x_k - x_{k-1}) \leq \sum_{k=1}^n f(t_k)(x_k - x_{k-1}) \leq \sum_{k=1}^n f(t_k^{**})(x_k - x_{k-1}).$$

In Riemann sum notation these inequalities are: $L(P, f) \leq R(P, T, f) \leq U(P, f)$.   □

We use Examples 3.2.21 and 3.2.34 to illustrate this proposition.

Examples 3.2.42 Illustrate the inequalities (3.2.5) of Proposition 3.2.41 in each case.

(1) Let A be the area of the region S bounded by the parabola $y = f(x) = x^2$, the x–axis and $x = 1$. Use the partition $P = \left\{0, \frac{1}{4}, \frac{1}{2}, \frac{3}{4}, 1\right\}$ of the interval $[0, 1]$ and $T = \left\{\frac{1}{8}, \frac{3}{8}, \frac{5}{8}, \frac{7}{8}\right\}$.

Solution In Example 3.2.21 (1) we computed the Riemann sum $R(P, T, f) = \frac{21}{64}$ to estimate A. In Example 3.2.34 (2) we computed $L(P, f) = \frac{7}{32}$ and $U(P, f) = \frac{15}{32}$. Inequality (3.2.5) in this case is: $\frac{7}{32} \leq \frac{21}{64} \leq \frac{15}{32}$.

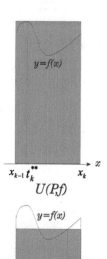

$U(P,f)$

$R(P,T,f)$

$L(P,f)$

Figure 3.2.40
Three Choices of R_k

(2) Let A' be the area of the region S' bounded by the graph of $g(x) = \cos x$, for $0 \le x \le \frac{\pi}{2}$, the x–axis and the y–axis . Use the partition $P' = \left\{ 0, \frac{\pi}{6}, \frac{\pi}{4}, \frac{\pi}{3}, \frac{\pi}{2} \right\}$ of the interval $\left[0, \frac{\pi}{2} \right]$ and $T' = \left\{ \frac{\pi}{12}, \frac{5\pi}{24}, \frac{7\pi}{24}, \frac{5\pi}{12} \right\}$.

Solution In Example 3.2.21 (1) we computed the Riemann sum $R(P', T', g) = 1.01$ to estimate A'. In Example 3.2.34 (2) we computed $L(P', g) = 0.77$ and $U(P', f) = 1.20$. Inequality (3.2.5) in this case is: $0.77 \le 1.01 \le 1.20$. □

Summary

The reader should be able to write sums in sigma notation and know how to manipulate sigma notation using the properties of Proposition 3.2.4. She should be able to approximate areas by Riemann sums. In particular, she should know how to approximate an area with a specified accuracy by using lower and upper Riemann sums.

Basic Exercises

1. Describe each sum in sigma notation.
(a) A is the sum of the first twenty positive odd numbers.
(b) B is the sum of the cubes of the integers of absolute value less than fifty.
(c) C is the sum of the first thirty positive powers of x whose exponents are divisible by three.
(d) D is the sum of the derivatives of the first forty positive powers of $\sin x$.
(e) Consider squares with two adjacent sides on the coordinate axes and opposite vertices at the origin and on the line x+y=n. E is the sum of the areas of these squares for the first ninety even positive integers n.
(f) F is the sum of the areas of the circles inscribed in the squares of (e).

2. Evaluate each sum.
(a) $\sum_{n=2}^{5}(2n+6)$ (b) $\sum_{n=0}^{4}(n^2-3)$ (c) $\sum_{n=1}^{3}\sqrt{n+5}$
(d) $\sum_{n=0}^{3}(-1)^n \frac{n}{n+1}$ (e) $\sum_{n=-1}^{4}\sin\frac{n\pi}{2}$ (f) $\sum_{n=0}^{4}\frac{x^n}{n+1}$
(g) $\sum_{n=0}^{3}\frac{x^{2n+3}}{n^2+1}$ (h) $\sum_{n=1}^{4}(-1)^n n^x$ (i) $\sum_{n=0}^{5}\frac{x^2}{3}$

3. Express each expression as a single sum.
(a) $5\sum_{n=0}^{9}n^2 - 3\sum_{n=0}^{9}4n + 2\sum_{n=0}^{9}7$ (b) $8\sum_{n=1}^{102}2^n + 9\sum_{k=1}^{102}3^k - \sum_{t=1}^{102}(-1)^t$
(c) $3\sum_{n=0}^{6}x^{4n} + 8\sum_{n=0}^{6}x^{2n} - 2\sum_{n=0}^{6}4$ (d) $x\sum_{n=1}^{42}\frac{x^{2n}}{n} - x^2\sum_{k=1}^{42}\frac{x^{3k}}{k+1} + x^3\sum_{t=1}^{42}\frac{x^{4t}}{t+2}$

4. Express each sum in sigma notation.
(a) $1 + 8 + 27 + 64$ (b) $8 - 18 + 32 - 50 + 72$
(c) $\arctan 3 + \arctan 5 + \arctan 7 + \arctan 9 + \arctan 11$
(d) $\sin\left(\frac{2\pi}{3}\right) - \sin\left(\frac{4\pi}{3}\right) + \sin\left(\frac{6\pi}{3}\right) - \sin\left(\frac{8\pi}{3}\right)$
(e) $x^2 + 2x^5 + 3x^8$ (f) $\frac{x^3}{2} - \frac{x^6}{4} + \frac{x^9}{12} - \frac{x^{12}}{48}$

5. Evaluate each sum.
(a) $\sum_{k=0}^{19}k$ (b) $\sum_{k=1}^{30}k^2$ (c) $\sum_{k=1}^{15}(3k-5)$
(d) $\sum_{k=1}^{24}(12k^2+9)$ (e) $\sum_{k=1}^{18}(4k^2-3k-8)$ (f) $\sum_{k=0}^{21}(6k-k^2-7)$

6. Use sigma notation to factor each polynomial.
(a) $x^{2n+1}+1$ (b) $x^{4n}-1$ (c) $\sum_{k=0}^{2n+1}x^k$ (d) $\sum_{k=0}^{2n+1}(-1)^k x^k$

7. Jane Doe has a $100,000 mortgage. Each month she pays $\frac{1}{2}\%$ of the remaining principal as interest plus $1,000 to reduce the principal.
(a) Use sigma notation to denote the total amount I of interest paid.
(b) Calculate the value of I.

8. A lottery sells a thousand tickets. There is one \$1,000 winner, two \$900 winners, three \$800 winners, . . . , nine \$200 winners and ten \$100 winners.
(a) Use sigma notation to denote the average winnings W of all the lottery tickets.
(b) Calculate the value of W.

9. A bar graph has ten adjacent bars. The first bar is 1 cm wide and 2 cm high. Each bar is twice as high and three times as wide as the preceding one. Use sigma notation to write the sum of the areas of these bars.

10. Twenty four identical cubic blocks of ice with side one meter long are placed in an empty vat. Each block melts so that it maintains its cubic shape with the length of its side decreasing by one centimeter each hour. One block is removed each hour. Ignoring the contraction of melting ice, use sigma notation to describe the amount of water in the vat one day later.

11. Find 3 partitions $P = \{x_0, \ldots, x_n\}$ which meet each of the given specifications.
   (a) Partition $[0, 6]$ with $n = 4$.   **(b)** Partition $[-3, 5]$ with $n = 10$.
   (c) Partition $[-4, 4]$ with $n = 5$.   **(d)** Partition $[-4, 10]$ with $n = 7$.

12. Find the mesh of each of your partitions in Exercise 11.

13. Find the regular partition of each interval into n subintervals.
   (a) $[0, 3]$, $n = 4$. **(b)** $[2, 5]$, $n = 9$ **(c)** $[-10, -1]$, $n = 6$ **(d)** $[-4, 10]$, $n = 7$

14. Find the mesh of each partition in Exercise 13.

Use a calculator to approximate your answers to the remaining exercises to three significant figures.

15. **(i)** Compute each Riemann sum $R(P, T, f)$.
   (ii) Sketch the rectangles that are used in each computation.
  (a) $P = \{0, \frac{1}{4}, \frac{1}{3}, \frac{3}{4}, 1\}$, $T = \{\frac{1}{10}, \frac{3}{10}, \frac{1}{2}, \frac{4}{5}\}$, $f(x) = x^2$
  (b) $P = \{-2, -1, 0, 1, 2, 3, 4\}$, $T = \{-2, -\frac{1}{2}, 1, \frac{5}{4}, \frac{5}{2}, 3\}$, $f(x) = x^3$
  (c) $P = \{-1, 1, 3, 4, 5\}$, $T = \{-1, 2, \frac{7}{2}, 5\}$, $f(x) = \frac{x}{1+x^2}$
  (d) $P = \{-2, -1, 0, 1, 2\}$, $T = \{-1, -\frac{1}{2}, 1, \frac{3}{2}\}$, $f(x) = \frac{x+2}{x^2-9}$
  (e) $P = \{0, 1, 4, 5, 7, 9\}$, $T = \{\frac{1}{4}, 4, 4, \frac{25}{4}, 9\}$, $f(x) = \sqrt{x}$
  (f) $P = \{1, 3, 5, 7, 8\}$, $T = \{1, 5, 5, 8\}$, $f(x) = x^{2/3}$
  (g) $P = \{0, \frac{\pi}{4}, \frac{\pi}{3}, \frac{\pi}{2}, \frac{5\pi}{6}, \pi\}$, $T = \{\frac{\pi}{6}, \frac{\pi}{4}, \frac{\pi}{2}, \frac{3\pi}{4}, \pi\}$, $f(x) = \sin^2 x$
  (h) $P = \{-\frac{\pi}{3}, -\frac{\pi}{4}, 0, \frac{\pi}{6}, \frac{\pi}{4}\}$, $T = \{-\frac{\pi}{3}, -\frac{\pi}{6}, 0, \frac{\pi}{4}\}$, $f(x) = \tan x$
  (i) $P = \{0, \frac{\pi}{6}, \frac{\pi}{4}, \frac{\pi}{3}, \frac{\pi}{2}\}$, $T = \{0, \frac{\pi}{6}, \frac{\pi}{3}, \frac{\pi}{3}\}$, $f(x) = \left\{ \begin{array}{ll} \frac{1-\cos x}{x^2} & \text{if } x \neq 0 \\ \frac{1}{2} & \text{if } x = 0 \end{array} \right\}$
  (j) $P = \{-1, 0, 1, 3, 6, 8\}$, $T = \{-1, 0, 1, 5, 7\}$, $f(x) = \arctan x$

16. Find each lower Riemann sum $L(P, f)$ and upper Riemann sum $U(P, f)$.
  (a) $f(x) = x^3$, $P = \{1, 2, 3, 4\}$ **(b)** $f(x) = \frac{1}{x}$, $P = \{2, 4, 5, 8\}$
  (c) $f(x) = \cot x$, $P = \{\frac{\pi}{6}, \frac{\pi}{4}, \frac{\pi}{3}, \frac{\pi}{2}, \frac{2\pi}{3}, \frac{5\pi}{6}\}$ **(d)** $f(x) = \arctan x$, $P = \{-3, -1, 0, 2, 4\}$
  (e) $f(x) = x^2$, $P = \{-5, -3, -1, 0, 2, 4\}$ **(f)** $f(x) = 1 - x^4$, $P = \{-4, -1, 0, 1, 3\}$
  (g) $f(x) = \frac{1}{1+x^2}$, $P = \{-2, -1, 0, 2, 3\}$ **(h)** $f(x) = \sec x$, $P = \{-\frac{\pi}{3}, -\frac{\pi}{4}, 0, \frac{\pi}{6}, \frac{\pi}{4}\}$

17. For each partition $P = \{x_0, \ldots, x_n\}$, let t_k be the midpoint of the interval $[x_{k-1}, x_k]$, $1 \leq k \leq n$. Define $T = \{t_1, \ldots, t_n\}$. Calculate the Riemann sum $R(P, T, f)$.
  (a) $[a, b] = [0, 5]$, $P = \{0, 1, 3, 4, 5\}$ and $f(x) = x^3$.
  (b) $[a, b] = [-4, 6]$, $P = \{-4, -3, -1, 0, 2, 5, 6\}$ and $f(x) = x^4$.
  (c) $[a, b] = [1, 4]$, $P = \{1, \frac{5}{4}, \frac{7}{4}, 2, \frac{5}{2}, 3, \frac{10}{3}, 4\}$ and $f(x) = x^2 - 1$.
  (d) $[a, b] = [-2, 3]$, $P = \{-2, -\frac{4}{3}, -\frac{1}{2}, 0, \frac{3}{4}, 2, 3\}$ and $f(x) = x^2 - x - 3$.
  (e) P is the regular partition of $[2, 8]$ into three subintervals and $f(x) = \frac{1}{x}$.
  (f) P is the regular partition of $[-10, 0]$ into five subintervals and $f(x) = \sqrt{x^2 + 1}$.
  (g) P is the regular partition of $[-2, 1]$ into six subintervals and $f(x) = \frac{x}{x+4}$.

(h) P is the regular partition of $[0, \pi]$ into three subintervals and $f(x) = \sin x$.
(i) P is the regular partition of $\left[0, \frac{\pi}{4}\right]$ into four subintervals and $f(x) = \tan x$.
(j) P is the regular partition of $\left[-\frac{\pi}{3}, \frac{\pi}{3}\right]$ into five subintervals and $f(x) = \sec x$.

18. For each part of Exercise 17, calculate $L(P, f)$ and $U(P, f)$.

19. Verify $L(P, f) \leq R(P, T, f) \leq U(P, f)$ for each part of Exercise 17.

20. In Exercises 17 (a),(b),(e),(f),(h),(i),(j) apply Prop. 3.2.32 to approximate the area of the corresponding region and give a bound on the error of the approximation.

21. Use the given P and T to estimate the area of each region.
(a) S_1 is the region bounded by the graph of $f(x) = x^2 - 4$ for $-2 \leq x \leq 2$ and the x–axis. P is the regular partition of $[-2, 2]$ into 4 subintervals and $T = \{-2, -1, 1, 2\}$.
(b) S_2 is the region bounded by the graph of $g(x) = \cos x$ for $\frac{\pi}{2} \leq x \leq \frac{3\pi}{2}$ and the x–axis. $P = \left\{\frac{\pi}{2}, \frac{3\pi}{4}, \pi, \frac{3\pi}{2}\right\}$ and $T = \left\{\frac{3\pi}{4}, \pi, \frac{5\pi}{4}\right\}$.
(c) S_3 is the region bounded by the graph of $h(x) = x^3$ for $-6 \leq x \leq -1$ and the x–axis. P is the regular partition of $[-6, -1]$ into five subintervals and $T = T^*$.
(d) S_4 is the region bounded by the graph of $k(x) = -\frac{1}{1+x^2}$ for $-2 \leq x \leq 3$ and the x–axis. P is the regular partition of $[-2, 3]$ into five subintervals and $T = T^{**}$.
(e) S_5 is the region bounded by the graph of $m(x) = x^2 - 6x + 8$ for $0 \leq x \leq 6$ and the x–axis. $P = \{0, 1, 2, 3, 4, 5, 6\}$ and $T = \{0, 1, 3, 3, 5, 6\}$.

22. Use formulas (3.2.4) to estimate the area of each region with error less than E.
(a) S_1 is bounded by the graph of $y = x^2$, for $0 \leq x \leq 1$, and $E = .01$.
(b) S_2 is bounded by the graph of $y = 6x$, for $2 \leq x \leq 9$, and $E = .001$.
(c) S_3 is bounded by the graph of $y = x^2 + x + 1$, for $0 \leq x \leq 2$, and $E = .003$.
(d) S_4 is bounded by the graph of $y = 15 - 2x$, for $2 \leq x \leq 5$, and $E = .005$.
(e) S_5 is bounded by the graph of $y = 6x^2 - 8x$, for $3 \leq x \leq 5$, and $E = .04$.
(f) S_6 is bounded by the graph of $y = 3x^2 + 5$, for $-4 \leq x \leq 4$, and $E = .02$.
(g) S_7 is bounded by the graph of $y = 7 + 4x - x^2$, for $2 \leq x \leq 4$, and $E = .03$.
(h) S_8 is bounded by the graph of $y = 10 - x^2$, for $-3 \leq x \leq 3$, and $E = .002$.

23. Let A be the area bounded by the graph of $y = f(x) = x^5$, the x–axis, $x = -1$ and $x = +1$.
(a) Let P_{2n} be the regular partition of the interval $[-1, 1]$ into $2n$ subintervals. Let T_{2n} be the set consisting of $\pm\frac{k}{n}$ for $1 \leq k \leq n$. Show that $R(P_{2n}, T_{2n}, f) = 0$.
(b) Does your computation in (a) indicate that A is approximately zero?

Challenging Problems

1. Let f be a continuous function with domain $[-a, a]$. Let P_n and Q_n be the regular partitions of the intervals $[-a, a]$ and $[0, a]$ into n subintervals.
(a) Compute $L(P_{2n}, f) + U(P_{2n}, f)$ when f is odd.
(b) Show $L(P_{2n}, f) = 2L(Q_n, f)$ and $U(P_{2n}, f) = 2U(Q_n, f)$ when f is even.

2. Let f be a continuous function with domain $[a, b]$.
(a) Let $P \subset Q$ be two partitions of the interval $[a, b]$. Show $L(P, f) \geq L(Q, f)$ and $U(Q, f) \leq U(P, f)$.
(b) Let P, Q be any two partition of the interval $[a, b]$. Show $L(P, f) \leq U(Q, f)$.

Appendix

We introduce the method of *induction* to establish the validity of a sequence of statements. Then we use induction to prove the formula for the sum of the squares of the first n positive integers given in Proposition 3.2.4(e). Recall that a natural number is a positive integer. The set of all natural numbers $\{1, 2, 3, \ldots\}$ is denoted N.

Proposition 3.2.43 *Let $S \subset N$. Assume:* **(a)** $1 \in S$; **(b)** *if $m \in S$, then $m+1 \in S$. Then $S = N$.*

Proof Assume S does not equal N. Let k be the smallest natural number which is not in S. By (a), $k > 1$. Hence $k-1$ is a natural number. Since $k-1 < k$, we have $k-1 \in S$. Apply (b) with $m = k-1$: we conclude $k \in S$ which is false. The assumption that S does not equal N led to this false conclusion. Hence this assumption must be false, i.e. $S = N$.   □

The use of Proposition 3.2.43 to prove a sequence of statements is called *induction*.

Corollary 3.2.44 (Induction) *Consider a sequence of statements, one statement $A(n)$ for each natural number n. Assume that:*

    (a) $A(1)$ *is true;*   **(b)** *if $m \in N$ and $A(m)$ is true, then $A(m+1)$ is true. Then all the statements $A(n)$ are true.*

Proof Let $S \subset N$ be the set of those n for which $A(n)$ is true. By (a), $1 \in S$. By (b), if $m \in S$ then $m+1 \in S$. By Prop. 3.2.43, $S = N$, i.e. every $A(n)$ is true.   □

This corollary agrees with our intuition. We know that $A(1)$ is true by (a). Apply (b) with $m = 1$ to conclude that $A(2)$ is true. Apply (b) again with $m = 2$ to conclude that $A(3)$ is true. Clearly we can continue in this way. For any natural number $n > 0$, apply (b) $n-1$ times to deduce that $A(n)$ is true. We illustrate the use of induction by proving Proposition 3.2.4 (e) and giving an alternate proof of Proposition 3.2.4 (d).

Proposition 3.2.4 **(d)** $\displaystyle\sum_{k=1}^{n} k = \frac{n(n+1)}{2}$ **(e)** $\displaystyle\sum_{k=1}^{n} k^2 = \frac{n(n+1)(2n+1)}{6}$

Proof **(d)** Let $A(n)$ be the statement: $\sum_{k=1}^{n} k = \frac{n(n+1)}{2}$. $A(1)$ says: $\sum_{k=1}^{1} k$ equals $\frac{1(2)}{2}$. Since both expressions equal one, $A(1)$ is true. Assume $A(m)$ is true. Then

$$
\begin{aligned}
\sum_{k=1}^{m+1} k &= 1 + 2 + 3 + \cdots + (m-1) + m + (m+1) \\[2mm]
&= \sum_{k=1}^{m} k + (m+1) = \frac{m(m+1)}{2} + m + 1 &&\text{[by A(m)]} \\[2mm]
&= \frac{m^2 + m + 2(m+1)}{2} = \frac{m^2 + 3m + 2}{2} = \frac{(m+1)(m+2)}{2}.
\end{aligned}
$$

Thus $A(m+1)$ is true. By induction, all the $A(n)$ are true.

(e) Let $B(n)$ be the statement: $\sum_{k=1}^{n} k^2 = \frac{n(n+1)(2n+1)}{6}$. $B(1)$ says: $\sum_{k=1}^{1} k^2$ equals $\frac{1(2)(3)}{6}$. Since both expressions equal one, $B(1)$ is true. Assume $B(m)$ is true. Then

$$
\begin{aligned}
\sum_{k=1}^{m+1} k^2 &= 1^2 + 2^2 + 3^2 + \cdots + (m-1)^2 + m^2 + (m+1)^2 \\[2mm]
&= \sum_{k=1}^{m} k^2 + (m+1)^2 = \frac{m(m+1)(2m+1)}{6} + (m+1)^2 &&\text{[by B(m)]} \\[2mm]
&= \frac{(m+1)(2m^2 + m) + 6(m+1)^2}{6} = \frac{(m+1)(2m^2 + 7m + 6)}{6} \\[2mm]
&= \frac{(m+1)(m+2)(2m+3)}{6} = \frac{(m+1)(m+2)\left[2(m+1)+1\right]}{6}.
\end{aligned}
$$

Thus $B(m+1)$ is true. By induction, all the $B(n)$ are true.   □

Exercises

Use induction to show that each statement is valid for every natural number n.

1. $\cos n\pi = (-1)^n$

2. $\sum_{k=1}^{n} k^3 = \frac{n^2(n+1)^2}{4}$

3. $3^{2n} - 1$ is divisible by eight.

4. $D(x^n) = nx^{n-1}$

5. $x^n - 1 = (x-1)\sum_{k=0}^{n-1} x^k$

6. $D[f(x)^n] = nf(x)^{n-1}Df(x)$

7. $2^n > n$

8. $\sum_{k=1}^{n} (2k-1) = n^2$

9. $\lim_{x \to c} [f(x)^n] = \left[\lim_{x \to c} f(x)\right]^n$

10. $\sum_{k=0}^{n} x^k = \frac{1-x^{n+1}}{1-x}$

11. $\sum_{k=1}^{n} \frac{1}{k(k+1)} = \frac{n}{n+1}$

12. $\sum_{k=1}^{n} k(k+1) = \frac{n(n+1)(n+2)}{3}$

3.3 Definite Integrals

In Section 2, we estimated areas by Riemann sums. In this section, we develop a procedure to compute the exact values of areas. In the first subsection, we define area, alias the definite integral, as the limit of Riemann sums. We state the theorem that the area of the region S in Figure 3.3.1, with f a continuous function, always exists. The proof of this deep result is given in Section 11. It is not practical to compute the exact values of areas directly from the definition. However, we use this definition in the second subsection to derive several properties of the definite integral. These properties are used in Section 4 to prove the Fundamental Theorems of Calculus. These theorems provide an easy method to evaluate definite integrals. The appendix to this section contains a proof of the additive property of definite integrals as well as a proof of the existence of definite integrals of piecewise monotone continuous functions.

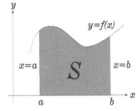

Figure 3.3.1
The Region S

Definition of the Definite Integral

Consider the region S bounded by the graph of $y = f(x)$ on top, by the x–axis on the bottom, the line $x = a$ on the left and the line $x = b$ on the right. See Figure 3.3.1. We approximate the area A of S as in Section 2. Let $P = \{x_0, \ldots, x_n\}$ be a partition which subdivides the interval $[a, b]$ into the n subintervals $[x_{k-1}, x_k]$ for $1 \le k \le n$. The lines $x = x_k$, for $0 \le k \le n$, subdivide S into the union of strips: $S = S_1 \cup \cdots \cup S_n$ as in Figure 3.3.2. Approximate each strip S_k by a rectangle R_k of width $x_k - x_{k-1}$ and length $f(t_k)$ as in Figure 3.3.3. The set $T = \{t_1, \ldots, t_n\}$ determines the lengths of the rectangles we are using. The Riemann sum $R(P, T, f)$, defined as the sum of the areas of these n rectangles, approximates the area A:

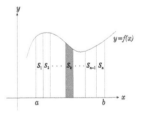

Figure 3.3.2
Subdividing S

$$A \approx R(P, T, f) = \sum_{k=1}^{n} f(t_k)(x_k - x_{k-1}). \qquad (3.3.1)$$

Recall the distinction between the Riemann sum $R(P, T, f)$ and Area $(R_1 \cup \cdots \cup R_n)$. Since $R(P, T, f)$ uses $f(t_k)(x_k - x_{k-1})$ as the area of the rectangle R_k, it computes the area of R_k as negative when $f(t_k) < 0$. That is, $R(P, T, f)$ computes the areas of rectangles above the x–axis as positive and the areas of rectangles below the x–axis as negative. However, when we compute $A = \text{Area}(S)$, we consider the areas of all regions to be positive. In particular, the approximation $A \approx R(P, T, f)$ of (3.3.1) refers to the case when the entire graph of f lies above the x–axis.

Figure 3.3.3
Approximating S_k by
the Rectangle R_k

Assume f is a continuous function, and fix a partition $P = \{x_0, \ldots, x_n\}$ of $[a, b]$. Let $f(t_k^*)$ be the minimum value of $f(x)$ on $[x_{k-1}, x_k]$ with $T^* = \{t_1^*, \ldots, t_n^*\}$, and let $f(t_k^{**})$ the maximum value of $f(x)$ on $[x_{k-1}, x_k]$ with $T^{**} = \{t_1^{**}, \ldots, t_n^{**}\}$. Recall from Proposition 3.2.41 that the lower Riemann sum $L(P, f) = R(P, T^*, f)$ is the smallest of the estimates $R(P, T, f)$ of A while the upper Riemann sum $U(P, f) = R(P, T^{**}, f)$ is the largest of these estimates. That is, for every choice of $T = \{t_1, \ldots, t_n\}$:

$$L(P, f) \leq R(P, T, f) \leq U(P, f). \tag{3.3.2}$$

The following example illustrates how the values of the lower and upper Riemann sums determine the area of the region S.

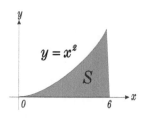

Figure 3.3.4

The Region S

Motivating Example 3.3.5 Compute the area A of the region S in Figure 3.3.4 bounded by the parabola $y = f(x) = x^2$, the x–axis and the line $x = 6$.

Solution   Let P_n be the regular partition of the interval $[0, 6]$ into n subintervals, each of width $\frac{6}{n}$. That is, P_n is the set of $\frac{6k}{n}$ for $0 \leq k \leq n$. The lower Riemann sum $L(P_n, f)$ is the area of the union L_n of the rectangles depicted in Fig. 3.3.6. The upper Riemann sum $U(P_n, f)$ is the area of the union U_n of the rectangles depicted in Fig. 3.3.7. Since $L_n \subset S \subset U_n$, we have $L(P_n, f) \leq A \leq U(P_n, f)$. In Example 3.2.38 we computed the values of $L(P_n, f)$ and $U(P_n, f)$:

$$72 - \frac{108}{n} + \frac{36}{n^2} = L(P_n, f) \leq A \leq U(P_n, f) = 72 + \frac{108}{n} + \frac{36}{n^2}.$$

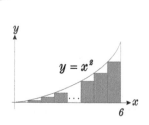

Figure 3.3.6

The Region L_n

Since only the number 72 satisfies these inequalities for every positive integer n, we conclude that $A = 72$.   □

The situation we encountered in this example should be true for every continuous function f whose graph lies above the x–axis. That is, let S be the region bounded by the graph of f, the x–axis and the lines $x = a$, $x = b$. Let $P = \{x_0, \ldots, x_n\}$ be a partition of $[a, b]$. The lower Riemann sum $L(P, f)$ is the area of a union of rectangles L which is contained in S while the upper Riemann sum $U(P, f)$ is the area of a union of rectangles U which contains S. Hence $L \subset S \subset U$ and the area A of S satisfies:

$$L(P, f) \leq A \leq U(P, f) \tag{3.3.3}$$

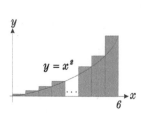

Figure 3.3.7

The Region U_n

Recall that mesh P is the largest of the widths $x_k - x_{k-1}$ of the rectangles R_k, $1 \leq k \leq n$. The smaller the mesh of P, the smaller the widths of all the rectangles R_k, and the better each R_k, for any choice of t_k, approximates the strip S_k. Thus the smaller the mesh of P, the better the Riemann sum $R(P, T, f)$ approximates the area A. Hence the smaller the mesh of P, the smaller the difference between the two Riemann sums $L(P, f)$ and $U(P, f)$. Thus, we expect that only one number A can satisfy inequalities (3.3.3) for all partitions P of $[a, b]$.

So far, our discussion has been based upon intuition. We assumed that we had a nice function f so that the meaning of the area A bounded by the graph of f, the x–axis and the lines $x = a$, $x = b$ is intuitively clear. However, there are complicated curves where our intuition cannot decide whether the area A is defined. For example, consider the function

$$f(x) = x^2 \sin^2 \frac{1}{x} \text{ if } 0 < x \leq \frac{1}{\pi}, \qquad f(0) = 0. \tag{3.3.4}$$

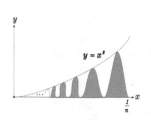

Figure 3.3.8

$f(x) = x^2 \sin^2 \frac{1}{x}$

The region R determined by f, depicted in Figure 3.3.8, consists of an infinite number of pieces of decreasing area. The sum of the areas of these pieces is an infinite sum which we have not defined.[1] Thus, we require a precise definition of A. Our intuitive

---

[1] Infinite sums will be defined in Section 5.3.

understanding of A led us to conclude that A is the unique number which satisfies the inequalities of (3.3.3) for every partition P. We base our definition of the definite integral upon this observation.

Definition 3.3.9 *Let f be a continuous function with domain $[a, b]$, $a \le b$.*
(a) *If there is a unique number I such that*

$$L(P, f) \le I \le U(P, f)$$

for every partition P of $[a, b]$, then we say that the definite integral of f on the interval $[a, b]$ exists and equals I. The usual notation is:

$$I = \int_a^b f(x) \; dx.$$

$f(x)$ *is called the integrand of I, a is called the lower bound of I and b is called the upper bound of I.*
(b) *Define $\int_b^a f(x) \; dx = -\int_a^b f(x) \; dx.$*

Note Let S be the region bounded by the graph f, the x–axis and the lines $x = a$, $x = b$. When the graph of f lies above the x–axis, each Riemann sum is a sum of positive numbers which approximates the area A of S. In this case, we define the area A as the definite integral $I = \int_a^b f(x) \; dx$. However, when the graph of f is not entirely above the x–axis, each Riemann sum is a sum of positive and negative numbers. Thus, the area A and the definite integral I are different in this case.

Although it is difficult to compute areas directly from this definition, we can use geometry in simple examples to determine the values of definite integrals.

Examples 3.3.12 (1) Evaluate $\int_{-2}^5 4 \; dx.$

> **Solution** This definite integral is the area of the rectangle of length 7 and width 4 depicted in Figure 3.3.10. Hence $\int_{-2}^5 4 \; dx = (7)(4) = 28.$

(2) Evaluate $\int_0^6 6 - x \; dx.$

> **Solution** This definite integral is the area of the right triangle depicted in Figure 3.3.11 which is bounded by the line $y = 6 - x$, the x–axis and the y–axis. Since each leg of this right triangle has length six, $\int_0^6 6 - x \; dx = \frac{1}{2}(6)^2 = 18.$

(3) Evaluate $\int_2^8 3 - 4x \; dx.$

> **Solution** This integral refers to the trapezoid T of Figure 3.3.13 bounded by the line $y = 3 - 4x$, the x–axis and the lines $x = 2$, $x = 8$. Since this region lies entirely under the x–axis the definite integral is the negative of the area of T. The bases of the trapezoid T have lengths 5 and 29 while its height is 6. Hence T has area $\frac{1}{2}(6)(5 + 29) = 102$, and $\int_2^8 3 - 4x \; dx = -102.$

(4) Evaluate $\int_6^0 x^2 \; dx.$

> **Solution** In Example 3.3.5 we showed that $\int_0^6 x^2 \; dx = 72$. Hence $\int_6^0 x^2 \; dx = -\int_0^6 x^2 \; dx = -72.$

(5) Evaluate $\int_{-\pi}^\pi \sin x \; dx.$

> **Solution** The function $g(x) = \sin x$ is an odd function, i.e. $g(-x) = -g(x)$. Therefore, the top region T in Fig. 3.3.14 is congruent the bottom region B.

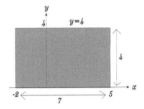

Figure 3.3.10
$\int_{-2}^5 4 \; dx$

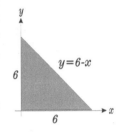

Figure 3.3.11
$\int_0^6 6 - x \; dx$

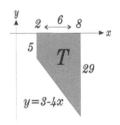

Figure 3.3.13
$\int_2^8 3 - 4x \; dx$

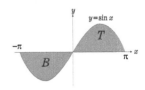

Figure 3.3.14
$\int_{-\pi}^\pi \sin x \; dx$

Thus Area $(B) =$ Area (T). Since the region B is below the x–axis, it is computed as negative by the definite integral. Therefore,

$$\int_{-\pi}^{\pi} \sin x \; dx = \text{Area}\,(T) - \text{Area}\,(B) = 0.$$

(6)   Evaluate Evaluate $\int_3^3 x^4 \; dx$.

Solution This definite integral represents the area of the line segment L of Fig. 3.3.15. Since L is one dimensional, it has zero area. Hence $\int_3^3 x^4 \; dx = 0$. $\square$

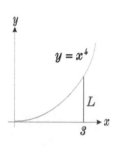

Figure 3.3.15
$\int_3^3 x^4 \; dx$

The following theorem says that the area bounded by the graph of any continuous function, the x–axis and the lines $x = a$, $x = b$ exists.

Theorem 3.3.16 *Let f be a function which is continuous on the closed interval $[a, b]$. Then the definite integral $\int_a^b f(x) \; dx$ exists.*

This theorem is a deep result, and we give its proof in Section 11. In the appendix to this section, we give a short proof of the special case of this theorem when f is piecewise monotone. Most of the examples we study are of this type. Although Theorem 3.3.16 tells us that definite integrals exist, it is very difficult to calculate their value directly from the definition. We use Theorem 3.3.16 to show that the area discussed above, between the graph of the function of (3.3.4) and the x–axis exists.

Example 3.3.17 Let $f(x) = \left\{ \begin{array}{ll} x^2 \sin^2 \frac{1}{x} & \text{if } x \neq 0 \\ 0 & \text{if } x = 0 \end{array} \right\}$. Show $\int_0^1 f(x) \; dx$ exists.

Solution Note $0 \leq x^2 \sin^2 \frac{1}{x} \leq x^2$ for all x. Since $\lim_{x \to 0} x^2 = 0$, it follows from the Pinching Theorem that $\lim_{x \to 0} f(x) = \lim_{x \to 0} x^2 \sin^2 \frac{1}{x} = 0 = f(0)$. Hence f is continuous at $x = 0$. Clearly f is continuous at all nonzero x. Thus f is a continuous function, and $\int_0^1 f(x) \; dx$ exists by Theorem 3.3.16. $\square$

Properties of the Definite Integral

Although the definition of definite integrals is not practical for computing their values, it is useful for establishing their properties. These properties are used in the next section to prove the Fundamental Theorems of Calculus which give an easy method to compute definite integrals.

The first property notes that the definite integral computes area above the x–axis as positive. See Figure 3.3.18.

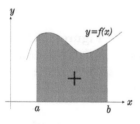

Figure 3.3.18
Property 1

Property 1 *Assume the function f is continuous on the interval $[a, b]$ with $a \leq b$. If $f(x) \geq 0$ for $x \in [a, b]$, then $\int_a^b f(x) \; dx \geq 0$.*

Proof Let $P = \{x_0, \ldots, x_n\}$ be a partition of $[a, b]$. Let the minimum value of f on $[x_{k-1}, x_k]$ occur at $x = t_k^*$ for $1 \leq k \leq n$. By hypothesis, each $f(t_k^*) \geq 0$. Hence

$$\int_a^b f(x) \; dx \geq L(P, f) = \sum_{k=1}^{n} f(t_k^*)(x_k - x_{k-1}) \geq 0. \qquad \square$$

The next result is used in the proof of Properties 3 and 7. When we multiply a function f by -1, the regions determined by f on one side of the x–axis become congruent regions determined by $-f$ on the other side of the x–axis. Hence the definite integral of $-f$ is the negative of the definite integral of f. See Figure 3.3.19.

Property 2 *Assume the function $f(x)$ is continuous for x between a, b. Then*

$$\int_a^b -f(x)\ dx = -\int_a^b f(x)\ dx.$$

Proof First, assume $a \le b$. Let $P = \{x_0, \dots, x_n\}$ be a partition of $[a, b]$ with $T^* = \{t_1^*, \dots, t_n^*\}$ and $T^{**} = \{t_1^{**}, \dots, t_n^{**}\}$. For $x \in [x_{k-1}, x_k]$: we have $f(t_k^*) \le f(x) \le f(t_k^{**})$ and thus $-f(t_k^*) \ge -f(x) \ge -f(t_k^{**})$. Therefore $L(P, -f)$ is computed using the $-f(t_k^{**})$ and $U(P, -f)$ is computed using the $-f(t_k^*)$. Hence

$$-L(P, f) = -\sum_{k=1}^{n} f(t_k^*)(x_k - x_{k-1}) = \sum_{k=1}^{n} -f(t_k^*)(x_k - x_{k-1}) = U(P, -f)$$

$$-U(P, f) = -\sum_{k=1}^{n} f(t_k^{**})(x_k - x_{k-1}) = \sum_{k=1}^{n} -f(t_k^{**})(x_k - x_{k-1}) = L(P, -f).$$

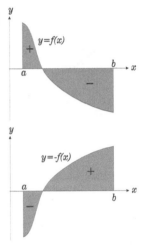

Multiply the inequalities $L(P, f) \le \int_a^b f(x)\ dx \le U(P, f)$ by -1:

$$L(P, -f) = -U(P, f) \le -\int_a^b f(x)\ dx \le -L(P, f) = U(P, -f)$$

for every partition P of $[a, b]$. By Definition 3.3.9, $\int_a^b -f(x)\ dx$ is the unique number with this property. Hence $\int_a^b -f(x)\ dx = -\int_a^b f(x)\ dx$.

Figure 3.3.19
Property 2

Now assume $b < a$. By the preceding case:

$$\int_a^b -f(x)\ dx = -\int_b^a -f(x)\ dx = \int_b^a f(x)\ dx = -\int_a^b f(x)\ dx. \qquad \square$$

The third property formalizes an observation of the preceding subsection: the definite integral computes area beneath the x–axis as negative. See Figure 3.3.20.

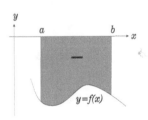

Property 3 *Assume the function f is continuous on the interval $[a, b]$ with $a \le b$. If $f(x) \le 0$ for $x \in [a, b]$, then $\int_a^b f(x)\ dx \le 0$.*

Proof Since $-f(x) \ge 0$ for $x \in [a, b]$, apply Property 1: $\int_a^b -f(x) \ge 0$. By Property 2, $\int_a^b f(x)\ dx = -\int_a^b -f(x)\ dx \le 0$. $\qquad \square$

The following property generalizes the observation made in Example 3.3.12 (6): the area of a line segment is zero. See Figure 3.3.21.

Figure 3.3.20
Property 3

Property 4 *Let f be a function which is continuous at $x = a$. Then*

$$\int_a^a f(x)\ dx = 0.$$

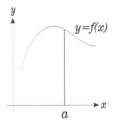

Proof If $P = \{x_0, \dots, x_n\}$ is a partition of $[a, a]$, then $a = x_0 = \cdots = x_n$. Hence each $x_k - x_{k-1} = 0$, and every Riemann sum is zero. Thus

$$0 = L(P, f) \le \int_a^a f(x)\ dx \le U(P, f) = 0,$$

Figure 3.3.21
Property 4

and $\int_a^a f(x)\ dx = 0$. $\qquad \square$

The fifth property generalizes Example 3.3.12 (1): the integral of the constant function $y = c$ over the interval $[a, b]$ is the area of a rectangle of dimensions c and $b - a$.

Property 5 *If $a, b, c \in \Re$, then*

$$\int_a^b c \; dx = c(b - a).$$

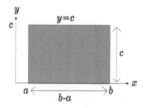

Figure 3.3.22
Property 5

Proof First, assume $a \le b$. Let $P = \{x_0, \ldots, x_n\}$ be a partition of $[a, b]$ with $T = \{t_1, \ldots, t_n\}$ a selection of points of the subintervals of $[a, b]$. Then

$$
\begin{aligned}
R(P, T, f) &= \sum_{k=1}^{n} f(t_k)(x_k - x_{k-1}) = \sum_{k=1}^{n} c(x_k - x_{k-1}) = c\sum_{k=1}^{n} (x_k - x_{k-1}) \\
&= c\left[(x_1 - x_0) + (x_2 - x_1) + \cdots + (x_{n-1} - x_{n-2}) + (x_n - x_{n-1})\right]
\end{aligned}
$$

Note x_0 and x_n only appear once in this sum. However, every other summand appears twice: once with a plus sign and once with a minus sign. These terms cancel, and $R(P, T, f) = c(x_n - x_0) = c(b - a)$. Thus

$$c(b - a) = L(P, f) \le \int_a^b c \; dx \le U(P, f) = c(b - a), \quad \text{and} \quad \int_a^b c \; dx = c(b - a).$$

Now assume $b < a$. By the preceding case:

$$\int_a^b c \; dx = -\int_b^a c \; dx = -c(a - b) = c(b - a). \qquad \square$$

For positive valued functions, larger functions describe larger areas. Property 6 says that this observation generalizes to definite integrals: larger functions have larger definite integrals. See Figure 3.3.23.

Property 6 *Assume the functions f, g are continuous on the interval $[a, b]$ with $a \le b$. If $f(x) \le g(x)$ for $x \in [a, b]$, then*

$$\int_a^b f(x) \; dx \le \int_a^b g(x) \; dx.$$

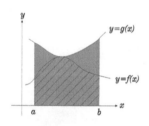

Figure 3.3.23
Property 6

Proof Let $P = \{x_0, \ldots, x_n\}$ be a partition of $[a, b]$. Let the maximum value of f on $[x_{k-1}, x_k]$ occur at $x = t_k^{**}$ for $1 \le k \le n$, and define $T^{**} = \{t_1^{**}, \ldots, t_n^{**}\}$. By hypothesis, $f(t_k^{**}) \le g(t_k^{**})$. Hence

$$\int_a^b f(x) \; dx \le U(P, f) = R(P, T^{**}, f) \le R(P, T^{**}, g) \le U(P, g).$$

Since $\int_a^b g(x) \; dx$ is the largest number less than or equal to every $U(P, g)$ for every partition P of $[a, b]$, it follows that $\int_a^b f(x) \; dx \le \int_a^b g(x) \; dx$. $\qquad \square$

The definite integral of $f(x)$ computes area above the x–axis as positive and area under the x–axis as negative. Therefore, its value is less than the area determined by the graph of $|f|$ where the areas of all regions are computed as positive. See Figure 3.3.24 where the region determined by $y = f(x)$ is filled with slanted lines, and the region determined by $y = |f(x)|$ is shaded.

Property 7 *If the function f is continuous on the interval $[a, b]$ with $a \le b$, then*

$$\left| \int_a^b f(x) \; dx \right| \le \int_a^b |f(x)| \; dx.$$

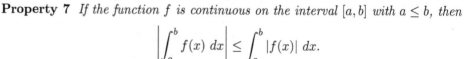

Proof Since $|f(x)|$ equals either $f(x)$ or $-f(x)$, it follows that

$$-|f(x)| \ \leq \ f(x) \ \leq \ |f(x)| \quad \text{and}$$

$$-\int_a^b |f(x)| \ dx \ = \ \int_a^b -|f(x)| \ dx \ \leq \ \int_a^b f(x) \ dx \ \leq \ \int_a^b |f(x)| \ dx \quad (3.3.5)$$

by Properties 2 and 6. Multiply these inequalities by -1:

$$\int_a^b |f(x)| \ dx \geq -\int_a^b f(x) \ dx \geq -\int_a^b |f(x)| \ dx. \qquad (3.3.6)$$

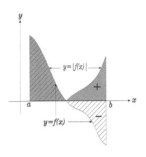

Figure 3.3.24
Property 7

Note that $|\int_a^b f(x) \ dx|$ is either $\int_a^b f(x) \ dx$ or $-\int_a^b f(x) \ dx$. Use (3.3.5) in the first case and (3.3.6) in the second case to conclude that $|\int_a^b f(x) \ dx| \leq \int_a^b |f(x)| \ dx$. $\qquad \square$

Example 3.3.25 Illustrate Property 7 for the function $f(x) = x$ with domain $[-2, 4]$.

Solution Observe from Figure 3.3.26 that the region bounded by the graph of f, the x–axis and the lines $x = -2$, $x = 4$ consists of two triangles: one in the fourth quadrant of area $\frac{1}{2}(2)(2) = 2$ and the other in the first quadrant of area $\frac{1}{2}(4)(4) = 8$. The definite integral computes the area of the former triangle to be negative since it lies beneath the x–axis while it computes the area of the latter triangle to be positive because it lies above the x–axis. Hence

$$\int_{-2}^4 f(x) \ dx = -2 + 8 = 6.$$

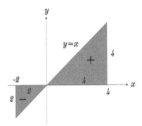

Figure 3.3.26
$\int_{-2}^4 x \ dx$

Observe from Figure 3.3.27 that the region bounded by the graph of $|f|$, the x–axis and the lines $x = -2$, $x = 4$ consists of two triangles: one in the second quadrant of area $\frac{1}{2}(2)(2) = 2$ and the other in the first quadrant of area $\frac{1}{2}(4)(4) = 8$. The definite integral computes the area of both these triangles as positive because they lie above the x–axis. Hence

$$\int_{-2}^4 |f(x)| \ dx = 2 + 8 = 10.$$

In this example, the inequality $|\int_{-2}^4 f(x) \ dx| \leq \int_{-2}^4 |f(x)| \ dx$ is $6 \leq 10$. $\qquad \square$

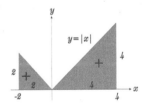

Figure 3.3.27
$\int_{-2}^4 |x| \ dx$

The next property generalizes Example 3.3.12 (5): the integral of an odd function over $[-a, a]$ is zero. The graph of an odd function determines congruent regions above and below the x–axis. The definite integral computes the areas of the former regions as positive and the areas of the latter regions as negative. Hence their sum is zero. For now, we only use this property to analyze examples, not to prove theorems. Therefore we postpone its verification to Section 6 where we give a simple proof.

Property 8 *Let f be a continuous odd function on the interval $[-a, a]$. Then*

$$\int_{-a}^a f(x) \ dx = 0.$$

Intuitively, the area of a region which consists of the union of two disjoint pieces is the sum of the areas of the two pieces. This result is still valid if the areas intersect on a common edge. This observation for the union of the two areas $\int_a^b f(x) \ dx$ and $\int_b^c f(x) \ dx$ in Figure 3.3.28 is called the additive property of definite integrals. We give a rigorous proof of this property in the appendix to this section.

Property 9 (Additive Property) *Assume the function f is continuous on the interval $[A, B]$ where A is the smallest of a, b, c and B is the largest of these numbers. Then*

$$\int_a^c f(x)\ dx = \int_a^b f(x)\ dx + \int_b^c f(x)\ dx. \tag{3.3.7}$$

Observe that the additive property does not require that $a < b < c$. It is valid for any three numbers a, b, c.

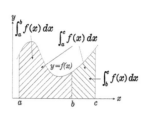

Figure 3.3.28
Property 9

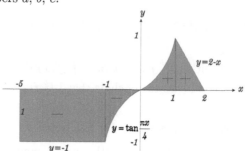

Figure 3.3.29 The Region of Example 3.3.30

Example 3.3.30 Evaluate $\int_{-5}^{2} f(x)\ dx$ where

$$f(x) = \left\{ \begin{array}{rl} -1 & \text{for } -5 \le x < -1 \\ \tan\frac{\pi x}{4} & \text{for } -1 \le x \le 1 \\ 2 - x & \text{for } 1 < x \le 2 \end{array} \right\}.$$

Solution Note the function f, whose graph is depicted in Figure 3.3.29, is continuous on the interval $[-5, 2]$. By the additive property:

$$\int_{-5}^{2} f(x)\ dx = \int_{-5}^{-1} -1\ dx + \int_{-1}^{1} \tan\frac{\pi x}{4}\ dx + \int_{1}^{2} 2 - x\ dx .$$

By Property 5: $\int_{-5}^{-1} -1\ dx = (-1)\left[-1 - (-5)\right] = -4$. Note $\int_{-1}^{1} \tan\frac{\pi x}{4}\ dx$ is the integral of an odd function on the interval $[-1, 1]$ which is zero by Property 8. Also, $\int_{1}^{2} 2 - x\ dx$ is the area of a right triangle above the x–axis with both legs of length one. This triangle has area $\frac{1}{2}(1)(1) = \frac{1}{2}$. Hence

$$\int_{-5}^{2} f(x)\ dx = -4 + 0 + \frac{1}{2} = -\frac{7}{2}. \qquad \square$$

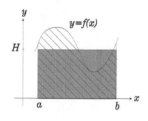

Figure 3.3.31
Average Value H

If $\int_a^b f(x)\ dx$ exists, then it equals the area of the rectangle in Figure 3.3.31 with one side the interval $[a, b]$ and width equal to the *average value* H of $f(x)$ on the interval $[a, b]$. That is, $\int_a^b f(x)\ dx = H(b - a)$, and

$$H = \frac{\int_a^b f(x)\ dx}{b - a}. \tag{3.3.8}$$

The mean value theorem for definite integrals states that a continuous function f takes on its average value on $[a, b]$. When $b < a$, $[a, b]$ denotes the closed interval $[b, a]$.

Property 10 (Mean Value Theorem for Definite Integrals) *Assume $f(x)$ is continuous on $[a, b]$ with $a \ne b$. Then there is at least one number $c \in [a, b]$ such that*

$$f(c) = \frac{\int_a^b f(x)\ dx}{b - a}. \tag{3.3.9}$$

Proof First, assume $a < b$. By the Maximum Value Theorem, f has a minimum value of $f(m)$ and a maximum value $f(M)$ on $[a, b]$ with $m, M \in [a, b]$. That is,

$$f(m) \leq f(x) \leq f(M)$$

for $x \in [a, b]$. By Properties 5 and 6:

$$f(m)(b - a) = \int_a^b f(m) \, dx \leq \int_a^b f(x) \, dx \leq \int_a^b f(M) \, dx = f(M)(b - a). \quad (3.3.10)$$

Divide these inequalities by $b - a$:

$$f(m) \leq \frac{\int_a^b f(x) \, dx}{b - a} \leq f(M).$$

By the Intermediate Value Theorem, there is at least one number c between m and M such that:

$$f(c) = \frac{\int_a^b f(x) \, dx}{b - a}.$$

Now assume $b < a$. By the preceding case there is $c \in [b, a]$ with

$$f(c) = \frac{\int_b^a f(x) \, dx}{a - b} = \frac{-\int_a^b f(x) \, dx}{a - b} = \frac{\int_a^b f(x) \, dx}{b - a}. \qquad \square$$

Examples 3.3.32 Find the average value of each function. Then verify the mean value theorem for definite integrals.

(1) $f(x) = 3x$ on the interval $[0, 6]$.

Solution Note $\int_0^6 3x \, dx$ is the area of the triangle with legs of length 6 and 18 in Figure 3.3.33. Therefore, $\int_0^6 3x \, dx = \frac{1}{2}(6)(18) = 54$, and the average value H of f on $[0, 6]$ is given by:

$$H = \frac{\int_0^6 3x \, dx}{6 - 0} = \frac{54}{6} = 9.$$

We verify the mean value theorem: $f(c) = 3c = 9$ has solution $c = 3 \in [0, 6]$.

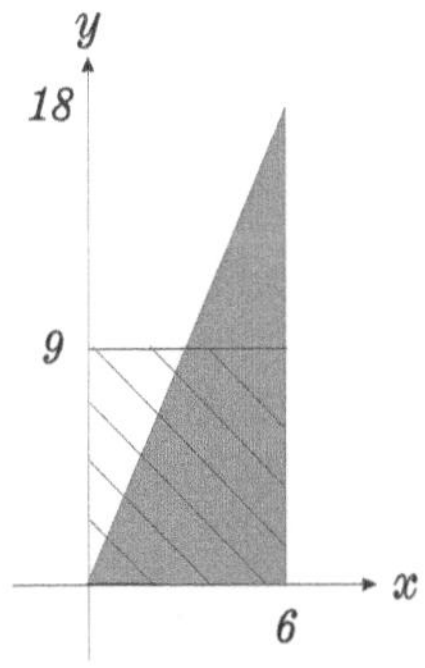

(2) $g(x) = \sin x$ on the interval $[-\pi, \pi]$.

Solution Since $\sin x$ is an odd function, $\int_{-\pi}^{\pi} \sin x \, dx = 0$ by Property 8. Hence the average value H of g on $[-\pi, \pi]$ is given by:

$$H = \frac{\int_{-\pi}^{\pi} \sin x \, dx}{\pi - (-\pi)} = \frac{0}{2\pi} = 0.$$

We verify the mean value theorem: $g(c) = \sin c = 0$ has solutions $c = -\pi, 0, \pi$ in $[-\pi, \pi]$. $\qquad \square$

Figure 3.3.33
$f(x) = 3x$

Historical Remarks

Our definition of area originates from the methods developed by three Greek geometers to study circles ca 430 BCE. Antiphon attempted to compute the area of a circle by approximating it with a sequence of inscribed polygons. Bryson of Heraclea extended this approach by making a simultaneous analysis of a sequence of inscribed polygons and a sequence of circumscribed polygons. Hippocrates of Chios showed that the area of a circle is proportional to the square of its radius by approximating a circle with a sequence of inscribed polygons. The mathematically rigorous formulation of using inscribed polygons to compute area is called the *method of exhaustion*. It is unclear whether this method was first formulated and applied by Hippocrates or by Eudoxus of Cnidus seventy years later. Archimedes' *method of compression*, ca. 250 BCE, generalizes this approach by using both inscribed and circumscribed polygons to compute areas.

We would formulate the method of compression to compute the area A of a given region S as follows. Find a sequence of polygons with areas L_n, which are contained inside S, and a second sequence of polygons with areas U_n, which contain S. If $\lim_{n \to \infty} (U_n - L_n) = 0$, then the common limit A of the L_n and the U_n is the area of S. The Greeks, however, did not use limits to formulate the method of compression because they considered limit arguments to be merely intuitive. Rather they used a double *reductio ad absurdum* argument. First, they used a sequence of inscribed polygons which approximate S to show that the assumption that the area of S is greater than A leads to a contradiction. Then they used a sequence of circumscribed polygons which approximate S to show that the assumption that the area of S is less than A also leads to a contradiction. It follows that the area of S equals A. This method reached its culmination with Archimedes' computations of areas bounded by ellipses, parabolas and spirals. However, these meticulously rigorous arguments of the Greeks were too clumsy and complicated to deal with more sophisticated problems.

In the seventeenth century, the tedious reductio ad absurdum arguments of the Greeks to compute the area A of a region S was replaced by the computation of A as the limit of areas of polygons which approximate S or by computing A as an infinite sum of infinitesmaly small areas. These intuitive procedures lacked rigor. However, in the seventeenth century rigor was viewed as an obstacle to progress which was relegated to philosophers who had nothing better to do with their time! For example, in 1609 Johann Kepler proved his laws of planetary motion by dividing a region into an infinite number of triangular *infinitesmal pieces*. In 1635, Bonaventura Cavalieri conjectured the value of $\int_0^b x^k \, dx$, for k a positive integer. He proved this conjecture for $k \leq 9$ by considering a region as a union of line segments and the area of the region as the sum of the lengths of these segments. One year later, John Pierre de Fermat, Blaise Pascal and Gilles Persone de Roberval used lower and upper Riemann sums, together with clever combinatorial arguments, to verify Cavalieri's conjecture for all positive integers k. By 1655, John Wallis, Fermat and Evangelista Torricelli had computed $\int_0^b x^{p/q} \, dx$ for positive integers p, q. Then Toricelli extended this computation to negative p/q.

Summary

The reader should know the definition of the definite integral. She should understand the interpretation of the ten properties of definite integrals in terms of areas. In particular, she should be able to find the average value H of a function f on a given closed interval and know how the mean value theorem for definite integrals interprets H in terms of the area under the graph of f.

Basic Exercises

1. Determine the value of each integral. In most cases, it is helpful to first sketch the corresponding area.

(a) $\int_1^9 6 \, dx$ (b) $\int_0^6 4x \, dx$ (c) $\int_2^2 \frac{1}{x} \, dx$ (d) $\int_0^8 2x + 3 \, dx$

(e) $\int_2^8 -5\,dx$ (f) $\int_{-5}^5 x^3\,dx$ (g) $\int_{-3}^3 \sqrt{9-x^2}\,dx$ (h) $\int_4^{-3} 2\,dx$ (i) $\int_{-5}^0 -3x\,dx$

(j) $\int_\pi^\pi \cos x\,dx$ (k) $\int_{-6}^6 4x^5 - 7x\,dx$ (l) $\int_{-4}^6 3x + 6\,dx$ (m) $\int_{-\pi/4}^{\pi/4} \tan x\,dx$

(n) $\int_0^{-3} 2x\,dx$ (o) $\int_{-6}^{-1} 4x - 1\,dx$ (p) $\int_{-4}^0 \sqrt{16-x^2}\,dx$ (q) $\int_{-\pi}^\pi x^2 \sin\frac{1}{x}\,dx$

(r) $\int_0^7 6 - 2x\,dx$ (s) $\int_{-2}^2 -\sqrt{4-x^2}\,dx$ (t) $\int_0^6 4x + 3\,dx$ (u) $\int_0^\pi \cos x\,dx$

2. Let P_n denote the regular partition of the interval $[a, b]$.
   (i) Calculate $L(P_n, f)$.   (ii) Calculate $U(P_n, f)$.
   (iii) Compute the value of each integral from its definition.

(a) $\int_2^6 7\,dx$ (b) $\int_{-3}^2 -5\,dx$ (c) $\int_0^4 3x\,dx$ (d) $\int_0^5 2x + 9\,dx$

(e) $\int_1^7 6x\,dx$ (f) $\int_2^8 5x - 4\,dx$ (g) $\int_0^4 x^2\,dx$ (h) $\int_{-3}^0 6x^2\,dx$

(i) $\int_0^2 3x^2 + 6x\,dx$ (j) $\int_0^4 5 + 2x^2\,dx$ (k) $\int_0^1 x^2 + x + 1\,dx$ (l) $\int_{-6}^0 2 - 4x + 3x^2\,dx$

3. Justify that each integral exists:

(a) $\int_0^1 f(x)\,dx$ where $f(x) = \begin{cases} x\sin\frac{1}{x} & \text{if } x \neq 0 \\ 0 & \text{if } x = 0 \end{cases}$;

(b) $\int_{-\pi}^\pi g(x)\,dx$ where $g(x) = \begin{cases} \frac{\sin x}{x} & \text{if } x \neq 0 \\ 1 & \text{if } x = 0 \end{cases}$;

(c) $\int_0^2 h(x)\,dx$ where $h(x) = \begin{cases} \frac{\sqrt{x+3}-2}{x-1} & \text{if } x \neq 1 \\ \frac{1}{4} & \text{if } x = 1 \end{cases}$.

4. Assume the function f is continuous on the interval $[a, b]$, and $c \in \Re$.
Show $\int_a^b cf(x)\,dx = c\int_a^b f(x)\,dx$.
Hint: First prove the case $c > 0$. Then deduce the case $c < 0$.

5. Use the additive property of the definite integral to evaluate each integral:

(a) $\int_0^6 f(x)\,dx$ where $f(x) = \begin{cases} 4x & \text{if } 0 \leq x \leq 2 \\ 3x + 2 & \text{if } 2 < x \leq 6 \end{cases}$;

(b) $\int_1^9 g(x)\,dx$ where $g(x) = \begin{cases} 2x + 1 & \text{if } 1 \leq x \leq 5 \\ 11 & \text{if } 5 < x \leq 9 \end{cases}$;

(c) $\int_{-3}^5 h(x)\,dx$ where $h(x) = \begin{cases} 5x & \text{if } 0 \leq x \leq 5 \\ 0 & \text{otherwise} \end{cases}$;

(d) $\int_{-1}^8 j(x)\,dx$ where $j(x) = \begin{cases} \sqrt{1-x^2} & \text{if } -1 \leq x \leq 0 \\ 1 & \text{if } x > 0 \end{cases}$;

(e) $\int_{-3}^5 k(x)\,dx$ where $k(x) = \begin{cases} +4 & \text{if } x \geq 2 \\ 2x & \text{if } -1 < x < 2 \\ -2 & \text{if } x \leq -1 \end{cases}$.

6. Find the average value of each function on the given interval:
   (a) $f(x) = 3$ on the interval $[-1, 7]$;
   (b) $k(x) = 2x^5 - 4x^3 + 7x$ on the interval $[-6, 6]$;
   (c) $g(x) = 6x$ on the interval $[0, 3]$;
   (d) $h(x) = 2x - 9$ on the interval $[0, 4]$;
   (e) $j(x) = \begin{cases} 5x - 3 & \text{if } x < 2 \\ 3x + 1 & \text{if } x \geq 2 \end{cases}$ on the interval $[1, 8]$;
   (f) $m(x) = \tan x$ on the interval $[-\pi/3, \pi/3]$.

7. In each case, verify the mean value theorem for definite integrals by finding all numbers $c \in [a, b]$ such that $f(c) = \frac{\int_a^b f(x)\,dx}{b-a}$.

(a) $f(x) = 5x$ on $[0, 6]$;   (b) $g(x) = |x|$ on $[-2, 6]$;
(c) $h(x) = 4x - 8$ on $[1, 5]$;   (d) $j(x) = x^5 - x$ on $[-2, 2]$;
(e) $k(x) = x^3 - 4x$ on $[-7, 7]$;   (f) $m(x) = \cos x$ on $[-5\pi/2, 7\pi/2]$.

8. Let G be a continuous odd function with domain $[-a, a]$.
(a) Let P_{2n} be the regular partition of $[-a, a]$ into $2n$ subintervals of width $\frac{2a}{n}$. Define the selection of points T_{2n} from these subintervals by choosing t_k arbitrarily for $1 \leq k \leq n$ and $t_k = -t_{2n-k+1}$ for $n+1 \leq k \leq 2n$. Calculate $R(P_{2n}, T_{2n}, G)$.
(b) Show $\int_{-a}^{a} G(x)\, dx = 0$.

Challenging Problems

1. Use the formulas

$$\sum_{k=1}^{n} k = \frac{n(n+1)}{2}, \qquad \sum_{k=1}^{n} k^2 = \frac{n(n+1)(2n+1)}{6}, \qquad \sum_{k=1}^{n} k^3 = \frac{n^2(n+1)^2}{4}$$

to compute the value of each integral from its definition.

(a) $\int_0^1 x^3\, dx$ (b) $\int_1^5 6x^3 - 2\, dx$ (c) $\int_{-2}^3 3x^3 + 6x^2 + 7\, dx$

2. **(a)** Let f be a continuous function defined on the interval $[a, b]$ with $f(x) \geq 0$ for $x \in [a, b]$. Show that if there is $c \in [a, b]$ with $f(c) > 0$, then $\int_a^b f(x)\, dx > 0$.
(b) Let f, g be continuous functions defined on the interval $[a, b]$ such that $f(x) \leq g(x)$ for $x \in [a, b]$. If there is $c \in [a, b]$ with $f(c) < g(c)$, show $\int_a^b f(x)\, dx < \int_a^b g(x)\, dx$.

Appendix

In this appendix we give a rigorous proof of Property 10, the additive property of definite integrals. Then we prove a special case of Theorem 3.3.16: the definite integral of a piecewise monotone continuous function exists.

Given a partition P of $[a, b]$, we construct a new partition P' of smaller mesh by subdividing each interval $[x_{k-1}, x_k]$ into subintervals. The Riemann sums $R(P', T', f)$ should approximate $\int_a^b f(x)\, dx$ better than the Riemann sums $R(P, T, f)$. We introduce terminology for this construction.

Definition 3.3.34 *Let P and P' be partitions of the interval $[a, b]$. If $P \subset P'$, we call P' a refinement of P.*

By definition, $\int_a^b f(x)\, dx$ is the unique number between each pair of lower and upper Riemann sums. The following proposition notes that this pair of Riemann sums get closer as the mesh of the partition decreases.

Proposition 3.3.35 *If P' is a refinement of the partition P of $[a, b]$, then*

$$\text{mesh } P' \leq \text{mesh } P. \tag{3.3.11}$$

Assume f is a function which is continuous on the interval $[a, b]$. Then

$$L(P, f) \leq L(P', f) \leq U(P', f) \leq U(P, f). \tag{3.3.12}$$

Proof Construct the partition P' from the partition P by adding one point at a time. Therefore, it suffices to consider the case where P' contains exactly one more point $\overline{x}$ than P. Say $x_{k-1} < \overline{x} < x_k$. In computing the mesh of P' we compare the widths of the same small intervals as when computing the mesh of P except that the interval

$[x_{k-1}, x_k]$ is replaced by the two smaller intervals $[x_{k-1}, \overline{x}]$ and $[\overline{x}, x_k]$. Therefore, the mesh of P' is less than or equal to the mesh of P. The only difference between the rectangles used to compute $L(P, f)$ and $L(P', f)$ is that the shaded rectangle R_k in Figure 3.3.36 is replaced by the two striped rectangles R'_k and R''_k. Thus, the summand

$$L_k = f(x^*_k)(x_k - x_{k-1})$$

of $L(P, f)$ representing the area of R_k is replaced by the summands

$$L'_k = f(x^{*'}_k)(\overline{x} - x_{k-1}) + f(x^{*''}_k)(x_k - \overline{x})$$

of $L(P', f)$. Note that $f(x^*_k)$ is the minimum value of f on $[x_{k-1}, x_k]$, while $f(x^{*'}_k)$ is the minimum value of f on $[x_{k-1}, \overline{x}]$ and $f(x^{*''}_k)$ is the minimum value of f on $[\overline{x}, x_k]$. Hence $f(x^*_k)$ is less than or equal to both $f(x^{*'}_k)$ and $f(x^{*''}_k)$. It follows that $L_k \leq L'_k$ and $L(P, f) \leq L(P', f)$. We leave the proof of $U(P', f) \leq U(P, f)$ for the exercises. $\square$

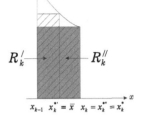

Figure 3.3.36
$L(P', f)$

The following lemma will be used in the proof of the additive property.

Lemma 3.3.37 *Let f be a continuous function with domain $[a, b]$. Assume $\int_a^b f(x)\, dx$ exists. If $\epsilon > 0$, then there is a partition P of $[a, b]$ such that*

$$U(P, f) - L(P, f) < \epsilon.$$

Proof By definition, $\int_a^b f(x)\, dx$ is the unique number between $L(P, f)$ and $U(P, f)$ for all partitions P of $[a, b]$. Hence $\int_a^b f(x)\, dx - \frac{\epsilon}{2}$ can not be larger than every $L(P, f)$, and $\int_a^b f(x)\, dx + \frac{\epsilon}{2}$ can not be smaller than every $U(P, f)$. Thus there are partitions P_1 and P_2 of $[a, b]$ such that:

$$\int_a^b f(x)\, dx - \frac{\epsilon}{2} < L(P_1, f) \qquad \text{and} \qquad U(P_2, f) < \int_a^b f(x)\, dx + \frac{\epsilon}{2}.$$

Hence $\quad \int_a^b f(x)\, dx - L(P_1, f) < \frac{\epsilon}{2} \qquad \text{and} \qquad U(P_2, f) - \int_a^b f(x)\, dx < \frac{\epsilon}{2}.$

Let $P = P_1 \cup P_2$. Since P is a refinement of both P_1 and P_2,

$$L(P_1, f) \leq L(P, f) \leq \int_a^b f(x)\, dx \quad \text{and} \quad \int_a^b f(x)\, dx \leq U(P, f) \leq U(P_2, f).$$

Therefore, $\quad \int_a^b f(x)\, dx - L(P, f) \quad \leq \quad \int_a^b f(x)\, dx - L(P_1, f) < \frac{\epsilon}{2} \qquad (3.3.13)$

and $\qquad U(P, f) - \int_a^b f(x)\, dx \quad \leq \quad U(P_2, f) - \int_a^b f(x)\, dx < \frac{\epsilon}{2}. \qquad (3.3.14)$

Add inequalities (3.3.13) and (3.3.14) to obtain: $U(P, f) - L(P, f) < \epsilon$. $\qquad \square$

We are now ready to prove the additive property.

Theorem 3.3.38 (Additive Property) *Assume the function f is continuous on the interval $[A, B]$ where A is the smallest of a, b, c and B is the largest of these numbers. If $\int_a^b f(x)\, dx$ and $\int_b^c f(x)\, dx$ exist, then $\int_a^c f(x)\, dx$ exists with*

$$\int_a^c f(x)\, dx = \int_a^b f(x)\, dx + \int_b^c f(x)\, dx.$$

Proof We first prove the case $a < b < c$. The other cases follow easily from this one. We prove that $\int_a^b f(x)\, dx + \int_b^c f(x)\, dx$ is the unique number between all pairs of lower and upper Riemann sums of f on $[a, c]$. Then it follows from the definition of $\int_a^c f(x)\, dx$ that this integral exists and equals $\int_a^b f(x)\, dx + \int_b^c f(x)\, dx$. We use the following two observations.

(1) If P_1 is a partition of $[a, b]$ and P_2 is a partition of $[b, c]$, then $P_1 \cup P_2$ is a partition of $[a, c]$. Moreover,

$$L(P_1 \cup P_2, f) \;=\; L(P_1, f) + L(P_2, f), \qquad (3.3.15)$$
$$U(P_1 \cup P_2, f) \;=\; U(P_1, f) + U(P_2, f). \qquad (3.3.16)$$

(2) If P is a partition of $[a, c]$, then $P \cup \{b\} = P' \cup P''$ where P' is a partition of $[a, b]$ and P'' is a partition of $[b, c]$. Furthermore,

$$L(P, f) \le L(P \cup \{b\}, f) \;=\; L(P', f) + L(P'', f),$$
$$U(P', f) + U(P'', f) \;=\; U(P \cup \{b\}, f) \le U(P, f).$$

It follows from these inequalities that if P is any partition of $[a, c]$, then

$$L(P, f) \le L(P', f) + L(P'', f) \le \int_a^b f(x)\, dx + \int_b^c f(x)\, dx \le U(P', f) + U(P'', f) \le U(P,$$

We show $\int_a^b f(x)\, dx + \int_b^c f(x)\, dx$ is the only number with this property. Suppose that A is a number such that

$$L(P, f) \le A \le U(P, f)$$

for all partitions P of $[a, c]$. If P_1 is any partition of $[a, b]$ and P_2 is any partition of $[b, c]$, then by (3.3.15) and (3.3.16):

$$L(P_1, f) + L(P_2, f) = L(P_1 \cup P_2) \le A \le U(P_1 \cup P_2) = U(P_1, f) + U(P_2, f).$$

We also know that

$$L(P_1, f) + L(P_2, f) \le \int_a^b f(x)\, dx + \int_b^c f(x)\, dx \le U(P_1, f) + U(P_2, f).$$

Thus

$$\left| A - \left(\int_a^b f(x)\, dx + \int_b^c f(x)\, dx \right) \right| \le [U(P_1, f) + U(P_2, f)] - [L(P_1, f) + L(P_2, f)].$$
$$(3.3.17)$$

Given $\epsilon > 0$, the preceding lemma says we can choose P_1 and P_2 so that

$$U(P_1, f) - L(P_1, f) < \frac{\epsilon}{2} \quad \text{and} \quad U(P_2, f) - L(P_2, f) < \frac{\epsilon}{2}.$$

Combine these two inequalities with (3.3.17):

$$\left| A - \left(\int_a^b f(x)\, dx + \int_b^c f(x)\, dx \right) \right| < \frac{\epsilon}{2} + \frac{\epsilon}{2} = \epsilon.$$

Since this inequality is true for every positive number ϵ, $A - \int_a^b f(x)\, dx + \int_b^c f(x)\, dx = 0$, and $A = \int_a^b f(x)\, dx + \int_b^c f(x)\, dx$. Thus $\int_a^b f(x)\, dx + \int_b^c f(x)\, dx$ is the unique number between $L(P, f)$ and $U(P, f)$ for every partition P of $[a, c]$. Therefore, $\int_a^c f(x)\, dx$ exists, and $\int_a^c f(x)\, dx$ equals $\int_a^b f(x)\, dx + \int_b^c f(x)\, dx$.

There are five other cases to consider. We leave the proof that $\int_a^c f(x)\,dx$ exists in these cases as an exercise. We verify the formula of the additive property when $b < a < c$. By the above case:

$$\int_b^c f(x)\,dx \;=\; \int_b^a f(x)\,dx + \int_a^c f(x)\,dx \;=\; -\int_a^b f(x)\,dx + \int_a^c f(x)\,dx$$

$$\int_a^c f(x)\,dx \;=\; \int_a^b f(x)\,dx + \int_b^c f(x)\,dx.$$

This formula for each of the other four cases is proved similarly, and we leave the verification as an exercise. □

Note We will use the additive property, in the case $a < b < c$, to show the integral of a piecewise monotone continuous function exists. That is why we did not assume $\int_a^c f(x)\,dx$ exists in the statement of this theorem.

The proof of the existence of the definite integral of a piecewise monotone continuous function uses the following criterion for the existence of a definite integral.

Lemma 3.3.39 *Let f be a continuous function with domain $[a,b]$. Assume for every number $\epsilon > 0$, there is a partition P_ϵ of $[a,b]$ such that*

$$U(P_\epsilon) - L(P_\epsilon) < \epsilon.$$

Then the definite integral $\int_a^b f(x)\,dx$ exists.

Proof[2] Let N and Q be two partitions of the closed interval $[a,b]$. Since $N \cup Q$ is a refinement of both N and Q,

$$L(N) \le L(N \cup Q) \le U(N \cup Q) \le U(Q).$$

Hence the set S_L of all lower Riemann sums $L(N)$ is bounded above by any upper Riemann sum $U(Q)$. Since the real numbers are complete, the set S_L has a least upper bound B. If P is any partition of $[a,b]$, then $L(P) \le B$ because B is an upper bound of S_L. In addition, $B \le U(P)$ because $U(P)$ is an upper bound of S_L while B is the least upper bound of S_L. Hence there is at least one number B which lies between all lower Riemann sums and all upper Riemann sums of f. To show that the definite integral exists, we must show there can not be more than one such number. Assume both B_1 and B_2 have this property. Say $B_1 \le B_2$. Then for every $\epsilon > 0$,

$$L(P_\epsilon, f) \le B_1 \le B_2 \le U(P_\epsilon, f) \quad \text{and}$$
$$B_2 - B_1 \le U(P_\epsilon, f) - L(P_\epsilon, f) < \epsilon.$$

Since this inequality is true for all $\epsilon > 0$, the only possibility is $B_1 = B_2$. □

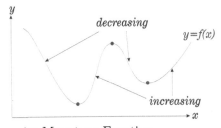

Figure 3.3.40 Piecewise Monotone Function

---

[2]This proof uses the concept of completeness which is exposited in Section 1.10.

Most nice functions, such as the one in Figure 3.3.40, have graphs which can be decomposed into a finite number of pieces on which f is either increasing or decreasing. We call such a function *piecewise monotone*. We use a partition of the domain of f to describe the decomposition of the graph of f into these pieces.

Definition 3.3.41 (a) *A function f with domain an interval I is called monotone if f is either increasing on I or decreasing on I.*
(b) *A function f with domain $[a,b]$ is called piecewise monotone if there is a partition $P = \{u_0, \ldots, u_n\}$ of the interval $[a,b]$ such that f restricted to the subinterval (u_{k-1}, u_k) is monotone for $1 \le k \le n$.*

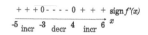

Figure 3.3.43
$f(x) = 2x^3 - 3x^2 - 72x$

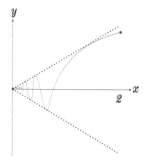

Figure 3.3.44
$g(x) = x \sin \frac{1}{x}$

Examples 3.3.42 (1) Show the function $f(x) = 2x^3 - 3x^2 - 72x$ with domain $[-5, 6]$ is piecewise monotone.

Solution Note that $f'(x) = 6x^2 - 6x - 72 = 6(x^2 - x - 12) = 6(x-4)(x+3)$. Hence f is increasing for $x \le -3$ or $x \ge 4$ where $f'(x) \ge 0$ while f is decreasing for $-3 \le x \le 4$ where $f'(x) \le 0$. See Figure 3.3.43. Let P be the partition $\{-5, -3, 4, 6\}$ of the domain $[-5, 6]$ of f. Then f is increasing on the first subinterval $[-5, -3]$ and on the third subinterval $[4, 6]$ while f is decreasing on the second subinterval $[-3, 4]$. Therefore, f is piecewise monotone.

(2) Show the function $g(x) = \left\{ \begin{array}{ccc} x \sin \frac{1}{x} & \text{if} & 0 < x \le 2 \\ 0 & \text{if} & x = 0 \end{array} \right\}$ with domain $[0, 2]$ is not piecewise monotone.

Solution The graph of g in Figure 3.3.44 consists of an infinite number of pieces where g is increasing and an infinite number of pieces where g is decreasing. Hence g is not a piecewise monotone function.   □

We use the criterion of Lemma 3.3.39 to prove that the definite integral of a piecewise monotone continuous function exists.

Theorem 3.3.45 *Let f be a piecewise monotone continuous function with domain $[a, b]$. Then $\int_a^b f(x)\, dx$ exists.*

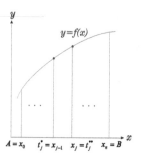

Figure 3.3.46
f Increasing on $[A, B]$

Proof Let $P = \{u_0, \ldots, u_n\}$ be a partition of the interval $[a, b]$ such that f restricted to each subinterval is monotone. We prove that $\int_{u_{k-1}}^{u_k} f(x)\, dx$ exists for $1 \le k \le n$. It then follows from the additive property, Theorem 3.3.38, that $\int_a^b f(x)\, dx$ exists. Let $A = u_{k-1}$ and $B = u_k$. Consider the case where f is increasing on $[A, B]$. If $P' = \{x_0, \ldots, x_N\}$ is any partition of $[A, B]$, then the minimum value of f on any $[x_{j-1}, x_j]$ occurs at the left endpoint x_{j-1}, while the maximum of f occurs at the right endpoint x_j. See Figure 3.3.46. That is, $t_j^* = x_{j-1}$ and $t_j^{**} = x_j$. Then

$$U(P', f) - L(P', f) = \sum_{j=1}^{n} f(x_j)(x_j - x_{j-1}) - \sum_{j=1}^{n} f(x_{j-1})(x_j - x_{j-1})$$

$$= \sum_{j=1}^{n} [f(x_j) - f(x_{j-1})](x_j - x_{j-1})$$

Thus $U(P', f) - L(P', f)$ is the sum of the areas of the shaded rectangles which lie over the interval $[A, B]$ in Figure 3.3.47. Project these rectangles to the left onto the y–axis. Their widths are at most mesh P'. Hence these projected rectangles lie inside the rectangle of length $f(B) - f(A)$ and width mesh P' depicted in Figure 3.3.47. Therefore

$$U(P', f) - L(P', f) \le [f(B) - f(A)] \cdot [\text{mesh } P']. \qquad (3.3.18)$$

Given $\epsilon > 0$, let P_ϵ be a partition of $[A, B]$ with mesh $P_\epsilon < \frac{\epsilon}{f(B)-f(A)}$. By (3.3.18), $U(P_\epsilon, f) - L(P_\epsilon, f) < \epsilon$. By Lemma 3.3.39, $\int_A^B f(x)\,dx$ exists. If f is decreasing on $[A, B]$, a similar argument applies. We leave the proof of this case as an exercise. $\square$

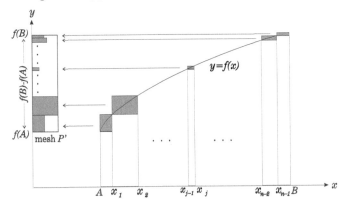

Figure 3.3.47 $U(P', f) - L(P', f)$

Exercises

1. Verify inequalities (3.3.11) and (3.3.12) of Proposition 3.3.35 in each case.
(a) $P = \{0, 2, 4, 6, 8\}$, $P' = \{0, 1, 2, 3, 4, 5, 6, 7, 8\}$ and $f(x) = x^2$.
(b) $P = \{1, 3, 5, 7\}$, $P' = \{1, 2, 3, 4, 5, 6, 7\}$ and $f(x) = \frac{1}{x}$.
(c) $P = \{0, \frac{\pi}{3}, \frac{\pi}{2}, \pi\}$, $P' = \{0, \frac{\pi}{4}, \frac{\pi}{3}, \frac{\pi}{2}, \frac{2\pi}{3}, \frac{3\pi}{4}, \frac{5\pi}{6}, \pi\}$ and $f(x) = \sin x$.

2. Let f be continuous on the interval $[a, b]$. Let P be a partition of $[a, b]$ with P' a refinement of P. Complete the proof of Prop. 3.3.35 by showing $U(P', f) \leq U(P, f)$.

3. **(a)** State and prove the formulas of the four cases of the additive property which are omitted from our proof of Theorem 3.3.38.
(b) Prove the existence of $\int_a^c f(x)\,dx$ in Thm. 3.3.38 in the cases other than $a \leq b \leq c$.

4. Show that each function is piecewise monotone.
(a) $f(x) = x^3 - 12x$ with domain $[-4, 3]$.
(b) $g(x) = 3x^5 - 65x^3 + 540x - 500$ with domain $[-5, 7]$.
(c) $h(x) = \sin^2 x$ with domain $[0, 4\pi]$.

5. Complete the proof of Theorem 3.3.45 by showing $\int_A^B f(x)\,dx$ exists when f is decreasing on $[A, B]$.

3.4 Fundamental Theorems of Calculus

Differential calculus computes slopes of tangent lines, while integral calculus computes areas. It seems that these two endeavors are unrelated. Newton and Leibniz, however, discovered the amazing fact that differentiation and integration are inverse operations. This deep result is called the Fundamental Theorem of Calculus. An immediate application is a simple method for computing integrals. If we want to compute the integral of f, we only need to determine a function F such that $DF = f$. Then the integral of f is the integral of the derivative of F which is F. That is, we show that $\int_a^b f(x)\,dx$ is merely $F(b) - F(a)$. We present this theorem in the first subsection and apply it to compute areas in the second subsection.

The Two Fundamental Theorems of Calculus

We divide the Fundamental Theorem of Calculus into two parts. The First Fundamental Theorem says that the derivative of the integral of a function equals the original function. The Second Fundamental Theorem says that the integral of the derivative of a function equals the original function. We apply the latter theorem to compute definite integrals.

The definite integral of the function f over the interval $[c, x]$, with x a variable, is a number which depends on the value of x, i.e. a function $F(x)$. The First Fundamental Theorem of Calculus computes the derivative of this function.

Motivating Example 3.4.1 Let $f(x) = 4x$. Find the derivative of the function

$$F(x) = \int_0^x f(t)\ dt = \int_0^x 4t\ dt.$$

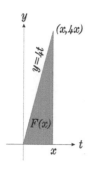

Figure 3.4.2
$F(x)$ for $x > 0$

Solution When x is positive, $F(x)$ is the area of the right triangle in Figure 3.4.2 with legs of length x and $4x$. Therefore,

$$F(x) = \int_0^x 4t\ dt = \frac{1}{2}(x)(4x) = 2x^2.$$

When x is negative, we first interchange the bounds of the integral that defines $F(x)$:

$$F(x) = \int_0^x 4t\ dt = -\int_x^0 4t\ dt$$

The latter integral refers to the right triangle in Figure 3.4.3 with legs of length $-x$ and $-4x$. Since this triangle lies below the x–axis, the definite integral computes its area as negative:

$$F(x) = -\int_x^0 4t\ dt = -\left[-\frac{1}{2}(-x)(-4x)\right] = 2x^2.$$

Figure 3.4.3
$F(x)$ for $x < 0$

Thus $F(x) = 2x^2$ for all $x \in \Re$. Note that this conclusion is a consequence of our definition of $\int_b^a f(x)\ dx$ as $-\int_a^b f(x)\ dx$ when $a < b$. For all $x \in \Re$:

$$F'(x) = \frac{d}{dx}\left(2x^2\right) = 4x = f(x). \qquad \square$$

The conclusion of the preceding example is that $f(x)$ equals the derivative of its integral $F(x)$. This is no coincidence and is true in general. That is, let $y = f(t)$ be a continuous function. The area

$$F(x) = \int_c^x f(t)\ dt,$$

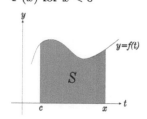

Figure 3.4.4
$F(x) = \text{Area}(S)$

of the region S bounded by the graph of $y = f(t)$, the t–axis and the vertical lines $t = c$, $t = x$ is depicted in Figure 3.4.4. The area $F(x)$ of S is a function of the variable x. That is, the vertical line $t = x$ which forms the right boundary of this area varies, and the value of the area of S depends on the position of this line. The First Fundamental Theorem of Calculus says that the derivative of the integral $F(x)$ equals $f(x)$. That is, the derivative of the integral of f equals f.

Theorem 3.4.5 (First Fundamental Theorem of Calculus) *Let $y = f(t)$ be a continuous function with domain the interval $[a, b]$, $a < b$. Let $a \le c \le b$. Define*

$$F(x) = \int_c^x f(t)\ dt$$

for $x \in [a, b]$. Then $F'(x) = f(x)$ for $x \in (a, b)$, $F'_+(a) = f(a)$ and $F'_-(b) = f(b)$.

Proof Use the definition of the derivative to compute $F'(x)$ for $x \in (a,b)$:

$$F'(x) = \lim_{h \to 0} \frac{F(x+h) - F(x)}{h} = \lim_{h \to 0} \frac{1}{h} \left[\int_c^{x+h} f(t) \, dt - \int_c^x f(t) \, dt \right]$$

$$F'(x) = \lim_{h \to 0} \frac{1}{h} \int_x^{x+h} f(t) \, dt \tag{3.4.1}$$

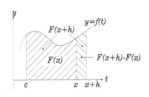

Figure 3.4.6
$F(x+h) - F(x)$

by the additive property of definite integrals. See Figure 3.4.6. In the limits above, x is fixed while h is a variable. Apply the mean value theorem for definite integrals to $y = f(t)$ on the interval from x to $x + h$: there is a number e_h between x and $x + h$ such that

$$\int_x^{x+h} f(t) \, dt = f(e_h) \left[(x+h) - x \right] = h f(e_h).$$

See Figure 3.4.7 for the case $h > 0$. Substitute this value of the integral into (3.4.1):

$$F'(x) = \lim_{h \to 0} \frac{1}{h} [h f(e_h)] = \lim_{h \to 0} f(e_h).$$

Note that e_h depends on the value of the variable h. Since e_h is a number between x and $x + h$, e_h approaches x as h approaches zero. Hence

$$F'(x) = \lim_{e_h \to x} f(e_h) = f(x)$$

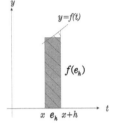

because f is continuous. The computations of $F'_+(a)$ and $F'_-(b)$ are done similarly, with the appropriate one–sided limits replacing the two sided limits above. $\square$

Figure 3.4.7
Mean Value Theorem

We apply the First Fundamental Theorem of Calculus to compute the derivatives of various integrals.

Examples 3.4.8 **(1)** Find the derivative of $F(x) = \int_0^x \tan \pi t \, dt$.

Solution Apply the First Fundamental Theorem of Calculus to $f(x) = \tan \pi x$:
$F'(x) = f(x) = \tan \pi x$.

(2) Find the derivative of $G(x) = \int_x^4 \sqrt{t^2 - 1} \, dt$.

Solution We cannot apply the First Fundamental Theorem of Calculus directly to $G(x)$, as defined above, because the variable x is in the lower bound of the integral. Hence we switch the bounds of this integral and then apply the First Fundamental Theorem of Calculus to $g(x) = \sqrt{x^2 - 1}$:

$$G'(x) = \frac{d}{dx} \left(\int_x^4 \sqrt{t^2 - 1} \, dt \right) = \frac{d}{dx} \left(- \int_4^x \sqrt{t^2 - 1} \, dt \right) = -\sqrt{x^2 - 1}.$$

(3) Find the derivative of $H(x) = \int_3^{x^4} \frac{t}{t^2+1} \, dt$.

Solution Let $u = x^4$. By the chain rule and the First Fundamental Theorem of Calculus:

$$H'(x) = \frac{d}{du} \left[\int_3^u \frac{t}{t^2 + 1} \, dt \right] \cdot \frac{du}{dx} = \frac{u}{u^2 + 1} \, 4x^3 = \frac{x^4}{(x^4)^2 + 1} \, 4x^3 = \frac{4x^7}{x^8 + 1}.$$

(4) Find the derivative of $K(x) = \int_{x^2}^{x^3} \sec t \, dt$.

Solution By the additive property of definite integrals:

$$K(x) = \int_{x^2}^0 \sec t \, dt + \int_0^{x^3} \sec t \, dt = - \int_0^{x^2} \sec t \, dt + \int_0^{x^3} \sec t \, dt.$$

Let $u = x^2$ and $v = x^3$. By the chain rule and the First Fundamental Theorem of Calculus:

$$\begin{aligned} K'(x) &= \frac{d}{du}\left[-\int_0^u \sec t \; dt\right] \cdot \frac{du}{dx} + \frac{d}{dv}\left[\int_0^v \sec t \; dt\right] \cdot \frac{dv}{dx} \\ &= (-\sec u)(2x) + (\sec v)(3x^2) = -2x\sec(x^2) + 3x^2\sec(x^3). \quad \square \end{aligned}$$

The Second Fundamental Theorem of Calculus says that the integral of the derivative of $G(x)$ equals $G(b) - G(a)$. That is, the operation of integration undoes the action of differentiation. Combining the two Fundamental Theorems of Calculus, we see that differentiation and integration are inverse operations.

Theorem 3.4.9 (Second Fundamental Theorem of Calculus) *Let $f(x)$ and $G(x)$ be continuous functions with domain $[a,b]$, $a < b$. If $G'(x) = f(x)$ for $x \in (a,b)$, then*

$$\int_a^b f(x) \; dx = G(b) - G(a).$$

Proof By the First Fundamental Theorem of Calculus the function

$$F(x) = \int_a^x f(t) \; dt$$

also has the property: $F'(x) = f(x)$ for $x \in (a,b)$. Thus, F and G have the same derivative. By Corollary 2.6.24, F and G are equal up to a constant:

$$G(x) = F(x) + C$$

for $x \in (a,b)$. The function F is differentiable on $[a,b]$ by the First Fundamental Theorem of Calculus. By Corollary 2.2.25, F is continuous at $x = a$ and at $x = b$ while G is continuous at $x = a$ and at $x = b$ by hypothesis. Hence

$$\begin{aligned} G(a) &= \lim_{x \to a^+} G(x) = \lim_{x \to a^+} F(x) + C = F(a) + C, \qquad (3.4.2) \\ G(b) &= \lim_{x \to b^-} G(x) = \lim_{x \to b^-} F(x) + C = F(b) + C. \qquad (3.4.3) \end{aligned}$$

We find the value of the constant C from (3.4.2):

$$G(a) = F(a) + C = \int_a^a f(t) \; dt + C = 0 + C = C.$$

Substitute $C = G(a)$ and the definition of $F(b)$ into (3.4.3):

$$\begin{aligned} G(b) &= F(b) + C = \int_a^b f(t) \; dt + G(a) \quad \text{and} \\ \int_a^b f(t) \; dt &= G(b) - G(a). \qquad \square \end{aligned}$$

Notation $G(b) - G(a)$ is written $G(x) \big|_a^b$.

Observe that by the First Fundamental Theorem of Calculus, there is always a function G whose derivative equals f: we can take $G(x) = \int_a^x f(t) \; dt$. In the following examples, we use our knowledge of differentiation to guess an explicit function G whose derivative equals f. Then we apply the Second Fundamental Theorem of Calculus to evaluate $\int_a^b f(x) \; dx$ as $G(b) - G(a)$.

Examples 3.4.10 (1) Evaluate $\int_1^4 x^2 \, dx$.

Solution Since $D(x^3) = 3x^2$, it follows that $D\left(\frac{1}{3}x^3\right) = x^2$. By the Second Fundamental Theorem of Calculus:

$$\int_1^4 x^2 \, dx = \frac{1}{3}x^3 \Big|_1^4 = \frac{4^3}{3} - \frac{1^3}{3} = 21.$$

(2) Evaluate $\int_0^9 \sqrt{x} \, dx$.

Solution Since $D\left(x^{3/2}\right) = \frac{3}{2}x^{1/2} = \frac{3}{2}\sqrt{x}$, it follows that $D\left(\frac{2}{3}x^{3/2}\right) = \sqrt{x}$. By the Second Fundamental Theorem of Calculus:

$$\int_0^9 \sqrt{x} \, dx = \frac{2}{3}x^{3/2} \Big|_0^9 = \frac{2}{3}9^{3/2} - \frac{2}{3}0^{3/2} = 18.$$

(3) Evaluate $\int_0^\pi \sin x \, dx$.

Solution Since $D(\cos x) = -\sin x$, it follows that $D(-\cos x) = \sin x$. By the Second Fundamental Theorem of Calculus:

$$\int_0^\pi \sin x \, dx = -\cos x \Big|_0^\pi = -\cos \pi - (-\cos 0) = -(-1) - (-1) = 2. \qquad \square$$

Computing Area

Now that we can compute the values of definite integrals, we turn to the problem of computing area. We begin by determining the area of the region bounded by $y = f(x)$, the lines $x = a$, $x = b$ and the x–axis. Then we find the area of a region bounded by two curves. We conclude by calculating areas using integration with respect to y.

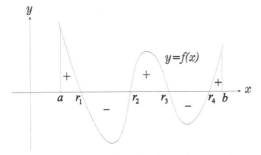

Figure 3.4.11 Computing Area with a Definite Integral

The areas of all regions are positive. However, the definite integral computes area above the x–axis as positive and area below the x–axis as negative. Therefore, to compute the area A bounded by the graph of f, the lines $x = a$, $x = b$ and the x–axis, we first compute the roots $r_1 \leq \cdots \leq r_n$ of f in $[a, b]$. The set $P = \{r_0, r_1, \ldots r_n, r_{n+1}\}$, with $r_0 = a$ and $r_{n+1} = b$, is a partition of $[a, b]$. The graph of f restricted to each of the subintervals $[r_{k-1}, r_k]$ lies entirely on one side of the x–axis. See Figure 3.4.11. Hence the area between the graph of f and the interval $[r_{k-1}, r_k]$ equals $\mid \int_{r_{k-1}}^{r_k} f(x) \, dx \mid$. Thus the entire area A is given by:

$$A = \left| \int_a^{r_1} f(x) \, dx \right| + \left| \int_{r_1}^{r_2} f(x) \, dx \right| + \cdots + \left| \int_{r_{n-1}}^{r_n} f(x) \, dx \right| + \left| \int_{r_n}^b f(x) \, dx \right|. \quad (3.4.4)$$

We illustrate this procedure with two examples. To evaluate the definite integrals which arise in computing areas, note that for $n \in \Re$:

$$x^n = D\left(\frac{x^{n+1}}{n+1}\right) \quad \text{when } n \neq -1. \quad (3.4.5)$$

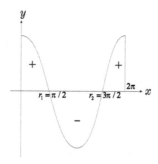

$f(x) = \cos x$ on the Interval $[0,2\pi]$

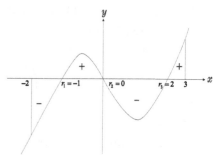

$g(x) = x^3 - x^2 - 2x$ on the Interval $[-1,2]$

Figure 3.4.12 Examples 3.4.13

Examples 3.4.13 (1) Find the area A of the region R between the graph of the function $f(x) = \cos x$ and the interval $[0, 2\pi]$ on the x–axis.

Solution The roots of $f(x) = \cos x$ between 0 and 2π are $r_1 = \frac{\pi}{2}$ and $r_2 = \frac{3\pi}{2}$. The region R is depicted in the left diagram of Figure 3.4.12. By (3.4.4):

$$A = \left| \int_0^{\pi/2} \cos x \; dx \right| + \left| \int_{\pi/2}^{3\pi/2} \cos x \; dx \right| + \left| \int_{3\pi/2}^{2\pi} \cos x \; dx \right|$$

$$= \left| \sin x \big|_0^{\pi/2} \right| + \left| \sin x \big|_{\pi/2}^{3\pi/2} \right| + \left| \sin x \big|_{3\pi/2}^{2\pi} \right| = |1 - 0| + |-1 - 1| + |0 - (-1)|$$

Alternatively, we can use the symmetry of this region to compute:

$$A = 4 \int_0^{\pi/2} \cos x \; dx = 4 \sin x \big|_0^{\pi/2} = 4(1) - 4(0) = 4.$$

(2) Find the area A of the region S between the graph of the function $g(x) = x^3 - x^2 - 2x$ and the interval $[-2, 3]$ on the x–axis.

Solution The roots of $g(x)$ are found by solving $g(x) = 0$:

$$0 = g(x) = x^3 - x^2 - 2x = x(x^2 - x - 2) = x(x - 2)(x + 1).$$

Thus, the roots of g are $r_1 = -1$, $r_2 = 0$ and $r_3 = 2$. The region S is depicted in the right diagram of Figure 3.4.12. By (3.4.4),

$$A = \left| \int_{-2}^{-1} x^3 - x^2 - 2x \; dx \right| + \left| \int_{-1}^0 x^3 - x^2 - 2x \; dx \right| + \left| \int_0^2 x^3 - x^2 - 2x \; dx \right|$$

$$+ \left| \int_2^3 x^3 - x^2 - 2x \; dx \right|$$

$$= \left| \frac{1}{4}x^4 - \frac{1}{3}x^3 - x^2 \big|_{-2}^{-1} \right| + \left| \frac{1}{4}x^4 - \frac{1}{3}x^3 - x^2 \big|_{-1}^0 \right| + \left| \frac{1}{4}x^4 - \frac{1}{3}x^3 - x^2 \big|_0^2 \right|$$

$$+ \left| \frac{1}{4}x^4 - \frac{1}{3}x^3 - x^2 \big|_2^3 \right|$$

$$= \left| -\frac{37}{12} \right| + \left| \frac{5}{12} \right| + \left| -\frac{8}{3} \right| + \left| \frac{59}{12} \right| = \frac{133}{12}. \qquad \square$$

The following proposition computes the area of a region between two curves, such as the area of the region S in the left diagram of Figure 3.4.14.

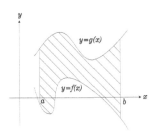

The Region S Between $y=f(x)$ and $y=g(x)$

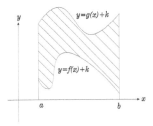

The Region S Shifted Upwards

Figure 3.4.14 Area of a Region Between Two Curves

Proposition 3.4.15 *Let f, g be continuous functions with domain $[a, b]$. Assume*

$$g(x) \geq f(x)$$

for $x \in [a, b]$. Let A be the area of the region S which is bounded by the graph of g on top, the graph of f on the bottom and the lines $x = a$, $x = b$ on the sides. Then

$$A = \int_a^b g(x) - f(x) \; dx. \qquad (3.4.6)$$

Proof If necessary, shift region S upwards by k units so that it lies above the x–axis. Note $f(x)$ is replaced by $f(x)+k$ and $g(x)$ is replaced by $g(x)+k$. In the right diagram of Figure 3.4.14, the shaded area A equals the area between the graph of $y = g(x) + k$ and the x–axis minus the area between the graph of $y = f(x) + k$ and the x–axis:

$$A = \int_a^b g(x) + k \; dx - \int_a^b f(x) + k \; dx. \qquad (3.4.7)$$

If $F'(x) = f(x)$ and $G'(x) = g(x)$, then $D\left[F(x) + kx\right] = f(x) + k$, $D\left[G(x) + kx\right] = g(x)+k$ and $D\left[G(x) - F(x)\right] = g(x) - f(x)$. Apply the Second Fundamental Theorem of Calculus to (3.4.7):

$$A = \left[G(x) + kx\right] \Big|_a^b - \left[F(x) + kx\right] \Big|_a^b = \left[G(x) - F(x)\right] \Big|_a^b = \int_a^b g(x) - f(x) \; dx. \quad \Box$$

We apply this proposition to compute areas between specific curves. It is important to sketch the curves to determine which curve is the top and which curve is the bottom. Also, we must determine where the two curves cross because the curve which is the top on one side of a crossing point may be the bottom curve on the other side. We compute the areas between adjacent crossing points separately and add them.

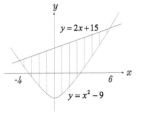

Examples 3.4.17 (1) Find the area A_1 of the region S_1 between the line $y = 2x+15$ and the parabola $y = x^2 - 9$.

Solution The line and parabola intersect when

$$
\begin{aligned}
2x + 15 &= x^2 - 9 \\
0 &= x^2 - 2x - 24 = (x + 4)(x - 6) \\
x &= -4 \quad \text{and} \quad x = 6.
\end{aligned}
$$

Figure 3.4.16
The Region S_1

From the graphs of this line and this parabola in Figure 3.4.16, we see that the line lies above the parabola throughout the interval $[-4, 6]$. By (3.4.6):

$$
\begin{aligned}
A_1 &= \int_{-4}^6 (2x + 15) - (x^2 - 9) \; dx = \int_{-4}^6 2x - x^2 + 24 \; dx \\
&= x^2 - \frac{1}{3}x^3 + 24x \Big|_{-4}^6 = 108 - \left(-\frac{176}{3}\right) = \frac{500}{3}.
\end{aligned}
$$

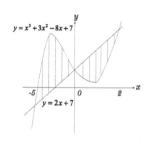

Figure 3.4.18
The Region S_2

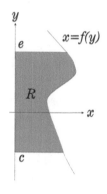

Figure 3.4.19
Area Using y

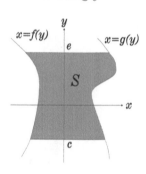

Figure 3.4.20
Area Between 2 Curves

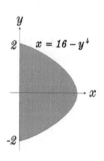

Figure 3.4.21
The Region S_1

(2) Find the area A_2 of the region S_2 between the graph of $f(x) = x^3 + 3x^2 - 8x + 7$ and the line $y = 2x + 7$.

Solution The cubic and the line intersect when

$$
\begin{aligned}
2x + 7 &= x^3 + 3x^2 - 8x + 7 \\
0 &= (x^3 + 3x^2 - 8x + 7) - (2x + 7) = x^3 + 3x^2 - 10x \\
0 &= x(x^2 + 3x - 10) = x(x + 5)(x - 2) \\
x &= -5, \quad x = 0 \quad \text{and} \quad x = 2.
\end{aligned}
$$

From the graphs of these functions in Figure 3.4.18, we see that the cubic lies above the line on the interval $[-5, 0]$ while the line lies above the cubic on the interval $[0, 2]$. By (3.4.6),

$$
\begin{aligned}
A_2 &= \int_{-5}^{0} (x^3 + 3x^2 - 8x + 7) - (2x + 7)\ dx \\
&\quad + \int_{0}^{2} (2x + 7) - (x^3 + 3x^2 - 8x + 7)\ dx \\
&= \int_{-5}^{0} x^3 + 3x^2 - 10x\ dx + \int_{0}^{2} 10x - x^3 - 3x^2\ dx \\
&= \left(\frac{1}{4}x^4 + x^3 - 5x^2 \right) \Big|_{-5}^{0} + \left(5x^2 - \frac{1}{4}x^4 - x^3 \right) \Big|_{0}^{2} = \frac{375}{4} + 8 = \frac{407}{4}. \ \square
\end{aligned}
$$

Sometimes it is convenient to compute the area of a region by using y as the independent variable instead of x. For example, consider the region R in Figure 3.4.19 bounded by the curve $x = f(y)$ on the right, the horizontal lines $y = c$, $y = e$ and the y–axis. Turn the paper sideways, so that the y–axis becomes horizontal and the x–axis points upwards. We see that the area A of R is given by:

$$
A = \int_{c}^{e} f(y)\ dy. \tag{3.4.8}
$$

Now consider the area S in Figure 3.4.20 bounded by the curve $x = f(y)$ on the left, $x = g(y)$ on the right and the horizontal lines $y = c$, $y = e$. Assume that $f(y) \leq g(y)$ for $y \in [c, e]$. When the paper is turned sideways, the curve $x = g(y)$ becomes the "top curve" while the curve $x = f(y)$ becomes the "bottom curve." Thus, the area A of S is given by:

$$
A = \int_{c}^{e} g(y) - f(y)\ dy \tag{3.4.9}
$$

where the graph of g lies to the right of the graph of f for $c \leq y \leq e$. As the following examples show, an integration problem may be easy in terms of one variable and complicated in terms of the other one. Remember that it is essential to first sketch the region. That sketch shows how to set up definite integral(s) to compute its area A using either x or y. Choose the simpler option to evaluate A.

Examples 3.4.22 (1) Find the area A_1 of the region S_1 bounded by the curve $x = 16 - y^4$ and the y–axis.

Solution The curve $x = 16 - y^4$ intersects the y–axis $x = 0$ when $0 = 16 - y^4$, i.e. when $y = \pm 2$. See Figure 3.4.21. By (3.4.8) the area of this region is:

$$
A_1 = \int_{-2}^{2} 16 - y^4\ dy = 16y - \frac{1}{5}y^5 \Big|_{-2}^{2} = \left(32 - \frac{32}{5} \right) - \left(-32 + \frac{32}{5} \right) = \frac{256}{5}.
$$

(2) Find the area A_2 of the region S_2 bounded by the curve $x = y^2 - 3$ and the line $2y = x - 5$.

Solution This region is sketched in Figure 3.4.23. To find the two points of intersection of the line $x = 2y + 5$ and the curve $x = y^2 - 3$, solve

$$\begin{aligned} y^2 - 3 &= 2y + 5, \\ 0 &= y^2 - 2y - 8 = (y + 2)(y - 4). \end{aligned}$$

Thus $y = -2$ and $y = 4$. The two points of intersection are $(1, -2)$ and $(13, 4)$. To find the area of S_2, using the variable x, divide the region into two parts by the line $x = 1$. Both regions are bounded by $y = \sqrt{x + 3}$ on top. However, the left region is bounded by $y = -\sqrt{x + 3}$ on the bottom, while the right region is bounded by $y = \frac{x-5}{2}$ on the bottom. Hence

$$A_2 = \int_{-3}^{1} \sqrt{x + 3} - (-\sqrt{x + 3})\ dx + \int_{1}^{13} \sqrt{x + 3} - (x - 5)/2\ dx.$$

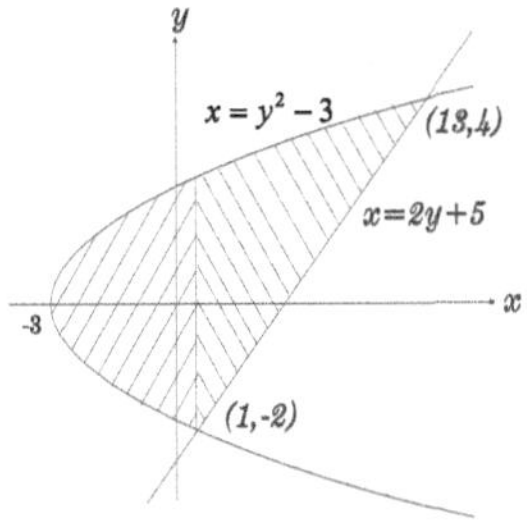

Figure 3.4.23
The Region S_2

This integral is difficult to evaluate since it would take a clever guess to find a function whose derivative is $\sqrt{x + 3}$. If we use the variable y to compute this area, the region S_2 is bounded by the line $x = 2y + 5$ on the right and by the parabola $x = y^2 - 3$ on the left. By (3.4.9):

$$A_2 = \int_{-2}^{4} (2y + 5) - (y^2 - 3)\ dy = \int_{-2}^{4} 2y - y^2 + 8\ dy.$$

Clearly it is easier to evaluate this integral to find the area A_2:

$$A_2 = \int_{-2}^{4} 2y - y^2 + 8\ dy = y^2 - \frac{1}{3}y^3 + 8y\Big|_{-2}^{4} = \frac{80}{3} - \left(\frac{-28}{3}\right) = 36.$$

(3) Find the area A_3 of the region S_3 bounded by the curves $x = y^2 - 1$ and $x = 1 - y^2$.

Solution The region R_3 between the two parabolas is sketched in Figure 3.4.24. We find the two points of intersection of these parabolas by solving

$$\begin{aligned} y^2 - 1 &= 1 - y^2, \\ 2y^2 &= 2. \end{aligned}$$

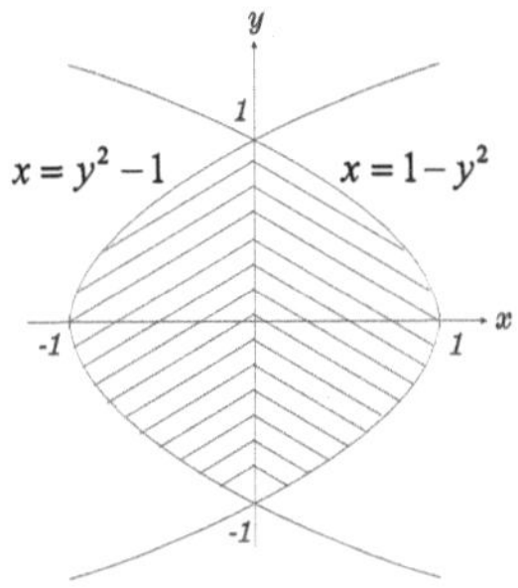

Figure 3.4.24
The Region S_3

Thus $y = \pm 1$, and the two points of intersection are $(0, -1)$ and $(0, 1)$. To find the area of S_3 with the variable x, use the y–axis to divide S_3 into two regions. To the left of the y–axis, the top of the region is given by $y = \sqrt{1 + x}$ and the bottom is given by $y = -\sqrt{1 + x}$. To the right of the y–axis, the top of the region is given by $y = \sqrt{1 - x}$ and the bottom is given by $y = -\sqrt{1 - x}$. Hence

$$\begin{aligned} A_3 &= \int_{-1}^{0} \sqrt{1 + x} - (-\sqrt{1 + x})\ dx + \int_{0}^{1} \sqrt{1 - x} - (-\sqrt{1 - x})\ dx \\ &= \int_{-1}^{0} 2\sqrt{1 + x}\ dx + \int_{0}^{1} 2\sqrt{1 - x}\ dx. \end{aligned}$$

Again, it would take a clever guess to find functions whose derivatives are $\sqrt{1 + x}$ and $\sqrt{1 - x}$. If we use the variable y to compute this area, the region S_3 is bounded by the parabola $x = 1 - y^2$ on the right and by the parabola $x = y^2 - 1$ on the left. By (3.4.9):

$$A_3 = \int_{-1}^{1} (1 - y^2) - (y^2 - 1)\ dy = \int_{-1}^{1} 2 - 2y^2\ dy.$$

Clearly, it is easier to evaluate the latter integral:

$$A_3 = \int_{-1}^{1} 2 - 2y^2\ dy = 2y - \frac{2}{3}y^3\Big|_{-1}^{1} = \frac{4}{3} - \left(-\frac{4}{3}\right) = \frac{8}{3}.$$

□

Historical Remarks

Isaac Newton and Gottfried Wilhelm Leibniz are credited with the discovery of calculus, because they discovered the Fundamental Theorem of Calculus and applied it to make computations.

The ideas underlying the Fundamental Theorem of Calculus were known before Newton and Leibniz. However, they were not formulated in a useful manner, nor were they used in computations. In 1638, Galileo showed that if an object travels along a coordinate line with position $s(t)$ and velocity $v(t) = s'(t)$ at time t, then the distance it travels from time $t = 0$ to time $t = b$ is the area under the curve $y = v(t)$. That is,

$$\int_0^b s'(t) \, dt = s(b) - s(0).$$

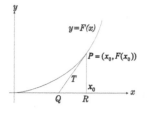

Figure 3.4.25
Barrow's Construction

By the 1660s, variations of the Fundamental Theorems of Calculus were formulated by William Neil, Isaac Barrow, James Gregory and Evangelista Torricelli. Neil studied the problem of finding a function with a given arc length. His major innovation was the introduction of the function $F(x) = \int_0^x f(t) \, dt$ to study the function $f(t)$. Barrow's study of the function F contained the following result. Given a point $P = (x_0, F(x_0))$ on the graph of $y = F(x)$, let Q be the point on the x–axis which is a distance $\frac{F(x_0)}{f(x_0)}$ from the point R on the x–axis with coordinate x_0. Then the line T joining the points P and Q is tangent to the graph of $y = F(x)$. See Figure 3.4.25. Barrow did not proceed any further. However, observe that the slope of T is

$$F'(x_0) = \frac{RP}{RQ} = \frac{F(x_0)}{F(x_0)/f(x_0)} = f(x_0).$$

This is the First Fundamental Theorem of Calculus. Newton probably attended Barrow's lectures on this subject at Cambridge.

Newton's approach to the Fundamental Theorem of Calculus is essentially the one presented in this section. He views the graph of f as the path traced out by a moving point $P = (x, y)$ whose velocity has horizontal component $\dot{x} = 1$ and vertical component $\dot{y} = f(x)$. The area $F(x) = \int_a^x f(x) \, dx$ becomes a function of time. He interprets the Fundamental Theorem of Calculus as saying: the rate of change of the area $F(x)$ equals the rate of change of the vertical component y of P. That is, $\dot{F} = \dot{y}$. By the chain rule, this is equivalent to the First Fundamental Theorem of Calculus:

$$f(x) = \frac{dy}{dt} = \frac{dF}{dt} = \frac{dF}{dx} \cdot \frac{dx}{dt} = \frac{dF}{dx} \cdot 1 = \frac{dF}{dx}.$$

Newton used this theorem to evaluate integrals and make other computations. Newton probably had formulated most of his basic ideas on calculus by 1666, and they were circulated in unpublished manuscripts of 1669 and 1671. His first published account appears in *Principia Mathematica*, published in 1687.

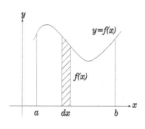

Figure 3.4.26
Leibniz's Infinitesmal
Rectangles

The preceding approaches to the Fundamental Theorems of Calculus all begin with a variation of the First Fundamental Theorem. Leibniz, however, found a direct approach to the Second Fundamental Theorem. He considers the area under the graph of $y = f(x)$ from $x = a$ to $x = b$ as the infinite sum of the areas of infinitesmal rectangles of area $f(x) \cdot dx$. See Figure 3.4.26. That is why Leibniz denotes this area as $\int_a^b f(x) \, dx$ where the integration symbol is a deformed S which represents an infinite sum. If $f = \frac{dF}{dx}$ then this area, $\int_a^b f(x) \, dx$, is the infinite sum of the

$$f(x) \cdot dx = \frac{dF}{dx} \cdot dx = dF$$

which add up to the change in F from $x = a$ to $x = b$, i.e. $F(b) - F(a)$. Leibniz used this point of view to make a large variety of computations. He developed this approach to calculus in the 1670s and published it in 1684.

Summary

The reader should understand the statements and proofs of the two Fundamental Theorems of Calculus. She should know how to apply the Second Fundamental Theorem of Calculus to compute definite integrals. In addition, she should be able to calculate areas, using the variable x or y, by evaluating the appropriate definite integrals.

Basic Exercises

1. Let $F(x) = \int_0^x 3t\, dt$. In each case, sketch the area corresponding to $F(x)$. Then use geometry to find the value of $F(x)$.
   - (a) $F(2)$
   - (b) $F(6)$
   - (c) $F(-4)$
   - (d) $F(0)$

2. Use geometry to compute $F(x) = \int_0^x f(t)\, dt$. Then verify that $F'(x) = f(x)$.
   - (a) $f(x) = 3x$
   - (b) $f(x) = -5x$
   - (c) $f(x) = 8$
   - (d) $f(x) = -4$
   - (e) $f(x) = 6x + 12$
   - (f) $f(x) = 2x - 8$
   - (g) $f(x) = 24 - 3x$
   - (h) $f(x) = -4x - 20$

3. Find the derivative of each function.
   - (a) $F(x) = \int_0^x \sqrt{t^3 + 8}\, dt$
   - (b) $G(x) = \int_1^x \tan(2t + 1)\, dt$
   - (c) $H(x) = \int_{-3}^x (t^2 + 4)^{3/4}\, dt$
   - (d) $I(x) = \int_x^5 \frac{1}{t}\, dt$
   - (e) $J(x) = \int_x^{-1} |\sin t|\, dt$
   - (f) $K(x) = \int_0^{\sqrt{x}} \frac{1}{1 + t^2}\, dt$
   - (g) $L(x) = \int_{\cos x}^1 \arctan t\, dt$
   - (h) $M(x) = \int_0^{\arcsin x} \sec t\, dt$
   - (i) $N(x) = \int_x^{x^2} \cot \sqrt{t}\, dt$
   - (j) $P(x) = \int_{2x+3}^{5x-4} \sin^4 t\, dt$
   - (k) $Q(x) = \int_{\sqrt[4]{x}}^{\sqrt[3]{x}} (t^2 + 1)^{100}\, dt$
   - (l) $R(x) = \int_{\cot x}^{\tan x} \sqrt[5]{3t + 4}\, dt$

4. Evaluate each definite integral.
   - (a) $\int_0^3 24x^3 - 2\, dx$
   - (b) $\int_{-2}^{-1} 6x^5 - 8x^3\, dx$
   - (c) $\int_0^{16} 3x^{5/4} + 2x^{3/2} + 7\, dx$
   - (d) $\int_1^9 \sqrt{x}\, dx$
   - (e) $\int_{-3}^1 |x^3 + 8|\, dx$
   - (f) $\int_{\pi/4}^{\pi/3} \csc^2 x\, dx$
   - (g) $\int_{-\pi/4}^{\pi/6} \sec x \tan x\, dx$
   - (h) $\int_{1/2}^{\sqrt{2}/2} \frac{6}{\sqrt{1-x^2}}\, dx$
   - (i) $\int_{-\pi/4}^{\pi/2} |\sin x|\, dx$
   - (j) $\int_0^{\sqrt{3}} \frac{4}{1+x^2}\, dx$
   - (k) $\int_0^1 18x(1 + x^2)^8\, dx$
   - (l) $\int_1^2 \frac{6}{x\sqrt{x^2-1}}\, dx$

5. Find the area between the graph of each function and the x–axis on the given interval.
   - (a) $f(x) = x^2 - 8$ on $[0, 2]$
   - (b) $g(x) = x^2 - x - 6$ on $[-3, 4]$
   - (c) $h(x) = x^4 - 1$ on $[-2, 3]$
   - (d) $i(x) = x^3 - 3x^2 + 2x$ on $[-1, 3]$
   - (e) $j(x) = (x - 1)(x - 2)(x - 3)$ on $[0, 4]$
   - (f) $k(x) = (x^2 - 1)(x^2 - 4)$ on $[-3, 4]$
   - (g) $m(x) = x^3 + 1$ on $[-2, 1]$
   - (h) $n(x) = \sin x$ on $\left[\frac{\pi}{4}, \frac{11\pi}{6}\right]$
   - (i) $p(x) = \tan x \sec x$ on $\left[-\frac{\pi}{3}, \frac{\pi}{4}\right]$
   - (j) $q(x) = \cos x$ on $\left[-\frac{2\pi}{3}, \frac{9\pi}{4}\right]$

6. In each case, find the area of the region between the graphs of f and g.
   - (a) $f(x) = x^2$ and $g(x) = 3x + 10$.
   - (b) $f(x) = x^2 - x - 5$ and $g(x) = 11 - 5x - x^2$.
   - (c) $f(x) = \sqrt{x}$ and $g(x) = x$.
   - (d) $f(x) = x^3$ and $g(x) = 4x$.
   - (e) $f(x) = x^3 + x^2 - 5x + 2$ and $g(x) = 4x^2 + 5x + 2$.
   - (f) $f(x) = x^3 + 2x^2 + 4x + 7$ and $g(x) = 3x^2 + 6x + 7$.
   - (g) $f(x) = \sin x$ and $g(x) = \frac{2x}{\pi}$.
   - (h) $f(x) = \sec^2 x$ and $g(x) = 1 + \frac{4x}{\pi}$ both with domain $\left(-\frac{\pi}{2}, \frac{\pi}{2}\right)$.
   - (i) $f(x) = \cos x$ and $g(x) = \sin x$ both with domain $[-2\pi, 2\pi]$.
   - (j) $f(x) = \sec^2 x$ and $g(x) = 8 - \sec^2 x$ both with domain $\left(-\frac{\pi}{2}, \frac{\pi}{2}\right)$.

7. In each case, find the area enclosed by the given curves:

(a) $x = \sqrt[3]{y}$, $y = 0$ and $y = 8$; (b) $x = \sec^2 y$, $y = 0$ and $y = \frac{\pi}{3}$;

(c) $x = y^2 - 1$ and the y–axis; (d) $x = y^2 - y - 6$ and $y = x - 2$;

(e) $x = y^2 - 2$ and $x = 6 - y^2$; (f) $x = y^3 - y^2 - 5y + 1$ and $y = x - 1$

(g) $y = \arcsin x$ $(0 \le x \le 1)$ and $y = \frac{\pi x}{2}$; (h) $y = \sqrt{\frac{2-x}{x}}$ $(0 < x \le 2)$, $y = x$ and

8. Find the average value of each function on the given interval:

(a) $f(x) = 9 - x^2$ on $[-3, 3]$; (b) $g(x) = 2x^3 - 4x$ on $[1, 3]$;

(c) $i(x) = \sqrt{x}$ on $[4, 9]$; (d) $j(x) = \cos x$ on $[-\frac{\pi}{2}, \frac{\pi}{2}]$;

(e) $k(x) = \frac{1}{1+x^2}$ on $[0, 1]$; (f) $m(x) = \sec x \tan x$ on $[\frac{\pi}{4}, \frac{\pi}{3}]$.

9. Is the following argument correct? $\int_{-1}^{1} \frac{2}{x^3}\, dx = -\frac{1}{x^2}\Big|_{-1}^{1} = -1 - (-1) = 0$. Justify your answer.

10. Let f, g be continuous functions with domain $[a, b]$. Let $h, k \in \Re$. Show $\int_a^b hf(x) + kg(x)\, dx = h \int_a^b f(x)\, dx + k \int_a^b g(x)\, dx$.

11. Complete the proof of the First Fundamental Theorem of Calculus by showing $F'_+(a) = f(a)$ and $F'_-(b) = f(b)$.

Challenging Problems

1. Evaluate each definite integral.

(a) $\int_0^1 (x^2 + 1)^3\, dx$ (b) $\int_0^4 \frac{5x^3 + 2x^2 + 1}{\sqrt{x}}\, dx$ (c) $\int_0^1 \frac{1}{9+x^2}\, dx$ (d) $\int_0^1 \frac{1}{\sqrt{4-x^2}}\, dx$

(e) $\int_1^3 \sqrt{5x + 3}\, dx$ (f) $\int_0^1 \sin \pi x\, dx$ (g) $\int 0^{\pi/4} \tan^2 x\, dx$

(h) $\int_0^{\pi/3} \sin^2 x\, dx$ (i) $\int_0^{\pi/6} \sin x \cos x\, dx$ (j) $\int_0^{\pi/4} \sin 9x \sin 5x\, dx$

3.5   Indefinite Integrals

To evaluate $\int_a^b f(x)\, dx$ by the Fundamental Theorem of Calculus, we must find a function $F(x)$ whose derivative is $f(x)$. We call $F(x)$ the *indefinite integral* of $f(x)$. It is often a difficult task to find an explicit function $F(x)$. The cases for which we know $F(x)$ are summarized in the first subsection. The second subsection describes several algebraic tricks which are sometimes helpful. In the last subsection we show how inverse trigonometric functions can be used to evaluate certain integrals.

Definition and Examples

After defining the indefinite integral, we rephrase the basic derivative formulas as formulas for indefinite integrals. The term *indefinite integral* of $f(x)$ denotes all functions $F(x)$ whose derivative is $f(x)$. If $F'(x) = f(x)$, then we also have

$$D[F(x) + C] = F'(x) + 0 = f(x)$$

for every number C. Moreover, by Corollary 2.6.24, if two functions $F(x)$ and $G(x)$ both have derivative $f(x)$, then F and G must differ by a constant, i.e. $G(x) = F(x) + C$ for some number C. Thus $F(x) + C$ gives *all functions* with derivative $f(x)$.

Definition 3.5.1 *When $f(x) = F'(x)$ for all x in Domain f, we call $F(x)$ the indefinite integral of the function $f(x)$ and write*

$$\int f(x)\, dx = F(x) + C. \qquad (3.5.1)$$

We call $f(x)$ the integrand of the integral $\int f(x)\, dx$ and C the constant of integration.

Notes (1) If $\int f(x)\,dx = F(x) + C$, then $\int_a^b f(x)\,dx = F(x)\,|_a^b = F(b) - F(a)$.
(2) Equation 3.5.1 can be rewritten as: $\int F'(x)\,dx = F(x) + C$.

We rephrase the first step in each calculation of a definite integral in Examples 3.4.10 as the computation of an indefinite integral.

Examples 3.5.2 (1) $\int x^2\,dx = \frac{1}{3}x^3 + C$ because $D\left(\frac{1}{3}x^3\right) = x^2$.

(2) $\int \sqrt{x}\,dx = \frac{2}{3}x^{3/2} + C$ because $D\left(\frac{2}{3}x^{3/2}\right) = x^{1/2} = \sqrt{x}$.

(3) $\int \sin x\,dx = -\cos x + C$ because $D(-\cos x) = \sin x$. $\qquad\qquad\square$

Every derivative formula $F'(x) = f(x)$ can be interpreted as the indefinite integral $\int f(x)\,dx = F(x) + C$. The following proposition summarizes the indefinite integrals which are determined by the basic derivative formulas.

Proposition 3.5.3 *Let n be any rational number other than -1.*

(a) $\int x^n\,dx = \frac{x^{n+1}}{n+1} + C.$ $\qquad$ (b) $\int \sin x\,dx = -\cos x + C$

(c) $\int \cos x\,dx = \sin x + C$ $\qquad$ (d) $\int \sec^2 x\,dx = \tan x + C$

(e) $\int \csc^2 x\,dx = -\cot x + C$ $\qquad$ (f) $\int \sec x \tan x\,dx = \sec x + C$

(g) $\int \csc x \cot x\,dx = -\csc x + C$ $\qquad$ (h) $\int \frac{1}{\sqrt{1-x^2}}\,dx = \arcsin x + C$

(i) $\int \frac{1}{1+x^2}\,dx = \arctan x + C$ $\qquad$ (j) $\int \frac{1}{|x|\sqrt{x^2-1}}\,dx = \operatorname{arcsec} x + C$

Proof (a) If $n \neq -1$ then $D(x^{n+1}) = (n+1)x^n$. Hence $D\left(\frac{x^{n+1}}{n+1}\right) = x^n$.
(b) $D(\cos x) = -\sin x$, so $D(-\cos x) = \sin x$.
(c), (d), (f) $D(\sin x) = \cos x$, $D(\tan x) = \sec^2 x$ and $D(\sec x) = \sec x \tan x$.
(e) $D(\cot x) = -\csc^2 x$, so $D(-\cot x) = \csc^2 x$.
(g) $D(\csc x) = -\csc x \cot x$, so $D(-\csc x) = \csc x \cot x$.
(h)–(j) $D(\arcsin x) = \frac{1}{\sqrt{1-x^2}}$, $D(\arctan x) = \frac{1}{1+x^2}$ and $D(\operatorname{arcsec} x) = \frac{1}{|x|\sqrt{x^2-1}}$. $\qquad\square$

Observe that the Second Fundamental Theorem of Calculus does not determine the values of all integrals. For example, there is no function we know whose derivative equals $\frac{1}{x}$. We will study $\int \frac{1}{x}\,dx$ in Chapter 4. Moreover, given a complicated function we cannot hope to guess its indefinite integral, and we need methods to compute its integral. These methods are the subject of the remainder of this section, Section 6 as well as much of Chapter 4.

Examples 3.5.4 Differentiate each function, and interpret the computation as an indefinite integral.

(1) $f(x) = \frac{x}{x^2+1}$.

Solution By the quotient rule, $f'(x) = \frac{(1)(x^2+1)-x(2x)}{(x^2+1)^2} = \frac{1-x^2}{(x^2+1)^2}$. Hence $\int \frac{1-x^2}{(x^2+1)^2}\,dx = \frac{x}{x^2+1} + C$.

(2) $g(x) = \sin^3 x$.

Solution By the chain rule: $g'(x) = 3\sin^2 x D(\sin x) = 3\sin^2 x \cos x$. Hence $\int 3\sin^2 x \cos x\,dx = \sin^3 x + C$.

(3) $h(x) = \arcsin \frac{x+1}{3}$.

   Solution By the chain rule:

$$h'(x) \;=\; \frac{1}{\sqrt{1 - \left(\frac{x+1}{3}\right)^2}} \frac{d}{dx}\left(\frac{x+1}{3}\right) = \frac{1}{\sqrt{1 - \frac{(x+1)^2}{9}}} \frac{1}{3} = \frac{1}{\sqrt{9 - (x+1)^2}}$$

$$= \frac{1}{\sqrt{8 - x^2 - 2x}}.$$

   Hence $\int \frac{1}{\sqrt{8-x^2-2x}}\,dx = \arcsin \frac{x+1}{3} + C.$ □

   Since differentiation is linear, its inverse operation, integration, is also linear. That is, we use the Second Fundamental Theorem of Calculus to deduce the linear property of integrals from the linear property of derivatives.

Proposition 3.5.5 (Linear Property of Integrals)
(a) *If f, g are continuous functions and h, $k \in \mathfrak{R}$, then*

$$\int hf(x) + kg(x)\,dx = h \int f(x)\,dx + k \int g(x)\,dx.$$

(b) *Let f, g be continuous on the interval $[a, b]$ with $h, k \in \mathfrak{R}$. Then*

$$\int_a^b hf(x) + kg(x)\,dx = h \int_a^b f(x)\,dx + k \int_a^b g(x)\,dx.$$

Proof **(a)** Let $DF(x) = f(x)$ and $DG(x) = g(x)$. Since differentiation is linear:

$$D\left[hF(x) + kG(x)\right] = hDF(x) + kDG(x) \;=\; hf(x) + kg(x).$$

Hence $\displaystyle\int hf(x) + kg(x)\,dx = hF(x) + kG(x) + C \;=\; h\int f(x)\,dx + k\int g(x)\,dx.$

(b) In the preceding notation,

$$\int_a^b hf(x) + kg(x)\,dx \;=\; \left[hF(x) + kG(x)\right]\big|_a^b = \left[hF(b) + kG(b)\right] - \left[hF(a) + kG(a)\right]$$

$$= \; h\left[F(b) - F(a)\right] + k\left[G(b) - G(a)\right]$$

$$= \; h\int_a^b f(x)\,dx + k\int_a^b g(x)\,dx.$$ □

Examples 3.5.6 (1) Evaluate $\int 5\sin x + 8\cos x\,dx$.

   Solution By the linear property of indefinite integrals:

$$\int 5\sin x + 8\cos x\,dx = 5\int \sin x\,dx + 8\int \cos x\,dx = -5\cos x + 8\sin x + C.$$

(2) Evaluate $\int_1^2 12x^3 - 6x^2 + 7\,dx$.

   Solution By the linear property of definite integrals:

$$\int_1^2 12x^3 - 6x^2 + 7\,dx \;=\; 12\int_1^2 x^3\,dx - 6\int_1^2 x^2\,dx + 7\int_1^2 1\,dx$$

$$= \; 12\left(\frac{x^4}{4}\bigg|_1^2\right) - 6\left(\frac{x^3}{3}\bigg|_1^2\right) + 7\left(x\big|_1^2\right)$$

$$= \; 12\left(4 - \frac{1}{4}\right) - 6\left(\frac{8}{3} - \frac{1}{3}\right) + 7(2 - 1) = 38. \quad □$$

Algebraic Manipulations

Consider an indefinite integral where the integrand involves products or quotients. Carrying out these operations may rewrite the integrand in a form which is routine to integrate.

Examples 3.5.7 (1) Evaluate $\int_0^2 (x+2)^3 \, dx$.

Solution Multiply out the integrand:

$$\int_0^2 (x+2)^3 \, dx \;=\; \int_0^2 x^3 + 6x^2 + 12x + 8 \, dx = \left. \frac{x^4}{4} + 2x^3 + 6x^2 + 8x \right|_0^2$$

$$= \;(4 + 16 + 24 + 16) - 0 = 60.$$

(2) Find $\int \frac{(x^2+3)(6x^2-1)}{x^2} \, dx$.

Solution Multiply out $(x^2+3)(6x^2-1)$ and then divide it by x^2:

$$\int \frac{(x^2+3)(6x^2-1)}{x^2} \, dx \;=\; \int \frac{6x^4 + 17x^2 - 3}{x^2} \, dx = \int 6x^2 + 17 - 3x^{-2} \, dx$$

$$= \; 2x^3 + 17x + 3x^{-1} + C. \qquad \square$$

If the integrand involves roots, rewrite them as fractional powers. Then carrying out products or quotients may change the integrand into a form we can integrate.

Examples 3.5.8 (1) Find $\int \left(\sqrt[3]{x} + \sqrt[4]{x} \right)^2 \, dx$.

Solution Write the roots as fractional powers and square their sum:

$$\int \left(\sqrt[3]{x} + \sqrt[4]{x} \right)^2 \, dx \;=\; \int \left(x^{1/3} + x^{1/4} \right)^2 \, dx = \int x^{2/3} + 2x^{7/12} + x^{1/2} \, dx$$

$$= \; \frac{3}{5} x^{5/3} + \frac{24}{19} x^{19/12} + \frac{2}{3} x^{3/2} + C$$

(2) Evaluate $\int_1^4 \frac{x^3 - 4x + 5}{\sqrt{x}} \, dx$.

Solution Write $\sqrt{x}$ as $x^{1/2}$, and divide it into the numerator:

$$\int_1^4 \frac{x^3 - 4x + 5}{\sqrt{x}} \, dx \;=\; \int_1^4 \frac{x^3 - 4x + 5}{x^{1/2}} \, dx = \int_1^4 x^{5/2} - 4x^{1/2} + 5x^{-1/2} \, dx$$

$$= \; \left. \frac{2}{7} x^{7/2} - \frac{8}{3} x^{3/2} + 10x^{1/2} \right|_1^4$$

$$= \; \left[\frac{2}{7}(128) - \frac{8}{3}(8) + 20 \right] - \left[\frac{2}{7} - \frac{8}{3} + 10 \right] = \frac{580}{21}. \qquad \square$$

Inverse Trigonometric Functions

We use the chain rule to generalize the indefinite integrals of Proposition 3.5.3(h),(i) whose values are inverse trigonometric functions.

By Proposition 3.5.3(h): $\int \frac{1}{\sqrt{1-x^2}} \, dx = \arcsin x + C$ because $D(\arcsin x) = \frac{1}{\sqrt{1-x^2}}$. Observe that by the chain rule:

$$\frac{d}{dx} \left(\arcsin \frac{x}{a} \right) = \frac{1}{\sqrt{1 - \left(\frac{x}{a}\right)^2}} D\left(\frac{x}{a}\right) = \frac{1}{\sqrt{1 - \frac{x^2}{a^2}}} \cdot \frac{1}{a} = \frac{1}{\sqrt{1 - \frac{x^2}{a^2}}} \cdot \frac{1}{\sqrt{a^2}} = \frac{1}{\sqrt{a^2 - x^2}}.$$

Hence

$$\int \frac{1}{\sqrt{a^2 - x^2}} \, dx \;=\; \arcsin \frac{x}{a} + C. \qquad (3.5.2)$$

Examples 3.5.9 (1) Evaluate $\int_0^1 \frac{1}{\sqrt{4-x^2}} \, dx$.

Solution Apply (3.5.2) with $a = 2$:

$$\int_0^1 \frac{1}{\sqrt{4 - x^2}} \, dx = \arcsin \frac{x}{2} \Big|_0^1 = \arcsin \frac{1}{2} - \arcsin 0 = \frac{\pi}{6} - 0 = \frac{\pi}{6}.$$

(2) Find $\int \frac{1}{\sqrt{8-4x^2}} \, dx$.

Solution Factor 4 from the square root to write the integral in a form where (3.5.2) applies:

$$\int \frac{1}{\sqrt{8 - 4x^2}} \, dx = \frac{1}{2} \int \frac{1}{\sqrt{2 - x^2}} \, dx = \frac{1}{2} \arcsin \frac{x}{\sqrt{2}} + C. \qquad \square$$

By Proposition 3.5.3(i): $\int \frac{1}{1+x^2} \, dx = \arctan x + C$ because $D(\arctan x) = \frac{1}{1+x^2}$. By the chain rule:

$$\frac{d}{dx} \left(\arctan \frac{x}{a} \right) = \frac{1}{1 + (x/a)^2} \frac{d}{dx} \left(\frac{x}{a} \right) = \frac{1}{1 + x^2/a^2} \cdot \frac{1}{a} = \frac{1}{1 + x^2/a^2} \cdot \frac{a}{a^2} = a \cdot \frac{1}{a^2 + }$$

Hence

$$\int \frac{1}{a^2 + x^2} \, dx = \frac{1}{a} \arctan \frac{x}{a} + C. \tag{3.5}$$

Examples 3.5.10 (1) Evaluate $\int_0^5 \frac{1}{x^2+25} \, dx$.

Solution Apply (3.5.3) with $a = 5$:

$$\int_0^5 \frac{1}{x^2 + 25} \, dx = \frac{1}{5} \arctan \frac{x}{5} \Big|_0^5 = \frac{1}{5} \arctan 1 - \frac{1}{5} \arctan 0 = \frac{1}{5} \cdot \frac{\pi}{4} - 0 = \frac{\pi}{20}.$$

(2) Find $\int \frac{1}{9x^2+4} \, dx$.

Solution Factor 9 from the denominator so that (3.5.3) applies:

$$\int \frac{1}{9x^2 + 4} \, dx = \frac{1}{9} \int \frac{1}{(2/3)^2 + x^2} \, dx = \frac{1}{9} \cdot \frac{1}{2/3} \arctan \frac{x}{2/3} + C = \frac{1}{6} \arctan \frac{3x}{2} + C. \ \square$$

Summary

The reader should be able to use her knowledge of differentiation to integrate simple indefinite integrals. She should recognize that performing multiplications and divisions in the integrand often results in rewriting the integrand as a function which can be integrated. She should also be able to use formulas (3.5.2) and (3.5.3) to integrate certain functions in terms of arcsin and arctan.

Basic Exercises

1. Differentiate each function, and interpret the computation as an indefinite integral.

(a) $f(x) = \frac{3x-5}{8x+7}$ **(b)** $g(x) = \sqrt{x^2 + 4x + 1}$ **(c)** $h(x) = (x^3 + 7)^{20}$

(d) $i(x) = \cos^5 x$ **(e)** $j(x) = \cot^6 x$ **(f)** $k(x) = \arctan \frac{3x-2}{7}$

Evaluate each of the following 36 indefinite integrals.

2. $\int 5x^7 \, dx$

3. $\int 3x^9 - 4x^6 - 5 \, dx$

4. $\int \frac{6}{x^3} + \frac{12}{x^2} \, dx$

5. $\int 3\sqrt{x} - \frac{5}{\sqrt{x}} \, dx$

6. $\int \frac{4}{x^{3/5}} - \frac{8}{x^{5/3}} \, dx$

7. $\int \frac{3}{(1-x)^2} \, dx$

8. $\int 4\sec^2 x - 1 \, dx$

9. $\int 6\csc x \cot x \, dx$

10. $\int 6\sin x - 4\cos x \, dx$

11. $\int \frac{5}{1+x^2} \, dx$

12. $\int (x^2 - 3)^2 \, dx$

13. $\int \frac{5x^3 - 3x^2 - x + 5}{x^6} \, dx$

14. $\int (2x + 5)^3 \, dx$

15. $\int \frac{1}{1+4x^2} \, dx$

16. $\int \frac{x^2 - 7x + 3}{\sqrt{x}} \, dx$

17. $\int \frac{(x+3)^2}{\sqrt{x}} \, dx$

18. $\int (3x - 2)(5x + 1) \, dx$

19. $\int \frac{1}{25+9x^2} \, dx$

20. $\int \frac{5x^2 - 8x + 9}{\sqrt{x}} \, dx$

21. $\int \frac{1}{\sqrt{4-x^2}} \, dx$

22. $\int \frac{(\sqrt{x}+2)(\sqrt{x}-3)}{\sqrt[3]{x}} \, dx$

23. $\int \frac{1}{16+9x^2} \, dx$

24. $\int \frac{(\sqrt{x}-2)^2}{x^3} \, dx$

25. $\int \frac{8}{81+x^2} \, dx$

26. $\int \frac{(x^2-1)^3}{\sqrt[4]{x}} \, dx$

27. $\int \frac{12}{25x^2+1} \, dx$

28. $\int \frac{(x^2-3)(x^2+4)}{x^6} \, dx$

29. $\int \frac{1}{\sqrt{1-9x^2}} \, dx$

30. $\int \frac{(2x+1)^3}{x^5} \, dx$

31. $\int \frac{1}{\sqrt{36-x^2}} \, dx$

32. $\int \frac{(x^2+4)(3x-2)}{x\sqrt{x}} \, dx$

33. $\int \frac{1}{\sqrt{4-16x^2}} \, dx$

34. $\int \frac{1}{\sqrt{9-49x^2}} \, dx$

35. $\int \frac{(x^2-4)^3}{x+2} \, dx$

36. $\int \frac{(x-1)^2}{\sqrt{x}-1} \, dx$

37. $\int \frac{x-1}{\sqrt[3]{x}-1} \, dx$

Evaluate each of the following 8 definite integrals.

38. $\int_1^3 12x^3 - 6x^2 + 5 \, dx$

39. $\int_1^{256} \frac{8}{x^{3/4}} - \frac{9}{x^{5/8}} \, dx$

40. $\int_{\pi/4}^{\pi/3} 6\sec x \tan x + 4\csc x \cot x \, dx$

41. $\int_1^9 \frac{x^2 - 4x + 1}{\sqrt{x}} \, dx$

42. $\int_1^4 \frac{(x+1)^3}{x^{3/2}} \, dx$

43. $\int_{-1}^2 (x^2 - 1)^2 \, dx$

44. $\int_0^{3/4} \frac{12}{\sqrt{9-4x^2}} \, dx$

45. $\int_0^{6/5} \frac{1}{36+25x^2} \, dx$

46. Find $\int \frac{1}{|x|\sqrt{x^2-a^2}} \, dx$.

47. Find the average value of each function on the given interval.

(a) $f(x) = (x+1)^3$ on $[2,3]$. (b) $g(x) = \frac{(2x-1)(3x+2)}{\sqrt{x}}$ on $[1,4]$.

(c) $h(x) = \frac{1}{\sqrt{36-x^2}}$ on $[0,3]$. (d) $k(x) = \frac{1}{16+x^2}$ on $[0,4]$.

(e) $m(x) = \frac{(x^2+2)^2}{x^4}$ on $[1,3]$. (f) $n(x) = \frac{1}{9+4x^2}$ on $[0, \frac{3}{2}]$.

48. Find the area between the curve $y = (x+1)^3$ and the line $y = x+1$.

49. Find the area in the first quadrant bounded by $y = \frac{(x^2-2)^2}{\sqrt{x}}$ and $y = x\sqrt{x}$.

50. Find the area between the curve $y = \frac{1}{\sqrt{9-x^2}}$ and the line $y = 1$.

51. Find the area between the curves $y = \frac{13}{25+4x^2}$ and $y = 2x^2$.

Challenging Problems

Evaluate each integral.

1. $\int \frac{1}{x^2 + 2x + 5} \, dx$

2. $\int \frac{x^3 + 3x^2 + 3x + 1}{x + 3x^{2/3} + 3x^{1/3} + 1} \, dx$

3. $\int \frac{1}{(\sin^4 x - \cos^4 x)^2} \, dx$

3.6   Change of Variables

Integration is the inverse operation of differentiation. Hence every differentiation formula can be interpreted as an integration formula. In the first subsection, we apply this procedure to the chain rule and obtain the substitution method for changing variables in an indefinite integral. This often allows us to write a complicated integral as a simpler integral in terms of a new variable. We apply this method to evaluate definite integrals in the second subsection. We also use a change of variables to prove that the definite integral of an odd function over the interval $[-a, a]$ is zero. In the last subsection, we use trigonometric identities and the substitution method to evaluate a variety of trigonometric integrals.

Substitution Method

We establish the formula for changing variables in an indefinite integral and apply it to evaluate several complicated integrals. Then we apply the substitution method to extend the integration formulas with values arcsin and arctan which were derived in the preceding section.

Recall that $\int g(x)\,dx$ refers to all functions of x whose derivative with respect to x is $g(x)$. The notation $\int \cdots dx$ is one symbol. The dx in this symbol indicates the name of the variable we are using. The following example illustrates the procedure for changing variables in an indefinite integral.

Motivating Example 3.6.1 Let $u = g(x)$ be a differentiable function. Compute the derivatives of $g(x)^{n+1} = u^{n+1}$ with respect to x and u. Interpret each differentiation formula as an indefinite integral.

Solution By the chain rule:

$$\frac{d}{dx}\left[g(x)^{n+1}\right] = \frac{d}{dx}\left[u^{n+1}\right] = \frac{d}{du}\left(u^{n+1}\right)\frac{du}{dx} = (n+1)u^n\frac{du}{dx}$$

$$= (n+1)g(x)^n\frac{du}{dx}$$

$$\frac{d}{dx}\left[\frac{1}{n+1}g(x)^{n+1}\right] = g(x)^n\frac{du}{dx}.$$

Interpret the function on the left as the integral of its derivative:

$$\int g(x)^n\frac{du}{dx}\,dx = \frac{1}{n+1}g(x)^{n+1} + C. \qquad (3.6.1)$$

On the other hand,
$$\frac{d}{du}\left[u^{n+1}\right] = (n+1)u^n$$

$$\frac{d}{du}\left[\frac{1}{n+1}u^{n+1}\right] = u^n.$$

Interpret the function on the left as the integral of its derivative:

$$\int u^n\,du = \frac{1}{n+1}u^{n+1} + C. \qquad (3.6.2)$$

Since $u = g(x)$, the integrals in (3.6.1) and (3.6.2) are equal:

$$\int g(x)^n\frac{du}{dx}\,dx = \int u^n\,du \quad \text{with } u = g(x). \qquad (3.6.3)$$

□

Equation (3.6.3) appears to substitute du for the expression $\frac{du}{dx}\,dx$ in the first integral. This is merely an illusion: the derivative $\frac{du}{dx}$ is not du divided by dx nor is dx in the symbol $\int \cdots dx$ an expression that can be replaced by another expression of equal value.[3] Nevertheless, the formula obtained by making this type of substitution is valid in general, not just in the example above.

Proposition 3.6.2 [Integration by Substitution] *Let $y = f(u)$ be a continuous function, and let $u = g(x)$ be a differentiable function with $f \circ g$ defined. Then*

$$\int f(g(x)) \frac{du}{dx} \, dx = \int f(u) \, du. \qquad (3.6.4)$$

Proof Let $F(u) = \int_a^u f(t) \, dt$. By the First Fundamental Theorem of Calculus, $F(u)$ is differentiable with $F'(u) = f(u)$. By the definition of the indefinite integral:

$$\int f(u) \, du = \int F'(u) \, du = F(u) + C \;=\; F(g(x)) + C = \int \frac{d}{dx} \left[F(g(x)) \right] \, dx.$$

Apply the chain rule: $\quad \int f(u) \, du \;=\; \int F'(g(x)) g'(x) \, dx = \int f(g(x)) \frac{du}{dx} \, dx.$ $\square$

As explained above, formula (3.6.4) does not multiply a fraction $\frac{du}{dx}$ by dx to replace the product by du. Nevertheless, this viewpoint is helpful in remembering the formula. In particular, we write $du = g'(x) \, dx$ to indicate that we intend to apply (3.6.4) to replace $\int \cdots g'(x) \, dx$ by $\int \cdots du$. In the examples below, we use equation (3.6.4) to rewrite a complicated integral in the variable x as a simpler integral in the variable u which we can integrate. The most challenging step in applying (3.6.4) is to decide how to define $u = g(x)$.

Examples 3.6.3 (1) Evaluate $\int \frac{x}{(3x^2+5)^2} \, dx$.

Solution Let $u = 3x^2 + 5$. Then $\frac{du}{dx} = 6x$ and $du = 6x \, dx$. By (3.6.4):

$$\int \frac{x}{(3x^2 + 5)^2} \, dx \;=\; \frac{1}{6} \int \frac{1}{(3x^2 + 5)^2} \, 6x \, dx = \frac{1}{6} \int \frac{1}{u^2} \, du = \frac{1}{6} \int u^{-2} \, du$$

$$=\; \frac{1}{6} \cdot \frac{u^{-1}}{-1} + C = -\frac{1}{18x^2 + 30} + C.$$

(2) Evaluate $\int \cos(4x - 7) \, dx$.

Solution Let $u = 4x - 7$. Then $\frac{du}{dx} = 4$ and $du = 4 \, dx$. By (3.6.4):

$$\int \cos(4x - 7) \, dx \;=\; \frac{1}{4} \int \cos(4x - 7) \, 4 \, dx = \frac{1}{4} \int \cos u \, du = \frac{1}{4} \sin u + C$$

$$=\; \frac{1}{4} \sin(4x - 7) + C.$$

(3) Evaluate $\int x\sqrt{5x + 3} \, dx$.

Solution Let $u = 5x + 3$. Then $x = \frac{u-3}{5}$, $\frac{du}{dx} = 5$ and $du = 5 \, dx$. By (3.6.4):

$$\int x\sqrt{5x + 3} \, dx \;=\; \frac{1}{5} \int x(5x + 3)^{1/2} \, 5 \, dx = \frac{1}{5} \int \frac{u - 3}{5} u^{1/2} \, du$$

$$=\; \frac{1}{25} \int u^{3/2} - 3u^{1/2} \, du = \frac{1}{25} \left(\frac{u^{5/2}}{5/2} - 3\frac{u^{3/2}}{3/2} \right) + C$$

$$=\; \frac{2}{125}(5x + 3)^{5/2} - \frac{2}{25}(5x + 3)^{3/2} + C.$$

---

[3]A reader who studied differentials in Section 2.11 can interpret $du = g'(x) \, dx$ as the correct mathematical statement that the differential du is the product of $g'(x)$ times the differential dx. However, it is impossible to interpret the replacement of $g'(x) \, dx$ by du in $\int \cdots g'(x) \, dx$ as the substitution of one algebraic expression for another one of equal value.

(4) Evaluate $\int \tan^2 x \sec^2 x \, dx$.

  Solution Let $u = \tan x$. Then $\frac{du}{dx} = \sec^2 x$ and $du = \sec^2 x \, dx$. By (3.6.4):

$$\int \tan^2 x \sec^2 x \, dx = \int u^2 \, du = \frac{u^3}{3} + C = \frac{1}{3}\tan^3 x + C. \qquad \square$$

We use substitutions to generalize the integrals of Section 5 whose values are given in terms of arcsin and arctan. The first of these integrals is:

$$\int \frac{1}{\sqrt{a^2 - x^2}} \, dx = \arcsin \frac{x}{a} + C. \tag{3.5.2}$$

We generalize this formula to integrate $\int \frac{1}{\sqrt{ax^2+bx+c}} \, dx$ when $a < 0$ and $b^2 - 4ac > 0$. In this case we complete the square and make a linear substitution to reduce the integral to the above form.

Examples 3.6.4 (1) Evaluate $\int \frac{1}{\sqrt{7+6x-x^2}} \, dx$.

  Solution Begin by completing the square:

$$\int \frac{1}{\sqrt{7 + 6x - x^2}} \, dx = \int \frac{1}{\sqrt{7 - (x^2 - 6x)}} \, dx = \int \frac{1}{\sqrt{16 - (x-3)^2}} \, dx.$$

Let $u = x - 3$. Then $\frac{du}{dx} = 1$ and $du = dx$. Change variables by (3.6.4) and apply (3.5.2):

$$\int \frac{1}{\sqrt{7 + 6x - x^2}} \, dx = \int \frac{1}{\sqrt{4^2 - u^2}} \, du = \arcsin \frac{u}{4} + C = \arcsin \frac{x-3}{4} + C.$$

(2) Evaluate $\int \frac{1}{\sqrt{9+5x-4x^2}} \, dx$.

  Solution First, complete the square:

$$\int \frac{1}{\sqrt{9 + 5x - 4x^2}} \, dx = \int \frac{1}{\sqrt{9 - (4x^2 - 5x)}} \, dx = \int \frac{1}{\sqrt{9 + \frac{25}{16} - \left(2x - \frac{5}{4}\right)^2}} \, d$$

$$= \int \frac{1}{\sqrt{\frac{169}{16} - \left(2x - \frac{5}{4}\right)^2}} \, dx$$

Let $u = 2x - \frac{5}{4}$. Then $\frac{du}{dx} = 2$ and $du = 2 \, dx$. Change variables by (3.6.4) and apply (3.5.2):

$$\int \frac{1}{\sqrt{7 + 5x - 4x^2}} \, dx \;=\; \frac{1}{2}\int \frac{1}{\sqrt{\frac{169}{16} - \left(2x - \frac{5}{4}\right)^2}} \, 2 \, dx = \frac{1}{2}\int \frac{1}{\sqrt{\left(\frac{13}{4}\right)^2 - u^2}} \, du$$

$$= \frac{1}{2}\arcsin \frac{u}{13/4} + C = \frac{1}{2}\arcsin \frac{4\left(2x - \frac{5}{4}\right)}{13} + C$$

$$= \frac{1}{2}\arcsin \frac{8x - 5}{13} + C \qquad \square$$

Recall the integration formula with value arctan from Section 5:

$$\int \frac{1}{a^2 + x^2} \, dx = \frac{1}{a}\arctan \frac{x}{a} + C. \tag{3.5.3}$$

This integration formula generalizes to integrate $\int \frac{1}{ax^2+bx+c} \, dx$ when $b^2 - 4ac < 0$. First, we complete the square. Then we make a linear substitution to reduce the integral to the above form.

Examples 3.6.5 (1) Evaluate $\int \frac{1}{x^2+6x+25}\,dx$.

Solution Begin by completing the square:

$$\int \frac{1}{x^2+6x+25}\,dx = \int \frac{1}{16+(x+3)^2}\,dx.$$

Let $u = x + 3$. Then $\frac{du}{dx} = 1$ and $du = dx$. Change variables by (3.6.4) and apply (3.5.3):

$$\int \frac{1}{x^2+6x+25}\,dx = \int \frac{1}{4^2+u^2}\,du = \frac{1}{4}\arctan\frac{u}{4} + C = \frac{1}{4}\arctan\frac{x+3}{4} + C.$$

(2) Evaluate $\int \frac{1}{4x^2+8x+41}\,dx$.

Solution Again, start by completing the square:

$$\int \frac{1}{4x^2+16x+41}\,dx = \int \frac{1}{(2x+4)^2+25}\,dx.$$

Let $u = 2x + 4$. Then $\frac{du}{dx} = 2$ and $du = 2\,dx$. Change variables by (3.6.4) and apply (3.5.3):

$$\int \frac{1}{4x^2+16x+41}\,dx = \frac{1}{2}\int \frac{1}{(2x+4)^2+25}\,2\,dx = \frac{1}{2}\int \frac{1}{u^2+25}\,du$$

$$= \frac{1}{2}\cdot\frac{1}{5}\arctan\frac{u}{5} + C = \frac{1}{10}\arctan\frac{2x+4}{5} + C. \qquad \square$$

Substitution in Definite Integrals

Consider the definite integral $\int_a^b f(g(x))g'(x)\,dx$ where we use the substitution $u = g(x)$ to compute its value. That is, we change variables from x to u and evaluate $\int f(u)\,du$ as $h(u) + C$. It is not necessary to write $h(u)$ as $h(g(x))$ to evaluate the original definite integral as $h(g(b)) - h(g(a))$. A simpler procedure is to replace the bounds $x = a$, $x = b$ in the original integral by the corresponding bounds $u = g(a)$, $u = g(b)$ in the integral $\int f(u)\,du$:

$$\int_{x=a}^{x=b} f(g(x))g'(x)\,dx = \int_{u=g(a)}^{u=g(b)} f(u)\,du. \qquad (3.6.5)$$

This procedure is more efficient because it does not require rewriting $h(u)$ as $h(g(x))$. Moreover, even if u appears several times in the function $h(u)$, we only need to compute $g(a)$ and $g(b)$ once.

Examples 3.6.6 (1) Evaluate $\int_0^5 x^2 \sin(\pi x^3)\,dx$.

Solution Let $u = \pi x^3$. Then $\frac{du}{dx} = 3\pi x^2$ and $du = 3\pi x^2\,dx$. In addition, when $x = 0$, $u = \pi 0^3 = 0$, while when $x = 5$, $u = \pi 5^3 = 125\pi$. By (3.6.5):

$$\int_0^5 x^2 \sin(\pi x^3)\,dx = \frac{1}{3\pi}\int_{x=0}^{x=5} \sin(\pi x^3)\,3\pi x^2\,dx = \frac{1}{3\pi}\int_{u=0}^{u=125\pi} \sin u\,du$$

$$= -\frac{1}{3\pi}\cos u\,\Big|_0^{125\pi} = -\frac{1}{3\pi}(\cos 125\pi - \cos 0)$$

$$= -\frac{1}{3\pi}(-1 - 1) = \frac{2}{3\pi}.$$

(2) Evaluate $\int_1^2 \frac{x^2}{\sqrt{3x+1}}\ dx$.

Solution Let $u = 3x + 1$. Then $x = \frac{u-1}{3}$, $\frac{du}{dx} = 3$ and $du = 3\ dx$. Also, when $x = 1$, $u = 4$ while when $x = 2$, $u = 7$. By (3.6.5):

$$
\begin{aligned}
\int_1^2 \frac{x^2}{\sqrt{3x+1}}\ dx &= \frac{1}{3}\int_{x=1}^{x=2} \frac{x^2}{\sqrt{3x+1}}\ 3\ dx = \frac{1}{3}\int_{u=4}^{u=7} \frac{[(u-1)/3]^2}{\sqrt{u}}\ du \\
&= \frac{1}{27}\int_4^7 \frac{u^2 - 2u + 1}{u^{1/2}}\ du = \frac{1}{27}\int_4^7 u^{3/2} - 2u^{1/2} + u^{-1/2}\ du \\
&= \frac{1}{27}\left(\frac{u^{5/2}}{5/2} - 2\frac{u^{3/2}}{3/2} + \frac{u^{1/2}}{1/2} \right)\Big|_4^7 \\
&= \frac{1}{27}\left[\left(\frac{2}{5}\cdot 7^{5/2} - \frac{4}{3}\cdot 7^{3/2} + 2\cdot 7^{1/2} \right) \right. \\
&\qquad\qquad \left. - \left(\frac{2}{5}\cdot 4^{5/2} - \frac{4}{3}\cdot 4^{3/2} + 2\cdot 4^{1/2} \right) \right] \\
&= \frac{184\sqrt{7} - 92}{405}
\end{aligned}
$$

(3) Evaluate $\int_6^{14} \frac{1}{x^2 - 12x + 100}\ dx$.

Solution Begin by completing the square:

$$
\int_6^{14} \frac{1}{x^2 - 12x + 100}\ dx = \int_{x=6}^{x=14} \frac{1}{(x-6)^2 + 64}\ dx.
$$

Now let $u = x - 6$ with $\frac{du}{dx} = 1$ and $du = dx$. Note that when $x = 6$, $u = 0$ and when $x = 14$, $u = 8$. Apply (3.5.3):

$$
\begin{aligned}
\int_0^2 \frac{1}{x^2 - 12x + 100}\ dx &= \int_{u=0}^{u=8} \frac{1}{u^2 + 64}\ dx = \frac{1}{8}\arctan \frac{u}{8}\Big|_0^8 \\
&= \frac{1}{8}\arctan 1 - \frac{1}{8}\arctan 0 = \frac{1}{8}\cdot\frac{\pi}{4} - 0 = \frac{\pi}{32}. \qquad \square
\end{aligned}
$$

We give an easy proof of Property 8 of definite integrals which was stated in Section 3.3: the integral of an odd function over the interval $[-a, a]$ is zero.

Property 8 Let f be an odd function which is continuous on $[-a, a]$. Then

$$
\int_{-a}^a f(x)\ dx = 0.
$$

Proof By the additive property:

$$
\int_{-a}^a f(x)\ dx = \int_{-a}^0 f(x)\ dx + \int_0^a f(x)\ dx.
$$

In the first integral, make the substitution $u = -x$ with $\frac{du}{dx} = -1$ and $du = -dx$. Since f is an odd function $f(-x) = -f(x)$. Thus

$$
\begin{aligned}
\int_{-a}^a f(x)\ dx &= \int_{x=-a}^{x=0} f(-x)\ -dx + \int_0^a f(x)\ dx = \int_{u=a}^{u=0} f(u)\ du + \int_0^a f(x)\ dx \\
&= -\int_0^a f(u)\ du + \int_0^a f(x)\ dx = 0. \qquad \square
\end{aligned}
$$

Trigonometric Identities

We use trigonometric identities to evaluate $\int \sin^m x \cos^n x \, dx$ and to evaluate integrals of products of $\sin Ax$ and $\cos Bx$. We break up our considerations of $\int \sin^m x \cos^n x \, dx$ into two cases: the case where at least one of m, n is an odd integer and the case where both m, n are even integers.

I. $\int \sin^m x \, \cos^n x \, dx$ (m or n an odd integer)

Consider the case where the exponent of $\sin x$ is an odd integer: $m = 2h + 1$. Use the Pythagorean identity

$$\sin^2 x + \cos^2 x = 1 \qquad (3.6.6)$$

to substitute $\sin^2 x = 1 - \cos^2 x$ in the given integral:

$$\int \sin^{2h+1} x \, \cos^n x \, dx = \int \left(\sin^2 x \right)^h \sin x \, \cos^n x \, dx = \int \left(1 - \cos^2 x \right)^h \cos^n x \, \sin x \, dx.$$

Make the substitution $u = \cos x$ with $du = -\sin x \, dx$:

$$
\begin{aligned}
\int \sin^{2h+1} x \, \cos^n x \, dx &= -\int \left(1 - \cos^2 x \right)^h \cos^n x \, (-\sin x) \, dx \\
&= -\int \left(1 - u^2 \right)^h u^n \, du. \qquad (3.6.7)
\end{aligned}
$$

Now multiply out the integrand and integrate.

Consider the case where the exponent of $\cos x$ is an odd integer: $n = 2k + 1$. Use the Pythagorean identity (3.6.6) to substitute $\cos^2 x = 1 - \sin^2 x$ in the given integral:

$$\int \sin^m x \, \cos^{2k+1} x \, dx = \int \sin^m x \, \left(\cos^2 x \right)^k \cos x \, dx = \int \sin^m x \, \left(1 - \sin^2 x \right)^k \cos x \, dx.$$

Make the substitution $v = \sin x$ with $dv = \cos x \, dx$:

$$\int \sin^m x \, \cos^{2k+1} x \, dx = \int v^m \left(1 - v^2 \right)^k \, dv. \qquad (3.6.8)$$

Now multiply out the integrand and integrate.

In the examples below we use the procedures described above rather than substituting into (3.6.7) and (3.6.8) directly which would require memorizing these formulas.

Examples 3.6.7 (1) Evaluate $\int \sin^3 x \, \cos^4 x \, dx$.

Solution Since the power of $\sin x$ is odd, let $u = \cos x$ with $du = -\sin x \, dx$:

$$
\begin{aligned}
\int \sin^3 x \, \cos^4 x \, dx &= \int (\sin^2 x) \sin x \, \cos^4 x \, dx \\
&= -\int (1 - \cos^2 x) \cos^4 x (-\sin x) \, dx = -\int (1 - u^2) u^4 \, du \\
&= -\int u^4 - u^6 \, du = -\frac{u^5}{5} + \frac{u^7}{7} + C = -\frac{\cos^5 x}{5} + \frac{\cos^7 x}{7} + C.
\end{aligned}
$$

(2) Evaluate $\int \sin^5 x \cos^5 x \, dx$.

Solution I Since the power of $\cos x$ is odd, let $v = \sin x$ with $dv = \cos x \, dx$:

$$\int \sin^5 x \cos^5 x \, dx = \int \sin^5 x (\cos^2 x)^2 \cos x \, dx = \int \sin^5 x (1 - \sin^2 x)^2 \cos x \, dx$$

$$= \int v^5 (1 - v^2)^2 \, dv = \int v^5 - 2v^7 + v^9 \, dv$$

$$= \frac{v^6}{6} - 2\frac{v^8}{8} + \frac{v^{10}}{10} + C$$

$$= \frac{1}{6} \sin^6 x - \frac{1}{4} \sin^8 x + \frac{1}{10} \sin^{10} x + C.$$

Solution II Since the power of $\sin x$ is odd, let $u = \cos x$ with $du = -\sin x \, dx$:

$$\int \sin^5 x \cos^5 x \, dx \;=\; \int \left(\sin^2 x\right)^2 \sin x \cos^5 x \, dx$$

$$= -\int \left(1 - \cos^2 x\right)^2 \cos^5 x(-\sin x) \, dx = -\int \left(1 - u^2\right)^2 u^5$$

$$= -\int u^5 - 2u^7 + u^9 \, du = -\left(\frac{u^6}{6} - 2\frac{u^8}{8} + \frac{u^{10}}{10}\right) + C$$

$$= -\frac{1}{6} \cos^6 x + \frac{1}{4} \cos^8 x - \frac{1}{10} \cos^{10} x + C.$$

The answers given by these two solutions must be equal up to a constant because they both have the same derivative $\sin^5 x \cos^5 x$. This fact is not obvious, and it would be difficult to verify it directly using trigonometric identities.   □

II. $\int \sin^m x \, \cos^n x \, dx$ (m and n even integers)

Write $m = 2h$ and $n = 2k$. To integrate $\int \sin^{2h} x \, \cos^{2k} x \, dx$ use the identities

$$\sin^2 x = \frac{1}{2}(1 - \cos 2x) \quad \text{and} \quad \cos^2 x = \frac{1}{2}(1 + \cos 2x) \qquad (3.6.9)$$

to rewrite this integral as a simpler integral in terms of the angle $2x$:

$$\int \sin^{2h} x \, \cos^{2k} x \, dx \;=\; \int \left(\sin^2 x\right)^h \left(\cos^2 x\right)^k \, dx$$

$$= \int \left[\frac{1}{2}(1 - \cos 2x)\right]^h \left[\frac{1}{2}(1 + \cos 2x)\right]^k \, dx. \qquad (3.6.10)$$

Note that the sum of the exponents in the new integral is $h + k$, half the sum of the exponents in the original integral. Now multiply out the integrand, and integrate each summand by the method of **I** or **II**.

Examples 3.6.8 (1) Evaluate $\int \sin^2 x \, dx$.

   Solution By the first trigonometric identity of (3.6.9),

$$\int \sin^2 x \, dx = \frac{1}{2} \int 1 - \cos 2x \, dx = \frac{1}{2}\left(x - \frac{1}{2}\sin 2x\right) + C = \frac{x}{2} - \frac{1}{4}\sin 2x + C.$$

(2) Evaluate $\int \sin^2 x \, \cos^4 x \, dx$.

   Solution By the second trigonometric identity of (3.6.9),

$$\int \sin^2 x \, \cos^4 x \, dx \;=\; \int \sin^2 x(\cos^2 x)^2 \, dx \;=\; \int \frac{1}{2}(1 - \cos 2x)\frac{1}{4}(1 + \cos 2x)^2 \, d$$

$$= \frac{1}{8} \int (1 - \cos 2x)(1 + 2\cos 2x + \cos^2 2x) \, dx$$

$$= \frac{1}{8} \int 1 + \cos 2x - \cos^2 2x - \cos^3 2x \, dx.$$

Use the second trigonometric identity of (3.6.9), with x replaced by $2x$, to evaluate the integral of the third summand. Substitute $u = \sin 2x$ with $du = 2\cos 2x\, dx$ to evaluate the integral of the last summand.

$$
\begin{aligned}
\int \sin^2 x \, \cos^4 x \, dx \;=\; & \frac{1}{8}\left[x + \frac{1}{2}\sin 2x - \frac{1}{2}\int 1 + \cos 4x \, dx \right. \\
& \left. - \frac{1}{2}\int (1 - \sin^2 2x)(2\cos 2x)\, dx \right] \\[4pt]
=\; & \frac{x}{8} + \frac{1}{16}\sin 2x - \frac{x}{16} - \frac{1}{64}\sin 4x - \frac{1}{16}\int 1 - u^2 \, du \\[4pt]
=\; & \frac{x}{16} + \frac{1}{16}\sin 2x - \frac{1}{64}\sin 4x - \frac{1}{16}\left(u - \frac{u^3}{3} \right) + C \\[4pt]
=\; & \frac{x}{16} + \frac{1}{16}\sin 2x - \frac{1}{64}\sin 4x - \frac{1}{16}\sin 2x + \frac{1}{48}\sin^3 2x + C \\[4pt]
=\; & \frac{x}{16} - \frac{1}{64}\sin 4x + \frac{1}{48}\sin^3 2x + C \qquad\qquad \square
\end{aligned}
$$

III. $\int \sin Ax \, \sin Bx \, dx$, $\int \cos Ax \, \cos Bx \, dx$, $\int \sin Ax \, \cos Bx \, dx$

These integrals are very important in the study of Fourier series. They are easily evaluated by using the following trigonometric identities.

Lemma 3.6.9 (a) $\sin\theta \sin\phi = \frac{1}{2}\left[\cos(\theta - \phi) - \cos(\theta + \phi)\right]$

(b) $\cos\theta \cos\phi = \frac{1}{2}\left[\cos(\theta - \phi) + \cos(\theta + \phi)\right]$

(c) $\sin\theta \cos\phi = \frac{1}{2}\left[\sin(\theta - \phi) + \sin(\theta + \phi)\right]$

Proof (a) Subtract the following two identities to obtain twice identity (a):

$$
\begin{aligned}
\cos(\theta - \phi) &= \cos\theta \cos\phi + \sin\theta \sin\phi \\
\cos(\theta + \phi) &= \cos\theta \cos\phi - \sin\theta \sin\phi.
\end{aligned}
$$

(b) Add the two identities above, to obtain twice identity (b).
(c) Add the following two identities to obtain twice identity (c):

$$
\begin{aligned}
\sin(\theta - \phi) &= \sin\theta \cos\phi - \cos\theta \sin\phi \\
\sin(\theta + \phi) &= \sin\theta \cos\phi + \cos\theta \sin\phi. \qquad\qquad \square
\end{aligned}
$$

We use Lemma 3.6.9 to evaluate integrals of products of $\sin mx$ and $\cos nx$.

Examples 3.6.10 (1) Evaluate $\int \sin 8x \, \sin 2x \, dx$.
Solution By Lemma 3.6.9(a):

$$
\int \sin 8x \, \sin 2x \, dx = \frac{1}{2}\int \cos 6x - \cos 10x \, dx = \frac{1}{12}\sin 6x - \frac{1}{20}\sin 10x + C.
$$

(2) Evaluate $\int_0^\pi \sin 3x \, \cos 5x \, dx$.
Solution By Lemma 3.6.9(c):

$$
\begin{aligned}
\int_0^\pi \sin 3x \, \cos 5x \, dx \;=\; & \frac{1}{2}\int_0^\pi -\sin 2x + \sin 8x \, dx = \left. \frac{1}{4}\cos 2x - \frac{1}{16}\cos 8x \right|_0^\pi \\[4pt]
=\; & \left(\frac{1}{4} - \frac{1}{16} \right) - \left(\frac{1}{4} - \frac{1}{16} \right) = 0. \qquad\qquad \square
\end{aligned}
$$

Summary

The reader should be able to define a new variable to change a given integral to a simpler integral in the new variable. In the case of a definite integral, she should know how to determine the bounds of the new integral. She should understand how to use substitutions and the identities for $\cos 2x$ to evaluate integrals of the form $\int \sin^m x \, \cos^n x \, dx$. She should also know how to use the trigonometric identities of Lemma 3.6.9 to integrate products of $\sin Ax$ and $\cos Bx$.

Basic Exercises

Evaluate each of these 51 integrals.

1. $\int \cos(5x+3) \, dx$
2. $\int x\sqrt{4x+1} \, dx$
3. $\int (8x^2+3)^{10} x \, dx$

4. $\int (9x+7)^{4/5} \, dx$
5. $\int \frac{x^2}{(5x^3+7)^3} \, dx$
6. $\int x^3 \sqrt{5x^4+6} \, dx$

7. $\int \frac{x^2}{\sqrt{3x+5}} \, dx$
8. $\int \cos^2 x \, dx$
9. $\int \tan x \, \sec^5 x \, dx$

10. $\int \sin^4 x \, \cos^3 x \, dx$
11. $\int \sin^8 x \cos^5 x \, dx$
12. $\int \sin^4 x \, \cos^2 x \, dx$

13. $\int \cos 6x \, \cos 8x \, dx$
14. $\int \sin^4 x \, \cos^7 x \, dx$
15. $\int \sin^3 x \, dx$

16. $\int \frac{1}{x^2+6x+25} \, dx$
17. $\int \cos^4 x \, dx$
18. $\int \sin^5 x \, \cos^6 x \, dx$

19. $\int \frac{3x^2-5}{(8x+2)^6} \, dx$
20. $\int \cos^5 x \, dx$
21. $\int \sin 3x \, \cos 4x \, dx$

22. $\int \frac{3x+7}{\sqrt{4x+9}} \, dx$
23. $\int \frac{\cos x}{\sqrt{5\sin x+3}} \, dx$
24. $\int \frac{8x^3+3x}{\sqrt[3]{4x^2-3}} \, dx$

25. $\int \sec^2 x (5\tan x + 8)^3 \, dx$
26. $\int \frac{1}{\sqrt{8-x^2-2x}} \, dx$
27. $\int \sin^6 6x \, dx$

28. $\int \frac{1}{x^2+4x+13} \, dx$
29. $\int \sin 7x \, \cos 5x \, dx$
30. $\int \sin^4 x \, \cos^4 x \, dx$

31. $\int \frac{\arctan x}{1+x^2} \, dx$
32. $\int \sin^7 2x \, \cos^5 2x \, dx$
33. $\int \frac{1}{4x^2+20x+41} \, dx$

34. $\int \frac{\cot x \csc x}{(1+\csc x)^2} \, dx$
35. $\int \frac{x}{\sqrt{2-x^4-6x^2}} \, dx$
36. $\int \csc^3 5x \cot 5x \, dx$

37. $\int \sin^7 x \, dx$
38. $\int \sin^2 3x \, \cos^6 3x \, dx$
39. $\int \csc^2 x \sqrt{5 \cot x + 2} \, dx$

40. $\int \frac{\csc x}{\sin x} \, dx$
41. $\int x \sin^4(x^2+1) \, dx$
42. $\int \frac{x^4+x^2}{\sqrt[3]{3x^5+5x^3+1}} \, dx$

43. $\int \frac{x}{4+x^4} \, dx$
44. $\int \frac{\sin(5\sqrt{x}) \, \sin(4\sqrt{x})}{\sqrt{x}} \, dx$
45. $\int \frac{\sin x - \cos x}{(\sin x + \cos x)^2} \, dx$

46. $\int \csc^2 x \cot^6 x \, dx$
47. $\int \frac{\arcsin x}{\sqrt{1-x^2}} \, dx$
48. $\int \frac{x^2}{\sin^2 \sqrt{x}+\cos^2 \sqrt{x}} \, dx$

49. $\int \frac{\tan x \, \sec^2 x}{\sqrt{\tan^2 x+1}} \, dx$
50. $\int \frac{1}{\sqrt{5-9x^2-12x}} \, dx$
51. $\int \frac{\cos x}{\sin^2 x+2\sin x+5} \, dx$

Evaluate each definite integral by making a substitution $u = g(x)$. Determine the appropriate bounds in the integral with respect to u, and use these bounds to evaluate the original integral.

52. $\int_1^3 \frac{x}{\sqrt{x^2+1}} \, dx$
53. $\int_0^{\pi/2} \sin x \sqrt{3+\cos x} \, dx$
54. $\int_0^{\pi/3} \sin^3 x \, \cos^4 x \, dx$

55. $\int_2^5 (2x-1)^{100} \, dx$
56. $\int_0^{\pi/4} \frac{\sec^2 x}{(1+3\tan x)^2} \, dx$
57. $\int_0^3 x^3 \sqrt{x+1} \, dx$

58. $\int_3^4 x^2 (x^3-10)^{2/3} \, dx$
59. $\int_1^4 \frac{1}{x^2+4x+10} \, dx$
60. $\int_0^{\pi/3} \tan x \, \sec^5 x \, dx$

61. $\int_0^{\pi/4} \frac{\sin 2x}{1+\cos^4 x} \, dx$
62. $\int_3^6 \frac{1}{\sqrt{27+6x-x^2}} \, dx$
63. $\int_0^{\pi/2} \sin 3x \, \cos 5x \, dx$

Find the average values of each of these six functions.

64. $f(x) = x\sqrt{5x-6}$ on $[2,3]$
65. $g(x) = \frac{\sin \pi \sqrt{x}}{\sqrt{x}}$ on $\left[\frac{1}{4}, 1\right]$

66. $h(x) = \frac{\arctan \sqrt{x}}{\sqrt{x}+x\sqrt{x}}$ on $[0,1]$
67. $j(x) = \tan^3 x \sec^4 x$ on $\left[0, \frac{\pi}{4}\right]$

68. $k(x) = \cos^8 x$ on $\left[0, \frac{\pi}{3}\right]$
69. $m(x) = \cos 4x \cos 7x$ on $\left[-\frac{\pi}{4}, \frac{\pi}{6}\right]$

70. Find the area between the graphs of $f(x) = (3x - 2)^{10}$ and $g(x) = 3x - 2$.

71. Find the area bounded by the graphs of $f(x) = x\sqrt{x^2 + 1}$ and $g(x) = 4x$.

72. Find the area between the graphs of $f(x) = \sec^4 x$ $(-\frac{\pi}{2} < x < \frac{\pi}{2})$ and $g(x) = 32 - \sec^4 x$.

73. Find the area between the graphs of $f(x) = \sin^2 x$ and $g(x) = \cos^2 x$, both with domain $[0, 2\pi]$.

74. Find the area between the curves $x = (2y + 1)^{100} + 15$ and $x = 8(2y + 1)^{50}$.

75. Find the area between the graphs of $f(x) = \sin^5 x$ and $g(x) = \sin^3 x$, both with domain $[0, \pi]$.

76. Find the area between the curves $x = \frac{y-1}{\sqrt{3y^2 - 6y + 9}}$ and $x = \frac{27y - 27}{(3y^2 - 6y + 9)^2}$.

77. Find the area between the graphs of $f(x) = \tan^2 x \sec^2 x$ and $g(x) = 2 \tan^2 x$, both with domain $\left(-\frac{\pi}{2}, \frac{\pi}{2}\right)$.

78. Find the area between the curves $x = \frac{17y}{1+y^4}$ and $y = x$.

79. Let $f(x)$ be an even function which is continuous on $[-a, a]$. Show that $\int_{-a}^{a} f(x) \, dx = 2 \int_{0}^{a} f(x) \, dx$.

Challenging Problems

1. Verify the formula for integration by substitution directly from the definition of the definite integral as outlined below. Let f be a continuous function, and let $u = g(x)$ be an increasing differentiable function.
(a) Let $P = \{x_0, \ldots, x_n\}$ be a partition of $[a, b]$. Show $g(P) = \{g(x_0), \ldots, g(x_n)\}$ is a partition of $[g(a), g(b)]$.
(b) Let $T = \{t_1, \ldots, t_n\}$ with $x_{k-1} \le t_k \le x_k$ for $1 \le k \le n$.
Show $g(T) = \{g(t_1), \ldots, g(t_n)\}$ satisfies $g(x_{k-1}) \le g(t_k) \le g(x_k)$ for $1 \le k \le n$.
(c) Show every Riemann sum $R(Q, S, f)$ for the integral $\int_{g(a)}^{g(b)} f(u) \, du$ is of the form $R(g(P), g(T), f)$.
(d) Show there is $t'_k \in (x_k, x_{k-1})$, for $1 \le k \le n$, such that

$$R(g(P), g(T), f) = f(g(t_1))g'(t'_1)(x_1 - x_0) + \cdots + f(g(t_n))g'(t'_n)(x_n - x_{n-1}).$$

(e) Show for each partition P of $[a, b]$, there is a choice of T such that

$$R(g(P), g(T), f) = R(P, T, (f \circ g)g').$$

(f) Use (a)–(e) to prove $\int_{g(a)}^{g(b)} f(u) \, du = \int_{a}^{b} f(g(x))g'(x) \, dx$.

3.7   Computation of Volume

In this section, we study three ways to use the definite integral to compute volumes. The first subsection introduces the *cross–section method*. This method is the basis of the more general procedure of computing volumes by iterated integrals presented in Chapter 8. The second subsection applies the cross-section method to compute volumes of solids of revolution by the disc method. The last subsection provides an alternate procedure for computing these volumes by the shell method.

Cross–Section Method for Computing Volume

The cross–section method computes the volume of a solid as the integral of its cross–sectional area. We begin with an example of this method, present the general method and then apply it to other examples.

Motivating Example 3.7.1 The Great Pyramid of Giza was built by the Pharaoh Cheops in 2560 B.C.E. With a height of 146 meters, it was the tallest building in the world for 4,000 years. Its base is a square of side 230 meters. Find the volume V of this pyramid.

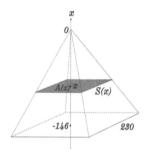

Figure 3.7.2
The Great Pyramid

Solution Construct a vertical x–axis through the center of the pyramid with the top of the pyramid at the origin and its bottom at $x = -146$. The horizontal plane through each $x \in [-146, 0]$ intersects the pyramid in a square having side of length $S(x)$. See Figure 3.7.2. Let P_n be the regular partition of the interval $[-146, 0]$ into n subintervals, each of width $\frac{146}{n}$. That is, P_n consists of $-\frac{146k}{n}$ for $0 \le k \le n$. The horizontal planes through these numbers on the x–axis cut the pyramid into n slices. Approximate the slice from $x = -\frac{146k}{n}$ to $x = -\frac{146(k-1)}{n}$ by a thin box of height $\frac{146}{n}$ with a square base having sides of length $S\left(-\frac{146k}{n}\right)$. The volume V_k of this thin box is the product of its dimensions:

$$V_k = S\left(-\frac{146k}{n}\right)^2 \frac{146}{n}$$

The volume of the pyramid is approximately the sum of volumes of these n thin boxes:

$$V \approx \sum_{k=1}^{n} V_k \approx \sum_{k=1}^{n} S\left(-\frac{146k}{n}\right)^2 \frac{146}{n}.$$

Since each subinterval of the partition P_n has width $\frac{146}{n}$, we can interpret the preceding sum as the Riemann sum

$$R(P_n, T_n, A) = \sum_{k=1}^{n} A(t_k) \frac{146}{n}$$

of the function $A(x) = S(x)^2$ with T_n the set of $t_k = -\frac{146k}{n}$, for $1 \le k \le n$. Note $A(x)$ is the area of the horizontal cross–section of this pyramid at the point x on the x–axis. As n gets larger and larger, the slices of the pyramid get thinner and thinner and our approximations of V get better and better. Hence

$$V = \lim_{n \to \infty} R(P_n, T_n, A) = \int_{-146}^{0} A(x)\, dx. \qquad (3.7.1)$$

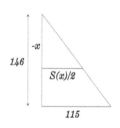

Figure 3.7.3
Finding $S(x)$

To continue, we need to find a formula for $A(x) = S(x)^2$ as a specific function of x. We find $S(x)$ from the similar triangles in the vertical cross–section of the right half of the pyramid depicted in Figure 3.7.3:

$$\frac{S(x)/2}{115} = \frac{-x}{146} \quad \text{and} \quad S(x) = -\frac{230}{146}x.$$

$$A(x) = S(x)^2 = \frac{230^2}{146^2}x^2.$$

By (3.7.1) : $V = \int_{-146}^{0} \frac{230^2}{146^2}x^2\, dx = \frac{230^2}{146^2}\int_{-146}^{0} x^2\, dx = \frac{230^2}{146^2} \left.\frac{x^3}{3}\right|_{-146}^{0}$

$$= 0 - \frac{230^2}{146^2}\frac{(-146)^3}{3} = \frac{146(230^2)}{3} \approx 2{,}574{,}467 \text{ meters}^3. \qquad \square$$

We generalize the procedure of the preceding example to compute the volume V of a solid S. That is, we slice S into n slices, approximate the volume of each slice and add these n approximations to approximate V. Define $A(x)$ as the cross-sectional area of the region obtained by slicing the solid S with the plane C_x perpendicular to the x–axis at the point x. See Figure 3.7.6. Assume that the entire solid S lies between C_a and C_b. To specify how to slice S into n slices, use the regular partition $P = \{x_0, \ldots, x_n\}$ of the interval $[a, b]$ into n subintervals of width Δx. The points of this partition determine vertical planes C_{x_k}, for $0 \leq k \leq n$, which cut the solid S into n slices of thickness Δx. See Figure 3.7.4. Choose a point t_k from each subinterval $[x_{k-1}, x_k]$, and denote the set of these selections by T. Let S_k be the slice of S between $C_{x_{k-1}}$ and C_{x_k} depicted in Figure 3.7.5. S_k is approximately a solid with constant cross–sectional area $A(t_k)$ and thickness Δx which has volume $A(t_k)\Delta x$. The volume V of S is estimated by the sum of these approximations of the volumes of the n slices:

Figure 3.7.4
Slicing S

$$V \approx \sum_{k=1}^{n} A(t_k)\Delta x = R(P, T, A).$$

Since $\int_a^b A(x)\, dx$ is the unique number approximated by these Riemann sums, we must have $V = \int_a^b A(x)\, dx$. We state this conclusion and give a more detailed proof.

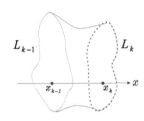

Figure 3.7.5
The Slice S_k

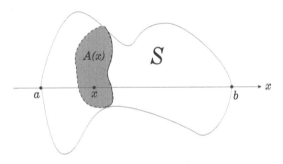

Figure 3.7.6 Cross-section of S with area $A(x)$

Proposition 3.7.7 *Let C_x be the plane perpendicular to the x–axis which passes through the point x on the x–axis. Let S be a solid of volume V which lies between the planes C_a and C_b. Define $A(x)$ as the area of the region given by the intersection of S with C_x. Assume $A(x)$ is a continuous function on the interval $[a, b]$. Then*

$$V = \int_a^b A(x)\, dx. \qquad (3.7.2)$$

Proof By Theorem 3.3.16, $\int_a^b A(x)\, dx$ is the unique number which lies between every lower Riemann sum $L(P, A)$ and every upper Riemann sum $U(P, A)$. Recall the definition of these Riemann sums. $P = \{x_0, \ldots, x_n\}$ is a partition of $[a, b]$. Let $x = t_k^*$ and $x = t_k^{**}$ be the locations of the minimum and maximum values of $A(x)$ on $[x_{k-1}, x_k]$. Then

$$L(P, A) = \sum_{k=1}^{n} A(t_k^*)\, (x_k - x_{k-1}), \qquad U(P, A) = \sum_{k=1}^{n} A(t_k^{**})\, (x_k - x_{k-1}).$$

Let S_k denote the slice of the solid S defined above, and let V_k denote its volume. Then $A(t_k^*)$ is the smallest cross-sectional area of S_k, and $A(t_k^{**})$ is the largest cross-sectional area of S_k. Since S_k has thickness $x_k - x_{k-1}$,

$$A(t_k^*)\, (x_k - x_{k-1}) \leq V_k \leq A(t_k^{**})\, (x_k - x_{k-1}).$$

Sum these inequalities for $1 \leq k \leq n$:

$$\sum_{k=1}^{n} A(t_k^*)\,(x_k - x_{k-1}) \leq \sum_{k=1}^{n} V_k \leq \sum_{k=1}^{n} A(t_k^{**})\,(x_k - x_{k-1})$$
$$L(P, A) \leq V \leq U(P, A).$$

However, $\int_a^b A(x)\,dx$ is the only number which satisfies these inequalities for every partition P. Hence $V = \int_a^b A(x)\,dx$.   □

We apply Proposition 3.7.7 to compute several volumes.

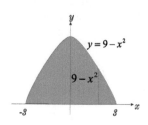

Examples 3.7.8 **(1)** The base of the solid S is the region bounded by the parabola $y = 9 - x^2$ and the x–axis. Each cross–section of S perpendicular to the x–axis is a square with one side in the xy–plane. Find the volume V of S.

Solution The base of the solid S is the region sketched in Figure 3.7.9. The cross-sectional area of S at x is the area of the square with side of length $9 - x^2$, i.e. $A(x) = (9 - x^2)^2$. By (3.7.2):

$$V = \int_{-3}^{3} (9 - x^2)^2\,dx = \int_{-3}^{3} 81 - 18x^2 + x^4\,dx = 81x - 6x^3 + \frac{x^5}{5}\Big|_{-3}^{3} = \frac{1296}{5}.$$

Figure 3.7.9
Base of S

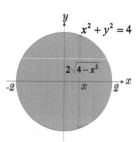

(2) Find the volume V of a pit whose top is the circle $x^2 + y^2 = 4$ and whose vertical cross–sections perpendicular to the x–axis are equilateral triangles.

Solution The equilateral triangle which is the cross–section of this pit with the vertical plane C_x has sides of length $2\sqrt{4 - x^2}$ as depicted in Fig. 3.7.10. Hence the cross–sectional area is $A(x) = (2\sqrt{4 - x^2})^2 \frac{\sqrt{3}}{4} = \sqrt{3}(4 - x^2)$. By (3.7.2):

$$V = \sqrt{3} \int_{-2}^{2} 4 - x^2\,dx = 4\sqrt{3}x - \frac{\sqrt{3}x^3}{3}\Big|_{-2}^{2} = \frac{32\sqrt{3}}{3}.$$

Figure 3.7.10
Top of the Pit

(3) Approximate the volume V of blueberries in a turnover with a flat bottom which is lying on the xy–plane where x and y are measured in inches. The crust has negligible thickness on the bottom of the pastry. Each cross–section of this turnover perpendicular to the x–axis, for $0 \leq x \leq 1$, is a semi–circle having diameter the line segment joining $(x, 0)$ with $(x, \sin \pi x)$. In each of these cross–sections, the top crust of the pastry forms half an annulus of thickness $1/8$ inch.

Solution The bottom of this turnover is shown in the left diagram of Fig. 3.7.11. The radius of the semi–circular cross–section of the blueberry filling determined by the vertical plane C_x is depicted in the right diagram of Fig. 3.7.11. It equals $\frac{1}{2}\left(\sin \pi x - \frac{1}{4}\right)$. Thus,

$$A(x) = \frac{1}{2}\pi \left[\frac{1}{2}\left(\sin \pi x - \frac{1}{4}\right)\right]^2 = \frac{\pi}{8}\left(\sin^2 \pi x - \frac{1}{2}\sin \pi x + \frac{1}{16}\right).$$

The left bound of integration is the solution of $\sin \pi x - \frac{1}{4} = 0$, or $\sin \pi x = \frac{1}{4}$, with πx an angle near zero. Since $\lim_{\theta \to 0} \dfrac{\sin \theta}{\theta} = 1$, use the approximation $\theta \approx \sin \theta$ for θ near zero. Therefore, the left bound of integration is approximately $\pi x = \frac{1}{4}$ or $x = \frac{1}{4\pi}$. (Alternatively, use a calculator to compute $\pi x = \arcsin \frac{1}{4} \approx 0.25 = \frac{1}{4}$.) By symmetry, we compute twice the volume of blueberries in the left half of the turnover where the right bound of integration is $\pi x = \frac{\pi}{2}$ or $x = \frac{1}{2}$. By (3.7.2):

$$
\begin{aligned}
V &\approx 2 \int_{1/4\pi}^{1/2} \frac{\pi}{8} \left(\sin^2 \pi x - \frac{1}{2} \sin \pi x + \frac{1}{16} \right) \, dx \\
&= \frac{\pi}{4} \int_{1/4\pi}^{1/2} \frac{1 - \cos 2\pi x}{2} - \frac{1}{2} \sin \pi x + \frac{1}{16} \, dx \\
&= \int_{1/4\pi}^{1/2} \frac{9\pi}{64} - \frac{\pi}{8} \cos 2\pi x - \frac{\pi}{8} \sin \pi x \, dx \\
&= \frac{9\pi}{64} x - \frac{1}{16} \sin 2\pi x + \frac{1}{8} \cos \pi x \Big|_{1/4\pi}^{1/2} \\
&= \left(\frac{9\pi}{128} - 0 + 0 \right) - \left(\frac{9}{256} - \frac{1}{16} \sin \frac{1}{2} + \frac{1}{8} \cos \frac{1}{4} \right) \approx 0.095 \text{ in}^3. \qquad \square
\end{aligned}
$$

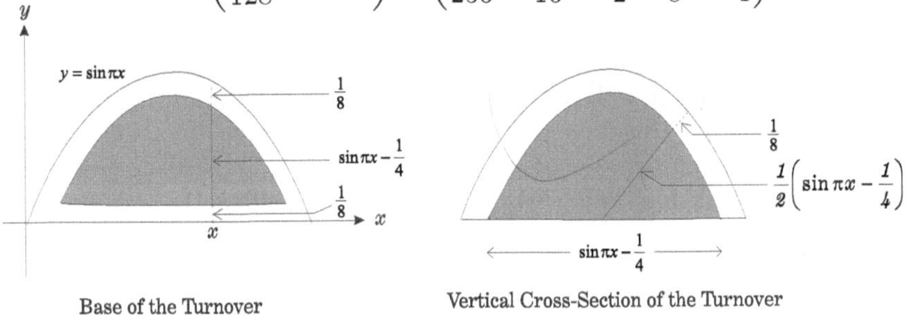

Base of the Turnover | Vertical Cross-Section of the Turnover

Figure 3.7.11 Blueberry Turnover

Solids of Revolution – the Disc Method

We apply the cross-section method to compute volumes of solids of revolution. First we consider solids of revolution with no holes whose cross–sections are discs. Then we study solids of revolution with holes whose cross–sections are annuli.

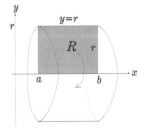

Figure 3.7.12
A Cylinder

A solid of revolution S is obtained by taking a region R in the xy–plane and revolving it around a fixed axis L. Think of R as spinning so fast that the solid S can be seen as the blur of the spinning region. The resulting solid looks as if it has been constructed on a lathe. The simplest case of this construction occurs when L is the x–axis and the region R is bounded by the x–axis, the vertical lines $x = a$, $x = b$ and the graph of $y = f(x)$ which lies entirely above the x–axis. See Figure 3.7.16.

Examples 3.7.14 In each example, consider the region R bounded by the x–axis, the vertical lines $x = a$, $x = b$ and the graph of $y = f(x)$. Describe the solid of revolution S obtained by revolving the region R around the x–axis.

(1) $f(x) = r$ with domain $[a, b]$ and $r > 0$.

Solution The region R is the rectangle depicted in Figure 3.7.12. That diagram shows that the solid of revolution S is a cylinder of radius r and height $b - a$.

(2) $f(x) = kx$ with domain $[0, b]$ and $k > 0$.

Solution The region R is the triangle depicted in Figure 3.7.13. That diagram shows that the solid of revolution S is a cone of radius kb and height b.

(3) $f(x) = \sqrt{r^2 - x^2}$ with domain $[-r, r]$ and $r > 0$.

Solution The region R is the semi–circle depicted in Figure 3.7.15. That diagram shows that the solid of revolution S is a sphere of radius r. $\qquad \square$

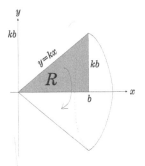

Figure 3.7.13
A Cone

We use the cross–section method of the preceding subsection to compute the volume of a solid of revolution. The resulting formula is called the *disc method* because the cross–sections of a solid of revolution are discs.

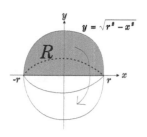

Figure 3.7.15
A Sphere

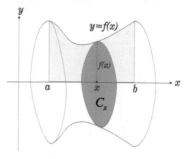

Figure 3.7.16 Cross-Section of a Solid of Revolution

Corollary 3.7.17 [Disc Method]   *Let $y = f(x)$ be a continuous function with domain $[a, b]$ such that $f(x) \geq 0$ for $x \in [a, b]$. Let R be the region bounded by the graph of f, the x–axis and the lines $x = a$, $x = b$. Let S be the solid of revolution obtained by revolving the region R around the x–axis. Then the volume V of S is given by:*

$$V = \pi \int_a^b f(x)^2 \ dx.$$

Proof Apply Proposition 3.7.7 to compute the volume of S. The cross-section obtained by slicing S with the vertical plane C_x perpendicular to the x-axis at the point x is a disc. See Figure 3.7.16. This disc is determined by rotating the line segment of points (x, y), $0 \leq y \leq f(x)$, around the x–axis. Hence it has radius $f(x)$ and area $A(x) = \pi f(x)^2$. By Proposition 3.7.7, $V = \int_a^b A(x) \ dx = \int_a^b \pi f(x)^2 \ dx.$ □

We apply Corollary 3.7.17 to compute the volumes of several solids of revolution.

Examples 3.7.18 (1)   Find the volume V of a cylinder of radius r and height H.

    Solution Consider V as the solid of revolution obtained by revolving the rectangle bounded by the lines $y = 0$, $y = r$, $x = 0$, $x = H$ around the x–axis as in Example 3.7.14 (1). Apply Corollary 3.7.17 with $f(x) = r$:

$$V = \pi \int_0^H r^2 \ dx = \pi r^2 x \big|_0^H = \pi r^2 H.$$

(2) Find the volume V of a cone of radius r and height H.

    Solution Consider V as the solid of revolution obtained by revolving the triangle bounded by the lines $y = 0$, $y = \frac{rx}{H}$, $x = H$ around the x–axis as in Example 3.7.14 (2) with $k = \frac{r}{H}$. Apply Corollary 3.7.17 with $f(x) = \frac{r}{H}x$:

$$V = \pi \int_0^H \left(\frac{r}{H}x\right)^2 \ dx = \frac{\pi r^2}{H^2} \int_0^H x^2 \ dx = \frac{\pi r^2}{H^2} \frac{x^3}{3}\Big|_0^H = \frac{\pi r^2}{H^2} \frac{H^3}{3} = \frac{1}{3}\pi r^2 H.$$

(3) Find the volume V of a sphere of radius r.

    Solution Consider V as the solid of revolution obtained by revolving the semi–circle bounded by $y = 0$ and $y = \sqrt{r^2 - x^2}$ around the x–axis as in

Example 3.7.14 (3). Apply Corollary 3.7.17 with $f(x) = \sqrt{r^2 - x^2}$:

$$V = \pi \int_{-r}^{r} \sqrt{r^2 - x^2}^2 \, dx = \pi \int_{-r}^{r} r^2 - x^2 \, dx = \pi \left(r^2 x - \frac{x^3}{3} \right) \Big|_{-r}^{r}$$

$$= \pi \left(r^3 - \frac{r^3}{3} \right) - \pi \left(-r^3 + \frac{r^3}{3} \right) = \frac{4}{3} \pi r^3.$$

(4) Find the volume V of the solid obtained by revolving the region bounded by the x–axis and the graph of $f(x) = \sin x$, $0 \leq x \leq \pi$, around the x–axis.

Solution By Corollary 3.7.17:

$$V = \pi \int_{0}^{\pi} \sin^2 x \, dx = \pi \int_{0}^{\pi} \frac{1 - \cos 2x}{2} \, dx = \frac{\pi}{2} \left(x - \frac{1}{2} \sin 2x \right) \Big|_{0}^{\pi}$$

$$= \frac{\pi}{2} [(\pi - 0) - (0 - 0)] = \frac{\pi^2}{2}. \qquad \square$$

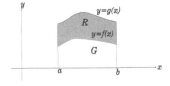
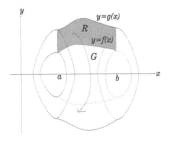

Figure 3.7.19 Solid of Revolution with a Hole

If there is a gap G between the region R and the axis of rotation L then the resulting solid of revolution S will have a hole. See Figure 3.7.19. Rotating the region G around the axis L produces the *hole*, while rotating the region $R \cup G$ around the axis L produces the *total solid* obtained from S by filling in the hole. Therefore, the volume V of S is given by:

$$V = volume\ of\ total\ solid \quad - \quad volume\ of\ the\ hole. \qquad (3.7.3)$$

We make this statement precise.

Corollary 3.7.20 *Let f and g be continuous functions with domain $[a, b]$ such that $0 \leq f(x) \leq g(x)$ for $x \in [a, b]$. Let R be the region bounded by the graphs of f and g as well as the vertical lines $x = a$, $x = b$. The volume V of the solid S obtained by rotating the region R around the x–axis is given by:*

$$V = \pi \int_{a}^{b} g(x)^2 - f(x)^2 \, dx. \qquad (3.7.4)$$

Proof By Corollary 3.7.17, the volume of the total solid is $\pi \int_a^b g(x)^2 \, dx$ and the volume of the hole is $\pi \int_a^b f(x)^2 \, dx$. By (3.7.3):

$$V = \pi \int_{a}^{b} g(x)^2 \, dx - \pi \int_{a}^{b} f(x)^2 \, dx = \pi \int_{a}^{b} g(x)^2 - f(x)^2 \, dx. \qquad \square$$

Warnings **(1)** The solid S obtained by revolving the region between the graphs of f and g around the x–axis is different than the solid S' obtained by revolving the region between the graph of $g - f$ and $y = 0$ around the x-axis. In particular, S and S' have different volumes. The volume V of S is given by (3.7.4) while the volume V' of S' is given by:

$$V' = \pi \int_a^b [g(x) - f(x)]^2 \ dx = \pi \int_a^b g(x)^2 - 2f(x)g(x) + f(x)^2 \ dx.$$

(2) When the region R is revolved around two different axes, we obtain two different solids with different volumes.

Examples 3.7.21 (1) R is the region bounded by the graphs of $f(x) = \frac{x}{2}$ and $g(x) = \sqrt{x}$.
(a) Find the volume V of the solid S obtained by revolving the region R around the x-axis.
(b) Find the volume V' of the solid S' obtained by revolving the region R around the y-axis.

Solution The region R is sketched in Figure 3.7.22. Observe that S is a bowl with a conical hole, while S' is a cone with a curved hole.
(a) The line $y = \frac{x}{2}$ and the parabola $y = \sqrt{x}$ intersect when $\sqrt{x} = \frac{x}{2}$. Squaring: $x = \frac{x^2}{4}$. Hence $x = 0$ or $x = 4$. By Corollary 3.7.20:

$$V = \pi \int_0^4 \sqrt{x}^2 - \left(\frac{x}{2}\right)^2 \ dx = \pi \int_0^4 x - \frac{x^2}{4} \ dx = \frac{\pi x^2}{2} - \frac{\pi x^3}{12}\Big|_0^4 = \frac{8\pi}{3}.$$

(b) Using y as the independent variable, the graph of f is given by $x = 2y$, and the graph of g is given by $x = y^2$. They intersect when $x = 0$ or $x = 4$, i.e. when $y = 0$ or $y = 2$. By Corollary 3.7.20:

$$V' = \pi \int_0^2 (2y)^2 - (y^2)^2 \ dy = \pi \int_0^2 4y^2 - y^4 \ dy = \frac{4\pi y^3}{3} - \frac{\pi y^5}{5}\Big|_0^2 = \frac{64\pi}{15}.$$

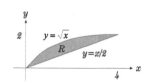

Figure 3.7.22
Example 1

(2) A pearl of radius R has a cylindrical hole of radius r drilled through its center with $r < R$. Find the volume V of the pearl.

Solution Let A be the region bounded by $g(x) = \sqrt{R^2 - x^2}$ and by $f(x) = r$. Revolving A around the x-axis produces this pearl. See Figure 3.7.23. The semi-circle $y = g(x)$ and the line $y = f(x)$ intersect when

$$r = \sqrt{R^2 - x^2}, \quad r^2 = R^2 - x^2 \quad \text{and} \quad x = \pm\sqrt{R^2 - r^2}.$$

By Corollary 3.7.20:

$$
\begin{aligned}
V &= \pi \int_{-\sqrt{R^2-r^2}}^{\sqrt{R^2-r^2}} \sqrt{R^2 - x^2}^2 - r^2 \ dx = \pi \int_{-\sqrt{R^2-r^2}}^{\sqrt{R^2-r^2}} R^2 - x^2 - r^2 \ dx \\
&= \pi(R^2 - r^2)x - \frac{\pi x^3}{3}\Big|_{-\sqrt{R^2-r^2}}^{\sqrt{R^2-r^2}} = 2\left[\pi\left(R^2 - r^2\right)^{3/2} - \frac{\pi}{3}\left(R^2 - r^2\right)^{3/2}\right] \\
&= \frac{4\pi}{3}\left(R^2 - r^2\right)^{3/2}. \qquad \qquad \square
\end{aligned}
$$

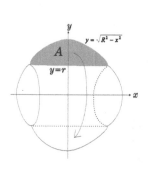

Figure 3.7.23
Pearl with Hole

Solids of Revolution – the Shell Method

The shell method is an alternate to the disc method to compute the volume of a solid of revolution. Sometimes one of these methods produces an integral which is easier to evaluate than the integral produced by the other method. In addition, the shell method applies directly to solids with holes.

Motivating Example 3.7.24 The bottom of the vase S depicted in Figure 3.7.25 is the surface obtained by rotating the curve $y = g(x) = \frac{8x^4}{1+x^4}$ inches, for $0 \le x \le 1$, around the y–axis. The vase has vertical sides and its top edge is a circle at height 10 inches. Find the volume V of this vase.

Solution This vase is obtained by revolving the shaded region A in Figure 3.7.25 around the y–axis. A is bounded by the y–axis, $y = g(x)$, $x = 1$ and $y = 10$ Let P_n be the regular partition of the interval $[0,1]$ into n subintervals of width $\frac{1}{n}$. The points $\frac{k}{n}$, $0 \le k \le n$, of P_n determine the lines $x = \frac{k}{n}$ which divide A into n strips $A_1, \ldots, A_n$ of width $\frac{1}{n}$. See Figure 3.7.26. The solids S_k obtained by rotating the regions S_k around the y–axis divide S into n concentric pieces $S_1, \ldots, S_n$. Approximate A_k by the rectangle R_k of width $x_k - x_{k-1}$ and length $10 - g(x_k)$ depicted in Figure 3.7.27. Then S_k is approximated by the cylindrical shell obtained by rotating the rectangle R_k around the y–axis. This shell is the cylinder of radius x_k with a cylindrical hole of radius x_{k-1}. Both cylinders have height $10 - g(x_k)$. Hence the volume V_k of S_k is approximately the volume of this cylindrical shell, the volume of the outside cylinder minus the volume of the hole:

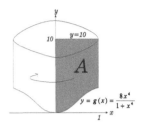

Figure 3.7.25
The Vase S

$$
\begin{aligned}
V_k &\approx \pi x_k^2 \left[10 - g(x_k)\right] - \pi x_{k-1}^2 \left[10 - g(x_k)\right] \\
&= \pi \left[10 - g(x_k)\right] \left(x_k^2 - x_{k-1}^2\right) = \pi \left[10 - g(x_k)\right] \left(x_k + x_{k-1}\right) \left(x_k - x_{k-1}\right) \\
&\approx \pi \left[10 - g(x_k)\right] \left(2x_k\right) \left(x_k - x_{k-1}\right) = 2\pi x_k \left[10 - g(x_k)\right] \left(x_k - x_{k-1}\right)
\end{aligned}
$$

V is approximately the sum of the volumes of these n cylindrical shells:

$$
V = \sum_{k=1}^{n} V_k \approx \sum_{k=1}^{n} 2\pi x_k \left[10 - g(x_k)\right] \left(x_k - x_{k-1}\right) = R(P_n, T_n, f)
$$

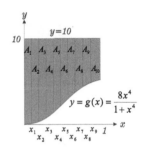

Figure 3.7.26
The n Strips $(n = 10)$

where $f(x) = 2\pi x \left[10 - g(x)\right]$ and $T_n = \{x_1, \ldots, x_n\}$. As n gets larger, the cylindrical shells get thinner and our approximations of V get better. Hence

$$
V = \lim_{n \to \infty} R(P_n, T_n, f) = \int_0^1 f(x)\, dx
$$

$$
V = \int_0^1 2\pi x \left[10 - g(x)\right]\, dx \tag{3.7.5}
$$

$$
V = \int_0^1 2\pi x \left(10 - \frac{8x^4}{1 + x^4}\right) dx.
$$

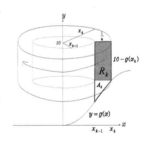

Figure 3.7.27
Cylindrical Shell S_k

To evaluate this integral, let $u = x^2$ with $du = 2x\, dx$:

$$
\begin{aligned}
V &= \int_0^1 20\pi x - \frac{8\pi x^4}{1 + x^4}\, 2x\, dx = 10\pi x^2 \big|_0^1 - 8\pi \int_0^1 \frac{u^2}{1 + u^2}\, du \\
&= 10\pi - 8\pi \int_0^1 1 - \frac{1}{1 + u^2}\, du = 10\pi - 8\pi \left(u - \arctan u \big|_0^1\right) \\
&= 10\pi - 8\pi(1 - \arctan 1) = 10\pi - 8\pi \left(1 - \frac{\pi}{4}\right) = 2\pi(\pi + 1) \text{ inches}^3. \quad \square
\end{aligned}
$$

Let A be any region bounded below by a curve $y = g(x)$ and bounded above by a curve $y = h(x)$. The procedure of the preceding example applies to the solid S obtained by revolving A around the y–axis. Formula (3.7.5) generalizes to give the volume of S with 10 replaced by the function $h(x)$.

Theorem 3.7.28 [Shell Method] *Let g, h be continuous functions with domain $[a, b]$ such that $0 \le a \le b$ and $g(x) \le h(x)$ for $x \in [a, b]$. Let A be the region in the xy–plane* *bounded by the graphs of g, h and the vertical lines $x = a$, $x = b$. Let S be the solid of revolution obtained by revolving A around the y–axis. Then the volume V of S is given by:*

$$V = 2\pi \int_a^b x\,[h(x) - g(x)]\ dx. \tag{3.7.6}$$

Proof There are serious technical problems that must be overcome to give a rigorous derivation of this formula. We present a proof in Section 12. We present one step of this proof here: the general case of this theorem is a consequence of the special case when the graph of g is the x–axis. Then we give a proof of this special case when h is an increasing function on $[a, b]$.

Assume (3.7.6) is true when $g(x) = 0$, i.e. the graph of g is the x–axis:

$$V = 2\pi \int_a^b xh(x)\ dx. \tag{3.7.7}$$

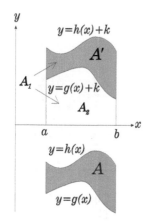

We show that the general case of formula (3.7.6) follows. Let k be a large positive number so that the graph of $y = g(x) + k$ lies entirely above the x–axis. See Figure 3.7.29. Let V_1 be the volume of the solid S_1 obtained by rotating the region A_1 between $y = h(x) + k$ and $y = 0$ around the y–axis. Let V_2 be the volume of the solid S_2 obtained by rotating the region A_2 between $y = g(x) + k$ and $y = 0$ around the y–axis. Let A' be the set of points of A_1 which are not in A_2. Note A' is the region A translated k units upwards. Therefore the solid S' obtained by rotating the region A' around the y–axis is congruent to the solid S and has volume V. The volume V of S' is the volume V_1 of S_1 minus the volume V_2 of S_2, and formula (3.7.7) applies to both S_1 and S_2. Thus

Figure 3.7.29
Reduction to $g(x) = 0$

$$V = V_1 - V_2 = 2\pi \int_a^b x\,[h(x) + k]\ dx - 2\pi \int_a^b x\,[g(x) + k]\ dx = 2\pi \int_a^b x\,[h(x) - g(x)]\ dx$$

as asserted in (3.7.6).

Next, we prove (3.7.7) when h is increasing on $[a, b]$. Let $P = \{x_0, \ldots, x_n\}$ be a partition of the interval $[a, b]$. The lines $x = x_k$, $0 \le k \le n$, divide A into n strips $A_1, \ldots, A_n$. Let S_k be the shell obtained by rotating A_k around the y–axis. Then $S_1, \ldots, S_n$ divide S into n shells. We approximate the volume V_k of S_k in two ways.

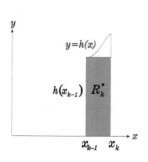

To make the first approximation of V_k, let R_k^* be the rectangle bounded by $x = x_{k-1}$, $x = x_k$, $y = h(x_{k-1})$ and the x–axis. See Figure 3.7.30. R_k^* is contained in A_k, and the volume V_k of S_k is greater than or equal to the volume V_k^* of the cylindrical shell S_k^*, depicted in Figure 3.7.31, obtained by rotating R_k^* around the y–axis. S_k^* is a cylinder of radius x_k and height $h(x_{k-1})$ with a cylindrical hole of radius x_{k-1} and height $h(x_{k-1})$. Hence

Figure 3.7.30
Rectangle R_k^*

$$V_k \ge V_k^* = \pi x_k^2 h(x_{k-1}) - \pi x_{k-1}^2 h(x_{k-1}) = \pi h(x_{k-1}) \left(x_k^2 - x_{k-1}^2\right)$$
$$V_k \ge \pi h(x_{k-1}) \left(x_k + x_{k-1}\right) \left(x_k - x_{k-1}\right) \ge \pi h(x_{k-1}) \left(2x_{k-1}\right) \left(x_k - x_{k-1}\right).$$

Since h is increasing it has its minimum value on $[x_{k-1}, x_k]$ at $x = x_{k-1}$. Thus

$$V = \sum_{k=1}^n V_k \ge \sum_{k=1}^n 2\pi x_{k-1} h(x_{k-1}) \left(x_k - x_{k-1}\right) = L(P, f) \tag{3.7.8}$$

where $f(x) = 2\pi x h(x)$.

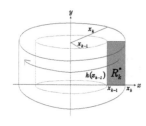

Figure 3.7.31
Cylindrical Shell S_k^*

To make the second approximation of V_k, let R_k^{**} be the rectangle bounded by $x = x_{k-1}$, $x = x_k$, $y = h(x_k)$ and the x–axis. See Figure 3.7.32. R_k^{**} contains A_k

and the volume V_k of S_k is less than or equal to the volume V_k^{**} of the cylindrical shell S_k^{**}, depicted in Fig. 3.7.33, obtained by rotating R_k^{**} around the y–axis. S_k^{**} is the cylinder of radius x_k and height $h(x_k)$ with a cylindrical hole of radius x_{k-1} and height $h(x_k)$. Hence

$$V_k \;\leq\; V_k^{**} = \pi x_k^2 h(x_k) - \pi x_{k-1}^2 h(x_k) = \pi h(x_k)\left(x_k^2 - x_{k-1}^2\right)$$
$$V_k \;\leq\; \pi h(x_k)\left(x_k + x_{k-1}\right)\left(x_k - x_{k-1}\right) \leq \pi h(x_k)\left(2x_k\right)\left(x_k - x_{k-1}\right).$$

Since h is increasing on $[x_{k-1}, x_k]$ it has its maximum value at $x = x_k$. Thus

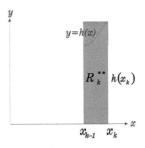

Figure 3.7.32
Rectangle R_k^{**}

$$V = \sum_{k=1}^{n} V_k \leq \sum_{k=1}^{n} 2\pi x_k h(x_k)\left(x_k - x_{k-1}\right) = U(P, f). \qquad (3.7.9)$$

By (3.7.8) and (3.7.9):

$$L(P, f) \leq V \leq U(P, f)$$

for every partition P of $[a, b]$. However, $\int_a^b f(x)\,dx$ is the only number with this property. Hence $V = \int_a^b f(x)\,dx = \int_a^b 2\pi x h(x)\,dx$. $\qquad\square$

We give several examples of computing volumes by the shell method.

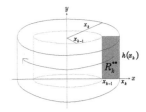

Figure 3.7.33
Cylindrical Shell S_k^{**}

Examples 3.7.34 (1) The roof of a concert hall is the surface of revolution obtained by revolving the graph of $y = \frac{1}{\sqrt{1+x^2}}$ feet, $0 \leq x \leq 1{,}000$, around the y–axis. Find the volume V of this concert hall.

Solution Apply the shell method (3.7.6) with $g(x) = 0$ and $h(x) = \frac{1}{\sqrt{1+x^2}}$:

$$V = 2\pi \int_0^{1{,}000} x \cdot \frac{1}{\sqrt{1+x^2}}\,dx.$$

To evaluate this integral, make the substitution $u = 1 + x^2$ with $du = 2x\,dx$:

$$V \;=\; \pi \int_{x=0}^{x=1{,}000} \frac{1}{\sqrt{1+x^2}} 2x\,dx = \pi \int_{u=1}^{u=1{,}000{,}001} \frac{1}{\sqrt{u}}\,du = \pi \int_1^{1{,}000{,}001} u^{-1/2}\,du$$

$$=\; 2\pi u^{1/2}\Big|_1^{1{,}000{,}001} \;=\; 2\pi\sqrt{1{,}000{,}001} - 2\pi \;\approx\; 6280 \text{ ft}^3.$$

(2) A glass egg is described as the solid of revolution obtained by revolving the ellipse $\frac{x^2}{a^2} + \frac{y^2}{b^2} = 1$ around the y–axis. This egg is packed in a cylindrical box of radius a and height $2b$. Find the volume V of sawdust required to fill the space between the egg and the box.

Solution By symmetry V is twice the volume of sawdust required to fill the top half of the box. The latter volume is the volume of the surface of revolution obtained by revolving the region in Figure 3.7.35 around the y–axis. Note the equation of the top half of the egg is the positive square–root obtained by solving the equation of the ellipse for y:

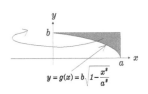

Figure 3.7.35
Glass Egg

$$y = b\sqrt{1 - \frac{x^2}{a^2}}.$$

Apply the shell method (3.7.6) with $g(x) = b\sqrt{1 - \frac{x^2}{a^2}}$ and $h(x) = b$:

$$V = 2\left[2\pi \int_0^a x\left(b - b\sqrt{1 - \frac{x^2}{a^2}}\right) dx\right] = 4\pi b \int_0^a x\,dx - 4\pi b \int_0^a x\sqrt{1 - \frac{x^2}{a^2}}\,dx$$

To evaluate the second integral, make the substitution $u = 1 - \frac{x^2}{a^2}$ with $du = -\frac{2}{a^2}x$

$$
\begin{aligned}
V &= 2\pi bx^2 \big|_0^a + 2\pi ba^2 \int_{x=0}^{x=a} \sqrt{1 - \frac{x^2}{a^2}} \left(-\frac{2}{a^2}x\right) \, dx = 2\pi ba^2 + 2\pi a^2 b \int_{u=1}^{u=0} u \\
&= 2\pi a^2 b + 2\pi a^2 b \frac{u^{3/2}}{3/2} \bigg|_1^0 = 2\pi a^2 b - \frac{4\pi a^2 b}{3} = \frac{2\pi a^2 b}{3}.
\end{aligned}
$$

(3) Find the volume V of the solid of revolution obtained by rotating the region R bounded by the graph of $y = \sqrt{\arcsin x}$, the x–axis and the line $x = 1$ around the x–axis.

Solution The region R is depicted in Figure 3.7.36. If we try to find V by the disc method (Corollary 3.7.17) we obtain the integral

$$
V = \pi \int_0^1 \arcsin x \, dx
$$

which we do not know how to evaluate. However, we can find this volume using the shell method. First, we solve $y = \sqrt{\arcsin x}$ for x:

$$
y^2 = \arcsin x \quad \text{and} \quad x = \sin(y^2).
$$

Now apply the shell method, viewing the region R as bounded by the line $x = 1$ on top and the graph of $x = \sin(y^2)$ on the bottom. See Figure 3.7.37. Note that when $x = 1$, $y = \sqrt{\arcsin 1} = \sqrt{\frac{\pi}{2}}$. By the shell method (3.7.6):

$$
V = 2\pi \int_0^{\sqrt{\pi/2}} y \left[1 - \sin(y^2)\right] \, dy = 2\pi \int_0^{\sqrt{\pi/2}} y \, dy - 2\pi \int_0^{\sqrt{\pi/2}} y \sin(y^2) \, dy.
$$

To evaluate the second integral, use the substitution $u = y^2$ with $du = 2y \, dy$:

$$
\begin{aligned}
V &= \pi y^2 \big|_0^{\sqrt{\pi/2}} - \pi \int_{y=0}^{y=\sqrt{\pi/2}} \sin(y^2) \, (2y) \, dy = \frac{\pi^2}{2} - \pi \int_{u=0}^{u=\pi/2} \sin u \, du \\
&= \frac{\pi^2}{2} + \pi \cos u \bigg|_0^{\pi/2} = \frac{\pi^2}{2} - \pi. \qquad \square
\end{aligned}
$$

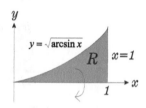

Figure 3.7.36
V by the Disc Method

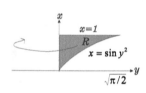

Figure 3.7.37
V by the Shell Method

Summary

The reader should be able to compute the volume of a solid from its cross-sectional areas. She should also know how to calculate volumes of solids of revolution by the disc method and by the shell method.

Basic Exercises

1. The base of each solid below is the region in the xy–plane bounded by the x–axis, the graph of $y = \sqrt{x}$ and the line $x = 3$. Find the volume of each solid.
(a) Each cross–section of S_1 perpendicular to the x–axis is a square with one side in the xy–plane.
(b) Each cross–section of S_2 perpendicular to the x–axis is a semi–circle with diameter in the xy–plane.
(c) Each cross–section of S_3 perpendicular to the x–axis is an equilateral triangle with one side in the xy–plane.
(d) Each cross–section of S_4 perpendicular to the y–axis is a square with one side in the xy–plane.

(e) Each cross-section of S_5 perpendicular to the y–axis is an isosceles right triangles with hypotenuse in the xy–plane.

2. The intersection of each solid below with the xy–plane is the region bounded by the lines $y = 2x$, $y = 6 - x$ and the x–axis. Find the volume of each solid.
(a) Each cross–section of S_1 perpendicular to the x–axis is a rectangle of length ten with one side in the xy–plane.
(b) Each cross–section of S_2 perpendicular to the x–axis is a circle with a diameter in the xy–plane.
(c) Each cross–section of S_3 perpendicular to the x–axis is an isosceles right triangle with hypotenuse in the xy–plane.
(d) Each cross–section of S_4 perpendicular to the y–axis is a square with one side in the xy–plane.
(e) Each cross–section of S_5 perpendicular to the y–axis is a semi–circle with diameter in the xy–plane.

3. The base of each solid below is the region in the xy–plane bounded by the parabolas $y = x^2 + 3$ and $y = 11 - x^2$. Find the volume of each solid.
(a) Each cross–section of S_1 perpendicular to the x–axis is a rectangle with one side in the xy–plane which is four times as long as the other side.
(b) Each cross–section of S_2 perpendicular to the x–axis is a semi–circle with diameter in the xy–plane.
(c) Each cross–section of S_3 perpendicular to the x–axis is a right triangle, with an angle of $\frac{\pi}{6}$, whose hypotenuse lies in the xy–plane.
(d) Each cross–section of S_4 perpendicular to the y–axis is a square with one side in the xy–plane.
(e) Each cross–section of S_5 perpendicular to the y–axis is an equilateral triangle with one side in the xy–plane.

4. The base of each solid below is the region in the xy–plane inside the ellipse

$$\frac{x^2}{a^2} + \frac{y^2}{b^2} = 1.$$

Find the volume of each solid.
(a) Each cross–section of S_1 perpendicular to the x–axis is a square with one side in the xy–plane.
(b) Each cross–section of S_2 perpendicular to the x–axis is a semi–circle with diameter in the xy–plane.
(c) Each cross–section of S_3 perpendicular to the x–axis is an equilateral triangle with one side in the xy–plane.
(d) Each cross–section of S_4 perpendicular to the y–axis is a square with one side in the xy–plane.
(e) Each cross–section of S_5 perpendicular to the y–axis is an isosceles right triangle with hypotenuse in the xy–plane.

5. (a) Find the volume of the pyramid of height H with base a square whose sides have length S.
(b) Find the volume of the pyramid of height H with base an equilateral triangle whose sides have length S.

6. Use the disc method to find the volume of each solid of revolution obtained by revolving the given region around the x–axis.
(a) R_1 is bounded by the graph of $f(x) = x^3$, the x–axis and the line $x = 2$.
(b) R_2 is bounded by the graph of $g(x) = \sqrt{x - 3}$, the x–axis and the line $x = 5$.
(c) R_3 is bounded by the graph of $h(x) = \cos x$, $-\frac{\pi}{2} \le x \le \frac{\pi}{2}$, and the x–axis.
(d) R_4 is the region in the first quadrant bounded by the graph of $i(x) = \sec x$, $x \in \left[0, \frac{\pi}{3}\right]$ and the line $x = \pi/3$.

(e) R_5 is bounded by the graph of $j(x) = x^2$ and the line $y = 4$.

(f) R_6 is bounded by the graphs of $k(x) = x^2 + x + 2$ and $m(x) = 2x + 4$.

(g) R_7 is bounded by the graphs of $n(x) = \csc x$, $0 < x < \pi$, and the line $y = 2$.

(h) R_8 is bounded by the graphs of $p(x) = x^3 + 10$ and $q(x) = 4x + 10$.

(i) R_9 is bounded by the graphs of $r(x) = 1 + \sin x$ and $s(x) = 1 + \cos x$, both with domain $[0, 2\pi]$.

(j) R_{10} is bounded by the x–axis as well as the semi–circles $t(x) = \sqrt{9 - x^2}$ and $u(x) = \sqrt{4 - x^2}$.

7. Use the disc method to find the volume of each solid of revolution obtained by revolving the given region around the y–axis.

(a) R_1 is bounded by the line $y = 5x$, the y–axis and the line $y = 10$.

(b) R_2 is bounded by the graph of $f(x) = \sqrt{x}$, the y–axis and the line $y = 5$.

(c) R_3 is bounded by the lines $y = 3x$, $y = 15 - 2x$ and the y–axis.

(d) R_4 is bounded by the curves $x = y^2 + 8$ and $x = 16 - y^2$.

(e) R_5 is the region in the first quadrant bounded by $y = x^3$ and the line $y = 9x$.

8. Use the shell method to find the volume of each solid of revolution obtained by revolving the given region around the y–axis.

(a) R_1 is bounded by the graph of $f(x) = x^2 - x$ and the x–axis.

(b) R_2 is the region in the first quadrant bounded by the graph of $g(x) = \sqrt{1 + x^2}$ and the line $x = 3$.

(c) R_3 is bounded by the graph of $h(x) = \sqrt{x - 4}$, the x–axis and the line $x = 20$.

(d) R_4 is bounded by the graph of $i(x) = x^2$ and the line $y = 6x$.

(e) R_5 is bounded by the graph of $j(x) = x^3 - 5x^2 + 7x$ and the line $y = 3x$.

(f) R_6 is bounded by the hyperbola $x^2 - y^2 = 144$ and the line $y = 2x - 21$.

(g) R_7 is bounded by the ellipse $\frac{x^2}{a^2} + \frac{y^2}{b^2} = 1$.

(h) R_8 is bounded by $y = \sin(x^2)$ and $y = \cos(x^2)$, both with domain $[0, 2]$.

(i) R_9 is bounded by $x = y^2$ and the line $x = 9$.

(j) R_{10} is bounded by $x = y^2$ and the line $y = x - 2$.

9. Use the shell method to find the volume of each solid of revolution obtained by revolving the given region around the x–axis.

(a) R_1 is bounded by the graph of $f(x) = x^2$ and the line $y = 3x$.

(b) R_2 is bounded by the graphs of $g(x) = x^2 + 1$ and $h(x) = 9 - x^2$.

(c) R_3 is bounded by the curve $y^2 - x^2 = 9$ and the line $y = 5$.

(d) R_4 is bounded by the y–axis and the lines $y = 2x + 1$, $y = 9 - 2x$.

(e) R_5 is bounded by graph $i(x) = \sqrt{\operatorname{arcsec} \sqrt{x}}$, $x \in [1, 4]$, the x–axis and $x = 4$.

Challenging Problems

1. Find the volume of blueberries in the turnover of Example 3.7.8 (3) if the bottom crust of the turnover is also $\frac{1}{8}$ inch thick.

2. (a) R is the region bounded by the graph of the continuous function f on the top, the line $y = c$ on the bottom, the line $x = a$ on the left and the line $x = b$ on the right. Assume R does not intersect the horizontal line $y = c$. S is the solid obtained by revolving the region R around $y = c$. Show the volume V of S is given by:

$$V = \pi \int_a^b [f(x) - c]^2 \ dx.$$

(b) Let f, g be continuous functions with $f(x) \le g(x)$ for $x \in [a, b]$. R is the region bounded by the graph of g on the top, the graph of f on the bottom, the line $x = a$ on the left and the line $x = b$ on the right. Assume R does not intersect the horizontal line $y = c$. S is the solid obtained by revolving the region R around $y = c$. Show the volume V of S is given by:

$$V = \pi \int_a^b [g(x) - c]^2 - [f(x) - c]^2 \ dx.$$

3. Use the formulas of Problem 2 to find the volume of each solid of revolution obtained by revolving the given region around the line L.
(a) R_1 is bounded by the graph of $f(x) = 16 - x^2$ and $y = 7$. L is the line $y = 2$.
(b) R_2 is bounded by the graph of $g(x) = x^3$, $x \geq 0$, and $y = 4x$. L is the line $y = -4$.
(c) R_3 is bounded by graph $h(x) = \cos x$, $0 \leq x \leq 2\pi$, and $y = 1$. L is the line $y = 1$.
(d) R_4 is bounded by the graph of $i(x) = \frac{1}{1+x^2}$, the x–axis, the y–axis and $x = 2$. L is the line $y = 3$.
(e) R_5 is bounded by $x = y^2$ and $x = 9$. L is the line $x = -4$.
(f) R_6 is bounded by the graph of $j(x) = x^2$ and $y = x + 2$. L is the line $x = -3$.

4. Assume g, h are continuous functions with $g(x) \leq h(x)$ for $x \in [a, b]$. Let R be the region bounded by the graph of h on the top, the graph of g on the bottom, the line $x = a$ on the left and the line $x = b$ on the right. Assume the vertical line $x = c$ lies to the left of the region R. Let S be the solid obtained by revolving the region R around $x = c$. Show the volume V of S is given by:

$$V = 2\pi \int_a^b (x - c) [h(x) - g(x)] \ dx.$$

5. Use the formula of Problem 4 to find the volume of each solid of revolution obtained by revolving the given region around the line L.
(a) R_1 is bounded by the graph of $f(x) = (x - 8)^2$ and $y = 4$. L is the line $x = 1$.
(b) R_2 is bounded by $y = 3x + 2$, $y = 2x - 1$, $x = 5$ and $x = 11$. L is the line $x = 2$.
(c) R_3 is bounded by $y = x^2 - 20x + 25$ and $y = 5 - 6x - x^2$. L is the line $x = -1$.
(d) R_4 is bounded by $x = y^2 - 4$ and $y = x - 2$. L is the line $y = 2$.
(e) R_5 is bounded by the parabola $y = x^2$ and $y = x + 20$. L is the line $y = -2$.

6. $y = f(x)$ is a continuous one–to–one function with domain $[a, b]$ whose graph lies in the first quadrant. Assume $f(a) = a$ and $f(b) = b$. Let R be the region bounded by the graphs of $y = f(x)$ and $y = f^{-1}(x)$. S is the solid obtained by revolving R around the x–axis, and T is the solid obtained by revolving R around the y–axis. Show S and T have the same volume.

Applications

3.8   Approximating Area

Prerequisite: Section 3.4

There are many definite integrals $\int_a^b f(x)\,dx$ which cannot be evaluated as explicit decimals. The problem may be that there is no formula, in terms of the functions we know, for an $F(x)$ with $F'(x) = f(x)$. Even if there is an exact expression for this integral, it may involve numbers such as π or trigonometric functions of angles, whose exact values are not known. Alternatively, the function f may result from data of an experiment so that there is no formula for $f(x)$. Two methods are presented for approximating the area $\int_a^b f(x)\,dx$. First we present rectangle rules, and then we study Simpson's rule. We illustrate them by approximating the integrals

$$\int_{-1}^{1} \frac{2}{1+x^2}\,dx = 2\arctan x\big|_{-1}^{1} = \pi \quad \text{and} \quad L(2) = \int_{1}^{2} \frac{1}{x}\,dx.$$

We also present formulas which give bounds on the errors of our approximations.[4]

Rectangle Rules

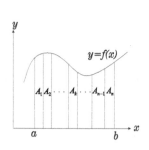

We divide the given region of area A into narrow strips. A rectangle rule approximates each narrow strip by a rectangle. We give six methods for defining the lengths of these rectangles which determine six rectangle rules for approximating A. After illustrating these six rules, we present formulas which bound the errors of the estimates of A produced by two of them.

Figure 3.8.1
The n Strips

Consider a region of area $A = \int_a^b f(x)\,dx$ bounded by the curve $y = f(x)$, the lines $x = a$, $x = b$ and the x–axis. Use a partition $P = \{x_0, \ldots, x_n\}$ of the interval $[a, b]$ to divide the given region into n strips. Recall this partition is called regular if all n widths $x_k - x_{k-1}$ of the subintervals $[x_{k-1}, x_k]$ equal $\frac{b-a}{n}$. The vertical lines $x = x_k$, for $0 \le k \le n$, divide the given region into n vertical strips of areas $A_1, \ldots, A_n$ as in Fig. 3.8.1. The key idea of this section is to approximate each of these narrow strips by a region whose area we can calculate. A rectangle rule approximates the kth strip by a rectangle of width $x_k - x_{k-1}$ and length L_k. We study 6 ways of selecting the L_k.

Left Endpoint Rule   Let $L_k = f(x_{k-1})$, the value of f on the left endpoint of the interval $[x_{k-1}, x_k]$.

Right Endpoint Rule   Let $L_k = f(x_k)$, the value of f on the right endpoint of the interval $[x_{k-1}, x_k]$.

Midpoint Rule   Let $L_k = f(\frac{x_{k-1}+x_k}{2})$, the value of f on the midpoint of $[x_{k-1}, x_k]$.

Trapezoid Rule   Let $L_k = \frac{f(x_{k-1})+f(x_k)}{2}$, the average value of f on the endpoints of the interval $[x_{k-1}, x_k]$.

Lower Riemann Sum   Let $L_k = f(t_k^*)$, the minimum value of f on $[x_{k-1}, x_k]$.

Upper Riemann Sum   Let $L_k = f(t_k^{**})$, the maximum value of f on $[x_{k-1}, x_k]$.

---

[4] We only consider the error produced by replacing the given region by another region. We do not consider the accumulated rounding–off error which occurs in calculating the approximation.

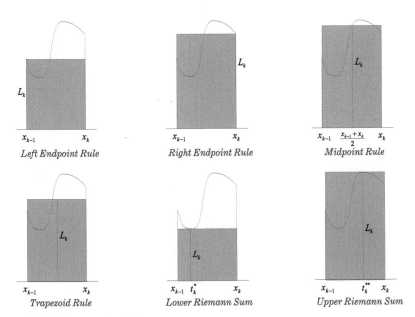

Figure 3.8.2 Rectangle Rules

These rectangles are illustrated in Figure 3.8.2. The trapezoid rule receives its name from the observation that the area of the rectangle of width $x_k - x_{k-1}$ and length $\frac{f(x_{k-1})+f(x_k)}{2}$ is the same as the area of the trapezoid in Figure 3.8.3 with height $x_k - x_{k-1}$ and bases of lengths $f(x_{k-1})$, $f(x_k)$. Note that if the function f is increasing, then the left endpoint rule coincides with the lower Riemann sum and the right endpoint rule coincides with the upper Riemann sum. If the function f is decreasing then the left endpoint rule coincides with the upper Riemann sum while the right endpoint rule coincides with the lower Riemann sum.

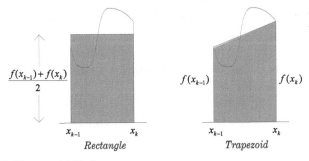

Figure 3.8.3 Trapezoid Rule

In each rectangle rule, the sum of the areas of the n rectangles approximates $\int_a^b f(x)\, dx$.

Proposition 3.8.4 [Rectangle Rules] *Let f be a continuous function with domain $[a, b]$. Let $P = \{x_0, \ldots, x_n\}$ be a partition of the interval $[a, b]$. For each $1 \le k \le n$, let $f(t_k^*)$ denote the minimum value of f on $[x_{k-1}, x_k]$ and let $f(t_k^{**})$ denote its maximum value. Choose L_k from one row of the following table.*

Left Endpoint Rule	$L_k = f(x_{k-1})$
Right Endpoint Rule	$L_k = f(x_k)$

Midpoint Rule	$L_k = f\left(\frac{x_{k-1}+x_k}{2}\right)$
Trapezoid Rule	$L_k = \frac{f(x_{k-1})+f(x_k)}{2}$
Lower Riemann Sum	$L_k = f(t_k^*)$
Upper Riemann Sum	$L_k = f(t_k^{**})$

Then $\displaystyle\int_a^b f(x)\,dx \approx \sum_{k=1}^n L_k(x_k-x_{k-1}) = L_1(x_1-x_0)+\cdots+L_n(x_n-x_{n-1}).$ (3.8.1)

If P is the regular partition of $[a,b]$ into n subintervals then

$$\int_a^b f(x)\,dx \approx \frac{b-a}{n}\sum_{k=1}^n L_k = \frac{b-a}{n}\left[L_1+\cdots+L_n\right]. \qquad (3.8.2)$$

Notes **(1)** The approximation produced by the trapezoid rule for a regular partition can be rewritten as

$$\int_a^b f(x)\,dx \approx \frac{b-a}{2n}\left[f(x_0)+2f(x_1)+2f(x_2)+\cdots+2f(x_{n-2})+2f(x_{n-1})+f(x_n)\right].$$
$$(3.8.3)$$

(2) We only use the hypothesis that f is continuous to define the lower and upper Riemann sums. The left endpoint, right endpoint, midpoint and trapezoid rules are defined even when f is not continuous.

(3) In Section 2, we used a seventh rectangle rule to estimate $\int_a^b f(x)\,dx$: the average of the lower and upper Riemann sums defined by taking $L_k = \frac{f(t_k^*)+f(t_k^{**})}{2}$.

We apply these rectangles rules to estimate π and $L(2)$.

Examples 3.8.5 (1) Estimate $\pi = \int_{-1}^1 \frac{2}{1+x^2}\,dx$ by each of the six rectangle rules using the regular partition of $[-1,1]$ into four subintervals.

Solution We use the partition $P = \left\{-1,\ -\frac{1}{2},\ 0,\ \frac{1}{2},\ 1\right\}$ with $f(x) = \frac{2}{1+x^2}$. The values of the L_k are displayed in the following table.

	Left Endpoint Rule	Right Endpoint Rule	Midpoint Rule	Trapezoid Rule	Lower Riemann Sum	Upper Riemann Sum
L_1	1	8/5	32/25	13/10	1	8/5
L_2	8/5	2	32/17	9/5	8/5	2
L_3	2	8/5	32/17	9/5	8/5	2
L_4	8/5	1	32/25	13/10	1	8/5

In all cases : $\displaystyle\int_{-1}^1 \frac{1}{1+x^2}\,dx \approx \frac{1}{2}\left[L_1+L_2+L_3+L_4\right].$

Use the values of L_k from the above table, to approximate $\int_{-1}^1 \frac{2}{1+x^2}\,dx$ by the six rectangle rules.

Left Endpoint Rule $\displaystyle\int_{-1}^1 \frac{2}{1+x^2}\,dx \approx \frac{1}{2}\left[1+\frac{8}{5}+2+\frac{8}{5}\right] = 3.10$

Right Endpoint Rule $\int_{-1}^{1} \frac{2}{1+x^2}\,dx \approx \frac{1}{2}\left[\frac{8}{5} + 2 + \frac{8}{5} + 1\right] = 3.10$

Midpoint Rule $\int_{-1}^{1} \frac{2}{1+x^2}\,dx \approx \frac{1}{2}\left[\frac{32}{25} + \frac{32}{17} + \frac{32}{17} + \frac{32}{25}\right] \approx 3.16$

Trapezoid Rule $\int_{-1}^{1} \frac{2}{1+x^2}\,dx \approx \frac{1}{2}\left[\frac{13}{10} + \frac{9}{5} + \frac{9}{5} + \frac{13}{10}\right] \approx 3.10$

Lower Riemann Sum $\int_{-1}^{1} \frac{2}{1+x^2}\,dx \approx \frac{1}{2}\left[1 + \frac{8}{5} + \frac{8}{5} + 1\right] = 2.60$

Upper Riemann Sum $\int_{-1}^{1} \frac{2}{1+x^2}\,dx \approx \frac{1}{2}\left[\frac{8}{5} + 2 + 2 + \frac{8}{5}\right] = 3.60$

The rectangles used to make these approximations are depicted in Figure 3.8.6.

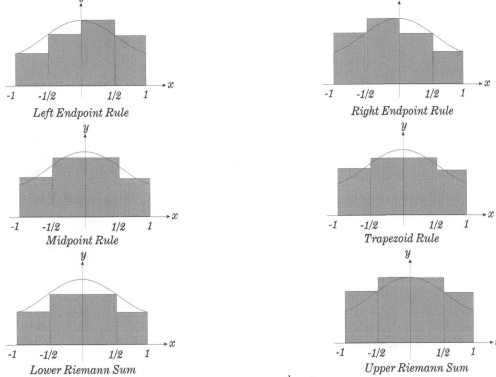

Figure 3.8.6 Rectangle Rules to Approximate $\int_{-1}^{1} \frac{2}{1+x^2}\,dx$

(2) Estimate $L(2) = \int_{1}^{2} \frac{1}{x}\,dx$ by each of the six rectangle rules using the regular partition of $[1,2]$ into five subintervals.

Solution We use the partition $P = \left\{1, \frac{6}{5}, \frac{7}{5}, \frac{8}{5}, \frac{9}{5}, 2\right\}$ with $f(x) = \frac{1}{x}$. The values of the L_k are displayed in the following table.

	Left Endpoint Rule	Right Endpoint Rule	Midpoint Rule	Trapezoid Rule	Lower Riemann Sum	Upper Riemann Sum
L_1	1	5/6	10/11	11/12	5/6	1
L_2	5/6	5/7	10/13	65/84	5/7	5/6
L_3	5/7	5/8	2/3	75/112	5/8	5/7
L_4	5/8	5/9	10/17	85/144	5/9	5/8
L_5	5/9	1/2	10/19	19/36	1/2	5/9

In all cases : $\qquad \int_{1}^{2} \frac{1}{x}\,dx \approx \frac{1}{5}\left[L_1 + L_2 + L_3 + L_4 + L_5\right].$

Since f is a decreasing function, the values of the L_k for the right endpoint rule and the lower Riemann sum coincide. Also, the values of the L_k for the left endpoint rule and the upper Riemann sum coincide. Use the values of the L_k from the above table, to approximate $\int_1^2 \frac{1}{x}\, dx$ by the six rectangle rules.

Left Endpoint Rule and

Upper Riemann Sum $\quad \int_1^2 \frac{1}{x}\, dx \quad \approx \frac{1}{5}\left[1 + \frac{5}{6} + \frac{5}{7} + \frac{5}{8} + \frac{5}{9} \right] \approx .746$

Right Endpoint Rule and

Lower Riemann Sum $\quad \int_1^2 \frac{1}{x}\, dx \quad \approx \frac{1}{5}\left[\frac{5}{6} + \frac{5}{7} + \frac{5}{8} + \frac{5}{9} + \frac{1}{2} \right] \approx .646$

Midpoint Rule $\quad \int_1^2 \frac{1}{x}\, dx \quad \approx \frac{1}{5}\left[\frac{10}{11} + \frac{10}{13} + \frac{2}{3} + \frac{10}{17} + \frac{10}{19} \right] \approx .692$

Trapezoid Rule $\quad \int_1^2 \frac{1}{x}\, dx \quad \approx \frac{1}{5}\left[\frac{11}{12} + \frac{65}{84} + \frac{75}{112} + \frac{85}{144} + \frac{19}{36} \right] \approx .$

The rectangles used to make these approximations are depicted in Fig. 3.8.7. □

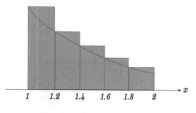

Left Endpoint Rule and Upper Riemann Sum

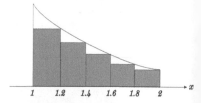

Right Endpoint Rule and Lower Riemann Sum

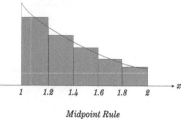

Midpoint Rule

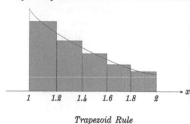

Trapezoid Rule

Figure 3.8.7 Rectangle Rules to Approximate $\int_1^2 \frac{1}{x}\, dx$

We make four observations on the approximations given by these six rectangle rules for a fixed partition P and continuous function f.

(A) The lower Riemann sum always gives the smallest of these six approximations while the upper Riemann sum always gives the largest approximation.

Justification The left endpoint, right endpoint and midpoint rules produce approximations of $\int_a^b f(x)\, dx$ which are Riemann sums. Every Riemann sum lies between the the lower and upper Riemann sums $L(P, f)$ and $U(P, f)$. The approximation of the trapezoid rule is the average of the approximations of the left endpoint and right endpoint rectangle rules. The average of two numbers between $L(P, f)$ and $U(P, f)$ is also between $L(P, f)$ and $U(P, f)$.

(B) The error in the approximation given by one of these rectangle rules is less than the difference between the lower and upper Riemann sums $U(P, f) - L(P, f)$.

Justification Each of these rectangle rule approximations R as well as $\int_a^b f(x)\, dx$ lie between $L(P, f)$ and $U(P, f)$. Hence $U(P, f) - L(P, f)$ is greater than or equal to the difference between R and $\int_a^b f(x)\, dx$.

(C) If f is either increasing on $[a, b]$ or decreasing on $[a, b]$, then the error in the approximation of $\int_a^b f(x)\, dx$ by any of these six rectangle rules is less than or equal to $|f(b) - f(a)|\, \mathrm{mesh}(P)$.

Justification Assume f is increasing. Recall $\mathrm{mesh}(P)$ is the width of the largest subinterval of P. By (B), the error in any of these approximations is less than

$$U(P, f) - L(P, f) = \sum_{k=1}^{n} f(x_k)(x_k - x_{k-1}) \;-\; \sum_{k=1}^{n} f(x_{k-1})(x_k - x_{k-1})$$

$$= \sum_{k=1}^{n} [f(x_k) - f(x_{k-1})](x_k - x_{k-1}) \;\leq\; \sum_{k=1}^{n} [f(x_k) - f(x_{k-1})]\, \mathrm{mesh}(P)$$

$$= [f(x_n) - f(x_0)]\, \mathrm{mesh}(P) \;=\; [f(b) - f(a)]\, \mathrm{mesh}(P).$$

Note all but two of the summands in the right sum of the second line above cancel because each of the $f(x_k)$, $1 \leq k \leq n-1$, appears twice: once with a plus sign and once with a minus sign.

The analogous argument, when f is decreasing, is given as an exercise.

(D) We expect the midpoint and trapezoid rules to give the most accurate approximations of these six rectangle rules. Moreover, we shall see below that the estimate produced by the midpoint rule is twice as accurate as the estimate produced by the trapezoid rule.

Justification When the mesh of P is small, f is usually increasing or decreasing on each subinterval $[x_{k-1}, x_k]$. Then the length of each rectangle used in the midpoint or trapezoid rules is between the minimum and maximum values of f on this subinterval. This produces more accurate approximations than using either the minimum or maximum value of f for the length of this rectangle. □

We apply these observations to the approximations of Examples 3.8.5.

Examples 3.8.8 (1) Check the above four observations for the six approximations of $\pi = \int_{-1}^{1} \frac{2}{1+x^2}\, dx$ made in Example 3.8.5 (1).

(A) The six approximations of π are 2.60, 3.10, 3.10, 3.10, 3.16 and 3.60. The lower Riemann sum 2.60 is the smallest of these approximations while the upper Riemann sum 3.60 is the largest one.

(B) The error of each of these estimates of π is less than $3.60 - 2.60 = 1.00$

(C) The error of each of these approximations on either $[-1, 0]$ or $[0, 1]$ is less than $(2 - 1)\frac{1}{2} = \frac{1}{2}$. Hence the error in each of these approximations of π is less than $\frac{1}{2} + \frac{1}{2} = 1$.

(D) The midpoint rule approximation 3.16 for π is the best of these estimates.

(2) Check the above four observations for the six approximations of $L(2) = \int_1^2 \frac{1}{x}\, dx$ made in Example 3.8.5 (2).

(A) The approximations $L(2)$ are .646, .692, .696 and .746. The lower Riemann sum .646 is the smallest of these approximations while the upper Riemann sum .746 is the largest one.

(B) The error of each of these estimates of $L(2)$ is less than $.746 - .646 = .1$

(C) The error of each of these estimates of $L(2)$ is less than $\left(1 - \frac{1}{2}\right)\frac{1}{5} = .1$

(D) $L(2) \approx .693$ to the nearest thousandth. Therefore the midpoint rule approximation .692 is the best of these estimates. □

We state, without proof, a theorem which gives bounds on the errors of the approximations produced by the midpoint and trapezoid rules.

Theorem 3.8.9 *Let f be a function whose domain contains $[a, b]$. Let P_n be the regular partition of $[a, b]$ into n subintervals. Assume there is a number K such that $f''(x)$ exists and*

$$|f''(x)| \leq K \text{ for } x \in [a, b].$$

(a) *If the midpoint rule produces the approximation M_n for $\int_a^b f(x) \, dx$, then*

$$\left| \int_a^b f(x) \, dx - M_n \right| < \frac{(b-a)^3 K}{24 n^2}. \tag{3.8.4}$$

(b) *If the trapezoid rule produces the approximation T_n for $\int_a^b f(x) \, dx$, then*

$$\left| \int_a^b f(x) \, dx - T_n \right| < \frac{(b-a)^3 K}{12 n^2}. \tag{3.8.5}$$

Note this theorem indicates that the midpoint rule is likely to produce an approximation which is twice as accurate as the approximation produced by the trapezoid rule. We determine the accuracy of our approximations of π and $L(2)$ by the midpoint and trapezoid rules.

Examples 3.8.10 (1)   Determine bounds on the errors in the approximations of $\pi = \int_{-1}^1 \frac{2}{1+x^2} \, dx$ by the midpoint and trapezoid rules in Example 3.8.5 (1).

Solution   Let $f(x) = \frac{2}{1+x^2} = 2(1+x^2)^{-1}$. Then $f'(x) = -4x(1+x^2)^{-2}$. By the product rule:

$$f''(x) = -4(1+x^2)^{-2} + 16x^2(1+x^2)^{-3} = \frac{-4(1+x^2) + 16x^2}{(1+x^2)^3} = \frac{12x^2 - 4}{(1+x^2)^3}.$$

Note that the maximum value of $|12x^2 - 4|$ on $[-1, 1]$ is 8 while the minimum value of $(1+x^2)^3$ on $[-1, 1]$ is 1. Hence $|f''(x)| \leq 8$ for $x \in [-1, 1]$, and we take $K = 8$. Since we are using the regular partition of $[-1, 1]$ into four subintervals, $n = 4$. By (3.8.4), the error of the approximation $M_4 = 3.16$ of π by the midpoint rule is less than

$$\frac{[1 - (-1)]^3 \, 8}{24 \cdot 4^2} = \frac{1}{6}.$$

That is, $|\pi - 3.16| < \frac{1}{6}$. Similarly (3.8.5) says that the error in the approximation $T_4 = 3.10$ of π produced by the trapezoid rule is less than $\frac{1}{3}$, i.e $|\pi - 3.10| < \frac{1}{3}$.

(2)   Determine bounds on the errors in the approximations of $L(2) = \int_1^2 \frac{1}{x} \, dx$ by the midpoint and trapezoid rules in Example 3.8.5 (2).

Solution   Let $f(x) = \frac{1}{x} = x^{-1}$. Then $f'(x) = -x^{-2}$ and $f''(x) = 2x^{-3} = \frac{2}{x^3}$. Hence the maximum value of $|f''(x)|$ on the interval $[1, 2]$ is $K = 2$. For the regular partition of $[1, 2]$ into five subintervals, $n = 5$. By (3.8.4), the error of the approximation $M_5 = .692$ of $L(2)$ by the midpoint rule is less than

$$\frac{(2 - 1)^3 (2)}{24 \cdot 5^2} = \frac{1}{300} \approx .003$$

That is, $|L(2) - .692| < .003$. Similarly (3.8.5) says that the error in the approximation $T_5 = .696$ of $L(2)$ produced by the trapezoid rule is less than $.007$, i.e. $|L(2) - .696| < .007$.   □

Simpson's Rule

Simpson's rule approximates the area $\int_a^b f(x)\,dx$ of a region R by dividing R into $2m$ vertical strips. Each pair of adjacent strips is approximated by replacing its top curve $y = f(x)$ by a segment of a parabola. We derive the formula for approximating an area by Simpson's Rule and give a bound on the error of this approximation.

Let R be the region bounded by $y = f(x)$, $x = a$, $x = b$ and the x–axis. Let $P = \{x_0, \dots, x_{2m}\}$ be the regular partition of $[a, b]$ into $2m$ subintervals. The vertical lines $x = x_k$, for $0 \le k \le 2m$, divide the region R into $2m$ strips. The pair of strips over $[x_{2k-2}, x_{2k}]$ is approximated by a simpler vertical strip in which the top curve $y = f(x)$ is replaced by a parabola $y = q_k(x)$. This parabola is defined by requiring it to pass through three points on the graph of f: the left point $(x_{2k-2}, f(x_{2k-2}))$, the middle point $(x_{2k-1}, f(x_{2k-1}))$, and the right point $(x_{2k}, f(x_{2k}))$. See Figure 3.8.11.

A *quadratic function* q refers to a degree two polynomial $q(x) = Ax^2 + Bx + C$. The graph of a quadratic function is a parabola. The following lemma shows how to define a quadratic function so that its graph passes through three specified points.

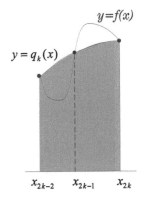

Figure 3.8.11
Simpson's Rule

Lemma 3.8.12 *Let* (a_1, b_1), (a_2, b_2), (a_3, b_3) *be three points with* a_1, a_2, a_3 *distinct numbers. Then the quadratic function*

$$q(x) = \frac{b_1(x - a_2)(x - a_3)}{(a_1 - a_2)(a_1 - a_3)} + \frac{b_2(x - a_1)(x - a_3)}{(a_2 - a_1)(a_2 - a_3)} + \frac{b_3(x - a_1)(x - a_2)}{(a_3 - a_1)(a_3 - a_2)}.$$

satisfies $q(a_1) = b_1$, $q(a_2) = b_2$ *and* $q(a_3) = b_3$.

Proof Clearly the above formula for $q(x)$ defines a quadratic function. When a_1 is substituted for x in the formula for $q(x)$, the first summand becomes b_1, while the last two summands are zero. Thus $q(a_1) = b_1$. Similarly $q(a_2) = b_2$ from the middle summand, and $q(a_3) = b_3$ from the last summand. $\square$

In fact, the quadratic function $q(x)$ of Lemma 3.8.12 is uniquely determined by requiring $q(a_k) = b_k$ for $1 \le k \le 3$. However, we do not need this fact and relegate its proof to a Challenging Problem. The next lemma computes the area of a strip whose top curve is a segment of a parabola.

Lemma 3.8.13 *For any quadratic function* $q(x)$,

$$\int_a^b q(x)\,dx = \frac{b - a}{6}\left[q(a) + 4q\left(\frac{a + b}{2}\right) + q(b)\right].$$

Proof Write $q(x) = \alpha x^2 + \beta x + \gamma$. Let $u = x - \frac{a+b}{2}$ with $du = dx$. If $h = \frac{b-a}{2}$, then $x = u + h + a$. For any constant K, $y = Ku$ is an odd function, and $\int_{-h}^h Ku\,du = 0$. Thus our integral becomes:[5]

$$\int_{x=a}^{x=b} q(x)\,dx = \int_{u=-h}^{u=h} \alpha(u + h + a)^2 + \beta(u + h + a) + \gamma\,du$$

$$= \int_{-h}^h \alpha u^2 + \alpha(h + a)^2 + \beta(h + a) + \gamma\,du$$

$$+ \int_{-h}^h [2\alpha(h + a) + \beta]\,u\,du$$

---

[5] We use *integration by substitution* from Section 3.6 in the first step of this proof.

$$
\begin{aligned}
&= \left. \alpha \frac{u^3}{3} + \alpha(h+a)^2 u + \beta(h+a)u + \gamma u \right|_{-h}^{h} + 0 \\
&= 2\frac{\alpha h^3}{3} + 2\alpha(h+a)^2 h + 2\beta(h+a)h + 2\gamma h \\
&= \frac{h}{3} \left\{ \frac{\alpha(b-a)^2}{2} + \frac{\alpha(b+a)^2}{2} + \beta(b+a) + 2\gamma \right. \\
&\qquad\qquad \left. + \left[\alpha(b+a)^2 + 2\beta(b+a) + 4\gamma\right] \right\} \\
&= \frac{h}{3} \left[\alpha b^2 + \alpha a^2 + \beta b + \beta a + 2\gamma + 4q(h+a)\right] \\
&= \frac{b-a}{6} \left[q(b) + q(a) + 4q\left(\frac{a+b}{2}\right)\right].
\end{aligned}
$$

$\square$

Return to the strategy of approximating $\int_a^b f(x) \, dx$ as described at the beginning of this subsection. We divide the region R into $2m$ vertical strips and replaced the top curve $y = f(x)$ of each adjacent pairs of strips by the parabola passing through the left, middle and right points of this curve. The equation of this parabola is given by Lemma 3.8.12 and the area of the strip it determines is computed in Lemma 3.8.13. The sum the areas of these parabolic strips approximates $\int_a^b f(x) \, dx$. The resulting formula is called Simpson's rule.

Theorem 3.8.14 [Simpson's Rule] *Let $P = \{x_0, \ldots, x_{2m}\}$ be the regular partition of the interval $[a, b]$ into $2m$ subintervals of width $\Delta x = \frac{b-a}{2m}$. Then*

$$
\begin{aligned}
\int_a^b f(x) \, dx \;\approx\; &\frac{\Delta x}{3} \left[f(x_0) + 4f(x_1) + 2f(x_2) + 4f(x_3) + 2f(x_4) \right. \\
&\left. + \cdots + 2f(x_{2m-4}) + 4f(x_{2m-3}) + 2f(x_{2m-2}) + 4f(x_{2m-1}) + f(x_{2m}) \right]
\end{aligned}
\tag{3.8.}
$$

Proof We approximate the area $\int_{x_{2k-2}}^{x_{2k}} f(x) \, dx$ of the kth pair of vertical strips. Let $q_k(x)$ be the quadratic function of Lemma 3.8.12 which passes through the points $(x_{2k-2}, f(x_{2k-2}))$, $(x_{2k-1}, f(x_{2k-1}))$ and $(x_{2k}, f(x_{2k}))$. By Lemma 3.8.13:

$$
\begin{aligned}
\int_{x_{2k-2}}^{x_{2k}} f(x) \, dx \;&\approx\; \int_{x_{2k-2}}^{x_{2k}} q_k(x) \, dx = \frac{2\Delta x}{6} \left[q_k(x_{2k-2}) + 4q_k(x_{2k-1}) + q_k(x_{2k})\right] \\
&= \frac{\Delta x}{3} \left[f(x_{2k-2}) + 4f(x_{2k-1}) + f(x_{2k})\right]
\end{aligned}
$$

by the definition of the quadratic function q_k. Sum these m approximations on the intervals $[x_{2k-2}, x_{2k}]$ for $1 \le k \le m$:

$$
\begin{aligned}
\int_a^b f(x) \, dx \;\approx\; &\frac{\Delta x}{3} \left\{ [f(x_0) + 4f(x_1) + f(x_2)] + [f(x_2) + 4f(x_3) + f(x_4)] + \cdots + \right. \\
&\left. [f(x_{2m-4}) + 4f(x_{2m-3}) + f(x_{2m-2})] + [f(x_{2m-2}) + 4f(x_{2m-1}) + f(x_{2m})] \right\} \\
=\; &\frac{\Delta x}{3} \left[f(x_0) + 4f(x_1) + 2f(x_2) + 4f(x_3) + 2f(x_4) \right. \\
&\left. + \cdots + 2f(x_{2m-4}) + 4f(x_{2m-3}) + 2f(x_{2m-2}) + 4f(x_{2m-1}) + f(x_{2m}) \right]
\end{aligned}
$$

We apply Simpson's rule to approximate π and $L(2)$.

Examples 3.8.15 (1) Estimate $\pi = \int_{-1}^1 \frac{2}{1+x^2} \, dx$ by Simpson's rule, using the regular partition of $[-1, 1]$ into eight subintervals.

Solution Use $f(x) = \frac{2}{1+x^2}$ and the partition $P = \left\{-1, -\frac{3}{4}, -\frac{1}{2}, -\frac{1}{4}, 0, \frac{1}{4}, \frac{1}{2}, \frac{3}{4}, 1\right\}$ with $\Delta x = \frac{1}{4}$. By Simpson's rule:

$$\int_{-1}^{1} \frac{2}{1+x^2}\, dx \approx \frac{1/4}{3}\left[f(-1) + 4f\left(-\frac{3}{4}\right) + 2f\left(-\frac{1}{2}\right) + 4f\left(-\frac{1}{4}\right)\right.$$
$$\left. +2f(0) + 4f\left(\frac{1}{4}\right) + 2f\left(\frac{1}{2}\right) + 4f\left(\frac{3}{4}\right) + f(1)\right]$$
$$\approx\ 3.14157$$

(2) Estimate $L(2) = \int_{1}^{2} \frac{1}{x}\, dx$ by Simpson's rule, using the regular partition of $[1, 2]$ into ten subintervals.

Solution Use the partition $P = \{1, 1.1, 1.2, 1.3, 1.4, 1.5, 1.6, 1.7, 1.8, 1.9, 2\}$ with $\Delta x = 0.1$. Let $f(x) = \frac{1}{x}$. By Simpson's rule:

$$\int_{1}^{2} \frac{1}{x}\, dx \approx \frac{0.1}{3}\left[f(1) + 4f(1.1) + 2f(1.2) + 4f(1.3) + 2f(1.4)\right.$$
$$\left. +4f(1.5) + 2f(1.6) + 4f(1.7) + 2f(1.8) + 4f(1.9) + f(2)\right]$$
$$\approx\ .693150 \qquad\qquad \square$$

We state, without proof, a theorem which gives bounds on the errors of the approximations produced by Simpson's rule.

Theorem 3.8.16 *Let f be a function whose domain contains $[a, b]$. Assume there is a number M such that $f^{(4)}(x)$ exists and*

$$|f^{(4)}(x)| \leq M \ \text{ for } x \in [a, b].$$

If Simpson's rule is applied to the regular partition of $[a, b]$ into $n = 2m$ subintervals to produce the approximation S_n for $\int_a^b f(x)\, dx$, then

$$\left| \int_a^b f(x)\, dx - S_n \right| < \frac{(b-a)^5 M}{180 n^4}. \qquad (3.8.7)$$

We determine the accuracy of our approximations of π and $L(2)$ by Simpson's rule in Examples 3.8.15.

Examples 3.8.17 (1) Determine a bound on the error in the approximation 3.14157 of $\pi = \int_{-1}^{1} \frac{2}{1+x^2}\, dx$ produced by Simpson's rule using the regular partition of $[-1, 1]$ into eight subintervals.

Solution Let $f(x) = \frac{2}{1+x^2} = 2(1+x^2)^{-1}$. We showed in Example 3.8.10 (1) that $f^{(2)}(x) = \frac{12x^2 - 4}{(1+x^2)^3}$. By the quotient rule:

$$f^{(3)}(x) = \frac{24x(1+x^2)^3 - (12x^2 - 4)(6x)(1+x^2)^2}{(1+x^2)^6} = \frac{48(x - x^3)}{(1+x^2)^4}.$$

Apply the quotient rule again:

$$f^{(4)}(x) = 48\frac{(1-3x^2)(1+x^2)^4 - (x - x^3)(8x)(1+x^2)^3}{(1+x^2)^8} = 48\frac{5x^4 - 10x^2 + 1}{(1+x^2)^5}.$$

Rewrite $f^{(4)}(x)$ in terms of $u = x^2$. We need to find the minimum and maximum values of

$$g(u) = 48\frac{5u^2 - 10u + 1}{(1+u)^5}$$

for $0 \le u \le 1$. By the quotient rule:

$$
\begin{aligned}
g'(u) &= 48\frac{(10u-10)(1+u)^5 - (5u^2-10u+1)5(1+u)^4}{(1+u)^{10}} \\
&= 240\frac{2(u-1)(u+1)-(5u^2-10u+1)}{(1+u)^6} = -240\frac{3u^2-10u+3}{(1+u)^6} \\
&= -240\frac{(3u-1)(u-3)}{(1+u)^6}.
\end{aligned}
$$

The only critical point of g, for $0 \le u \le 1$, is the root $\frac{1}{3}$ of its numerator. Thus the maximum and minimum values of $f^{(4)}(x) = g(u)$, for $0 \le u \le 1$, occur at $g(0) = 48$, at $g\left(\frac{1}{3}\right) = -\frac{81}{4}$ or at $g(1) = -6$. Therefore the maximum value of $|f^{(4)}(x)|$ is $M = 48$. Since we are using the regular partition of $[-1, 1]$ into eight subintervals, $n = 8$. By (3.8.7), the error of the approximation $S_8 = 3.14157$ produced by Simpson's rule is less than

$$
\frac{[1-(-1)]^5\,48}{180 \cdot 8^4} = \frac{1}{480} \approx .00208
$$

That is, $|\pi - 3.14157| < .00208$. In fact, $\pi = 3.14159$ to five decimal places.

(2) Find a bound on the error in the approximation $.693150$ of $L(2) = \int_1^2 \frac{1}{x}\,dx$ produced by Simpson's rule using the regular partition of $[1, 2]$ into ten subintervals.

Solution Let $f(x) = \frac{1}{x} = x^{-1}$. Then $f^{(4)}(x) = \frac{24}{x^5}$, and $|f^{(4)}(x)| \le 24$ for $x \in [1, 2]$. Thus choose $M = 24$. Since we are using the regular partition of $[1, 2]$ into ten subintervals, $n = 10$. By (3.8.7), the error of the approximation $S_{10} = .693150$ produced by Simpson's rule is less than

$$
\frac{(2-1)^5(24)}{180 \cdot 10^4} \approx .000013.
$$

Hence $|L(2) - .693150| < .000013$. In fact, $L(2) = .693147$ to 6 decimal places. □

Historical Remarks

The approximation of a region by a union of rectangles and triangles to estimate its area is an ancient method which was common practice in Egypt and Babylonia by the second millenium BCE. For example, in 1400 BCE Rameses II divided the farmland of Egypt into rectangular plots to assess property tax. However, the annual flooding of the Nile often washed away portions of these plots. Therefore, surveyors were appointed to determine the corresponding reduction in taxes.

More sophisticated forms of estimating area approximate the boundary of a region with the graph of a polynomial. For example, Lemma 3.8.12 approximates a curve with the graph of a quadratic polynomial, and the resulting formula for estimating area is Simpson's rule. The general form of Lemma 3.8.12, which gives a formula for the coefficients of the polynomial of degree n which passes through $n+1$ given points, was stated without proof by James Gregory in 1670. Isaac Newton also discovered the formula in 1676, published it in 1687 and gave a proof in *Methodus Differentialis* in 1711. Newton and his students applied this result to approximate a definite integral by replacing the graph of the integrand by arcs of polynomials. For example, *Methodus Differentialis* contains the following generalization of Lemma 3.8.13. Let $x_k = x_0 + k\Delta x$ for $k = 1, 2, 3$. Then the integral of the unique cubic polynomial $C(x)$ which passes through the four points (x_0, y_0), (x_1, y_1), (x_2, y_2), (x_3, y_3) is given by:

$$
\int_{x_0}^{x_3} C(x)\,dx = \frac{3\Delta x}{8}\,(y_0 + 3y_1 + 3y_2 + y_3) \tag{3.8.8}
$$

This formula is called the *Newton–Cotes three–eighths rule*. Lemma 3.8.13 for the integral of the unique quadratic polynomial which passes through three points was probably also known to Newton as it was used by two of his students, Roger Cotes and James Stirling. It was rediscovered by Thomas Simpson in 1743 and its application (3.8.6) to approximate definite integrals is erroneously called Simpson's rule.

Summary

The reader should be able to approximate definite integrals with the six rectangle rules and Simpson's rule. In addition she should be able to determine a bound for the error of the approximations produced by the midpoint rule, the trapezoid rule and Simpson's rule.

Basic Exercises

Estimate each definite integral in Exercises 1–10 with the given partition using:
  (a) the left endpoint rule;   (b) the right endpoint rule;
  (c) the midpoint rule;   (d) the trapezoid rule;
  (e) the lower Riemann sum;   (f) the upper Riemann sum;
  (g) Simpson's rule.
Determine bounds on the error of the approximations you made using:
  (h) the midpoint rule;   (i) the trapezoid rule;   (j) Simpson's rule.

1. $L(3) = \int_1^3 \frac{1}{x}\, dx$ with P the regular partition of $[1,3]$ into 6 subintervals.

2. $\int_0^2 \frac{x}{x+1}\, dx$ with P the regular partition of $[0,2]$ into 4 subintervals.

3. $\pi = \int_0^{1/2} \frac{6}{\sqrt{1-x^2}}\, dx$ with P the regular partition of $[0, \frac{1}{2}]$ into 4 subintervals.

4. $\int_0^{\pi/4} \sec x \, dx$ with P the regular partition of $[0, \frac{\pi}{4}]$ into 4 subintervals.

5. $\int_0^1 \frac{x}{x^2+1}\, dx$ with P the regular partition of $[0,1]$ into 10 subintervals.

6. $\int_0^2 \sqrt{1+x^2}\, dx$ with P the regular partition of $[0,2]$ into 8 subintervals.

7. $\int_0^1 \arctan x \, dx$ with P the regular partition of $[0,1]$ into 10 subintervals.

8. $\int_0^1 \sqrt{1+x^3}\, dx$ with P the regular partition of $[0,1]$ into 4 subintervals.

9. $\int_{2\pi}^{3\pi} \frac{\sin x}{x}\, dx$ with P the regular partition of $[2\pi, 3\pi]$ into 6 subintervals.

10. $\int_0^1 \sin(\pi x^2)\, dx$ with P the regular partition of $[0,1]$ into 4 subintervals.

In each of Exercises 11–15 estimate the given definite integral using:
  (a) the trapezoid rule;   (b) Simpsons's rule.

11. Estimate $\int_1^3 f(x)\, dx$ using the data given in the following table.

x	1	2	3
$f(x)$	3.2	5.1	8.4

12. Estimate $\int_0^1 g(x)\, dx$ using the data given in the following table.

x	0	.25	.5	.75	1
$g(x)$	1.1	1.9	2.1	1.8	1.4

13. Estimate $\int_2^{3.2} h(x)\, dx$ using the data given in the following table.

x	2	2.2	2.4	2.6	2.8	3.0	3.2
$h(x)$	−4.3	−4.1	−3.7	−3.6	−3.8	−4.2	−4.5

14. Estimate $\int_{-3}^{-1} k(x)\,dx$ using the data given in the following table.

x	−3	−2.7	−2.4	−2.1	−1.8	−1.6	−1.4	−1.2	−1
$k(x)$	1.4	1.5	1.7	2.0	2.4	2.9	3.3	3.5	3.6

15. Estimate $\int_0^1 m(x)\,dx$ using the data given in the following table.

x	0	.1	.2	.3	.4	.5	.6	.7	.8	.9	1
$m(x)$	2.4	2.3	2.1	1.8	1.7	1.9	2.2	2.5	2.9	3.2	3.3

16. Show that if f is decreasing on the interval $[a, b]$, then the error in the approximation of $\int_a^b f(x)\,dx$ from any of the six rectangle rules is less than $|f(b)-f(a)|$ mesh (P).

Challenging Problems

1. Show the quadratic function $q(x)$ in Lemma 3.8.12 is the *unique* quadratic function which satisfies $q(a_k) = b_k$ for $1 \le k \le 3$.

2. **(a)** Generalize Lemma 3.8.12 to the following situation. Suppose we are given $n + 1$ points in the plane $(a_0, b_0), \ldots, (a_n, b_n)$ with the numbers $a_0, \ldots, a_n$ distinct. Define a specific polynomial $Q(x)$ of degree n such that $Q(a_k) = b_k$ for $0 \le k \le n$.
(b) Show that the polynomial $Q(x)$ that you defined in (a) is the unique polynomial of degree n with the property that $Q(a_k) = b_k$ for $0 \le k \le n$.

3. **(a)** Show Simpson's rule gives the exact value of the integral $\int_a^b Ax^2 + Bx + C\,dx$ of a quadratic function.
(b) Show Simpson's rule gives the exact value of the integral $\int_a^b Ax^3 + Bx^2 + Cx + D\,dx$ of a cubic function.

4. Show that Simpson's rule is a rectangle rule.

5. Using a regular partition of the interval $[a, b]$, let M be the approximation of $\int_a^b f(x)\,dx$ produced by the midpoint rule and let T be the approximation of $\int_a^b f(x)\,dx$ produced by the trapezoid rule. Show that if the graph of f lies above the x-axis and f is concave up on the interval $[a, b]$, then $M \le \int_a^b f(x)\,dx \le T$.

6. Prove the Newton–Cotes three–eighths rule (3.8.8).

3.9   Applications to Physics

Prerequisite: Sections 2.9 and 3.6

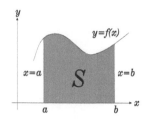

Figure 3.9.1
Region S

The definite integral $\int_a^b f(x)\,dx$ is defined to compute the area of the region S in Figure 3.9.1 bounded by the graph of $y = f(x)$ and the lines $x = a$, $x = b$, $y = 0$. We can view the definite integral as an invariant of continuous functions $y = f(x)$ with domain $[a, b]$. Many invariants of functions, other than area, that are studied in the sciences and social sciences are also given by definite integrals. In particular, when an invariant on the constant function $f(x) = k$ has value $k(b-a)$, then we try to identify the invariant as a definite integral. We illustrate this procedure by identifying several invariants from physics as definite integrals: distance traveled, mass of an object, work done by a force and force on a submerged plate.

Distance Traveled

Consider an object moving on a coordinate line. Recall from Section 2.9 that the velocity $v(t)$ of the object at time t is defined so that $|v(t)|$ is its speed while the sign of $v(t)$ denotes its direction of motion. That is, if $v(t) > 0$ the object is moving in the direction of the coordinate line, while if $v(t) < 0$ the object is moving in the opposite direction. If we know $v(t)$, we show how to compute the distance D traveled by this object from time $t = a$ to time $t = b$. Note if the object has constant speed k meters per second, then the distance traveled in $b - a$ seconds is $D = k(b - a)$. The following example illustrates the computation of D when $v(t)$ is not constant.

Motivating Example 3.9.2 A car travels west, accelerating from rest at $-6t$ m/sec^2. How far does the car travel in 4 seconds?

Solution Let $v(t)$ and $a(t)$ denote the velocity and the acceleration of the car at time t seconds. We are given that $v'(t) = a(t) = -6t$. Hence

$$v(t) = \int v'(t) \, dt = \int -6t \, dt = -3t^2 + C.$$

Since the car starts at rest, $v(0) = 0$, i.e. $C = 0$. Thus $v(t) = -3t^2$. To estimate the distance D traveled from $t = 0$ to $t = 4$, consider the regular partition $P_n = \{t_0, \ldots, t_n\}$ of $[0, 4]$ with $t_k = \frac{4}{k}$ for $0 \le k \le n$. For time t between t_{k-1} and t_k, approximate the motion of the car by considering its velocity to have constant value $v(t_k) = -3t_k^2$. In this approximation, the speed of the car has constant value $|v(t_k)| = 3t_k^2$ meters/second. Hence the car travels approximately $3t_k^2 (t_k - t_{k-1})$ meters from $t = t_{k-1}$ to $t = t_k$. Thus,

$$D \approx \sum_{k=1}^{n} 3t_k^2 (t_k - t_{k-1}) = \sum_{k=1}^{n} |v(t_k)| (t_k - t_{k-1}),$$

the Riemann sum $R(P_n, T_n, |v|)$ with $T_n = \{t_1, \ldots, t_n\}$. The approximations

$$D \approx R(P_n, T_n, |v|) \quad \text{and} \quad R(P_n, T_n, |v|) \approx \int_0^4 |v(t)| \, dt \tag{3.9.1}$$

become better as n gets larger. Hence

$$D = \int_0^4 |v(t)| \, dt = \int_0^4 |-3t^2| \, dt = \int_0^4 3t^2 \, dt = t^3 \big|_0^4 = 64 \text{ meters.} \qquad \square$$

The argument used to compute the distance traveled by this car applies to compute the distance traveled by any object moving in a straight line. We give an intuitive justification of the following proposition. A rigorous proof is given in Section 12.

Proposition 3.9.3 *An object moves along a coordinate line with continuous velocity $v(t)$ at time t. The distance D traveled by this object from time $t = a$ to time $t = b$ is*

$$D = \int_a^b |v(t)| \, dt. \tag{3.9.2}$$

Intuitive Justification To estimate D, consider the regular partition $P_n = \{t_0, \ldots, t_n\}$ of $[a, b]$ into n subintervals of width $\frac{b-a}{n}$. For time t between t_{k-1} and t_k approximate the motion of the object by considering its speed to be constant with value

$0 \to 2 \to 3 \to 6$

$(t-2)(t-3)$

$|v(t_k)|$. Hence this object travels approximately $|v(t_k)|\,(t_k - t_{k-1})$ units from $t = t_{k-1}$ to $t = t_k$. Thus

$$D \approx \sum_{k=1}^{n} |v(t_k)|\,(t_k - t_{k-1}) = R(P_n, T_n, |v|)$$

with $T_n = \{t_1, \ldots, t_n\}$. The approximations $D \approx R(P_n, T_n, |v|)$ and $R(P_n, T_n, |v|) \approx \int_a^b |v(t)|\,dt$ become better as n gets larger. Hence $D = \int_a^b |v(t)|\,dt$. $\qquad\square$

To apply (3.9.2) to compute the distance traveled, note that

$$|v(t)| = \left\{ \begin{array}{ll} v(t) & \text{if } v(t) \geq 0 \\ -v(t) & \text{if } v(t) \leq 0 \end{array} \right\}.$$

Since v is continuous, $v(t)$ changes sign when $v(t) = 0$. Thus we find the roots of $v(t)$. Then we use the additive property to write the integral of $|v(t)|$ as a sum of integrals over subintervals of $[a, b]$ so that $v(t) \geq 0$ or $v(t) \leq 0$ on each subinterval.

$t^2 - 3t - 2t + 6$

$t(t-3) - 2(t-3)$

$t^2 - 6t + t + 6$

$t(t+6) + 1(t-6)$

Examples 3.9.4 (1) An object travels on a coordinate line from time $t = 0$ to time $t = 5$ seconds with velocity $v(t) = 2t - 6$ feet per second. Find the distance D traveled by this object.

Solution Since $v(t) = 2(t - 3)$, it has one root at $t = 3$. We have $v(t) \leq 0$ for $t \leq 3$ and $v(t) \geq 0$ for $t \geq 3$. Hence

$$\begin{aligned} D &= \int_0^5 |v(t)|\,dt = \int_0^3 -v(t)\,dt + \int_3^5 v(t)\,dt \\ &= \int_0^3 6 - 2t\,dt + \int_3^5 2t - 6\,dt = (6t - t^2)\big|_0^3 + (t^2 - 6t)\big|_3^5 \\ &= (9 - 0) + (-5 + 9) = 13 \text{ feet.} \end{aligned}$$

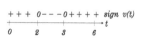

$+++\; 0\;---\;0\;++++\;$ *sign $v(t)$*

$0 \qquad 2 \qquad 3 \qquad 6$

Figure 3.9.5
Velocity in Example 2

(2) An object begins traveling on a coordinate line at time $t = 0$ with velocity $v(t) = t^2 - 5t + 6$ meters per minute at t minutes. Find the distance D traveled by this object during the first 6 minutes.

Solution Solve $v(t) = 0$ to determine the roots of $v(t)$:

$$0 = v(t) = t^2 - 5t + 6 = (t - 3)(t - 2)$$

when $t = 2$ or $t = 3$. The sign of $v(t)$ is given in Figure 3.9.5. Thus

$$\begin{aligned} D &= \int_0^6 |v(t)|\,dt = \int_0^2 v(t)\,dt + \int_2^3 -v(t)\,dt + \int_3^6 v(t)\,dt \\ &= \int_0^2 t^2 - 5t + 6\,dt + \int_2^3 5t - t^2 - 6\,dt + \int_3^6 t^2 - 5t + 6\,dt \\ &= \left(\frac{t^3}{3} - \frac{5t^2}{2} + 6t \right)\Big|_0^2 + \left(\frac{5t^2}{2} - \frac{t^3}{3} - 6t \right)\Big|_2^3 + \left(\frac{t^3}{3} - \frac{5t^2}{2} + 6t \right)\Big|_3^6 \\ &= \left(\frac{14}{3} - 0 \right) + \left(-\frac{9}{2} + \frac{14}{3} \right) + \left(18 - \frac{9}{2} \right) = \frac{55}{3} \text{ meters.} \end{aligned}$$

(3) One end of a spring is attached to a wall, and an object is attached to its other end. The velocity of this object is found to be $v(t) = 2\sin 5\pi t$ feet/sec. Find the distance D traveled by this object from time $t = 0$ to $t = 10$ sec.

Solution This spring's motion is periodic. Its motion from time to $t = 0$ to $t = \frac{2}{5}$ (when the angle $5\pi t$ equals 2π) is repeated 25 times from $t = 0$ to $t = 10$.

On the interval $[0, 2/5]$, the function $v(t) = 2\sin 5\pi t$ is positive for $0 < t < 1/5$ and negative for $1/5 < t < 2/5$. Hence

$$D = 25\int_0^{2/5} |v(t)|\ dt = 25\int_0^{1/5} v(t)\ dt + 25\int_{1/5}^{2/5} -v(t)\ dt$$

$$= 25\int_0^{1/5} 2\sin 5\pi t\ dt + 25\int_{1/5}^{2/5} -2\sin 5\pi t\ dt$$

$$= -\frac{10}{\pi}\cos 5\pi t\Big|_0^{1/5} + \frac{10}{\pi}\cos 5\pi t\Big|_{1/5}^{2/5} = \left(\frac{10}{\pi} + \frac{10}{\pi}\right) + \left(\frac{10}{\pi} + \frac{10}{\pi}\right) = \frac{40}{\pi}\ \text{ft.}\ \square$$

$$-\frac{2}{5\pi}\cos 5\pi t$$

Mass

Consider a rod of variable density with mass M. The density of the rod at any point is defined as its mass per unit length. Place the rod on the x–axis. Say the rod lies between $x = a$ and $x = b$. See Figure 3.9.6. Let $\rho(x)$ denote the density[6] of the rod at the point x. If $\rho(x) = k$ is constant, then the mass of the rod is its mass per unit length k times its length $b - a$, i.e. $M = k(b - a)$. When $\rho(x)$ is not constant, the following proposition identifies M as the integral of $\rho(x)$ over $[a, b]$. This proposition is proved in Section 12, but we give an intuitive justification here.

Figure 3.9.6
A Rod

Proposition 3.9.7 *A rod of mass M lies on the interval $[a, b]$ of the x–axis with continuous density $\rho(x)$. Then*

$$M = \int_a^b \rho(x)\ dx. \tag{3.9.3}$$

Intuitive Justification To estimate M, consider the regular partition $P_n = \{x_0, \ldots, x_n\}$ of $[a, b]$ into n subintervals of width $\frac{b-a}{n}$. The piece of the rod between x_{k-1} and x_k has density approximately $\rho(x_k)$ and mass approximately $\rho(x_k)(x_k - x_{k-1})$. Hence

$$M \approx \sum_{k=1}^n \rho(x_k)\,(x_k - x_{k-1}) = R(P_n, T_n, \rho)$$

with $T_n = \{x_1, \ldots, x_n\}$. The approximations $M \approx R(P_n, T_n, \rho)$ and $R(P_n, T_n, \rho) \approx \int_a^b \rho(x)\ dx$ become better as n gets larger. Hence $M = \int_a^b \rho(x)\ dx$. $\square$

Examples 3.9.8 (1) Find the mass M of the rod with density $\rho(x) = 12x$ kg/m for $0 \le x \le 3$ meters.

Solution $M = \int_0^3 \rho(x)\ dx = \int_0^3 12x\ dx = 6x^2\Big|_0^3 = 54$ kg.

(2) Find the mass M of the rod with density $\rho(x) = \sin^2 \pi x$ lbs/ft for $0 \le x \le 2$ ft.

Solution $M = \int_0^2 \rho(x)\ dx = \int_0^2 \sin^2 \pi x\ dx = \int_0^2 \frac{1 - \cos 2\pi x}{2}\ dx.$
Make the substitution $u = 2\pi x$ with $du = 2\pi\ dx$:

$$M = \frac{1}{4\pi}\int_{x=0}^{x=2} (1 - \cos 2\pi x)2\pi\ dx = \frac{1}{4\pi}\int_{u=0}^{u=4\pi} 1 - \cos u\ du$$

$$= \frac{1}{4\pi}(u - \sin u)\Big|_0^{4\pi} = \frac{1}{4\pi}(4\pi - 0) = 1\ \text{pound.} \qquad \square$$

---

[6] ρ is the Greek letter *rho* which is the standard notation for density.

Work

Consider a force acting on an object which moves along the x–axis. This force does work on the object. The work done by a constant force $F(x) = k$ acting on an object which moves a distance of s units is defined as ks. We want to extend this definition to define the work W done by a variable force $F(x)$ as the object moves from $x = a$ to $x = b$. To estimate W, consider the regular partition $P_n = \{x_0, \ldots, x_n\}$ of $[a, b]$ into n subintervals of width $\frac{b-a}{n}$. Choose $t_k \in [x_{k-1}, x_k]$, for $1 \leq k \leq n$, and let $T_n = \{t_1, \ldots, t_n\}$. As the object moves from $x = x_{k-1}$ to $x = x_k$, the force acting on it is approximately $F(t_k)$ and work done is approximately $F(t_k)(x_k - x_{k-1})$. Hence

$$W \approx \sum_{k=1}^{n} F(t_k)\,(x_k - x_{k-1}) = R(P_n, T_n, F)$$

with $T_n = \{x_1, \ldots, x_n\}$. The approximations $W \approx R(P_n, T_n, F)$ and $R(P_n, T_n, F) \approx \int_a^b F(x)\,dx$ become better as n gets larger. Hence we define the work W done by the force $F(x)$ as the object moves from $x = a$ to $x = b$ as:

$$W = \int_a^b F(x)\,dx. \tag{3.9.4}$$

In the metric system force is measured in *Newtons*. One Newton is the force required to accelerate an object of mass one kilogram at the rate of one meter/second^2. The work done by a force of one Newton in moving an object one meter is called a *joule*. In the British system both mass and force are measured in *pounds*. That is, these two different units have the same name! A one pound force accelerates a one pound mass at the rate of 32.174 feet/second^2, the acceleration of gravity. The work done by a one pound force in moving an object one foot is called a *foot-pound*.

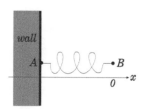

Figure 3.9.10
Equilibrium Position

Examples 3.9.9 (1) One end A of the spring in Figure 3.9.10 is attached to a wall. Hooke's Law says that the force exerted by this spring is proportional to the distance the other end B is stretched from its equilibrium position. Consider a spring where this constant of proportionality is 4 when distance is measured in feet and force is measured in pounds. Find the work W done in stretching this spring 2 feet from its equilibrium position.

Solution Place an x–axis along the spring with the origin at B when the spring is in its equilibrium position. The force of this spring is $F(x) = 4x$, and

$$W = \int_0^2 4x\,dx = 2x^2\Big|_0^2 = 8 \text{ foot} - \text{pounds}.$$

(2) The force exerted by the Earth on an object is given by

$$F = \frac{GmM}{s^2}$$

where m is the mass of the object, $M = 5.97 \times 10^{24}$ kilograms is the mass of the Earth, s is the distance of the object from the center of the Earth and $G = 6.673 \times 10^{-11}$ is the universal gravitation constant. The radius of the Earth is 6.37×10^6 meters. Find the work W done by a rocket in raising a 1,000 kilogram satellite 200 kilometers from the surface of the Earth.

Solution Note that 200 km equals 0.2×10^6 meters. Hence

$$W = \int_{6.37 \times 10^6}^{6.57 \times 10^6} \frac{(6.673 \times 10^{-11})(1,000)(5.97 \times 10^{24})}{s^2}\,ds$$

$$= 3.98 \times 10^{17} \int_{6.37 \times 10^6}^{6.57 \times 10^6} \frac{1}{s^2} \, ds = 3.98 \times 10^{17} \left(-\frac{1}{s} \right) \Big|_{6.37 \times 10^6}^{6.57 \times 10^6}$$

$$= 3.98 \times 10^{17} \left(-\frac{1}{6.57 \times 10^6} + \frac{1}{6.37 \times 10^6} \right) \approx 1.90 \text{ billion joules.} \qquad \Box$$

Force on a Submerged Object

A fluid exerts pressure on submerged objects. The amount of this pressure depends on the weight of the fluid rather than its mass. We digress to clarify the distinction between these two concepts. The *mass* of an object is the quantity of matter it contains while the *weight* of an object is the force exerted on it by gravity. There are two corresponding concepts of density. The *density* of an object refers to its mass per unit volume while its *weight–density* refers to its weight per unit volume. Thus the mass and density of an object are the same on the surface of the Earth as they are in outer space. On the other hand, the weight and weight–density of an object are less at a higher altitude and are zero in outer space. Since weight is a force, it is measured in units of force. In the metric system mass is measured in kilograms, and weight is measured in Newtons. Thus density is measured in kilograms per cubic meter, and weight–density is measured in Newtons per cubic meter. In the British system the unit of mass and the unit of weight are both called pounds. Hence the units of density and weight–density are both called pounds per cubic foot. At a place on the Earth's surface where the gravitational constant $g = 9.80665$ m/sec$^2 = 32.174$ lb/sec^2, a mass of one kilogram weighs 9.80665 Newtons and a one pound mass weighs one pound.

Return to a fluid which exerts pressure on a submerged object. At a constant depth h, the pressure $P(h)$ is equal in all directions. Since pressure is defined as the force per unit area, we calculate the value of $P(h)$ by submerging a flat plate of area A in a horizontal position at depth h. The force on this plate will equal $AP(h)$. Experimentation shows that the force on this plate equals the weight of the fluid above the plate. The volume of the fluid above the plate equals Ah, so the weight of this fluid of weight–density ρ is $Ah\rho$. Therefore $AP(h) = Ah\rho$ and

$$P(h) = h\rho. \tag{3.9.5}$$

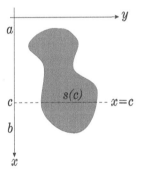

Figure 3.9.11
Vertical Plate

Now we compute the force F on a flat plate which is submerged in a vertical position. Place the x and y axes in the plane of this plate with the x–axis pointing downwards and the y–axis on the surface of the fluid. Assume that the plate lies between $x = a$ and $x = b$. For each number c, let $s(c)$ denote the length of the line segment in Figure 3.9.11 which is the intersection of the line $x = c$ with the plate. Assume s is a continuous function. To estimate F, let $P_n = \{x_0, \ldots, x_n\}$ be the regular partition of $[a, b]$ into n subintervals of width $\frac{b-a}{n}$. The lines $x = x_k$, for $0 \leq k \leq n$, divide the plate into n thin vertical strips. Choose $t_k \in [x_{k-1}, x_k]$, for $1 \leq k \leq n$, with $T_n = \{t_1, \ldots, t_n\}$. Approximate the kth vertical strip V_k of the plate by a horizontal rectangle H_k at depth t_k of length $s(t_k)$ and width $x_k - x_{k-1}$. The force of the water on V_k is approximately the force of the water on H_k. The latter force is the volume of the water above H_k (the area $s(t_k)(x_k - x_{k-1})$ of H_k times its depth t_k) times the weight density ρ of the fluid. Hence

$$F \approx \sum_{k=1}^{n} \rho t_k s(t_k)(x_k - x_{k-1}) = R(P_n, T_n, \rho x s(x)).$$

As n gets larger, the approximation of F by $R(P_n, T_n, \rho x s(x))$ and the approximation of $\int_a^b \rho x s(x) \, dx$ by $R(P_n, T_n, \rho x s(x))$ get better. Hence

$$F = \int_a^b \rho x s(x) \, dx. \tag{3.9.6}$$

A rigorous derivation of this formula is given in Section 12.

In our examples the fluid is water which has weight–density 9,800 Newtons per cubic meter or 62.5 pounds per cubic foot.

Examples 3.9.12 (1) Find the force on a circular horizontal plate of radius 2 meters which is submerged 8 meters under water.

Solution By (3.9.5), the pressure on the plate equals $(8)(9,800) = 78,400\,\text{N/m}^2$. Thus, the force on this plate equals the area of the plate $\pi(2)^2\,\text{m}^2$ times the pressure $78,400\,\text{N/m}^2$ which is $313,600\pi$ Newtons.

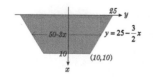

Figure 3.9.13
Vertical Dam

(2) Find the force on a vertical dam whose face has the shape of an isosceles trapezoid which is 50 feet across at the surface of the water and 20 feet across at the bottom of the river, 10 feet below the surface of the water.

Solution This dam is sketched in Figure 3.9.13 with the x–axis at the center of the plate. The right edge of the dam passes through the points $(0,25)$ and $(10,10)$. Hence it has equation $y = 25 - \frac{3}{2}x$. Therefore the width $s(x)$ of the dam at depth x is $s(x) = 2y = 50 - 3x$ feet. By (3.9.6), the force on this dam is

$$\int_0^{10} 62.5x(50-3x)\,dx = 1562.5x^2 - 62.5x^3\Big|_0^{10} = 156,250 - 62,500 = 93,750\,\text{lbs.}\,\square$$

Summary

Given the velocity of an object, the reader should be able to compute the distance the object travels. Given the density of a rod, she should be able to calculate its mass. She should also be able to find the work done by a force which acts on a moving object. Finally, she should be able to compute the force on a flat plate which is submerged in either a horizontal or vertical position.

Basic Exercises

Distance Traveled

1. In each case, an object is traveling on a coordinate line between two given times with a given velocity. Find the distance traveled by the object.
    (a) $v(t) = 3t^2 + 7$ miles per hour from $t = 0$ to $t = 3$ hours.
    (b) $v(t) = 4t - 8$ km. per hour from $t = 0$ to $t = 5$ hours.
    (c) $v(t) = 6 - 3t$ km. per hour from $t = -1$ to $t = 4$ hours.
    (d) $v(t) = t^2 - 6t + 8$ miles per hour from $t = 0$ to $t = 6$ hours.
    (e) $v(t) = t^2 + t - 6$ miles per hour from $t = -4$ to $t = 3$ hours.
    (f) $v(t) = t^2 - t + 3$ km. per hour from $t = -2$ to $t = 4$ hours.
    (g) $v(t) = t^3 - 1$ km. per hour from $t = 0$ to $t = 2$ hours.
    (h) $v(t) = t^6 - t^4$ miles per hour from $t = -2$ hours to $t = 3$ hours.
    (i) $v(t) = 2\sin 3\pi t$ miles per hour from $t = 0$ to $t = 2$ hours.
    (j) $v(t) = \frac{t}{(t^2-16)^2}$ km per hour from $t = -1$ to $t = 1$ hours.

2. One end of a spring is attached to a wall, and the other end is attached to an object. The velocity of this object is $v(t) = 7\cos 6\pi t$ centimeters per second. Find the distance traveled by this object from time $t = 0$ to $t = 5$ seconds.

Information for Exercises 3 to 6

Gravity acts on an object which moves along a vertical coordinate line pointing upwards. If no other forces act on this object, its velocity is $v(t) = v(0) - 32t$ ft/sec.

3. John throws a ball vertically upwards at 96 feet per second and catches it when it comes down. How far has this ball traveled?

4. A rock is kicked upwards from the edge of a 200 foot high cliff at the rate of 90 ft/sec. How far has the rock traveled when it lands at the bottom of the cliff?

5. An arrow is shot vertically upwards into the air. What should its initial velocity be so that it reaches a maximum height of exactly 150 feet?

6. A rubber ball is dropped from a window 64 feet above the ground. Each bounce is two thirds as high as the preceding one. Find the distance traveled by this ball during the first 3 seconds.

Mass

7. Find the mass of a rod with density $\rho(x) = 6x^2$ lbs/ft for $0 \le x \le 4$ ft.

8. Find the mass of a pole with density $\rho(x) = \sec^2 \frac{\pi x}{4}$ kg/m for $0 \le x \le 1$ meters.

9. Find the mass of a rod with density $\rho(x) = \cos^4(\pi x)$ kg/m for $0 \le x \le 6$ meters.

10. Find the mass of a pole with density $\rho(x) = \frac{8}{9+x^2}$ lbs/ft for $0 \le x \le 3$ ft.

11. Find the mass of a rod with density $\rho(x) = \frac{1}{\sqrt{2-x^2}}$ lbs/ft for $0 \le x \le 1$ ft.

Work

12. How much work is done when a constant force of 3 Newtons moves an object 4 m?

13. How much work is done when the force $F(x) = 6x^2 + 7$ lbs moves an object from $x = 0$ to $x = 2$ ft?

14. How much work is done by the force $F(x) = 5\cos^2 \pi x$ lbs in moving an object from $x = 0$ to $x = 3$ ft?

15. How much work is done by the force $F(x) = \frac{1}{x^2+4}$ Newtons in moving an object from $x = 0$ to $x = 2$ m?

16. The force in stretching a spring x feet from its equilibrium position is $F(x) = 5x$. Find the work done in stretching this spring 6 inches from its equilibrium position.

17. It requires 10 foot–pounds of work to stretch a spring one foot from its equilibrium position. How much work is required to stretch this spring three feet?

18. Find the work done by a motor which raises a 500 kilogram elevator 300 meters to the top floor of a skyscraper.

19. Find the work done by a 20,000 kilogram helicopter which rises 1,000 meters vertically from the ground.

20. A rocket weighing 8,000 kg rises vertically from the surface of the Earth to an altitude of 1,500 km above the Earth's surface. As the rocket burns fuel, its mass decreases in such a way that at a height of s km the mass of the rocket is given by $8,000 - 5s$ kg. Find an integral which gives the work done by this rocket.

Force on a Submerged Object

21. (a) Find the force on a square, whose side is 2 meters long, which is submerged horizontally 3 meters under water.
(b) Find the force on this square when it is in a vertical position with its top edge 3 meters under water.

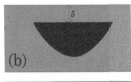

22. (a) Find the force on a parabolic plate bounded by $y = x^2$, $-3 \le x \le 3$, and $y = 9$ which is submerged horizontally 5 meters under water. (Both x and y are measured in meters.)
(b) Find the force on this plate when it is in a vertical position with its straight edge 5 meters under water.

Exercise 22

(c) Find the force on this plate if it is turned upside down from its position in (b) so that its straight edge is 14 meters under water.

23. (a) A cylindrical drum with no top, of height 3 meters and radius 2 meters, is half filled with water and capped. If the drum is turned to a horizontal position, what is the force on the cap?
(b) If the drum were completely filled before being capped and turned, what would the force be on the cap?

24. (a) The cross–section of a trough has the shape of an equilateral triangle of side 3 feet. The top of this cross–section is a horizontal side of this triangle. Find the force on each end of the trough when the trough is filled with water.
(b) Find the force on each end of this trough if it is filled with water to half its height.

25. A 6 cm deep gutter has the cross–section of an isosceles trapezoid which is 8 cm wide at the top and 4 cm wide at the bottom. Find the force on each end of this gutter when it filled with water.

26. A trough has vertical cross–sections equal to the bottom half of an ellipse which has a vertical axis of 4 feet and a horizontal axis of 6 feet. If the trough is filled to half its height with water, find the force of the water on each end of the trough.

27. The vertical face of a dam has the shape of a semi–circle of radius 16 meters with its diameter on the surface of the water. Find the force on this dam.

28. Fifty pounds of dirt are placed in a wooden box measuring 5 feet long, 3 feet wide and 2 feet high. If this box is placed in water, how deep will it sink?

29. A rectangular swimming pool of length 50 feet and width 20 feet is 4 feet deep at one end and 10 feet deep at the other end. The floor of the pool is a slanted rectangle. Find the force on each of the four vertical sides of this pool.

30. A dam, in the shape of an isosceles trapezoid, is to be constructed at a spot where a river is 30 meters wide at the top, 15 meters wide at the bottom and 5 meters deep. The face of the dam facing the river will be vertical. The thickness of the dam in meters at depth x meters will be one millionth the force of the water (measured in Newtons) on that part of the dam at depth less than or equal to x. Find the thickness of the bottom of this dam.

Challenging Problems

1. Consider a tank and a vertical x–axis pointing downwards with the top of the tank at $x = 0$. The cross-sectional area of this tank at depth x is $A(x)$. The bottom of this tank is at $x = b$ and the tank is filled with liquid of weight–density ρ to height $x = a$. Let W denote the work required to pump the liquid in the tank out over the top of the tank. Show that $W = \int_a^b \rho x A(x)\ dx$.

2. A chain hangs from the top of a building. Its weight–density at distance x from the top of the building is $\rho(x)$. Let W denote the work required to lift the part of the chain from $x = a$ to $x = b$ to the top of the building. Show that $W = \int_a^b x\rho(x)\ dx$.

3. The flat face of a dam makes an acute angle θ with the vertical. Let $s(x)$ be cross-sectional width of the face of the dam at depth x. Show that the force F on the section of this dam between depth $x = a$ and $x = b$ is given by $F = \int_a^b \rho x s(x) \sec\theta\ dx$ where ρ is the weight–density of water.

3.10 Center of Mass

Prerequisite: Sections 3.7 and 3.9[7]

Newton's second law of motion says that when a force is applied to an object, the object accelerates in the direction of the force. However, when the object is attached to a pivot and a force is applied, the object rotates. The amount of rotation produced by this force depends on the magnitude and direction of the force, the distance between the point of application of the force and the pivot, as well as the mass of the object. We study the special case where the force of gravity acts on an object. The concept of *moment* is introduced to measure the resulting rotation. We determine the point of placement of a pivot which will balance the object. This point is called the *center of mass* of the object. In the first subsection, we study the center of mass of a finite set of discrete point masses while in the second section we compute the center of mass of a rod of variable density. In the third subsection, we determine the *centroid* of a region R. It is defined as the center of mass of the plate with shape R and constant density. We conclude with a theorem of Pappus which gives a simple method to calculate the volume of a solid of revolution S from the centroid and area of the region which generates S.

Center of Mass of Discrete Objects

Consider a seesaw which is balanced in a horizontal position. The support upon which the seesaw pivots is called a *fulcrum*. Place a number line on the seesaw. When a mass is placed on one side of the seesaw, gravity acts on this mass to create a rotational force which makes that side of the seesaw rotate downwards. See Figure 3.10.1. Experimentation validates the following two principles.

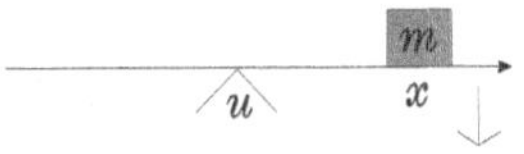

Principle A *The rotational force produced by a mass is proportional to the magnitude of the mass.*

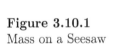

Figure 3.10.1
Mass on a Seesaw

Principle B *The rotational force produced by a mass is proportional to the distance between the mass and the fulcrum.*

Based on these two principles we define the moment of this mass which, up to a constant of proportionality, gives the rotational force that it exerts on the seesaw.

Definition 3.10.2 *Consider a number line with a fulcrum at coordinate u. An object of mass m at coordinate x has moment $X_u = m(x - u)$ about $x = u$. The moment of this object refers to its moment $M = X_0 = mx$ about $x = 0$.*

Observe that the sign of the moment indicates the side of the fulcrum on which the mass has been placed. If the moment is positive, the mass is on the side of the fulcrum in the direction that the number line is pointing. If the moment is negative, the mass is on the other side. For example, the moments of the masses m_1, m_2 in Figure 3.10.4 are positive while the moments of the masses m_3, m_4, m_5 are negative.

Example 3.10.3 A fulcrum is located at $x = 2$ cm on a number line. Determine the moment of each of the following masses:
(a) 4 gm located at 7 cm; (b) 3 gm located at 2 cm; (c) 7 gm located at −6 cm.

Solution (a) The moment is $X_2 = (4)(7 - 2) = 20$ gm–cm.
(b) The moment is $X_2 = 3(2 - 2) = 0$ gm–cm.
(c) The moment is $X_2 = 7(-6 - 2) = -56$ gm–cm. □

---

[7] Only the subsection *Mass* from Section 3.9 is required as a prerequisite.

If several masses are placed on a seesaw, as in Figure 3.10.4, experimentation shows that the net rotational force on the seesaw is the sum of the rotational forces of the individual masses. We use this observation to extend the definition of moment to this situation.

Figure 3.10.4 A Set of Five Masses on a Seesaw

Definition 3.10.5 *Consider a number line with a fulcrum at coordinate u and an object of mass m_k at coordinate x_k for $1 \leq k \leq t$. The moment of the set of masses $\{m_1, \ldots, m_t\}$ about $x = u$ is*

$$X_u = m_1(x_1 - u) + \cdots + m_t(x_t - u).$$

The moment of this set of masses refers to its moment M about $x = 0$:

$$M = X_0 = m_1 x_1 + \cdots + m_t x_t.$$

Example 3.10.6 A fulcrum is located at $x = -3$ ft on a number line. Find the moment of the following set of masses:

$$\begin{array}{lll}
2 \text{ lbs at } x = -10 \text{ ft}; & 5 \text{ lbs at } x = -4 \text{ ft}; & 3 \text{ lbs at } x = 0 \text{ ft}; \\
1 \text{ lbs at } x = 1 \text{ ft}; & 6 \text{ lbs } at \ x = 3 \text{ ft}; & 4 \text{ lbs at } x = 2 \text{ ft}.
\end{array}$$

Solution The moment of this set of masses is:

$$\begin{aligned}
X_{-3} &= 2\left[-10 - (-3)\right] + 5\left[-4 - (-3)\right] + 3\left[0 - (-3)\right] + 1\left[1 - (-3)\right] \\
&\quad + 6\left[3 - (-3)\right] + 4\left[2 - (-3)\right] \\
&= -14 - 5 + 9 + 4 + 36 + 20 = 50 \text{ ft} - \text{lbs}. \qquad \square
\end{aligned}$$

Suppose a set of masses S is attached to a seesaw with a movable fulcrum. A basic problem is to find the position of the fulcrum which balances the seesaw in a horizontal position. This position is called the *center of mass* of S.

Proposition 3.10.7 *Let S be the set of masses which consists of an object of mass m_k at coordinate x_k, for $1 \leq k \leq t$. Define the center of mass of S by:*

$$\overline{x} = \frac{m_1 x_1 + \cdots + m_t x_t}{m_1 + \cdots + m_t} = \frac{M}{m}$$

where M is the moment of these masses and m is their total mass. Then the moment of S about $x = \overline{x}$ equals zero.

Proof Note that $m\overline{x} = M$. Hence the moment of S about $\overline{x}$ is given by:

$$\begin{aligned}
X_{\overline{x}} &= \left[m_1(x_1 - \overline{x}) + \cdots + m_t(x_t - \overline{x})\right]\frac{m}{m} = m_1\frac{mx_1 - M}{m} + \cdots + m_t\frac{mx_t - M}{m} \\
&= \frac{(m_1 m x_1 - m_1 M) + \cdots + (m_t m x_t - m_t M)}{m}.
\end{aligned}$$

If $i \neq j$, each $m_i m_j x_k$ occurs twice in the numerator of this fraction: as a summand of $m_i m x_k$ and of $m_j m x_k$. Each $-m_i m_j x_k$ also occurs twice: as a summand of $-m_i M$ and of $-m_j M$. These four summands of the numerator cancel. If $i = j$, each $m_i^2 x_k$ occurs once in the numerator as a summand of $m_i m x_k$. Each $-m_i^2 x_k$ occurs once as a summand of $-m_i M$. These two summands cancel. Hence the numerator of the above fraction is zero, and $X_{\overline{x}} = 0$. $\qquad \square$

Example 3.10.8 Find the center of mass of the set of masses S of Example 3.10.6.

Solution The total mass of S is: $m = 2 + 5 + 3 + 1 + 6 + 4 = 21$ lbs.
The moment of S is: $M = 2(-10) + 5(-4) + 3(0) + 1(1) + 6(3) + 4(2) = -13$ ft $-$ lbs.
Therefore, the center of mass is $\overline{x} = \frac{M}{m} = -\frac{13}{21}$ ft.   □

We show that the moment about $x = u$ of a set of masses, of total mass m, is the same as the moment about $x = u$ of the point pass m located at $x = \overline{x}$.

Corollary 3.10.9 *Consider the set S of masses which consists of an object of mass m_k at coordinate x_k for $1 \le k \le t$. Let m denote the total mass of S, and let $\overline{x}$ denote its center of mass. Then for any u:*

$$X_u = m(\overline{x} - u).$$

Proof Since $m\overline{x} = M$,

$$
\begin{aligned}
m(\overline{x} - u) &= M - mu = (m_1 x_1 + \cdots + m_t x_t) - (m_1 + \cdots + m_t)u \\
&= m_1(x_1 - u) + \cdots + m_t(x_t - u) = X_u.
\end{aligned}
$$
     □

It is easy to compute moments by using this corollary.

Example 3.10.10 Find the moment about $x = 5$ of the set of masses S of Example 3.10.6.

Solution In Example 3.10.8, we found the total mass of S to be 21 lbs and the center of mass to be $\overline{x} = -\frac{13}{21}$ ft. By Corollary 3.10.9, $X_5 = 21\left(-\frac{13}{21} - 5\right) = -118$ ft $-$ lbs. □

Center of Mass of a Rod

We generalize the concepts of moment and center of mass to a rod R of variable density which lies on a seesaw. Place the x–axis on the seesaw. The rod lies on the interval $[a, b]$ of this axis and the fulcrum of the seesaw is at $x = u$. See Figure 3.10.11. Let $\rho(x)$ denote the density of this rod at x. We want to define the moment X_u about $x = u$ of R. Let $P = \{x_0, \ldots, x_n\}$ be a partition of $[a, b]$ with $x_{k-1} \le t_k \le x_k$, for $1 \le k \le n$. Write $T = \{t_1, \ldots, t_n\}$. Estimate the mass of the section of the rod R_k on the interval $[x_{k-1}, x_k]$ by considering its density to have constant value $\rho(t_k)$: the mass of R_k is approximately $\rho(t_k)(x_k - x_{k-1})$. Estimate the moment of R_k about $x = u$ by considering its mass to be concentrated at t_k. By Definition 3.10.2, the moment of R_k is approximately $(t_k - u)\rho(t_k)(x_k - x_{k-1})$. The moment of the entire rod R is approximately the sum of these n approximations of the moments of the R_k:

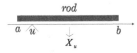

Figure 3.10.11
Moment of a Rod

$$X_u \approx \sum_{k=1}^{n} (t_k - u)\rho(t_k)(x_k - x_{k-1}) = R(P, T, (x - u)\rho(x)).$$

This sum is an approximation of both X_u and $\int_a^b (x - u)\rho(x)\, dx$. Hence we define X_u as this integral.

Definition 3.10.12 *Let R be a rod on the interval $[a, b]$ of the x–axis. Assume its density $\rho(x)$ is continuous on $[a, b]$. Define the moment X_u of R about $x = u$ by:*

$$X_u = \int_a^b (x - u)\rho(x)\, dx.$$

The moment M of this rod is its moment about $x = 0$:

$$M = \int_a^b x\rho(x)\, dx.$$

Example 3.10.13 Compute the moment about $x = 2$ cm of the rod R of density $\rho(x) = \frac{1}{\sqrt{64-x^2}}$ gm/cm, for $0 \le x \le 4$ cm.

Solution The moment of this rod about $x = 2$ is:

$$X_2 = \int_0^4 \frac{x-2}{\sqrt{64-x^2}} \, dx = \int_0^4 \frac{x}{\sqrt{64-x^2}} \, dx - 2 \int_0^4 \frac{1}{\sqrt{64-x^2}} \, dx.$$

Let $t = 64 - x^2$ with $dt = -2x \, dx$. Then

$$X_2 = -\frac{1}{2} \int_{64}^{48} t^{-1/2} \, dt - 2 \arcsin \frac{x}{8} \Big|_0^4 = -t^{1/2} \Big|_{64}^{48} - 2 \arcsin \frac{1}{2} = 8 - 4\sqrt{3} - \frac{\pi}{3} \text{ gm} - \text{cm}.$$

If a rod is attached to a seesaw with a movable fulcrum, the *center of mass* of the rod is the position of the fulcrum for which the seesaw balances horizontally.

Proposition 3.10.14 *Let R be a rod of continuous density $\rho(x)$, for $a \le x \le b$, with mass m and moment M. Then there is a unique number $\bar{x}$ for which the moment of the rod about $\bar{x}$ equals zero. In particular,*

$$\bar{x} = \frac{M}{m} = \frac{\int_a^b x\rho(x) \, dx}{\int_a^b \rho(x) \, dx}.$$

Proof We solve the equation which says the moment of R about $x = \bar{x}$ equals zero:

$$X_{\bar{x}} = \int_a^b (x - \bar{x})\rho(x) \, dx \quad = \quad 0.$$

By linearity of the integral : $\displaystyle\int_a^b x\rho(x) \, dx - \bar{x} \int_a^b \rho(x) \, dx \quad = \quad 0.$

This equation can be rewritten as $M - \bar{x}m = 0$. Therefore $\bar{x} = \frac{M}{m}$ as asserted.   □

Example 3.10.15 Compute the center of mass of the rod of Example 3.10.13.

Solution The moment of this rod is

$$M = \int_0^4 x\rho(x) \, dx = \int_0^4 \frac{x}{\sqrt{64-x^2}} \, dx.$$

Let $t = 64 - x^2$ with $dt = -2x \, dx$. Then

$$M = -\frac{1}{2} \int_{64}^{48} t^{-1/2} \, dt = -t^{1/2} \Big|_{64}^{48} = 8 - 4\sqrt{3} \text{ gm} - \text{cm}.$$

The mass of this rod is:

$$m = \int_0^4 \rho(x) \quad = \quad \int_0^4 \frac{1}{\sqrt{64-x^2}} \, dx = \arcsin \frac{x}{8} \Big|_0^4 = \arcsin \frac{1}{2} - \arcsin 0 = \frac{\pi}{6} \text{ gm}.$$

Thus, $\bar{x} = \dfrac{M}{m} \quad = \quad \dfrac{8 - 4\sqrt{3}}{\pi/6} = \dfrac{48 - 24\sqrt{3}}{\pi}$ cm.   □

As with a set of masses, the moment about $x = u$ of a rod of mass m is the same as the moment about $x = u$ of a point mass m located at $x = \bar{x}$.

Corollary 3.10.16 *Let R be a rod of continuous density $\rho(x)$, for $a \le x \le b$, with mass m and center of mass $\bar{x}$. Then*

$$X_u = m(\bar{x} - u).$$

Proof Since $\bar{x} = \frac{M}{m}$,

$$m(\bar{x} - u) = M - um = \int_a^b x\rho(x)\,dx - u\int_a^b \rho(x)\,dx = \int_a^b (x-u)\rho(x)\,dx = X_u. \qquad \square$$

It is easy to compute moments by using this corollary.

Example 3.10.17 Compute the moment of the rod of Example 3.10.13 about $x = 3$.

Solution In Example 3.10.15 we found the mass of this rod to be $\frac{\pi}{6}$ gm and its center of mass to be at $\frac{48 - 24\sqrt{3}}{\pi}$ cm. By Corollary 3.10.16,

$$X_3 = \frac{\pi}{6}\left(\frac{48 - 24\sqrt{3}}{\pi} - 3\right) = 8 - 4\sqrt{3} - \frac{\pi}{2} \text{ gm} - \text{cm}. \qquad \square$$

Centroid of a Region

We generalize the concept of center of mass to a two dimensional plate P of constant density. The center of mass of P only depends on the geometrical shape S of P and is called the centroid of S. We conclude with Pappus's Theorem which gives a formula for the volume of a solid of revolution in terms of the area of the region S being revolved and the position of its centroid.

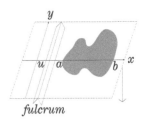

Figure 3.10.18
Moment About $x = u$

We begin by generalizing the concept of moment to a plate. Recall that the moment of a rod R about a point Q is a measure of the rotational force which results from placing R on a seesaw with fulcrum at Q. In our case, replace the seesaw by a board coordinatized as the xy–plane. Assume this board is balanced in a horizontal position on an edge (fulcrum). Now place a plate P, of variable density, on this board. The board will rotate. If the fulcrum lies along the line $x = u$, as in Figure 3.10.18, [or along the line $y = v$ as in Figure 3.10.19] we want to define the moment $X_u(P)$ [or $Y_v(P)$] to measure this rotational force. The derivation of formulas for $X_u(P)$ and $Y_v(P)$ in this situation requires the use of double integrals, and we return to this problem in Chapter 8. However, using our present knowledge, we can analyze the case when the plate P has constant density ρ_0.

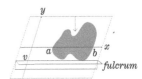

Figure 3.10.19
Moment About $y = v$

We give an intuitive argument to motivate the definitions of the moments $X_u(P)$ and $Y_v(P)$ of a plate P of constant density ρ_0. We restrict our analysis to the case where the shape S of P is most convenient. That is, assume S is bounded by the x–axis, the lines $x = a$, $x = b$ and the graph of a continuous function f with $f(x) \geq 0$ for $a \leq x \leq b$. Let $P = \{x_0, \ldots, x_n\}$ be a partition of $[a,b]$, and select $t_k \in [x_{k-1}, x_k]$ for $1 \leq k \leq n$. The partition P divides the region S into n vertical strips, and we approximate the moment of the strip which lies between $x = x_{k-1}$ and $x = x_k$ by the moment of the rectangle R_k with height $f(t_k)$ depicted in Figure 3.10.20. The mass of R_k is its area $f(t_k)(x_k - x_{k-1})$ times its density ρ_0. Its moment about any line is estimated by the moment of the point mass which concentrates the mass $\rho_0 f(t_k)(x_k - x_{k-1})$ of R_k at the center $\left(\frac{x_{k-1}+x_k}{2}, \frac{f(t_k)}{2}\right)$ of this rectangle. Hence the moment of the rectangle R_k about the line $x = u$ is approximately

$$\left(\frac{x_{k-1} + x_k}{2} - u\right)\rho_0 f(t_k)(x_k - x_{k-1}).$$

The moment of the rectangle R_k about $y = v$ is approximately

$$\left(\frac{f(t_k)}{2} - v\right)\rho_0 f(t_k)(x_k - x_{k-1}).$$

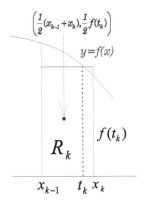

Figure 3.10.20
Approximating a Strip
by a Rectangle

Therefore the moment $X_u(P)$ of the entire plate P about $x = u$ is approximately

$$\sum_{k=1}^{n} \left(\frac{x_{k-1} + x_k}{2} - u \right) \rho_0 f(t_k)(x_k - x_{k-1}).$$

This sum is also an approximation of $\int_a^b (\frac{x+x}{2} - u)\rho_0 f(x)\, dx$. Thus we define $X_u(P) = \int_a^b (x - u)\rho_0 f(x)\, dx$ below. The moment $Y_v(P)$ of the entire plate P about $y = v$ is approximately

$$\sum_{k=1}^{n} \left(\frac{f(t_k)}{2} - v \right) \rho_0 f(t_k)(x_k - x_{k-1}).$$

This sum is also an approximation of $\int_a^b (\frac{f(x)}{2} - v)\rho_0 f(x)\, dx$. Thus we define $Y_v(P) = \int_a^b \left(\frac{f(x)}{2} - v \right) \rho_0 f(x)\, dx$ below.

Definition 3.10.21 *Let P be a plate of constant density ρ_0 which has the shape of the region S in the xy–plane bounded by the graph of the continuous function f, the x–axis and the lines $x = a$, $x = b$. Assume the graph of f lies entirely above the x–axis. Define the moment $X_u(P)$ of P about the vertical line $x = u$ and the moment $Y_v(P)$ of P about the horizontal line $y = v$ by:*

$$X_u(P) = \rho_0 \int_a^b (x - u)f(x)\, dx, \quad Y_v(P) = \frac{\rho_0}{2} \int_a^b f(x)\,[f(x) - 2v]\,\ dx.$$

Example 3.10.22 Let S be the region bounded by the x–axis and the parabola $y = 9 - x^2$. Find the moments about the lines $x = 2$ and $y = 1$ of the plate P of constant density 30 lbs/ft^2 which has the shape of the region S.

Solution This parabola intersects the x–axis when $9 - x^2 = 0$, i.e. at $x = \pm 3$. See Figure 3.10.23. By Definition 3.10.21, the moment of P about the line $x = 2$ is:

$$\begin{aligned}
X_2(P) &= \int_{-3}^{3} 30(x - 2)(9 - x^2)\, dx = 30 \int_{-3}^{3} -x^3 + 2x^2 + 9x - 18\, dx \\
&= 30 \left(-\frac{x^4}{4} + \frac{2x^3}{3} + \frac{9x^2}{2} - 18x \right) \Big|_{-3}^{3} = -2160 \text{ ft} - \text{lbs}.
\end{aligned}$$

By Definition 3.10.21, the moment of P about the line $y = 1$ is:

$$\begin{aligned}
Y_1(P) &= 15 \int_{-3}^{3} (9 - x^2)(9 - x^2 - 2)\, dx = 15 \int_{-3}^{3} x^4 - 16x^2 + 63\, dx \\
&= 3x^5 - 80x^3 + 945x \Big|_{-3}^{3} = 2808 \text{ ft} - \text{lbs}. \quad \square
\end{aligned}$$

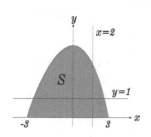

Figure 3.10.23
Moments of P

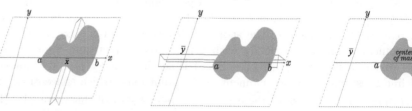

Figure 3.10.24 Center of Mass of a Plate

Consider a plate P whose shape is the region S. There is a unique line $x = \overline{x}$ where a fulcrum balances this plate. Similarly, there is a unique line $y = \overline{y}$ where a fulcrum balances P. Then the plate P balances on a pin placed at the point $(\overline{x}, \overline{y})$. See Figure 3.10.24. This point is the *center of mass* of the plate P. When P has constant density ρ_0, its center of mass is also called the *centroid* of the region S. We show that the centroid depends only on the geometry of the region S and not on the value of ρ_0.

Proposition 3.10.25 *Let P be a plate with constant density ρ_0 whose shape is the region S of area A. Assume S is bounded by the x–axis, the lines $x = a$, $x = b$ and the graph of the continuous function f which lies entirely above the x–axis. Then there is a unique point $(\overline{x}, \overline{y})$ such that the moments of P about the lines $x = \overline{x}$ and $y = \overline{y}$ are both zero. In particular,*

$$\overline{x} = \frac{X_0(P)}{\rho_0 A} = \frac{\int_a^b x f(x)\,dx}{\int_a^b f(x)\,dx}, \qquad \overline{y} = \frac{Y_0(P)}{\rho_0 A} = \frac{\int_a^b f(x)^2\,dx}{2\int_a^b f(x)\,dx}.$$

The center of mass of P is defined as $(\overline{x}, \overline{y})$ which only depends on S and not on ρ_0.

Proof We solve the following equations for $\overline{x}$ and $\overline{y}$:

$$0 = X_{\overline{x}}(P) = \int_a^b \rho_0(x - \overline{x})f(x)\,dx \text{ and } 0 = Y_{\overline{y}}(P) = \int_a^b \frac{\rho_0}{2}f(x)\,[f(x) - 2\overline{y}]\,dx.$$

These equations can be rewritten as:

$$\overline{x}\rho_0 \int_a^b f(x)\,dx = \rho_0 \int_a^b x f(x)\,dx \text{ and } \overline{y}\rho_0 \int_a^b f(x)\,dx = \frac{\rho_0}{2}\int_a^b f(x)^2\,dx.$$

Divide by $\rho_0 \int_a^b f(x)\,dx$ to produce the required formulas for $\overline{x}$ and $\overline{y}$. $\qquad\square$

We use moments and centers of mass of plates to define corresponding geometric properties of regions.

Definition 3.10.27 *Let S be a region in the xy–plane of area A. Let P be the plate of constant density one which has the shape S. Define the moments of S as:*

$$X_u(S) = X_u(P), \qquad Y_v(S) = Y_v(P).$$

The moment M_x of S refers to its moment about the y–axis, and the moment M_y of S refers its moment about the x–axis:

$$M_x = X_0(S), \qquad M_y = Y_0(S).$$

The center of mass $(\overline{x}, \overline{y})$ of P is called the centroid of S:

$$\overline{x} = \frac{M_x}{A}, \qquad \overline{y} = \frac{M_y}{A}.$$

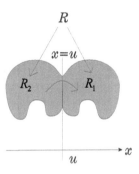

Figure 3.10.26
Symmetry of R with
Respect to $x = u$

Consider a region R that is symmetric with respect to the line $x = u$. That is, R consists of two congruent pieces: R_1 to the right of this line and R_2 to the left of this line. Moreover, reflection about the line $x = u$ transforms R_2 to R_1. See Figure 3.10.26. This transformation changes the sign of the moment about $x = u$: the moment of R_2 about $x = u$ is the negative of the moment of R_1 about $x = u$. By the additive property of the integral for $X_u(R)$, the moment of $R = R_1 \cup R_2$ about $x = u$ equals zero. Thus, $\overline{x} = u$. Similarly, if the region S is symmetric with respect to the line $y = v$, as in Figure 3.10.28, then $\overline{y} = v$.

Example 3.10.29 Find the centroid of the region S described in Example 3.10.22.

Solution Since S is symmetric about the y–axis ($x = 0$), we have $\overline{x} = 0$. The moment of S about the x–axis is

$$M_y = \frac{1}{2}\int_{-3}^{3}\left(9 - x^2\right)^2\,dx = \frac{1}{2}\int_{-3}^{3} 81 - 18x^2 + x^4\,dx$$

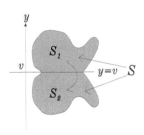

Figure 3.10.28
Symmetry of S with
Respect to $y = v$

$$= \frac{1}{2}\left(81x - 6x^3 + \frac{x^5}{5}\right)\Big|_{-3}^{3} = \frac{648}{5}.$$

The area A of S is : $A = \int_{-3}^{3} 9 - x^2 \, dx = 9x - \frac{x^3}{3}\Big|_{-3}^{3} = 36$

Hence $\qquad\qquad \overline{y} = \frac{M_y}{A} = \frac{648/5}{36} = \frac{18}{5}.$

Thus the centroid of S is $\left(0, \frac{18}{5}\right)$.

As with rods, the moment of a plate of mass m is the same as the moment of a point mass m located at its center of mass.

Corollary 3.10.30 *Let P be a plate of constant density and mass m whose shape is the region S. The region S is bounded by the x-axis, the lines $x = a$, $x = b$ and the graph of a continuous function f which lies entirely above the x-axis. Let $(\overline{x}, \overline{y})$ be the centroid of S. Then*

$$X_u(P) = m(\overline{x} - u) \quad and \quad Y_v(P) = m(\overline{y} - v).$$

Proof Let A denote the area of S, and let ρ_0 denote the density of P. Then $m = A\rho_0$. By Definitions 3.10.21 and 3.10.27:

$$X_u(P) = \int_a^b \rho_0(x-u)f(x) \, dx = \rho_0 \int_a^b xf(x) \, dx - u\rho_0 \int_a^b f(x) \, dx$$
$$= \rho_0 M_x - u\rho_0 A = \rho_0 A\overline{x} - u\rho_0 A = m\overline{x} - um,$$
$$Y_v(P) = \frac{\rho_0}{2}\int_a^b f(x)\left[f(x) - 2v\right] \, dx = \frac{\rho_0}{2}\int_a^b f(x)^2 \, dx - \rho_0 v \int_a^b f(x) \, dx$$
$$= \rho_0 M_y - \rho_0 vA = \rho_0 A\overline{y} - v\rho_0 A = m\overline{y} - vm. \qquad \square$$

Example 3.10.31 Find the moments, about the lines $x = -1$ and $y = 4$, of the plate P of shape S described in Example 3.10.22.

Solution In Example 3.10.29, we found the area of the region S to be 36 and its centroid to be $\left(0, \frac{18}{5}\right)$. The constant density of P is given as 30 lbs per square foot. Hence the mass of S is $(36)(30) = 1080$ lbs. By Corollary 3.10.30:

$$X_{-1}(P) = 1080\,(0 - (-1)) = 1080 \text{ ft} - \text{lbs},$$
$$Y_4(P) = 1080\left(\frac{18}{5} - 4\right) = -512 \text{ ft} - \text{lbs}. \qquad \square$$

We extend the concepts of moment and center of mass to plates with more general shapes. These extensions follow from Definition 3.10.21 and the principle that a moment of two disjoint plates is the sum of the moments of the individual plates. We call this principle the *additive property* of moments.

Corollary 3.10.32 *Let P be a plate of constant density ρ_0 and mass m which has the shape of the region S of area A. Assume S is bounded on top by the graph of the continuous function g, on the bottom by the graph of the continuous function f and on the sides by the lines $x = a$, $x = b$.*

(a) *The moments of P about the lines $x = u$ and $y = v$ are:*

$$X_u(P) = \rho_0 \int_a^b (x-u)\left[g(x) - f(x)\right] \, dx \quad and$$
$$Y_v(P) = \frac{\rho_0}{2}\int_a^b g(x)\left[g(x) - 2v\right] - f(x)\left[f(x) - 2v\right] \, dx.$$

(b) *The coordinates of the centroid $(\overline{x}, \overline{y})$ of S are:*

$$\overline{x} = \frac{M_x}{A} = \frac{\int_a^b x \left[g(x) - f(x)\right] \, dx}{\int_a^b g(x) - f(x) \, dx} \quad and \quad \overline{y} = \frac{M_y}{A} = \frac{\int_a^b g(x)^2 - f(x)^2 \, dx}{2 \int_a^b g(x) - f(x) \, dx}.$$

(c) *The moments of P are also given by:*

$$X_u(P) = m(\overline{x} - u) \quad and \quad Y_v(P) = m(\overline{y} - v).$$

Proof Let S' be the translation of S upwards by k units so that this translation of the graph of f on the interval $[a, b]$ lies entirely above the x–axis. Replacing S by S' replaces $f(x)$ by $f(x) + k$ and replaces $g(x)$ by $g(x) + k$. S and S' are two positions of the same plate with the same horizontal position and vertical positions which differ by k units. P refers to the case where the plate is placed at S while P' refers to the case where the plate is placed at S'. Then

$$X_u(P) = X_u(P') \quad and \quad Y_v(P) = Y_{v+k}(P').$$

Let R be the plate of constant density ρ_0 bounded by the graph of $y = f(x) + k$, the x–axis and the lines $x = a$, $x = b$.

(a) By the additive property of moments:

$$X_u(P) = X_u(P') = X_u(P' \cup R) - X_u(R).$$

Definition 3.10.21 applies to both $P' \cup R$ and R:

$$
\begin{aligned}
X_u(P) &= \int_a^b \rho_0 (x - u) \left[g(x) + k\right] \, dx - \int_a^b \rho_0 (x - u) \left[f(x) + k\right] \, dx \\
&= \rho_0 \int_a^b (x - u) \left[g(x) - f(x)\right] \, dx.
\end{aligned}
$$

By the additive property of moments and Definition 3.10.21:

$$
\begin{aligned}
Y_v(P) &= Y_{v+k}(P') = Y_{v+k}(P' \cup R) - Y_{v+k}(R) \\
&= \frac{\rho_0}{2} \int_a^b \left[g(x) + k\right] \left[g(x) + k - 2(v + k)\right] \, dx \\
&\quad - \frac{\rho_0}{2} \int_a^b \left[f(x) + k\right] \left[f(x) + k - 2(v + k)\right] \, dx \\
&= \frac{\rho_0}{2} \int_a^b g(x) \left[g(x) - 2v\right] - f(x) \left[f(x) - 2v\right] \, dx.
\end{aligned}
$$

(b), (c) The proofs are analogous to the proofs of Prop. 3.10.25 and Corollary 3.10.30 and are given as exercises. □

Symmetry considerations also apply to simplify the determination of the centroids of these more general plates. If the shape of P is the region S and S is symmetric with respect to $x = u$, then $\overline{x} = u$. If S is symmetric with respect to $y = v$, then $\overline{y} = v$.

Examples 3.10.33 (1) Find the centroid of a circle.

Solution The centroid of a circle is its center because a circle is symmetric with respect to its horizontal diameter and its vertical diameter.

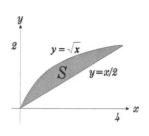

Figure 3.10.34
Plate P

(2) Let P be the plate of constant density 1 gm/cm^2 whose shape is the region S bounded by $y = \sqrt{x}$ and $y = \frac{x}{2}$. Find the moments of P with respect to $x = 1$ and $y = 3$.

Solution This curve and this line intersect when $\sqrt{x} = \frac{x}{2}$. Squaring this equation: $x = \frac{x^2}{4}$ or $x^2 - 4x = 0$. Hence $x = 0$ or $x = 4$. See Figure 3.10.34. By Corollary 3.10.32(a):

$$X_1(P) = \int_0^4 (x-1) \left[\sqrt{x} - \frac{x}{2} \right] dx = \int_0^4 x^{3/2} - x^{1/2} - \frac{x^2}{2} + \frac{x}{2} \, dx$$

$$= \frac{x^{5/2}}{5/2} - \frac{x^{3/2}}{3/2} - \frac{x^3}{6} + \frac{x^2}{4} \Big|_0^4 = \frac{4}{5} \text{ gm} - \text{cm}$$

$$Y_3(P) = \frac{1}{2} \int_0^4 \sqrt{x} \left[\sqrt{x} - 6 \right] - \frac{x}{2} \left[\frac{x}{2} - 6 \right] dx = \int_0^4 2x - 3x^{1/2} - \frac{x^2}{8} \, dx$$

$$= x^2 - 2x^{3/2} - \frac{x^3}{24} \Big|_0^4 = -\frac{8}{3} \text{ gm} - \text{cm}.$$

(3) Find the centroid of the region S of Example 2.

Solution The area of the region S is:

$$A = \int_0^4 \sqrt{x} - \frac{x}{2} \, dx = \frac{2x^{3/2}}{3} - \frac{x^2}{4} \Big|_0^4 = \frac{4}{3} \text{ cm}^2.$$

By Corollary 3.10.32(b):

$$\bar{x} = \frac{M_x}{A} = \frac{\int_0^4 x \left[\sqrt{x} - x/2 \right] \, dx}{4/3} = \frac{3}{4} \int_0^4 x^{3/2} - \frac{x^2}{2} \, dx = \frac{3x^{5/2}}{10} - \frac{x^3}{8} \Big|_0^4$$

$$= \frac{8}{5} \text{ gm} - \text{cm}$$

$$\bar{y} = \frac{M_y}{A} = \frac{\frac{1}{2} \int_0^4 x - x^2/4 \, dx}{4/3} = \frac{3x^2}{16} - \frac{x^3}{32} \Big|_0^4 = 1 \text{ gm} - \text{cm}.$$

Hence the centroid of S is $\left(\frac{8}{5}, 1 \right)$.

(4) Find the moments of the plate P of Example 2 about the lines $x = 2$ and $y = -2$.

Solution Since P has density one, its mass m equals its area: $m = \frac{4}{3}$ gm. By Corollary 3.10.32(c):

$$X_2(S) = m(\bar{x} - 2) = \frac{4}{3} \left(\frac{8}{5} - 2 \right) = -\frac{8}{15} \text{ gm} - \text{cm},$$

$$Y_{-2}(S) = m[\bar{y} - (-2)] = \frac{4}{3}(1+2) = 4 \text{ gm} - \text{cm}. \qquad \square$$

Pappus's Theorem gives a formula for the volume of a solid of revolution in terms of the area and centroid of the region which generates the solid.

Corollary 3.10.35 [Pappus's Theorem] *Let f and g be continuous functions with domain $[a, b]$. Let R be the region bounded by the graph of g on the top, the graph of f on the bottom and the lines $x = a$, $x = b$ on the sides. Let S be the solid of revolution obtained by rotating the region R around the line $x = u$ where $u \leq a$. Then the volume V of S equals the area A of R times the circumference of the circle C obtained by rotating the centroid of R around the line $x = u$:*

$$V = 2\pi (\bar{x} - u) A.$$

Proof By the shell method, the volume of S is given by:

$$
\begin{aligned}
V &= 2\pi \int_a^b (x-u)\left[g(x)-f(x)\right]\,dx \\
&= 2\pi \left\{ \int_a^b x\left[g(x)-f(x)\right]\,dx - u\int_a^b \left[g(x)-f(x)\right]\,dx \right\} \\
&= 2\pi\left(A\bar{x} - uA\right) = 2\pi\left(\bar{x}-u\right)A.
\end{aligned}
$$

Note the circle C has radius $\bar{x}-u$ and circumference $2\pi(\bar{x}-u)$. $\qquad\square$

Example 3.10.36 Let $0 \le r \le a$ and $0 \le k$. Find the volume V of the torus (doughnut) obtained by revolving the circle $(x-a)^2 + (y-b)^2 = r^2$ around the line $x = -k$.

Solution The circle being rotated has radius r and area πr^2. By symmetry, its centroid is its center (a, b) which is $a+k$ units from the axis of rotation. See Fig. 3.10.37. By Pappus's Thm. the volume of the torus is: $V = 2\pi(a+k)(\pi r^2) = 2\pi^2(a+k)r^2$.$\square$

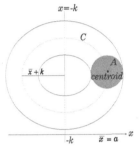

Figure 3.10.37 Torus

Historical Remarks

The concepts of moment and center of mass were well known to the Greeks at the time of Euclid, ca. 300 BCE. For example, they identified the centroid of a triangle as the intersection point of its three medians at a point $\frac{2}{3}$ of the distance from each vertex. They understood the concept of moment and its application to levers. However, it was Archimedes (287 to 212 BCE) who developed a mathematical theory of moments, proving theorems in mechanics from basic axioms. He applied this theory to construct powerful machines. In addition, he used centers of mass to compute areas and volumes.

This is how Archimedes uses moments to compute the area $A(R)$ of the region R. He selects a second region S of known area $A(S)$ and known centroid $(\bar{x}_S, \bar{y}_S)$. Thinking of R and S as thin plates of density one, he places them on the same interval $[a, b]$. For each $c \in [a, b]$, let R_c and S_c be the lengths of the cross–sections of R and S perpendicular to the x–axis at $x = c$. See Figure 3.10.38. The region S is selected so that there is a number k with

$$ xS_x = kR_x $$

for each $x \in [a, b]$. Think of the xy–plane as a board with fulcrum along the y–axis. If the entire mass $A(R)$ of R is placed at the point $x = -k$ on the x–axis, it has moment

$$ -kA(R) = -k\int_a^b R_x\,dx. $$

The region S has moment $\quad A(S)\bar{x}_S = \int_a^b xS_x\,dx = \int_a^b kR_x\,dx.$

Thus, the total moment of the region S and the point mass $A(R)$ at $x = -k$ equals zero. That is, $kA(R) = A(S)\bar{x}_S$. Hence

$$ A(R) = \frac{A(S)\bar{x}_S}{k}. \tag{3.10.1} $$

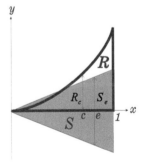

Figure 3.10.38
$xS_x = kR_x$

Archimedes called this formula the *mechanical method*. For example, let R be the region bounded by the parabola $y = x^2$, the x–axis and the line $x = 1$. Let S be the triangle bounded by the lines $y = \frac{x}{2}$, $y = -\frac{x}{2}$ and $x = 1$. See Figure 3.10.38. The triangle S has area $A(S) = \frac{1}{2}$ and centroid $(\bar{x}, \bar{y}) = \left(\frac{2}{3}, 0\right)$. Since $R_x = x^2$ and $S_x = x$, we have $xS_x = kR_x$ with $k = 1$. By (3.10.1):

$$ A(R) = \frac{(1/2)(2/3)}{1} = \frac{1}{3}. $$

Since Archimedes did not have calculus at his disposal, he could not give a rigorous argument that the condition $xS_x = kR_x$ implies equation (3.10.1). Therefore, he used the mechanical method to discover formulas for areas and volumes. Then he would give rigorous proofs of these formulas using standard Greek methods such as the *method of exhaustion* described in the Historical Remarks of Section 2. Archimedes used the mechanical method to compute a number of complicated areas. In addition he generalized this method to compute volumes of solids of revolution.

Pappus of Alexandria discovered Corollary 3.10.35 in about 300 CE. The simple geometric formula it gives for the volume of a solid of revolution is a significant application of calculus.

Summary

The concept of moment for a point mass, a set of point masses, a rod of variable density or a region of constant density should be understood. The reader should know how to compute the moments and centers of mass of these objects. She should also be able to apply Pappus's Theorem to compute the volume of a solid of revolution.

Basic Exercises

1. Find the moment of each of the following masses about $x = u$:
  (a) $u = 5$ m, $m = 3$ kg, $x = 2$ m;   **(b)** $u = -4$ ft, $m = 6$ lbs, $x = 1$ ft;
  (c) $u = -8$ cm, $m = 2$ gm, $x = -4$ cm;   **(d)** $u = 7$ in, $m = 4$ oz, $x = 10$ in;
  (e) $u = 0$ cm, $m = 5$ kg, $x = -3$ cm;   **(f)** $u = -5$ ft, $m = 3$ lbs, $x = 0$.

2. Find the moment of each of the following sets of masses about $x = u$.
  (a) $u = 0$ ft, $m_1 = 3$ lbs , $x_1 = -2$ ft, $m_2 = 4$ lbs, $x_2 = 8$ ft, $m_3 = 2$ lbs, $x_3 = -3$ ft.
  (b) $u = 5$ ft, $m_1 = 4$ lbs , $x_1 = 9$ ft, $m_2 = 3$ lbs, $x_2 = 0$ ft, $m_3 = 5$ lbs, $x_3 = -1$ ft, $m_4 = 2$ lbs $x_4 = -2$ ft.
  (c) $u = -3$ cm, $m_1 = 2$ gm, $x_1 = -2$ cm, $m_2 = 5$ gm, $x_2 = 1$ cm, $m_3 = 4$ gm, $x_3 = -7$ cm, $m_4 = 3$ gm, $x_4 = -10$ cm.
  (d) $u = 2$ cm, $m_1 = 5$ gm, $x_1 = 4$ cm, $m_2 = 3$ gm, $x_2 = -4$ cm, $m_3 = 2$ gm, $x_3 = -1$ cm, $m_4 = 6$ gm, $x_4 = 3$ cm, $m_5 = 4$ gm, $x_5 = -3$ cm.
  (e) $u = -4$ m, $m_1 = 2$ kg, $x_1 = -8$ m, $m_2 = 5$ kg, $x_2 = -6$ m, $m_3 = 4$ kg, $x_3 = 2$ m, $m_4 = 3$ kg, $x_4 = 5$ m, $m_5 = 9$ kg, $x_5 = -4$ m.

3. Find the center of mass of each set of masses in Exercise 2.

4. Use Corollary 3.10.9 to recompute each of the moments in Exercise 2.

5. Find the moment of each rod of variable density $\rho(x)$ about $x = u$.
  (a) $u = -1$ m and $\rho(x) = x^2 + x + 1$ kg/m for $0 \le x \le 2$ m.
  (b) $u = 4$ cm and $\rho(x) = (2x + 3)^{10}$ gm/cm for $-1 \le x \le 3$ cm.
  (c) $u = 2$ ft and $\rho(x) = \sqrt{3x + 1}$ lbs/ft for $0 \le x \le 8$ ft.
  (d) $u = 0$ and $\rho(x) = 1/\sqrt{25 - x^2}$ lbs/ft for $1 \le x \le 4$ ft.
  (e) $u = -3$ cm and $\rho(x) = x\sqrt{x + 10}$ gm/cm for $-6 \le x \le 2$ cm.
  (f) $u = 2$ feet and $\rho(x) = x^2/(2x + 5)^{3/2}$ lbs/ft for $-2 \le x \le 2$ ft.
  (g) $u = 0$ and $\rho(x) = \sec^2(\pi x^2/12)$ oz/in for $0 \le x \le 2$ in.
  (h) $u = 0$ and $\rho(x) = \sin(\pi x^2)$ kg/m for $0 \le x \le 1$ m.

6. Find the center of mass of each rod in Exercises 5(a)–(f).

7. Use Corollary 3.10.16 to recompute each of the moments in Exercise 5.

8. Find the moments of each plate P of constant density ρ_0 about $x = u$ and about $y = v$.
  (a) $\rho_0 = 4$, $u = -2$, $v = 4$; P is a rectangle with vertices $(2, 0)$, $(7, 0)$, $(2, 3)$, $(7, 3)$.

(b) $\rho_0 = 7$, $u = -1$, $v = -1$ and P is the upper semi-circle with radius r and center the origin.

(c) $\rho_0 = 2$, $u = 2$, $v = 1$ and P is the region bounded by $y = x^2$, the x–axis and the lines $x = -3$, $x = 2$.

(d) $\rho_0 = 5$, $u = 1$, $v = -2$ and P is the region bounded by $y = \sqrt{5x+1}$, the x–axis and $x = 7$.

(e) $\rho_0 = 10$, $u = 4$, $v = -3$ and P is the region bounded by $y = (2x+1)^{-3/2}$, the x–axis, the y–axis and the line $x = 4$.

(f) $\rho_0 = 8$, $u = -5$, $v = 4$ and P is the region bounded by $y = (3x+1)^{2/3}$, the x–axis and the line $x = 1$.

(g) $\rho_0 = 1$, $u = 2$, $v = 3$; P is the region bounded by $y = 3x$, $y = 20 - x$, the x–axis.

(h) $\rho_0 = 3$, $u = -3$, $v = 2$ and P is the region bounded by $y = |x|$, the x–axis and the lines $x = -5$, $x = 4$.

9. Find the center of mass of each plate in Exercise 8.

10. Use Corollary 3.10.30 to recompute each of the moments in Exercise 8.

11. Find the moments of each plate P of constant density ρ_0 about $x = u$ and about $y = v$.

(a) $\rho_0 = 9$, $u = 2$ and $v = 1$ with P the region bounded by $y = x^2$ and $y = x + 2$.

(b) $\rho_0 = 2$, $u = 0$ and $v = 0$ with P the region enclosed by the circle $(x-8)^2 + (y-5)^2 = 16$.

(c) $\rho_0 = 5$, $u = 2$ and $v = 3$ with P the region bounded by $y = 5x - x^2$, $y = 9x$ and $x = 2$.

(d) $\rho_0 = 3$, $u = -1$ and $v = 5$ with P the region bounded by $y = (x-3)^2$ and $y = 9 + 6x - x^2$.

(e) $\rho_0 = 12$, $u = 2$ and $v = 1$ with P the region bounded by $y = \sqrt{x}$, $y = \sqrt{6-x}$ and the y–axis.

(f) $\rho_0 = 8$, $u = -2$ and $v = -4$ with P the region bounded by $y = 6x - x^2 - 5$, $y = 2x - 1$ and $x = 5$.

(g) $\rho_0 = 10$, $u = 3$ and $v = 5$ with P the region bounded by $y = x^2 + 2x$ and $y = 6x - x^2 + 16$.

(h) $\rho_0 = 15$, $u = 0$ and $v = 0$ with P the region bounded by $y = \sqrt{4 - (x-2)^2}$ and $y = x - \frac{x^2}{4}$.

12. Find the center of mass of each plate in Exercise 11.

13. Use Corollary 3.10.32(c) to recompute each of the moments in Excersce 11.

14. Let $(\overline{x}, \overline{y})$ be the centroid of the region S bounded by the continuous function $x = h(y)$ on the right, the y–axis on the left and the lines $y = c$, $y = e$. Show that

$$\overline{x} = \frac{\int_c^e h(y)^2 \, dy}{2 \int_c^e h(y) \, dy}, \qquad \overline{y} = \frac{\int_c^e y h(y) \, dy}{\int_c^e h(y) \, dy}.$$

15. Find the centroid of each region.

(a) S_1 is the rectangle with vertices $(-3, -5)$, $(-3, 4)$, $(2, -5)$, $(2, 4)$.

(b) S_2 is the ellipse $\frac{(x-4)^2}{9} + \frac{(y+8)^2}{25} = 1$.

(c) S_3 is the triangle with vertices $(1, 2)$, $(5, 3)$, $(7, 1)$.

(d) S_4 is the region bounded by the parabola $x = y^2$ and the line $y = x - 2$.

(e) S_5 is the region bounded by the parabolas $y = x^2 + 6$ and $y = 24 - x^2$.

(f) S_6 is the region bounded by the curve $y = x^3 + x^2 + 1$ and the line $y = 2x + 1$.

(g) S_7 is the region bounded by $y = \sin x$ and $y = \cos x$, both with domain $[0, 2\pi]$.

(h) S_8 is the region bounded by $y = |x - 3|$ and $y = 21 - |4 - x|$.

16. Consider the region S in Figure 3.10.28 which is symmetric with respect to the line $y = v$. Show that $\bar{y} = v$.

17. (a) Prove Corollary 3.10.32(b).   **(b)** Prove Corollary 3.10.32(c).

18. Use Pappus's Theorem to compute the volume of each solid:
(a) a sphere of radius R;
(b) a cylinder of radius R and height H;
(c) a cone of radius R and height H;
(d) a frustrum of a cone of height H and radii r, R;
(e) the ice cream used to fill a cone of radius R and height H and top it with a
hemisphere of ice cream of radius R.

19. Revolve each region in Exercise 11 about the x–axis to obtain a solid of revolution
Use Pappus's Theorem to find the volume of each solid.

20. Revolve each region in Exercise 11 about the y–axis to obtain a solid of revolution
Use Pappus's Theorem to find the volume of each solid.

Challenging Problems

1. Use calculus to show that the three medians of a triangle intersect in a point, two
thirds of the distance from each vertex, which is the centroid of the triangle.

2. Let P be a plate with variable density $\rho(x, y)$ at the point (x, y) and shape the
region S in the xy–plane.
(a) Assume S is bounded by the x–axis, the lines $x = a$, $x = b$ and the graphs of the
continuous functions $y = f(x)$ and $y = g(x)$ with $f(x) \leq g(x)$ for $x \in [a, b]$. Find the
x–coordinate of the center of mass of P by crushing P into a rod on the x–axis.
(b) Assume S is bounded by the y–axis, the lines $y = c$, $y = e$ and the graphs of the
continuous functions $x = h(y)$ and $x = k(y)$ with $h(y) \leq k(y)$ for $y \in [c, e]$. Find the
y–coordinate of the center of mass of P by crushing P into a rod on the y–axis.

3.   Let R be the region in the first quadrant bounded above by the graph of the
continuous function f, below by the x–axis and on the sides by the lines $x = a$, $x = b$
with $0 \leq a \leq b$.
(a) Define and then locate the center of mass of the solid of revolution obtained by
rotating R around the x–axis.
(b) Define and then locate the center of mass of the solid of revolution obtained by
rotating R around the y–axis.

Theory

3.11 Existence of Definite Integrals

Prerequisites: Section 1.12 and the Appendix to Section 3.3

We prove that the definite integral of a continuous function exists. This fact was stated in Theorem 3.3.16 and has been used throughout this chapter. The proof of this theorem, in the first subsection, uses a strong property of continuous functions on closed intervals called *uniform continuity*. In the second subsection, the definition of the definite integral is generalized to bounded functions which may not be continuous. This definition is applied to compute the definite integrals of functions which have a finite number of jump discontinuities.

Integrals of Continuous Functions

The goal of this section is to prove Theorem 3.3.16: the definite integral of a continuous function exists. First, we review the definition of continuity and Lemma 3.3.39. This lemma gives a criterion for establishing the existence of a definite integral. The application of this lemma to prove Theorem 3.3.16 raises a technical problem. We resolve this problem by introducing the concept of uniform continuity and showing that a continuous function on a closed interval has this property.

Let $y = f(x)$ be a function with domain the interval J. By definition, f is continuous at $x = c \in J$ if $\lim_{x \to c} f(x) = f(c)$. Recall that $d(A, B) = |B - A|$ denotes the distance between A and B on the real number line. See Figure 3.11.1. We used this notation to rephrase the definition of this limit in (1.10.3): $\lim_{x \to c} f(x) = f(c)$ if for every $\epsilon > 0$ there is a corresponding $\delta > 0$ such that:

if $x \in J$ and $d(x, c) < \delta$, then $d(f(x), f(c)) < \epsilon$.

Figure 3.11.1
Distance Between A, B

We recall Lemma 3.3.39, from the appendix to Section 3, which we will use to show $\int_a^b f(x)\, dx$ exists. First, we introduce the notation. f is a function which is continuous on $[a, b]$. The integral $\int_a^b f(x)\, dx$ is approximated by Riemann sums. For each partition $P = \{x_0, \ldots, x_n\}$ of $[a, b]$, let $f(t_k^*)$ and $f(t_k^{**})$ be the minimum and maximum values of $f(x)$ on $[x_{k-1}, x_k]$. The lower Riemann sum

$$L(P, f) = \sum_{k=1}^{n} f(t_k^*)(x_k - x_{k-1}),$$

is the smallest of all Riemann sums while the upper Riemann sum

$$U(P, f) = \sum_{k=1}^{n} f(t_k^{**})(x_k - x_{k-1})$$

is the largest of all Riemann sums. In addition, every lower Riemann sum $L(P, f)$ is less than or equal to every upper Riemann sum $U(Q, f)$. That is, all the lower Riemann sums lie to the left of all the upper Riemann sums on the real number line. By definition of the definite integral, if there is a unique number I between all lower and upper Riemann sums, then $\int_a^b f(x)\, dx$ exists and has value I. Lemma 3.3.39 rephrases this condition to say: $\int_a^b f(x)\, dx$ exists when there is no gap on the real number line between the lower Riemann sums on the left and the upper Riemann sums on the right. See Figure 3.11.2.

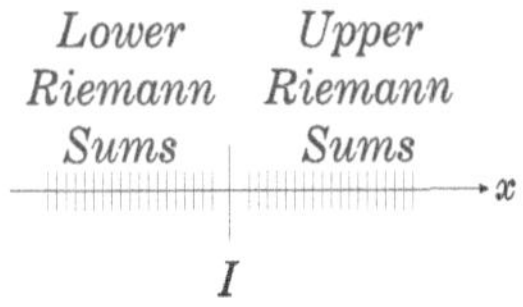

Lower Riemann Sums   *Upper Riemann Sums*

I

Figure 3.11.2
Lower and Upper Riemann Sums

Lemma 3.3.39 *Let f be a continuous function with domain $[a, b]$. Assume for every number $e > 0$, there is a partition P_e of $[a, b]$ such that*

$$U(P_e) - L(P_e) < e.$$

Then the definite integral $\int_a^b f(x)\, dx$ exists.

We use this lemma to prove Theorem 3.3.16.

Theorem 3.3.16 *Let f be a function which is continuous on the closed interval $[a, b]$. Then the definite integral $\int_a^b f(x)\, dx$ exists.*

Proof Let $P = \{x_0, \ldots, x_n\}$ be a partition of the interval $[a, b]$. Then

$$
\begin{aligned}
U(P, f) - L(P, f) &= \sum_{k=1}^n f(t_k^{**})(x_k - x_{k-1}) - \sum_{k=1}^n f(t_k^*)(x_k - x_{k-1}) \\
&= \sum_{k=1}^n \left[f(t_k^{**}) - f(t_k^*) \right] (x_k - x_{k-1}) \\
&= \sum_{k=1}^n d\left(f(t_k^*), f(t_k^{**}) \right) (x_k - x_{k-1}) \qquad (3.11.1)
\end{aligned}
$$

Let ϵ be any positive number. By the continuity of f at $x = t_k^*$, there is $\delta_k > 0$ such that if $d(t_k^*, t_k^{**}) < \delta_k$ then

$$d\left(f(t_k^*), f(t_k^{**}) \right) < \epsilon \quad \text{for} \ \ 1 \le k \le n. \qquad (3.11.2)$$

There is a subtle difficulty here. First we chose the partition $P = \{x_0, \ldots, x_n\}$ with small mesh, say mesh $P < \delta$. The choice of P determines the locations of the numbers t_k^*, t_k^{**} in each subinterval $[x_{k-1}, x_k]$ of $[a, b]$. Then t_k^* determines the value of δ_k. We know $d(t_k^*, t_k^{**}) \le d(x_{k-1}, x_k) < \delta$, but we need $d(t_k^*, t_k^{**}) < \delta_k$ to establish (3.11.2).

Problem How do we know that $\delta \le \delta_k$ for $1 \le k \le n$?

We complete the proof, assuming that we can solve this problem. Apply the inequalities (3.11.2) to the summands of (3.11.1):

$$U(P, f) - L(P, f) < \sum_{k=1}^n \epsilon(x_k - x_{k-1}) = \epsilon \sum_{k=1}^n (x_k - x_{k-1}) = \epsilon(b - a)$$

because the sum of the lengths of the n subintervals $[x_{k-1}, x_k]$ of $[a, b]$ is the length $b - a$ of the entire interval $[a, b]$. Now apply Lemma 3.3.39. Let $e > 0$ be given. Define $\epsilon = \frac{e}{b-a}$. By the above computation, if the mesh of P is sufficiently small, then

$$U(P, f) - L(P, f) < \epsilon(b - a) = \frac{e}{b - a}(b - a) = e.$$

By Lemma 3.3.39, $\int_a^b f(x)\, dx$ exists. $\square$

There is a special kind of function, called *uniformly continuous*, where we can easily solve the problem encountered in the proof of Theorem 3.3.16. When f is uniformly continuous and $\epsilon > 0$ is given, we can choose the same value of δ in the definition of continuity of f at $x = c$ for every $c \in [a, b]$. If P is a partition of $[a, b]$ with mesh less than δ, then all the δ_k in the proof above will have the same value δ. Thus we have a complete proof of Theorem 3.3.16 when f is uniformly continuous on $[a, b]$.

Definition 3.11.3 f *is a function whose domain contains the interval J. We say f is uniformly continuous on J if for every $\epsilon > 0$ there is a number $\delta > 0$ such that*

$$\text{if } x, c \in J \text{ and } d(x, c) < \delta, \text{ then } d(f(x), f(c)) < \epsilon.$$

To complete the proof of Theorem 3.3.16 we must show that a continuous function f with domain a closed interval $[a, b]$ is uniformly continuous on $[a, b]$. This fact is a consequence of the Bolzano–Weierstrass Theorem (Theorem 1.12.6). Recall this theorem says that a closed and bounded set, such as $[a, b]$, is compact, i.e. every sequence of numbers in $[a, b]$ has a subsequence which converges to a number in $[a, b]$.

Theorem 3.11.4 *Let f be a continuous function whose domain includes the closed interval $[a, b]$. Then f is uniformly continuous on $[a, b]$.*

Proof Assume this theorem is false. Then there is a number $\epsilon > 0$ such that for each positive integer n there are $x_n, c_n \in [a, b]$ with

$$d(x_n, c_n) < \frac{1}{n} \quad \text{and} \quad d(f(x_n), f(c_n)) \geq \epsilon. \tag{3.11.3}$$

By the Bolzano–Weierstrass Thm. the sequence S of the x_n has a convergent subsequence with limit $P \in [a, b]$. Since f is continuous at P, there is $\delta_P > 0$ such that:

$$\text{if } d(x, P) < \delta_P, \quad \text{then } d(f(x), f(P)) < \frac{\epsilon}{2}. \tag{3.11.4}$$

Since P is the limit of a subsequence of S, there are an infinite number of $x_n \in \left(P - \frac{\delta_P}{2}, P + \frac{\delta_P}{2}\right)$ with $n > \frac{2}{\delta_P}$, i.e. $\frac{1}{n} < \frac{\delta_P}{2}$. For each of these x_n:

$$d(f(x_n), f(P)) < \frac{\epsilon}{2} \tag{3.11.5}$$

since $d(P, x_n) < \delta_P$. By the triangle inequality (Proposition 1.10.3(c)):

$$d(P, c_n) \leq d(P, x_n) \quad + \quad d(x_n, c_n) < \frac{\delta_P}{2} + \frac{1}{n} < \frac{\delta_P}{2} + \frac{\delta_P}{2} = \delta_P.$$

By (3.11.4): $\quad d(f(P), f(c_n)) \quad < \quad \frac{\epsilon}{2}.$ $\hspace{3cm}$ (3.11.6)

Apply the triangle inequality:

$$d(f(x_n), f(c_n)) \quad \leq \quad d(f(x_n), f(P)) + d(f(P), f(c_n)) < \frac{\epsilon}{2} + \frac{\epsilon}{2} = \epsilon$$

by (3.11.5) and (3.11.6). This contradicts (3.11.3). Therefore, our initial assumption that this theorem is false is not correct, i.e. this theorem is true. $\hspace{1cm}$ □

A continuous function on an interval J which is not closed may be uniformly continuous, as in Example 1 below, or may not be uniformly continuous, as in Example 2. For each $\epsilon > 0$ in that example, we have to choose smaller values of δ in the definition of $\lim_{x \to c} f(x) = f(c)$, as c gets closer to the missing endpoint of J.

Examples 3.11.5 (1) Show $f(x) = 3x + 2$ is uniformly continuous on $J = (0, 1)$.

Solution Let $\epsilon > 0$. If $d(x, c) < \frac{\epsilon}{3}$, then

$$d(f(x), f(c)) = d(3x + 2, 3c + 2) = |(3x + 2) - (3c + 2)| = 3|x - c| = 3d(x, c) < \epsilon.$$

Thus take $\delta = \frac{\epsilon}{3}$ to show f is uniformly continuous.

(2) (a) Show $g(x) = \frac{1}{x}$ is not uniformly continuous on the interval $J = (0, 1)$.
(b) Find an explicit formula for $\delta_{c, \epsilon}$, as a function of c and ϵ, such that if $x, c \in (0, 1)$ and $d(x, c) < \delta_{c, \epsilon}$, then $d(f(x), f(c)) < \epsilon$.

Solution (a) Let $\epsilon > 0$ be given. g is uniformly continuous on $(0, 1)$ if there is a $\delta > 0$ such that:

$$\text{if } x, c \in (0, 1) \text{ with } d(x, c) < \delta, \text{ then } d(g(x), g(c)) = d\left(\frac{1}{x}, \frac{1}{c}\right) < \epsilon. \quad (3.11.7)$$

Note $c - \delta > 0$ because $\lim_{x \to 0^+} \frac{1}{x} = \infty$. However, when $c = \frac{\delta}{2}$ we have $c - \delta = -\frac{\delta}{2} < 0$, a contradiction. Thus there is no single number $\delta > 0$ which makes (3.11.7) true for every $c \in (0, 1)$. Hence g is not uniformly continuous on $(0, 1)$.

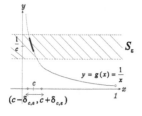

Figure 3.11.6

Choosing $\delta_{c,\epsilon}$

(b) Let $\epsilon > 0$ be given. Since g is a continuous function, all the $\delta_{c,\epsilon}$ exist. Our problem is to find an explicit formula for $\delta_{c,\epsilon}$. If $x \in (c - \delta_{c,\epsilon}, c + \delta_{c,\epsilon})$, then

$$g(c + \delta_{c,\epsilon}) = \frac{1}{c + \delta_{c,\epsilon}} < g(x) = \frac{1}{x} < g(c - \delta_{c,\epsilon}) = \frac{1}{c - \delta_{c,\epsilon}}$$

since g is decreasing. It suffices to find a number $\delta_{c,\epsilon}$ with $0 < \delta_{c,\epsilon} < c$ and

$$\frac{1}{c} - \epsilon < \frac{1}{c + \delta_{c,\epsilon}} < \frac{1}{c - \delta_{c,\epsilon}} < \frac{1}{c} + \epsilon. \quad (3.11.8)$$

Then $\frac{1}{c} - \epsilon < \frac{1}{x} < \frac{1}{c} + \epsilon$ and $d\left(\frac{1}{x}, \frac{1}{c}\right) < \epsilon$ as required. To establish (3.11.8), assume $\epsilon c < 1$. (If not, $\epsilon \geq \frac{1}{c} = g(c)$, and take $\delta_{c,\epsilon} = \delta_{c,1/2c}$.) We show $\delta_{c,\epsilon} = \frac{\epsilon c^2}{2}$ satisfies the inequalities of (3.11.8). See Figure 3.11.6.

$$\frac{1}{c - \delta_{c,\epsilon}} = \frac{1}{c - \epsilon c^2 / 2} = \frac{1}{c} + \frac{\epsilon c / 2}{c(1 - \epsilon c / 2)} = \frac{1}{c} + \frac{\epsilon}{2 - \epsilon c} < \frac{1}{c} + \epsilon \quad (3.11.9)$$

since $2 - \epsilon c > 1$. Similarly

$$\frac{1}{c + \delta_{c,\epsilon}} = \frac{1}{c + \epsilon c^2 / 2} = \frac{1}{c} - \frac{\epsilon c / 2}{c(1 + \epsilon c / 2)} = \frac{1}{c} - \frac{\epsilon}{2 + \epsilon c} > \frac{1}{c} - \epsilon. \quad (3.11.10)$$

Combine inequalities (3.11.9) and (3.11.10) to establish (3.11.8). $\qquad \square$

Integrals of Bounded Functions

The definition of the definite integral $\int_a^b f(x)\, dx$, given in Section 3, only applies to functions f which are continuous on $[a, b]$. In this subsection, we generalize the definition of $\int_a^b f(x)\, dx$ to functions f which may not be continuous on $[a, b]$ but are bounded there. That is, there is a number M with $|f(x)| \leq M$ for $x \in [a, b]$. We apply this definition to show $\int_a^b f(x)\, dx$ exists when f is continuous, except for a finite number of jump discontinuities.

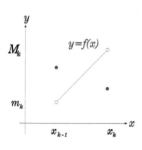

Figure 3.11.7

Bounded Function

Recall the definite integral of a continuous function f on $[a, b]$ is defined as the unique number which lies between all lower Riemann sums $L(P, f)$ and all upper Riemann sums $U(P, f)$. If $P = \{x_0, \ldots, x_n\}$ is a partition of $[a, b]$, then $L(P, f) = \sum_{k=1}^{n} f(t_k^*)(x_k - x_{k-1})$ where $f(t_k^*)$ is the minimum value of f on the interval $[x_{k-1}, x_k]$. Also, $U(P, f) = \sum_{k=1}^{n} f(t_k^{**})(x_k - x_{k-1})$ where $f(t_k^{**})$ is the maximum value of f on $[x_{k-1}, x_k]$. When f is not continuous, it may not have a minimum or maximum value on the closed interval $[x_{k-1}, x_k]$ as in Figure 3.11.7. However, by the Completeness Theorem 1.10.34, if f is bounded on $[a, b]$ then it has a greatest lower bound m_k and a least upper bound M_k on $[x_{k-1}, x_k]$. We generalize the definition of the lower and upper Riemann sums to bounded functions by using m_k instead of $f(t_k^*)$ and M_k instead of $f(t_k^{**})$. This allows us to define the definite integral of a bounded function.

Definition 3.11.8 *Let f be a function which is bounded on the closed interval $[a, b]$, and let $P = \{x_0, \ldots, x_n\}$ a partition of $[a, b]$.*
(a) *Let m_k be the greatest lower bound of f on the interval $[x_{k-1}, x_k]$. Define the lower Riemann sum by:*

$$L(P, f) = \sum_{k=1}^{n} m_k(x_k - x_{k-1}) \,.$$

(b) *Let M_k be the least upper bound of f on the interval $[x_{k-1}, x_k]$. Define the upper Riemann sum by:*

$$U(P, f) = \sum_{k=1}^{n} M_k(x_k - x_{k-1}) \,.$$

The following lemma says that, for a continuous function, the preceding definitions agree with the definitions of lower and upper Riemann sums given in Section 2.

Lemma 3.11.9 *Let f be a function which is continuous on $[a, b]$. Let $P = \{x_0, \ldots, x_n\}$ be a partition of $[a, b]$. Then Definition 3.2.28 and Definition 3.11.8 of the lower and upper Riemann sums agree:*

$$L(P, f) = \sum_{k=1}^{n} f(t_k^*)(x_k - x_{k-1}) = \sum_{k=1}^{n} m_k(x_k - x_{k-1})$$

$$U(P, f) = \sum_{k=1}^{n} f(t_k^{**})(x_k - x_{k-1}) = \sum_{k=1}^{n} M_k(x_k - x_{k-1}) \,.$$

Proof Since f is continuous on $[x_{k-1}, x_k]$, the minimum value $f(t_k^*)$ of f is the greatest lower bound m_k of f. Also, the maximum value $f(t_k^{**})$ of f is the least upper bound M_k of f. □

In Proposition 3.2.5 we showed that when f is continuous, $L(P, f)$ is the smallest of the Riemann sums $R(P, T, f)$ and $U(P, f)$ is the largest of these Riemann sums. This fact generalizes to Riemann sums of bounded functions.

Proposition 3.11.10 *Let f be a function which is bounded on $[a, b]$, and let $R(P, T, f)$ be a Riemann sum of f on $[a, b]$. Then*

$$L(P, f) \leq R(P, T, f) \leq U(P, f).$$

Proof Write the partition P of $[a, b]$ as $P = \{x_0, \ldots, x_n\}$. Write $T = \{t_1, \ldots, t_n\}$ with $x_{k-1} \leq t_k \leq x_k$ for $1 \leq k \leq n$. Then $m_k \leq f(t_k) \leq M_k$, and

$$L(P, f) = \sum_{k=1}^{n} m_k(x_k - x_{k-1}) \leq \sum_{k=1}^{n} f(t_k)(x_k - x_{k-1}) = R(P, T, f)$$

$$\leq \sum_{k=1}^{n} M_k(x_k - x_{k-1}) = U(P, f). \quad\square$$

Let P and P' be partitions of $[a, b]$. Recall P' is a refinement of P if $P' \subset P$. In Proposition 3.3.35 we showed that when we refine a partition, the lower Riemann sum of a continuous function becomes larger and the upper Riemman sum becomes smaller. This fact generalizes to Riemann sums of bounded functions.

Proposition 3.11.11 *Let P' be a refinement of the partition P of $[a, b]$. If f is a function which is bounded on $[a, b]$, then*

$$L(P, f) \leq L(P', f) \leq U(P', f) \leq U(P, f).$$

Proof Construct P' from the partition P by adding one point at a time. Thus it suffices to consider the case where P' contains exactly one more point $\overline{x}$ than P. Say $x_{k-1} < \overline{x} < x_k$. The only difference between $L(P, f)$ and $L(P', f)$ is that the summand

$$L_k = m_k(x_k - x_{k-1})$$

of $L(P, f)$ is replaced by the summands

$$L'_k = m'_k(\overline{x} - x_{k-1}) + m''_k(x_k - \overline{x})$$

of $L(P', f)$. Here m_k is the greatest lower bound of f on $[x_{k-1}, x_k]$, m'_k is the greatest lower bound of f on $[x_{k-1}, \overline{x}]$, and m''_k is the greatest lower bound of f on $[\overline{x}, x_k]$. Note m_k is the smaller of m'_k and m''_k. Hence $L_k \leq L'_k$ and $L(P, f) \leq L(P', f)$. Similarly $U(P', f) \leq U(P, f)$. $\quad\square$

The definition of the definite integral generalizes from continuous functions to bounded functions.

Definition 3.11.12 *Let f be a function which is bounded on $[a, b]$ where $a \leq b$.*
(a) *If there is a unique number A such that*

$$L(P, f) \leq A \leq U(P, f)$$

for all partitions P of $[a, b]$, we say the definite integral of f on the interval $[a, b]$ exists and equals A. Write $A = \int_a^b f(x)\, dx$.
(b) *Define $\int_b^a f(x)\, dx = -\int_a^b f(x)\, dx$.*

Note If f is continuous on $[a, b]$, it follows from Lemma 3.11.9 that the definition of $\int_a^b f(x)\, dx$ by Definition 3.3.9 and the preceding definition of $\int_a^b f(x)\, dx$ agree.

We want to establish the additive property for definite integrals of bounded functions. The proof of the additive property for definite integrals of continuous functions, Thm. 3.3.38, uses Lemma 3.3.37. The proofs of Lemma 3.3.37 and Thm. 3.3.38 do not use the definitions of $L(P, f)$ and $U(P, f)$. Rather, their proofs use Prop. 3.2.5 and 3.3.35 for continuous functions which we generalized as Prop. 3.11.10 and 3.11.11 for bounded functions. Hence the additive property holds for bounded functions.

Proposition 3.11.13 (Additive Property) *Assume the function f is bounded on the interval $[A, B]$ where A is the smallest of a, b, c and B is the largest of these numbers. If $\int_a^b f(x)\, dx$ and $\int_b^c f(x)\, dx$ exist, then $\int_a^c f(x)\, dx$ exists with*

$$\int_a^c f(x)\, dx \;=\; \int_a^b f(x)\, dx \;+\; \int_b^c f(x)\, dx.$$

We apply the definition of the definite integral of a bounded function to an important class of examples: functions with a finite number of jump discontinuities. See Figure 3.11.17. In this case, we partition $[a, b]$ at its points of discontinuity into a finite union of subintervals such that f is continuous on the *interior* of each subinterval. The following lemma says that the existence and value of a definite integral do not depend on the values of the function at the endpoints of the interval.

Lemma 3.11.14 *Let F be a continuous function with domain $[a,b]$. Let p and q be any two numbers. Define the function f with domain $[a,b]$ by:*

$$f(x) = \left\{ \begin{array}{ll} p & if\ x = a \\ F(x) & if\ a < x < b \\ q & if\ x = b \end{array} \right\}.$$

Then $\int_a^b f(x)\ dx$ exists, and $\int_a^b f(x)\ dx = \int_a^b F(x)\ dx$.

Proof The area given by $\int_a^b f(x)\ dx$ is depicted in Figure 3.11.15. Let $P = \{x_0, \ldots, x_n\}$ be any partition of the interval $[a,b]$. Let $F(t_k^*)$ denote the minimum value of F, and let $F(t_k^{**})$ denote the maximum value of F on the subinterval $[x_{k-1}, x_k]$. Let $F(m)$ denote the minimum value of F, and let $F(M)$ denote the maximum value of F on the interval $[a,b]$. If $x \in [a,b]$, then $F(m) \le f(x) \le F(M)$. Hence for $r \in \Re$:

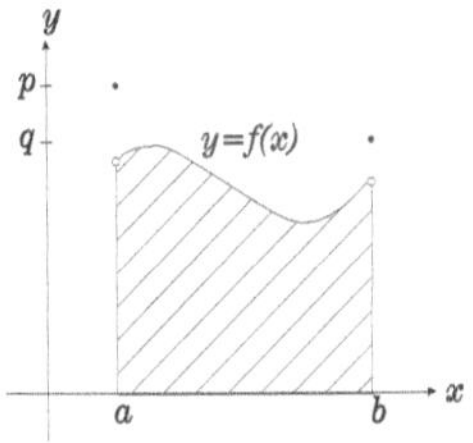

$$d(F(x), r) \le d(F(m), r)\ if\ F(x) \le r\ \ while$$
$$d(F(x), r) \le d(F(M), r)\ if\ r \le F(x).$$
In all cases:
$$d(F(x), r) \le d(F(m), r) + d(F(M), r). \tag{3.11.11}$$

Figure 3.11.15 $\int_a^b f(x)\ dx$

The only differences between $L(P, F)$ and $L(P, f)$ are the areas of the first and last rectangles. The height of the first rectangle used to compute $L(P, f)$ is either $F(t_1^*)$ or p while the height of the nth rectangle is either $F(t_n^*)$ or q. Thus,

$$\begin{aligned} d(L(P,F), L(P,f)) &= |L(P,F) - L(P,f)| \\ &\le |F(t_1^*) - p|(x_1 - x_0) + |F(t_n^*) - q|(x_n - x_{n-1}) \\ &\le d(F(t_1^*), p)(\text{mesh } P) + d(F(t_n^*), q)(\text{mesh } P) \\ &= [d(F(t_1^*), p) + d(F(t_n^*), q)]\,(\text{mesh } P) \\ &\le [d(F(m), p) + d(F(M), p) + d(F(m), q) + d(F(M), q)]\,(\text{mesh } P) \end{aligned}$$

using (3.11.11) with $r = p$ and with $r = q$. Similarly,

$$\begin{aligned} d(U(P,F), U(P,f)) &\le [d(F(t_1^{**}), p) + d(F(t_n^{**}), q)]\,(\text{mesh } P) \\ &\le [d(F(m), p) + d(F(M), p) + d(F(m), q) + d(F(M), q)]\,(\text{mesh } P). \end{aligned}$$

Let $\epsilon > 0$ be given. Then every partition P of $[a,b]$ with

$$\text{mesh } P < \frac{\epsilon/3}{d(F(m), p) + d(F(M), p) + d(F(m), q) + d(F(M), q)}$$

satisfies $d(L(P,F), L(P,f)) < \frac{\epsilon}{3}$ and $d(U(P,F), U(P,f)) < \frac{\epsilon}{3}$. $\qquad (3.11.12)$

If, in addition, $\text{mesh}(P)$ is small enough so that $U(P,F) - L(P,F) < \frac{\epsilon}{3}$, then

$$\begin{aligned} d(U(P,f), L(P,f)) &\le d(U(P,f), U(P,F)) + d(U(P,F), L(P,F)) + d(L(P,F), L(P,f)) \\ &< \frac{\epsilon}{3} + \frac{\epsilon}{3} + \frac{\epsilon}{3} = \epsilon \end{aligned}$$

by the triangle inequality. Hence there is a unique number which lies between all lower and upper Riemann sums of f, and $\int_a^b f(x)\ dx$ exists. By (3.11.12), this unique number is the same unique number which lies between all lower and upper Riemann sums of F. Hence $\int_a^b f(x)\ dx = \int_a^b F(x)\ dx$. $\qquad \square$

We are now ready to prove the existence of the definite integral of a function with a finite number of jump discontinuities. The proof uses the additive property of definite integrals and the preceding lemma.

Theorem 3.11.16 *Assume the function f is continuous at each point of the interval* $[a, b]$ *except for a finite number of jump discontinuities. Then $\int_a^b f(x)\,dx$ exists.*

Proof Let $P = \{x_0, \ldots, x_n\}$ be the partition of $[a, b]$ at the points of discontinuity of f, i.e. the jump discontinuities of f occur at $x_1, \ldots, x_{n-1}$ and possibly at x_0, x_n. Then P divides $[a, b]$ into subintervals $[x_{k-1}, x_k]$, $1 \le k \le n$, with f continuous on the interiors (x_{k-1}, x_k) of these subintervals. By Example 1.7.14 (3), the one–sided limits of f exist at each point where f has a jump discontinuity. Hence the one–sided limits at the endpoints of the subintervals $[x_{k-1}, x_k]$ exist:

$$\lim_{x \to x_{k-1}+} f(x) = A_{k-1}, \qquad \lim_{x \to x_k-} f(x) = B_k.$$

Define functions F_k with domain $[x_{k-1}, x_k]$ by: $F_k(x) = \begin{cases} A_{k-1} & \text{if } x = x_{k-1} \\ f(x) & \text{if } x_{k-1} < x < x_k \\ B_k & \text{if } x = x_k \end{cases}$

Then F_k is continuous on the interval $[x_{k-1}, x_k]$, and $\int_{x_{k-1}}^{x_k} F_k(x)\,dx$ exists. By the preceding lemma, $\int_{x_{k-1}}^{x_k} f(x)\,dx$ exists, and $\int_{x_{k-1}}^{x_k} F_k(x)\,dx = \int_{x_{k-1}}^{x_k} f(x)\,dx$. By the additive property, $\int_a^b f(x)\,dx$ exists with

$$\int_a^b f(x)\,dx \;=\; \int_a^{x_1} f(x)\,dx + \int_{x_1}^{x_2} f(x)\,dx \tag{3.11.13}$$

$$+ \cdots + \int_{x_{n-2}}^{x_{n-1}} f(x)\,dx + \int_{x_{n-1}}^b f(x)\,dx. \qquad \square$$

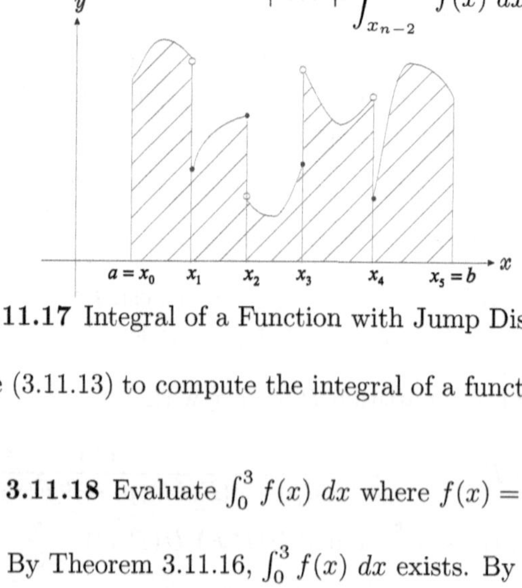

Figure 3.11.17 Integral of a Function with Jump Discontinuities

We use (3.11.13) to compute the integral of a function with jump discontinuities.

Example 3.11.18 Evaluate $\int_0^3 f(x)\,dx$ where $f(x) = \begin{Bmatrix} 4x + 3 & \text{if } 0 \le x \le 1 \\ 6x^2 & \text{if } 1 < x < 2 \\ 15 - 2x & \text{if } 2 \le x \le 3 \end{Bmatrix}$.

Solution By Theorem 3.11.16, $\int_0^3 f(x)\,dx$ exists. By (3.11.13),

$$\int_0^3 f(x)\,dx \;=\; \int_0^1 f(x)\,dx + \int_1^2 f(x)\,dx + \int_2^3 f(x)\,dx$$

$$=\; \int_0^1 4x + 3\,dx + \int_1^2 6x^2\,dx + \int_2^3 15 - 2x\,dx$$

$$=\; (2x^2 + 3x)\big|_0^1 + 2x^3\big|_1^2 + (15x - x^2)\big|_2^3 = 5 + 14 + 10 = 29. \qquad \square$$

Summary

The reader should understand uniform continuity and its use in the proof of the existence of the definite integral of a continuous function. She should know the definition of the definite integral of a bounded function and be able to use the additive property to compute the definite integral of a function with jump discontinuities.

Basic Exercises

1. Determine whether each function is uniformly continuous on the given interval J.
(a) $f(x) = 3 - 5x$ with $J = (3, 8]$.
(b) $g(x) = \tan x$ with $J = [0, \pi/4]$.
(c) $h(x) = \tan x$ with $J = (0, \pi/2)$.
(d) $k(x) = \sqrt{x^2 + 1}$ with $J = [1, 4]$.
(e) $m(1) = 4$ and $m(x) = \frac{x^2 - 4}{x^2 - 1}$ for $x \neq 1$ with $J = [0, 2]$.
(f) $n(0) = 0$ and $n(x) = x \sin(1/x)$ for $x \neq 0$ with $J = [0, \pi]$.
(g) $p(x) = 9x + 2$ with $J = \Re$.
(h) $q(x) = x^2$ with $J = [0, \infty)$.

2. Show if f is a continuous function with domain (a, b) such that $\lim\limits_{x \to a^+} f(x)$ and $\lim\limits_{x \to b^-} f(x)$ exist and are elements of $\Re$, then f is uniformly continuous on (a, b).

3. Show if f is not continuous on $[a, b]$, then f is not uniformly continuous on $[a, b]$.

4. Let f be continuous on $(a, b]$ with a vertical asymptote at $x = a$. Show f is not uniformly continuous on $(a, b]$.

5. (a) Let f be continuous on $[a, \infty)$ with a horizontal asymptote on the right. Show f is uniformly continuous on $[a, \infty)$.
(b) Let g be continuous on $\Re$ with horizontal asymptotes on both the left and the right. Show f is uniformly continuous on $\Re$.

6. Let f and g be uniformly continuous on the interval J.
(a) Show $f + g$ is uniformly continuous on J.
(b) If $k \in \Re$, show kf is uniformly continuous on J.
(c) Show fg is uniformly continuous on J.
(d) If $f \circ g$ is defined, show $f \circ g$ is uniformly continuous on J.
(e) If $0 \notin \text{Range } g$, is $\frac{f}{g}$ uniformly continuous on J? Justify your answer.

7. Evaluate each integral:
(a) $\int_{-2}^{4} \frac{x}{|x|} \, dx$;
(b) $\int_{0}^{4} \frac{x^2 - 4}{|x - 2|} \, dx$;
(c) $\int_{0}^{3} \sqrt[3]{\frac{|x-1|}{x^2 - 1}} \, dx$;

(d) $\int_{0}^{8} f(x) \, dx$ where $f(x) = \begin{cases} 3x + 1 & \text{if } x \leq 1 \\ 9 - 5x & \text{if } 1 < x \leq 4 \\ 6x + 2 & \text{if } x > 4 \end{cases}$;

(e) $\int_{-5}^{6} g(x) \, dx$ where $g(x) = \begin{cases} x^2 & \text{if } x < -3 \\ 10 - x^2 & \text{if } -3 \leq x \leq 0 \\ \sqrt{x} & \text{if } 0 < x < 4 \\ 9\sqrt{3x - 8} & \text{if } x \geq 4 \end{cases}$.

8. Complete the proof of Proposition 3.11.11 by showing $U(P', f) \leq U(P, f)$.

9. Assume the function f is bounded on the interval $[a, b]$.
(a) If $f(x) \geq 0$, show $\int_{a}^{b} f(x) \, dx \geq 0$. (b) If $f(x) \leq 0$, show $\int_{a}^{b} f(x) \, dx \leq 0$.

10. Assume f is bounded on the interval $[a, b]$. Show $\int_{a}^{b} kf(x) \, dx = k \int_{a}^{b} f(x) \, dx$.

11. Assume the functions f and g are bounded on the interval $[a, b]$ with $f(x) \leq g(x)$ for $x \in [a, b]$. Show $\int_{a}^{b} f(x) \, dx \leq \int_{a}^{b} g(x) \, dx$.

12. Assume f is bounded on the interval $[a, b]$. Show $\left| \int_{a}^{b} f(x) \, dx \right| \leq \int_{a}^{b} |f(x)| \, dx$.

13. By Definition 3.3.41(b), a function f with doman $[a, b]$ is called piecewise monotone if there is a partition $P = \{u_0, \ldots, u_n\}$ of $[a, b]$ such that f restricted to (u_{k-1}, u_k) is either increasing or decreasing for $1 \leq k \leq n$.
(a) Give an example of a continuous function which is not piecewise monotone.
(b) Use the ideas of the proofs of Theorems 3.3.45 and 3.11.16 to show that if f is a bounded piecewise monotone function with domain $[a, b]$, then $\int_{a}^{b} f(x) \, dx$ exists.

Challenging Problems

1. Each function is uniformly continuous on the given interval J. Find a formula for δ as a function of ϵ such that if $x, c \in J$ and $|x - c| < \delta$, then $|f(x) - f(c)| < \epsilon$.

    (a) $f(x) = 8x$ with $J = [1, \infty)$.   **(b)** $g(x) = 7x + 2$ with $J = \Re$.

    (c) $h(x) = |4x - 3|$ with $J = \Re$.   **(d)** $k(x) = x^2$ with $J = [0, 2]$.

    (e) $m(x) = \sqrt{x}$ with $J = [0, 3]$.   **(f)** $n(x) = \frac{1}{1+x^2}$ with $J = \Re$.

2. Each function is not uniformly continuous on the given interval J. Find a formula for δ as a function of ϵ and c such that if $x, c \in J$ and $|x - c| < \delta$, then $|f(x) - f(c)| < \epsilon$.

    (a) $f(x) = \frac{1}{x-1}$ with $J = (0, 1)$.   **(b)** $g(x) = \frac{1}{x+3}$ with $J = (-3, 0)$.

    (c) $h(x) = x^2$ with $J = \Re$.   **(d)** $k(x) = \frac{1}{x^2-1}$ with $J = (-1, 1)$.

3. Show $f(x) = \sqrt{x}$ is uniformly continuous on the interval $[0, \infty)$.

4. Let f and g be bounded functions with domain $[a, b]$ such that $\int_a^b f(x)\, dx$ exists Show if $f(x) = g(x)$ for all $x \in [a, b]$, other than a finite set of numbers, then $\int_a^b g(x)\, dx$ exists and $\int_a^b g(x)\, dx = \int_a^b f(x)\, dx$.

5. Determine whether $\int_0^1 f(x)\, dx$ exists where $f(x) = \left\{ \begin{array}{ll} 1 & \text{if } x \text{ is rational} \\ 0 & \text{if } x \text{ is irrational} \end{array} \right\}$.

6. Define the function g with domain $\Re$ by $g(x) = \left\{ \begin{array}{ll} \sin \frac{1}{x} & \text{if } x \neq 0 \\ 0 & \text{if } x = 0 \end{array} \right\}$.

(a) Determine whether $\int_0^1 xg(x)\, dx$ exists. **(b)** Determine whether $\int_0^1 g(x)\, dx$ exists.

3.12   Integral Invariants

Prerequisite: Sections 3.7 and 3.11[8]

In Sections 7, 9 and 10 we used intuitive methods to represent volume, distance, mass, work, force on a submerged object and moments as definite integrals. In this section we develop rigorous methods to justify these formulas. In the first subsection, we introduce an integral invariant as an abstraction of area. An integral invariant assigns a number to a continuous function on a closed interval. Moreover, it has three of the basic properties of area. We use these properties to prove that an integral invariant is given by a definite integral. Then we use integral invariants to identify the simple invariants of Sections 7, 9 and 10 as definite integrals. The second subsection introduces continuous functions of several variables which are used in the following two subsections. In the third subsection, we define a generalized integral invariant to identify the more complicated invariants of Sections 7, 9 and 10 as definite integrals. The proof of the theorem which identifies a generalized integral invariant as a definite integral is given in the last subsection.

Integral Invariants

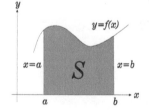

Figure 3.12.1

Area $A_a^b(f)$

In Section 2, we studied the area $A_a^b(f)$ of the region S, depicted in Figure 3.12.1, bounded by the graph of the continuous function f, the vertical lines $x = a$, $x = b$ and the x–axis. This area is an invariant of the function f and the interval $[a, b]$. In Section 3, we used several key properties of $A_a^b(f)$ to identify $A_a^b(f)$ with the definite integral $\int_a^b f(x)\, dx$. We define an integral invariant $I_a^b(f)$ to be an invariant of a

---

[8]Sections 3.9 and 3.10 are helpful, but not necessary, to appreciate some of the applications of this section. Section 3.11 is only required to read the last subsection.

continuous function f on an interval $[a, b]$ which has these key properties. We mimic the arguments of Section 3 to identify $I_a^b(f)$ with $\int_a^b f(x)\, dx$. We use integral invariants to identify volume as the integral of the cross-sectional area, distance as the integral of speed, mass as the integral of density and work as the integral of force. We begin by defining an invariant.

Definition 3.12.2 *An invariant is a function I which assigns the number $I_a^b(f)$ to every function f which is continuous on the closed interval $[a, b]$.*

There are many invariants in addition to the area invariant $A_a^b(f)$.

Examples 3.12.3 Let f, v, ρ, F, A be continuous functions with domain $[a, b]$.

(1) The arclength invariant $L_a^b(f)$ is the length of the curve $y = f(x)$ for $a \le x \le b$.

(2) The height invariant $H_a^b(f)$ is the change in height, $f(b) - f(a)$, of the graph of f on the interval $[a, b]$.

(3) The distance invariant $D_a^b(v)$ is the distance traveled by an object moving along the x–axis with velocity $v(t)$ from time $t = a$ to time $t = b$.

(4) The mass invariant $M_a^b(\rho)$ is the mass of the rod with density $\rho(x)$ which lies on the interval $[a, b]$ of the x–axis.

(5) The center of mass invariant $C_a^b(\rho)$ gives the point of the rod of Example 4 where it balances on a pin.

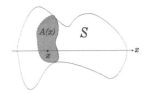

(6) The work invariant $W_a^b(F)$ is the work done by a force $F(x)$ on an object which moves along the x–axis form $x = a$ to $x = b$.

Figure 3.12.4
Cross-Sectional Area

(7) Let $A(x)$ be the cross–sectional area of the solid S at the point x on the x–axis. See Figure 3.12.4. That is, $A(x)$ is the area of the intersection of S with the plane P_x through x perpendicular to the x–axis. The volume invariant $V_a^b(A)$ is the volume of the slice of S, depicted in Figure 3.12.5, which lies between the planes P_a and P_b. □

There is no reason to expect that an invariant $I_a^b(f)$ will be given by the definite integral of f unless it has the fundamental properties of the area invariant $A_a^b(f)$. Such an invariant is called an integral invariant.

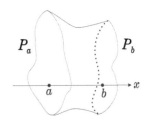

Figure 3.12.5
Slice of Volume $V_a^b(A)$

Definition 3.12.6 *An invariant $I_a^b(f)$ is called an integral invariant if it satisfies the following three properties.*

Increasing Property *If $f_1(x) \le f_2(x)$ for $x \in [a, b]$, then $I_a^b(f_1) \le I_a^b(f_2)$.*

Additive Property *If $a \le b \le c$ and $[a, c]$ is contained in the domain of f, then*

$$I_a^c(f) = I_a^b(f) + I_b^c(f).$$

Multiplicative Property *Let $C_k(x) = k$ be a constant function. Then*

$$I_a^b(C_k) = k(b - a).$$

We determine which of the invariants of Example 3.12.3 are integral invariants.

Examples 3.12.7 (1)   Show the area $A_a^b(f)$ of Fig. 3.12.1 is an integral invariant.

Solution The increasing property for area says: if region R_1 is contained in region R_2, then area R_1 is less than or equal to area R_2. The additive property for area says: the area of the union of two regions, which intersect in a line, is the sum of the areas of the two regions. The multiplicative property for area says: the area of a rectangle of dimensions k and $b - a$ is $k(b - a)$. Clearly all three of these properties are valid, and $A_a^b(f)$ is an integral invariant.

(2)   Show the arclength invariant $L_a^b(f)$ is not an integral invariant.

Solution The arclength invariant satisfies the additive property: the length of the union of two curves, which intersect in a point, is the sum of the lengths of the two curves. However, the increasing property does not hold: $x \leq 2$ for $0 \leq x \leq 1$, but the length of $f_1(x) = x$ on $[0, 1]$ is $\sqrt{2}$ while the length of $f_2(x) = 2$ on $[0, 1]$ is only 1. The multiplicative property also fails: $L_a^b(C_k)$ is $b - a$, not $k(b - a)$. Thus $L_a^b(f)$ is not an integral invariant.

(3)   Show the height invariant $H_a^b(f)$ is not an integral invariant.

Solution Height satisfies none of the three properties of an integral invariant. We leave the details as an exercise.

(4)   Show the distance invariant $D_a^b(|v|)$, considered as an invariant of the speed $|v(t)|$, is an integral invariant.

Solution The distance invariant satisfies the increasing property because an object which moves faster travels a greater distance. It satisfies the additive property because the distance traveled from time $t = a$ to time $t = c$ is the sum of the distance traveled from time $t = a$ to time $t = b$ and the distance traveled from time $t = b$ to time $t = c$. It satisfies the multiplicative property because an object moving with constant speed k from time $t = a$ to time $t = b$ travels $k(b - a)$ units. Thus distance $D_a^b(|v|)$ is an integral invariant of the speed $|v(t)|$. We leave it as an exercise to show the distance $D_a^b(v)$, as an invariant of the velocity $v(t)$, is not an integral invariant.

(5)   Show the mass invariant $M_a^b(\rho)$ is an integral invariant.

Solution Mass satisfies the increasing property: a rod which is denser has greater mass. It satisfies the additive property: if a rod is cut into two pieces, the mass of the entire rod is the sum of the masses of the individual pieces. It has the multiplicative property: a rod of constant density k has mass equal to k times the length of the rod. Thus $M_a^b(\rho)$ is an integral invariant.

(6)   Show the center of mass invariant $C_a^b(\rho)$ is not an integral invariant.

Solution The center of mass satisfies none of the three properties of an integral invariant. We leave the details as an exercise.

(7)   Show the work invariant $W_a^b(F)$ is an integral invariant.

Solution Work satisfies the increasing property: a stronger force does more work. It satisfies the additive property: the work done when an object moves from $x = a$ to $x = c$ is the sum of the work done in moving from $x = a$ to $x = b$ and the work done in moving from $x = b$ to $x = c$. It satisfies the multiplicative property: the work done by a constant force $F(x) = k$, when an object moves $b - a$ units, is $k(b - a)$. Thus $W_a^b(F)$ is an integral invariant.

(8)   Show the volume invariant $V_a^b(A)$ is an integral invariant.

Solution Volume satisfies the increasing property: a solid which has greater cross-sectional area at each point has greater volume. It satisfies the additive property: if a solid is cut into two pieces, the volume of the entire solid is the sum of the volumes of the pieces. It satisfies the multiplicative property: a solid of constant cross–sectional area has volume equal to the cross–sectional area times the length of the solid. Thus $V_a^b(A)$ is an integral invariant. $\square$

By its definition, an integral invariant $I_a^b(f)$ has similar properties to the area invariant $A_a^b(f)$. However, there is an even stronger resemblance. The following theorem says that an integral invariant $I_a^b(f)$ equals the definite integral $\int_a^b f(x)\, dx$. The reason is: the increasing, additive and multiplicative properties of $A_a^b(f)$ are all that are used to show $A_a^b(f)$ lies between the lower and upper Riemann sums of every partition of the interval $[a, b]$. Hence the same argument can be used to show $I_a^b(f)$ also lies between all lower and upper Riemann sums. However, $\int_a^b f(x)\, dx$ is the unique number with this property.

Theorem 3.12.8 [Integral Invariant Theorem] *If $I_a^b(f)$ is an integral invariant, then*

$$I_a^b(f) = \int_a^b f(x)\, dx.$$

Proof We show $I_a^b(f)$ lies between all lower and upper Riemann sums in the same manner that we demonstrated this fact for area in (3.2.2). Let $P = \{x_0, \ldots, x_n\}$ be a partition of the interval $[a, b]$. By the additive property, the value of $I_a^b(f)$ is the sum of the values of I on the n subintervals of the partition P:

$$I_a^b(f) = I_{x_0}^{x_1}(f) + \cdots + I_{x_{n-1}}^{x_n}(f). \tag{3.12.1}$$

By the Maximum Value Theorem, we can choose t_k^* and t_k^{**} in each interval $[x_{k-1}, x_k]$ so that f restricted to this interval has its minimum value at $x = t_k^*$ and its maximum value at $x = t_k^{**}$. That is,

$$f(t_k^*) \leq f(x) \leq f(t_k^{**}) \text{ for } x \in [x_{k-1}, x_k].$$
Then $I_{x_{k-1}}^{x_k}(C_{f(t_k^*)}) \leq I_{x_{k-1}}^{x_k}(f) \leq I_{x_{k-1}}^{x_k}(C_{f(t_k^{**})})$ by the increasing property;
$$f(t_k^*)(x_k - x_{k-1}) \leq I_{x_{k-1}}^{x_k}(f) \leq f(t_k^{**})(x_k - x_{k-1}) \text{ by the multiplicative property.}$$

Sum the preceding inequalities for $1 \leq k \leq n$:

$$\sum_{k=1}^{n} f(t_k^*)(x_k - x_{k-1}) \leq \sum_{k=1}^{n} I_{x_{k-1}}^{x_k}(f) \leq \sum_{k=1}^{n} f(t_k^{**})(x_k - x_{k-1}).$$

The sum on the left is the definition of the lower Riemann sum $L(P, f)$ while the sum on the right is the definition of the upper Riemann sum $U(P, f)$. By (3.12.1), the middle sum equals $I_a^b(f)$. Thus the preceding inequalities can be rewritten as:

$$L(P, f) \leq I_a^b(f) \leq U(P, f).$$

Since f is a continuous function on the interval $[a, b]$, $\int_a^b f(x)\, dx$ exists by Thm. 3.3.16. Hence $\int_a^b f(x)\, dx$ is the unique number which lies between the lower Riemann sum $L(P, f)$ and the upper Riemann sum $U(P, f)$ for every partition P. Therefore, the only possible value of $I_a^b(f)$ is $\int_a^b f(x)\, dx$. $\square$

By the Integral Invariant Theorem, the integral invariants of Example 3.12.7 are given by definite integrals.

Corollary 3.12.9 (a) *If* $f(x)$ *is continuous on* $[a, b]$, *then the area invariant satisfies* $A_a^b(f) = \int_a^b f(x)\ dx$.

(b) *If the velocity* $v(t)$ *of an object moving on the* x–*axis is continuous on the tim interval* $[a, b]$, *then the distance invariant satisfies:* $D_a^b(|v|) = \int_a^b v(t)\ dt$.

(c) *If the density* $\rho(x)$ *of a rod is continuous on* $[a, b]$, *then the mass invariant satsifies* $M_a^b(\rho) = \int_a^b \rho(x)\ dx$.

(d) *If the force* $F(x)$ *is continuous on* $[a, b]$, *then the work invariant satisfies:* $W_a^b(F) = \int_a^b F(x)\ dx$.

(e) *If the cross-sectional area* $A(x)$ *of a solid* S *is continuous on* $[a, b]$, *then th volume invariant satisfies:* $V_a^b(A) = \int_a^b A(x)\ dx$.

Functions of Several Variables

In the next subsection, we define generalized integral invariants to identify complicatec invariants as definite integrals. One of the properties of a generalized integral invarian uses a continuous function of several variables. These functions will be studied in detai in Section 7.3. We digress to exposit the basic facts about these functions which wil be used in the remainder of this section.

Definition 3.12.10 *Let* I *be an interval. The box* I^m *denotes the set of all* $\mathbf{x} = (x_1, \ldots, x_m)$ *with* $x_1, \ldots, x_m \in I$. *A real valued function* $g(x_1, \ldots, x_m)$ *of* m *variables with domain* I^m *assigns a number* $g(\mathbf{x}) = g(x_1, \ldots, x_m) \in \Re$ *to every* $\mathbf{x} \in I$.

Examples 3.12.11 (1) $f(x) = 5x^3 - 19$ is a real valued function of 1 variable ($m =$

(2) $g(x_1, x_2) = \sin(x_1 x_2 + 1)$ is a real valued function of 2 variables ($m = 2$).

(3) $h(x_1, x_2, x_3) = x_1^2 - x_1 x_2 x_3 + 8$ is a real valued function of 3 variables ($m = 3$).

We want to define when a real valued function $g(x_1, \ldots, x_m)$ of m variables is continuous. Recall that when $m = 1$, we defined $g(x)$ to be continuous at $x = c$ if $\lim_{x \to c} g(x) = g(c)$. We begin the generalization of this definition to $m > 1$ by defining the limit of $g(x_1, \ldots, x_m)$ as $(x_1, \ldots, x_m)$ approaches $(c_1, \ldots, c_m)$ to equal L if $g(x_1, \ldots, x_m)$ is arbitrarily close to L when each x_k is sufficiently close to c_k.

Definition 3.12.12 *Let an interval* I *and* $m \geq 1$ *be given. Let* $\mathbf{c} \in I^m$.
(a) *The punctured box* $I_\delta(\mathbf{c})$ *denotes the set of all* $\mathbf{x} \in I^m$, *other than* $\mathbf{c}$, *such that*

$$x_k \in I \quad and \quad d(x_k, c_k) < \delta \quad for \ 1 \leq k \leq m.$$

(b) *Assume* $g(\mathbf{x}) = g(x_1, \ldots, x_m) \in \Re$ *is a real valued function of* m *variables with domain* I^m. *We say*

$$\lim_{\mathbf{x} \to \mathbf{c}} g(\mathbf{x}) = L$$

if given any $\epsilon > 0$, *there is a corresponding* $\delta > 0$ *such that* $\mathbf{x} \in I_\delta(\mathbf{c})$ *implies that* $d(g(\mathbf{x}), L) < \epsilon$, *i.e.* $g(\mathbf{x}) \in (L - \epsilon, L + \epsilon)$.

Observe that when $m = 1$, this limit is the same as the usual limit $\lim_{x \to c} g(x) = L$. When $m = 2$, the set $I_\delta(c_1, c_2)$ is the square of side 2δ with its center $\mathbf{c} = (c_1, c_2)$ removed. See Figure 3.12.13. When $m = 3$, the set $I_\delta(\mathbf{c})$ is the cube of side 2δ with its center $\mathbf{c}$ removed. The following proposition gives six basic properties of these limits which are used to compute limits in concrete examples. The first five of these properties are proved in exactly the same way that the corresponding properties of functions of one variable were proved in Section 1.6. We postpone the details to Section 7.3 and only prove the sixth property here.

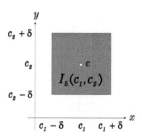

Figure 3.12.13
$I_\delta(c_1, c_2)$ with $m = 2$

Proposition 3.12.14 *Let I be an interval, $\mathbf{c} \in I^m$ and $f(\mathbf{x})$, $g(\mathbf{x})$ functions of m variables with domain I^m.*

(1) *If $\lim_{\mathbf{x} \to \mathbf{c}} f(\mathbf{x}) = L$ and $\lim_{\mathbf{x} \to \mathbf{c}} f(\mathbf{x}) = L'$, then $L = L'$.*

(2) *If $\lim_{\mathbf{x} \to \mathbf{c}} f(\mathbf{x}) = A$ and $\lim_{\mathbf{x} \to \mathbf{c}} g(\mathbf{x}) = B$, then $\lim_{\mathbf{x} \to \mathbf{c}} [f(\mathbf{x}) + g(\mathbf{x})] = A + B$.*

(3) *If $\lim_{\mathbf{x} \to \mathbf{c}} f(\mathbf{x}) = A$ and $\lim_{\mathbf{x} \to \mathbf{c}} g(\mathbf{x}) = B$, then $\lim_{\mathbf{x} \to \mathbf{c}} [f(\mathbf{x})g(\mathbf{x})] = AB$.*

(4) *If $\lim_{\mathbf{x} \to \mathbf{c}} f(\mathbf{x}) = A$ and $\lim_{t \to \mathbf{c}} g(\mathbf{x}) = B$ with $B \neq 0$, then $\lim_{\mathbf{x} \to \mathbf{c}} \dfrac{f(\mathbf{x})}{g(\mathbf{x})} = \dfrac{A}{B}$.*

(5) *Let J be an open interval containing B with $h : J \to \Re$ continuous at B. If $\lim_{\mathbf{x} \to \mathbf{c}} g(\mathbf{x}) = B$, then $\lim_{\mathbf{x} \to \mathbf{c}} h(g(\mathbf{x})) = h\left(\lim_{\mathbf{x} \to \mathbf{c}} g(\mathbf{x})\right) = h(B) = L$.*

(6) *Let $k : [a, b] \to \Re$, let $1 \leq i \leq m$ be fixed and assume $\lim_{x \to c_i} k(x) = L$. Define $K(\mathbf{x}) = K(x_1, \ldots, x_m) = k(x_i)$ for $\mathbf{x} \in I^m$. Then $\lim_{\mathbf{x} \to \mathbf{c}} K(\mathbf{x}) = L$.*

Proof (6) Given $\epsilon > 0$, there is $\delta > 0$ such that $x \in I_\delta(c_i)$ implies $k(x) \in (L - \epsilon, L + \epsilon)$. If $\mathbf{x} \in I_\delta(\mathbf{c})$, then $x_i \in I_\delta(c_i)$ and $K(\mathbf{x}) = k(x_i) \in (L - \epsilon, L + \epsilon)$. Hence $\lim_{\mathbf{x} \to \mathbf{c}} K(\mathbf{x}) = L$. $\square$

The following examples use limit properties to compute explicit limits.

Examples 3.12.15 (1) Let $\mathbf{x} = (x_1, x_2)$ and $\mathbf{c} = (3, -2)$. Evaluate $\lim_{\mathbf{x} \to \mathbf{c}} \dfrac{5x_1 x_2^3}{4x_1 - 2x_2 + 1}$.

Solution By Property 6:

$$\lim_{\mathbf{x} \to \mathbf{c}} 5x_1 = 15, \quad \lim_{\mathbf{x} \to \mathbf{c}} x_2^3 = -8, \quad \lim_{\mathbf{x} \to \mathbf{c}} 4x_1 = 12 \quad \text{and} \quad \lim_{\mathbf{x} \to \mathbf{c}} (-2x_2 + 1) = 5.$$

By Property 3, $\lim_{\mathbf{x} \to \mathbf{c}} 5x_1 x_2^3 = -120$. By Property 2, $\lim_{\mathbf{x} \to \mathbf{c}} (4x_1 - 2x_2 + 1) = 17$. By Property 4, $\lim_{\mathbf{x} \to \mathbf{c}} \dfrac{5x_1 x_2^3}{4x_1 - 2x_2 + 1} = -\dfrac{120}{17}$.

(2) Let $\mathbf{x} = (x_1, x_2, x_3)$ and $\mathbf{c} = (1, 5, 4)$. Evaluate $\lim_{\mathbf{x} \to \mathbf{c}} \ln\left(1 + x_1^3 x_2^2 x_3\right)$.

Solution By Property 6:

$$\lim_{\mathbf{x} \to \mathbf{c}} 1 = 1, \quad \lim_{\mathbf{x} \to \mathbf{c}} x_1^3 = 1, \quad \lim_{\mathbf{x} \to \mathbf{c}} x_2^2 = 25 \quad \text{and} \quad \lim_{\mathbf{x} \to \mathbf{c}} x_3 = 4.$$

By Property 3, $\lim_{\mathbf{x} \to \mathbf{c}} x_1^3 x_2^2 x_3 = 100$. By Property 2, $\lim_{\mathbf{x} \to \mathbf{c}} (1 + x_1^3 x_2^2 x_3) = 101$. By Property 5, $\lim_{\mathbf{x} \to \mathbf{c}} \ln\left(1 + x_1^3 x_2^2 x_3\right) = \ln\left[\lim_{\mathbf{x} \to \mathbf{c}} \left(1 + x_1^3 x_2^2 x_3\right)\right] = \ln 101$. $\square$

Just as in the case of a function of one variable, the function $f(x_1, \ldots, x_m)$ is continuous at $(c_1, \ldots, c_m)$ if the limit and the value of the function at $(c_1, \ldots, c_m)$ agree.

Definition 3.12.16 *Let I be an interval and $f(\mathbf{x})$ a function of m variables with domain I^m.*
(a) *f is continuous at $\mathbf{c} \in I^m$ if $\lim_{\mathbf{x} \to \mathbf{c}} f(\mathbf{x}) = f(\mathbf{c})$.*
(b) *f is continuous on I^m if f is continuous at every $\mathbf{c} \in I^m$.*

Observe when $m = 1$ this definition of continuity agrees with the usual one. The following properties of continuous functions are immediate consequences of Prop. 3.12.14.

Corollary 3.12.17 *Let I be an interval and $f(\mathbf{x})$, $g(\mathbf{x})$ functions of m variables wit domain I^m.*

(1) *If f, g are continuous on I^m, then $f + g$ is continuous on I^m.*

(2) *If f, g are continuous on I^m, then fg is continuous on I^m.*

(3) *If f, g are continuous on I^m and $0 \notin \operatorname{Range} g$, then $\frac{f}{g}$ is continuous on I^m.*

(4) *Let J be an open interval containing $\operatorname{Range} g$, and let $h : J \to \Re$. If g i continuous on I^m and h is continuous on J, then $h \circ g$ is continuous on I^m.*

(5) *Let $k : I \to \Re$ be continuous, and let $1 \le i \le m$ be fixed. Define $K(\mathbf{x}) = K(x_1, \ldots, x_m) = k(x_i)$ for $\mathbf{x} \in I^m$. Then K is continuous on I^m.*

The following examples use the preceding corollary to deduce that a functio $f(x_1, \ldots, x_m)$, constructed from continuous functions of one variable, is continuous.

Examples 3.12.18 (1) Show $F(x_1, x_2) = \frac{7x_1^3 x_2^5}{x_1^2 + x_2^2 + 1}$ is continuous on $\Re^2$.

  Solution The functions

$$f(x_1, x_2) = 7x_1^2, \quad g(x_1, x_2) = x_2^5, \quad h(x_1, x_2) = x_1^2 \text{ and } i(x_1, x_2) = x_2^2 + 1$$

are continuous on $\Re^2$ by Corollary 3.12.17 (5). Then $(fg)(x_1, x_2) = 7x_1^2 x_2^5$ i continuous on $\Re^2$ by Corollary 3.12.17 (2), and $(h + i)(x_1, x_2) = x_1^2 + x_2^2 + 1$ i continuous on $\Re^2$ by Corollary 3.12.17 (1). By Corollary 3.12.17 (3), $F(x_1, x_2) = \left(\frac{fg}{h+i}\right)(x_1, x_2)$ is continuous on $\Re^2$.

(2) Show $G(x_1, x_2, x_3) = \sin(x_1 + x_2^2 + x_3^3)$ is continuous on $\Re^3$.

  Solution The functions

$$f(x_1, x_2, x_3) = x_1, \quad g(x_1, x_2, x_3) = x_2^2 \text{ and } h(x_1, x_2, x_3) = x_3^3$$

are continuous on $\Re^3$ by Corollary 3.12.17 (5). By Corollary 3.12.17 (1), $(f + g + h)(x_1, x_2, x_3) = x_1 + x_2^2 + x_3^3$ is continuous on $\Re^3$. Let $k(x) = \sin x$. By Cor. 3.12.17 (4), $G(x_1, x_2, x_3) = k\left((f + g + h)(x_1, x_2, x_3)\right)$ is continuous on $\Re^3$.

(3) Show if $f(x)$ is a continuous function on the interval I, then the function $H(x_1, x_2, x_3) = (x_1 + x_2)f(x_3)$ is continuous on I^3.

  Solution The functions

$$g(x_1, x_2, x_3) = x_1, \quad h(x_1, x_2, x_3) = x_2 \text{ and } i(x_1, x_2, x_3) = f(x_3)$$

are continuous on I^3 by Corollary 3.12.17 (5). By Corollary 3.12.17 (1), $(g + h)(x_1, x_2, x_3) = x_1 + x_2$ is continuous on I^3. By Corollary 3.12.17 (2), $H(x_1, x_2, x_3) = [(g + h)i](x_1, x_2, x_3)$ is continuous on I^3.

(4) Show if $f(x)$, $g(x)$, $h(x)$ are continuous functions on the interval I, then

$$K(x_1, x_2, x_3) = \sqrt{f(x_1)^2 + g(x_2)^2 + h(x_3)^2}$$

is continuous on I^3.

  Solution The functions

$$F(x_1, x_2, x_3) = f(x_1)^2, \quad G(x_1, x_2, x_3) = g(x_2)^2 \text{ and } H(x_1, x_2, x_3) = h(x_3)^2$$

are continuous on I^3 by Corollary 3.12.17 (5). Then

$$(F + G + H)(x_1, x_2, x_3) = f(x_1)^2 + g(x_2)^2 + h(x_3)^2$$

is continuous on I^3 by Cor. 3.12.17 (1). Let $p(x) = \sqrt{x}$. By Cor. 3.12.17 (4), $K(x_1, x_2, x_3) = p\left((F + G + H)(x_1, x_2, x_3)\right)$ is continuous on I^3. □

Generalized Integral Invariants

Generalized integral invariants are used to identify invariants $A_a^b(f)$ which are more complicated than just the integral of f. Generalized integral invariants are additive. However, the increasing and multiplicative properties of an integral invariant are replaced by the more general bounded property. We state the Generalized Integral Invariant Theorem which identifies a generalized integral invariant as a definite integral. Its proof is given in the next subsection. We use generalized integral invariants to derive the formulas that compute volume by the shell method, find the force on a submerged object and compute moments of rods of variable density.

Motivating Example 3.12.19 Let f be a continuous positive function with domain $I = [B, C]$, $0 \leq B \leq C$. Assume $A_a^b(f)$ is defined for $B \leq a \leq b \leq C$ and is an additive invariant which satisfies the property:

if $m \leq xf(x) \leq M$ for $a \leq x \leq b$, then $m(b - a) \leq A_a^b(f) \leq M(b - a)$.   (3.12.2)

Try to apply the method used to prove the Integral Invariant Theorem to identify the invariant $A_a^b(f)$ as $\int_a^b xf(x)\, dx$.

Solution Let $P = \{x_0, \ldots, x_n\}$ be a partition of $[a, b]$. Let $f(t_k^*)$ and $f(t_k^{**})$ be the minimum and maximum values of f on the interval $[x_{k-1}, x_k]$. For $x \in [x_{k-1}, x_k]$,

$$x_{k-1} f(t_k^*) \leq xf(x) \leq x_k f(t_k^{**})$$

Apply property (3.12.2) with $[a, b] = [x_{k-1}, x_k]$, $m = x_{k-1} f(t_k^*)$ and $M = x_k f(t_k^{**})$:

$$x_{k-1} f(t_k^*)(x_k - x_{k-1}) \leq A_{x_{k-1}}^{x_k}(f) \leq x_k f(t_k^{**})(x_k - x_{k-1}).$$ (3.12.3)

Sum these inequalities for $1 \leq k \leq n$, and use the additive property of I:

$$\sum_{k=1}^n x_{k-1} f(t_k^*)(x_k - x_{k-1}) \leq \sum_{k=1}^n A_{x_{k-1}}^{x_k}(f) = A_a^b(f) \leq \sum_{k=1}^n x_k f(t_k^{**})(x_k - x_{k-1}).$$

(3.12.4)

This presents a technical problem. The left sum in (3.12.4) is not a Riemann sum of $xf(x)$, because the expression $x_{k-1} f(t_k^*)$ involves two numbers in the interval $[x_{k-1}, x_k]$, namely t_k^* and x_{k-1}, while only one such number is allowed. For this reason, this expression is not a summand of any Riemann sum $R(P, T, xf(x))$. Similarly, the right sum in (3.12.4) is not a Riemann sum of $xf(x)$. In particular, we can not interpret (3.12.4) as saying $L(P, xf(x)) \leq A_a^b(f) \leq U(P, xf(x))$. If we knew these inequalities, we could conclude that $A_a^b(f)$ equals $\int_a^b xf(x)\, dx$. $\square$

We define a generalized integral invariant as an additive invariant which satisfies inequalities like those of (3.12.3) in the preceding example. We call these inequalities the *bounded property*. We study the invariant $A_a^b(f)$ for a fixed continuous function f and denote $A_a^b(f)$ as A_a^b.

Definition 3.12.20 *Assume I is a closed interval such that for every closed interval $[a, b]$ contained in I there is a number A_a^b which has the following properties.*

Additive Property *If $a \leq b \leq c$ and the interval $[a, c]$ is contained in I, then*

$$A_a^c = A_a^b + A_b^c.$$

Bounded Property *There is a continuous function $g(\mathbf{t})$ on I^m such that for each closed interval $[a, b]$ contained in I there are $\mathbf{r}, \mathbf{s} \in [a, b]^m$ with*

$$g(\mathbf{r})(b - a) \leq A_a^b \leq g(\mathbf{s})(b - a).$$

A_a^b is called a generalized integral invariant, and g is called its bounding function.

In Example 3.12.19, we tried to identify an invariant $A_a^b = A_a^b(f)$ as $\int_a^b x f(x)\, dx$ for a continuous positive function f and $0 \le B \le a \le b \le C$. We show A_a^b is a generalized integral invariant with $I = [B, C]$. A_a^b is assumed to be additive. Let

$$g(x_1, x_2) = x_2 f(x_1).$$

g is continuous by Corollary 3.12.17 (2), (5). Now (3.12.3) can be rewritten as:

$$g(t_k^*, x_{k-1})(x_k - x_{k-1}) \le A_{x_{k-1}}^{x_k} \le g(t_k^{**}, x_k)(x_k - x_{k-1}).$$

This inequality is valid for all $B \le x_{k-1} \le x_k \le C$. Hence A_a^b satisfies the bounded property and is a generalized integral invariant with $m = 2$ and bounding function g. The following theorem identifies a generalized integral invariant as a definite integral.

Theorem 3.12.21 [Generalized Integral Invariant Theorem] *Let A_a^b be a generalized integral invariant defined when $a \le b$ are in the closed interval I. Let g be the bounding function of this invariant. Then $G(x) = g(x, \ldots, x)$ is continuous on I, and*

$$A_a^b = \int_a^b G(x)\, dx = \int_a^b g(x, \ldots, x)\, dx.$$

In Example 3.12.19, $g(x_1, x_2) = x_2 f(x_1)$ and $G(x) = g(x, x) = x f(x)$ which is continuous. The Generalized Integral Invariant Theorem identifies A_a^b as $\int_a^b G(x)\, dx = \int_a^b x f(x)\, dx$. We use the Generalized Integral Invariant Theorem to derive the formula that computes the volume of a solid of revolution by the shell method.

Theorem 3.7.28 [Shell Method] *Let g, h be continuous functions with domain $[a, b]$ such that $0 \le a \le b$ and $g(x) \le h(x)$ for $x \in [a, b]$. Let R be the region in the xy–plane bounded by the graphs of g, h and the lines $x = a$, $x = b$. Let S be the solid of revolution obtained by revolving R around the y–axis. Then the volume V of S is given by:*

$$V = 2\pi \int_a^b x\left[h(x) - g(x)\right]\, dx. \tag{3.12.5}$$

Proof In Section 3.7, we showed the general case of this theorem follows from the case $g(x) = 0$. Thus it suffices to derive (3.12.5) in this case:

$$V = 2\pi \int_a^b x h(x)\, dx. \tag{3.12.6}$$

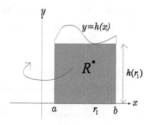
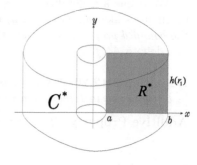

Figure 3.12.22 A Lower Bound of V_a^b

Let V_a^b denote the volume described in the statement of the theorem. Let $h(r_1)$ be the minimum value of h on $[a, b]$. The region R contains the rectangle R^* of width $b - a$ and length $h(r_1)$ shown in the left diagram in Figure 3.12.22. Hence S contains the cylindrical shell C^* obtained by revolving the rectangle R^* around the y–axis. Note

C^* is the cylinder of radius b and height $h(r_1)$ with the cylindrical hole of radius a and height $h(r_1)$ depicted in the right diagram of Figure 3.12.22. Thus,

$$V_a^b \geq \text{Volume } C^* = \pi b^2 h(r_1) - \pi a^2 h(r_1) = \pi h(r_1)(b+a)(b-a). \qquad (3.12.7)$$

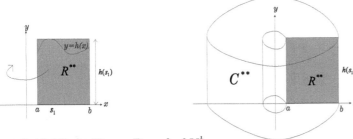

Figure 3.12.23 An Upper Bound of V_a^b

Let $h(s_1)$ be the maximum value of h on $[a, b]$. The region R is contained in the rectangle R^{**} of width $b - a$ and length $h(s_1)$ shown in the left diagram of Figure 3.12.23. Therefore, S is contained in the cylindrical shell C^{**} obtained by revolving the rectangle R^{**} around the y–axis. Note C^{**} is the cylinder of radius b and height $h(s_1)$ with the cylindrical hole of radius a and height $h(s_1)$ depicted in the right diagram of Figure 3.12.23. Thus,

$$V_a^b \leq \text{Volume } C^{**} = \pi b^2 h(s_1) - \pi a^2 h(s_1) = \pi h(s_1)(b+a)(b-a). \qquad (3.12.8)$$

Define a bounding function g by:

$$g(t_1, t_2, t_3) = \pi (t_2 + t_3) h(t_1).$$

Since h is continuous, it follows that g is continuous. By (3.12.7) and (3.12.8):

$$g(r_1, a, b)(b-a) \leq V_a^b \leq g(s_1, a, b)(b-a).$$

Thus, V_a^b satisfies the bounded property. V_a^b satisfies the additive property because the volume of a solid which consists of two pieces is the sum of the volumes of the individual pieces. Hence V_a^b is a generalized integral invariant. By the Generalized Integral Invariant Theorem:

$$V = \int_a^b g(x, x, x) \, dx = \int_a^b \pi(x + x)h(x) \, dx = 2\pi \int_a^b x h(x) \, dx$$

which verifies (3.12.6). $\square$

In the next application, we use the Generalized Integral Invariant Theorem to derive the formula for the force on a submerged vertical plate.

Proposition 3.12.24 *A plate is submerged in a vertical position in a fluid of weight density ρ. Let $s(c)$ denote the length of the line segment of this plate at depth c depicted in Figure 3.12.25. Assume s is a continuous function. Then the force F_a^b of the fluid on the segment of this plate which lies between depth a and depth b is given by:*

$$F_a^b = \int_a^b \rho x s(x) \, dx.$$

Proof Place the x–axis in the plane of the plate so that it points downwards with its origin at the surface of the fluid. Place a y–axis on the surface of the fluid so that

it lies in the plane of the submerged plate. Then the plate lies in the xy–plane. Se
Figure 3.12.25. The invariant F is additive because if we make a horizontal cut in th
plate then the force on the original plate is the sum of the forces on the two pieces
We must find a bounding function for F_a^b. Let $s(t^*)$ denote the minimum value c
$s(x)$, and let $s(t^{**})$ denote the maximal value of $s(x)$ on the interval $[a,b]$. The are
A of the plate equals $\int_a^b s(x)\,dx$. By (3.3.10):

$$s(t^*)(b-a) \le A \le s(t^{**})(b-a). \tag{3.12.9}$$

By (3.9.5), the pressure of the fluid $P(x)$ on this plate at depth x is $P(x) = x\rho$ whic
is an increasing function of x. Thus for $a \le x \le b$:

$$a\rho = P(a) \le P(x) \le P(b) = b\rho.$$

If two forces act on a plate with the pressure of the first force less than that of th
second force at each point of the plate, then the first force is less than the second one
Also, a force which exerts constant pressure on a plate has value the pressure of the
force times the area of the plate. Hence the force F_a^b satisfies:

$$a\rho A = P(a)A \le \quad F_a^b \quad \le P(b)A = b\rho A$$
$$a\rho s(t^*)(b-a) \le a\rho A \le \quad F_a^b \quad \le b\rho A \le b\rho s(t^{**})(b-a).$$

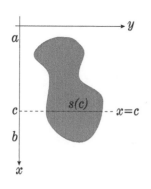

Figure 3.12.25
Submerged Plate

by (3.12.9). Define the continuous function

$$g(t_1, t_2) = \rho s(t_1)t_2.$$

Rewrite the preceding inequalities as:

$$g(t^*, a)(b-a) \le F_a^b \le g(t^{**}, b)(b-a).$$

Then g is a bounding function for F_a^b, and F_a^b is a generalized integral invariant. By
the Generalized Integral Invariant Theorem:

$$F_a^b = \int_a^b g(x, x)\,dx = \int_a^b \rho x s(x)\,dx. \qquad \square$$

Let R be a rod which lies on the interval $I = [a, b]$ of the x–axis with density $\rho(x)$.
In Section 10, we motivated the definition $X_u(R) = \int_a^b (x-u)\rho(x)\,dx$ of the moment
of R about a fulcrum at $x = u$. However, if we make three simple assumptions about
moments, we can use a generalized integral invariant to deduce this formula for $X_u(R)$.
For $a \le c \le e \le b$, let R_c^e denote the section of the rod R on the interval $[c, e]$ with
$X_u(R_c^e)$ its moment about a fulcrum at $x = u$. This invariant is additive because if
we cut a rod into two pieces, the moment of the entire rod is the sum of the moments
of the pieces.

Additive Axiom of Moments Let the rod R lie on the interval $[a, b]$.
If $a \le c \le e \le f \le b$, then $X_u(R_c^f) = X_u(R_c^e) + X_u(R_e^f)$.

The monotone property of moments states that the heavier the rod, the greater
the absolute value of its moment.

Monotone Axiom of Moments Assume the rod R of density $\rho_R(x)$ and the rod
S of density $\rho_s(x)$ both lie on the interval $[a, b]$. If $\rho_R(x) \le \rho_S(x)$ for $a \le x \le b$, then

$$X_u(R_a^b) \le X_u(S_a^b) \text{ if } u \le a \text{ and } X_u(S_a^b) \le X_u(R_a^b) \text{ if } b \le u.$$

Note When $u \le a$, both $X_u(R_a^b)$ and $X_u(S_a^b)$ are positive. However, when $b \le u$
both $X_u(R_a^b)$ and $X_u(S_a^b)$ are negative.

The third axiom of moments follows from the following observations.

- If $u \leq a$, then $X_u(R_a^b)$ is positive. Its moment decreases if we concentrate all its mass at the point $x = a$ which is closest to the fulcrum while its moment increases if we concentrate all its mass at the point $x = b$ which is furthest from the fulcrum.

- If $b \leq u$, then $X_u(R_a^b)$ is negative. Its negative moment decreases if we concentrate all its mass at the point $x = a$ which is furthest from the fulcrum. Its negative moment increases if we concentrate all its mass at the point $x = b$ which is closest to the fulcrum.

- If $a \leq u \leq b$, then $R_a^b = R_a^u \cup R_u^b$. Moving the mass of R_a^u away from the fulcrum to $x = a$ decreases its negative moment. Moving the mass of R_u^b to $x = a$ changes its moment from positive to negative. Hence moving all the mass of R_a^b to $x = a$ decreases its moment.

On the other hand, moving the mass of R_u^b away from the fulcrum to $x = b$ increases its positive moment. Moving the mass of R_a^u to $x = b$ changes its moment from negative to positive. Hence moving all the mass of R_a^b to $x = b$ increases its moment.

Thus in all cases moving the mass of R_a^b to $x = a$ decreases its moment while moving its mass to $x = b$ increases its moment. Recall that the moment of a point mass m at $x = c$ about a fulcrum at $x = u$ is $m(c - u)$. The concentration axiom quantifies the preceding observation.

Concentration Axiom of Moments Let m_a^b denote the mass of the rod R_a^b. Then

$$(a - u)m_a^b \leq X_u(R_a^b) \leq (b - u)m_a^b.$$

We use the Generalized Integral Invariant Theorem to identify an invariant which satisfies the additive, monotone and concentration axioms as a definite integral.

Theorem 3.12.26 *Let R be a rod which lies on the interval $I = [a, b]$ with continuous density $\rho(x)$. Assume $X_u(R_a^b)$ is an invariant that satisfies the additive, monotone and concentration axioms. Then*

$$X_u(R_a^b) = \int_a^b (x - u)\rho(x) \, dx \ . \tag{3.12.10}$$

Proof We break up the proof into three cases: **(I)** $u \leq a$, **(II)** $b \leq u$ and **(III)** $a \leq u \leq b$. For u fixed, let M_a^b denote $X_u(R_a^b)$.

Case I: $u \leq a$

We show M_a^b is a generalized integral invariant. Since M_a^b satisfies the additive axiom, it suffices to verify the bounded property. Let $\rho(t^*)$ be the minimum value of ρ and let $\rho(t^{**})$ be the maximum value of ρ on the closed interval $[a, b]$. Let R_2 be a second rod of constant density $\rho_2(x) = \rho(t^*)$ for $a \leq x \leq b$, and let R_3 be a third rod of constant density $\rho_3(x) = \rho(t^{**})$ for $a \leq x \leq b$. Since $u \leq a$, the monotone axiom applies:

$$X_u(R_2) \leq M_a^b \leq X_u(R_3).$$

The mass of R_2 is $\rho(t^*)(b - a)$, while the mass of R_3 is $\rho(t^{**})(b - a)$. By the concentration axiom:

$$(a - u)\rho(t^*)(b - a) \leq X_u(R_2) \leq M_a^b \leq X_u(R_3) \leq (b - u)\rho(t^{**})(b - a).$$

Since ρ is continuous, the function

$$g(x_1, x_2) = (x_2 - u)\rho(x_1)$$

is also continuous. The preceding inequalities can be rewritten as:

$$g(t^*, a)(b - a) \leq M_a^b \leq g(t^{**}, b)(b - a).$$

Thus M_a^b is a generalized integral invariant with bounding function g. By the Generalized Integral Invariant Theorem:

$$X_u(R_a^b) = M_a^b = \int_a^b g(x, x) \ dx = \int_a^b (x - u)\rho(x) \ dx \ .$$

Case II: $b \leq u$

The identification of M_a^b in this case is analogous to the argument of Case I. We leave the details as an exercise.

Case III: $a \leq u \leq b$

Observe $X_u(R_a^u) = \int_a^u (x - u)\rho(x) \ dx$ by Case II, while $X_u(R_u^b) = \int_u^b (x - u)\rho(x) \ dx$ by Case I. By the additive axiom:

$$X_u(R_a^b) = X_u(R_a^u) + X_u(R_u^b) = \int_a^u (x-u)\rho(x) \ dx + \int_u^b (x-u)\rho(x) \ dx = \int_a^b (x-u)\rho(x) \ dx$$

Proof of the Generalized Integral Invariant Theorem

In the preceding subsection, we defined a generalized integral invariant A_a^b as an invariant which satisfies the additive and bounded properties. We stated the Generalized Integral Invariant Theorem which identifies a generalized integral invariant as a definite integral. This subsection is devoted to proving that theorem. If g is the bounding function of A_a^b, the bounded property says that there are $\mathbf{r}, \mathbf{s} \in [a, b]^m$ with

$$g(\mathbf{r})(b - a) \leq A_a^b \leq g(\mathbf{s})(b - a).$$

Hence each of the lower and upper estimates of A_a^b, produced by a generalized integral invariant, uses m points of $[a, b]$ while a Riemann sum can only use one point of $[a, b]$. Thus the lower and upper estimates of A_a^b produced by a generalized integral invariant *are not* Riemann sums. Our strategy to identify A_a^b as a definite integral is to compare these lower and upper estimates of A_a^b with new estimates, which *are* Riemann sums.

The proof, given in Section 11, of the existence of the definite integral of a continuous function f on $[a, b]$ uses the theorem that says f is uniformly continuous. So too, the identification of a generalized integral invariant with a definite integral uses the fact that a continuous function on $[a, b]^m$ is uniformly continuous. Before establishing this fact, we simplify our notation by defining the distance between two m–tuples.

Definition 3.12.27 *Define the distance between two point* $\mathbf{s}$ *and* $\mathbf{t}$ *of* $\Re^m$ *by*

$$d(\mathbf{s}, \mathbf{t}) = \max \left\{ |t_1 - s_1|, \ldots, |t_m - s_m| \right\}.$$

Example 3.12.28 The distance between $\mathbf{s} = (4, -2, 7)$ and $\mathbf{t} = (-1, 0, 3)$ is $d(\mathbf{s}, \mathbf{t}) = \max \left\{ |-1 - 4|, \ |0 - (-2)|, \ |3 - 7| \right\} = \max \left\{ 5, \ 2, 4 \right\} = 5.$ $\square$

The distance function we defined on $\Re^m$ differs from the usual one for $m \geq 2$. Nevertheless, it has the usual properties of distance.

Proposition 3.12.29 *Let* $\mathbf{s}, \mathbf{t}, \mathbf{u} \in \Re^m$. *Then*

(a) $d(\mathbf{s}, \mathbf{t}) = d(\mathbf{t}, \mathbf{s})$;

(b) $d(\mathbf{s}, \mathbf{t}) = 0$ *if and only if* $\mathbf{s} = \mathbf{t}$;

(c) $d(\mathbf{s}, \mathbf{u}) \leq d(\mathbf{s}, \mathbf{t}) + d(\mathbf{t}, \mathbf{u})$ (Triangle Inequality).

Proof

(a) $d(\mathbf{s}, \mathbf{t}) = \max\{|t_1 - s_1|, \ldots, |t_m - s_m|\} = \max\{|s_1 - t_1|, \ldots, |s_m - t_m|\} = d(\mathbf{t}, \mathbf{s})$.

(b) $d(\mathbf{s}, \mathbf{t}) = \max\{|t_1 - s_1|, \ldots, |t_m - s_m|\} = 0$ if and only if $|t_i - s_i| = 0$ for $1 \le i \le m$. This occurs if and only if $t_i = s_i$ for $1 \le i \le m$, i.e. $\mathbf{s} = \mathbf{t}$.

(c) We use the triangle inequality for the absolute value:

$$
\begin{aligned}
d(\mathbf{s}, \mathbf{u}) &= \max\{|u_1 - s_1|, \ldots, |u_m - s_m|\} \\
&= \max\{|(u_1 - t_1) + (t_1 - s_1)|, \ldots, |(u_m - t_m) + (t_m - s_m)|\} \\
&\le \max\{|u_1 - t_1| + |t_1 - s_1|, \ldots, |u_m - t_m| + |t_m - s_m|\} \\
&\le \max\{|u_1 - t_1|, \ldots, |u_m - t_m|\} + \max\{|t_1 - s_1|, \ldots, |t_m - s_m|\} \\
&= d(\mathbf{t}, \mathbf{u}) + d(\mathbf{s}, \mathbf{t}). \qquad \square
\end{aligned}
$$

We use the distance function to restate the definition of continuity for a function of m variables.

Lemma 3.12.30 *Let I be an interval with $\mathbf{c} \in I^m$. Let $g(\mathbf{x})$ be a function of m variables with domain I^m. Then g is continuous at $\mathbf{x} = \mathbf{c}$ if and only if given any $\epsilon > 0$, there is a corresponding $\delta > 0$ such that*

$$\mathbf{x} \in I^m \text{ with } d(\mathbf{c}, \mathbf{x}) < \delta \text{ implies } d(g(\mathbf{c}), g(\mathbf{x})) < \epsilon.$$

Poof Recall the definition of the continuity of g at $\mathbf{x} = \mathbf{c}$. The punctured box $I_\delta(\mathbf{c})$ consists of those $\mathbf{x} \in I$, other than $\mathbf{c}$, with $|x_k - c_k| < \delta$ for $1 \le k \le m$. That is, $I_\delta(\mathbf{c})$ consists of those $\mathbf{x} \in I$, other than $\mathbf{c}$, with $d(\mathbf{x}, \mathbf{c}) < \delta$. We defined g to be continuous at $\mathbf{x} = \mathbf{c}$ if $\lim_{\mathbf{x} \to \mathbf{c}} g(\mathbf{x}) = g(\mathbf{c})$. Equivalently, given $\epsilon > 0$, there is a corresponding $\delta > 0$ such that if $\mathbf{x} \in I_\delta(\mathbf{c})$, then $d(g(\mathbf{c}), g(\mathbf{x})) < \epsilon$, i.e. if $d(\mathbf{x}, \mathbf{c}) < \delta$ and $\mathbf{x} \ne \mathbf{c}$, then $d(g(\mathbf{c}), g(\mathbf{x})) < \epsilon$. When $\mathbf{x} = \mathbf{c}$, we have $d(g(\mathbf{c}), g(\mathbf{x})) = 0 < \epsilon$. $\qquad \square$

We use this formulation of continuity to define uniform continuity for a function of m variables. Just as in the case $m = 1$, which we studied in Section 11, a uniformly continuous function is a continuous function where we can choose the same value of δ for every $\mathbf{c}$.

Definition 3.12.31 *Let I be an interval with $g(\mathbf{x})$ a function of m variables having domain I^m. We say g is uniformly continuous on I^m if for every $\epsilon > 0$, there is a corresponding $\delta > 0$ such that*

$$\mathbf{c}, \mathbf{x} \in I^m \text{ with } d(\mathbf{c}, \mathbf{x}) < \delta \text{ implies } d(g(\mathbf{c}), g(\mathbf{x})) < \epsilon.$$

The proof that a continuous function $g(\mathbf{x})$ on $[a, b]^m$ is uniformly continuous is a generalization of the proof of Theorem 3.11.4 where we showed a continuous function of one variable on a closed interval is uniformly continuous.

Theorem 3.12.32 *Let $I = [a, b]$ be a closed interval. If $g(\mathbf{x})$ is a function of m variables which is continuous on I^m, then g is uniformly continuous on I^m.*

Proof Assume this theorem is false. Then there are $\epsilon > 0$ and $\mathbf{c}_n, \mathbf{x}_n \in I^m$ with

$$\mathbf{x}_n = (x_{n1}, \ldots, x_{nm}), \quad d(\mathbf{c}_n, \mathbf{x}_n) < \frac{1}{n} \quad \text{and} \quad d(g(\mathbf{c}_n), g(\mathbf{x}_n)) \ge \epsilon \qquad (3.12.11)$$

for every positive integer n. Let S denote the sequence of these $\mathbf{x}_n$. By the Bolzano–Weierstrass Theorem, the sequence of first coordinates of the m–tuples in S has a convergent subsequence S_1' with limit $L_1 \in I$. Let S_1 denote the subsequence of S corresponding to S_1'. For $2 \le k \le m$ we show there is a sequence S_k and a number $L_k \in I$ such that:

- S_k is a subsequence of S_{k-1};
- the kth coordinates of the sequence S_k converge to L_k.

Assume $2 \le k \le m$, and assume we have found S_i and L_i for $1 \le i < k$. We show how to find S_k and L_k. By the Bolzano–Weierstrass Theorem, the sequence of kth coordinates of the m-tuples in S_{k-1} has a convergent subsequence S_k' with limit $L_k \in I$. Let S_k denote the subsequence of S_{k-1} corresponding to S_k'. Apply this procedure for $k = 2$, $k = 3$, $\dots$, $k = m$ to construct the sequences $S_2, \dots, S_m$.

Let $\mathbf{L} = (L_1, \dots, L_m) \in I^m$. Since g is continuous at $\mathbf{L}$ there is a $\delta > 0$ such that

$$\text{if } \mathbf{x} \in I^m \text{ and } d(\mathbf{x}, \mathbf{L}) < \delta, \text{ then } d(g(\mathbf{x}), g(\mathbf{L})) < \frac{\epsilon}{2}. \tag{3.12.12}$$

Since the first coordinates of the sequence S_m converge to L_1, there is a subsequence T_1 of S_m consisting of those $\mathbf{x}_n \in S_m$ with $d(x_{n1}, L_1) < \frac{\delta}{2}$ and $n > \frac{2}{\delta}$. Hence $\frac{1}{n} < \frac{\delta}{2}$.

For $2 \le k \le m$, we construct a sequence T_k such that:

- T_k is a subsequence of T_{k-1};
- if $\mathbf{x}_n \in T_k$, then $d(x_{nk}, L_k) < \frac{\delta}{2}$;
- if $\mathbf{x}_n \in T_k$, then $\frac{1}{n} < \frac{\delta}{2}$.

Assume $2 \le k \le m$, and assume T_i has been constructed for $1 \le i < k$. The sequence of kth coordinates of the sequence T_{k-1} converges to L_k. Let T_k be the subsequence of T_{k-1} consisting of those $\mathbf{x}_n$ with $d(\mathbf{x}_{nk}, L_k) < \frac{\delta}{2}$. Since each $x_n \in T_k$ is in T_1, we have $n > \frac{2}{\delta}$, and $\frac{1}{n} < \frac{\delta}{2}$. Apply this procedure for $k = 2$, $k = 3$, $\dots$, $k = m$ to construct the sequences $T_2, \dots, T_m$. If $\mathbf{x}_n \in T_m$, then

$$d(\mathbf{x}_n, \mathbf{L}) = \max\left\{|L_1 - x_{n1}|, \dots, |L_m - x_{nm}|\right\} < \max\left\{\frac{\delta}{2}, \dots, \frac{\delta}{2}\right\} = \frac{\delta}{2}. \tag{3.12.13}$$

Use the triangle inequality and apply (3.12.11), (3.12.13):

$$d(\mathbf{c}_n, \mathbf{L}) \le d(\mathbf{c}_n, \mathbf{x}_n) + d(\mathbf{x}_n, \mathbf{L}) < \frac{1}{n} + \frac{\delta}{2} < \frac{\delta}{2} + \frac{\delta}{2} = \delta. \tag{3.12.14}$$

By (3.12.13) and (3.12.14), the inequality of (3.12.12) applies to both $\mathbf{x}_n$ and $\mathbf{c}_n$:

$$d(g(\mathbf{x}_n), g(\mathbf{L})) < \frac{\epsilon}{2} \quad \text{and} \quad d(g(\mathbf{c}_n), g(\mathbf{L})) < \frac{\epsilon}{2}.$$

By the triangle inequality:

$$d(g(\mathbf{c}_n), g(\mathbf{x}_n)) \le d(g(\mathbf{c}_n), g(\mathbf{L})) + d(g(\mathbf{L}), g(\mathbf{x}_n)) < \frac{\epsilon}{2} + \frac{\epsilon}{2} = \epsilon.$$

This contradicts the choices of $\mathbf{x}_n$ and $\mathbf{c}_n$. Hence our initial assumption that g is not uniformly continuous is not correct, i.e. g is uniformly continuous on I^m. $\square$

We show the function $G(x)$ defined by the bounding function of a generalized integral invariant is continuous.

Lemma 3.12.33 *Let A_a^b be a generalized integral invariant defined for a, b in the closed interval I. Let $g(t_1, \dots, t_m)$ be the bounding function of this invariant. Then the function $G(x) = g(x, \dots, x)$ is continuous on I.*

Proof Let $c \in I$, and let $\epsilon > 0$. Since g is continuous at $\mathbf{c} = (c, \ldots, c)$, there is $\delta > 0$ such that if $\mathbf{x} \in I^m$ and $d(\mathbf{x}, \mathbf{c}) < \delta$, then $d(g(\mathbf{x}), g(\mathbf{c})) < \epsilon$. Let $x \in I$ with $d(x, c) < \delta$. Define $\mathbf{x} = (x, \ldots, x)$. Then

$$d(\mathbf{x}, \mathbf{c}) = \max \{|x - c|, \ldots, |x - c|\} = |x - c| < \delta.$$

Hence $d(G(x), G(c)) = d(g(\mathbf{x}), g(\mathbf{c})) < \epsilon$, and G is continuous at each $c \in I$. $\square$

Recall that $\int_a^b f(x)\, dx$ is approximated by the Riemann sums

$$R(P, T, f) = \sum_{k=1}^{n} f(t_k)(x_k - x_{k-1})$$

where $P = \{x_0, \ldots, x_n\}$ is a partition of $[a, b]$ and $T = \{t_1, \ldots, t_n\}$ with $t_k \in [x_{k-1}, x_k]$ for $1 \le k \le n$. Note T is a choice of one point from each of the subintervals $[x_{k-1}, x_k]$ determined by P. Generalized integral invariants A_a^b produce generalized Riemann sums to estimate A_a^b where $f(t_k)$ is replaced by a function of m numbers selected from $[x_{k-1}, x_k]$.

Definition 3.12.34 *Let $g(\mathbf{x})$ be a function of m variables with domain $[a, b]^m$. Let $P = \{x_0, \ldots, x_n\}$ be a partition of $[a, b]$. Let $\mathbf{T} = \{\mathbf{t}_1, \ldots, \mathbf{t}_n\}$ with $\mathbf{t}_k \in [x_{k-1}, x_k]^m$, i.e. $\mathbf{t}_k = (t_{k1}, \ldots, t_{km})$ is a selection of m points of $[x_{k-1}, x_k]$. The generalized Riemann sum determined by P and $\mathbf{T}$ is defined by:*

$$R(P, \mathbf{T}, g) = \sum_{k=1}^{n} g(\mathbf{t}_k)(x_k - x_{k-1}).$$

Example 3.12.35 Evaluate the generalized Riemann sum $R(P, \mathbf{T}, g)$ for the partition $P = \{0, 2, 3, 6\}$ of $[0, 6]$, $\mathbf{T} = \left\{(0, 2, 1), \left(3, 2, \frac{3}{2}\right), (6, 3, 4)\right\}$ and $g(x_1, x_2, x_3) = x_1 + x_2 x_3$.

Solution By the preceding definition:

$$\begin{aligned} R(P, \mathbf{T}, g) &= g(0, 2, 1)(2 - 0) + g\left(3, 2, \frac{3}{2}\right)(3 - 2) + g(6, 3, 4)(6 - 3) \\ &= (2)(2) + (6)(1) + (18)(3) = 64. \end{aligned}$$

$\square$

We use uniform continuity to show the generalized Riemann sums which estimate a generalized integral invariant A_a^b also approximate $\int_a^b G(x)\, dx$. Recall P' is called a refinement of the partition P of $[a, b]$ if $P \subset P'$.

Lemma 3.12.36 *Let A_a^b be a generalized integral invariant defined for a, b in the closed interval I. Let $g(t_1, \ldots, t_m)$ be the bounding function of this invariant with $G(x) = g(x, \ldots, x)$. Given $\epsilon > 0$, there is a corresponding partition P_ϵ of $[a, b]$ such that if $P = \{x_0, \ldots, x_n\}$ is a partition of $[a, b]$ which is a refinement of P_ϵ, then*

$$d\left(\int_a^b G(x)\, dx,\ R(P, \mathbf{T}, g)\right) < \epsilon$$

for every $\mathbf{T} = \{\mathbf{t}_1, \ldots, \mathbf{t}_n\}$ with $\mathbf{t}_k = (t_{k1}, \ldots, t_{km}) \in [x_{k-1}, x_k]^m$ for $1 \le k \le n$.

Proof By Theorem 3.12.32, g is uniformly continuous on I^m. Hence there is $\delta > 0$ such that:

$$\text{if } \mathbf{u}, \mathbf{v} \in I^m \text{ and } d(\mathbf{u}, \mathbf{v}) < \delta, \text{ then } d(g(\mathbf{u}), g(\mathbf{v})) < \frac{\epsilon}{2(b - a)}. \tag{3.12.15}$$

Let $P = \{x_0, \ldots, x_n\}$ be a partition of $[a, b]$ of mesh less than δ. Let $\mathbf{T} = \{\mathbf{t}_1, \ldots, \mathbf{t}_n\}$ with $\mathbf{t}_k \in [x_{k-1}, x_k]^m$ for $1 \leq k \leq n$. Define $T = \{t_{11}, \ldots, t_{n1}\}$. The Riemann sum $R(P, T, G)$ approximates $\int_a^b G(x)\, dx$, and the generalized Riemann sum $R(P, \mathbf{T}, g)$ approximates A_a^b:

$$R(P, \mathbf{T}, g) = \sum_{k=1}^{n} g(\mathbf{t}_k)(x_k - x_{k-1})$$

$$R(P, T, G) = \sum_{k=1}^{n} G(t_{k1})(x_k - x_{k-1}) = \sum_{k=1}^{n} g(t_{k1}, \ldots, t_{k1})(x_k - x_{k-1})$$

$$= \sum_{k=1}^{n} g(\mathbf{t}_{k1})(x_k - x_{k-1})$$

where $\mathbf{t}_{k1} = (t_{k1}, \ldots, t_{k1})$. Note

$$d(\mathbf{t}_k, \mathbf{t}_{k1}) = \max\left\{|t_{k1} - t_{k1}|, \ldots, |t_{k1} - t_{km}|\right\} \leq x_k - x_{k-1} \leq \mathrm{mesh}(P) < \delta.$$

The distance between the above sums on the number line is:

$$d(R(P, T, G), R(P, \mathbf{T}, g)) = |R(P, \mathbf{T}, g) - R(P, T, G)|$$

$$= \left| \sum_{k=1}^{n} [g(\mathbf{t}_k) - g(\mathbf{t}_{k1})] [x_k - x_{k-1}] \right| \leq \sum_{k=1}^{n} |g(\mathbf{t}_k) - g(\mathbf{t}_{k1})| (x_k - x_{k-1})$$

$$= \sum_{k=1}^{n} d(g(\mathbf{t}_k), g(\mathbf{t}_{k1})) (x_k - x_{k-1}) < \sum_{k=1}^{n} \frac{\epsilon}{2(b-a)} (x_k - x_{k-1}) \qquad \text{by (3.12.15)}$$

$$= \frac{\epsilon}{2(b-a)} \sum_{k=1}^{n} (x_k - x_{k-1}) = \frac{\epsilon}{2(b-a)}(b - a) = \frac{\epsilon}{2}$$

since the sum of the lengths $x_k - x_{k-1}$ of the n subintervals $[x_{k-1}, x_k]$ is the length $b - a$ of $[a, b]$. Since G is continuous, $\int_a^b G(x)\, dx$ exists. By the triangle inequality:

$$d\left(R(P, \mathbf{t}, g), \int_a^b G(x)\, dx\right) \leq d\left(R(P, \mathbf{t}, g), R(P, T, G)\right) + d\left(R(P, T, G), \int_a^b G(x)\, dx\right)$$

$$\leq \frac{\epsilon}{2} + d\left(R(P, T, G), \int_a^b G(x)\, dx\right). \qquad (3.12)$$

By Theorem 3.3.37, there is a partition P' of $[a, b]$ such that $U(P', G) - L(P', G) < \frac{\epsilon}{2}$. Let P'' be a refinement of P' with $R(P'', T'', G)$ a Riemann sum of G. Since both $R(P'', T'', G)$ and $\int_a^b G(x)\, dx$ are between $L(P'', G)$ and $U(P'', G)$,

$$d\left(R(P'', T'', G), \int_a^b G(x)\, dx\right) \leq d(L(P'', G), U(P'', G))$$

$$\leq d(L(P', G), U(P', G)) < \frac{\epsilon}{2}. \qquad (3.12.17)$$

Let P_ϵ be a partition of mesh less than δ which is a refinement of P'. If P is a refinement of P_ϵ, then by (3.12.16) and (3.12.17):

$$d\left(R(P, \mathbf{t}, g), \int_a^b G(x)\, dx\right) < \frac{\epsilon}{2} + \frac{\epsilon}{2} = \epsilon. \qquad \square$$

Now we can identify a generalized integral invariant as a definite integral.

Theorem 3.12.21 [Generalized Integral Invariant Theorem] *Let A_a^b be a generalized integral invariant defined when $a \leq b$ are in the closed interval I. Let g be the bounding function of this invariant. Then $G(x) = g(x, \ldots, x)$ is continuous on I, and*

$$A_a^b = \int_a^b G(x) \, dx = \int_a^b g(x, \ldots, x) \, dx.$$

Proof $G(x)$ is continuous on I by Lemma 3.12.33. Let $\epsilon > 0$. By Lemma 3.12.36, there is a partition $P_{\epsilon/3}$ of $[a, b]$ such that if $P = \{x_0, \ldots, x_n\}$ is a partition of $[a, b]$ which is a refinement of $P_{\epsilon/3}$, then

$$d\left(R(P, \mathbf{T}, g), \int_a^b G(x) \, dx \right) < \frac{\epsilon}{3}. \tag{3.12.18}$$

for any $\mathbf{T} = \{\mathbf{t}_1, \ldots, \mathbf{t}_n\}$ with $\mathbf{t}_k \in [x_{k-1}, x_k]^m$, $1 \leq k \leq n$. By the additive property:

$$A_a^b = A_{x_0}^{x_1} + \cdots + A_{x_{n-1}}^{x_n}.$$

Apply the bounded property to each interval $[x_{k-1}, x_k]$ to find $\mathbf{r}_k, \mathbf{s}_k \in [x_{k-1}, x_k]$ with

$$g(\mathbf{r}_k)(x_k - x_{k-1}) \leq A_{x_{k-1}}^{x_k} \leq g(\mathbf{s}_k)(x_k - x_{k-1}).$$

Add these inequalities, for $1 \leq k \leq n$, to obtain lower and upper estimates of A_a^b which we denote as $L'(P)$ and $U'(P)$:

$$L'(P) = \sum_{k=1}^n g(\mathbf{r}_k)(x_k - x_{k-1}) \leq \sum_{k=1}^n A_{x_{k-1}}^{x_k} = A_a^b \leq \sum_{k=1}^n g(\mathbf{s}_k)(x_k - x_{k-1}) = U'(P)$$

$$\tag{3.12.19}$$

Apply (3.12.18) with $\mathbf{t}_k = \mathbf{r}_k$ for $1 \leq k \leq n$ and again with $\mathbf{t}_k = \mathbf{s}_k$ for $1 \leq k \leq n$:

$$d\left(L'(P), \int_a^b G(x) \, dx \right) < \frac{\epsilon}{3} \quad \text{and} \quad d\left(U'(P), \int_a^b G(x) \, dx \right) < \frac{\epsilon}{3}. \tag{3.12.20}$$

By the triangle inequality and (3.12.20):

$$\begin{aligned}
d\left(L'(P), U'(P) \right) &\leq d\left(L'(P), \int_a^b G(x) \, dx \right) + d\left(\int_a^b G(x) \, dx, U'(P) \right) \\
&< \frac{\epsilon}{3} + \frac{\epsilon}{3} = \frac{2\epsilon}{3}. \tag{3.12.21}
\end{aligned}$$

By (3.12.19), $d\left(A_a^b, L'(P) \right) \leq d\left(U'(P), L'(P) \right)$. Apply the triangle inequality:

$$\begin{aligned}
d\left(A_a^b, \int_a^b G(x) \, dx \right) &\leq d\left(A_a^b, L'(P) \right) + d\left(L'(P), \int_a^b G(x) \, dx \right) \\
&\leq d\left(U'(P), L'(P) \right) + d\left(L'(P), \int_a^b G(x) \, dx \right) < \frac{2\epsilon}{3} + \frac{\epsilon}{3} = \epsilon.
\end{aligned}$$

by (3.12.21) and (3.12.20). Since the preceding inequality is true for every $\epsilon > 0$, it follows that $d\left(A_a^b, \int_a^b G(x) \, dx \right) = 0$, and $A_a^b = \int_a^b G(x) \, dx$. $\qquad \square$

Summary

The reader should understand the definition of an integral invariant as well as the proof of the Integral Invariant Thm. She should be able to apply this theorem to identify an invariant as a definite integral by verifying the increasing, additive and multiplicative properties. She should understand the concepts of limit, continuity and uniform continuity for functions of m variables. She should know the definition of a generalized integral invariant and be able to apply the Generalized Integral Invariant Thm. to identify an invariant as a definite integral by verifying the additive and bounded properties. Furthermore, she should understand how uniform continuity and generalized Riemann sums are used to prove the Generalized Integral Invariant Thm.

Basic Exercises

1. Show the height invariant $H_a^b(f)$ of Example 3.12.3 (2) does not satisfy any of the properties of an integral invariant.

2. Show the distance invariant $D_a^b(v)$ of Example 3.12.3 (3) is not an integral invariant of the velocity $v(t)$.

3. Show the center of mass invariant $C_a^b(\rho)$ of Example 3.12.3 (5) does not satisfy any of the three properties of an integral invariant.

4. Use Proposition 3.12.14 to evaluate each limit.

(a) $\lim\limits_{t \to c} \dfrac{t_1 - t_2^2}{1 + 5t_1 t_2}$ with $\mathbf{c} = (-4, 6)$;   (b) $\lim\limits_{t \to c} \tan(t_1^2 + t_2^2 + t_3^2)$ with $\mathbf{c} = (-1, 3, 2)$

(c) $\lim\limits_{t \to c} \sqrt{\dfrac{t_1^4 + t_2^2 t_3^2 + 1}{t_1 t_2 + t_2 t_3 + 9}}$, $\mathbf{c} = (-2, 0, 1)$; (d) $\lim\limits_{t \to c} \sec(t_1 t_2) \cos t_1 \sin t_2$, $\mathbf{c} = (\frac{\pi}{4}, -\frac{\pi}{6})$;

(e) $\lim\limits_{t \to c} 1$ with $\mathbf{c} = (6, -8, -9, 7)$;   (f) $\lim\limits_{t \to c} \arctan \dfrac{t_1 t_2 - 7}{1 + t_1^2 t_2^2}$ with $\mathbf{c} = (-1, 3)$.

5. Use Corollary 3.12.17 to show each function is continuous:

(a) $f(t_1, t_2) = 7t_1^3 t_2^4 + 8t_1^5 + 4$ on $\Re^2$;   (b) $g(t_1, t_2, t_3) = \sin(t_1^2 + t_2^4 + t_3^6 + 1)$ on $\Re^3$;

(c) $h(t_1, t_2) = \frac{\sin t_1 + \cos t_2}{1 + t_1^2 t_2^2}$ on $\Re^2$;   (d) $i(t_1, t_2, t_3, t_4) = \sqrt{t_1 t_3} \tan \pi(t_2 + t_4)$ on $[0, \frac{1}{5}]^4$;

(e) $j(t_1, t_2) = F(t_1 + t_2) G(t_1 - t_2)$ on $\Re^2$ where $F(x)$, $G(x)$ are continuous on $\Re$;

(f) $k(t_1, t_2, t_3) = \sqrt{t_1 F(t_2)^2 G(t_3)^4}$ on $[0, 1]^3$ where $F(x)$, $G(x)$ are continuous on $\Re$.

6. Let f be a continuous function with domain $\Re$. Find a bounding function for each generalized integral invariant.

(a) $A_a^b = \int_a^b x^2 f(x)\, dx$ (b) $B_a^b = \int_a^b f(x)^2\, dx$

(c) $C_a^b = \int_a^b \dfrac{f(x)}{1 + f(x)^2}\, dx$ (d) $D_a^b = \int_a^b \dfrac{x f(x)}{1 + f(x)^2}\, dx$

(e) $E_a^b = \int_a^b x \arctan f(x)\, dx$ (f) $F_a^b = \int_a^b \sqrt{x^2 + f(x)^2}\, dx$

7. Complete the proof of Theorem 3.12.26 by proving the case $b \le u$.

8. Compute the distance $d(\mathbf{x}, \mathbf{y})$ in each case:

(a) $\mathbf{x} = (4, -1)$ and $\mathbf{y} = (3, 2)$;   (b) $\mathbf{x} = (-1, 2, 1)$ and $\mathbf{y} = (3, -1, 4)$;

(c) $\mathbf{x} = (4, -3, 7, -2)$ and $\mathbf{y} = (6, -1, 4, 3)$;

(d) $\mathbf{x} = (5, 0, 6, -4, -8)$ and $\mathbf{y} = (3, -2, 7, -5, -6)$.

9. If $\mathbf{c} = (c_1, \ldots, c_m) \in I^m$ and $k \in \Re$, define $k\mathbf{c} = (kc_1, \ldots, kc_m)$. Show if $\mathbf{x}$, $\mathbf{y} \in I^m$ and $r \in \Re$, then $d(r\mathbf{x}, r\mathbf{y}) = |r| d(\mathbf{x}, \mathbf{y})$.

10. (a) Let $\mathbf{c} = (c_1, c_2)$ be a fixed point in the plane, and let k be a fixed positive number. Describe the set of points $\mathbf{x} = (x_1, x_2)$ in the plane such that $d(\mathbf{x}, \mathbf{c}) = k$.

(b) Let $\mathbf{c} = (c_1, c_2, c_3)$ be a fixed point in $\Re^3$, and let k be a fixed positive number. Describe the set of points $\mathbf{x} \in \Re^3$ such that $d(\mathbf{x}, \mathbf{c}) = k$.

11. Show $d((x_1, y_1)), (x_2, y_2))$ is not the usual distance between the points (x_1, y_1) and (x_2, y_2) in the plane.

12. Determine whether each of these functions is uniformly continuous.

(a) $f(x_1, x_2) = \sin(x_1 x_2)$ on $[0, \pi]^2$.

(b) $g(x_1, x_2, x_3) = \frac{1}{x_1 x_2 x_3}$ on $[-1, 1]^3$.

(c) $h(x_1, x_2) = \frac{1}{x_1 + x_2}$ on $[1, \infty)^2$.

(d) $k(x_1, x_2, x_3, x_4) = \frac{x_1 x_4}{1 + x_2 x_3}$ on $(0, 1)^4$.

(e) $m(x_1, x_2) = x_1 x_2 \sin \frac{1}{x_1 x_2}$ on $(0, 1)^2$.

(f) $n(x_1, x_2, x_3) = \frac{x_1 x_2 x_3}{3 + x_1 + x_2 + x_3}$ on $(-1, 1)^3$.

13. Compute each generalized Riemann sum $R(P, \mathbf{T}, g)$.

(a) $P = \{0, 2, 3, 6, 9\}$, $\mathbf{T} = \{(0, 1), (3, 2), (5, 4), (8, 7)\}$ and $g(x_1, x_2) = x_1 x_2$.

(b) $P = \{2, 4, 5, 8\}$, $\mathbf{T} = \{(4, 2, 3), (4, 5, 4), (7, 5, 6)\}$ and $g(x_1, x_2, x_3) = x_1^2 + x_2^2 + x_3^2$.

(c) $P = \{-6, -2, 1, 3\}$, $\mathbf{T} = \{(-2, -4, -3, -5), (1, 0, -1, 1), (3, 2, 1, 3)\}$ and $g(x_1, x_2, x_3, x_4) = x_1 x_4 + x_2 x_3$.

(d) $P = \{1, 3, 6, 9\}$, $\mathbf{T} = \{(3, 1), (4, 5), (9, 7)\}$ and $g(x_1, x_2) = \frac{x_1}{1 + x_2}$.

(e) $P = \{0, 2, 12\}$, $\mathbf{T} = \{(2, 1, 2), (12, 3, 4)\}$ and $g(x_1, x_2, x_3) = \sqrt{x_1^2 + x_2^2 + x_3^2}$.

Challenging Problems

1. Prove Proposition 3.12.14.

2. Show $g(t_1, t_2) = \left\{ \begin{array}{cc} \frac{t_1 t_2}{t_1^2 + t_2^2} & \text{for } (t_1, t_2) \neq (0, 0) \\ 0 & \text{for } (t_1, t_2) = (0, 0) \end{array} \right\}$ is not continuous on $[0, 1]^2$.

3.13   Review Exercises for Chapter 3

Decide if each of these 40 statements is True or False. Justify your answers

1. $\sum_{k=2}^{50}(k^2 - 2k + 1) = \sum_{n=0}^{48}(n^2 + 2n + 1)$

2. $\sum_{k=1}^{100}(6k^2 - 4k - 99) = 2,000,000$

3. Let $f(x) = \frac{1}{1+x^2}$ with P the regular partition of $[0,1]$ into 20 subintervals. Th upper Riemann sum $U(P, f)$ equals $\frac{1}{20}\sum_{k=1}^{20}\frac{1}{1+k^2/400}$.

4. Let f be a continuous function with domain $[a, b]$. The Riemann sum $R(P, T, f$ approximates the area bounded by the graph of $y = f(x)$, the x–axis and the line $x = a$, $x = b$.

5. Let f be continuous on $[5, 9]$ with P a partition of $[5, 9]$. If $L(P, f) = -3.8$ and $U(P, f) = -3.4$, then $\int_9^5 f(x)\, dx \approx -3.6$ with error less than .2.

6. Let f be continuous on $[a, b]$ with P a partition of $[a, b]$. If $L(P, f) = 2.47$ and $U(P, f) = 2.53$. Then the error in approximating $\int_a^b f(x)\, dx$ by any Riemann sum $R(P, T, f)$ is less than .06.

7. $\int_0^1 x \sin\frac{1}{x}\, dx$ does not exist.

8. Let $f(x) = \left\{ \begin{array}{ll} 1 & \text{if } x \text{ is rational} \\ 0 & \text{if } x \text{ is irrational} \end{array} \right\}$. Then $\int_0^1 f(x)\, dx$ does not exist.

9. $\int_0^\pi \cos x\, dx$ is the area bounded by the graph of $y = \cos x$, the x–axis, the y–axis and the line $x = \pi$.

10. $\int_{-\pi}^\pi x \sin(x^3)\, dx = 0$.   **11.** $\int_0^{100\pi} |\sin x|\, dx \leq \int_0^{100\pi} \sin^2 x\, dx$

12. If f is continuous on $[0, 1]$, then there is $0 \leq c \leq 1$ with $f(c) = \int_0^1 f(x)\, dx$.

13. Let f be a continuous one–to–one function with domain $[a, b]$. Then $\int_a^b f(x)\, dx$ is the area bounded by the graph of $y = f^{-1}(x)$, the y–axis and the lines $y = a$, $y = b$.

14. $\int_0^{\pi/2} \cos^m x \sin^n x\, dx \leq \frac{\pi}{2}$ for all positive integers m and n.

15. $\frac{d}{dx}\int_0^x \cot\theta\, d\theta = \cot\theta$.

16. Let a and k be positive numbers. Then
(a) $\frac{d}{dx}\int_{ka}^{kx}\frac{1}{t}\, dt = \frac{d}{dx}\int_a^x \frac{1}{t}\, dt$ **(b)** $\int_{ka}^{kx}\frac{1}{t}\, dt = \int_a^x \frac{1}{t}\, dt$

17. $\int_0^{\pi/7} \tan^4 x\, dx = \frac{1}{3}\tan^3\frac{\pi}{7} - \tan\frac{\pi}{7} + \frac{\pi}{7}$ **18.** $\int \frac{1}{x}\, dx = -\frac{1}{x^2} + C$.

19.   The area between the curve $y = x^3 - x^2 - 2x$ and the x–axis is given by $\left|\int_{-1}^2 x^3 - x^2 - 2x\, dx\right|$.

20. The area between the curves $y = x^4$ and $y = x^2$ is $\int_{-1}^1 x^4 - x^2\, dx$.

21. If f is a continuous function, then $\int f(x)\, dx = \int_a^x f(t)\, dt + C$.

22. $\int_0^1 \arcsin x\, dx + \int_0^{\pi/2} \sin x\, dx$ equals the area of the rectangle with sides of length 1 and $\frac{\pi}{2}$.

23. $\int \tan x \sec^2 x\, dx$ is both $\frac{1}{2}\tan^2 x + C$ and $\frac{1}{2}\sec^2 x + C$.

24. $\int \frac{1}{x^2+16}\, dx = \arctan\frac{x}{4} + C$ **25.** $1 + \int \frac{1}{x}\, dx = 2 + \int \frac{1}{x}\, dx$

26. $30\int \frac{x^2-1}{\sqrt{x}-1}\, dx = \sqrt{x}(12x^2 + 15x\sqrt{x} + 20x + 30\sqrt{x}) + C$

27. $\int \cos^2 x \, dx + \int \sin^2 x \, dx = x + C$

28. Let $g(x)$ have a continuous derivative. If $u = \frac{x}{g(x)}$, then $\int \frac{g(x) - xg'(x)}{x^2} \, dx = \int \frac{1}{u^2} \, du$.

29. If $\theta = \arcsin x$, then $\int \sqrt{1 - x^2} \, dx = \int \cos^2 \theta \, d\theta$.

30. If $\theta = \arctan x$, then $\int \sqrt{1 + x^2} \, dx = \int \sec^2 \theta \, d\theta$.

31. $\int \sin^4 x \, dx + \int \cos^4 x \, dx = x - \frac{1}{2} \int \sin^2 2x \, dx$.

32. If $u = 1 - x$, then $\int_0^1 (1 - x)^{100} \, dx = \int_0^1 u^{100} \, du$

33. $\int \frac{1}{x^2 + 6x + 5} \, dx = \frac{1}{2} \arctan \frac{x + 3}{2} + C$

34. The cross–section of the solid S perpendicular to the x–axis is a square whose side has length $\sin x$ for $0 \le x \le \pi$. Then the volume of S is $\pi - \int_0^\pi \cos^2 y \, dy$.

35. Let R be the region bounded by the curves $y = 8 - x^2$ and $y = x^2$. Let V be the volume of the solid obtained by revolving R around the x–axis. Then $V = \pi \int_{-2}^2 \left(8 - 2x^2 \right)^2 \, dx$.

36. Let f be a continuous odd function with domain $\mathfrak{R}$. Let R be the region bounded by the graph of $y = f(x)$, the x–axis and the lines $x = -a$, $x = a$. Then the volume of the solid of revolution obtained by revolving the region R around the x–axis is zero.

37. Let R be the region bounded by the curve $y = \sin x$, $0 \le x \le \pi$, and the x–axis. Let V be the volume of the solid obtained by revolving R around the x–axis. Then $V = 2\pi \int_0^\pi x \sin x \, dx$.

38. No polyhedron is a solid of revolution.

39. Let R be a region in the first quadrant. Let S be the solid obtained by revolving R around the x–axis, and let T be the solid obtained by revolving R around the y–axis. If S and T are congruent solids, then they must be spheres.

40. Let $f(x)$ and $g(x)$ satisfy the hypotheses of Rolle's Theorem on the interval $[a, b]$. If $g(x)$ never has value zero on $[a, b]$, then $\int_a^b \frac{f'(x)g(x) - f(x)g'(x)}{g(x)^2} \, dx = 0$.

Solve each of the following problems.

41. Find all positive integers n such that $\left(\sum_{k=1}^n k \right)^2 = \sum_{k=1}^n k^2$.

42. Find $\lim_{n \to \infty} \frac{\sum_{k=1}^{n^2} k^2}{\sum_{k=1}^{n^3} k}$.

43. Find $\lim_{n \to \infty} \sum_{k=1}^n \frac{k}{n^2} \sin \pi \frac{k^2}{n^2}$.

44. Use the average of a lower and an upper Riemann sum to estimate $\int_0^2 x^2 + x \, dx$ with error less than .01.

45. Which is larger: $\int_{-\pi/3}^{\pi/4} \sqrt{\tan^6 x} \, dx$ or $\left| \int_{-\pi/3}^{\pi/4} \tan^3 x \, dx \right|$?

46. Find $\frac{d}{dx} \left(\int_{\arctan x}^{\arcsin x} \tan t \, dt \right)$. Simplify your answer.

47. Find the area bounded by $y = \sqrt{x + 1}$ and $x = \pm\sqrt{4y + 1}$.

48. Find the area bounded by $x = \frac{192}{4 + y^2}$ and $x = y^2$.

49. Find the area bounded by the graphs of the functions $f(\theta) = \sin^4 \theta$ and $g(\theta) = \cos^4 \theta$, both with domain $[0, \pi]$.

50. Make the substitution $\theta = \arctan x$ in $\int_1^{\sqrt{3}} \frac{x^2}{\sqrt{1 + x^2}} \, dx$. **Do not integrate.**

51. Integrals compute area above the x–axis as positive. The graph of $y = |x|$ lies

entirely above the x–axis. Note that $\int |x| \, dx = -\frac{x^2}{2} + C$ for $x \leq 0$. Why is thi
integral negative?

52. **(a)** Show the area of a quadrilateral with perpendicular diagonals of lengths
and t is $\frac{1}{2}st$.
(b) Find the volume of the pyramid whose base has vertices $(1,0)$, $(0,4)$, $(-5,0)$
$(0,-6)$ and has its top vertex 8 units above the origin.

53. Let A be the part of the curve $x = 4 - y^2$ which lies in the first and fourth
quadrants. Let B be the part of the curve $y = 1 - x^2$ which lies in the first and secon
quadrants. A peach has the shape of the curve A revolved around the y–axis whil
its pit has the shape of the curve B revolved around the x–axis. Find the volume o
edible fruit in this peach that remains when its pit is removed.

54. The bottom of a large bowl has the shape of the curve $y = \sqrt{1 + x^6}$ revolvec
around the y–axis where x and y are measured in cm. This bowl is filling with rai
water whose depth is rising at 3 cm/hr. Find the rate of increase of the volume o
water in this bowl when the water is 2 cm deep.

55. A spherical apple of radius 5 cm has its center at the origin. A cylindrical core
of radius 3 cm is removed with the origin at the center of the core. Find the volume
of the cored apple.

Evaluate each of the following integrals.

56. $\int \frac{1+x\sqrt{x}}{1+\sqrt{x}} \, dx$ **57.** $\int_{-2}^{3} |x^2 - 1| \, dx$ **58.** $\int \frac{(1+\tan x)^3}{\sqrt{\sin^3 x \cos x}} \, dx$

59. $\int \frac{x^2}{\sqrt{3x+2}} \, dx$ **60.** $\int \frac{\sin x - \cos x}{1 + \sin 2x} \, dx$ **61.** $\int \frac{\cos x}{9 + \sin^2 x} \, dx$

62. $\int \frac{\sin x \cos x}{\sqrt{4 - \sin^4 x}} \, dx$ **63.** $\int (\sin \theta + \cos \theta)^3 \, d\theta$ **64.** $\int \frac{(\sqrt[3]{x} + \sqrt[4]{x})^2}{\sqrt{x}} \, dx$

65. $\int_{7\pi/4}^{9\pi/4} (\sin x + \tan x)^{101} \, dx$ **66.** $\int_0^{\pi} |\sin x - \cos x| \, dx$ **67.** $\int_{-\pi}^{\pi} \theta \cos \theta \, d\theta$

The following exercise is for those who have studied Section 3.8.

68. Estimate $\int_1^3 \frac{1}{x} \, dx$ with error less than .01 using:
    (a) the midpoint rule;   **(b)** Simpson's rule.

The following 2 exercises are for those who have studied Section 3.9.

69. Find the distance traveled by the object which moves from $x = 0$ at $t = 0$ to
$x = 4$ cm at $t = 4$ seconds with acceleration $a(t) = 6t - 4$ cm/sec^2.

70. How much work is done in lifting a 4 pound rock 10 feet?

Chapter 4

Logarithms and Exponentials

4.1 Introduction

Exponential and logarithm functions play a central and unique role in calculus. In particular, they allow us to evaluate many integrals, such as $\int \frac{1}{x}\, dx$, whose values can not be expressed in terms of other functions. After establishing the properties of the exponential and logarithm functions, the basic techniques of integration are developed. The optional sections of this chapter are devoted to the study of differential equations. The problem is to find all functions which satisfy a given equation relating the function and its derivatives. We establish methods to solve a variety of these equations. We also prove that, under reasonable conditions, solutions always exist.

The *Concepts* portion commences in Section 2 with definitions of the logarithm and exponential functions. Our definitions produce differentiable functions with the usual properties. Applications of logarithms and exponentials to integration, graphing, logarithmic differentiation and limits are given in Section 3. The next five sections are devoted to methods of integration. Section 4 studies integration by parts, a formula that often simplifies the integral of a product. In Section 5, we integrate powers of trigonometric functions. In Section 6, trigonometric substitutions are used to integrate powers of quadratic functions. In Section 7, we use the method of partial fractions to integrate rational functions. In Section 8, substitutions are used to integrate rational functions of trigonometric functions, exponential functions and fractional powers of x. In Section 9, we define and evaluate improper integrals which represent areas of unbounded regions.

The *Applications* portion begins in Section 10 with the study of hyperbolic functions and their inverses. We show these functions arise in the descriptions of hanging cables and of motions of objects falling under gravity subject to air resistance. Section 11 is devoted to solving several basic varieties of first order differential equations. Section 12 shows how to solve linear second order differential equations. The importance of differential equations is their use in the sciences and social sciences. Often a principle which is discovered there to describe the behavior of a phenomenon can be phrased as a differential equation. We give applications of differential equations to exponential and logistic growth, continuously compounded interest, rates of chemical reactions, mixing problems, price prediction and harmonic motion.

The *Theory* portion begins in Section 13 with the the study of fixed points of functions and operators. Fixed point theory is used in Section 14 to prove an existence theorem for solutions of first order differential equations.

The flow chart in Figure 4.1.1 indicates the dependence of the sections of th
Applications and *Theory* portions of this chapter on preceding sections.

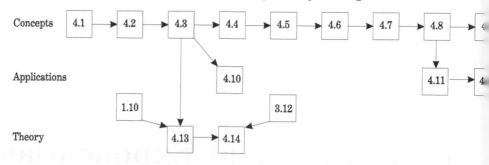

Figure 4.1.1 The Sections of Chapter 4

4.2 Definitions and Properties

In this section, we study the exponential function b^x and its inverse function, the
logarithm function $\log_b y$ for $b > 0$. (That is, $x = \log_b y$ if $y = b^x$.) Definitions
of these functions are given which make them differentiable. The basic properties of
logarithms and exponentials are established from their definitions. Several applications
of logarithms are presented in the next section.

Recall from high school that if $b > 0$, then b^t is defined for t a rational number:

$$\textbf{(1)}\ b^0 = 1; \qquad\qquad\qquad \textbf{(2)}\ b^n = b \cdots b \ \ (n\ \text{times});$$

$$\textbf{(3)}\ b^{-n} = \tfrac{1}{b^n}; \qquad\qquad\qquad \textbf{(4)}\ b^{p/q} = \left(\sqrt[q]{b}\right)^p$$

where n, p, q are integers with n, $q > 0$. However, we do not have a definition of b^t for
t an irrational number. Thus, we can only define $y = \log_b t$ if $t = b^y$ with y a rational
number. For example, $\log_2 0.125 = \log_2 2^{-3} = -3$, $\log_2 0.250 = \log_2 2^{-2} = -2$, but
$\log_2 0.200$ is not defined. There are a variety of physical quantities, such as bacteria
population, whose value $P(t)$, at time t, is given by the exponential function $P(t) = b^t$.
Even though the rate of population growth $P'(t)$ clearly exists, this only suggests that
there should be a differentiable exponential function $y = b^t$. We do not even have
a rigorous argument that there is a continuous function $y = b^t$ which extends the
above definitions. The focus of this section is to rectify this situation by defining the
functions $y = b^t$ and $y = \log_b x$ for all real numbers t, x with $x > 0$.

For $b > 0$, we follow an indirect procedure to define the logarithm function $\log_b y$
and the exponential function b^x. It consists of four steps.

Step I We introduce a new function $\ln x$ as the name for area under a hyperbola.
The properties of $\ln x$ are the same as the properties of $\log_b x$.

Step II We show the function $x = \ln y$ is one–to–one, and call its inverse function
$y = \exp x$. The properties of $\exp x$ are the same as the properties of b^x.

Step III We define $b^x = \exp(x \ln b)$. The function b^x has the usual properties of
an exponential function and satisfies conditions (1)–(4) above. In addition, we
show there is a number e between 2 and 3 such that $\exp x = e^x$.

Step IV We show the function $x = b^y$ is one–to–one, and define $y = \log_b x$ as its
inverse function. The function $\log_b x$ has the usual properties of a logarithm
function, and $\ln x = \log_e x$.

Step I: The Function $y = \ln x$

Recall : $\qquad \int x^n \, dx = \dfrac{x^{n+1}}{n+1} + C$ for $n \neq -1$.

When $n = -1$, we do not know of any function whose derivative is $\frac{1}{x}$. That is, we cannot express $\int \frac{1}{x} \, dx$ in terms of the functions we know. However, the First Fundamental Theorem of Calculus says that the area under the hyperbola $y = \frac{1}{t}$ in Figure 4.2.1, with a variable upper bound x, has derivative $\frac{1}{x}$:

$$\frac{d}{dx}\left(\int_1^x \frac{1}{t} \, dt \right) = \frac{1}{x}.$$

We denote this integral as $\ln x$. It is only defined for $x > 0$ because the hyperbola $y = \frac{1}{t}$ has a vertical asymptote at $t = 0$.

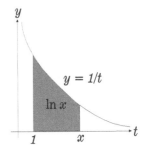

Definition 4.2.2 *For $x > 0$, define*

$$\ln x = \int_1^x \frac{1}{t} \, dt.$$

Figure 4.2.1
$\ln x = \int_1^x \frac{1}{t} \, dt$

The function $\ln x$ is called the *natural logarithm* function. Most read it as *log x*, although some read it as *lawn x*. We will see in Step IV that $\ln x$ is a logarithm function in the usual sense. Our first project is to sketch the graph of $y = \ln x$. With this goal in mind, we make ten observations.

(1) The function $\ln x$ has domain $(0, \infty)$ by its definition.

(2) **(i)** If $0 < x < 1$, then $\ln x < 0$. **(ii)** $\ln 1 = 0$ **(iii)** If $x > 1$, then $\ln x > 0$.
   Justification For $x > 1$, $\int_1^x \frac{1}{t} \, dt$ represents area above the x–axis and is therefore a positive number. If $0 < x < 1$, then $\ln x = \int_1^x \frac{1}{t} \, dt = -\int_x^1 \frac{1}{t} \, dt < 0$. Clearly, $\ln 1 = \int_1^1 \frac{1}{t} \, dt = 0$.

(3) The function $\ln x$ is differentiable, hence continuous. In particular,

$$D(\ln x) = \frac{1}{x}. \qquad (4.2.1)$$

   Justification As noted above, $\frac{d}{dx}(\ln x) = \frac{d}{dx}(\int_1^x \frac{1}{t} \, dt) = \frac{1}{x}$ by the First Fundamental Theorem of Calculus.

(4) 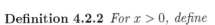 The function $\ln x$ is increasing.
   Justification $D(\ln x) = \frac{1}{x} > 0$ since $x > 0$. By Corollary 2.6.29(a), the function $y = \ln x$ is increasing.

(5) The graph of $\ln x$ is concave down.
   Justification $D^2(\ln x) = D(\frac{1}{x}) = -\frac{1}{x^2} < 0$. By Theorem 2.7.23(b), the function $y = \ln x$ is concave down.

(6) If $x > 0$ and q is a rational number, then $\ln(x^q) = q \ln x$.
   Justification Let $u = x^q$. By the chain rule

$$\frac{d}{dx} \ln(x^q) = \frac{d}{du}(\ln u) \frac{du}{dx} = \frac{1}{u} q x^{q-1} = \frac{q x^{q-1}}{x^q} = \frac{q}{x} = \frac{d}{dx}(q \ln x).$$

   By Corollary 2.6.24 the two functions $\ln(x^q)$ and $q \ln x$, with the same derivative, differ by a constant, i.e.

$$\ln(x^q) = q \ln x + C.$$

   Substitute $x = 1$ into this equation: $0 = \ln 1 = q \ln 1 + C = 0 + C = C.$

(7) $\ln 3 > 1$

Justification Let $f(x) = \frac{1}{x}$. We compute the lower Riemann sum $L(P, f)$ fo
$\ln 3 = \int_1^3 \frac{1}{t}\, dt$ using the partition $P = \left\{1, \frac{5}{4}, \frac{3}{2}, \frac{7}{4}, 2, \frac{9}{4}, \frac{5}{2}, \frac{11}{4}, 3\right\}$ of $[1, 3$
See Figure 4.2.3.

$$\ln 3 = \int_1^3 \frac{1}{t}\, dt \geq L(P, f)$$

$$= f\left(\frac{5}{4}\right)\frac{1}{4} + f\left(\frac{3}{2}\right)\frac{1}{4} + f\left(\frac{7}{4}\right)\frac{1}{4} + f(2)\frac{1}{4} + f\left(\frac{9}{4}\right)\frac{1}{4}$$

$$+ f\left(\frac{5}{2}\right)\frac{1}{4} + f\left(\frac{11}{4}\right)\frac{1}{4} + f(3)\frac{1}{4}$$

$$= \frac{1}{4}\left[\frac{4}{5} + \frac{2}{3} + \frac{4}{7} + \frac{1}{2} + \frac{4}{9} + \frac{2}{5} + \frac{4}{11} + \frac{1}{3}\right]$$

$$> \frac{1}{4}(.8 + .66 + .57 + .5 + .44 + .4 + .36 + .33) = \frac{4.06}{4} > 1.$$

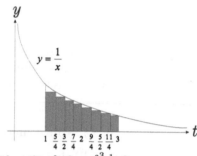

Figure 4.2.3 Estimating $\ln 3 = \int_1^3 \frac{1}{t}\, dt$

(8) $\displaystyle\lim_{x \to \infty} \ln x = \infty$

Justification By Property 6, $\ln 3^n = n \ln 3$ for all integers n. By Property 7,
$\ln 3 > 1$. Since $\ln x$ is an increasing function,

$$\lim_{x \to \infty} \ln x = \lim_{n \to \infty} \ln(3^n) = \lim_{n \to \infty} n \ln 3 = \infty.$$

(9) $\ln x$ has the y-axis as a vertical asymptote on the left.

Justification By Property 6, $\ln 3^{-n} = -n \ln 3$ for all integers n. By Property 7,
$\ln 3 > 1$. Since $\ln x$ is an increasing function,

$$\lim_{x \to 0^+} \ln x = \lim_{n \to \infty} \ln\frac{1}{3^n} = \lim_{n \to \infty} \ln 3^{-n} = \lim_{n \to \infty} -n \ln 3 = -\infty.$$

Hence $y = \ln x$ has the vertical asymptote $x = 0$ on the left.

(10) The range of $\ln x$ equals $\Re$.

Justification Let $y \in \Re$. We show y is in the range of $\ln$. By Property 8,
there is a number $B > y$ in the range of $\ln$, say $\ln b = B$. By Property 9, there
is a number $A < y$ in the range of $\ln$, say $\ln a = A$. Since $\ln$ is an increasing
function, $a < b$. By the Intermediate Value Theorem, there is $c \in [a, b]$ with
$\ln c = y$. Thus the range of $\ln$ is $\Re$.

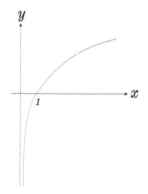

Figure 4.2.4
$y = \ln x$

From these ten observations we sketch the graph of $y = \ln x$ in Figure 4.2.4. The
examples below illustrate how the function $\ln x$ is used in calculus.

Examples 4.2.5 (1) Find $\int_2^7 \frac{1}{x}\, dx$.

Solution $\int_2^7 \frac{1}{x}\, dx = \ln x\big|_2^7 = \ln 7 - \ln 2 = \ln\frac{7}{2}.$

(2) Find $\int \frac{1}{9x+7}\, dx$.

Solution Let $u = 9x + 7$ with $du = 9\, dx$. Then

$$\int \frac{1}{9x+7}\, dx = \frac{1}{9}\int \frac{1}{u}\, du = \frac{1}{9}\ln u + C = \frac{1}{9}\ln(9x+7) + C.$$

(3) Find $D\left[\ln\left(x^2 + 1\right)\right]$.

Solution Let $u = x^2 + 1$. By the chain rule

$$\frac{d}{dx}\ln\left(x^2 + 1\right) = \frac{d}{du}(\ln u)\cdot\frac{d}{dx}\left(x^2 + 1\right) = \frac{1}{u}\, 2x = \frac{2x}{x^2 + 1}. \qquad \square$$

We show the function $\ln x$ has the same properties as a logarithm function.

Proposition 4.2.6 *Let x and y be positive numbers.*
(a) $\ln(xy) = \ln x + \ln y$
(b) $\ln\left(x^q\right) = q\ln x$ *for q a rational number.*
(c) $\ln\left(\frac{x}{y}\right) = \ln x - \ln y$

Proof **(a)** Think of x as a variable and y as a constant. By the chain rule:

$$\frac{d}{dx}\ln(xy) \;=\; \frac{1}{xy}\frac{d}{dx}(xy) = \frac{1}{xy}y = \frac{1}{x}.$$

Also, $\quad \dfrac{d}{dx}(\ln x + \ln y) \;=\; \dfrac{1}{x} + 0 = \dfrac{1}{x}.$

Thus, we have two functions with the same derivative. By Corollary 2.6.24 these functions differ by a constant:

$$\ln(xy) = \ln x + \ln y + C.$$

When $x = 1$ this equation says: $\ln y = 0 + \ln y + C$. Hence $C = 0$.
(b) We proved this formula as Property 6 above.
(c) By (a) and (b),

$$\ln\frac{x}{y} = \ln\left(x\cdot\frac{1}{y}\right) = \ln x + \ln\frac{1}{y} = \ln x + \ln\left(y^{-1}\right) = \ln x - \ln y. \qquad \square$$

Step II: The Function $y = \exp x$

Since the function $\ln$ is increasing, it is one–to–one. Therefore it has an inverse function which we call $\exp$. We deduce the properties of $\exp$ from the properties of $\ln$.

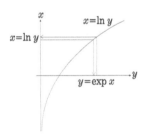

Figure 4.2.7
Definition of exp

Definition 4.2.8 *Define the exponential function $y = \exp x$ as the inverse function of $x = \ln y$. That is,*

$$y = \exp x \quad \text{if and only if} \quad x = \ln y.$$

The relationship between the function $x = \ln y$ and its inverse function $y = \exp x$ is shown in Figures 4.2.7. The logarithm undoes the action of the exponential function as depicted in Figure 4.2.9. That is, if we start with $x \in \Re$ and take its exponential, then application of the logarithm gives the original number x:

$$\ln(\exp x) = x. \qquad (4.2.2)$$

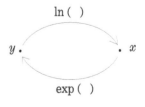

Figure 4.2.9
exp as the Inverse of ln

Similarly, the exponential function undoes the action of the logarithm. That is, if we start with $y > 0$ and take its logarithm, then application of the exponential function gives the original number y:

$$\exp(\ln y) = y. \qquad (4.2.3)$$

We establish the properties of the function $y = \exp x$ which are needed to sketch its graph. These properties are derived from the corresponding properties of its inverse function $x = \ln y$.

(1) The function $\exp x$ has domain $\Re$.

> **Justification** The domain of $y = \exp x$ equals the range of its inverse function $x = \ln y$ which, by Property 10 of ln, is $\Re$.

(2) The function $\exp x$ has range $(0, \infty)$. That is, $\exp x > 0$ for all $x \in \Re$.

> **Justification** The range of $y = \exp x$ equals the domain of its inverse function $x = \ln y$ which, by Property 1 of ln, is $(0, \infty)$.

(3) $\exp 0 = 1$.

> **Justification** By Property 2(ii) of ln: $\ln 1 = 0$. Hence $\exp 0 = 1$.

(4) The function $\exp x$ is differentiable:

$$D(\exp x) = \exp x \quad \text{and} \quad \int \exp x \, dx = \exp x + C. \qquad (4.2.4)$$

> **Justification** Let $y = \exp x$. Use the chain rule to take the derivative of the equation $x = \ln(\exp x)$:

$$\frac{d}{dx}(x) = \frac{d}{dx}\ln(\exp x) = \frac{d}{dy}(\ln y) \cdot \frac{d}{dx}(\exp x)$$

$$1 = \frac{1}{y}\frac{d}{dx}(\exp x)$$

$$\frac{d}{dx}(\exp x) = y = \exp x.$$

By the Second Fundamental Theorem of Calculus, $\int \exp x \, dx = \exp x + C$.

(5) The function $\exp x$ is increasing for $x \in \Re$.

> **Justification** $D(\exp x) = \exp x > 0$ for $x \in \Re$. By Corollary 2.6.29(a), the function $\exp x$ is increasing.

(6) The function $\exp x$ is concave up for $x \in \Re$.

> **Justification** $D^2(\exp x) = D(\exp x) = \exp x > 0$ for $x \in \Re$. By Theorem 2.7.23 $\exp x$ is concave up.

(7) The graph of $\exp x$ has the x–axis as a horizontal asymptote on the left.

> **Justification** Let $x = \ln y$. By Property 9 of ln: x approaches $-\infty$ as y approaches zero from the right. Equivalently, $y = \exp x$ approaches zero as x approaches $-\infty$.

(8) $\lim\limits_{x \to \infty} \exp x = \infty$.

> **Justification** Let $x = \ln y$. By Property 8 of ln: x approaches ∞ as y approaches ∞. Hence $y = \exp x$ approaches ∞ as x approaches ∞.

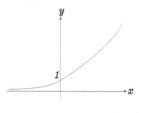

Figure 4.2.10
Graph of $y = \exp x$

We use these observations to sketch the graph of $y = \exp x$ in Figure 4.2.10. The following examples apply Property 4 of exp to compute derivatives and integrals.

Examples 4.2.11 (1) Find $D[\exp(\sin x)]$.

    Solution Let $y = \sin x$. By the chain rule

$$\frac{d}{dx}[\exp(\sin x)] = \frac{d}{dy}(\exp y) \cdot \frac{d}{dx}(\sin x) = \exp y \cos x = \exp(\sin x) \cos x.$$

(2) Find $\int x \exp(-x^2)\, dx$.

    Solution Let $u = -x^2$ with $du = -2x\, dx$. Then

$$\int x \exp(-x^2)\, dx = -\frac{1}{2}\int \exp u \, du = -\frac{1}{2}\exp u + C = -\frac{1}{2}\exp(-x^2) + C. \quad \square$$

The function exp has the properties of an exponential function.

Proposition 4.2.12 *Let $x, y \in \Re$.*

(a) $(\exp x)(\exp y) = \exp(x + y)$

(b) $\frac{\exp x}{\exp y} = \exp(x - y)$

(c) $(\exp x)^q = \exp(qx)$ *for q a rational number.*

Proof (a) Both sides of formula (a) have the same logarithm:

$$\ln[(\exp x)(\exp y)] = \ln(\exp x) + \ln(\exp y) = x + y = \ln[\exp(x + y)].$$

Since the function ln is one–to–one, $(\exp x)(\exp y) = \exp(x + y)$.
(b) Both sides of formula (b) have the same logarithm:

$$\ln\left(\frac{\exp x}{\exp y}\right) = \ln(\exp x) - \ln(\exp y) = x - y = \ln[\exp(x - y)].$$

Since the function ln is one–to–one, $\frac{\exp x}{\exp y} = \exp(x - y)$.
(c) Both sides of formula (c) have the same logarithm:

$$\ln[(\exp x)^q] = q\ln(\exp x) = qx = \ln[\exp(qx)].$$

Since the function ln is one–to–one, $(\exp x)^q = \exp(qx)$. $\square$

Step III: The Function b^x

We know how to define b^q for $b > 0$ and q a rational number. In this subsection, we define b^x for all numbers x, including irrational ones, in a manner that allows us to compute its derivative and establish its properties.

    Observe that when q is a rational number, $\ln(b^q) = q\ln b = \ln[\exp(q\ln b)]$. Since ln is a one–to–one function, $b^q = \exp(q\ln b)$. The expression $\exp(x\ln b)$ is defined even when x is irrational, and we use it to define b^x.

Definition 4.2.13 *Let b be a positive real number, and let x be any real number. Define the exponential function*

$$b^x = \exp(x\ln b). \tag{4.2.5}$$

We show b^x, as defined by (4.2.5), has the usual properties of an exponential function. In addition, we extend the exponential properties of ln and exp from rational powers to arbitrary powers. These results follow from the properties of ln and exp established in the preceding two subsections.

Proposition 4.2.14 Let $x, y, b, r \in \Re$ with $b > 0$. In each of the following statement the expressions b^u are defined by (4.2.5).

(a) $b^0 = 1$ and $b^1 = b$.   (b) $b^x b^y = b^{x+y}$.

(c) $\frac{b^x}{b^y} = b^{x-y}$.   (d) $(b^x)^r = b^{rx}$.

(e) $\ln(x^r) = r \ln x$ for $x > 0$.   (f) $(\exp x)^r = \exp(rx)$.

Proof (a) $b^0 = \exp(0 \ln b) = \exp 0 = 1$ and $b^1 = \exp(1 \ln b) = \exp(\ln b) = b$.

(b) $b^x b^y = \exp(x \ln b) \exp(y \ln b) = \exp(x \ln b + y \ln b) = \exp[(x + y) \ln b] = b^{x+y}$.

(c) $\frac{b^x}{b^y} = \frac{\exp(x \ln b)}{\exp(y \ln b)} = \exp(x \ln b - y \ln b) = \exp[(x - y) \ln b] = b^{x-y}$.

(d) $(b^x)^r = \exp[r \ln(b^x)] = \exp[r \ln(\exp(x \ln b))] = \exp(rx \ln b) = b^{rx}$.

(e) $\ln(x^r) = \ln[\exp(r \ln x)] = r \ln x$.

(f) $(\exp x)^r = \exp[r \ln(\exp x)] = \exp(rx)$ □

Observe when q is a rational number, we have two definitions of b^q: the usual one given at the beginning of this section as well as the one given by (4.2.5). We use properties of the exponential function to show that these two definitions of b^q agree.

Corollary 4.2.15 When q is a rational number, the exponential function b^q defined by (4.2.5) agrees with the usual definition.

Proof The usual definition of b^q is given by the following four cases, as listed at the beginning of this section. Note all expressions b^u below are defined by (4.2.5).

(1) $b^0 = 1$ by Proposition 4.2.14(a).

(2) If n is a positive integer, then $b^n = b^{1+\cdots+1} = b^1 \cdots b^1 = b \cdots b$ (n times) by Proposition 4.2.14(a),(b).

(3) For any rational numbe q: $\frac{1}{b^q} = \frac{b^0}{b^q} = b^{0-q} = b^{-q}$ by Proposition 4.2.14(a),(c).

(4) Let p, q be integers with $q > 0$. We show $b^{p/q} = \left(\sqrt[q]{b}\right)^p$. Observe $b^{1/q}$ multiplied with itself q times is $\left(b^{1/q}\right)^q = b^{q/q} = b^1 = b$ by Proposition 4.2.14 (a),(d) and (1). Hence $b^{1/q} = \sqrt[q]{b}$. By Prop. 4.2.14(d), $b^{p/q} = \left(b^{1/q}\right)^p = \left(\sqrt[q]{b}\right)^p$.   □

Observe $b^x = \exp(x \ln b)$ is a continuous function. Thus we approximate b^r by using the first few digits of the decimal expansion $r = r_0.r_1 \cdots r_n \cdots$

$$b^r = \lim_{n \to \infty} b^{r_0.r_1 \cdots r_n} \approx b^{r_0.r_1 \cdots r_n} \tag{4.2.6}$$

Examples 4.2.16 (1) Approximate 5^π.

Solution The decimal expansion of π begins as $\pi = 3.14159 \cdots$. Hence

$$5^3 = 125, \qquad 5^{3.1} \approx 146.827, \qquad 5^{3.14} \approx 156.591,$$
$$5^{3.141} \approx 156.842, \qquad 5^{3.1415} \approx 156.969, \qquad 5^{3.14159} \approx 156.992$$

are increasingly better approximations of 5^π.

(2) Approximate $3^{\sqrt{2}}$.

Solution The decimal expansion of $\sqrt{2}$ begins as $\sqrt{2} = 1.4142 \cdots$. Hence

$$3^1 = 3, \quad 3^{1.4} \approx 4.656, \quad 3^{1.41} \approx 4.707, \quad 3^{1.414} \approx 4.728, \quad 3^{1.4142} \approx 4.729$$

are increasingly better approximations of $3^{\sqrt{2}}$.   □

We have two new families of functions: the exponential functions b^x, for $x > 0$, and the power functions x^r, for r irrational. We compute their derivatives and integrals.

Proposition 4.2.17 (a) *If b is a positive number, then*

$$D\left(b^x\right) = (\ln b)\, b^x \quad and \quad \int b^x\, dx = \frac{b^x}{\ln b} + C.$$

(b) *If $r \in \mathbb{R}$, then*

$$D\left(x^r\right) = r x^{r-1} \quad and \quad \int x^r\, dx = \frac{x^{r+1}}{r+1} + C \ \ if \ r \neq -1.$$

Proof (a) Let $u = x \ln b$. Then $b^x = \exp(x \ln b) = \exp u$. By the chain rule:

$$\frac{d}{dx}\left(b^x\right) = \frac{d}{du}(\exp u) \cdot \frac{du}{dx} = (\exp u)(\ln b) = b^x \ln b.$$

By the 2nd Fundamental Thm. of Calculus: $\int b^x\, dx = \frac{1}{\ln b} \int b^x \ln b\, dx = \frac{1}{\ln b} b^x + C.$
(b) Let $u = r \ln x$. Then $x^r = \exp(r \ln x) = \exp u$. By the chain rule:

$$\frac{d}{dx}\left(x^r\right) = \frac{d}{du}(\exp u) \cdot \frac{du}{dx} = (\exp u)\frac{r}{x} = x^r \frac{r}{x} = r x^{r-1}.$$

If $r \neq -1$, then $D\left(\frac{x^{r+1}}{r+1}\right) = x^r$. By the Second Fundamental Theorem of Calculus,
$\int x^r\, dx = \frac{x^{r+1}}{r+1} + C.$ ☐
Warning Be careful of the distinction between power functions and exponential functions. $f(x) = x^3$ is very different from $g(x) = 3^x$. In particular, $f'(x) = 3x^2$ while $g'(x) = 3^x \ln 3$. Note $g'(x)$ is *not* equal to $x3^{x-1}$.

Examples 4.2.18 (1) Find $D(x^\pi)$.
   Solution By Proposition 4.2.17(b): $D(x^\pi) = \pi x^{\pi-1}$.

(2) Find $D(8^{\sin x})$.
   Solution By the chain rule:

$$\frac{d}{dx}\left(8^{\sin x}\right) = (\ln 8) 8^{\sin x} \frac{d}{dx}(\sin x) = (\ln 8) 8^{\sin x} \cos x.$$

(3) Evaluate $\int \frac{5^{\ln x}}{x}\, dx$.
   Solution Let $u = \ln x$ with $du = \frac{1}{x}\, dx$. Then

$$\int \frac{5^{\ln x}}{x}\, dx = \int 5^u\, du = \frac{5^u}{\ln 5} + C = \frac{5^{\ln x}}{\ln 5} + C.$$

(4) Evaluate $\int \frac{x^{\sqrt 2}}{x + x^{1+\sqrt 8}}\, dx$.
   Solution Let $u = x^{\sqrt 2}$ with $du = \sqrt 2 x^{\sqrt 2 - 1}\, dx$. Then

$$\begin{aligned}
\int \frac{x^{\sqrt 2}}{x + x^{1+\sqrt 8}}\, dx &= \int \frac{x^{\sqrt 2}}{x + x x^{2\sqrt 2}}\, dx = \frac{1}{\sqrt 2} \int \frac{\sqrt 2 x^{\sqrt 2 - 1}}{1 + (x^{\sqrt 2})^2}\, dx \\
&= \frac{1}{\sqrt 2} \int \frac{1}{1 + u^2}\, du = \frac{1}{\sqrt 2} \arctan u + C \\
&= \frac{1}{\sqrt 2} \arctan\left(x^{\sqrt 2}\right) + C \qquad\qquad ☐
\end{aligned}$$

We assemble information to sketch the graph of $y = b^x$ for a positive number $b \neq 1$.

(1)   Domain $b^x = \Re$.

   Justification $b^x = \exp(x \ln b)$ is defined for every number x.

(2)   Range $b^x = (0, \infty)$, and the graph of $y = b^x$ lies entirely above the x–axis.

   Justification The range of $b^x = \exp(x \ln b)$ equals Range $\exp x$ which is $(0, \infty)$

(3) $b^0 = 1$.

   Justification This is Proposition 4.2.14(a).

(4)   If $b > 1$, then b^x is increasing for $x \in \Re$. If $b < 1$, then b^x is decreasing for $x \in \Re$

   Justification $D(b^x) = (\ln b)b^x$ for $x \in \Re$. Since b^x is always positive, $D(b^x)$ has the same sign as $\ln b$ which is positive if $b > 1$ or negative if $b < 1$. By Corollary 2.6.29, b^x is increasing if $b > 1$ or decreasing if $b < 1$.

(5) b^x is concave up everywhere.

   Justification $D^2(b^x) = D\left[(\ln b)b^x\right] = (\ln b)^2 b^x > 0$ for $x \in \Re$.
   By Theorem 2.7.23(a) b^x is concave up.

(6)   If $b > 1$, then $\lim\limits_{x \to \infty} b^x = \infty$ and $\lim\limits_{x \to -\infty} b^x = 0$.

   Justification $\lim\limits_{x \to \infty} b^x = \lim\limits_{x \to \infty} \exp(x \ln b) = \infty$ as $\ln b > 0$ and $x \ln b$ approaches
   Similarly, $\lim\limits_{x \to -\infty} b^x = \lim\limits_{x \to -\infty} \exp(x \ln b) = 0$ since $x \ln b$ approaches $-\infty$.

(7)   If $b < 1$, then $\lim\limits_{x \to \infty} b^x = 0$ and $\lim\limits_{x \to -\infty} b^x = \infty$.

   Justification $\lim\limits_{x \to \infty} b^x = \lim\limits_{x \to \infty} \exp(x \ln b) = 0$ as $\ln b < 0$ and $x \ln b$ approaches
   Similarly, $\lim\limits_{x \to -\infty} b^x = \lim\limits_{x \to -\infty} \exp(x \ln b) = \infty$ since $x \ln b$ approaches ∞.

Using these observations, we sketch the graph of the function $y = b^x$ in Figure 4.2.19.

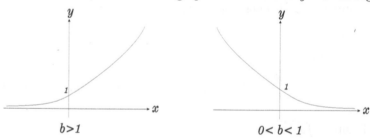

Figure 4.2.19 Graph of $y = b^x$

There is a special number e such that $\exp x$ equals the exponential function e^x.

Proposition 4.2.20 (a) *There is a unique number e such that $2 < e < 3$ and $\ln e = 1$.*
(b) $\exp x = e^x$ *for $x \in \Re$.*

Proof (a) Since 1 is a positive number and the range of $\ln x$ equals all positive numbers, there must be a number e with $\ln e = 1$. Since the function $\ln x$ is one–to–one, there is only one such number e. We showed in Property 7 of $\ln$ that $1 < \ln 3$. We show below that $\ln 2 < 1$. Since the function $\ln x$ is increasing and $\ln 2 < \ln e < \ln 3$, it follows that $2 < e < 3$. It remains to show $\ln 2 < 1$. Let $f(t) = \frac{1}{t}$, and consider the partition $P = \{1, \frac{3}{2}, 2\}$ of the interval $[1,2]$ in Figure 4.2.21. Then

$$\ln 2 = \int_1^2 \frac{1}{t}\, dt \leq U(P, f) = f(1)\frac{1}{2} + f\left(\frac{3}{2}\right)\frac{1}{2} = (1)\frac{1}{2} + \left(\frac{2}{3}\right)\frac{1}{2} = \frac{5}{6} < 1.$$

(b) $e^x = \exp(x \ln e) = \exp x$ since $\ln e = 1$. $\square$

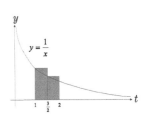

Step IV: The Function $\log_b x$

We define $y = \log_b x$, the logarithm function with base b, as the inverse of the exponential function $x = b^y$. We use the properties of exponential functions to derive the properties of logarithms. We also identify $\ln x$ as $\log_e x$.

Property 4 of the exponential function $x = b^y$ says it is increasing when $b > 1$ or decreasing when $0 < b < 1$. In both cases $x = b^y$ is one–to–one, and its inverse function is defined. We call the inverse function *the logarithm of x to the base b*.

Definition 4.2.22 *Let b be a positive number other than one. Define $y = \log_b x$, the logarithm of x to the base b, as the inverse function of $x = b^y$. That is,*

$$y = \log_b x \quad \text{if and only if} \quad x = b^y. \tag{4.2.7}$$

The definition of $y = \log_b x$ as the inverse function of $x = b^y$ is depicted in Figure 4.2.23. If we start with a real number y, take its exponential b^y and then apply the logarithm $\log_b(-)$, we obtain the original number:

$$\log_b (b^y) = y. \tag{4.2.8}$$

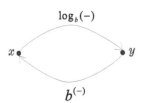

Similarly, if we start with a positive number x, take its logarithm $\log_b x$ and then apply the exponential $b^{(-)}$, we obtain the original number:

$$b^{\log_b x} = x. \tag{4.2.9}$$

Figure 4.2.23
Definition of $\log_b x$

We list the properties of the logarithm function $y = \log_b x$. The first four properties are derived from the corresponding properties of the exponential function $x = b^y$.

Proposition 4.2.24 *Let a, b, c, x, y be positive numbers with $a \neq 1$ and $b \neq 1$.*

(a) $\log_b(xy) = \log_b x + \log_b y$ **(b)** $\log_b\left(\frac{x}{y}\right) = \log_b x - \log_b y$

(c) $\log_b(x^r) = r\log_b x$ *for* $r \in \mathcal{R}$ **(d)** $\ln x = \log_e x$ **(e)** $\log_a c = (\log_a b)(\log_b c)$

Proof Let $p = \log_b x$ and $q = \log_b y$. Then $x = b^p$ and $y = b^q$.
(a) Note $xy = b^p b^q = b^{p+q}$. Hence $\log_b(xy) = p + q = \log_b x + \log_b y$.
(b) Since $\frac{x}{y} = \frac{b^p}{b^q} = b^{p-q}$, it follows that $\log_b \frac{x}{y} = p - q = \log_b x - \log_b y$.
(c) Since $x^r = (b^p)^r = b^{pr}$, it follows that $\log_b(x^r) = rp = r\log_b x$.
(d) $\log_e x$ is the inverse function of the exponential function e^x. However, $e^x = \exp x$ which is the inverse function of $\ln x$. Thus $\log_e x$ and $\ln x$ are inverse functions of the same function and must be equal.
(e) If $s = \log_a b$ and $t = \log_b c$, then $b = a^s$ and $c = b^t$. Hence $c = (a^s)^t = a^{st}$, and $\log_a c = st = (\log_a b)(\log_b c)$. $\square$

The following examples illustrate the computation of $\log_b x$.

Examples 4.2.25 (1) Find $\log_4$ of each number:

(a) 64   (b) $\frac{1}{16}$ (c) 8   (d) $\frac{1}{32}$ (e) $\sqrt{2}$

Solution (a) Since $64 = 4^3$, $\log_4 64 = 3$.

(b) Since $\frac{1}{16} = \frac{1}{4^2} = 4^{-2}$, $\log_4 \frac{1}{16} = -2$.

(c) Since $8 = 2^3 = (4^{1/2})^3 = 4^{3/2}$, $\log_4 8 = \frac{3}{2}$.

(d) Since $\frac{1}{32} = 2^{-5} = (4^{1/2})^{-5} = 4^{-5/2}$, $\log_4 \frac{1}{32} = -\frac{5}{2}$.

(e) Since $\sqrt{2} = 2^{1/2} = (4^{1/2})^{1/2} = 4^{1/4}$, $\log_4 \sqrt{2} = \frac{1}{4}$.

(2) Express each number in terms of $\log_{10} 4$ and $\log_{10} 9$:

(a) $\log_{10} 144$ (b) $\log_{10} 24$ (c) $\log_{10} \frac{27}{32}$

Solution (a) $\log_{10} 144 = \log_{10}(4^2 9) = \log_{10} 4^2 + \log_{10} 9 = 2\log_{10} 4 + \log_{10} 9$.

(b) $\log_{10} 24 = \log_{10} 2^3 3 = \log_{10} 4^{3/2} 9^{1/2} = \log_{10} 4^{3/2} + \log_{10} 9^{1/2}$
$= \frac{3}{2}\log_{10} 4 + \frac{1}{2}\log_{10} 9$.

(c) $\log_{10} \frac{27}{32} = \log_{10} \frac{3^3}{2^5} = \log_{10} \frac{9^{3/2}}{4^{5/2}} = \log_{10} 9^{3/2} - \log_{10} 4^{5/2}$
$= \frac{3}{2}\log_{10} 9 - \frac{5}{2}\log_{10} 4$.

We compute the derivatives of logarithm functions.

Corollary 4.2.26 *Let b and x be positive numbers with $b \neq 1$. Then*

$$\text{(a)} \ \log_b x = \frac{\ln x}{\ln b} \qquad\qquad \text{(b)} \ D\left(\log_b x\right) = \frac{1}{x \ln b}.$$

Proof (a) By Proposition 4.2.24(e), $\ln x = \log_e x = (\log_e b)(\log_b x) = (\ln b)(\log_b x)$.
(b) $D(\log_b x) = D\left(\frac{\ln x}{\ln b}\right) = \frac{1}{\ln b}D(\ln x) = \frac{1}{x \ln b}$.

Examples 4.2.27 (1) Find the derivative of $\log_{10}(x^2 + 1)$.

Solution By the chain rule:

$$\frac{d}{dx}\left[\log_{10}(x^2 + 1)\right] = \frac{1}{(\ln 10)(x^2 + 1)}\frac{d}{dx}(x^2 + 1) = \frac{2x}{(\ln 10)(x^2 + 1)}.$$

(2) Evaluate $\int \frac{1}{x \log_3 x} \, dx$.

Solution Let $u = \log_3 x$. Then $du = \frac{dx}{x \ln 3}$. Hence

$$\int \frac{1}{x \log_3 x} \, dx = (\ln 3) \int \frac{1}{u} \, du = (\ln 3) \ln u + C = (\ln 3) \ln(\log_3 x) + C.$$

The following six properties determine the graph of $y = \log_b x$ for $b > 0$, $b \neq 1$.

(1) The function $\log_b x$ has domain $(0, \infty)$ and range $\Re$.

Justification The domain of $y = \log_b x$ equals the range $(0, \infty)$ of its inverse function $x = b^y$. The range of $y = \log_b x$ equals the domain $\Re$ of its inverse function $x = b^y$.

(2) $\log_b 1 = 0$.

Justification $\log_b 1 = 0$ because $b^0 = 1$.

(3) If $b > 1$, then $\log_b x < 0$ for $0 < x < 1$ and $\log_b x > 0$ for $x > 1$.

If $b < 1$, then $\log_b x > 0$ for $0 < x < 1$ and $\log_b x < 0$ for $x > 1$.

Justification Note $\log_b x = \frac{\ln x}{\ln b}$. If $b > 1$, then $\ln b > 0$. Thus, $\log_b x$ and $\ln x$ have the same sign: positive for $x > 1$ and negative for $0 < x < 1$. If $b < 1$, then $\ln b < 0$. Thus, $\log_b x$ and $\ln x$ have opposite signs: $\log_b x$ is positive for $0 < x < 1$ and negative for $x > 1$.

(4) If $b > 1$, then $\log_b x$ is increasing. If $b < 1$, then $\log_b x$ is decreasing.

Justification Since $x > 0$, the sign of $D(\log_b x) = \frac{1}{x \ln b}$ is the same as the sign of $\ln b$. Thus $D(\log_b x)$ is positive if $b > 1$ or negative if $b < 1$. By Corollary 2.6.29, $\log_b x$ is increasing if $b > 1$ or decreasing if $b < 1$.

(5) If $b > 1$, then $\log_b x$ is concave down. If $b < 1$, then $\log_b x$ is concave up.

Justification $D^2(\log_b x) = D(\frac{1}{x \ln b}) = -\frac{1}{x^2 \ln b}$ which is negative if $b > 1$ or positive if $b < 1$. By Theorem 2.7.23, $\log_b x$ is concave down if $b > 1$ or concave up if $b < 1$.

(6) $\log_b x$ has the y–axis as a vertical asymptote.

Justification If $b > 1$, then $x = b^y > 0$ approaches 0 as y approaches $-\infty$. Therefore $y = \log_b x$ approaches $-\infty$ as x approaches zero from the right. If $b < 1$, then $x = b^y > 0$ approaches zero as y approaches ∞. Hence $y = \log_b x$ approaches ∞ as x approaches zero from the right.

We use the preceding properties to sketch the graph of $y = \log_b x$ in Figure 4.2.28.

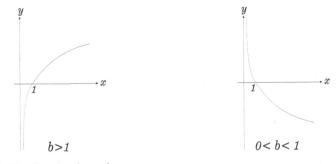

Figure 4.2.28 Graph of $y = \log_b x$

If $B > b > 1$, then

$$\log_B x = (\log_B b)(\log_b x) \quad \text{with} \quad 0 < \log_B b < 1.$$

Thus the graph of $y = \log_B x$ lies above the graph of $y = \log_b x$ for $0 < x < 1$ and lies below the graph of $y = \log_b x$ for $x > 1$. See the left diagram in Figure 4.2.29.

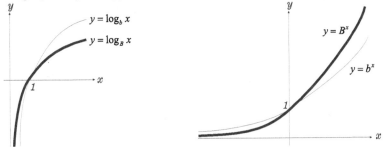

Figure 4.2.29 Graphs of Logarithm and Exponential Functions for Bases $B > b > 1$

Also, $B^x = \exp(x \ln B) = \exp(x \ln b)^{\frac{\ln B}{\ln b}} = (b^x)^{\frac{\ln B}{\ln b}}$ with $\dfrac{\ln B}{\ln b} > 1.$

Thus the graph of $y = B^x$ lies above the graph of $y = b^x$ for $x > 0$ and lies below the graph of $y = b^x$ for $x < 0$. See the right diagram in Figure 4.2.29.

Historical Remarks

At the end of the sixteenth century, 15 place tables of trigonometric functions were used in navigation and astronomy. The need for an easy way to multiply and divide these numbers led to the discovery of logarithms by John Napier in 1614 and by Jost Bürgi in 1620. They began with the observation that if $x = AR^m$ and $y = AR^n$, then

$$ xy = A \cdot AR^{m+n} \quad \text{and} \quad \frac{x}{y} = \frac{1}{A} \cdot AR^{m-n}. $$

Thus, a table of the numbers AR^t for various values of t allows multiplications and divisions to be replaced by the much easier computations of sums and differences. For such a table to be useful, the consecutive powers of R must be close to each other, i.e. R must be close to one. In addition, A should be a power of ten so that multiplication and division by A is merely a shift of the decimal point. Napier chose $A = 10^7$ and $R = 1 - 10^{-7}$, while Bürgi chose $A = 10^8$ and $R = 1 + 10^{-4}$. Napier constructed tables giving the values of his logarithms to six significant figures. He also introduced decimal notation to replace fractions, i.e. he wrote 37.2589 instead of $37\frac{2,589}{10,000}$.

Observe the relationship between the approach of Napier–Bürgi and modern logarithms. If $x = AR^y$, write $y = \text{Nlog } x$. Then $R^y = \frac{x}{A}$ and

$$ \text{Nlog } x = y = \log_R \frac{x}{A}. $$

Apply the Mean Value Theorem to the function $f(t) = \ln(1 + t)$ on the interval $[0, x]$: there is $c \in (0, x)$ such that $\ln(1 + x) - \ln 1 = \frac{1}{1+c}(x - 0)$. Thus, $\ln(1 + x) \approx x$ when $|x|$ is small. If $R = 1 \pm 10^{-p}$, then

$$ \pm 10^{-p} \text{Nlog } x = \pm 10^{-p} \log_R \frac{x}{A} = \pm 10^{-p} \frac{\ln(x/A)}{\ln(1 \pm 10^{-p})} \approx \frac{\pm 10^{-p}}{\pm 10^{-p}} \ln \frac{x}{A} = \ln \frac{x}{A}. $$

Note that $R = 1 - 10^{-p}$ with $A = 10^p$ for Napier while $R = 1 + 10^{-p}$ with $A = 10^{2p}$ for Bürgi. In fact, Napier was aware of the approximation

$$ \text{Nlog } x \approx -10^p \ln \frac{x}{A} = -A \ln \frac{x}{A}. $$

Since exponential notation and rational powers were not defined in his days, Napier gives the following definition of $\text{nlog} x$ which produces a continuous function defined for all numbers x. Consider an object moving from A to 0 on a number line so that if $x(t)$ is its position at time t, then its speed at time t equals $-x(t)$. Napier defines $\text{nlog } x(t) = At$. We leave the identification of $\text{nlog } x$ with $-A \ln \frac{x}{A}$ as a Challenging Problem. The close relationship between $\text{nlog } x$ and $\ln x$ gives rise to the name *Napierian logarithm* for $\ln x$. Some do not think that Napier should be credited with the discovery of $\ln x$ and call this function the *natural logarithm*.

In 1624, Henry Briggs constructed part of a 14 digit table of $\log_{10} x$ for x a positive integer less than $100,000$ which he called *improved logarithms*. This table, up to 10 digits, was completed by Adrian Vlacq in 1628. Today we call these logarithms *common logarithms*. These are useful for applications to base ten arithmetic because

$$ \log_{10}(10^n x) = n + \log_{10} x. $$

The close relationship between logarithms and area under the hyperbola $y = \frac{1}{x}$ was deduced from the following identity discovered by Gregory St. Vincent in 1647. In our notation, he showed that for $a, b, s > 0$:

$$\int_{sa}^{sb} \frac{1}{x}\, dx = \int_{a}^{b} \frac{1}{x}\, dx. \tag{4.2.10}$$

We verify this identity by rewriting it as: $\ln(sb) - \ln(sa) = \ln b - \ln a$. In 1649, Alfonso Antonio de Sarasa applied Gregory's identity to show that

$$L(x) = \int_{1}^{x} \frac{1}{t}\, dt$$

has the properties of a logarithm function:

$$L(1) = 0, \quad L(xy) = L(x) + L(y), \quad L\left(\frac{x}{y}\right) = L(x) - L(y).$$

For example,

$$L(xy) = \int_{1}^{xy} \frac{1}{t}\, dt = \int_{1}^{x} \frac{1}{t}\, dt + \int_{x}^{xy} \frac{1}{t}\, dt = \int_{1}^{x} \frac{1}{t}\, dt + \int_{1}^{y} \frac{1}{t}\, dt = L(x) + L(y).$$

In 1667, Isaac Newton studied the function

$$A(1 + x) = \int_{0}^{x} \frac{1}{1+t}\, dt.$$

He showed it has the logarithmic properties:

$$A(0) = 0, \quad A\left[(1+x)(1+y)\right] = A(1+x) + A(1+y), \quad A\left(\frac{1+x}{1+y}\right) = A(1+x) - A(1+y).$$

Of course, we know that $L(x) = \ln x$ and $A(1 + x) = \ln(1 + x)$.

The definition $y = \log_b x$ if $x = b^y$, the definition of e and the identification of $\ln x$ with $\log_e x$ were made by Leonhard Euler in 1728 although he first published these results in his calculus textbook of 1748. We used Euler's approach in this section.

Summary

The reader should know the definitions of the logarithm functions $\ln x$ and $\log_b x$ as well as the exponential functions $\exp x$ and b^x. She should understand their properties, including their derivatives and graphs. In particular, she should be able to compute derivatives and integrals involving these functions.

Basic Exercises

1. Evaluate each exponential.
(a) 2^{-3} (b) $27^{2/3}$ (c) $32^{-3/5}$ (d) $(\frac{1}{125})^{4/3}$ (e) $(64)^{-5/6}$

2. Use decimal expansions to the nearest thousandth to approximate each exponential.
(a) $3^{\sqrt{5}}$ (b) 2^{π} (c) $4^{-\sqrt{3}}$ (d) $6^{\sqrt{7}}$ (e) $5^{-\sqrt{2}}$

3. Evaluate each logarithm.
(a) $\log_4 64$ (b) $\log_8 32$ (c) $\log_{16} 8$ (d) $\log_{64} \frac{1}{16}$ (e) $\log_{1/27} 9$

4. Evaluate each natural logarithm.
(a) $\ln e$ (b) $\ln 1$ (c) $\ln(e^3)$ (d) $\ln \sqrt{e}$ (e) $\ln(1/e^2)$

5. Simplify each expression.
(a) $\log_5(5^\pi)$ (b) $e^{\ln 3}$ (c) $(\log_3 2)(\log_2 3)$ (d) $\frac{\log_{10} 7}{\log_2 7}$ (e) $e^{x \ln 8}$

6. Write each expression in terms of exp and ln. Simplify your answer.
(a) $5^{\sqrt{2}}$ (b) π^e (c) $\log_\pi e$ (d) $\log_{e^2} \pi$

(e) $\log_2 \frac{(3^9)(7^4)}{(5^8)(4^6)}$ (f) $\log_5 \sqrt{\frac{2^8}{(3^6)(7^3)}}$ (g) $\log_3 \frac{\sqrt{10^7}}{\sqrt[3]{8^5}}$ (h) $\log_4 \left[\frac{(4^7)(9^5)}{(8^4)(3^6)}\right]^{3/2}$

7. Simplify each expression.
(a) $\frac{\exp(4)\exp(3)}{\exp(6)}$ (b) $\sqrt{\exp(8)}$ (c) $5\ln 2 + 4\ln 3 - 3\ln$

(d) $\ln \sqrt{8} - \ln 16 + 3\ln 4$ (e) $\ln \sqrt{\exp(2)\exp(3)\exp(4)}$ (f) $\ln \left(\frac{\exp(4)^{3/2}\exp(27)^{2/}}{\exp(8)^{4/3}\exp(9)^{5/2}}\right)$

(g) $\exp(3\ln 4 - 2\ln 5)$ (h) $\sqrt{\exp(2\ln 6 - 3\ln 2)}$ (i) $\ln \sqrt{\exp \sqrt{\ln \sqrt{\exp(6}}}$

8. Give the definition of each expression.
(a) $\ln 4$ (b) $\exp 2$ (c) 3^π (d) $\log_3 10$

9. Use this table of approximations to estimate the natural logarithms below.

N	1	2	3	4	5	6	7	8	9	10
$\ln N$	0	.69	1.10	1.39	1.61	1.79	1.95	2.08	2.20	2.30

(a) $\ln 15$ (b) $\ln \frac{7}{9}$ (c) $\ln \sqrt{60}$ (d) $\ln 4.8$ (e) $\ln .108$

10. Estimate each natural logarithm $\ln x$ by computing the lower and upper Riemann sums of $\int_1^x \frac{1}{t}\, dt$ for the regular partition of $[1, x]$ into n subintervals.
(a) $\ln 5$ with $n = 4$ (b) $\ln 5$ with $n = 8$ (c) $\ln 4$ with $n = 6$
(d) $\ln(1/2)$ with $n = 4$ (e) $\ln(1/3)$ with $n = 5$ (f) $\ln(3/4)$ with $n = 3$

11. Find the domain of each function.
(a) $f(x) = \frac{3^x}{5^{-x}+1}$ (b) $g(x) = \frac{7^x - 4^x}{8^x - 2}$ (c) $h(x) = \frac{3^x}{5^x - 7^x}$
(d) $k(x) = \frac{2^x - 3^x}{4^{3x+2} - 8^{5x-10}}$ (e) $p(x) = \sqrt{1 - 3^x}$ (f) $q(x) = (2^x - 4)^{5/6}$

12. Find the derivative of each function.
(a) π^x (b) x^π (c) $\sqrt{3}^x$ (d) $x^{\sqrt{3}}$

13. Compute each derivative.

(a) $D\left(x^{\sqrt{5}-3}\right)$ (b) $D(\ln \sin x)$ (c) $D(x\ln x)$

(d) $D\left(\frac{\log_3 x}{\log_2 x}\right)$ (e) $D(10^x)$ (f) $D\left(e^{x^2}\right)$

(g) $D\left(\frac{e^x + e^{-x}}{e^x - e^{-x}}\right)$ (h) $D\left(5^x x^5\right)$ (i) $D\left[\log_2\left(3^x + 7^x\right)\right]$

(j) $D\left[\ln(\ln x)\right]$ (k) $D\left(e^{e^x}\right)$ (l) $D\left(8^{\cos e^x}\right)$

(m) $D\left(\log_x e\right)$ (n) $D\left[(\arcsin \ln x)(\arctan \ln x)\right]$ (o) $D\left(\sqrt{\log_2 x + \log_3 x}\right)$

14. Find the third derivative of each function.
(a) $y = \exp(-x^2)$ (b) $y = 2^{\cos x}$ (c) $y = \ln \tan x$ (d) $y = \sin \log_4 x$ (e) $y = \sqrt{\ln x}$

15. Find all the higher derivatives of each function.
(a) $f(x) = \exp(x)$ (b) $g(x) = 5^x$ (c) $h(x) = \ln x$ (d) $k(x) = \log_3 x$ (e) $m(x) = x^\pi$

16. Find the derivative $\frac{dy}{dx}$ of each implicitly defined function.
(a) $\ln(x + y) = 1 + \ln(x - y)$ (b) $\ln \sin(xy) = \sin \ln(xy)$
(c) $\arctan(\ln(x^2 + y^2)) = xy^2 + 1$ (d) $\log_6(1 + \sqrt{xy}) = 6^{x+y}$
(e) $\exp(\log_2 x + \log_4 y) = xy + 1$ (f) $\log_{10} \exp(x^2y^2) = \cot(x + y)$

17. Find the second derivative $\frac{d^2y}{dx^2}$ of each implicitly defined function.
(a) $\exp(x + y) = xy$ (b) $2^y = x + y$ (c) $\ln(x^2 + y^2) = y - 5$

18. Determine whether each function has a vertical tangent, cusp or vertical cusp at the indicated point.

(a) $y = \exp(x^{2/3})$ at $x = 0$ (b) $y = \sqrt[3]{\ln x}$ at $x = 1$ (c) $y = |\ln x|$ at $x = 1$

(d) $y = 2^{\left(x^{1/5}\right)}$ at $x = 0$ (e) $y = |\log_7(x+1)|$ at $x = 0$ (f) $y = \exp|x|$ at $x = 0$

19. (a) Since $\ln x = \log_e x$, we have $D(\ln x) = \frac{1}{x \ln e}$ by Corollary 4.2.26(b). However, $D(\ln x) = \frac{1}{x}$ by (4.2.1). Which answer is correct?

(b) Since $\exp(x) = e^x$, we have $D(\exp(x)) = (\ln e)e^x$ by Corollary 4.2.17(a). However, $D(\exp(x)) = \exp(x)$ by (4.2.4). Which answer is correct?

20. (a) Show $\ln(\log_2 x)$ and $\ln(\ln x)$ have the same derivative.

(b) Explain how these two different functions can have the same derivative.

21. Compute each integral.

(a) $\int x^{\pi - \sqrt{2}} \, dx$ (b) $\int e^{5x+1} \, dx$ (c) $\int x e^{-x^2} \, dx$ (d) $\int \pi^x \, dx$

(e) $\int_3^9 \frac{1}{x} \, dx$ (f) $\int \frac{1}{11x - 5} \, dx$ (g) $\int \frac{\ln x}{x} \, dx$ (h) $\int \frac{e^x}{\sqrt{1 - e^x}} \, dx$

(i) $\int (e^x + 1)^2 \, dx$ (j) $\int \frac{x}{x^2 + 1} \, dx$ (k) $\int \frac{\log_5 \sqrt{3x+1}}{3x+1} \, dx$ (l) $\int \frac{e^x}{e^x + 1} \, dx$

22. Sketch the graph of each function.

(a) $f(x) = \log_3 x$ (b) $g(x) = \log_{1/5} x$ (c) $h(x) = \ln(1 + x)$ (d) $i(x) = 4^x$

(e) $j(x) = 3^{-x}$ (f) $k(x) = e^{1+x}$ (g) $m(x) = e^{-5x}$ (h) $n(x) = \ln \sqrt{5^{3x}}$

23. Deduce the graph of each function by reflecting the graph of its inverse function about the line $y = x$.

(a) $y = \exp x$ (b) $y = \log_b x$ for $0 < b < 1$ (c) $y = \log_b x$ for $b > 1$

24. Evaluate each limit by identifying it as the definition of a derivative.

(a) $\lim\limits_{x \to 1} \dfrac{\ln x}{x - 1}$ (b) $\lim\limits_{x \to e} \dfrac{\ln x - 1}{x - e}$ (c) $\lim\limits_{x \to 0} \dfrac{\exp x - 1}{x}$ (d) $\lim\limits_{x \to 1} \dfrac{\exp x - e}{x - 1}$

(e) $\lim\limits_{x \to 0} \dfrac{5^x - 1}{x}$ (f) $\lim\limits_{x \to 1} \dfrac{7^x - 7}{x - 1}$ (g) $\lim\limits_{x \to 1} \dfrac{\log_3 x}{x - 1}$ (h) $\lim\limits_{x \to 2} \dfrac{\log_2 x - 1}{x - 2}$

25. Use the definition of x^x to find its derivative.

Challenging Problems

1. Show e^x is the only function which equals its own derivative.

2. In the notation of the Historical Note, show $\text{nlog } x = -A \ln \frac{x}{A}$.

3. Show $\text{nlog } x$ has the following properties:

(a) $\text{nlog}(xy) = \text{nlog } x + \text{nlog } y - A \ln A$;   (b) $\text{nlog}\left(\frac{x}{y}\right) = \text{nlog } x - \text{nlog } y + A \ln A$.

4. Prove Gregory's formula (4.2.10) directly as follows.

(a) Let $P = \{x_0, \ldots, x_n\}$ be a partiton of $[a, b]$ and let $T = \{t_1, \ldots, t_n\}$ with $x_{k-1} \le t_k \le x_k$ for $1 \le k \le n$. Show $\overline{P} = \{sx_0, \ldots, sx_n\}$ is a partition of $[sa, sb]$ and $\overline{T} = \{st_0, \ldots, st_n\}$ has the property $sx_{k-1} \le st_k \le sx_k$ for $1 \le k \le n$.

(b) Let $f(x) = \frac{1}{x}$. Show $L(P, f) = L(\overline{P}, f)$ and $U(P, f) = U(\overline{P}, f)$.

(c) Prove Gregory's formula: $\int_a^b \frac{1}{x} \, dx = \int_{sa}^{sb} \frac{1}{x} \, dx$.

5. Deduce de Sarasa's identity $L\left(\frac{x}{y}\right) = L(x) - L(y)$ from Gregory's formula (4.2.10).

6. Show Newton's function $A(1 + x) = \int_0^x \frac{1}{1+t} \, dt$ has the properties:

(a) $A[(1 + x)(1 + y)] = A(1 + x) + A(1 + y)$;   (b) $A\left(\frac{1+x}{1+y}\right) = A(1 + x) - A(1 + y)$.

4.3   Applications of Logarithms

In this section, we study several ways that logarithms are used in calculus. We begin b
showing how logarithms arise in the integration of several basic functions. In the sec
ond subsection, we introduce logarithmic differentiation. This method uses logarithm
to simplify the computation of derivatives of products, quotients and exponentials i
the same way that logarithms are used to simplify arithmetic computations. In th
third subsection, L'Hôpital's Rule is used to compute limits involving exponentials an
logarithms. In addition, we evaluate the logarithms of certain limits which can not b
evaluated directly. In the fourth subsection, we sketch the graphs of several function
which involve logarithms and exponentials. The methods of the third subsection ar
used to compute the limits which give the horizontal and vertical asymptotes.

Integration

By definition, $\int \frac{1}{x}\, dx = \ln x + C$ for $x > 0$. We begin by extending this formula t
all nonzero x. Examples of the resulting formula are presented. In particular, w
integrate $\tan x$, $\cot x$, $\sec x$ and $\csc x$.

To evaluate $\int \frac{1}{x}\, dx$ when $x < 0$, substitute $u = -x$. Note $u > 0$ when $x < 0$ an
$du = -dx$. Hence

$$\int \frac{1}{x}\, dx = -\int \frac{1}{x} - dx = -\int \frac{1}{-u}\, du = \int \frac{1}{u}\, du = \ln u + C = \ln(-x) + C.$$

Since $x < 0$, $-x = |x|$ and $\int \frac{1}{x}\, dx = \ln |x| + C$. When $x > 0$, $\int \frac{1}{x}\, dx = \ln x + C =$
$\ln |x| + C$. Combining these two equations:

$$\int \frac{1}{x}\, dx = \ln |x| + C \quad \text{for } x \neq 0. \tag{4.3.1}$$

We generalize this formula to integrate $\frac{f'(x)}{f(x)}$. Let $v = f(x)$ with $dv = f'(x)\, dx$. Ther

$$\int \frac{f'(x)}{f(x)}\, dx = \int \frac{1}{v}\, dv = \ln |v| + C = \ln |f(x)| + C. \tag{4.3.2}$$

We apply the First Fundamental Theorem of Calculus to (4.3.1) and (4.3.2) to obtair
formulas needed in the next subsection:

$$\frac{d}{dx}(\ln |x|) = \frac{1}{x} \quad \text{and} \quad \frac{d}{dx}(\ln |f(x)|) = \frac{f'(x)}{f(x)}. \tag{4.3.3}$$

Examples 4.3.1 (1) Find $\int \frac{e^x}{e^x + 1}\, dx$.

Solution Let $f(x) = e^x + 1$. Then $f'(x) = e^x$, and (4.3.2) applies:

$$\int \frac{e^x}{e^x + 1}\, dx = \ln |e^x + 1| + C = \ln(e^x + 1) + C$$

since $e^x + 1 > 1 > 0$ for all x.

(2) Find $\int \frac{1}{x \ln x}\, dx$.

Solution Let $f(x) = \ln x$. Since $f'(x) = \frac{1}{x}$, we can apply (4.3.2):

$$\int \frac{1}{x \ln x}\, dx = \int \frac{1/x}{\ln x}\, dx = \ln |\ln x| + C.$$

(3) Find $\int \frac{2x+7}{x^2+7x-3}\,dx$.

Solution Let $f(x) = x^2 + 7x - 3$. Since $f'(x) = 2x + 7$, (4.3.2) applies:

$$\int \frac{2x+7}{x^2+7x-3}\,dx = \ln|x^2 + 7x - 3| + C$$

(4) Find $\int \frac{6x+7}{x^2+4x+20}\,dx$.

Solution Complete the square of the quadratic polynomial:

$$\int \frac{6x+7}{x^2+4x+20}\,dx = \int \frac{6x+7}{(x+2)^2+16}\,dx \ .$$

Let $u = x + 2$ with $x = u - 2$ and $du = dx$:

$$\int \frac{6x+7}{x^2+4x+20}\,dx = \int \frac{6(u-2)+7}{u^2+16}\,du = \int \frac{6u-5}{u^2+16}\,du$$

$$= 3\int \frac{2u}{u^2+16}\,du - 5\int \frac{1}{u^2+4^2}\,du \ .$$

Apply (4.3.2) to the first integral, and (3.5.3) to the second one:

$$\int \frac{6x+7}{x^2+4x+20}\,dx = 3\ln(u^2+16) - \frac{5}{4}\arctan\frac{u}{4} + C$$

$$= 3\ln[(x+2)^2 + 16] - \frac{5}{4}\arctan\left(\frac{x+2}{4}\right) + C.$$

Note we do not need to take the absolute value of $u^2 + 16$ above because it is always positive. □

The integrals of $\sin x$ and $\cos x$ are immediate consequences of the Second Fundamental Theorem of Calculus. We use (4.3.2) to integrate the other trigonometric functions.

Proposition 4.3.2 (a) $\int \tan x\ dx = -\ln|\cos x| + C = \ln|\sec x| + C$
(b) $\int \cot x\ dx = \ln|\sin x| + C$
(c) $\int \sec x\ dx = \ln|\tan x + \sec x| + C$
(d) $\int \csc x\ dx = -\ln|\cot x + \csc x| + C$

Proof (a) $\int \tan x\ dx = -\int \frac{-\sin x}{\cos x}\,dx$. Thus (4.3.2) applies to this integral:

$$\int \tan x\ dx = -\ln|\cos x| + C = \ln\left|(\cos x)^{-1}\right| + C = \ln|\sec x| + C.$$

(b) Note $\int \cot x\ dx = \int \frac{\cos x}{\sin x}\,dx$. Apply (4.3.2) to this integral with $f(x) = \sin x$ to produce the formula in (b).
(c) We use the trick of multiplying $\sec x$ by $\frac{\sec x + \tan x}{\sec x + \tan x}$ to rewrite the integrand in the form required to apply (4.3.2):

$$\int \sec x\ dx = \int \frac{\sec x\,(\sec x + \tan x)}{\sec x + \tan x}\,dx = \int \frac{\sec^2 x + \sec x \tan x}{\tan x + \sec x}\,dx$$

$$= \int \frac{D(\tan x + \sec x)}{\tan x + \sec x}\,dx = \ln|\tan x + \sec x| + C.$$

(d) Use a trick similar to the preceding one: multiply $\csc x$ by $\frac{\csc x + \cot x}{\csc x + \cot x}$:

$$\int \csc x\ dx = \int \frac{\csc x\,(\csc x + \cot x)}{\csc x + \cot x}\,dx = -\int \frac{-\csc^2 x - \csc x \cot x}{\cot x + \csc x}\,dx$$

$$= -\int \frac{D(\cot x + \csc x)}{\cot x + \csc x}\,dx = -\ln|\cot x + \csc x| + C. \qquad \square$$

Logarithmic Differentiation

Logarithms were originally introduced as an arithmetic aid to turn products and quo tients into sums and differences. In the same way, we use them in calculus to conver derivatives of complicated products and quotients into derivatives of sums and differ ences. That is, we start with $y = f(x)$ and take the logarithm of this equation befor differentiating. When this method is appropriate, $\ln y = \ln f(x)$ is the sum of simpl terms of the form $\ln s(x)$. Differentiating this equation by the chain rule is easier tha differentiating $f(x)$ directly. That is, if $u = s(x)$ then by (4.3.3):

$$\frac{d}{dx}\ln|y| = \frac{d}{dy}(\ln|y|)\cdot\frac{dy}{dx} = \frac{1}{y}\frac{dy}{dx} = \frac{f'(x)}{f(x)} \tag{4.3.4}$$

$$\frac{d}{dx}\ln|s(x)| = \frac{d}{du}(\ln|u|)\cdot\frac{du}{dx} = \frac{1}{u}\frac{du}{dx} = \frac{s'(x)}{s(x)}. \tag{4.3.5}$$

Note the use of absolute value signs to include those x where y and/or $s(x)$ ar negative. This method of computing $f'(x)$ is called *logarithmic differentiation*.

Examples 4.3.3 (1) Find the derivative of $f(x) = x^5 \sin x \cos x$.

Solution Apply ln to the equation $|f(x)| = |x|^5|\sin x||\cos x|$:

$$\ln|f(x)| = \ln\left[|x|^5|\sin x||\cos x|\right] = \ln|x|^5 + \ln|\sin x| + \ln|\cos x| = 5\ln|x| + \ln|\sin|$$

Use (4.3.4) and (4.3.5) to differentiate this equation:

$$\frac{f'(x)}{f(x)} = \frac{5}{x} + \frac{\cos x}{\sin x} - \frac{\sin x}{\cos x}$$

$$f'(x) = f(x)\left[\frac{5}{x} + \cot x - \tan x\right] = x^5\sin x\cos x\left[\frac{5}{x} + \cot x - \tan x\right]$$

$$= 5x^4\sin x\cos x + x^5\cos^2 x - x^5\sin^2 x.$$

(2) Find the derivative of $g(x) = \sqrt{\frac{x^2+1}{x^4+1}}$.

Solution Apply ln to the equation $g(x) = \sqrt{\frac{x^2+1}{x^4+1}} = \left[\frac{x^2+1}{x^4+1}\right]^{1/2}$:

$$\ln g(x) = \ln\left[\frac{x^2+1}{x^4+1}\right]^{1/2} = \frac{1}{2}\ln\left[\frac{x^2+1}{x^4+1}\right] = \frac{1}{2}\left[\ln\left(x^2+1\right) - \ln\left(x^4+1\right)\right].$$

Use (4.3.4) and (4.3.5) to differentiate this equation:

$$\frac{g'(x)}{g(x)} = \frac{1}{2}\left[\frac{2x}{x^2+1} - \frac{4x^3}{x^4+1}\right]$$

$$g'(x) = g(x)\left[\frac{x}{x^2+1} - \frac{2x^3}{x^4+1}\right] = \sqrt{\frac{x^2+1}{x^4+1}}\left[\frac{x}{x^2+1} - \frac{2x^3}{x^4+1}\right]. \qquad \square$$

Logarithmic differentiation can also be used to compute the derivative of an expo nential function when both the base and exponent are functions of x.

Examples 4.3.4 (1) Let $f(x) = x^x$ with domain $(0, \infty)$. Find $f'(x)$.

Solution Apply ln to the equation $f(x) = x^x$:

$$\ln f(x) = \ln x^x = x\ln x.$$

Use (4.3.4) and the product rule to differentiate this equation:

$$\frac{f'(x)}{f(x)} = x\frac{1}{x} + (1)\ln x = 1 + \ln x$$
$$f'(x) = f(x)(1 + \ln x) = x^x(1 + \ln x) .$$

(2) Let $g(x) = (\sin x)^{\cos x}$ with domain $(0, \pi)$. Find $g'(x)$.

Solution Apply ln to the equation $g(x) = (\sin x)^{\cos x}$:

$$\ln g(x) = \ln[(\sin x)^{\cos x}] = (\cos x)\ln(\sin x) .$$

Use (4.3.4) and (4.3.5) to differentiate this equation:

$$\frac{g'(x)}{g(x)} = (-\sin x)\ln(\sin x) + (\cos x)\left(\frac{\cos x}{\sin x}\right)$$
$$g'(x) = g(x)\left[-(\sin x)\ln(\sin x) + \cos x \cot x\right]$$
$$= (\sin x)^{\cos x}\left[\cos x \cot x - (\sin x)\ln(\sin x)\right] . \qquad \square$$

Limits

In the first set of examples we use L'Hôpital's Rule to evaluate several limits directly.
Then we evaluate limits of $f(x)^{g(x)}$ indirectly by finding the logarithms of the limits.

Examples 4.3.5 (1) Let $p(x)$ be any polynomial of positive degree. Find $\lim\limits_{x \to \infty} \dfrac{\ln x}{p(x)}$.

Solution Substitution of $x = \infty$ into this limit produces the indeterminate
form $\frac{\infty}{\infty}$. Thus L'Hôpital's Rule applies:

$$\lim_{x \to \infty} \frac{\ln x}{p(x)} = \lim_{x \to \infty} \frac{D(\ln x)}{D(p(x))} = \lim_{x \to \infty} \frac{1/x}{p'(x)} = \lim_{x \to \infty} \frac{1}{xp'(x)} = 0.$$

This limit says that logarithmic growth is slower than polynomial growth.

(2) Find $\lim\limits_{x \to 0^+} x \ln x$.

Solution To apply L'Hôpital's Rule, rewrite this limit as a quotient:

$$\lim_{x \to 0^+} x \ln x = \lim_{x \to 0^+} \frac{\ln x}{1/x}.$$

Substitution of $x = 0^+$ into this limit produces the indeterminate form $\frac{-\infty}{\infty}$.
Thus L'Hôpital's Rule applies:

$$\lim_{x \to 0^+} x \ln x = \lim_{x \to 0^+} \frac{D(\ln x)}{D(1/x)} = \lim_{x \to 0^+} \frac{1/x}{-1/x^2} = \lim_{x \to 0^+} -x = 0.$$

(3) Let $p(x)$ be any polynomial. Find $\lim\limits_{x \to \infty} \dfrac{p(x)}{e^x}$.

Solution Let n be the degree of the polynomial $p(x)$. If $n \geq 1$, substitution
of $x = \infty$ into this limit produces the indeterminate form $\frac{\infty}{\infty}$. Thus L'Hôpital's
Rule applies:

$$\lim_{x \to \infty} \frac{p(x)}{e^x} = \lim_{x \to \infty} \frac{D(p(x))}{D(e^x)} = \lim_{x \to \infty} \frac{p'(x)}{e^x}.$$

Observe that $p'(x)$ is also a polynomial, but its degree is $n - 1$. For any $n \geq 0$,
apply L'Hôpital's Rule n times to the original limit to obtain the limit of a
fraction whose numerator is a constant K and whose denominator is e^x. Substi-
tution of $x = \infty$ into this fraction produces $\frac{K}{\infty} = 0$. Thus $\lim\limits_{x \to \infty} \dfrac{p(x)}{e^x} = 0$. This
limit says that exponential growth is faster than polynomial growth. $\qquad \square$

In the following examples, we consider the limit L of an indeterminate form when direct substitution produces 1^∞, 0^0 or ∞^0. Such limits may not exist, and when they do exist they may have any value. L'Hôpital's Rule does not apply directly to such limits. However, we can evaluate the logarithm A of this limit by L'Hôpital's Rule. Then $\ln L = A$ and $L = e^A$.

Examples 4.3.6 (1) Find $\lim\limits_{x\to\infty} x^{1/x}$.

Solution Substitution of $x = \infty$ into $x^{1/x}$ produces the indeterminate form ∞^0. Let $L = \lim\limits_{x\to\infty} x^{1/x}$. Then

$$\ln L = \ln\left[\lim_{x\to\infty} x^{1/x}\right] = \lim_{x\to\infty} \ln\left(x^{1/x}\right) = \lim_{x\to\infty} \frac{\ln x}{x} = 0$$

by Example 4.3.5 (1). Note we moved $\ln$ past the limit in the first step above because $\ln$ is a continuous function. Thus, $A = \ln L = 0$ and $L = e^A = e^0 = 1$.

(2) Find $\lim\limits_{x\to 0^+} x^x$.

Solution Substitution of $x = 0$ into x^x produces the indeterminate form 0^0. Let $L = \lim\limits_{x\to 0^+} x^x$. Then

$$\ln L = \ln\left[\lim_{x\to 0^+} x^x\right] = \lim_{x\to 0^+} \ln\left(x^x\right) = \lim_{x\to 0^+} x\ln x = 0$$

by Example 4.3.5 (2). Then $A = \ln L = 0$ and $L = e^A = e^0 = 1$.

The next example is so important that we state it as a proposition.

Proposition 4.3.7 (Euler's Limit) *For any $x \in \Re$:*

$$\lim_{k\to\infty} \left(1 + \frac{x}{k}\right)^k = e^x.$$

Proof Note k is the variable in this limit while x is a constant. Substitution of $k = \infty$ into $(1 + \frac{x}{k})^k$ gives the indeterminate form 1^∞. Let $L = \lim\limits_{k\to\infty} \left(1 + \frac{x}{k}\right)^k$. Then

$$\ln L = \ln\left[\lim_{k\to\infty}\left(1 + \frac{x}{k}\right)^k\right] = \lim_{k\to\infty} \ln\left[\left(1+\frac{x}{k}\right)^k\right] = \lim_{k\to\infty} k\ln\left(1+\frac{x}{k}\right) = \lim_{k\to\infty} \frac{\ln\left(1+\frac{x}{k}\right)}{1/k}$$

Substitution of $k = \infty$ into $\frac{\ln\left(1+\frac{x}{k}\right)}{1/k}$ produces the indeterminate form $\frac{0}{0}$. Hence L'Hôpital's Rule applies:

$$\ln L = \lim_{k\to\infty} \frac{\frac{d}{dk}\left[\ln\left(1+\frac{x}{k}\right)\right]}{\frac{d}{dk}\left(\frac{1}{k}\right)} = \lim_{k\to\infty} \frac{\frac{-x/k^2}{1+x/k}}{-1/k^2} = \lim_{k\to\infty} \frac{x}{1+\frac{x}{k}} = \frac{x}{1+0} = x.$$

Thus $\ln L = x$, and $L = e^x$.   □

Graphs

We use the methods of Section 2.7 to construct the graphs of functions which involve logarithms and exponentials.

Examples 4.3.8 (1) Sketch the graph of the function $f(x) = e^{-x}$ with domain $\Re$.

Solution We construct the graph of $f(x) = e^{-x}$ from the graph of its reciprocal $F(x) = e^x$. That is, (x, y) lies on the graph of f if and only if $\left(x, \frac{1}{y}\right)$ lies on the graph of F. In other words, the graph of f is the reflection of the graph of F about the line $y = 1$, compressed to lie between the x–axis and $y = 1$. See Figure 4.3.9.

(2) Sketch the graph of the function $g(x) = e^{x^2}$ with domain $\Re$.

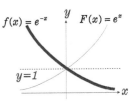

Solution We construct the graph of g from the following information.
- g has no x–intercepts because an exponential is always positive.
- The y–intercept of g is $g(0) = e^0 = 1$.
- g has no horizontal asymptotes since $g(x)$ has infinite limit when x approaches either ∞ or $-\infty$.
- g has no vertical asymptotes because g is continuous at all $x \in \Re$.
- By the chain rule: $g'(x) = 2xe^{x^2}$.
- g is decreasing for $x < 0$ because $g'(x)$ is negative there.
- g is increasing for $x > 0$ because $g'(x)$ is positive there.
- The critical point $x = 0$ is a local minimum by the First Derivative Test.
- By the product rule: $g''(x) = 2e^{x^2} + 4x^2e^{x^2} = e^{x^2}\left(2 + 4x^2\right) > 0$.
  Thus, the graph of g is concave up everywhere and has no inflection points.

Figure 4.3.9
$y = e^{-x}$ and $y = e^x$

The resulting graph of g is sketched in Figure 4.3.10. Note g is an even function, i.e. $g(-x) = g(x)$. Hence the graph of g is symmetric with respect to the y–axis.

(3) Sketch the graph of the function $h(x) = e^{-x^2}$ with domain $\Re$.

Solution Note $h(x) = e^{-x^2}$ is the reciprocal of $g(x) = e^{x^2}$. Hence the graph of h is the reflection of the graph of g about the line $y = 1$, compressed to lie between the x–axis and $y = 1$. See Figure 4.3.11. The concavity of the graph of h, however, is not clear from this method. By the chain and product rules:

$$h'(x) = -2xe^{-x^2}$$

$$h''(x) = -2e^{-x^2} + 4x^2e^{-x^2} = 4e^{-x^2}\left(-\frac{1}{2} + x^2\right)$$

$$= 4e^{-x^2}\left(x - \frac{1}{\sqrt{2}}\right)\left(x + \frac{1}{\sqrt{2}}\right)$$

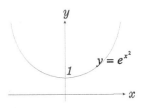

Figure 4.3.10
$y = e^{x^2}$

Thus $h''(x) > 0$ if either $x < -\frac{1}{\sqrt{2}}$ or $x > \frac{1}{\sqrt{2}}$, and the graph of h is concave up. If $-\frac{1}{\sqrt{2}} < x < \frac{1}{\sqrt{2}}$ then $h''(x) < 0$, and the graph of h is concave down. Thus the concavity changes at $x = \pm\frac{1}{\sqrt{2}}$ which are inflection points.

(4) Sketch the graph of the function $k(x) = \frac{\ln x}{x}$ with domain $(0, \infty)$.

Solution We construct the graph of k from the following observations.
- k has x–intercept $x = 1$ where $\ln x = 0$.
- k has the y–axis as a vertical asymptote because $\displaystyle\lim_{x \to 0^+} \frac{\ln x}{x} = \frac{-\infty}{+0} = -\infty$.
- k has the x–axis as a horizontal asymptote on the right since $\displaystyle\lim_{x \to \infty} \frac{\ln x}{x} = 0$ by Example 4.3.5 (1).
- By the quotient rule:

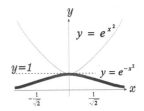

Figure 4.3.11
$y = e^{-x^2}$ and $y = e^{x^2}$

$$k'(x) = \frac{(1/x)x - (1)\ln x}{x^2} = \frac{1 - \ln x}{x^2}.$$

- k is increasing for $0 < x < e$ because $1 - \ln x > 0$ there.
- k is decreasing for $x > e$ because $1 - \ln x < 0$ there.

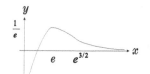

Figure 4.3.12

$$y = \frac{\ln x}{x}$$

- k has a local maximum at $x = e$ by the First Derivative Test.
- By the quotient rule:

$$k''(x) = \frac{(-1/x)(x^2) - (2x)(1 - \ln x)}{x^4} = \frac{2\ln x - 3}{x^3}.$$

Note $\ln x = \frac{3}{2}$ when $x = e^{3/2}$.

- When $x < e^{3/2}$, $k''(x) < 0$ and the graph of k is concave down. When $x > e^{3/2}$, $k''(x) > 0$ and the graph of k concave up. Thus k has an inflectio point at $x = e^{3/2}$.

The graph of k is sketched in Figure 4.3.12.

(5) Sketch the graph of the function $m(x) = x^{1/x}$ with domain $(0, \infty)$.

Solution We construct the graph of m from the following observations.

- $\displaystyle\lim_{x \to 0^+} m(x) = 0^{+\infty} = 0.$
- m has the horizontal asymptote $y = 1$ on the right because $\displaystyle\lim_{x \to \infty} x^{1/x} = 1$ b Example 4.3.6 (1).
- By logarithmic differentiation:

$$\frac{m'(x)}{m(x)} = \frac{d}{dx}\ln m(x) = \frac{d}{dx}\ln\left(x^{1/x}\right) = \frac{d}{dx}\left(\frac{\ln x}{x}\right)$$

$$m'(x) = m(x)\frac{d}{dx}\left(\frac{\ln x}{x}\right) = m(x)\frac{1 - \ln x}{x^2}.$$

- m is increasing for $0 < x < e$ because $m'(x) > 0$ there.
- m is decreasing for $x > e$ because $m'(x) < 0$ there.
- m has a local maximum at $x = e$ by the First Derivative Test.
- We show $\displaystyle\lim_{x \to 0^+} m'(x) = 0$. For x a small positive number:

$$1 - \ln x < -2\ln x \quad \text{and} \quad -\ln x < \frac{1}{x}. \tag{4.3.6}$$

The first inequality follows from $\displaystyle\lim_{x \to 0^+} \ln x = -\infty$. The second inequality follows from Example 4.3.5 (2): $\displaystyle\lim_{x \to 0^+} x\ln x = 0$. Combine the formula for $m'(x)$ with the inequalities of (4.3.6) when x is a small positive number:

$$0 < m'(x) < m(x)\left(\frac{-2\ln x}{x^2}\right) < m(x)\frac{2}{x^3} = 2x^{\frac{1}{x} - 3}$$

$$0 \le \lim_{x \to 0^+} m'(x) \le \lim_{x \to 0^+} 2x^{\frac{1}{x} - 3} = 0^{\infty} = 0.$$

By the Pinching Theorem, $\displaystyle\lim_{x \to 0^+} m'(x) = 0$.

- By the preceding limit, the graph of m is concave up near $x = 0$. The graph is concave down at the local maximum at $x = e$, and it is concave up for large values of x as it approaches its horizontal asymptote. Hence it has an inflection point between 0 and e and another inflection point at a number greater than e.

The graph of m is sketched in Figure 4.3.13. □

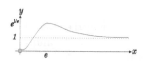

Figure 4.3.13

$$y = x^{1/x}$$

Summary

The reader should be able to use logarithmic differentiation to find the derivatives of products, quotients and exponentials. She should be able to compute the integral of expressions of the form $\frac{f'(x)}{f(x)}$. The application of L'Hôpital's Rule to evaluate logarithmic and exponential limits should be understood. In addition, she should know how to evaluate exponential indeterminate forms by computing their logarithms by L'Hôpital's Rule. She should be able to graph logarithmic and exponential functions by the methods of Section 2.7.

Basic Exercises

1. Use logarithmic differentiation to find the derivative of each function.

(a) $y = (x^2 + 1)(5x^3 - 2)(4x^5 + 7)$

(b) $y = \frac{(3x^4+5)(6x^8-3)}{(2x^2+5)(7x^6-8)}$

(c) $y = (5x + 4)^9(2x^2 + 1)^6(8x^3 - 7)^3$

(d) $y = \sqrt{(3x^2 - 1)^9(5x^3 + 2)^7}$

(e) $y = \sqrt[3]{\frac{(x^3+3)^4(5x^4-6)^8}{(2x^2-7)^2(6x^6-9)^5}}$

(f) $y = \sin^3 x \cos^4 x \tan^2 x$

(g) $y = \frac{(3\sin x+8)^5(5\tan x-9)^3}{(6\cos x-1)^2(4\cot x-3)^6}$

(h) $y = (3e^x + 7)^5(4e^{-x} + 1)^{10}(e^{3x} - 8)^4$

(i) $y = \frac{(2^x-1)^5(5^{-x}+7)^8}{(3^x+2)^4(7^{-x}-8)^3}$

(j) $y = \sqrt{\frac{[3+\ln(x^5)]^7[\ln(5x+2)]^3}{(\ln\sin x)^5}}$

(k) $y = x^{-x}$ (l) $y = (\sin x)^x$

(m) $y = x^{\sin x}$ (n) $y = (\arcsin x)^x$

(o) $y = x^{\arctan x}$ (p) $y = (\arctan x)^{\tan x}$

(q) $y = (\ln x)^x$ (r) $y = (\ln x)^{\ln x}$

2. Evaluate each integral.

(a) $\int \frac{x^4}{x^5+2}\, dx$

(b) $\int \frac{x^2+1}{2x^3+6x-5}\, dx$

(c) $\int \frac{\sin x}{3\cos x+7}\, dx$

(d) $\int \frac{\sec^2 x}{4\tan x+3}\, dx$

(e) $\int \frac{2^x}{2^x+5}\, dx$

(f) $\int \frac{1}{x\log_3 x}\, dx$

(g) $\int \frac{1}{\sqrt{x}(3\sqrt{x}+2)}\, dx$

(h) $\int \frac{x^2\sqrt{x}}{x^3\sqrt{x}+1}\, dx$

(i) $\int \frac{1}{x\ln x\ln\ln x}\, dx$

(j) $\int \frac{1}{\sqrt{1-x^2}\arcsin x}\, dx$

(k) $\int \frac{2^x-2^{-x}}{2^x+2^{-x}}\, dx$

(l) $\int \frac{\tan x}{\ln\cos x}\, dx$

3. Evaluate each integral as a logarithm plus an arctangent.

(a) $\int \frac{4x+7}{x^2-8x+17}\, dx$

(b) $\int \frac{2x+1}{x^2-4x+13}\, dx$

(c) $\int \frac{3x+2}{x^2+2x+5}\, dx$

(d) $\int \frac{5x-3}{x^2+6x+25}\, dx$

(e) $\int \frac{x-4}{2x^2-6x+27}\, dx$

(f) $\int \frac{x}{3x^2-18x+45}\, dx$

4. Evaluate each limit.

(a) $\lim\limits_{x\to\infty} \ln\left(\frac{x}{x+1}\right)$

(b) $\lim\limits_{x\to 0} \frac{3^x - 2^x}{x}$

(c) $\lim\limits_{x\to 0^+}\left(\frac{\ln x}{\cot x}\right)$

(d) $\lim\limits_{x\to 0^+} (\sin x)^x$

(e) $\lim\limits_{x\to \frac{\pi}{2}^-} (\cos x)^{\sec x}$

(f) $\lim\limits_{x\to\infty} (3e^x + 5)^{1/x}$

(g) $\lim\limits_{x\to\infty} (7x^3 - 4)^{1/\ln x}$

(h) $\lim\limits_{x\to\infty} (\ln x)^{1/x}$

(i) $\lim\limits_{x\to 1^-}\left[\frac{1}{x-1} - \ln(1-x)\right]$

(j) $\lim\limits_{x\to\infty} \sqrt{\exp\frac{\ln x}{x}}$

(k) $\lim\limits_{x\to\infty} \frac{\exp\left(1+\frac{1}{x}\right)^x}{\left(1-\frac{1}{x}\right)^x}$

(l) $\lim\limits_{x\to 0^+} (\exp\sin x)^{1/x}$

(m) $\lim\limits_{x\to 0^+} (1 - x)^{1/x}$

(n) $\lim\limits_{x\to\infty}\left(1+\frac{\ln k}{x}\right)^x$

(o) $\lim\limits_{x\to\infty}\left(2+\frac{1}{x}\right)^x$

(p) $\lim\limits_{x\to\infty}\left(\frac{x}{x+1}\right)^x$

(q) $\lim\limits_{x\to 0}(1+\sin x)^{1/x}$

(r) $\lim\limits_{x\to\infty}\left(1+\frac{1}{x^2}\right)^x$

5. Graph each function.

(a) $f(x) = xe^x$

(b) $g(x) = xe^{-x^2}$

(c) $h(x) = x^2e^{-x^2}$

(d) $i(x) = (\ln x)^2$ for $x > 0$

(e) $j(x) = x\ln x$ for $x > 0$

(f) $k(x) = 2^x + 2^{-x}$

(g) $m(x) = \frac{3^x+3^{-x}}{3^x-3^{-x}}$

(h) $n(x) = x^x$ for $x > 0$

(i) $p(x) = x^{-1/x}$ for $x > 0$

6. Find the area bounded by $y = \ln x$, $y = 1$, the x-axis and the y-axis.

7. Find the area bounded by $y = e^x$, $y = e^{-x}$ and $y = \frac{1}{e}$.

8. Find the area in the first quadrant bounded by $y = xe^{-x^2}$ and $y = \frac{x}{e}$.

9. Find the area bounded by $y = \frac{\ln x}{x}$ and $y = \frac{1}{e} + \frac{(e^2+1)(x-e)}{e^2-1}$.
Hint: these curves intersect at $x = e$ and at $x = \frac{1}{e}$.

10. Find the volume of the solid of revolution obtained by revolving the graph of $f(x) = \frac{\ln x}{\sqrt{x}}$, for $1 \leq x \leq e^2$, around the x-axis.

11. Find the volume of the solid of revolution obtained by revolving the graph of $g(x) = e^x$, for $0 \leq x \leq 1$, around the x-axis.

12. Find the volume of the solid of revolution obtained by revolving the graph of $g(x) = e^{-x^2}$, for $0 \leq x \leq 1$, around the y-axis.

The following three exercises are for those who have studied Section 3.9.

13. Find the distance traveled by each object with the given velocity.
(a) $v(t) = \frac{4t-6}{t^2-3t+3}$ miles per hour from $t = 0$ to $t = 4$ hours.
(b) $v(t) = \frac{t^2-4}{t^3-12t+20}$ miles per hour from $t = -3$ to $t = 3$ hours.
(c) $v(t) = \frac{\cos t}{2+\sin t}$ km. per hour from $t = 0$ to $t = 2\pi$ hours.
(d) $v(t) = \frac{1}{t\ln t}$ km. per hour from $t = 2$ to $t = 5$ hours.

14. Find the mass of a two meter long rod of density $\rho(x) = \frac{e^x}{e^x+e^{-x}}$ kg per meter for $0 \leq x \leq 2$ meters.

15. How much work is done when the force $F(x) = \frac{\sin x \cos x}{\cos^2 x - \sin^2 x}$ Newtons moves an object from $x = 0$ to $x = \frac{\pi}{6}$ meters?

The following two exercises are for those who have studied Section 3.10.

16. Find the center of mass of each rod of density $\rho(x)$.
(a) $\rho(x) = \frac{1}{x^2+9}$ kilograms per meter for $0 \leq x \leq 3$ meters.
(b) $\rho(x) = \frac{x}{x^2+1}$ pounds per foot for $0 \leq x \leq 2$ feet.
(c) $\rho(x) = \frac{1}{3x+5}$ pounds per foot for $1 \leq x \leq 4$ feet.

17. Find the centroid of the region R bounded by the graph of $y = \frac{1}{2x+1}$, the line $x = 3$, the x-axis and the y-axis.

4.4   Integration by Parts

The *Integration by parts* formula is obtained from the product rule of derivatives. This formula gives the value of a given integral in terms of another integral. In the examples of the first subsection, the new integral is easier to evaluate than the original one. In the second subsection, integrals are evaluated where the new integral is a multiple of the original one. In the third subsection, integration by parts is used to derive recursion formulas. These are iterated to evaluate integrals involving powers of certain functions.

Simple Examples

We derive the formula for integration by parts and apply it to evaluate a variety of indefinite and definite integrals. Consider the product rule for finding the derivative of the product of two differentiable functions:

$$\frac{d}{dx}[f(x)g(x)] = f'(x)g(x) + f(x)g'(x).$$

If f' and g' are continuous, then by the Second Fundamental Theorem of Calculus:

$$\int f'(x)g(x)\ dx + \int f(x)g'(x)\ dx = f(x)g(x) + C$$

We rewrite this equation to obtain the formula for integration by parts.

Proposition 4.4.1 [Integration by Parts] *Let $u = f(x)$ and $v = g(x)$ be continuously differentiable functions. Then*

$$\int f'(x)g(x)\ dx = f(x)g(x) - \int f(x)g'(x)\ dx \qquad (4.4.1)$$

In Leibniz notation, $du = f'(x)\ dx$, $dv = g'(x)\ dx$ and

$$\int v\ du = uv - \int u\ dv. \qquad (4.4.2)$$

The strategy to use (4.4.1) to evaluate the integral $\int H(x)\ dx$ is to write $H(x) = F(x)g(x)$ as a product in such a way that:

- we know the integral $f(x)$ of $F(x)$;

- the integral of $f(x)g'(x)$ is simpler than the original integral.

Then $f'(x) = F(x)$ and

$$\int H(x)\ dx = \int f'(x)g(x)\ dx = f(x)g(x) - \int f(x)g'(x)\ dx$$

is a simplification of the original integration problem. There are several situations where this strategy applies. In the following examples, the motivation for selecting the factor $g(x)$ is: *an integral involving $g'(x)$ is probably easier to integrate than an integral involving $g(x)$*. After $g(x)$ is selected, the function $f'(x) = \frac{H(x)}{g(x)}$ is determined. We use integration by parts if we can integrate $f'(x)$ to find $f(x)$.

Examples 4.4.2 (1) Find $\int xe^x\ dx$.

Solution We integrate by parts with $g(x) = x$ because its derivative $g'(x) = 1$ is simpler than $g(x)$. Then $f'(x) = e^x$, and $f(x) = e^x$. By (4.4.1):

$$\int xe^x\ dx = \int D(e^x)x\ dx = xe^x - \int e^x D(x)\ dx = xe^x - \int e^x\ dx = xe^x - e^x + C.$$

(2) Find $\int x\ln x\ dx$.

Solution We integrate by parts with $g(x) = \ln x$ because its derivative $g'(x) = \frac{1}{x}$ is simpler than $g(x)$. Then $f'(x) = x$, and $f(x) = \frac{1}{2}x^2$. By (4.4.1):

$$\begin{aligned}
\int x\ln x\ dx &= \int D\left(\frac{1}{2}x^2\right)\ln x\ dx = \frac{1}{2}x^2\ln x - \int \frac{1}{2}x^2 D(\ln x)\ dx \\
&= \frac{1}{2}x^2\ln x - \int \frac{1}{2}x^2 \cdot \frac{1}{x}\ dx = \frac{1}{2}x^2\ln x - \frac{1}{2}\int x\ dx \\
&= \frac{1}{2}x^2\ln x - \frac{1}{4}x^2 + C.
\end{aligned}$$

(3)   Find $\int x^2 \sin x \, dx$.

Solution We integrate by parts with $g(x) = x^2$ because its derivative $g'(x) = 2$ is simpler than $g(x)$. Then $f'(x) = \sin x$, and $f(x) = -\cos x$. By (4.4.1):

$$\int x^2 \sin x \, dx \;=\; \int D(-\cos x)x^2 \, dx = (-\cos x)x^2 - \int (-\cos x)D(x^2) \, dx$$

$$=\; -x^2 \cos x + \int 2x \cos x \, dx.$$

The new integral can not be evaluated immediately. However, it is similar to the original one but simpler: one factor of the integrand is a trigonometric function and the other factor is x instead of x^2. Thus we integrate by parts again: redefine $g(x) = 2x$, $f'(x) = \cos x$, and $f(x) = \sin x$. By (4.4.1):

$$\int x^2 \sin x \;=\; -x^2 \cos x + \int D(\sin x)(2x) \, dx$$

$$=\; -x^2 \cos x + (\sin x)(2x) - \int (\sin x)D(2x) \, dx$$

$$=\; -x^2 \cos x + 2x \sin x - \int 2 \sin x \, dx$$

$$=\; -x^2 \cos x + 2x \sin x + 2 \cos x + C.$$

Another context in which integration by parts may be helpful is when we cannot evaluate $\int g(x) \, dx$, but we can integrate many functions with a factor $g'(x)$. In this case, we use integration by parts with $f'(x) = 1$ and $f(x) = x$.

Examples 4.4.3 (1)   Find $\int \ln x \, dx$.

Solution Observe that $D(\ln x) = \frac{1}{x}$, and we can integrate many functions with $\frac{1}{x}$ as a factor. Thus we integrate by parts with $g(x) = \ln x$, $f'(x) = 1$ and $f(x) = x$. By (4.4.1):

$$\int \ln x \, dx \;=\; \int D(x) \ln x \, dx = x \ln x - \int x D(\ln x) \, dx = x \ln x - \int x \left(\frac{1}{x}\right)$$

$$=\; x \ln x - \int 1 \, dx = x \ln x - x + C.$$

(2)   Find $\int \arctan x \, dx$.

Solution Note $D(\arctan x) = \frac{1}{1+x^2}$, and we can integrate some functions with $\frac{1}{1+x^2}$ as a factor. Thus we integrate by parts with $g(x) = \arctan x$, $f'(x) = 1$ and $f(x) = x$. By (4.4.1):

$$\int \arctan \, dx \;=\; \int D(x) \arctan x \, dx = x \arctan x - \int x D(\arctan x) \, dx$$

$$=\; x \arctan x - \int x \frac{1}{1 + x^2} \, dx = x \arctan x - \frac{1}{2} \int \frac{2x}{1 + x^2} \, dx$$

$$=\; x \arctan x - \frac{1}{2} \int \frac{D(1 + x^2)}{1 + x^2} \, dx$$

$$=\; x \arctan x - \frac{1}{2} \ln \left(1 + x^2\right) + C. \qquad \square$$

When integration by parts is applied to a definite integral, we substitute the bounds into the first summand before completing the integration.

Corollary 4.4.4 *Let f and g be continuously differentiable functions. Then*

$$\int_a^b f'(x)g(x)\ dx = f(x)g(x)|_a^b - \int_a^b f(x)g'(x)\ dx.$$

Examples 4.4.5 Find $\int_0^1 \arcsin x\ dx$.

Solution Let $g(x) = \arcsin x$ with $f'(x) = 1$ and $f(x) = x$. By (4.4.1):

$$\int_0^1 \arcsin x\ dx = \int_0^1 D(x)\arcsin x\ dx = x\arcsin x|_0^1 - \int_0^1 xD(\arcsin x)\ dx$$

$$= \frac{\pi}{2} - \int_{x=0}^{x=1} x\frac{1}{\sqrt{1-x^2}}\ dx.$$

Make the substitution $u = 1 - x^2$ with $du = -2x\ dx$:

$$\int_0^1 \arcsin x\ dx = \frac{\pi}{2} + \frac{1}{2}\int_{u=1}^{u=0} \frac{1}{\sqrt{u}}\ dx = \frac{\pi}{2} + \frac{1}{2}\frac{u^{1/2}}{1/2}\Big|_1^0 = \frac{\pi}{2} - 1. \qquad \square$$

Circular Examples

Sometimes, after integration by parts, we obtain a multiple kI of the original integral I. When the coefficient k is not one, we consider the equation $I = \cdots + kI$ as a linear equation in I which can easily be solved for the value of I.

Examples 4.4.6 (1) Find $\int e^x \sin \pi x\ dx$.

Solution We integrate by parts. There are two factors in the integrand: e^x and $\sin \pi x$. It does not matter which one we designate as $g(x)$. Say $g(x) = \sin \pi x$ with $f'(x) = e^x$ and $f(x) = e^x$. Integration by parts gives:

$$\int e^x \sin \pi x\ dx = \int D(e^x)\sin \pi x\ dx = e^x \sin \pi x - \int e^x D(\sin \pi x)\ dx$$

$$= e^x \sin \pi x - \pi \int e^x \cos \pi x\ dx.$$

The new integral is analogous to the original one and is just as difficult. Let's integrate by parts a second time with $g(x) = \cos \pi x$, $f'(x) = e^x$ and $f(x) = e^x$:

$$\int e^x \sin \pi x\ dx = e^x \sin \pi x - \pi \int D(e^x)\cos \pi x\ dx$$

$$= e^x \sin \pi x - \pi \left[e^x \cos \pi x - \int e^x D(\cos \pi x)\ dx \right]$$

$$= e^x \sin \pi x - \pi e^x \cos \pi x + \pi \int e^x(-\pi \sin \pi x)\ dx$$

$$= e^x \sin \pi x - \pi e^x \cos \pi x - \pi^2 \int e^x \sin \pi x\ dx.$$

The integral on the right side of this equation is the original integral with coefficient $-\pi^2$. Bring this integral to the left side of the equation:

$$(1 + \pi^2) \int e^x \sin \pi x\ dx = e^x \sin \pi x - \pi e^x \cos \pi x + C \quad \text{and}$$

$$\int e^x \sin \pi x\ dx = \frac{1}{1 + \pi^2} e^x \sin \pi x - \frac{\pi}{1 + \pi^2} e^x \cos \pi x + C.$$

Handwritten margin notes:

quiz 4 only → $\int \frac{P(x)}{q(x)}$ where all roots of $q(x)$ are real

$= \ln|\sec x| + C$

$= -\ln|\cos x| + C$

$\int \tan x = -\ln|\cos x| + C$

$\int \sec x = \ln|\tan x + \sec x| + C$

$\int \csc x = \ln|\csc x + \cot x| + C$

$\int \csc x = -\ln|\sin x|$

$\int \cot x = \ln|\sin x|$

$x \sin x$

$f'(x)g(x) = f(x)g(x) - f(x)g'(x)$

$f(x)g(x) = f(x)g(x) - f(x)g'(x)$

$\int \sec x \frac{(-\cos x)}{x} x$

$\ln|\tan x + \sec x| + C$

Note the "C" in the last equation is $\frac{1}{1+\pi^2}$ times the "C" in the preceding lin̄ We use the same letter in both cases, however, because the "$+C$" in the value ̄ an indefinite integral merely denotes that adding any constant to the precedir̄ function does not change its derivative.

(2) Find $\int \sec^3 x \, dx$.

Solution We integrate by parts with $f'(x) = \sec^2 x$ because $f(x) = \tan x$ is simpler function. Then $g(x) = \sec x$, and integration by parts yields:

$$\int \sec^3 x \, dx = \int D(\tan x) \sec x \, dx = \tan x \sec x - \int \tan x D(\sec x) \, dx$$

$$= \tan x \sec x - \int \tan x(\tan x \sec x) \, dx$$

$$= \tan x \sec x - \int \tan^2 x \sec x \, dx.$$

Use the Pythagorean identity $\tan^2 x = \sec^2 x - 1$ to rewrite the latter integral̄

$$\int \sec^3 x \, dx = \tan x \sec x - \int (\sec^2 x - 1) \sec x \, dx$$

$$= \tan x \sec x - \int \sec^3 x \, dx + \int \sec x \, dx.$$

By Proposition 4.3.2(c), $\int \sec x \, dx = \ln|\sec x + \tan x| + C$. Note the new cop̄ of $\int \sec^3 x \, dx$ has coefficient -1, and we bring it to the left side of the equation

$$2\int \sec^3 x \, dx = \tan x \sec x + \ln|\sec x + \tan x| + C \quad \text{and}$$

$$\int \sec^3 x \, dx = \frac{1}{2}\tan x \sec x + \frac{1}{2}\ln|\sec x + \tan x| + C.$$

The integral $\int \sec^3 x \, dx$ will arise repeatedly in the following sections.

Recursion Formulas

A recursion formula expresses the value of an integral involving a positive integer N in terms of another integral of the same type involving an integer less than N. Ān integral of this type can be evaluated by the iterated application of this formula. Wē present two examples where integration by parts produces recursion formulas.

Examples 4.4.7 (1) **(a)** Find a recursion formula for $\int x^N e^x \, dx$ with $N \geq 1$.
(b) Use the recursion formula of (a) to evaluate $\int x^4 e^x \, dx$.
Solution (a) Integrate by parts with $g(x) = x^N$, $f'(x) = e^x$ and $f(x) = e^x$:

$$\int x^N e^x \, dx = \int x^N D(e^x) \, dx = x^N e^x - \int D(x^N) e^x \, dx = x^N e^x - N \int x^{N-1} e^x \, \text{̄}$$

(b) Apply the recursion formula of (a) with $N = 4$:

$$\int x^4 e^x \, dx = x^4 e^x - 4 \int x^3 e^x \, dx.$$

Apply this recursion formula again with $N = 3$:

$$\int x^4 e^x \, dx = x^4 e^x - 4x^3 e^x + 12 \int x^2 e^x \, dx.$$

Apply this recursion formula two more times, with $N = 2$ and $N = 1$:

$$\int x^4 e^x \, dx \;=\; x^4 e^x - 4x^3 e^x + 12x^2 e^x - 24 \int x e^x \, dx$$

$$= \; x^4 e^x - 4x^3 e^x + 12x^2 e^x - 24x e^x + 24 \int e^x \, dx$$

$$= \; x^4 e^x - 4x^3 e^x + 12x^2 e^x - 24x e^x + 24 e^x + C.$$

(2) **(a)** Find a recursion formula for $\int \sin^N x \, dx$ with $N \geq 2$.
(b) Use the recursion formula of (a) to evaluate $\int \sin^5 x \, dx$.

Solution (a) Integrate by parts with $g(x) = \sin^{N-1} x$, $f'(x) = \sin x$ and $f(x) = -\cos x$:

$$\int \sin^N x \, dx \;=\; \int D(-\cos x)\sin^{N-1} x \, dx$$

$$= \; (-\cos x)\sin^{N-1} x - \int (-\cos x)D(\sin^{N-1} x) \, dx$$

$$= \; -\sin^{N-1} x \cos x + \int (\cos x)\left[(N-1)\sin^{N-2} x \cos x\right] \, dx$$

$$= \; -\sin^{N-1} x \cos x + (N-1)\int \sin^{N-2} x \cos^2 x \, dx.$$

Apply the Pythagorean identity $\cos^2 x = 1 - \sin^2 x$:

$$\int \sin^N x \, dx \;=\; -\sin^{N-1} x \cos x + (N-1)\int \sin^{N-2} x \left(1 - \sin^2 x\right) \, dx$$

$$= \; -\sin^{N-1} x \cos x + (N-1)\int \sin^{N-2} x \, dx - (N-1)\int \sin^N x \, dx.$$

Bring the last summand on the right to the left side of the equation:

$$N \int \sin^N x \, dx \;=\; -\sin^{N-1} x \cos x + (N-1)\int \sin^{N-2} x \, dx$$

$$\int \sin^N x \, dx \;=\; -\frac{1}{N}\sin^{N-1} x \cos x + \frac{N-1}{N}\int \sin^{N-2} x \, dx. \quad (4.4.3)$$

(b) Apply the recursion formula of (a) with $N = 5$:

$$\int \sin^5 x \, dx = -\frac{1}{5}\sin^4 x \cos x + \frac{4}{5}\int \sin^3 x \, dx.$$

Apply this recursion formula again with $N = 3$:

$$\int \sin^5 x \, dx \;=\; -\frac{1}{5}\sin^4 x \cos x + \frac{4}{5}\left[-\frac{1}{3}\sin^2 x \cos x + \frac{2}{3}\int \sin x \, dx\right]$$

$$= \; -\frac{1}{5}\sin^4 x \cos x - \frac{4}{15}\sin^2 x \cos x - \frac{8}{15}\cos x + C. \qquad \square$$

Historical Remarks

Integration by parts was discovered independently by Isaac Newton and Gottfried Leibniz as a method for evaluating integrals. Newton's derivation is the one presented in this section. Leibniz derived the formula for integration by parts from his *transmutation method*. This method determines the relationship between the areas of two regions by comparing their infinitesmal slices. In the left diagram of Figure 4.4.8, L is

the tangent line to the graph of $y = f(x)$ at the point $P = (x, f(x))$ with Q' the inte
section of L with the y-axis. The length z of OQ' is a function of x: $z = g(x)$. Leibn
applies his transmutation method to compare the area $\int_a^b g(x)\,dx$ of the striped re
gion R with the area of the shaded region U bounded by the graph of $y = f(x)$, fo
$a \le x \le b$, and the line segments OA, OB. Divide the region U into infinitesma
thin sectors OPQ where $P = (x, f(x))$ and $Q = (x + dx, f(x) + dy)$ as in the midd
diagram of Figure 4.4.8. The secant line S determined by P and Q approximates th
tangent line L to the graph of f at P. Observe z equals the ordinate $y = f(x)$ of
minus the change in ordinate from Q' to P. This change in ordinate is the product of
the slope $\frac{dy}{dx}$ of L with the change x of the abscissa:

$$z = y - x\frac{dy}{dx}. \tag{4.4.4}$$

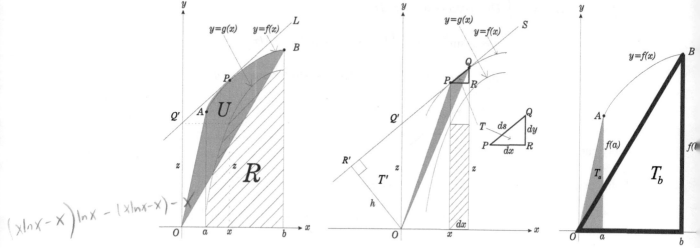

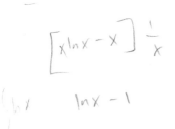

Figure 4.4.8 Leibniz's Transmutation Method

Approximate the area of sector OPQ by the area of triangle OPQ which has base ds
height h and area $\frac{1}{2}h\,ds$. Since triangle $T' = OQ'R'$ is similar to triangle $T = PQR$:

$$\frac{h}{dx} = \frac{z}{ds} \quad \text{and}$$

$$\text{Area } \triangle OPQ = \frac{1}{2}h\,ds = \frac{1}{2}z\,dx. \tag{4.4.5}$$

In the middle diagram of Figure 4.4.8, the striped rectangle of dimensions z, dx is an
infinitesmal slice of region R while the shaded triangle OPQ is an infinitesmal slice of
region U. By (4.4.5), the area $\frac{1}{2}h\,ds$ of this shaded triangle is half the area $z\,dx$ of
this striped rectangle. Sum the areas of these infinitesmal triangles and rectangles:

$$\text{Area }(U) = \sum \frac{1}{2}h\,ds = \frac{1}{2}\sum z\,dx = \frac{1}{2}\text{Area }(R) = \frac{1}{2}\int_a^b z\,dx. \tag{4.4.6}$$

This derivation of this formula is called the *transmutation method*.

The formula for integration by parts follows from (4.4.6) as follows. The right
diagram in Figure 4.4.8 shows:

$$\int_a^b f(x)\,dx + \text{area }\triangle T_a = \text{area }\triangle T_b + \text{Area }(U).$$

$$\int_a^b f(x)\ dx + \frac{1}{2}af(a) = \frac{1}{2}bf(b) + \frac{1}{2}\int_a^b z\ dx \qquad \text{by (4.4.6)}$$

$$\int_a^b f(x)\ dx = \frac{1}{2}[bf(b) - af(a)] + \frac{1}{2}\int_a^b y - x\frac{dy}{dx}\ dx \qquad \text{by (4.4.4)}$$

$$\int_a^b f(x)\ dx = \frac{1}{2}xf(x)\Big|_a^b + \frac{1}{2}\int_a^b y\ dx - \frac{1}{2}\int_{x=a}^{x=b} x\frac{dy}{dx}\ dx$$

$$\frac{1}{2}\int_a^b y\ dx = \frac{1}{2}xy\Big|_a^b - \frac{1}{2}\int_{y=f(a)}^{y=f(b)} x\ dy$$

as $y = f(x)$ and $dy = \frac{dy}{dx}dx$. Multiplying by 2 gives the integration by parts formula.

Summary

The reader should be able to use integration by parts to evaluate appropriate indefinite and definite integrals. In addition, she should know how to derive recursion formulas and apply these formulas to evaluate specific integrals.

Basic Exercises

Evaluate the following thirty integrals.

1. $\int x \sin x\ dx$ **2.** $\int x5^x\ dx$ **3.** $\int x^2 \cos x\ dx$ **4.** $\int x \log_{10} x\ dx$

5. $\int x^2 e^x\ dx$ **6.** $\int x^7 \ln x\ dx$ **7.** $\int \arcsin x\ dx$ **8.** $\int x^5 e^{x^3}\ dx$

9. $\int (\ln x)^2\ dx$ **10.** $\int e^x \cos x\ dx$ **11.** $\int e^{2x} \sin 5x\ dx$ **12.** $\int \ln(1 + x^2)\ dx$

13. $\int \sin(\ln x)\ dx$ **14.** $\int x \sec^2 x\ dx$ **15.** $\int x^2 \arcsin x\ dx$ **16.** $\int \log_{10} x\ dx$

17. $\int 3^x \cos \pi x\ dx$ **18.** $\int 6^x \sin x\ dx$ **19.** $\int \frac{\ln(\ln x)}{x}\ dx$ **20.** $\int (x + \sin x)^2\ dx$

21. $\int \sqrt{x} \sin \sqrt{x}\ dx$ **22.** $\int xe^x \sin x\ dx$ **23.** $\int e^x(x+1) \ln x\ dx$ **24.** $\int \csc^3 x\ dx$

25. $\int_3^8 \ln \sqrt{x+1}\ dx$ **26.** $\int_0^{\pi/6} x^2 \cos x\ dx$ **27.** $\int_0^1 x \arctan x\ dx$

28. $\int_{-1}^1 (e^x + x)^2\ dx$ **29.** $\int_0^{1/3} \sec^3 \pi x\ dx$ **30.** $\int_1^e \ln x\ dx$

31. Find recursion formulas for each integral.

         (a) $\int \cos^N x\ dx$ **(b)** $\int (\ln x)^N\ dx$ **(c)** $\int x^N \sin x\ dx$

         (d) $\int x^N \cos x\ dx$ **(e)** $\int \sec^N x\ dx$ **(f)** $\int x^M (\ln x)^N\ dx$

32. Find the area in the first quadrant bounded by $y = x \ln(1 + x^2)$ and $y = 2x$.

33. Find the area in the first quadrant bounded by $y = e^x \sin \pi x$ and $y = 2xe^x$.

34. Find the area bounded by $y = x^7 e^{-x^4}$ and $y = \frac{x^9}{e}$.

35. Find the area in the first quadrant bounded by $y = \arctan(4x)$ and $y = \pi x$.

36. Find the area bounded by $y = \arcsin x$, $y = \arcsin(1 - x)$ and $y = \frac{\pi}{2}$.

37. Find the volume of the solid of revolution obtained by rotating the region in the first quadrant bounded by $y = e^x$ and $x = 1$ around the y–axis.

38. Consider the region R bounded by $y = \ln x$, $y = 1$, the x–axis and the y–axis.
(a) Find the volume of the solid obtained by rotating the region R around the x–axis
(b) Find the volume of the solid obtained by rotating the region R around the y–axis

39. Consider the region bounded by $y = \sin x$ for $0 \le x \le \pi$ and the x–axis. Find the volume of the solid of revolution obtained by rotating this region around the y–axis

40. Find the volume of the solid of revolution obtained by rotating the region in the first quadrant bounded by $y = \sec^2 x$ and $x = \frac{\pi}{4}$ around the y–axis.

The following three exercises are for those who have studied Section 3.9.

41. Find the distance traveled by each object with the given velocity.
(a) $v(t) = t^3 2^t$ miles per hour from $t = -1$ to $t = 3$ hours.
(b) $v(t) = t^2 \arcsin t$ km per hour from $t = -1/2$ to $t = 1$ hours.
(c) $v(t) = t \cos t$ km per hour from $t = -\pi$ to $t = 3\pi/2$ hours.
(d) $v(t) = \ln(t - 1) + \ln(t + 1)$ km per hour from $t = 1.1$ to $t = 2$ hours.

42. Find the mass of a rod of density $\rho(x) = 5^x \sin 2x$ lbs per ft for $0 \le x \le \pi/2$ ft

43. How much work is done when the force $F(x) = \csc^3 x$ Newtons moves an object from $x = \frac{\pi}{4}$ to $x = \frac{2\pi}{3}$ meters?

The following two exercises are for those who have studied Section 3.10.

44. Find the center of mass of each rod of density $\rho(x)$.
(a) $\rho(x) = \log_{10} x$ pounds per foot for $1 \le x \le 100$ feet.
(b) $\rho(x) = 4^x$ pounds per foot for $0 \le x \le 3$ feet.
(c) $\rho(x) = \arctan x$ kilograms per meter for $0 \le x \le 1$ meters.
(d) $\rho(x) = x \cos x$ kilograms per meter for $0 \le x \le \pi/2$ meters.

45. Find the centroid of each region.
(a) Let R be the region bound by $y = \ln x$, the line $x = e$ and the x–axis.
(b) Let S be the region bounded by $y = 2^x$, the line $x = 3$, the x–axis and the y–axis
(c) Let T be the region bounded by $y = \sin \pi x$, for $0 \le x \le 1$, and the x–axis.
(d) Let U be the region bounded by $y = xe^x$, the line $x = 1$ and the x–axis.

The following four exercises are for those who have studied Section 4.10.

46. Evaluate each integral.

(a) $\int x^2 \sinh x \, dx$ (b) $\int e^x \cosh x \, dx$ (c) $\int \sinh^{-1} x \, dx$ (d) $\int \tanh^{-1} x \, dx$

47. Find the area bounded by $y = \operatorname{arcsech} x$, $y = \ln(5x)$ and the x–axis.

48. Find the area bounded by $y = x \sinh x$ and the line $y = \frac{12x}{5}$.

49. Consider the region R bounded by $y = \cosh x$ and the line $y = 2$.
(a) Find the volume of the solid obtained by rotating R around the x–axis.
(a) Find the volume of the solid obtained by rotating R around the y–axis.

Challenging Problems

1. (a) Assume $f''(x)$ exists and is continuous on the interval $[a, b]$. Show

$$f(b) - f(a) = f'(a)(b - a) - \int_a^b f''(x)(x - b) \, dx.$$

(b) Assume $f^{(3)}(x)$ exists and is continuous on the interval $[a, b]$. Show

$$f(b) - f(a) = f'(a)(b - a) + \frac{f''(a)}{2}(b - a)^2 + \frac{1}{2}\int_a^b f^{(3)}(x)(x - b)^2 \, dx.$$

2. (a) Let f and g be continuous functions with domain $[a, b]$. Use Cauchy's Mean Value Theorem to show there is at least one number $c \in (a, b)$ such that

$$f(c) \int_a^b g(x) \, dx = g(c) \int_a^b f(x) \, dx.$$

(b) Assume f and g are continuous functions with domain $[a, b]$ such that $g(x) \neq 0$ for all $x \in (a, b)$. Show there is at least one number $c \in (a, b)$ such that

$$\int_a^b f(x)g(x) \, dx = f(c) \int_a^b g(x) \, dx.$$

(c) Assume f is a continuous function and g is a continuously differentiable function, both with domain $[a, b]$. If $g'(x) \neq 0$ for all $x \in (a, b)$, show there is at least one number $c \in (a, b)$ such that:

$$\int_a^b f(x)g(x) \, dx = g(a) \int_a^c f(x) \, dx + g(b) \int_c^b f(x) \, dx.$$

3. (a) Use the recursion formula (4.4.3) to show:

$$\int_0^{\pi/2} \sin^{2n} x \, dx \;=\; \frac{2n-1}{2n} \cdot \frac{2n-3}{2n-2} \cdots \frac{3}{4} \cdot \frac{1}{2} \cdot \frac{\pi}{2} \quad \text{and}$$

$$\int_0^{\pi/2} \sin^{2n+1} x \, dx \;=\; \frac{2n}{2n+1} \cdot \frac{2n-2}{2n-1} \cdots \frac{4}{5} \cdot \frac{2}{3}.$$

(b) Start with the inequality $0 \leq \sin^{2n+1} x \leq \sin^{2n} x \leq \sin^{2n-1} x$ for $0 \leq x \leq \frac{\pi}{2}$. Use the recursion formula (4.4.3) to show:

$$1 \leq \frac{\int_0^{\pi/2} \sin^{2n} x \, dx}{\int_0^{\pi/2} \sin^{2n+1} x \, dx} \leq 1 + \frac{1}{2n}.$$

(c) Prove Wallis's formula:

$$\frac{\pi}{2} = \lim_{n \to \infty} \left[\frac{2^2}{3^2} \cdot \frac{4^2}{5^2} \cdots \frac{(2n)^2}{(2n+1)^2} \cdot (2n+1) \right].$$

Although π is not a rational number, Wallis's Formula describes π as a simple infinite product of rational numbers.

4.5 Integration of Trigonometric Functions

This section is devoted to integrating products of tangents and secants as well as products of cotangents and cosecants. The first two subsections present cases where a substitution transforms the integral into the integral of a polynomial. The remaining cases are evaluated in the third subsection by recursion formulas.

Integrating $\tan^m x \sec^n x$ and $\cot^m x \csc^n x$ with n even

When $m = 1$ and $n = 0$, evaluate these integrals by Proposition 4.3.2(a),(b):

$$\int \tan x \, dx = -\ln|\cos x| + C = \ln|\sec x| + C, \qquad \int \cot x \, dx = \ln|\sin x| + C.$$

If $m + n \geq 2$, make the substitution:

$$u = \tan x \text{ with } du = \sec^2 x \, dx \quad \text{or} \quad u = \cot x \text{ with } du = -\csc^2 x \, dx. \qquad (4.5.1)$$

To carry out this substitution, use the Pythagorean identity

$$\sec^2 x = \tan^2 x + 1 \quad \text{or} \quad \csc^2 x = \cot^2 x + 1 \qquad (4.5.2)$$

to write the integrand as a sum of terms of the form $\tan^k x \sec^2 x$ or $\cot^k x \csc^2 x$.

Examples 4.5.1 (1)   Evaluate $\int \tan^3 x \sec^6 x \, dx$.

Solution Note the exponent 6 of $\sec x$ is even. Use the left Pythagorean identity of (4.5.2) to prepare for making the substitution $u = \tan x$:

$$\int \tan^3 x \sec^6 x \, dx = \int \tan^3 x \left(\sec^2 x\right)^2 \sec^2 x \, dx = \int \tan^3 x \left(1 + \tan^2 x\right)^2 \sec^2 $$

Make the substitution $u = \tan x$ with $du = \sec^2 x \, dx$:

$$
\begin{aligned}
\int \tan^3 x \sec^6 x \, dx &= \int u^3 \left(1 + u^2\right)^2 \, du = \int u^7 + 2u^5 + u^3 \, du \\
&= \frac{u^8}{8} + \frac{u^6}{3} + \frac{u^4}{4} + C = \frac{1}{8} \tan^8 x + \frac{1}{3} \tan^6 x + \frac{1}{4} \tan^4 x +
\end{aligned}
$$

(2)   Evaluate $\int \tan^6 x \, dx$.

Solution Think of $\sec x$ as a factor of the integrand with even exponent zero. Use the left Pythagorean identity of (4.5.2) to prepare for making the substitution $u = \tan x$:

$$
\begin{aligned}
\int \tan^6 x \, dx &= \int \tan^4 x \left(\tan^2 x\right) \, dx = \int \tan^4 x \left(\sec^2 x - 1\right) \, dx \\
&= \int \tan^4 x \sec^2 x - \tan^2 x \left(\tan^2 x\right) \, dx \\
&= \int \tan^4 x \sec^2 x - \tan^2 x \left(\sec^2 x - 1\right) \, dx \\
&= \int \tan^4 x \sec^2 x - \tan^2 x \sec^2 x + \tan^2 x \, dx \\
&= \int \tan^4 x \sec^2 x - \tan^2 x \sec^2 x + \sec^2 x - 1 \, dx \\
&= \int \left(\tan^4 x - \tan^2 x + 1\right) \sec^2 x \, dx - \int 1 \, dx.
\end{aligned}
$$

Next make the substitution $u = \tan x$ with $du = \sec^2 x \, dx$:

$$
\begin{aligned}
\int \tan^6 x \, dx &= \int u^4 - u^2 + 1 \, du - x = \frac{1}{5} u^5 - \frac{1}{3} u^3 + u - x + C \\
&= \frac{1}{5} \tan^5 x - \frac{1}{3} \tan^3 x + \tan x - x + C.
\end{aligned}
$$

(3)   Evaluate $\int \sec^{10} x \, dx$.

Solution Note the exponent 10 of $\sec x$ is even. Use the left Pythagorean identity of (4.5.2) to prepare for making the substitution $u = \tan x$:

$$\int \sec^{10} x \, dx = \int \left(\sec^2 x\right)^4 \sec^2 x \, dx = \int \left(1 + \tan^2 x\right)^4 \sec^2 x \, dx.$$

Make the substitution $u = \tan x$ with $du = \sec^2 x \, dx$:

$$
\begin{aligned}
\int \sec^{10} x \, dx &= \int \left(1 + u^2\right)^4 \, du = \int u^8 + 4u^6 + 6u^4 + 4u^2 + 1 \, du \\
&= \frac{1}{9} u^9 + \frac{4}{7} u^7 + \frac{6}{5} u^5 + \frac{4}{3} u^3 + u + C \\
&= \frac{1}{9} \tan^9 x + \frac{4}{7} \tan^7 x + \frac{6}{5} \tan^5 x + \frac{4}{3} \tan^3 x + \tan x + C.
\end{aligned}
$$

(4) Evaluate $\int \cot^4 x \csc^4 x \, dx$.

Solution Note the exponent 4 of $\csc x$ is even. Use the right Pythagorean identity of (4.5.2) to prepare for making the substitution $u = \cot x$:

$$\int \cot^4 x \csc^4 x \, dx \;=\; \int \cot^4 x \left(\csc^2 x\right) \csc^2 x \, dx$$

$$=\; -\int \cot^4 x \left(1 + \cot^2 x\right) \left(-\csc^2 x\right) \, dx.$$

Now make the substitution $u = \cot x$ with $du = -\csc^2 x \, dx$:

$$\int \cot^4 x \csc^4 x \, dx \;=\; -\int u^4 \left(1 + u^2\right) \, du = -\int u^4 + u^6 \, du$$

$$=\; -\frac{1}{5}u^5 - \frac{1}{7}u^7 + C = -\frac{1}{5}\cot^5 x - \frac{1}{7}\cot^7 x + C. \qquad \square$$

Integrating $\tan^m x \sec^n x$ and $\cot^m x \csc^n x$ with m odd

To evaluate $\int \tan^m x \sec^n x \, dx$ when m is odd, make the substitution:

$$v = \sec x \text{ with } dv = \sec x \tan x \, dx \;\; \text{ or } \;\; v = \csc x \text{ with } dv = -\csc x \cot x \, dx. \quad (4.5.3)$$

Use one factor of $\tan x$ [or $\cot x$] for this substitution. Then use a Pythagorean identity of (4.5.2) to rewrite the integrand as a sum of terms of the form $\sec^k x(\sec x \tan x)$ [or $\csc^k x(\csc x \cot x)$]. When m is odd and n is even, either the substitution $u = \tan x$ or $v = \sec x$ [$u = \cot x$ or $v = \csc x$] will evaluate this integral. As in the second example below, the two correct answers look very different. However, since they have the same derivative, they must be equal up to a constant.

Examples 4.5.2 (1) Evaluate $\int \tan^5 x \sec^3 x \, dx$.

Solution Note the exponent 5 of $\tan x$ is odd. Use the left Pythagorean identity of (4.5.2) to prepare for making the substitution $v = \sec x$:

$$\int \tan^5 x \sec^3 x \, dx \;=\; \int \left(\tan^2 x\right)^2 \tan x \sec^3 x \, dx$$

$$=\; \int \left(\sec^2 x - 1\right)^2 \left(\sec^2 x\right) (\sec x \tan x) \, dx.$$

Now make the substitution $v = \sec x$ with $dv = \sec x \tan x \, dx$:

$$\int \tan^5 x \sec^3 x \, dx \;=\; \int \left(v^2 - 1\right)^2 v^2 \, dv = \int v^6 - 2v^4 + v^2 \, dv$$

$$=\; \frac{1}{7}v^7 - \frac{2}{5}v^5 + \frac{1}{3}v^3 + C = \frac{1}{7}\sec^7 x - \frac{2}{5}\sec^5 x + \frac{1}{3}\sec^3 x + C.$$

(2) Evaluate $\int \cot^7 x \csc^6 x \, dx$.

Solution 1 Note the exponent 7 of $\cot x$ is odd. Use the right Pythagorean identity of (4.5.2) to prepare for making the substitution $v = \csc x$:

$$\int \cot^7 x \csc^6 x \, dx \;=\; \int \left(\cot^2 x\right)^3 \cot x \csc^6 x \, dx$$

$$=\; -\int \left(\csc^2 x - 1\right)^3 \left(\csc^5 x\right) (-\csc x \cot x) \, dx.$$

Now make the substitution $v = \csc x$ with $dv = -\csc x \cot x\, dx$:

$$
\begin{aligned}
\int \cot^7 x \csc^6 x\, dx &= -\int \left(v^2 - 1\right)^3 v^5\, dv = -\int v^{11} - 3v^9 + 3v^7 - v^5\, dv \\
&= -\frac{1}{12}v^{12} + \frac{3}{10}v^{10} - \frac{3}{8}v^8 + \frac{1}{6}v^6 + C \\
&= -\frac{1}{12}\csc^{12} x + \frac{3}{10}\csc^{10} x - \frac{3}{8}\csc^8 x + \frac{1}{6}\csc^6 x + C.
\end{aligned}
$$

Solution 2 Note the exponent 6 of $\csc x$ is even. Use the right Pythagorean identity of (4.5.2) to prepare for making the substitution $u = \cot x$:

$$
\begin{aligned}
\int \cot^7 x \csc^6 x\, dx &= \int \cot^7 x(\csc^2 x)^2 \csc^2 x\, dx \\
&= -\int \cot^7 x(1 + \cot^2 x)^2(-\csc^2 x)\, dx.
\end{aligned}
$$

Now make the substitution $u = \cot x$ with $du = -\csc^2 x\, dx$:

$$
\begin{aligned}
\int \cot^7 x \csc^6 x\, dx &= -\int u^7(1 + u^2)^2\, du = -\int u^{11} + 2u^9 + u^7\, du \\
&= -\frac{1}{12}u^{12} - \frac{1}{5}u^{10} - \frac{1}{8}u^8 + C \\
&= -\frac{1}{12}\cot^{12} x - \frac{1}{5}\cot^{10} x - \frac{1}{8}\cot^8 x + C.
\end{aligned}
$$

Although the answers given by these two methods look very different, they must be equal up to a constant because they have the same derivative $\cot^7 x \csc^6 x$.

Integrating $\tan^m x \sec^n x$ and $\cot^m x \csc^n x$ with m even and n odd

These integrals can not be evaluated by the substitution $u = \tan x$ or $v = \sec x$ [$u = \cot x$ or $v = \csc x$]. The first step to evaluate these integrals is to use a Pythagorean identity of (4.5.2) to reduce the problem to integrating odd powers of $\sec x$ [odd powers of $\csc x$]. The latter integrals are evaluated by the following recursion formulas.

Proposition 4.5.3 *Let k be a positive integer.*

(a) $\displaystyle \int \sec^{2k+1} x\, dx = \frac{1}{2k}\tan x \sec^{2k-1} x + \frac{2k-1}{2k}\int \sec^{2k-1} x\, dx$ \qquad (4.5.4)

$\displaystyle \int \sec x\, dx = \ln|\sec x + \tan x| + C.$

(b) $\displaystyle \int \csc^{2k+1} x\, dx = -\frac{1}{2k}\cot x \csc^{2k-1} x + \frac{2k-1}{2k}\int \csc^{2k-1} x\, dx.$ \qquad (4.5.5)

$\displaystyle \int \csc x\, dx = -\ln|\csc x + \cot x| + C.$

Proof **(a)** We evaluated $\int \sec x\, dx$ in Proposition 4.3.2(c). To derive the recursion formula for $\int \sec^{2k+1} x\, dx$, integrate by parts noting that $\sec^2 x = D(\tan x)$:

$$
\begin{aligned}
\int \sec^{2k+1} x\, dx &= \int D(\tan x) \sec^{2k-1} x\, dx = \tan x \sec^{2k-1} x - \int \tan x\, D\left(\sec^{2k-1} x\right) \\
&= \tan x \sec^{2k-1} x - \int (\tan x)\left[(2k-1)(\sec^{2k-2} x)(\sec x \tan x)\right]\, dx \\
&= \tan x \sec^{2k-1} x - (2k-1)\int \tan^2 x \sec^{2k-1} x\, dx.
\end{aligned}
$$

Use the Pythagorean identity $\tan^2 x = \sec^2 x - 1$:

$$\int \sec^{2k+1} x \, dx \;=\; \tan x \sec^{2k-1} x - (2k-1) \int \left(\sec^2 x - 1\right) \sec^{2k-1} x \, dx$$

$$=\; \tan x \sec^{2k-1} x - (2k-1) \int \sec^{2k+1} x \, dx + (2k-1) \int \sec^{2k-1} x \, dx$$

Solve this linear equation for the integral $\int \sec^{2k+1} dx$ by bringing $(2k-1) \int \sec^{2k+1} x \, dx$ to the left side of the equation:

$$2k \int \sec^{2k+1} x \, dx = \tan x \sec^{2k-1} x + (2k-1) \int \sec^{2k-1} x \, dx.$$

Dividing by $2k$ produces the desired recursion formula.

(b) We evaluated $\int \csc x \, dx$ in Proposition 4.3.2(d). The derivation of this recursion formula is similar to the derivation of the recursion formula in (a). We leave the details as an exercise. $\qquad\square$

Examples 4.5.4 (1) Evaluate $\int \sec^5 x \, dx$.

Solution By recursion formula (4.5.4) with $k = 2$:

$$\int \sec^5 x \, dx = \frac{1}{4} \tan x \sec^3 x + \frac{3}{4} \int \sec^3 x \, dx.$$

Apply (4.5.4) again with $k = 1$:

$$\int \sec^5 x \, dx \;=\; \frac{1}{4} \tan x \sec^3 x + \frac{3}{4} \left[\frac{1}{2} \tan x \sec x + \frac{1}{2} \int \sec x \, dx \right]$$

$$=\; \frac{1}{4} \tan x \sec^3 x + \frac{3}{8} \tan x \sec x + \frac{3}{8} \ln |\sec x + \tan x| + C.$$

(2) Evaluate $\int \tan^4 x \sec^3 x \, dx$.

Solution Use the left Pythagorean identity of (4.5.2) to rewrite the integrand in terms of $\sec x$:

$$\int \tan^4 x \sec^3 x \, dx \;=\; \int \left(\sec^2 x - 1\right)^2 \sec^3 x \, dx$$

$$=\; \int \sec^7 \, dx - 2 \int \sec^5 x \, dx + \int \sec^3 x \, dx.$$

Apply recursion formula (4.5.4) with $k = 3$ to $\int \sec^7 \, dx$:

$$\int \tan^4 x \sec^3 x \, dx \;=\; \frac{1}{6} \tan x \sec^5 x + \left(\frac{5}{6} - 2\right) \int \sec^5 x \, dx + \int \sec^3 x \, dx$$

$$=\; \frac{1}{6} \tan x \sec^5 x - \frac{7}{6} \int \sec^5 x \, dx + \int \sec^3 x \, dx.$$

Apply (4.5.4) with $k = 2$ to $\int \sec^5 x \, dx$:

$$\int \tan^4 x \sec^3 x \, dx \;=\; \frac{1}{6} \tan x \sec^5 x - \frac{7}{24} \tan x \sec^3 x + \left(1 - \frac{7}{6}\frac{3}{4}\right) \int \sec^3 x \, dx$$

$$=\; \frac{1}{6} \tan x \sec^5 x - \frac{7}{24} \tan x \sec^3 x + \frac{1}{8} \int \sec^3 x \, dx.$$

Apply (4.5.4) again with $k = 1$:

$$\int \tan^4 x \sec^3 x \; dx = \frac{1}{6} \tan x \sec^5 x - \frac{7}{24} \tan x \sec^3 x + \frac{1}{16} \tan x \sec x$$

$$+ \frac{1}{16} \int \sec x \; dx$$

$$= \frac{1}{6} \tan x \sec^5 x - \frac{7}{24} \tan x \sec^3 x + \frac{1}{16} \tan x \sec x$$

$$+ \frac{1}{16} \ln |\tan x + \sec x| + C.$$

(3) Evaluate $\int \cot^4 x \csc x \; dx$.

Solution Use the right Pythagorean identity of (4.5.2) to rewrite the integrand in terms of $\csc x$:

$$\int \cot^4 x \csc x \; dx = \int \left(\csc^2 x - 1 \right)^2 \csc x \; dx$$

$$= \int \csc^5 x \; dx - 2 \int \csc^3 x \; dx + \int \csc x \; dx.$$

Apply recursion formula (4.5.5) with $k = 2$ to $\int \csc^5 x \; dx$:

$$\int \cot^4 x \csc x \; dx = -\frac{1}{4} \cot x \csc^3 x + \left(\frac{3}{4} - 2 \right) \int \csc^3 x \; dx + \int \csc x \; dx$$

$$= -\frac{1}{4} \cot x \csc^3 x - \frac{5}{4} \int \csc^3 x \; dx + \int \csc x \; dx.$$

Apply (4.5.5) with $k = 1$ to $\int \csc^3 x \; dx$:

$$\int \cot^4 x \csc x \; dx = -\frac{1}{4} \cot x \csc^3 x + \frac{5}{8} \cot x \csc x + \left(1 - \frac{5}{8} \right) \int \csc x \; dx$$

$$= -\frac{1}{4} \cot x \csc^3 x + \frac{5}{8} \cot x \csc x - \frac{3}{8} \ln |\csc x + \cot x| + C.$$

Summary

The reader should be able to evaluate integrals of the forms $\int \tan^m x \sec^n x \; dx$ and $\int \cot^m x \csc^n x \; dx$. When m is odd or n is even, she should make the appropriate substitution. When m is even and n is odd she should use the appropriate recursion formula.

Basic Exercises

Evaluate the following 22 integrals.

1. $\int \tan^2 x \; dx$
2. $\int \tan^2 x \sec x \; dx$
3. $\int \tan^3 x \; dx$
4. $\int \cot^3 x \csc^4 x \; dx$
5. $\int \cot^5 x \csc^5 x \; dx$
6. $\int \sec^4 x \; dx$
7. $\int \csc^5 x \; dx$
8. $\int \tan^2 x \sec^3 x \; dx$
9. $\int \tan^5 x \sec^5 x \; dx$
10. $\int \tan^4 x \sec^4 x \; dx$
11. $\int \sec^7 \pi x \; dx$
12. $\int \cot^2(3x - 7) \csc^3(3x - 7) \; dx$
13. $\int e^x \tan^7(e^x) \sec^8(e^x) \; dx$
14. $\int x \tan^4(x^2) \sec(x^2) \; dx$
15. $\int \frac{1}{x} \cot^6(\ln x) \csc^6(\ln x) \; dx$
16. $\int 2^x \tan^9(2^x) \sec(2^x) \; dx$
17. $\int \cot^9 \pi x \; dx$
18. $\int \sqrt{x} \tan^3(x^{3/2}) \sec^6(x^{3/2}) \; dx$
19. $\int x^3 \tan^6(x^4) \sec(x^4) \; dx$
20. $\int \frac{1}{\sqrt{x}} \tan^9 \sqrt{x} \; dx$
21. $\int 5^x \tan^6(5^x) \sec^8(5^x) \; dx$
22. $\int \frac{\sin^4 x}{\cos^{12} x} \; dx$

23. Find the area of the region bounded by the curves $y = \sec^4 x$, $y = \csc^4 x$, the x–axis, the y–axis and the line $x = \frac{\pi}{2}$.

24. Find the area of the region bounded by the curves $y = \tan^6 x$, $y = \cot^6 x$ for $0 \leq x \leq \frac{\pi}{2}$ and the x–axis.

25. Find the volume of the solid of revolution obtained by revolving the region bounded by the line $y = 2$ and the curve $y = \sec^2 x$, $-\frac{\pi}{2} < x < \frac{\pi}{2}$, around the x–axis.

26. Find the volume of the solid of revolution obtained by revolving the region in the first quadrant bounded by $y = \tan^3 x$ and $y = 1$ around the x–axis.

27. Derive the recursion formula of Proposition 4.5.3(b) for $\int \csc^{2k+1} x \, dx$.

The following three exercises are for those who have studied Section 3.9.

28. Find the distance traveled by each object with the given velocity.
(a) $v(t) = \cot^3 t$ miles per hour from $t = \frac{\pi}{4}$ to $t = \frac{5\pi}{6}$.
(b) $v(t) = \tan^3 t \sec^3 t$ miles per hour from $t = -\frac{\pi}{6}$ to $t = \frac{\pi}{3}$.

29. Find the mass of the object with density $\rho(x) = \tan^3 \pi x$ kg/m for $0 \leq x \leq \frac{1}{4}$ m.

30. How much work is done when the force $F(x) = \sec^6 x$ Newtons moves an object from $x = 0$ to $x = \frac{\pi}{4}$ meters?

The following two exercises are for those who have studied Section 3.10.

31. Find the center of mass of each rod of density $\rho(x)$.
(a) $\rho(x) = \tan^4 x$ kilograms per meter for $0 \leq x \leq \frac{\pi}{4}$ meters.
(b) $\rho(x) = \sec^4 x$ kilograms per meter for $0 \leq x \leq \frac{\pi}{3}$ meters.

32. Find the centroid of each region.
(a) Let R be the region bounded by $y = \tan^2 x$, the line $x = \frac{\pi}{4}$ and the x–axis.
(b) Let S be the region bounded by $y = \sec^2 x$, the lines $x = \pm\frac{\pi}{3}$ and the x–axis.

The following three exercises are for those who have studied Section 4.10.

33. Use hyperbolic functions to evaluate each integral.
(a) $\int \operatorname{sech}^3 x \tanh^3 x \, dx$ (b) $\int \operatorname{csch}^4 x \coth^4 x \, dx$ (c) $\int \operatorname{sech}^6 x \, dx$
(d) $\int \operatorname{csch} x \, dx$ (e) $\int \operatorname{sech} x \, dx$ (f) $\int \operatorname{csch}^3 x \, dx$

34. Find the area of the region bounded by $y = -1 + \cosh x$ and $y = 6 \operatorname{sech} x$.

35. Let R be the region bounded by $y = (e^2 + 1) \tanh^2 x$ and $y = e^2 - 1$.
(a) Find the area of R.
(b) Find the volume of the solid of revolution obtained by revolving the region R around the x–axis.

4.6 Trigonometric Reverse Substitutions

In Section 3.6, we changed variables to rewrite a complicated integral $\int f(g(x))g'(x) \, dx$ in the variable x as the simpler integral $\int f(u) \, du$ in the variable $u = g(x)$. The motivation is clear: it should be easier to evaluate a simple integral than a complicated one. There are exceptions, however, where the two integrals are of different types. We may not be able to evaluate a certain type of simple integral in x while we can evaluate a complicated integral in θ of another type. The substitution $x = h(\theta)$, with $dx = h'(\theta) \, d\theta$, transforms the simple integral $\int f(x) \, dx$ into the more complicated integral $\int f(h(\theta))h'(\theta) \, d\theta$ is called a *reverse substitution*. The three subsections consider simple algebraic integrals involving $a^2 - x^2$, $a^2 + x^2$ and $x^2 - a^2$ that we can not evaluate directly. In each case, an appropriate trigonometric reverse substitutio transforms the integral into a more complicated one involving powers of trigonometr functions. The latter integral is evaluated by a method of Section 3.6 or 4.5.

Integrands Involving $a^2 - x^2$

We evaluate integrals of algebraic expressions in $a^2 - x^2$. The reverse substitution

$$x = a\sin\theta \ \text{ with } \ a^2 - x^2 = a^2\cos^2\theta \quad \text{and} \quad dx = a\cos\theta \, d\theta \qquad (4.6.$$

produces equivalent integrals involving either powers of $\sin\theta$, $\cos\theta$ or powers of $\tan$ $\sec\theta$. Evaluate $\int \sin^m\theta\cos^n\theta \, d\theta$ by the methods of Section 3.6:

- if m is odd, change variables to $u = \cos\theta$ with $du = -\sin\theta \, d\theta$;

- if n is odd, change variables to $v = \sin\theta$ with $dv = \cos\theta \, d\theta$;

- if m and n are even, simplify the integrand with the trigonometric identities

$$\sin\theta\cos\theta = \frac{1}{2}\sin 2\theta, \ \ \sin^2\theta = \frac{1}{2}(1 - \cos 2\theta) \ \text{ and } \ \cos^2\theta = \frac{1}{2}(1 + \cos 2\theta).$$

Evaluate $\int \tan^m\theta\sec^n\theta \, d\theta$ by the methods of Section 4.5:

- if n is even, change variables to $u = \tan\theta$ with $du = \sec^2\theta \, d\theta$;

- if m is odd, change variables to $v = \sec\theta$ with $dv = \tan\theta\sec\theta \, d\theta$;

- if m is even and n is odd, the integrand can be rewritten in terms of odd power of $\sec\theta$ which are integrated by the recursion formula of Proposition 4.5.3(a):

$$\int \sec^{2k+1}\theta \, d\theta = \frac{1}{2k}\tan\theta\sec^{2k-1}\theta + \frac{2k-1}{2k}\int \sec^{2k-1}\theta \, d\theta.$$

Examples 4.6.1 (1) Evaluate $\int \frac{x^2}{\sqrt{4-x^2}} \, dx$.

Solution In this case $a^2 - x^2 = 4 - x^2$, and $a = 2$. Make the substitution $x = 2\sin\theta$ with $dx = 2\cos\theta \, d\theta$:

$$\int \frac{x^2}{\sqrt{4-x^2}} \, dx = \int \frac{4\sin^2\theta}{\sqrt{4 - 4\sin^2\theta}} 2\cos\theta \, d\theta = 4\int \sin^2\theta \, d\theta$$

$$= 4\int \frac{1}{2}(1 - \cos 2\theta) \, d\theta = 2\theta - \sin 2\theta + C.$$

It remains to rewrite this answer in terms of x. The first step is to substitute $\sin 2\theta = 2\sin\theta\cos\theta$ so that we have trigonometric functions of the angle θ:

$$\int \frac{x^2}{\sqrt{4-x^2}} \, dx = 2\theta - 2\sin\theta\cos\theta + C.$$

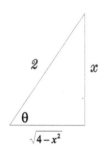

Figure 4.6.2
$\sin\theta = \frac{x}{2}$

Since $\sin\theta = \frac{x}{2}$, construct a right triangle in Figure 4.6.2 with angle θ by letting the opposite side have length x and the hypotenuse have length 2. By the Pythagorean Theorem, the adjacent side has length $\sqrt{4 - x^2}$. From this triangle, $\cos\theta = \frac{\sqrt{4-x^2}}{2}$. In addition, restate $\sin\theta = \frac{x}{2}$ as $\theta = \arcsin\frac{x}{2}$. Thus

$$\int \frac{x^2}{\sqrt{4-x^2}} \, dx = 2\arcsin\frac{x}{2} - 2\left(\frac{x}{2}\right)\left(\frac{\sqrt{4-x^2}}{2}\right) + C = 2\arcsin\frac{x}{2} - \frac{x}{2}\sqrt{4-x^2} + C$$

(2) Evaluate $\int \frac{x^4}{(1-x^2)^{9/2}}\,dx$.

Solution In this case $a^2 - x^2 = 1 - x^2$, and $a = 1$. Make the substitution $x = \sin\theta$ with $dx = \cos\theta\, d\theta$:

$$\begin{aligned}\int \frac{x^4}{(1-x^2)^{9/2}}\,dx &= \int \frac{\sin^4\theta}{\left(\sqrt{1-\sin^2\theta}\right)^9}\cos\theta\,d\theta = \int \frac{\sin^4\theta}{\cos^9\theta}\cos\theta\,d\theta \\ &= \int \frac{\sin^4\theta}{\cos^4\theta}\frac{1}{\cos^4\theta}\,d\theta = \int \tan^4\theta\sec^4\theta\,d\theta.\end{aligned}$$

Since the exponent of $\sec\theta$ is even, let $u = \tan\theta$ with $du = \sec^2\theta d\theta$:

$$\begin{aligned}\int \frac{x^4}{(1-x^2)^{9/2}}\,dx &= \int \tan^4\theta(1+\tan^2\theta)\sec^2\theta\,d\theta = \int u^4(1+u^2)\,du \\ &= \int u^4 + u^6\,du = \frac{u^5}{5} + \frac{u^7}{7} + C = \frac{1}{5}\tan^5\theta + \frac{1}{7}\tan^7\theta + C.\end{aligned}$$

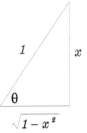

Figure 4.6.3
$\sin\theta = x$

Since $\sin\theta = x$, construct a right triangle in Figure 4.6.3 with angle θ where the opposite side has length x and the hypotenuse has length one. By the Pythagorean Theorem, the adjacent side has length $\sqrt{1-x^2}$. From this triangle, we see that $\tan\theta = \frac{x}{\sqrt{1-x^2}}$. Hence

$$\int \frac{x^4}{(1-x^2)^{9/2}}\,dx = \frac{x^5}{5(1-x^2)^{5/2}} + \frac{x^7}{7(1-x^2)^{7/2}} + C.$$

(3) Find the area A of the region enclosed by the ellipse $\frac{x^2}{a^2} + \frac{y^2}{b^2} = 1$.

Solution Solve the equation of the ellipse for y:

$$y = \pm\frac{b}{a}\sqrt{a^2 - x^2}.$$

Figure 4.6.4
The Ellipse $\frac{x^2}{a^2} + \frac{y^2}{b^2} = 1$

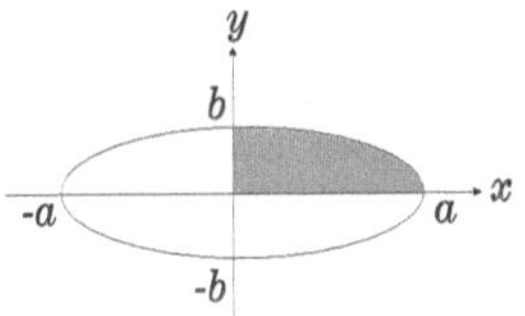

By symmetry, the area A of the entire ellipse is four times the area of the quarter ellipse which lies in the first quadrant. See Figure 4.6.4. Thus,

$$A = 4\int_{x=0}^{x=a} \frac{b}{a}\sqrt{a^2 - x^2}\,dx.$$

Let $x = a\sin\theta$ with $dx = a\cos\theta\,d\theta$. Observe $x = 0$ when $\theta = 0$ while $x = a$ when $\theta = \frac{\pi}{2}$. Thus

$$\begin{aligned}A &= \frac{4b}{a}\int_{\theta=0}^{\theta=\pi/2}\sqrt{a^2 - a^2\sin^2\theta}\,a\cos\theta\,d\theta = 4ab\int_0^{\pi/2}\cos^2\theta\,d\theta \\ &= 4ab\int_0^{\pi/2}\frac{1}{2}(1+\cos 2\theta)\,d\theta = 2ab\theta + ab\sin 2\theta\big|_0^{\pi/2} = \pi ab.\end{aligned}$$

□

Integrands Involving $a^2 + x^2$

We evaluate integrals of algebraic expressions in $a^2 + x^2$. The reverse substitution

$$x = a\tan\theta \text{ with } a^2 + x^2 = a^2\sec^2\theta \quad\text{and}\quad dx = a\sec^2\theta\,d\theta \qquad (4.6.2)$$

produces equivalent integrals involving powers of $\tan\theta$, $\sec\theta$ or $\sin\theta$, $\cos\theta$. These integrals are evaluated by the methods of Sections 3.6 and 4.5 which are summarized at the beginning of the first subsection. Integrals of this form will arise when we integrate rational functions in the next section.

Examples 4.6.5 (1) Evaluate $\int \frac{1}{(16+x^2)^2}\, dx$.

Solution In this case $a^2 + x^2 = 16 + x^2$, and $a = 4$. Make the substitution $x = 4\tan\theta$ with $dx = 4\sec^2\theta\, d\theta$:

$$
\int \frac{1}{(16+x^2)^2}\, dx = \int \frac{1}{(16+16\tan^2\theta)^2} 4\sec^2\theta\, d\theta = \frac{1}{64}\int \frac{1}{\sec^4\theta}\sec^2\theta\, d\theta
$$

$$
= \frac{1}{64}\int \cos^2\theta\, d\theta = \frac{1}{64}\int \frac{1}{2}(1+\cos 2\theta)\, d\theta
$$

$$
= \frac{1}{128}\theta + \frac{1}{256}\sin 2\theta + C = \frac{1}{128}\theta + \frac{1}{128}\sin\theta\cos\theta + C.
$$

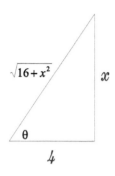

Figure 4.6.6
$\tan\theta = \frac{x}{4}$

Rewrite $\tan\theta = \frac{x}{4}$ as $\theta = \arctan\frac{x}{4}$. To write $\sin\theta$ and $\cos\theta$ in terms of x construct a right triangle with angle θ in Figure 4.6.6. Since $\tan\theta = \frac{x}{4}$, let the opposite side have length x and the adjacent side have length 4. By the Pythagorean Theorem, the hypotenuse has length $\sqrt{16+x^2}$. From this triangle

$$
\sin\theta = \frac{x}{\sqrt{16+x^2}} \quad\text{and}\quad \cos\theta = \frac{4}{\sqrt{16+x^2}}.
$$

$$
\text{Hence}\quad \int \frac{1}{(16+x^2)^2}\, dx = \frac{1}{128}\arctan\frac{x}{4} + \frac{1}{128}\frac{x}{\sqrt{16+x^2}}\frac{4}{\sqrt{16+x^2}} + C
$$

$$
= \frac{1}{128}\arctan\frac{x}{4} + \frac{x}{32(16+x^2)} + C.
$$

(2) Evaluate $\int \sqrt{x^2 - 8x + 25}\, dx$.

Solution We complete the square to recognize that this integral can be integrated by the method of this subsection:

$$
\int \sqrt{x^2 - 8x + 25}\, dx = \int \sqrt{(x-4)^2 + 9}\, dx.
$$

Now make the substitution $x - 4 = 3\tan\theta$ with $dx = 3\sec^2\theta\, d\theta$:

$$
\int \sqrt{x^2 - 8x + 25}\, dx = \int \sqrt{9\tan^2\theta + 9}\, (3\sec^2\theta)\, d\theta = 9\int \sec^3\theta\, d\theta.
$$

By Example 4.4.6(2) or by the recursion formula (4.5.4):

$$
\int \sqrt{x^2 - 8x + 25}\, dx = \frac{9}{2}\tan\theta\sec\theta + \frac{9}{2}\ln|\sec\theta + \tan\theta| + C.
$$

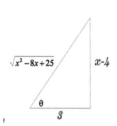

Figure 4.6.7
$\tan\theta = \frac{x-4}{3}$

To write this expression in terms of x, construct a right triangle with angle θ in Figure 4.6.7. Since $\tan\theta = \frac{x-4}{3}$, let the opposite side have length $x - 4$ and the adjacent side have length 3. By the Pythagorean Theorem, the hypotenuse has length $\sqrt{x^2 - 8x + 25}$. From this triangle, we see that $\sec\theta = \frac{\sqrt{x^2-8x+25}}{3}$. Thus

$$
\int \sqrt{x^2 - 8x + 25}\, dx
$$

$$
= \frac{9}{2}\left(\frac{x-4}{3}\right)\left(\frac{\sqrt{x^2-8x+25}}{3}\right) + \frac{9}{2}\ln\left|\frac{\sqrt{x^2-8x+25}}{3} + \frac{x-4}{3}\right| +
$$

$$
= \frac{1}{2}(x-4)\sqrt{x^2 - 8x + 25} + \frac{9}{2}\ln\left|\sqrt{x^2-8x+25} + x - 4\right| + C'
$$

where $C' = C - \frac{9}{2}\ln 3$.

Integrands Involving $x^2 - a^2$

To integrate algebraic expressions in $x^2 - a^2$, we use the reverse substitution

$$x = a \sec\theta \quad \text{with} \quad x^2 - a^2 = a^2 \tan^2\theta \quad \text{and} \quad dx = a \sec\theta \tan\theta \, d\theta.$$

The resulting integrals involve products of $\tan\theta$, $\sec\theta$ or $\sin\theta$, $\cos\theta$. These integrals are evaluated by the methods of Section 3.6 and 4.5 which are summarized at the beginning of the first subsection.

Examples 4.6.8 (1) Evaluate $\int \frac{1}{\sqrt{x^2-4}} \, dx$.

Solution In this case, $x^2 - a^2 = x^2 - 4$, and $a = 2$. Hence make the substitution $x = 2\sec\theta$ with $dx = 2\sec\theta\tan\theta\, d\theta$:

$$\int \frac{1}{\sqrt{x^2-4}} \, dx = \int \frac{1}{\sqrt{4\sec^2\theta - 4}} 2\sec\theta\tan\theta \, d\theta = \int \frac{1}{2\tan\theta} 2\sec\theta\tan\theta \, d\theta$$

$$= \int \sec\theta \, d\theta = \ln|\sec\theta + \tan\theta| + C.$$

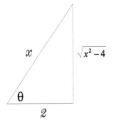

To write this expression in terms of x, construct a right triangle with angle θ in Figure 4.6.9. Since $\sec\theta = \frac{x}{2}$, let the adjacent side have length 2 and the hypotenuse have length x. By the Pythagorean Theorem the opposite side has length $\sqrt{x^2-4}$. From this triangle, we see that $\tan\theta = \frac{\sqrt{x^2-4}}{2}$. Thus

Figure 4.6.9
$\sec\theta = \frac{x}{2}$

$$\int \frac{1}{\sqrt{x^2-4}} \, dx = \ln\left|\frac{x}{2} + \frac{\sqrt{x^2-4}}{2}\right| + C = \ln\left|x + \sqrt{x^2-4}\right| + C', \quad (C' = C - \ln 2).$$

(2) Evaluate $\int \sqrt{x^2 + 14x + 13} \, dx$.

Solution Begin by completing the square in the quadratic polynomial to identify the appropriate reverse trigonometric substitution.

$$\int \sqrt{x^2 + 14x + 13} \, dx = \int \sqrt{(x+7)^2 - 36} \, dx.$$

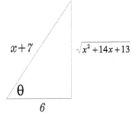

Make the substitution $x + 7 = 6\sec\theta$ with $dx = 6\sec\theta\tan\theta \, d\theta$:

$$\int \sqrt{x^2 + 14x + 13} \, dx = \int \sqrt{36\sec^2\theta - 36} \, (6\sec\theta\tan\theta) \, d\theta$$

$$= 36 \int (\tan\theta)(\sec\theta\tan\theta) \, d\theta = 36 \int \sec\theta\tan^2\theta \, d\theta$$

Figure 4.6.10
$\sec\theta = \frac{x+7}{6}$

$$= 36 \int \sec\theta \left(\sec^2\theta - 1\right) \, d\theta = 36 \int \sec^3\theta - \sec\theta \, d\theta$$

$$= 18 \left(\tan\theta\sec\theta + \ln|\sec\theta + \tan\theta|\right) - 36\ln|\sec\theta + \tan\theta| + C$$

$$= 18\tan\theta\sec\theta - 18\ln|\sec\theta + \tan\theta| + C$$

by Example 4.4.6(2). To write this expression in terms of x, construct a right triangle with angle θ in Figure 4.6.10. Since $\sec\theta = \frac{x+7}{6}$, let the hypotenuse of this triangle have length $x + 7$ and the adjacent side have length 6. By the Pythagorean Theorem the opposite side has length $\sqrt{x^2 + 14x + 13}$. From this triangle, we see that $\tan\theta = \frac{\sqrt{x^2+14x+13}}{6}$. Thus

$$\int \sqrt{x^2 + 14x + 13} \, dx$$

$$= 18 \left(\frac{\sqrt{x^2 + 14x + 13}}{6}\right)\left(\frac{x+7}{6}\right) - 18\ln\left|\frac{x+7}{6} + \frac{\sqrt{x^2 + 14x + 13}}{6}\right| + C$$

$$= \frac{1}{2}(x+7)\sqrt{x^2 + 14x + 13} - 18\ln\left|x + 7 + \sqrt{x^2 + 14x + 13}\right| + C' \quad (C' = C + 18\ln 6).$$

Summary

The reader should know how to use trigonometric reverse substitutions to evalua
integrals whose integrands involve:

$$a^2 - x^2 \quad (\text{use} \quad x = a \sin \theta),$$
$$a^2 + x^2 \quad (\text{use} \quad x = a \tan \theta),$$
$$x^2 - a^2 \quad (\text{use} \quad x = a \sec \theta).$$

Basic Exercises

Evaluate the following 21 integrals.

1. $\int \sqrt{x^2 + 9}\, dx$ **2.** $\int \sqrt{25 - x^2}\, dx$ **3.** $\int \frac{1}{(x^2+25)^3}\, dx$

4. $\int x^2 \sqrt{x^2 - 1}\, dx$ **5.** $\int \frac{x^2}{\sqrt{16-x^2}}\, dx$ **6.** $\int (x^2 + 2x + 5)^{3/2}\, dx$

7. $\int (x^2 + 6x)^{3/2}\, dx$ **8.** $\int \frac{1}{(x^2-4x+13)^2}\, dx$ **9.** $\int (x^2 - 8x + 9)^{5/2}\, dx$

10. $\int \frac{x^2}{(4-x^2)^3}\, dx$ **11.** $\int \frac{x}{\sqrt{x^2+6x+15}}\, dx$ **12.** $\int \frac{1}{x^{3/2}(x-6)^{3/2}}\, dx$

13. $\int \frac{1}{(9-2x-x^2)^{3/2}}\, dx$ **14.** $\int \frac{x^5}{\sqrt{x^4-9}}\, dx$ **15.** $\int \frac{1}{\cos^2 x (5 - \tan^2 x)^{5/2}}\, dx$

16. $\int \frac{e^x}{(e^{2x}+4e^x+8)^2}\, dx$ **17.** $\int \sqrt{\frac{x+4\sqrt{x}+1}{x}}\, dx$ **18.** $\int (2^x + 4^x)(5 - 2^{x+1} - 4^x)^{5/2}$

19. $\int \frac{x}{x^8 + 2x^4 + 1}\, dx$ **20.** $\int \frac{e^x - e^{-x}}{(e^{2x}+e^{-2x})^2}\, dx$ **21.** $\int \frac{1}{x\sqrt{(\ln x)^2 + 6 \ln x}}\, dx$

22. Find the area of the region bounded by the curve $x^2 - y^2 = 1$ and the line $y = -2$, $y = 2$.

23. Find the area of the region bounded by $y = \frac{1}{(1+x^2)^2}$ and $y = \frac{x^2}{(1+x^2)^2}$.

24. Find the area of the region bounded by $y = \frac{e^{x+1}}{\sqrt{2e^{2x}-e}}$, $y = e^x \sqrt{2e^{2x} - e}$ and $x =$

25. Find the volume of the solid of revolution obtained by revolving the region bounded by $y = \frac{1}{1+x^2}$, $y = 1$ and $x = \frac{1}{10}$ around the x–axis.

26. Find the volume of the solid of revolution obtained by revolving the region bonded by $y = \sqrt{x^4 - 1}$, $y = 0$ and $x = 5$ around the y–axis.

27. Find the volume of the solid of revolution obtained by revolving the region bounded by $y = \frac{6x}{\sqrt{1-x^2}}$, $y = 8$ and $x = 0$ around the y–axis.

The following three exercises are for those who have studied Section 3.9.

28. Find the distance traveled by each object with the given velocity.
(a) $v(t) = t^3 \sqrt{1 - t^2}$ kilometers per hour from $t = -1$ to $t = \frac{1}{2}$.
(b) $v(t) = \frac{1}{(1+t^2)^{5/2}}$ kilometers per hour from $t = -2$ to $t = 3$.

29. Find the mass of the two meter long rod of density $\rho(x) = x^2 \sqrt{1 + x^2}$ kilograms per meter for $0 \le x \le 2$ meters.

30. How much work is done when the force $F(x) = \frac{x^2}{\sqrt{16-x^2}}$ Newtons moves an object from $x = 0$ to $x = 2$ meters?

The following two exercises are for those who have studied Section 3.10.

31. Find the center of mass of each rod with density $\rho(x)$.
(a) $\rho(x) = \frac{1}{(x^2+9)^2}$ pounds per foot for $0 \le x \le 3$ feet.
(b) $\rho(x) = x^4 \sqrt{x^2 - 4}$ pounds per foot for $2 \le x \le 4$ feet.

32. Find the centroid of each region.

(a) S is the region bounded by $y = \frac{1}{1+x^2}$, $x = 3$, the x–axis and the y–axis.

(b) T is the region bounded by $y = (x^2 + 2x + 5)^{3/2}$, $x = 2$, the x-axis and the y–axis.

The following exercise is for those who have studied Section 4.10.

33. Use hyperbolic functions to evaluate each integral.

     (a) $\int \sqrt{1+x^2}\, dx$ **(b)** $\int \frac{1}{x^2-1}\, dx$ **(c)** $\int \frac{x^2}{\sqrt{1-x^2}}\, dx$

4.7   Integration of Rational Functions

Recall that a rational function $\frac{p(x)}{q(x)}$ is a quotient of polynomials $p(x)$ and $q(x)$. In this section, we show that every rational function can be integrated by using algebraic methods to simplify the integrand. This procedure rewrites the integrand as a sum of simple rational functions which we know how to integrate. The first subsection presents several examples of this procedure which motivate the statement of the four steps involved. The key step is the method of partial fractions. We divide the detailed study of this method into two cases: the second subsection considers the case where all the roots of $q(x)$ are real while the third subsection studies the case where $q(x)$ has complex roots.

The Procedure

We illustrate the strategy for integrating rational functions with three examples. For simplicity, the denominators of the integrands have no complex roots. These examples motivate the four steps used to integrate a rational function.

Motivating Example 4.7.1 (1)   Evaluate $\int \frac{x}{x^2-4}\, dx$.

    Solution   Observe that $x^2 - 4 = (x-2)(x+2)$. If we add the fractions $\frac{A}{x-2}$ and $\frac{B}{x+2}$, the common denominator will be $(x-2)(x+2)$:

$$\frac{A}{x-2} + \frac{B}{x+2} = \frac{A(x+2) + B(x-2)}{(x-2)(x+2)} = \frac{(A+B)x + 2(A-B)}{(x-2)(x+2)}.$$

    This sum equals the integrand $\frac{x}{(x-2)(x+2)}$ when the numerator $(A+B)x + 2(A-B)$ of the sum equals the numerator x of the integrand. This occurs when the coefficient $A + B$ of x in the sum equals one and the constant term $2(A - B)$ in the sum equals zero. By the latter condition, $A = B$. Then the former condition becomes: $2A = 2B = A + B = 1$, and $A = B = \frac{1}{2}$. Hence

$$\int \frac{x}{x^2-4}\, dx = \int \frac{1/2}{x-2} + \frac{1/2}{x+2}\, dx = \frac{1}{2}\ln|x-2| + \frac{1}{2}\ln|x+2| + C.$$

(2)   Evaluate $\int \frac{6x^3}{x^2-x-2}\, dx$.

    Solution   Simplify the integrand by dividing its denominator $x^2 - x - 2$ into its numerator $6x^3$ as in Figure 4.7.2: $6x^3 = (6x+6)(x^2-x-2) + 18x + 12$. Hence

$$\int \frac{6x^3}{x^2-x-2}\, dx = \int 6x + 6 + \frac{18x+12}{x^2-x-2}\, dx$$

$$= 3x^2 + 6x + \int \frac{18x+12}{(x-2)(x+1)}\, dx. \qquad (4.7.1)$$

    $$\begin{array}{r} 6x+6 \\ \hline x^2-x-2\,\big)\,6x^3 \\ 6x^3 - 6x^2 - 12x \\ \hline 6x^2 + 12x \\ 6x^2 - 6x - 12 \\ \hline 18x + 12 \end{array}$$

Figure 4.7.2
Polynomial Division

$A - B = 0$ $A = 1/2,\ B = 1/2$

$A + B = 1$

The new integrand is simplified as in the preceding example. The sum $\frac{A}{x-2} + \frac{B}{x+}$ has least common denominator $(x-2)(x+1)$:

$$\frac{A}{x-2} + \frac{B}{x+1} = \frac{A(x+1) + B(x-2)}{(x-2)(x+1)} = \frac{(A+B)x + (A-2B)}{(x-2)(x+1)}.$$

We want this sum to equal the new integrand $\frac{18x+12}{(x-2)(x+2)}$. This occurs when the coefficient $A + B$ of x in the sum equals the coefficient 18 of x in the integrand and the constant term $A - 2B$ in the sum equals the constant term 12 in the integrand. The equations

$$\begin{aligned} A + B &= 18 \\ A - 2B &= 12 \end{aligned}$$

have solution $A = 16$ and $B = 2$. Hence we rewrite (4.7.1) as:

$$\begin{aligned} \int \frac{6x^3}{x^2 - x - 2} \, dx &= 3x^2 + 6x + \int \frac{16}{x-2} + \frac{2}{x+1} \, dx \\ &= 3x^2 + 6x + 16 \ln|x-2| + 2 \ln|x+1| + C. \end{aligned}$$

(3) Evaluate $\int \frac{x^3 - x^2 - x - 1}{x^4 - x^3} \, dx$.

Solution Observe that $x^4 - x^3 = x^3(x-1)$. We want to add several simple fractions, with numbers as numerators, whose least common denominator is $x^3(x-1)$. Moreover the sum of these fractions should have a numerator which is a linear combination of 1, x, x^2 and x^3. Let's try

$$\begin{aligned} \frac{A}{x-1} + \frac{B}{x} + \frac{C}{x^2} + \frac{D}{x^3} &= \frac{Ax^3 + Bx^2(x-1) + Cx(x-1) + D(x-1)}{x^3(x-1)} \\ &= \frac{(A+B)x^3 + (C-B)x^2 + (D-C)x - D}{x^3(x-1)}. \end{aligned}$$

We want the numerator $n(x) = (A+B)x^3 + (C-B)x^2 + (D-C)x - D$ of this sum to equal the numerator $p(x) = x^3 - x^2 - x - 1$ of the integrand.
• The constant term $-D$ of $n(x)$ equals the constant term -1 of $p(x)$ when $D = 1$.
• The coefficient $D - C = 1 - C$ of x in $n(x)$ equals the coefficient -1 of x in $p(x)$ when $C = 2$.
• The coefficient $C - B = 2 - B$ of x^2 in $n(x)$ equals the coefficient -1 of x^2 in $p(x)$ when $B = 3$.
• The coefficient $A + B = A + 3$ of x^3 in $n(x)$ equals the coefficient 1 of x^3 in $p(x)$ when $A = -2$. Thus

$$\begin{aligned} \int \frac{x^3 - x^2 + x - 1}{x^4 - x^3} \, dx &= \int -\frac{2}{x-1} + \frac{3}{x} + \frac{2}{x^2} + \frac{1}{x^3} \, dx \\ &= -2 \ln|x-1| + 3 \ln|x| - \frac{2}{x} - \frac{1}{2x^2} + C. \end{aligned}$$

These examples indicate that there are four steps in integrating a rational function $f(x) = \frac{p(x)}{q(x)}$ when $q(x)$ has no complex roots.

Step I (Division) If the degree of the numerator $p(x)$ is greater than or equal to the degree of the denominator $q(x)$, divide $q(x)$ into $p(x)$:

$$p(x) = k(x)q(x) + r(x)$$

where $k(x)$ and $r(x)$ are polynomials with degree $r(x)$ less than degree $q(x)$. Divide the above equation by $q(x)$:

$$f(x) = \frac{p(x)}{q(x)} = k(x) + \frac{r(x)}{q(x)}.$$

Step II (Factorization) Since $q(x)$ has no complex roots, factor $q(x)$ into linear factors:

$$q(x) = c\,(x - a_1)^{s_1} \cdots (x - a_m)^{s_m}$$

where $a_1, \ldots, a_m \in \Re$ are distinct.

Step III (Partial Fractions) Write $\frac{r(x)}{q(x)}$ as the sum of the simple fractions

$$\frac{A_{i,1}}{x - a_i}, \quad \frac{A_{i,2}}{(x - a_i)^2}, \quad \ldots, \quad \frac{A_{i,s_i}}{(x - a_i)^{s_i}} \quad \text{for } 1 \leq i \leq m.$$

Step IV (Integration) Integrate the rational function $f(x)$, using its representation in Steps I and III as a sum of functions which we know how to integrate.

We skip Step I when the degree of the numerator of $f(x)$ is less than the degree of its denominator as in Examples 1 and 3 above. However, when the degree of the numerator is greater than or equal to the degree of the denominator, omission of Step I makes it impossible to carry out the method of partial fractions in Step III.

Step II is always possible in theory: the Fundamental Theorem of Algebra says that a polynomial with no complex roots can be factored into linear factors. There are algorithms for performing this factorization when the polynomial $q(x)$ has degree two, three or four. (This algorithm is an application of the *quadratic formula* when the degree of $q(x)$ equals two.) However, it can be shown that there is no algorithm for performing this factorization when the degree of $q(x)$ is greater than four.

Step III, the method of partial fractions, can be carried out in several ways which are described and illustrated in the next subsection.

The integrations in Step IV are straightforward:

$$\int \frac{1}{x - a}\, dx = \ln|x - a| + C \quad \text{and}$$

$$\int \frac{1}{(x - a)^f}\, dx = -\frac{1}{f - 1} \cdot \frac{1}{(x - a)^{f-1}} + C \quad \text{if } f \geq 2. \qquad (4.7.2)$$

Rational Functions $\frac{p(x)}{q(x)}$ where $q(x)$ has no Complex Roots

In the preceding subsection, we established a four step procedure to integrate a rational function whose denominator has no complex roots: division, factorization of the denominator, partial fractions and integration. We begin with a detailed strategy for carrying out the method of partial fractions. Then we illustrate the procedure for integrating these rational functions with two examples.

The method of partial fractions starts with a rational function

$$\frac{r(x)}{q(x)} = \frac{r(x)}{(x - a_1)^{s_1} \cdots (x - a_m)^{s_m}}$$

with degree $r(x)$ less than degree $q(x)$. This method writes $\frac{r(x)}{q(x)}$ as a sum of simp
rational functions:

$$\frac{r(x)}{q(x)} = \sum_{i=1}^{m} \left(\sum_{j=1}^{s_i} \frac{A_{i,j}}{(x - a_i)^j} \right) \tag{4.7.}$$

by finding the values of the numbers $A_{i,j}$. Note the numbers $a_1, \ldots, a_m$ are distin
and each factor $(x - a_i)^{s_i}$ of $q(x)$ produces the s_i summands

$$\frac{A_{i,1}}{x - a_i}, \quad \frac{A_{i,2}}{(x - a_i)^2}, \quad \ldots, \quad \frac{A_{i,s_i}}{(x - a_i)^{s_i}}$$

of $\frac{r(x)}{q(x)}$. If the polynomial $q(x)$ has degree n then $n = s_1 + \cdots + s_m$ and there a
n numbers $A_{i,j}$. To find the values of the $A_{i,j}$, add the fractions on the right sic
of equation (4.7.3) over their least common denominator $(x - a_1)^{s_1} \cdots (x - a_m)^{s_m}$
$q(x)$. This sum will have the form $\frac{h(x)}{q(x)}$. Equation (4.7.3) requires that $\frac{r(x)}{q(x)} = \frac{h(x)}{q(x)}$
Multiplying by $q(x)$, we want to solve the equation

$$r(x) = h(x) \tag{4.7.4}$$

for the n unknown numbers $A_{i,j}$. Note the coefficients of the polynomial $r(x)$ a
numbers, while the coefficients of the polynomial $h(x)$ are linear combinations of th
$A_{i,j}$. There are three alternative methods to solve for the values of the $A_{i,j}$.

Method 1   Since the polynomials $h(x)$ and $r(x)$ are equal, the coefficients of th
corresponding powers of x must be equal. Equating these coefficients produce
a system of n linear equations in the n unknowns $A_{i,j}$ which has a unique solutio
that can be found by methods from high school algebra.

Method 2   Since the polynomials $h(x)$ and $r(x)$ are equal, they must have equa
values when we substitute any number for x. Each such substitution produces
linear equation in the n unknowns $A_{i,j}$. Substituting n numbers for x produce
a linear system of n equations in n unknowns which we can solve. As we wi
see, substitution of $a_1, \ldots, a_m$ for x give the values of m of the $A_{i,j}$.

Method 3   This method is a variation of Method 2 when the values of some of th
exponents s_i are greater than one. First substitute $a_1, \ldots, a_m$ into (4.7.4) t
obtain the values of m of the $A_{i,j}$. Then substitute the values of these $A_{i,j}$ int
the corresponding m summands of (4.7.4), bring these summands to the left sid
of the equation, and divide the resulting equation by $(x - a_1) \cdots (x - a_m)$. Sub
stitute $x = a_i$ into this equation for those i with $s_i \geq 2$. Iterate this procedur
until the values of all the $A_{i,j}$ have been found.

We illustrate this procedure for integrating rational functions, using each of th
three methods of partial fractions. Of course, only one method needs to be used in
each example.

Examples 4.7.3 (1)   Evaluate $\int \frac{x^4 - 2}{x^4 - 5x^2 + 4} \, dx$.

Solution   In this example, the degree of the numerator and the denominator
are both four. Therefore Step I requires that we divide $x^4 - 5x^2 + 4$ into $x^4 - 2$:

$$x^4 - 2 = 1 \left(x^4 - 5x^2 + 4 \right) + \left(5x^2 - 6 \right).$$

Hence

$$\int \frac{x^4 - 2}{x^4 - 5x^2 + 4} \, dx = \int 1 + \frac{5x^2 - 6}{x^4 - 5x^2 + 4} \, dx = x + \int \frac{5x^2 - 6}{x^4 - 5x^2 + 4} \, dx. \tag{4.7.5}$$

Step II requires that we factor the denominator of the integrand:

$$x^4 - 5x^2 + 4 = \left(x^2 - 1\right)\left(x^2 - 4\right) = (x - 1)(x + 1)(x - 2)(x + 2).$$

In Step III the method of partial fractions finds numbers A, B, C, D with[1]:

$$\frac{5x^2 - 6}{x^4 - 5x^2 + 4} = \frac{A}{x - 1} + \frac{B}{x + 1} + \frac{C}{x - 2} + \frac{D}{x + 2}. \qquad (4.7.6)$$

We show how to find these four numbers by Methods 1 and 2. (Method 3 is not applicable here since none of the linear factors of the denominator are raised to a power greater than one.) Both methods begin by adding the fractions on the right side of equation (4.7.6):

$$\frac{5x^2 - 6}{(x - 1)(x + 1)(x - 2)(x + 2)} = \frac{\begin{array}{c} A(x + 1)(x - 2)(x + 2) + B(x - 1)(x - 2)(x + 2) \\ +C(x - 1)(x + 1)(x + 2) + D(x - 1)(x + 1)(x - 2) \end{array}}{(x - 1)(x + 1)(x - 2)(x + 2)}.$$

Multiply this equation by the denominator $(x - 1)(x + 1)(x - 2)(x + 2)$:

$$\begin{aligned} 5x^2 - 6 = {}& A(x + 1)(x - 2)(x + 2) + B(x - 1)(x - 2)(x + 2) \\ & + C(x - 1)(x + 1)(x + 2) + D(x - 1)(x + 1)(x - 2) \end{aligned} \qquad (4.7.7)$$

At this point, Methods 1 and 2 proceed differently.

Method 1

Do the algebra on the right side of (4.7.7) to determine the coefficient of each power of x:

$$\begin{aligned} 5x^2 - 6 \;=\; & (A + B + C + D)x^3 + (A - B + 2C - 2D)x^2 \\ & + (-4A - 4B - C - D)x + (-4A + 4B - 2C + 2D). \end{aligned}$$

Equate the coefficients of the corresponding powers of x to obtain the following linear system of four equations in four unknowns:

$$\begin{array}{rcrcrcrcrl} 0 & = & A & + & B & + & C & + & D & \text{(coefficients of } x^3) \\ 5 & = & A & - & B & + & 2C & - & 2D & \text{(coefficients of } x^2) \\ 0 & = & -4A & - & 4B & - & C & - & D & \text{(coefficients of } x) \\ -6 & = & -4A & + & 4B & - & 2C & + & 2D & \text{(constant terms)} \end{array}$$

Solve this linear system using your favorite high school method[2]:

$$A = \frac{1}{6}, \qquad B = -\frac{1}{6}, \qquad C = \frac{7}{6}, \qquad D = -\frac{7}{6}.$$

Method 2

We substitute the roots $x = 1$, $x = -1$, $x = 2$, $x = -2$ of $x^4 - 5x^2 + 4$ into equation (4.7.7).

Set $x = +1$ in (4.7.7): $-1 = -6A$ and $A = \frac{1}{6}$.

Set $x = -1$ in (4.7.7): $-1 = 6B$ and $B = -\frac{1}{6}$.

Set $x = +2$ in (4.7.7): $14 = 12C$ and $C = \frac{7}{6}$.

Set $x = -2$ in (4.7.7): $14 = -12D$ and $D = -\frac{7}{6}$.

Clearly this method is much easier than Method 1.

---

[1] In the notation of (4.7.3): $m = 4$, $s_1 = 1$, $a_1 = 1$, $A_{1,1} = A$, $s_2 = 1$, $a_2 = -1$, $A_{2,1} = B$, $s_3 = 1$, $a_3 = 2$, $A_{3,1} = C$, $s_4 = 1$, $a_4 = -2$ and $A_{4,1} = D$.

[2] The details of solving this linear system are tedious and have been omitted.

Having found the values of A, B, C, D by either Method 1 or Method 2, conclude with the integration of Step IV. By (4.7.5) and (4.7.6):

$$\int \frac{x^4 - 2}{x^4 - 5x^2 + 4}\, dx = x + \int \frac{1/6}{x - 1} - \frac{1/6}{x + 1} + \frac{7/6}{x - 2} - \frac{7/6}{x + 2}\, dx$$

$$= x + \frac{1}{6}\ln|x - 1| - \frac{1}{6}\ln|x + 1| + \frac{7}{6}\ln|x - 2| - \frac{7}{6}\ln|x + 2| + C.$$

(2) Evaluate $\int \frac{1}{x^5 - 2x^4 + x^3}\, dx$.

Solution In this example the numerator has degree zero and the denominator has degree five. Hence the polynomial division of Step I is not necessary. We proceed to Step II and factor the denominator of the integrand:

$$x^5 - 2x^4 + x^3 = x^3\left(x^2 - 2x + 1\right) = x^3 (x - 1)^2.$$

In Step III we use the method of partial fractions to write:

$$\frac{1}{x^5 - 2x^4 + x^3} = \frac{A}{x} + \frac{B}{x^2} + \frac{C}{x^3} + \frac{D}{x - 1} + \frac{E}{(x - 1)^2}. \qquad (4.7.8$$

Note x determines three summands because x^3 is a factor of the denominator while $x - 1$ determines two summands because $(x - 1)^2$ is a factor of the denominator.[3] We show how to use Methods 1, 2 and 3 to find A, B, C, D, E. Begin by adding the fractions on the right side of equation (4.7.8):

$$\frac{1}{x^3(x - 1)^2} = \frac{Ax^2(x - 1)^2 + Bx(x - 1)^2 + C(x - 1)^2 + Dx^3(x - 1) + Ex^3}{x^3(x - 1)^2}.$$

Multiply this equation by the denominator $x^3(x - 1)^2$:

$$1 = Ax^2(x - 1)^2 + Bx(x - 1)^2 + C(x - 1)^2 + Dx^3(x - 1) + Ex^3. \qquad (4.7.9$$

At this point the three methods proceed differently.

Method 1

Do the algebra on the right side of equation (4.7.9), grouping the coefficients of each power of x:

$$1 = (A + D)x^4 + (-2A + B - D + E)x^3 + (A - 2B + C)x^2 + (B - 2C)x + C$$

Equate the corresponding coefficients of the polynomials on the left and right sides of this equation to obtain a linear system of five equations in five unknowns:

$0 =$	A			$+\ D$	(coefficients of x^4)
$0 =$	$-2A\ +$	B		$-\ D\ +\ E$	(coefficients of x^3)
$0 =$	$A\ -$	$2B\ +$	C		(coefficients of x^2)
$0 =$		$B\ -$	$2C$		(coefficients of x)
$1 =$			C		(constant terms)

Use any high school method to solve this linear system:

$$A = 3, \quad B = 2, \quad C = 1, \quad D = -3, \quad E = 1.$$

---

[3] In the notation of (4.7.3): $m = 5$, $s_1 = 3$, $a_1 = 0$, $A_{1,1} = A$, $A_{1,2} = B$, $A_{1,3} = C$, $s_2 = 2$, $a_2 = 1$, $A_{2,1} = D$ and $A_{2,2} = E$.

Methods 2 and 3

Substitute the roots $x = 0$ and $x = 1$ of $x^5 - 2x^4 + x^3$ into equation (4.7.9).
Set $x = 0$ in (4.7.9): $1 = C$.
Set $x = 1$ in (4.7.9): $1 = E$.

Method 2

Substitute $x = -1$, $x = 2$ and $x = -2$ in (4.7.9) to obtain a linear system of three equations in the remaining three unknowns:

$$
\begin{array}{rcrcrcr}
-2 &=& 4A &-& 4B &+& 2D \\
-8 &=& 4A &+& 2B &+& 8D \\
0 &=& 36A &-& 18B &+& 24D
\end{array}
$$

Solve this linear system: $A = 3$, $B = 2$, $D = -3$.

Method 3

Substitute $C = 1$ and $E = 1$, as found above, into equation (4.7.9):

$$1 = Ax^2(x-1)^2 + Bx(x-1)^2 + (x-1)^2 + Dx^3(x-1) + x^3.$$

Bring the two summands which do not involve A, B, D to the left side of the equation and divide the resulting equation by $x(x-1)$:

$$
\begin{array}{rcl}
1 - x^3 - (x-1)^2 &=& Ax^2(x-1)^2 + Bx(x-1)^2 + Dx^3(x-1) \\
(x-1)\left[-(1+x+x^2) - (x-1)\right] &=& x(x-1)\left[Ax(x-1) + B(x-1) + Dx^2\right] \\
x(x-1)(-x-2) &=& x(x-1)\left[Ax(x-1) + B(x-1) + Dx^2\right] \\
-x - 2 &=& Ax(x-1) + B(x-1) + Dx^2 \quad (4.7.10)
\end{array}
$$

Set $x = 0$ in equation (4.7.10): $-2 = 0 - B + 0 = -B$ and $B = 2$.
Set $x = 1$ in equation (4.7.10): $-3 = 0 + 0 + D$ and $D = -3$.
Substitute the values $B = 2$, $D = -3$ into equation (4.7.10), bring the terms not involving A to the left side of the equation, and divide the resulting equation by $x(x-1)$:

$$
\begin{array}{rcl}
-x - 2 &=& Ax(x-1) + 2(x-1) - 3x^2 \\
-x - 2 - 2(x-1) + 3x^2 &=& Ax(x-1) \\
3x(x-1) &=& Ax(x-1) \\
3 &=& A
\end{array}
$$

Substitute $A = 3$, $B = 2$, $C = 1$, $D = -3$, $E = 1$ into equation (4.7.8), and perform the integration of Step IV:

$$
\begin{aligned}
\int \frac{1}{x^5 - 2x^4 + x^3}\, dx &= \int \frac{3}{x} + \frac{2}{x^2} + \frac{1}{x^3} - \frac{3}{x-1} + \frac{1}{(x-1)^2}\, dx \\
&= 3\ln|x| - \frac{2}{x} - \frac{1}{2x^2} - 3\ln|x-1| - \frac{1}{x-1} + C. \quad \square
\end{aligned}
$$

After studying the preceding examples, it is worthwhile to reread the description of the four steps for integrating a rational function at the end of the first subsection as well as the description of the three methods of partial fractions at the beginning of this subsection.

Rational Functions $\frac{p(x)}{q(x)}$ where $q(x)$ has Complex Roots

When $q(x)$ has complex roots, we integrate $f(x) = \frac{p(x)}{q(x)}$ by the same four step procedure as when all the roots are real. However, the factorization, partial fractions and integration are more difficult.

Step I (Division) If the degree of the numerator $p(x)$ is greater than or equal the degree of the denominator $q(x)$, divide $q(x)$ into $p(x)$ to write:

$$f(x) = \frac{p(x)}{q(x)} = k(x) + \frac{r(x)}{q(x)}.$$

where $k(x)$ and $r(x)$ are polynomials with degree $r(x)$ less than degree $q(x)$.

Step II (Factorization) Factor $q(x)$ into linear and quadratic factors:

$$q(x) = e\,(x - a_1)^{s_1} \cdots (x - a_m)^{s_m} \left(x^2 + b_1 x + c_1\right)^{t_1} \cdots \left(x^2 + b_n x + c_n\right)^{t_n}.$$

These linear and quadratic polynomials are distinct, and the quadratic polyne mials have no real roots.

Step III (Partial Fractions) Write $\frac{r(x)}{q(x)}$ as the sum of the simple fractions

$$\frac{A_{i,1}}{x - a_i}, \quad \frac{A_{i,2}}{(x - a_i)^2}, \quad \cdots, \quad \frac{A_{i,s_i}}{(x - a_i)^{s_i}},$$

$$\frac{B_{j,1} x + C_{j,1}}{x^2 + b_j x + c_j}, \quad \frac{B_{j,2} x + C_{j,2}}{(x^2 + b_j x + c_j)^2}, \quad \cdots, \quad \frac{B_{j,t_j} x + C_{j,t_j}}{\left(x^2 + b_j x + c_j\right)^{t_j}} \tag{4.7.11}$$

for $1 \le i \le m$ and $1 \le j \le n$.

Step IV (Integration) Integrate the rational function $f(x)$, using its representa tion in Steps I and III as a sum of functions which we know how to integrate.

Note that we skip the polynomial division of Step I when the degree of $p(x)$ is les than the degree of $q(x)$.

The Fundamental Theorem of Algebra says that it is always possible to factor th polynomial $q(x)$ as required in Step II. However, there is no algorithm for carryin out this factorization when the degree of $q(x)$ is greater than four. The linear factor correspond to the real roots of $q(x)$. The quadratic factors correspond to the comple roots of $q(x)$ which always occur in conjugate pairs. In our examples, this factorizatio will either be easy or given.

The method of partial fractions in Step III begins by summing the fractions of (4.7.11 over their common denominator $q(x)$. We want this sum $\frac{h(x)}{q(x)}$ to equal $\frac{r(x)}{q(x)}$, i.e.

$$r(x) = h(x).$$

The coefficients of $r(x)$ are numbers while the coefficients of $h(x)$ are linear combi nations of the A_i, B_j, C_j. Either Method 1 or 2 of the preceding subsection can b used to solve for the A_i, B_j, C_j. Let N be the degree of $q(x)$. In Method 1 we equate the corresponding coefficients of $r(x)$ and $h(x)$. This produces a system of N linear equations in the N unknowns A_i, B_j, C_j which we solve. In Method 2 we substitute N values of x to obtain N linear equations in the N unknowns A_i, B_j, C_j which we solve. We always choose $a_1, \ldots, a_m$ as m of these values of x to obtain the values of m of the A_i. Unfortunately, there are no real values of x which directly give the values of any of the B_j or C_j.

In Step IV we integrate each of the $\frac{A_i}{(x - a_i)^k}$ as in (4.7.2). The evaluation of the

$$\int \frac{Bx + C}{(x^2 + bx + c)^k} \, dx$$

is more complicated. First complete the square:

$$x^2 + bx + c = (x + \alpha)^2 + \beta^2$$

with $\alpha = \frac{b}{2}$ and $\beta^2 = c - \frac{b^2}{4}$. (The latter number is positive because this quadratic polynomial has no real roots). Substitute $u = x + \alpha$ with $x = u - \alpha$ and $du = dx$:

$$\int \frac{Bx + C}{(x^2 + bx + c)^k}\, dx = \int \frac{Bu + (C - B\alpha)}{(u^2 + \beta^2)^k}\, du$$

$$= \frac{B}{2} \int \frac{2u}{(u^2 + \beta^2)^k}\, du + (C - B\alpha) \int \frac{1}{(u^2 + \beta^2)^k}\, du.$$

The first integral is evaluated by the substitution $v = u^2 + \beta^2$ with $dv = 2u\, du$. To evaluate the second integral, make the reverse trigonometric substitution:

$$u = \beta \tan \theta \text{ with } du = \beta \sec^2 \theta\, d\theta \text{ and}$$
$$u^2 + \beta^2 = (\beta \tan \theta)^2 + \beta^2 = \beta^2 \sec^2 \theta.$$

The second integral is transformed into the integral of an even power of $\cos \theta$. In Section 3.6 we evaluated these integrals by iterated use of the trigonometric identity:

$$\cos^2 \phi = \frac{1}{2}(1 + \cos 2\phi)$$

for ϕ equal to appropriate multiples of θ. This rewrites the original integral as a sum of integrals of odd powers of various cosines. Each of these integrals $\int \cos^{2e+1} \psi\, d\psi$ is evaluated by the substitution:

$$w = \sin \psi \text{ with } \cos^2 \psi = 1 - \sin^2 \psi = 1 - w^2 \text{ and } dw = \cos \psi\, d\psi.$$

The resulting integrand is a polynomial in w.

This four step procedure for integrating rational functions whose denominators have complex roots is illustrated in the following examples.

Examples 4.7.4 (1) Evaluate $\int \frac{x^4 + 8x^3 + 17x^2}{x^4 + 10x^2 + 9}\, dx$.

Solution Perform the polynomial division of Step I:

$$x^4 + 8x^3 + 17x^2 = 1(x^4 + 10x^2 + 9) + 8x^3 + 7x^2 - 9 \text{ and}$$
$$\int \frac{x^4 + 8x^3 + 17x^2}{x^4 + 10x^2 + 9}\, dx = \int 1 + \frac{8x^3 + 7x^2 - 9}{x^4 + 10x^2 + 9}\, dx$$

$$= x + \int \frac{8x^3 + 7x^2 - 9}{x^4 + 10x^2 + 9}\, dx. \qquad (4.7.12)$$

Factor the denominator as required by Step 2:

$$x^4 + 10x^2 + 9 = (x^2 + 1)(x^2 + 9).$$

Now use partial fractions to write

$$\frac{8x^3 + 7x^2 - 9}{x^4 + 10x^2 + 9} = \frac{Ax + B}{x^2 + 1} + \frac{Cx + D}{x^2 + 9}.$$

We use Method 1 to solve for A, B, C, D. Add the two fractions on the right side of this equation:

$$\frac{8x^3 + 7x^2 - 9}{(x^2 + 1)(x^2 + 9)} = \frac{(Ax + B)(x^2 + 9) + (Cx + D)(x^2 + 1)}{(x^2 + 1)(x^2 + 9)}. \qquad (4.7.13)$$

Multiply this equation by $(x^2 + 1)(x^2 + 9)$:

$$
\begin{aligned}
8x^3 + 7x^2 - 9 &= (Ax + B)(x^2 + 9) + (Cx + D)(x^2 + 1) \\
&= (A + C)x^3 + (B + D)x^2 + (9A + C)x + (9B + D).
\end{aligned}
$$

Equate the corresponding coefficients of these polynomials to produce the linea. system:

$$
\begin{array}{lllllll}
8 &=& A &+& & C & & \text{(coefficients of } x^3) \\
7 &=& & & B &+ D & & \text{(coefficients of } x^2) \\
0 &=& 9A & & + C & & & \text{(coefficients of } x) \\
-9 &=& & & 9B &+ D & & \text{(constant terms)}
\end{array}
$$

Solve the first and third equations: $A = -1$, $C = 9$. Solve the second and fourt equations: $B = -2$, $D = 9$. By (4.7.12) and (4.7.13):

$$
\int \frac{x^4 + 8x^3 + 17x^2}{x^4 + 10x^2 + 9}\, dx = x + \int \frac{-x - 2}{x^2 + 1}\, dx + \int \frac{9x + 9}{x^2 + 9}\, dx
$$

$$
= x - \frac{1}{2} \int \frac{2x}{x^2 + 1}\, dx - 2 \int \frac{1}{x^2 + 1}\, dx + \frac{9}{2} \int \frac{2x}{x^2 + 9}\, dx + 9 \int \frac{1}{x^2 + 9}\, dx
$$

$$
= x - \frac{1}{2} \ln\left(x^2 + 1\right) - 2\arctan x + \frac{9}{2} \ln\left(x^2 + 9\right) + 3\arctan \frac{x}{3} + C.
$$

(2) Evaluate $\int \frac{64}{(x-1)(x^2+2x+5)^2}\, dx$.

Solution Since the degree of the numerator is zero, the polynomial division c Step 1 is not necessary. Since the denominator is already factored, Step 2 ha been done for us. Begin Step 3 by writing

$$
\frac{64}{(x-1)(x^2+2x+5)^2} = \frac{A}{x-1} + \frac{Bx+C}{x^2+2x+5} + \frac{Dx+E}{(x^2+2x+5)^2} \tag{4.7}
$$

$$
= \frac{A(x^2+2x+5)^2 + (Bx+C)(x-1)(x^2+2x+5) + (Dx+E)(x}{(x-1)(x^2+2x+5)^2}
$$

Note $x^2 + 2x + 5$ determines two summands because $(x^2 + 2x + 5)^2$ is a facto. of the denominator. Multiply the preceding equation by $(x-1)(x^2+2x+5)^2$:

$$
64 = A(x^2+2x+5)^2 + (Bx+C)(x-1)(x^2+2x+5) + (Dx+E)(x-1)
$$

Solve for A by substituting $x = 1$: $64 = 64A$ and $A = 1$. We find the values of B, C, D, E by Method 1. Collect the coefficients of each power of x on the right side of the preceding equation:

$$
\begin{aligned}
64 &= (1 + B)\, x^4 + (4 + B + C)\, x^3 + (14 + 3B + C + D)\, x^2 \\
&\quad + (20 - 5B + 3C - D + E)\, x + (25 - 5C - E).
\end{aligned}
$$

Equate the corresponding coefficients of the polynomials on the left and right sides of this equation to produce the linear system:

$$
\begin{array}{lllllllll}
0 &=& 1 &+& B & & & & \text{(coefficients of } x^4) \\
0 &=& 4 &+& B &+ C & & & \text{(coefficients of } x^3) \\
0 &=& 14 &+& 3B &+ C &+ D & & \text{(coefficients of } x^2) \\
0 &=& 20 &-& 5B &+ 3C &- D &+ E & \text{(coefficients of } x) \\
64 &=& 25 & & &- 5C & &- E & \text{(constant terms)}
\end{array}
$$

From the first equation, $B = -1$. From the second equation, $C = -3$. From the third equation, $D = -8$. From the last equation, $E = -24$. By (4.7.14):

$$\int \frac{64}{(x-1)(x^2+2x+5)^2}\,dx = \int \frac{1}{x-1} + \frac{-x-3}{x^2+2x+5} + \frac{-8x-24}{(x^2+2x+5)^2}\,dx$$

$$= \ln|x-1| + \int \frac{-x-3}{(x+1)^2+4} + \frac{-8x-24}{[(x+1)^2+4]^2}\,dx \ .$$

Substitute $u = x + 1$ with $x = u - 1$ and $dx = du$:

$$\int \frac{64}{(x-1)(x^2+2x+5)^2}\,dx = \ln|x-1| + \int \frac{-(u-1)-3}{u^2+4} + \frac{-8(u-1)-24}{(u^2+4)^2}\,du$$

$$= \ln|x-1| - \frac{1}{2}\int \frac{2u}{u^2+4} - 2\frac{1}{u^2+4}\left(-4\frac{2u}{(u^2+4)^2}\,du - 16\int \frac{1}{(u^2+4)^2}\,du\right.$$

$$= \ln|x-1| - \frac{1}{2}\ln(u^2+4) - \arctan\frac{u}{2} + \frac{4}{u^2+4} - 16\int \frac{1}{(u^2+4)^2}\,du. \qquad (4.7.15)$$

In the remaining integral, make the reverse trigonometric substitution $u = 2\tan\theta$ with $du = 2\sec^2\theta\,d\theta$ and $u^2 + 4 = 4\tan^2\theta + 4 = 4\sec^2\theta$:

$$\int \frac{1}{(u^2+4)^2}\,du = \int \frac{1}{(4\sec^2\theta)^2}2\sec^2\theta\,d\theta = \frac{1}{8}\int \frac{1}{\sec^2\theta}\,d\theta$$

$$= \frac{1}{8}\int \cos^2\theta\,d\theta = \frac{1}{8}\int \frac{1}{2}(1+\cos 2\theta)\,d\theta$$

$$= \frac{1}{16}\theta + \frac{1}{32}\sin 2\theta + C = \frac{1}{16}\theta + \frac{1}{16}\sin\theta\cos\theta + C.$$

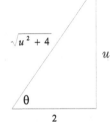

Figure 4.7.5

$\tan\theta = \frac{u}{2}$

Observe that $\tan\theta = \frac{u}{2}$. Construct a triangle in Figure 4.7.5 with angle θ by letting the opposite side have length u and the adjacent side have length 2. By the Pythagorean Theorem, the hypotenuse has length $\sqrt{u^2+4}$. Hence

$$\int \frac{1}{(u^2+4)^2}\,du = \frac{1}{16}\arctan\frac{u}{2} + \frac{1}{16}\frac{u}{\sqrt{u^2+4}}\frac{2}{\sqrt{u^2+4}} + C = \frac{1}{16}\arctan\frac{u}{2} + \frac{u}{8(u^2+4)} + C.$$

Substitute the value of this integral into (4.7.15), and replace u by $x + 1$:

$$\int \frac{64}{(x-1)(x^2+2x+5)^2}\,dx$$

$$= \ln|x-1| - \frac{1}{2}\ln(u^2+4) - \arctan\frac{u}{2} + \frac{4}{u^2+4} - \arctan\frac{u}{2} - \frac{2u}{u^2+4} + C$$

$$= \ln|x-1| - \frac{1}{2}\ln(u^2+4) - 2\arctan\frac{u}{2} + \frac{4-2u}{u^2+4} + C$$

$$= \ln|x-1| - \frac{1}{2}\ln(x^2+2x+5) - 2\arctan\frac{x+1}{2} + \frac{2-2x}{x^2+2x+5} + C \quad \square$$

Historical Remarks

Solving equations has always been recognized as a fundamental problem of mathematics. In Babylon ca. 1700 BCE, in Euclid's *Elements* ca. 300 BCE and in the Chinese *Jiuzhang Suanshu* ca. 200 BCE many quadratic equations were solved for a positive root. The methods used were geometric, originating from proofs of the Pythagorean theorem. (Yes, the Pythagorean Theorem was known over a thousand years before Pythagoras!) The quadratic formula for finding one root of a quadratic equation with real positive roots was found by the Indian mathematician Brahmuagupta in 628. The full formula for two positive roots was found by the Iraqi mathematician Muhammad

ibn Musa al-Khwarizmi in 825. Indian mathematicians were aware of negative num
bers, and Bhaskara II in 1150 used the full quadratic formula. However he discarde
negative solutions as being "incongruous" because "people do not approve a negati
absolute number."

The Persian mathematician Sharaf al-Din al Tusii ca. 1200 found the positi
roots of cubic equations of the form $x^3 - ax^2 + d = 0$ with $a, d > 0$ as follows. Th
minimum value of $f(x) = x^3 - ax^2$ occurs at $x = \frac{2a}{3}$. If $f\left(\frac{2a}{3}\right) > -d$, i.e. $d > \frac{4a}{2}$
there will be no solution. If $f\left(\frac{2a}{3}\right) = -d$, i.e. $d = \frac{4a^3}{27}$, there will be one solution.
$f\left(\frac{2a}{3}\right) < -d$, i.e. $d < \frac{4a^3}{27}$, there will be two solutions. See Figure 4.7.6. When the
are solutions, he finds them by geometric methods.

Algebraic methods for solving equations were developed in Italy. In 1344, Maestr
Dardi of Pisa published methods for solving 198 types of equations of degree at mos
four. An algebraic solution of the cubic $x^3 + cx = d$ was found by Scipione del Ferro i
1515, and a solution of the cubic $x^3 + bx^2 = d$ was found by Niccolo Tartaglia in 153
The solution to the general cubic equation, based on these methods, was publishe
by Gerolamo Cardano in 1545. He also published the solution to the general quart
equation which was found by his student Lodvico Ferrari. These solutions include
consideration of negative and complex roots. (For the next 250 years complex solutior
were described as "useless" or "impossible.")

In 1629, Albert Girard stated, without proof, the Fundamental Theorem of A
gebra: a polynomial of degree n has n roots (counting repetitions). In additior
he identified the coefficients of a polynomial as symmetric polynomials of its roots
In 1746, Jean Baptiste d'Alembert showed that complex roots of polynomials occu
in conjugate pairs. He also gave a proof of the Fundamental Theorem of Algebra
However it contained a lemma with an unsatisfactory proof. A complete proof c
d'Alembert's Lemma was given by V. A. Puiseux in 1850. Carl Friedrich Gauss pub
lished several proofs of the Fundamental Theorem of Algebra which used the geometri
interpretation of complex numbers. His 1799 proof, at the age of 22, was his Ph.D
thesis. It contained a gap which Gauss claimed was intuitively clear. However, i
was only in 1920 that the first rigorous proof of the point in question was given b
A. Ostrowski! Gauss' proof of 1816 relied on the Intermediate Value Theorem which
although intuitively clear, was only proved by Bolzano several years later.

For 250 years mathematicians searched for formulas for the roots of the genera
polynomials of degrees greater than four. In 1798, Paola Ruffini gave an incomplete
proof that the general polynomial of degree five is not *solvable by radicals*: there is n
general formula for its roots which involves only addition, subtraction, multiplication
division and roots. A complete proof, using ideas of Gauss, was given by Niels Henril
Abel in 1824 at the age of 22. Multiplication by x^{k-5} then shows that the genera
equation of degree k cannot be solved by radicals for all $k \geq 5$. In 1829, Evariste
Galois tried unsuccessfully to publish his theory of equations at the age of 17, three
years before his death in a duel. His manuscript contained three of the fundamenta
concepts of abstract algebra: groups, fields and the group associated to a polynomial
He showed how to use these concepts to determine which polynomial equations car
be solved by radicals. His ideas were not understood and largely ignored. It was not
until 1870 that Camille Jordan gave a clear exposition of Galois' ideas, which have
become the cornerstone for understanding the roots of polynomials.

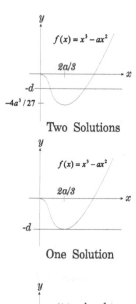

Two Solutions

One Solution

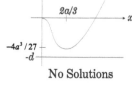

No Solutions

Figure 4.7.6
$x^3 - ax^2 = -d$ $(a, d > 0)$

Summary

The reader should know how to use the four step procedure to integrate rationa
functions: polynomial division, factorization of the denominator, the method of partia
fractions and integration.

Basic Exercises

1. Perform polynomial division, if necessary, to write each rational function as $k(x) + \frac{r(x)}{q(x)}$ with degree $r(x)$ less than degree $q(x)$.

(a) $\frac{x^2}{x^2-5x+8}$ (b) $\frac{x^3+5x-2}{x^5-4x^2+7}$ (c) $\frac{x^5-3x^4+8x-2}{x^4+1}$ (d) $\frac{x^2-5x+3}{x-2}$

(e) $\frac{x^4}{x^3-x-1}$ (f) $\frac{x}{x+6}$ (g) $\frac{x^{11}-7x^9-4x^8+7x^6+5x^2-3x+9}{x}$ (h) $\frac{x^6-3x^4+x-6}{x^3+x+1}$

(i) $\frac{x^7-x+8}{x^2+3x-2}$ (j) $\frac{x^4+x^2+1}{x^2-3x+5}$ (k) $\frac{x^8+1}{x^9+x^7+x^5+x^3+x+1}$ (l) $\frac{x^8-x^6+x^4-x^2+1}{x-1}$

2. Factor each polynomial into linear factors and quadratic factors with complex roots. Recall that integer roots of a monic polynomial with integer coefficients $p(x) = x^n + a_{n-1}x^{n-1} + \cdots + a_1 x + a_0$ are divisors of its constant term a_0.

(a) $x^2 - 2x - 15$ (b) $x^3 - 2x^2 - 5x + 6$ (c) $x^3 + 3x^2 + 3x + 1$

(d) $x^6 - 3x^4 + 3x^2 - 1$ (e) $x^3 - 1$ (f) $x^5 - 4x^4 + x^3 + 10x^2 - 4x - 8$

(g) $x^3 - 2x^2 - 5x + 6$ (h) $x^4 - 13x^2 + 36$ (i) $x^4 - 3x^3 - 3x^2 + 11x - 6$

(j) $x^4 + 5x^2 + 6$ (k) $x^3 - 2x + 4$ (l) $x^4 + 2x^3 + 3x^2 + 2x + 1$

3. Use the method of partial fractions to write each rational function as a sum of simple rational functions.

(a) $\frac{x}{x^2-9}$ (b) $\frac{1}{x^2-x-6}$ (c) $\frac{x-21}{x^2+3x-18}$

(d) $\frac{9x^3-16x^2+21x-20}{x^4-4x^3}$ (e) $\frac{x^4-9x^3+7x^2+48x+13}{(x-2)^2(x+1)^3}$ (f) $\frac{6x^5+24x^4+16x^3-22x^2+8x+8}{x^6+2x^5+4x^4}$

(g) $\frac{5x^2+2x+3}{x^3+x^2+x}$ (h) $\frac{6x^2-26x+32}{(x-3)(x^2-3x+4)}$ (i) $\frac{x^5+4x^4+37x^3+54x^2+300x+130}{(x^2+9)^2(x^2+25)}$

(j) $\frac{6x^3-2x^2+31x-12}{(x^2+2x+3)(x^2-x+5)}$ (k) $\frac{2x^3-2x^2-19x-16}{(x+2)^2(x^2-x-5)}$ (l) $\frac{2x^7-x^6+11x^5-9x^4+18x^3-19x^2-3x+4}{(x^2+1)^3(x^2+4)}$

4. Integrate each rational function.

(a) $\int \frac{3}{x-4}\,dx$ (b) $\int \frac{5}{(x+2)^3}\,dx$ (c) $\int \frac{7}{(x-4)^6}\,dx$ (d) $\int \frac{4}{x+3}\,dx$

(e) $\int \frac{6x+7}{x^2+9}\,dx$ (f) $\int \frac{4x-3}{x^2+25}\,dx$ (g) $\int \frac{2x+5}{x^2+2x+10}\,dx$ (h) $\int \frac{8x-7}{x^2+6x+25}\,dx$

(i) $\int \frac{2x+1}{(x^2+4)^2}\,dx$ (j) $\int \frac{4x-9}{(x^2+1)^3}\,dx$ (k) $\int \frac{6x+3}{(x^2+4x+13)^2}\,dx$ (l) $\int \frac{4x-8}{(x^2+6x+25)^3}\,dx$

Evaluate each of the following 20 integrals.

5. $\int \frac{x}{x^2-1}\,dx$ **6.** $\int \frac{3x-5}{x^2-3x-10}\,dx$ **7.** $\int \frac{x^4}{x^2-16}\,dx$

8. $\int \frac{x^2+3}{x^2-x-6}\,dx$ **9.** $\int \frac{6x+1}{x^3-x^2-x+1}\,dx$ **10.** $\int \frac{2x^3-x^2+5}{x^4+4x^3+3x^2-4x-4}\,dx$

11. $\int \frac{x^3+8x^2-2}{x^4-5x-6}\,dx$ **12.** $\int \frac{x^3+2x^2+4}{x^4-2x^3+2x-1}\,dx$ **13.** $\int \frac{x^4+10x^3+16x^2+x}{x^5+2x^4-2x^3-8x^2-7x-2}\,dx$

14. $\int \frac{x^2-5x+7}{x^4-6x^2+8}\,dx$ **15.** $\int \frac{x^2-4}{x^3+9x}\,dx$ **16.** $\int \frac{x^3}{x^3+x^2-x-1}\,dx$

17. $\int \frac{x^4+16}{x^4-16}\,dx$ **18.** $\int \frac{x^2-x+2}{x^3+4x^2+13x}\,dx$ **19.** $\int \frac{x^3+2x-3}{x^3+7x^2+31x+25}\,dx$

20. $\int \frac{x^2+3x-4}{x^4+13x^2+36}\,dx$ **21.** $\int \frac{x^3-2x^2+4}{x^5+8x^3+16x}\,dx$ **22.** $\int \frac{3x^4+19x^3+140x^2+308x+785}{(x+2)(x^2+4x+29)^2}\,dx$

23. $\int \frac{-2x^4+12x^3+6x^2+297x+141}{(x-4)(x^2+2x+17)^2}\,dx$ **24.** $\int \frac{x^8-2x^7+3x^6-3x^5+6x^4+x^3+9x^2-3x+1}{x^9+4x^7+6x^5+4x^3+x}\,dx$

25. Find the area of the region bounded by $y = \frac{80}{9-x^2}$ and $y = 12 - 2x$.

26. Find the area of the region bounded by $y = \frac{100}{(x^2+2x+2)^2}$ and $y = 7 - 3x$ for $1 \le x \le 2$.

27. Find the volume of the solid of revolution obtained by rotating the region bounded by $y = \frac{x}{x^2-9}$, the x–axis and the line $x = 2$ around the x–axis.

28. Consider the region R bounded by $y = \frac{x}{x^2+1}$, the x–axis and the line $x = 1$.
(a) Find the volume of the solid of revolution obtained by rotating the region R around the x–axis.
(b) Find the volume of the solid of revolution obtained by rotating the region R around the y–axis.

The following three exercises are for those who have studied Section 3.9

29. Find the distance traveled by each object with the given velocity $v(t)$.

(a) $v(t) = \frac{t^2-1}{t^4-81}$ miles per hour from $t = -2$ to $t = 2$ hours.

(b) $v(t) = \frac{t^4-16}{t^3-125}$ cm per second from $t = -3$ seconds to $t = 4$ seconds.

30. Find the mass of the two foot long rod of density $\rho(x) = \frac{x^2+3}{x^4+x^2}$ pounds per fo⸱ for $1 \leq x \leq 3$ feet.

31. How much work is done when the force $F(x) = \frac{7x^3+2x^2+31x+18}{x^4+10x^2+9}$ Newtons mov⸱ an object from $x = 0$ to $x = 2$ meters?

The following two exercises are for those who have studied Section 3.10.

32. Find the center of mass of each rod of density $\rho(x)$.

(a) $\rho(x) = \frac{x}{x^2-x-12}$ pounds per foot for $0 \leq x \leq 2$ feet.

(b) $\rho(x) = \frac{1}{x^4+x^3}$ pounds per foot for $1 \leq x \leq 4$ feet.

33. Find the centroid of each region.

(a) R is the region bounded by $y = \frac{x}{4-x^2}$, $x = 1$, the x–axis and the y–axis.

(b) S is the region bounded by $y = \frac{1}{x^3+9x}$, $x = 1$, $x = 2$ and the x–axis.

4.8   Rationalizing Substitutions

In the preceding section, the method of partial fractions was used to integrate ra⸱ tional functions $\frac{p(x)}{q(x)}$ where $p(x)$ and $q(x)$ are polynomials in x. These functions ar⸱ constructed from x using the arithmetic operations of addition, subtraction, mul⸱ tiplication and division. In this section, we integrate rational functions which ar⸱ constructed from either trigonometric functions or e^x or fractional powers of x b⸱ addition, subtraction, multiplication and division. In all three cases, an appropriat⸱ change of variables transforms the given integral into a rational function of a nev⸱ variable u which can be integrated by the method of partial fractions. A substitio⸱ which transforms an integral in x into the integral of a rational function of a nev⸱ variable u is called a *rationalizing substitution*.

Rational Functions of Trigonometric Functions

In this subsection, we integrate rational functions of trigonometric functions. Thes⸱ functions are constructed from the six trigonometric functions of θ using the operations of addition, subtraction, multiplication and division. When an integrand has this form the integral is evaluated in three steps.

Step 1   Write the integrand as a rational function of $\sin\theta$ and $\cos\theta$ by using the definitions of the other four trigonometric functions:

$$\tan\theta = \frac{\sin\theta}{\cos\theta}, \quad \cot\theta = \frac{\cos\theta}{\sin\theta}, \quad \sec\theta = \frac{1}{\cos\theta} \quad \text{and} \quad \csc\theta = \frac{1}{\sin\theta}.$$

Step 2   Make the rationalizing substitution given in the proposition below.

Step 3   Integrate the resulting rational function by the method of partial fractions.

Proposition 4.8.1 *The change of variables*

$$x = \tan\frac{\theta}{2}$$

is a rationalizing substitution which changes a rational function of $\sin\theta$ and $\cos\theta$ into a rational function of x. In particular,

$$d\theta = \frac{2}{1+x^2}\,dx, \qquad \sin\theta = \frac{2x}{1+x^2}, \qquad \cos\theta = \frac{1-x^2}{1+x^2}.$$

Proof $x = \tan\frac{\theta}{2}$ means $\frac{\theta}{2} = \arctan x$, and $\theta = 2\arctan x$. Hence

$$\frac{d\theta}{dx} = \frac{2}{1+x^2} \quad \text{and} \quad d\theta = \frac{2}{1+x^2}\,dx.$$

Construct a triangle with angle $\frac{\theta}{2}$ in Figure 4.8.2 by letting the opposite side have length x and the adjacent side have length 1. By the Pythagorean Theorem, the hypotenuse has length $\sqrt{1+x^2}$. From this triangle, we see that:

$$\sin\frac{\theta}{2} = \frac{x}{\sqrt{1+x^2}} \quad \text{and} \quad \cos\frac{\theta}{2} = \frac{1}{\sqrt{1+x^2}}.$$

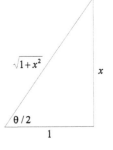

Figure 4.8.2
$x = \tan\frac{\theta}{2}$

By the double angle formulas:

$$\sin\theta = \sin 2\left(\frac{\theta}{2}\right) = 2\sin\frac{\theta}{2}\cos\frac{\theta}{2} = 2\frac{x}{\sqrt{1+x^2}}\frac{1}{\sqrt{1+x^2}} = \frac{2x}{1+x^2},$$

$$\cos\theta = \cos 2\left(\frac{\theta}{2}\right) = \cos^2\frac{\theta}{2} - \sin^2\frac{\theta}{2} = \frac{1}{1+x^2} - \frac{x^2}{1+x^2} = \frac{1-x^2}{1+x^2}. \qquad \square$$

The examples below illustrate the three step procedure for integrating rational functions of trigonometric functions.

Examples 4.8.3 (1) Evaluate $\int \frac{1}{\sin\theta+\cos\theta-1}\,d\theta$.

Solution Let $x = \tan\frac{\theta}{2}$. By Proposition 4.8.1:

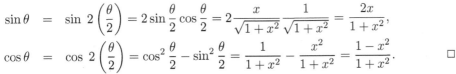

$$\int \frac{1}{\sin\theta+\cos\theta-1}\,d\theta = \int \frac{1}{\frac{2x}{1+x^2}+\frac{1-x^2}{1+x^2}-1}\cdot\frac{2}{1+x^2}\,dx$$

$$= \int \frac{2}{2x+(1-x^2)-(1+x^2)}\,dx = \int \frac{2}{2x-2x^2}\,dx = \int \frac{-1}{x^2-x}\,dx.$$

Note $x^2 - x = x(x-1)$. We use the method of partial fractions to write

$$\frac{-1}{x^2-x} = \frac{A}{x} + \frac{B}{x-1} = \frac{A(x-1)+Bx}{x^2-x}.$$

Multiply this equation by $x^2 - x$:

$$-1 = A(x-1) + Bx.$$

Substitute $x = 0$: $-1 = -A$ and $A = 1$. Substitute $x = 1$: $-1 = B$. Thus

$$\int \frac{1}{\sin\theta+\cos\theta-1}\,d\theta = \int \frac{1}{x} + \frac{-1}{x-1}\,dx = \ln|x| - \ln|x-1| + C$$

$$= \ln\left|\frac{x}{x-1}\right| + C = \ln\left|\frac{\tan\frac{\theta}{2}}{\tan\frac{\theta}{2}-1}\right| + C.$$

(2) Use the procedure of this subsection to evaluate $\int \sec\theta\,d\theta$.

Solution First, write the integrand in terms of $\sin\theta$ and $\cos\theta$:

$$\int \sec\theta \; d\theta = \int \frac{1}{\cos\theta} \; d\theta.$$

Let $x = \tan\frac{\theta}{2}$, and apply Proposition 4.8.1:

$$\int \sec\theta \; d\theta = \int \frac{1+x^2}{1-x^2} \cdot \frac{2}{1+x^2} \; dx = \int \frac{2}{1-x^2} \; dx.$$

Note $1 - x^2 = (1-x)(1+x)$. We use the method of partial fractions to write

$$\frac{2}{1-x^2} = \frac{A}{1-x} + \frac{B}{1+x} = \frac{A(1+x) + B(1-x)}{1-x^2}.$$

Multiply this equation by $1 - x^2$:

$$2 = A(1+x) + B(1-x).$$

Set $x = 1$: $2 = 2A$ and $A = 1$. Set $x = -1$: $2 = 2B$ and $B = 1$. Thus

$$\begin{aligned}
\int \sec\theta \; d\theta &= \int \frac{1}{1-x} + \frac{1}{1+x} \; dx = -\ln|1-x| + \ln|1+x| + C \\
&= \ln\left|\frac{1+x}{1-x}\right| + C = \ln\left|\frac{1+\tan\frac{\theta}{2}}{1-\tan\frac{\theta}{2}}\right| + C.
\end{aligned}$$

Of course this form of $\int \sec\theta \; d\theta$ must equal the usual form $\ln|\sec\theta + \tan\theta|$ u
to a constant. Verification of this statement is given as Exercise 34.

Rational Functions of e^x

In this subsection, we integrate rational functions of e^x. An integral of this type
evaluated by the rationalizing substitution $u = e^x$ which transforms it into the integra
of a rational function of u. The latter integral can be integrated by the method o
partial fractions. In particular, $\frac{du}{dx} = e^x$ and $du = e^x \; dx$. Hence $dx = \frac{1}{e^x} \; du = \frac{1}{u} \; du$
To summarize, we use the substitution:

$$u = e^x \quad \text{with} \quad dx = \frac{1}{u} \; du. \tag{4.8.1}$$

The following example illustrates this procedure.

Examples 4.8.4 Evaluate $\int \frac{e^x - e^{-x} + 3}{e^x - 1} \; dx$.

Solution Let $u = e^x$ with $dx = \frac{1}{u} \; du$:

$$\int \frac{e^x - e^{-x} + 3}{e^x - 1} \; dx = \int \frac{u - u^{-1} + 3}{u - 1} \cdot \frac{1}{u} \; du = \int \frac{u - \frac{1}{u} + 3}{u - 1} \cdot \frac{u}{u^2} \; du = \int \frac{u^2 - 1 + 3u}{u^2(u-1)} \; d$$

We use the method of partial fractions to write

$$\frac{u^2 + 3u - 1}{u^2(u-1)} = \frac{A}{u} + \frac{B}{u^2} + \frac{C}{u-1} = \frac{Au(u-1) + B(u-1) + Cu^2}{u^2(u-1)}.$$

Multiply this equation by $u^2(u-1)$:

$$u^2 + 3u - 1 = Au(u-1) + B(u-1) + Cu^2.$$

Set $u = 1$: $3 = C$. Set $u = 0$: $-1 = -B$ and $B = 1$. The preceding equation becomes:

$$u^2 + 3u - 1 = Au(u-1) + (u-1) + 3u^2$$
$$-2u^2 + 2u = -2u(u-1) = Au(u-1)$$

Divide this equation by $u(u-1)$: $-2 = A$. Thus

$$\int \frac{e^x - e^{-x} + 3}{e^x - 1} \, dx = \int \frac{-2}{u} + \frac{1}{u^2} + \frac{3}{u-1} \, du = -2\ln|u| - \frac{1}{u} + 3\ln|u-1| + C$$

$$= -2\ln e^x - \frac{1}{e^x} + 3\ln|e^x - 1| + C = -2x - e^{-x} + 3\ln|e^x - 1| + C. \qquad \square$$

Rational Functions of Fractional Powers of x

In this subsection, we integrate functions obtained by adding, subtracting, multiplying and dividing fractional powers of x. Let n be the least common multiple of the denominators of the fractional exponents which appear in the integrand. (This is the denominator we would use to add these fractions.) The rationalizing substitution $x = u^n$ transforms the given integral into the integral of a rational function of u:

$$x = u^n \quad \text{with} \quad dx = nu^{n-1} \, du. \tag{4.8.2}$$

Note $u \geq 0$ if n is even. We illustrate this procedure with two examples.

Examples 4.8.5 (1) Evaluate $\int \frac{\sqrt[3]{x}}{\sqrt[3]{x}+1} \, dx$.

Solution The only fractional power of x in the integrand is $x^{1/3}$. Since the denominator of the exponent is three, let $x = u^3$. Then $u = x^{1/3} = \sqrt[3]{x}$, and $dx = 3u^2 \, du$. Thus

$$\int \frac{\sqrt[3]{x}}{\sqrt[3]{x}+1} \, dx = \int \frac{u}{u+1} \cdot 3u^2 \, du = \int \frac{3u^3}{u+1} \, du$$

$$= \int 3u^2 - 3u + 3 - \frac{3}{u+1} \, du \qquad \text{[by polynomial division]}$$

$$= u^3 - \frac{3}{2}u^2 + 3u - 3\ln|u+1| + C$$

$$= x - \frac{3}{2}x^{2/3} + 3\sqrt[3]{x} - 3\ln|\sqrt[3]{x}+1| + C.$$

(2) Evaluate $\int \frac{\sqrt{x}}{x^{2/3}+x^{1/3}-2} \, dx$.

Solution The fractional powers of x in the integrand are $x^{1/2}$, $x^{2/3}$ and $x^{1/3}$. The denominators of the exponents are 2 and 3 which have least common multiple 6. Thus let $x = u^6$ with $u = x^{1/6}$ and $dx = 6u^5 \, du$. Then $\sqrt{x} = x^{1/2} = x^{3/6} = u^3$ while $x^{1/3} = x^{2/6} = u^2$ and $x^{2/3} = (x^{1/3})^2 = u^4$. Hence

$$\int \frac{\sqrt{x}}{x^{2/3} + x^{1/3} - 2} \, dx = \int \frac{u^3}{u^4 + u^2 - 2} \cdot 6u^5 \, du = \int \frac{6u^8}{u^4 + u^2 - 2} \, du$$

$$= \int 6u^4 - 6u^2 + 18 + \frac{-30u^2 + 36}{u^4 + u^2 - 2} \, du \tag{4.8.3}$$

by polynomial division. Next, factor the denominator of the integrand:

$$u^4 + u^2 - 2 = (u^2 - 1)(u^2 + 2) = (u - 1)(u + 1)(u^2 + 2).$$

We use the method of partial fractions to rewrite the fraction in the integran

$$\frac{-30u^2 + 36}{u^4 + u^2 - 2} = \frac{A}{u - 1} + \frac{B}{u + 1} + \frac{Cu + D}{u^2 + 2} \tag{4}$$

$$= \frac{A(u + 1)(u^2 + 2) + B(u - 1)(u^2 + 2) + (Cu + D)(u - 1)(u}{u^4 + u^2 - 2}$$

Multiply this equation by $u^4 + u^2 - 2$:

$$-30u^2 + 36 = A(u + 1)(u^2 + 2) + B(u - 1)(u^2 + 2) + (Cu + D)(u - 1)(u + 1)$$

Substitute $u = 1$: $6 = 6A$ and $A = 1$. Substitute $u = -1$: $6 = -6B$ an
$B = -1$. Substitute $u = 0$: $36 = 2 + 2 - D$ and $D = -32$. Substitute $u =$
$-84 = 18 - 6 + (2C - 32)(1)(3)$ and $C = 0$. By (4.8.3) and (4.8.4):

$$\int \frac{\sqrt{x}}{x^{2/3} + x^{1/3} - 2}\, dx = \int 6u^4 - 6u^2 + 18 + \frac{1}{u - 1} + \frac{-1}{u + 1} + \frac{-32}{u^2 + 2}\, du$$

$$= \frac{6}{5}u^5 - 2u^3 + 18u + \ln|u - 1| - \ln|u + 1| - \frac{32}{\sqrt{2}}\arctan\frac{u}{\sqrt{2}} + C$$

$$= \frac{6}{5}x^{5/6} - 2x^{1/2} + 18x^{1/6} + \ln\left|\frac{\sqrt[6]{x} - 1}{\sqrt[6]{x} + 1}\right| - 16\sqrt{2}\arctan\sqrt[6]{\frac{x}{8}} + C.$$

Summary

The reader should know how to use rationalizing substitutions to integrate ration
functions of trigonometric functions, rational functions of e^x and rational functions o
fractional powers of x.

Basic Exercises

Evaluate each integral.

1. $\int \frac{1}{\sin x + \cos x + 1}\, dx$

2. $\int \frac{1 + \sin x}{1 - \cos x}\, dx$

3. $\int \frac{\sin x + \cos x}{3\sin x - 4\cos x}\, dx$

4. $\int \frac{1 - \cos x}{1 + \cos^2 x}\, dx$

5. $\int \frac{\sin x}{1 + \csc x}\, dx$

6. $\int \frac{1}{7 + 3\cos x}\, dx$

7. $\int \frac{1}{4\csc x + 5}\, dx$

8. $\int \frac{\sec x + 2\tan x}{2 + \sec x}\, dx$

9. $\int \frac{\sin x + \cot x}{1 + \csc x}\, dx$

10. $\int \frac{2\csc x + 3\cot x}{3 + 4\cot x}\, dx$

11. $\int \frac{e^x + 1}{e^{2x} + 1}\, dx$

12. $\int \frac{1}{e^{2x} - 1}\, dx$

13. $\int \frac{1 + e^x}{1 - e^x}\, dx$

14. $\int \frac{1 - e^{-x}}{1 + e^x}\, dx$

15. $\int \frac{e^{2x} + 4e^x + 1}{e^{3x} - 1}\, dx$

16. $\int \frac{1}{1 + 2^{x+1} + 4^x}\, dx$

17. $\int \frac{27^x + 9^x + 3^x + 1}{9^x - 1}\, dx$

18. $\int \frac{1}{1 + 5^x + 25^x}\, dx$

19. $\int \frac{x}{1 + \sqrt{x}}\, dx$

20. $\int \frac{\sqrt[4]{x} - 1}{\sqrt[4]{x} + 1}\, dx$

21. $\int \frac{\sqrt{x}}{x - \sqrt[3]{x}}\, dx$

22. $\int \frac{\sqrt{x} + \sqrt[3]{x}}{1 + \sqrt[4]{x}}\, dx$

23. $\int \frac{x}{5 + \sqrt{x + 2}}\, dx$

24. $\int \frac{3 - \sqrt[3]{x + 5}}{6 + \sqrt[3]{x + 5}}\, dx$

25. $\int \frac{x^{2/3} + x^{5/6}}{\sqrt{x} + 1}\, dx$

26. $\int \frac{x^{2/3} + x^{-1/3}}{x^{2/3} - x^{-1/3}}\, dx$

27. $\int \frac{\sqrt[3]{x} + \sqrt{x}}{\sqrt[3]{x} - \sqrt{x}}\, dx$

28. Find the area of the region bounded by $y = \frac{2}{1 + \sin x}$ and $y = 3 - 2\sin x$, both with
domain $\left(-\frac{\pi}{2}, \pi\right)$.

29. Find the area of the region bounded by $y = \frac{1}{e^x + 2}$, $y = e^{-2x}$ and the y–axis.

30. Find the area of the region in the 1st quadrant bounded by $y = \frac{6}{1 + \sqrt[3]{x}}$ and $y = \sqrt[3]{x}$.

31. Find the volume of the solid of revolution obtained by rotating the region bounded
by $y = \frac{1}{2 + \cos x}$, the x–axis and the lines $x = 0$, $x = \frac{\pi}{2}$ around the x–axis.

32. Find the volume of the solid of revolution obtained by rotating the region bounded
by $y = \frac{1 - \sqrt{x}}{1 + \sqrt{x}}$, the x–axis and the line $x = 0$ around the y–axis.

33. Find the volume of the solid of revolution obtained by rotating the region bounded by $y = \frac{1}{1+e^x}$, the x–axis and the line $x = 0$ around the x–axis.

34. Show the forms $\ln|\sec\theta + \tan\theta| + C$ and $\ln\left|\frac{1+\tan\frac{\theta}{2}}{1-\tan\frac{\theta}{2}}\right| + C'$ of $\int \sec\theta\, d\theta$ are equal.

The following three exercises are for those who have studied Section 3.9.

35. Find the distance traveled by each object with the given velocity $v(t)$.

(a) $v(t) = \frac{1+2\sin t}{3+5\cos t}$ miles per hour from $t = -\frac{\pi}{2}$ to $t = \frac{\pi}{2}$ hours.

(b) $v(t) = \frac{e^{2t}-5e^t+6}{e^{3t}+1}$ kilometers per hour from $t = 0$ to $t = 1$ hours.

(c) $v(t) = \frac{\sqrt[3]{t^2}-3\sqrt[3]{t}+2}{t(\sqrt{t}+4\sqrt[3]{t}+5\sqrt[6]{t}+2)}$ kilometers per hour from $t = 1$ to $t = 2$ hours.

36. Find the mass of the rod of density $\rho(x) = \frac{\sin\pi x}{2+\cos\pi x}$ lbs per ft for $0 \leq x \leq \frac{1}{2}$ ft.

37. How much work is done when the force $F(x) = \frac{1}{e^{3x}+1}$ Newtons moves an object from $x = 0$ to $x = 2$ meters?

The following two exercises are for those who have studied Section 3.10.

38. Find the center of mass of the rod of variable density $\rho(x) = \frac{\sqrt[3]{x}}{1+\sqrt{x}}$ kilograms per meter for $0 \leq x \leq 1$ meter.

39. Find the centroid of the region bounded by $y = \frac{x^{1/4}}{x+x^{5/6}}$, the x–axis and the lines $x = 1$, $x = 64$.

4.9   Improper Integrals

So far, our computations of areas have been restricted to bounded regions R, i.e. the region R is contained in some rectangle. We defined the definite integral to compute the area of the region depicted in Figure 4.9.1 bounded by the x–axis, the vertical lines $x = a$, $x = b$ and the graph of a continuous function f on $[a, b]$. We call these integrals *proper integrals*. In this section, we study *improper integrals* which give the areas of unbounded regions. We will see that some unbounded regions have finite area while others have infinite area. The first subsection is devoted to defining and computing the area of a region which is unbounded on either the left or the right. The second subsection applies the Comparison Test to determine whether an unbounded region has finite area. In the third subsection, we find areas of regions which are unbounded on both the left and the right. The fourth subsection studies areas of regions which are unbounded in several places.

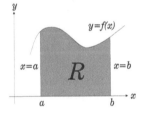

Figure 4.9.1
   The Region R

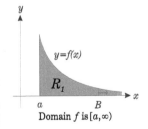

Figure 4.9.2
   Basic Region R_1

Areas of Basic Unbounded Regions

The area of an unbounded region R is estimated by the area of a bounded region R' which approximates R. We define the area of R as the limit of these estimates. We present examples that apply this definition directly to compute areas of unbounded regions. We conclude by showing that our definition gives the correct value for the area of a bounded region.

There are four basic types of unbounded regions.

(1)   The shaded region R_1 in Figure 4.9.2 is bounded by the graph of the continuous function f on $[a, \infty)$, the x–axis (its horizontal asymptote on the right) and the line $x = a$. It is estimated by the proper integral $\int_a^B f(x)\, dx$ for $B > a$.

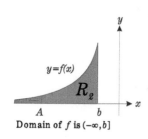

$y=f(x)$

R_2

Domain of f is $(-\infty, b]$

Figure 4.9.3
Basic Region R_2

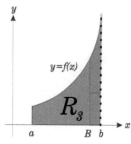

$y=f(x)$

R_3

Domain of f is $[a, b)$

Figure 4.9.4
Basic Region R_3

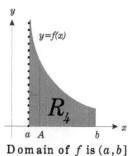

$y=f(x)$

R_4

Domain of f is $(a, b]$

Figure 4.9.6
Basic Region R_4

(2) The shaded region R_2 in Figure 4.9.3 is bounded by the graph of the continuo function f on $(-\infty, b]$, the x–axis (its horizontal asymptote on the left) and t line $x = b$. It is estimated by the proper integral $\int_A^b f(x)\, dx$ for $A < b$.

(3) The shaded region R_3 in Figure 4.9.4 is bounded by the x–axis, the graph of t continuous function f on $[a, b)$, its vertical asymptote $x = b$ and the line $x =$ It is estimated by the proper integral $\int_a^B f(x)\, dx$ for $a < B < b$.

(4) The shaded region R_4 in Figure 4.9.6 is bounded by the x–axis, the graph of t continuous function f on $(a, b]$, its vertical asymptote $x = a$ and the line $x =$ It is estimated by the proper integral $\int_A^b f(x)\, dx$ for $a < A < b$.

The closer B is to b, the better $\int_a^B f(x)\, dx$ estimates the area of the unbound region R_1 or R_3. Hence we define the area of R_1 or R_3 as the limit of the estimat $\int_a^B f(x)\, dx$ as B approaches b from the left. Similarly, the closer A is to a, the bett $\int_A^b f(x)\, dx$ estimates the area of the unbounded region R_2 or R_4. Hence we define t area of R_2 or R_4 as the limit of the estimates $\int_A^b f(x)\, dx$ as A approaches a from t right. These limits are indicated by the arrows in the four figures.

Definition 4.9.5 (a) *Let f be a continuous function with domain $[a, b)$ where* $-\infty < a < b \le \infty$. *Define the improper integral*

$$\int_a^b f(x)\, dx = \lim_{B \to b^-} \int_a^B f(x)\, dx.$$

(b) *Let f be a continuous function with domain $(a, b]$ where* $-\infty \le a < b < \infty$ *Define the improper integral*

$$\int_a^b f(x)\, dx = \lim_{A \to a^+} \int_A^b f(x)\, dx.$$

(c) *An improper integral converges if the above limit for its value exists and equa a number. On the other hand, an improper integral diverges if the above limit for it value does not exist or is infinite.*

We apply this definition to compute the values of several improper integrals. Th first two examples are especially important, as they will be used in the next subsectio to study more complicated integrals. The first example will also be used in Chapter to analyze certain important infinite series.

Examples 4.9.7 (1a) Evaluate $\int_1^\infty \frac{1}{x^p}\, dx$ when $0 < p < 1$.

Solution This integral gives the area of a region as in Figure 4.9.2 with $a = 1$ By Definition 4.9.5(a):

$$\int_1^\infty \frac{1}{x^p}\, dx = \lim_{B \to \infty} \int_1^B x^{-p}\, dx = \lim_{B \to \infty} \frac{x^{1-p}}{1-p}\Big|_1^B = \lim_{B \to \infty} \left(\frac{B^{1-p}}{1-p} - \frac{1}{1-p} \right).$$

Since $1 - p > 0$, this limit is infinite. Thus $\int_1^\infty \frac{1}{x^p}\, dx$ diverges when $0 < p < 1$.

(1b) Evaluate $\int_1^\infty \frac{1}{x^p}\, dx$ when $p > 1$.

Solution This integral also gives the area of a region as in Figure 4.9.2 wit $a = 1$. Note $p - 1 > 0$. By the computation in Example 1(a):

$$\int_1^\infty \frac{1}{x^p}\, dx = \lim_{B \to \infty} \left(\frac{B^{1-p}}{1-p} - \frac{1}{1-p} \right) = \lim_{B \to \infty} \left(\frac{1}{(1-p)B^{p-1}} + \frac{1}{p-1} \right) = \frac{1}{p-1}$$

Thus $\int_1^\infty \frac{1}{x^p}\, dx$ converges with value $\frac{1}{p-1}$ when $p > 1$.

(1c) Evaluate $\int_1^\infty \frac{1}{x}\,dx$.

Solution This integral also gives the area of a region as in Figure 4.9.2 with $a = 1$. By Definition 4.9.5(a):

$$\int_1^\infty \frac{1}{x}\,dx = \lim_{B \to \infty} \int_1^B \frac{1}{x}\,dx = \lim_{B \to \infty} \ln x\Big|_1^B = \lim_{B \to \infty} \ln B = \infty.$$

Thus $\int_1^\infty \frac{1}{x}\,dx$ diverges.

Summary

$$\int_1^\infty \frac{1}{x^p}\,dx \text{ diverges if } p \leq 1 \text{ and converges with value } \frac{1}{p-1} \text{ if } p > 1. \quad (4.9.1)$$

Note all these regions have the same appearance. We can only determine which have finite areas by evaluating the appropriate limits.

(2a) Evaluate $\int_0^1 \frac{1}{x^p}\,dx$ when $0 < p < 1$.

Solution The function $f(x) = \frac{1}{x^p}$ has the y–axis as a vertical asymptote. Hence this integral gives the area of a region as in Figure 4.9.6 with $a = 0$ and $b = 1$. By Definition 4.9.5(b):

$$\int_0^1 \frac{1}{x^p}\,dx = \lim_{A \to 0^+} \int_A^1 x^{-p}\,dx = \lim_{A \to 0^+} \frac{x^{1-p}}{1-p}\Big|_A^1 = \lim_{A \to 0^+} \left(\frac{1}{1-p} - \frac{A^{1-p}}{1-p} \right) = \frac{1}{1-p}$$

since $1 - p > 0$. Thus $\int_0^1 \frac{1}{x^p}\,dx$ converges with value $\frac{1}{1-p}$ when $0 < p < 1$.

(2b) Evaluate $\int_0^1 \frac{1}{x^p}\,dx$ when $p > 1$.

Solution This integral also gives the area of a region as in Figure 4.9.6 with $a = 0$ and $b = 1$. Note $p - 1 > 0$. By the computation in Example 2(a):

$$\int_0^1 \frac{1}{x^p}\,dx = \lim_{A \to 0^+} \left(\frac{1}{1-p} - \frac{A^{1-p}}{1-p} \right) = \lim_{A \to 0^+} \left(\frac{1}{1-p} + \frac{1}{(p-1)A^{p-1}} \right) = \infty.$$

Thus $\int_0^1 \frac{1}{x^p}\,dx$ diverges when $p > 1$.

(2c) Evaluate $\int_0^1 \frac{1}{x}\,dx$.

Solution This integral also gives the area of a region as in Figure 4.9.6 with $a = 0$ and $b = 1$. By Definition 4.9.5(b):

$$\int_0^1 \frac{1}{x}\,dx = \lim_{A \to 0^+} \int_A^1 \frac{1}{x}\,dx = \lim_{A \to 0^+} \ln x\Big|_A^1 = \lim_{A \to 0^+} -\ln A = -(-\infty) = \infty.$$

Thus $\int_0^1 \frac{1}{x}\,dx$ diverges.

Summary

$$\int_0^1 \frac{1}{x^p}\,dx \text{ diverges if } p \geq 1 \text{ and converges with value } \frac{1}{1-p} \text{ if } p < 1. \quad (4.9.2)$$

(3) Evaluate $\int_0^{\pi/2} \tan x\,dx$.

Solution The function $f(x) = \tan x$ has the line $x = \frac{\pi}{2}$ as a vertical asymptote. This integral is the area of a region as in Figure 4.9.4 with $a = 0$ and $b = \frac{\pi}{2}$. By Definition 4.9.5(a):

$$\int_0^{\pi/2} \tan x\,dx = \lim_{B \to \frac{\pi}{2}^-} \int_0^B \tan x\,dx = \lim_{B \to \frac{\pi}{2}^-} -\ln|\cos x|\Big|_0^B$$

$$= \lim_{B \to \frac{\pi}{2}^-} -\ln|\cos B| + \ln|\cos 0| = \lim_{B \to \frac{\pi}{2}^-} -\ln|\cos B|.$$

Note as B approaches $\frac{\pi}{2}$, $\cos B$ approaches zero and $\ln|\cos B|$ approaches $-c$
Thus the improper integral $\int_0^{\pi/2} \tan x \, dx$ diverges.

(4) Evaluate $\int_{-\infty}^0 e^x \, dx$.

Solution This integral gives the area of a region as in Figure 4.9.3 with $b =$
By Definition 4.9.5(b):

$$\int_{-\infty}^0 e^x \, dx = \lim_{A \to -\infty} \int_A^0 e^x \, dx = \lim_{A \to -\infty} e^x \big|_A^0 = \lim_{A \to -\infty} 1 - e^A = 1.$$

Thus $\int_{-\infty}^0 e^x \, dx$ converges and has value 1.

(5) Evaluate $\int_2^\infty \frac{1}{x \ln x} \, dx$.

Solution This integral is the area of a region as in Figure 4.9.2 with $a = 2$. $[$
Definition 4.9.5(a):

$$\int_2^\infty \frac{1}{x \ln x} \, dx = \lim_{B \to \infty} \int_2^B \frac{1}{x \ln x} \, dx.$$

Substitute $u = \ln x$ with $du = \frac{1}{x} \, dx$:

$$\int_{x=2}^{x=\infty} \frac{1}{x \ln x} \, dx = \lim_{B \to \infty} \int_{u=\ln 2}^{u=\ln B} \frac{1}{u} \, du = \lim_{B \to \infty} \ln u \big|_{\ln 2}^{\ln B}$$
$$= \lim_{B \to \infty} \ln \ln B - \ln \ln 2 = \infty.$$

Thus the improper integral $\int_2^\infty \frac{1}{x \ln x} \, dx$ diverges.

Note Definition 4.9.5 defines $\int_a^b f(x) \, dx$ for any function f which is continuous o
$[a, b)$ or $(a, b]$. In particular, it applies to a function f which is continuous on $[a, b$
For such a function, $\int_a^b f(x) \, dx$ is defined as an improper integral[4] by Definition 4.9.
and as a proper integral by Definition 3.3.9. The following proposition says that bot
these definitions of $\int_a^b f(x) \, dx$ have the same value.

Proposition 4.9.8 *Let f be a continuous function with domain $[a, b]$. Then*

$$\lim_{A \to a^+} \int_A^b f(x) \, dx = \lim_{B \to b^-} \int_a^B f(x) \, dx = \int_a^b f(x) \, dx. \qquad (4.9.3$$

Proof Note $\lim\limits_{A \to a^+} \int_A^b f(x) \, dx$, $\lim\limits_{B \to b^-} \int_a^B f(x) \, dx$, $\int_a^b f(x) \, dx$ in (4.9.3) is the defini
tion of $\int_a^b f(x) \, dx$ by Def. 4.9.5(b), Def. 4.9.5(a), Def. 3.3.9, respectively. We prov
the second equality and give verification of the first one as Exercise 60. Define

$$F(x) = \int_a^x f(t) \, dt$$

for $x \in [a, b]$. By the First Fundamental Theorem of Calculus, F is a differentiable
function on $[a, b]$ with $F'(x) = f(x)$ for $a < x < b$, $F'_+(a) = f(a)$ and $F'_-(b) = f(b)$
By Corollary 2.2.25, F is continuous on $[a, b]$. In particular, F is continuous at $x = b$

$$\int_a^b f(x) \, dx = F(b) = \lim_{B \to b^-} F(B) = \lim_{B \to b^-} \int_a^B f(x) \, dx. \qquad \square$$

---

[4]In the future, we will not call $\int_a^b f(x) \, dx$ an improper integral when f is continuous on $[a, b]$.

Comparison Test

Consider an improper integral I whose exact value is difficult or impossible to calculate. In some cases we can find a larger convergent improper integral. It follows that the smaller improper integral I must also be convergent. In other cases, we can find a divergent improper integral that is smaller than I. Then the larger improper integral I must also diverge. This procedure for analyzing I is called the *Comparison Test*.

There is a geometric method to determine that an improper integral I is less than another one J. Each improper integral gives the area of a region. If the region R is contained in the region S, then the area I of R will be less than or equal to the area J of S. See Figure 4.9.9. Say R is determined by the graph of f and S is determined by the graph of g. Then R is contained in S if $0 \le f(x) \le g(x)$ for all x.

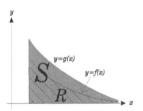

Figure 4.9.9
Comparison Test

Note there are two varieties of an improper integral $\int_a^b f(x)\ dx$ which represents the area of an unbounded region R. R may have $x = b$ as a vertical asymptote or have the x–axis as a horizontal asymptote on the right ($b = \infty$) as in Figures 4.9.4 and 4.9.2. In these cases, R is unbounded on the right, and the domain of f is $D = [a, b)$. In other cases, R may have $x = a$ as a vertical asymptote or have the x–axis as a horizontal asymptote on the left ($a = -\infty$) as in Figures 4.9.6 and 4.9.3. In these cases, the domain of f is is unbounded on the left, and the domain of f is $D = (a, b]$. We are now ready to state the Comparison Test.

Proposition 4.9.10 (Comparison Test)   *Let* $-\infty < a < b \le \infty$ *with* $D = [a, b)$, *or let* $-\infty \le a < b < \infty$ *with* $D = (a, b]$. *Assume* f *and* g *are continuous functions with domain* D *such that*

$$0 \le f(x) \le g(x) \quad \text{for } x \in D. \tag{4.9.4}$$

(a) *If* $\int_a^b g(x)\ dx$ *converges, then* $\int_a^b f(x)\ dx$ *also converges and* $\int_a^b f(x)\ dx \le \int_a^b g(x)\ dx$.

(b) *If* $\int_a^b f(x)\ dx$ *diverges, then* $\int_a^b g(x)\ dx$ *also diverges.*

Proof   Consider the case $D = [a, b)$. The other case is given as Exericse 61. Since the functions f and g are always non–negative, the limits which define their improper integrals exist as numbers or as $+\infty$. By (4.9.4):

$$0 \le \int_a^B f(x)\ dx \quad \le \quad \int_a^B g(x)\ dx$$

$$0 \le \lim_{B \to b^-} \int_a^B f(x)\ dx \quad \le \quad \lim_{B \to b^-} \int_a^B g(x)\ dx.$$

If the second limit is finite, then so is the first one. Moreover, the first limit will be less than or equal to the second one. This proves (a). On the other hand, if the first limit is infinite, then so is the second one. This proves (b).   □

The key step in applying the Comparison Test is the selection of a function similar to, but simpler than, the given integrand.

Examples 4.9.11 Determine whether each improper integral converges or diverges.

(1) $\int_0^1 \frac{1}{x(e^x - 1)}\ dx$

    Solution The analysis is a sequence of six steps.
- The simple function $\frac{1}{x}$ is similar to the integrand $\frac{1}{x(e^x - 1)}$.

- Note $\frac{1}{x} \leq \frac{1}{x(e^x-1)}$ when $0 < e^x - 1 \leq 1$, i.e. when $1 < e^x \leq 2$. Since $\ln$ is
  increasing function, the preceding inequality is equivalent to $\ln 1 < \ln e^x \leq \ln$
  i.e. $0 < x \leq \ln 2$. Thus

$$\frac{1}{x} \leq \frac{1}{x(e^x - 1)} \quad \text{for} \quad 0 < x \leq \ln 2.$$

- We showed in Example 4.9.7 (2) that $\int_0^1 \frac{1}{x} \, dx$ diverges.
- Then $\int_0^{\ln 2} \frac{1}{x} \, dx = \int_0^1 \frac{1}{x} \, dx - \int_{\ln 2}^1 \frac{1}{x} \, dx$ is a divergent improper integral min
  a proper integral which diverges.
- By the Comparison Test, $\int_0^{\ln 2} \frac{1}{x(e^x-1)} \, dx$ also diverges.
- Thus the given integral $\int_0^1 \frac{1}{x(e^x-1)} \, dx = \int_0^{\ln 2} \frac{1}{x(e^x-1)} \, dx + \int_{\ln 2}^1 \frac{1}{x(e^x-1)} \, dx$
  the sum of a divergent improper integral and a proper integral which diverges

(2) $\int_0^\infty e^{-x^2} \, dx$

Solution We analyze this integral in a sequence of five steps.
- The function e^{-x} is simpler than the integrand e^{-x^2}.
- Since $-x^2 \leq -x$ for $x \geq 1$ and exp is an increasing function,

$$0 < e^{-x^2} \leq e^{-x} \quad \text{for} \quad x \geq 1.$$

- We can easily evaluate the improper integral $\int_1^\infty e^{-x} \, dx$:

$$\int_1^\infty e^{-x} \, dx = \lim_{B \to \infty} \int_1^B e^{-x} \, dx = \lim_{B \to \infty} -e^{-x}\big|_1^B = \lim_{B \to \infty} \left(e^{-1} - e^{-B} \right) = \frac{1}{e}.$$

Thus $\int_1^\infty e^{-x} \, dx$ converges.
- By the Comparison Test, the improper integral $\int_1^\infty e^{-x^2} \, dx$ converges.
- The given integral $\int_0^\infty e^{-x^2} \, dx = \int_0^1 e^{-x^2} \, dx + \int_1^\infty e^{-x^2} \, dx$ is the sum of
  proper integral and a convergent improper integral which converges.

(3) $\int_1^\infty \frac{\sin x}{x^p}$ where $p > 1$.

Solution The analysis proceeds through a sequence of six steps.
- Note $\frac{|\sin x|}{x^p} \leq \frac{1}{x^p}$ since $0 \leq |\sin x| \leq 1$.
- By Example 4.9.7(1b), $\int_1^\infty \frac{1}{x^p} \, dx$ converges.
- By the Comparison Test, $\int_1^\infty \frac{|\sin x|}{x^p} \, dx$ also converges.
- When $\sin x \geq 0$, we have $\sin x + |\sin x| = 2|\sin x| \leq 2$. On the other hand
  when $\sin x \leq 0$, we have $\sin x + |\sin x| = 0$. Thus in all cases:

$$0 \leq \frac{\sin x + |\sin x|}{x^p} \leq \frac{2}{x^p}.$$

- By the Comparison Test, $\int_1^\infty \frac{\sin x + |\sin x|}{x^p} \, dx$ converges.
- The given integral,

$$\int_1^\infty \frac{\sin x}{x^p} \, dx = \int_1^\infty \frac{\sin x + |\sin x|}{x^p} \, dx - \int_1^\infty \frac{|\sin x|}{x^p} \, dx,$$

is the difference of two convergent improper integrals which converges.

Areas of Regions Unbounded on the Left and the Right

We have been studying improper integrals $\int_a^b f(x)\,dx$ which represent areas of regions R which are unbounded at one end. That is, R may be unbounded at $x = a$ with Domain $f = (a, b]$, or R may be unbounded at $x = b$ with Domain $f = [a, b)$. In this subsection, we study improper integrals $\int_a^b f(x)\,dx$ which represent the areas of regions which are unbounded at both $x = a$ and $x = b$, i.e. Domain $f = (a, b)$.

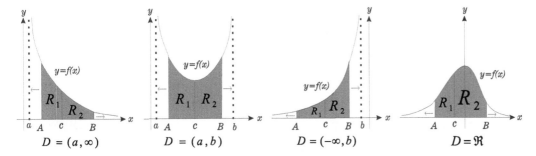

Figure 4.9.12 Regions which are Unbounded at Both Ends of the Interval D

Figure 4.9.12 illustrates the four types of regions, bounded by the graph of a function f with domain D, which are unbounded at both ends of D. From left to right:

(i)   when $D = (a, \infty)$ there may be a vertical asymptote at $x = a$ and the x–axis may be a horizontal asymptote on the right;

(ii)   when $D = (a, b)$ there may be vertical asymptotes at both $x = a$ and $x = b$;

(iii)   when $D = (-\infty, b)$ the x–axis may be a horizontal asymptote on the left and there may be a vertical asymptote at $x = b$;

(iv)   when $D = \Re = (-\infty, \infty)$ the x–axis may be a horizontal asymptote on both the left and the right.

The strategy to define the area of one of these four regions R is to select a line $x = c$ which divides R into two regions. The area of the subset R_1 of R to the left of $x = c$ is defined because it is only unbounded on the left. The area of the subset R_2 of R to the right of $x = c$ is defined because it is only unbounded on the right. We define the area of R as the sum of the areas of R_1 and R_2. See Figure 4.9.12.

Definition 4.9.13 *Let f be a continuous function having domain (a, b) with $-\infty \le a < c < b \le \infty$.*
(a) The improper integral $\int_a^b f(x)\,dx$ diverges if one or both of the integrals $\int_a^c f(x)\,dx$, $\int_c^b f(x)\,dx$ diverge.
(b)   The improper integral $\int_a^b f(x)\,dx$ converges if both $\int_a^c f(x)\,dx$ and $\int_c^b f(x)\,dx$ converge. In this case, define

$$\int_a^b f(x)\,dx = \int_a^c f(x)\,dx + \int_c^b f(x)\,dx.$$

There are two points which require clarification. We verify in the next proposition that the definition of $\int_a^b f(x)\,dx$ does not depend on the choice of c. We must also justify why we considered the area of the region near a separately from the area of the region near b. Why did we reject the alternate approach of defini $\int_a^b f(x)\ dx$ as one limit of areas of bounded regions. For example, why do we n define $\int_{-a}^a f(x)\ dx = \lim\limits_{A \to a^-} \int_{-A}^A f(x)\ dx$? The reason for rejecting this approach c be seen from the example $\int_{-\pi/2}^{\pi/2} \tan x\ dx$. By Example 4.9.7(3),

$$\int_{-\pi/2}^0 \tan x\ dx = -\infty, \qquad \int_0^{\pi/2} \tan x\ dx = +\infty \quad \text{while} \quad \int_{-A}^A \tan x\ dx = 0,$$

for $0 < A < \frac{\pi}{2}$ since $\tan x$ is an odd function. Definition 4.9.13(a), with $c = 0$, sa that $\int_{-\pi/2}^{\pi/2} \tan x\ dx$ diverges. To understand why we reject the alternate definition

$$\int_{-\pi/2}^{\pi/2} \tan x\ dx = \lim_{A \to \pi/2} \int_{-A}^A \tan x\ dx = \lim_{A \to \pi/2} 0 = 0,$$

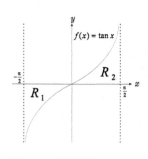

Figure 4.9.14
$\int_{-\pi/2}^{\pi/2} \tan x\ dx$

consider the region R in Figure 4.9.14 whose area determines the value of $\int_{-\pi/2}^{\pi/2} \tan x$ (R consists of two congruent pieces: R_1 in the third quadrant and R_2 in the first qua rant. R_1 has infinite area, lies under the x–axis and $\int_{-\pi/2}^0 \tan x\ dx = -\infty$. R_2 h area $\int_0^{\pi/2} \tan x\ dx = +\infty$. Hence

$$\int_{-\pi/2}^{\pi/2} \tan x\ dx = \int_{-\pi/2}^0 \tan x\ dx + \int_0^{\pi/2} \tan x\ dx = -\infty + \infty$$

which is not defined. Thus it is correct to say that $\int_{-\pi/2}^{\pi/2} \tan x\ dx$ diverges. It woul be wrong to say: $\int_{-\pi/2}^{\pi/2} \tan x\ dx = -\infty + \infty = 0$.

Proposition 4.9.15 *Let f be a continuous function having domain (a, b) with $-\infty \le a < c < e < b \le \infty$.*
(a) *$\int_a^c f(x)\ dx$ converges if and only if $\int_a^e f(x)\ dx$ converges.*
(b) *$\int_c^b f(x)\ dx$ converges if and only if $\int_e^b f(x)\ dx$ converges.*
(c) *If all of these integrals converge, then*

$$\int_a^c f(x)\ dx + \int_c^b f(x)\ dx = \int_a^e f(x)\ dx + \int_e^b f(x)\ dx.$$

Proof (a) Observe that

$$\int_a^e f(x)\ dx = \lim_{A \to a^+} \int_A^e f(x)\ dx = \lim_{A \to a^+} \int_A^c f(x)\ dx + \int_c^e f(x)\ dx$$

$$= \int_a^c f(x)\ dx + \int_c^e f(x)\ dx$$

where $\int_c^e f(x)\ dx$ is a proper integral. Thus $\int_a^c f(x)\ dx$ converges if and only i $\int_a^e f(x)\ dx$ converges.
(b) The proof of (b) is given as Exercise 62.
(c) Observe that

$$\int_a^c f(x)\ dx + \int_c^b f(x)\ dx = \lim_{A \to a^+} \int_A^c f(x)\ dx + \lim_{B \to b^-} \int_c^B f(x)\ dx$$

$$= \lim_{A \to a^+} \int_A^c f(x)\ dx + \int_c^e f(x)\ dx + \lim_{B \to b^-} \int_e^B f(x)\ dx$$

$$= \lim_{A \to a^+} \int_A^e f(x)\ dx + \lim_{B \to b^-} \int_e^B f(x)\ dx = \int_a^e f(x)\ dx + \int_e^b f(x)\ dx. \ \ \square$$

Examples 4.9.16 Determine whether each improper integral converges or diverges.

(1) $\int_0^\infty \frac{1}{x^p}\,dx$ for $p > 0$

Solution This integral represents the area of the left region in Figure 4.9.12 with $f(x) = \frac{1}{x^p}$ and $a = 0$. Apply Definition 4.9.13 with $c = 1$:

$$\int_0^\infty \frac{1}{x^p}\,dx = \int_0^1 \frac{1}{x^p}\,dx + \int_1^\infty \frac{1}{x^p}\,dx.$$

We evaluated these integrals in Examples 4.9.7(1), (2). If $p > 1$, the first integral diverges. If $p = 1$, both integrals diverge. If $p < 1$, the second integral diverges. Thus the improper integral $\int_0^\infty \frac{1}{x^p}\,dx$ diverges in all cases.

(2) $\int_{-\infty}^\infty e^{-x^2}\,dx$

Solution This integral represents the area of the right region in Figure 4.9.12 with $f(x) = e^{-x^2}$. Apply Definition 4.9.13 with $c = 0$:

$$\int_{-\infty}^\infty e^{-x^2}\,dx = \int_{-\infty}^0 e^{-x^2}\,dx + \int_0^\infty e^{-x^2}\,dx.$$

Since e^{-x^2} is an even function, the two integrals on the right are equal. By Example 4.9.11(2), the second integral is convergent. Hence the first integral also converges, and the improper integral $\int_{-\infty}^\infty e^{-x^2}\,dx$ converges.

(3) $\int_{-2}^2 \frac{1}{\sqrt[3]{4-x^2}}\,dx$

Solution This integral represents the area of the second region in Figure 4.9.12 with $f(x) = \frac{1}{\sqrt[3]{4-x^2}}$, $a = -2$ and $b = 2$. Apply Definition 4.9.13 with $c = -1$:

$$\int_{-2}^2 \frac{1}{\sqrt[3]{4-x^2}}\,dx = \int_{-2}^{-1} \frac{1}{\sqrt[3]{4-x^2}}\,dx + \int_{-1}^2 \frac{1}{\sqrt[3]{4-x^2}}\,dx. \qquad (4.9.5)$$

To analyze the first integral on the right, make the substitution $u = x + 2$ with $du = dx$. Observe that $4 - x^2 = 4u - u^2 = u(4 - u) \geq u$ for $0 \leq u \leq 1$. Hence

$$0 \leq \int_{x=-2}^{x=-1} \frac{1}{\sqrt[3]{4-x^2}}\,dx = \int_{u=0}^{u=1} \frac{1}{\sqrt[3]{4u-u^2}}\,du \leq \int_0^1 \frac{1}{\sqrt[3]{u}}\,du.$$

The right integral converges by Example 4.9.7(2a). By the Comparison Test, $\int_{-2}^{-1} \frac{1}{\sqrt[3]{4-x^2}}\,dx$ also converges. To analyze the second integral on the right side of (4.9.5), make the substitution $v = -x$ with $dv = -dx$:

$$\begin{aligned}
\int_{-1}^2 \frac{1}{\sqrt[3]{4-x^2}}\,dx &= \int_{-1}^1 \frac{1}{\sqrt[3]{4-x^2}}\,dx + \int_{x=1}^{x=2} \frac{1}{\sqrt[3]{4-x^2}}\,dx \\
&= \int_{-1}^1 \frac{1}{\sqrt[3]{4-x^2}}\,dx + \int_{v=-1}^{v=-2} -\frac{1}{\sqrt[3]{4-v^2}}\,dv \\
&= \int_{-1}^1 \frac{1}{\sqrt[3]{4-x^2}}\,dx + \int_{-2}^{-1} \frac{1}{\sqrt[3]{4-v^2}}\,dv
\end{aligned}$$

which is the sum of a proper integral and a convergent improper integral. Thus the improper integral $\int_{-1}^2 \frac{1}{\sqrt[3]{4-x^2}}\,dx$ converges. Since both integrals on the right side of (4.9.5) converge, the improper integral $\int_{-1}^2 \frac{1}{\sqrt[3]{4-x^2}}\,dx$ converges. $\qquad \square$

Areas of More General Unbounded Regions

The general improper integral $\int_a^b f(x)\, dx$ has an integrand f with several discontin-
ities. The partition $P = \{x_0, \ldots, x_n\}$ of $[a, b]$ designates the points of discontinuity

$$a = x_0 < x_1 < \cdots < x_{n-1} < x_n = b.$$

Note this notation includes the case where $a = -\infty$ or $b = +\infty$. Each integr
$\int_{x_{k-1}}^{x_k} f(x)\, dx$ represents a region which is only unbounded at the left and right en
of the interval (x_{k-1}, x_k). Hence it is defined by Definition 4.9.13. We define $\int_a^b f(x)\,$
as the sum of these integrals.

Definition 4.9.17 *Let $P = \{x_0, \ldots, x_n\}$ be a partition of the interval $[a, b]$. Assum*
the function $f(x)$ is continuous on each open interval (x_{k-1}, x_k), <u>for</u> $1 \le k \le n$.
(a) *If one or more of the improper integrals $\int_{x_{k-1}}^{x_k} f(x)\, dx$ diverges, then the improp*
integral $\int_a^b f(x)\, dx$ diverges.
(b) *If all the improper integrals $\int_{x_{k-1}}^{x_k} f(x)\, dx$ converge, then the improper integr*
$\int_a^b f(x)\, dx$ converges, and

$$\int_a^b f(x)\, dx = \int_{x_0}^{x_1} f(x)\, dx + \cdots + \int_{x_{n-1}}^{x_n} f(x)\, dx.$$

Examples 4.9.18 Determine whether each improper integral converges or diverges

(1) $\int_{-\infty}^{\infty} \frac{1}{x^3}\, dx$

> **Solution** The function $f(x) = \frac{1}{x^3}$ has a discontinuity at $x = 0$ where it has
> vertical asymptote. Hence $\int_{-\infty}^{\infty} \frac{1}{x^3}\, dx$ is defined by Definition 4.9.17 using th
> partition $\{-\infty, 0, \infty\}$ of $\Re = (-\infty, \infty)$:
>
> $$\int_{-\infty}^{\infty} \frac{1}{x^3}\, dx = \int_{-\infty}^{0} \frac{1}{x^3}\, dx + \int_{0}^{\infty} \frac{1}{x^3}\, dx.$$
>
> Apply Definition 4.9.13 with $c = 1$ to evaluate the second integral:
>
> $$\int_{0}^{\infty} \frac{1}{x^3}\, dx = \int_{0}^{1} \frac{1}{x^3}\, dx + \int_{1}^{\infty} \frac{1}{x^3}\, dx.$$

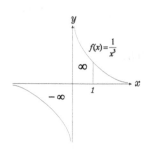

Figure 4.9.19
$\int_{-\infty}^{\infty} \frac{1}{x^3}\, dx$

> By Examples 4.9.7 (2b) and (1b), the first integral diverges while the sec
> ond integral converges. By Definition 4.9.13, $\int_0^{\infty} \frac{1}{x^3}\, dx$ diverges. Then b
> Definition 4.9.17(a), $\int_{-\infty}^{\infty} \frac{1}{x^3}\, dx$ also diverges. Observe this integral refers t
> the region depicted in Figure 4.9.19 which consists of two congruent regions
> one in the first quadrant with area $+\infty$ and another in the third quadrant wit
> area $-\infty$. The area of the total region, $\infty + -\infty$, is not defined. We *do not sa*
> that the area of the total region is zero.

(2) $\int_{-\infty}^{\infty} \frac{e^{-x^2}}{x^{2/3}}\, dx$

> **Solution** Observe $g(x) = \frac{e^{-x^2}}{x^{2/3}}$ has a discontinuity at $x = 0$ where it ha
> a vertical asymptote. See Figure 4.9.20. Hence $\int_{-\infty}^{\infty} \frac{e^{-x^2}}{x^{2/3}}\, dx$ is defined by
> Definition 4.9.17 using the partition $\{-\infty, 0, \infty\}$ of $\Re = (-\infty, \infty)$:
>
> $$\int_{-\infty}^{\infty} \frac{e^{-x^2}}{x^{2/3}}\, dx = \int_{-\infty}^{0} \frac{e^{-x^2}}{x^{2/3}}\, dx + \int_{0}^{\infty} \frac{e^{-x^2}}{x^{2/3}}\, dx = 2\int_{0}^{\infty} \frac{e^{-x^2}}{x^{2/3}}\, dx \qquad (4.9.6)$$

since $y = \frac{e^{-x^2}}{x^{2/3}}$ is an even function. Apply Definition 4.9.13 with $c = 1$:

$$\int_0^\infty \frac{e^{-x^2}}{x^{2/3}}\, dx = \int_0^1 \frac{e^{-x^2}}{x^{2/3}}\, dx + \int_1^\infty \frac{e^{-x^2}}{x^{2/3}}\, dx. \qquad (4.9.7)$$

To evaluate the first integral on the right, note $x^{2/3} \geq 1$ when $x \geq 1$. Hence

$$0 \leq \frac{e^{-x^2}}{x^{2/3}} \leq \frac{e^{-x^2}}{1} = e^{-x^2} \text{ for } x \geq 1.$$

By Example 4.9.11(2), $\int_1^\infty e^{-x^2}\, dx$ converges. By the Comparison Test, $\int_1^\infty \frac{e^{-x^2}}{x^{2/3}}\, dx$ also converges. To evaluate the second integral on the right side of (4.9.7), note $e^{-x^2} \leq 1$ for all x. Hence

$$0 \leq \frac{e^{-x^2}}{x^{2/3}} \leq \frac{1}{x^{2/3}}.$$

By Example 4.9.7 (2a), $\int_0^1 \frac{1}{x^{2/3}}\, dx$ converges because $\frac{2}{3} < 1$. By the Comparison Test, $\int_0^1 \frac{e^{-x^2}}{x^{2/3}}\, dx$ also converges. By (4.9.7), $\int_0^\infty \frac{e^{-x^2}}{x^{2/3}}\, dx$ converges. By (4.9.6), $\int_{-\infty}^\infty \frac{e^{-x^2}}{x^{2/3}}\, dx$ converges. $\square$

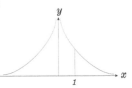

Figure 4.9.20

$y = \frac{e^{-x^2}}{x^{2/3}}$

Summary

The reader should understand the definition of the various types of improper integrals. She should be able to determine whether an improper integral converges or diverges, either directly from the appropriate definition or by using the Comparison Test. In particular she should know that:

$$\int_1^\infty \frac{1}{x^p}\, dx \left\{ \begin{array}{ll} \text{converges} & \text{if } p > 1 \\ \text{diverges} & \text{if } 0 < p \leq 1 \end{array} \right\} \text{ and } \int_0^1 \frac{1}{x^p}\, dx \left\{ \begin{array}{ll} \text{converges} & \text{if } 0 < p < 1 \\ \text{diverges} & \text{if } p \geq 1 \end{array} \right\}.$$

Basic Exercises

Determine whether each improper integral converges or diverges. If it converges, find its value.

1. $\int_2^\infty \frac{1}{x^3}\, dx$
2. $\int_0^{81} \frac{1}{\sqrt[4]{x}}\, dx$
3. $\int_0^\infty \frac{1}{\sqrt{x}}\, dx$
4. $\int_{-1}^8 \frac{1}{x^{2/3}}\, dx$

5. $\int_{-\infty}^\infty \frac{1}{1+x^2}\, dx$
6. $\int_0^1 \ln x\, dx$
7. $\int_{-\infty}^0 e^x\, dx$
8. $\int_0^{\pi/2} \cot x\, dx$

9. $\int_0^{\pi/2} \sec x\, dx$
10. $\int_0^\pi 2\arctan x\, dx$
11. $\int_0^1 \frac{1}{\sqrt{1-x^2}}\, dx$
12. $\int_0^5 \frac{1}{x^2-1}\, dx$

13. $\int_3^\infty \frac{x+1}{x^3-x}\, dx$
14. $\int_{-\infty}^\infty \frac{2x+1}{x^2+6x+25}\, dx$
15. $\int_0^\infty \frac{\sin x}{e^x}\, dx$
16. $\int_{-\infty}^\infty \frac{\cos x}{2^x}\, dx$

17. $\int_0^\pi \csc x\, dx$
18. $\int_0^\pi \sec x\, dx$
19. $\int_0^{5/2} \frac{1}{\sqrt{5-2x}}\, dx$
20. $\int_0^\infty xe^{-x}\, dx$

21. $\int_1^\infty \frac{\ln x}{x}\, dx$
22. $\int_2^\infty \frac{3x^2+x-1}{x^4-1}\, dx$
23. $\int_0^\infty \frac{1}{\sqrt{x}(x+3)}\, dx$
24. $\int_1^2 \frac{x}{\sqrt[4]{x^2-1}}\, dx$

25. $\int_1^\infty \frac{1}{x\sqrt{x^2-1}}\, dx$
26. $\int_{-\infty}^\infty \frac{x}{(x^2+1)^2}\, dx$
27. $\int_3^6 \frac{1}{6x-x^2}\, dx$
28. $\int_1^\infty \frac{1}{x\ln x}\, dx$

29. $\int_0^{\pi/2} \frac{\tan x}{\ln \cos x}\, dx$
30. $\int_{-\infty}^\infty xe^{-x^2}\, dx$
31. $\int_0^\infty \frac{1}{xe^x}\, dx$
32. $\int_{-\infty}^\infty \frac{x}{1-x^4}\, dx$

Determine whether each improper integral converges or diverges.

33. $\int_{-\infty}^\infty \frac{\cos x}{x^2}\, dx$
34. $\int_{-\infty}^\infty e^{-x^3}\, dx$
35. $\int_{-\infty}^\infty e^{-x^4}\, dx$
36. $\int_0^\infty \frac{1}{3x^4+2x^2+5}\, dx$

37. $\int_8^\infty \frac{x^3}{x^4-x^2-1}\, dx$
38. $\int_2^\infty \frac{1}{\sqrt{x}\ln x}\, dx$
39. $\int_{-\infty}^\infty 10^{-x^2}\, dx$
40. $\int_1^\infty \frac{1}{xe^x}\, dx$

41. $\int_0^{\pi/2} \sec^9 x\, dx$
42. $\int_1^\infty \frac{x^2}{\sqrt{x^8+x^2+1}}\, dx$
43. $\int_0^\infty \frac{1}{x+\sqrt{x}}\, dx$
44. $\int_0^1 \frac{\sin x}{x}\, dx$

45. $\int_0^1 \frac{\ln(x+1)}{x-1}\, dx$
46. $\int_0^2 \sqrt{\frac{x}{2-x}}\, dx$
47. $\int_0^1 \frac{\sin(e^x+1)}{\sqrt{x}}\, dx$
48. $\int_0^\infty \frac{1}{\sqrt{x}e^x}\, dx$

Determine the positive numbers p for which each improper integral converges.

49. $\int_1^\infty \frac{1}{x^p e^x}\, dx$ **50.** $\int_1^\infty \frac{x^p}{e^x}\, dx$ **51.** $\int_2^\infty \frac{1}{x(\ln x)^p}\, dx$

52. $\int_1^\infty \frac{\ln x}{x^p}\, dx$ **53.** $\int_0^1 x^p \sin(1/x)\, dx$ **54.** $\int_0^\pi \frac{\sin x}{x^p}\, dx$

55. Let p be a positive number. Consider the region R bounded by $y = \frac{1}{x^p}$ for $x \geq$ the line $x = 1$ and the x–axis.
(a) Let S be the solid of revolution obtained by revolving the region R around t. x–axis. Determine the values of p for which the volume of S is finite.
(b) For what values of p is the area of R infinite and the volume of S finite?
(c) Let T be the solid of revolution obtained by revolving the region R around t. y–axis. Determine the values of p for which the volume of T is finite.

56. Let p be a positive number. Consider the region R bounded by $y = \frac{1}{x^p}$, $x =$ the x–axis and the y–axis.
(a) Let S be the solid of revolution obtained by revolving the region R around t x–axis. Determine the values of p for which the volume of S is finite.
(b) For what values of p is the area of R finite and the volume of S infinite?
(c) Let T be the solid of revolution obtained by revolving the region R around t y–axis. Determine the values of p for which the volume of T is finite.

57. Let R be the region bounded by $y = \frac{1}{\sqrt{x}\ln x}$, the line $x = 2$ and the x–axis. Fin the volume of the solid of revolution obtained by rotating R around the x–axis.

58. Let R be the region bounded by $y = \frac{1}{\sqrt{1+x^2}}$ and the x–axis. Find the volume the solid of revolution obtained by rotating R around the x–axis.

59. Let R be the region bounded by $y = e^{-x^2}$ for $x \geq 1$, the line $x = 1$ and th x–axis. Find the volume of the solid of revolution obtained by rotating R around th y–axis.

60. Let f be a continuous function with domain $[a, b]$. Show

$$\lim_{A \to a+} \int_A^b f(x)\, dx = \int_a^b f(x)\, dx.$$

61. Prove Proposition 4.9.10 in the case $I = (a, b]$.

62. Prove Proposition 4.9.15(b).

The following three exercises are for those who have studied Section 3.9.

63. Find the distance traveled by each object with given velocity $v(t)$.
(a) $v(t) = \frac{1}{t(\ln t)^2}$ miles per hour for $t \geq e^3$ hours.
(b) $v(t) = \frac{1}{\sqrt{t-1}}$ miles per hour for $0 \leq t < 1$ hour.
(c) $v(t) = \frac{t}{e^{t^2}}$ kilometers per hour for all $t \in \Re$.
(d) $v(t) = \frac{\cos t}{e^t}$ kilometers per hour for $t \geq 0$.

64.   Find the mass of the rod of infinite length whose density is $\rho(x) = x^2 e^{-}$ kilograms per meter for $x \geq 0$ meters.

65. How much work is done when the force $F(x) = \frac{1}{\sqrt{4-x}}$ Newtons moves an objec from $x = 0$ to $x = 4$ meters?

The following two exercises are for those who have studied Section 3.10.

66. Find the center of mass of each rod of variable density $\rho(x)$.
(a) $\rho(x) = \frac{1}{1+x^3}$ kg/m for $1 \leq x < \infty$ m.   **(b)** $\rho(x) = 2^{-x}$ kg/m for $0 \leq x < \infty$ m
(c) $\rho(x) = \frac{1}{\sqrt{x}+\sqrt[3]{x}}$ lbs/ft for $0 < x \leq 64$ ft   **(d)** $\rho(x) = -\ln x$ lbs/ft for $0 < x \leq 1$ ft

67. Find the centroid of each region.
(a) Let R be the region bounded by $y = \frac{1}{x^4}$ for $x \geq 1$, the line $x = 1$ and the x–axis.
(b) Let S be the region bounded by $y = xe^{-x}$ for $x \geq 0$ and the x–axis.
(c) Let T be the region bounded by $y = \ln x$ for $0 < x \leq 1$, the x–axis and the y–axis.

The following two exercises are for those who have studied Section 4.10.

68. Determine whether each improper integral converges or diverges. If it converges, find its value.
(a) $\int_0^1 \coth x \; dx$ **(b)** $\int_{-\infty}^\infty \operatorname{sech} x \; dx$ **(c)** $\int_0^1 \operatorname{csch} x \; dx$ **(d)** $\int_0^1 \tanh^{-1} x \; dx$
(e) $\int_1^2 \coth^{-1} x \; dx$ **(f)** $\int_2^\infty \coth^{-1} x \; dx$ **(g)** $\int_0^1 \operatorname{sech}^{-1} x \; dx$ **(h)** $\int_0^1 \operatorname{csch}^{-1} x \; dx$

69. Determine whether each improper integral converges or diverges.
(a) $\int_1^\infty \frac{\tanh x}{x^3} \; dx$ **(b)** $\int_0^1 \frac{\operatorname{sech} x}{\sqrt{x}} \; dx$ **(c)** $\int_{-\infty}^\infty \sin^2 x \operatorname{sech} x \; dx$ **(d)** $\int_0^1 \frac{\coth x}{x} \; dx$

Challenging Problems

1. Show the definition of $\int_a^b f(x) \; dx$ in Def. 4.9.17 does not depend on the choice of a partition of $[a, b]$. That is, let $\{x_0, \ldots, x_m\}$ and $\{y_0, \ldots, y_n\}$ be two partitions of $[a, b]$ such that $f(x)$ is continuous on (x_{h-1}, x_h) for $1 \leq h \leq m$ and on (y_{k-1}, y_k) for $1 \leq k \leq n$. Show $\int_{x_{h-1}}^{x_h} f(x) \; dx$ converges for $1 \leq h \leq m$ if and only if $\int_{y_{k-1}}^{y_k} f(x) \; dx$ converges for $1 \leq k \leq n$. Moreover, $\sum_{h=1}^m \int_{x_{h-1}}^{x_h} f(x) \; dx = \sum_{k=1}^n \int_{x_{k-1}}^{x_k} f(x) \; dx$.

2. Show the following properties of definite integrals are also valid for the convergent improper integrals of Definition 4.9.17.
(a) $\int_a^b hf(x) + kg(x) \; dx = h \int_a^b f(x) \; dx + k \int_a^b g(x) \; dx$ where $h, k \in \mathbb{R}$.
(b) $\int_a^c f(x) \; dx + \int_c^b f(x) \; dx = \int_a^b f(x) \; dx$.

3. Let $f(x) \geq 0$ and $g(x) \geq 0$ be continuous functions with domain $[a, b)$. Assume
$$\lim_{x \to b^-} \frac{f(x)}{g(x)} = L \in \mathbb{R}.$$
(a) If $L \neq 0$, show $\int_a^b f(x) \; dx$ converges if and only if $\int_a^b g(x) \; dx$ converges.
(b) What is the relationship between $\int_a^b f(x) \; dx$ and $\int_a^b g(x) \; dx$ when $L = 0$?

4. Let $p(x)$ and $q(x)$ be polynomials. Use the Problem 3 to prove each statement.
(a) Assume $q(x)$ has no roots r with $r \geq a$. Then $\int_a^\infty \frac{p(x)}{q(x)} \; dx$ is convergent if and only if $1 + \deg p(x) < \deg q(x)$.
(b) Assume $p(a) \neq 0$ and $x = a$ is a root of $q(x)$ but $q(x)$ has no roots r with $a < r \leq b$. Then $\int_a^b \frac{p(x)}{q(x)} \; dx$ diverges.
(c) Let n be a positive integer. Assume $p(a) \neq 0$ and $x = a$ is a root of $q(x) = (x-a)^k q_1(x)$ where $q_1(x)$ has no roots r with $a \leq r \leq b$. Then $\int_a^b \sqrt[n]{\frac{p(x)}{q(x)}} \; dx$ converges if and only if $k < n$.

5. (a) If $\int_a^\infty f(x) \; dx$ converges and $\lim_{x \to \infty} f(x) \; dx = L$ exists, show that $L = 0$.
(b) Consider the continuous function $f(x)$ with domain $[1, \infty)$ whose graph on the interval $[n, n+1]$ is given in Figure 4.9.21 for every positive integer n.
(i) Show $\lim_{x \to \infty} f(x)$ does not exist. **(ii)** Show $\int_1^\infty f(x) \; dx$ converges.

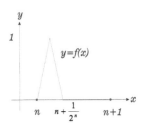

Figure 4.9.21
Problem 5

6. (a) Let $f(x)$ be a continuous function with domain $[a, b)$. If $\int_a^b |f(x)| \; dx$ is convergent, show $\int_a^b f(x) \; dx$ is also convergent.
(b) Give an example of a continuous function $f(x)$ with domain $[1, \infty)$ such that $\int_1^\infty f(x) \; dx$ is convergent, but $\int_1^\infty |f(x)| \; dx$ is divergent.

7. **(a)** Show $\int_0^\infty \frac{\sin x}{x}\, dx$ is a convergent improper integral.

Hint: consider $\int_{2n\pi}^{(2n+1)\pi} \frac{\sin x}{x}\, dx + \int_{(2n+1)\pi}^{(2n+2)\pi} \frac{\sin x}{x}\, dx$.

(b) Show $\int_0^\infty \frac{\sin kx}{x}\, dx$ converges for all nonzero numbers k and its value does n depend on k.

8. Let R be the region bounded by the graph of the continuous function f with doma $[a, \infty)$, the line $x = a$ and the x–axis. Let S be the solid of revolution obtained rotating R around the x–axis. Assume the area of R is finite and $\lim_{x \to \infty} f(x) = 0$. Sho the volume of S is finite.

9. Determine whether $\int_0^\infty \frac{1}{e^x \ln x}\, dx$ converges or diverges.

Applications

4.10 Hyperbolic Functions

Prerequisite: Section 4.3

Certain algebraic combinations of e^x and e^{-x} are called hyperbolic functions. These functions have very nice properties which are analogous to the properties of trigonometric functions. In the first subsection, we define the hyperbolic functions and establish their basic properties. In the second subsection, we study the inverse hyperbolic functions. As with inverse trigonometric functions, certain integrals can be expressed in terms of inverse hyperbolic functions. In the third subsection, we show that hyperbolic functions arise as solutions to the differential equations which describe the shape of a hanging cable and the motion of a falling object subject to air resistance.

Definitions and Basic Properties

We define the six hyperbolic functions. Their names are derived from the names of the six trigonometric functions by adding an "h". The definitions of $\tanh x$, $\coth x$, $\operatorname{sech} x$, $\operatorname{csch} x$ in terms of $\sinh x$, $\cosh x$ are direct analogues of the definitions of the corresponding trigonometric functions. The basic properties of the hyperbolic functions are derived directly from their definitions. We conclude by sketching their graphs.

Definition 4.10.1 *For $x \in \Re$:*

$$\sinh x = \frac{e^x - e^{-x}}{2} \qquad \cosh x = \frac{e^x + e^{-x}}{2}$$

$$\tanh x = \frac{\sinh x}{\cosh x} = \frac{e^x - e^{-x}}{e^x + e^{-x}} \qquad \coth x = \frac{\cosh x}{\sinh x} = \frac{e^x + e^{-x}}{e^x - e^{-x}}, \quad x \neq 0,$$

$$\operatorname{sech} x = \frac{1}{\cosh x} = \frac{2}{e^x + e^{-x}} \qquad \operatorname{csch} x = \frac{1}{\sinh x} = \frac{2}{e^x - e^{-x}}, \quad x \neq 0.$$

Hyperbolic functions satisfy seven types of identities which are almost identical to those satisfied by trigonometric functions. Property (h) below motivates the terminology *circular functions* for trigonometric functions and *hyperbolic functions* for the functions of this section. The first six sets identities below follow directly from the above definitions. We prove the first identity of each type and relegate the verification of the remaining identities to the exercises.

Proposition 4.10.2 *Hyperbolic functions have the following properties.*

(a) $\cosh$ *and* sech *are even functions:*

$$\cosh(-x) = \cosh x, \qquad \operatorname{sech}(-x) = \operatorname{sech} x.$$

(b) $\sinh$, $\tanh$, $\coth$ *and* csch *are odd functions:*

$$\sinh(-x) = -\sinh x, \qquad \tanh(-x) = -\tanh x,$$
$$\coth(-x) = -\coth x, \qquad \operatorname{csch}(-x) = -\operatorname{csch} x.$$

(c) *These functions satisfy the hyperbolic identities:*

$$\cosh^2 x - \sinh^2 x = 1$$
$$1 - \tanh^2 x = \operatorname{sech}^2 x$$
$$\coth^2 x - 1 = \operatorname{csch}^2 x.$$

(d) *The functions* $\sinh$ *and* $\cosh$ *satisfy addition formulas:*

$$\sinh(x+y) = \sinh x \,\cosh y + \cosh x \,\sinh y$$
$$\cosh(x+y) = \cosh x \,\cosh y + \sinh x \,\sinh y.$$

(e) *The functions* $\sinh$ *and* $\cosh$ *satisfy double angle formulas:*

$$\sinh 2x = 2 \sinh x \,\cosh x$$
$$\cosh 2x = \cosh^2 x + \sinh^2 x = 2\cosh^2 x - 1 = 1 + 2\sinh^2 x.$$

(f) *The derivatives of these functions are:*

$$D(\sinh x) = \cosh x \qquad\qquad D(\coth x) = -\operatorname{csch}^2 x$$
$$D(\cosh x) = \sinh x \qquad\qquad D(\operatorname{sech} x) = -\operatorname{sech} x \,\tanh x$$
$$D(\tanh x) = \operatorname{sech}^2 x \qquad\qquad D(\operatorname{csch} x) = -\operatorname{csch} x \,\coth x.$$

(g) *These functions satisfy the integration formulas:*

$$\int \cosh x \, dx = \sinh x + C \qquad\qquad \int \operatorname{csch}^2 x = -\coth x + C$$
$$\int \sinh x \, dx = \cosh x + C \qquad\qquad \int \operatorname{sech} x \,\tanh x \, dx = -\operatorname{sech} x + C$$
$$\int \operatorname{sech}^2 x \, dx = \tanh x + C \qquad\qquad \int \operatorname{csch} x \,\coth x \, dx = -\operatorname{csch} x + C.$$

(h) *The circular sector in Figure 4.10.3 determined by the point* $(\cos t, \sin t)$ *on th*
   unit circle $x^2 + y^2 = 1$ *has area* $\frac{t}{2}$.
   The hyperbolic sector in Figure 4.10.4 determined by the point $(\cosh t, \sinh t)$ *o*
   the unit hyperbola $x^2 - y^2 = 1$ *has area* $\frac{t}{2}$.

Proof (a) We verify $\cosh$ is an even function:

$$\cosh(-x) = \frac{e^{-x} + e^{-(-x)}}{2} = \frac{e^{-x} + e^x}{2} = \cosh x.$$

(b) We prove $\sinh$ is an odd function:

$$\sinh(-x) = \frac{e^{-x} - e^{-(-x)}}{2} = \frac{e^{-x} - e^x}{2} = -\sinh x.$$

(c) We verify the first identity:

$$\cosh^2 x - \sinh^2 x = \frac{(e^x + e^{-x})^2}{4} - \frac{(e^x - e^{-x})^2}{4} = \frac{(e^{2x} + 2 + e^{-2x}) - (e^{2x} - 2 + e^{-2x})}{4}$$

(d) We prove the first identity.

$$\sinh x \cosh y + \cosh x \sinh y = \frac{e^x - e^{-x}}{2} \frac{e^y + e^{-y}}{2} + \frac{e^x + e^{-x}}{2} \frac{e^y - e^{-y}}{2}$$
$$= \frac{e^{x+y} + e^{x-y} - e^{y-x} - e^{-x-y}}{4} + \frac{e^{x+y} - e^{x-y} + e^{y-x} - e^{-x-y}}{4}$$
$$= \frac{2e^{x+y} - 2e^{-x-y}}{4} = \sinh(x+y)$$

(e) The first identity follows from (d):

$$\sinh 2x = \sinh x \cosh x + \cosh x \sinh x = 2 \sinh x \cosh x.$$

(f) We verify the first formula:

$$D(\sinh x) = D \left(\frac{e^x - e^{-x}}{2} \right) = \frac{e^x - (-e^{-x})}{2} = \frac{e^x + e^{-x}}{2} = \cosh x.$$

(g) These integral formulas follow from the derivative formulas in (f) by the Second Fundamental Theorem of Calculus.

(h) The shaded circular sector in Figure 4.10.3 has angle t radians while the entire circle has angle 2π radians. Hence the area of the sector equals $\frac{t}{2\pi}$ of the area $\pi(1)^2$ of the entire circle: $\frac{t}{2\pi} \cdot \pi(1)^2 = \frac{t}{2}$.

The area $A(t)$ of the shaded hyperbolic sector in Figure 4.10.4 is the area of the right triangle with legs of length $\cosh t$, $\sinh t$ minus the area under the hyperbola from $x = 1$ to $x = \cosh t$:

$$A(t) = \frac{1}{2} \cosh t \sinh t - \int_1^{\cosh t} \sqrt{x^2 - 1} \, dx.$$

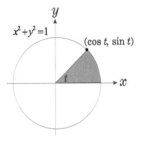

Figure 4.10.3
Circular Sector

We use (f) to differentiate this equation. As in Section 3.4, the chain rule and the First Fundamental Theorem of Calculus give the derivative of the second summand:

$$A'(t) = \frac{1}{2} \cosh^2 t + \frac{1}{2} \sinh^2 t - \sqrt{\cosh^2 t - 1} \; \sinh t = \frac{1}{2} \cosh^2 t + \frac{1}{2} \sinh^2 t - \sinh^2 t$$

$$= \frac{1}{2} \cosh^2 t - \frac{1}{2} \sinh^2 t = \frac{1}{2}.$$

Hence $A(t) = \int A'(t) \, dt = \int \frac{1}{2} \, dt = \frac{t}{2} + C$. When $t = 0$: $(\cosh 0, \sinh 0) = (1, 0)$, the hyperbolic sector degenerates to a line segment and $C = A(0) = 0$. Thus $A(t) = \frac{t}{2}$. □

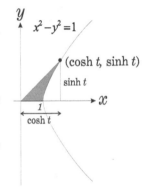

The following examples illustrate how the properties of hyperbolic functions are used in differentiation and integration problems.

Examples 4.10.5 (1) Find the derivative of $f(x) = \cosh \sqrt{1 - x^2}$.

Figure 4.10.4
Hyperbolic Sector

    Solution By the chain rule:

$$f'(x) = \sinh \sqrt{1 - x^2} \, \frac{d}{dx} \left(\sqrt{1 - x^2} \right) = \sinh \sqrt{1 - x^2} \cdot \frac{-2x}{2\sqrt{1 - x^2}}$$

$$= -\frac{x \sinh(\sqrt{1 - x^2})}{\sqrt{1 - x^2}}.$$

(2) Evaluate $\int \sinh^2 x \, dx$.

    Solution By the double angle formula, $\sinh^2 x = \frac{\cosh 2x - 1}{2}$. Hence

$$\int \sinh^2 x \, dx = \int \frac{\cosh 2x - 1}{2} \, dx = \frac{1}{2} \left(\frac{1}{2} \sinh 2x - x \right) + C.$$

(3) Evaluate $\int \tanh^6 x \, \mathrm{sech}^4 x \, dx$.

    Solution Let $u = \tanh x$ with $du = \mathrm{sech}^2 x \, dx$. By the hyperbolic identity:

$$\int \tanh^6 x \, \mathrm{sech}^4 x \, dx = \int \tanh^6 x \left(1 - \tanh^2 x \right) \mathrm{sech}^2 x \, dx$$

$$= \int u^6 \left(1 - u^2 \right) \, du = \int u^6 - u^8 \, du = \frac{u^7}{7} - \frac{u^9}{9} + C$$

$$= \frac{1}{7} \tanh^7 x - \frac{1}{9} \tanh^9 x + C. \qquad \square$$

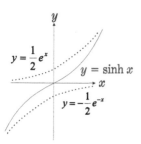

Figure 4.10.6
Graph of $y = \sinh x$

We sketch the graphs of the hyperbolic functions. $f(x) = \sinh x$ has domain Its derivative $f'(x) = \cosh x > 0$ for $x \in \Re$, so the function f is increasing. Th the origin is the only x–intercept and the y–intercept. Note $f''(x) = \sinh x$ which negative for $x < 0$ and positive for $x > 0$. Thus the graph of f is concave down $x < 0$, concave up for $x > 0$ and has an inflection point at $x = 0$. Write

$$f(x) = \frac{1}{2}e^x - \frac{1}{2}e^{-x}.$$

As x approaches $-\infty$, the first summand goes to zero through positive numbers. Her the graph of f has the curve $y = -\frac{1}{2}e^{-x}$ as a curved asymptote on the left, which approaches from the top. On the other hand, as x approaches ∞ the second summa above goes to zero through negative numbers. Hence the graph of f has the cur $y = \frac{1}{2}e^x$ as a curved asymptote on the right, which it approaches from the bottom. particular, $\displaystyle\lim_{x \to -\infty} f(x) = \lim_{x \to -\infty} -\frac{1}{2}e^{-x} = -\infty$ while $\displaystyle\lim_{x \to \infty} f(x) = \lim_{x \to \infty} \frac{1}{2}e^x = \infty$. the Intermediate Value Theorem, the range of f is $\Re$. The graph of $f(x) = \sinh x$ sketched in Figure 4.10.6.

The function $g(x) = \cosh x$ also has domain $\Re$. Its derivative $g'(x) = \sinh x$ negative for $x < 0$ and positive for $x > 0$. Thus g is decreasing for $x < 0$, increasi for $x > 0$ and has an absolute minimum at $x = 0$. As $g(0) = 1$, the graph of g has x–intercept. Since $g''(x) = \cosh x > 0$ for $x \in \Re$, the graph of g is concave up and h no inflection points. Write

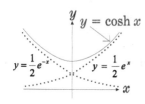

Figure 4.10.7
Graph of $y = \cosh x$

$$g(x) = \frac{1}{2}e^x + \frac{1}{2}e^{-x}.$$

As x approaches $-\infty$, the first summand goes to zero through positive number Hence the graph of f has the curve $y = \frac{1}{2}e^{-x}$ as a curved asymptote on the lef which it approaches from the top. On the other hand, as x approaches ∞ the secon summand above goes to zero through positive numbers. Hence the graph of f h the curve $y = \frac{1}{2}e^x$ as a curved asymptote on the right, which it approaches from th top. In particular, $\displaystyle\lim_{x \to -\infty} g(x) = \lim_{x \to -\infty} \frac{1}{2}e^{-x} = \infty$, and $\displaystyle\lim_{x \to \infty} g(x) = \lim_{x \to \infty} \frac{1}{2}e^x = \infty$ By the Intermediate Value Theorem, $g(x)$ takes on all values greater than or equal t its absolute minimium value $g(0) = 1$. Thus the range of g is $[1, \infty)$. The graph $g(x) = \cosh x$ is sketched in Figure 4.10.7.

The function $h(x) = \tanh x = \frac{\sinh x}{\cosh x}$ has domain $\Re$ because $\cosh x$ is alway positive. In addition $h'(x) = \text{sech}^2 x > 0$ for all x, so the function h is increas ing. Thus the only x–intercept and the y–intercept of h occurs at the origin. Not $h''(x) = -2\text{sech}^2 x \tanh x$ is positive for $x < 0$ and negative for $x > 0$. Thus the grap of h is concave up for $x < 0$, concave down for $x > 0$ and has an inflection point a $x = 0$. Observe

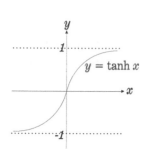

Figure 4.10.8
Graph of $y = \tanh x$

$$\lim_{x \to \infty} \tanh x = \lim_{x \to \infty} \frac{e^x - e^{-x}}{e^x + e^{-x}} \cdot \frac{e^{-x}}{e^{-x}} = \lim_{x \to \infty} \frac{1 - e^{-2x}}{1 + e^{-2x}} = 1 \quad \text{and}$$

$$\lim_{x \to -\infty} \tanh x = \lim_{x \to -\infty} \frac{e^x - e^{-x}}{e^x + e^{-x}} \cdot \frac{e^x}{e^x} = \lim_{x \to -\infty} \frac{e^{2x} - 1}{e^{2x} + 1} = -1.$$

Thus the graph of h has the line $y = 1$ as a horizontal asymptote on the right and th line $y = -1$ as a horizontal asymptote on the left. Hence range of h is $(-1, 1)$. Th graph of $h(x) = \tanh x$ is sketched in Figure 4.10.8.

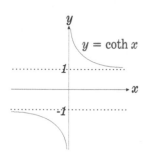

Figure 4.10.9
Graph of $y = \coth x$

The graph of $k(x) = \coth x = \frac{1}{\tanh x}$ consists of the points $\left(x, \frac{1}{y}\right)$ for (x, y) on th graph of $h(x) = \tanh x$. The domain of k is $(-\infty, 0) \cup (0, \infty)$, the numbers where $\tanh x \neq 0$. Since $\tanh 0 = 0$, the graph of k has the y–axis as a vertical asymptote As x approaches infinity, $h(x) = \frac{1}{\tanh x}$ approaches one. Hence the graph of k has th line $y = 1$ as a horizontal asymptote on the right. As x approaches $-\infty$, $h(x) = \frac{1}{\tanh x}$ approaches -1. Hence the graph of k has the line $y = -1$ as a horizontal asymptote on the left. The graph of k decreases from its horizontal asymptote $x = -1$ to its vertical asymptote at the y–axis. Then it decreases from the y–axis to its horizontal asymptote $x = 1$. Thus the range of k is $(-\infty, -1) \cup (1, \infty)$. The graph of k is sketched in Figure 4.10.9.

The graph of $m(x) = \operatorname{sech} x = \frac{1}{\cosh x}$ consists of the points $\left(x, \frac{1}{y}\right)$ for (x, y) on the graph of $g(x) = \cosh x$. The domain of m is $\Re$ because $\cosh x$ is never zero. Since $\cosh x$ approaches infinity when x approaches either $-\infty$ or $+\infty$, the graph of m has the x–axis as a horizontal asymptote on both the left and right. The graph of m increases from the x–axis to its absolute maximum at $(0, 1)$ and then decreases to the x–axis. Hence the range of m is $(0, 1]$. The graph of m is sketched in Figure 4.10.10.

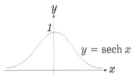

Figure 4.10.10
Graph of $y = \operatorname{sech} x$

The graph of $n(x) = \operatorname{csch} x = \frac{1}{\sinh x}$ consists of $\left(x, \frac{1}{y}\right)$ for (x, y) on the graph of $f(x) = \sinh x$. The domain of n is $(-\infty, 0) \cup (0, \infty)$, the numbers where $\sinh x \neq 0$. Since $\sinh 0 = 0$, the graph of n has the y–axis as a vertical asymptote. The graph of $\sinh x$ approaches $-\infty$ when x approaches minus infinity and approaches ∞ when x approaches infinity. Hence the graph of n has the x–axis as a horizontal asymptote on both the left and the right. The graph of n decreases from the x–axis, its horizontal asymptote on the left, to its vertical asymptote, the y–axis. Then it decreases from the y–axis to the x–axis, its horizontal asymptote on the right. Thus the range of n consists of all nonzero numbers. The graph of n is sketched in Figure 4.10.11.

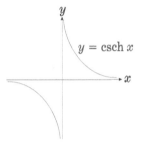

Inverse Hyperbolic Functions

We define the inverse functions of the hyperbolic functions, determine their graphs and compute their derivatives. As with inverse trigonometric functions, the corresponding integration formulas are useful. We conclude by deriving identities which express the inverse hyperbolic functions of x as natural logarithms of algebraic expressions in x.

Figure 4.10.11
Graph of $y = \operatorname{csch} x$

Recall $y = f^{-1}(x)$ is the inverse function of $x = f(y)$ means

$$y = f^{-1}(x) \text{ if and only if } x = f(y).$$

Using this notation:

$$f^{-1}(f(y)) = f^{-1}(x) = y \quad \text{and} \quad f(f^{-1}(x)) = f(y) = x.$$

When $f(y_1) = f(y_2)$, then $f^{-1}(x)$ equals both y_1 and y_2. Therefore, the definition of f^{-1} defines a function only when $f(y_1) = f(y_2)$ implies $y_1 = y_2$. A function f with this property is called one–to–one. Note that an increasing function is one–to–one. Hence $\sinh$ and $\tanh$ are one–to–one, and the functions $\sinh^{-1}$ and $\tanh^{-1}$ exist. Similarly, a decreasing function is one–to–one. Hence $\coth$ and csch are one–to–one, and the functions $\tanh^{-1}$ and csch^{-1} exist.

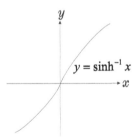

We begin by studying $y = f^{-1}(x) = \sinh^{-1} x$, the inverse function of $x = f(y) = \sinh y$. The domain of $\sinh^{-1}$ equals the range of $\sinh$ which is $\Re$. The range of $\sinh^{-1}$ equals the domain of $\sinh$ which is also $\Re$. Thus,

Figure 4.10.12
Inverse of $y = \sinh x$

$$\sinh^{-1}(\sinh y) = y \quad \text{and} \quad \sinh\left(\sinh^{-1} x\right) = x \qquad (4.10.1)$$

for $x, y \in \Re$. The graph of $f^{-1}(x) = \sinh^{-1} x$ in Figure 4.10.12 is obtained by reflecting the graph of $f(x) = \sinh x$ about the line $y = x$.

Consider $y = h^{-1}(x) = \tanh^{-1} x$, the inverse function of $x = h(y) = \tanh y$. The domain of $\tanh^{-1}$ equals the range of $\tanh$ which is $(-1, 1)$. The range of $\tanh^{-1}$

equals the domain of tanh which is $\Re$. Thus

$$\tanh^{-1}(\tanh y) = y \quad \text{and} \quad \tanh\left(\tanh^{-1} x\right) = x \qquad (4.10$$

for $-1 < x < 1$ and $y \in \Re$. The graph of $h^{-1}(x) = \tanh^{-1} x$ in Figure 4.10.13
obtained by reflecting the graph of $h(x) = \tanh x$ about the line $y = x$.

Consider $y = k^{-1}(x) = \coth^{-1} x$, the inverse function of $x = k(y) = \coth y$. T
domain of $\coth^{-1}$ equals the range of coth which is $(-\infty, -1) \cup (1, \infty)$. The range
$\coth^{-1}$ equals the domain of coth which is the set of nonzero numbers. Thus

$$\coth^{-1}(\coth y) = y \quad \text{and} \quad \coth\left(\coth^{-1} x\right) = x \qquad (4.10$$

for $|x| > 1$ and $y \neq 0$. The graph of $k^{-1}(x) = \coth^{-1} x$ in Figure 4.10.14 is obtain
by reflecting the graph of $k(x) = \coth x$ about the line $y = x$.

Consider $y = n^{-1}(x) = \operatorname{csch}^{-1} x$, the inverse function of $x = n(y) = \operatorname{csch} y$. T
domain of csch^{-1} equals the range of csch, the set of nonzero numbers. The range
csch^{-1} equals the domain of csch which is also the set of nonzero numbers. Thus

$$\operatorname{csch}^{-1}(\operatorname{csch} y) = y \quad \text{and} \quad \operatorname{csch}\left(\operatorname{csch}^{-1} x\right) = x \qquad (4.10.$$

for $x, y \neq 0$. The graph of $n^{-1}(x) = \operatorname{csch}^{-1} x$ in Figure 4.10.15 is obtained by reflecti
the graph of $n(x) = \operatorname{csch} x$ about the line $y = x$.

The remaining two hyperbolic functions cosh and sech are even functions, i.
$\cosh(-x) = \cosh x$ and $\operatorname{sech}(-x) = \operatorname{sech} x$. Hence cosh and sech are not one–to–on
However, if we restrict their domains from $\Re$ to $[0, \infty)$, the resulting functions a
one–to–one. See the left diagrams in Figures 4.10.16 and 4.10.17. We define $\cosh^{-}$
and sech^{-1} as the inverse functions of these restrictions.

Consider the one–to–one function $x = G(y) = \cosh y$ with domain $[0, \infty)$. The
$y = G^{-1}(x) = \cosh^{-1} x$ has range $[0, \infty)$. The domain of $\cosh^{-1}$ equals the range
G which is $[1, \infty)$. Thus for $x \geq 1$ and $y \geq 0$:

$$\cosh^{-1}(\cosh y) = y \quad \text{and} \quad \cosh\left(\cosh^{-1} x\right) = x. \qquad (4.10.$$

The graph of $G^{-1}(x) = \cosh^{-1} x$, in the right diagram of Figure 4.10.16, is obtaine
by reflecting the graph of $G(x) = \cosh x$ about the line $y = x$.

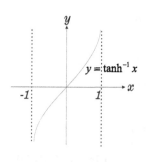

Figure 4.10.13
Inverse of $y = \tanh x$

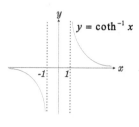

Figure 4.10.14
Inverse of $y = \coth x$

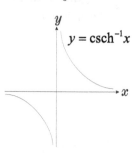

Figure 4.10.15
Inverse of $y = \operatorname{csch} x$

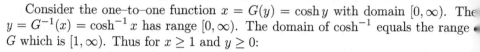
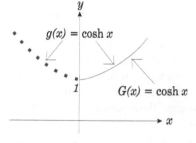

Figure 4.10.16 The Inverse Function of $y = \cosh x$

Now consider the one–to–one function $x = M(y) = \operatorname{sech} y$ with domain $[0, \infty)$
Then $y = M^{-1}(x) = \operatorname{sech}^{-1} x$ has range $[0, \infty)$. The domain of sech^{-1} equals th
range of sech which is $(0, 1]$. Thus for $0 < x \leq 1$ and $y \geq 0$:

$$\operatorname{sech}^{-1}(\operatorname{sech} y) = y \quad \text{and} \quad \operatorname{sech}\left(\operatorname{sech}^{-1} x\right) = x. \qquad (4.10.6$$

The graph of $M^{-1}(x) = \operatorname{sech}^{-1} x$, in the right diagram of Figure 4.10.17, is obtained
by reflecting the graph of $M(x) = \operatorname{sech} x$ about the line $y = x$.

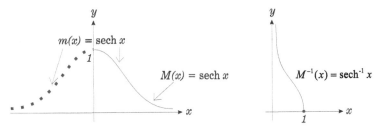

Figure 4.10.17 The Inverse Function of $y = \operatorname{sech} x$

Examples 3 and 5 below illustrate that one must be careful when evaluating $\cosh^{-1}(\cosh y)$ and $\operatorname{sech}^{-1}(\operatorname{sech} y)$ for y a negative number.

Examples 4.10.18 (1) $\tanh\left(\tanh^{-1}\frac{1}{7}\right) = \frac{1}{7}$.

(2) $\sinh^{-1}(\sinh 19) = 19$.

(3) $\cosh^{-1}(\cosh -57) = \cosh^{-1}(\cosh 57) = 57$ because -57 is not in the domain of $G(y) = \cosh y$ whose inverse is $y = \cosh^{-1} x$.

(4) $\coth^{-1}(\coth -83) = -83$.

(5) $\operatorname{sech}^{-1}(\operatorname{sech} -26) = \operatorname{sech}^{-1}(\operatorname{sech} 26) = 26$ because -26 is not in the domain of $M(y) = \operatorname{sech} y$ whose inverse is $y = \operatorname{sech}^{-1} x$.   □

By Theorem 2.5.4 each of the six inverse hyperbolic functions $y = F^{-1}(x)$ is differentiable. Moreover,

$$D(F^{-1}(x)) = \frac{1}{DF(F^{-1}(x))} = \frac{1}{DF(y)}. \qquad (4.10.7)$$

Proposition 4.10.19 *The inverse hyperbolic functions have the following derivatives:*

(a) $D\left(\sinh^{-1}\right)(x) = \frac{1}{\sqrt{x^2+1}}$ *for* $x \in \Re$; **(d)** $D\left(\coth^{-1}\right)(x) = \frac{1}{1-x^2}$ *for* $|x| > 1$;

(b) $D\left(\cosh^{-1}\right)(x) = \frac{1}{\sqrt{x^2-1}}$ *for* $x > 1$; **(e)** $D\left(\operatorname{sech}^{-1}\right)(x) = -\frac{1}{x\sqrt{1-x^2}}$ *for* $0 < x < 1$;

(c) $D\left(\tanh^{-1}\right)(x) = \frac{1}{1-x^2}$ *for* $|x| < 1$; **(f)** $D\left(\operatorname{csch}^{-1}\right)(x) = -\frac{1}{|x|\sqrt{1+x^2}}$ *for* $x \neq 0$.

Proof We prove (a), (b), (f) and leave the proofs of (c), (d), (e) as exercises.

(a) $y = f^{-1}(x) = \sinh^{-1} x$ is the inverse function of $x = f(y) = \sinh y$. By (4.10.7):

$$D(\sinh^{-1}(x)) = \frac{1}{D(\sinh y)} = \frac{1}{\cosh y}.$$

By the hyperbolic identity, $\cosh^2 y = 1 + \sinh^2 y = 1 + x^2$. Since $\cosh y$ is always positive, $\cosh y = +\sqrt{1 + x^2}$, and $D\left(\sinh^{-1}\right)(x) = \frac{1}{\sqrt{1+x^2}}$.

(b) $y = G^{-1}(x) = \cosh^{-1} x$ is the inverse function of $x = G(y) = \cosh y$. By (4.10.7):

$$D(\cosh^{-1}(x)) = \frac{1}{D(\cosh y)} = \frac{1}{\sinh y}.$$

By the hyperbolic identity, $\sinh^2 y = \cosh^2 y - 1 = x^2 - 1$. The domain of $G(y) = \cosh y$ is $[0, \infty)$ where $\sinh y$ is always positive. Hence $\sinh y = +\sqrt{x^2 - 1}$, and $D\left(\cosh^{-1}\right)(x) = \frac{1}{\sqrt{x^2-1}}$.

(f) $y = n^{-1}(x) = \operatorname{csch}^{-1} x$ is the inverse function of $x = n(y) = \operatorname{csch} y$. By (4.10.7):

$$D(\operatorname{csch}^{-1}(x)) = \frac{1}{D(\operatorname{csch} y)} = \frac{1}{-\operatorname{csch} y \,\coth y}.$$

By the hyperbolic identity, $\coth^2 y = 1 + \operatorname{csch}^2 y = 1 + x^2$. For $y \neq 0$, $\coth y$ a $\operatorname{csch} y$ are both positive or both negative. Hence their product is always positive, a $D\left(\operatorname{csch}^{-1}\right)(x) = -\frac{1}{|x|\sqrt{1+x^2}}$.

By the Fundamental Theorem of Calculus, the differentiation formulas of Prop sition 4.10.19 translate into integration formulas.

Corollary 4.10.20 (a) $\int \frac{1}{\sqrt{x^2+1}}\,dx = \sinh^{-1} x + C$ *for* $x \in \Re$.

(b) $\int \frac{1}{\sqrt{x^2-1}}\,dx = \cosh^{-1} x + C$ *for* $x > 1$.

(c) $\int \frac{1}{1-x^2}\,dx = \tanh^{-1} x + C$ *for* $|x| < 1$.

(d) $\int \frac{1}{1-x^2}\,dx = \coth^{-1} x + C$ *for* $|x| > 1$.

(e) $\int \frac{1}{x\sqrt{1-x^2}}\,dx = -\operatorname{sech}^{-1} x + C$ *for* $0 < x < 1$.

(f) $\int \frac{1}{|x|\sqrt{1+x^2}}\,dx = -\operatorname{csch}^{-1} x + C$ *for* $x \neq 0$.

The following examples illustrate the use of inverse hyperbolic functions in diffe entiation and integration problems.

Examples 4.10.21 (1) Find the derivative of $y = \cosh^{-1}(\sec x)$.

Solution By the chain rule,

$$\frac{d}{dx}\left[\cosh^{-1}(\sec x)\right] = \frac{1}{\sqrt{\sec^2 x - 1}} \cdot \frac{d}{dx}(\sec x) = \frac{1}{\tan x} \cdot \sec x \tan x = \sec x.$$

Thus we have an alternate form of the value of $\int \sec x \, dx$:

$$\int \sec x \, dx = \cosh^{-1}(\sec x) + C. \qquad (4.10.8$$

(2) Evaluate $\int \frac{1}{x\sqrt{9-x^2}}\,dx$.

Solution Let $x = 3u$ with $dx = 3\,du$. Then

$$\int \frac{1}{x\sqrt{9-x^2}}\,dx = \int \frac{1}{3u\sqrt{9-9u^2}}\,3\,du = \frac{1}{3}\int \frac{1}{u\sqrt{1-u^2}}\,du$$

$$= -\frac{1}{3}\operatorname{sech}^{-1} u + C = -\frac{1}{3}\operatorname{sech}^{-1}\frac{x}{3} + C.$$

(3) Evaluate $\int \frac{e^x}{\sqrt{1+e^{2x}}}\,dx$.

Solution Let $u = e^x$ with $du = e^x \, dx$. Then

$$\int \frac{e^x}{\sqrt{1+e^{2x}}}\,dx = \int \frac{1}{\sqrt{1+u^2}}\,du = \sinh^{-1} u + C = \sinh^{-1}(e^x) + C. \qquad \square$$

The hyperbolic functions are not new functions, merely convenient algebraic com binations of e^x and e^{-x}. So too, the inverse hyperbolic functions are not new func tions. The following corollary expresses them as logarithms of algebraic expressions Nevertheless, the inverse hyperbolic functions are useful because they give simple for mulations of the values of the integrals of Corollary 4.10.20.

Corollary 4.10.22 (a) $\sinh^{-1} x = \ln\left(x + \sqrt{x^2 + 1}\right)$ *for* $x \in \Re$.

(b) $\cosh^{-1} x = \ln(x + \sqrt{x^2 - 1})$ *for* $x \geq 1$.

(c) $\tanh^{-1} x = \frac{1}{2}\ln\left(\frac{1+x}{1-x}\right)$ *for* $|x| < 1$. **(d)** $\coth^{-1} x = \frac{1}{2}\ln\left(\frac{x+1}{x-1}\right)$ *for* $|x| > 1$.

(e) $\operatorname{sech}^{-1} x = \ln\left(\frac{1+\sqrt{1-x^2}}{x}\right)$ *for* $0 < x \leq 1$. **(f)** $\operatorname{csch}^{-1} x = \ln\left(\frac{1}{x} + \frac{\sqrt{1+x^2}}{|x|}\right)$ *for* $x \neq 0$.

Proof There are three methods to verify each of these formulas. We illustrate these methods in proving (a), (b), (c). The proofs of (d), (e), (f) are given as exercises.

(a) If $y = \sinh^{-1} x$, then $x = \sinh y = \frac{e^y - e^{-y}}{2}$. We solve this equation for y. First, multiply the equation by $2e^y$: $2xe^y = e^{2y} - 1$. This is a quadratic equation in e^y:

$$\left(e^y\right)^2 - 2xe^y - 1 = 0.$$

Solve for e^y by the quadratic formula:

$$e^y = \frac{2x \pm \sqrt{4x^2 + 4}}{2} = x + \sqrt{x^2 + 1}$$

since e^y is always positive. Hence $y = \ln(e^y) = \ln(x + \sqrt{x^2 + 1})$.

(b) We evaluate $\int \frac{1}{\sqrt{x^2-1}}\, dx$ in two different ways. The two values of this indefinite integral must be equal up to a constant. First, use the trigonometric substitution $x = \sec\theta$. Then $\sqrt{x^2 - 1} = \tan\theta$ and $dx = \sec\theta\tan\theta\, d\theta$. Thus

$$\int \frac{1}{\sqrt{x^2-1}}\, dx = \int \frac{\sec\theta\tan\theta}{\tan\theta}\, d\theta = \int \sec\theta\, d\theta = \ln|\sec\theta + \tan\theta| + C = \ln\left(x + \sqrt{x^2-1}\right) + C$$

By Corollary 4.10.20(b), $\int \frac{1}{\sqrt{x^2-1}}\, dx = \cosh^{-1} x + C$. Thus $\cosh^{-1} x = \ln\left(x + \sqrt{x^2-1}\right) + C$. Substitute $x = 1$: $0 = \cosh^{-1} 1 = \ln\left(1 + \sqrt{0}\right) + C = C$.

(c) We verify that both sides of the asserted equation have the same derivative. By the chain and quotient rules:

$$\frac{d}{dx}\left[\frac{1}{2}\ln\left(\frac{1+x}{1-x}\right)\right] = \frac{1}{2}\frac{1}{\frac{1+x}{1-x}}\frac{(1)(1-x) - (-1)(1+x)}{(1-x)^2}$$

$$= \frac{1}{2}\frac{2}{(1+x)(1-x)} = \frac{1}{1-x^2} = \frac{d}{dx}\left(\tanh^{-1} x\right)$$

by Proposition 4.10.19(c). Hence $\tanh^{-1} x = \frac{1}{2}\ln\left(\frac{1+x}{1-x}\right) + C$. Set $x = 0$:
$0 = \tanh^{-1} 0 = \frac{1}{2}\ln\frac{1+0}{1-0} + C = C$. $\qquad\square$

Applications of Hyperbolic Functions

We study two types of differential equations whose solutions are expressed in terms of hyperbolic functions. First, we solve the differential equation which describes the position of a hanging cable. Then we solve the differential equation which describes the motion of a falling body subject to air resistance.

Consider the hanging cable of Figure 4.10.23 which is supported at both of its ends. Assume the cable has constant density ρ. Let τ denote the tension of the cable at its lowest point P. Using elementary physics, it can be shown that the equation of this cable $y = f(x)$ satisfies the differential equation:

$$y'' = \frac{\rho}{\tau}\sqrt{1 + (y')^2}.$$

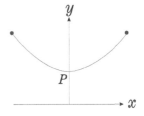

Figure 4.10.23
Hanging Cable

Position the x–axis so that the lowest point of the cable occurs at $x = 0$. Then $y = f(x)$ has a local minimum at $x = 0$ and $y'(0) = 0$. The following proposition gives the solution of this initial value problem.

Proposition 4.10.24 *The initial value problem*

$$y'' = k\sqrt{1 + (y')^2} \quad \text{with} \quad y'(0) = 0$$

has the unique solution

$$y = \frac{1}{k}\cosh kx + y(0) - \frac{1}{k}.$$

Proof Let $u = y'$. Restate this differential equation as:

$$u' = k\sqrt{1 + u^2} \quad \text{with} \quad u(0) = 0.$$

Integrate the equation $\frac{1}{\sqrt{1+u^2}}\frac{du}{dx} = k$ with respect to x:

$$\int \frac{1}{\sqrt{1+u^2}}\frac{du}{dx}\,dx \;=\; \int k\,dx \;=\; kx + C$$

$$\sinh^{-1} u \;=\; \int \frac{1}{\sqrt{1+u^2}}\,du \;=\; kx + C.$$

Substitute $x = 0$: $C = \sinh^{-1}u(0) = \sinh^{-1}y'(0) = \sinh^{-1}0 = 0$. Thus

$$\frac{dy}{dx} \;=\; u = \sinh(kx) \quad \text{and} \quad y = \int \sinh(kx)\,dx = \frac{1}{k}\cosh(kx) + B.$$

Substitute $x = 0$: $y(0) = \frac{1}{k}\cosh 0 + B = \frac{1}{k} + B$ and $B = y(0) - \frac{1}{k}$.

The curve $y = A + B\cosh Cx$ of Figure 4.10.23, which describes the shape of hanging cable, is called a *catenary*.

Examples 4.10.25 A telephone wire of density 5 kilograms per meter hangs betwee the tops of two poles of equal height which are 30 meters apart.
(a) What is the relationship between the distance s that the wire dips at its lowe point P and the tension τ at P?
(b) If the tension at P is 100 Newtons, how far does the cable dip?
(c) If the cable dips 5 meters, estimate the tension at P.

Solution (a) Let P be the origin of our coordinate system. Then $y(0) = 0$. The tele phone poles are at $x = \pm 15$. By Prop. 4.10.24 with $k = \frac{5}{\tau}$, the equation of the wire i

$$y = \frac{\tau}{5}\cosh\frac{5x}{\tau} - \frac{\tau}{5} = \frac{\tau}{5}\left(\cosh\frac{5x}{\tau} - 1\right).$$

The amount s that the cable dips is the height of the right telephone pole at $x = 1\vphantom{5}$

$$s = y(15) = \frac{\tau}{5}\left(\cosh\frac{75}{\tau} - 1\right). \tag{4.10.9}$$

(b) By the preceding equation, $s = 20\,(\cosh .75 - 1) \approx 5.9$ meters.
(c) By (4.10.9):

$$5 = \frac{\tau}{5}\left(\cosh\frac{75}{\tau} - 1\right) \quad \text{and} \quad 1 + \frac{25}{\tau} = \cosh\frac{75}{\tau}.$$

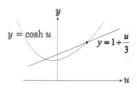

$y = \cosh u$
$y = 1 + \dfrac{u}{3}$

Figure 4.10.26
Solving $1 + \frac{u}{3} = \cosh u$

Let $u = \frac{75}{\tau}$. We need to solve $1 + \frac{u}{3} = \cosh u$ for $u > 0$. The graphs in Figure 4.10.2 show there is a unique solution. We use trial and error to estimate this solution.

u	$\cosh u$	$1 + \frac{u}{3}$
1	1.5431	1.3333
.7	1.2552	1.2333
.6	1.1855	1.2000
.65	1.2188	1.2167
.64	1.2119	1.2133
.645	1.2153	1.2150

Thus $u = .64$ approximates the solution, and $\tau = \frac{75}{u} \approx 117$ Newtons.   □

We will use the differential equation of the following proposition to describe the motion of a falling object subject to air resistance when the air resistance is proportional to the square of the velocity of the object.

Proposition 4.10.27 *Let p, $q > 0$. The differential equation*

$$y'' = q^2(y')^2 - p^2$$

has the solution

$$y = B - \frac{1}{q^2} \ln \cosh (pqx - C) \quad with$$

$$y' = \frac{p}{q} \tanh (C - pqx) \quad and \quad \lim_{x \to \infty} y'(x) = -\frac{p}{q}. \qquad (4.10.10)$$

Proof Let $u = \frac{q}{p}y'$ with $y' = \frac{p}{q}u$. Rewrite our differential equation as $\frac{p}{q}u' = p^2u^2 - p^2 = -p^2(1 - u^2)$ or $\frac{1}{1-u^2}\frac{du}{dx} = -pq$. Integrate this equation with respect to x:

$$C - pqx = \int -pq \, dx = \int \frac{1}{1-u^2}\frac{du}{dx} \, dx = \int \frac{1}{1-u^2} \, du = \tanh^{-1} u$$

$$y' = \frac{p}{q}u = \frac{p}{q} \tanh (C - pqx).$$

Since tanh has the horizontal asymptote $y = -1$ on the left, $\lim_{x \to \infty} y'(x) = -\frac{p}{q}$.
Let $z = \cosh (C - pqx)$ with $dz = -pq \sinh (C - pqx) \, dx$. Then

$$y = \frac{p}{q} \int \tanh(C - pqx) \, dx = -\frac{1}{q^2} \int \frac{-pq \sinh (C - pqx)}{\cosh (C - pqx)} \, dx = -\frac{1}{q^2} \int \frac{1}{z} \, dz$$

$$= -\frac{1}{q^2} \ln |z| + B = -\frac{1}{q^2} \ln \cosh (C - pqx) + B \qquad □$$

We apply Proposition 4.10.27 to describe the motion of a falling object which is subject to air resistance.

Examples 4.10.28 Consider an object of mass m which is falling from rest with the air resistance proportional to the square of its velocity. Find the velocity and position of this object at time t. Also find the limiting velocity as t approaches infinity.

Solution Let the coordinate axis point upwards with origin at the initial position of the object. Let $s(t)$, $v(t) = s'(t)$, $a(t) = s''(t)$ denote the position, velocity, acceleration, respectively, of this object at time t. Then $s(0) = v(0) = 0$. Gravity accelerates this object downwards with constant acceleration $-g$ and constant force $-mg$. Air resistance is an upwards force of magnitude $kv(t)^2$, $k > 0$. By Newton's second law of motion, the net force $kv(t)^2 - mg$ on this object equals its mass times its acceleration:

$$ma(t) = kv(t)^2 - mg \quad \text{or} \quad s''(t) = \frac{k}{m}[s'(t)]^2 - g.$$

Apply Proposition 4.10.27 with $p = \sqrt{g}$ and $q = \sqrt{\frac{k}{m}}$:

$$v(t) = s'(t) = \frac{p}{q} \tanh(C - pqt) = \sqrt{\frac{g}{k/m}} \tanh \left(C - t\sqrt{\frac{gk}{m}} \right).$$

Moreover, $0 = v(0) = \sqrt{\frac{gm}{k}} \tanh C$, and $C = 0$. Hence

$$v(t) = s'(t) = -\sqrt{\frac{gm}{k}} \tanh\left(t\sqrt{\frac{gk}{m}}\right).$$

By (4.10.10), the limiting velocity as t approaches infinity is:

$$\lim_{t \to \infty} v(t) = -\frac{p}{q} = -\sqrt{\frac{gm}{k}}.$$

By Proposition 4.10.27:

$$s(t) = B - \frac{1}{q^2} \ln \cosh (pqt - C) = B - \frac{m}{k} \ln \cosh \left(t\sqrt{\frac{gk}{m}}\right).$$

Note $0 = s(0) = B - \frac{m}{k} \ln \cosh 0 = B - \frac{m}{k} \ln 1 = B$. Hence

$$s(t) = -\frac{m}{k} \ln \cosh \left(t\sqrt{\frac{gk}{m}}\right).$$

Historical Remarks

Galileo Galilei studied a hanging cable and conjectured that its shape is a parabol
In 1646 Christiaan Huygens showed that Galileo's conjecture is false. The proble
of finding this curve was raised again by Jakob Bernoulli in 1690. Within a ye
solutions were found by Jakob Bernoulli, Johann Bernoulli, Huygens and Gottfri
Wilhelm Leibniz. The latter named the solution a *catenary*.

Hyperbolic functions next arose in 1748 when Leonhard Euler found infinite seri
and infinite product representations for $\frac{e^x - e^{-x}}{2}$ and $\frac{e^x + e^{-x}}{2}$. In 1768, Johann Heinric
Lambert introduced the current names for the hyperbolic functions and establishe
their basic properties. These functions were of sufficient significance in applicatior
that Christof Gudermann published a seven place table of the values of $\log_{10} \sinh$
and $\log_{10} \cosh x$ in 1832.

Summary

The definitions of the hyperbolic functions as well as their graphs and basic propertie
should be understood. In particular, the reader should be able to evaluate derivative
and integrals which involve hyperbolic functions. In addition, the graphs and basi
properties of the inverse hyperbolic functions should be known. The reader should b
able to identify the values of certain integrals in terms of inverse hyperbolic function:
She should also understand how hyperbolic functions are useful in the solution c
certain differential equations.

Basic Exercises

1. Approximate each of the 6 hyperbolic functions at x to 3 significant figures when
(a) $x = 0$;   **(b)** $x = 1$;   **(c)** $x = -1$;   **(d)** $x = 2$;   **(e)** $x = \ln 2$.

2. In each case below, use the given value of one of the hyperbolic functions t
determine the values of the other five.
    (a) $\sinh x = -2$ **(b)** $\cosh y = 4\ (y > 0)$ **(c)** $\tanh z = \frac{1}{3}$
    (d) $\coth t = -5$ **(e)** $\operatorname{sech} u = \frac{2}{5}\ (u < 0)$ **(f)** $\operatorname{csch} v = 3$

3. Show $\operatorname{sech} (-x) = \operatorname{sech} x$.

4. Verify each of these identities:
(a) $\tanh(-x) = -\tanh x$; **(b)** $\coth(-x) = -\coth x$; **(c)** $\operatorname{csch}(-x) = -\operatorname{csch} x$.

5. Show: **(a)** $1 - \tanh^2 x = \operatorname{sech}^2 x$; **(b)** $\coth^2 x - 1 = \operatorname{csch}^2 x$.

6. Prove each of the following identities:
(a) $\cosh(x + y) = \cosh x \cosh y + \sinh x \sinh y$.
(b) $\sinh(x - y) = \sinh x \cosh y - \cosh x \sinh y$.
(c) $\cosh(x - y) = \cosh x \cosh y - \sinh x \sinh y$.
(d) $\cosh 2x = \cosh^2 x + \sinh^2 x = 2\cosh^2 x - 1 = 1 + 2\sinh^2 x$.

7. **(a)** Find an identity for $\tanh(x + y)$ in terms of $\tanh x$ and $\tanh y$.
(b) Find an identity for $\tanh(x - y)$ in terms of $\tanh x$ and $\tanh y$.
(c) Find an identity for $\tanh(2x)$ in terms of $\tanh x$.

8. **(a)** Find an identity for $\sinh x \cosh y$ in terms of $\sinh(x + y)$ and $\sinh(x - y)$.
(b) Find an identity for $\sinh x \sinh y$ in terms of $\cosh(x + y)$ and $\cosh(x - y)$.
(c) Find an identity for $\cosh x \cosh y$ in terms of $\cosh(x + y)$ and $\cosh(x - y)$.

9. Show that for all $t \in \Re$, the point $(\cosh t, \sinh t)$ lies on the right branch of the hyperbola $x^2 - y^2 = 1$ while $(-\cosh t, \sinh t)$ lies on the left branch of the hyperbola.

10. Verify the formulas for $D(\cosh x)$, $D(\tanh x)$, $D(\coth x)$, $D(\operatorname{sech} x)$, $D(\operatorname{csch} x)$ stated in Proposition 4.10.2(f).

11. Find the derivative of each function.
(a) $y = \tanh(5x + 1)$ **(b)** $y = \sqrt{1 + \cosh x}$ **(c)** $y = \frac{x^2 + \tanh x}{x^2 - \coth x}$
(d) $y = \operatorname{sech} x \tanh x$ **(e)** $y = \sin \operatorname{csch} x$ **(f)** $y = \ln \sinh x$
(g) $y = \cosh(x + \tan x)$ **(h)** $y = \coth \sqrt{5 - \sin x}$ **(i)** $y = \tanh \ln x$

12. Evaluate each integral.
(a) $\int \tanh x \, dx$ **(b)** $\int \coth x \, dx$ **(c)** $\int \operatorname{sech} x \, dx$
(d) $\int \operatorname{csch} x \, dx$ **(e)** $\int \frac{\cosh x}{\sinh^3 x} \, dx$ **(f)** $\int \sinh^3 x \cosh^3 x \, dx$
(g) $\int \sinh x \cosh^6 x \, dx$ **(h)** $\int \coth^5 x \operatorname{csch}^2 x \, dx$ **(i)** $\int \tanh^3 x \operatorname{sech}^3 x \, dx$
(j) $\int \sinh^2 x \, dx$ **(k)** $\int \frac{1}{(e^x - e^{-x})^2} \, dx$ **(l)** $\int \frac{1}{(e^x + e^{-x})^4} \, dx$

13. Let T be the tangent line to the hyperbola $x^2 - y^2 = 1$ at the point $(\cosh t, \sinh t)$. Show T intersects the y–axis at the point $(0, -\operatorname{csch} t)$.

14. Derive the formulas for $D(\tanh^{-1})(x)$, $D(\coth^{-1})(x)$, $D(\operatorname{sech}^{-1})(x)$ of Prop. 4.10.19.

15. Verify the formulas for $\coth^{-1} x$, $\operatorname{sech}^{-1} x$, $\operatorname{csch}^{-1} x$ stated in Corollary 4.10.22.

16. Prove each identity:
(a) $\coth^{-1} x = \tanh^{-1}\left(\frac{1}{x}\right)$; **(b)** $\operatorname{sech}^{-1} x = \cosh^{-1}\left(\frac{1}{x}\right)$; **(c)** $\operatorname{csch}^{-1} x = \sinh^{-1}\left(\frac{1}{x}\right)$.

17. Evaluate each expression.
(a) $\sinh(\sinh^{-1} 24)$;   **(b)** $\tanh^{-1}(\tanh 8)$;   **(c)** $\cosh^{-1}(\cosh(-3))$;
(d) $\operatorname{sech}^{-1}(\operatorname{sech}(-11))$;   **(e)** $\coth\left(\tanh^{-1} \frac{2}{7}\right)$;   **(f)** $\operatorname{csch}\left(\cosh^{-1} 5\right)$;
(g) $\sinh\left(\coth^{-1}(-2)\right)$;   **(h)** $\tanh\left(\sinh^{-1}(-9)\right)$;   **(i)** $\tanh^{-1}(\operatorname{sech} \ln 2)$.

18. Use Corollary 4.10.22 to estimate each expression to three significant figures.
(a) $\sinh^{-1} 5$ **(b)** $\cosh^{-1} 12$ **(c)** $\tanh^{-1} \frac{3}{7}$
(d) $\coth^{-1}(-4)$ **(e)** $\operatorname{sech}^{-1} \frac{4}{9}$ **(f)** $\operatorname{csch}^{-1}(-10)$

19. Find the derivative of each function.
(a) $y = \sinh^{-1} \sqrt{x^2 - 1}$ **(b)** $y = \frac{\sinh^{-1} x}{\cosh^{-1} x}$ **(c)** $y = \sqrt{1 + (\operatorname{sech}^{-1} x)^2}$
(d) $y = \frac{1}{x + \tanh^{-1} x}$ **(e)** $y = \tanh^{-1}(\cos x)$ **(f)** $y = \left(\operatorname{sech}^{-1} x\right)^3$
(g) $y = \cosh^{-1}(\tan x)$ **(h)** $y = \operatorname{csch}^{-1}(\sinh x + \coth x)$ **(i)** $y = \ln\left(\cosh^{-1} x + \sinh^{-1} x\right)$

20. Evaluate each integral.

(a) $\int \frac{1}{4-9x^2}\, dx$

(b) $\int \frac{1}{x\sqrt{25-4x^2}}\, dx$

(c) $\int \frac{1}{x\sqrt{9+16x^2}}\, dx$

(d) $\int \frac{\cos x}{\sqrt{4\sin^2 x-1}}\, dx$

(e) $\int \frac{e^x}{\sqrt{e^{2x}-9}}\, dx$

(f) $\int \frac{1}{\sqrt{1+4^x}}\, dx$

(g) $\int \frac{1}{\cos^2 x\sqrt{\tan^2 x-1}}\, dx$

(h) $\int \frac{1}{\sqrt{x^2-x}}\, dx$

(i) $\int \frac{1}{\sqrt{x^2+6x}}\, dx$

21. A wire of density 3 kilograms per meter hangs between the tops of two poles equal height which are 50 meters apart. If the tension at the lowest point of this w. equals 80 Newtons, how far does the lowest point of the wire dip?

22. A rope of density 4 kilograms per meter hangs between the tops of two po of equal height which are 40 meters apart. If the lowest point P of this rope dips meters, estimate the tension at P.

23. A ball weighing 2 pounds is kicked upwards at a velocity of 16 feet per secon Assume air resistance equals one fifth of the square of the velocity of the ball.
(a) How long does it take for the ball to reach its highest point?
(b) How high does the ball go?
(c) How long does it take until the ball returns to the ground?
(d) If the ball fell into a very deep hole, what would be its limiting velocity?

24. A rock weighing 5 kilograms is thrown upwards from the top edge of a 40 met cliff at 3 meters per second. Assume air resistance equals one fourth of the square the velocity of the rock.
(a) How long does it take the rock to reach its highest point?
(b) How long does it take the rock to reach the bottom of the cliff?
(c) If the rock were to fall into a very deep hole, what would be its limiting velocit

The following three exercises are for those who have studied Section 3.9

25. Find the distance traveled by each object with velocity $v(t)$.
(a) $v(t) = \frac{\sinh t}{1+\cosh t}$ miles per hour from $t = -1$ to $t = 2$ hours.
(b) $v(t) = t\cosh(t^2)$ miles per hour from $t = -2$ to $t = 3$ hours.
(c) $v(t) = \frac{2^t}{2^t\sqrt{100-4^t}}$ km. per hour from $t = 0$ to $t = 3$ hours.

26. Find the mass of a rod of density $\rho(x) = \sinh^2 x \cosh^4 x$ lbs/ft for $0 \le x \le 2$ ft.

27. How much work is done when the force $F(x) = \frac{x^2}{\sqrt{x^6+1}}$ Newtons moves an objec from $x = 0$ to $x = 2$ meters?

Challenging Problems

1. (a) Show $x = g(\theta) = \sinh^{-1}(\tan\theta)$, for $\theta \in \left(-\frac{\pi}{2}, \frac{\pi}{2}\right)$, is a one–to–one function The inverse function $\theta = gd(x) = \arctan(\sinh x)$ is called the *gudermannian* of x.
(b) Show $D(gd)(x) = \frac{1}{\sec\theta}$.   **(c)** Show $\int \sec\theta\, d\theta = g(\theta) + C$.
(d) Show $\theta = gd(x)$ has the horizontal asymptotes $\theta = -\frac{\pi}{2}$ on the left and $\theta = \frac{\pi}{2}$ o the right.
(e) Graph the function $\theta = gd(x)$.

2. A small town in India of population $15,000$ is infested with 500 cobras. The deat rate of people each year is .0002 times the number of cobras while the death rate o cobras is .0013 times the number of people.
(a) Show the number of people $P(t)$ in this town at time t satisfies the differentia equation $P'' = 2.6 \times 10^{-7}P$.
(b) Solve for $P(t)$. **(c)** Solve for the number of cobras $C(t)$ in this town at time t
(d) Estimate how long it will take until no cobras remain in this town.

4.11 First Order Differential Equations

Prerequisite: Section 4.8

A *differential equation* is an equation which involves a function and its derivatives. Solving a differential equation means finding functions which satisfy the equation. The *general solution* describes all solutions of the equation. The *order* of a differential equation is the highest derivative which appears there. This section is devoted to solving first order differential equations. A differential equation is a common way in which calculus arises in applications. Often a principle can be established empirically about a phenomenon. This principle may state a relationship between quantities and their rates of change which can be formulated as a differential equation. Solving this differential equation produces a model to describe the phenomenon.

In the first subsection, we study the differential equation $y' = ky$ for exponential growth. It describes a quantity y whose rate of change is proportional to the amount present. Applications are made to bacteria growth, exponential decay and continuously compounded interest. In the second subsection, we study separable differential equations $y' = f(x)g(y)$ which can be solved easily by integration. Applications are made to mixing problems, logistic growth and chemical reactions. In the third subsection, we solve linear differential equations $y' + p(x)y = q(x)$. Applications are made to general mixing problems and price prediction.

The general solution of a first order differential equation usually involves a constant of integration C. If we specify a value $y(x_0) = y_0$ of the solution, then we have an *initial value problem* and can determine the value of C.

Examples 4.11.1 (1) The differential equation $y' = 6x^2$ has order one. By the Second Fundamental Theorem of Calculus, the general solution is:

$$y = \int y' \, dx = \int 6x^2 \, dx = 2x^3 + C.$$

Say we are asked to solve the initial value problem with $y(1) = -3$. Then $-3 = y(1) = 2 + C$, $C = -5$, and the unique solution is $y = 2x^3 - 5$.

(2) The differential equation $xy'' + y' = 5$ has order two. Note it can be rewritten as $\frac{d}{dx}(xy') = 5$. Then

$$xy' = \int \frac{d}{dx}(xy') \, dx = \int 5 \, dx = 5x + C \quad \text{and} \quad y' = 5 + \frac{C}{x}.$$

Integrate a second time:

$$y = \int y' \, dx = \int 5 + \frac{C}{x} \, dx = 5x + C \ln|x| + B.$$

(3) The differential equation $y^{(4)} = y$ has order four. Observe that $y = e^x$, $y = e^{-x}$, $y = \sin x$ and $y = \cos x$ are all solutions. It can be shown that the general solution of this differential equation is $y = Ae^x + Be^{-x} + C \sin x + D \cos x$. □

Exponential Growth

Exponential growth refers to the amount y of a quantity which changes at a rate proportional to the amount present. That is, $y' = ky$. We solve this differential equation and give applications to bacteria growth, radioactive decay and continuously compounded interest.

Proposition 4.11.2 *The differential equation* $y' = ky$ *has the general solution*

$$y = y(0)e^{kx}.$$

Proof Multiply the differential equation $0 = y' - ky$ by e^{-kx}:

$$0 = e^{-kx}y' - kye^{-kx} = \frac{d}{dx}\left(e^{-kx}y\right).$$

By Corollary 2.6.23, only constant functions have zero derivative. Hence $e^{-kx}y =$
and $y = Ce^{kx}$. Substitute $x = 0$: $y(0) = Ce^0 = C$ and $y = y(0)e^{kx}$.

Observe the constant k in the differential equation $y' = ky$ may be positive
negative. In applications, y is usually positive. When k is positive, y is an increasi
function of x, and the solution $y = y(0)e^{kx}$ describes exponential growth. See Exa
ples 1, 3 below. When k is negative, y is a decreasing function of x, and the soluti
$y = y(0)e^{kx}$ describes exponential decay. See Example 2 below.

Examples 4.11.3 (1) Bacteria grows at a rate proportional to the number preser
1,000 bacteria are placed in a petrie dish filled with agar. Twenty four hou
later there are 10,000 bacteria in this dish.
(a) How many bacteria will be in this dish after one week?
(b) How long will it take until there are 1,000,000 bacteria in this dish?

Solution Let $y(t)$ denote the number of bacteria in the dish after t days. The
$y' = ky$ and $y = y(0)e^{kt} = 1,000e^{kt}$. To determine the value of k, substitu
$t = 1$: $10,000 = 1,000e^k$ and $e^k = 10$. Thus

$$y = 1,000\left(e^k\right)^t = 1,000\left(10^t\right) = 10^{t+3}.$$

(a) $y(7) = 10^{10}$, and there will be 10 billion bacteria in this dish after one wee.
(b) We solve $10^6 = y(t) = 10^{t+3}$. Then $6 = t + 3$, and $t = 3$ days.

(2) A radioactive substance decays at a rate proportional to the amount present.
6 gram lump of this substance is found to have decayed to 2 grams after 4 hour
(a) How much of this substance will be left after 24 hours?
(b) How long will it take until there is only 1 gram of this substance remaining
(c) What is the half–life[5] of this substance?

Solution Let $y(t)$ be the number of grams of this substance remaining after
hours. Then $y' = -ky$ and $y = y(0)e^{-kt} = 6e^{-kt}$. To determine the value of k
substitute $t = 4$: $2 = y(4) = 6e^{-4k}$ and $e^{-4k} = \frac{1}{3} = 3^{-1}$. Then

$$y = 6e^{-kt} = 6\left(e^{-4k}\right)^{t/4} = 6\left(3^{-t/4}\right).$$

(a) $y(24) = 6\left(3^{-24/4}\right) = \frac{6}{3^6} = \frac{2}{243}$ grams.
(b) Solve $1 = y(t) = 6\left(3^{-t/4}\right)$ by taking the logarithm to the base three:

$$0 = \log_3 1 = \log_3\left[6\left(3^{-t/4}\right)\right] = \log_3 6 - \frac{t}{4}.$$

Hence $t = 4\log_3 6 \approx 6.5$ hours.
(c) We solve for t with $y(t) = \frac{1}{2}y(0)$. That is, $6\left(3^{-t/4}\right) = 3$. Then $3^{-t/4} = \frac{1}{2}$
$-\frac{t}{4} = \log_3\left(\frac{1}{2}\right) = -\log_3 2$ and $t = 4\log_3 2 = \log_3 16 \approx 2.5$ hours.

---

[5] The half–life of a substance refers to the time required for half of the substance to decay.

(3) A $10,000 certificate of deposit collects interest which is compounded continuously at the rate of 5%.

(a) How much will this certificate be worth after 3 years?

(b) How long will it take until this certificate is worth $12,000?

Solution Let $A(t)$ denote the balance of this account after t years. We begin by interpreting the meaning of continuous compounding at R percent. If the interest were compounded n times each year, then each nth of a year the account would receive interest of $A\frac{R}{100n}$. Thus over a time period of $\Delta t = \frac{1}{n}$ we have $\Delta A = A\frac{R}{100n} = \frac{R}{100}A\Delta t$, and $\frac{\Delta A}{\Delta t} = \frac{R}{100}A$. Take the limit as Δt approaches zero:

$$\frac{dA}{dt} = \frac{R}{100}A.$$

This differential equation describes the rate of change of the value of this certificate under continuously compounded interest of R percent. The solution of this differential equation is $A = A(0)e^{\frac{R}{100}t}$. In the case $R = 5$ and $A(0) = 10,000$ we have $A(t) = 10,000e^{.05t}$.

(a) $A(3) = 10,000e^{.15} \approx \$11,618.34$.

(b) We solve $12,000 = A(t) = 10,000e^{.05t}$. Then $e^{.05t} = 1.2$ and $.05t = \ln 1.2$. Hence $t = 20\ln 1.2 \approx 3.6$ years. $\square$

Separable Equations

A differential equation of the form

$$y' = f(x)g(y)$$

is called *separable*. Separate the function of x and the function of y in this equation to rewrite it as $f(x) = \frac{1}{g(y(x))}\frac{dy}{dx}$. Now integrate with respect to x:

$$\int f(x)\,dx = \int \frac{1}{g(y(x))}\frac{dy}{dx}\,dx.$$

Make the substitution $y = y(x)$ with $dy = \frac{dy}{dx}\,dx$:

$$\int f(x)\,dx = \int \frac{1}{g(y)}\,dy.$$

After solving two separable equations, we give applications to mixing problems, logistic growth and rates of chemical reactions.

Examples 4.11.4 (1) Solve the differential equation $y' = xy$.

Solution Note $y = 0$ is the *trivial solution* of this differential equation. When $y \neq 0$, rewrite this differential equation as $x = \frac{y'(x)}{y(x)} = \frac{1}{y(x)}\frac{dy}{dx}$. Then

$$\frac{x^2}{2} + C = \int x\,dx = \int \frac{1}{y(x)}\frac{dy}{dx}\,dx = \int \frac{1}{y}\,dy = \ln|y|$$

$$|y| = e^{C+\frac{x^2}{2}}$$

$$y = \pm e^{C+\frac{x^2}{2}} = \pm e^C e^{\frac{x^2}{2}} = Ae^{\frac{x^2}{2}}$$

where A is any real number. (The values $A = \pm e^C$ are all nonzero numbers, and $A = 0$ gives the trivial solution.)

(2)   Solve the initial value problem $y' = 6x^2(y^2 + 1)$ with $y(0) = 1$.

Solution Rewrite this equation as $6x^2 = \frac{1}{y(x)^2+1}\frac{dy}{dx}$. Then

$$\int 6x^2 \, dx = \int \frac{1}{y(x)^2 + 1}\frac{dy}{dx}\, dx = \int \frac{1}{y^2 + 1}\, dy$$

$$2x^3 + C = \arctan y$$

$$y = \tan\left(2x^3 + C\right).$$

Now $1 = y(0) = \tan C$, and $C = \frac{\pi}{4} + n\pi$ for n an integer. Hence the gene: solution of this initial value problem is:

$$y = \tan\left(2x^3 + \frac{\pi}{4} + n\pi\right) = \tan\left(2x^3 + \frac{\pi}{4}\right).$$

Mixing Problems

Separable differential equations arise in *mixing problems* where a container is fill with a solution of a substance S. Fluid leaves the container at a constant rate and simultaneously replaced by a solution of S with a different concentration. We assur the container is stirred so that the concentration of S is uniform throughout. To discu: this situation, it is important to understand the relationship between the volume A S in a volume V of this solution and the concentration of S. Note $\frac{A}{V}$ is the fractic of the solution which consists of S. Multiplying by 100 gives the percentage of S the solution, which is called the *concentration* C of S:

$$C = 100\frac{A}{V} \quad \text{or} \quad A = V\frac{C}{100}. \tag{4.11.}$$

The basic principle to set up a differential equation to describe this situation is:

$$\text{Rate of Change of } S = \text{Rate of } S \text{ In} - \text{Rate of } S \text{ Out}.$$

Examples 4.11.5 (1) A 1,000 liter vat of 5% brine (salt water) develops a leak .4 liters per minute. The vat is kept full by adding 2% brine to replace the bri lost through the leak. Assume the vat rotates so that the concentration of sa in the vat is uniform. How long does it take until the concentration of salt the vat is reduced to 3%?

Solution Let $A(t)$ denote the number of liters of salt in the vat t minutes afte the leak develops. The volume of salt that enters the vat each minute is given b the second equation of (4.11.1): the volume .4 liters of solution that enters time its concentration of 2% divided by 100. Thus salt enters the vat at .008 liter per minute. The rate at which salt leaves the tank each minute is the volume c salt in the .4 liters that leaks from the 1,000 liter vat that contains $A(t)$ liter of salt: $\frac{.4}{1,000}A(t) = .0004A(t)$ liters per minute. Then

$$A'(t) = \text{Rate of Salt In} - \text{Rate of Salt Out} = .008 - .0004A(t).$$

To find $A(t)$, we solve this differential equation by separation of variables: $1 = \frac{A'(t)}{.008 - .0004A(t)}$ and

$$\int 1 \, dt = \int \frac{A'(t)}{.008 - .0004A(t)}\, dt = \int \frac{1}{.008 - .0004A}\, dA$$

$$t + K_1 = = -2500\ln|.008 - .0004A|$$

$$\ln|.008 - .0004A(t)| = -\frac{t}{2500} + K_2$$

$$|.0004A(t) - .008| = e^{-\frac{t}{2500} + K_2}$$

$$.0004A(t) - .008 = \pm e^{-\frac{t}{2500} + K_2} = \pm e^{K_2} e^{-\frac{t}{2500}}$$

$$A(t) = 20 \pm 2500 e^{K_2} e^{-\frac{t}{2500}} = 20 + K e^{-\frac{t}{2500}}.$$

At time $t = 0$ the concentration of salt in the vat is 5%. Then $A(0)$ equals 5% of the 1,000 liter vat which is 50 liters. Substitute $t = 0$ in the equation for $A(t)$: $50 = A(0) = 20 + Ke^0 = 20 + K$ and $K = 30$. Thus

$$A(t) = 20 + 30 e^{-\frac{t}{2500}}.$$

We solve for the time t when the concentration equals 3%. By the second equation in (4.11.1): $A(t) = 1,000 \frac{3}{100} = 30$ liters:

$$30 = A(t) = 20 + 30 e^{-\frac{t}{2500}}$$

$$e^{-\frac{t}{2500}} = \frac{1}{3}$$

$$-\frac{t}{2500} = \ln \frac{1}{3} = -\ln 3$$

$$t = 2500 \ln 3 \approx 2747 \text{ minutes.}$$

Thus approximately 45 hours and 47 minutes after the leak starts the concentration of salt in the vat will be 3%. Note that in the long run $A(t)$ approaches 20 liters and $C(t)$ approaches 2%. See Figure 4.11.6.

(2) A 100,000 gallon swimming pool contains more than 2.5% chlorine. The pool filter draws 40 gallons per minute of pool water, filters it and returns the filtered water to the pool. This filter only transmits the square root of the amount of chlorine in the water which passes through it each minute. Assuming the concentration of chlorine in the pool is uniform, find the concentration of chlorine in the pool in the long run.

Solution Let $A(t)$ denote the number of gallons of chlorine in the pool after t minutes. The volume of chlorine which leaves the pool and passes through the filter each minute is $A(t) \frac{40}{100,000} = \frac{A(t)}{2500}$ gallons. The filter returns $\sqrt{\frac{A(t)}{2500}} = \frac{\sqrt{A(t)}}{50}$ gallons of chlorine per minute to the pool. Thus

$$A'(t) = \text{ Rate of Chlorine In } - \text{ Rate of Chlorine Out } = \frac{\sqrt{A(t)}}{50} - \frac{A(t)}{2500}.$$

Note for this problem to be realistic, $A'(0)$ must be negative, i.e. $A(0) > 2500$. This is why we require the initial concentration of chlorine to be greater than 2.5%. We solve this differential equation by separation of variables:

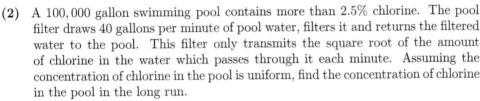

$$1 = \frac{A'(t)}{\frac{\sqrt{A(t)}}{50} - \frac{A(t)}{2500}} = \frac{2500 A'(t)}{50\sqrt{A(t)} - A(t)} \quad \text{and}$$

$$t + K_1 = \int 1 \, dt = \int \frac{2500 A'(t)}{50\sqrt{A(t)} - A(t)} \, dt = \int \frac{2500}{\sqrt{A}(50 - \sqrt{A})} \, dA.$$

Let $u = 50 - \sqrt{A}$ with $du = -\frac{1}{2\sqrt{A}} \, dA$. Then

$$t + K_1 = -5000 \int \frac{1}{50 - \sqrt{A}} \left(-\frac{1}{2\sqrt{A}} \right) dA = -5000 \int \frac{1}{u} \, du = -5000 \ln |u|$$

$$\ln |u| = -\frac{t}{5000} + K_2$$

$$50 - \sqrt{A} = u = \pm e^{-\frac{t}{5000} + K_2} = \pm e^{K_2} e^{-\frac{t}{5000}} = -K e^{-\frac{t}{5000}}$$

$$A(t) = \left(50 + K e^{-\frac{t}{5000}} \right)^2.$$

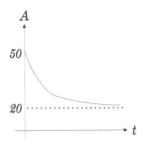

$$A(t) = 20 + 30 e^{-t/2500}$$

Figure 4.11.6
Example 1

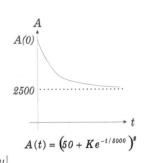

$$A(t) = \left(50 + K e^{-t/5000} \right)^2$$

Figure 4.11.7
Example 2

Observe $K > 0$ because $A(0) > 2500$. Then

$$\lim_{t \to \infty} A(t) = \lim_{t \to \infty} \left(50 + K e^{-\frac{t}{5000}}\right)^2 = (50 + 0)^2 = 2500.$$

In the long run there will be 2500 gallons of chlorine in the $100,000$ gallon p with concentration $100\frac{2500}{100,000} = 2.5\%$. Note this long term concentration chlorine does not depend on the initial concentration. See Figure 4.11.7.

Logistic Growth

The differential equation

$$y' = ky, \quad k > 0, \tag{4.11}$$

which describes exponential growth is reasonable when there is sufficient food a enough space in the environment to support the growth. As t gets large, one or bo of these assumptions will no longer be valid. For example, bacteria may exhaust t supply of agar in a petrie dish, and fish may become overcrowded in a pond. take these factors into account, the differential equation (4.11.2) must be modified subtracting a summand from ky which is small when y is small and becomes lar when y is large. For example,

$$y' = ky - by^2 \text{ for } b, k > 0 \tag{4.11.}$$

meets these criteria when $\frac{k}{b}$ is large. The separable differential equation (4.11.3) called the *logistic equation*, or the *Verhulst equation*, of population growth.

Example 4.11.8 One thousand bacteria are placed in a petrie dish containing aga The number of bacteria $y(t)$ in this dish after t hours is found to satisfy the equatic

$$y'(t) = 2y(t) - .00001y(t)^2.$$

Find the number of bacteria in the petrie dish in the long run.

Solution Observe $k = 2$ and $b = .00001$. Then $\frac{k}{b} = \frac{2}{.00001} = 200,000$ is ver large. Thus the assumption of the logistic equation is satisfied. Solve this equation t separation of variables:

$$1 = \frac{y'(t)}{2y(t) - .00001y(t)^2}$$

$$\int 1 \, dt = \int \frac{y'(t)}{2y(t) - .00001y(t)^2} \, dt = \int \frac{1}{2y - .00001y^2} \, dy$$

$$t + C = \int \frac{-100,000}{y^2 - 200,000y} \, dy = \int \frac{-100,000}{y(y - 200,000)} \, dy. \tag{4.11.4}$$

Use the method of partial fractions to evaluate this integral. Write

$$\frac{-100,000}{y^2 - 200,000y} = \frac{A}{y} + \frac{B}{y - 200,000} = \frac{A(y - 200,000) + By}{y^2 - 200,000y}.$$

Multiply this equation by $y^2 - 200,000y$:

$$-100,000 = A(y - 200,000) + By.$$

Setting $y = 0$ gives $A = \frac{1}{2}$ while setting $y = 200,000$ gives $B = -\frac{1}{2}$. By (4.11.4):

$$t + C = \frac{1}{2} \int \frac{1}{y} - \frac{1}{y - 200,000} \, dy = \frac{1}{2} \left(\ln|y| - \ln|y - 200,000|\right) = \ln \left|\frac{y}{y - 200,000}\right|^{1/2}$$

Setting $t = 0$ gives $C = \ln|\frac{1,000}{-199,000}|^{1/2} \approx -2.647$. Then

$$e^{t-2.647} = e^{t+C} = \left| \frac{y}{y - 200,000} \right|^{1/2}$$

Square this equation:

$$e^{2t-5.294} = -\frac{y(t)}{y(t) - 200,000}.$$

We place a minus sign on the right side of the preceding equation, because $e^{2t-5.294}$ is always positive and $\frac{y}{y-200,000}$ is negative for $t = 0$, hence for all $t > 0$ with $y(t) < 200,000$. Cross–multiply to solve this equation for $y(t)$:

$$
\begin{aligned}
e^{2t-5.294} y(t) - 200,000 e^{2t-5.294} &= -y(t) \\
y(t)\left(e^{2t-5.294} + 1\right) &= 200,000 e^{2t-5.294} \\
y(t) = \frac{200,000 e^{2t-5.294}}{e^{2t-5.294} + 1} &= \frac{200,000}{1 + e^{5.294-2t}}.
\end{aligned}
$$

Thus $\lim_{t\to\infty} y(t) = \lim_{t\to\infty} \frac{200,000}{1 + e^{5.294-2t}} = \frac{200,000}{1 + 0} = 200,000$ bacteria. $\square$

There is an important variation of the logistic equation which applies to the situation of *constant harvesting*. For example, say fish are grown in a pond and harvested at a constant rate of h fish per day. Then the number of fish $y(t)$ in the pond on day t satisfies the differential equation:

$$y' = ky - by^2 - h \quad \text{for } b, h, k > 0. \tag{4.11.5}$$

We retain the hypothesis that $\frac{k}{b}$ is large.

Example 4.11.9 Thirty five hundred trout are placed in a pond. The intention is to harvest h trout per day with $h \geq 22$. Let $y(t)$ denote the number of trout in the pond on day t. The rate of growth of the trout satisfies the differential equation

$$y' = .02y - .000004y^2 - h.$$

(a) Find the relationship between the population of trout in the pond and the harvesting rate h.
(b) What is the largest value of h for which fish can be harvested indefinitely?

Solution **(a)** Note $k = .02$ and $b = .000004$. Then $\frac{k}{b} = \frac{.02}{.000004} = 5,000$ is large. Thus the assumption of the logistic equation is satisfied. We solve this separable differential equation:

$$
\begin{aligned}
1 &= \frac{y'(t)}{.02y(t) - .000004y(t)^2 - h} = \frac{250,000y'(t)}{5,000y(t) - y(t)^2 - 250,000h} \\
\int 1 \, dt &= \int \frac{250,000y'(t)}{5,000y(t) - y(t)^2 - 250,000h} \, dt \\
t + C &= \int \frac{-250,000}{y^2 - 5,000y + 250,000h} \, dy. \tag{4.11.6}
\end{aligned}
$$

We use the method of partial fractions. First, consider the case $h < 25$. Write

$$
\begin{aligned}
\frac{-250,000}{y^2 - 5,000y + 250,000h} &= \frac{-250,000}{(y - 2500)^2 - 250,000(25 - h)} \\
&= \frac{-250,000}{(y - 2500 - 500\sqrt{25 - h})(y - 2500 + 500\sqrt{25 - h})}
\end{aligned}
$$

$$= \frac{A}{y - 2500 - 500\sqrt{25 - h}} + \frac{B}{y - 2500 + 500\sqrt{25 - h}}$$

$$= \frac{A(y - 2500 + 500\sqrt{25 - h}) + B(y - 2500 - 500\sqrt{25 - h}}{y^2 - 5,000y + 250,000h}$$

Multiply this equation by $y^2 - 5,000y + 250,000h$:

$$-250,000 = A(y - 2500 + 500\sqrt{25 - h}) + B(y - 2500 - 500\sqrt{25 - h}).$$

Setting $y = 2500 + 500\sqrt{25 - h}$ gives $A = -\frac{250}{\sqrt{25-h}}$, while setting
$y = 2500 - 500\sqrt{25 - h}$ gives $B = \frac{250}{\sqrt{25-h}}$. By (4.11.6),

$$t + C = \frac{250}{\sqrt{25 - h}} \int -\frac{1}{y - 2500 - 500\sqrt{25 - h}} + \frac{1}{y - 2500 + 500\sqrt{25 -}}$$

$$\frac{\sqrt{25 - h}}{250}(t + C) = -\ln|y - 2500 - 500\sqrt{25 - h}| + \ln|y - 2500 + 500\sqrt{25 - h}|$$

$$\frac{\sqrt{25 - h}}{250}(C + t) = \ln\left|\frac{y - 2500 + 500\sqrt{25 - h}}{y - 2500 - 500\sqrt{25 - h}}\right| \tag{4}$$

$$\pm e^{\frac{\sqrt{25-h}}{250}C} e^{\frac{\sqrt{25-h}}{250}t} = \pm e^{\frac{\sqrt{25-h}}{250}(C+t)} = \frac{y - 2500 + 500\sqrt{25 - h}}{y - 2500 - 500\sqrt{25 - h}} \tag{4}$$

Set $t = 0$ in (4.11.7). Since $y(0) = 3500$,

$$\frac{\sqrt{25 - h}}{250}C = \ln\left|\frac{2 + \sqrt{25 - h}}{2 - \sqrt{25 - h}}\right| \quad \text{and} \quad Q = e^{\frac{\sqrt{25-h}}{250}C} = \left|\frac{2 + \sqrt{25 - h}}{2 - \sqrt{25 - h}}\right|.$$

By (4.11.8),

$$\pm Q e^{\frac{\sqrt{25-h}}{250}t} = \frac{y - 2500 + 500\sqrt{25 - h}}{y - 2500 - 500\sqrt{25 - h}}.$$

Solve this equation for y:

$$y(t) = \frac{\pm Q e^{\frac{\sqrt{25-h}}{250}t}(2500 + 500\sqrt{25 - h}) - 2500 + 500\sqrt{25 - h}}{\pm Q e^{\frac{\sqrt{25-h}}{250}t} - 1} \quad \text{when } h < 25. \tag{4.11.9}$$

Now consider the case $h = 25$. From (4.11.6):

$$t + C = \int \frac{-250,000}{y^2 - 5,000y + 6,250,000}\, dy = \int \frac{-250,000}{(y - 2500)^2}\, dy = \frac{250,000}{y - 2500}.$$

Set $t = 0$: $C = \frac{250,000}{3,500-2,500} = 250$. Solve the preceding equation for y:

$$y(t) = 2500 + \frac{250,000}{t + 250} \quad \text{when } h = 25. \tag{4.11.10}$$

Consider the last case, $h > 25$. By (4.11.6):

$$t+C = -250,000 \int \frac{1}{(y - 2500)^2 + 250,000(h - 25)}\, dy = -\frac{500}{\sqrt{h - 25}} \arctan \frac{y - 2500}{500\sqrt{h - 25}}.$$

Set $t = 0$. Since $y(0) = 3500$, $C = -\frac{500}{\sqrt{h-25}} \arctan \frac{2}{\sqrt{h-25}}$. Thus

$$t - \frac{500}{\sqrt{h - 25}} \arctan \frac{2}{\sqrt{h - 25}} = -\frac{500}{\sqrt{h - 25}} \arctan \frac{y - 2500}{500\sqrt{h - 25}} \quad \text{when } h > 25. \tag{4.11.11}$$

(b) By (4.11.11), if $h > 25$ then $y = 0$ when

$$t = \frac{500}{\sqrt{h - 25}} \left(\arctan \frac{5}{\sqrt{h - 25}} + \arctan \frac{2}{\sqrt{h - 25}} \right)$$

and no fish remain in the pond at that time. If $h < 25$, then by (4.11.9):

$$\lim_{t \to \infty} y(t) = 2500 + 500\sqrt{25 - h} \text{ fish.}$$

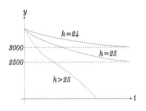

Figure 4.11.10
Fish Harvesting

The largest value of h in this case is $h = 24$ for which there will be 3000 fish in the pond in the long run. If $h = 25$, then by (4.11.10) there will be 2500 fish in the pond in the long run. Thus 25 fish should be harvested each day to obtain the maximum yield which is sustainable in the long run. See Figure 4.11.10.   □

Rates of Chemical Reactions

Consider a reaction in which the chemicals $A_1, \ldots, A_s$ in a homogeneous gas or solution react to produce the chemicals $B_1, \ldots, B_u$:

$$\alpha_1 A_1 + \cdots + \alpha_s A_s \to \beta_1 B_1 + \cdots + \beta_u B_u. \tag{4.11.12}$$

This equation means that α_1 molecules of chemical A_1, ... , α_s molecules of chemical A_s react to produce β_1 molecules of chemical B_1, ... , β_u molecules of chemical B_u. Recall that a mole denotes 6.023×10^{23} molecules. Thus the above equation implies that α_1 moles of chemical A_1, ... , α_s moles of chemical A_s react to produce β_1 moles of chemical B_1, ... , β_u moles of chemical B_u. Let $[C]$ denote the concentration of the chemical C measured in moles per liter where $[C]$ is a function of time t measured in seconds. Let $y(t)$ denote the concentration of the products of this reaction at time t:

$$y = [B_1] + \cdots + [B_u].$$

Experimentation shows that there are positive integers $m_1, \ldots, m_s$, k such that the rate of this reaction y' is given by:

$$y' = k \, [A_1]^{m_1} \cdots [A_s]^{m_s}. \tag{4.11.13}$$

The values of the exponents $m_1, \ldots, m_s$ and the *rate constant* k are determined by experimentation. In particular, they cannot be determined from the equation (4.11.12) which describes the reaction. The sum of the exponents $m_1 + \cdots + m_s$ is called the *order* of the reaction. Choose a fixed number i with $1 \le i \le s$. For each occurrence of this reaction, α_i molecules A_i are used in the production of $\beta = \beta_1 + \cdots + \beta_u$ molecules of the products. Thus

$$y' = -\frac{\beta}{\alpha_i} \, [A_i]'.$$

Integrate this equation:

$$y = -\frac{\beta}{\alpha_i} \, [A_i] + C.$$

Let a_i denote the concentration $[A_i]$ of A_i at $t = 0$. When $t = 0$ the preceding equation becomes: $0 = y(0) = -\frac{\beta}{\alpha_i} a_i + C$ and $C = \frac{\beta a_i}{\alpha_i}$. Thus

$$\alpha_i y = -\beta \, [A_i] + \beta a_i \quad \text{and} \quad [A_i] = a_i - \frac{\alpha_i}{\beta} y.$$

Substitute these values of $[A_1], \ldots, [A_s]$ into (4.11.13):

$$y' = k \left(a_1 - \frac{\alpha_1}{\beta} y \right)^{m_1} \cdots \left(a_s - \frac{\alpha_s}{\beta} y \right)^{m_s}. \tag{4.11.14}$$

Solving this separable differential equation gives the concentration y of the products of this reaction as a function of time.

Examples 4.11.11 (1) A container is filled with the gas dinitrogen pentoxide wh
decomposes into nitrogen dioxide and oxygen in the first order reaction

$$2N_2O_5 \rightarrow 4NO_2 + O_2.$$

The initial concentration of N_2O_5 is .3 moles per liter. If the rate constant
6×10^{-4}, find the total concentration y of NO_2 and O_2 produced by this reacti
as a function of time t.

Solution In this reaction $k = 6 \times 10^{-4}$, $s = 1$, $u = 2$, $\alpha_1 = 2$, $a_1 = .3$, $m_1 =$
$\beta_1 = 4$, $\beta_2 = 1$ and $\beta = 4 + 1 = 5$. Equation (4.11.14) for this reaction is:

$$y' = 6 \times 10^{-4} \left(.3 - \frac{2}{5}y \right).$$

Then $6 \times 10^{-4} = \frac{y'(t)}{.3 - .4y(t)}$. Integrate this equation:

$$\int 6 \times 10^{-4} \, dt = \int \frac{y'(t)}{.3 - .4y(t)} \, dt = \int \frac{1}{.3 - .4y} \, dy$$

$$6 \times 10^{-4}t + C = -2.5 \ln |.3 - .4y|$$

$$\ln |.3 - .4y| = -.4C - 2.4 \times 10^{-4}t$$

$$.3 - .4y = \pm e^{-.4C} e^{-2.4 \times 10^{-4}t} = A e^{-2.4 \times 10^{-4}t}.$$

Substitute $t = 0$: $.3 - 0 = Ae^0$ and $A = .3$. Hence $.3 - .4y = .3e^{-2.4 \times 10^{-4}t}$, an

$$y = .75 - .75e^{-2.4 \times 10^{-4}t}.$$

The limit of y, as t approaches infinity, is .75 moles. Alternatively, every 2 mol
of N_2O_5 eventually produces $\beta = 5$ moles of products, and $\frac{5}{2}(.3) = .75$ moles.

(2) Half a mole of the corrosive gas nitrosyl chloride is accidentally released into
room. It decomposes into nitric oxide and chlorine in the second order reactic

$$2NOCl \rightarrow 2NO + Cl_2$$

with rate constant .02. How long will it take until the amount of nitrosyl chloric
is reduced to .01 moles?

Solution In this reaction $k = .02$, $s = 1$, $u = 2$, $\alpha_1 = 2$, $a_1 = .5$, $m_1 = $
$\beta_1 = 2$, $\beta_2 = 1$ and $\beta = 2 + 1 = 3$. Equation (4.11.14) for this reaction is:

$$y' = .02 \left(.5 - \frac{2}{3}y \right)^2 = \frac{1}{1800}(3 - 4y)^2.$$

Then $\frac{1}{1800} = \frac{y'(t)}{[3 - 4y(t)]^2}$. Integrate this equation:

$$\int \frac{1}{1800} \, dt = \int \frac{y'(t)}{[3 - 4y(t)]^2} \, dt = \int \frac{1}{(3 - 4y)^2} \, dy$$

$$\frac{t}{1800} + C = \frac{1/4}{3 - 4y}$$

$$3 - 4y = \frac{1/4}{\frac{t}{1800} + C} = \frac{450}{t + 1800C}.$$

Substitute $t = 0$: $3 - 0 = \frac{450}{1800C} = \frac{1}{4C}$ and $C = \frac{1}{12}$. Hence $3 - 4y = \frac{450}{t+150}$, and

$$y = \frac{3}{4} - \frac{225}{2t + 300}.$$

The amount of $NOCl$ in the room will be .01 moles when $y = \frac{3}{2}(.5 - .01) = .735$:

$$
\begin{aligned}
.735 &= .750 - \frac{225}{2t + 300} \\[2mm]
\frac{225}{2t + 300} &= .015 \\[2mm]
2t + 300 &= 15000 \\[2mm]
t &= 7350 \text{ seconds} = 2 \text{ hours } 2 \text{ minutes } 30 \text{ seconds.} \qquad \square
\end{aligned}
$$

Linear Equations

A *linear differential equation* has the form

$$y' + p(x)y = q(x). \tag{4.11.15}$$

We show how to solve this equation and then give applications to general mixing problems and price prediction.

A linear differential equation is solved by generalizing the method used to solve the differential equation $y' - ky = 0$ for exponential growth. In that case, we multiplied the equation by e^{-kx} and recognized $e^{-kx}y' - ke^{-kx}y$ as the derivative of $e^{-kx}y$. The exponent $-kx$ of e was chosen because $D(-kx) = -k$. Similarly, multiply the linear differential equation (4.11.15) by $e^{P(x)}$:

$$e^{P(x)}y' + p(x)e^{P(x)}y = q(x)e^{P(x)}.$$

We choose $P(x)$ so that the left side of this equation is the derivative of $e^{P(x)}y$. This occurs when $P'(x) = p(x)$. Therefore define

$$P(x) = \int p(x)\, dx.$$

By the First Fundamental Theorem of Calculus, $P'(x) = p(x)$ as required. We call the function $e^{P(x)}$ an *integrating factor* for this differential equation. Then

$$
\begin{aligned}
\frac{d}{dx}\left(e^{P(x)}y\right) &= e^{P(x)}y' + p(x)e^{P(x)}y = q(x)e^{P(x)} \\[2mm]
e^{P(x)}y &= \int q(x)e^{P(x)}\, dx \\[2mm]
y &= e^{-P(x)}\int q(x)e^{P(x)}\, dx.
\end{aligned}
$$

The following proposition summarizes our conclusions.

Proposition 4.11.12 *Let $p(x)$, $q(x)$ be continuous functions. The linear differential equation*

$$y' + p(x)y = q(x)$$

has the general solution

$$y = e^{-P(x)}\int q(x)e^{P(x)}\, dx \quad where \quad P(x) = \int p(x)\, dx.$$

Examples 4.11.13 (1) Solve the differential equation $y' + \frac{y}{x} = 12x^2$.

Solution Since $\int \frac{1}{x}\, dx = \ln x + B$, an integrating factor for this different equation is $e^{\ln x} = x$. Then

$$
\begin{aligned}
xy' + y &= 12x^3 \\
\frac{d}{dx}(xy) &= xy' + y = 12x^3 \\
xy &= \int 12x^3\, dx = 3x^4 + C \\
y &= 3x^3 + \frac{C}{x}.
\end{aligned}
$$

(2) Solve the differential equation $y' \cos^2 x = y + \tan x$.

Solution Divide this equation by $\cos^2 x$:

$$
y' - y \sec^2 x = \tan x\ \sec^2 x.
$$

Since $\int - \sec^2 x\, dx = -\tan x + B$, use $e^{-\tan x}$ as an integrating factor:

$$
\begin{aligned}
e^{-\tan x} y' - y(\sec^2 x)e^{-\tan x} &= (\tan x)(\sec^2 x)e^{-\tan x} \\
\frac{d}{dx}\left(e^{-\tan x} y\right) &= (\tan x)(\sec^2 x)e^{-\tan x} \\
e^{-\tan x} y &= \int (-\tan x)e^{-\tan x}(-\sec^2 x)\, dx = \int ue^u\, du
\end{aligned}
$$

where $u = -\tan x$ with $du = -\sec^2 x\, dx$. Now integrate by parts:

$$
\begin{aligned}
e^{-\tan x} y &= \int uD(e^u)\, du = ue^u - \int D(u)e^u\, du = ue^u - \int e^u\, du \\
e^{-\tan x} y &= ue^u - e^u + C = -(\tan x)e^{-\tan x} - e^{-\tan x} + C \\
y &= -\tan x - 1 + Ce^{\tan x}.
\end{aligned}
$$

General Mixing Problems

Although simple mixing problems are modeled by separable differential equation more complicated mixing problems give rise to linear differential equations which a not separable. The method of setting up the differential equation remains the sam the rate of change of the substance in a container equals the rate at which the substance enters minus the rate at which the substance leaves.

Example 4.11.14 At the beginning of the rainy season in Brazil, a billion gallo pond contains .3% pollutants. Rain containing .5% pollutants begins to fall on th pond at the rate of 10,000 gallons per hour. Water exits this pond through a strean at the rate of 8,000 gallons per hour. Find the level of pollution in this pond whe the rainy season ends 150 days later.

Solution Let t denote time measured in hours where $t = 0$ is the start of the rain season. Let $A(t)$ be the number of gallons of pollutants in this pond at time t. Th volume of pollutants in 10,000 gallons of rain is $10,000(.005) = 50$ gallons. Thus th rate at which pollutants enter this pond is 50 gallons per hour. The volume of thi pond at time t is $10^9 + 10,000t - 8,000t = 10^9 + 2,000t$. Hence 8,000 gallons of pon water at time t contain $A(t)\frac{8,000}{10^9 + 2,000t}$ gallons of pollutants which is the rate at whicl pollutants leave this pond per hour. Then

$$
A'(t) = \text{Rate In} - \text{Rate Out} = 50 - \frac{8,000A(t)}{10^9 + 2,000t} \quad \text{and}
$$

$$
A'(t) + \frac{8,000}{10^9 + 2,000t}A(t) = 50. \tag{4.11.16}
$$

This is a linear differential equation. Since $\int \frac{8,000}{10^9+2,000t}\,dt = 4\ln\left(10^9+2,000t\right)+B$, an integrating factor for this differential equation is:

$$e^{4\ln\left(10^9+2,000t\right)} = \left[e^{\ln\left(10^9+2,000t\right)}\right]^4 = \left(10^9+2,000t\right)^4.$$

Multiply the differential equation (4.11.16) by this integrating factor:

$$A'(t)\left(10^9+2,000t\right)^4 \;+\; 8,000A(t)\left(10^9+2,000t\right)^3 \;=\; 50\left(10^9+2,000t\right)^4$$

$$\frac{d}{dt}\left[A(t)\left(10^9+2,000t\right)^4\right] \;=\; 50\left(10^9+2,000t\right)^4$$

$$A(t)\left(10^9+2,000t\right)^4 \;=\; 50\int\left(10^9+2,000t\right)^4 dt \;=\; \frac{1}{200}\left(10^9+2,000t\right)^5+C$$

$$A(t) \;=\; \frac{1}{200}\left(10^9+2,000t\right)+C\left(10^9+2,000t\right)^{-4}$$

$$A(t) \;=\; 5\left(10^6+2t\right)+C\times10^{-12}\left(10^6+2t\right)^{-4}.$$

Observe that $A(0) = .003\times10^9 = 3\times10^6$ gallons. Set $t=0$ in the preceding equation: $3\times10^6 = 5\times10^6 + C\times10^{-36}$ and $C = -2\times10^{42}$. Then

$$A(t) = 5\left(10^6+2t\right)-2\times10^{30}\left(10^6+2t\right)^{-4}.$$

Since 150 days $=3600$ hours, the amount of pollutants at the end of the rainy season is:

$$A(3600) = 5\left(10^6+7200\right)-2\times10^{30}\left(10^6+7200\right)^{-4}\approx 3.09\times10^6 \text{ gallons.}$$

The volume of the pond after 150 days is $10^9+(10,000-8,000)(3600) = 1.0072\times10^9$ gallons. Thus the level of pollution after 150 days is $\frac{3.09\times10^6}{1.0072\times10^9}\times100\approx .31\%$.   □

Price Prediction

We want to predict a product's price $P(t)$ which is determined by supply and demand. We restrict our considerations to the case in which the rate of change of $P(t)$ is proportional to the difference between the demand D and the supply S:

$$P'(t) = k(D-S).$$

The demand D and the supply S depend on the time t and the price P. Usually we are given D and S as specific functions of t and P. Consider the special case where

$$D(t) = m(t)P+a(t) \quad \text{and} \quad S(t) = n(t)P+b(t).$$

Then we have the linear differential equation:

$$P'(t) = k\left[m(t)P(t)+a(t)\right]-k\left[n(t)P(t)+b(t)\right] = k\left[m(t)-n(t)\right]P(t)+k\left[a(t)-b(t)\right].$$

Examples 4.11.15 (1) The demand for minivans in Carpool City is given by

$$D = 10,600 - .02P \text{ vans}$$

where P is the price of a minivan in dollars. The supply of minivans is given by

$$S = 9,800 + .03P \text{ vans.}$$

If the rate of change per day of the price is 4% of the difference between the demand and supply, find the price of minivans in the long run.

Solution We solve the differential equation

$$P'(t) = .04\left[(10,600 - .02P(t)) - (9,800 + .03P(t))\right] = 32 - .002P(t).$$

where t is time measured in days. This is a separable differential equation:

$$1 = \frac{P'(t)}{32 - .002P(t)}$$

$$\int 1\, dt = \int \frac{P'(t)}{32 - .002P(t)}\, dt = \int \frac{1}{32 - .002P}\, dP$$

$$t + C = -500\ln|32 - .002P|$$

$$32 - .002P(t) = \pm e^{-\frac{t+C}{500}}$$

$$P(t) = 16,000 \pm 500 e^{-\frac{C}{500}} e^{-\frac{t}{500}} = 16,000 + Ae^{-\frac{t}{500}} \text{ dollars.}$$

In the long run the term $Ae^{-\frac{t}{500}}$ goes to zero, and the price of a minivan w approach \$16,000.

(2) The demand for tomatoes in Pizza Village is given by

$$D = 8,001 - .04P \text{ bushels}$$

where P is the price per bushel in dollars. The supply of tomatoes varies wi the seasons and is given by

$$S = 7,999 + .02P + 500\sin\frac{\pi t}{180} \text{ bushels}$$

where t is the time from May 1st measured in days. The rate of change of t price of tomatoes is 1% of the difference between the demand and supply. Fi the price of tomatoes under these conditions in the long run.

Solution We solve the differential equation

$$P'(t) = .01\left[(8,001 - .04P(t)) - \left(7,999 + .02P(t) + 500\sin\frac{\pi}{18}\right.\right.$$

$$P'(t) = .02 - .0006P(t) - 5\sin\frac{\pi t}{180}$$

$$P'(t) + .0006P(t) = .02 - 5\sin\frac{\pi t}{180}.$$

This is a linear differential equation. Since $\int .0006\, dt = .0006t + B$, an integratin factor is $e^{.0006t}$. Multiply the differential equation by $e^{.0006t}$:

$$\frac{d}{dt}\left(Pe^{.0006t}\right) = P'(t)e^{.0006t} + .0006P(t)e^{.0006t} = .02e^{.0006t} - 5e^{.0006t}\sin\frac{\pi t}{180}$$

$$Pe^{.0006t} = \frac{100}{3}e^{.0006t} - 5\int e^{.0006t}\sin\frac{\pi t}{180}\, dt. \qquad (4.11.1\text{?})$$

Integrate by parts twice, using a calculator to estimate the coefficients:

$$\int e^{.0006t}\sin\frac{\pi t}{180}\, dt = 1667\int D\left(e^{.0006t}\right)\sin\frac{\pi t}{180}\, dt$$

$$= 1667 e^{.0006t}\sin\frac{\pi t}{180} - 1667\int e^{.0006t} D\left(\sin\frac{\pi t}{180}\right)\, dt$$

$$= 1667 e^{.0006t}\sin\frac{\pi t}{180} - 29.09\int e^{.0006t}\cos\frac{\pi t}{180}\, dt$$

$$= 1667 e^{.0006t}\sin\frac{\pi t}{180} - 48,480\int D\left(e^{.0006t}\right)\cos\frac{\pi t}{180}\, dt$$

$$= 1667e^{.0006t} \sin \frac{\pi t}{180} - 48,480e^{.0006t} \cos \frac{\pi t}{180} + 48,480 \int e^{.0006t} D\left(\cos \frac{\pi t}{180}\right) dt$$

$$= 1667e^{.0006t} \sin \frac{\pi t}{180} - 48,480e^{.0006t} \cos \frac{\pi t}{180} - 846.1 \int e^{.0006t} \sin \frac{\pi t}{180} dt$$

$$847.1 \int e^{.0006t} \sin \frac{\pi t}{180} dt = 1667e^{.0006t} \sin \frac{\pi t}{180} - 48,480e^{.0006t} \cos \frac{\pi t}{180} + C$$

$$\int e^{.0006t} \sin \frac{\pi t}{180} dt = 1.968e^{.0006t} \sin \frac{\pi t}{180} dt - 57.23e^{.0006t} \cos \frac{\pi t}{180} + C.$$

Substitute the value of this integral into (4.11.17):

$$P(t)e^{.0006t} = 33.33e^{.0006t} - 9.840e^{.0006t} \sin \frac{\pi t}{180} + 286.2e^{.0006t} \cos \frac{\pi t}{180} + C,$$

$$P(t) = 33.33 - 9.840 \sin \frac{\pi t}{180} + 286.2 \cos \frac{\pi t}{180} + Ce^{-.0006t}.$$

In the long run, the term $Ce^{-.0006t}$ approaches zero and

$$P(t) \approx 33.33 - 9.840 \sin \frac{\pi t}{180} + 286.2 \cos \frac{\pi t}{180}.$$

Observe that in the long run the price of tomatoes becomes periodic, matching the periodic supply function. □

Historical Remarks

By the Second Fundamental Theorem of Calculus, every computation of an integral $\int f(x) \, dx = F(x) + C$ is equivalent to solving the differential equation $y' = f(x)$. Thus the history of differential equations began at the end of the seventeenth century with the discovery of the Fundamental Theorems of Calculus by Newton and Leibniz. In addition, Newton's study of mechanics included formulating physical principles as differential equations and solving them. He also made abstract studies of differential equations of the forms $y' = f(x)$, $y' = g(y)$ and $y' = h(x, y)$. He originated the method of finding power series solutions to these equations, which we will study in Sections 5.6 and 5.7. In the early 1690s, Leibniz developed the methods for solving separable and linear differential equations which were presented in this section.

Beginning in the late 1680s, the Swiss brothers Jakob and Johann Bernoulli, as well as the latter's son Daniel, used calculus to solve problems in mechanics. They formulated these problems in terms of differential equations and developed methods for solving them. We illustrate their work with the *brachistochrone problem*[6]. The problem originated with Galileo Galilei in 1638 who showed that a bead starting at rest at P would slide down a frictionless wire in the shape of a polygonal path to Q more quickly than along a straight line from P to Q. See Figure 4.11.16. Taking the limit, he concluded that the bead would slide more quickly along a circular arc than along a straight line or any polygonal path. In addition, he claimed that a circular arc is the curve along which the bead travels from P to Q in the shortest possible time. In 1696, Johann Bernoulli publicly posed the question of finding the path which gives the shortest time. The problem was solved by Newton and Leibniz as well as each of the Bernoulli brothers. The question of priority started a nasty argument between the two brothers. Johann, in fact, was such a jealous person that he threw his son Daniel out of the house for winning a prize, that he coveted, from the French Academy!

Figure 4.11.16
Sliding Bead

We present Johann Bernoulli's solution of the brachistochrone problem. His derivation of the differential equation is based upon the derivation of Snell's Law using

---

[6] The word brachistochrone comes from the Greek words *brachistos*, shortest, and *chronos*, time.

Fermat's principle of least time. In 1621, Snell used experimental evidence to deme strate that when light passes from one medium to another, it changes direction w $\frac{\sin\theta_1}{\sin\theta_2} = k$. See Figure 4.11.17. The constant k was identified in 1637 by Re Descartes as the quotient $\frac{v_1}{v_2}$ of the velocities of light in the two media. In 1657 F mat proved this law from the assumption light travels in the path which takes it fr P to Q in the least possible time. See Problem 4 of Section 2.12. Bernoulli rewr this equation as $\frac{\sin\theta_1}{v_1} = \frac{\sin\theta_2}{v_2}$. He deduced that if light passes through a medium variable density it will satisfy the equation

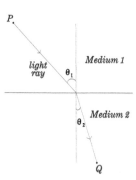

Figure 4.11.17
Refraction of Light

$$\frac{\sin\theta}{v} = C \tag{4.11.1}$$

where v is the velocity of light and θ is the angle between the tangent to the pa of the light and the vertical. Return to the brachistochrone problem. Introdu coordinates so that the starting position P of the bead is the origin. By the princi of conservation of energy, the kinetic energy $\frac{1}{2}mv^2$ of a bead of mass m must equ its loss in potential energy mgy. Hence $v = \sqrt{2gy}$ depends only on the depth y of t bead below its original position. Bernoulli concluded that a bead moving from P to in the least possible time would move along the same curve as a beam of light passi through a medium of decreasing opacity with velocity $\sqrt{2gy}$ at depth y. Thus t path of this bead will satisfy (4.11.18). From Figure 4.11.18, the slope of the tange to the bead's path is $y' = \tan(\pi/2 - \theta) = \cot\theta$. Then

$$\sin\theta = \frac{1}{\csc\theta} = \frac{1}{\sqrt{1+\cot^2\theta}} = \frac{1}{\sqrt{1+(y')^2}}$$

Thus (4.11.18) can be rewritten as

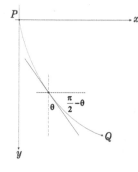

Figure 4.11.18
Brachistochrone Problem

$$\frac{1}{\sqrt{1+(y')^2}} = \sin\theta = Cv = C\sqrt{2gy} \quad \text{or} \quad y\left[1+(y')^2\right] = B.$$

Rewrite this differential equation as the separable equation $y' = \sqrt{\frac{B-y}{y}}$. Then

$$x = \int 1\, dx = \int \sqrt{\frac{y(x)}{B-y(x)}}\, y'(x)\, dx = \int \sqrt{\frac{y}{B-y}}\, dy. \tag{4.11.1}$$

Make the substitution $\tan\phi = \sqrt{\frac{y}{B-y}}$. Then

$$\tan^2\phi = \frac{y}{B-y}$$
$$y(1+\tan^2\phi) = B\tan^2\phi$$
$$y = \frac{B\tan^2\phi}{\sec^2\phi} = B\sin^2\phi. \tag{4.11.20}$$

Thus $dy = 2B\sin\phi\cos\phi\, d\phi$ and by (4.11.19):

$$x = \int (\tan\phi)(2B\sin\phi\cos\phi)\, d\phi = 2B\int \sin^2\phi\, d\phi = B\int 1 - \cos 2\phi\, d\phi$$
$$x = \frac{B}{2}(2\phi - \sin 2\phi) + C = \frac{B}{2}(2\phi - \sin 2\phi) \tag{4.11.21}$$

since $x = \phi = 0$ when $y = 0$. We shall see in Section 6.6 that equations (4.11.20) an (4.11.21) describe a curve called a *cycloid*. Note this path is not an arc of a circle, a conjectured by Galileo.

Summary

The reader should be able to solve separable and linear first order differential equations. She should be able to formulate problems dealing with growth, mixing, chemical reactions and price prediction as differential equations. In addition, she should be able to carry out similar analyses in other contexts.

Basic Exercises

1. Find the general solution of each differential equation.
   (a) $y' = x \ln x$ (b) $y' = 5y^2$ (c) $xy' = \ln x$
   (d) $y' = 6x^2 y$ (e) $(y')^2 = x^2 + 1$ (f) $y' = -e^y$
   (g) $y' + 3y = 2x + 1$ (h) $y' = \sin x \cos y$ (i) $xy' + y = e^x$
   (j) $y' = xy + x + y + 1$ (k) $y' + y \sin x = 12x^3 e^{\cos x}$ (l) $xy' + y^2 = 1$

2. Solve each initial value problem.
   (a) $y' = -6y$, $y(0) = 3$ (b) $y' = x \sin x$, $y(\pi) = 3$
   (c) $y' = \pi y$, $y(1) = -2$ (d) $y' = y \tan x$, $y(\pi/4) = 5$
   (e) $\sqrt{x} y' = x^2 + 1$, $y(1) = 4$ (f) $y' + 5y = x$, $y(0) = 4$
   (g) $y' = e^{x+y}$, $y(5) = 2$ (h) $x^2 y' + 8xy = \ln x$, $y(1) = 0$
   (i) $yy' - 3y^2 = xy$, $y(2) = -1$ (j) $3y' = \frac{1}{y^2} - \frac{y}{x}$, $y(2) = 3$
   (k) $\frac{x}{yy'} = x^2 + 1$, $y(-1) = 4$ (l) $\frac{x}{y+y'} = e^x$, $y(0) = 3$

3. Five thousand bacteria are placed in a dish of agar. Twenty four hours later there are 25,000 bacteria in this dish.
   (a) How many bacteria will be in this dish at the end of three days?
   (b) How long did it take for the number of bacteria to double?

4. Bacteria are placed in a dish. Five hours later there are 8,000 bacteria present and seven hours later there are 15,000 bacteria present.
   (a) How many bacteria were initially in this dish?
   (b) How many bacteria will be present after 12 hours?

5. A certain strain of bacteria is found to double every three hours. 15,000 of these bacteria are placed in a dish.
   (a) How many bacteria will be present after 10 hours?
   (b) In how long will there be 25,000 bacteria in this dish?

6. A radioactive substance of 10 ounces decays to 8 ounces in twelve hours.
   (a) How much of this substance will be left after 15 hours?
   (b) How long will it take until there is only one ounce of this substance remaining?

7. A radioactive substance is placed on a table. After 6 hours there are 12 grams remaining, while after 15 hours there are 10 grams remaining.
   (a) How many grams of this substance were placed on the table?
   (b) How much of this substance will remain after 24 hours?
   (c) How long will it take until there are only 5 grams of this substance remaining?

8. Consider a radioactive substance R. Show there is a fixed time t_H called the *half-life of R* such that any mass of R decays to half its mass in time t_H.

9. All living organisms contain carbon of which one ten thousandth is the radioactive isotope ^{14}C. As long as an organism is alive, it maintains the level of ^{14}C amongst all of its carbon at one in ten thousand. However, when an organism dies the ^{14}C begins to decay with a half-life of 5700 years.
   (a) An Egyptian mummy is found to have three-fourths of its original ^{14}C. How long ago did the mummy die?
   (b) An archaeologist discovers a parchment scroll in 1998 which is known to have been written in 450 BCE. What percent of the original ^{14}C remains in this parchment?

10. A five thousand dollar certificate of deposit earns 6% interest, compound continuously, for five years. How much will this certificate be worth at maturity?

11. What rate of interest, compounded continuously, must a $8,000 certificate deposit earn to be worth $20,000 after ten years?

12. How long does it take money to double if it is compounded continuously at 5

13. A 5,000 gallon tank of 3% brine begins to leak at one quart per hour. T leaking brine is simultaneously replaced with pure water. Find the concentration brine in this tank one day later.

14. A pond is created by placing a dam on a stream. The volume of this pond kept constant at one million liters because the thousand liters of water per hour whi enter from the stream cause an equal volume of water to flow over the top of the da The pond contains 4 parts per million (ppm) of PCBs and the stream contains 6 pp of PCBs. What will the concentration of PCBs in this pond be 30 days later?

15. A hundred gallon fish tank contains 2% oxygen. One fourth of a gallon water per minute passes through an aerator which is attached to this tank. The wat which the aerator returns to the tank each minute has one tenth of the square ro of the gallons of oxygen which were in the water that entered the aerator. Find t concentration of oxygen in this tank in the long run.

16. A large vat contains 1,000 liters of 4% brine. A solution of 3% brine is add at the rate of two liters per minute while brine is drawn out of the vat at the rate $1\frac{1}{2}$ liters per minute. Find the concentration of brine in this vat after ten hours?

17. A fruit canning company has a large vat of syrup which contains 500 gallons 15% sugar. The canning process uses two gallons per minute of this syrup. If a syru of 20% sugar is added at the rate of 2.5 gallons per minute, find the concentration sugar in this vat after four hours.

18. Water is poured into a vat containing 50 liters of 5% brine at the rate of tw liters per minute. Brine flows from this vat at two liters per minute into a second v which contains 200 liters of 3% brine. Brine from the second vat flows through a ta at the rate of two liters per minute.
(a) What is the concentration of each of these vats after t minutes?
(b) What is the maximum concentration of salt in the second vat?

19. The number of bacteria $y(t)$ in a petrie dish is found to satisfy the differenti equation $y' = 3y - .0002y^2$. If 10,000 bacteria are placed in this dish, find the numb of bacteria in this dish in the long run.

20. The number of bees $y(t)$ in a hive is found to satisfy the differential equatic $y' = 1.5y - .003y^2$. If there are initially one hundred bees in this hive, find th population of bees in this hive in the long run.

21. 80 fish are placed in a pond. If h fish are harvested per day, the number of fis $y(t)$ in this pond on day t satisfies the differential equation $y' = .03y - .00005y^2 - h$ Find the largest value of h for which fish can be harvested indefinitely.

22. Fifty thousand yeast bacteria are placed in a growth medium. If h bacteria ar harvested each day, the number of bacteria in this medium after t days satisfies th differential equation $y' = 4y - .0002y^2 - h$. Find the largest value of h for which yeas can be harvested indefinitely.

23. There are 500 moose in a national forest. h permits per year are issued, each o which allows the bearer to hunt two moose. The population $y(t)$ of moose after t year is found to satisfy the differential equation $y' = 1.5y - .004y^2 - h$. Find the larges value of h for which the moose population of this forest will survive indefinitely.

24. A store finds that the demand for its designer jeans is $D = 800 - .05P$ jeans where P is the price of a pair of jeans in dollars. The supply of jeans is $S = 780 + .02P$ jeans. If the rate of change of the price is 2% of the difference between the demand and supply, find the price of jeans in the long run.

25. A new computer chip, superior to all of its competitors, has been developed. The supply of this chip is $S = 500 + .01P$ chips where P is the price of each chip. The demand for this chip after t days is $D = 510 + .0001t$ chips. The initial price of this chip is $10, and the rate of change of the price is 3% of the difference between the demand and supply. Find the price of this chip after 30 days.

26. The demand for feed corn in Ohio is $D = 81,276 - .08P$ bushels where P is the price of corn per bushel. Let time t be measured in days starting on May first. The supply of feed corn is $S = 81,264 + .04P - 100\cos\frac{\pi t}{180}$ bushels. If the rate of change of the price of corn is 2% of the difference between the demand and supply, find the approximate price of feed corn in the long run.

27. The compound SO_2Cl_2 decomposes into sulphur dioxide and chlorine in the first order reaction $SO_2Cl_2 \rightarrow SO_2 + Cl_2$. The initial concentration of SO_2Cl_2 is .4 moles per liter and the rate constant is $k = 2 \times 10^{-5}$. When will the concentration of SO_2Cl_2 be reduced to .2 moles per liter?

28. The gas hydrogen iodide decomposes into hydrogen and iodine in the second order reaction $2HI \rightarrow H_2 + I_2$. The initial concentration of HI is .05 moles per liter and the rate constant is $k = .08$. When will the concentration of HI be reduced to .01 moles per liter?

29. Nitrogen monoxide combines with oxygen to form nitrogen dioxide in the third order reaction $2NO + O_2 \rightarrow 2NO_2$ with reaction rate $7 \times 10^9 [NO]^2 [O_2]$. The initial concentration of NO is .01 moles per liter, and the initial concentration of O_2 is .03 moles per liter. How long does it take to produce .002 moles of NO_2?

30. The half–life of a reaction for the decomposition of the compound C is the time required for half of C to disappear. Let k be the rate constant of this reaction.
(a) Show the half–life of the decomposition of C in a 1st order reaction is $\frac{\beta \ln 2}{k\alpha_1}$.
(b) Show the half–life of the decomposition of C in a 2nd order reaction is $\frac{\beta}{k a_1 \alpha_1}$.

31. A chemical S is decomposing through a first order reaction. Let $c(t)$ denote the concentration of S at time t in moles per liter.
(a) Find the value of the constant h such that $\ln\frac{c(t)}{c(0)} = ht$.
(b) Show $c(t) = c(0)e^{ht}$.

32. A chemical S is decomposing through a second order reaction. Let $c(t)$ denote the concentration of S at time t in moles per liter.
(a) Find the value of the constant h such that $\frac{c(t) - c(0)}{c(0)c(t)} = ht$.
(b) Show $c(t) = \frac{c(0)}{1 - hc(0)t}$.

33. An object is placed in a medium of constant temperature. Newton's law of cooling states that the rate of change of the temperature T of this object is proportional to the difference between its temperature and the temperature T_0 of the surrounding medium. That is, $T' = k(T_0 - T)$ with $k > 0$.
(a) Solve for T in terms of the time t and the constants T_0, k, $T(0)$.
(b) A metal bar of temperature $1000°F$ is removed from an oven and placed in a large room of temperature $70°F$. If the bar cools to $900°F$ in ten seconds, how long will it take the bar to cool to $100°F$?

34.   A two pound rock is kicked upwards with an initial velocity of 128 feet
second. Assume the air resistance on this rock is one hundredth of its velocity.
Newton's second law of motion: $2a(t) = -32 - .01v(t)$.
(a) Find the velocity $v(t)$ of the rock at time t.
(b) What is the maximum height of the rock?
(c) How long does it take for the rock to return to the ground?

35. An epidemic is spreading in a population of N people. Let x denote the fract
of well people, and let y denote the fraction of sick people. The rate at which
epidemic spreads is proportional to the product of x and y: $y' = kxy$.
(a) Solve for y as a function of the time t and the constants k, N, $y(0)$.
(b) If the epidemic spreads according to this model, what occurs in the long run?

36. By Lambert's law of absorption, the fraction of sunlight absorbed by each th
horizontal layer of the ocean is proportional to its thickness. If half the sunlight
absorbed at a depth of 30 m, what percent of the sunlight remains at a depth of 50 r

The following two problems are for those who have studied Section 4.10

37. Find the general solution of each differential equation.
(a) $y' = x\operatorname{sech} y$ **(b)** $y' = 2y + \sinh x$ **(c)** $y' = e^x \tanh^3 y$ **(d)** $xy' = 12x \cosh x -$

38. Solve each initial value problem.   **(a)** $y' = 8x^3 \coth y$, $y(0) = \ln 5$
(b) $y'\operatorname{csch} x = \cosh x - y$, $y(0) = 2$ **(c)** $y' = e^x \sinh y$, $y(0) = 1$

Challenging Problems

1. (a) If $g(x, y)$ can be written as a function of $\frac{y}{x}$, we call g a *homogeneous functio*
Assume if $g(x, y)$ is defined, then $g(tx, ty)$ is also defined for $t \neq 0$. Show $g(x, y)$
homogeneous if and only if $g(tx, ty) = g(x, y)$ for $t \neq 0$.
(b) Show each function is homogeneous and write it as a function of $\frac{y}{x}$.

  (i) $g(x, y) = \frac{x^2 - y^2}{x^2 + y^2}$ **(ii)** $h(x, y) = 5 - \frac{xy^3}{x^4 + x^2 y^2}$
  (iii) $j(x, y) = \frac{x + y}{\sqrt{x^2 + y^2}}$ **(iv)** $k(x, y) = 2\ln|x - y| - \ln(x^2 + y^2)$

2.   **(a)** A *homogeneous* differential equation has the form $y' = f\left(\frac{y}{x}\right)$. Show t
substitution $u = \frac{y}{x}$ produces the separable differential equation $xu' = f(u) - u$.
(b) Solve each homogeneous differential equation.

  (i) $2x^2 y' = x^2 + y^2$ **(ii)** $y' = \frac{x+y}{x-y}$ **(iii)** $xy' = y + \sqrt{x^2 + y^2}$
  (iv) $y' = \frac{x}{y} + \frac{y}{x}$ **(v)** $x^2 y' = x^2 - xy + y^2$

3. The differential equation $y' + p(x)y = q(x)y^n$ is called a *Bernoulli equation*.
(a) Solve the Bernoulli equation when $n = 0$.
(b) Solve the Bernoulli equation when $n = 1$.
(c) Show if $n \neq 0, 1$ then the change of variables $u = y^{1-n}$ transforms the Bernoul
equation into the linear equation $u' + (1 - n)p(x)u = (1 - n)q(x)$.
(d) Solve each Bernoulli equation.

  (i) $y' = 2y - 6y^2$ **(ii)** $y' = 5y + y^3 \sin x$ **(iii)** $xy' = 4y + 7\frac{x}{y}$
  (iv) $xy' = 8xy^{2/5} + 3y$ **(v)** $y^2 y' = 12y^3 + e^{-x}$ **(vi)** $xyy' = y^2 + x^2 \ln x$

4. According to Torricelli's Law, the velocity $v(t)$ of water that flows through a hol
in the bottom of a tank equals the velocity it would have if it fell through air fror
rest at the surface of the water to the bottom of the tank.
(a) Let $y(t)$ denote the depth of water in the tank at time t. Show $v = \sqrt{2gy}$.
(b) If the hole has area A and the cross-sectional area of the tank at height y is $A(y)$
show $A(y)\frac{dy}{dt} = -A\sqrt{2gy}$.
(c) A hemispherical sink of radius 15 cm. is full of water. The plug is removed fror
a circular hole of radius 2 cm. at the bottom of the sink. How long does it take fo
all the water to flow out this hole?

4.12   Second Order Differential Equations

Prerequisite: Section 4.11

This section is devoted to solving various types of second order differential equations. In the first subsection, we consider examples where the substitution $u = y'$ transforms the second order differential equation in y into a first order differential equation in u. In the second subsection, we solve linear homogeneous equations with constant coefficients: $y'' + by' + cy = 0$. In the third subsection, we use the method of variation of parameters and the method of undetermined coefficients to solve linear nonhomogeneous equations with constant coefficients: $y'' + by' + cy = f(x)$. The methods of these subsections generalize to solve higher order linear differential equations. The fourth subsection models various types of harmonic motion by second order linear differential equations with constant coefficients.

Simple Examples

We solve certain second order differential equations by replacing the dependent variable $y(x)$ with the new variable $u(x) = y'(x)$. We solve the resulting first order differential equation for u. Then we integrate u to find y. The general solution will involve two constants of integration. However, an *initial value problem*, where we are given the values $y(x_0)$ and $y'(x_0)$, will have a unique solution. We illustrate this procedure with three types of examples.

(1)   The differential equation $y'' = f(x)$ is solved by integrating twice. First find $u(x) = y'(x) = \int f(x)\,dx$, and then find $y(x) = \int u(x)\,dx$.

(2)   To solve the differential equation $y'' = g(x, y')$ substitute $u(x) = y'(x)$. Solve the first order equation $u'(x) = g(x, u)$ for $u(x)$. Then $y(x) = \int u(x)\,dx$.

(3)   To solve the differential equation $y'' = h(y, y')$ substitute $u = y'(x)$. Then

$$\frac{du}{dy}u = \frac{du}{dy}\frac{dy}{dx} = \frac{du}{dx} = y'' = h(y, y') = h(y, u). \tag{4.12.1}$$

Solve this first order equation for $u = y'(x)$ as a function of y. Then $y(x)$ is the solution of this separable first order differential equation.

Examples 4.12.1 (1) (a) Solve the differential equation $y'' = 12x^2$.
  (b) Solve the initial value problem $y'' = 12x^2$ with $y(0) = 3$ and $y'(0) = -4$.

  Solution (a) Integrate twice:

$$y'(x) = \int y''(x)\,dx = \int 12x^2\,dx = 4x^3 + C$$

$$y(x) = \int y'(x)\,dx = \int 4x^3 + C\,dx = x^4 + Cx + B.$$

  (b) $-4 = y'(0) = 4(0^3) + C = C$, and $3 = y(0) = 0^4 + C(0) + B = B$. Thus the solution to the initial value problem is $y = x^4 - 4x + 3$.

(2) **(a)** Solve the differential equation $xy'' \ln x = y'$.
  (b) Solve the initial value problem $xy'' \ln x = y'$ with $y(e) = 2$ and $y'(e) = 5$.

  Solution (a) Let $u = y'$. Then $xu' \ln x = u$. Clearly $u = 0$ is one solution. When $u \neq 0$, integrate this separable differential equation:

$$\int \frac{1}{x \ln x}\,dx = \int \frac{1}{u}\frac{du}{dx}\,dx = \int \frac{1}{u}\,du$$

$$\ln|\ln x| + C \;=\; \ln|u|$$
$$|u| \;=\; e^{\ln|u|} \;=\; e^{\ln|\ln x| + C} \;=\; e^C e^{\ln|\ln x|} \;=\; A|\ln x|$$

for $A \geq 0$ (including the solution $u = 0$). Hence $y' = u = A\ln x$ for $A \in$
Integrate by parts to find y:

$$y \;=\; \int y'\,dx = \int A\ln x\,dx = A\int D(x)\ln x\,dx = Ax\ln x - A\int x D(\ln x)$$

$$=\; Ax\ln x - A\int 1\,dx = Ax\ln x - Ax + B.$$

(b) $5 = y'(e) = A\ln e = A$, and $2 = y(e) = 5e\ln e - 5e + B = B$. Thus t
solution to the initial value problem is $y = 5x\ln x - 5x + 2$.

(3) **(a)** Solve the differential equation $y'y'' = 1$.
(b) Solve the initial value problem $y'y'' = 1$ with $y(1) = -1$ and $y'(1) = 2$.

Solution   **(a)** Rewrite this equation as $y'' = \frac{1}{y'}$, and let $u = y'$. In this ca
the first order differential equation (4.12.1) is $\frac{du}{dy}u = \frac{1}{u}$. Integrate this separal
differential equation:

$$\int 1\,dy \;=\; \int u^2\frac{du}{dy}\,dy = \int u^2\,du$$

$$y + C \;=\; \frac{u^3}{3}$$

$$y' \;=\; u = (3y + 3C)^{1/3}.$$

Integrate this separable differential equation:

$$\int 1\,dx \;=\; \int \frac{1}{(3y + 3C)^{1/3}}\frac{dy}{dx}\,dx = \int \frac{1}{(3y + 3C)^{1/3}}\,dy$$

$$x + B' \;=\; \frac{1}{2}(3y + 3C)^{2/3}$$

$$(2x + B)^{3/2} \;=\; 3y + 3C \qquad\qquad [\text{where } B = 2B']$$

$$y \;=\; \frac{1}{3}(2x + B)^{3/2} - C.$$

(b) $2 = y'(1) = [3y(1) + 3C]^{1/3} = [3(-1) + 3C]^{1/3}$. Hence $3C - 3 = 8$ and $C =$
Also, $-1 = y(1) = \frac{1}{3}(2 + B)^{3/2} - \frac{11}{3}$. Then $8 = (2 + B)^{3/2}$, $2 + B = 8^{2/3} = 4$ an
$B = 2$. Thus the initial value problem has solution $y = \frac{1}{3}(2x + 2)^{3/2} - \frac{11}{3}$.

Homogeneous Equations with Constant Coefficients

This subsection is devoted to solving the second order homogeneous linear differentia
equation with constant coefficients:

$$y'' + by' + cy = 0. \tag{4.12.2}$$

This equation is called *homogeneous* because its right side is zero. It is called *linec*
because its left side is a linear combination of y'', y' and y. It has *constant coefficien*
$a, b, c \in \Re$. We find the general solution and show that every associated initial valu
problem has a unique solution.

Reformulate the differential equation (4.12.2) by using the symbol D for the firs
derivative and D^2 for the second derivative:

$$(D^2 + bD + c)(y) = 0.$$

The following proposition says that linear combinations of solutions of this equation are also solutions.

Proposition 4.12.2 (Superposition Principle) *If u and v are solutions of the linear homogeneous differential equation with constant coefficients $(D^2 + bD + c)(y) = 0$ and $h, k \in \mathfrak{R}$, then $hu + kv$ is also a solution.*

Proof We know $(D^2 + bD + c)(u) = 0$ and $(D^2 + bD + c)(v) = 0$. Since the derivatives D and D^2 are linear:

$$
\begin{aligned}
(D^2 + bD + c)(hu + kv) &= h(D^2 + bD + c)(u) + k(D^2 + bD + c)(v) \\
&= (h)(0) + (k)(0) = 0. \qquad \square
\end{aligned}
$$

The polynomial $p^2 + bp + c$ is called the *characteristic polynomial* of the equation $y'' + by' + cy = 0$. We divide our discussion into the cases when the roots r, s of the characteristic polynomial are real and when they are complex. If these roots are real, $p^2 + bp + c = (p - s)(p - r)$ and

$$0 = (D^2 + bD + c)(y) = (D - s)(D - r)(y). \tag{4.12.3}$$

We know the first order equation $0 = (D - r)(y) = y' - ry$ has a solution $y = e^{rx}$, and the equation $0 = (D - s)(y) = y' - sy$ has a solution $y = e^{sx}$. Then

$$
\begin{aligned}
(D - s)(D - r)(e^{rx}) &= (D - s)(0) = 0 \quad \text{and} \\
(D - s)(D - r)(e^{sx}) &= (D - r)(D - s)(e^{sx}) = (D - r)(0) = 0.
\end{aligned}
$$

By the superposition principle, $Ae^{rx} + Be^{sx}$ is a solution of (4.12.3) for $A, B \in \mathfrak{R}$. We show this is the general solution when $r \neq s$ and then consider the case $r = s$.

Proposition 4.12.3 *Let $r, s \in \mathfrak{R}$. Consider the linear homogeneous second order differential equation*

$$(D - s)(D - r)(y) = 0.$$

(a) *If $r \neq s$, the general solution is*

$$y = Ae^{rx} + Be^{sx} \quad \text{for } A, B \in \mathfrak{R}.$$

(b) *If $r = s$, the general solution is*

$$y = Ae^{rx} + Cxe^{rx} \quad \text{for } A, C \in \mathfrak{R}.$$

Proof We know $y = e^{rx}$ is one solution. Let $u = ye^{-rx}$. By the product rule:

$$
\begin{aligned}
0 = (D - s)(D - r)(y) &= (D - s)(D - r)(ue^{rx}) = (D - s)[u(D - r)(e^{rx}) + e^{rx}Du] \\
&= (D - s)(u'e^{rx}) \qquad \text{[since } (D - r)(e^{rx}) = 0] \\
D(u'e^{rx}) &= su'e^{rx}.
\end{aligned}
$$

Let $v = u'e^{rx}$. Then $Dv = D(u'e^{rx}) = su'e^{rx} = sv$. The general solution is $v = Ce^{sx}$. Hence

$$u' = ve^{-rx} = Ce^{sx}e^{-rx} = Ce^{(s-r)x}. \tag{4.12.4}$$

(a) Since $s - r \neq 0$, integration of (4.12.4) gives:

$$
\begin{aligned}
ye^{-rx} &= u = \int u' \, dx = \int Ce^{(s-r)x} \, dx = Be^{(s-r)x} + A \text{ [where } B = C/(s - r)]. \\
y &= Be^{sx} + Ae^{rx}
\end{aligned}
$$

(b) Since $s - r = 0$, equation (4.12.4) becomes $u' = C$ and

$$ye^{-rx} \;=\; u \;=\; \int u'\,dx \;=\; \int C\,dx \;=\; Cx + A,$$

$$y \;=\; Cxe^{rx} + Ae^{rx}.$$

$\quad$ ⌐

Notes **(1)** When $r = 0$, $s \neq 0$ the general solution of $y'' - sy' = 0$ is $y = A + Be$
(2) When $r = s = 0$, the general solution of $y'' = 0$ is $y = A + Cx$.

Examples 4.12.4 **(1)** Solve the linear differential equation $y'' + 2y' - 15y = 0$.

$\quad$ **Solution** The characteristic polynomial is $p^2 + 2p - 15 = (p+5)(p-3)$ with ro
$\quad$ -5 and $+3$. By Proposition 4.12.3(a), the general solution of this different
$\quad$ equation is $y = Ae^{-5x} + Be^{3x}$.

(2) Solve the linear differential equation $y'' - 8y' + 16y = 0$.

$\quad$ **Solution** The characteristic polynomial is $p^2 - 8p + 16 = (p-4)^2$ with repeat
$\quad$ root 4. By Proposition 4.12.3(b), the general solution of this differential equati
$\quad$ is $y = Ae^{4x} + Cxe^{4x}$.

$\quad$ Now consider the case when the characteristic polynomial of the linear homog
neous differential equation with constant coefficients $y'' + by' + cy = 0$ has comp
roots. Complete the square: $p^2 + bp + c = (p - r)^2 + k^2$ with $r = -\frac{b}{2}$ and $k^2 = c - $
Hence the differential equation can be written as:

$$\left[(D - r)^2 + k^2 \right](y) = 0. \tag{4.12.}$$

Let $u = e^{-rx}y$. Then

$$
\begin{aligned}
(D - r)(y) \;&=\; (D - r)(ue^{rx}) = u'e^{rx} + u(D - r)\,(e^{rx}) = u'e^{rx} &&\text{(4.12.}\\
-k^2 u e^{rx} \;&=\; -k^2 y = (D - r)^2(y) &&\text{[by (4.12.5}\\
-k^2 u e^{rx} \;&=\; (D - r)\,(u'e^{rx}) &&\text{[by (4.12.}\\
-k^2 u e^{rx} \;&=\; u''e^{rx} + u'(D - r)\,(e^{rx}) = u''e^{rx}\\
u'' \;&=\; -k^2 u. &&\text{(4.12.}
\end{aligned}
$$

We solve this differential equation for u in the following proposition. Then $y = ue$
is the general solution of the original differential equation (4.12.5).

Proposition 4.12.5 *Let $k \neq 0$. The differential equation*

$$y'' = -k^2 y$$

has the general solution

$$y = A\sin kx + B\cos kx \quad \text{for } A, B \in \Re.$$

Proof Clearly all the functions $y = A\sin kx + B\cos kx$ are solutions of the differenti
equation. Let $y = f(x)$ be any nonzero solution. Let $B = f(0)$ and $A = \frac{f'(0)}{k}$. Defin
$F(x) = A\sin kx + B\cos kx$. We show that $F(x) = f(x)$. By the quotient rule:

$$\frac{d}{dx}\left[\frac{F(x)}{f(x)} \right] = \frac{F'(x)f(x) - F(x)f'(x)}{f(x)^2}. \tag{4.12.8}$$

By the product rule, the derivative of the numerator of this fraction is

$$[F''(x)f(x) + F'(x)f'(x)] - [F'(x)f'(x) + F(x)f''(x)]$$
$$= F''(x)f(x) - F(x)f''(x) = -k^2 F(x)f(x) + k^2 F(x)f(x) = 0.$$

Hence the numerator of the fraction on the right hand side of (4.12.8) is a constant. Substitute $x = 0$ into this numerator: $(kA)(B) - (B)(kA) = 0$. Thus $\frac{F(x)}{f(x)}$ has zero derivative and must be a constant. Since $F(0) = f(0) = B$, it follows that $\frac{F(x)}{f(x)} = \frac{F(0)}{f(0)} = 1$ and $F(x) = f(x)$. □

We use this proposition to complete the solution of a second order linear homogeneous differential equation with constant coefficients when its characteristic polynomial has complex roots.

Corollary 4.12.6 *Let $k \neq 0$. The differential equation*

$$\left[(D - r)^2 + k^2\right](y) = 0$$

has the general solution

$$y = Ae^{rx} \sin kx + Be^{rx} \cos kx \quad \text{for } A, B \in \Re.$$

Proof Let $u = e^{-rx}y$. By (4.12.7), $u'' = -k^2 u$. By Proposition 4.12.5:

$$e^{-rx}y = u = A \sin kx + B \cos kx$$
$$y = Ae^{rx} \sin kx + Be^{rx} \cos kx. \qquad \square$$

Examples 4.12.7 (1) Solve the differential equation $y'' + 4y = 0$.

Solution The characteristic polynomial of this differential equation is $p^2 + 2^2$. By Proposition 4.12.5, the general solution of this differential equation is $y = A \cos 2x + B \sin 2x$.

(2) Solve the differential equation $y'' + 6y' + 25y = 0$.

Solution The characteristic polynomial of this differential equation is $p^2 + 6p + 25 = (p + 3)^2 + 4^2$. By Corollary 4.12.6, the general solution of this differential equation is $y = Ae^{-3x} \sin 4x + Be^{-3x} \cos 4x$. □

To solve initial value problems for these differential equations, we use the following criterion to determine when a linear system of two equations in two unknowns has a unique solution.

Proposition 4.12.8 *Consider two linear equations in the unknowns x, y:*

$$ax + by = h$$
$$cx + dy = k$$

If $ad - bc \neq 0$, then this linear system has the unique solution:

$$x = \frac{dh - bk}{ad - bc} \qquad y = \frac{ak - ch}{ad - bc}.$$

Proof Consider d times the first equation minus b times the second equation:

$$(ad - bc)x = dh - bk.$$

If $ad - bc \neq 0$, then $x = \frac{dh - bk}{ad - bc}$. Similarly, consider c times the first equation minus a times the second equation:

$$(bc - ad)y = ch - ak.$$

If $ad - bc \neq 0$, then $y = \frac{ch - ak}{bc - ad} = \frac{ak - ch}{ad - bc}$. □

The expressions $ad - bc$ which will occur when we solve initial value problems are called *Wronskians*.

Definition 4.12.9 *The Wronskian of the differentiable functions $u(x)$ and $v(x)$ i*

$$W(u, v) = u(x)v'(x) - u'(x)v(x).$$

We compute the Wronskian for the functions which occur as solutions to homo neous linear second order differential equations with constant coefficients.

Lemma 4.12.10 *Let k, r, s be nonzero numbers with $r \neq s$. Then*

(a) $W(1, x) = 1;$ (d) $W(e^{rx}, xe^{rx}) = e^{2rx};$

(b) $W(1, e^{rx}) = re^{rx};$ (e) $W(\cos kx, \sin kx) = k;$

(c) $W(e^{rx}, e^{sx}) = (s - r)e^{(r+s)x};$ (f) $W(e^{rx}\cos kx, e^{rx}\sin kx) = ke$

In all cases the Wronskian is never zero.

Proof (a) $W(1, x) = 1D(x) - D(1)x = 1 - 0 = 1.$

(b) $W(1, e^{rx}) = 1D(e^{rx}) - D(1)e^{rx} = re^{rx} - 0 = re^{rx}.$

(c) $W(e^{rx}, e^{sx}) = e^{rx}D(e^{sx}) - D(e^{rx})e^{sx} = se^{rx+sx} - re^{rx+sx} = (s - r)e^{(r+s)x}.$

(d) $W(e^{rx}, xe^{rx}) = e^{rx}D(xe^{rx}) - D(e^{rx})xe^{rx} = (e^{2rx} + rxe^{2rx}) - rxe^{2rx} = e^{2rx}.$

(e) $W(\cos kx, \sin kx) = \cos kx D(\sin kx) - D(\cos kx)\sin kx = k\cos^2 kx + k\sin^2 kx =$

(f) $W(e^{rx}\cos kx, e^{rx}\sin kx) = e^{rx}\cos kx D(e^{rx}\sin kx) - D(e^{rx}\cos kx)e^{rx}\sin kx$

$= (e^{rx}\cos kx)(re^{rx}\sin kx + ke^{rx}\cos kx) - (re^{rx}\cos kx - ke^{rx}\sin kx)(e^{rx}\sin kx)$

$= ke^{2rx}\cos^2 kx + ke^{2rx}\sin^2 kx = ke^{2rx}.$

We apply the preceding two results to conclude that an initial value problem sociated to a homogeneous second order linear differential equation with consta coefficients has a unique solution.

Corollary 4.12.11 *The following initial value problem has a unique solution.*

$$y'' + by' + cy = 0, \quad y(x_0) = h, \quad y'(x_0) = k$$

Proof Let $y(x) = Au(x) + Bv(x)$ be the general solution of this differential equatio Then $y'(x) = Ay'(x) + Bv'(x)$. We want to solve the following linear system:

$$Au(x_0) + Bv(x_0) = h$$
$$Au'(x_0) + Bv'(x_0) = k.$$

Here A, B are unknowns while $u(x_0)$, $u'(x_0)$, $v(x_0)$, $v'(x_0)$, h, k are known number By the preceding lemma, $u(x_0)v'(x_0) - u'(x_0)v(x_0) = W(u(x_0), v(x_0))$ is nonzero. E Proposition 4.12.8, the above linear system has a unique solution for A, B.

Examples 4.12.12 Solve the initial value problem

$$y'' - y' - 12y = 0, \quad y(0) = 10, \quad y'(0) = 12.$$

Solution The characteristic polynomial is $p^2 - p - 12 = (p - 4)(p + 3)$. The genera solution of this differential equation is $y = Ae^{4x} + Be^{-3x}$. Then $y' = 4Ae^{4x} - 3Be^{-3}$

We require : $10 = y(0) = A + B,$

                     $12 = y'(0) = 4A - 3B.$

The unique solution of this linear system is $A = 6$ and $B = 4$. Hence the initial valu problem has the unique solution $y = 6e^{4x} + 4e^{-3x}$.

Nonhomogeneous Equations with Constant Coefficients

This subsection is devoted to solving second order linear differential equations with constant coefficients:

$$y'' + by' + cy = f(x). \tag{4.12.9}$$

This equation is called *nonhomogeneous* when $f(x) \neq 0$. We use the method of variation of parameters to find the general solution. Then we introduce the method of undetermined coefficients, an easier procedure to solve this equation in many cases.

By the following proposition, the general solution of (4.12.9) is the sum of a particular solution and the *complementary solution*, the general solution of the corresponding homogeneous equation.

Proposition 4.12.13 *Let* $y(x) = Au(x) + Bv(x)$ *be the general solution of the homogeneous differential equation with constant coefficients*

$$y'' + by' + cy = 0. \tag{4.12.10}$$

Let $y_p(x)$ *be one particular solution of the nonhomogeneous differential equation*

$$y'' + by' + cy = f(x). \tag{4.12.11}$$

Then the general solution of this nonhomogeneous differential equation is:

$$y(x) = y_p(x) + Au(x) + Bv(x). \tag{4.12.12}$$

Proof Observe

$$
\begin{aligned}
\left(D^2 + bD + c\right)(y_p + Au + Bv) &= \left(D^2 + bD + c\right)(y_p) + \left(D^2 + bD + c\right)(Au + Bv) \\
&= f(x) + 0 = f(x).
\end{aligned}
$$

Thus all the functions of (4.12.12) are solutions of the nonhomogeneous equation (4.12.11). Let $y(x)$ be any solution of this nonhomogeneous equation. Then

$$\left(D^2 + bD + c\right)(y - y_p) = \left(D^2 + bD + c\right)(y_p) - \left(D^2 + bD + c\right)(y) = f(x) - f(x) = 0.$$

Hence $y(x) - y_p(x)$ is a solution of the homogeneous equation (4.12.10), and there are $A, B \in \Re$ with $y(x) - y_p(x) = Au(x) + Bv(x)$. Then $y(x) = y_p(x) + Au(x) + Bv(x)$. Thus every solution of the nonhomogeneous equation (4.12.11) is given by (4.12.12). $\square$

The *method of variation of parameters* gives a particular solution $y(x)$ of the nonhomogeneous equation (4.12.11) of the form

$$y(x) = \alpha(x)u(x) + \beta(x)v(x) \tag{4.12.13}$$

where $Au(x) + Bv(x)$, for $A, B \in \Re$, is the general solution of the homogeneous equation (4.12.10). We find functions $\alpha(x)$ and $\beta(x)$ so that the $y(x)$ of (4.12.13) is a solution of the nonhomogeneous equation (4.12.11). First, we find $y'(x)$ and $y''(x)$. By the product rule:

$$y'(x) = \alpha'(x)u(x) + \alpha(x)u'(x) + \beta'(x)v(x) + \beta(x)v'(x).$$

To simplify the computation of $y''(x)$, we find a solution of (4.12.11) where

$$
\begin{aligned}
0 &= \alpha'(x)u(x) + \beta'(x)v(x). \tag{4.12.14} \\
\text{Then} \quad y'(x) &= \alpha(x)u'(x) + \beta(x)v'(x) \quad \text{and} \tag{4.12.15} \\
y''(x) &= \alpha'(x)u'(x) + \alpha(x)u''(x) + \beta'(x)v'(x) + \beta(x)v''(x) \tag{4.12.16}
\end{aligned}
$$

by the product rule. Note $y(x)$ is a solution of (4.12.11) if

$$f(x) = \left(D^2 + bD + c\right)[y(x)] = D^2 y(x) + bDy(x) + cy(x).$$

By (4.12.13), (4.12.15) and (4.12.16), we require:

$$
\begin{aligned}
f(x) &= [\alpha'(x)u'(x) + \alpha(x)u''(x) + \beta'(x)v'(x) + \beta(x)v''(x)] + b[\alpha(x)u'(x) + \beta(x) \\
&\qquad\qquad + c[\alpha(x)u(x) + \beta(x) \\
&= \alpha(x)\left(D^2 + bD + c\right)[u(x)] + \beta(x)\left(D^2 + bD + c\right)[v(x)] + \alpha'(x)u'(x) + \beta' \\
&= \alpha'(x)u'(x) + \beta'(x)v'(x).
\end{aligned}
$$

To find a particular solution of the nonhomogeneous equation (4.12.11) we solve preceding equation as well as equation (4.12.14). That is, we solve for the unknow $\alpha'(x)$, $\beta'(x)$ in the linear system:

$$
\begin{aligned}
\alpha'(x)u(x) + \beta'(x)v(x) &= 0 \\
\alpha'(x)u'(x) + \beta'(x)v'(x) &= f(x).
\end{aligned}
$$

By Lemma 4.12.10, the Wronskian $W(u(x), v(x)) = u(x)v'(x) - u'(x)v(x) \neq 0$ for .
By Proposition 4.12.8, the linear system above has the unique solution:

$$\alpha'(x) = \frac{-v(x)f(x)}{W(u(x), v(x))}, \qquad \beta'(x) = \frac{u(x)f(x)}{W(u(x), v(x))}$$

Now integrate these equations to find $\alpha(x)$ and $\beta(x)$.

Proposition 4.12.14 [Variation of Parameters] *Let f be a continuous function w. domain the interval I. Let $Au(x) + Bv(x)$, for $A, B \in \Re$ be the general solution of t homogeneous equation $y'' + by' + cy = 0$. Then the nonhomogeneous linear different. equation with constant coefficients $y'' + by' + cy = f(x)$ has the general solution*

$$y(x) = \alpha(x)u(x) + \beta(x)v(x) + Au(x) + Bv(x)$$

for $x \in I$ where

$$\int \frac{-v(x)f(x)}{W(u(x), v(x))}\,dx = \alpha(x) + C, \qquad \int \frac{u(x)f(x)}{W(u(x), v(x))}\,dx = \beta(x) + C. \qquad (4.12.1$$

Examples 4.12.15 (1)   Solve the differential equation $y'' + 3y' + 2y = 6x - 5$.

Solution The general solution of the homogeneous equation

$$0 = y'' + 3y' + 2y = (D^2 + 3D + 2)(y) = (D+1)(D+2)(y)$$

is $y = Ae^{-x} + Be^{-2x}$. Let $u(x) = e^{-x}$ and $v(x) = e^{-2x}$. By Lemma 4.12.10(c $W(u(x), v(x)) = W(e^{-x}, e^{-2x}) = -e^{-3x}$. To find $\alpha(x)$ by (4.12.17), integra by parts:

$$
\begin{aligned}
\alpha(x) &= \int \frac{-v(x)f(x)}{W(u(x), v(x))}\,dx = \int \frac{-e^{-2x}(6x - 5)}{-e^{-3x}}\,dx = \int e^x(6x - 5)\,dx \\
&= \int D\left(e^x\right)(6x - 5)\,dx = e^x(6x - 5) - \int e^x D(6x - 5)\,dx \\
&= e^x(6x - 5) - \int 6e^x\,dx = e^x(6x - 5) - 6e^x + C = e^x(6x - 11) + C.
\end{aligned}
$$

Similarly, integrate by parts to find $\beta(x)$ by (4.12.17):

$$
\begin{aligned}
\beta(x) &= \int \frac{u(x)f(x)}{W(u(x),v(x))}\,dx = \int \frac{e^{-x}(6x-5)}{-e^{-3x}}\,dx = -\int e^{2x}(6x-5)\,dx\\
&= -\frac{1}{2}\int D\left(e^{2x}\right)(6x-5)\,dx = -\frac{1}{2}e^{2x}(6x-5)+\frac{1}{2}\int e^{2x}D(6x-5)\,dx\\
&= -\frac{1}{2}e^{2x}(6x-5)+\frac{1}{2}\int 6e^{2x}\,dx = -\frac{1}{2}e^{2x}(6x-5)+\frac{3}{2}e^{2x}+C\\
&= e^{2x}(4-3x)+C.
\end{aligned}
$$

By (4.12.13), a particular solution of the given nonhomogeneous equation is

$$
\begin{aligned}
y_p(x) &= \alpha(x)u(x)+\beta(x)v(x) = [e^x(6x-11)]\,e^{-x}+\left[e^{2x}(4-3x)\right]e^{-2x}\\
&= (6x-11)+(4-3x) = 3x-7.
\end{aligned}
$$

By Proposition 4.12.14, the general solution of this nonhomogeneous equation is

$$
y(x) = 3x-7+Ae^{-x}+Be^{-2x}.
$$

(2) Solve the differential equation $y''+y = \sin^2 x$.

Solution The general solution of the homogeneous equation

$$
0 = y''+y = (D^2+1)(y)
$$

is $y = A\cos x+B\sin x$. Let $u(x) = \cos x$ and $v(x) = \sin x$. By Lemma 4.12.10(e), $W(u(x),v(x)) = W(\cos x,\sin x) = 1$. By (4.12.17),

$$
\alpha(x) = \int \frac{-v(x)f(x)}{W(u(x),v(x))}\,dx = \int \frac{-(\sin x)(\sin^2 x)}{1}\,dx = \int (1-\cos^2 x)(-\sin x)\,dx.
$$

Let $u = \cos x$ with $du = -\sin x\,dx$. Then

$$
\alpha(x) = \int 1-u^2\,du = u-\frac{u^3}{3}+C = \cos x-\frac{1}{3}\cos^3 x+C.
$$

Similarly,

$$
\begin{aligned}
\beta(x) &= \int \frac{u(x)f(x)}{W(u(x),v(x))}\,dx = \int \frac{(\cos x)(\sin^2 x)}{1}\,dx\\
&= \int \sin^2 x\cos x\,dx = \frac{1}{3}\sin^3 x+C.
\end{aligned}
$$

By (4.12.13), a particular solution of the given nonhomogeneous equation is

$$
\begin{aligned}
y_p(x) &= \alpha(x)u(x)+\beta(x)v(x) = \left(\cos x-\frac{1}{3}\cos^3 x\right)\cos x+\left(\frac{1}{3}\sin^3 x\right)\sin x\\
&= \cos^2 x-\frac{1}{3}\cos^4 x+\frac{1}{3}\sin^4 x\\
&= \cos^2 x-\frac{1}{3}\left(\cos^2 x+\sin^2 x\right)\left(\cos^2 x-\sin^2 x\right)\\
&= \cos^2 x-\frac{1}{3}\left(\cos^2 x-\sin^2 x\right) = \frac{2}{3}\cos^2 x+\frac{1}{3}\sin^2 x = \frac{1}{3}\cos^2 x+\frac{1}{3}.
\end{aligned}
$$

By Proposition 4.12.14, the general solution of this nonhomogeneous equation is

$$
y(x) = \frac{1}{3}\cos^2 x+\frac{1}{3}+A\cos x+B\sin x. \qquad \square
$$

Example 4.12.15 (1) indicates that the method of variation of parameters m be making the search for a particular solution unnecessarily complicated. Our f simplification says that we can consider each summand of $f(x)$ separately.

Proposition 4.12.16 Let $b, c, h, k \in \Re$. If $y_1(x)$ is a particular solution of $(D^2 + bD + c)y = f_1(x)$ and $y_2(x)$ is a particular solution of $(D^2 + bD + c)y = f_2$ then $hy_1(x) + ky_2(x)$ is a particular solution of $(D^2 + bD + c)y = hf_1(x) + kf_2(x)$

Proof Since the derivatives D and D^2 are linear,

$$
\begin{aligned}
\left(D^2 + bD + c\right)(hy_1 + ky_2) &= h\left(D^2 + bD + c\right)(y_1) + k\left(D^2 + bD + c\right)(y_2 \\
&= hf_1(x) + kf_2(x).
\end{aligned}
$$

Consider the linear nonhomogeneous differential equation with constant coefficient

$$y'' + by' + c = f(x) \quad \text{where} \quad f(x) = Q(x)e^{kx}\sin mx + R(x)e^{kx}\cos m$$
$$Q(x) = A_n x^n + \cdots + A_1 x + A_0 \quad \text{and} \quad R(x) = B_n x^n + \cdots + B_1 x + B_0. \quad (4.12.1$$

We want to find a particular solution of this differential equation of the form

$$y_p(x) = q(x)e^{kx}\sin mx + r(x)e^{kx}\cos mx \qquad (4.1$$
where $\quad q(x) = a_n x^n + \cdots + a_1 x + a_0$ and $r(x) = b_n x^n + \cdots + b_1 x + b_0.$

Observe what occurs when we apply the operator $D^2 + bD + c$ to the simple functio from which $y_p(x)$ is constructed. Let $a_0, \ldots, a_n, b, c, k, m \in \Re$. Then

$$
\begin{aligned}
\left(D^2 + bD + c\right)(a_n x^n + \cdots + a_1 x + a_0) &= G_n x^n + \cdots + G_1 x + G_0 \\
\left(D^2 + bD + c\right)\left(e^{kx}\right) &= Ae^{kx} \\
\left(D^2 + bD + c\right)(\sin mx) &= B\sin mx + C\cos mx \\
\left(D^2 + bD + c\right)(\cos mx) &= E\sin mx + F\cos mx
\end{aligned}
$$

where $A, B, C, E, F, G_0, \ldots, G_n \in \Re$. The preceding computations and the produ rule indicate that $(D^2 + bD + c)[y_p(x)]$ has the form $Q_0(x)e^{kx}\sin mx + R_0(x)e^{kx}\cos m$ where $Q_0(x)$ and $R_0(x)$ are polynomials of degree at most n. Thus $y_p(x)$ is a solutic of the differential equaiton (4.12.18) if we can find numbers $a_0, \ldots a_n, b_0, \ldots b_n$ terms of the given numbers $b, c, k, m, A_0, \ldots, A_n, B_0, \ldots, B_n$ such that:

$$
\begin{aligned}
\left(D^2 + bD + c\right)[y_p(x)] &= \left(D^2 + bD + c\right)\left[q(x)e^{kx}\sin mx + r(x)e^{kx}\cos mx\right] \\
&= Q(x)e^{kx}\sin mx + R(x)e^{kx}\cos mx, \\
&= (A_n x^n + \cdots + A_1 x + A_0)\,e^{kx}\sin mx \\
&\quad + (B_n x^n + \cdots + B_1 x + B_0)\,e^{kx}\cos mx. \qquad (4.12.2($$

It turns out that as long as k is not a root of the characteristic polynomial $p^2 + bp +$ there is a $y_p(x)$, as in (4.12.19), which is a particular solution of the differentia equation $(D^2 + bD + c)(y) = f(x)$. Finding $y_p(x)$ in this way is called the *method c undetermined coefficients*.

Note (1) Both summands of (4.12.19) must be included even if $Q(x)$ or $R(x)$ is zerc
(2) If $m = 0$ then $\sin mx = 0$, and we take $q(x) = 0$.

Examples 4.12.17 (1) Use the method of undetermined coefficients to solve th differential equation $y'' - 3y' + 2y = 170e^{5x}\sin x.$

Solution In the notation of (4.12.18), $n = 0$, $k = 5$ and $m = 1$. Hence we look for a particular solution of the form

$$y_p(x) = ae^{5x} \sin x + be^{5x} \cos x.$$
$$\text{Then } \left(D^2 - 3D + 2\right)[y_p(x)] = (11a - 7b)e^{5x} \sin x + (7a + 11b)e^{5x} \cos x.$$

Hence we must select a and b with:

$$11a - 7b = 170$$
$$7a + 11b = 0.$$

The unique solution is: $a = 11$ and $b = -7$. Thus $y_p(x) = 11e^{5x} \sin x - 7e^{5x} \cos x$ is a particular solution. The characteristic polynomial of the associated homogeneous equation is $p^2 - 3p + 2 = (p - 2)(p - 1)$. Hence the complementary solution is $y = Ae^x + Be^{2x}$. Thus the general solution of the given nonhomogeneous equation is:

$$y = y_p(x) + Ae^x + Be^{2x} = 11e^{5x} \sin x - 7e^{5x} \cos x + Ae^x + Be^{2x}.$$

(2) Use the method of undetermined coefficients to solve $y'' + 4y = x^2 e^{2x}$.

Solution In the notation of (4.12.18), $n = 2$, $k = 2$ and $m = 0$. Hence we look for a particular solution of the form

$$y_p(x) = \left(ax^2 + bx + c\right)e^{2x} \cos 0x = ax^2 e^{2x} + bxe^{2x} + ce^{2x}.$$
$$\text{Then } \left(D^2 + 4\right)[y_p(x)] = 8ax^2 e^{2x} + (8a + 8b)xe^{2x} + (2a + 4b + 8c)e^{2x}.$$

We must select a, b, c so that:

$$8a = 1$$
$$8a + 8b = 0$$
$$2a + 4b + 8c = 0.$$

Then $a = \frac{1}{8}$, $b = -\frac{1}{8}$, $c = \frac{1}{32}$. Hence $y_p(x) = \frac{1}{8}x^2 e^{2x} - \frac{1}{8}xe^{2x} + \frac{1}{32}e^{2x}$ is a particular solution. The characteristic polynomial of the associated homogeneous equation is $p^2 + 2^2$. Hence the complementary solution is $y = A\sin 2x + B\cos 2x$. Thus the general solution of the given nonhomogeneous equation is:

$$y = y_p(x) + A\sin 2x + B\cos 2x = \frac{1}{8}x^2 e^{2x} - \frac{1}{8}xe^{2x} + \frac{1}{32}e^{2x} + A\sin 2x + B\cos 2x. \quad \square$$

Harmonic Motion

The motion of an object attached to a spring is called *harmonic motion*. It is described by a second order linear differential equation with constant coefficients. We use the methods of this section to solve this equation and determine the motion of the object. In addition to the basic case, we discuss *damped harmonic motion* where there is a force resisting the motion of the object as well as *forced harmonic motion* where an external force acts on the object.

Simple Harmonic Motion

Consider an object of mass m attached to the end of a spring. The spring is stretched from its equilibrium position E and released. This situation, depicted in the upper left picture of Figure 4.12.19, is called *free undamped harmonic motion* or *simple harmonic motion*.[7] Experiments show that the spring exerts a force on the object

---

[7]The adjective *free* indicates that there is no external force acting on the object while the adjective *undamped* means that there is no force resisting its motion.

towards E which is proportional to the distance between the object and E. T
physical principle is called *Hooke's Law*. Construct a horizontal coordinate line w
the equilibrium position E at the origin. Let $s(t)$ denote the position of the object
time t. By Newton's second law of motion, the force on the object equals its mass
times its acceleration $s''(t)$. Hence Hooke's Law can be stated as $ms''(t) = -k^2 s$
Note the constant of proportionality $-k^2$ is negative because the spring is pulling
object to the left ($s''(t) < 0$) when it is to the right of E ($s(t) > 0$) or pulling it to
right ($s''(t) > 0$) when it is to the left of E ($s(t) < 0$). Thus Hooke's Law is given
a homogeneous linear differential equation with constant coefficients:

$$s''(t) + \left(\frac{k}{\sqrt{m}}\right)^2 s(t) = 0. \tag{4.12.}$$

To simplify the discussion, assume the object is initially at rest, i.e $s'(0) = v(0) = $
The general solution of this differential equation is:

$$s(t) = s(0) \cos \frac{kt}{\sqrt{m}}. \tag{4.12.}$$

Figure 4.12.18
Simple Harmonic
Motion

The graph of $s(t)$, in the case $s(0) > 0$, is depicted in Figure 4.12.18. Note the moti
of this object is periodic with period $\frac{2\pi\sqrt{m}}{k}$.

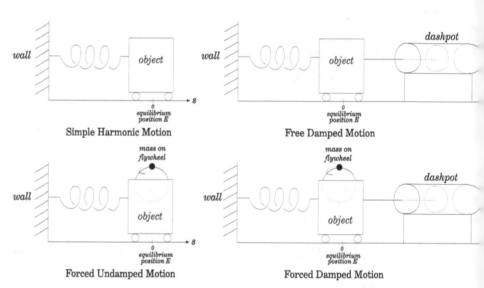

Figure 4.12.19 Harmonic Motion

Damped Free Harmonic Motion

Consider the situation of simple harmonic motion with an additional complication
Suppose there is a force $-cv = -cs'(t)$, with $c > 0$, slowing the movement of the
object. This force may be caused by air resistance or by attaching the object to
piston inside a fluid filled cylinder called a *dashpot*. See the upper right picture
Figure 4.12.19. This situation is called *damped free harmonic motion*. By Newton
second law of motion, the force on the object is $ms''(t) = -cs'(t) - k^2 s(t)$. This ca
be written as a linear homogeneous differential equation with constant coefficients:

$$s''(t) + \frac{c}{m} s'(t) + \frac{k^2}{m} s(t) = 0. \tag{4.12.23}$$

As before, assume the object starts at rest: $s'(0) = 0$. By the quadratic formula, th characteristic polynomial $x^2 + \frac{c}{m}x + \frac{k^2}{m}$ has roots

$$\frac{-c \pm \sqrt{c^2 - 4mk^2}}{2m}.$$

There are three cases.

Case I: Overdamped Free Harmonic Motion $(c^2 - 4mk^2 > 0)$

In this case, the characteristic polynomial has two negative real roots $-p, -q$. The general solution of the differential equation (4.12.23) is $s(t) = Ae^{-pt} + Be^{-qt}$. The values of A and B are obtained by solving the linear system:

$$\begin{aligned} s(0) &= A + B \\ 0 = s'(0) &= -pA - qB. \end{aligned}$$

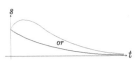

Figure 4.12.20
Overdamped Free
Harmonic Motion

Thus $A = \frac{q}{q-p}s(0)$, $B = -\frac{p}{q-p}s(0)$ and

$$s(t) = \frac{s(0)}{q-p}\left(qe^{-pt} - pe^{-qt}\right). \tag{4.12.24}$$

The graph of $s(t)$, when $q > p$ and $s(0) > 0$, is shown in Figure 4.12.20.

Case II: Critically Damped Free Harmonic Motion $(c^2 - 4mk^2 = 0)$

In this case, the characteristic polynomial has one repeated negative real root $-\frac{c}{2m}$. Thus the general solution of the differential equation (4.12.23) is $s(t) = (A+Bt)e^{-\frac{ct}{2m}}$. Then $s(0) = A$, and $0 = v(0) = -s(0)\frac{c}{2m} + B$. Hence $B = s(0)\frac{c}{2m}$ and

$$s(t) = s(0)\left(1 + \frac{ct}{2m}\right)e^{-\frac{ct}{2m}}. \tag{4.12.25}$$

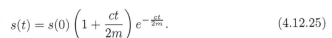

Figure 4.12.21
Critically Damped
Free Harmonic Motion

The two possible graphs of this motion, when $s(0) > 0$, are depicted in Figure 4.12.21.

Case III: Underdamped Free Harmonic Motion $(c^2 - 4mk^2 < 0)$

In this case, the characteristic polynomial has complex roots. Let $\omega = \frac{\sqrt{4mk^2-c^2}}{2m}$. The differential equation (4.12.23) has the general solution

$$s(t) = e^{-\frac{ct}{2m}}\left[A\sin\omega t + B\cos\omega t\right].$$

Let $G = \sqrt{A^2 + B^2}$ and $\tan\theta = \frac{A}{B}$. From Fig. 4.12.22, $\sin\theta = \frac{A}{G}$ and $\cos\theta = \frac{B}{G}$. Then

$$\begin{aligned} s(t) &= Ge^{-\frac{ct}{2m}}\left[\frac{A}{G}\sin\omega t + \frac{B}{G}\cos\omega t\right] = Ge^{-\frac{ct}{2m}}\left[\sin\theta\sin\omega t + \cos\theta\cos\omega t\right] \\ s(t) &= Ge^{-\frac{ct}{2m}}\cos\left(\omega t - \theta\right) \end{aligned} \tag{4.12.26}$$

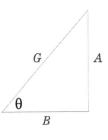

Figure 4.12.22
The Angle θ

The values of θ and G are determined by solving

$$\begin{aligned} s(0) &= G\cos\theta \\ 0 = v(0) &= \frac{-cG}{2m}\cos\theta + G\omega\sin\theta. \end{aligned}$$

By the second equation $\tan\theta = \frac{\sin\theta}{\cos\theta} = \frac{c}{2\omega m}$. Then

$$\theta = \arctan\frac{c}{2\omega m}, \tag{4.12.27}$$

$$G = s(0)\sec\theta = s(0)\sqrt{\tan^2\theta + 1} = s(0)\frac{\sqrt{c^2 + 4\omega^2 m^2}}{2\omega m}. \tag{4.12.28}$$

The graph of this motion when $s(0) > 0$ is given in Figure 4.12.23.

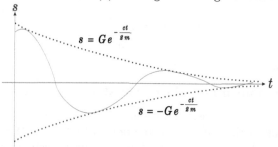

Figure 4.12.23 Underdamped Free Harmonic Motion

Forced Harmonic Motion

In the case of *forced harmonic motion* we have one of the above types of harmo
motion as well as an external force which acts on the object. We restrict our analy
to a periodic external force given by

$$F(t) = F_0 \cos \lambda t.$$

This situation occurs when the object is on a cart which contains a flywheel. T
flywheel consists of a mass which rotates about a point directly above the center
the cart. See the two bottom pictures of Figure 4.12.19. The differential equati
given by Newton's second law of motion is a linear nonhomogeneous equation wi
constant coefficients:

$$ms''(t) + cs'(t) + k^2 s(t) = F_0 \cos \lambda t. \qquad (4.12.2$$

To simplify our computations, assume

$$s(0) = 0 \quad \text{and} \quad v(0) = s'(0) = 0.$$

First we analyze the case with no damping and then the case with damping.

Forced Undamped Harmonic Motion

In the undamped case $c = 0$, and the differential equation (4.12.29) becomes:

$$ms''(t) + k^2 s(t) = F_0 \cos \lambda t \quad \text{or} \quad \left(D^2 + \frac{k^2}{m} \right) s(t) = \frac{F_0}{m} \cos \lambda t. \qquad (4.12.3$$

The associated homogeneous equation $\left(D^2 + \frac{k^2}{m} \right) s(t) = 0$, with characteristic pol
nomial $x^2 + \frac{k^2}{m}$, has the general solution

$$s(t) = A \sin \omega t + B \cos \omega t \quad \text{with} \quad \omega = \frac{k}{\sqrt{m}}. \qquad (4.12.3$$

ω is called the *natural frequency* of this motion. It remains to find a particular solutic
of the nonhomogeneous equation. There are two cases.

Case I: Forced Undamped Harmonic Motion with $\lambda \neq \omega$

Since $D^2(\cos \lambda t) = -\lambda^2 \cos \lambda t$, there is a particular solution of the nonhomogeneou
equation (4.12.30) of the form $s = C \cos \lambda t$. Note

$$\left(D^2 + \frac{k^2}{m} \right)(C \cos \lambda t) = C \left(\frac{k^2}{m} - \lambda^2 \right) \cos \lambda t = C \left(\omega^2 - \lambda^2 \right) \cos \lambda t.$$

This is a solution to (4.12.30) when $\frac{F_0}{m} = C(\omega^2 - \lambda^2)$, i.e. $C = \frac{F_0}{m(\omega^2 - \lambda^2)}$. The general solution of the nonhomogeneous equation (4.12.30) is this particular solution plus the complementary solution (4.12.31):

$$s(t) = \frac{F_0}{m(\omega^2 - \lambda^2)} \cos \lambda t + A \sin \omega t + B \cos \omega t.$$

Our initial value problem is: $0 = s(0) = \frac{F_0}{m(\omega^2 - \lambda^2)} + B$ and $0 = s'(0) = A\omega$. Hence $A = 0$ and $B = -\frac{F_0}{m(\omega^2 - \lambda^2)}$. Thus the solution of our initial value problem is:

$$s(t) = \frac{F_0}{m(\omega^2 - \lambda^2)} \left[\cos \lambda t - \cos \omega t\right]. \tag{4.12.32}$$

By the trigonometric identity $2 \sin \theta \sin \phi = \cos(\theta - \phi) - \cos(\theta + \phi)$:

$$s(t) = \frac{2F_0}{m(\omega^2 - \lambda^2)} \sin \frac{1}{2}(\lambda + \omega) t \sin \frac{1}{2}(\omega - \lambda) t. \tag{4.12.33}$$

Equation (4.12.32) says that this motion is the difference of two periodic motions: one with the natural frequency ω minus another with the frequency λ of the external force. Equation (4.12.33) implies that when λ approaches ω, the amplitude of this motion increases to infinity. This phenomenon is called *resonance*. In addition, when λ is close to ω, the number $\frac{1}{2}(\lambda + \omega)$ is large compared to $\frac{1}{2}(\omega - \lambda)$. Write

$$s(t) = \alpha(t) \sin \frac{1}{2}(\lambda + \omega) t \quad \text{where} \quad \alpha(t) = \frac{2F_0}{m(\omega^2 - \lambda^2)} \sin \frac{1}{2}(\omega - \lambda) t. \tag{4.12.34}$$

The graph of $s(t)$ in Figure 4.12.24 is a rapidly varying oscillation with period $\frac{4\pi}{\lambda + \omega}$ having slowly varying periodic amplitude $\alpha(t)$ with period $\frac{4\pi}{|\omega - \lambda|}$. This phenomenon is called *beats*. It occurs when two musical instruments, which are not tuned to match each other, play the same note. The two sound waves produced have similar, but different, frequencies.

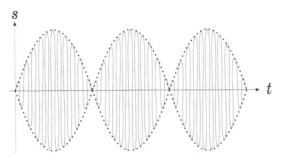

Figure 4.12.24 The Phenomenon of Beats

Case II: Forced Undamped Harmonic Motion with $\lambda = \omega$

Observe $D^2(t \sin \omega t) = D(\sin \omega t + \omega t \cos \omega t) = 2\omega \cos \omega t - \omega^2 t \sin \omega t$. Hence

$$\left(D^2 + \frac{k^2}{m}\right)(Ct \sin \omega t) = \left(D^2 + \omega^2\right)(Ct \sin \omega t) = 2C\omega \cos \omega t$$

as $\omega^2 = \frac{k^2}{m}$. Hence a partiuclar solution of the nonhomogeneous equation (4.12.30) is

$$s(t) = \frac{F_0}{2m\omega} t \sin \omega t. \tag{4.12.35}$$

In fact, this is the solution of our initial value problem because it satisfies $s(0)$
$s'(0) = 0$. Thus when the frequency of the external force coincides with the natu
frequency, we have a periodic motion with a linearly increasing amplitude.   T
phenomenon is also called *resonance*. See Figure 4.12.25.

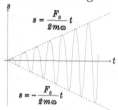

Figure 4.12.25 The Phenomenon of Resonance

Forced Harmonic Motion with Damping

In the case of *forced harmonic motion with damping* all the constants in the different
equation (4.12.29) are nonzero. We find a particular solution of the form

$$s_p(t) = A \sin \lambda t + B \cos \lambda t.$$

Observe $D[s_p(t)] = A\lambda \cos \lambda t - B\lambda \sin \lambda t$ and $D^2[s_p(t)] = -A\lambda^2 \sin \lambda t - B\lambda^2 \cos$
Then (4.12.29) becomes:

$$
\begin{aligned}
F_0 \cos \lambda t &= \left(mD^2 + cD + k^2 \right) [s_p(t)] \\
&= \left(k^2 A - c\lambda B - m\lambda^2 A \right) \sin \lambda t + \left(k^2 B + c\lambda A - m\lambda^2 B \right) \cos \lambda t.
\end{aligned}
$$

We have a solution when we select A and B as solutions of the linear system:

$$
\begin{aligned}
(k^2 - m\lambda^2)A - c\lambda B &= 0 \\
c\lambda A + (k^2 - m\lambda^2)B &= F_0
\end{aligned}
$$

By Proposition 4.12.8, the unique solution is $A = \frac{c\lambda F_0}{(k^2 - m\lambda^2)^2 + c^2\lambda^2}$ and $B = \frac{F_0(k^2 - m}{(k^2 - m\lambda^2)^2}$

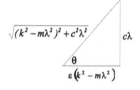

Figure 4.12.26
The Angle θ

Let $\qquad \epsilon = 1$ if $k^2 - m\lambda^2 > 0$ or $\epsilon = -1$ if $k^2 - m\lambda^2 < 0.$ $\qquad$ (4.12.

Then $s_p(t) = \dfrac{\epsilon F_0}{\sqrt{(k^2 - m^2)^2 + c^2\lambda^2}} \left(\dfrac{\epsilon c\lambda}{\sqrt{(k^2 - m\lambda^2)^2 + c^2\lambda^2}} \sin \lambda t \right.$

$$\left. + \dfrac{\epsilon(k^2 - m\lambda^2)}{\sqrt{(k^2 - m\lambda^2)^2 + c^2\lambda^2}} \cos \lambda t \right) \qquad (4.12.$$

Let $\qquad \alpha = \dfrac{\epsilon F_0}{\sqrt{(k^2 - m\lambda^2)^2 + c^2\lambda^2}}$ and $\theta = \arctan \dfrac{c\lambda}{\epsilon(k^2 - m\lambda^2)}.$ $\qquad$ (4.12.

The right triangle in Fig. 4.12.26 has angle θ with $\tan \theta = \frac{c\lambda}{\epsilon(k^2 - m\lambda^2)}$. It shows: the co
efficients of $\sin \lambda t$ and $\cos \lambda t$ in (4.12.37) are $\epsilon \sin \theta = \sin(\epsilon\theta)$ and $\cos \theta = \cos(\epsilon\theta)$. Thi

$$s_p(t) = \alpha[\sin(\epsilon\theta)\sin \lambda t + \cos(\epsilon\theta)\cos \lambda t] = \alpha \cos(\lambda t - \epsilon\theta). \qquad (4.12.39$$

The associated homogeneous equation of (4.12.29) describes damped free harmoni
motion. As we showed above, there are three cases: the overdamped case, the criticall
damped case and the underdamped case. Denote the complementary solution as $s_c(t$
Observe that in all cases $\lim_{t\to\infty} s_c(t) = 0$. The general solution $s(t) = s_c(t) + s_p(t)$
the nonhomogeneous equation (4.12.29) is:

$$s(t) = s_c(t) + \alpha\cos(\lambda t - \epsilon\theta) \text{ and } \lim_{t\to\infty}[s(t) - \alpha\cos(\lambda t - \epsilon\theta)] = 0. \qquad (4.12.4C$$

Thus, as t increases, $s(t)$ approaches the periodic oscillation $s_p(t) = \alpha \cos(\lambda t - \epsilon\theta)$ which is called the *steady periodic solution* of the initial value problem. The amplitude α of this oscillation for fixed F_0, c, k, m depends on λ and has a maximum value of $\frac{F_0 \sqrt{m}}{ck}$ when $\lambda = \frac{k}{\sqrt{m}}$. This value of λ is called the *resonant frequency*. When a bridge is forced to vibrate near this frequency, the amplitude of the vibrations may become large enough to destroy the bridge. This phenomenon occurred on the Broughton Bridge in Manchester, England which was destroyed by the resonance of marching soldiers in 1831. It also caused the destruction of the Tacoma Narrows Bridge in 1940 from the resonance caused by a storm. Today a model of a proposed bridge is tested in a wind tunnel to insure that resonance does not occur.

Examples 4.12.27 Find the position function and the steady periodic solution for the object that moves from rest at the origin with forced damped harmonic motion where $m = 2$, $c = 8$, $k^2 = 26$, $F_0 = 6630$, $\lambda = 6$ in the notation of (4.12.29).

Solution We are asked to solve the initial value problem

$$2s'' + 8s' + 26s = 6630 \cos 6t \ \text{ with } \ s(0) = s'(0) = 0.$$

The associated homogeneous equation $(D^2 + 4D + 13)y = 0$ has characteristic polynomial is $x^2 + 4x + 13 = (x + 2)^2 + 3^2$. Hence the complementary solution is:

$$s_c(t) = he^{-2t} \sin 3t + ke^{-2t} \cos 3t.$$

Note $\epsilon = -1$ because $k^2 - m\lambda^2 = -46 < 0$. Then $\alpha = \dfrac{\epsilon F_0}{\sqrt{(k^2 - m\lambda^2)^2 + c^2\lambda^2}} = \dfrac{-6630}{\sqrt{4420}} = -3\sqrt{1105}$. Also $\theta = \arctan \dfrac{c\lambda}{\epsilon(k^2 - m\lambda^2)} = \arctan \frac{24}{23}$. By (4.12.39) the steady periodic solution is:

$$s_p(t) = \alpha \cos(\lambda t - \epsilon\theta) = -3\sqrt{1105} \cos(6t + \theta).$$

By (4.12.40), the general solution of this nonhomogeneous differential equation is:

$$s(t) = -3\sqrt{1105} \cos(6t + \theta) + he^{-2t} \sin 3t + ke^{-2t} \cos 3t.$$

The initial value problem for the position function is: $0 = s(0) = -69 + k$ and $0 = s'(0) = 18(24) + 3h - 2k$. Hence $h = -98$, $k = 69$ and the position function is:

$$s(t) = -3\sqrt{1105} \cos(6t + \theta) - 98e^{-2t} \sin 3t + 69e^{-2t} \cos 3t.$$

The graph of $s(t)$ is sketched in Figure 4.12.28. Note the graph of $s(t)$ approaches the graph of $s_p(t)$ as t gets large. □

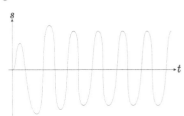

Figure 4.12.28 Forced Damped Harmonic Motion

Historical Remarks

The first strategy to solve nontrivial higher order differential equations was to use a change of variables so that the differential equation, rewritten in terms of the new

variable, has order one less than the original equation. This trick was used by Joha
Bernoulli in 1700 to solve the differential equation

$$y + a_1 xy^{(1)} + a_2 x^2 y^{(2)} + \cdots + a_n x^n y^{(n)} = 0.$$

In 1712, Count Jacopo Riccati solved differential equations of the form $f(y, y', y'') =$
by this strategy using the change of variables $u = y'$.

In 1739, Leonhard Euler solved linear homogeneous differential equations w
constant coefficients using the roots of their characteristic polynomials. In 1740,
solved nonhomogeneous equations of this type by using a sequence of integrat
factors, each of which reduced the order by one. The method of variation of paramet
was discovered by Joseph Louis Lagrange in 1774. The Wronskian is named after
Polish mathematician Hoene Wronski who introduced this concept in 1821.

Summary

The reader should be able to solve differential equations of the form $y'' = f($
$y'' = g(x, y')$ and $y'' = h(y, y')$. She should know how to solve second order hom
geneous linear differential equations with constant coefficients using the roots of t
characteristic polynomial. She should be able to solve nonhomogeneous linear diff
ential equations with constant coefficients by the methods of variation of paramet
and undetermined coefficients. The solutions of the differential equations for sim
harmonic motion, damped harmonic motion and forced harmonic motion should
understood.

Basic Exercises

1. Solve each differential equation.
(a) $y'' = \sin \pi x$ (b) $y'' = x(y')^3$ (c) $xy'' - y' = 12$ (d) $y'' = y$
(e) $y^2 y'' - y(y')^2 = 6y^2$ (f) $x^3 y'' + 2x^2 y' = 1$ (g) $xy'' = x^2 + 1$ (h) $y'' = ($

2. Solve each initial value problem.
(a) $y'' = xe^x$, $y(0) = 2$, $y'(0) = -1$ (b) $y'' = 2y'$, $y(0) = -3$, $y'(0) = 8$
(c) $y'y'' = 6$, $y(1) = -1$, $y'(1) = 4$ (d) $y'' + y' = 3y^2 y'$, $y(0) = 2$, $y'(0) =$
(e) $y'' + 3y' = x$, $y(0) = -2$, $y'(0) = 2$ (f) $(x^2 + 1)y'' = 1$, $y(0) = 3$, $y'(0) = 5$

3. Solve each homogeneous linear differential equation.
(a) $y'' + y' - 12y = 0$ (b) $y'' - 4y' - 21y = 0$ (c) $y'' + 9y = 0$
(d) $y'' + 4y' + 20y = 0$ (e) $y'' + 6y' + 9y = 0$ (f) $y'' - 2y' - 15y = 0$
(g) $y'' - 8y' + 16y = 0$ (h) $y'' - 6y' + 45y = 0$ (i) $y'' + 10y' + 34y = 0$

4. Solve each initial value problem.
(a) $y'' - 7y' + 12y = 0$, $y(0) = 3$, $y'(0) = -4$
(b) $y'' - 4y' + 4y = 0$, $y(0) = -2$, $y'(0) = 6$
(c) $y'' - 2y' + y = 0$, $y(1) = 4e$, $y'(1) = -5e$
(d) $y'' + 49y = 0$, $y(0) = 6$, $y'(0) = 8$
(e) $y'' + 8y' + 80y = 0$, $y(\pi) = 2$, $y'(\pi) = 0$
(f) $y'' - 12y' + 85y = 0$, $y\left(\frac{\pi}{2}\right) = -1$, $y'\left(\frac{\pi}{2}\right) = 3$

5. Compute each Wronskian.
(a) $W(x^2 + 4, x^4 - 5)$ (b) $W(\tan x, \sec x)$ (c) $W(e^x, \ln x)$ (d) $W(x^2 + 1, \arctan x$

6. Use the method of undetermined coefficients to solve each differential equation.
(a) $y'' - 3y' - 28y = 56x + 34$ (b) $y'' - 9y = e^{4x}$
(c) $y'' + 6y' + 9y = \cos \pi x$ (d) $y'' - 4y' + 20y = 34xe^{3x}$
(e) $y'' + 2y' - 24y = 200x \sin 2x$ (f) $y'' + 8y' + 32y = 290e^{-x} \cos 3x$
(g) $y'' - y' - 12y = 170xe^x \cos x$ (h) $y'' - 2y' + 10y = 10x^2 - 3x + 5$
(i) $y'' + y' = 10x^2 e^{2x} \sin x$ (j) $y'' + 4y = 10xe^{-x} \cos x$

7. Use the method of variation of parameters to solve each differential equation.
 (a) $y'' + 3y' - 10y = e^{5x}$ (b) $y'' + y = \sin^3 x$ (c) $y'' - 10y' + 25y = e^{5x} \sec^2 x$
 (d) $y'' + 4y = \cos^4 x$ (e) $y'' - 4y' + 4y = e^{-2x}$ (f) $y'' + y' - 2y = e^x \sin x$

8. Show that if y, u, v are differentiable functions, then $W(yu, yv) = y^2 W(u, v)$.

9. Find the position function $s(t)$ for undamped free harmonic motion in the general case where $s(0)$ and $v(0)$ are any two real numbers.

10. An object moves with damped free harmonic motion and $v(0) = 0$. Find the position function $s(t)$ and determine whether the motion is overdamped, critically damped or underdamped. In the latter case, write $s(t)$ in the form $Ge^{-Ht} \cos(\omega t - \theta)$.
 (a) $m = 2$, $c = 12$, $k = 3$ (b) $m = 6$, $c = 5$, $k = 8$
 (c) $m = 9$, $c = 24$, $k = 4$ (d) $m = 3$, $c = 4$, $k = 2$
 (e) $m = 9$, $c = 30$, $k = 5$ (f) $m = 4$, $c = 25$, $k = 5$

11. An object moves with undamped forced harmonic motion and $s(0) = v(0) = 0$. Find the position function $s(t)$.
 (a) $m = 2$, $k = 6$, $F(t) = 8 \cos \pi t$ (b) $m = 3$, $k = 4$, $F(t) = 2 \cos 5t$
 (c) $m = 4$, $k = 8$, $F(t) = 6 \cos 4t$ (d) $m = 1$, $k = 3$, $F(t) = 5 \cos 3t$

12. Find the position function $s(t)$ for forced undamped harmonic motion with the forcing function $F(t) = F_0 \cos \lambda t$ when $s(0)$ and $v(0)$ are any two real numbers.

13. Find the position function $s(t)$ for forced undamped harmonic motion with the forcing function $F(t) = F_0 \sin \lambda t$ when $s(0) = v(0) = 0$.

14. An object moves with damped forced harmonic motion and $s(0) = v(0) = 0$. Find the position function $s(t)$. Express the steady periodic solution in the form $\alpha \cos(\lambda t - \epsilon \theta)$.
 (a) $m = 1$, $c = 10$, $k = 5$, $F(t) = 2 \cos 3t$
 (b) $m = 2$, $c = 9$, $k = 3$, $F(t) = 6 \cos 2t$
 (c) $m = 3$, $c = 6$, $k = 12$, $F(t) = 24 \cos 4t$
 (d) $m = 4$, $c = 32$, $k = 10$, $F(t) = 12 \cos t$

The following four problems are for those who have studied Section 4.10.

15. Solve each differential equation.
 (a) $y'' = 100 \sinh 5x$ (b) $y'' = (\operatorname{sech} y)(\tanh y)y'$, $y'(0) = -1$
 (c) $(y'')^2 = 4y' (\sinh x + \cosh 2x)^2$, $y'(0) = 1$

16. Compute each Wronskian.
 (a) $W(\sinh kx, \cosh kx)$ (b) $W(e^{rx} \sinh kx, e^{rx} \cosh kx)$ (c) $W(\tanh kx, \coth kx)$

17. Show the general solution of the differential equation $y'' = k^2 y$ is given by $y = A \cosh kx + B \sinh kx$.

18. Solve each linear differential equation.
 (a) $y'' - 3y' - 28y = 11e^x \cosh x$ (b) $y'' + 2y' - 24y = 60 \sinh 4x$
 (c) $y'' - 4y = \sinh 2x$ (d) $y'' - 16y = 100\operatorname{csch}^2 4x$

Challenging Problems

1. Consider the linear differential equation

$$y'' + b(x)y' + c(x)y = 0$$

where $b(x)$ and $c(x)$ are continuous functions.
 (a) (Abel's Identity) If u and v are solutions of this differential equation, show there is a constant k such that $W(u, v) = ke^{-\int b(x)\, dx}$.

(b) Show if u, v are solutions of this differential equation, then their Wronsk
$W(u, v)$ is never zero or always zero.

2. (a) (Reduction of Order) Let $u(x)$ be a solution of the linear differential equat

$$y'' + b(x)y' + c(x)y = 0$$

where $b(x)$, $c(x)$ are continuous functions. If $z = \frac{y}{u}$, show z' satisfies the first or
linear differential equation

$$z'' + \left[b(x) + 2\frac{u'(x)}{u(x)} \right] z' = 0.$$

(b) The differential equation $(1 - x^2)y'' - 2xy' + 2y = 0$ has one solution $y = x$. U
reduction of order to find all the solutions of this differential equation.

3. (a) Find the position function $s(t)$ for overdamped free harmonic motion in t
case where $s(0)$ and $v(0)$ are arbitrary numbers.
(b) Show in this case the object can pass through its equilibrium position at m
once. Under what conditions does this occur?

4. (a) Find the position function $s(t)$ for critically damped free harmonic moti
in the case where $s(0)$ and $v(0)$ are arbitrary numbers.
(b) Show in this case the object can pass through its equilibrium position at m
once. Under what conditions does this occur?
(c) Under what conditions does $s(t)$, $t > 0$, have a local extremum? Find the locati
of this extremum.

5. (a) Find the position function $s(t)$ for underdamped free harmonic motion in t
case where $s(0)$ and $v(0)$ are arbitrary numbers.
(b) Show if the local maxima occur at $t_1 < \cdots < t_n < \cdots$, then $t_n - t_{n-1}$ is
constant. Find the value of this constant.

6. (a) Find the steady periodic solution of forced damped harmonic motion wi
forcing function $F(t) = F_1 \sin \lambda t$ when $s(0) = v(0) = 0$. Express your answer in t
form $\alpha \sin(\lambda t - \epsilon \theta)$.
(b) Find the steady periodic solution of forced damped harmonic motion with forci
function $F(t) = F_0 \cos \lambda t + F_1 \sin \lambda t$ when $s(0) = v(0) = 0$. Express your answer
the form $\alpha \cos(\lambda t - \epsilon \theta)$.

Theory

4.13 Fixed Point Theory

Prerequisites: Sections 1.10 and 4.3

Let $f : S \to S$ be a function from a set S to itself. We call $p \in S$ a *fixed point* of f if $f(p) = p$. Some functions have fixed points while others do not. In the first subsection, we show that a distance shrinking function, called a *contraction*, has a unique fixed point. The remainder of this section generalizes this result from the context of functions defined on numbers to the context of operators defined on functions. In the second subsection, we generalize several topological concepts from sets of numbers to sets of functions. We use this viewpoint in the third subsection to prove that an operator, which is a contraction, has a unique fixed point. In the next section this result is applied to show that an initial value problem which satisfies a Lipschitz condition has a unique solution.

Contractions

The examples below demonstrate that some continuous functions have fixed points while others do not. We show that a continuous function from a closed interval to itself always has a fixed point. Then we define a contraction, show it has a unique fixed point p and give a method for finding p.

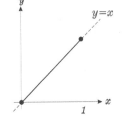

Figure 4.13.1
$f(x) = x$

Examples 4.13.2 (1) The function $f(x) = x$ with domain $[0, 1]$ has every number $p \in [0, 1]$ as a fixed point.

(2) Each function $g_n(x) = x^n$ with domain $[0, 1]$, for $n \geq 2$, has 2 fixed points: 0 and 1.

(3) Each function $h_n(x) = x^n$ with domain $(0, 1)$, for $n \geq 2$, has no fixed points. □

A fixed point p of a function f is a solution of the equation $f(x) = x$. Therefore fixed points of f correspond to the intersection of the graph of f with the line $y = x$. The analyses of the functions of Example 4.13.2 by this method are depicted in Figures 4.13.1, 4.13.3 and 4.13.5.

The following corollary of the Intermediate Value Theorem provides many examples of continuous functions with fixed points.

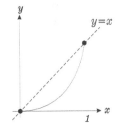

Figure 4.13.3
$g_n(x) = x^n, \ n \geq 2$

Corollary 4.13.4 *Let f be a continuous function with domain the closed interval $[a, b]$ such that the range of f is contained in $[a, b]$. Then f has a fixed point.*

Proof A fixed point of f is a solution of the equation $g(x) = f(x) - x = 0$. Observe $g(a) = f(a) - a$ and $g(b) = f(b) - b$. Since the range of f is contained in $[a, b]$, $g(a) \geq 0$ and $g(b) \leq 0$. By the Intermediate Value Theorem there is at least one number $c \in [a, b]$ with $g(c) = 0$. Each of these numbers c is a fixed point of f. □

The first example below gives an application of this corollary. Examples 2 and 3 show if the domain of f is not a closed interval or the range of f is not contained in $[a, b]$, then f may not have a fixed point.

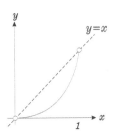

Figure 4.13.5
$h_n(x) = x^n, \ n \geq 2$

Examples 4.13.6 (1) Consider the continuous function $f(x) = 1 - \sin x$ with domain the closed interval $[0, \pi]$. Since the range of f is $[0, 1]$, which is contained in $[0, \pi]$, f has a fixed point by Corollary 4.13.4.

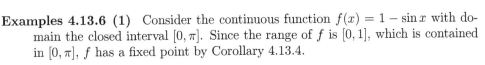

(2) Consider the continuous function $g(x) = x^2$ with domain $\left(0, \frac{1}{2}\right]$. The range of the interval $\left(0, \frac{1}{4}\right]$ which is contained in the domain of g. However, the equat $x = x^2$ has no solution $x \in \left(0, \frac{1}{2}\right]$. Hence g has no fixed point. Corollary 4.1 does not apply because the domain $\left(0, \frac{1}{2}\right]$ of g is not a closed interval.

(3) Consider the continuous function $h(x) = \frac{1}{x}$ with domain the closed interval $[2.$ The equation $x = \frac{1}{x}$ has no solution for $x \in [2, 3]$. Hence h has no fixed po Corollary 4.13.4 does not apply because the range of h is $\left[\frac{1}{3}, \frac{1}{2}\right]$ which is : contained in the domain $[2, 3]$ of h.

There are functions $f : I \to I$, called *contractions*, which always have unique fi points as long as the interval I is a closed set. This statement applies even if f is continuous and I is not bounded.

Definition 4.13.7 *Let $f : I \to I$ be a function with domain an interval I. We cal a contraction if there is a number $0 < c < 1$ such that*

$$d(f(x), f(y)) < c \cdot d(x, y)$$

for all $x, y \in I$ with $x \neq y$. We call c the constant of the contraction f.

The following corollary of the Mean Value Theorem identifies many contractio Recall the derivative at an endpoint of an interval refers to the appropriate one–sid derivative.

Corollary 4.13.8 *Let $f : I \to I$ be a continuous function with I an interval. Assu there is a number B such that $f'(x)$ exists and $|f'(x)| \leq B$ for each $x \in I$. Then f a contraction if and only if B can be chosen with $B < 1$.*

Proof Assume $B < 1$. Choose c with $B < c < 1$. We show f is a contractic Let $x, y \in I$ with $x < y$. By the Mean Value Theorem, there is $k \in (x, y)$ such th $f(y) - f(x) = f'(k)(y - x)$. Then f is a contraction because

$$d(f(x), f(y)) = |f(y) - f(x)| = |f'(k)| \cdot |y - x| \leq B \cdot |y - x| = B \cdot d(x, y) < c \cdot d(x,$$

Now assume f is a contraction with constant $c < 1$. Select B as the least upp bound of the bounded set U of numbers $|f'(x)|$ for $x \in I$. We show $B < 1$. Obser there are tangent lines to the graph of f whose slopes have absolute value arbitrari close to B. Hence there are secant lines S of the graph of f whose slopes m_S ha absolute value arbitrarily close to B. If S is determined by the points $(x, f(x))$ a $(y, f(y))$ on the graph of f, with $x \neq y$, then

$$m_S = \frac{f(y) - f(x)}{y - x} \quad \text{and}$$

$$|m_S| d(y, x) = |m_S| \cdot |y - x| = |f(y) - f(x)| = d(f(y), f(x)) < c \cdot d(y, x).$$

Hence $|m_S| < c$. It follows that c is an upper bound of the set U. Hence $B \leq c < 1$

We apply this corollary to give examples of contractions.

Examples 4.13.9 Show each function is a contraction.

(1) $f(x) = \frac{1}{2} \cos x$ with domain $\Re$.

> **Solution** The range of f is $\left[-\frac{1}{2}, \frac{1}{2}\right] \subset \Re$. Now $f'(x) = -\frac{1}{2} \sin x$ and $\left|-\frac{1}{2} \sin x\right| \leq \frac{1}{2} < 1$ for $x \in \Re$. By Corollary 4.13.8, f is a contraction.

(2) $g(x) = 2 + \ln x$ with domain $[2, \infty)$.

Solution Since g is an increasing function, $g(x) \geq g(2) = 2 + \ln 2 > 2$ for $x \in [2, \infty)$. Hence the range of g is contained in $[2, \infty)$. Observe that $g'(x) = \frac{1}{x} \leq \frac{1}{2} < 1$ for $x \in [2, \infty)$. By Corollary 4.13.8, g is a contraction.

(3) $h(x) = \frac{|x|}{2}$ with domain $\Re$.

Solution Since $h'(0)$ does not exist, we cannot apply Corollary 4.13.8 to this example. However, if x and y have the same sign, then

$$d(h(x), h(y)) = |h(y) - h(x)| = \frac{1}{2}|y - x| = \frac{1}{2}d(x, y).$$

If x and y have opposite signs, then

$$d(h(x), h(y)) = |h(y) - h(x)| = \frac{1}{2}|y + x| < \frac{1}{2}|y - x| = \frac{1}{2}d(x, y).$$

Thus in all cases $d(h(x), h(y)) < \frac{3}{4}d(x, y)$, and h is a contraction. $\square$

Recall that a sequence $\{x_n\}$ of numbers is called a Cauchy sequence if for every $\epsilon > 0$ there is a positive integer N such that if $m, n \geq N$ then $d(x_m, x_n) < \epsilon$. In Theorem 1.10.38 we showed that every Cauchy sequence converges. We prove that a contraction has a unique fixed point. The fixed point is identified by constructing a Cauchy sequence. This sequence must converge, and its limit is the unique fixed point of the contraction. Note that an interval which is a closed set has the form $\Re$, $[a, b]$, $[a, \infty)$ or $(-\infty, b]$.

Theorem 4.13.10 *Let $f : I \to I$ be a contraction with I an interval which is a closed set.*
(a) *Then f is continuous.*
(b) *Let f^n denote the composite of f with itself n times, and let x be any number in I. Then f has a unique fixed point p given by:*

$$p = \lim_{n \to \infty} f^n(x). \qquad (4.13.1)$$

Proof Let c be the constant of the contraction f.
(a) If $x, y \in I$, $\epsilon > 0$ and $d(x, y) < \epsilon$, then

$$d(f(x), f(y)) < c \cdot d(x, y) < c\epsilon < \epsilon.$$

Hence f is continuous at x.
(b) We show $\{f^n(x)\}$ is a Cauchy sequence. Let $\epsilon > 0$. Since $\lim_{n \to \infty} c^n = 0$, find a positive integer N such that $c^N < \frac{\epsilon(1-c)}{d(x, f(x))}$. If $n > m \geq N$ then

$$d(f^m(x), f^{m+1}(x)) < cd(f^{m-1}(x), f^m(x)) < c^2 d(f^{m-2}(x), f^{m-1}(x)) < \cdots < c^m d(x, f(x))$$

By the triangle inequality, Proposition 1.10.3(c):

$$
\begin{aligned}
d(f^m(x), f^n(x)) &\leq d(f^m(x), f^{m+1}(x)) + d(f^{m+1}(x), f^{m+2}(x)) + \cdots + d(f^{n-1}(x), f^n(x)) \\
&< c^m d(x, f(x)) + c^{m+1}d(x, f(x)) + \cdots + c^{n-1}d(x, f(x)) \\
&= d(x, f(x))\left[c^m + c^{m+1} + \cdots + c^{n-1}\right] \\
&= d(x, f(x))\frac{c^m - c^n}{1 - c} \leq d(x, f(x))\frac{c^m}{1-c} \leq d(x, f(x))\frac{c^N}{1-c} < \epsilon.
\end{aligned}
$$

Thus $\{f^n(x)\}$ is a Cauchy sequence and converges to a number p. Since I is a closed set, $p \in I$. Since f is continuous,

$$f(p) = f\left(\lim_{n \to \infty} f^n(x)\right) = \lim_{n \to \infty} f^{n+1}(x) = \lim_{m \to \infty} f^m(x) = p \text{ where } m = n + 1.$$

Hence p is a fixed point of f. Let q be any fixed point of f. If $p \neq q$, then

$$d(p, q) = d(f(p), f(q)) < c \cdot d(p, q).$$

Divide by $d(p, q)$: $1 < c$, a contradiction. Thus $p = q$, and f has a unique fixed po

We apply this theorem to Examples 4.13.9.

Examples 4.13.12 Estimate the unique fixed point of each contraction.

(1) $f(x) = \frac{1}{2} \cos x$ with domain $\Re$.

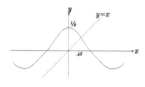

Figure 4.13.11
$f(x) = \frac{1}{2} \cos x$

Solution We construct the first 6 terms of the sequence $\{f^n(0)\}$ to estim
the fixed point.

$$f(0) = .5000 \qquad\qquad f^2(0) = f(0.5000) = 0.4388$$
$$f^3(0) = f(0.4388) = 0.4526 \quad f^4(0) = f(0.4526) = 0.4497$$
$$f^5(0) = f(0.4497) = 0.4503 \quad f^6(0) = f(0.4503) = 0.4502$$

Hence the unique fixed point p of f is $p \approx 0.45$. See Figure 4.13.11.

(2) $g(x) = 2 + \ln x$ with domain $[2, \infty)$.

Solution We construct the first 7 terms of the sequence $\{g^n(2)\}$ to estim
the fixed point.

$$g(2) = 2.693 \qquad g^2(2) = g(2.693) = 2.991 \quad g^3(2) = g(2.991) = 3.0$$
$$g^4(2) = g(3.096) = 3.130 \quad g^5(2) = g(3.130) = 3.141 \quad g^6(2) = g(3.141) = 3.1$$
$$g^7(2) = g(3.145) = 3.146$$

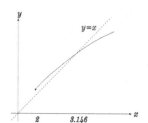

Figure 4.13.13
$g(x) = 2 + \ln x$

Hence the unique fixed point p of g occurs at $p \approx 3.146$. See Figure 4.13.13.

(3) $h(x) = \frac{|x|}{2}$ with domain $\Re$.

Solution Observe for any number x:

$$\lim_{n \to \infty} h^n(x) = \lim_{n \to \infty} \frac{|x|}{2^n} = |x| \lim_{n \to \infty} \frac{1}{2^n} = 0$$

Hence the unique fixed point p of h is $p = 0$. See Figure 4.13.14.

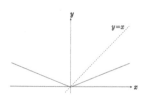

Figure 4.13.14
$h(x) = \frac{|x|}{2}$

Say f itself is not a contraction, but the composition of f with itself several tim
is a contraction. In this case, f also has a unique fixed point.

Corollary 4.13.15 *Let I be an interval which is a closed set. Let $f : I \to I$ su
that the composite of f with itself n times is a contraction. Then f has a unique fix
point p. For any number $x \in I$:*

$$p = \lim_{k \to \infty} f^k(x).$$

Proof By Theorem 4.13.10, f^n has a unique fixed point $p = f^n(p)$. Apply th
equation k times:

$$f(p) = f\left[(f^n)^k(p)\right] = f^{nk+1}(p) = f^{nk}(f(p)) = (f^n)^k(f(p)).$$

Take the limit of this equation as k goes to infinity:

$$f(p) = \lim_{k \to \infty} f(p) = \lim_{k \to \infty} (f^n)^k (f(p)) = p$$

by (4.13.1) applied to f^n with $x = f(p)$. Hence p is a fixed point of f. Note that a fixed point of f must be a fixed point of f^n, and f^n has a unique fixed point. Thus f cannot have more than one fixed point. For $x \in I$ and $0 \leq s \leq n - 1$:

$$p = \lim_{m \to \infty} (f^n)^m(f^s(x)) = \lim_{m \to \infty} f^{mn+s}(x).$$

Since every positive integer q can be written as $q = mn + s$, with $0 \leq s \leq n - 1$, $\lim_{q \to \infty} f^q(x)$ exists and equals p. □

Example 4.13.16 Let $f(x) = \cos x$ with domain $\Re$.
(a) Show f is not a contraction.
(b) Show f^2 is a contraction.
(c) Estimate the unique fixed point of f.

Solution (a) Note $f'(x) = -\sin x$, and $f'\left(\frac{3\pi}{2}\right) = 1$. By Corollary 4.13.8, f is not a contraction.
(b) Apply Corollary 4.13.8 to $f^2(x) = \cos(\cos x)$. By the chain rule: $D(f^2)(x) = -\sin(\cos x)(-\sin x) = \sin(\cos x)\sin x$. Observe that

$$-\frac{\pi}{3} < -1 \leq \cos x \leq 1 < \frac{\pi}{3}.$$

Since the sin function is increasing on the interval $\left[-\frac{\pi}{3}, \frac{\pi}{3}\right]$, we have $|\sin(\cos x)| \leq |\sin\frac{\pi}{3}| = \frac{\sqrt{3}}{2}$. By Corollary 4.13.8, f^2 is a contraction because

$$|D(f^2)(x)| = |\sin(\cos x)| \cdot |\sin x| \leq \frac{\sqrt{3}}{2} \cdot 1 < 1.$$

(c) By Corollary 4.13.15 f has a unique fixed point p. We calculate the first 21 terms of the sequence $\{f^n(0)\}$ to estimate p:

$f(0) = 1.0000$ $f^2(0) = \cos 1.000 = .5403$ $f^3(0) = \cos.5403 = .8576$
$f^4(0) = \cos.8576 = .6543$ $f^5(0) = \cos.6543 = .7935$ $f^6(0) = \cos.7935 = .7014$
$f^7(0) = \cos.7014 = .7640$ $f^8(0) = \cos.7640 = .7221$ $f^9(0) = \cos.7221 = .7504$
$f^{10}(0) = \cos.7504 = .7314$ $f^{11}(0) = \cos.7314 = .7442$ $f^{12}(0) = \cos.7442 = .7356$
$f^{13}(0) = \cos.7356 = .7414$ $f^{14}(0) = \cos.7414 = .7375$ $f^{15}(0) = \cos.7375 = .7401$
$f^{16}(0) = \cos.7401 = .7384$ $f^{17}(0) = \cos.7384 = .7396$ $f^{18}(0) = \cos.7396 = .7388$
$f^{19}(0) = \cos.7388 = .7393$ $f^{20}(0) = \cos.7393 = .7389$ $f^{21}(0) = \cos.7389 = .7392$

Thus $p \approx .739$. □

Topology of Functions

We study the topological properties of the set of continuous functions with domain $[a, b]$. Our goal is to show that this set of functions is complete in the sense that every Cauchy sequence converges. We begin by defining distance, convergent sequences and Cauchy sequences for functions.

The distance $d(f, g)$ between two continuous functions f and g with domain $[a, b]$ is defined as the length of the longest vertical line segment L_x between the corresponding points $(x, f(x))$ on the graph of f and $(x, g(x))$ on the graph of g. See Figure 4.13.18.

Definition 4.13.17 *Let f, g be continuous functions with domain the closed interval $[a, b]$. The distance $d(f, g)$ between f and g is defined as the maximum value of*

$$D(x) = d(f(x), g(x)) = |g(x) - f(x)| \quad \text{for } x \in [a, b].$$

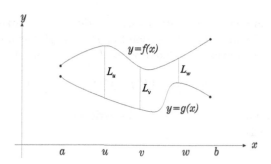

Figure 4.13.18 Distance Between f and g

Note the continuous function D with domain $[a, b]$ has a maximum value by
Maximum Value Theorem.

Examples 4.13.20 (1) Let $f(x) = x$ and $g(x) = x^2$ with domain $[0, 1]$. Find $d(f,$

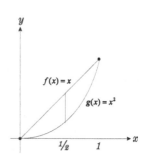

Figure 4.13.19
$d(f, g)$

   Solution We compute the maximum value of

$$D(x) = d(f(x), g(x)) = |g(x) - f(x)| = |x^2 - x| = x - x^2.$$

   for $x \in [0, 1]$. Since $D'(x) = 1 - 2x$, $D(x)$ has a critical point at $x = \frac{1}{2}$. N
   $D''(x) = -2 < 0$. By the Second Derivative Test, $D(x)$ has its maximum va
   at $x = \frac{1}{2}$. Hence $d(f, g) = D\left(\frac{1}{2}\right) = \frac{1}{4}$. See Figure 4.13.19.

(2) Let $h(x) = \sin x$ and $k(x) = \cos x$ with domain $[0, \pi]$. Find $d(h, k)$.

   Solution For $x \in [0, \pi]$,

$$
\begin{aligned}
D(x) &= d(h(x), k(x)) = |k(x) - h(x)| = |\cos x - \sin x| \\
&= \left\{
\begin{array}{ll}
\cos x - \sin x & \text{if } 0 \le x \le \pi/4 \\
\sin x - \cos x & \text{if } \pi/4 < x \le \pi
\end{array}
\right\}.
\end{aligned}
$$

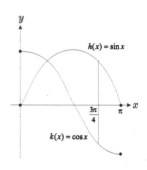

Figure 4.13.21
$d(h, k)$

   Compute $D'(x)$, for $x \in (0, \pi)$, to determine the critical points of $D(x)$:

$$
D'(x) = \left\{
\begin{array}{ll}
-\sin x - \cos x & \text{if } 0 < x < \pi/4 \\
DNE & \text{if } x = \pi/4 \\
\cos x + \sin x & \text{if } \pi/4 < x < \pi
\end{array}
\right\}.
$$

   Thus $D(x)$ has critical points at $x = \frac{\pi}{4}$, where $D'(x)$ does not exist, and
   $x = \frac{3\pi}{4}$ where $D'(x) = 0$. The maximum value of D occurs at one of the
   critical points or at one of the endpoints of the interval $[0, \pi]$. Since $D(0) =$
   $D\left(\frac{\pi}{4}\right) = 0$, $D\left(\frac{3\pi}{4}\right) = \sqrt{2}$ and $D(\pi) = 1$, the maximum value of $D(x)$ is $\sqrt{ }$
   Thus $d(h, k) = \sqrt{2}$. See Figure 4.13.21.

Distance between functions has the usual properties of distance.

Proposition 4.13.22 *Let f, g, h be continuous functions with domain $[a, b]$. Ther*
(a) $d(f, g) = d(g, f)$;
(b) $d(f, g) \ge 0$ with $d(f, g) = 0$ if and only if $f = g$;
(c) [Triangle Inequality] $d(f, h) \le d(f, g) + d(g, h)$.

Proof (a) $d(f, g)$ is the maximum value of the function $D(x) = |g(x) - f(x)|$ c
$[a, b]$. This function is the same as the function $E(x) = |f(x) - g(x)|$ whose maximu
value on $[a, b]$ is $d(g, f)$.

(b) $d(f, g)$ is the maximum value of the non–negative numbers $D(x) = |g(x) - f(x)|$ for $x \in [a, b]$. Hence $d(f, g) \geq 0$. When $f = g$, $D(x) = 0$ for $x \in [a, b]$ and $d(f, f) = 0$. If $d(f, g) = 0$, then the maximum value of the function $D(x) = |g(x) - f(x)|$ on $[a, b]$ is zero. Hence $g(x) - f(x) = 0$ for $x \in [a, b]$, and $f = g$.
(c) Let $x \in [a, b]$. By the triangle inequality for numbers:

$$D(x) = d(f(x), h(x)) \leq d(f(x), g(x)) + d(g(x), h(x)) \leq d(f, g) + d(g, h).$$

for $x \in [a, b]$. Hence the maximum value $d(f, h)$ of $D(x)$ on $[a, b]$ is less than or equal to the number $d(f, g) + d(g, h)$. □

We used the completeness property of $\Re$ to construct fixed points: every Cauchy sequence of numbers converges. Just as a sequence of numbers is a list of numbers, so too a *sequence $\{f_n\}$ of functions* is a list of functions:

$$f_1, \ f_2, \ f_3, \ \ldots, \ f_n, \ \ldots$$

Convergence of a sequence of functions and a Cauchy sequence of functions are defined in the same way as for sequences of numbers in Section 1.10. Our goal is to prove the completeness property for functions: a Cauchy sequence of continuous functions converges to a continuous function.

Definition 4.13.23 *Let $\{f_n\}$ be a sequence of functions, each with domain $[a, b]$.*
(a) *The sequence $\{f_n\}$ converges to the function f with domain $[a, b]$ if for every $\epsilon > 0$ there is a positive integer N such that if $n \geq N$, then $d(f_n, f) < \epsilon$.*
(b) *A sequence of functions which does not converge, is called divergent.*
(c) *The sequence of functions $\{f_n\}$ is a Cauchy sequence if for every $\epsilon > 0$ there is a positive integer N such that if $m, n \geq N$ then $d(f_m, f_n) < \epsilon$.*

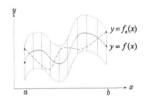

Observe that $d(f_n, f) < \epsilon$ if and only if the graph of f_n lies inside the strip $S_\epsilon(f)$ of width 2ϵ around the graph of f. See Figure 4.13.24. We show that a convergent sequence of functions evaluated at a point produces a convergent sequence of numbers. Similarly, a Cauchy sequence of functions evaluated at a point produces a Cauchy sequence of numbers.

Figure 4.13.24
Graph of f_n Lies in the Strip $S_\epsilon(f)$

Proposition 4.13.25 *Let $\{f_n\}$ be a sequence of functions, each with domain $[a, b]$.*
(a) *If the sequence of functions $\{f_n\}$ converges to the function f, then the sequence of numbers $\{f_n(x)\}$ converges to the number $f(x)$ for each $x \in [a, b]$.*
(b) *If the sequence of functions $\{f_n\}$ is a Cauchy sequence, then the sequence of numbers $\{f_n(x)\}$ is a Cauchy sequence for each $x \in [a, b]$.*

Proof (a) If $\epsilon > 0$, there is a positive integer N such that $n \geq N$ implies $d(f_n, f) < \epsilon$. If $x \in [a, b]$, then for $n \geq N$:

$$d(f_n(x), f(x)) \leq d(f_n, f) < \epsilon.$$

Hence the sequence of numbers $\{f_n(x)\}$ converges to $f(x)$.
(b) If $\epsilon > 0$, there is a positive integer N such that $m, n \geq N$ implies $d(f_m, f_n) < \epsilon$. If $x \in [a, b]$, then for $m, n \geq N$:

$$d(f_m(x), f_n(x)) \leq d(f_m, f_n) < \epsilon.$$

Hence the sequence of numbers $\{f_n(x)\}$ is a Cauchy sequence. □

The converse of this proposition is false. As the next example shows, a divergent sequence of functions may produce a convergent sequence of numbers when evaluated at each point of $[a, b]$. Also a sequence of functions which is not a Cauchy sequence may produce a Cauchy sequence of numbers when evaluated at each point of $[a, b]$.

Examples 4.13.26 (1) Let $f_n(x) = x^n$ with domain $[0,1]$ for $n \geq 1$.

(a) Show the sequence of numbers $\{f_n(x)\}$ converges for each $x \in [0,1]$.

(b) Show the sequence of functions $\{f_n\}$ diverges.

(c) Show the sequence of numbers $\{f_n(x)\}$ is a Cauchy sequence for each x

(d) Show the sequence of functions $\{f_n\}$ is not a Cauchy sequence.

Solution (a) If $0 \leq x < 1$, the sequence $\{x^n\}$ converges to 0. If $x = 1$, sequence $\{x^n\} = \{1\}$ converges to 1. See Figure 4.13.27.

(b) By the proof of (a), if the sequence $\{f_n\}$ converges then it can only conve to the function f defined by:

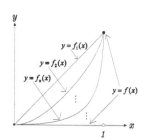

Figure 4.13.27
Example 1(a)

$$f(x) = \left\{ \begin{array}{ll} 0 & \text{if } 0 \leq x < 1 \\ 1 & \text{if } x = 1 \end{array} \right\}.$$

Take $\epsilon = \frac{1}{3}$ and let n be a positive integer. Choose x with $\sqrt[n]{1/3} < x < 1$. Th $d(f_n(x), f(x)) = |x^n - 0| = x^n > \frac{1}{3}$. Thus $d(f_n, f) > \frac{1}{3} = \epsilon$, and the seque $\{f_n\}$ does not converge to f. Note in Figure 4.13.28 that the graph of $y = f_n$ does not lie in the strip $S_{1/3}(f)$.

(c) Since every convergent sequence of numbers is a Cauchy sequence, convergent sequence $\{f_n(x)\}$ is a Cauchy sequence for each $x \in [0,1]$.

(d) Let $\epsilon = \frac{1}{3}$, and let s be a positive integer. We show below how to cho $x \in [0,1]$ and a positive integer $t > s$ such that $x^s > \frac{2}{3}$ and $x^t < \frac{1}{3}$. Then

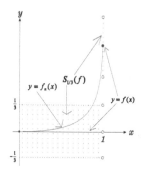

Figure 4.13.28
Example 1(b)

$$d(f_s, f_t) \geq d(f_s(x), f_t(x)) = d(x^s, x^t) > d\left(\frac{1}{3}, \frac{2}{3}\right) = \frac{1}{3} = \epsilon.$$

Hence $\{f_n\}$ is not a Cauchy sequence.

We show how to select x and t. Choose x with $\sqrt[s]{2/3} < x < 1$. Then $x^s >$ as required. Select t so that $t > -\frac{\ln 3}{\ln x}$. Since $\ln x < 0$, $\ln x^t = t \ln x < -\ln 3$ $\ln \frac{1}{3}$. Since $\ln$ is an increasing function, $x^t < \frac{1}{3}$.

(2) Let $g_n(x) = x^n$ with domain $\left[0, \frac{1}{2}\right]$ for $n \geq 1$. Show the sequence $\{g_n\}$ has lim the zero function, $g(x) = 0$ for $x \in \left[0, \frac{1}{2}\right]$.

Solution Given $\epsilon > 0$, choose a positive integer N with $\frac{1}{2^N} < \epsilon$. If $n \geq N$ a $x \in \left[0, \frac{1}{2}\right]$, then

$$d(g_n(x), g(x)) = d(x^n, 0) = x^n \leq \frac{1}{2^n} \leq \frac{1}{2^N}.$$

Hence $d(g_n, g) \leq \frac{1}{2^N} < \epsilon$, and the sequence $\{g_n\}$ converges to g.

(3) Let $b > 0$, and let $h_n(x) = \left(1 + \frac{x}{n}\right)^n$ with domain $[0,b]$ for $n \geq 1$. Show t sequence $\{h_n\}$ has limit $h(x) = e^x$.

Solution Exercise 17 gives a guide through the derivation of the inequalitie

$$\left(1 + \frac{x}{n}\right)^n \leq e^x \leq \left(1 + \frac{x}{n}\right)^{n+x} \quad \text{for } x \geq 0. \tag{4.13.}$$

Then for $n > x$:

$$
\begin{aligned}
d(h_n(x), h(x)) &= |h(x) - h_n(x)| = \left| e^x - \left(1 + \frac{x}{n}\right)^n \right| \\
&\leq \left(1 + \frac{x}{n}\right)^{n+x} - \left(1 + \frac{x}{n}\right)^n = \left(1 + \frac{x}{n}\right)^n \left[\left(1 + \frac{x}{n}\right)^x - 1\right] \\
&\leq e^x \left[\left(1 + \frac{x}{n}\right)^x - 1\right] \tag{4.13.}
\end{aligned}
$$

Observe $\ln(1 + x) \leq x$ for $x \geq 0$. [Let $f(x) = x - \ln(1 + x)$. For $x \geq 0$, $f'(x) = 1 - \frac{1}{1+x} \geq 0$ and f is increasing. Since $f(0) = 0$, it follows that $f(x) \geq 0$ for $x \geq 0$.] Let $0 < x < b < n$. By logarithmic differentiation:

$$\frac{d}{dx}\left[\ln\left(1 + \frac{x}{n}\right)^x\right] = \frac{d}{dx}\left[x\ln\left(1 + \frac{x}{n}\right)\right]$$

$$\frac{1}{(1 + x/n)^x}\frac{d}{dx}\left[\left(1 + \frac{x}{n}\right)^x\right] = \ln\left(1 + \frac{x}{n}\right) + \frac{x}{x + n} < \frac{x}{n} + \frac{x}{0 + n}$$

$$\frac{d}{dx}\left[\left(1 + \frac{x}{n}\right)^x\right] < \left(1 + \frac{x}{n}\right)^x\left[\frac{x}{n} + \frac{x}{n}\right] \leq \left(1 + \frac{x}{n}\right)^n\frac{2x}{n}$$

$$< e^x\frac{2x}{n} < e^b\frac{2b}{n}.$$

For $0 < x \leq b$, apply the Mean Value Theorem to $g(t) = \left(1 + \frac{t}{n}\right)^t$ on the interval $[0, x]$: there is $0 < c < x$ such that

$$\left(1 + \frac{x}{n}\right)^x - 1 = g'(c)(x - 0) < e^b\frac{2b}{n} \cdot b = \frac{2b^2e^b}{n}. \tag{4.13.4}$$

Given $\epsilon > 0$, choose a positive integer N with $\frac{2b^2e^{2b}}{N} < \epsilon$. If $n \geq N > b$, then by (4.13.3) and (4.13.4):

$$d(h_n(x), h(x)) < e^x \cdot \frac{2b^2e^b}{n} \leq \frac{2b^2e^{2b}}{N} \quad \text{and} \quad d(h_n(0), h(0)) = 0.$$

Hence $d(h_n, h) \leq \frac{2b^2e^{2b}}{N} < \epsilon$, and the sequence $\{h_n\}$ converges to h. □

In Example 1 above, the divergent sequence of continuous functions $\{f_n\}$ converges at each $x \in [0, 1]$ to $f(x)$, but the function f is not continuous. We show that when a sequence of continuous functions converges, the limit must be continuous. We also generalize Theorem 1.10.38: a sequence of continuous functions converges if and only if it is a Cauchy sequence.

Theorem 4.13.29 [Completeness Theorem] *Let f be a function and let $\{f_n\}$ be a sequence of continuous functions, all with domain $[a, b]$.*
(a) *If the sequence $\{f_n\}$ converges to f, then f is continuous.*
(b) *$\{f_n\}$ is a Cauchy sequence if and only if it converges.*

Proof (a) We show f is continuous at $c \in [a, b]$. Let $\epsilon > 0$. Since the sequence $\{f_n\}$ converges to f, there is a positive integer n such that $d(f_n, f) < \frac{\epsilon}{3}$. Since the function f_n is continuous, there is $\delta > 0$ such that if $x \in [a, b]$ and $d(c, x) < \delta$ then $d(f_n(c), f_n(x)) < \frac{\epsilon}{3}$. Let $x \in [a, b]$ with $d(x, c) < \delta$. By the triangle inequality:

$$\begin{aligned} d(f(c), f(x)) &\leq d(f(c), f_n(c)) + d(f_n(c), f_n(x)) + d(f_n(x), f(x)) \\ &\leq d(f, f_n) + d(f_n(c), f_n(x)) + d(f_n, f) < \frac{\epsilon}{3} + \frac{\epsilon}{3} + \frac{\epsilon}{3} = \epsilon. \end{aligned}$$

Thus f is continuous at $x = c$.
(b) Assume the sequence $\{f_n\}$ converges to f. Given $\epsilon > 0$, there is a positive integer n such that if $k \geq n$, then $d(f_k, f) < \frac{\epsilon}{4}$. If $s, t \geq n$ and $x \in [a, b]$, then

$$d(f_s(x), f_t(x)) \leq d(f_s(x), f(x)) + d(f(x), f_t(x)) \leq d(f_s, f) + d(f, f_t) < \frac{\epsilon}{4} + \frac{\epsilon}{4} = \frac{\epsilon}{2}$$

by the triangle inequality. Hence $d(f_s, f_t) \leq \frac{\epsilon}{2} < \epsilon$, and $\{f_n\}$ is a Cauchy sequence.

Let $\{f_n\}$ be a Cauchy sequence of functions. If $x \in [a, b]$, the sequence $\{f_n($ is a Cauchy sequence of numbers. By Theorem 1.10.38 it converges to a number call $f(x)$. This defines a function f with domain $[a, b]$. We show the sequence $\{$ converges to f. Assume this sequence $\{f_n\}$ does not converge to f. We show assumption leads to a contradiction. Find $\epsilon' > 0$ such that for every positive inte s there is another positive integer $t_s > s$ with $\epsilon' \leq d(f, f_{t_s})$. Let $\epsilon = \frac{\epsilon'}{2}$. Then ther $x_s \in [a, b]$ such that $\epsilon < d(f_{t_s}(x_s), f(x_s))$. Since the sequence $\{f_n(x_s)\}$ converges $f(x_s)$, find a positive integer $u_s > t_s$ such that $d(f_{u_s}(x_s), f(x_s)) < \frac{\epsilon}{2}$. Thus

$$\epsilon < d(f_{t_s}(x_s), f(x_s)) \leq d(f_{t_s}(x_s), f_{u_s}(x_s)) + d(f_{u_s}(x_s), f(x_s)) < d(f_{t_s}(x_s), f_{u_s}(x_s))\cdot$$

by the triangle inequality. Hence $\frac{\epsilon}{2} < d(f_{t_s}(x_s), f_{u_s}(x_s))$, and

$$\frac{\epsilon}{2} < d(f_{t_s}, f_{u_s}). \qquad (4.13$$

Thus for each positive integer s we can find $u_s > t_s > s$ such that (4.13.5) holds. T contradicts the hypothesis that $\{f_n\}$ is a Cauchy sequence. Hence the assumpt that the sequence $\{f_n\}$ does not converge to f must be false, i.e. the sequence $\{$ converges to f.

The Completeness Theorem generalizes to the following set of functions $C_M[a,$

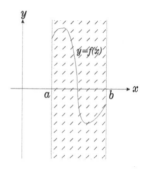

Figure 4.13.30

$R_\infty[a, b]$

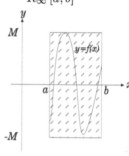

Figure 4.13.32

$R_M[a, b]$

Definition 4.13.31 (a) *Let $C_\infty[a, b]$ denote the set of all continuous functions u domain $[a, b]$.*
(b) *If M is a positive number, let $C_M[a, b]$ denote the set of all continuous functi with domain $[a, b]$ and range contained in $[-M, M]$.*

Observe that the graph of a function in $C_\infty[a, b]$ is contained in the vertical st $R_\infty[a, b]$ of Figure 4.13.30 between the lines $x = a$ and $x = b$. If M is a positi number, then $C_M[a, b]$ consists of all continuous functions with domain $[a, b]$ who graphs are contained in the rectangle $R_M[a, b]$ of Figure 4.13.32 bounded by $x =$ $x = b$, $y = M$ and $y = -M$. We generalize the Completeness Theorem to $C_M[a, b]$

Corollary 4.13.33 *Each $C_M[a, b]$ is complete in the sense that if $\{f_n\}$ is a Cauc sequence of functions in $C_M[a, b]$, then $\{f_n\}$ converges to a function $f \in C_M[a, b]$.*

Proof By the Completeness Theorem, the Cauchy sequence $\{f_n\}$ in $C_M[a, b]$ co verges to a function $f \in C_\infty[a, b]$. We must show $f \in C_M[a, b]$. Now $f_n \in C_M[a,$ means $f_n(x) \in [-M, M]$ for each $x \in [a, b]$ and each integer $n > 0$. The sequen of numbers $\{f_n(x)\}$ in $[-M, M]$ converges to $f(x)$. Since $[-M, M]$ is a closed s $f(x) \in [-M, M]$. Hence $f \in C_M[a, b]$.

Operators

Theorem 4.13.10 says that a function which is a contraction on a closed interval $[a,$ has a unique fixed point. We generalize this theorem by replacing the interval $[a, b]$ the function space $C_M[a, b]$ and replacing a function from $[a, b]$ to itself by an operat on $C_M[a, b]$. That is, we show that an operator which is a contraction on the set functions $C_M[a, b]$ has a unique fixed point.

Definition 4.13.34 *An operator on $C_M[a, b]$ is a function $T : C_M[a, b] \to C_M[a, b]$*

Examples 4.13.35 (1) Let $M \geq 1$, and let S be the operator on $C_M[-\pi, \pi]$ define by $S(f)(x) = \sin x$ for $f \in C_M[-\pi, \pi]$ and $x \in [-\pi, \pi]$. This is a *constal operator* since it assigns every function f to the sine function.

(2) Let T be the operator on $C_\infty[-2,5]$ defined by $T(f)(x) = x^3 f(x) + e^x$ for each $f \in C_\infty[-2,5]$ and $x \in [-2,5]$. Since specific functions, such as $m(x) = x^3$ and $b(x) = e^x$, play the role of constants, T is a generalization of a linear function $y = mx + b$ to operators.

(3) If g is a specific continuous function with domain $\Re$, define the operator G on $C_\infty[a,b]$ by $G(f) = g \circ f$ for each $f \in C_\infty[a,b]$. For example, a polynomial $g(x) = k_n x^n + k_{n-1} x^{n-1} + \cdots + k_1 x + k_0$ defines the operator G on $C_\infty[a,b]$ by

$$G(f)(x) = k_n f(x)^n + \cdots + k_1 f(x) + k_0 \quad \text{for } f \in C_\infty[a,b] \quad \text{and} \quad x \in [a,b].$$

If the Range of g is contained in the interval $[-M, M]$, then G is also an operator on $C_M[a,b]$. □

We generalize the definitions of continuity, contraction and fixed point from functions on $[a,b]$ to operators on $C_M[a,b]$.

Definition 4.13.36 (a) *An operator T on $C_M[a,b]$ is continuous at $f \in C_M[a,b]$ if for every $\epsilon > 0$ there is a corresponding $\delta > 0$ such that:*

$$\text{if } g \in C_M[a,b] \quad \text{and} \quad d(f,g) < \delta \quad \text{then} \quad d(T(f), T(g)) < \epsilon.$$

(b) *An operator T on $C_M[a,b]$ is continuous if it is continuous at every $f \in C_M[a,b]$.*
(c) *An operator T on $C_M[a,b]$ is a contraction if there is a number $0 < c < 1$ with*

$$d(T(f), T(g)) < c \cdot d(f,g) \quad \text{for all } f, g \in C_M[a,b] \quad \text{with } f \neq g.$$

(d) *A function $h \in C_M[a,b]$ is a fixed point of an operator T on $C_M[a,b]$ if $T(h) = h$.*

The definition of continuity of the operator T at f has the following geometric interpretation. For every strip $S_\epsilon(T(f))$ around the graph of $T(f)$, there is a strip $S_\delta(f)$ around the graph of f such that if the graph of g is contained in the strip $S_\delta(f)$, then the graph of $T(g)$ is contained in the strip $S_\epsilon(T(f))$. See Figure 4.13.37.

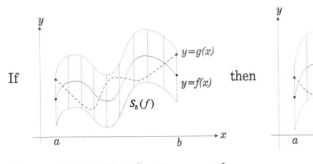

Figure 4.13.37 T is Continuous at f

Examples 4.13.38 (1) Let $M \geq 1$. Show the operator S on $C_M[-\pi, \pi]$ defined by $S(f)(x) = \sin x$, for $f \in C_M[-\pi, \pi]$ and $x \in [-\pi, \pi]$, is continuous.

Solution Given $\epsilon > 0$, choose any $\delta > 0$. If $f, g \in C_M[-\pi, \pi]$ and $d(f,g) < \delta$, then $d(S(f)(x), S(g)(x)) = d(\sin x, \sin x) = 0$. Hence $d(S(f), S(g)) = 0 < \epsilon$, and S is continuous.

(2) Show the operator T on $C_\infty[-2,5]$ defined by $T(f)(x) = x^3 f(x) + e^x$, for $f \in C_\infty[-2,5]$ and $x \in [-2,5]$, is continuous.

Solution Note if $f, g \in C_\infty[-2, 5]$ and $d(f, g) < \delta$ then for $x \in [-2, 5]$:

$$
\begin{aligned}
d(T(f)(x), T(g)(x)) &= d(x^3 f(x) + e^x, x^3 g(x) + e^x) \\
&= \left| [x^3 f(x) + e^x] - [x^3 g(x) + e^x] \right| = \left| x^3 f(x) - x^3 g(\right. \\
&= |x|^3 |f(x) - g(x)| \leq 125|f(x) - g(x)| < 125\delta. \quad (4.1
\end{aligned}
$$

If $\epsilon > 0$, choose $\delta = \frac{\epsilon}{250}$. By (4.13.6), if $d(f, g) < \delta$ and $x \in [-2, 5]$, t
$d(T(f)(x), T(g)(x)) < \frac{\epsilon}{2}$. Hence $d(T(f), T(g)) \leq \frac{\epsilon}{2} < \epsilon$, and T is continuous

(3) Assume a function g with domain $\Re$ and a number $K > 0$ are given such tha

$$
d(g(x), g(y)) \leq K \cdot d(x, y) \text{ for every } x, y \in \Re.
$$

Define the operator G on $C_\infty[a, b]$ by $G(f) = g \circ f$.
(a) Show G is a continuous operator.
(b) If $K < 1$, show G is a contraction.

Solution (a) Observe if $f, h \in C_\infty[a, b]$, then

$$
d(G(f)(x), G(h)(x)) = d(g(f(x)), g(h(x))) \leq K \cdot d(f(x), h(x)) \leq K \cdot
$$

Hence
$$
d(G(f), G(h)) \leq K \cdot d(f, h). \quad (4
$$

Given $\epsilon > 0$, let $\delta = \frac{\epsilon}{2K}$. If $d(f, h) < \delta$, then $d(G(f), G(h)) \leq K\delta = \frac{\epsilon}{2} <$
Hence G is continuous.
(b) Select c between K and one. By (4.13.7),

$$
d(G(f), G(h)) \leq K \cdot d(f, h) < c \cdot d(f, h)
$$

for every $f, h \in C_\infty[a, b]$. Thus G is a contraction.

(4) Define the operator H on $C_M[a, b]$ by:

$$
H(f)(x) = \frac{1}{2} e^{-x^2} f(x) \text{ for } f \in C_M[a, b] \text{ and } x \in [a, b].
$$

(a) Show H is a contraction.   **(b)** Find all the fixed points of H.

Solution (a) Observe that the maximum value of the function e^{-x^2} is c
Hence if $f, g \in C_M[a, b]$ and $x \in [a, b]$, then

$$
d(H(f)(x), H(g)(x)) = \left| \frac{1}{2} e^{-x^2} g(x) - \frac{1}{2} e^{-x^2} f(x) \right| = \frac{1}{2} e^{-x^2} |g(x) - f(x)| \leq \frac{1}{2} d
$$

Therefore $d(H(f), H(g)) \leq \frac{1}{2} d(f, g) < \frac{3}{4} d(f, g)$, and H is a contraction.
(b) $f \in C_M[a, b]$ is a fixed point of H if and only if $H(f)(x) = f(x)$ for
$x \in [a, b]$. That is, we require $\frac{1}{2} e^{-x^2} f(x) = f(x)$ for all $x \in [a, b]$. If $f(x) \neq$
then $e^{-x^2} = 2$, an impossibility. Hence $f(x) = 0$ for all $x \in [a, b]$, i.e. the z
function is the only fixed point of H.

The proofs of Theorem 4.13.10 and Corollary 4.13.15 generalize directly to operato

Theorem 4.13.39 *Let T be a contraction on $C_M[a, b]$.*
(a) *Then T is continuous.*
(b) *Let T^k denote the composite of T with itself k times. Then T has a unique fi
point $h \in C_M[a, b]$ defined by*

$$
h(x) = \lim_{n \to \infty} T^n(f)(x) \text{ for } x \in [a, b] \quad (4.13
$$

where f is any function in $C_M[a, b]$.

Proof Since T is a contraction, there is $0 < c < 1$ such that $d(T(p), T(q)) < c \cdot d(p, q)$ for all $p, q \in C_M[a, b]$.

(a) Let $f \in C_M[a, b]$ and $\epsilon > 0$. If $d(f, g) < \epsilon$ then $d(T(f), T(g)) < c \cdot d(f, g) < c\epsilon < \epsilon$. Hence T is continuous at f.

(b) We show $\{T^n(f)\}$ is a Cauchy sequence. Let $\epsilon > 0$. Since $\lim_{n \to \infty} c^n = 0$, find a positive integer N such that $c^N < \frac{\epsilon(1-c)}{d(f, T(f))}$. If $m \geq N$ then

$$d(T^m(f), T^{m+1}(f)) < cd(T^{m-1}(f), T^m(f)) < c^2 d(T^{m-2}(f), T^{m-1}(f)) < \cdots < c^m d(f, T(f)).$$

Let $n > m \geq N$. By the triangle inequality,

$$
\begin{aligned}
d(T^m(f), T^n(f)) &\leq d(T^m(f), T^{m+1}(f)) + d(T^{m+1}(f), T^{m+2}(f)) + \cdots + d(T^{n-1}(f), T^n(f)) \\
&< c^m d(f, T(f)) + c^{m+1} d(f, T(f)) + \cdots + c^{n-1} d(f, T(f)) \\
&= d(f, T(f)) \left[c^m + c^{m+1} + \cdots + c^{n-1} \right] \\
&= d(f, T(f)) \frac{c^m - c^n}{1 - c} \leq d(f, T(f)) \frac{c^m}{1 - c} \leq d(f, T(f)) \frac{c^N}{1 - c} < \epsilon.
\end{aligned}
$$

Thus $\{T^n(f)\}$ is a Cauchy sequence in $C_M[a, b]$. Since $C_M[a, b]$ is complete, $\{T^n(f)\}$ converges to a function $h \in C_M[a, b]$. By Exercise 18:

$$T(h) = T\left(\lim_{n \to \infty} T^n(f) \right) = \lim_{n \to \infty} T^{n+1}(f) = \lim_{m \to \infty} T^m(f) = h \quad \text{where } m = n + 1.$$

Hence h is a fixed point of T. Let $k \in C_M[a, b]$ be any fixed point of T. Then

$$d(h, k) = d(T(h), T(k)) < c \cdot d(h, k).$$

Either $d(h, k) = 0$ or $1 < c$. Since $c < 1$, we must have $d(h, k) = 0$, and $h = k$. Thus T has a unique fixed point. $\square$

The operators T that arise from initial value problems in the next section are often not contractions, while the composition of T with itself several times is a contraction. The following corollary shows that an operator with this property has a unique fixed point.

Corollary 4.13.40 *Let T be an operator on $C_M[a, b]$ such that the composite T^n is a contraction. Then T has a unique fixed point $h \in C_M[a, b]$ defined by:*

$$h(x) = \lim_{m \to \infty} T^m(f)(x) \quad \text{for } x \in [a, b]$$

where f is any function in $C_M[a, b]$.

Proof By Theorem 4.13.39, T^n has a fixed point $h \in C_M[a, b]$. Hence $h = T^n(h)$. Apply this equation k times:

$$T(h) = T\left[(T^n)^k(h) \right] = T^{kn+1}(h) = T^{nk}(T(h)) = (T^n)^k(T(h)).$$

Take the limit of this equation as k goes to infinity:

$$T(h) = \lim_{k \to \infty} T(h) = \lim_{k \to \infty} (T^n)^k (T(h)) = h$$

using (4.13.8) applied to T^n with $f = T(h)$. Hence h is a fixed point of T. A fixed point of T must be a fixed point of T^n, which has a unique fixed point. Thus T cannot have more than one fixed point. For $f \in C_M[a, b]$ and $0 \leq s \leq n - 1$:

$$h = \lim_{m \to \infty} (T^n)^m (T^s(f)) = \lim_{m \to \infty} T^{mn+s}(f).$$

Since every positive integer q can be written as $q = mn + s$, with $0 \le s \le n$ -
$\lim\limits_{q \to \infty} T^q(f)$ exists and equals h.

Summary

The reader should be able to identify a contraction and construct a sequence wh
converges to its unique fixed point. She should be able to compute the dista
between two functions. She should know how to determine whether a given seque
of functions converges and how to compute its limit. She should be able to determ
whether a given operator is continuous and whether it is a contraction. The proof t
an operator, which is a contraction, has a unique fixed point should be understoo

Basic Exercises

1. Find all the fixed points of each function.
  (a) $f(x) = \sin x$ with domain $\Re$.   **(b)** $g(x) = x^2 - 2x - 10$ with domain $\Re$
  (c) $h(x) = x^3$ with domain $\Re$.   **(d)** $k(x) = 1/x$ with domain $(0, \infty)$.
  (e) $m(x) = \arctan x$ with domain $\Re$.   **(f)** $n(x) = x\sqrt{2-x}$ with domain $(-\infty,$

2. Use Corollary 4.13.4 to show that each function has a fixed point.
  (a) $f(x) = x^4 - 1$ with domain $[-1, 1]$.   **(b)** $g(x) = \frac{x^3}{3} - x^2 + 1$ with domain
  (c) $h(x) = \frac{1}{x^4 + x^2 + 1}$ with domain $\left[\frac{1}{21}, 2\right]$.   **(d)** $k(x) = \cos x$ with domain $[-\pi/2,$
  (e) $m(x) = e^{-x}$ with domain $[0, 1]$.   **(f)** $n(x) = \arctan x$ with domain $[-\pi$

3. Show each function is a contraction and estimate its unique fixed point.
  (a) $f(x) = \frac{1}{4}(x^3 + 1)$ with domain $[0, 1]$.   **(b)** $g(x) = \frac{x^3}{6} - \frac{1}{2}$ with domain $[-1,$
  (c) $h(x) = \frac{1}{1+x^2}$ with domain $\left[\frac{1}{2}, 1\right]$.   **(d)** $j(x) = \frac{3}{4}(x + \frac{1}{x})$ with domain $[1,$
  (e) $k(x) = \frac{1}{2}(1 + \sin x)$ with domain $\Re$.   **(f)** $m(x) = \frac{1}{3}e^{-x}$ with domain $[0, \infty)$
  (g) $n(x) = \ln(2 + x)$ with domain $[0, \infty)$.   **(h)** $p(x) = \frac{1}{5}|x^2 - 1|$ with domain $[-$

4. **(a)** Let $f : I \to I$ be a function with I an interval. Show:
if $|f(y) - f(x)| < |y - x|$ for all $x, y \in I$, then f has at most one fixed point.
  (b) In addition, if $I = [a, b]$ and f is continuous, show f has a unique fixed point.

5. Let $f : [a, b] \to [a, b]$ be a continuous function which is differentiable on (a, b). U
Exercise 4 to show: if $|f'(x)| < 1$ for all $x \in (a, b)$, then f has a unique fixed point.

6. Show the function $f(x) = \frac{1}{x} + x$ with domain $[1, \infty)$ has no fixed point. Why do
this not contradict Theorem 4.13.10?

7. Show each function is not a contraction but that the function composed with its
is a contraction. Then estimate the unique fixed point of the function.
  (a) $f(x) = \cos(x + \frac{\pi}{10})$ with domain $\Re$.   **(b)** $g(x) = \frac{\pi}{2} + \sin x$ with domain $\Re$.
  (c) $h(x) = e^{-x}$ with domain $[0, \infty)$.   **(d)** $j(x) = \frac{1}{4}|x^2 - 1|$ with domain $[-2, 2$
  (e) $k(x) = \frac{2}{x} + \frac{x}{10}$ with domain $[1, \infty)$.   **(f)** $m(x) = 2 + \arctan x$ with domain $[-$

8. Let $f(x) = \sqrt{x}$ with domain $[0.1, \infty)$.
  (a) Show f and f^2 are not contractions.   **(b)** Show f^3 is a contraction.
  (c) Find the unique fixed point of f.

9. **(a)** Find $d(f, g)$ where $f(x) = x^2 - x + 1$, $g(x) = x - x^2 - 1$ both with domain
  (b) Find $d(h, j)$ where $h(x) = 5x^4 - 10x^3 + 20x^2 + 30x + 5$ and
$j(x) = 2x^4 + 10x^3 - 16x^2 + 30x - 15$, both with domain $[0, 4]$.
  (c) Find $d(k, m)$ where $k(x) = e^x$ and $m(x) = e^{-x}$, both with domain $[-1, 1]$.
  (d) Find $d(n, p)$ where $n(x) = \ln(x + 1)$, $p(x) = \ln(x - 1)$, both with domain $[2, 1($
  (e) Find $d(q, r)$ where $q(x) = \sin \pi x$ and $r(x) = \cos \pi x$, both with domain $\left[-\frac{1}{2}, \frac{1}{2}\right]$

10. Verify the triangle inequality, $d(f, h) \le d(f, g) + d(g, h)$, in the case whe
$f(x) = x^2$, $g(x) = x + 1$ and $h(x) = 1 - x^2$, all with domain $[-1, 1]$.

11.   Determine whether each sequence of functions converges or diverges. If the sequence converges, find its limit.

(a) $f_n(x) = x^n$, for $n \geq 1$, each with domain $[-1, 0]$.

(b) $g_n(x) = x^n$, for $n \geq 1$, each with domain $\left[-\frac{1}{2}, 0\right]$.

(c) $h_n(x) = \frac{x^n}{n}$ for $n \geq 1$, each with domain $[-1, 1]$.

(d) $j_n(x) = \sqrt[n]{x}$ for $n \geq 1$, each with domain $[0, 1]$.

(e) $k_n(x) = \sin(\pi x^n)$ for $n \geq 1$, each with domain $[0, 1]$.

(f) $p_n(x) = \cos\left(\frac{\pi x^n}{2}\right)$ for $n \geq 1$, each with domain $[0, 1]$.

(g) $q_n(x) = \sin^n x$ for $n \geq 1$, each with domain $[0, \pi]$.

(h) $r_n(x) = (1 + \frac{x}{n^2})^{n^2}$ for $n \geq 1$, each with domain $[0, 2]$.

(i) $s_n(x) = \ln(1 + x^n)$ for $n \geq 1$, each with domain $[0, 1]$.

(j) $t_n(x) = \ln \sqrt[n]{1 + \sqrt[n]{x}}$ for $n \geq 1$, each with domain $[0, 1]$.

12. Determine whether each sequence of functions in Example 11 is a Cauchy sequence.

13. Show the limit of a sequence of functions is unique, i.e. if f_n, f, g are functions with domain $[a, b]$, $\lim\limits_{n \to \infty} f_n = f$ and $\lim\limits_{n \to \infty} f_n = g$, then $f = g$.

14. Show each operator is continuous.

(a) $P(f)(x) = e^x$ for each $f \in C_\infty[0, 1]$ and $x \in [0, 1]$.

(b) $Q(f)(x) = f(x) \sin x - \cos x$ for each $f \in C_\infty[0, 1]$ and $x \in [0, 1]$.

(c) $R(f)(x) = \arctan f(x)$ for each $f \in C_{\frac{\pi}{2}}[0, 1]$ and $x \in [0, 1]$.

(d) $S(f)(x) = e^{-f(x)^2}$ for each $f \in C_1[0, 1]$ and $x \in [0, 1]$.

(e) $T(f)(x) = \sin \pi f(x)$ for each $f \in C_1[0, 1]$ and $x \in [0, 1]$.

(f) $U(f)(x) = x^2 \sin f(x) + x^3 \cos f(x)$ for each $f \in C_\infty[0, 1]$ and $x \in [0, 1]$.

(g) $V(f)(x) = f\left(\frac{x}{2}\right)$ for each $f \in C_M[0, 1]$ and $x \in [0, 1]$.

(h) $W(f)(x) = f(1 - x)$ for each $f \in C_M[0, 1]$ and $x \in [0, 1]$.

15. Show each operator is a contraction on $C_M[0, 1]$.

(a) $S(f)(x) = \frac{f(x)}{3 + x^2}$ for each $f \in C_M[0, 1]$ and $x \in [0, 1]$.

(b) $T(f)(x) = f(x) \arctan x$ for each $f \in C_M[0, 1]$ and $x \in [0, 1]$.

(c) $U(f)(x) = \frac{2}{3} \sin f(x)$ for each $f \in C_M[0, 1]$, $M \geq 1$, and $x \in [0, 1]$.

(d) $V(f)(x) = \frac{1}{8M^2} e^{-f(x)^2}$ for each $f \in C_M[0, 1]$, $M \geq 1$, and $x \in [0, 1]$.

(e) $W(f)(x) = \frac{4}{5} \arctan f(x)$ for each $f \in C_M[0, 1]$ and $x \in [0, 1]$.

16. Find the unique fixed point of each operator in Exercise 15.

17. Let x and n be two positive numbers. Derive the inequality (4.13.2) as follows.

(a) Use the inequality $\frac{1}{t} \leq 1$, for $1 \leq t$, to show $\ln\left(1 + \frac{x}{n}\right) = \int_1^{1 + \frac{x}{n}} \frac{1}{t} \, dt \leq \frac{x}{n}$.

(b) Deduce that $1 + \frac{x}{n} \leq e^{\frac{x}{n}}$ and $\left(1 + \frac{x}{n}\right)^n \leq e^x$.

(c) Use $\frac{1}{1 + \frac{x}{n}} \leq \frac{1}{t}$, for $1 \leq t \leq 1 + \frac{x}{n}$, to show $\frac{x}{x + n} \leq \int_1^{1 + \frac{x}{n}} \frac{1}{t} \, dt = \ln\left(1 + \frac{x}{n}\right)$.

(d) Deduce that $e^{\frac{x}{x+n}} \leq 1 + \frac{x}{n}$ and $e^x \leq \left(1 + \frac{x}{n}\right)^{x+n}$.

18. Let T be a continuous operator on $C_M[a, b]$. Show if $h, f_n \in C_M[a, b]$, for $n \geq 1$, with $\lim\limits_{n \to \infty} f_n = h$, then $T\left(\lim\limits_{n \to \infty} f_n\right) = \lim\limits_{n \to \infty} T(f_n)$.

The following two exercises are for those who have studied Section 4.10.

19. Show each function is a contraction and estimate its unique fixed point.

(a) $f(x) = \frac{1}{2}(1 + \tanh x)$ with domain $\Re$.   **(b)** $g(x) = \frac{1}{3}(x - \text{sech } x)$ with domain $\Re$.

20. Show each operator is a contraction on $C_M [0, 1]$.

(a) $S(f)(x) = \frac{3}{4} f(x) \tanh x$ for each $f \in C_M [0, 1]$ and $x \in [0, 1]$.

(b) $T(f)(x) = \frac{1}{2}\text{sech}\, f(x)$ for each $f \in C_M [0, 1]$ and $x \in [0, 1]$.

(c) $U(f)(x) = \frac{\cosh f(x)}{1 + 2 \cosh f(x)}$ for each $f \in C_M [0, 1]$ and $x \in [0, 1]$.

Challenging Problems

1. Let f be a contraction with domain $[a, b]$. Assume f is differentiable on $(a$ Show there is a number B such that $|f'(x)| \leq B$ for all $x \in (a, b)$.

2. Show the distance between points in the plane satisfies the three properties Proposition 4.13.22.

3. Let S and T be continuous operators on $C_\infty [a, b]$.
(a) Define the operator $S+T$ on $C_\infty [a, b]$ by $[(S+T)(f)](x) = [S(f)](x) + [T(f)]$ for $f \in C_\infty [a, b]$ and $x \in [a, b]$. Show the operator $S + T$ is continuous.
(b) Let $k \in C_\infty [a, b]$. Define the operator kS on $C_\infty [a, b]$ by $[(kS)(f)](x)$ $k(x) [S(f)](x)$ for $f \in C_\infty [a, b]$ and $x \in [a, b]$. Show the operator kS is conti ous.
(c) Define the operator ST on $C_\infty [a, b]$ by $[(ST)(f)](x) = [S(f)](x) \cdot [T(f)](x)$ $f \in C_\infty [a, b]$ and $x \in [a, b]$. Show the operator ST is continuous.

4. Prove the following generalization of Theorem 4.13.10. Let $s > 0$ with $f : (a - s, a + s) \to \Re$ a contraction, i.e. there is $0 < k < 1$ such that $d(f(x), f(y))$ $k \cdot d(x, y)$ for $x, y \in (a - s, a + s)$ with $x \neq y$. Assume $d(a, f(a)) < s(1 - k)$. The has a unique fixed point in the interval $(a - s, a + s)$.

5. Let $f : [a, b] \to \Re$ be differentiable with $f'(x)$ never zero. Define $F : (a, b) \to \Re$

$$F(x) = x - \frac{f(x)}{f'(x)}.$$

(a) Assume there is a number $0 < k < 1$ such that for all $x \in [a, b]$:

$$\left| \frac{f(x) f''(x)}{f'(x)^2} \right| < k.$$

Show F is a contraction.
(b) In addition, assume there are numbers $x_0 \in (a, b)$ and $s > 0$ such that

$$\left| \frac{f(x_0)}{f'(x_0)} \right| < s(1 - k).$$

Define the Newton–Raphson sequence $x_n = F(x_{n-1})$ for $n \geq 1$. Use Problem 4 to sho that $R = \lim_{n \to \infty} x_n$ exists and is the unique root of f on the interval $[a, b] \cap (x_0 - s, x_0 +$

4.14 Solution of Initial Value Problems

Prerequisite: Sections 3.12 and 4.13

Let $x_0 \in [a, b]$, and let f be a continuous function on $R_M [a, b]$. An *initial value proble* (IVP) says: find all differentiable functions $y = y(x)$ having domain $[a, b]$ with

$$y' = f(x, y) \quad \text{and} \quad y(x_0) = y_0 \tag{4.14.}$$

In the first subsection, we associate an operator F on $C_M [a, b]$ to this IVP so tha solutions of the IVP are precisely the fixed points of F. By Theorem 4.13.39, when is a contraction, it has a unique fixed point which is the unique solution of the IVP. In the second subsection, we show that if f satisfies a Lipschitz condition, then the IVP has a unique solution.

Reduction of the IVP to Finding a Fixed Point

We show solving an IVP is equivalent to solving an associated integral equation. We associate an operator F to this integral equation and show solving the IVP is equivalent to finding a fixed point of F. When F is a contraction, we apply Theorem 4.13.39 to find the fixed point of F as the limit of the sequence $\{F^n(g)\}$ for any $g \in C_M[a, b]$.

Recall that $R_\infty[a, b]$ denotes the vertical strip of Figure 4.13.30 between the lines $x = a$ and $x = b$. If M is a positive number, then $R_M[a, b]$ denotes the rectangle depicted in Figure 4.13.32 bounded by the lines $x = a$, $x = b$, $y = -M$ and $y = M$. We begin by establishing two properties of a continuous function $f(x, y)$ defined on $R_M[a, b]$ in the case $f(x, y) = s(x)t(y)$.

Proposition 4.14.1 *Let $M > 0$ be a number with $t : [-M, M] \to \Re$, or let $M = \infty$ with $t : \Re \to \Re$. Let $s : [a, b] \to \Re$. Assume s and t are continuous. Then the function*

$$f(x, y) = s(x)t(y)$$

with domain $R_M[a, b]$ has the following properties.

(1) *If $g \in C_M[a, b]$, then the function $G(x) = f(x, g(x))$ is continuous on $[a, b]$.*

(2) *If M is a positive number, then the function f is bounded, i.e. there is $B \in \Re$ such that $|f(x, y)| \leq B$ for $(x, y) \in R_M[a, b]$.*

Proof **(1)** Observe $G(x) = s(x)t(g(x))$ which is clearly continuous.
(2) Let B_s be the maximum value of $|s(x)|$ for $x \in [a, b]$, and let B_t be the maximum value of $|t(y)|$ for $y \in [-M, M]$. Then $|f(x, y)| = |s(x)|\,|t(y)| \leq B_s B_t$ for $(x, y) \in R_M[a, b]$. Thus Property 2 is valid with $B = B_s B_t$. $\quad\square$

Let $f(x, y)$ be a function with domain $R_M[a, b]$. In Section 3.12, we defined $f(x, y)$ to be continuous at the point (c, d) if $f(x, y)$ is near $f(c, d)$ whenever (x, y) is near (c, d). We called $f(x, y)$ continuous if f is continuous at each point of its domain. These continuous functions include products of continuous functions $f(x, y) = s(x)t(y)$ as well as other functions such as $f(x, y) = \sin(xy)$. In Section 7.3, we will show that all continuous functions $f(x, y)$ satisfy properties (1) and (2) of Proposition 4.14.1. Thus we will use these two properties of continuous functions in the proofs of this section. However, the examples in this section will be restricted to products of continuous functions $f(x, y) = s(x)t(y)$ for which we have established these properties.

Observe it suffices to solve the IVP (4.14.1) when $x_0 = y_0 = 0$. The general case can be reduced to this special case by the change of variables $\overline{x} = x - x_0$ and $\overline{y} = y - y_0$. Then $a - x_0 \leq 0 \leq b - x_0$. The resulting IVP on the interval $[a - x_0, b - x_0]$ is:

$$\overline{y}' = f(\overline{x} + x_0, \overline{y} + y_0) \quad \text{and} \quad \overline{y}(0) = 0$$

We show solving this IVP is equivalent to solving an integral equation.

Proposition 4.14.2 *Let $f(x, y)$ be a continuous function on $R_M[a, b]$ with M a positive number or infinity. Then the initial value problem*

$$y'(x) = f(x, y(x)) \quad \text{and} \quad y(0) = 0 \tag{4.14.2}$$

has the same solutions $y(x)$, with domain $[a, b]$, as the integral equation

$$y(x) = \int_0^x f(t, y(t))\, dt. \tag{4.14.3}$$

Proof First change the name of the variable x in (4.14.2) to t: $y'(t) = f(t, y)$
Apply the integral $\int_0^x \cdots \, dt$ to this equation:

$$\int_0^x y'(t) \, dt = \int_0^x f(t, y(t)) \, dt.$$

By the Second Fundamental Theorem of Calculus, the left integral equals $y(t)|_0^x =$
$y(x) - y(0) = y(x)$. Thus a solution of the IVP satisfies the integral equation.

Now let $y(x)$ be a solution of the integral equation. Differentiate (4.14.3):

$$y'(x) = \frac{d}{dx} \left(\int_0^x f(t, y(t)) \, dt \right) = f(x, y(x)) \ \text{ and } \ y(0) = \int_0^0 f(t, y(t)) \, dt = 0$$

by the First Fundamental Theorem of Calculus. Hence $y(x)$ is a solution of the I
Thus the IVP and its associated integral equation have the same set of solutions.

To solve the integral equation (4.14.3), interpret its right side as an operato
applied to the function $y(t)$.

Definition 4.14.3 *Define the operator F on $C_\infty[a, b]$ associated to the initial va
problem $y(x) = f'(x, y(x))$ with $y(0) = 0$ by:*

$$F(g)(x) = \int_0^x f(t, g(t)) \, dt \ \text{ for } g \in C_\infty[a, b].$$

Note solutions of the integral equation (4.14.3) are the same as the fixed points
the operator F. We rephrase the preceding proposition in terms of this operator.

Corollary 4.14.4 *$y(x)$ is a solution of the IVP $y(x) = f'(x, y(x))$ with $y(0) = 0$
and only if $y(x)$ is a fixed point of the operator F associated to this IVP.*

If the operator F is a contraction, Theorem 4.13.39 says that F has a unique fix
point and the IVP (4.14.2) has a unique solution. We apply this procedure to a simp
example. First, we digress to introduce the factorial construction. Define n *factor*
as the product of the first n positive integers:

$$n! = n(n-1)(n-2)\cdots(3)(2)(1). \tag{4.14.}$$

Thus $1! = 1$, $2! = 2(1) = 2$, $3! = 3(2)(1) = 6$, etc. Observe that $n!$ has the sar
factors as $(n-1)!$ as well as the factor n:

$$n! = n(n-1)! \tag{4.14.}$$

Example 4.14.5 Let $0 < A < 1$. Solve the initial value problem

$$y' = y + 1 \ \text{ with } \ y(0) = 0 \ \text{ on the interval } [-A, A]. \tag{4.14.}$$

Solution In this example, $f(x, y) = y + 1$. The associated integral equation is:

$$y(x) = \int_0^x y(t) + 1 \, dt.$$

The operator F on $C_\infty[-A, A]$ associated to this IVP is:

$$F(y)(x) = \int_0^x y(t) + 1 \, dt = x + \int_0^x y(t) \, dt \ \text{ for } y \in C_\infty[-A, A].$$

If $y, z \in C_\infty [-A, A]$, then

$$
d(F(y)(x), F(z)(x)) = \left| \left[x + \int_0^x z(t) \, dt \right] - \left[x + \int_0^x y(t) \, dt \right] \right|
$$

$$
= \left| \int_0^x z(t) \, dt - \int_0^x y(t) \, dt \right| = \left| \int_0^x z(t) - y(t) \, dt \right| \leq \left| \int_0^x |z(t) - y(t)| \, dt \right|
$$

$$
= \left| \int_0^x d(y(t), z(t)) \, dt \right| \leq \left| \int_0^x d(y, z) \, dt \right| = d(y, z) \left| \int_0^x 1 \, dt \right| = d(y, z)|x| \leq A \cdot d(y, z).
$$

Thus
$$
d(F(y), F(z)) \leq A \cdot d(y, z).
$$

Since $A < 1$, the operator F is a contraction. By Theorem 4.13.39, F has a unique fixed point given by $\lim_{n \to \infty} F^n(g)$ for any $g \in C_\infty [-A, A]$. Take g to be the constant function $g(x) = 0$. Then

$$
\begin{aligned}
F(g)(x) &= x + \int_0^x 0 \, dt = x \\[2mm]
F^2(g)(x) &= x + \int_0^x t \, dt = x + \left. \frac{t^2}{2} \right|_0^x = x + \frac{x^2}{2} \\[2mm]
F^3(g)(x) &= x + \int_0^x t + \frac{t^2}{2} \, dt = x + \left. \left(\frac{t^2}{2} + \frac{t^3}{3 \cdot 2} \right) \right|_0^x = x + \frac{x^2}{2} + \frac{x^3}{2 \cdot 3}.
\end{aligned}
$$

Clearly $F^n(g)(x)$ is a polynomial of degree n in which the lowest degree monomial summand is x. For $k \geq 2$, the coefficient of x^k equals the coefficient of x^{k-1} divided by k. By (4.14.5) the coefficient of x^k is $\frac{1}{k!}$, and

$$
F^n(g)(x) = x + \frac{x^2}{2!} + \cdots + \frac{x^n}{n!}.
$$

Thus the given IVP has the unique solution on the interval $[-A, A]$:

$$
h(x) = \lim_{n \to \infty} \left(\frac{x}{1!} + \frac{x^2}{2!} + \cdots + \frac{x^n}{n!} \right).
$$

Observe $D(e^x - 1) = e^x = (e^x - 1) + 1$ and $e^0 - 1 = 0$. Hence $y = e^x - 1$ is another description of the unique solution of the given IVP. Therefore $e^x - 1 = h(x)$, and

$$
e^x = \lim_{n \to \infty} \left(1 + \frac{x}{1!} + \frac{x^2}{2!} + \cdots + \frac{x^n}{n!} \right) \tag{4.14.7}
$$

for $-1 < x < 1$.[8] $\square$

Solution of the IVP when f Satisfies a Lipschitz Condition

We introduce Lipschitz conditions for a function $f(x, y)$. When $f(x, y)$ satisfies a Lipschitz condition, we show the IVP $y' = f(x, y)$ with $y(0) = 0$ has an associated operator F for which some F^n is a contraction. By Corollary 4.13.40, F has a unique fixed point which is the unique solution of the given IVP.

Definition 4.14.6 *Let $f(x, y)$ have domain $R_M [a, b]$ where $M > 0$ is either a number or ∞. We say $f(x, y)$ satisfies the Lipschitz condition with Lipschitz constant K if*

$$
d(f(x, y_1), f(x, y_2)) \leq K d(y_1, y_2) \text{ for every } (x, y_1), (x, y_2) \in R_M [a, b].
$$

---

[8] In Example 4.14.13 we show this equation is true for all $x \in \Re$.

In the first example below, $f(x, y)$ satisfies a Lipschitz condition on a vertical s $R_\infty [a, b]$. In the second example, $g(x, y)$ only satisfies a Lipschitz condition when values of y are restricted to an interval $[-M, M]$. In the third example, $h(x, y)$, v domain the rectangle $R_4 [a, b]$, does not satisfy any Lipschitz condition.

Examples 4.14.7 (1) The function $f(x, y) = y + 1$ of Example 4.14.5 satisfies Lipschitz condition with $K = 1$, $[a, b]$ arbitrary and $M = \infty$:

$$d(f(x, y_1), f(x, y_2)) = d(y_1+1, y_2+1) = d(y_1, y_2) \quad \text{for } (x, y_1),\ (x, y_2) \in R_\infty\ [a$$

(2) Let $g(x, y) = y^2$. Then

$$d(g(x, y_1), g(x, y_2)) = \left| y_2^2 - y_1^2 \right| = |y_1 + y_2|\, |y_2 - y_1| = |y_1 + y_2|\ d\,(y_1, y_2).$$

Thus $g(x, y)$ does not satisfy a Lipschitz condition on any vertical strip R_∞ [a However if M is a positive number, then on the rectangle $R_M [a, b]$:

$$d(g(x, y_1), g(x, y_2)) \leq 2M\ d\,(y_1, y_2).$$

Thus $g(x, y)$ satisfies the Lipschitz condition with $K = 2M$ on the rectar $R_M [a, b]$ for every interval $[a, b]$.

(3) Let $h(x, y) = 2\sqrt{y + 4}$ for $(x, y) \in R_4 [a, b]$ with $a < 0 < b$. Then

$$
\begin{aligned}
d(h(x, y_1), h(x, y_2)) &= \left| 2\sqrt{y_2 + 4} - 2\sqrt{y_1 + 4} \right| \\
&= 2\left| \sqrt{y_2 + 4} - \sqrt{y_1 + 4} \right| \cdot \frac{\sqrt{y_1 + 4} + \sqrt{y_2 + 4}}{\sqrt{y_1 + 4} + \sqrt{y_2 + 4}} \\
&= \frac{2|y_2 - y_1|}{\sqrt{y_1 + 4} + \sqrt{y_2 + 4}} = \frac{2d(y_1, y_2)}{\sqrt{y_1 + 4} + \sqrt{y_2 + 4}}
\end{aligned}
$$

In particular if $y_1 = y$ and $y_2 = -4$:

$$d(h(x, y), h(x, -4)) = \frac{2}{\sqrt{y + 4}} d(y, -4).$$

Since there is no number K for which $\frac{2}{\sqrt{y+4}} \leq K$ for all $y \in (-4, 4]$, the functi h does not satisfy a Lipschitz condition on the rectangle $R_4 [a, b]$.

We show if the function $f(x, y)$ in the IVP (4.14.2) is bounded on $[a, b]$ and satisf a certain Lipschitz condition, then its associated operator is a contraction and the IN has a unique solution.

Proposition 4.14.8 *Let $M > 0$ be a number or ∞ with $a \leq 0 \leq b$. Assume* **(i)**–**(iv**
(i) *The function $f(x, y)$ is continuous on $R_M [a, b]$.*
(ii) *$f(x, y)$ satisfies the Lipschitz condition with constant K.*
(iii) *If $M \neq \infty$, then f is bounded:[9] there is $B \in \Re$ with $|f(x, y)| \leq B$ for $(x, y) \in R$*
(iv) *If $M \neq \infty$ and m the larger of $|a|$ and $|b|$, then $Bm \leq M$.*
Let F be the operator on $C_M [a, b]$ associated to f:

$$F(g)(x) = \int_0^x f(t, g(t))\ dt \quad \text{for } g \in C_M [a, b] \ \text{and } x \in [a, b].$$

---

[9] The existence of a bound B of $|f(x, y)|$ on the rectangle $R_M [a, b]$ is the Maximum Value Theore for functions of two variables which is stated in Section 7.3 and proved in Section 7.10.

(a) *Then F is an operator on $C_M[a, b]$.*
(b) *If $Km < 1$, then F is a contraction on $C_M[a, b]$.*
(c) *The following conditions are equivalent.*
  (1) *$y(x)$ is a solution of the initial value problem on $[a, b]$:*

$$y'(x) = f(x, y(x)) \quad and \quad y(0) = 0.$$

  (2) *$y(x)$ satisfies the integral equation:*

$$y(x) = \int_0^x f(t, y(t)) \, dt \quad for \ x \in [a, b].$$

  (3) *$y(x)$ is a fixed point of the operator F.*

Proof (a) By the First Fundamental Theorem of Calculus, if $g \in C_M[a, b]$ then $F(g)$ is differentiable hence continuous. Hence F is an operator on $C_\infty[a, b]$. Assume M is a positive number, and $g \in C_M[a, b]$. Then

$$|F(g)(x)| = \left| \int_0^x f(t, g(t)) \, dt \right| \leq \left| \int_0^x |f(t, g(t))| \, dt \right| \leq \left| \int_0^x B \, dt \right| = B|x|.$$

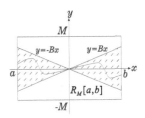

Figure 4.14.9
Graph of $y = F(g)(x)$

Thus the graph of $F(g)$ lies in the wedge shaped region between the lines $y = Bx$ and $y = -Bx$. See Figure 4.14.9. By (iv), $B|a| \leq M$ and $B|b| \leq M$. Hence this wedge shaped region lies inside $R_M[a, b]$. Thus F is an operator on $C_M[a, b]$.
(b) If $y, z \in C_M[a, b]$, then

$$d(F(y)(x), F(z)(x)) = \left| \int_0^x f(t, z(t)) \, dt - \int_0^x f(t, y(t)) \, dt \right|$$

$$= \left| \int_0^x f(t, z(t)) - f(t, y(t)) \, dt \right| \leq \left| \int_0^x |f(t, z(t)) - f(t, y(t))| \, dt \right|$$

$$= \left| \int_0^x d(f(t, z(t)), f(t, y(t))) \, dt \right| \leq \left| \int_0^x K \cdot d(y(t), z(t)) \, dt \right|$$

$$\leq \left| \int_0^x K \cdot d(y, z) \, dt \right| = K \cdot d(y, z) \left| \int_0^x 1 \, dt \right| = K \cdot d(y, z) \cdot |x| \leq K \cdot d(y, z) \cdot m$$

Hence $d(F(y), F(z)) \leq Km \cdot d(y, z)$. Since $Km < 1$, F is a contraction.
(c) This is a restatement of Proposition 4.14.2 and Corollary 4.14.4. □

Note Proposition 4.14.8(b) only says the operator F associated to the given IVP is a contraction when $Km < 1$. However, we show that some iterated composite of F with itself is always a contraction.

Proposition 4.14.10 *Let $f(x, y)$ satisfy the hypotheses of the preceding proposition. Let F be the operator on $C_M[a, b]$ associated to f. If n is a large positive integer, then F^n is a contraction.*

Proof Let K be the Lipschitz constant of f. We show:

$$d(F^n(y_1), F^n(y_2)) \leq \frac{[K(b-a)]^n}{n!} d(y_1, y_2) \quad for \ y_1, y_2 \in C_M[a, b]. \qquad (4.14.8)$$

Let $x \in [a, b]$. By the definition of the operator F:

$$d(F(y_1)(x), F(y_2)(x)) = |F(y_2)(x) - F(y_1)(x)|$$

$$= \left| \int_0^x f(t, y_2(t))\, dt - \int_0^x f(t, y_1(t))\, dt \right| \le \left| \int_0^x |f(t, y_2(t)) - f(t, y_1(t))| \right.$$

$$= \left| \int_0^x d\left(f(t, y_1(t)), f(t, y_2(t))\right)\, dt \right| \le \left| \int_0^x K \cdot d(y_1(t), y_2(t))\, dt \right|$$

$$= K \left| \int_0^x d(y_1(t), y_2(t))\, dt \right| \tag{4.1}$$

$$\le K \left| \int_0^x d(y_1, y_2)\, dt \right| = K \cdot d(y_1, y_2) \left| \int_0^x 1\, dt \right| = K \cdot d(y_1, y_2) \cdot |x|$$

This is the case $n = 1$ of the following inequality for $y_1, y_2 \in C_M[a, b]$ and $x \in [a,$

$$d\left(F^n(y_1)(x), F^n(y_2)(x)\right) \le \frac{(K|x|)^n}{n!}\, d(y_1, y_2) \tag{4.14.}$$

We show (4.14.10) is true for all integers $n > 0$. Assume $n \ge 2$ and (4.14.10) is t
for $n - 1$. Apply (4.14.9) with y_1 replaced by $F^{n-1}y_1$ and y_2 replaced by $F^{n-1}y_2$:

$$d(F^n(y_1)(x), F^n(y_2)(x)) \le K \left| \int_0^x d(F^{n-1}(y_1)(t), F^{n-1}(y_2)(t))\, dt \right|$$

$$\le K \left| \int_0^x \frac{(K|t|)^{n-1}}{(n-1)!} d(y_1, y_2)\, dt \right| = \frac{K^n d(y_1, y_2)}{(n-1)!} \left| \int_0^x |t|^{n-1}\, dt \right|$$

$$= \frac{K^n d(y_1, y_2)}{(n-1)!} \left(\frac{|t|^n}{n} \Big|_0^x \right) = \frac{K^n d(y_1, y_2)}{n!} |x|^n$$

and (4.14.10) is true for n. Thus (4.14.10) must be true for all integers $n > 0$. Si
$|x| \le b - a$, (4.14.8) follows from (4.14.10). By (4.14.8), F^n is a contraction if

$$\frac{[K(b - a)]^n}{n!} < 1. \tag{4.14.}$$

By the following lemma, this inequality is true when n is large.

Lemma 4.14.11 *For every $c \in \Re$,* $\quad \lim_{n \to \infty} \dfrac{c^n}{n!} = 0.$

Proof To verify this limit, choose an integer $k > |c|$. For $n > k$:

$$\left| \frac{c^n}{n!} \right| = \frac{|c|}{1} \frac{|c|}{2} \cdots \frac{|c|}{k-1} \frac{|c|}{k} \frac{|c|}{k+1} \cdots \frac{|c|}{n}$$

$$\le \frac{|c|}{1} \frac{|c|}{2} \cdots \frac{|c|}{k-1} \frac{|c|}{k} \frac{|c|}{k} \cdots \frac{|c|}{k} = \frac{|c|^{k-1}}{(k-1)!} \left(\frac{|c|}{k} \right)^{n-k+1}.$$

The number $\frac{|c|^{k-1}}{(k-1)!}$ does not depend on n and $\lim\limits_{n \to \infty} \left(\dfrac{|c|}{k} \right)^{n-k+1} = 0$ because $\frac{|c|}{k} <$

Hence $\lim\limits_{n \to \infty} \dfrac{c^n}{n!} = 0.$

Inequality (4.14.11) with $n = 1$ is convenient to show the operator F is a contractio

Corollary 4.14.12 *Let $y' = f(x, y)$ with $y(0) = 0$ be an IVP which satisfies t
hypotheses of Proposition 4.14.8. In particular, $f(x, y)$ satisfies the Lipschitz conditi
with constant K. Let F be the operator on $C_M[a, b]$ associated to f. If $K(b - a) <$
then F is a contraction.*

After giving an application of this corollary, we return to the IVP which we solved for $|x| < 1$ in Example 4.14.5. We use Proposition 4.14.8 to show the solution we obtained there is valid for all $x \in \Re$.

Examples 4.14.13 (1) Consider the IVP $40y' = 3x^2y + 8$ with $y(0) = 0$. Show the operator F associated to this IVP is a contraction on $C_\infty[-1, 2]$.

Solution In this IVP, $f(x, y) = \frac{3}{40}x^2y + \frac{1}{5}$. Hence

$$
\begin{aligned}
d(f(x, y_1), f(x, y_2)) &= \left|\left(\frac{3}{40}x^2y_2 + \frac{1}{5}\right) - \left(\frac{3}{40}x^2y_1 + \frac{1}{5}\right)\right| \\
&= \left|\frac{3}{40}x^2y_2 - \frac{3}{40}x^2y_1\right| = \frac{3}{40}x^2\,|y_2 - y_1|\,.
\end{aligned}
$$

On $[-1, 2]$ we have $x^2 \le 4$. Hence

$$
d(f(x, y_1), f(x, y_2)) \le \frac{3}{40}(4)\,|y_2 - y_1| = \frac{3}{10}d(y_1, y_2).
$$

Thus $f(x, y)$ satisfies the Lipschitz condition with constant $K = \frac{3}{10}$. Since $\frac{3}{10}[2 - (-1)] = \frac{9}{10} < 1$, F is a contraction on $C_\infty[-1, 2]$ by Corollary 4.14.12.

(2) Solve the IVP $y' = y + 1$ and $y(0) = 0$ on the interval $[-A, A]$ with $A > 0$.

Solution In Example 4.14.7 (1) we showed the function $f(x, y) = y + 1$ satisfies the Lipschitz condition with constant $K = 1$. By Proposition 4.14.10, the iterated composite of the associated operator F^n is a contraction on $C_\infty[-A, A]$ for n large. For example, if $A = 3$ then $\frac{6^n}{n!} < 1$ for $n \ge 14$, i.e. (4.14.11) is satisfied and F^n is a contraction. Thus Corollary 4.13.40 applies to F for every $A > 0$. Hence the IVP has solution $h = \lim_{n \to \infty} F^n(g)$ where $g(x) = 0$ is the constant function. By the computation in Example 4.14.5, this solution is:

$$
h(x) = \lim_{n \to \infty}\left(\frac{x}{1!} + \frac{x^2}{2!} + \cdots + \frac{x^n}{n!}\right) \quad \text{for all } x \in \Re.
$$

That is, the solution of this IVP for $A < 1$ derived in Example 4.14.5 is also valid when $A \ge 1$. This IVP is also has the solution $y = e^x - 1$. Since the solution of this IVP is unique, these two solutions are the same:

$$
e^x = \lim_{n \to \infty}\left(1 + \frac{x}{1!} + \frac{x^2}{2!} + \cdots + \frac{x^n}{n!}\right) \quad \text{for all } x \in \Re. \qquad \Box
$$

The next theorem combines several of our propositions to conclude that every IVP, $y' = f(x, y)$ with $y(0) = 0$, has a unique solution on a sufficiently small interval when $f(x, y)$ is continuous, bounded and satisfies a Lipschitz condition.

Theorem 4.14.14 *Let $M > 0$ be a number or ∞, and let $a \le 0 \le b$. Assume:*
(1) $f(x, y)$ is a continuous function with domain $R_M[a, b]$;
(2) $f(x, y)$ satisfies a Lipschitz condition on $R_M[a, b]$;
(3) if $M \in \Re$, then $|f(x, y)|$ is bounded by B on $R_M[a, b]$.

$$
\text{Define } \alpha = \left\{\begin{array}{ll} a & \text{if } M = \infty \\ \max\left\{a, -\frac{M}{B}\right\} & \text{if } M < \infty \end{array}\right\}, \quad \beta = \left\{\begin{array}{ll} b & \text{if } M = \infty \\ \min\left\{b, \frac{M}{B}\right\} & \text{if } M < \infty \end{array}\right\}.
$$

Then there is a unique solution $y = h(x)$, on the interval $[\alpha, \beta]$, of the IVP:

$$
y'(x) = f(x, y(x)) \quad \text{and} \quad y(0) = 0 \tag{4.14.12}
$$

Let F be the operator associated to this IVP. If g is any element of $C_M[\alpha, \beta]$, then

$$
h = \lim_{n \to \infty} F^n(g). \tag{4.14.13}
$$

Proof Note α and β are defined so that $B \max\{|\alpha|, |\beta|\} \leq M$. By Prop. 4.14.8
the operator F associated to the IVP on $[\alpha, \beta]$ is an operator on $C_M[\alpha, \beta]$.
Prop. 4.14.10, some iterated composite F^n of F is a contraction. By Corollary 4.13
F has a unique fixed point h given by (4.14.13). By Prop. 4.14.8(c), the original in
value problem has h as its unique solution.

The following example illustrates that when the function f in the IVP (4.14
does not satisfy a Lipschitz condition on any rectangle $R_M[a, b]$, then the IVP
have no solution or more than one solution.

Example 4.14.15 Solve the IVP $y' = \frac{y^2}{x^2}$ and $y(0) = k$ on the interval $[0, 1]$.

Solution Solve this separable differential equation by integrating $\frac{1}{x^2} = \frac{y'}{y^2}$:

$$-\frac{1}{x} = \int \frac{1}{x^2}\, dx = \int \frac{1}{y^2}\frac{dy}{dx}\, dx = \int y^{-2}\, dy = -\frac{1}{y} + C$$

$$y = \frac{x}{Cx+1}. \tag{4.14.}$$

Now $y(0) = 0$ for any value of C. Thus if $k \neq 0$, this IVP has no solution. Howeve
$k = 0$ then (4.14.14) is a solution of this IVP for all values of C, and this IVP has
infinite number of solutions. Observe that the function $f(x, y) = \frac{y^2}{x^2}$ does not sati
a Lipschitz condition on any rectangle $R_M[0, 1]$. In fact, the function $f(x, y)$ is
even defined when $x = 0$.

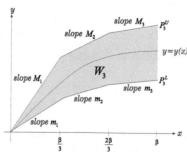

Figure 4.14.16 The Cauchy–Lipschitz Method for $n = 3$

Historical Remarks

The first method for demonstrating the existence of a unique solution of the IV
$y' = f(x, y)$ with $y(0) = 0$ originated with Leonhard Euler in 1768. It was develop
by Augustin Louis Cauchy in his lectures during the 1820s and refined by Rud
Lipschitz in 1876. We sketch a simplified version of this method using the notation
Theorem 4.14.14. Divide $R_M[0, \beta]$ into n strips $S_k = R_M\left[\frac{(k-1)\beta}{n}, \frac{k\beta}{n}\right]$ for $1 \leq k \leq$
Let m_k be the minimum value of $f(x, y)$ on S_k, and let M_k be the maximum val
of $f(x, y)$ on S_k . Then the slopes of the tangent lines of the graph of a soluti
$y = y(x)$ of this initial value problem must lie between m_k and M_k on the interv
$I_k = \left[\frac{(k-1)\beta}{n}, \frac{k\beta}{n}\right]$. Construct a polygonal path P_n^L, beginning at the origin, whi
consists of a segment of slope m_1 over I_1, a segment of slope m_2 over I_2, ..., a segme
of slope m_n over I_n. The graph of $y = y(x)$ on $[0, \beta]$ must lie above P_n^L. Constru
a second polygonal path P_n^U, beginning at the origin, which consists of a segment
slope M_1 over I_1, a segment of slope M_2 over I_2, ..., a segment of slope M_n over I
The graph of $y = y(x)$ on $[0, \beta]$ must lie below P_n^U. Thus the graph of a soluti
$y = y(x)$ of the IVP on $[0, \beta]$ lies in the wedge W_n between the polygonal paths P
and P_n^U. The case $n = 3$ is illustrated in Figure 4.14.16. As n increases, the wedg

W_n shrink. There is a unique curve contained in all the W_n which is the graph of the unique solution of the IVP.

The second method for proving the existence of a unique solution of this IVP is the one we presented. It originated in 1838 with Joseph Liouville's analysis of heat flow in a one dimensional conductor. He first reduced the problem to solving the equivalent integral equation. Then he used the integral operator to construct a sequence of successively better approximations of the solution. This method was generalized to all first order initial value problems by Emile Picard in 1893. In 1922, Stefan Banach proved the fixed point theorem for contractions and reformulated this method in the form exposited in this section. Since then, fixed point theory has become an important method for proving existence theorems for initial value problems of both ordinary and partial differential equations.

Summary

The reader should know how to solve the IVP $y' = f(x, y)$, $y(0) = 0$ for a continuous function $f(x, y)$ which satisfies a Lipschitz condition on $R_M [a, b]$. In particular, she should be able to determine the interval $[\alpha, \beta]$ for which the associated operator F is an operator on $C_M [\alpha, \beta]$. Then she should find the solution of the IVP as $\lim_{n \to \infty} F^n(g)$ for any $g \in C_M [\alpha, \beta]$.

Basic Exercises

1. Rewrite each IVP in terms of new variables $\overline{x}, \overline{y}$ so that the initial value condition becomes $\overline{y}(0) = 0$.
(a) $y' = xy$ and $y(1) = 2$. (b) $y' = e^{x+y}$ and $y(4) = -1$.
(c) $y' = (x^2 - 1)(y^2 + 1)$, $y(2) = 3$. (d) $y' = \sin(x + y) \cos(x - y)$, $y(\pi/2) = -1$.

2. Find the associated integral equation of each IVP.
(a) $y' = x^2 + y^2$ and $y(0) = 0$. (b) $y' = \ln(x + y + 1)$ and $y(0) = 0$.
(c) $y' = \tan(xy)$ and $y(0) = 0$. (d) $y' = \sqrt{4 - x^2 - y^2}$ and $y(0) = 0$.

3. Find the associated operator of each IVP in Exercise 2.

4. (i) Show the operator F associated to each IVP is a contraction.
(ii) Apply Theorem 4.13.39 to approximate the solution of the IVP by $F^n(g)$ where $g(x) = 0$ is the constant function.
  (a) $5y' = y + 2$ and $y(0) = 0$ on $R_\infty [-2, 2]$ for all n.
  (b) $3y' = x + y$ and $y(0) = 0$ on $R_\infty [-1, 1]$ for all n.
  (c) $7y' = x^2 + y$ and $y(0) = 0$ on $R_2 [-3, 3]$ for all n.
  (d) $4y' = x^2 y + xy^2 + 1$ and $y(0) = 0$ on $R_1 [-1, 1]$ for $n = 3$.
  (e) $y' = \sqrt{x + y + 11}$ and $y(0) = 0$ on $R_1 [-1, 1]$ for $n = 1$.
  (f) $y' = \frac{1}{x + y + 7}$ and $y(0) = 0$ on $R_2 [-1, 1]$ for $n = 1$.

5. Determine whether each function satisfies a Lipschitz condition. If it does, determine a value of the Lipschitz constant K.
(a) $f(x, y) = x^2 y$ on $R_\infty [-1, 1]$. (b) $g(x, y) = y \sin x$ on $R_\infty [-\pi, \pi]$.
(c) $h(x, y) = y^2$ on $R_1 [-1, 1]$. (d) $m(x, y) = x^2 y^3$ on $R_2 [-\frac{1}{2}, \frac{1}{2}]$.
(e) $p(x, y) = x\sqrt{y + 3}$ on $R_2 [-3, 3]$. (f) $q(x, y) = x\sqrt{y + 3}$ on $R_3 [-3, 3]$.

6. For each function in Exercise 5 which satisfies a Lipschitz condition, let F denote the associated operator. Determine the least integer n for which F^n is a contraction.

7. (i) Find the operator F associated with each IVP on $[-1, 1]$.
(ii) By Theorem 4.14.14 this IVP has a unique solution. Approximate this solution by $F^3(g)$ where $g(x) = 0$ is the constant function.

(a) $y' = y^2 - 1$ and $y(0) = 0$.   (b) $y' = xy^2 + 1$ and $y(0) = 0$.
(c) $y' = y^3 + 5$ and $y(0) = 0$.   (d) $y' = ye^x - 1$ and $y(0) = 0$.
(e) $y' = y\sin x + 1$ and $y(0) = 0$.   (f) $y' = x^2 + y^2$ and $y(0) = 0$.

8. (i) Find $\alpha < 0 < \beta$ so that each IVP has a unique solution on $R_M\,[\alpha, \beta]$.
(ii) Find the unique solution of this IVP.
   (a) $y' = ky$, $M = 3$ and $y(0) = 1$ (b) $y' = x + y$, $M = 2$ and $y(0) =$
   (c) $y' = x(x + y)$, $M = \infty$ and $y(0) = 2$. (d) $y' = 2x(1 + y)$, $M = 3$ and $y($
   (e) $y' = 4xy$, $M = 5$, and $y(0) = 4$.   (f) $y' = 6x^2y$, $M = 1$ and $y(0) =$

9.   Determine whether Theorem 4.14.14 (with an appropriate change of variab
applies to each IVP on $R_1\,[-1, 1]$. Find the number of solutions of the IVP.
   (a) $y' = \frac{y}{x}$ and $y(0) = k$.   (b) $y' = \frac{y}{x^2+1}$ and $y(0) = k$.
   (c) $y' = \frac{y}{x+1}$ and $y(0) = k$.   (d) $y' = \frac{y^2}{(x+1)^2}$ and $y(0) = k$.
   (e) $y' = \frac{y}{x^3}$ and $y(0) = k$.   (f) $y' = x\frac{\sin y}{y}$ and $y(0) = k$.

10.   Consider the IVP: $y' = \sqrt{y+1}$, $y(0) = 0$. This separable differential equat
has solution $2\sqrt{y+1} = x + C$. Solving for y, we have $y = \frac{1}{4}(x + C)^2 - 1$. T
$0 = y(0) = \frac{1}{4}C^2 - 1$ and $C = \pm 2$. Thus we have **two solutions** of this IV
$y = \frac{1}{4}(x+2)^2 - 1$ and $y = \frac{1}{4}(x-2)^2 - 1$. Does this example contradict Thm. 4.14
which says this IVP has a **unique solution**?

Challenging Problems

1.   Assume $f(x, y)$ is a continuous function which satisfies a Lipschitz condition
$R_\infty\,[a, b]$ with $a \le 0 \le b$. Consider the integral equation

$$y(x) = \phi(x) + \int_0^x f(t, y(t))\, dt.$$

(a) If ϕ is a differentiable function with domain $[a, b]$, find an IVP which has
same solutions as this integral equation. Conclude that the integral equation ha
unique solution on $[a, b]$.
(b) If ϕ is a continuous function with domain $[a, b]$, show this integral equation
a unique solution.

2.   Let $f(x, y)$ be a continuous function which satisfies a Lipschitz condition
$R_\infty\,[a, b]$, $a \le 0 \le b$. Let $\lambda : [a, b] \to [a, b]$ be a function. Consider the integral equ

$$y(x) = \int_0^{\lambda(x)} f(t, y(t))\, dt.$$

(a) If λ is continuously differentiable, find an IVP which has the same solutions as
integral equation. Conclude that the integral equation has a unique solution on $[a,$
(b) If λ is continuous, show this integral equation has a unique solution.

3. Let $\psi : [a, b] \to [a, b]$ be continuous with $a \le 0 \le b$. Assume $f(x, y)$ is continuo
and satisfies a Lipschitz condition on $R_M\,[a, b]$. In the notation of Theorem 4.14.
show the following integral equation has a unique solution on $[\alpha, \beta]$.

$$y(x) = \int_0^x f(\psi(t), y(t))\, dt.$$

4.   Let $\phi\,[a, b] \to \Re$, $\lambda : [a, b] \to [a, b]$, $\psi : [a, b] \to [a, b]$ be continuous with $a \le 0 \le$
Assume $f(x, y)$ is continuous and satisfies a Lipschitz condition on $R_\infty\,[a, b]$. Shc
the following integral equation has a unique solution on $[a, b]$.

$$y(x) = \phi(x) + \int_0^{\lambda(x)} f(\psi(t), y(t))\, dt.$$

4.15 Review Exercises for Chapter 4

Decide whether each of the following 30 statements is True or False. Justify your answers.

1. The definition of $\exp x$ is: $\exp x = e^x$.

2. The definition of b^2 is b times b.

3. $\int_{2a}^{2b} \frac{1}{x}\, dx = \int_{3a}^{3b} \frac{1}{x}\, dx$ for all $0 < a < b$.

4. $\frac{d^2}{dx^2}(x \ln x) = \frac{d}{dx}(\ln x)$

5. $D(x^x)$ is the sum of $x \cdot x^{x-1}$ and $(\ln x)x^x$.

6. $\int \sec x\, dx = \int \frac{1}{\cos x}\, dx = \ln|\cos x| + C$

7. $\lim\limits_{x \to 0^+} x^{x/\sin x} = 0$

8. Let $f(x)$ be a continuous function with domain an interval. Then the graph of $y = \exp[-f(x)]$ has no vertical asymptotes.

9. $\int_a^b f'(x)g(x)\, dx = f(x)g(x)\big|_a^b + \int_b^a f(x)g'(x)\, dx$

10. The value of $\int e^x \cos x\, dx$ can be obtained by integrating by parts twice: first let $f'(x) = e^x$ with $g(x) = \cos x$ and then let $G(x) = e^x$ with $F'(x) = \sin x$.

11. Iterated application of the recursion formula for $\int x^n e^x\, dx$ derived in Example 4.4.7 (1a) allows us to evaluate this integral for every integer n.

12. $\int \tan^8 x\, dx$ can be evaluated with the substitution $u = \tan x$.

13. $\int \cot^7 x \, \csc^7 x\, dx$ can be evaluated with the substitution $u = \cot x$ and with the substitution $v = \csc x$.

14. $\int \tan^8 x \sec^9 x\, dx$ can not be evaluated with the substitution $u = \tan x$ nor with the substitution $v = \sec x$.

15. $\int \frac{x^2+4}{\sqrt{9-x^2}}\, dx$ can be evaluated with the substitution $x = 3\sin\theta$ and with the substitution $x = 2\tan\theta$.

16. $\int \sqrt{x^2 + 6x + 5}\, dx$ can be evaluated with the substitution $x + 3 = 2\tan\theta$.

17. $\int \frac{x^2}{\sqrt{1-x^2}}\, dx$ can be integrated with the substitution $u = \arcsin x$.

18. Long division is not necessary to evaluate $\int \frac{x^4}{5x^4 - x^3 + 7}\, dx$ by the method of partial fractions.

19. A polynomial of degree five must have a real root.

20. Assume all roots of the polynomial $q(x)$ are real. Substitution of each of these roots for x once in the method of partial fractions produces the values of all the unknowns.

21. $\int \frac{\tan\theta}{\sec\theta + \csc\theta + 1}\, d\theta$ can be integrated by the rationalizing substitution $x = \tan\frac{\theta}{2}$.

22. $\int \frac{8^x - 4^x}{16^x + 32^x + 1}\, dx$ can be integrated by the rationalizing substitution $u = 2^x$.

23. $\int \frac{\sqrt{x}\,\sqrt[3]{x}}{\sqrt[9]{x} + \sqrt[4]{x}}\, dx$ can be integrated by the rationalizing substitution $x = u^{18}$.

24. $\int_{-2}^2 \frac{1}{x^5}\, dx = 0$.

25. $\int_0^\infty \frac{1}{x^p}\, dx$ diverges for every $p > 0$.

26. $\int_0^\infty \frac{1}{x^p e^x}\, dx$ converges for every $p > 0$.

27. Let f be a decreasing continuous function with positive values having dom $[0, \infty)$. If $\int_0^\infty f(x)\, dx$ converges, then $\int_0^\infty f(x)^2\, dx$ also converges.

The next 3 questions are for those who have studied Section 4.11.

28. If $y'(t) = ky(t)$ and $k \neq 0$, then $y(t)$ is one–to–one.

29. The differential equation $y' = xy + x + y + 1$ is linear but not separable.

30. Every solution of the differential equation $y' - x^2 y^2 = 1 + x^4$ is an increas function.

Solve each of the following problems.

31. Let $f(x) = \int_2^{\ln \ln x} \frac{1}{t \ln t}\, dt$.
(a) Find $f'(x)$.   (b) Find $f(x)$.

32. Evaluate $\int_1^3 \frac{1}{\log_2 x}\, dx$.

33. Show the function $f(x) = \ln \sqrt{1 + e^x}$, with domain $\Re$, is one–to–one.

34. Simplify $\frac{\log_{10} x}{\ln x}$.

35. Find the derivative of $y = \tan(x^x)$.

36. Solve for x: $e^x = \lim_{k \to \infty} \left(1 + \frac{x^2}{k}\right)^k$.

37. Sketch the graph of $f(x) = \frac{2^x - 2^{-x}}{2^x + 2^{-x}}$.

38. Find a recursion formula for $\int \tan^N x\, dx$ with $N \geq 3$.

39. Find the volume of the surface of revolution obtained by revolving the cur $y = \frac{1}{x}$, for $x \geq 1$, around the x–axis.

40. Find the volume of the surface of revolution obtained by revolving the cur $y = \frac{1}{1+x^2}$, for $x \in \Re$, around the x–axis.

41. Find the area in the first quadrant bounded by the curves $y = \frac{5}{\sqrt{x^2-4}}$, $y = \frac{x}{x^2}$ and the line $x = 5$.

42. Find the area bounded by the graphs of the functions $f(x) = \pi \tan^4 x$ a $g(x) = 4x \sin 2x$, both with domain $[0, \frac{\pi}{2})$.

43. Let S be the surface of revolution obtained by revolving the region bound by the curve $y = \frac{\sin x}{x}$, the x–axis and the y–axis around the x–axis. Show that t volume V of S is finite.

44. Let S be the surface of revolution obtained by revolving the region bounded the curve $y = \frac{f(x)}{x^2}$, the line $x = 1$ and the x–axis around the y–axis. Assume f a continuous decreasing function with $f(1) = 1$ and $\lim_{x \to \infty} f(x) = 0$. Find a speci function $f(x)$ for which the volume V of S is infinite.

45. Evaluate $L = \lim_{x \to 0} \frac{\ln(1+x)}{x}$ without using L'Hopital's Rule.

The next 2 questions are for those who have studied Section 4.11.

46. Bacteria are placed in a petrie dish containing agar at noon. At 3:00 p.m. the are 20 bacteria in the dish while at 4:00 p.m. there are 400 bacteria there. How mar bacteria will be in the dish at 9:00 p.m.?

47. A $1,000$ liter vat of 3% brine and a 500 liter vat of 5% brine are connected by two tubes. Brine flows from the large vat to the smaller one through the top tube at 2 liters per minute while brine flows from the small vat to the larger one through the bottom tube at 2 liters per minute. What is the concentration of salt in each vat in the long run?

Evaluate each integral.

48. $\int \sec x \csc x \, dx$

49. $\int \frac{\ln x}{x^2} \, dx$

50. $\int x^2 \cos x \, dx$

51. $\int \tan^6 x \, dx$

52. $\int \sec^6 x \, dx$

53. $\int \frac{x^2}{\sqrt{6x - x^2 + 16}} \, dx$

54. $\int \frac{1}{(\cos^2 x)\sqrt{4 + \tan^2 x}} \, dx$

55. $\int \frac{\sqrt{(\ln x)^2 - 1}}{x} \, dx$

56. $\int \frac{x^3}{x^3 - 1} \, dx$

57. $\int \frac{1}{(x^3 - x)^3} \, dx$

58. $\int \frac{1}{(x^3 + 4x^2 + 13x)^2} \, dx$

59. $\int \frac{\tan \theta}{1 - \cot \theta} \, d\theta$

60. $\int \frac{\sqrt{e^x}}{1 + e^x} \, dx$

61. $\int \frac{x^{1/3}}{x^{1/2} + x^{1/6}} \, dx$

62. $\int \frac{\ln \tan x}{\sin x \cos x} \, dx$

63. $\int (\ln x)^3 \, dx$

64. $\int \frac{1 + \arctan x}{(1 + x^2)(1 + \arctan x)^2} \, dx$

65. $\int \frac{\sqrt{x}}{x + 1} \, dx$

Determine whether each of these improper integrals converges or diverges.

66. $\int_0^\infty \frac{1}{\sqrt{x} e^x} \, dx$

67. $\int_{-\infty}^\infty \frac{1}{x^4 - 1} \, dx$

68. $\int_0^\infty \frac{1}{e^x - 1} \, dx$

Solve each differential equation.

69. $y' = e^{x+y}$

70. $(y' - x)^2 + (y - 1)^2 = (y' - 1)^2 + (y + x)^2 + (x - 1)^2$

Chapter 5

Infinite Series

5.1 Introduction

Infinite series provide a new approach to represent functions. This approach is used to evaluate integrals and solve differential equations. It is motivated by the observation that if the function $f(x)$ is approximated by the polynomial $p(x)$ on the interval $[a, b]$, then the integral of the polynomial $\int_a^b p(x)\, dx$ approximates the integral of the function $\int_a^b f(x)\, dx$. This procedure is useful when it is impossible to evaluate $\int_a^b f(x)\, dx$ or when we only require an approximate value of this integral for a specific application. This approach is implemented as follows: suppose we have numbers c_n, for $n \geq 0$, such that for each $n \geq 0$ and $x \in [a, b]$

$$f(x) \approx c_0 + c_1 x + c_2 x^2 + \cdots + c_n x^n. \tag{5.1.1}$$

In addition, assume the errors of these approximations approaches zero as n increases. In this case, write

$$f(x) = c_0 + c_1 x + c_2 x^2 + \cdots + c_n x^n + \cdots \quad \text{for } x \in [a, b]. \tag{5.1.2}$$

This infinite sum generalizes a polynomial and is called a *power series representation* of $f(x)$ on the interval $[a, b]$. We will see that power series are just as easy to manipulate as the polynomials they generalize. In particular, the integral of $f(x)$ is obtained by integrating its power series term by term:

$$\int_a^b f(x)\, dx = \left(c_0 x + c_1 \frac{x^2}{2} + c_2 \frac{x^3}{3} + \cdots + c_n \frac{x^{n+1}}{n+1} + \cdots \right) \Bigg|_a^b.$$

Power series and their use in calculus are the unifying themes of this chapter.

The *Concepts* portion of this chapter begins with the study of sequences in Sections 2 and 3. The basic properties of sequences are deduced in Section 2, and we compute limits of sequences in Section 3. In Section 4, we use sequences to define infinite series just as we use the sequence of approximations (5.1.1) to define the series (5.1.2). In particular, we study geometric power series and their integrals. In Section 5, we show how specific power series, *Taylor series in* x, are constructed to represent a smooth function near $x = 0$. Applications are given to evaluate limits and to approximate functions and integrals. In Section 6, we extend this construction from power series in x to power series in $x - c$ to approximate functions near $x = c$. Section 7 is devoted to the study of power series. In particular, we show a power series in $x - c$ *converges* (has a finite sum) for values of x in an interval $(c - R, c + R)$, for some $0 \leq R \leq \infty$. called the *radius of convergence* of the power series. We also show that the deriva of a power series is the infinite sum of the derivatives of its summands, while integral of a power series is the infinite sum of the integrals of its summands. particular, we use power series to evaluate and approximate integrals.

The *Applications* portion of this chapter consists of Section 8 where we find po series solutions of differential equations. The differential equation $y' - f(x) = 0$ solution $y = \int f(x)\, dx$ which we represented as a power series in Section 7. Section 8, we find power series solutions of first order linear differential equati $y' + g(x)y = 0$ and second order linear differential equations $y'' + g(x)y' + h(x)y =$ This procedure is useful when the differential equation can not be solved or when ϵ an approximation of the solution is required.

The *Theory* portion of this chapter begins with the study of numerical infi series in Section 9. Procedures are introduced to determine whether an infinite se converges. These procedures are applied to determine whether a power series, w finite radius of convergence R, converges at $x = c - R$ and at $x = c + R$. In Section we define complex numbers and establish their basic algebraic and geometric prop ties. Applications are made to find complex roots of polynomials and to simplify partial fractions computation in the integration of a rational function whose deno nator has complex roots. In Section 11, we show the basic properties of power se in a real valued variable x generalize to power series in a complex valued variable We use complex power series to extend the definitions of e^z, $\log z$, $\sin z$ and $\cos z$ complex numbers z. Functions which can be represented by complex power se are called *analytic functions*. These functions have interesting properties and pla fundamental role in mathematics. We continue their study in Sections 7.13 and 9.

The flow chart in Figure 5.1.1 shows the dependence of sections of the *Applicati* and *Theory* portions of this chapter on preceding sections.

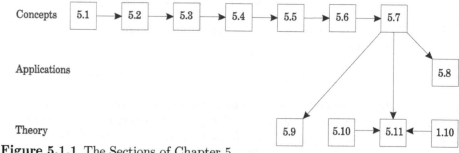

Figure 5.1.1 The Sections of Chapter 5

5.2   Sequences

This section is devoted to the study of sequences or infinite lists. Although the co cepts of this section originate with sequences of numbers, our main interest is in th application to sequences of functions. In the first subsection, we define sequences a determine whether a given sequence is increasing or decreasing. In the second subse tion, we determine whether a given sequence is bounded. In the third subsection, study sequences of numbers defined by a function.

Increasing and Decreasing Sequences

We begin by defining a sequence and presenting various examples. We shall see th some sequences are increasing, some are decreasing and others are neither.

Motivating Example 5.2.1 Consider the infinite list of numbers

$$\frac{1}{2}, \ \frac{2}{3}, \ \frac{3}{4}, \ \frac{4}{5}, \ \frac{5}{6}, \ \cdots \qquad\qquad (5.2.1)$$

These fractions have a pattern: each denominator is one larger than its numerator. If we denote its numerator by n, then its denominator is $n+1$. Thus we describe this list as $\left\{\frac{n}{n+1}\right\}$. The list is generated by letting n run through the positive integers. Note $\frac{n}{n+1}$ is the nth number of the list. This notation has advantages over the description (5.2.1). Clearly it is shorter and more precise. In addition, if the formula for the nth number of the list were complicated, it would be difficult to discover this formula from the first few numbers of the list.   □

We give a formal definition of a sequence as an infinite list of numbers a_n. The sequence of Example 5.2.1 started with $n = 1$. However, there are contexts where we want to start with other values of n such as $n = 0$.

Definition 5.2.2 (a) *A sequence of numbers* $\{a_n\}$ *is an infinite list of numbers:*

$$a_k, \ a_{k+1}, \ a_{k+2}, \ \ldots, \ a_n, \ \ldots$$

(b) *A sequence of functions* $\{f_n\}$ *is an infinite list of functions with the same domain D:*

$$f_k(x), \ f_{k+1}(x), \ f_{k+2}(x), \ \ldots, \ f_n(x), \ \ldots$$

Notes (1) The notation $\{a_n\}$ does not tell us the starting value k of n. We must be told the value of k or be able to determine its value from the context.
(2) A sequence of functions $\{f_n\}$ produces a sequence of numbers $\{f_n(x)\}$ for each fixed value of $x \in D$.
(3) A sequence of numbers can be depicted as dots on the number line. For example, the sequence of Example 5.2.1 is displayed in Figure 5.2.3.

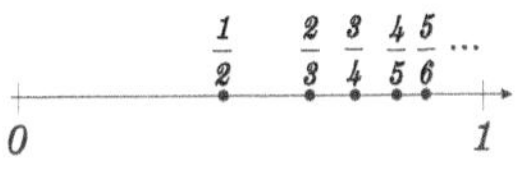

Figure 5.2.3
The Sequence $\left\{\frac{n}{n+1}\right\}$

Examples 5.2.4 (1) The sequence $\{(-1)^n\}$, for $n \geq 1$, is the infinite list

$$-1, \ +1, \ -1, \ +1, \ -1, \ +1, \ \cdots$$

(2) We take the sequence $\left\{\frac{1}{\ln n}\right\}$ to begin with $n = 2$ because $\ln 1 = 0$. This sequence is the infinite list

$$\frac{1}{\ln 2}, \ \frac{1}{\ln 3}, \ \frac{1}{\ln 4}, \ \frac{1}{\ln 5}, \ \frac{1}{\ln 6}, \ \cdots$$

(3) The sequence of functions $\{x^n\}$, for $n \geq 0$, is the infinite list

$$1, \ x, \ x^2, \ x^3, \ x^4, \ x^5, \ \cdots$$

When $x = 3$, we obtain the sequence of numbers $\{3^n\}$, for $n \geq 0$, which is the infinite list $1, \ 3, \ 9, \ 27, \ 81, \ldots$. When $x = -4$, we obtain the sequence of numbers $\{(-4)^n\}$, for $n \geq 0$, which is the infinite list $1, \ -4, \ 16, \ -64, \ 256, \ldots$.□

Many important sequences involve the factorial construction.

Definition 5.2.5 *Define 0 factorial as $0! = 1$. For $n \geq 1$, define n factorial as the product of the first n positive integers:*

$$n! = n(n-1)\cdots(3)(2)(1).$$

In particular, $1! = 1$, $2! = 2(1) = 2$, $3! = 3(2)(1) = 6$ and $4! = 4(3)(2)(1) =$ Observe that $(n+1)!$, being the product of the first $n+1$ positive integers, has same factors as $n!$ as well as the additional factor $n+1$:

$$(n+1)! = (n+1) \cdot n! \quad \text{for } n \geq 0. \tag{5.}$$

Examples 5.2.6 (1) The sequence $\left\{ \frac{n^2}{(n!)^2} \right\}$, for $n \geq 1$, is the infinite list

$$\frac{1^2}{1^2}, \quad \frac{2^2}{(1^2)(2^2)}, \quad \frac{3^2}{(1^2)(2^2)(3^2)}, \quad \frac{4^2}{(1^2)(2^2)(3^2)(4^2)}, \quad \frac{5^2}{(1^2)(2^2)(3^2)(4^2)(5^2)}, \quad \cdot$$

which is the same as the sequence

$$\frac{1}{1}, \quad \frac{1}{1^2}, \quad \frac{1}{(1^2)(2^2)}, \quad \frac{1}{(1^2)(2^2)(3^2)}, \quad \frac{1}{(1^2)(2^2)(3^2)(4^2)},$$

Alternatively, $\left\{ \frac{n^2}{(n!)^2} \right\} = \left\{ \frac{n^2}{[n(n-1)!]^2} \right\} = \left\{ \frac{n^2}{n^2(n-1)!^2} \right\} = \left\{ \frac{1}{(n-1)!^2} \right\}$

(2) The sequence $\left\{ \frac{x^n}{n!} \right\}$, for $n \geq 0$, is the infinite list

$$1, \quad \frac{x}{1}, \quad \frac{x^2}{2!}, \quad \frac{x^3}{3!}, \quad \frac{x^4}{4!}, \quad \frac{x^5}{5!}, \quad \cdots$$

Recall the function $f(x)$ is called *increasing* if its values get larger as x gets larg and the function $g(x)$ is called *decreasing* if its values get smaller as x gets larger. introduce the same terminology for sequences.

Definition 5.2.7 (a) *The sequence $\{a_n\}$ is increasing if $a_n \leq a_{n+1}$ for all n.*
(b) *The sequence $\{b_n\}$ is decreasing if $b_n \geq b_{n+1}$ for all n.*

Note the dots of an increasing sequence on the number line move to the right wh the dots of an decreasing sequence move to the left. See Figure 5.2.8. In partic lar, Figure 5.2.3 indicates that the sequence $\left\{ \frac{n}{n+1} \right\}$ of Example 5.2.1 is increasi One way to determine whether a sequence $\{a_n\}$ of positive numbers is increasing decreasing is to evaluate the quotient $\frac{a_{n+1}}{a_n}$.

- If $\frac{a_{n+1}}{a_n} \geq 1$ for all n, then $a_{n+1} \geq a_n$ and the sequence is increasing.

- If $\frac{a_{n+1}}{a_n} \leq 1$ for all n, then $a_{n+1} \leq a_n$ and the sequence is decreasing.

Increasing Sequence   Decreasing Sequence

Figure 5.2.8

Examples 5.2.9 Determine whether each sequence is increasing or decreasing.

(1) $a_n = n^2$ for $n \geq 0$.

Solution This sequence is the list of numbers $0, \ 1, \ 4, \ 9, \ 16, \ \ldots$ Since $\frac{a_{n+1}}{a_n} = \frac{(n+1)^2}{n^2} = \frac{n^2+2n+1}{n^2} = 1 + \frac{2n+1}{n^2} \geq 1$, this sequence is increasing.

(2) $b_n = (-1)^n n^2$ for $n \geq 0$.

Solution This sequence is the list of numbers $0, -1, 4, -9, 16, \ldots$
This sequence oscillates between positive and negative numbers. Therefore, it is neither increasing nor decreasing.

(3) $c_n = \frac{n}{2^n}$ for $n \geq 1$.

Solution This sequence is the list of numbers: $\frac{1}{2}, \frac{2}{4}, \frac{3}{8}, \frac{4}{16}, \frac{5}{32}, \ldots$
Observe for $n \geq 1$:

$$\frac{c_{n+1}}{c_n} = \frac{(n+1)/2^{n+1}}{n/2^n} = \frac{n+1}{n} \frac{2^n}{2^{n+1}} = \frac{n+1}{2n} \leq 1.$$

Hence $c_{n+1} \leq c_n$ for $n \geq 1$, and this sequence is decreasing.

(4) $d_n = \frac{n^2+1}{n!}$ for $n \geq 1$.

Solution Observe

$$
\begin{aligned}
\frac{d_{n+1}}{d_n} &= \frac{\left[(n+1)^2 + 1\right]/(n+1)!}{(n^2+1)/n!} = \frac{(n+1)^2 + 1}{n^2+1} \cdot \frac{n!}{(n+1)!} \\
&= \frac{n^2 + 2n + 2}{n^2 + 1} \cdot \frac{n!}{(n+1)n!} = \frac{n^2 + 2n + 2}{n^2 + 1} \cdot \frac{1}{n+1} \\
&= \frac{(n^2 + n + 1) + (n+1)}{(n^2 + n + 1) + n^3}.
\end{aligned}
$$

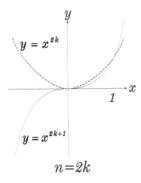

When $n \geq 2$ this quotient is less than one because $n + 1 < n^3$. Hence the sequence $\{d_n\}$, for $n \geq 2$, is decreasing. However, the given sequence $\{d_n\}$, for $n \geq 1$, is not decreasing because $d_1 = 2$ and $d_2 = \frac{5}{2}$.

(5) Let $e_n = \frac{n^n}{n!}$ for $n \geq 1$.

Solution Observe for $n \geq 1$:

$$
\begin{aligned}
\frac{e_{n+1}}{e_n} &= \frac{(n+1)^{n+1}/(n+1)!}{n^n/n!} = \frac{(n+1)^{n+1}}{n^n} \cdot \frac{n!}{(n+1)!} \\
&= \frac{(n+1)(n+1)^n}{n^n} \cdot \frac{n!}{(n+1)n!} = \left(\frac{n+1}{n}\right)^n = \left(1 + \frac{1}{n}\right)^n \geq 2
\end{aligned}
$$

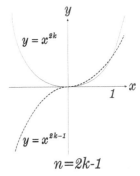

(Clearly the preceding quotient is greater than one. The reason it is greater than two, is outlined in Exercise 8.) Thus $e_{n+1} \geq 2e_n \geq e_n$, and the sequence $\{e_n\}$ is increasing.

(6) $f_n(x) = x^n$ for $n \geq 0$.

Solution The sequence $\{f_n(x)\}$ is the list of functions $1, \; x, \; x^2, \; x^3, \ldots$
Observe for $n \geq 0$:

$$\frac{f_{n+1}(x)}{f_n(x)} = \frac{x^{n+1}}{x^n} = x \qquad (5.2.3)$$

Figure 5.2.10
The Sequence $\{x^n\}$

- By (5.2.3), this sequence is decreasing if $0 \leq x \leq 1$. This is indicated in the first quadrants of the graphs in Figure 5.2.10 to the left of $x = 1$ where the solid curve $y = f_{n+1}(x)$ is below the dotted curve $y = f_n(x)$. For example, when $x = \frac{1}{2}$, the decreasing sequence $\left\{f_n\left(\frac{1}{2}\right)\right\}$ is: $1, \; \frac{1}{2}, \; \frac{1}{4}, \; \frac{1}{8}, \; \frac{1}{16}, \; \frac{1}{32}, \; \ldots$
- By (5.2.3), this sequence is increasing if $x \geq 1$. This is seen in the first quadrants of the graphs in Figure 5.2.10 to the right of $x = 1$ where the solid curve $y = f_{n+1}(x)$ is above the dotted curve $y = f_n(x)$. For example, when $x = 2$, the increasing sequence $\{f_n(2)\}$ is: $1, \; 2, \; 4, \; 8, \; 16, \; 32, \; \ldots$

- If $x < 0$, this sequence oscillates between positive and negative numb This is depicted by the left sides of the graphs in Figure 5.2.10. Note solid curve $y = f_{n+1}(x)$ is above the dotted curve $y = f_n(x)$ for n odd below the dotted curve $y = f_n(x)$ for n even. Hence $\{f_n(x)\}$ is neither creasing nor decreasing. For example, when $x = -2$, the sequence $\{f_n(-2)$
  $$1, \quad -2, \quad 4, \quad -8, \quad 16, \quad -32, \quad \ldots$$

(7) Let $g_n(x) = \frac{x^n}{n!}$ for $n \geq 0$.

Solution The sequence $\{g_n(x)\}$ is the list of functions:

$$1, \quad x, \quad \frac{x^2}{2}, \quad \frac{x^3}{6}, \quad \frac{x^4}{24}, \quad \ldots$$

Observe for $n \geq 0$:

$$\frac{g_{n+1}(x)}{g_n(x)} = \frac{x^{n+1}/(n+1)!}{x^n/n!} = \frac{x^{n+1}}{x^n} \cdot \frac{n!}{(n+1)!} = x \frac{n!}{(n+1)n!} = \frac{x}{n+1} \qquad (5.\!$$

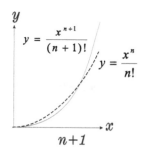

$$y = \frac{x^{n+1}}{(n+1)!}$$
$$y = \frac{x^n}{n!}$$

Figure 5.2.11
The Sequence $\left\{\frac{x^n}{n!}\right\}$

- If $0 \leq x \leq 1$, then $x \leq n+1$ and the sequence $\{g_n(x)\}$ is decreasing. Figure 5.2.11 where the solid curve $y = g_{n+1}(x)$ lies below the dotted cu $y = g_n(x)$.
- If $x > 1$, then (5.2.4) implies that the sequence increases at first and t decreases, i.e. the sequence is neither increasing nor decreasing. This is depic in Figure 5.2.11 where the solid curve $y = g_{n+1}(x)$ lies above the dotted cu $y = g_n(x)$ when $x > n+1$ and below the dotted curve $y = g_n(x)$ when $x < n+$
- If $x < 0$, the sequence $\{g_n(x)\}$ is neither increasing nor decreasing beca the $g_n(x)$ alternate between positive and negative numbers.

Bounded and Unbounded Sequences

We define a bounded sequence. Then we examine the sequences studied in the p ceding section and see that some of them are bounded while others are unbounded

We call a sequence $\{a_n\}$ bounded above if all the points a_n on the number l are to the left of some number U. We call $\{a_n\}$ bounded below if all the points on the number line are to the right of some number L. We call $\{a_n\}$ bounded if it is bounded above and below, i.e. we can find two numbers L and U such that all t points a_n on the number line are between L and U. See Figure 5.2.12. For examp the sequence $\left\{\frac{n}{n+1}\right\}$ of Example 5.2.1 is bounded because $0 \leq \frac{n}{n+1} \leq 1$, for $n \geq 1$, we can take $L = 0$ and $U = 1$.

$\ldots\ a_{k+8}\ a_{k+7}\ a_{k+6}\ a_{k+5}\ a_{k+4}\ a_{k+3}\ a_{k+2}\ a_{k+1}\ a_k\quad U$

A Sequence which is Bounded Above

$L\quad a_k\ a_{k+1}\ a_{k+2}\ a_{k+3}\ a_{k+4}\ a_{k+5}\ a_{k+6}\ a_{k+7}\ a_{k+8}\ \ldots$

A Sequence which is Bounded Below

$L\ \ldots\ a_{k+8}\ a_{k+7}\ a_{k+6}\ a_{k+5}\ a_{k+4}\ a_{k+3}\ a_{k+2}\ a_{k+1}\ a_k\ U$

A Bounded Sequence

$\ldots\ a_{k+8}\ a_{k+6}\ a_{k+4}\ a_{k+2}\ a_k\ a_{k+1}\ a_{k+3}\ a_{k+5}\ a_{k+7}\ \ldots$

An Unbounded Sequence

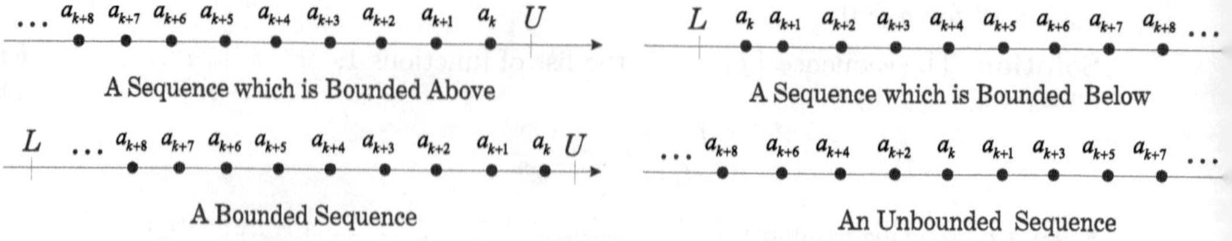

Figure 5.2.12

Definition 5.2.13 (a) *The sequence $\{a_n\}$ is bounded above if there is a number such that $a_n \leq U$ for all n. U is called an upper bound of the sequence.*
(b) *The sequence $\{a_n\}$ is bounded below if there is a number L such that $L \leq a_n$ f all n. L is called a lower bound of the sequence.*
(c) *The sequence $\{a_n\}$ is bounded if it is bounded below and bounded above.*

(d) *The sequence $\{a_n\}$ is unbounded if it is not bounded.*

Notes (1) Lower bounds are not unique. If the sequence $\{a_n\}$ has the lower bound L, then every number smaller than L is also a lower bound.
(2) Upper bounds are also not unique. If the sequence $\{a_n\}$ has the upper bound U, then every number larger than U is also an upper bound.
(3) The sequence $\{a_n\}$ is bounded if and only if there is a number B such that $|a_n| \leq B$ for all n. (If B exists, take $L = -B$ and $U = B$. If L and U exist, take B to be the larger of $|L|$, $|U|$.)
(4) An increasing sequence is bounded below by its first term, while a decreasing sequence is bounded above by its first term.

Examples 5.2.14 Determine whether each sequence is bounded or unbounded.

(1) $a_n = n^2$ for $n \geq 0$.

> **Solution** This increasing sequence, studied in Example 5.2.9 (1), is bounded below by its first term $L = 0$ and is not bounded above. Hence it is unbounded.

(2) $b_n = (-1)^n n^2$ for $n \geq 0$.

> **Solution** This sequence, studied in Example 5.2.9 (2), is neither bounded below nor bounded above. Hence it is unbounded.

(3) $c_n = \frac{n}{2^n}$ for $n \geq 1$.

> **Solution** This decreasing sequence, studied in Example 5.2.9 (3), is bounded above by $U = c_1 = \frac{1}{2}$. Since all the c_n are positive, this sequence is bounded below by $L = 0$. Thus the sequence $\{c_n\}$ is bounded.

(4) $d_n = \frac{n^2+1}{n!}$ for $n \geq 1$.

> **Solution** In Example 5.2.9 (4), we showed $\{d_n\}$ is decreasing for $n \geq 2$. Since $d_1 = 2$ and $d_2 = \frac{5}{2}$, this sequence is bounded above by $U = \frac{5}{2}$. Since the $d_n > 0$, this sequence is bounded below by $L = 0$. Therefore $\{d_n\}$ is bounded.

(5) $e_n = \frac{n^n}{n!}$ for $n \geq 1$.

> **Solution** Note this sequence of positive numbers is bounded below by $L = 0$. In Example 5.2.9 (5), we showed this sequence is increasing and satisfies $e_{n+1} \geq 2e_n$. Hence it is not bounded above and is unbounded.

(6) $f_n(x) = x^n$ for $n \geq 0$.

> **Solution** We studied this sequence in Example 5.2.9 (6).
> • Let $0 \leq x \leq 1$. For these x, we showed this sequence is decreasing. Hence it is bounded above by $U = x^0 = 1$ and is bounded below by $L = 0$. Thus this sequence is bounded.
> • Let $x > 1$. For these x, we showed this sequence is increasing. Thus it is bounded below by $L = x^0 = 1$. Since $\ln x^n = n \ln x$ and $\ln x > 0$, this sequence is not bounded above and is unbounded.
> • Let $-1 \leq x < 0$. Observe $|f_n(x)| = f_n(|x|)$. By the first case, the sequence $\{|f_n(x)|\}$ is bounded below by 0 and bounded above by 1. Hence the sequence $\{f_n(x)\}$ is bounded below by $L = -1$, bounded above by $U = 1$ and is bounded.
> • Let $x < -1$. Again, $|f_n(x)| = f_n(|x|)$. By the second case, the sequence $\{|f_n(x)|\}$ is not bounded above. Hence the sequence $\{f_n(x)\}$ is unbounded.

(7) Let $g_n(x) = \frac{x^n}{n!}$ for $n \geq 0$.

> **Solution** We studied this sequence in Example 5.2.9 (7).
> • Let $0 \leq x \leq 1$. We showed this sequence is decreasing. Hence it is bounded

above by $U = g_0(x) = 1$. Since $\{g_n(x)\}$ is a sequence of positive numbers,
bounded below by $L = 0$. Thus this sequence is bounded.
- Let $x > 1$. We showed this sequence increases at first and then decreases
  $n \geq x - 1$. Hence it is bounded above. Since $\{g_n(x)\}$ is a sequence of posi
  numbers, it is bounded below by $L = 0$. Thus this sequence is bounded.
- Let $x < 0$. Since $|g_n(x)| = g_n(|x|)$, the sequence $\{|g_n(x)|\}$ is bounded by
  first two cases. Hence the sequence $\{g_n(x)\}$ is bounded for $x < 0$.

Combine these three cases: the sequence $\{g_n(x)\}$ is bounded for all $x \in \Re$.

Sequences Defined by Functions

Often we can identify a sequence $\{a_n\}$ as the values of a continuous function: $a_n = f$
If we can establish a property of the function f, we may be able to conclude that
sequence $\{a_n\}$ also has this property.

Motivating Example 5.2.1 (continued) The sequence $\left\{\frac{n}{n+1}\right\}$, for $n \geq 1$, is defi
by the function $f(x) = \frac{x}{x+1}$ with domain $[1, \infty)$. The sequence $\left\{\frac{n}{n+1}\right\}$ corresponds
the heights of the points $(n, f(n))$ on the graph of f. See Figure 5.2.15.

By the quotient rule, $f'(x) = \frac{(1)(x+1) - x(1)}{(x+1)^2} = \frac{1}{(x+1)^2} \geq 0$. Hence $y = f(x)$ is
increasing function, i.e. $f(x)$ get larger as the number x increases. Hence $f(n) = \frac{1}{}$
get larger as the integer n increases, i.e. $\left\{\frac{n}{n+1}\right\}$ is an increasing sequence.

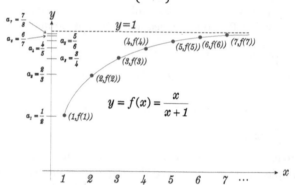

Figure 5.2.15 A Sequence Defined by a Function

The procedure above generalizes to a sequence $\{a_n\}$, $n \geq k$, defined by a functi
$y = f(x)$ with domain $[k, \infty)$, i.e. $f(n) = a_n$ for $n \geq k$. Assume f is differentiable.
- If $f'(x) \geq 0$, then f is increasing and the sequence $\{a_n\}$ must also be increasing
- If $f'(x) \leq 0$, then f is decreasing and the sequence $\{a_n\}$ must also be decreasing
- When the function f is neither increasing nor decreasing, it is possible that the
  sequence $\{a_n\}$ is increasing or decreasing. See Example 3 below.

Examples 5.2.16 Determine whether each sequence is increasing or decreasing.

(1) $a_n = \frac{\ln n}{n}$, for $n \geq 3$.

Solution This sequence is the list of numbers: $\frac{\ln 3}{3}$, $\frac{\ln 4}{4}$, $\frac{\ln 5}{5}$, $\frac{\ln 6}{6}$, $\ldots$
Let $f(x) = \frac{\ln x}{x}$. Then $f(n) = a_n$ for $n \geq 3$. By the quotient rule:

$$f'(x) = \frac{\frac{1}{x} \cdot x - \ln x \cdot 1}{x^2} = \frac{1 - \ln x}{x^2} < 0 \text{ for } x \geq 3.$$

Therefore the function f is decreasing for $x \geq 3$. Hence $a_n = f(n) \geq f(n+1) = a_{n+1}$ for $n \geq 3$, and the sequence $\{a_n\}$ is decreasing.

(2) $b_n = \frac{n}{2^n}$, for $n \geq 2$.

Solution Define $g(x) = \frac{x}{2^x}$. Then $g(n) = b_n$ for $n \geq 2$. By the quotient rule:

$$g'(x) = \frac{1 \cdot 2^x - x(\ln 2)2^x}{(2^x)^2} = \frac{1 - x \ln 2}{2^x} \leq 0 \text{ for } x \geq 2.$$

Therefore the the function g is decreasing for $x \geq 2$. Hence $b_n = g(n) \geq g(n+1) = b_{n+1}$ for $n \geq 2$, and the sequence $\{b_n\}$ is decreasing.

(3) $c_n = \sin \pi n$, for $n \geq 0$.

Solution Let $h(x) = \sin \pi x$. Then $c_n = h(n)$ for $n \geq 0$. Observe the function h is neither increasing nor decreasing. See Figure 5.2.20. The sequence $\{c_n\}$ is the constant sequence of zeros since $\sin \pi n = 0$ for every integer n. Thus this sequence is both increasing and decreasing. $\square$

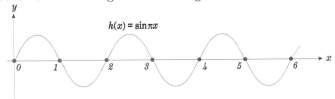

Figure 5.2.20 The Sequence $\{\sin \pi n\}$ and the Function $h(x) = \sin \pi x$

Motivating Example 5.2.1 (continued) Return to the sequence $\{a_n\} = \left\{ \frac{n}{n+1} \right\}$, $n \geq 1$, defined by the function $f(x) = \frac{x}{x+1}$ with domain $[1, \infty)$. Observe the increasing function $f(x)$ has the line $y = 1$ as a horizontal asymptote on the right. See Figure 5.2.15. Hence $f(x) < 1$ for $x \geq 1$. In particular, all the $f(n) = \frac{n}{n+1}$ are less than one. Hence the increasing sequence $\left\{ \frac{n}{n+1} \right\}$ has its first term $L = \frac{1}{2}$ has a lower bound and $U = 1$ as an upper bound. Thus the sequence $\left\{ \frac{n}{n+1} \right\}$ is bounded. $\square$

The preceding observations generalize to any sequence defined by a function. We call a function *bounded above* if its graph lies below a horizontal line $y = U$. See Figure 5.2.17. We call a function *bounded below* if its graph lies above a horizontal line $y = L$. See Figure 5.2.18. We call a function *bounded* if its graph lies between two horizontal lines $y = L$ and $y = U$. See Figure 5.2.19. A function which is not bounded is called *unbounded*. For example, $f(x) = x \sin \frac{1}{x}$, for $x > 0$ is unbounded. See Figure 5.2.21. Let $\{a_n\}$, $n \geq k$, be a sequence which is defined by the function f with domain $[k, \infty)$, i.e. $a_n = f(n)$ for $n \geq k$.

- If f is bounded above, then $\{a_n\}$ is also bounded above.
- If f is bounded below, then $\{a_n\}$ is also bounded below.
- If f is bounded, then $\{a_n\}$ is also bounded.
- If f is unbounded, it is possible that $\{a_n\}$ is bounded. See Example 3 below.

Examples 5.2.22 Determine whether each sequence is bounded or unbounded.

(1) $a_n = \frac{\ln n}{n}$, for $n \geq 3$.

Solution This sequence is defined by the function $f(x) = \frac{\ln x}{x}$ with domain $[3, \infty)$. In Example 5.2.16 (1) we showed this function, with positive values, is decreasing. Hence f is bounded above by $f(3) = \frac{\ln 3}{3}$ and is bounded below by zero. Hence the sequence $\left\{ \frac{\ln n}{n} \right\}$ is bounded above by $\frac{\ln 3}{3}$ and bounded below by zero. Thus $\{a_n\}$ is a bounded sequence.

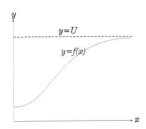

Figure 5.2.17
f is Bounded Above

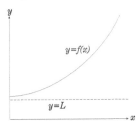

Figure 5.2.18
f is Bounded Below

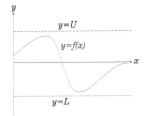

Figure 5.2.19
f is Bounded

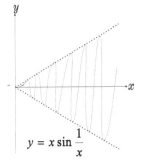

Figure 5.2.21
f is Unbounded

(2) $b_n = \frac{e^n}{n}$, for $n \geq 1$.

Solution This sequence is defined by the function $g(x) = \frac{e^x}{x}$ with dom $[1, \infty)$. By the quotient rule $g'(x) = \frac{xe^x - e^x}{x^2} = \frac{e^x(x-1)}{x^2} \geq 0$. Hence $g(x)$ i increasing function, and $\{b_n\}$ is an increasing sequence which is bounded be by $b_1 = e$. By L'Hôpital's Rule, $\lim_{x \to \infty} \frac{e^x}{x} = \lim_{x \to \infty} \frac{e^x}{1} = \infty$. Hence the seque $\{b_n\}$ is not bounded above and is unbounded.

(3) $c_n = n \sin \pi n$, for $n \geq 0$.

Solution This sequence is defined by the function $h(x) = x \sin \pi x$ with dom $[0, \infty)$. This function, whose graph is depicted in Figure 5.2.21, is unbounc Nevertheless, the sequence of $c_n = h(n) = n \sin \pi n = 0$ is bounded.

Summary

The reader should be able to determine whether a sequence of numbers or monom is increasing, decreasing, bounded above or bounded below. When the sequence is fined by a function she should know how to determine these properties of the seque from the corresponding properties of the function.

Basic Exercises

1. Describe each infinite list as a sequence.
    (a) 1, 8, 27, 64, 125, ...   **(b)** -1, $\sqrt{2}$, $-\sqrt{3}$, 2, $-\sqrt{5}$, ...
    (c) 1, 1, $\frac{1}{2}$, $\frac{1}{6}$, $\frac{1}{24}$, ...   **(d)** -1, 2, -24, 720, $-40,320$, ...
    (e) $\frac{x}{2}$, $\frac{x^2}{3}$, $\frac{x^3}{4}$, $\frac{x^4}{5}$, ...   **(f)** 1, $-\frac{1}{x}$, $\frac{2}{x^2}$, $-\frac{6}{x^3}$, $\frac{24}{x^4}$, ...

2. Write the first five terms of each sequence where $n \geq 1$.
    (a) $\left\{\frac{1}{n^2}\right\}$ **(b)** $\{(-1)^n(3n-1)\}$ **(c)** $\left\{\frac{1}{n!}\right\}$ **(d)** $\left\{\frac{n!}{(2n)!}\right\}$
    (e) $\left\{\frac{\cos(n\pi)}{n}\right\}$ **(f)** $\left\{\frac{x^n}{3n}\right\}$ **(g)** $\left\{\sin\frac{x^{2n}}{n}\right\}$ **(h)** $\left\{\frac{\ln(x^n)}{n!}\right\}$

3. Find the function that defines each sequence.
    (a) $\{n^3 - 2n + 6\}$ for $n \geq 1$ **(b)** $\left\{\frac{e^n}{\ln n}\right\}$ for $n \geq 2$ **(c)** $\left\{\frac{\sin n}{\arctan n}\right\}$ for $n \geq 1$
    (d) $\{\sqrt{n^2 - n - 6}\}$ for $n \geq 3$ **(e)** $\{\sqrt[n]{n}\}$ for $n \geq 1$ **(f)** $\left\{\left(1 + \frac{1}{n}\right)^n\right\}$ for $n \geq$

4. Write the first five terms of each sequence of functions, with $n \geq 0$, at the specifi value of x.
    (a) $\{x^{2n}\}$ at $x = -1$ **(b)** $\left\{\frac{x^n}{n!}\right\}$ at $x = 3$ **(c)** $\{n^x\}$ at $x = 2$
    (d) $\left\{\frac{\ln(n+1)}{x^n}\right\}$ at $x = -2$ **(e)** $\left\{\frac{x^{2n}}{(2n)!}\right\}$ at $x = -1$ **(f)** $\left\{\frac{x^{3/2}}{n+1}\right\}$ at $x = 4$

5. Verify each formula for $n \geq 2$.
    (a) $\frac{(2n+2)!}{(2n)!} = 4n^2 + 6n + 2$ **(b)** $\frac{n!}{(2n)!} = \frac{1}{(2^n)(2n-1)(2n-3)\cdots(3)(1)}$
    (c) $\frac{(n!)^2}{n^2} = [(n-1)!]^2$ **(d)** $\frac{(n-1)!}{k!(n-k-1)!} + \frac{(n-1)!}{(k-1)!(n-k)!} = \frac{n!}{k!(n-k)!}$ for $1 \leq k <$

6. Determine whether each sequence, for $n \geq 1$, is increasing, decreasing, bound below or bounded above.
    (a) $\{\sqrt{n}\}$ **(b)** $\{(-1)^n\sqrt{n}\}$ **(c)** $\left\{\frac{\sin n}{n}\right\}$
    (d) $\{n\sin(1/n)\}$ **(e)** $\left\{2^{-\frac{n}{n^2+1}}\right\}$ **(f)** $\left\{\sin\frac{n\pi}{2}\right\}$
    (g) $\{\arctan(n^2 + 1)\}$ **(h)** $\left\{(-1)^{2n}\frac{3n+1}{e^n+1}\right\}$ **(i)** $\left\{\ln\left(\frac{n}{3n+1}\right)\right\}$
    (j) $\left\{(-1)^{n^2}\frac{n^3+1}{5n^3+2n+1}\right\}$ **(k)** $\left\{\frac{n\ln n}{n^2+1}\right\}$ **(l)** $\left\{\frac{(2n)!}{(n!)^2}\right\}$
    (m) $\left\{\frac{n^2}{7n^2+3}\cos n\pi\right\}$ **(n)** $\{e^{\tan n}\sin n\pi\}$ **(o)** $\left\{\frac{1+(-1)^n}{\sqrt{n^3+5}}\right\}$

7. Determine the values of x for which each sequence of functions, for $n \geq 1$, is increasing, decreasing, bounded below or bounded above.

(a) $\left\{ \frac{x^n}{n} \right\}$ (b) $\left\{ \frac{x^{2n+1}}{n!} \right\}$ (c) $\left\{ \sqrt{1 + \frac{x^2}{n}} \right\}$ (d) $\left\{ \sqrt[3]{1 - \frac{x^2}{n}} \right\}$

(e) $\left\{ \ln \left(\frac{1+x^2}{1+x^{2n}} \right) \right\}$ (f) $\left\{ \frac{3x^4 + 5x^2 + 9}{n! + 1} \right\}$ (g) $\left\{ \sqrt[n]{x^2} \right\}$ (h) $\{ \arctan(x^n) \}$

(i) $\left\{ \sin \left(\frac{x}{n} \right) \right\}$ (j) $\{ e^{nx} \}$ (k) $\left\{ \frac{x^{2n}}{(2n)!} \right\}$ (l) $\left\{ \frac{x^n}{(2n)!} \right\}$

8. Let $f(x) = \left(1 + \frac{1}{x} \right)^x$. Verify each statement.

(a) $f'(x) = \left(1 + \frac{1}{x} \right)^x \left[\ln \left(1 + \frac{1}{x} \right) - \frac{1}{x+1} \right]$ (b) $D \left[\ln \left(1 + \frac{1}{x} \right) - \frac{1}{x+1} \right] \leq 0$ for $x \geq 1$.

(c) $\lim_{x \to \infty} \left[\ln \left(1 + \frac{1}{x} \right) - \frac{1}{x+1} \right] = 0$ (d) $f'(x) \geq 0$ for $x \geq 1$. (e) $f(x) \geq 2$ for $x \geq 1$.

Challenging Problem

1. Determine the values of x for which the sequence $\left\{ \left(1 + \frac{x}{n} \right)^n \right\}$ is increasing, and determine the values of x for which this sequence is decreasing.

5.3 Convergence of Sequences

In this section, we study limits of sequences. In the first subsection, we define and compute the limits of simple sequences. In the second subsection, we derive limit properties and use them to compute the limits of more complicated sequences. In the last subsection, we interpret the completeness of $\Re$ in terms of sequences and use this viewpoint to prove a strong form of the Intermediate Value Theorem.

Limits of Sequences

We define the limit of a sequence and compute simple limits directly from this definition. When the sequence $\{a_n\}$ is defined by the function g we use the limit of $g(x)$, as x goes to infinity, to determine the limit of the sequence $\{a_n\}$.

Recall the function f has limit L, as x goes to infinity, if the values of $f(x)$ get close to L as x gets large. This phenomenon also occurs with sequences.

Motivating Example 5.3.1 In Section 5.2, we showed $\left\{ \frac{n}{n+1} \right\}$ is an increasing sequence which is bounded above by one. Note $1 - \frac{n}{n+1} = \frac{n+1}{n+1} - \frac{n}{n+1} = \frac{1}{n+1}$. Hence as n increases, the values of $\frac{n}{n+1}$ get close to one. Thus it is reasonable to say that the sequence $\left\{ \frac{n}{n+1} \right\}$ has limit one. □

In general, we want a precise definition which says that the sequence $\{a_n\}$ has limit L if the values of a_n get close to L as n increases. We model our definition on the definition of $\lim_{x \to \infty} f(x) = L$. By Definition 1.5.20(a), $\lim_{x \to \infty} f(x) = L$ if given any interval $(L - \epsilon, L + \epsilon)$ of numbers close to L, the values of $f(x)$ lie in this interval for x sufficiently large. Analogously, we define the limit of the sequence $\{a_n\}$ to be L if given any interval $(L - \epsilon, L + \epsilon)$ of numbers close to L, the numbers a_n lie in this interval for n sufficiently large. A sequence that has a limit is called *convergent* while a sequence with no limit is called *divergent*.

Definition 5.3.2 (a) *The sequence $\{a_n\}$ converges to the number L if for every $\epsilon > 0$, there is a corresponding positive integer N_ϵ such that $a_n \in (L - \epsilon, L + \epsilon)$ when*

$n \geq N_\epsilon$. L is called the *limit of the sequence*, and we write $\lim_{n \to \infty} a_n = L$.

(b) When the sequence $\{a_n\}$ does not converge, we say it *diverges*.

A convergent sequence of numbers with limit L corresponds to a sequence of po on the real number line which eventually enter every interval $(L - \epsilon, L + \epsilon)$ and n leave. See Figure 5.3.3 where $N_\epsilon = 5$. Observe that a convergent sequence only h finite number of terms which are not in an interval $(L - \epsilon, L + \epsilon)$. Thus a converg sequence is bounded. Equivalently, an unbounded sequence must diverge.

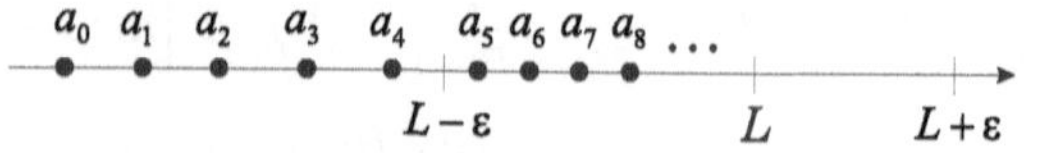

Figure 5.3.3 A Convergent Sequence

Examples 5.3.4 Determine whether each sequence converges or diverges. Find the limit of each convergent sequence.

(1) Let $a_n = \frac{1}{n}$ for $n \geq 1$.

Solution This sequence is the list of decreasing numbers $1, \frac{1}{2}, \frac{1}{3}, \frac{1}{4}, \ldots, \frac{1}{n}$, which approach zero. To establish that this sequence has limit zero, let interval $(-\epsilon, \epsilon)$ be given. Select a positive integer $N_\epsilon > \frac{1}{\epsilon}$. Then $\frac{1}{N_\epsilon} < \epsilon$. $n \geq N_\epsilon$, then $0 - \epsilon < 0 < \frac{1}{n} \leq \frac{1}{N_\epsilon} < 0 + \epsilon$. Hence the sequence $\{\frac{1}{n}\}$ conver with $\lim_{n \to \infty} \frac{1}{n} = 0$.

(2) Let $b_n = \frac{n}{n+1}$ for $n \geq 1$.

Solution In Example 5.3.1 we saw that this sequence of increasing numb approaches one. To establish that this sequence has limit one, say the inter $(1 - \epsilon, 1 + \epsilon)$ is given. Select a positive integer $N_\epsilon > \frac{1}{\epsilon}$. If $n \geq N_\epsilon$, th $\frac{1}{n+1} < \frac{1}{n} \leq \frac{1}{N_\epsilon} < \epsilon$, and

$$1 + \epsilon > 1 > \frac{n}{n+1} = 1 - \frac{1}{n+1} > 1 - \epsilon.$$

Hence the sequence $\left\{\frac{n}{n+1}\right\}$ converges with $\lim_{n \to \infty} \frac{n}{n+1} = 1$.

(3) Let $c_n = n^2$ for $n \geq 0$.

Solution Since this sequence is unbounded, it must diverge.

(4) Let $d_n = (-1)^n$ for $n \geq 0$.

Solution This sequence alternates between $+1$ and -1. We show that neith of these numbers is the limit of this sequence. Take $\epsilon = \frac{1}{2}$. For any large the odd terms $d_{2n+1} = -1$ of this sequence, with $2n + 1 \geq N$, are not in t interval $(1 - \epsilon, 1 + \epsilon) = \left(\frac{1}{2}, \frac{3}{2}\right)$. Hence $\{d_n\}$ does not have limit 1. Also t even terms $d_{2n} = 1$ of this sequence, with $2n \geq N$, are not in the interv $(-1 - \epsilon, -1 + \epsilon) = \left(-\frac{3}{2}, -\frac{1}{2}\right)$. Hence $\{d_n\}$ does not have limit -1. Therefore t sequence $\{d_n\}$ diverges.

Consider the sequence $\{a_n\}$, with $n \geq k$, defined by the function f with doma $[k, \infty)$, i.e. $f(n) = a_n$ for $n \geq k$. Equivalently, $\{a_n\}$ is the sequence of the heights of tl points $(n, f(n))$ on the graph of f. We may be able to use L'Hôpital's Rule to compu $\lim_{x \to \infty} f(x) = L$. Then the graph of $y = f(x)$ has $y = L$ as a horizontal asymptote the right, and the heights of the points $(n, f(n))$ approach L as n increases. Th

$\lim\limits_{n\to\infty} a_n = \lim\limits_{n\to\infty} f(n) = L$ and the sequence $\{a_n\}$ converges. See Figure 5.3.5.

Warning! Example 3 below shows when $\lim\limits_{x\to\infty} f(x)$ does not exist, it is still possible that the sequence $\{a_n\}$ converges.

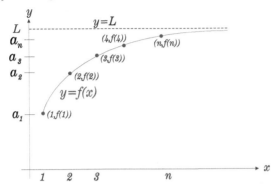

Figure 5.3.5 A Sequence Defined by a Function

Examples 5.3.6 In Examples 1–3, determine whether the sequence converges or diverges. Find the limit of each convergent sequence.

(1) $a_n = \frac{\ln n}{n}$, for $n \geq 3$.

Solution This sequence is defined by the function $f(x) = \frac{\ln x}{x}$, i.e. $f(n) = \frac{\ln n}{n} = a_n$ for $n \geq 3$. By L'Hôpital's Rule:

$$\lim_{x\to\infty} h(x) = \lim_{x\to\infty} \frac{D(\ln x)}{D(x)} = \lim_{x\to\infty} \frac{1/x}{1} = 0.$$

Therefore the sequence $\{a_n\}$ converges with limit zero.

(2) $b_n = \frac{n}{2^n}$, for $n \geq 1$.

Solution This sequence is defined by the function $g(x) = \frac{x}{2^x}$ for $x \geq 1$, i.e. $g(n) = \frac{n}{2^n} = b_n$ for $n \geq 1$. By L'Hôpital's Rule:

$$\lim_{x\to\infty} f(x) = \lim_{x\to\infty} \frac{D(x)}{D(2^x)} = \lim_{x\to\infty} \frac{1}{2^x \ln 2} = 0.$$

Hence the sequence $\{b_n\}$ converges with limit zero.

(3) $c_n = \sin \pi n$, for $n \geq 0$.

Solution This sequence is defined by the function $h(x) = \sin \pi x$ for $x \geq 0$, i.e. $h(n) = \sin \pi n = c_n$ for $n \geq 0$. From the graph of $y = h(x)$ in Figure 5.2.20, it is clear that $y = h(x)$ does not have a horizontal asymptote on the right, i.e. $\lim\limits_{x\to\infty} h(x)$ does not exist. Note the sequence $\{c_n\}$ is the constant sequence of zeros since $\sin \pi n = 0$ for every integer n. Hence $\{c_n\}$ converges with limit zero despite the fact that $\lim\limits_{x\to\infty} h(x)$ does not exist.

(4) Let $k_n(x) = x^n$, for $n \geq 0$. Determine the values of x for which $\{k_n(x)\}$ is a convergent sequence and the values of x for which $\{k_n(x)\}$ is a divergent sequence. When $\{k_n(x)\}$ converges, find its limit.

Solution • If $x = 1$, then $\{k_n(1)\}$ is the constant sequence of ones which converges with limit one.
• If $x = -1$, then $\{k_n(-1)\} = \{(-1)^n\}$ which we showed to be divergent in Example 5.3.4 (4).

- If $x < -1$ or $x > 1$, we showed in Example 5.2.14 (6) that this sequen⋯ unbounded. Hence it diverges for these values of x.
- Let $-1 < x < 1$. Think of x as a constant. Let $b = |x|$, and define $p(u) =$ for $u \geq 0$. Then $p(n) = |x^n| = |k_n(x)|$ for $n \geq 0$. In constructing the gr⋯ of $y = p(u) = b^u$, for $0 < b < 1$, in Figure 4.2.19 of Section 4.2, we sho⋯ $\lim_{u \to \infty} p(u) = 0$. Hence the sequence $\{k_n(x)\}$ converges to zero for $-1 < x <$

Summary The sequence $\{k_n(x)\}$ diverges for $x \leq -1$ or $x > 1$, converge⋯ one for $x = 1$ and converges to zero for $-1 < x < 1$.

Properties of Limits

Recall the strategy used in Section 1.6 to compute limits of functions. We only ⋯ the definition of the limit to compute the limits of the simplest functions. Then⋯ used limit properties to compute other limits. Analogously, limits of sequences ⋯ similar properties to limits of functions. These limit properties are used to comp⋯ the limits of complicated sequences.

In Example 5.3.4 (4) we studied the sequence $\{(-1)^n\}$. The even terms have ⋯ as a limit while the odd terms have -1 as a limit. Therefore, there is no one num⋯ which all the terms of this sequence are approaching. Hence the sequence $\{(-1)^n\}$ d⋯ not converge. The argument of that example generalizes to show that a converg⋯ sequence can not have more than one number as a limit. We leave the details of proof for Exercise 60.

Proposition 5.3.7 *The limit of a convergent sequence is unique.*

Recall the Pinching Theorem for limits of functions from Section 1.6 $f(x) \leq g(x) \leq h(x)$ for $x \geq k$ and $\lim_{x \to \infty} f(x) = \lim_{x \to \infty} h(x) = L$, then $\lim_{x \to \infty} g(x) =$ Limits of sequences have the analagous property.

Theorem 5.3.8 (Pinching Theorem) *Let $\{a_n\}$, $\{c_n\}$, for $n \geq k$, be convergent* *quences with* $\lim_{n \to \infty} a_n = \lim_{n \to \infty} c_n = L$. *If* $a_n \leq b_n \leq c_n$ *for* $n \geq k$, *then* $\{b_n\}$ *i⋯ convergent sequence with* $\lim_{n \to \infty} b_n = L$.

Proof Let $\epsilon > 0$. Find positive integers N_1 and N_2 so that if $n \geq N_1$ then a_n⋯ $(L - \epsilon, L + \epsilon)$, and if $n \geq N_2$ then $c_n \in (L - \epsilon, L + \epsilon)$. Let N be the larger of N_1 a⋯ N_2. If $n \geq N$ then a_n and c_n are both in $(L - \epsilon, L + \epsilon)$. Since $a_n \leq b_n \leq c_n$, t⋯ number b_n must also be in $(L - \epsilon, L + \epsilon)$. Thus the sequence $\{b_n\}$ converges to L.

The Pinching Theorem is used to demonstrate the convergence of a complicat⋯ sequence by trapping it between two simpler sequences with the same limit.

Examples 5.3.9 Determine whether each sequence converges or diverges. Find the limit of each convergent sequence.

(1) $a_n = \frac{n^2+1}{n!}$ for $n \geq 1$.

Solution Observe that for $n \geq 3$:

$$(n-1)(n-2)(n-3) = n^3 - 6n^2 + 11n - 6 = (n^2 + 1) + (n^3 - 7n^2 + 11n - 7$$

For $n \geq 7$ we have $n^3 - 7n^2 + 11n - 7 = (n^3 - 7n^2) + (11n - 7) \geq 0$ becau⋯ $n^3 \geq 7n^2$ and $11n \geq 7$. Hence for $n \geq 7$:

$$n^2 + 1 \ \leq \ (n-1)(n-2)(n-3) \leq (n-1)!$$

$$\text{and } 0 \leq a_n \leq \frac{(n-1)!}{n!} = \frac{(n-1)!}{n(n-1)!} = \frac{1}{n}.$$

Since $\lim\limits_{n\to\infty} 0 = \lim\limits_{n\to\infty} \frac{1}{n} = 0$, it follows from the Pinching Theorem that the sequence $\{a_n\}$ converges with limit zero.

(2) $b_n(x) = \frac{x^n}{n!}$ for $n \geq 0$.

Solution Let $c_m(x) = \frac{|x|^m}{m!}$ for $m \geq 0$. We showed in Example 5.2.9 (7) that

$$0 \leq c_m(x) = \frac{|x|}{m} c_{m-1}(x) \tag{5.3.1}$$

for $m \geq 1$. If N be an integer greater than $|x|$. For $n \geq N$ apply (5.3.1) with $m = n$, $m = n-1$, . . . , $m = N+2$, $m = N+1$:

$$0 \leq c_n(x) = \frac{|x|}{n} c_{n-1}(x) = \frac{|x|^2}{n(n-1)} c_{n-2}(x) = \cdots = \frac{|x|^{n-N}}{n(n-1)\cdots(N+1)} c_N(x)$$

$$0 \leq c_n(x) \leq \left(\frac{|x|}{N+1}\right)^{n-N} c_N(x). \tag{5.3.2}$$

Since $\frac{|x|}{N+1} < 1$, we have

$$\lim_{n\to\infty}\left[\left(\frac{|x|}{N+1}\right)^{n-N} c_N(x)\right] = c_N(x)\left(\lim_{n\to\infty}\frac{|x|}{N+1}\right)^{n-N} = 0. \tag{5.3.3}$$

By (5.3.2) and (5.3.3), the Pinching Theorem says that the sequence $\{c_n(x)\}$, for $n \geq N$, converges with limit 0. Hence $\{c_n(x)\}$, for $n \geq 0$, converges with limit 0. Note $b_n(x)$ is either $c_n(x)$ or $-c_n(x)$. Hence $-c_n(x) \leq b_n(x) \leq c_n(x)$. Apply the Pinching Theorem again to conclude that the sequence $\{b_n(x)\}$ also converges with limit zero. $\square$

Limits of sequences have the same algebraic properties as the limits of functions which were established in Section 1.6:

    (a) the limit of a sum is the sum of the limits;

    (b) the limit of a product is the product of the limits;

    (c) the limit of a quotient is the quotient of the limits.

We list these properties in the next proposition and leave the proof as Problem 1.

Proposition 5.3.10 *Assume $\{a_n\}$ and $\{b_n\}$ are two convergent sequences, for $n \geq k$, with $\lim\limits_{n\to\infty} a_n = A$ and $\lim\limits_{n\to\infty} b_n = B$. Then*

- *$\{a_n + b_n\}$ is a convergent sequence with limit $A + B$;*
- *$\{c\, a_n\}$ is a convergent sequence with limit cA;*
- *$\{a_n b_n\}$ is a convergent sequence with limit AB;*
- *$\left\{\frac{a_n}{b_n}\right\}$ converges with limit $\frac{A}{B}$ when $B \neq 0$ and $b_n \neq 0$ for $n \geq k$.*

Example 5.3.11 The sequence $\left\{\frac{2n}{n+1}\right\}$ converges with limit 2, and the sequence $\left\{\frac{6n^2}{2n^2+5}\right\}$ converges with limit 3. Therefore:

- the sequence $\left\{\frac{2n}{n+1} + \frac{6n^2}{2n^2+5}\right\} = \left\{\frac{10n^3+6n^2+10n}{(n+1)(2n^2+5)}\right\}$ converges with limit $2 + 3 = 5$;

- the sequence $\left\{ 7 \cdot \frac{2n}{n+1} \right\} = \left\{ \frac{14n}{n+1} \right\}$ converges with limit $7 \cdot 2 = 14$;

- the sequence $\left\{ \frac{2n}{n+1} \cdot \frac{6n^2}{2n^2+5} \right\} = \left\{ \frac{12n^3}{(n+1)(2n^2+5)} \right\}$ converges with limit $2 \cdot 3 = 6$;

- the sequence $\left\{ \frac{2n}{n+1} \div \frac{6n^2}{2n^2+5} \right\} = \left\{ \frac{2n^2+5}{3n^2+3n} \right\}$ converges with limit $2 \div 3 = \frac{2}{3}$.

We evaluate the limit of a composite function, when the function f is continu as $\lim_{x\to\infty} f(g(x)) = f\left[\lim_{x\to\infty} g(x) \right]$. Sequences have the analogous property:

$$\lim_{n\to\infty} f(a_n) = f\left(\lim_{n\to\infty} a_n \right). \tag{5.}$$

The notation $t \in \Re$ indicates that t is a number, not one of the symbols ∞ or $-\infty$

Proposition 5.3.12 *Let f be a continuous function whose domain is an interva Let each of A, L be a number or one of the symbols ∞, $-\infty$. Assume one of the following three conditions is true.*

    (1) $A \in I$ *with* $f(A) = L$.

    (2) A *is the left endpoint of I with* $\lim_{x\to A^+} f(x) = L$.

    (3) A *is the right endpoint of I with* $\lim_{x\to A^-} f(x) = L$.

Assume the interval I contains all the numbers of the sequence $\{a_n\}$.

(a) *If $\lim_{n\to\infty} a_n = A$ and $L \in \Re$, then $\{f(a_n)\}$ converges with limit L.*

(b) *If $\lim_{n\to\infty} f(a_n) = L$, $A \in \Re$ and f is one-to-one, then $\{a_n\}$ converges with limit*

Notes **(1)** When condition 1 applies, this proposition is exactly (5.3.4). W condition 2 or 3 applies, the limits in (5.3.4) can have value ∞ or $-\infty$.

(2) Condition 2 applies when $I = (a, b)$ with $A = a$ and $L = \lim_{x\to a^+} f(x)$. We m have $a = -\infty$ or f may have a vertical asymptote at $x = a$.

(3) Condition 3 applies when $I = (a, b)$ with $A = b$ and $L = \lim_{x\to b^-} f(x)$. We m have $b = \infty$ or f may have a vertical asymptote at $x = b$.

Proof **(a)** Condition (1) implies $\lim_{x\to A} f(x) = L$. Let $\epsilon > 0$. By condition ((2) or (3) find $\delta > 0$ such that $f[I \cap (A - \delta, A + \delta)] \subset (L - \epsilon, L + \epsilon)$. Since sequence $\{a_n\}$ converges to A, there is a positive integer N_δ such that if $n \geq N_\delta$ th $a_n \in (A - \delta, A + \delta)$ and $f(a_n) \in (L - \epsilon, L + \epsilon)$. Hence $\{f(a_n)\}$ converges to L. **(b)** Apply (a) to the sequence $\{f(a_n)\}$ and the function f^{-1}.

We use Proposition 5.3.12(a) to evaluate the limit of the sequence $\{f(a_n)\}$ as f the limit of the simpler sequence $\{a_n\}$. We use Proposition 5.3.12(b) with $f(x) = \ln$ to evaluate the limit of a sequence $\{b_n\}$ which contains exponentials. The new sequer $\{\ln(b_n)\}$ may be simpler than the original sequence $\{b_n\}$. In this case, we determi the limit of the original sequence from the limit of the new one.

Examples 5.3.13 Determine whether each sequence, for $n \geq 1$, in Examples 1 converges or diverges. If it converges, find its limit.

(1) $a_n = \cos\frac{\pi n}{n+1}$.

    Solution By Example 5.3.4(2), the sequence $\left\{ \frac{n}{n+1} \right\}$ converges with limit or

    By Proposition 5.3.10, the sequence $\left\{ \frac{\pi n}{n+1} \right\}$ converges. By Proposition 5.3.12(

the sequence $\{a_n\} = \left\{ \cos \frac{\pi n}{n+1} \right\}$ converges and

$$\lim_{n \to \infty} \cos \frac{\pi n}{n+1} = \cos \left(\lim_{n \to \infty} \frac{\pi n}{n+1} \right) = \cos \left(\pi \lim_{n \to \infty} \frac{n}{n+1} \right) = \cos \pi(1) = -1.$$

(2) $b_n = \frac{1}{n^p}$ with $p > 0$.

Solution Apply Proposition 5.3.12(b) with the one–to–one function $f(x) = \ln x$:

$$\{f(b_n)\} = \left\{ \ln \left(\frac{1}{n^p} \right) \right\} = \{\ln (n^{-p})\} = \{-p \ln n\}$$

is a convergent sequence with limit $-\infty$ which also equals $\lim_{x \to 0^+} \ln x$. Hence Condition 2 is satisfied with $A = 0$ and $L = -\infty$. By Proposition 5.3.12(b), the sequence $\{b_n\}$ converges with limit zero.

(3) $c_n = \sqrt[n]{n}$.

Solution Apply Proposition 5.3.12(b) with the one–to–one function $f(x) = \ln x$:

$$\{f(c_n)\} = \{\ln \sqrt[n]{n}\} = \{\ln n^{1/n}\} = \left\{ \frac{\ln n}{n} \right\}.$$

This is a convergent sequence with limit zero because $\lim_{x \to \infty} \frac{\ln x}{x} = 0$ by L'Hôpital's Rule. Thus $\{f(c_n)\}$ converges with limit $0 = \ln 1 = f(1)$. Hence Condition 1 is satisfied with $A = 1$ and $L = 0$. By Proposition 5.3.12(b), the sequence $\{\sqrt[n]{n}\}$ converges with limit 1.

(4) $d_n(x) = \cos \frac{\pi x}{n^2+1}$ for $x \in \Re$.

Solution Apply Proposition 5.3.12(a) with $f(u) = \cos u$. For any $x \in \Re$:

$$\lim_{n \to \infty} d_n(x) = \lim_{n \to \infty} \cos \frac{\pi x}{n^2 + 1} = \cos \left(\lim_{n \to \infty} \frac{\pi x}{n^2 + 1} \right) = \cos 0 = 1.$$

Hence the sequence $\{d_n(x)\}$ converges with limit one for all $x \in \Re$.

(5) $e_n(x) = \sqrt[n]{x}$ for $x \geq 0$.

Solution If $x = 0$, then $\{e_n(0)\}$ is the constant sequence of zeros which converges to zero. Assume $x > 0$. Apply Proposition 5.3.12(b) with the one–to–one function $f(u) = \ln u$. For any $x > 0$, the sequence

$$\{f(d_n(x))\} = \{\ln \sqrt[n]{x}\} = \left\{ \frac{\ln x}{n} \right\}$$

converges with limit $0 = \ln 1 = f(1)$. Hence Condition 1 holds with $A = 1$ and $L = 0$. By Prop. 5.3.12(b), the sequence $\{\sqrt[n]{x}\}$ converges with limit 1 when $x > 0$.

(6) $g_n(x) = \left(1 + \frac{x}{n}\right)^n$ for $x \in \Re$.

Solution Apply Proposition 5.3.12(b) with the one–to–one function $f(u) = \ln u$:

$$\{f(g_n(x))\} = \left\{ \ln \left(1 + \frac{x}{n}\right)^n \right\} = \left\{ n \ln \left(1 + \frac{x}{n}\right) \right\}.$$

Note x is a fixed number. Let $t \geq 1$ be a variable. Then

$$\lim_{t \to \infty} t \ln \left(1 + \frac{x}{t}\right) = \lim_{t \to \infty} \frac{\ln \left(1 + \frac{x}{t}\right)}{t^{-1}}$$

By L'Hôpital's Rule:

$$\lim_{t\to\infty} t\ln\left(1+\frac{x}{t}\right) = \lim_{t\to\infty}\frac{\frac{d}{dt}\ln\left(1+\frac{x}{t}\right)}{\frac{d}{dt}\left(t^{-1}\right)} = \lim_{t\to\infty}\frac{\frac{-x/t^2}{1+\frac{x}{t}}}{-t^{-2}} = \lim_{t\to\infty}\frac{x}{1+\frac{x}{t}} = x.$$

Hence the sequence $\{f(g_n(x))\}$ converges with limit $x = \ln e^x = f(e^x)$. 1 Condition 1 is satisfied with $A = e^x$ and $L = x$. By Proposition 5.3.12(b), sequence $\{g_n(x)\}$ converges with limit e^x.

(7)   The sequence $\{(-1)^n\}$ diverges. Nevertheless, when $f(x) = x^2$, the seque $\{f((-1)^n)\} = \{1\}$ converges to one. Even though Condition 1 is satisfied $A = L = 1$, this example does not violate Proposition 5.3.12(b) because f is a one–to–one function.

Completeness of $\Re$

The real number line $\Re$ is complete in the sense that it has no holes. We state property in terms of sequences. Then we use this viewpoint to prove a corollar the Intermediate Value Theorem which describes the range of a continuous funct whose domain is an interval.

Motivating Example 5.3.14 There is a number $\sqrt{2}$ whose square equals two. 1

$$\sqrt{2} = 1.d_1 d_2 d_3 \ldots$$

be its infinite decimal expansion. Let a_n be the first $n + 1$ digits of this decimal:

$$a_n = 1.d_1 d_2 d_3 \ldots d_n$$

- Each $a_n = 1 + \sum_{k=1}^{n}\frac{d_k}{10^k}$ is a rational number.
- Note $a_{n+1} = a_n + d_{n+1} \times 10^{-(n+1)}$. Hence $\{a_n\}$ is an increasing sequence.
- Since $(a_n)^2 < 2$ for $n \geq 1$, the sequence $\{a_n\}$ is bounded above.
- Every number greater than or equal to $\sqrt{2}$ is an upper bound of this sequence.
- Since $|a_n - \sqrt{2}| < 10^{-n}$, the smallest upper bound of this sequence is $\sqrt{2}$.
- The sequence $\{a_n\}$ converges with limit $\sqrt{2}$.

Conclusion The increasing sequence of rational numbers $\{a_n\}$ is bounded above a converges with limit the smallest of all its upper bounds $\sqrt{2}$.

The construction of the preceding example applies to any number $r \in \Re$. That the decimal expansion of r produces an increasing sequence of rational numbers whi is bounded above and has limit the smallest of all its upper bounds r.

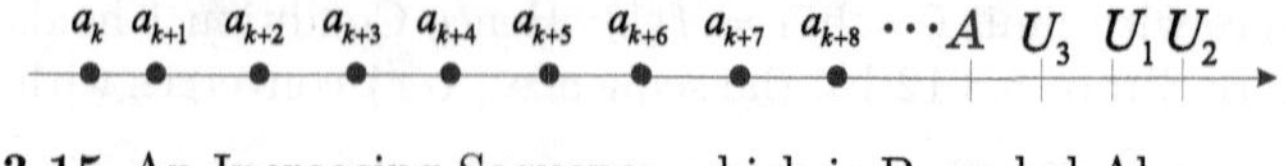

Figure 5.3.15 An Increasing Sequence which is Bounded Above

In general, let $\{a_n\}$ be an increasing sequence which is bounded above. Then $\{a$ has many upper bounds. For example, if U is one upper bound then every numb greater than U is also an upper bound. However, amongst all the upper bounds the must be a smallest one A, the *least upper bound* of the sequence. If not, there wou be a hole in the real number line to the right of all the points a_n and to the left of the upper bounds U of the sequence. It can be shown that the real number line $\Re$ h no holes. This property of the real numbers is called *completeness*. See Figure 5.3.1

Theorem 5.3.16 (Completeness Theorem) *Every increasing sequence of real numbers which is bounded above, has a least upper bound.*

The preceding result was stated as Theorem 1.10.34 and proved in Section 1.10.

Corollary 5.3.17 *Let $\{a_n\}$ be an increasing sequence which is bounded above. Then $\{a_n\}$ converges with limit its least upper bound.*

Proof By the Completeness Theorem, this sequence has a least upper bound A. Given $\epsilon > 0$, $A - \epsilon$ is not an upper bound of the sequence $\{a_n\}$. Hence there is an $a_N > A - \epsilon$. Then $a_n \geq a_N > A - \epsilon$ for $n \geq N$. Since A is an upper bound of the sequence, $a_n \leq A < A + \epsilon$ for $n \geq N$. Thus $a_n \in (A - \epsilon, A + \epsilon)$ for $n \geq N$, and the sequence $\{a_n\}$ converges with limit A. □

Similarly a decreasing sequence $\{b_n\}$ which is bounded below has many lower bounds L. However, amongst all the lower bounds there is a greatest one B which is to the left of all the points b_n and to the right of all the lower bounds of the sequence. B is called the *greatest lower bound* of the sequence. See Figure 5.3.18.

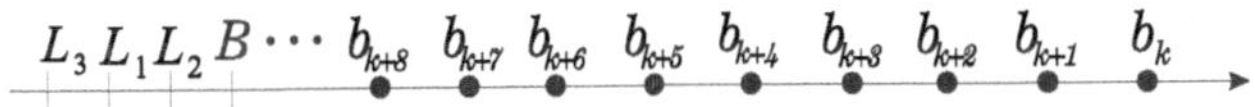

$$L_3\, L_1 L_2\, B \cdots b_{k+8}\ b_{k+7}\ b_{k+6}\ b_{k+5}\ b_{k+4}\ b_{k+3}\ b_{k+2}\ b_{k+1}\ b_k$$

Figure 5.3.18 A Decreasing Sequence which is Bounded Below

Corollary 5.3.19 *Let $\{b_n\}$ be a decreasing sequence which is bounded below. Then the sequence $\{b_n\}$ converges with limit its greatest lower bound.*

The proof of this corollary is analogous to the proof of Corollary 5.3.17 and is given as Exercise 68.

Examples 5.3.20 (1) Find the least upper bound and greatest lower bound of the sequence of $a_n = (-1)^n \frac{n}{n+1}$, for $n \geq 0$.

Solution The even terms of this sequence are $\left\{ \frac{2n}{2n+1} \right\}$, for $n \geq 0$. This is an increasing sequence with limit one which is its least upper bound. Since the odd terms of $\{a_n\}$ are negative, the least upper bound of $\{a_n\}$ is $+1$. The odd terms of $\{a_n\}$ are $\left\{ -\frac{2n-1}{2n} \right\}$, for $n \geq 1$. This is a decreasing sequence with limit -1 which is its greatest lower bound. Since the even terms of $\{a_n\}$ are positive, the greatest lower bound of $\{a_n\}$ is -1.

(2) For each $x \in \Re$, determine whether the sequence of $b_n(x) = \frac{x}{2^n}$, for $n \geq 0$, converges. When it converges, find its limit.

Solution Observe that $b_{n+1}(x) = \frac{1}{2} b_n(x)$ for all $x \in \Re$. Let $x \geq 0$. Then the sequence $\{b_n(x)\}$ is decreasing and is bounded below by zero. Therefore $\{b_n(x)\}$ converges to its greatest lower bound which is zero. Let $x \leq 0$. Then $\{b_n(x)\}$ is increasing and is bounded above by zero. Therefore $\{b_n(x)\}$ converges to its least upper bound which is zero. Thus in all cases, $\lim_{n \to \infty} b_n(x) = 0$. □

In Section 1.7 we stated a reformulation of the Intermediate Value Theorem. We prove this reformulation using the concepts of greatest lower bound and least upper bound.

Corollary 1.7.31 *Let f be a continuous function whose domain is an interval I. Then the range R of f is also an interval.*

Proof Let a be the left endpoint of I, and let b be the right endpoint of I.

Let $a_n \in I$, $n \geq 1$, be a sequence with limit a, and let $b_n \in I$, $n \geq 1$, be a sequ$\bullet$ with limit b. See Figure 5.3.21.

Apply the Maximum Value Theorem: let m_n be the minimum value of f on $[a_n,$ and let M_n be the maximum value of f on $[a_n, b_n]$.

Note I is the union of the intervals $[a_n, b_n]$, and R is the union of the intervals $[m_n,$ I If $\{m_n\}$ is bounded below, let m be the greatest lower bound of the sequence $\{m$ If $\{m_n\}$ is not bounded below, let $m = -\infty$.

If $\{M_n\}$ is bounded above, let M be the least upper bound of the sequence $\{M_n$ If $\{M_n\}$ is not bounded above, let $M = \infty$.

We show $R \subset [m, M]$. Let $f(x) \in R$ with $x \in I$. Then $x \in [a_n, b_n]$ for som Hence $m \leq m_n \leq f(x) \leq M_n \leq M$, and $f(x) \in [m, M]$. Thus $R \subset [m, M]$.

We show $(m, M) \subset R$. Let $k \in (m, M)$. Since $m < k$, it follows that k is n$\bullet$ lower bound of $\{m_n\}$. Hence there is an m_s with $m_s < k$. Since $k < M$, it foll that k is not an upper bound of $\{M_n\}$. Hence there is an M_t with $k < M_t$. T $m \leq m_s < k < M_t \leq M$ and $m_s, M_t \in R$. By the Intermediate Value Theor $k \in R$. Therefore, $(m, M) \subset R$.

We have shown $(m, M) \subset R \subset [m, M]$. Thus R is an interval with left endpoin$\bullet$ and right endpoint M.

Figure 5.3.21 The Interval I

Summary

The reader should be able to determine whether a sequence of numbers or function$\bullet$ convergent or divergent. If it converges, she should know how to find its limit dire$\bullet$ or by one of several methods. If the sequence is defined by a function f she should f the limit of the sequence by applying L'Hôpital's rule to f. She should also be a$\bullet$ to evaluate limits of sequences by the Pinching Theorem, by applying a function by using the completeness property of $\Re$. In addition, she should know the follow$\bullet$ limits which we established in the examples of this section.

$$\lim_{n \to \infty} \frac{\ln n}{n} = 0 \tag{5.3}$$

$$\lim_{n \to \infty} \frac{1}{n^p} = 0 \quad \text{for } p > 0 \tag{5.3}$$

$$\lim_{n \to \infty} \sqrt[n]{n} = 1 \tag{5.3}$$

$$\lim_{n \to \infty} x^n = 0 \quad \text{for } |x| < 1 \tag{5.3}$$

$$\lim_{n \to \infty} \sqrt[n]{x} = 1 \quad \text{for } x > 0 \tag{5.3}$$

$$\lim_{n \to \infty} \frac{x^n}{n!} = 0 \quad \text{for } x \in \Re \tag{5.3.1}$$

$$\lim_{n \to \infty} \left(1 + \frac{x}{n}\right)^n = e^x \quad \text{for } x \in \Re \tag{5.3.1}$$

Basic Exercises

1. Use the definition of convergence to determine whether each sequence, with $n \geq$ converges or diverges. If it converges, find its limit.

(a) $\left\{\frac{1}{n^2}\right\}$ (b) $\left\{\frac{2n}{3n+1}\right\}$ (c) $\{\sqrt{n}\}$ (d) $\{2^{-n}\}$

(e) $\left\{(-1)^{n^2}\right\}$ (f) $\left\{(-1)^{n(n+1)}\right\}$ (g) $\left\{(-1)^n \frac{1}{n}\right\}$ (h) $\left\{(-1)^n \frac{n}{n+1}\right\}$

2. Show each sequence, with $n \geq 2$, is defined by a function. Find the limit of the function as x approaches infinity. Then determine whether the sequence converges or diverges. If it converges, find its limit.

(a) $\left\{ \frac{n}{e^n} \right\}$ (b) $\left\{ \frac{\ln n}{e^n} \right\}$ (c) $\left\{ \frac{\arctan(2n+1)}{\arctan(5n+3)} \right\}$ (d) $\left\{ \frac{n}{\ln n} \right\}$

(e) $\{ \cos \pi n \}$ (f) $\{ \tan \pi n \}$ (g) $\left\{ \frac{n^2}{n^2-1} - \frac{n}{2n+1} \right\}$ (h) $\left\{ n \sin \frac{1}{n} \right\}$

3. Use the Pinching Theorem to determine whether each sequence converges or diverges. If it converges, find its limit.

(a) $\left\{ \frac{\sin n}{n} \right\}$ (b) $\left\{ (-1)^{n(n+1)/2} \frac{n}{n^2+1} \right\}$ (c) $\left\{ \frac{\arctan n}{n} \right\}$ (d) $\left\{ \frac{x^n}{(2n+1)!} \right\}$ (e) $\left\{ (-1)^n \frac{x^n}{(2n)!} \right\}$

4. Assume $\lim_{n \to \infty} a_n = 5$ and $\lim_{n \to \infty} b_n = 10$. Find the limit of each sequence.

(a) $\{ 3a_n - 4b_n \}$ (b) $\{ a_n^3 b_n^2 \}$ (c) $\left\{ \frac{a_n}{b_n^3} \right\}$ (d) $\{ \sqrt{a_n b_n} \}$ (e) $\{ \ln(b_n - a_n) \}$

Determine whether each sequence in exercises 5 to 31 converges or diverges. If the sequence converges, find its limit.

5. $\{ \ln n \}$ **6.** $\{ (-1)^n \ln n \}$ **7.** $\left\{ \frac{\cos n}{n} \right\}$

8. $\{ n \tan(1/n) \}$ **9.** $\left\{ 2^{-\frac{n}{n^2+1}} \right\}$ **10.** $\left\{ \sin \frac{n\pi}{2} \right\}$

11. $\{ \arctan(n^2+1) \}$ **12.** $\left\{ (-1)^n \frac{3n+1}{e^n+1} \right\}$ **13.** $\left\{ \ln \left(\frac{n}{3n+1} \right) \right\}$

14. $\left\{ (-1)^{n^2} \frac{n^3+1}{5n^3+2n+1} \right\}$ **15.** $\left\{ \frac{n \ln n}{n^2+1} \right\}$ **16.** $\left\{ \frac{(2n)!}{(n!)^2} \right\}$

17. $\left\{ \frac{n^2}{7n^2+3} \cos n\pi \right\}$ **18.** $\{ e^{\tan n} \sin n\pi \}$ **19.** $\left\{ \frac{1+(-1)^n}{\sqrt{n^3+5}} \right\}$

20. $\left\{ \frac{n^3-2n+7}{n!} \right\}$ **21.** $\left\{ \frac{\cos \pi (n^2+3n-5)}{n^2} \right\}$ **22.** $\left\{ \cos \frac{\pi(n^2+3n-5)}{n^2} \right\}$

23. $\left\{ \frac{\ln n}{n!} \right\}$ **24.** $\left\{ \frac{n}{5n+2} - \frac{n}{3n-4} \right\}$ **25.** $\left\{ (-1)^n \frac{n^3+4n-6}{2n^3+8n^2+7} \right\}$

26. $\{ \arctan \sqrt[n]{n} \}$ **27.** $\{ \arctan(n^n) \}$ **28.** $\left\{ ne^{1/n} \sin \frac{\pi}{n} \right\}$

29. $\left\{ \frac{1+(-1)^n}{\cos(1/n)} \right\}$ **30.** $\{ \sqrt{n^2+1} - \sqrt{n^2-1} \}$ **31.** $\left\{ \frac{e^n+e^{-n}}{e^n-e^{-n}} \right\}$

For each sequence in exercises 32 to 55, determine the values of x for which it converges, and find the limit of the sequence for these values of x.

32. $\left\{ \frac{x^n}{n} \right\}$ **33.** $\left\{ \frac{x^{2n+1}}{n!} \right\}$ **34.** $\left\{ \sqrt{1 + \frac{x^2}{n}} \right\}$ **35.** $\left\{ \sqrt[3]{1 - \frac{x^2}{n}} \right\}$

36. $\left\{ \ln \frac{1+x^2}{1+x^{2n}} \right\}$ **37.** $\left\{ \frac{3x^4+5x^2+9}{n!+1} \right\}$ **38.** $\{ \sqrt[n]{x^2} \}$ **39.** $\{ \arctan(x^n) \}$

40. $\left\{ \sin \frac{x}{n} \right\}$ **41.** $\{ e^{nx} \}$ **42.** $\left\{ \frac{x^n}{n^2+1} \right\}$ **43.** $\{ e^{x/n} \}$

44. $\left\{ \cos \frac{x^n}{n!} \right\}$ **45.** $\{ x^{-n} \}$ **46.** $\{ \arctan(nx) \}$ **47.** $\left\{ \frac{\sin \pi x}{n} \right\}$

48. $\left\{ \ln \frac{x^{2n}}{2^n+1} \right\}$ **49.** $\{ n^x \}$ **50.** $\left\{ \frac{x^n}{x^{2n}+1} \right\}$ **51.** $\left\{ \frac{x^n n!}{n^2+1} \right\}$

52. $\left\{ \left(1 + \frac{x}{2n} \right)^n \right\}$ **53.** $\left\{ \left(2 + \frac{x}{n} \right)^n \right\}$ **54.** $\left\{ (-1)^n \frac{x^{2n}}{(2n)!} \right\}$ **55.** $\left\{ \frac{x^{2n}}{n!} \right\}$

56. What does Proposition 5.3.12(b) say about each sequence $\{ a_n \}$?

(a) $\lim_{n \to \infty} e^{a_n} = 3$ (b) $\lim_{n \to \infty} a_n^2 = 9$ (c) $\lim_{n \to \infty} \sqrt{a_n} = 4$

(d) $\lim_{n \to \infty} \ln a_n = 2$ (e) $\lim_{n \to \infty} a_n^3 = 64$ (f) $\lim_{n \to \infty} \cos a_n = 1$

57. Find the least upper bound and greatest lower bound of each sequence where $n \geq 1$.

(a) $\left\{ \frac{3^n}{3^n+1} \right\}$ (b) $\left\{ \frac{1-2^n}{2^n+1} \right\}$ (c) $\left\{ \frac{n^2+1}{2n-1} \right\}$ (d) $\left\{ (-1)^n \frac{n}{2n-1} \right\}$

(e) $\left\{ \frac{\ln n}{n} \right\}$ (f) $\left\{ \frac{n}{2^{n/2}} \right\}$ (g) $\left\{ \sin \frac{\pi n}{2} \right\}$ (h) $\{ (-1)^n \arctan n \}$

58. A sequence $\{ a_n \}$ is defined *recursively* if we are given its first p terms $a_1, \ldots, a_p$

and a formula for computing the nth term from the preceding p terms. Write the six terms of each recursive sequence.

(a) $a_1 = 2$ and $a_n = 2a_{n-1} + 1$ for $n \geq 2$.

(b) $a_1 = -1$ and $a_n = -\frac{a_{n-1}}{n}$ for $n \geq 2$.

(c) $a_1 = 0$ and $a_n = a_{n-1}^2 + 1$ for $n \geq 2$.

(d) $a_1 = a_2 = 1$ and $a_n = a_{n-1} + a_{n-2}$ for $n \geq 3$.

(e) $a_1 = 1$, $a_2 = 2$ and $a_n = a_{n-1}a_{n-2}$ for $n \geq 3$.

(f) $a_1 = 0$, $a_2 = 1$ and $a_n = 2a_{n-1} - a_{n-2} + 3$ for $n \geq 3$.

(g) $a_1 = 1$, $a_2 = 2$, $a_3 = 3$ and $a_n = a_{n-1} + a_{n-2} - a_{n-3}$ for $n \geq 4$.

59. In the notation of the preceding exercise, determine whether each recursi defined sequence converges or diverges. If the sequence converges find its limit.

(a) $a_1 = 1$ and $a_{n+1} = \frac{a_n}{n}$ for $n \geq 1$.   (b) $a_1 = 3$ and $a_{n+1} = \sqrt[n]{a_n}$ for $n \geq 1$.

(c) $a_1 = 4$ and $a_{n+1} = \frac{na_n}{n+1}$ for $n \geq 1$.   (d) $a_1 = 2$ and $a_{n+1} = \frac{1}{a_n}$ for $n \geq 1$.

(e) $a_1 = 1$, $a_2 = 3$ and $a_{n+1} = \frac{a_{n-1}+a_n}{n}$ for $n \geq 2$.

(f) $a_1 = 2$, $a_2 = 1$ and $a_{n+1} = \frac{a_{n-1}}{a_n}$ for $n \geq 2$.

60. Prove Proposition 5.3.7: the limit of a convergent sequence is unique.

61. In the notation of Proposition 5.3.12, show if $\lim\limits_{n \to \infty} a_n = A$ and $L = \pm\infty$, t the sequence $\{f(a_n)\}$ diverges.

62. Let $\{a_n\}$ be an increasing sequence which is bounded above with least up bound L. Show that for every $\epsilon > 0$ there is some $a_n \in (L - \epsilon, L]$.

63. Let $\{a_n\}$ be an increasing sequence which is bounded above with least up bound L. If $L = a_n$ for some n, what can be said about the set S of all the points of this sequence?

64. Give an example of a sequence $\{a_n\}$ which converges with limit L such that neither the least upper bound nor the greatest lower bound of the sequence.

65. Give an example of a continuous function whose range is not an interval.

66. Give an example of a continuous function whose domain is not an interval a whose range is an interval.

67. (a) Show that the set of rational numbers is not complete. That is, give example of a specific increasing sequence $\{a_n\}$ of rational numbers that is bound above but does not have a least upper bound which is a rational number.
(b) Show that the set of positive real numbers is not complete.

68. Prove Corollary 5.3.19: a decreasing sequence $\{b_n\}$ which is bounded bel converges with limit its greatest lower bound.

Challenging Problems

1. Prove Proposition 5.3.10.

2. Show the sequence $\{a_n\}$ converges if and only if the sequences $\{a_{2n}\}$ and $\{a_{2n-}$ both converge and have the same limit.

3. Determine the values of x for which the sequence $\left\{\frac{x^n(2n)!}{(n!)^2}\right\}$ converges. Find t limit of this sequence for these values of x.

4. A subsequence of $\{a_n\}$ is a sequence $a_{k_1}, \ldots, a_{k_n}, \ldots$ with $k_1 < k_2 < \cdots < k_n$.
(a) Show a convergent sequence has a convergent subsequence which is either incre ing or decreasing.
(b) Show a bounded sequence has a convergent subsequence which is either increasi or decreasing.

5.4   Infinite Sums

This section is devoted to defining infinite sums (also called infinite series), establishing their properties and presenting examples. Infinite sums, unlike finite sums, are not always defined. We call those that are defined *convergent* and those that are not defined *divergent*. In the first subsection, we define convergent infinite sums and present examples. The second subsection is devoted to the nth Term Test which identifies certain infinite sums as divergent. The third subsection studies geometric series in detail. In particular, we integrate geometric power series to produce new series, including infinite series with sums $\ln 2$ and $\frac{\pi}{4}$. In the fourth subsection, we show convergent infinite series fall into two categories. *Absolutely convergent* series do not depend on cancellation between their positive and negative summands for convergence. On the other hand, *conditionally convergent* series only converge because of the cancellation between their positive and negative summands. To make our notation concise, we denote sums with the sigma notation introduced in Section 3.2.

Convergent Series

The basic idea to define the infinite sum

$$S = \sum_{k=0}^{\infty} a_k = a_0 + a_1 + a_2 + \cdots + a_k + \cdots$$

is that S is approximated by the finite sums

$$S_n = \sum_{k=0}^{n} a_k = a_0 + a_1 + a_2 + \cdots + a_{n-1} + a_n.$$

The larger the value of n, the better S_n should approximate S.

Motivating Example 5.4.1 Consider the infinite sum

$$S = \sum_{k=0}^{\infty} \frac{9}{10^k} = 9 + \frac{9}{10} + \frac{9}{100} + \frac{9}{1000} + \cdots + \frac{9}{10^n} + \cdots$$

S is approximated by

$$S_n = \sum_{k=0}^{\infty} \frac{9}{10^k} = 9 + \frac{9}{10} + \frac{9}{100} + \frac{9}{1000} + \cdots + \frac{9}{10^n}$$

for n large. Write S_n as a decimal:

$$S_n = 9.99\ldots 9$$

with n nines to the right of the decimal point. The larger the value of n, the closer S_n is to ten. In particular, $10 - S_n = 0.00\ldots 01$, with $n-1$ zeroes to the right of the decimal point. That is, $S_n = 10 - 0.00\ldots 01 = 10 - 10^{-n} = 10 - \frac{1}{10^n}$. Hence

$$S = \lim_{n \to \infty} S_n = \lim_{n \to \infty} \left(10 - \frac{1}{10^n} \right) = 10.$$

Thus the infinite sum $\sum_{k=0}^{\infty} \frac{9}{10^k}$ should be defined to have value 10.   □

Depending on the example, we start labelling the summands a_k of an infinite sum at different values of k, say at $k = p$. In the example above, we started with $a_0 = 9$ and $p = 0$. If $a_k = \frac{1}{\ln k}$ we might start with $a_2 = \frac{1}{\ln 2}$ and $p = 2$. Let S_n denote the sum of the terms from a_p through a_n. We define the sum of all the a_k to be the limit S of the S_n, as n goes to infinity. Equivalently, S is the limit of the sequence $\{S_n\}$.

Definition 5.4.2 *Let p be an integer with $a_k \in \mathfrak{R}$ for $k \geq p$.*

(a) *The nth partial sum, for $n \geq p$, is defined as the finite sum*

$$S_n = \sum_{k=p}^{n} a_k = a_p + a_{p+1} + \cdots + a_{n-1} + a_n.$$

(b) *If the sequence $\{S_n\}$ converges with limit $S \in \mathfrak{R}$, then the infinite series*

$$\sum_{k=p}^{\infty} a_k = a_p + a_{p+1} + a_{p+2} + \cdots$$

is said to converge with sum S. Write $S = \sum_{k=p}^{\infty} a_k$.

(c) *If the sequence $\{S_n\}$ diverges, then the infinite series $\sum_{k=p}^{\infty} a_k$ is said to dive*

Note $S_{n+1} = S_n + a_{n+1}$. If all the $a_k > 0$, the partial sums $\{S_n\}$ form an increasing quence. Either this sequence is bounded above with limit $S \in \mathfrak{R}$ or this sequence is bounded with limit ∞. That is, $\sum_{k=1}^{\infty} a_k$ converges or it diverges with $\sum_{k=1}^{\infty} a_k =$ When some of the a_k are positive and others are negative, $\sum_{k=1}^{\infty} a_k$ may conve If the series diverges there are three possibilities: $\lim_{n \to \infty} S_n = \infty$, $\lim_{n \to \infty} S_n = -\infty$ $\lim_{n \to \infty} S_n$ does not exist. The examples below illustrate these phenomena.

Examples 5.4.3 (1) Determine whether the series $\sum_{k=1}^{\infty} \dfrac{1}{2^k}$ converges or diverges

Solution Observe that

$$S_n = \sum_{k=1}^{n} \frac{1}{2^k} = \frac{1}{2} + \frac{1}{4} + \cdots + \frac{1}{2^{n-1}} + \frac{1}{2^n}.$$

Therefore

$$\begin{aligned} S_n + \frac{1}{2^n} &= \frac{1}{2} + \frac{1}{4} + \cdots + \frac{1}{2^{n-3}} + \frac{1}{2^{n-2}} + \frac{1}{2^{n-1}} + \frac{2}{2^n} \\ &= \frac{1}{2} + \frac{1}{4} + \cdots + \frac{1}{2^{n-3}} + \frac{1}{2^{n-2}} + \frac{2}{2^{n-1}} \\ &= \frac{1}{2} + \frac{1}{4} + \cdots + \frac{1}{2^{n-3}} + \frac{2}{2^{n-2}} = \cdots = \frac{2}{2} = 1. \end{aligned}$$

The sequence of partial sums $\{S_n\} = \{1 - \frac{1}{2^n}\}$ converges with limit one. He the infinite series $\sum_{k=1}^{\infty} \frac{1}{2^k}$ converges with sum one:

$$\sum_{k=1}^{\infty} \frac{1}{2^k} = 1.$$

This is an example of a series whose terms are all positive where the increas sequence of its partial sums converges to the number 1.

(2) Determine whether the telescoping series $\sum_{k=1}^{\infty} \dfrac{1}{k(k+1)}$ converges or diverges.

Solution Observe that

$$\frac{1}{k(k+1)} = \frac{1}{k} - \frac{1}{k+1}.$$

Therefore

$$\begin{aligned} S_n &= \frac{1}{1(2)} + \frac{1}{2(3)} + \frac{1}{3(4)} + \cdots + \frac{1}{(n-1)n} + \frac{1}{n(n+1)} \\ &= \left[\frac{1}{1} - \frac{1}{2}\right] + \left[\frac{1}{2} - \frac{1}{3}\right] + \left[\frac{1}{3} - \frac{1}{4}\right] + \cdots + \left[\frac{1}{n-1} - \frac{1}{n}\right] + \left[\frac{1}{n} - \frac{1}{n+1}\right] \\ &= 1 - \frac{1}{n+1}. \end{aligned}$$

Notice how all the summands in S_n, except for the first and last ones, appear twice: once with a plus sign and once with a minus sign. The resulting cancellation reduces the long sum for S_n into a short one, just as a long pocket telescope of several sections compactifies by sliding the sections together. The sequence of partial sums $\{S_n\} = \left\{1 - \frac{1}{n+1}\right\}$ converges with limit one, and the series $\sum_{k=1}^{\infty} \frac{1}{k(k+1)}$ converges with sum one:

$$\sum_{k=1}^{\infty} \frac{1}{k(k+1)} = 1.$$

This is another example of a series whose terms are all positive where the increasing sequence of its partial sums converges to the number 1.

(3) Determine whether the series $\sum_{k=0}^{\infty} k^2$ converges or diverges.

Solution Observe that

$$S_n = 0^2 + 1^2 + 2^2 + \cdots + (n-1)^2 + n^2 \geq n^2 \geq n.$$

Therefore the sequence of partial sums $\{S_n\}$ is unbounded and diverges. Hence the series $\sum_{k=0}^{\infty} k^2$ diverges. This is an example of a series whose terms are all positive where the increasing sequence of its partial sums diverges with limit ∞.

(4) Determine whether the harmonic series $\sum_{k=1}^{\infty} \frac{1}{k}$ converges or diverges.

Solution We show $S_{2^n} \geq \frac{n+1}{2}$. Hence the increasing sequence of partial sums $\{S_n\}$ is unbounded and diverges. Thus the harmonic series diverges. Observe

$$
\begin{aligned}
S_1 &= 1 \geq \frac{1}{2} \\
S_2 &= 1 + \frac{1}{2} \geq \frac{1}{2} + \frac{1}{2} = \frac{2}{2} \\
S_4 &= 1 + \frac{1}{2} + \left[\frac{1}{3} + \frac{1}{4}\right] \geq \frac{1}{2} + \frac{1}{2} + 2\left(\frac{1}{4}\right) = \frac{1}{2} + \frac{1}{2} + \frac{1}{2} = \frac{3}{2} \\
S_8 &= S_4 + \left[\frac{1}{5} + \frac{1}{6} + \frac{1}{7} + \frac{1}{8}\right] \geq \frac{3}{2} + 4\left(\frac{1}{8}\right) = \frac{3}{2} + \frac{1}{2} = \frac{4}{2}.
\end{aligned}
$$

Similarly S_{2^n} equals $S_{2^{n-1}}$ plus 2^{n-1} terms greater than or equal to $\frac{1}{2^n}$. Hence

$$S_{2^n} \geq S_{2^{n-1}} + 2^{n-1}\left(\frac{1}{2^n}\right) = S_{2^{n-1}} + \frac{1}{2}.$$

It follows that $S_{2^n} \geq \frac{n+1}{2}$, for all $n \geq 0$, as asserted. This is another example of a series whose terms are all positive where the increasing sequence of its partial sums diverges with limit ∞.

(5) Let $a_{2k} = -\frac{1}{2^k}$ and $a_{2k+1} = k$ for $k \geq 1$. Determine whether the series $\sum_{k=2}^{\infty} a_k$ converges or diverges.

Solution By Example 1, the series of the even summands $\sum_{k=1}^{\infty} -\frac{1}{2^k}$ converges with sum -1. Hence the limit of the partial sums of this series is the sum of its odd summands minus one. However, the series of the odd summands $\sum_{k=1}^{\infty} k$ diverges with sum ∞. Hence the given series diverges. It is an example of a series with positive and negative summands whose partial sums have limit ∞.

(6) Let $b_{2k} = \frac{1}{2^k}$ and $b_{2k+1} = -k$ for $k \geq 0$. Determine whether the series $\sum_{k=}^{\infty}$ converges or diverges.

Solution This series is the negative of the series in the preceding example therefore diverges. This is an example of a series with positive and nega summands whose partial sums have limit $-\infty$.

(7) Determine whether the series $\sum_{k=1}^{\infty}(-1)^k$ converges or diverges.

Solution Observe the even summands of this series equal one, while the summands of this series equal minus one. Therefore S_{2n} is the sum of n ones n minus ones, i.e. $S_{2n} = 0$. On the other hand, S_{2n+1} is the sum of n ones $n+1$ minus ones, i.e. $S_{2n+1} = -1$. Thus the sequence $\{S_n\}$ oscillates betw -1 and 0, hence diverges. Therefore the series $\sum_{k=1}^{\infty}(-1)^k$ diverges. Th an example of an infinite series with positive and negative summands where limit of its partial sums does not exist.

n^{th} Term Test

In the subject of infinite series, the word *test* indicates a method to determine whe a given series converges or diverges. The nth Term Test identifies certain series t diverge. We also show that sums of convergent series satisfy a linear property.

Comparison of Examples 5.4.3 (1) and (4) is interesting. Both series are sim in the sense that the summands are decreasing positive numbers whose limit is z Nevertheless, one series converges while the other diverges. On the other hand, if summands of a series do not have limit zero, as in Examples 5.4.3 (3), (5), (6), then the nth Term Test says the series diverges.

Proposition 5.4.4 (n^{th} Term Test) *If* $\lim_{n \to \infty} a_n \neq 0$, *the series* $\sum_{k=p}^{\infty} a_k$ *diverges*

Proof We show that if the series $\sum_{k=p}^{\infty} b_k$ converges with sum S, then $\lim_{n \to \infty} b_n =$ This statement is equivalent to this proposition. Observe the summands of

$$S_n = (b_1 + b_2 + \cdots + b_{n-1}) + b_n$$

consist of the summands of S_{n-1} as well as b_n, i.e. $S_n = S_{n-1} + b_n$. Then

$$\lim_{n \to \infty} b_n = \lim_{n \to \infty} (S_n - S_{n-1}) = \lim_{n \to \infty} S_n - \lim_{n \to \infty} S_{n-1} = S - S = 0.$$

Warning! The nth Term Test can only be used to conclude that certain ser diverge. It can never be used to conclude that a series converges. Moreover, th are some series, such as the harmonic series, which diverge even though they have property that their n^{th} term has limit zero.

Examples 5.4.5 (1) What does the nth Term Test say about the series $\sum_{k=1}^{\infty} \dfrac{k}{\sqrt{4k^2}}$

Solution Since $\lim_{k \to \infty} \dfrac{k}{\sqrt{4k^2 + 1}} = \dfrac{1}{2}$ which is not zero, the nth Term Test sa that the series $\sum_{k=1}^{\infty} \frac{k}{\sqrt{4k^2+1}}$ diverges.

(2) What does the nth Term Test say about the series $\sum_{k=1}^{\infty} \dfrac{1}{\sqrt{4k^2 + 1}}$?

Solution Since $\lim_{k \to \infty} \dfrac{1}{\sqrt{4k^2 + 1}} = 0$, the nth Term Test tells us nothing abo the series $\sum_{k=1}^{\infty} \frac{1}{\sqrt{4k^2+1}}$. In fact, we will see in Section 9 that this series diverg

(3) What does the nth Term Test say about the series $\displaystyle\sum_{k=0}^{\infty}(-1)^k\frac{(2k)!}{(k!)^2}$?

Solution Let $a_k=(-1)^k\frac{(2k)!}{(k!)^2}$ denote the kth term of this series. Observe that

$$
\begin{aligned}
\frac{|a_{k+1}|}{|a_k|} &= \frac{[2(k+1)]!/[(k+1)!]^2}{(2k)!/(k!)^2} = \frac{(2k+2)!}{(2k)!}\left[\frac{k!}{(k+1)!}\right]^2 \\
&= \frac{(2k+2)(2k+1)(2k)!}{(2k)!}\left[\frac{k!}{(k+1)k!}\right]^2 \\
&= \frac{(2k+2)(2k+1)}{(k+1)^2} = \frac{2(2k+1)}{k+1} = 1 + \frac{3k+1}{k+1} \geq 1
\end{aligned}
$$

for $k \geq 0$. Thus $\left\{\left|(-1)^k\frac{(2k)!}{(k!)^2}\right|\right\}$ is an increasing sequence which can not converge to zero. Hence the sequence $\left\{(-1)^k\frac{(2k)!}{(k!)^2}\right\}$ diverges. By the nth Term Test, the series $\sum_{k=0}^{\infty}(-1)^k\frac{(2k)!}{(k!)^2}$ diverges.   □

Proposition 3.2.4(a),(b) says that finite sums are linear. It follows that infinite sums are also linear.

Proposition 5.4.6 *Let $\sum_{k=p}^{\infty}a_k$ and $\sum_{k=p}^{\infty}b_k$ be convergent infinite series. Then*

(a) *the series $\sum_{k=p}^{\infty}(a_k+b_k)$ is convergent with*

$$\sum_{k=p}^{\infty}(a_k+b_k)=\sum_{k=p}^{\infty}a_k+\sum_{k=p}^{\infty}b_k;$$

(b) *for $c\in\Re$, the series $\sum_{k=p}^{\infty}ca_k$ is convergent with*

$$\sum_{k=p}^{\infty}ca_k=c\sum_{k=p}^{\infty}a_k.$$

Proof Let S_n denote the nth partial sum of $\sum_{k=p}^{\infty}a_k$ with $S=\lim_{n\to\infty}S_n=\sum_{k=p}^{\infty}a_k$. Let T_n denote the nth partial sum of $\sum_{k=p}^{\infty}b_k$ with $T=\lim_{n\to\infty}T_n=\sum_{k=p}^{\infty}b_k$.

(a) By Proposition 3.2.4(a), the nth partial sum of $\sum_{k=p}^{\infty}(a_k+b_k)$ equals S_n+T_n:

$$\sum_{k=p}^{n}(a_k+b_k)=\sum_{k=p}^{n}a_k+\sum_{k=p}^{n}b_k=S_n+T_n.$$

Then $\displaystyle\sum_{k=p}^{\infty}(a_k+b_k)=\lim_{n\to\infty}(S_n+T_n)=\lim_{n\to\infty}S_n+\lim_{n\to\infty}T_n=S+T=\sum_{k=p}^{\infty}a_k+\sum_{k=p}^{\infty}b_k.$

(b) By Proposition 3.2.4(b), the nth partial sum of $\sum_{k=p}^{\infty}ca_k$ equals cS_n:

$$\sum_{k=p}^{n}ca_k=c\sum_{k=p}^{n}a_k=cS_n.$$

Then $\displaystyle\sum_{k=p}^{\infty}ca_k=\lim_{n\to\infty}cS_n=c\lim_{n\to\infty}S_n=cS=c\sum_{k=p}^{\infty}a_k.$ □

This proposition rewrites a complicated series as a linear combination of simpler ones.

Examples 5.4.7 Evaluate $\displaystyle\sum_{k=1}^{\infty} \frac{2^k + 3k^2 + 3k}{2^{k+1}k(k+1)}$.

Solution This series is a linear combination of the series of Examples 5.4.3 (1),

$$
\begin{aligned}
\sum_{k=1}^{\infty} \frac{2^k + 3k^2 + 3k}{2^{k+1}k(k+1)} &= \sum_{k=1}^{\infty} \frac{2^k}{2^{k+1}k(k+1)} + \sum_{k=1}^{\infty} \frac{3k^2 + 3k}{2^{k+1}k(k+1)} \\
&= \sum_{k=1}^{\infty} \frac{1}{2k(k+1)} + \sum_{k=1}^{\infty} \frac{3}{2^{k+1}} \\
&= \frac{1}{2}\sum_{k=1}^{\infty} \frac{1}{k(k+1)} + \frac{3}{2}\sum_{k=1}^{\infty} \frac{1}{2^k} = \frac{1}{2}(1) + \frac{3}{2}(1) = 2.
\end{aligned}
$$

Geometric Series

Geometric series are basic examples of infinite series which are simple to analyze.
use them to study geometric power series, infinite polynomials which are geome
series. In particular, we integrate these power series to obtain new interesting se
The use of power series in calculus is the central theme of this chapter.

A *geometric series* begins with an *initial term* a. Each of the other term
obtained by multiplying the preceding term by the *ratio* R. Equivalently, the kth t
of a geometric series is aR^k for $k \geq 0$:

$$
\sum_{k=0}^{\infty} aR^k = a + aR + aR^2 + \cdots + aR^k + \cdots
$$

We derive a formula for the partial sum S_n of this series by subtracting RS_n from

$$
\begin{aligned}
S_n &= a + aR + aR^2 + \cdots + aR^{n-1} + aR^n \\
RS_n &= aR + aR^2 + \cdots + aR^{n-1} + aR^n + aR^{n+1} \\
S_n - RS_n &= a - aR^{n+1}
\end{aligned}
$$

Thus $(1 - R)S_n = a - aR^{n+1}$ and $S_n = \frac{a - aR^{n+1}}{1-R}$ when $R \neq 1$.

Proposition 5.4.8 *Consider the geometric series $\sum_{k=0}^{\infty} aR^k$.*
(a) *If $R \neq 1$, the nth partial sum S_n of this series is:*

$$
S_n = \frac{a - aR^{n+1}}{1 - R} \tag{5.4}
$$

(b) *If $a \neq 0$, this geometric series diverges when $R \leq -1$ or $R \geq +1$.*
(c) *If $-1 < R < 1$, this geometric series converges with sum*

$$
S = \frac{a}{1 - R}. \tag{5.4}
$$

Proof (a) We computed this formula for S_n above.
(b) If $R \geq 1$ and $a > 0$, then $\lim_{n\to\infty} aR^n = \infty$. If $R \geq 1$ and $a < 0$, then $\lim_{n\to\infty} aR^n = -$
If $R \leq -1$, then $\lim_{n\to\infty} aR^n$ does not exist. In all these cases, $\lim_{n\to\infty} aR^n$ does not equ
zero, and the geometric series diverges by the nth Term Test.
(c) Since $|R| < 1$, $\lim_{n\to\infty} R^n = 0$. By the formula for S_n in (a), $\lim_{n\to\infty} S_n = \frac{a}{1 - }$
Hence this geometric series converges with sum $\frac{a}{1-R}$.

Note Example 5.4.1 is a convergent geometric series with ratio $\frac{1}{10}$, Example 5.4.3 (1) is a convergent geometric series with ratio $\frac{1}{2}$ and Example 5.4.3 (7) is a divergent geometric series with ratio -1. We present two additional examples.

Examples 5.4.9 (1) Determine whether the series $\sum_{k=0}^{\infty} \frac{25}{3^k}$ converges or diverges.

Solution This is a geometric series with initial term $a = 25$ and ratio $R = \frac{1}{3}$. Since $|R| = \frac{1}{3} < 1$, this series converges with sum $\frac{a}{1-R} = \frac{25}{1-\frac{1}{3}} = \frac{75}{2}$.

(2) Determine whether the series $\sum_{k=1}^{\infty} (-1)^k \frac{5^k}{4^{2k+1}}$ converges or diverges.

Solution Observe $(-1)^k \frac{5^k}{4^{2k+1}} = \frac{1}{4} \left(-\frac{5}{16} \right)^k$. Thus we have a geometric series with initial term $a = -\frac{5}{64}$ and ratio $R = -\frac{5}{16}$. Since $|R| = \frac{5}{16} < 1$, this series converges with sum $\frac{a}{1-R} = \frac{-5/64}{1-(-5/16)} = -\frac{5}{84}$. □

A rational number $\frac{p}{q}$ is written as a decimal by performing the long division of dividing q into p. At each step the remainder is one of the integers 0, 1, 2, $\ldots$, $q-2$, $q-1$. After at most q steps the remainder repeats, and we obtain a repeating decimal. Ignoring the placement of the decimal point, the quotient consists of s digits $c_1, \ldots, c_s$ followed by t repeating digits $d_1, \ldots, d_t$:

$$\frac{p}{q} = c_1 \cdots c_s d_1 \cdots d_t d_1 \cdots d_t d_1 \cdots d_t \cdots = c_1 \cdots c_s \overline{d_1 \cdots d_t}$$

The notation $\overline{d_1 \cdots d_t}$ means that these digits are repeated indefinitely. Conversely, as illustrated by Example 2 below, every repeating decimal is a convergent geometric series. Its sum, given by (5.4.2), represents the repeating decimal as a rational number.

Examples 5.4.10 (1) Write the rational number $\frac{36}{11}$ as a repeating decimal.

Solution Divide 11 into 36:

$$
\begin{array}{r}
3.27 \\
11\overline{)36} \\
\underline{33} \\
30 \\
\underline{22} \\
80 \\
\underline{77} \\
3
\end{array}
$$

Since the remainder 3 has appeared again, the pattern of the last two divisions will repeat indefinitely producing the iterated digits 27 in the quotient. Thus $\frac{36}{11}$ equals the repeating decimal $3.\overline{27}$.

(2) Write the repeating decimal $4.9\overline{253}$ as a rational number.

Solution Observe that

$$4.9\overline{253} = 4.9 + 253 \times 10^{-4} + 253 \times 10^{-7} + 253 \times 10^{-10} + \cdots + 253 \times 10^{-3n-1} + \cdots$$

Therefore $4.9\overline{253}$ equals 4.9 plus a convergent geometric series with initial term $.0253$ and ratio 10^{-3}. By (5.4.2):

$$4.9\overline{253} = 4.9 + \frac{.0253}{1 - 10^{-3}} = \frac{49}{10} + \frac{253}{9,990} = \frac{(49)(999) + 253}{9,990} = \frac{49,204}{9,990} = \frac{24,602}{4,995}. \quad □$$

The infinite series which interest us the most, because of their relevance to calc[are *power series in* x:

$$\sum_{k=0}^{\infty} a_k x^k = a_0 + a_1 x + a_2 x^2 + \cdots + a_n x^n + \cdots$$

As written above, x is a symbol. However, we can substitute any real number for obtain an infinite series of numbers. The resulting series of numbers may converg diverge depending on the value of x. The following examples analyze several geom[power series, i.e. power series which are geometric series.

Examples 5.4.11 (1) For each value of x, determine whether the power s[$\sum_{k=0}^{\infty} \frac{x^k}{6^k}$ converges or diverges.

Solution Since $\frac{x^k}{6^k} = \left(\frac{x}{6}\right)^k$, this power series is a geometric series with in[term $a = 1$ and ratio $R = \frac{x}{6}$. This series diverges when $|R| = \frac{|x|}{6} \geq 1$, i.e. w[$x \leq -6$ or $x \geq 6$. When $-6 < x < 6$, $|R| = \frac{|x|}{6} < 1$ and this series conve[with sum $\frac{a}{1-R} = \frac{1}{1-\frac{x}{6}} = \frac{6}{6-x}$:

$$\sum_{k=0}^{\infty} \frac{x^k}{6^k} = \frac{6}{6-x} \quad \text{for } -6 < x < 6.$$

(2) Write $\frac{x^5}{7+3x^2}$ as a power series in x. Specify the values of x for which [representation is valid.

Solution Observe $\frac{x^5}{7+3x^2} = \frac{x^5/7}{1-(-3x^2/7)}$ which is the sum of the geometric se[with initial term $\frac{x^5}{7}$ and ratio $-\frac{3x^2}{7}$. Hence

$$\frac{x^5}{7+3x^2} = \sum_{k=0}^{\infty} \frac{x^5}{7}\left(-\frac{3x^2}{7}\right)^k = \sum_{k=0}^{\infty}(-1)^k \frac{3^k x^{2k+5}}{7^{k+1}}$$

for $\left|-\frac{3x^2}{7}\right| < 1$, i.e for $-\sqrt{\frac{7}{3}} < x < \sqrt{\frac{7}{3}}$.

(3) **(a)** For each value of x, determine whether the series $\sum_{k=0}^{\infty}(-1)^k \frac{(3x+2)^k}{5^{k+1}}$ c[verges or diverges.
(b) Write this series as a power series in a variable y of the form $y = x + c$[

Solution **(a)** Since $(-1)^k \frac{(3x+2)^k}{5^{k+1}} = \frac{1}{5}\left(-\frac{3x+2}{5}\right)^k$, we have a geometric se[with initial term $a = \frac{1}{5}$ and ratio $R = -\frac{3x+2}{5}$. Observe $|R| = \left|\frac{3x+2}{5}\right| \geq 1$ w[$\frac{3x+2}{5} \leq -1$ or $\frac{3x+2}{5} \geq 1$. Thus this series diverges when $x \leq -\frac{7}{3}$ or $x \geq 1$. N[$|R| < 1$ when $-1 < \frac{3x+2}{5} < 1$. Hence this series converges when $-\frac{7}{3} < x < 1$. [this case, its sum is $\frac{a}{1-R} = \frac{1/5}{1+\frac{3x+2}{5}} = \frac{1}{3x+7}$. Thus

$$\sum_{k=0}^{\infty}(-1)^k \frac{(3x+2)^k}{5^{k+1}} = \frac{1}{3x+7} \quad \text{for } -\frac{7}{3} < x < 1.$$

(b) Note

$$\sum_{k=0}^{\infty}(-1)^k \frac{(3x+2)^k}{5^{k+1}} = \sum_{k=0}^{\infty}(-1)^k \frac{3^k (x+2/3)^k}{5^{k+1}} = \sum_{k=0}^{\infty}(-1)^k \frac{3^k}{5^{k+1}} y^k$$

where $y = x + \frac{2}{3}$. Thus $\sum_{k=0}^{\infty}(-1)^k \frac{(3x+2)^k}{5^{k+1}}$ is a power series in $y = x + \frac{2}{3}$.

We use geometric series to preview how power series are used in calculus. Consider a geometric series with initial term 1, ratio R and sum S. By (5.4.1), the nth partial sum $S_n = \frac{1-R^{n+1}}{1-R} = \frac{1}{1-R} - \frac{R^{n+1}}{1-R}$ and

$$\frac{1}{1-R} = S_n + \frac{R^{n+1}}{1-R} = 1 + R + \cdots + R^n + \frac{R^{n+1}}{1-R}.$$

If $R = -t^p$, the preceding equation becomes:

$$\frac{1}{1+t^p} = 1 - t^p + \cdots + (-1)^n t^{np} + (-1)^{n+1} \frac{t^{np+p}}{1+t^p}. \tag{5.4.3}$$

If $t^p \neq -1$ for t between 0 and x, integrate this equation over the interval from 0 to x:

$$\int_0^x \frac{1}{1+t^p} \, dt = \int_0^x 1 - t^p + t^{2p} + \cdots + (-1)^n t^{np} \, dt + \int_0^x (-1)^{n+1} \frac{t^{np+p}}{1+t^p} \, dt$$

$$= \left(t - \frac{t^{p+1}}{p+1} + \frac{t^{2p+1}}{2p+1} + \cdots + (-1)^n \frac{t^{np+1}}{np+1} \right) \Big|_0^x + (-1)^{n+1} \int_0^x \frac{t^{np+p}}{1+t^p} \, dt$$

$$= x - \frac{x^{p+1}}{p+1} + \frac{x^{2p+1}}{2p+1} + \cdots + (-1)^n \frac{x^{np+1}}{np+1} + (-1)^{n+1} \int_0^x \frac{t^{np+p}}{1+t^p} \, dt. \tag{5.4.4}$$

When the integral I_n on the right side of (5.4.4) is small, we have an approximation of $\int_0^x \frac{1}{1+t^p} \, dt$. Moreover, if the limit of the I_n, as n goes to infinity, equals zero then we have a power series representation of $\int_0^x \frac{1}{1+t^p} \, dt$. We summarizes these conclusions.

Proposition 5.4.12 *Let* $x, p \in \Re$. *Assume* $t^p \neq -1$ *for* t *between 0 and* x. *Define*

$$I_n = \int_0^x \frac{t^{np+p}}{1+t^p} \, dt.$$

(a) $|I_n|$ *is the error in the approximation:*

$$\int_0^x \frac{1}{1+t^p} \, dt \approx x - \frac{x^{p+1}}{p+1} + \frac{x^{2p+1}}{2p+1} + \cdots + (-1)^n \frac{x^{np+1}}{np+1} \tag{5.4.5}$$

(b) *If* $\lim\limits_{n \to \infty} I_n = 0$, *then the value of this integral is given by the power series:*

$$\int_0^x \frac{1}{1+t^p} \, dt = \sum_{k=0}^{\infty} (-1)^k \frac{x^{kp+1}}{kp+1} \tag{5.4.6}$$

The difficulty in applying this proposition is to identify the x for which $\lim\limits_{n \to \infty} I_n = 0$.

Examples 5.4.13 (1) (a) Estimate $\int_0^1 \frac{1}{1+t^8} \, dt$ with error less than .03.
(b) Find a convergent infinite series whose sum is $\int_0^1 \frac{1}{1+t^8} \, dt$.
Solution (a) Apply (5.4.5) with $x = 1$ and $p = 8$:

$$\int_0^1 \frac{1}{1+t^8} \, dt \approx 1 - \frac{1}{9} + \frac{1}{17} + \cdots + (-1)^n \frac{1}{8n+1}$$

with error $I_n = \int_0^1 \frac{t^{8n+8}}{1+t^8} \, dt$. Since $1 \leq 1 + t^8$ for $0 \leq t \leq 1$:

$$I_n = \int_0^1 \frac{t^{8n+8}}{1+t^8} \, dt \leq \int_0^1 t^{8n+8} \, dt = \frac{t^{8n+9}}{8n+9} \Big|_0^1 = \frac{1}{8n+9} \tag{5.4.7}$$

When $n = 4$, we have the approximation

$$\int_0^1 \frac{1}{1+t^8}\, dt \approx 1 - \frac{1}{9} + \frac{1}{17} - \frac{1}{25} + \frac{1}{33} \approx 0.94$$

with error $|I_4|$ less than $\frac{1}{8(4)+9} = \frac{1}{41} < .03$.

(b) By (5.4.7), $0 \le I_n \le \frac{1}{8n+9}$, and $\displaystyle\lim_{n\to\infty} \frac{1}{8n+9} = 0$. By the Pinching Theo $\displaystyle\lim_{n\to\infty} I_n = 0$, and by (5.4.6):

$$\int_0^1 \frac{1}{1+t^8}\, dt = \sum_{k=0}^\infty (-1)^k \frac{1}{8k+1}.$$

(2) Find a power series representation for $\ln(1+x)$, and specify the values of x which this representation is valid.

Solution For $x > -1$,

$$\int_0^x \frac{1}{1+t}\, dt = \ln(1+t)\big|_0^x = \ln(1+x) - \ln 1 = \ln(1+x).$$

Apply (5.4.5) with $x > -1$ and $p = 1$:

$$\ln(1+x) = \int_0^x \frac{1}{1+t}\, dt \approx x - \frac{x^2}{2} + \frac{x^3}{3} + \cdots + (-1)^n \frac{x^{n+1}}{n+1}$$

with error $I_n = \int_0^x \frac{t^{n+1}}{1+t}\, dt$. If $x \ge 0$, then

$$|I_n| = \left|\int_0^x \frac{t^{n+1}}{1+t}\, dt\right| \le \int_0^x t^{n+1}\, dt = \frac{t^{n+2}}{n+2}\bigg|_0^x = \frac{x^{n+2}}{n+2}. \tag{5.}$$

Let $-1 < x < 0$. For $x \le t \le 0$, $|t| = -t$ and $1+t \ge 1+x \ge 0$. Thus

$$
\begin{aligned}
|I_n| &= \left|\int_0^x \frac{t^{n+1}}{1+t}\, dt\right| = \int_x^0 \frac{(-t)^{n+1}}{1+t}\, dt \le \int_x^0 \frac{(-1)^{n+1} t^{n+1}}{1+x}\, dt \\
&= \frac{(-1)^{n+1}}{1+x}\int_x^0 t^{n+1}\, dt = \frac{(-1)^{n+1}}{1+x}\frac{t^{n+2}}{n+2}\bigg|_x^0 = -\frac{(-1)^{n+1}}{1+x}\frac{x^{n+2}}{n+2} \\
&= \frac{(-x)^{n+2}}{(1+x)(n+2)} = \frac{|x|^{n+2}}{(1+x)(n+2)}. \tag{5.4}
\end{aligned}
$$

By (5.4.8) and (5.4.9):

$$\lim_{n\to\infty} |I_n| = \lim_{n\to\infty}\left|\int_0^x \frac{t^{n+1}}{1+t}\, dt\right| = 0 \text{ for } -1 < x \le 1.$$

By (5.4.6) : $\displaystyle \ln(1+x) = \sum_{n=1}^\infty (-1)^{n-1}\frac{x^n}{n} \text{ for } -1 < x \le 1. \tag{5.4.}$

In particular, when $x = 1$:

$$\ln 2 = \sum_{n=1}^\infty (-1)^{n-1}\frac{1}{n} = 1 - \frac{1}{2} + \frac{1}{3} + \cdots + (-1)^n \frac{1}{n+1} + \cdots \tag{5.4.1}$$

Observe the power series of (5.4.10) for $x = -1$ is $\sum_{n=1}^\infty -\frac{1}{n} = -\sum_{n=1}^\infty \frac{1}{n}$, t negative of the divergent harmonic series. We use the nth Term Test to sh the power series (5.4.10) diverges for $|x| > 1$. By L'Hôpital's Rule:

$$\lim_{n\to\infty} \frac{|x|^n}{n} = \lim_{n\to\infty} \frac{|x|^n \ln|x|}{1} = \infty.$$

Thus $\lim_{n\to\infty}(-1)^{n-1}\dfrac{x^n}{n} = -\infty$ for $x < -1$ and $\lim_{n\to\infty}(-1)^{n-1}\dfrac{x^n}{n}$ DNE for $x > 1$. By the nth Term Test, the power series $\sum_{n=1}^{\infty}(-1)^{n-1}\dfrac{x^n}{n}$ diverges for $|x| > 1$.

Conclusion For $-1 < x \le 1$, $\ln(1+x)$ equals the power series (5.4.10). That power series diverges for $x \le -1$ or $x > 1$.

(3) Find a power series representation for $\arctan x$, and specify the values of x for which this representation is valid.

Solution For all $x \in \Re$,

$$\int_0^x \frac{1}{1+t^2}\, dt = \arctan t\big|_0^x = \arctan x - \arctan 0 = \arctan x.$$

Apply (5.4.5) with $p = 2$:

$$\arctan x = \int_0^x \frac{1}{1+t^2}\, dt \approx x - \frac{x^3}{3} + \frac{x^5}{5} + \cdots + (-1)^n \frac{x^{2n+1}}{2n+1}$$

with error $I_n = \int_0^x \frac{t^{2n+2}}{1+t^2}\, dt$. Moreover,

$$0 \le |I_n| = \left|\int_0^x \frac{t^{2n+2}}{1+t^2}\, dt\right| \le \left|\int_0^x t^{2n+2}\, dt\right| = \frac{|t^{2n+3}|}{2n+3}\bigg|_0^x = \frac{|x|^{2n+3}}{2n+3}.$$

Hence $\lim_{n\to\infty}|I_n| = \lim_{n\to\infty}\left|\int_0^x \frac{t^{2n+2}}{1+t^2}\, dt\right| = 0$ for $-1 \le x \le 1$

by the Pinching Theorem. By (5.4.6):

$$\arctan x = \sum_{n=0}^{\infty}(-1)^n \frac{x^{2n+1}}{2n+1} \quad \text{for } -1 \le x \le 1. \tag{5.4.12}$$

In particular, when $x = 1$:

$$\frac{\pi}{4} = \arctan 1 = \sum_{n=0}^{\infty}(-1)^n \frac{1}{2n+1} = 1 - \frac{1}{3} + \frac{1}{5} + \cdots + (-1)^n \frac{1}{2n+1} + \cdots \tag{5.4.13}$$

We use the nth Term Test to show the power series $\sum_{n=0}^{\infty}(-1)^n \frac{x^{2n+1}}{2n+1}$ diverges for $|x| > 1$. By L'Hôpital's Rule:

$$\lim_{n\to\infty} \frac{|x|^{2n+1}}{2n+1} = \lim_{n\to\infty} \frac{2|x|^{2n+1}\ln|x|}{2} = \infty.$$

Hence $\lim_{n\to\infty}(-1)^n \frac{x^{2n+1}}{2n+1}$ does not exist for $x < -1$ or $x > 1$. By the nth Term Test, the power series $\sum_{n=0}^{\infty}(-1)^n \frac{x^{2n+1}}{2n+1}$ diverges for $|x| > 1$.

Conclusion For $-1 \le x \le 1$, $\arctan x$ equals the power series (5.4.12). That power series diverges for $x < -1$ or $x > 1$. $\square$

Absolute and Conditional Convergence

We study an interesting phenomenon of convergent series with sum S that have infinitely many positive and negative terms. In some examples, S is the sum of the positive summands minus the sum of the negative summands. We call these series *absolutely convergent*. In other examples, the positive summands have sum ∞ while the negative summands have sum $-\infty$. The partial sums S_n, after cancellation between the positive and negative summands, have a finite limit S. We call these series *conditionally convergent*. The following example illustrates this type of series.

Motivating Example 5.4.14 Consider the harmonic series and the series for l

$$1 + \frac{1}{2} + \frac{1}{3} + \frac{1}{4} + \cdots + \frac{1}{k} + \cdots \ = \ \sum_{k=1}^{\infty} \frac{1}{k},$$

$$\ln 2 \ = \ 1 - \frac{1}{2} + \frac{1}{3} - \frac{1}{4} + \cdots + (-1)^{k-1}\frac{1}{k} + \cdots \ = \ \sum_{k=1}^{\infty} (-1)^{k-1}\frac{1}{k}.$$

By Example 5.4.3(4), the harmonic series has sum $+\infty$ and diverges. The sec series converges with sum $\ln 2$ by Example 5.4.13(2). Each pair of consecutive pos and negative terms in this series combine to produce a number which is so small these combinations add up to $\ln 2$. We show the positive summands of the conver series for $\ln 2$ have sum $+\infty$, and the negative summands have sum $-\infty$.

$$1 + \frac{1}{3} + \frac{1}{5} + \cdots + \frac{1}{2n-1} + \cdots \ \geq \ \frac{1}{2} + \frac{1}{4} + \frac{1}{6} + \cdots + \frac{1}{2n} + \cdots$$

$$\geq \ \frac{1}{2}\left(1 + \frac{1}{2} + \frac{1}{3} + \cdots + \frac{1}{n} + \cdots\right) = \infty$$

$$-\frac{1}{2} - \frac{1}{4} - \frac{1}{6} - \cdots - \frac{1}{2n} - \cdots \ = \ -\frac{1}{2}\left(1 + \frac{1}{2} + \frac{1}{3} + \cdots + \frac{1}{n} + \cdots\right) = -\infty.$$

An absolutely convergent series $\sum_{k=p}^{\infty} b_k$, unlike the series above with sum $\ln 2$, not rely on cancellation between its positive and negative summands to converge. next proposition gives the following criterion to identify this type of series. Cha the signs of all the summands to make them positive. If the resulting series $\sum_{k=p}^{\infty}$ converges, then the original series $\sum_{k=p}^{\infty} b_k$ converges but with a different sum.

Proposition 5.4.15 *If the series* $\displaystyle\sum_{k=p}^{\infty} |b_k|$ *converges, the series* $\sum_{k=p}^{\infty} b_k$ *also con*

Proof Let S_n denote the nth partial sum of the series $\sum_{k=p}^{\infty} |b_k|$, and let T_n den the nth partial sum of the series $\sum_{k=p}^{\infty} (b_k + |b_k|)$. Both series are sums of posi terms. Hence the sequences $\{S_n\}$ and $\{T_n\}$ are increasing. For $k \geq p$:

$$0 \ \leq \ b_k + |b_k| \ \leq \ 2\,|b_k|$$

Thus for $n \geq p$: $\qquad\qquad\qquad 0 \ \leq \ T_n \ \leq \ 2S_n.$

Since the sequence $\{2S_n\}$ converges, it is bounded above. Therefore, $\{T_n\}$ is increasing sequence which is bounded above and hence converges. That is, the se $\sum_{k=p}^{\infty} (b_k + |b_k|)$ converges. Then

$$\sum_{k=p}^{\infty} b_k = \sum_{k=p}^{\infty} (b_k + |b_k|) - \sum_{k=p}^{\infty} |b_k|,$$

being the difference of two convergent series, also converges.

The following definition distinguishes between the two types of convergent seri An absolutely convergent series does not depend on cancellation to converge wh a conditionally convergent series only converges because of cancellation between positive and negative summands.

Definition 5.4.16 (a) *The series* $\sum_{k=p}^{\infty} a_k$ *is absolutely convergent if the ser* $\sum_{k=p}^{\infty} |a_k|$ *converges.*
(b) *The series* $\sum_{k=p}^{\infty} b_k$ *is conditionally convergent if it converges and the ser* $\sum_{k=p}^{\infty} |b_k|$ *diverges.*

We restate Proposition 5.4.15 in this terminology.

Corollary 5.4.17 *An absolutely convergent series is convergent.*

Let $\sum_{k=p}^{\infty} a_k$ be a series where the a_k are all positive: $\sum_{k=p}^{\infty} |a_k| = \sum_{k=p}^{\infty} a_k$. In this case, convergence and absolute convergence are synonymous. A series of this type is never conditionally convergent. Since $S_{n+1} = S_n + a_{n+1}$, the partial sums form an increasing sequence $\{S_n\}$ of positive numbers. There are two possibilities:

(1) The increasing sequence of partial sums is bounded above and converges to its least upper bound S. The series $\sum_{k=p}^{\infty} a_k$ is absolutely convergent with sum S.

(2) The increasing sequence of partial sums is unbounded, i.e $\lim_{n \to \infty} S_n = \infty$. The series $\sum_{k=p}^{\infty} a_k$ diverges.

To summarize, each infinite series $\sum_{k=p}^{\infty} a_k$ is of one of three types:

   (1) absolutely convergent,   (2) conditionally convergent   or   (3) divergent.

In an example, test for absolute convergence first: does $\sum_{k=p}^{\infty} |a_k|$ converge or diverge? If absolute convergence fails, test for conditional convergence: does $\sum_{k=p}^{\infty} a_k$ converge or diverge? If conditional convergence also fails, we have a divergent series. Sometimes we show directly that a series diverges without considering the possibilities of absolute or conditional convergence.

Examples 5.4.18 (1) Determine whether the series $\sum_{k=1}^{\infty} (-1)^k \frac{1}{k}$ is absolutely convergent, conditionally convergent or divergent.

   Solution Test for absolute convergence by considering the series $\sum_{k=1}^{\infty} \left| (-1)^k \frac{1}{k} \right|$ $= \sum_{k=1}^{\infty} \frac{1}{k}$. This is the harmonic series which diverges. Therefore the original series is not absolutely convergent. The original series however is convergent with sum $-\ln 2$. Hence the series $\sum_{k=1}^{\infty} (-1)^k \frac{1}{k}$ is conditionally convergent.

(2) Determine whether the series $\sum_{k=1}^{\infty} (-1)^k \frac{5}{k(k+1)}$ is absolutely convergent, conditionally convergent or divergent.

   Solution Test for absolute convergence by considering the series $\sum_{k=1}^{\infty} \left| (-1)^k \frac{5}{k(k+1)} \right|$ $= 5 \sum_{k=1}^{\infty} \frac{1}{k(k+1)}$. This is the telescoping series which converges with sum 5. Therefore the original series $\sum_{k=1}^{\infty} (-1)^k \frac{5}{k(k+1)}$ is absolutely convergent.

(3) Determine whether the series $\sum_{k=1}^{\infty} (-1)^k \frac{k!}{8^k}$ is absolutely convergent, conditionally convergent or divergent.

   Solution Since $\lim_{k \to \infty} \frac{k!}{8^k} = \infty \neq 0$, this series diverges by the nth Term Test.

(4) For each value of x, determine whether the power series $\sum_{k=0}^{\infty} (-1)^{k(k+1)/2} \frac{x^k}{3^k}$ is absolutely convergent, conditionally convergent or divergent.

   Solution Test for absolute convergence by considering the series

$$\sum_{k=0}^{\infty} \left| (-1)^{k(k+1)/2} \frac{x^k}{3^k} \right| = \sum_{k=0}^{\infty} \left(\frac{|x|}{3} \right)^k .$$

This is a geometric series with ratio $\frac{|x|}{3}$ which converges for $|x| < 3$ and diverges for $|x| \geq 3$. Thus the given power series is absolutely convergent for $-3 < x < 3$. When $|x| \geq 3$, $\lim_{k \to \infty} (-1)^{k(k+1)/2} \frac{x^k}{3^k}$ does not exist. By the nth Term Test, the given power series diverges for $x \leq -3$ or $x \geq 3$.   □

Historical Remarks

About 450 BCE, the Greek philosopher Zeno of Elea posed a paradox. He asse
that a runner will never finish a race because after completing the first half, he
still run the second half. After finishing three quarters of the race, he must still
the remaining quarter. Since this process continues indefinitely, the runner will n
finish the entire race. Aristotle refuted this paradox 100 years later by observing
both finite intervals of time and distance are infinitely divisible: in each subinterv:
time the runner covers a corresponding subinterval of distance. Therefore the rur
can complete the race in a finite amount of time. From our point of view, Zeno
questioning how the infinite geometric series

$$\frac{1}{2} + \frac{1}{4} + \cdots + \frac{1}{2^n} + \cdots$$

can have a finite sum. Although Greek mathematicians understood that a geome
series can have a finite sum, they were disturbed by the process of forming an infi
sum and avoided the issue. For example, as part of Archimedes's computation of
area of a parabolic segment he encountered the geometric series $1 + \frac{1}{4} + \cdots + \frac{1}{4^n} +$
To deal with this sum, he proved the identity

$$1 + \frac{1}{4} + \cdots + \frac{1}{4^n} + \frac{1}{3}\frac{1}{4^n} = \frac{4}{3}.$$

Then he used reductio ad absurdum arguments (proofs by contradiction) to show
sum S of this series is neither less than $\frac{4}{3}$ nor greater than $\frac{4}{3}$. Therefore $S = \frac{4}{3}$.

A variety of infinite series were studied in Europe in the middle ages beginn
in the early fourteenth century at Merton College in Oxford, England. For exam
Richard Swineshead proved that

$$\frac{1}{2} + \frac{2}{4} + \frac{3}{8} + \cdots + \frac{n}{2^n} + \cdots = 2.$$

We describe his proof in Problem 2. In 1350 the French mathematician Nicole Ores
generalized Archimedes' geometric series. He showed for $k > 1$ an integer, the geom
ric series with initial term $\frac{a}{k}$ and ratio $1 - \frac{1}{k}$ has sum a. He also devised a geome
method to sum series. For example, he sums Swineshead's series as follows: consi
the area under the infinite staircase where each step has height one and the nth s
has width $\frac{1}{2^n}$. Calculate this area in two different ways: either divide the staircase i
vertical rectangles or divide it into horizontal rectangles. See Figure 5.4.19. The
vertical rectangle has area $(n)\left(\frac{1}{2^n}\right) = \frac{n}{2^n}$ while the nth horizontal rectangle has a
$(1)\left(\frac{1}{2^n}\right) = \frac{1}{2^n}$. Thus Swineshead's series has the same sum as the geometric ser
with initial term 1 and ratio $\frac{1}{2}$. The latter series has sum 2.

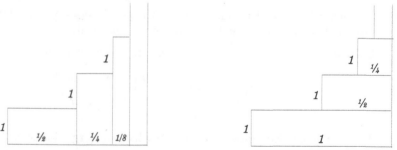

Figure 5.4.19 Oresme's Summation of Swineshead's Series

Summary

The reader should understand the meaning of an infinite series being convergent, divergent, absolutely convergent or conditionally convergent. Given a numerical series or power series, she should be able to determine which of these terms apply. In particular, she should know how to apply the nth Term Test to show certain series diverge. She should also know how to use geometric series to approximate certain integrals. These are several important series we analyzed:

$$(\text{Telescoping Series}) \qquad \sum_{k=1}^{\infty} \frac{1}{k(k+1)} = 1; \qquad\qquad (5.4.14)$$

$$(\text{Harmonic Series}) \qquad \sum_{k=1}^{\infty} \frac{1}{k} \text{ diverges}; \qquad\qquad (5.4.15)$$

$$(\text{Geometric Series}) \qquad \sum_{k=0}^{\infty} aR^k = \frac{a}{1-R} \qquad \text{for } -1 < R < 1; \qquad (5.4.16)$$

$$\ln(1+x) = \sum_{k=1}^{\infty} (-1)^{k-1} \frac{1}{k} x^k \qquad \text{for } -1 < x \le 1; \qquad (5.4.17)$$

$$\arctan x = \sum_{k=0}^{\infty} (-1)^k \frac{1}{2k+1} x^{2k+1} \quad \text{for } -1 \le x \le 1. \qquad (5.4.18)$$

Basic Exercises

1. Find S_5 for each series.

(a) $\displaystyle\sum_{k=0}^{\infty} k^3$ (b) $\displaystyle\sum_{k=1}^{\infty} (-1)^k k$ (c) $\displaystyle\sum_{k=2}^{\infty} \frac{k}{2^k}$ (d) $\displaystyle\sum_{k=3}^{\infty} (-1)^k \frac{k!}{(2k-4)!}$ (e) $\displaystyle\sum_{k=4}^{\infty} \frac{1}{k!}$

2. What does the nth Term Test say about each series?

(a) $\displaystyle\sum_{k=1}^{\infty} \frac{k^2}{9k^2+1}$ (b) $\displaystyle\sum_{k=1}^{\infty} \frac{k^2}{7k^3+1}$ (c) $\displaystyle\sum_{k=1}^{\infty} (-1)^k \frac{1}{(1+1/k)^k}$

(d) $\displaystyle\sum_{k=1}^{\infty} \frac{1}{k}$ (e) $\displaystyle\sum_{k=1}^{\infty} (-1)^k \frac{1}{k}$ (f) $\displaystyle\sum_{k=1}^{\infty} \frac{1}{2^k}$

3. Determine whether each series converges or diverges. If it converges, find its sum.

(a) $\displaystyle\sum_{k=3}^{\infty} \frac{4}{5^k}$ (b) $\displaystyle\sum_{k=1}^{\infty} \frac{k}{10k+7}$ (c) $\displaystyle\sum_{k=1}^{\infty} (-1)^{k(k+1)}$ (d) $\displaystyle\sum_{k=1}^{\infty} (-1)^{\frac{1}{2}k(k+1)}$

(e) $\displaystyle\sum_{k=0}^{\infty} \frac{1}{(k+5)(k+6)}$ (f) $\displaystyle\sum_{k=1}^{\infty} \frac{k!}{2^k}$ (g) $\displaystyle\sum_{k=1}^{\infty} \frac{1}{5k-1}$ (h) $\displaystyle\sum_{k=0}^{\infty} (-1)^k \frac{2^k}{3^k}$

(i) $\displaystyle\sum_{k=0}^{\infty} (-1)^k \frac{3^k}{2^k}$ (j) $\displaystyle\sum_{k=1}^{\infty} \frac{1}{k(k+2)}$ (k) $\displaystyle\sum_{k=1}^{\infty} k \sin(1/k)$ (l) $\displaystyle\sum_{k=1}^{\infty} \frac{5^k - 3^k}{10^k}$

(m) $\displaystyle\sum_{k=1}^{\infty} \frac{4^k - 1}{5^k}$ (n) $\displaystyle\sum_{k=0}^{\infty} \frac{1}{8k+1}$ (o) $\displaystyle\sum_{k=0}^{\infty} \frac{2^{k+3}}{7^{k-1}}$ (p) $\displaystyle\sum_{k=1}^{\infty} \frac{1}{k^2+3k}$

(q) $\displaystyle\sum_{k=2}^{\infty} (-1)^k \frac{1}{k^2 - k}$ (r) $\displaystyle\sum_{k=1}^{\infty} \frac{2^k + 7^k}{5^k}$ (s) $\displaystyle\sum_{k=0}^{\infty} \frac{5+(-1)^k}{k+3}$ (t) $\displaystyle\sum_{k=1}^{\infty} \frac{6+(-1)^k}{k^2+3k+2}$

(u) $\displaystyle\sum_{k=0}^{\infty} \frac{4-(-1)^k}{3^k}$ (v) $\displaystyle\sum_{k=1}^{\infty} \frac{8k+2^{k+3}}{k2^k}$ (w) $\displaystyle\sum_{k=1}^{\infty} \ln\left(\frac{k}{k+1}\right)$ (x) $\displaystyle\sum_{k=1}^{\infty} \left(\frac{k}{k+1}\right)^k$

4. Write each rational number as a repeating decimal.

(a) $\frac{2}{13}$ (b) $4\frac{8}{11}$ (c) $-\frac{4}{15}$ (d) $\frac{19}{7}$ (e) $-7\frac{5}{12}$

5. Write each repeating decimal as a rational number.

(a) $0.\overline{43}$ (b) $3.28\overline{74}$ (c) $10.416\overline{999}$ (d) $-9.37\overline{815}$ (e) $20.41\overline{2936}$

6. For every value of x, determine whether each power series converges or dive Find the sum for those values of x for which the series converges.

(a) $\sum_{n=0}^{\infty} \frac{x^n}{7^n}$

(b) $\sum_{n=0}^{\infty}(-1)^n \frac{x^{4n+3}}{3^{2n+1}}$

(c) $\sum_{n=1}^{\infty} \frac{x^n}{n}$

(d) $\sum_{n=1}^{\infty}(-1)^n \frac{x^{2n}}{n}$

(e) $\sum_{n=0}^{\infty} \frac{x^{2n+1}}{2n+1}$

(f) $\sum_{n=0}^{\infty} \frac{x^{4n+3}}{2n+1}$

(g) $\sum_{n=1}^{\infty}(-1)^n \frac{x^{4n}}{2n-1}$

(h) $\sum_{n=1}^{\infty}(-1)^{n+1} \frac{(n+2^n)x^n}{n2^n}$

(i) $\sum_{n=1}^{\infty}(-1)^n \frac{3n-1}{2n^2-n}x^2$

7. Write each function as a power series in x. Specify the values of x for which representation is valid.

(a) $\dfrac{x}{1-x^4}$ (b) $\dfrac{x^6}{1+x^5}$ (c) $\dfrac{x^3}{1+x^2}$ (d) $\dfrac{x^4}{1-x^3}$

(e) $\dfrac{x}{1-7x^3}$ (f) $\dfrac{x^2}{1+4x^5}$ (g) $\dfrac{7x^4}{16-9x^4}$ (h) $\dfrac{6x^5}{8+5x^3}$

8. Write each series as a power series in a variable y of the form $y = x+c$. Deter the values of x for which each series converges, and find its sum for those x.

(a) $\sum_{k=0}^{\infty} \dfrac{(x+5)^k}{7^k}$

(b) $\sum_{k=1}^{\infty} \dfrac{(x-3)^{2k}}{4^{k+1}}$

(c) $\sum_{k=0}^{\infty} \dfrac{(x+1)^{3k+2}}{5^{2k+1}}$

(d) $\sum_{k=0}^{\infty}(-1)^k \dfrac{(x-5)^{4k+}}{16^{3k+1}}$

(e) $\sum_{k=2}^{\infty} \dfrac{(2x+5)^k}{8^k}$

(f) $\sum_{k=1}^{\infty} \dfrac{(4x-3)^{3k}}{27^{5k}}$

(g) $\sum_{k=0}^{\infty} \dfrac{(2x+7)^{5k+3}}{3^{4k+1}}$

(h) $\sum_{k=0}^{\infty}(-1)^k \dfrac{(3x+5)}{4^{3k}}$

9. Add the first N terms of the appropriate power series to estimate each inte with error less than E. Indicate the value of N that you use.

(a) $\int_0^1 \frac{1}{1+t^4}\, dt$, $E = .05$ (b) $\int_0^2 \frac{1}{8+t^3}\, dt$, $E = .02$ (c) $\int_0^{1/2} \frac{1}{1-t^5}\, dt$, $E = .0000$

(d) $\int_0^2 \frac{1}{64+t^6}\, dt$, $E = .002$ (e) $\int_0^{.9} \frac{1}{1-t^8}\, dt$, $E = .006$

10. Derive a power series representation for each function. State the values of x which each representation is valid.

(a) $\arctan(x^2)$ (b) $\ln(1-x)$ (c) $\int_0^x \frac{1}{1-t^2}\, dt$ (d) $\int_0^x \frac{1}{1+t^3}\, dt$ (e) $\int_0^x$

(f) $\int_0^x \frac{t}{1-t^5}\, dt$ (g) $\int_0^x \frac{t^5}{16-t^4}\, dt$ (h) $\int_0^{x^2} \frac{t}{27-t^3}\, dt$ (i) $\ln\frac{1+x^2}{1-x^2}$ (j) $\ln(2$

11. Determine whether each series is absolutely convergent, conditionally converg or divergent.

(a) $\sum_{k=0}^{\infty}(-1)^k \dfrac{1}{3^k}$

(b) $\sum_{k=0}^{\infty}(-1)^k \dfrac{2^k}{k^5+1}$

(c) $\sum_{k=0}^{\infty}(-1)^k \dfrac{1}{2k+1}$

(d) $\sum_{k=0}^{\infty}(-1)^k \dfrac{k}{5k+7}$

(e) $\sum_{k=2}^{\infty} \left|\dfrac{\cos \pi k}{k^2-1}\right|$

(f) $\sum_{k=0}^{\infty}(-1)^{k(k+1)(2k+1)/6} \dfrac{3}{4^2}$

(g) $\sum_{k=1}^{\infty} \dfrac{2^k+(-1)^k 2^{2k+1}}{3^{k+2}}$

(h) $\sum_{k=1}^{\infty} \dfrac{\cos \pi k}{k}$

(i) $\sum_{k=1}^{\infty}(-1)^k \dfrac{1}{k-k\cos(1/k)}$

(j) $\sum_{k=1}^{\infty}(-1)^k \dfrac{1}{3k-2}$

(k) $\sum_{k=1}^{\infty} \dfrac{1+(-1)^k 2}{5^k}$

(l) $\sum_{k=1}^{\infty}(-1)^{k(k+1)/2} \dfrac{1}{k^2+k}$

12. For each value of x, determine whether the power series is absolutely converge conditionally convergent or divergent.

(a) $\sum_{n=0}^{\infty} \dfrac{x^{5n+2}}{32^n}$

(b) $\sum_{n=1}^{\infty}(-1)^n \dfrac{x^{4n}}{8n}$

(c) $\sum_{n=0}^{\infty}(-1)^{n(n+1)(2n+1)/6} \dfrac{x^{8n+6}}{81^{2n+1}}$

(d) $\sum_{n=0}^{\infty} \dfrac{x^{6n+1}}{6n+1}$

(e) $\sum_{n=0}^{\infty} \dfrac{(1-2(-1)^n)x^n}{5^{3n+2}}$

(f) $\sum_{n=1}^{\infty}(-1)^{n(n+1)/2} \dfrac{x^{n^2+n}}{n^2+n}$

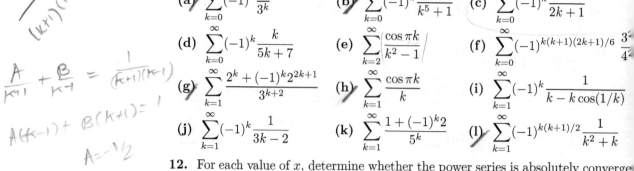

$\dfrac{A}{k+1} + \dfrac{B}{k-1} = \dfrac{1}{(k+1)(k-1)}$

$A(k-1) + B(k+1) = 1$

$A = -\frac{1}{2}$

$B = \frac{1}{2}$

13. A ball is dropped from four feet above the ground. Each bounce of the ball is two–thirds as high as the previous bounce. If this ball keeps bouncing indefinitely, find the total distance traveled by the ball.

14. A diver jumps two feet up from a diving board. He keeps jumping on the diving board until he is at least ten feet above the board before doing a front tuck with a triple twist. If each jump is 20% higher than the previous one, how many jumps must the diver make? Approximate the vertical distance traveled by the diver in his upward jumps before beginning his dive.

15. A $10,000 savings account earns 5% annual interest compounded monthly for ten years. Each month the account balance is entered in the passbook. Approximate the sum of these entries.

16.   **(a)** Show if $\sum_{k=1}^{\infty} a_k$ is a convergent series and $\sum_{k=1}^{\infty} b_k$ is a divergent series, then $\sum_{k=1}^{\infty} (a_k + b_k)$ is a divergent series.
(b) Give an example of two divergent series $\sum_{k=1}^{\infty} a_k$ and $\sum_{k=1}^{\infty} b_k$ such that $\sum_{k=1}^{\infty} (a_k + b_k)$ converges.

17.   **(a)** Show if $\sum_{k=1}^{\infty} a_k$ converges, with $a_k \neq 0$ for $k \geq 1$, then $\sum_{k=1}^{\infty} \frac{1}{a_k}$ diverges.
(b) Given an example of a divergent series $\sum_{k=1}^{\infty} b_k$ such that $\sum_{k=1}^{\infty} \frac{1}{b_k}$ also diverges.

18. Show if $a_k > m > 0$ for $k \geq 1$, then the series $\sum_{k=1}^{\infty} a_k$ diverges.

19. Show the series $\sum_{k=1}^{\infty} a_k$ converges if and only if the series $\sum_{k=N}^{\infty} a_k$ converges for every positive integer N.

20.   **(a)** Show if $\sum_{k=p}^{\infty} a_k$ and $\sum_{k=p}^{\infty} b_k$ are absolutely convergent, then $\sum_{k=p}^{\infty} (a_k + b_k)$ is also absolutely convergent.
(b) Show if $\sum_{k=p}^{\infty} a_k$ is absolutely convergent and $\sum_{k=p}^{\infty} b_k$ is conditionally convergent, then $\sum_{k=p}^{\infty} (a_k + b_k)$ is conditionally convergent.
(c) Give an example of two conditionally convergent series $\sum_{k=p}^{\infty} a_k$ and $\sum_{k=p}^{\infty} b_k$ such that $\sum_{k=p}^{\infty} (a_k + b_k)$ is absolutely convergent.

21. Show $\sum_{n=1}^{\infty} a_n^2$ is absolutely convergent if $\sum_{n=1}^{\infty} a_n$ is absolutely convergent.

Challenging Problems

1. Find the first time after 3 : 00 that the minute and hour hands of a clock coincide.

2.   **(a)** Show for $n \geq 1$: $\sum_{k=1}^{\infty} \frac{k}{2^k} = \sum_{k=1}^{n} \frac{2}{2^k} + \sum_{k=n+1}^{\infty} \frac{k-n}{2^k}$.
(b) Derive the sum of Swineshead's series: $\sum_{k=1}^{\infty} \frac{k}{2^k} = 2$.

5.5   Taylor Series at $x = 0$

The Taylor series of $f(x)$ is a power series representation of $f(x)$, derived from information about the higher derivatives of f. In the first subsection, we construct Taylor polynomials $P_n(x)$ to approximate a smooth function $f(x)$ for x near zero. The larger the value of n and the closer x is to zero, the better the approximation. In the second subsection, we take the limit of the Taylor polynomials of $f(x)$ to define a power series, called the *Taylor series of* $f(x)$ *at* $x = 0$, which equals $f(x)$ for x near zero. We concentrate on several basic examples: the Taylor series of e^x, $\sin x$, $\cos x$ and binomial series. The technical problem is to bound the size of the remainder $R_n(x) = f(x) - P_n(x)$. There are two approaches. The derivative form of the remainder is used in this subsection while the integral form of the remainder is derived in the appendix. The third subsection presents two applications of Taylor series. First,

we use Taylor series to evaluate indeterminate forms. Then we use the integrals o
Taylor polynomials $P_n(x)$ to approximate the integral of $f(x)$. The appendix of
section verifies the convergence of binomial series which are introduced in the se
subsection and used in the third subsection.

Taylor Polynomials

In this subsection, we approximate $f(x)$ by polynomials $P_n[f(x)]$ of degree n for
$n \geq 0$. We derive a formula for the coefficients of these polynomials and com
these polynomials for several examples. There are two conditions we require on
approximations $f(x) \approx P_n[f(x)]$.

Condition 1 The larger the value of n, the better the approximation.

Condition 2 The closer x is to zero, the better the approximation.

Write $P_n[f(x)] = a_0 + a_1 x + a_2 x^2 + \cdots + a_{n-1} x^{n-1} + a_n x^n$. Both $P_{n-1}[f(x)]$
$a_0 + a_1 x + a_2 x^2 + \cdots + a_{n-1} x^{n-1}$ are degree $n-1$ polynomials that approxim
$f(x)$. We assume these two polynomials are equal. Equivalently, we define $P_n[f$
from $P_{n-1}[f(x)]$ by choosing a_n and letting $P_n[f(x)] = P_{n-1}[f(x)] + a_n x^n$.
Conditions 1 and 2, we choose a_n so that $P_n[f(x)]$ is a better approximation of
for x near zero than $P_{n-1}[f(x)]$. Thus x^k has the same coefficient a_k in every $P_n[f$
for $n \geq k$. Hence for all $n \geq 0$, write

$$P_n[f(x)] = \sum_{k=0}^{n} a_k x^k = a_0 + a_1 x + a_2 x^2 + \cdots + a_{n-1} x^{n-1} + a_n x^n.$$

We illustrate the procedure for identifying the coefficients a_k of these polynomial

Motivating Example 5.5.1 Approximate $f(x) = e^x$ by degree n polynomials P_n
to satisfy Conditions 1 and 2 above.

Solution Note $y = P_0[e^x] = a_0$ is a horizontal line. By Condition 2, this line sho
pass through $(0, f(0))$. Hence we define

$$P_0(x) = f(0) = e^0 = 1 \text{ and } a_0 = 1.$$

By Condition 2, $y = P_1[e^x] = a_0 + a_1 x$ is the line which best approximates $y =$
near $x = 0$. By Condition 1, this line should be a better approximation than
horizontal line $y = 1$. The tangent line of $y = e^x$ at $x = 0$ has this property. N
$f'(x) = e^x$. Then the tangent line $y = P_1[e^x]$ has equation

$$P_1[e^x] = f(0) + f'(0)x = e^0 + e^0 x = 1 + x \text{ and } a_1 = 1.$$

By Conditions 1 and 2, $y = P_2[e^x] = 1 + x + a_2 x^2$ is a parabola which approxima
$y = f(x) = e^x$ for x near zero better than the tangent line $y = 1 + x$. Then y
$P_2'[e^x] = a_1 + 2a_2 x$ is a line which approximates $y = f'(x) = e^x$ near $x = 0$. Take t
line to be the tangent line $y = P_1[e^x] = 1 + x$, i.e. $P_2'[e^x] = 1 + x$. Then

$$P_2[e^x] = \int P_2'[e^x] \, dx = \int 1 + x \, dx = a_0 + x + \frac{x^2}{2} = 1 + x + \frac{x^2}{2} \text{ and } a_2 = \frac{1}{2}.$$

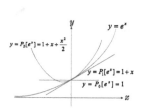

Figure 5.5.2

Taylor Polynomials

Observe in Figure 5.5.2 that $y = P_0[e^x]$, $y = P_1[e^x]$, $y = P_2[e^x]$ are increasin
better approximations of $y = e^x$ for x near zero. The procedure we used to find $P_2[$
generalizes to find $P_t[e^x]$ if we know $P_{t-1}[e^x]$. By Condition 2, $P_t'[e^x]$ is a degree t
polynomial which approximates $f'(x) = e^x$ for x near zero. Take $P_t'[e^x] = P_{t-1}[e^x$

Then $P_t[e^x] = \int P_t'[e^x] \, dx = \int P_{t-1}[e^x] \, dx$

$$= \int a_0 + a_1 x + a_2 x^2 + \cdots + a_{t-1} x^{t-1} \ dx$$

$$= a_0 + \frac{a_0 x}{1} + \frac{a_1 x^2}{2} + \frac{a_2 x^3}{3} + \cdots + \frac{a_{t-1} x^t}{t}.$$

Since a_k is the coefficient of x^k in $P_t [e^x]$, we have

$$a_k = \frac{a_{k-1}}{k} \ \text{ for } 1 \le k \le t. \tag{5.5.1}$$

Iterate (5.5.1) k times to find the values of the a_k:

$$a_k \ = \ \frac{a_{k-1}}{k} = \frac{a_{k-2}}{k(k-1)} = \frac{a_{k-3}}{k(k-1)(k-2)} = \cdots = \frac{a_1}{k(k-1)(k-2) \cdots (2)}$$

$$= \ \frac{a_0}{k(k-1)(k-2) \cdots (2)(1)} = \frac{1}{k!}.$$

Thus $P_n [e^x] \ = \ 1 + \dfrac{x}{1!} + \dfrac{x^2}{2!} + \dfrac{x^3}{3!} + \cdots + \dfrac{x^n}{n!} = \displaystyle\sum_{k=0}^{n} \dfrac{x^k}{k!}.$ □

Since all the derivatives of a polynomial exist, we restrict our considerations to functions which have this property.

Definition 5.5.3 *A function f is called smooth at $x = c$ if there is an open interval I containing c such that all the higher derivatives $f^{(n)}(x)$ exist for $x \in I$.*

The derivation of the coefficients of the polynomials $P_n [e^x]$ which approximate e^x generalizes to find the coefficients of the polynomials $P_n [f(x)]$ which approximate any smooth function $f(x)$. Let $a_{k,f}$ denote the coefficient of x^k in $P_n [f(x)]$. Then

$$P_n [f(x)] = a_{0,f} + a_{1,f} x + a_{2,f} x^2 + \cdots + a_{n,f} x^n.$$

By Condition 2, $P_t' [f(x)]$ and $P_{t-1} [f'(x)]$ are degree $t - 1$ polynomials which approximate $f'(x)$ for x near zero. Take $P_t' [f(x)] = P_{t-1} [f'(x)]$. Then

$$P_t [f(x)] \ = \ \int P_t' [f(x)] \ dx = \int P_{t-1} [f'(x)] \ dx$$

$$= \ \int a_{0,f'} + a_{1,f'} x + a_{2,f'} x^2 + \cdots + a_{t-1,f'} x^{t-1} \ dx$$

$$= \ a_{0,f} + \frac{a_{0,f'} x}{1} + \frac{a_{1,f'} x^2}{2} + \frac{a_{2,f'} x^3}{3} + \cdots + \frac{a_{t-1,f'} x^t}{t}.$$

Since $a_{k,f}$ is the coefficient of x^k in $P_t [f(x)]$, we have

$$a_{k,f} = \frac{a_{k-1,f'}}{k} \ \text{ for } 1 \le k \le t. \tag{5.5.2}$$

By Condition 2, we want $P_t [f(x)] (0)$, which is $a_{0,f}$, to equal $f(0)$:

$$a_{0,f} = f(0). \tag{5.5.3}$$

To find the coefficient $a_{k,f}$ of x^k in $P_n [f(x)]$, for $n \ge k \ge 1$, apply (5.5.2) k times: first with $t = n$, then with $t = n - 1$, then with $t = n - 2$, ..., then with $t = n - (k-1)$ and then with $t = n - k$:

$$a_{k,f} \ = \ \frac{a_{k-1,f'}}{k} = \frac{a_{k-2,f^{(2)}}}{k(k-1)} = \frac{a_{k-3,f^{(3)}}}{k(k-1)(k-2)} = \cdots = \frac{a_{1,f^{(k-1)}}}{k(k-1)(k-2) \cdots (2)}$$

$$= \ \frac{a_{0,f^{(k)}}}{k(k-1)(k-2) \cdots (2)(1)} = \frac{a_{0,f^{(k)}}}{k!} = \frac{f^{(k)}(0)}{k!}$$

by (5.5.3). This discussion motivates the following definition of the Taylor polynomials $P_n [f(x)]$. We use the convention $f^{(0)}(x) = f(x)$ in this definition.

Definition 5.5.4 *Let $f(x)$ be a function which is smooth at $x = 0$. For each poε integer n, define the nth Taylor polynomial of $f(x)$ at $x = 0$ by:*

$$P_n\left[f(x)\right] = f(0) + \frac{f^{(1)}(0)}{1!}x + \frac{f^{(2)}(0)}{2!}x^2 + \frac{f^{(3)}(0)}{3!}x^3 + \cdots + \frac{f^{(n)}(0)}{n!}x^n = \sum_{k=0}^{n} \frac{f^{(k)}(0}{k!}$$

Note These polynomials are also called the Maclaurin polynomials of $f(x)$.

The Taylor polynomial $P_n\left[f(x)\right]$ of degree n has zero k^{th} derivative for $k > n$. show that the approximation $P_n\left[f(x)\right] \approx f(x)$ implies that $P_n\left[f(x)\right]$ has the same derivatives as $f(x)$, at $x = 0$, for $k \le n$.

Proposition 5.5.5 *Let $f(x)$ be a function which is smooth at $x = 0$. For $0 \le k$ the kth derivatives of $f(x)$ and $P_n(x)$ have the same value at $x = 0$.*

Proof $\frac{d^k}{dx^k}\left(P_n\left[f(x)\right]\right)(0)$ is the constant term of the polynomial $\frac{d^k}{dx^k}\left(P_n\left[f(x)\right]\right)$. ferentiating a polynomial reduces each power of x by one. Hence the constant t of $\frac{d^k}{dx^k}\left(P_n\left[f(x)\right]\right)$ is:

$$\frac{d^k}{dx^k}\left[\frac{f^{(k)}(0)}{k!}x^k\right] = \frac{f^{(k)}(0)}{k!}\frac{d^k}{dx^k}\left[x^k\right] = \frac{f^{(k)}(0)}{k!}\frac{d^{k-1}}{dx^{k-1}}\left[kx^{k-1}\right]$$

$$= \frac{f^{(k)}(0)}{k!}\frac{d^{k-2}}{dx^{k-2}}\left[k(k-1)x^{k-2}\right] = \frac{f^{(k)}(0)}{k!}\frac{d^{k-3}}{dx^{k-3}}\left[k(k-1)(k-2)x^{k-3}\right] = \cdots$$

$$= \frac{f^{(k)}(0)}{k!}\frac{d}{dx}\left[k(k-1)\cdots(2)x\right] = \frac{f^{(k)}(0)}{k!}\left[k(k-1)\cdots(2)(1)\right] = \frac{f^{(k)}(0)}{k!}(k!) = f^{(k)}$$

In Example 5.5.1 we computed the Taylor polynomials of e^x at $x = 0$. We compute the Taylor polynomials at $x = 0$ of the functions $\sin x$ and $(1+x)^p$.

Examples 5.5.6 (1) Find the Taylor polynomials of $g(x) = \sin x$ at $x = 0$.
    Solution Observe that

$$g(x) = \sin x, \quad g^{(1)}(x) = \cos x, \quad g^{(2)}(x) = -\sin x, \quad g^{(3)}(x) = -\cos x, \quad g^{(4)}(x)$$

Thus the higher derivatives of $\sin x$ consist of four functions which repeat:

$$g^{(4k)}(x) = \sin x, \quad g^{(4k+1)}(x) = \cos x, \quad g^{(4k+2)}(x) = -\sin x, \quad g^{(4k+3)}(x) = -$$

for $k \ge 0$. Substitute $x = 0$:

$$g^{(4k)}(0) = \sin 0 = 0, \qquad\qquad g^{(4k+1)}(0) = \cos 0 = 1,$$
$$g^{(4k+2)}(0) = -\sin 0 = 0, \qquad\qquad g^{(4k+3)}(0) = -\cos 0 = -1.$$

Thus $P_{2n-1}\left[\sin x\right] = P_{2n}\left[\sin x\right]$

$$= x - \frac{x^3}{3!} + \frac{x^5}{5!} - \frac{x^7}{7!} + \cdots + (-1)^{n-1}\frac{x^{2n-1}}{(2n-1)!} = \sum_{k=0}^{n-1}(-1)^k\frac{x^{2k+1}}{(2k+1}$$

(2) Find the Taylor polynomials of $h(x) = (1+x)^p$ at $x = 0$ for p a nonzero consta
    Solution Observe that

$$h^{(1)}(x) = p(1+x)^{p-1}, \qquad\qquad\qquad h^{(2)}(x) = p(p-1)(1+x)^{p-2},$$
$$h^{(3)}(x) = p(p-1)(p-2)(1+x)^{p-3}, \ldots,$$
$$h^{(n)}(x) = p(p-1)(p-2)\cdots(p-n+1)(1+x)^{p-n}.$$

Substitute $x = 0$:

$$h(0) = 1, \qquad h^{(1)}(0) = p, \qquad h^{(2)}(0) = p(p-1),$$
$$h^{(3)}(0) = p(p-1)(p-2), \quad \ldots \quad , h^{(n)}(0) = p(p-1)(p-2)\cdots(p-n+1).$$

$$\text{Therefore} \quad P_n\left[(1+x)^p\right] = 1 + px + \frac{p(p-1)}{2!}x^2 + \frac{p(p-1)(p-2)}{3!}x^3$$
$$+ \cdots + \frac{p(p-1)(p-2)\cdots(p-n+1)}{n!}x^n. \quad \square$$

We introduce notation for the coefficients of the Taylor polynomials of $(1+x)^p$:

$$\begin{pmatrix} s \\ 0 \end{pmatrix} = 1, \qquad \begin{pmatrix} s \\ t \end{pmatrix} = \frac{s(s-1)(s-2)\cdots(s-t+1)}{t!} \qquad (5.5.4)$$

for $s \in \Re$ and t a positive integer. These numbers are called *generalized binomial coefficients*. In this notation:

$$h^{(n)}(x) = n!\begin{pmatrix} p \\ n \end{pmatrix}(1+x)^{p-n} \quad \text{and} \qquad (5.5.5)$$

$$P_n\left[(1+x)^p\right] = \sum_{k=0}^{n}\begin{pmatrix} p \\ k \end{pmatrix}x^k. \qquad (5.5.6)$$

Observe when p is a positive integer, the binomial coefficients $\begin{pmatrix} p \\ k \end{pmatrix}$ for $k > p$ have the factor $p - j$ for $j = p$ and are zero. Hence when p is a positive integer:

$$(1+x)^p = P_p\left[(1+x)^p\right]$$
$$= 1 + \begin{pmatrix} p \\ 1 \end{pmatrix}x + \cdots + \begin{pmatrix} p \\ k \end{pmatrix}x^k + \cdots + \begin{pmatrix} p \\ p-1 \end{pmatrix}x^{p-1} + x^p. \qquad (5.5.7)$$

Taylor Series

The Taylor series $T(x)$ of a smooth function $f(x)$ is the limit of its Taylor polynomials. Criteria are derived to determine the values of x for which $f(x) = T(x)$. We apply these criteria to establish that e^x, $\sin x$ and $\cos x$ equal their Taylor series for all x. We also state that $(1+x)^p$ equals its binomial series for $|x| < 1$. Verification of this fact is given in the appendix to this section.

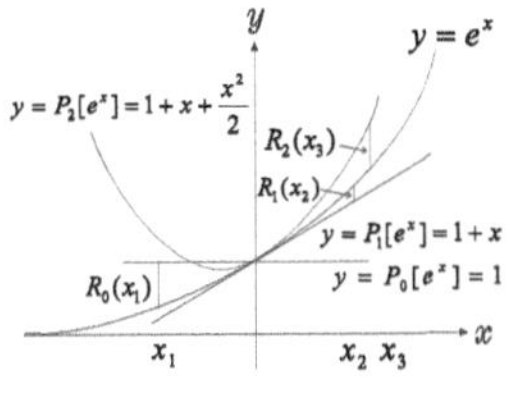

When $f(x)$ is smooth at $x = 0$, $f(x)$ is approximated by its Taylor polynomials:[1]

$$f(x) \approx P_n(x) = \sum_{k=0}^{n}\frac{f^{(k)}(0)}{k!}x^k.$$

Figure 5.5.7
Remainders of $f(x) = e^x$

The error $R_n(x) = f(x) - P_n(x)$ in this approximation is called the nth *remainder* of $f(x)$. These remainders for $f(x) = e^x$ are the lengths of the vertical line segments depicted in Figure 5.5.7.

Definition 5.5.8 *Let $f(x)$ be a function which is smooth at $x = 0$. The nth remainder of $f(x)$ at $x = 0$ is:*

$$R_n(x) = f(x) - P_n(x). \qquad (5.5.8)$$

---

[1]From now on, we denote the Taylor polynomial $P_n\left[f(x)\right]$ as $P_n(x)$ with the appropriate function $f(x)$ understood from the context.

The Taylor series of $f(x)$ at $x = 0$ is the power series

$$\sum_{k=0}^{\infty} \frac{f^{(k)}(0)}{k!} x^k.$$ (5-

Notes **(1)** The partial sums of the Taylor series (5.5.9) of $f(x)$ are the T
polynomials $P_n(x)$ of $f(x)$.
(2) For those x with $\lim_{n \to \infty} R_n(x) = 0$,

$$f(x) = \sum_{k=0}^{\infty} \frac{f^{(k)}(0)}{k!} x^k.$$ (5.5

For x near zero, the approximations $f(x) \approx P_n(x)$ get better as n gets la
i.e. the limit of the errors $R_n(x)$ of these approximations is zero. Thus we ex
the Taylor series of $f(x)$ to converge with sum $f(x)$ for x near zero. We establi
formula for the remainder $R_n(x)$, called its *derivative form*, which we use to deter
the values of x for which $\lim_{n \to \infty} R_n(x) = 0$.

We begin with a generalization of the Mean Value Theorem for Integrals (3.
which we use to derive the derivative form of the remainder $R_n(x)$.

Theorem 5.5.9 (Generalized Mean Value Theorem for Integrals) *Let $u(x)$ and*
be continuous functions with domain $[a, b]$. Assume either $v(x) \geq 0$ for all $x \in [$
or $v(x) \leq 0$ for all $x \in [a, b]$. Then there is at least one number $c \in [a, b]$ such tha

$$\int_a^b u(x)v(x)\, dx = u(c) \int_a^b v(x)\, dx.$$ (5.5

Proof Consider the case $v(x) \geq 0$ for all $x \in [a, b]$. By the Maximum Value Theor
the continuous function $u(x)$ has a minimum value $m = u(A)$ and a maximum va
$M = u(B)$ on $[a, b]$. That is, $m \leq u(x) \leq M$ and hence $mv(x) \leq u(x)v(x) \leq Mv$
for $x \in [a, b]$. Thus

$$m \int_a^b v(x)\, dx \leq \int_a^b u(x)v(x)\, dx \leq M \int_a^b v(x)\, dx.$$

If $v(x) = 0$ for all $x \in [a, b]$, then (5.5.11) is merely $0 = 0$. If $v(x) \neq 0$ for so
$x \in [a, b]$, then $\int_a^b v(x)\, dx > 0$, and

$$m \leq \frac{\int_a^b u(x)v(x)\, dx}{\int_a^b v(x)\, dx} \leq M.$$

By the Intermediate Value Theorem, there is a number c between A and B with

$$u(c) = \frac{\int_a^b u(x)v(x)\, dx}{\int_a^b v(x)\, dx}.$$

Since A and B are in $[a, b]$, c is also in $[a, b]$ and we have verified (5.5.11).
We leave the proof of (5.5.11), when $v(x) \leq 0$ for all $x \in [a, b]$, as Exercise 15.

To show the partial sums $P_n(x)$ of the Taylor series have limit $f(x)$, we use t
following formula for the error $R_n(x)$ of the approximation $f(x) \approx P_n(x)$.

Proposition 5.5.10 [Derivative Form of the Remainder] *Let f be a smooth function on an open interval I which contains zero. If $x \in I$, there is a number $t(x)$ between 0 and x such that:*

$$R_n(x) = \frac{f^{(n+1)}(t(x))}{(n+1)!}x^{n+1}. \tag{5.5.12}$$

Proof Apply the Mean Value Theorem to the function $f^{(n)}(x)$ on the interval with endpoints zero and x: there is a number $t_1(x)$ between 0 and x such that

$$f^{(n+1)}(t_1(x)) = \frac{f^{(n)}(x) - f^{(n)}(0)}{x - 0} \quad \text{or}$$

$$f^{(n)}(x) = f^{(n)}(0) + f^{(n+1)}(t_1(x))x.$$

Integrate this equation with respect to x:

$$\int_0^X f^{(n)}(x)\,dx = \int_0^X f^{(n)}(0)\,dx + \int_0^X f^{(n+1)}(t_1(x))x\,dx.$$

Integrate the first two integrals, and apply the Generalized Mean Value Theorem for Integrals to the third integral: there is $t_2(X)$ between 0 and X such that

$$f^{(n-1)}(X) - f^{(n-1)}(0) = f^{(n)}(0)X + f^{(n+1)}(t_2(X))\int_0^X x\,dx$$

$$f^{(n-1)}(X) = f^{(n-1)}(0) + f^{(n)}(0)X + f^{(n+1)}(t_2(X))\frac{X^2}{2}.$$

This equation is valid for all $X \in I$. Change the name of the variable in this equation from X to x, integrate with respect to x, and apply the Generalized Mean Value Theorem for Integrals to the last integral:

$$\int_0^X f^{(n-1)}(x)\,dx = \int_0^X f^{(n-1)}(0) + f^{(n)}(0)x\,dx + \int_0^X f^{(n+1)}(t_2(x))\frac{x^2}{2}\,dx$$

$$f^{(n-2)}(X) - f^{(n-2)}(0) = f^{(n-1)}(0)X + f^{(n)}(0)\frac{X^2}{2} + f^{(n+1)}(t_3(X))\int_0^X \frac{x^2}{2}\,dx$$

$$f^{(n-2)}(X) = f^{(n-2)}(0) + f^{(n-1)}(0)X + f^{(n)}(0)\frac{X^2}{2} + f^{(n+1)}(t_3(X))\frac{X^3}{3(2)}$$

for some number $t_3(X)$ between 0 and X. Iterate this argument an additional $n-2$ times to produce formula (5.5.12). □

We find the derivative form of the remainder for the functions studied in the examples of the first subsection.

Examples 5.5.11 (1) Write $f(x) = P_n(x) + R_n(x)$ when $f(x) = e^x$.

Solution Note $f^{(n+1)}(x) = e^x$. The formula $P_n(x) = \sum_{k=0}^{n} \frac{x^k}{k!}$ was derived in Example 5.5.1. By Proposition 5.5.10 there is a number $t(x)$ between 0 and x such that:

$$e^x = \sum_{k=0}^{n} \frac{x^k}{k!} + \frac{e^{t(x)}}{(n+1)!}x^{n+1}.$$

(2) Write $g(x) = P_{2n}(x) + R_{2n}(x)$ when $g(x) = \sin x$.

Solution In Example 5.5.6(1) we computed $g^{(2n+1)}(x) = (-1)^n \cos x$, and derived the formula $P_{2n}(x) = \sum_{k=0}^{n-1}(-1)^k \frac{x^{2k+1}}{(2k+1)!}$. By Proposition 5.5.10 there is a number $t(x)$ between 0 and x such that:

$$\sin x = \sum_{k=0}^{n-1}(-1)^k \frac{x^{2k+1}}{(2k+1)!} + (-1)^n \frac{\cos t(x)}{(2n+1)!}x^{2n+1}.$$

(3) Write $h(x) = P_n(x) + R_n(x)$ when $h(x) = (1+x)^p$ with p a nonzero consta

Solution In Example 5.5.6(2) we showed that

$$h^{(n+1)}(x) = p(p-1)\cdots(p-n)(1+x)^{p-n-1}$$

and derived the formula $P_n(x) = \sum_{k=0}^{n} \binom{p}{k} x^k$. By Proposition 5.5.10 t
is a number $t(x)$ between 0 and x such that:

$$
\begin{aligned}
(1+x)^p &= \sum_{k=0}^{n} \binom{p}{k} x^k + \frac{p(p-1)\cdots(p-n)\left[1+t(x)\right]^{p-n-1}}{(n+1)!} x^{n+1} \\
&= \sum_{k=0}^{n} \binom{p}{k} x^k + \binom{p}{n+1} \left[1+t(x)\right]^{p-n-1} x^{n+1}.
\end{aligned}
$$

The following corollary of Proposition 5.5.10 gives a bound on the size of
remainder $R_n(x)$.

Corollary 5.5.12 *Let f be a smooth function on an interval I which contains*
Fix $x \in I$ and a positive integer n. Assume there is a number $M_{n+1}(x)$ such tha
all numbers t between zero and x:

$$\left| f^{(n+1)}(t) \right| \le M_{n+1}(x)$$

Then the nth remainder $R_n(x)$ of $f(x)$ at $x = 0$ satisfies:

$$|R_n(x)| \le \frac{M_{n+1}(x)}{(n+1)!} |x|^{n+1}. \tag{5.5}$$

This corollary is used to show $\lim\limits_{n\to\infty} R_n(x) = 0$ for certain values of x near z
For these values of x, the partial sums $P_n(x)$ of the Taylor series have limit $f(x)$,
the Taylor series has sum $f(x)$, i.e. $f(x) = \sum_{k=0}^{\infty} \frac{f^{(k)}(0)}{k!} x^k$. We establish the Ta
series of the functions studied in the examples of the first subsection. We use the li
established in Example 5.3.9 (2):

$$\lim_{m\to\infty} \frac{u^m}{m!} = 0 \quad \text{for all } u \in \Re. \tag{5.3}$$

Proposition 5.5.13 (a) *For all $x \in \Re$:*

$$e^x = \sum_{k=0}^{\infty} \frac{x^k}{k!}. \tag{5.5.}$$

(b) *For all $x \in \Re$:*

$$\sin x = \sum_{k=0}^{\infty} (-1)^k \frac{x^{2k+1}}{(2k+1)!}. \tag{5.5.}$$

(c) *For all $x \in \Re$:*

$$\cos x = \sum_{k=0}^{\infty} (-1)^k \frac{x^{2k}}{(2k)!}. \tag{5.5.}$$

(d) *If **either** $p > 0$ with $-1 < x \le 1$ **or** $p < 0$ with $-1 < x < 1$:*

$$(1+x)^p = \sum_{k=0}^{\infty} \binom{p}{k} x^k. \tag{5.5.}$$

These Taylor series are called binomial series.

Proof **(a)** Let $f(x) = e^x$. For $n \geq 0$, the function $f^{(n+1)}(x) = e^x$ is increasing. Hence for t between 0 and x:

$$\left| f^{(n+1)}(t) \right| \leq \left\{ \begin{array}{ll} e^x & \text{if } 0 \leq x \\ e^0 = 1 & \text{if } x \leq 0 \end{array} \right\}.$$

Apply Corollary 5.5.12 with $M_{n+1}(x) = e^x$ if $x \geq 0$ or $M_{n+1}(x) = 1$ if $x \leq 0$:

$$|R_n(x)| \leq \frac{x^{n+1}}{(n+1)!} e^x \quad \text{if } x \geq 0 \tag{5.5.18}$$

$$|R_n(x)| \leq \frac{|x|^{n+1}}{(n+1)!} \quad \text{if } x \leq 0 \tag{5.5.19}$$

In both cases, $\lim_{n\to\infty} |R_n(x)| = 0$ by (5.3.10), and e^x is the limit of its Taylor polynomials $P_n(x)$. By Example 5.5.1, $P_n(x) = \sum_{k=0}^{n} \frac{x^k}{k!}$. Hence $e^x = \sum_{k=0}^{\infty} \frac{x^k}{k!}$ for all $x \in \Re$.

(b) Let $g(x) = \sin x$. In Example 5.5.6 (1) we computed $g^{(2n+1)}(x) = (-1)^n \cos x$. Hence $|g^{(2n+1)}(x)| \leq 1$. Apply Corollary 5.5.12 for any $x \in \Re$ with $M_{2n+1}(x) = 1$:

$$|R_{2n}(x)| \leq \frac{|x|^{2n+1}}{(2n+1)!}. \tag{5.5.20}$$

In Example 5.5.6 (1) we showed $P_{2n-1}(x) = P_{2n}(x) = \sum_{k=0}^{n-1} (-1)^k \frac{x^{2k+1}}{(2k+1)!}$. Hence $R_{2n-1}(x) = R_{2n}(x)$, and $\lim_{n\to\infty} R_n(x) = 0$ by (5.3.10). Thus $\sin x$ is the limit of its Taylor polynomials: $\sin x = \sum_{k=0}^{\infty} (-1)^k \frac{x^{2k+1}}{(2k+1)!}$ for all $x \in \Re$.

(c) The derivation of this Taylor series is analogous to the derivation of the Taylor series of $\sin x$. The details are left as Exercise 7(a). An alternate derivation is given in Example 5.7.13 (1).

(d) We establish these Taylor series in the appendix of this section. $\square$

Examples 5.5.14 (1) Approximate e with error less than .01.

Solution Let $f(x) = e^x$. By (5.5.18): $|R_5(1)| \leq \frac{e}{6!} = \frac{e}{720} < \frac{3}{720} = \frac{1}{240} < .005$. Thus the approximation $f(1) = e \approx P_5(1)$ has error less than .005. By (5.5.14):

$$e \approx \frac{1^0}{0!} + \frac{1^1}{1!} + \frac{1^2}{2!} + \frac{1^3}{3!} + \frac{1^4}{4!} + \frac{1^5}{5!} \approx 1 + 1 + .500 + .1667 + .0417 + .0083 \approx 2.717$$

(2) Approximate the sine of one radian with error less than .00001.

Solution By (5.5.20): $|R_8(1)| \leq \frac{1}{9!} = \frac{1}{362,880} < .0000028$. Thus the approximation $\sin 1 \approx P_8(1)$ has error less than .0000028. By (5.5.15):

$$\sin 1 \approx \frac{1^1}{1!} - \frac{1^3}{3!} + \frac{1^5}{5!} - \frac{1^7}{7!} \approx 1 - .166667 + .008333 - .000198 \approx .841468$$

(3) Use a binomial series to estimate $\sqrt{2}$ with error less than .04.

Solution Consider the binomial series of $h(x) = (1+x)^{1/2}$. Then $h(1) = \sqrt{2}$. By Example 5.5.11 (3) with $p = \frac{1}{2}$, $h^{(k)}(x) = k! \binom{1/2}{k}$. By Proposition 5.5.10, there is a number $0 \leq t \leq 1$ such that:

$$|R_3(1)| = \left| \binom{1/2}{4} \right| (1+t)^{\frac{1}{2}-4} (1)^4 = \left| \binom{1/2}{4} \right| \frac{1}{(1+t)^{7/2}}$$

$$\leq \left| \binom{1/2}{4} \right| = \left| \frac{\frac{1}{2} \left(-\frac{1}{2}\right) \left(-\frac{3}{2}\right) \left(-\frac{5}{2}\right)}{4!} \right| = \frac{5}{128} < .04.$$

Thus the approximation $\sqrt{2} \approx P_3(1)$ has error less than .04. By (5.5.17):

$$\sqrt{2} \approx \binom{1/2}{0} 1^0 + \binom{1/2}{1} 1^1 + \binom{1/2}{2} 1^2 + \binom{1/2}{3} 1^3 = 1 + \frac{1}{2} - \frac{1}{8} + \frac{1}{16}$$

(4) Find the Taylor series at $x = 0$ of $q(x) = \left\{ \begin{array}{ll} e^{-1/x^2} & \text{if } x \neq 0 \\ 0 & \text{if } x = 0 \end{array} \right\}.$

Solution Let $t = \frac{1}{x^2}$. Then $q(x)$ is continuous at $x = 0$ because

$$\lim_{x \to 0} q(x) = \lim_{t \to \infty} e^{-t} = 0 = q(0).$$

We show $q^{(k)}(0) = 0$ for $k \geq 1$. Assume $k \geq 0$ and we have shown $q^{(k)}(0)$
Observe $q^{(k)}(x)$ has the form $\frac{n_k(x)}{d_k(x)} e^{-1/x^2}$ for $x \neq 0$ where $n_k(x)$ and $d_k(x)$
polynomials. Let $u = \frac{1}{h}$. Then

$$q^{(k+1)}(0) = \lim_{h \to 0} \frac{q^{(k)}(h) - q^{(k)}(0)}{h} = \lim_{h \to 0} \frac{n_k(h) e^{-1/h^2}}{h d_k(h)}$$

$$q_+^{(k+1)}(0) = \lim_{u \to \infty} \frac{n_k(1/u) e^{-u^2}}{\frac{1}{u} d_k(1/u)} = \lim_{u \to \infty} \frac{u n_k(1/u)}{e^{u^2} d_k(1/u)} = \lim_{u \to \infty} \frac{N_k(u)}{e^{u^2} D_k(u)}$$

where $N_k(u)$ and $D_k(u)$ are polynomials in u. If $N_k(u)$ has degree m, then ap
L'Hopital's Rule m times to show that the preceding limit equals zero. Simila
$q_-^{(k+1)}(0) = 0$. Hence $q^{(k+1)}(0) = 0$, the Taylor series $T(x)$ of $q(x)$ at $x = 0$
all zero coefficients and $T(x) = 0$. Thus $q(x) = T(x)$ only when $x = 0$, des
the facts that $q(x)$ is smooth at all $x \in \Re$ and $T(x)$ converges for all $x \in \Re$.

Applications

We illustrate how Taylor series are used as an alternate to L'Hopital's Rule to evalu
indeterminate forms. Then we approximate definite integrals by the integrals of
Taylor polynomials of their integrands.

Examples 5.5.15 (1) Evaluate $\displaystyle\lim_{x \to 0} \frac{\sin x - x}{2e^x - 2 - 2x - x^2}$.

Solution We use (5.5.14) and (5.5.15) to find power series representation
the numerator and denominator of the given fraction:

$$\sin x - x = \sum_{k=0}^{\infty} (-1)^k \frac{x^{2k+1}}{(2k+1)!} - x = \sum_{k=1}^{\infty} (-1)^k \frac{x^{2k+1}}{(2k+1)!},$$

$$2e^x - 2 - 2x - x^2 = 2 \sum_{k=0}^{\infty} \frac{x^k}{k!} - 2 - 2x - x^2 = \sum_{k=3}^{\infty} \frac{2x^k}{k!}.$$

Thus we can represent the given fraction as a quotient of power series. Af
dividing numerator and denominator by their common factor x^3, we substit
$x = 0$ to evaluate the limit.

$$\lim_{x \to 0} \frac{\sin x - x}{2e^x - 2 - 2x - x^2} = \lim_{x \to 0} \frac{\displaystyle\sum_{k=1}^{\infty} (-1)^k \frac{x^{2k+1}}{(2k+1)!}}{\displaystyle\sum_{k=3}^{\infty} \frac{2x^k}{k!}}$$

$$
\begin{aligned}
&= \lim_{x \to 0} \frac{-\dfrac{x^3}{6} + x^5 \displaystyle\sum_{k=2}^{\infty} (-1)^k \dfrac{x^{2k-4}}{(2k+1)!}}{\dfrac{2x^3}{6} + x^4 \displaystyle\sum_{k=4}^{\infty} \dfrac{2x^{k-4}}{k!}} = \lim_{x \to 0} \frac{-\dfrac{1}{6} + x^2 \displaystyle\sum_{k=2}^{\infty} (-1)^k \dfrac{x^{2k-4}}{(2k+1)!}}{\dfrac{1}{3} + x \displaystyle\sum_{k=4}^{\infty} \dfrac{2x^{k-4}}{k!}} \\
&= \frac{-\dfrac{1}{6} + 0}{\dfrac{1}{3} + 0} = -\frac{1}{2}.
\end{aligned}
$$

(2) Evaluate $\displaystyle\lim_{x \to 0} \frac{4\sqrt[4]{1+x^3} - 4 - x^3}{\ln(1+x^3) - x^3}$.

Solution We use (5.5.17), with $p = \frac{1}{4}$ and x replaced by x^3, to find a power series representation of the numerator, and use (5.4.10), with x replaced by x^3, to find a power series representation of the denominator.

$$
4\sqrt[4]{1+x^3} - 4 - x^3 = 4\sum_{k=0}^{\infty} \binom{1/4}{k} x^{3k} - 4 - x^3 = 4\sum_{k=2}^{\infty} \binom{1/4}{k} x^{3k},
$$

$$
\ln(1+x^3) - x^3 = \sum_{k=1}^{\infty} (-1)^{k-1} \frac{x^{3k}}{k} - x^3 = \sum_{k=2}^{\infty} (-1)^{k-1} \frac{x^{3k}}{k}.
$$

Now we can represent the given fraction as a quotient of power series. After dividing numerator and denominator by their common factor x^6, we substitute $x = 0$ to evaluate the limit.

$$
\lim_{x \to 0} \frac{4\sqrt[4]{1+x^3} - 4 - x^3}{\ln(1+x^3) - x^3} = \lim_{x \to 0} \frac{4\displaystyle\sum_{k=2}^{\infty} \binom{1/4}{k} x^{3k}}{\displaystyle\sum_{k=2}^{\infty} (-1)^{k-1} \frac{x^{3k}}{k}}
$$

$$
= \lim_{x \to 0} \frac{-\dfrac{3x^6}{8} + 4x^9 \displaystyle\sum_{k=3}^{\infty} \binom{1/4}{k} x^{3k-9}}{-\dfrac{x^6}{2} + x^9 \displaystyle\sum_{k=3}^{\infty} (-1)^{k-1} \frac{x^{3k-9}}{k}} = \lim_{x \to 0} \frac{-\dfrac{3}{8} + 4x^3 \displaystyle\sum_{k=3}^{\infty} \binom{1/4}{k} x^{3k-9}}{-\dfrac{1}{2} + x^3 \displaystyle\sum_{k=3}^{\infty} (-1)^{k-1} \frac{x^{3k-9}}{k}}
$$

$$
= \frac{-\dfrac{3}{8} + 0}{-\dfrac{1}{2} + 0} = \frac{3}{4}. \qquad \square
$$

For x near zero, a smooth function $f(x)$ is approximated by its Taylor polynomial $P_n(x)$ with error $R_n(x)$. Hence $\int_0^b f(x)\, dx$ is approximated by the integral of its Taylor polynomial $\int_0^b P_n(x)\, dx$ with error $\int_0^b R_n(x)\, dx$.

Examples 5.5.16 (1) Estimate $\int_0^{1/2} e^{-t^2}\, dt$ with error less than .00001.

Solution Substitute $x = -t^2$ in the formula for the nth Taylor polynomial of e^x in Example 5.5.11 (1):

$$
e^{-t^2} = P_n(-t^2) + R_n(-t^2) = \sum_{k=0}^{n} (-1)^k \frac{t^{2k}}{k!} + R_n(-t^2).
$$

Note $P_n(-t^2)$ refers to the nth Taylor polynomial of e^x with $x = -t^2$ which is the $2n^{\text{th}}$ Taylor polynomial of e^{-t^2}. $R_n(-t^2)$ refers to the nth remainder of e^x

with $x = -t^2$. Then

$$\int_0^{1/2} e^{-t^2}\, dt \;=\; \int_0^{1/2} \sum_{k=0}^{n} (-1)^k \frac{t^{2k}}{k!} + R_n(-t^2)\, dt$$

$$= \sum_{k=0}^{n} (-1)^k \frac{t^{2k+1}}{k!\,(2k+1)} \Bigg|_0^{1/2} + \int_0^{1/2} R_n(-t^2)\, dt$$

$$= \sum_{k=0}^{n} (-1)^k \frac{1}{k!\,(2k+1)2^{2k+1}} + \int_0^{1/2} R_n(-t^2)\, dt. \qquad (5.5$$

By (5.5.19) with $x = -t^2$:

$$\big|R_n(-t^2)\big| \;\leq\; \frac{t^{2(n+1)}}{(n+1)!} = \frac{t^{2n+2}}{(n+1)!} \quad \text{for } 0 \leq t \leq$$

Hence

$$\left|\int_0^{1/2} R_n(-t^2)\, dt\right| \;\leq\; \int_0^{1/2} \big|R_n(-t^2)\big|\, dt \leq \int_0^{1/2} \frac{t^{2n+2}}{(n+1}$$

$$= \frac{t^{2n+3}}{(n+1)!\,(2n+3)} \Bigg|_0^{1/2} \;=\; \frac{1}{(n+1)!\,(2n+3)2^{2n+3}}.$$

Thus for $n = 3$:

$$\left|\int_0^{1/2} R_3(-t^2)\, dt\right| \;\leq\; \frac{1}{4!\,(9)2^9} = \frac{1}{110,592} < .00001$$

By (5.5.21) with $n = 3$:

$$\int_0^{1/2} e^{-t^2}\, dt \;\approx\; \frac{1}{2} - \frac{1}{1!\,(3)2^3} + \frac{1}{2!\,(5)2^5} - \frac{1}{3!\,(7)2^7} = \frac{1}{2} - \frac{1}{24} + \frac{1}{320} - \frac{1}{5}$$

$$\approx\; .46127 \quad \text{with error less than } .00001.$$

(2) Estimate $\int_0^{1/3} \sqrt{1+t^4}\, dt$ with error less than .000001.

Solution Substitute $x = t^4$ in the formula for the nth Taylor polynomia $(1+x)^{1/2}$ given in Example 5.5.6 (3):

$$(1+t^4)^{1/2} = P_n(t^4) + R_n(t^4) = \sum_{k=0}^{n} \binom{1/2}{k} t^{4k} + R_n(t^4).$$

Note $P_n(t^4)$ refers to the nth Taylor polynomial of $(1+x)^{1/2}$ with $x = t^4$ wh is the $4n^{\text{th}}$ Taylor polynomial of $\sqrt{1+t^4}$. $R_n(t^4)$ refers to the nth remainde $(1+x)^{1/2}$ with $x = t^4$. Then

$$\int_0^{1/3} \sqrt{1+t^4}\, dt \;=\; \int_0^{1/3} \sum_{k=0}^{n} \binom{1/2}{k} t^{4k} + R_n(t^4)\, dt$$

$$= \sum_{k=0}^{n} \binom{1/2}{k} \frac{t^{4k+1}}{4k+1} \Bigg|_0^{1/3} + \int_0^{1/3} R_n(t^4)\, dt$$

$$= \sum_{k=0}^{n} \binom{1/2}{k} \frac{1}{(4k+1)3^{4k+1}} + \int_0^{1/3} R_n(t^4)\, dt. \qquad (5.5.$$

Let $0 \leq t \leq \frac{1}{3}$. By Example 5.5.11 (3) with $x = t^4$, there is $0 \leq T \leq t$ such th

$$\big|R_n(t^4)\big| \leq \left|\binom{1/2}{n+1} (1+T)^{\frac{1}{2}-n-1} \left(t^4\right)^{n+1}\right| \;\leq\; \left|\binom{1/2}{n+1}\right| (1)t^{4n+4}.$$

Hence $\left| \int_0^{1/3} R_n(t^4)\, dt \right| \le \int_0^{1/3} |R_n(t^4)|\ dt \ \le\ \int_0^{1/3} \left| \binom{1/2}{n+1} \right| t^{4n+4}\, dt$

$$\le \left| \binom{1/2}{n+1} \right| \frac{t^{4n+5}}{4n+5} \Big|_0^{1/3} \ =\ \left| \binom{1/2}{n+1} \right| \frac{1}{3^{4n+5}(4n+5)}.$$

Thus for $n = 1$:

$$\left| \int_0^{1/3} R_1(t)\, dt \right| \le \left| \binom{1/2}{2} \right| \frac{1}{3^9(9)} = \frac{1}{8} \cdot \frac{1}{177,147} \ =\ \frac{1}{1,417,176} < .000001.$$

By (5.5.22): $\displaystyle \int_0^{1/3} \sqrt{1 + t^4}\, dt \approx \binom{1/2}{0} \frac{1}{(1)3^1} \ +\ \binom{1/2}{1} \frac{1}{(5)3^5} \approx .333745$

with error less than .000001. $\qquad\qquad\qquad\qquad\qquad\qquad\qquad\qquad\qquad\qquad\qquad$ $\square$

Historical Remarks

The binomial series for $(1+x)^p$ with p a positive integer was discovered independently several times. In the early formulations, the coefficient $\binom{p}{k}$ of x^k in the binomial expansion of $(1+x)^p$ was defined by an algorithm which we call *Pascal's triangle*: the binomial coefficients $\binom{p}{k}$ form a triangle in which the number $\binom{p}{k}$ is obtained by adding the adjacent pair of numbers $\binom{p-1}{k-1}$ and $\binom{p-1}{k}$ from the preceding row.

$$
\begin{array}{ccccccccccccc}
& & & & & & 1 & & & & & & \\
& & & & & 1 & & 1 & & & & & \\
& & & & 1 & & 2 & & 1 & & & & \\
& & & 1 & & 3 & & 3 & & 1 & & & \\
& & 1 & & 4 & & 6 & & 4 & & 1 & & \\
& 1 & & 5 & & 10 & & 10 & & 5 & & 1 & \\
1 & & 6 & & 15 & & 20 & & 15 & & 6 & & 1
\end{array}
$$

This triangle was first discovered by the Chinese mathematician Jia Xian and the Iraqi mathematician Abu Bakr al-Karaji in the eleventh century. It was rediscovered in Europe by the Germans Petrus Apianus in 1527 and Michael Stifel in 1544. In 1654, Blaise Pascal introduced mathematical induction to give a formal proof that this triangle produces the binomial coefficients. He also derived the formula

$$\binom{p}{k} = \frac{p(p-1)\cdots(p-k+1)}{k!}$$

and applied binomial coefficients to solve problems in probability.

The discovery of the binomial series for $(1+x)^p$, with p a rational number, by Isaac Newton in 1665 was a stunning accomplishment. He first considered the case $p = \frac{n}{2}$ for n an integer. Using nonrigorous interpolation arguments, modeled upon John Wallis's 1665 derivation of an infinite product for π, he determined the coefficients of the power series representation of $\int_0^x (1 - t^2)^{n/2}\, dt$. He then differentiated this power series to obtain the values of the coefficients

$$\binom{n/2}{k} = \frac{\frac{n}{2}\left(\frac{n}{2} - 1\right)\cdots\left(\frac{n}{2} - k + 1\right)}{k!}$$

of the binomial series for $(1+x)^{n/2}$. He then guessed the values

$$\binom{n/q}{k} = \frac{\frac{n}{q}\left(\frac{n}{q} - 1\right)\cdots\left(\frac{n}{q} - k + 1\right)}{k!}$$

of the binomial coefficients in the power series expansion of $(1+x)^{n/q}$ for n and q
integers. Newton verified the conjectured values of these coefficients in several sp
binomial series which he calculated by other methods. The first rigorous derivati
the binomial series was given by Niels Henrik Abel in 1826.

Summary

The reader should be able to find the Taylor polynomials and Taylor series of a g
function $f(x)$. She should know how to bound the remainder terms to deter
when the Taylor series converges to $f(x)$. She should be able to apply Taylor s
to evaluate indeterminate forms. She should also know how to use the integra
Taylor polynomials to estimate $\int_0^b f(x)\,dx$. In particular she should know the follo
Taylor series which we have derived.

$$\ln(1+x) = \sum_{k=1}^{\infty}(-1)^{k-1}\frac{x^k}{k} \qquad \text{for } -1 < x \le 1. \qquad (5.5$$

$$\arctan x = \sum_{k=0}^{\infty}(-1)^k\frac{x^{2k+1}}{2k+1} \qquad \text{for } -1 \le x \le 1. \qquad (5.5$$

$$e^x = \sum_{k=0}^{\infty}\frac{x^k}{k!} \qquad \text{for } x \in \mathbb{R}. \qquad (5.5$$

$$\sin x = \sum_{k=0}^{\infty}(-1)^k\frac{x^{2k+1}}{(2k+1)!} \qquad \text{for } x \in \mathbb{R}. \qquad (5.5$$

$$\cos x = \sum_{k=0}^{\infty}(-1)^k\frac{x^{2k}}{(2k)!} \qquad \text{for } x \in \mathbb{R}. \qquad (5.5$$

$$(1+x)^p = \sum_{k=0}^{\infty}\binom{p}{k}x^k \qquad \text{for } p > 0,\; -1 < x \le 1$$

$$\text{or } p < 0,\; -1 < x < 1. \quad (5.5$$

Basic Exercises

1. Find the Taylor polynomials $P_0(x)$, $P_1(x)$, $P_2(x)$, $P_3(x)$, $P_4(x)$ for each functi
   (a) $f(x) = \frac{1}{1+x}$ (b) $g(x) = e^{-x^2}$ (c) $h(x) = \ln(1+x^2)$
   (d) $i(x) = \sqrt{1+x^4}$ (e) $j(x) = \arcsin x$ (f) $k(x) = \arctan x$
   (g) $m(x) = \tan x$ (h) $n(x) = \sec x$ (i) $p(x) = \frac{1}{x^3+3x^2+3x+1}$
   (j) $q(x) = \cos x$ (k) $r(x) = \sqrt[3]{1-x}$ (l) $s(x) = (1+x)^e$

2. Describe the Taylor polynomials $P_n(x)$ of a polynomial $f(x)$.

3. Use Definition 5.5.4 to show $\frac{d}{dx}\left(P_n\left[f(x)\right]\right) = P_{n-1}\left[f'(x)\right]$ for $n \ge 1$.

4. Evaluate each generalized binomial coefficient.
   (a) $\binom{5}{3}$ (b) $\binom{2/3}{2}$ (c) $\binom{-4}{3}$ (d) $\binom{\pi}{4}$ (e) $\binom{-\frac{1}{2}}{3}$ (f) $\binom{e}{0}$

5. Let n and k be integers with $0 \le k \le n$. Show $\binom{n}{k} = \frac{n!}{k!(n-k)!}$.

6. Verify Pascal's triangle: for positive integers k and n, with $1 \le k \le n-1$:
$$\binom{n}{k} = \binom{n-1}{k-1} + \binom{n-1}{k}.$$

7. For each function f and interval I:

(i) find $f^{(k)}(x)$ for $k \geq 0$ and $x \in I$;

(ii) find the Taylor polynomials $P_n(x)$ for $n \geq 0$;

(iii) find an upper bound for the nth remainder $|R_n(x)|$ for $x \in I$;

(iv) show $\lim_{n \to \infty} R_n(x) = 0$ for $x \in I$;

(v) find the Taylor series of $f(x)$ at $x = 0$.

(a) $f(x) = \cos x$ and $I = \Re$

(b) $f(x) = e^{-3x}$ and $I = \Re$

(c) $f(x) = \frac{1}{2}(e^x - e^{-x})$ and $I = \Re$

(d) $f(x) = \frac{1}{2}(e^x + e^{-x})$ and $I = \Re$

(e) $f(x) = 5^x$ and $I = \Re$

(f) $f(x) = \sin \pi x$ and $I = \Re$

(g) $f(x) = \cos(4x - \pi)$ and $I = \Re$

(h) $f(x) = \sqrt{9 - x}$ and $I = (-9, 0]$

(i) $f(x) = \frac{1}{\sqrt[4]{1+2x}}$ and $I = [0, \frac{1}{2}]$

(j) $f(x) = (3 + 5x)^\pi$ and $I = [0, \frac{3}{5}]$

8. Use the Taylor series previously derived in this section to find the Taylor series of each function in Exercises 7(b)-(j).

9. Use the appropriate Taylor polynomial $P_n(x)$ to estimate each expression with error less than .01. Use Corollary 5.5.12 to determine the value of n.

(a) $\sqrt{e}$
(b) $\frac{1}{e}$
(c) $\sin .5$
(d) $\sin \frac{3\pi}{7}$ using $\pi \approx 3.14$

(e) $\cos .2$
(f) $\sqrt[3]{9}$
(g) $\frac{1}{\sqrt[4]{1.4}}$
(h) $42^{.2}$

10. Let $q(x) = \left\{ \begin{array}{ll} e^{-1/x^2} & \text{for } x \neq 0 \\ 0 & \text{for } x = 0 \end{array} \right\}$. Complete Example 5.5.14 (4) by showing $q_-^{(k)}(0) = 0$.

11. Use Taylor series to evaluate each limit.

(a) $\lim_{x \to 0} \frac{1 - e^x}{\sqrt{1 + x} - 1}$

(b) $\lim_{x \to 0} \frac{\sin x - x}{\cos x - 1}$

(c) $\lim_{x \to 0} \frac{\arctan x - x}{2 \ln(1 + x) - 2x + x^2}$

(d) $\lim_{x \to 0} \frac{3\sqrt[3]{1 + x^4} - 3 - x^4}{2 \cos(x^2) - 2 + x^4}$

(e) $\lim_{x \to 0} \frac{e^{-x^2/2} - \cos x}{\sin(x^2) - x^2}$

(f) $\lim_{x \to 0} \frac{\sin x - \arctan x}{e^x + \ln(1 - x) - 1}$

(g) $\lim_{x \to 0} \frac{e^x + e^{-x} - 2}{\sqrt{1 + x^2} - 1}$

(h) $\lim_{x \to 0} \frac{2 \cos x - \ln(1 - x^2) - 2}{e^{x^2} - x^2 - 1}$

(i) $\lim_{x \to 0} \frac{\sqrt[4]{1 + x^3} + \sin x - x - 1}{x \cos x - x}$

(j) $\lim_{x \to 0} \frac{x + x e^{(-x^2)} - 2x \cos x}{5x + x e^{(-x^2)} - 6 \sin x}$

12. Integrate an appropriate Taylor polynomial $P_n(x)$ to estimate each definite integral with error less than .01. Use Corollary 5.5.12 to determine the value of n.

(a) $\int_0^1 e^{t^2} \, dt$

(b) $\int_0^1 e^{-t^4} \, dt$

(c) $\int_0^{1/4} e^{\sqrt{t}} \, dt$

(d) $\int_0^{1/2} \frac{\sin t}{t} \, dt$

(e) $\int_0^{4/5} \cos(t^3) \, dt$

(f) $\int_0^{3/4} \cos \sqrt{t} \, dt$

(g) $\int_0^1 \sqrt{t} \sin \sqrt{t} \, dt$

(h) $\int_0^{.9} \sqrt[3]{1 + t^2} \, dt$

(i) $\int_0^{.85} \frac{1}{\sqrt{1 + t^3}} \, dt$

(j) $\int_0^{3/4} \frac{t}{1 + t^8} \, dt$

(k) $\int_0^{1/2} \frac{1 - \cos t}{t^2} \, dt$

(l) $\int_0^1 \frac{e^t - \sin t - 1}{t^2} \, dt$

13. Find the number c given by the Generalized Mean Value Theorem so that $\int_a^b u(x) v(x) \, dx = u(c) \int_a^b v(x) \, dx$ in each case:

(a) $\int_0^2 e^x x^2 \, dx$ with $u(x) = e^x$;

(b) $\int_0^\pi x \sin x \, dx$ with $u(x) = \sin x$;

(c) $\int_1^e x \ln x \, dx$ with $u(x) = \ln x$;

(d) $\int_0^1 \frac{x}{1 + x^2} \, dx$ with $u(x) = x$.

14. Show the Mean Value Theorem for Integrals is a special case of the Generalized Mean Value Theorem.

15. Prove the Generalized Mean Value Theorem in the case $v(x) \leq 0$ for all $x \in [a, b]$.

16. Let $f(x)$ be a smooth function at $x = 0$.

(a) Show $f(x) = \sum_{k=0}^\infty \frac{f^{(2k)}(0)}{(2k)!} x^{2k}$ if and only if $f(x)$ is an even function.

(b) Show $f(x) = \sum_{k=0}^\infty \frac{f^{(2k+1)}(0)}{(2k+1)!} x^{2k+1}$ if and only if $f(x)$ is an odd function.

The following four exercises are for those who have studied Section 4.

17. Find the Taylor for $f(x) = \sinh x$. Show it converges to $\sinh x$ for all $x \in \Re$

18. Find the Taylor for $g(x) = \cosh x$. Show it converges to $\cosh x$ for all $x \in \Re$

19. Use Taylor series to evaluate each limit.

(a) $\displaystyle\lim_{x \to 0} \frac{x - \sinh x}{x - \sin x}$ **(b)** $\displaystyle\lim_{x \to 0} \frac{1 - \cosh x}{1 - \cos x}$ **(c)** $\displaystyle\lim_{x \to 0} \frac{x^2 - x \sinh x}{2 + x^2 - 2 \cosh x}$

20. Integrate an appropriate Taylor polynomial $P_n(x)$ to estimate each def integral with error less than .01. Use Corollary 5.5.12 to determine the value of

(a) $\int_0^1 \cosh \sqrt{t}\, dt$ **(b)** $\int_0^1 \frac{\sinh t}{t}\, dt$ **(c)** $\int_0^1 \sqrt{t}\sinh\sqrt{t}\, dt$

Challenging Problems

1. Show for any positive integers k, m and n with $k \leq m + n$:

$$\binom{m+n}{k} = \sum_{i=0}^{k} \binom{m}{i}\binom{n}{k-i}.$$

2. **(a)** Let $S_n = \sum_{k=0}^{n} \frac{1}{k!}$ be the nth partial sum of the series for e. Show $0 < n!(e - S_n) < \frac{1}{n}$.
(b) Show e is irrational.
Hint: If e were the rational number $\frac{m}{n}$, show $n!(e - S_n)$ would be a positive int between zero and one.

Appendix: Binomial Series

In Proposition 5.5.13 we stated that the binomial series, the Taylor series of $h(x$ $(1 + x)^p$, converges to $h(x)$ for x between -1 and $+1$. In this appendix we prove assertion. This requires showing that the remainder terms have limit zero:

$$\lim_{n \to \infty} R_n(x) = 0. \tag{5.5}$$

For x between 0 and 1, we use the derivative form of the remainder, Corollary 5.5 to establish (5.5.29). For $-1 < x < 0$, we derive and apply the integral form of remainder to establish (5.5.29).

Let $P_n(x)$ denote the nth Taylor polynomial of the smooth function $f(x)$. Re the nth remainder term $R_n(x) = f(x) - P_n(x)$ is the error in the approximat $f(x) \approx P_n(x)$. We need a bound on these remainder terms to show their limit, n goes to infinity, is zero, i.e. $f(x)$ equals its Taylor series. In the examples of second subsection, we used the derivative form of the remainder to establish (5.5. For other examples, however, it is easier to use the following integral form.

Proposition 5.5.17 [Integral Form of the Remainder] *Let f be a function which smooth on an open interval I which contains zero. Then for $x \in I$ and $n \geq 1$, the remainder term is given by:*

$$R_n(x) = \frac{1}{n!} \int_0^x f^{(n+1)}(t)(x - t)^n\, dt \tag{5.5.}$$

Proof Begin with the case $n = 1$. By the Second Fundamental Theorem of Calcul $\int_0^x f^{(1)}(t)\, dt = f(x) - f(0)$. Integrate by parts with t the variable and x constant:

$$f(x) = f(0) + \int_0^x f^{(1)}(t)\, dt = f(0) + \int_0^x D(t - x)f^{(1)}(t)\, dt$$

$$= f(0) + (t-x)f^{(1)}(t)\Big|_0^x - \int_0^x (t-x)f^{(2)}(t)\,dt$$

$$= f(0) + xf^{(1)}(0) + \int_0^x (x-t)f^{(2)}(t)\,dt = P_1(x) + \int_0^x (x-t)f^{(2)}(t)\,dt$$

Then $R_1(x) = f(x) - P_1(x) = \int_0^x (x-t)f^{(2)}(t)\,dt$ which establishes (5.5.30) for $n=1$.

Now assume (5.5.30) is true for $n=N$. Under this assumption, we show (5.5.30) is also true for $n = N+1$. Integrate (5.5.30), for $n=N$, by parts:

$$\begin{aligned}
R_N(x) &= \frac{1}{N!}\int_0^x f^{(N+1)}(t)(x-t)^N dt = \frac{1}{N!}\int_0^x f^{(N+1)}(t)D\left[-\frac{(x-t)^{N+1}}{N+1}\right]dt \\
&= -\frac{1}{N!}f^{(N+1)}(t)\frac{(x-t)^{N+1}}{N+1}\Big|_0^x - \frac{1}{N!}\int_0^x f^{(N+2)}(t)\left[-\frac{(x-t)^{N+1}}{N+1}\right]dt \\
&= \frac{f^{(N+1)}(0)}{(N+1)!}x^{N+1} + \frac{1}{(N+1)!}\int_0^x f^{(N+2)}(t)(x-t)^{N+1}\,dt.
\end{aligned}$$

$$\begin{aligned}
\text{Then }\ f(x) &= P_N(x) + R_N(x) \\
&= P_N(x) + \frac{f^{(N+1)}(0)}{(N+1)!}x^{N+1} + \frac{1}{(N+1)!}\int_0^x f^{(N+2)}(t)(x-t)^{N+1}\,dt \\
&= P_{N+1}(x) + \frac{1}{(N+1)!}\int_0^x f^{(N+2)}(t)(x-t)^{N+1}\,dt \quad \text{and}
\end{aligned}$$

$$R_{N+1}(x) = f(x) - P_{N+1}(x) = \frac{1}{(N+1)!}\int_0^x f^{(N+2)}(t)(x-t)^{N+1}\,dt.$$

This establishes (5.5.30) for $n = N+1$. By mathematical induction, (5.5.30) is true for all values of n. □

We now have the methods to show $h(x) = (1+x)^p$ equals its binomial series for x between -1 and $+1$.

Proposition 5.5.13(d) If **either** $p > 0$ *with* $-1 < x \le 1$ **or** $p < 0$ *with* $-1 < x < 1$:

$$(1+x)^p = \sum_{k=0}^\infty \binom{p}{k}x^k.$$

Proof Recall the formulas for $h^{(n+1)}(x)$ and $P_n(x)$ derived in Example 5.5.6(3):

$$h^{(n+1)}(x) = (n+1)!\binom{p}{n+1}(1+x)^{p-(n+1)} \quad \text{and}\quad P_n(x) = \sum_{k=0}^n \binom{p}{k}x^k.$$

Case I Either $p > 0$ with $0 < x \le 1$ or $p < 0$ with $0 < x < 1$

We use the derivative form of the remainder. Let $0 \le t \le x$. If $p \le n$, then

$$\begin{aligned}
\left|h^{(n+1)}(t)\right| &= \left|\binom{p}{n+1}\frac{(n+1)!}{(1+t)^{n+1-p}}\right| \le \left|\binom{p}{n+1}\frac{(n+1)!}{(1+0)^{n+1-p}}\right| \\
&= (n+1)!\left|\binom{p}{n+1}\right|.
\end{aligned}$$

By Corollary 5.5.12,

$$|R_n(x)| \le \frac{(n+1)!\left|\binom{p}{n+1}\right|}{(n+1)!}x^{n+1} = \left|\binom{p}{n+1}\right|x^{n+1} \qquad (5.5.31)$$

$$= \ |p||p-1|\cdots|p-n|\frac{x^{n+1}}{(n+1)!}$$

$$= \ |p|x \cdot \frac{|p-1|x}{1} \cdot \frac{|p-2|x}{2} \cdots \frac{|p-n|x}{n} \cdot \frac{1}{n+1}$$

$$= \ |p|x \cdot \left|1-\frac{p}{1}\right|x \cdot \left|1-\frac{p}{2}\right|x \cdots \left|1-\frac{p}{n}\right|x \cdot \frac{1}{n+1}.$$

Select a positive integer $N > p$ such that $N > \frac{-px}{1-x}$ if $p < 0$. (By hypothesis, x when $p < 0$.) The reason for these conditions will become apparent below. For n

$$|R_n(x)| \le |p|x \cdot \left|1-\frac{p}{1}\right|x \cdot \left|1-\frac{p}{2}\right|x \cdots \left|1-\frac{p}{N}\right|x \cdot \left|1-\frac{p}{N+1}\right|x \cdots \left|1-\frac{p}{n}\right|x \cdot \frac{}{n}$$

The product of the first $N+2$ terms above is the constant

$$A = |p|x \cdot \left|1-\frac{p}{1}\right|x \cdot \left|1-\frac{p}{2}\right|x \cdots \left|1-\frac{p}{N}\right|x$$

which is independent of n. Then

$$|R_n(x)| \le \begin{cases} \dfrac{A}{n+1}\left[\left(1-\dfrac{p}{N+1}\right)x\right]^{n-N} & \text{if } p < 0 \\[4mm] \dfrac{Ax^{n-N}}{n+1} & \text{if } p > 0 \end{cases} . \qquad (5.5$$

If $p > 0$, then $0 \le x \le 1$ and $\lim\limits_{n\to\infty} |R_n(x)| = 0$. If $p < 0$, we chose N such $N+1 > N > \frac{-px}{1-x}$. Then

$$\begin{aligned} (N+1)(1-x) &> -px \\ N+1 &> (N+1)x - px = (N+1-p)x \\ 1 &> \left(1-\frac{p}{N+1}\right)x. \end{aligned}$$

Thus when $p < 0$, then $0 \le x < 1$ and it follows from (5.5.32) that $\lim\limits_{n\to\infty} |R_n(x)| =$

Case II $\ -1 < x < 0$

We use the integral form of the remainder.

- If $p > 0$, let q be an integer larger than p.
- If $p < 0$, let q be a positive integer with $\left(1-\frac{p}{q}\right)|x| < 1$.

Then for $-1 < t \le 0$ and $n > q$:

$$\left|\frac{h^{(n+1)}(t)}{(n+1)!}\right| = \left|\binom{p}{n+1}(1+t)^{p-(n+1)}\right| = |p|\frac{|p-1|}{1}\frac{|p-2|}{2}\cdots\frac{|p-n|}{n}\frac{(1+t)^p}{n+}$$

$$\left|\frac{h^{(n+1)}(t)}{n!}\right| = |p||p-1|\left|1-\frac{p}{2}\right|\left|1-\frac{p}{3}\right|\cdots\left|1-\frac{p}{q-1}\right|\left|1-\frac{p}{q}\right|\cdots\left|1-\frac{p}{n}\right|(1+t)^p$$

The product of the first q terms above is the constant

$$C = |p||p-1|\left|1-\frac{p}{2}\right|\left|1-\frac{p}{3}\right|\cdots\left|1-\frac{p}{q-1}\right|$$

which does not depend on n or t. Thus

$$\left|\frac{h^{(n+1)}(t)}{n!}\right| = C\left|1-\frac{p}{q}\right|\cdots\left|1-\frac{p}{n}\right|(1+t)^{p-n-1}$$

$$\le \begin{cases} C(1+t)^{p-n-1} & \text{if } p > 0 \\[3mm] C\left(1-\dfrac{p}{q}\right)^{n-q+1}(1+t)^{p-n-1} & \text{if } p < 0 \end{cases} . \qquad (5.5.$$

Consider the integral form of the remainder:

$$|R_n(x)| = \frac{1}{n!}\left|\int_0^x h^{(n+1)}(t)(x-t)^n \, dt\right| \leq \int_x^0 \left|\frac{h^{(n+1)}(t)}{n!}\right|(t-x)^n \, dt. \qquad (5.5.34)$$

First consider the case $p > 0$. By (5.5.33) and (5.5.34):

$$|R_n(x)| \leq \int_x^0 C(1+t)^{p-n-1}(t-x)^n \, dt.$$

By the Mean Value Theorem for integrals there is an x_n with $x \leq x_n \leq 0$ such that:

$$|R_n(x)| \leq C(1+x_n)^{p-n-1}(x_n - x)^n|x|.$$

Note $1 + x_n \geq 1 + x > 0$. Hence

$$|R_n(x)| \leq C\left(\frac{x_n - x}{1 + x_n}\right)^{n-p}\frac{(x_n - x)^p|x|}{1 + x_n} \leq C\left(\frac{x_n + |x|}{1 + x_n}\right)^{n-p}\frac{|x|^{p+1}}{1 + x}.$$

Since $0 < |x| < 1$ and $x_n < 0$, we have $x_n + |x| < |x|x_n + |x| = |x|(x_n + 1)$. Thus

$$0 \leq \lim_{n\to\infty}|R_n(x)| \leq \lim_{n\to\infty} C\left(\frac{|x|(x_n + 1)}{1 + x_n}\right)^{n-p}\frac{|x|^{p+1}}{1 + x}$$

$$= \lim_{n\to\infty} C|x|^{n-p}\frac{|x|^{p+1}}{1 + x} = \frac{C}{1 + x}\lim_{n\to\infty}|x|^{n+1} = 0.$$

Hence $\lim_{n\to\infty}|R_n(x)| = 0$ for $p > 0$ and $-1 < x < 0$. Now consider the case $p < 0$. By (5.5.33) and (5.5.34):

$$|R_n(x)| \leq \int_x^0 C\left(1 - \frac{p}{q}\right)^{n-q+1}(1+t)^{p-n-1}(t-x)^n \, dt.$$

By the Mean Value Theorem for integrals there is an x_n with $x \leq x_n \leq 0$ such that:

$$|R_n(x)| \leq C\left(1 - \frac{p}{q}\right)^{n-q+1}(1+x_n)^{p-n-1}(x_n - x)^n|x|.$$

Since $1 - \frac{p}{q} \geq 1$ and $1 + x_n \geq 1 + x > 0$,

$$|R_n(x)| \leq C\left(1 - \frac{p}{q}\right)^{n-q+1}\left(\frac{x_n - x}{1 + x_n}\right)^n\frac{|x|}{(1 + x_n)^{1-p}}$$

$$\leq C\left(1 - \frac{p}{q}\right)^{n+1}\left(\frac{x_n + |x|}{1 + x_n}\right)^n\frac{|x|}{(1 + x)^{1-p}}.$$

As above, $x_n + |x| < |x|(x_n + 1)$. By the choice of q, we have $\left(1 - \frac{p}{q}\right)|x| < 1$. Therefore

$$0 \leq |R_n(x)| \leq C\left(1 - \frac{p}{q}\right)^{n+1}|x|^n\frac{|x|}{(1 + x)^{1-p}}$$

and $\lim_{n\to\infty} C\left(1 - \frac{p}{q}\right)^{n+1}|x|^n\frac{|x|}{(1 + x)^{1-p}} = \frac{C}{(1 + x)^{1-p}}\lim_{n\to\infty}\left[\left(1 - \frac{p}{q}\right)|x|\right]^{n+1} = 0.$

By the Pinching Theorem, $\lim_{n\to\infty} R_n(x) = 0$ for $p < 0$ and $-1 < x < 0$. $\qquad\square$

Exercise

State the integral form of the remainder $R_n(x)$ for each function.

(a) $f(x) = e^x$ (b) $g(x) = \sin x$ (c) $h(x) = \cos x$

(d) $k(x) = \ln(1 + x)$ (e) $m(x) = \sqrt{1 + x}$ (f) $k(x) = \frac{1}{(1+x)^3}$

5.6   General Taylor Series

In Section 5, we studied the Taylor polynomials $P_n(x)$ of a function $f(x)$ whi
smooth at $x = 0$. $P_n(x)$ is a degree n polynomial in x which approximates
best for x near 0. In the first subsection, we extend this construction from 0 to
$c \in \Re$ when $f(x)$ is smooth at $x = c$. That is, we approximate $f(x)$ by a degr
polynomial $P_n(x, c)$ in $x - c$ which approximates $f(x)$ best for x near c. In the se
subsection, we take the limit of these polynomials to define the Taylor series of $f($.
$x = c$. In the third subsection, we give applications to evaluate limits as x approa
c and to estimate definite integrals $\int_c^b f(x)\, dx$. Before proceeding with details of t
constructions, we present an example that uses a representation of $f(x)$ as a p
series in $x - c$ to estimate $f(x)$ for x close to c.

Motivating Example 5.6.1 Estimate $\sqrt{4.01}$ as the sum of the first three term
an infinite series.

Solution We write $\sqrt{x}$ as a power series in $x - 4$. Its partial sums, with $x = $
give good estimates of $\sqrt{4.01}$ because $x - 4 = .01$ is close to zero. Write

$$\sqrt{x} = \sqrt{4 + (x-4)} = 2\sqrt{1 + \frac{x-4}{4}} = 2\left(1 + \frac{x-4}{4}\right)^{1/2}.$$

Apply the binomial series for $(1+u)^{1/2}$ with $u = \frac{x-4}{4}$ and $-1 < u \le 1$:

$$\sqrt{x} = 2\sum_{k=0}^{\infty} \binom{1/2}{k}\left(\frac{x-4}{4}\right)^k = \sum_{k=0}^{\infty} \binom{1/2}{k}\frac{(x-4)^k}{2^{2k-1}}. \qquad (5.$$

This is valid for $-1 < \frac{x-4}{4} \le 1$, i.e $0 < x \le 8$. In particular, when $x = 4.01$:

$$\sqrt{4.01} = \sum_{k=0}^{\infty} \binom{1/2}{k}\frac{(.01)^k}{2^{2k-1}}.$$

We estimate $\sqrt{4.01}$ by the sum of the first three terms of this series:

$$\sqrt{4.01} \approx 2 + \frac{1}{2}\times\frac{.01}{2} - \frac{1}{8}\times\frac{.0001}{8} = 2 + .0025 - .0000015625 = 2.0024984375$$

In fact, $\sqrt{2} = 2.0024984395$ to ten decimal places!

Taylor Polynomials in $x - c$

Recall the Taylor polynomial $P_n(x)$ of the smooth function $f(x)$ at $x = 0$ is the deg
n polynomial in x given by $P_n(x) = \sum_{k=0}^{n}\frac{f^{(k)}(0)}{k!}x^k$ which approximates $f(x)$ best
x near 0. When $f(x)$ is smooth at $c \in \Re$ we define a degree n polynomial in $x - c$

$$P_n(x,c) = a_0 + a_1(x-c) + a_2(x-c)^2 + \cdots + a_{n-1}(x-c)^{n-1} + a_n(x-c)^n = \sum_{k=0}^{n} a_k(x-$$

which approximates $f(x)$ best for x near c. Change variables to $u = x - c$:

$$P_n(x,c) = a_0 + a_1 u + a_2 u^2 + \cdots + a_n u^n$$

should approximate $f(x) = f(u+c)$ for $u = x - c$ near 0. Hence we take $P_n(x,c)$
be the nth Taylor polynomial $P_n(u)$ of $g(u) = f(u+c)$ at $u = 0$. Since $\frac{du}{dx} = 1$,

$$g^{(k)}(u) = \frac{d^k}{du^k}f(u+c) = f^{(k)}(u+c) \qquad (5.6$$

by the chain rule. Then

$$P_n(x,c) = P_n(u) = \sum_{k=0}^{n} \frac{g^{(k)}(0)}{k!} u^k = \sum_{k=0}^{n} \frac{f^{(k)}(0+c)}{k!} u^k = \sum_{k=0}^{n} \frac{f^{(k)}(c)}{k!} (x-c)^k. \quad (5.6.3)$$

Thus the Taylor polynomials in $x-c$ are merely translations of the Taylor polynomials in u which we studied in Section 5.

Definition 5.6.2 *Let $f(x)$ be a smooth function at $x = c$. Define the nth Taylor polynomial $P_n(x,c)$ of $f(x)$ at $x = c$ as*

$$\begin{aligned} P_n(x,c) &= f(c) + f^{(1)}(c)(x-c) + \frac{f^{(2)}(c)}{2!}(x-c)^2 + \cdots + \frac{f^{(n)}(c)}{n!}(x-c)^n \\ &= \sum_{k=0}^{n} \frac{f^{(k)}(c)}{k!}(x-c)^k. \end{aligned}$$

Since the polynomial $P_n(x,c)$ has degree n, its kth derivatives are zero for $k > n$. However, Proposition 5.5.5 generalizes from $x = 0$ to $x = c$: the approximation $P_n(x,c) \approx f(x)$ for x near c implies that $P_n(x,c)$ and $f(x)$ have the same k^{th} derivatives at $x = c$ for $k \le n$.

Proposition 5.6.3 *Let f be a smooth function at $x = c$. For $0 \le k \le n$, the kth derivatives of $f(x)$ and $P_n(x,c)$ have the same value at $x = c$.*

Proof For $0 \le k \le n$, the kth derivative of $P_n(x,c)$, at $x = c$, is the constant term of $P_n^{(k)}(x,c)$, written as a polynomial in $x-c$. This is the kth derivative of $\frac{f^{(k)}(c)}{k!}(x-c)^k$ which is $k! \frac{f^{(k)}(c)}{k!} = f^{(k)}(c)$. $\qquad\square$

This proposition supports our expectation that $P_n(x,c)$ approximates $f(x)$ for n large and x near c. For example, $P_1(x,c) = f(c) + f'(c)(x-c)$ is the tangent line to the graph of $y = f(x)$ at $x = c$.

Examples 5.6.4 (1) Find the third Taylor polynomial $P_3(x,e)$ of $f(x) = \ln x$.

Solution Compute the first three derivatives of f:

$$f^{(1)}(x) = \frac{1}{x}, \qquad f^{(2)}(x) = -\frac{1}{x^2}, \qquad f^{(3)}(x) = \frac{2}{x^3}.$$

$$\begin{aligned} \text{Hence } P_3(x,e) &= f(e) + \frac{f^{(1)}(e)}{1!}(x-e) + \frac{f^{(2)}(e)}{2!}(x-e)^2 + \frac{f^{(3)}(e)}{3!}(x-e)^3 \\ &= 1 + \frac{1}{e}(x-e) - \frac{1}{2e^2}(x-e)^2 + \frac{1}{3e^3}(x-e)^3 . \end{aligned}$$

(2) Write $g(x) = x^4 - 5x^3 - 2x^2 + 7x + 8$ as a polynomial in $x + 1$.

Solution $g(x)$ and its Taylor polynomial $P_4(x,-1)$ must be equal because they are both degree four polynomials with the same higher derivatives at $x = -1$. To calculate the coefficients of $P_4(x,-1)$, compute the derivatives of g:

$$\begin{aligned} g^{(1)}(x) &= 4x^3 - 15x^2 - 4x + 7, & g^{(2)}(x) &= 12x^2 - 30x - 4, \\ g^{(3)}(x) &= 24x - 30, & g^{(4)}(x) &= 24 . \end{aligned}$$

Thus $g(-1) = 5, g'(-1) = -8, g^{(2)}(-1) = 38, g^{(3)}(-1) = -54, g^{(4)}(-1) = 24$.

$$\text{Hence } g(x) \;=\; P_4(x,-1) = g(-1) + \frac{g^{(1)}(-1)}{1!}(x+1) + \frac{g^{(2)}(-1)}{2!}(x+1)^2$$

$$+ \frac{g^{(3)}(-1)}{3!}(x+1)^3 + \frac{g^{(4)}(-1)}{4!}(x+1$$

$$= \; 5 - 8(x+1) + 19(x+1)^2 - 9(x+1)^3 + (x+1)^4 \,.$$

Taylor Series at $x = c$

The Taylor polynomials $P_n(x,c)$ of $f(x)$ at $x = c$ are the partial sums of the p
series $\sum_{k=0}^{\infty} \frac{f^{(k)}(c)}{k!}(x-c)^k$, called the Taylor series of $f(x)$ at $x = c$. This is an exa
of a *power series in $x - c$*:

$$\sum_{k=0}^{\infty} a_k(x-c)^k = a_0 + a_1(x-c) + a_2(x-c)^2 + \cdots + a_n(x-c)^n + \cdots \qquad (5$$

There are two ways to determine where $f(x)$ equals its Taylor series. We can s
the derivative form of the remainders $R_n(x,c) = f(x) - P_n(x,c)$ have limit zero
many examples, however, it is easier to establish this Taylor series as the transla
of a Taylor series at $x = 0$ which we derived in Section 5.

To show the errors $R_n(x,c) = f(x) - P_n(x,c)$ of the approximations $f(x) \approx P_n($
have limit zero, we need a formula for their value. In the following proposit
we translate the descriptions of $R_n(x) = R_n(x,0)$, given in Proposition 5.5.10
Corollary 5.5.12, to obtain corresponding descriptions of $R_n(x,c)$.

Proposition 5.6.5 *Let f be a smooth function on an open interval I which cont
Define the nth remainder of $f(x)$ at $x = c$ by:*

$$R_n(x,c) = f(x) - P_n(x,c).$$

(a) *If $x \in I$, there is a number $t(x)$ between c and x such that*

$$R_n(x,c) = \frac{f^{(n+1)}(t(x))}{(n+1)!}(x-c)^{n+1}. \qquad (5.$$

(b) *If $\left| f^{(n+1)}(t) \right| \le M_{n+1}(x)$ for all numbers t between x and c, then*

$$|R_n(x,c)| \le \frac{M_{n+1}(x)}{(n+1)!}|x-c|^{n+1}. \qquad (5.$$

Proof (a) In the discussion before Definition 5.6.2, we defined $u = x - c$ and
$g(u) = f(u+c)$. In (5.6.2) and (5.6.3) we showed:
   (1) $g^{(k)}(u) = f^{(k)}(u+c)$;
   (2) $P_n(x,c)$ is the nth Taylor polynomial $P_n(u)$ of $g(u)$.
Let $R_n(u) = g(u) - P_n(u)$ be the nth remainder of $g(u)$ at $u = 0$. By (2):

$$R_n(x,c) = f(x) - P_n(x,c) = f(u+c) - P_n(u) = g(u) - P_n(u) = R_n(u).$$

By Proposition 5.5.10 there is a number $T(u)$ between u and 0 with:

$$R_n(x,c) = R_n(u) = \frac{g^{(n+1)}(T(u))}{(n+1)!}u^{n+1} = \frac{f^{(n+1)}(T(u)+c)}{(n+1)!}(x-c)^{n+1}$$

by (1). Let $t(x) = T(u) + c$ which lies between $u + c = x$ and $0 + c = c$.
(b) This is an immediate consequence of (a).

Proposition 5.6.5(b) is used to show $\lim_{n\to\infty} R_n(x,c) = 0$ for certain values of x near c. For these values of x, we have the power series representation:

$$f(x) = \sum_{k=0}^{\infty} \frac{f^{(k)}(c)}{k!}(x-c)^k. \tag{5.6.7}$$

This power series is called the *Taylor series* of $f(x)$ at $x = c$. We can often establish the validity of (5.6.7) by translating a known Taylor series from $x = 0$ to $x = c$.

Examples 5.6.6 (1) **(a)** Find the Taylor series $P(x)$ of $f(x) = \frac{1}{x}$ at $x = 2$.
(b) Show $f(x)$ equals this Taylor series $P(x)$ for $1 < x < 3$.
Solution **(a)** Write $f(x) = x^{-1}$ and observe that for $k \geq 0$:

$$f^{(k)}(x) = (-1)^k k! x^{-k-1} = (-1)^k \frac{k!}{x^{k+1}}.$$

Hence $f^{(k)}(2) = (-1)^k \frac{k!}{2^{k+1}}$ and

$$P(x) = \sum_{k=0}^{\infty} (-1)^k \frac{k!}{2^{k+1}} \frac{(x-2)^k}{k!} = \sum_{k=0}^{\infty} (-1)^k \frac{1}{2^{k+1}}(x-2)^k.$$

(b) For $1 < x < 3$,

$$\left| f^{(k)}(x) \right| = \frac{k!}{x^{k+1}} \leq k!$$

Apply Proposition 5.6.5(b), for $1 < x < 3$, with $M_{n+1}(x) = (n+1)!$:

$$0 \leq |R_n(x,2)| \leq \frac{(n+1)!}{(n+1)!}|x-2|^{n+1} = |x-2|^{n+1}.$$

Since $-1 < x - 2 < 1$, we have $\lim_{n\to\infty} |x-2|^{n+1} = 0$. By the Pinching Theorem, $\lim_{n\to\infty} R_n(x,2) = 0$. Hence

$$\frac{1}{x} = \sum_{k=0}^{\infty} (-1)^k \frac{1}{2^{k+1}}(x-2)^k \quad \text{for } 1 < x < 3. \tag{5.6.8}$$

(2) **(a)** Find the Taylor series of $f(x) = e^x$ at $x = 1$.
(b) Assuming $e \approx 2.7183$, approximate $e^{.8}$ with error less than .004.

Solution **(a)** By Proposition 5.5.13(a), $e^u = \sum_{k=0}^{\infty} \frac{u^k}{k!}$ for $u \in \Re$. Substitute $u = x - 1$:

$$e^{x-1} = \sum_{k=0}^{\infty} \frac{(x-1)^k}{k!} \quad \text{for } x \in \Re.$$

Multiply the preceding equation by e:

$$e^x = \sum_{k=0}^{\infty} \frac{e(x-1)^k}{k!} \quad \text{for } x \in \Re.$$

(b) Since $f^{(k)}(x) = e^x \leq e$ for $x \leq 1$, we take $M_{n+1}(.8) = e$ in (5.6.6):

$$|R_n(.8,1)| \leq \frac{e}{(n+1)!}(.2)^{n+1}.$$

When $n = 2$, $|R_2(.8,1)| \leq \frac{e(.2)^3}{3!} < .004$. Hence

$$e^{.8} \approx P_2(.8,1) = e + e(.8-1) + \frac{e(.8-1)^2}{2} \approx 2.229$$

with error less than .004.

(3) **(a)** Find the Taylor series of $g(x) = \cos x$ at $x = \frac{\pi}{2}$.

   (b) Assuming $\pi \approx 3.1416$, estimate $\cos \frac{5\pi}{12}$ to the nearest .0002.

Solution **(a)** By Proposition 5.5.13(b), $\sin u = \sum_{k=0}^{\infty}(-1)^k \frac{u^{2k+1}}{(2k+1)!}$ for u
For $x \in \Re$, let $u = x - \frac{\pi}{2}$. Then

$$g(x) = \cos x = -\sin\left(x - \frac{\pi}{2}\right) = -\sum_{k=0}^{\infty}(-1)^k \frac{\left(x - \frac{\pi}{2}\right)^{2k+1}}{(2k+1)!} = \sum_{k=0}^{\infty}(-1)^{k+1} \frac{(x - }{(2}$$

(b) For $k \geq 1$, $g^{(k)}(x)$ is either $\pm\cos x$ or $\pm\sin x$. Hence $|g^{(k)}(x)| \leq 1$ for x
and we take $M_{2n+2}(x) = 1$ in (5.6.6):

$$\left| R_{2n+1}\left(\frac{5\pi}{12}, \frac{\pi}{2}\right) \right| \leq \frac{\left(\frac{\pi}{12}\right)^{2n+2}}{(2n+2)!}.$$

When $n = 1$:
$$\left| R_3\left(\frac{5\pi}{12}, \frac{\pi}{2}\right) \right| \leq \frac{\left(\frac{\pi}{12}\right)^4}{4!} < .0002 \text{ and}$$

$$\cos\frac{5\pi}{12} \approx \frac{\pi}{12} - \frac{\left(\frac{\pi}{12}\right)^3}{6} \approx .2588$$

with error less than .0002.

Applications

We begin with two examples that use Taylor series at $x = c$ to evaluate limits
approaches c. Then we give an example that uses a Taylor series at $x = c$ to estin
a definite integral of the form $\int_c^b f(x)\, dx$.

   In Section 5, we evaluated limits of fractions, as x approaches 0, by represen
the numerator and denominator as power series in x. The special feature of x is th
becomes zero when $x = 0$. After canceling the common factors of x in the numer
and denominator we substituted $x = 0$ to evaluate the limit. We evaluate the li
of a fraction, as x approaches c, by representing the numerator and denominato.
power series in $x - c$. Note $x - c$ has the feature that it becomes zero when $x =$
After canceling the common factors of $x - c$ in the numerator and denominator
substitute $x = c$ to evaluate the limit.

Examples 5.6.7 (1) Evaluate $\lim_{x \to \pi} \frac{1 + \cos x}{2\sqrt{\pi x} - \pi - x}$.

Solution Direct substitution of $x = \pi$ into the above fraction produces
indeterminate form $\frac{0}{0}$. We represent the numerator and denominator as po
series in $x - \pi$. Observe that $\cos(x - \pi) = \cos x \cos \pi + \sin x \sin \pi = -\cos x$.
use the Taylor series at $u = 0$ of $\cos u$ given in (5.5.16), with $u = x - \pi$, to w.
$1 + \cos x$ as a power series in $x - \pi$:

$$1 + \cos x \;=\; 1 - \cos(x - \pi) = 1 - \cos u = 1 - \sum_{k=0}^{\infty}(-1)^k \frac{1}{(2k)!}u^{2k}$$

$$=\; 1 - \sum_{k=0}^{\infty}(-1)^k \frac{1}{(2k)!}(x - \pi)^{2k} = 1 - \left[1 + \sum_{k=1}^{\infty}(-1)^k \frac{1}{(2k)!}(x - \pi)\right.$$

$$=\; \sum_{k=1}^{\infty}(-1)^{k+1} \frac{1}{(2k)!}(x - \pi)^{2k}. \tag{5.}$$

We use the binomial series $(1 + v)^{1/2}$, with $v = \frac{x - \pi}{\pi}$, to write $2\sqrt{\pi x} - \pi - x$ as a power series in $x - \pi$:

$$
\begin{aligned}
2\sqrt{\pi x} - \pi - x &= 2\pi \left(1 + \frac{x - \pi}{\pi}\right)^{1/2} - \pi - x = 2\pi(1 + v)^{1/2} - \pi - x \\
&= 2\pi \sum_{k=0}^{\infty} \binom{1/2}{k} v^k - \pi - x = 2\pi \sum_{k=0}^{\infty} \binom{1/2}{k} \left(\frac{x - \pi}{\pi}\right)^k - \pi - x \\
&= 2\pi \left[1 + \frac{1}{2}\left(\frac{x - \pi}{\pi}\right) + \sum_{k=2}^{\infty} \binom{1/2}{k}\left(\frac{x - \pi}{\pi}\right)^k\right] - \pi - x \\
&= 2\pi \sum_{k=2}^{\infty} \binom{1/2}{k} \frac{(x - \pi)^k}{\pi^k}.
\end{aligned}
$$

Now we can write the given fraction as a quotient of power series in $x - \pi$. After dividing numerator and denominator by their common fact $(x - \pi)^2$, we substitute $x = \pi$ to evaluate the limit.

$$
\lim_{x \to \pi} \frac{1 + \cos x}{2\sqrt{\pi x} - \pi - x} = \lim_{x \to \pi} \frac{\displaystyle\sum_{k=1}^{\infty} (-1)^{k+1} \frac{1}{(2k)!}(x - \pi)^{2k}}{2\pi \displaystyle\sum_{k=2}^{\infty} \binom{1/2}{k} \frac{(x - \pi)^k}{\pi^k}}
$$

$$
= \lim_{x \to \pi} \frac{\dfrac{1}{2}(x - \pi)^2 + (x - \pi)^4 \displaystyle\sum_{k=2}^{\infty} (-1)^{k+1} \frac{1}{(2k)!}(x - \pi)^{2k-4}}{-\dfrac{1}{4\pi}(x - \pi)^2 + 2\pi(x - \pi)^3 \displaystyle\sum_{k=3}^{\infty} \binom{1/2}{k} \frac{(x - \pi)^{k-3}}{\pi^k}}
$$

$$
= \lim_{x \to \pi} \frac{\dfrac{1}{2} + (x - \pi)^2 \displaystyle\sum_{k=2}^{\infty} (-1)^{k+1} \frac{1}{(2k)!}(x - \pi)^{2k-4}}{-\dfrac{1}{4\pi} + 2\pi(x - \pi) \displaystyle\sum_{k=3}^{\infty} \binom{1/2}{k} \frac{(x - \pi)^{k-3}}{\pi^k}} = \frac{\frac{1}{2} + 0}{-\frac{1}{4\pi} + 0} = -2\pi.
$$

(2) Evaluate $\lim_{x \to 1} \dfrac{8\sqrt{x} - 3 - 6x + x^2}{2\ln x + 3 - 4x + x^2}$.

Solution Direct substitution of $x = 1$ into the above fraction produces the indeterminate form $\frac{0}{0}$. We write the numerator and denominator of this fraction as Taylor series at $x = 1$. Recall the Taylor series (5.4.10) for $\ln(1 + u)$ and the binomial series for $\sqrt{1 + u} = (1 + u)^{1/2}$:

$$
\ln(1 + u) = \sum_{k=1}^{\infty} (-1)^{k-1} \frac{u^k}{k} \quad \text{and} \quad \sqrt{1 + u} = \sum_{k=0}^{\infty} \binom{1/2}{k} u^k
$$

for $-1 < u \le 1$. Substitute $u = x - 1$. Then for $0 < x \le 2$:

$$
\begin{aligned}
\ln x &= \sum_{k=1}^{\infty} (-1)^{k-1} \frac{(x - 1)^k}{k} = x - 1 - \frac{(x - 1)^2}{2} + \frac{(x - 1)^3}{3} + \sum_{k=4}^{\infty} (-1)^{k-1} \frac{(x - 1)^k}{k} \\
&= -\frac{3}{2} + 2x - \frac{x^2}{2} + \frac{(x - 1)^3}{3} + (x - 1)^4 \sum_{k=4}^{\infty} (-1)^{k-1} \frac{(x - 1)^{k-4}}{k} \\
\sqrt{x} &= \sum_{k=0}^{\infty} \binom{1/2}{k} (x - 1)^k
\end{aligned}
$$

$$= 1 + \frac{1}{2}(x-1) - \frac{1}{8}(x-1)^2 + \frac{1}{16}(x-1)^3 + \sum_{k=4}^{\infty} \binom{1/2}{k}(x-1)^k$$

$$= \frac{3}{8} + \frac{3}{4}x - \frac{1}{8}x^2 + \frac{1}{16}(x-1)^3 + (x-1)^4 \sum_{k=4}^{\infty} \binom{1/2}{k}(x-1)^{k-4}$$

Substitution of these power series into the given limit, writes that limit quotient of power series in $x - 1$. After dividing numerator and denominate their common factor $(x-1)^3$, we substitute $x = 1$ to evaluate the limit.

$$\lim_{x \to 1} \frac{8\sqrt{x} - 3 - 6x + x^2}{2\ln x + 3 - 4x + x^2}$$

$$= \lim_{x \to 1} \frac{8\left[\frac{3}{8} + \frac{3}{4}x - \frac{1}{8}x^2 + \frac{1}{16}(x-1)^3 + \sum_{k=4}^{\infty}\binom{1/2}{k}(x-1)^k\right] - 3 - 6x \cdot}{2\left[-\frac{3}{2} + 2x - \frac{x^2}{2} + \frac{(x-1)^3}{3} + \sum_{k=4}^{\infty}(-1)^{k-1}\frac{(x-1)^k}{k}\right] + 3 - 4x + x^2}$$

$$= \lim_{x \to 1} \frac{\frac{1}{2}(x-1)^3 + 8(x-1)^4 \sum_{k=4}^{\infty}\binom{1/2}{k}(x-1)^{k-4}}{\frac{2(x-1)^3}{3} + 2(x-1)^4 \sum_{k=4}^{\infty}(-1)^{k-1}\frac{(x-1)^{k-4}}{k}}$$

$$= \lim_{x \to 1} \frac{\frac{1}{2} + 8(x-1)\sum_{k=4}^{\infty}\binom{1/2}{k}(x-1)^{k-4}}{\frac{2}{3} + 2(x-1)\sum_{k=4}^{\infty}(-1)^{k-1}\frac{(x-1)^{k-4}}{k}} = \frac{\frac{1}{2}+0}{\frac{2}{3}+0} = \frac{3}{4}.$$

The Taylor polynomial $P_n(x,c)$ approximates $f(x)$ for x near c with error $R_n($ Hence $\int_c^b P_n(x,c)\,dx$ approximates $\int_c^b f(x)\,dx$ with error $\int_c^b R_n(x,c)\,dx$.

Example 5.6.8 Estimate the integral $\int_4^6 \frac{1}{\sqrt[3]{t^2-8t+24}}\,dt$ with error less than .004.

Solution Since the lower bound of this definite integral is 4 we write the integr as a power series in $x - 4$. Observe that

$$\frac{1}{\sqrt[3]{t^2-8t+24}} = (t^2-8t+24)^{-1/3} = \left[8 + (t-4)^2\right]^{-1/3}$$

$$= 8^{-1/3}\left[1 + (t-4)^2/8\right]^{-1/3} = \frac{1}{2}\left[1 + (t-4)^2/8\right]^{-1/3}.$$

Substitute $u = \frac{(t-4)^2}{8}$ in the binomial series $(1+u)^{-1/3} = \sum_{k=0}^{\infty}\binom{-1/3}{k}u^k$:

$$\frac{1}{\sqrt[3]{t^2-8t+24}} = \frac{1}{2}\left[1 + \frac{(t-4)^2}{8}\right]^{-1/3} = \frac{1}{2}\sum_{k=0}^{\infty}\binom{-1/3}{k}\frac{(t-4)^{2k}}{8^k} \tag{5.6.}$$

for $0 \le u = \frac{(t-4)^2}{8} < 1$, i.e. $4 - \sqrt{8} < t < 4 + \sqrt{8}$. Hence the representation (5.6.10 valid for $4 \le t \le 6$. Let $R_n(u)$ denote the nth remainder term of the binomial se for $(1+u)^{-1/3}$. Then

$$\int_4^6 \frac{1}{\sqrt[3]{t^2-8t+24}}\,dt = \int_4^6 \frac{1}{2}\sum_{k=0}^{n}\binom{-1/3}{k}\frac{(t-4)^{2k}}{8^k} + \frac{1}{2}R_n\left(\frac{(t-4)^2}{8}\right)\,dt$$

$$= \frac{1}{2}\sum_{k=0}^{n}\binom{-1/3}{k}\frac{(t-4)^{2k+1}}{(2k+1)8^k}\Big|_4^6 + \int_4^6 \frac{1}{2}R_n\left(\frac{(t-4)^2}{8}\right)\,dt$$

$$= \frac{1}{2}\sum_{k=0}^{n}\binom{-1/3}{k}\frac{2^{2k+1}}{(2k+1)8^k} + \int_4^6 \frac{1}{2}R_n\left(\frac{(t-4)^2}{8}\right)\,dt$$

$$= \sum_{k=0}^{n} \binom{-1/3}{k} \frac{1}{(2k+1)2^k} + \int_4^6 \frac{1}{2} R_n \left(\frac{(t-4)^2}{8} \right) dt. \tag{5.6.11}$$

Let $h(u) = (1+u)^{-1/3}$ for $0 \le u < 1$. By (5.5.5):

$$\left| h^{(n+1)}(u) \right| = (n+1)! \left| \binom{-1/3}{n+1} (1+u)^{1/3-n-1} \right|$$

$$\le (n+1)! \left| \binom{-1/3}{n+1} \right| (1+0)^{-n-2/3} = (n+1)! \left| \binom{-1/3}{n+1} \right|$$

Apply Cor. 5.5.12 with $M_{n+1}(u) = (n+1)! \left| \binom{-1/3}{n+1} \right|$, $u = \frac{(t-4)^2}{8}$ and $4 \le t \le 6$:

$$\left| \frac{1}{2} R_n \left(\frac{(t-4)^2}{8} \right) \right| \le \frac{1}{2} \frac{M_{n+1}(u)}{(n+1)!} \left| \frac{(t-4)^2}{8} \right|^{n+1} = \frac{1}{2} \left| \binom{-1/3}{n+1} \right| \frac{(t-4)^{2(n+1)}}{8^{n+1}}$$

$$\le \left| \binom{-1/3}{n+1} \right| \frac{(t-4)^{2n+2}}{2^{3n+4}}.$$

It follows that $\left| \int_4^6 \frac{1}{2} R_n \left(\frac{(t-4)^2}{8} \right) dt \right| \le \int_4^6 \left| \frac{1}{2} R_n \left(\frac{(t-4)^2}{8} \right) \right| dt$

$$\le \int_4^6 \left| \binom{-1/3}{n+1} \right| \frac{(t-4)^{2n+2}}{2^{3n+4}} dt = \left| \binom{-1/3}{n+1} \right| \frac{(t-4)^{2n+3}}{2^{3n+4}(2n+3)} \Big|_4^6$$

$$= \left| \binom{-1/3}{n+1} \right| \frac{2^{2n+3}}{2^{3n+4}(2n+3)} = \left| \binom{-1/3}{n+1} \right| \frac{1}{2^{n+1}(2n+3)}.$$

For n = 2 : $\left| \frac{1}{2} \int_4^6 R_2 \left(\frac{(t-4)^2}{8} \right) dt \right| \le \left| \binom{-1/3}{3} \right| \frac{1}{2^3(7)} = \frac{1}{324} < .004.$

By (5.6.11) : $\int_4^6 \frac{1}{\sqrt[3]{t^2 - 8t + 24}} dt \approx \binom{-1/3}{0} + \binom{-1/3}{1} \frac{1}{(3)2^1} + \binom{-1/3}{2} \frac{1}{(5)2^2}$

$$= 1 - \frac{1}{18} + \frac{1}{90} \approx .9556$$

with error less than .004. $\qquad \square$

Historical Remarks

The first terms of the Taylor series of a function were calculated by James Gregory in 1671. He never communicated his method but did indicate that it involved finding the higher derivatives of the function. Isaac Newton calculated the Taylor series (5.5.15) for $\sin x$ by inverting the Taylor series for $\arcsin x$. (He obtained the latter series by observing that $\theta = \arcsin x$ is twice the area of a sector of radius one and angle θ. From Figure 5.6.9, the area of this sector equals $\int_0^x \sqrt{1-t^2} \, dt - \frac{1}{2} x \sqrt{1-x^2}$. Newton used binomial series to evaluate this expression as a power series.) He then derived the Taylor series (5.5.16) of $\cos x$ by computing $\sqrt{1 - \sin^2 x}$. Newton was the first to state the general formula (5.6.7) for the Taylor series of a function in 1691. It was first published by Newton's student Brook Taylor in 1715. He was accused of plagiarism by Johann Bernoulli, who in 1694 had published the series:

$$\int_0^x u(x) \, dv(x) = \sum_{k=1}^{\infty} (-1)^{k-1} \frac{v(x)^k}{k!} \frac{d^{k-1} u(x)}{dx^{k-1}}. \tag{5.6.12}$$

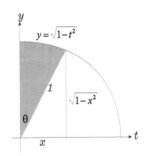

Figure 5.6.9
Newton's Taylor Series for $\arcsin x$

The derivation of Bernoulli's formula and the deduction of Taylor's series from (5.6.12) are given as Problem 1. The case $c = 0$ of Taylor's series was also published by another

of Newton's students, Colin Maclaurin, in 1742 who gave credit to Taylor for the r
Nevertheless, Maclaurin's name is usually associated with this series! He used
series to prove the following generalization of the Second Derivative Test.

> Let f be a smooth function at $x = c$. Assume $f^{(k)}(c) = 0$ for $1 \le k \le n$,
> and $f^{(n+1)}(c) \ne 0$. When n is odd, f has a local minimum at $x = c$ if
> $f^{(n+1)}(c) > 0$ while f has a local maximum at $x = c$ if $f^{(n+1)}(c) < 0$.
> When n is even, f does not have a local extremum at $x = c$.

The verification of this result is given as Problem 2. The Taylor series for e
published by Leonhard Euler in 1748. The remainder $R_n(x, c)$ in approximating
by the Taylor polynomial $P_n(x, c)$ was first studied by Joseph Louis Lagrange in
who derived the derivative form (5.5.12) of $R_n(x, c)$. The integral form, present
the appendix of Section 5, was discovered by Augustin Cauchy in 1823.

Summary

The reader should be able to find the Taylor polynomials $P_n(x, c)$ and the T
series of $f(x)$ at $x = c$. She should be able to translate Taylor series at $x =$
construct Taylor series at $x = c$. She should also know how to bound the rema
terms $R_n(x, c)$ to determine when this Taylor series converges to $f(x)$. She s
be able to apply Taylor series to evaluate indeterminate forms. She should also
how to use the integrals of the Taylor polynomials of $f(x)$ to estimate $\int_c^b f(x)\,d$

Basic Exercises

1. Find the indicated Taylor polynomial for each function:

(a) $P_3(x, 1)$ for $f(x) = 2^x$;   (b) $P_4(x, 1)$ for $g(x) = \ln x$;

(c) $P_5(x, 2)$ for $h(x) = \frac{1}{x}$;   (d) $P_4(x, 4)$ for $j(x) = \sqrt{x}$;

(e) $P_3\left(x, \frac{\pi}{4}\right)$ for $k(x) = \tan x$;   (f) $P_3\left(x, \frac{\pi}{2}\right)$ for $m(x) = \csc x$;

(g) $P_3\left(x, \frac{1}{2}\right)$ for $n(x) = \arcsin x$;   (h) $P_3(x, 1)$ for $p(x) = \arctan x$;

(i) $P_5(x, \pi)$ for $q(x) = \sin^2 x$;   (j) $P_3(x, 8)$ for $r(x) = x^{2/3}$.

2. Write each polynomial as a polynomial in $x - c$ for the indicated value of c:

(a) $f(x) = x^3 + 5x^2 - 2x + 4$ with $c = 2$;   (b) $g(x) = x^4 + x^2 + 1$ with $c =$
(c) $h(x) = x^3 - 6x^2 + 3x + 7$ with $c = 3$;   (d) $k(x) = x^4 - 8x + 5$ with $c =$

3.   Find the Taylor series of each function at $x = c$. Justify that the Taylor s
converges to the given function for $x \in I$.

(a) $f(x) = 3^x$ at $c = 2$, $I = \Re$ (b) $g(x) = e^{2x-7}$ $c = 4$, $I = \Re$
(c) $h(x) = \sin x$, $c = \pi$ $I = \Re$ (d) $j(x) = \cos \pi x$, $c = 3$, $I = \Re$
(e) $k(x) = \sqrt[3]{x}$, $c = 1$, $I = (0, 2]$ (f) $m(x) = \ln x$, $c > 0$, $I = (0, 2c]$
(g) $p(x) = \frac{1}{\sqrt{x^2 - 4x + 13}}$, $c = 2$ $I = [2, 5)$ (h) $q(x) = \frac{1}{7+x}$, $c = -2$, $I = [-2,$
(i) $r(x) = \frac{1}{\sqrt{x}}$, $c = 9$, $I = (0, 18)$ (j) $s(x) = \arctan(1 + x)$, $c = -1$, $I = [-2$

4.   Use the appropriate Taylor polynomial $P_n(x, c)$ to estimate each of the follo
with error less than .01. Use Proposition 5.6.5 to determine the value of n.

(a) $3^{2.1}$ given $\ln 3 \approx 1.099$ (b) $e^{3.2}$ given $e \approx 2.718$ (c) $\sin \frac{9\pi}{10}$ given $\pi \approx 3$

(d) $\cos \frac{17\pi}{5}$ given $\pi \approx 3.142$ (e) $\sqrt[3]{1.2}$ (f) $\ln 6.3$ given $\ln 6 \approx 1$

(g) $p(2.3)$ with $p(x) = \frac{1}{\sqrt{x^2 - 4x + 13}}$ (h) $q(-1.4)$ with $q(x) = \frac{1}{7+x}$ (i) $\sqrt{15}$.

5. Use Taylor series to evaluate each limit.

(a) $\lim\limits_{x \to 1} \dfrac{e^x - e}{x - 1}$ (b) $\lim\limits_{x \to \pi} \dfrac{1 + \cos x}{\sin(x - \pi)^2}$ (c) $\lim\limits_{x \to e^+} \dfrac{\ln x - 1}{\cos\sqrt{x - e} - 1}$ (d) $\lim\limits_{x \to 64} \dfrac{\sqrt[3]{x} - 4}{\sqrt{x} - 8}$

(e) $\lim\limits_{x \to \frac{\pi}{2}} \dfrac{1 - \sin x}{\cos x}$ (f) $\lim\limits_{x \to \pi} \dfrac{e^{x/\pi} + e\cos x}{\ln x - \ln \pi}$ (g) $\lim\limits_{x \to 1} \dfrac{x^\pi + \cos \pi x}{e^x - e x^e}$ (h) $\lim\limits_{x \to 1} \dfrac{x^{3/2} - x^{2/3}}{e - e^x}$

6. Integrate the appropriate Taylor polynomial $P_n(t, c)$ to estimate each integral with error less than .01. Use Proposition 5.6.5 to determine the value of n.

(a) $\int_1^{1.8} \sqrt[4]{t^2 - 2t + 5}\, dt$ (b) $\int_3^{3.4} \dfrac{1}{(t^2 - 6t + 10)^7}\, dt$ (c) $\int_{-4}^{-3.1} \dfrac{1}{\sqrt[5]{t^2 + 8t + 48}}\, dt$

(d) $\int_1^{1.3} \dfrac{\sin \pi t}{t - 1}\, dt$ using $\pi \approx 3.1416$ (e) $\int_1^2 e^{t^2 - t}\, dt$ using $e \approx 2.718$

The following two exercises are for those who have studied Section 4.10.

7. Find the Taylor series of each function at $x = c$. Justify that the Taylor series converges to the given function for $x \in \Re$.

(a) $f(x) = \sinh x$, $c = 1$ (b) $g(x) = \cosh x$, $c = 1$ (c) $h(x) = \sinh x + \cosh x$, $c = 2$

8. Integrate the appropriate Taylor polynomial $P_3(t, 1)$ to estimate $\int_1^2 \dfrac{\sinh t + \cosh t - e}{t - 1}\, dt$.

Challenging Problems

1. (a) Use integration by parts to derive Bernoulli's series (5.6.12).
(b) Deduce Taylor's series at $x = 0$ from Bernoulli's series (5.6.12) with $v(x) = x$ and $u(x) = f^{(n)}(x)$ for $n \geq 0$.

2. (a) Prove Maclaurin's generalization of the Second Derivative Test.
(b) Apply this result to identify each of the following local extrema:

(i) $f(x) = \sin(x^4)$ at $x = 0$;   (ii) $g(x) = e^{-x^6}$ at $x = 0$;

(iii) $h(x) = \ln(1 + x^5)$ at $x = 0$;   (iv) $k(x) = \cos(x^7)$ at $x = 0$.

3. For those who studied the appendix to Section 5, prove the following general statement of the integral form of the remainder.

Let f be a function which is smooth on an open interval I which contains c. Let $P_n(x, c)$ be the nth Taylor polynomial of $f(x)$ at $x = c$ with $R_n(x, c) = f(x) - P_n(x, c)$ the nth remainder. Then for all $x \in I$: $R_n(x, c) = \frac{1}{n!} \int_c^x f^{(n+1)}(t)(x - t)^n\, dt$.

5.7   Power Series

We have constructed various examples of power series: geometric series and their integrals in Section 4 as well as Taylor series in Sections 5, 6. In this section, we study general power series and derive their fundamental properties. Before proceeding, consider several power series we derived in the preceding three sections.

Motivating Example 5.7.1 Recall these power series from the indicated references.

$$e^x = \sum_{k=0}^{\infty} \frac{(x - 0)^k}{k!} \qquad \text{for } |x - 0| < \infty \qquad (5.5.14)$$

$$\ln(1 + x) = \sum_{k=1}^{\infty} (-1)^{k-1} \frac{(x - 0)^k}{k} \qquad \text{for } |x - 0| < 1 \text{ or } x = 1 \qquad (5.4.10)$$

$$\arctan x = \sum_{k=0}^{\infty} (-1)^k \frac{(x - 0)^{2k+1}}{2k + 1} \qquad \text{for } |x - 0| < 1 \text{ or } x = \pm 1 \qquad (5.4.12)$$

$$\sqrt{x} = \sum_{k=0}^{\infty} \binom{1/2}{k} \frac{(x-4)^k}{2^{2k-1}} \qquad \text{for } |x-4| < 4 \text{ or } x = 8 \qquad (5.6$$

$$\cos x = \sum_{k=0}^{\infty} (-1)^{k+1} \frac{(x-\pi)^{2k}}{(2k)!} \qquad \text{for } |x-\pi| < \infty \qquad (5.6$$

$$\frac{1}{\sqrt[3]{x^2 - 8x + 24}} = \sum_{k=0}^{\infty} \binom{-1/3}{k} \frac{(x-4)^{2k}}{2^{3k+1}} \qquad \text{for } |x-4| < \sqrt{8} \qquad (5.6.$$

For each of these power series $\sum_{k=0}^{\infty} a_k (x-c)^k$ there is an $R \geq 0$ such that:

- R is either a nonnegative number or ∞;
- the power series converges for $|x-c| < R$, i.e for $x \in (c - R, c + R)$;
- the power series diverges for $|x-c| > R$, i.e for $x \in (-\infty, c - R) \cup (c + R, c$
- there is no pattern of what happens when $x = c - R$ or $x = c + R$.

We call R the *radius of convergence* of the power series. In particular,

$$c = 0 \text{ and } R = \infty \text{ in } (5.5.14), \qquad c = 0 \text{ and } R = 1 \text{ in } (5.4.10),$$
$$c = 0 \text{ and } R = 1 \text{ in } (5.4.12), \qquad c = 4 \text{ and } R = 4 \text{ in } (5.6.1),$$
$$c = \pi \text{ and } R = \infty \text{ in } (5.6.9), \qquad c = 4 \text{ and } R = \sqrt{8} \text{ in } (5.6.10).$$

The observations deduced in the example above are valid for all power series the first subsection, we derive the Ratio Test. In the second section, we apply Ratio Test to show every power series $P(x)$ in $x - c$ has a radius of convergence R the third subsection, we show the derivative of the power series $P(x)$ equals the of the derivatives of its summands for $|x - c| < R$. For these x, the definite int of $P(x)$ equals the sum of the integrals of its summands. In the fourth subsec we show power series can be multiplied by a distributive law to obtain their Ca product. In the appendix we derive the Root Test, an alternative to the Ratio to find the radius of convergence of a power series.

Ratio Test

In this subsection, we study the Ratio Test. We begin with the Comparison which deduces the convergence of a given series by comparing it with a known Then we apply the Comparison Test to derive the Ratio Test.

Proposition 5.7.2 (Comparison Test) *Let $\sum_{k=p}^{\infty} a_k$ and $\sum_{k=q}^{\infty} b_k$ be two infi series. Assume there is a positive integer N such that for $k \geq N$:*

$$0 \leq |a_k| \leq b_k.$$

If the series $\sum_{k=q}^{\infty} b_k$ converges, then the series $\sum_{k=p}^{\infty} a_k$ is absolutely convergent.

Proof The partial sums of the series $\sum_{k=N}^{\infty} |a_k|$ are less than or equal to the responding partial sums of the series $S = \sum_{k=N}^{\infty} b_k$. The latter partial sums form increasing sequence with limit S. Therefore, the partial sums of the series $\sum_{k=N}^{\infty}$ form an increasing sequence which is bounded above by S. This sequence conver to its least upper bound, and the series $\sum_{k=N}^{\infty} a_k$ is absolutely convergent. Hence series $\sum_{k=p}^{\infty} a_k = a_p + \cdots + a_{N-1} + \sum_{k=N}^{\infty} a_k$ is also absolutely convergent.

Examples 5.7.3 Show the series $\sum_{k=1}^{\infty} (-1)^k \frac{1}{2^k + 101}$ is absolutely convergent.

Solution $\sum_{k=1}^{\infty} \frac{1}{2^k}$ is a geometric series with ratio $\frac{1}{2}$ which converges since $\frac{1}{2} < 1$

$$\left| (-1)^k \frac{1}{2^k + 101} \right| = \frac{1}{2^k + 101} < \frac{1}{2^k} \quad \text{for } k \geq 1.$$

By the Comparison Test, the series $\sum_{k=1}^{\infty}(-1)^k \frac{1}{2^k+101}$ is absolutely convergent. □

Consider the geometric series $\sum_{k=0}^{\infty} aR^k$, with $a \neq 0$ and ratio $R = \frac{aR^{k+1}}{aR^k} > 0$. This geometric series is absolutely convergent for $|R| < 1$ and divergent for $|R| > 1$. The Ratio Test gives the same conclusion about any series $\sum_{k=p}^{\infty} a_k$ with $R = \lim_{k\to\infty} \frac{a_{k+1}}{a_k}$. The derivation of the Ratio Test uses the Comparison Test to compare the series $\sum_{k=p}^{\infty} a_k$ with a geometric series having ratio R.

Proposition 5.7.4 (Ratio Test) *Let $\sum_{k=p}^{\infty} a_k$ be an infinite series such that the limit*

$$\lim_{k\to\infty} \left| \frac{a_{k+1}}{a_k} \right| = R$$

exists. There are three cases:

   (a) *if $R < 1$, the series is absolutely convergent;*
   (b) *if $R > 1$, the series is divergent;*
   (c) *if $R = 1$, there is no conclusion.*

Proof (a) Choose $R < S < 1$. For k large enough, say $k \geq N$, we have $\frac{|a_{k+1}|}{|a_k|} < S$ and $|a_{k+1}| < S|a_k|$. For $k \geq N$:

$$|a_k| < S|a_{k-1}| < S^2|a_{k-2}| < \cdots < S^{k-N}|a_N|$$

The geometric series $\sum_{k=N}^{\infty} |a_N|S^{k-N}$ has ratio S with $0 < S < 1$ and therefore converges. By the Comparison Test, the series $\sum_{k=p}^{\infty} a_k$ is absolutely convergent.

(b) For k large enough, say $k \geq N$, $\frac{|a_{k+1}|}{|a_k|} > 1$ and $|a_{k+1}| > |a_k| > 0$. Therefore $\lim_{k\to\infty} a_k$ can not equal zero, and the series $\sum_{k=p}^{\infty} a_k$ diverges by the nth Term Test.

(c) When $R = 1$, the series may converge or diverge. For example, the divergent harmonic series $\sum_{k=1}^{\infty} \frac{1}{k}$ has $R = \lim_{k\to\infty} \frac{\frac{1}{k+1}}{\frac{1}{k}} = \lim_{k\to\infty} \frac{k}{k+1} = 1$ while the convergent telescoping series $\sum_{k=1}^{\infty} \frac{1}{k(k+1)}$ also has $R = \lim_{k\to\infty} \frac{\frac{1}{(k+1)(k+2)}}{\frac{1}{k(k+1)}} = \lim_{k\to\infty} \frac{k}{k+2} = 1$. □

The Ratio Test can be applied to numerical series and to power series. To evaluate the limit for R, group corresponding parts of $\frac{a_{k+1}}{a_k}$ as separate fractions.

Examples 5.7.5 (1) Determine whether $\sum_{k=1}^{\infty} \frac{(k!)^2}{(2k)!}$ converges or diverges.

   Solution Apply the Ratio Test:

$$\begin{aligned} R &= \lim_{k\to\infty} \frac{[(k+1)!]^2 / [2(k+1)]!}{(k!)^2/(2k)!} = \lim_{k\to\infty} \left[\frac{(k+1)!}{k!} \right]^2 \frac{(2k)!}{(2k+2)!} \\ &= \lim_{k\to\infty} \left[\frac{(k+1)(k!)}{(k!)} \right]^2 \frac{(2k)!}{(2k+2)(2k+1)(2k)!} = \lim_{k\to\infty} (k+1)^2 \frac{1}{2(k+1)(2k+1)} \\ &= \lim_{k\to\infty} \frac{k+1}{4k+2} = \frac{1}{4} < 1. \end{aligned}$$

By the Ratio Test, the series $\sum_{k=1}^{\infty} \frac{(k!)^2}{(2k)!}$ converges.

(2)   Apply the Ratio Test to the power series $\sum_{k=0}^{\infty} \frac{kx^k}{3^k+1}$.

Solution The limit R of the Ratio Test is given by:

$$R = \lim_{k \to \infty} \frac{(k+1)|x|^{k+1}/(3^{k+1}+1)}{k|x|^k/(3^k+1)} = \lim_{k \to \infty} \frac{k+1}{k} \frac{|x|^{k+1}}{|x|^k} \frac{3^k+1}{3^{k+1}+1}$$

$$= \lim_{k \to \infty} \frac{k+1}{k}|x|\frac{1+3^{-k}}{3+3^{-k}} = 1|x|\frac{1+0}{3+0} = \frac{|x|}{3}.$$

By the Ratio Test this power series is absolutely convergent if $R = \frac{|x|}{3} < 1$ a divergent if $R = \frac{|x|}{3} > 1$. That is, we have absolute convergence for $-3 < x$ and divergence for $x < -3$ or $x > 3$. In the terminology of Example 5.7.1 radius of convergence of this power series is 3.

Radius of Convergence

In Examples 5.7.1 and 5.7.5 we gave seven examples of the radius of convergen of a power series. We use the Ratio Test to show every power series has a radi convergence. Then we give additional examples.

Proposition 5.7.6 *Let $P(x) = \sum_{n=0}^{\infty} a_n(x-c)^n$ be a power series. Then the* $0 \le R \le \infty$, *called the radius of convergence of $P(x)$, such that:*

$$P(x) \text{ converges absolutely if } |x-c| < R \text{ and } P(x) \text{ diverges if } |x-c| > R.$$

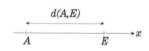

d(A,E)

$A \qquad\qquad E \qquad \to x$

Figure 5.7.7
Distance Between A, E

Proof We use the notation $d(A,E) = |A-E|$ to denote the distance betwee and E on the real number line. See Figure 5.7.7. The proofs of the following statements are given below.

(1) If $P(u)$ converges and $d(w,c) < d(u,c)$ then $P(w)$ is absolutely converge

(2) If $P(u)$ diverges and $d(w,c) > d(u,c)$ then $P(w)$ also diverges.

We show our proposition is a consequence of these two statements. Let S be the of numbers $d(u,c)$ for which $P(u)$ is absolutely convergent.

Case I: there is no number x for which $P(x)$ diverges.

By (1), $S = [0,\infty)$ and $R = \infty$.

Case II: there is a number x for which $P(x)$ diverges.

By (2), $d(x,c)$ is an upper bound of S. Let R be the least upper bound of S.
• If $d(y,c) > R$, then $P(y)$ diverges. If $P(y)$ were to converge, choose z with $R < d(z,c) < d(y,c)$. By (1), $P(z)$ would be absolutely convergent. Then $d(z,c)$ with $d(z,c) > R$ which would contradict the fact that R is an upper bound of S.
• If $d(y,c) < R$, choose z with $d(y,c) < d(z,c) < R$. If $P(z)$ diverges then by $d(z,c)$ is an upper bound of S. This contradicts the fact that R is the least up bound of S. Hence $P(z)$ converges. By (1), $P(y)$ is absolutely convergent.

Proof of (1) Let $Q = \frac{d(w,c)}{d(u,c)} = |\frac{w-c}{u-c}| < 1$. Since the series $P(u)$ converges, the li of its nth term must be zero. Thus select a positive integer N so that if $k \ge N$, t $|a_k(u-c)^k| < 1$. Then for $k \ge N$:

$$|a_k(w-c)^k| = |a_k(u-c)^k|\left|\frac{w-c}{u-c}\right|^k < \left|\frac{w-c}{u-c}\right|^k = Q^k.$$

The geometric series $\sum_{k=N}^{\infty} Q^k$ has ratio Q with $0 < Q < 1$ and hence converges. the Comparison Test, the series $\sum_{k=0}^{\infty} a_k(w-c)^k$ is absolutely convergent.

Proof of (2) Since the series $\sum_{k=0}^{\infty} a_k(u-c)^k$ diverges, the series $\sum_{k=0}^{\infty} a_k(w-c)^k$ can not converge by (1). Hence it diverges. □

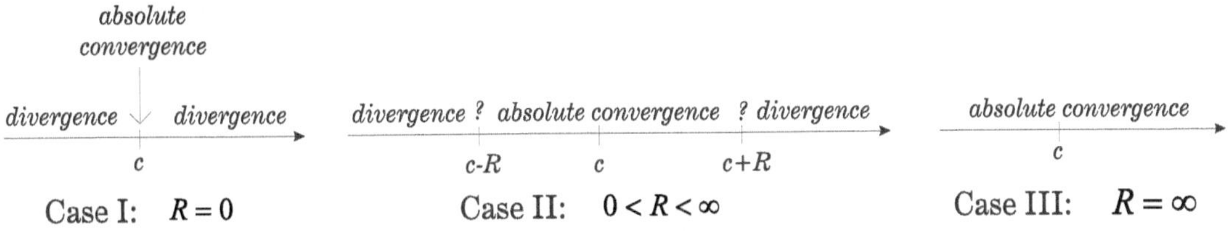

Figure 5.7.8 Radius of Convergence R of a Power Series in $x - c$

There are three cases of the radius of convergence of a power series in $x - c$ which are illustrated in Figure 5.7.8. We explain each case and give examples.

Case I: $R = 0$

In this case the power series converges absolutely for $x = c$ and diverges for $x \neq c$.

Example 5.7.9 Find the radius of convergence R of the power series $\sum_{k=0}^{\infty} k! x^k$.

Solution Apply the Ratio Test:

$$\lim_{k \to \infty} \frac{(k+1)!|x|^{k+1}}{k!|x|^k} = \lim_{k \to \infty} \frac{(k+1)!}{k!} \frac{|x|^{k+1}}{|x|^k} = \lim_{k \to \infty} (k+1)|x| = \left\{ \begin{array}{ll} 0 & \text{if } x = 0 \\ \infty & \text{if } x \neq 0 \end{array} \right\}.$$

Thus $\sum_{k=0}^{\infty} k! x^k$ converges absolutely when $x = 0$ and diverges if $x \neq 0$. The radius of convergence of this power series is $R = 0$. □

Case II: $0 < R < \infty$

In this case the power series $P(x)$ converges absolutely for $|x - c| < R$ and diverges for $|x - c| > R$. In other words:

- $P(x)$ is absolutely convergent for $x \in (c - R, c + R)$;
- $P(x)$ is divergent for $x \in (-\infty, c - R) \cup (c + R, \infty)$.

Anything is possible at $x = c - R$ and at $x = c + R$. We will study this phenomenon in Section 9. We studied examples of this case in (5.4.10), (5.4.12), (5.6.1), (5.6.8), (5.6.10) and Example 5.7.5 (2). Here are two more examples.

Examples 5.7.10 (1) Find the radius of convergence R of the power series $\sum_{k=1}^{\infty} \frac{k! x^k}{k^k}$.

Solution Apply the Ratio Test:

$$\lim_{k \to \infty} \frac{(k+1)!|x|^{k+1}/(k+1)^{k+1}}{k!|x|^k/k^k} = \lim_{k \to \infty} \frac{(k+1)!}{k!} \frac{|x|^{k+1}}{|x|^k} \frac{k^k}{(k+1)^{k+1}}$$

$$= \lim_{k \to \infty} \frac{(k+1)k!}{k!} |x| \left(\frac{k}{k+1} \right)^k \frac{1}{k+1} = \lim_{k \to \infty} \frac{|x|}{\left(1 + \frac{1}{k} \right)^k} = \frac{|x|}{e}$$

by (5.3.11). Thus $\sum_{k=1}^{\infty} \frac{k! x^k}{k^k}$ converges absolutely when $\frac{|x|}{e} < 1$ and diverges when $\frac{|x|}{e} > 1$. That is, this power series converges absolutely when $x \in (-e, e)$ and diverges when $x \in (-\infty, -e) \cup (e, \infty)$. The radius of convergence is $R = e$.

(2) Find the radius of convergence R of the power series $\sum_{k=0}^{\infty} \frac{2^k}{k^2+1}(x-3)^k$.

Solution Apply the Ratio Test:

$$\lim_{k \to \infty} \frac{2^{k+1}|x-3|^{k+1}/\left[(k+1)^2+1\right]}{2^k|x-3|^k/(k^2+1)} = \lim_{k \to \infty} \frac{2^{k+1}}{2^k} \frac{|x-3|^{k+1}}{|x-3|^k} \frac{k^2+1}{k^2+2k+2}$$

$$= 2|x-3|(1) = 2|x-3|.$$

When $2|x-3| < 1$ this power series converges absolutely, and when $2|x-3|$
it diverges, i.e. this power series is absolutely convergent when $x \in \left(\frac{5}{2}, \frac{7}{2}\right)$
divergent when $x \in \left(-\infty, \frac{5}{2}\right) \cup \left(\frac{7}{2}, \infty\right)$. The radius of convergence is $R =$

Case III: $R = \infty$

In this case, the power series converges absolutely for all $x \in \Re$. We saw exampl
this phenomenon in (5.5.14), (5.5.15) and (5.5.16). Here is another one.

Example 5.7.11 Find the radius of convergence R of the power series $\sum_{k=0}^{\infty} \frac{(k+}{}$

Solution Apply the Ratio Test:

$$
\lim_{k \to \infty} \frac{(k+2)|x|^{k+1}/(k+1)!}{(k+1)|x|^k/k!} = \lim_{k \to \infty} \frac{k+2}{k+1} \frac{|x|^{k+1}}{|x|^k} \frac{k!}{(k+1)!}
$$

$$
= \lim_{k \to \infty} \frac{k+2}{k+1}|x| \frac{k!}{(k+1)k!} = \lim_{k \to \infty} \frac{k+2}{k+1}|x| \lim_{k \to \infty} \frac{1}{k+1} = (1)(|x|)(0) = 0.
$$

Thus $\sum_{k=0}^{\infty} \frac{(k+1)x^k}{k!}$ converges absolutely for all $x \in \Re$, and $R = \infty$.

A function and its Taylor series are two individual functions. Specifically, let
be the Taylor series of $f(x)$ at $x = c$, and let R be the radius of convergence of T
Since $T(x)$ is only defined for $x \in (c - R, c + R)$, it only makes sense to ask whe
$f(x)$ equals $T(x)$ for $|x - c| < R$. Here are two extreme examples.

• When $f(x) = e^x$, we have $T(x) = \sum_{k=0}^{\infty} \frac{x^k}{k!}$ with radius of convergence $R = \infty$.
(5.5.14), $f(x) = T(x)$ for all $x \in \Re$.

• In Example 5.5.14(4), we showed the Taylor series of $f(x) = \begin{cases} e^{-1/x^2} & \text{if } x \neq \\ 0 & \text{if } x = \end{cases}$
is $T(x) = 0$ with radius of convergence $R = \infty$. Although both $f(x)$ and $T(x)$
defined for all $x \in \Re$, we have $f(x) = T(x)$ only when $x = 0$. In this example,
and $T(x)$ are totally different functions.

Taylor series can be used to define functions. For example, the functions $f(x$
$\sum_{k=1}^{\infty} \frac{k!x^k}{k^k}$, with domain $(-e, e)$, and $g(x) = \sum_{k=0}^{\infty} \frac{2^k}{k^2+1}(x-3)^k$, with domain $\left(\frac{5}{2}\right.$
of Example 5.7.10 are new functions that we can not describe in any other mann

Differentiation of Power Series

An important property of power series is that they can be differentiated and integra
term by term. Thus we can easily establish a power series representation for
derivative or integral of a power series. We also show that every power series is
Taylor series of its sum.

The derivative of a finite sum of functions is the sum of the derivatives of
summands. A power series is an infinite sum of monomials. We show its derivativ
the sum of the derivatives of these monomials.

Theorem 5.7.12 *Let $P(x) = \sum_{k=0}^{\infty} a_k(x - c)^k$ be a power series with radius of c
vergence R. Then for $|x - c| < R$, the power series $P_1(x) = \sum_{k=1}^{\infty} ka_k(x - c)^{k-1}$
absolutely convergent and*

$$
\frac{d}{dx}\left[\sum_{k=0}^{\infty} a_k(x - c)^k\right] = \sum_{k=1}^{\infty} ka_k(x - c)^{k-1}.
$$

Proof We show the general case of this theorem follows from the case $c = 0$. Let $u = x - c$ with $\frac{du}{dx} = 1$. For $|x - c| = |u| < R$, apply the chain rule:

$$\frac{d}{dx}[P(x)] = \frac{dP}{du}\frac{du}{dx} = \frac{d}{du}\left[\sum_{k=0}^{\infty} a_k u^k\right] = \sum_{k=1}^{\infty} k a_k u^{k-1} = \sum_{k=1}^{\infty} k a_k (x - c)^{k-1}$$

by the case $c = 0$. Thus it suffices to prove this theorem when $c = 0$.

We begin by showing the formal derivative $P_1(x)$ of $P(x)$ is absolutely convergent for $|x| < R$. Clearly this is true for $x = 0$. Assume $0 < |x| < R$. Choose $|x| < |y| < R$. Since the series $\sum_{k=0}^{\infty} a_k y^k$ is convergent, the limit of its summands is zero. Thus there is a positive integer N such that if $k \geq N$, then $|a_k y^k| < 1$. For $k \geq N$:

$$|k a_k x^k| = k|a_k y^k|\left|\frac{x}{y}\right|^k < k\left|\frac{x}{y}\right|^k = k S^k \tag{5.7.1}$$

where $0 < S = \left|\frac{x}{y}\right| < 1$. Apply the Ratio Test to the series $\sum_{k=N}^{\infty} k S^k$:

$$\lim_{k\to\infty} \frac{(k+1)S^{k+1}}{k S^k} = \lim_{k\to\infty} \frac{k+1}{k}S = S < 1.$$

Thus $\sum_{n=N}^{\infty} k S^k$ is convergent. By (5.7.1), $\sum_{k=1}^{\infty} k a_k x^k$ is absolutely convergent by the Comparison Test. Therefore $P_1(x) = \sum_{k=1}^{\infty} k a_k x^{k-1} = \frac{1}{x}\sum_{k=1}^{\infty} k a_k x^k$ and its formal derivative $P_2(x) = \sum_{k=2}^{\infty} k(k-1) a_k x^{k-2}$ are absolutely convergent for $|x| < R$.

Let $|x_0| < R$. We show $P'(x_0) = P_1(x_0)$. Select T with $|x_0| < T < R$. Let $|x| < T$. By the definition of the derivative:

$$P'(x_0) - P_1(x_0) = \lim_{x\to x_0} \frac{P(x) - P(x_0)}{x - x_0} - P_1(x_0)$$

$$= \lim_{x\to x_0} \sum_{k=0}^{\infty} \left[\frac{a_k x^k - a_k x_0^k}{x - x_0} - k a_k x_0^{k-1}\right]$$

$$= \lim_{x\to x_0} \sum_{k=2}^{\infty} a_k \left[\frac{(x - x_0)(x^{k-1} + x^{k-2}x_0 + \cdots + x x_0^{k-2} + x_0^{k-1})}{x - x_0} - k x_0^{k-1}\right]$$

$$= \lim_{x\to x_0} \sum_{k=2}^{\infty} a_k \left[x^{k-1} + x^{k-2}x_0 + \cdots + x x_0^{k-2} - (k-1)x_0^{k-1}\right] \tag{5.7.2}$$

To continue, we use the following computation.

$$(x - x_0)\ \left[1x^{k-2} + 2x_0 x^{k-3} + \cdots + (k-2)x_0^{k-3}x + (k-1)x_0^{k-2}\right]$$

$$= \left[x^{k-1} - x_0 x^{k-2}\right] + \left[2x_0 x^{k-2} - 2x_0^2 x^{k-3}\right] + \left[3x_0^2 x^{k-3} - 3x_0^3 x^{k-4}\right]$$

$$+ \cdots + \left[(k-2)x_0^{k-3}x^2 - (k-2)x_0^{k-2}x\right] + \left[(k-1)x_0^{k-2}x - (k-1)x_0^{k-1}\right]$$

$$= x^{k-1} + x_0 x^{k-2} + x_0^2 x^{k-3} + \cdots + x_0^{k-3}x^2 + x_0^{k-2}x - (k-1)x_0^{k-1}$$

Apply this identity to (5.7.2):

$$|P'(x_0) - P_1(x_0)|$$

$$= \left|\lim_{x\to x_0} \sum_{k=2}^{\infty} a_k (x - x_0)\left[x^{k-2} + 2x_0 x^{k-3} + \cdots + (k-2)x_0^{k-3}x + (k-1)x_0^{k-2}\right]\right|$$

$$\leq \lim_{x\to x_0} \sum_{k=2}^{\infty} |a_k|\,|x - x_0|\left[|x|^{k-2} + 2|x_0||x|^{k-3}\right.$$

$$\left. + \cdots + (k-2)|x_0|^{k-3}|x| + (k-1)|x_0|^{k-2}\right]$$

$$\leq \lim_{x\to x_0} |x - x_0| \sum_{k=2}^{\infty} |a_k| T^{k-2}\left[1 + 2 + \cdots + (k-2) + (k-1)\right]$$

since $|x_0|$ and $|x|$ are less than T. By Proposition 3.2.4(d):

$$|P'(x_0) - P_1(x_0)| \leq \lim_{x \to x_0} |x - x_0| \sum_{k=2}^{\infty} |a_k| T^{k-2} \frac{k(k-1)}{2}$$

$$\leq \lim_{x \to x_0} \frac{|x - x_0|}{2} \sum_{k=2}^{\infty} k(k-1) |a_k| T^{k-2} \qquad ($$

The latter series converges because $P_2(T) = \sum_{k=2}^{\infty} k(k-1)a_k T^{k-2}$ is absolutely vergent. By (5.7.3), $|P'(x_0) - P_1(x_0)| = 0$. Thus $P'(x_0)$ exists and equals $P_1(x_0$

We apply this theorem to give an easy derivation of the Taylor series of $\cos x$

Examples 5.7.13 (1) Establish the Taylor series $\cos x = \sum_{k=0}^{\infty} (-1)^k \frac{x^{2k}}{(2k)!}$ for x

Solution By Proposition 5.5.13(b):

$$\sin x = \sum_{k=0}^{\infty} (-1)^k \frac{x^{2k+1}}{(2k+1)!} \quad \text{for } x \in \Re.$$

By Theorem 5.7.12, $\cos x = \frac{d}{dx}(\sin x) = \frac{d}{dx}\left[\sum_{k=0}^{\infty} (-1)^k \frac{x^{2k+1}}{(2k+1)!} \right]$

$$= \sum_{k=0}^{\infty} (-1)^k \frac{(2k+1)x^{2k}}{(2k+1)(2k)!} = \sum_{k=0}^{\infty} (-1)^k \frac{x^{2k}}{(2k)!} \quad \text{fo}$$

(2)   Use the Taylor series of $g(x) = \cos x$ at $x = \frac{\pi}{2}$, derived in Example 5.6.6 (3 find the Taylor series of $\sin x$ at $x = \frac{\pi}{2}$.

Solution By Example 5.6.6 (3), $g(x) = \cos x = \sum_{k=0}^{\infty} (-1)^{k+1} \frac{\left(x - \frac{\pi}{2}\right)^{2k+1}}{(2k+1)!}$ $x \in \Re$. By Theorem 5.7.12:

$$\sin x = -g'(x) = -\frac{d}{dx}\left[\sum_{k=0}^{\infty} (-1)^{k+1} \frac{\left(x - \frac{\pi}{2}\right)^{2k+1}}{(2k+1)!} \right]$$

$$= -\sum_{k=0}^{\infty} (-1)^{k+1} \frac{(2k+1)\left(x - \frac{\pi}{2}\right)^{2k}}{(2k+1)(2k)!} = \sum_{k=0}^{\infty} (-1)^k \frac{\left(x - \frac{\pi}{2}\right)^{2k}}{(2k)!} \quad \text{for } x \in$$

Recall, not every smooth function equals its Taylor series. For example, the fu tion $f(x) = e^{-1/x^2}$ of Example 5.5.14 (4) only equals its Taylor series at $c = 0$ w $x = 0$. Nevertheless, we show every power series equals its Taylor series.

Corollary 5.7.14 Let $P(x) = \sum_{k=0}^{\infty} a_k(x-c)^k$ be a power series with radius of vergence $R > 0$. Then

$$a_n = \frac{P^{(n)}(c)}{n!} \quad \text{for } n \geq 0.$$

Proof By Theorem 5.7.12, for $|x - c| < R$:

$$P^{(n)}(x) = \sum_{k=n}^{\infty} k(k-1)\cdots(k-n+1)a_k(x-c)^{k-n}$$

$$= n(n-1)\cdots(2)(1)a_n + \sum_{k=n+1}^{\infty} k(k-1)\cdots(k-n+1)a_k(x-c)^{k-n}$$

$$= n!a_n + (x - c)\sum_{k=n+1}^{\infty} k(k-1)\cdots(k-n+1)a_k(x-c)^{k-n-1}.$$

Take $x = c$: we have $P^{(n)}(c) = n!a_n$ as asserted. □

This corollary can be used to compute the higher derivatives of a function from its power series representation.

Examples 5.7.15 (1) Find the higher derivatives of $f(x) = \arctan x$ at $x = 0$.

Solution In Example 5.4.13 (3), we derived the power series representation $\arctan x = \sum_{k=0}^{\infty}(-1)^k \frac{x^{2k+1}}{2k+1}$ for $-1 \le x \le 1$. Let $k \ge 0$. It follows from Corollary 5.7.14 that $f^{(2k)}(0) = 0$ and

$$\frac{f^{(2k+1)}(0)}{(2k+1)!} = (-1)^k\frac{1}{2k+1}$$

$$f^{(2k+1)}(0) = (-1)^k\frac{(2k+1)!}{2k+1} = (-1)^k\frac{(2k+1)(2k)!}{2k+1} = (-1)^k(2k)! \,.$$

(2) Find the values of the higher derivatives of $g(x) = e^{-x^4}$ at $x = 0$.

Solution We know $e^u = \sum_{k=0}^{\infty}\frac{u^k}{k!}$ for $u \in \Re$. Let $u = -x^4$. Then $e^{-x^4} = \sum_{k=0}^{\infty}(-1)^k\frac{x^{4k}}{k!}$. Therefore $g^{(n)}(0) = 0$ if n is not divisible by four while

$$\frac{g^{(4k)}(0)}{(4k)!} = (-1)^k\frac{1}{k!} \quad \text{and} \quad g^{(4k)}(0) = (-1)^k\frac{(4k)!}{k!} \quad \text{for } k \ge 0. \quad \square$$

By the Fundamental Theorems of Calculus, integration is the inverse operation of differentiation. Thus we can integrate a power series term by term. That is, the integral of an infinite sum of monomials is the sum of the integrals of these monomials.

Corollary 5.7.16 *Let* $P(x) = \sum_{k=0}^{\infty} a_k(x-c)^k$ *be a power series with radius of convergence* R. *Then* $\sum_{k=0}^{\infty}\frac{a_k}{k+1}(x-c)^{k+1}$ *converges absolutely for* $|x-c| < R$ *and*

$$\int \sum_{k=0}^{\infty} a_k(x-c)^k \, dx = \sum_{k=0}^{\infty}\frac{a_k}{k+1}(x-c)^{k+1} + B.$$

Proof Observe for $k \ge 0$:

$$\frac{|a_k||x-c|^{k+1}}{k+1} \le |a_k||x-c|^{k+1}.$$

When $|x-c| < R$, the series $\sum_{k=0}^{\infty} a_k(x-c)^k$ is absolutely convergent, and the series $|x-c|\sum_{k=0}^{\infty}|a_k||x-c|^k = \sum_{k=0}^{\infty}|a_k||x-c|^{k+1}$ converges. By the Comparison Test, the series $\sum_{k=0}^{\infty}\frac{a_k(x-c)^{k+1}}{k+1}$ is absolutely convergent. By Theorem 5.7.12:

$$\frac{d}{dx}\left(\sum_{k=0}^{\infty}\frac{a_k(x-c)^{k+1}}{k+1}\right) = \sum_{k=0}^{\infty} a_k(x-c)^k.$$

for $|x-c| < R$. By the Second Fundamental Theorem of Calculus, $\int \sum_{k=0}^{\infty} a_k(x-c)^k \, dx = \sum_{k=0}^{\infty}\frac{a_k}{k+1}(x-c)^{k+1} + B$. □

We use this corollary to represent a definite integral as an infinite series. The partial sums of this series approximate the integral. An analysis of the remainder, as in Examples 5.4.13, 5.5.16 and 5.6.7, determines the accuracy of these approximations.

Examples 5.7.17 (1)   Evaluate $\int_0^1 \cos \sqrt{t}\, dt$ as an infinite series.

Solution $\cos x = \sum_{k=0}^{\infty} (-1)^k \frac{x^{2k}}{(2k)!}$ for $x \in \Re$. Let $x = \sqrt{t} = t^{1/2}$. Then

$$
\int_0^1 \cos \sqrt{t}\, dt \;=\; \int_0^1 \sum_{k=0}^{\infty} (-1)^k \frac{t^k}{(2k)!}\, dt = \sum_{k=0}^{\infty} (-1)^k \frac{t^{k+1}}{(k+1)(2k)!} \Bigg|_0^1
$$

$$
= \; \sum_{k=0}^{\infty} (-1)^k \frac{1}{(k+1)(2k)!}.
$$

(2)   Find a power series representation of $\arcsin x$.

Solution   For $-1 < u < 1$, we have the binomial series
$(1+u)^{-1/2} = \sum_{k=0}^{\infty} \binom{-1/2}{k} u^k$. Let $u = -t^2$:

$$
(1-t^2)^{-1/2} = \sum_{k=0}^{\infty} \binom{-1/2}{k} (-1)^k t^{2k} \quad \text{for } -1 < t < 1.
$$

For $-1 < x < 1$, integrate this series over the interval with endpoints 0 and

$$
\arcsin x \;=\; \int_0^x \frac{1}{\sqrt{1-t^2}}\, dt = \int_0^x (1-t^2)^{-1/2}\, dt = \int_0^x \sum_{k=0}^{\infty} (-1)^k \binom{-1/2}{k}
$$

$$
= \; \sum_{k=0}^{\infty} (-1)^k \binom{-1/2}{k} \frac{t^{2k+1}}{2k+1} \Bigg|_0^x = \sum_{k=0}^{\infty} (-1)^k \binom{-1/2}{k} \frac{x^{2k+1}}{2k+1}.
$$

Cauchy Product

The product of two finite sums is evaluated by the distributive law. This proceed extends to products of infinite sums. That is, the series constructed by the distributive property from two absolutely convergent series, called the *Cauchy product*, is absolutely convergent series whose sum is the product of the sums of the original series. In particular, we multiply power series in this way. The Cauchy product be used in Sections 8 and 11.

Theorem 5.7.18   *Assume the series* $A = \sum_{k=0}^{\infty} a_k$ *and* $B = \sum_{k=0}^{\infty} b_k$ *are absolute convergent. Then the Cauchy product*

$$
\sum_{n=0}^{\infty} \left(\sum_{k=0}^{n} a_k b_{n-k} \right) = \sum_{n=0}^{\infty} (a_0 b_n + a_1 b_{n-1} + \cdots + a_{n-1} b_1 + a_n b_0)
$$

is absolutely convergent with sum AB.

Proof   Let $c_n = a_0 b_n + a_1 b_{n-1} + \cdots + a_{n-1} b_1 + a_n b_0$, and let

$$
A'_n = \sum_{k=0}^{n} |a_k|, \qquad B'_n = \sum_{k=0}^{n} |b_k|, \qquad C'_n = \sum_{k=0}^{n} |c_k|
$$

denote the partial sums of the three series

$$
A' = \sum_{k=0}^{\infty} |a_k|, \qquad B' = \sum_{k=0}^{\infty} |b_k|, \qquad \sum_{k=0}^{\infty} |c_k|.
$$

Since the series $\sum_{k=0}^{\infty} a_k$ and $\sum_{k=0}^{\infty} b_k$ are absolutely convergent, they converge. Note

$$C_n' \leq \sum_{k=0}^{n} (|a_0||b_k| + |a_1||b_{k-1}| + \cdots + |a_{k-1}||b_1| + |a_k||b_0|)$$

$$\leq \sum_{k=0}^{n} \sum_{m=0}^{n} |a_m||b_k| = A_n' B_n' \leq A' B'.$$

Thus the increasing sequence $\{C_n'\}$ is bounded above by $A'B'$ and converges to its least upper bound. Hence the series $\sum_{n=0}^{\infty} c_n = \sum_{n=0}^{\infty} (\sum_{k=0}^{n} a_k b_{n-k})$ is absolutely convergent. Let C denote its sum. It remains to show $C = AB$. Observe we have also shown that the series $\sum_{k=0}^{\infty} c_k''$ is convergent where

$$c_k'' = |a_0||b_k| + |a_1||b_{k-1}| + \cdots + |a_{k-1}||b_1| + |a_k||b_0|.$$

Let $A_n = \sum_{k=0}^{n} a_k$, $B_n = \sum_{k=0}^{n} b_k$, $C_n = \sum_{k=0}^{n} c_k$ denote the partial sums. Then

$$|A_n B_n - C_n| = \left| \sum a_s b_t \right| \leq \sum |a_s b_t|$$

where the sums on the right are taken over all those s and t for which $0 \leq s \leq n$, $0 \leq t \leq n$ and $s + t > n$. Therefore

$$|A_n B_n - C_n| \leq \sum_{k=n+1}^{\infty} \sum_{m=0}^{n} |a_m b_{k-m}| = \sum_{k=n+1}^{\infty} c_k''.$$

The latter sum is the difference between the sum of the convergent series $\sum_{k=0}^{\infty} c_k''$ and its nth partial sum which has limit zero as n approaches infinity. Hence $|AB - C| = \lim_{n \to \infty} |A_n B_n - C_n| = 0$, and $AB = C$ as asserted. □

The following corollary computes the Cauchy product of two power series.

Corollary 5.7.19 *Let $P(x) = \sum_{k=0}^{\infty} a_k (x - c)^k$ and $Q(x) = \sum_{k=0}^{\infty} b_k (x - c)^k$ be two power series whose radii of convergence are at least R. Then the Cauchy product of these power series has radius of convergence at least R and for $|x - c| < R$:*

$$P(x)Q(x) = \sum_{n=0}^{\infty} \left(\sum_{k=0}^{n} a_k b_{n-k} \right) (x - c)^n$$

Proof Theorem 5.7.18 applies to the Cauchy product of $P(x)$ and $Q(x)$ because these two power series are absolutely convergent for $|x - c| < R$. □

Example 5.7.20 Find the Taylor series of $f(x) = e^x \sin x$ at $x = 0$ and $f^{(n)}(0)$, $n \geq 0$.

Solution By (5.5.14) and (5.5.15):

$$e^x = \sum_{k=0}^{\infty} \frac{x^k}{k!} \quad \text{and} \quad \sin x = \sum_{k=0}^{\infty} (-1)^k \frac{x^{2k+1}}{(2k+1)!} \quad \text{for } x \in \Re.$$

By Corollary 5.7.19, the Cauchy product of these two series is absolutely convergent and has sum $e^x \sin x$ for $x \in \Re$:

$$e^x \sin x = \sum_{n=0}^{\infty} \left(\sum_{k=0}^{[(n-1)/2]} (-1)^k \frac{1}{(2k+1)!(n-2k-1)!} \right) x^n.$$

The notation $[(n-1)/2]$ denotes the largest integer less than or equal to $\frac{n-1}{2}$. By Corollary 5.7.14, this power series is the Taylor series of $e^x \sin x$ at $x = 0$, and

$$f^{(n)}(0) = \sum_{k=0}^{[(n-1)/2]} (-1)^k \frac{n!}{(2k+1)!(n-2k-1)!} = \sum_{k=0}^{[(n-1)/2]} (-1)^k \binom{n}{2k+1} \quad \text{for } n \geq 0. \ \square$$

Historical Remarks

The study of power series began in southwest India at the end of the fourt[
century. Motivated by astronomical calculations, Madhava obtained the Taylor [
of $\sin x$, $\cos x$ and $\arctan x$. The *Yuktibhasa* written by Jyesthadeva 200 years [
gives the geometric and algebraic arguments used by Madhava. His derivation [
series for $\arctan x$ is outlined in Problem 5.

The study of power series in England in the 1660s played a central role i[
development of calculus. Isaac Newton discovered the binomial series in 166[
1667, he integrated the geometric series $\frac{1}{1+t} = 1 - t + t^2 + \cdots + (-1)^n t^n + \cdots$ fr[
to x to obtain the series

$$A(1+x) = x - \frac{x^2}{2} + \frac{x^3}{3} + \cdots + (-1)^{n-1}\frac{x^n}{n} + \cdots.$$

Newton noted that $A(1+x)$ has logarithmic properties, as discussed in the histc
remarks to Section 4.2. This series was obtained independently by Nicolas Merc
in 1668 who recognized the sum as $\ln(1+x)$. In 1669, Newton published the T[
series of $\sin x$, $\cos x$ and $\arcsin x$. By 1671, James Gregory had calculated the T[
series of several functions including $\arcsin x$, $\arctan x$ and $\tan x$. Gottfried Wil[
Leibniz calculated these series independently in 1673. Both Newton and Leibniz [
power series as a fundamental tool in their development of calculus. In fact, it [
only through the term by term differentiation of power series that they could com[
the derivatives of transcendental functions.

The use of power series increased dramatically in the 18th century through[
work of the Bernoulli brothers and Leonhard Euler. They manipulated power s[
as if they were polynomials and paid little attention to the rigor of their calculati[
Between 1689 and 1704, the Bernoulli brothers published papers that illustrate the [
of power series to represent derivatives, integrals and solutions of differential equati[
In the 1730s, Euler began publishing many interesting results on infinite series [
infinite products. In 1748, he published the Taylor series of e^x as part of his exposi[
of logarithm and exponential functions which we presented in Section 4.2. He also [
power series to extend the domain of a function from real values to complex va[
(See Section 11.) The need for rigor in using infinite series was known but igno[
since no one had a reasonable approach. For example, Jakob Bernoulli rediscov[
Oresme's proof (ca 1350) that the harmonic series diverges. Neither the Berno[
nor Euler could decide what to conclude about the sum of the series $\sum_{k=0}^{\infty}(-1)^k$.[

In 1772, Joseph Louis Lagrange wrote a calculus text with the intent of remo[
the intuitive concepts of fluxions, infinitesmals, differentials and limits from the [
ject. He proposed to reduce calculus to the algebraic manipulation of power se[
For example, he defined the derivative $f'(c)$ as the coefficient of $x - c$ in the po[
series representation of $f(x)$ in powers of $x - c$. His methods apply to a function w[
is representable by a power series with a positive radius of convergence at each p[
of its domain. However, both Lagrange and Euler incorrectly assumed the Ta[
series at $x = c$ of a smooth function $f(x)$ equals $f(x)$ on an open interval contair[
c. In 1823, Augustin-Louis Cauchy produced the counterexample $f(x) = e^{-1/x^2}$, v[
$c = 0$, to Lagrange's assertion. Cauchy, however, made a fundamental error of [
own in 1821. He gave a fallacious proof that a convergent infinite series of continu[
functions $f_n(x)$ has a continuous sum. Niels Henrik Abel produced a counterex[
ple in 1826 and showed that if the series $\sum_{n=1}^{\infty} f_n(x)$ is *uniformly convergent*, t[
its sum is continuous. However he did not identify this concept. Moreover neit[
Abel nor Cauchy realized that this hypothesis was also necessary for computing [
derivative of a convergent series by term by term differentiation. It was not until 1[
that the German mathematician Karl Weierstrass gave a correct proof of this theor[

under the hypothesis of uniform convergence. An outline of this proof is presented in Problems 6 and 7.

Summary

The reader should be able to apply the Comparison and Ratio Tests to numerical series. She should also know how to use the Ratio Test to determine the radius of convergence of a power series. The derivative of a power series should be determined by differentiating it term by term while its integral should be found by integrating it term by term. The reader should also be able to compute the Cauchy product of two absolutely convergent series.

Basic Exercises

1. Determine whether each series converges or diverges.

(a) $\sum_{k=1}^{\infty} \frac{1}{3^k+5}$

(b) $\sum_{k=1}^{\infty} \frac{k^5+1}{k!}$

(c) $\sum_{k=1}^{\infty} (-1)^k \frac{k^3+5k-2}{k^k}$

(d) $\sum_{k=1}^{\infty} \frac{(3k)!}{k!(2k)!}$

(e) $\sum_{k=1}^{\infty} \frac{\sin k}{5^k}$

(f) $\sum_{k=1}^{\infty} (-1)^k \frac{k!5^k}{(2k+1)!}$

(g) $\sum_{k=1}^{\infty} \frac{1}{k^2}$

(h) $\sum_{k=1}^{\infty} (-1)^k \frac{1}{k^3}$

(i) $\sum_{k=1}^{\infty} \frac{k!}{(2)(4)\cdots(2k)}$

(j) $\sum_{k=1}^{\infty} \frac{(2k)!}{(1)(3)\cdots(2k+1)}$

(k) $\sum_{k=1}^{\infty} \frac{2^k}{k!\sqrt{k}}$

(l) $\sum_{k=1}^{\infty} \frac{1}{k^2+k+12}$

2. (a) Assume $0 \le a_k \le b_k$ for $k \ge p$. Show if $\sum_{k=p}^{\infty} a_k$ diverges, then $\sum_{k=p}^{\infty} b_k$ diverges.
(b) Determine whether each series converges or diverges.

(i) $\sum_{k=1}^{\infty} \frac{1}{\sqrt{k}}$

(ii) $\sum_{k=0}^{\infty} \frac{1}{1+\left(\frac{3}{2}\right)^k}$

(iii) $\sum_{k=2}^{\infty} \frac{1}{\ln k}$

(iv) $\sum_{k=1}^{\infty} \frac{1}{k+2^k}$

(v) $\sum_{k=1}^{\infty} \frac{1}{k+\left(\frac{1}{2}\right)^k}$

(vi) $\sum_{k=1}^{\infty} \frac{1}{k \sin^2 k}$

3. Show $\sum_{k=1}^{\infty} \frac{1}{k^p}$ converges for $p > 1$ and diverges for $p \le 1$.
Hint: compare these series with the telescoping and harmonic series.

4. Consider the geometric series $\sum_{k=0}^{\infty} aR^k$ with $a \ne 0$.
• By the Ratio Test, this series converges absolutely if $-1 < R < 1$ and diverges if $R < -1$ or $R > 1$.
• When $R = 1$, $\sum_{k=0}^{\infty} aR^k = \sum_{k=0}^{\infty} a$ which diverges by the nth Term Test.
• When $R = -1$, $\sum_{k=0}^{\infty} aR^k = \sum_{k=0}^{\infty} (-1)^k a$ which diverges by the nth Term Test.
Is this a correct derivation of the conditions under which a geometric series converges?

5. Find the radius of convergence of each power series.

(a) $\sum_{k=0}^{\infty} \frac{x^{2k}}{3^k+1}$

(b) $\sum_{k=0}^{\infty} \frac{x^k}{k^3+1}$

(c) $\sum_{k=0}^{\infty} (-1)^k \frac{k!x^k}{(2k)!}$

(d) $\sum_{k=2}^{\infty} \frac{2^{3k+1}x^k}{k \ln k}$

(e) $\sum_{k=0}^{\infty} (-1)^k \frac{(4^k-3)x^{2k}}{k^4+3}$

(f) $\sum_{k=0}^{\infty} (-1)^k \frac{k!x^k}{10^k}$

(g) $\sum_{k=0}^{\infty} \frac{k^{k/2}x^k}{k!}$

(h) $\sum_{k=0}^{\infty} (-1)^k \frac{kx^k}{10^k}$

(i) $\sum_{k=1}^{\infty} (-1)^k \frac{(x-2)^k}{5^k\sqrt{k}}$

(j) $\sum_{k=1}^{\infty} \frac{(2k)!}{k!k^k}(x+1)^{3k}$

(k) $\sum_{k=0}^{\infty} \frac{k!(x-3)^k}{1(3)\cdots(2k+1)}$

(l) $\sum_{k=0}^{\infty} \frac{9^k+k}{k^3+k+1}(x+4)^{2k}$

6. What conclusions can you deduce about the radius of convergence R of each power series $P(x)$ in $x - c$?
(a) $c = 0$, $P(-5)$ converges and $P(10)$ diverges.
(b) $c = 2$, $P(-5)$ converges and $P(10)$ diverges.
(c) $c = -3$, $P(2)$ converges and $P(-9)$ diverges.
(d) $c = -4$, $P(-1)$ converges, $P(-6)$ converges and $P(0)$ diverges.
(e) $c = 0$ and $P(n)$ converges for every positive integer n.
(f) $c = 0$ and $P(n)$ diverges for every positive integer n.
(g) $c = 0$ and $P(q)$ diverges for every nonzero rational number q.
(h) $c = 0$ and $P\left((-1)^n n^2\right)$ converges for every positive integer n.

7. Give an example of a power series $P(x)$ in x with radius of convergence $0 < R$ which meets the criterion of each case.

(a) $P(-R)$ and $P(R)$ both converge.   (b) $P(-R)$ and $P(R)$ both diverge

(c) $P(-R)$ converges and $P(R)$ diverges.   (d) $P(-R)$ diverges and $P(R)$ con

8. Let n be a positive integer. Show for every infinite series of the form $P($ $\sum_{k=0}^{\infty} a_k (\sqrt[n]{x})^k$ there is an $0 \leq R \leq \infty$ such that $P(x)$ is absolutely converger $0 \leq x < R$ and divergent for $x > R$.

9. Consider the power series $\sum_{k=0}^{\infty} a_k x^k$. If $\lim_{k\to\infty} |a_{k+1}/a_k| = L$, what is the ra of convergence R of this power series?

10. What is the difference between $\cos\sqrt{x}$ and its Taylor series $T(x) = \sum_{k=0}^{\infty} (-1$

11. Evaluate each limit: (a) $\lim_{k\to\infty} \dfrac{c^k}{k!}$;   (b) $\lim_{k\to\infty} \dfrac{k!}{k^k}$.

12. Find the power series representation of the derivative of each power series. the interval on which your computation is valid.

(a) $\sum_{k=0}^{\infty} (-1)^k \frac{x^k}{k!}$ (b) $\sum_{k=0}^{\infty} \frac{x^{2k}}{(2k)!}$ (c) $\sum_{k=1}^{\infty} \frac{x^k}{k^k}$

(d) $\sum_{k=0}^{\infty} (-1)^k \frac{x^k}{(k!)^2}$ (e) $\sum_{k=0}^{\infty} \frac{k}{k^2+1}(x+2)^k$ (f) $\sum_{k=1}^{\infty} (-1)^k \frac{3^k}{\sqrt{k}}(x-5)$

(g) $\sum_{k=1}^{\infty} \frac{\ln k}{k}(x+6)^k$ (h) $\sum_{k=0}^{\infty} \frac{e^k}{(k!)^2}(x-3)^{k^2}$ (i) $\sum_{k=0}^{\infty} \binom{-2}{k} \frac{x^{3k}}{8^k}$

13. Use power series to evaluate the derivatives $f^{(n)}(c)$, $n \geq 0$, for each function

(a) $f(x) = e^{-x^2}$ and $c = 0$;   (b) $f(x) = \sqrt{x}\sin\sqrt{x}$ and $c = 0$;

(c) $f(x) = e^{x^2-6x}$ and $c = 3$;   (d) $f(x) = \arcsin x$ and $c = 0$;

(e) $f(x) = \sqrt{1+x^4}$ and $c = 0$;   (f) $f(x) = \frac{x}{x^2+8x+25}$ and $c = -4$.

14. Find a power series representation of each integral.

(a) $\int_0^{1/2} e^{-x^3}\, dx$ (b) $\int_0^{1/3} \sin(x^2)\, dx$ (c) $\int_0^{1/2} \frac{1}{\sqrt{1-x^5}}$ C

(d) $\int_0^{2/3} x^2 \cos\sqrt{x}\, dx$ (e) $\int_0^{1/4} \sqrt{x}\arctan\sqrt{x}\, dx$ (f) $\int_0^{1/2} \ln(1+x$

15. Use Theorem 5.7.12 to find the Taylor series of $\sin x$ from that of $\cos x$.

16. Assume e^x has a representation as a power series in x. Use Theorem 5.7.1 find the coefficients of this power series.

17. Use Corollary 5.7.16 to find a representation of each function as a power seri

(a) $\ln(1-x)$ (b) $\arctan x$ (c) $\cos x$

18. Use the Cauchy product to find 5 terms of the Taylor series at $x = c$ of each

(a) $e^x \cos x$ at $x = 0$;   (b) $e^x \arctan x$ at $x = 0$;

(c) $(\sin x)\sqrt{1+x^2}$ at $x = 0$;   (d) $\frac{\cos x}{1+x^4}$ at $x = 0$;

(e) $(\ln x)e^{2x-x^2}$ at $x = 1$;   (f) $(\sin x)\ln(2\pi x - x^2)$ at $x = \pi$.

The following four exercises are for those who have studied Section 4.1

19. Use power series to evaluate the derivatives $f^{(n)}(0)$, $n \geq 0$, for each function.

(a) $f(x) = \cosh\sqrt{x}$ (b) $f(x) = \sinh(x^4)$ (c) $f(x) = \cosh(x^3)$

20. Find a power series representation of each integral.

(a) $\int_0^1 \cosh\sqrt{x}\, dx$ (b) $\int_0^{1/2} \frac{\sinh x}{x}\, dx$ (c) $\int_0^{1/3} \sqrt{x}\sinh\sqrt{x}\, dx$

21. Use Thm. 5.7.12 to find the Taylor series of $\cosh x$ at $x = 0$ from that of $\sinh$

22. Use the Cauchy product to find the Taylor series at $x = 0$ of each function.

(a) $e^x \sinh x$ (b) $(\sinh x)(\cosh x)$ (c) $(\cosh x)\ln(1+x)$

Challenging Problems

1. **(a)** Let $P(x) = \sum_{k=0}^{\infty} a_k x^k$ have radius of convergence at least R. If $a_0 \neq 0$, show that there is a positive number S such that

$$\frac{1}{P(x)} = \sum_{k=0}^{\infty} b_k x^k \quad \text{for } |x| < S \text{ where}$$

$$b_0 = \frac{1}{a_0} \quad \text{and} \quad b_k = -\frac{1}{a_0}(a_1 b_{k-1} + a_2 b_{k-2} + \cdots + a_k b_0) \quad \text{for } k \geq 1$$

(b) Find the first four nonzero summands of each quotient of Taylor series at $x = 0$:
  (i) $\sec x = \frac{1}{\cos x}$ with $R = \frac{\pi}{2}$; **(ii)** $\tan x = \frac{\sin x}{\cos x}$ with $R = \frac{\pi}{2}$;
  (iii) $\frac{e^x}{e^x + 1}$ with $R = \infty$; **(iv)** $\frac{\ln(1+x^2)}{(1+x^2)^3}$ with $R = 1$.

2. **(a)** The Bernoulli numbers B_k, $k \geq 0$, are defined by $\frac{x}{e^x - 1} = \sum_{k=0}^{\infty} \frac{B_k x^k}{k!}$. Compute the first eight Bernoulli numbers.
(b) Show $\frac{x}{e^x - 1} + \frac{x}{2}$ is an even function, and conclude that $B_{2k+1} = 0$ for $k \geq 1$.

3. **(a)** Use the trigonometric identity for $\tan(A + B)$ to verify Machin's identity:

$$\frac{\pi}{4} = 4\arctan\frac{1}{5} - \arctan\frac{1}{239}.$$

(b) Add the first six terms of the series for $4\arctan\frac{1}{5}$, and subtract the first two terms of the series for $\arctan\frac{1}{239}$ to estimate π to seven decimal places.

4. Assume the power series $P(x) = \sum_{n=0}^{\infty} a_n (x - c)^n$ has radius of convergence R and converges for $x = R$. Show $\lim_{x \to R^-} P(x)$ exists and equals $P(R)$.

5. Fill in the details in Madhava's derivation of the Taylor series for $\arctan x$.
(a) Let BC be a small arc of a circle with center O and radius one. Let A be a point on the circle. Let B_0 and C_0 denote the intersection of OB and OC with the tangent line T to the circle at A. See Figure 5.7.21. Show the length of the arc BC is approximately $\frac{B_0 C_0}{1 + (AB_0)^2}$.
(b) Let θ denote the angle AOC. Then $x = AC_0 = \tan\theta$. Divide the line segment AC_0 into n equal segments, apply the formula of (a) to each one and sum to obtain:

$$\theta = \arctan x \approx \sum_{k=0}^{n-1} \frac{x/n}{1 + \left(\frac{kx}{n}\right)^2}.$$

(c) Take the limit of the preceding formula as n approaches infinity, expand each $\frac{x/n}{1 + \left(\frac{kx}{n}\right)^2}$ as a geometric series and simplify to conclude that for $0 \leq x \leq 1$:

$$\arctan x = \lim_{n \to \infty} \left[x + \sum_{p=1}^{\infty} (-1)^p \frac{x^{2p+1}}{n^{2p+1}} \sum_{k=1}^{n-1} k^{2p} \right].$$

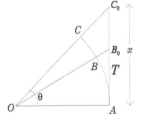

Figure 5.7.21
Problem 5

(d) Show for q a positive integer: $n \sum_{k=1}^{n} k^{q-1} = \sum_{k=1}^{n} k^q + \sum_{h=1}^{n-1} \sum_{k=1}^{h} k^{q-1}$.

(e) Deduce $\lim_{n \to \infty} \frac{1}{n^{q+1}} \sum_{k=1}^{n} k^q = \frac{1}{q+1}$.

(f) Combine (c) and (e) to conclude that for $0 \leq x \leq 1$: $\arctan x = \sum_{p=0}^{\infty} (-1)^p \frac{x^{2p+1}}{2p+1}$.

6. A sequence of functions $\{f_n\}$, each with domain an interval I, is said to *con* *uniformly* to the function f with domain I if for all $\epsilon > 0$, there is a positive in N such that if $n \geq N$ then $f_n(x) \in (f(x) - \epsilon, f(x) + \epsilon)$ for all $n \geq N$ and all x Note the important condition that N is independent of x.

(a) If the sequence of continuous functions $\{f_n\}$ converges uniformly to f o interval I, show f is continuous.

(b) Let $f_n(x) = x^n$ for $x \in [0, 1]$ and $n \geq 1$. Identify the limit f of this conve sequence. Show that the sequence $\{f_n\}$ does not converge uniformly to f.

7. A convergent infinite series of functions $f(x) = \sum_{n=0}^{\infty} f_n(x)$ is *uniformly conve* if its sequence of partial sums converges uniformly to f.

(a) Show the sum of a uniformly convergent series of continuous functions o interval I is also continuous on I.

(b) Show the sum $f(x)$ of a uniformly convergent series of differentiable func $\sum_{n=0}^{\infty} f_n(x)$ on an open interval I is differentiable on I. Moreover, $f'(x) = \sum_{n=0}^{\infty}$ for $x \in I$.

(c) Let $f(x)$ be the sum of a uniformly convergent series of continuous func $\sum_{n=0}^{\infty} f_n(x)$ on an open interval I. Show $\int f(x)\, dx = \sum_{n=0}^{\infty} \int f_n(x)\, dx$ on I.

(d) Show a power series with radius of convergence R at $x = c$ converges unifo to its sum on any closed interval $[a, b]$ contained in $(c - R, c + R)$.

Appendix: Root Test

The Root Test determines whether a series converges based on the value of the of the kth root of its kth term. It is an alternative to the Ratio Test for finding radius of convergence of a power series. However, it is less practical than the R Test, because the limit which requires evaluation is often a difficult one.

Proposition 5.7.22 (Root Test) *Let* $\sum_{k=p}^{\infty} a_k$ *be an infinite series such that the*

$$\lim_{k \to \infty} \sqrt[k]{|a_k|} = R$$

exists. There are three cases:

    (a) *if $R < 1$, the series is absolutely convergent;*

    (b) *if $R > 1$, the series is divergent;*

    (c) *if $R = 1$, there is no conclusion.*

Proof **(a)** Choose $R < S < 1$. For k large enough, say $k \geq N$, $\sqrt[k]{|a_k|} < S$ $|a_k| < S^k$. The geometric series $\sum_{k=N}^{\infty} S^k$ has ratio S, with $0 < S < 1$, and there converges. By the Comparison Test, the series $\sum_{k=p}^{\infty} a_k$ is absolutely convergent.

(b) For k large enough, say $k \geq N$, $\sqrt[k]{|a_k|} > 1$ and $|a_k| > 1$. Hence $\lim_{k \to \infty} a_k \neq 0$, the series $\sum_{k=p}^{\infty} a_k$ diverges by the nth Term Test.

(c) When $R = 1$, the series may converge or diverge. For example, the telesco series $\sum_{k=1}^{\infty} \frac{1}{k(k+1)}$ converges and $\lim_{k \to \infty} \sqrt[k]{\frac{1}{k(k+1)}} = 1$. On the other hand, the se $\sum_{k=1}^{\infty} 1$ diverges and $\lim_{k \to \infty} \sqrt[k]{1} = 1$.

The Root Test works best with series defined by polynomials and powers. I not practical for series which involve factorials. We often use $\lim_{k \to \infty} \sqrt[k]{k} = 1$, derived Example 5.3.13 (2).

Examples 5.7.23 (1) Determine whether $\sum_{k=1}^{\infty} \frac{k^4+k^2+1}{k^k}$ converges or diverges.

Solution The limit R of the Root Test is given by:

$$
\begin{aligned}
R &= \lim_{k\to\infty} \sqrt[k]{\frac{k^4+k^2+1}{k^k}} = \lim_{k\to\infty} \frac{(k^4+k^2+1)^{1/k}}{(k^k)^{1/k}} \\
&\le \lim_{k\to\infty} \frac{(2k^4)^{1/k}}{k} = \lim_{k\to\infty} \frac{2^{1/k}(\sqrt[k]{k})^4}{k} = \frac{(1)(1)^4}{\infty} = 0.
\end{aligned}
$$

Thus $R = 0$, and the series $\sum_{k=1}^{\infty} \frac{k^4+k^2+1}{k^k}$ converges by the Root Test.

(2) Determine whether $\sum_{k=1}^{\infty} \frac{3^k}{k^3}$ converges or diverges.

Solution The limit R of the Root Test is given by:

$$
R = \lim_{k\to\infty} \sqrt[k]{\frac{3^k}{k^3}} = \lim_{k\to\infty} \frac{(3^k)^{1/k}}{(\sqrt[k]{k})^3} = \frac{3}{(1)^3} = 3.
$$

Since $R = 3 > 1$, this series diverges by the Root Test. $\qquad\square$

The Root Test can be used to find the radius of convergence of some power series.

Example 5.7.24 Find the radius of convergence of the power series $\sum_{k=1}^{\infty} \frac{kx^{2k}}{5^{2k-1}}$.

Solution The limit R of the Root Test applied to this power series is given by:

$$
R = \lim_{k\to\infty} \sqrt[k]{\frac{k|x|^{2k}}{5^{2k-1}}} = \lim_{k\to\infty} \frac{\sqrt[k]{k}(|x|^{2k})^{1/k}}{(5^{2k-1})^{1/k}} = \lim_{k\to\infty} \frac{\sqrt[k]{k}|x|^2}{5^{2-\frac{1}{k}}} = \frac{1|x|^2}{5^2} = \frac{|x|^2}{25}.
$$

By the Root Test we have absolute convergence if $R = \frac{|x|^2}{25} < 1$ and divergence if $R = \frac{|x|^2}{25} > 1$. That is, we have absolute convergence for $-5 < x < 5$ and divergence for $x < -5$ or $x > 5$. Hence this power series has radius of convergence 5. $\qquad\square$

Exercises

1. What does the Root Test say about each series?

(a) $\sum_{k=1}^{\infty}(-1)^k \left(\frac{2k+7}{3k+5}\right)^k$ (b) $\sum_{k=1}^{\infty}(-1)^k \frac{2^{k^2}}{k^k}$ (c) $\sum_{k=1}^{\infty} \frac{1}{k}$

(d) $\sum_{k=1}^{\infty}(-1)^k \frac{k^2+1}{k^k}$ (e) $\sum_{k=1}^{\infty} \frac{(3k+1)^{2k}}{(7k+4)^k}$ (f) $\sum_{k=1}^{\infty} \frac{k}{\ln k}$

(g) $\sum_{k=1}^{\infty} \frac{10^k}{k^4+1}$ (h) $\sum_{k=1}^{\infty} \frac{2k-1}{3k^4+k^2+1}$ (i) $\sum_{k=1}^{\infty} \frac{k!}{(2k)^k}$

2. Use the Root Test to find the radius of convergence R of each power series.

(a) $\sum_{k=0}^{\infty} \frac{x^k}{2^k+1}$ (b) $\sum_{k=0}^{\infty}(-1)^k \frac{x^k}{k^2+1}$ (c) $\sum_{k=0}^{\infty} \frac{x^k}{k^k+1}$

(d) $\sum_{k=0}^{\infty} \frac{x^{2k}}{9^k-2}$ (e) $\sum_{k=0}^{\infty}(-1)^k \frac{x^{3k}}{k8^k}$ (f) $\sum_{k=0}^{\infty} \frac{kx^{4k}}{3^{8k+1}}$

3. If the Root Test determines that the series $\sum_{k=1}^{\infty} a_k$ diverges, what can be said about the series $\sum_{k=1}^{\infty}(-1)^k a_k$?

4. Consider the power series $\sum_{k=0}^{\infty} a_k x^k$. If $\lim_{k\to\infty} \sqrt[k]{|a_k|} = L$, what is the radius of convergence of this power series?

Applications

5.8   Power Series Solutions of Differential Equati

Prerequisite: Section 5.7

The simple differential equation $y' = f(x)$ is solved by integration: $y = \int f(x)\, dx$ have seen that when we can not evaluate this integral as an explicit function, w integrate the Taylor series of $f(x)$ term by term to obtain a power series represent. of the integral. The following example indicates that this method extends to ol power series solutions of other differential equations.

Motivating Example 5.8.1 Solve the differential equation $y' - 2y = 0$.

Solution We find the solutions of this differential equation, represented by p series in x with radius of convergence $R > 0$:

$$y = \sum_{k=0}^{\infty} a_k x^k \ \text{ with } \ y' = \sum_{k=1}^{\infty} k a_k x^{k-1} = \sum_{t=1}^{\infty} t a_t x^{t-1} \ \text{ for } x \in (-R, R)$$

by Theorem 5.7.12. We will see below why we change the name of the dummy var. k to t. Substitute these power series into our differential equation:

$$\sum_{t=1}^{\infty} t a_t x^{t-1} - 2 \sum_{k=0}^{\infty} a_k x^k = 0$$

Change the dummy variable in the first sum to $k = t - 1$, i.e. $t = k + 1$. Then use distributive property on the second sum and subtract the two sums:

$$0 = \sum_{k=0}^{\infty} (k+1) a_{k+1} x^k - \sum_{k=0}^{\infty} 2 a_k x^k = \sum_{k=0}^{\infty} \left[(k+1) a_{k+1} x^k - 2 a_k x^k \right]$$

$$0 = \sum_{k=0}^{\infty} \left[(k+1) a_{k+1} - 2 a_k \right] x^k.$$

A power series, like a polynomial, can only be zero when all of its coefficients are z

$$(k+1) a_{k+1} - 2 a_k = 0$$
$$(k+1) a_{k+1} = 2 a_k$$
$$a_{k+1} = \frac{2 a_k}{k+1} \quad \text{for } k \geq 0.$$

This equation is called a *recursion formula*. We apply it for $k = 0$, then for $k =$ $k = 2$ and $k = 3$. We do not simplify the arithmetic to discover the pattern.

$$k = 0: \quad a_1 = \frac{2 a_0}{1} = \frac{2}{1} a_0$$

$$k = 1: \quad a_2 = \frac{2 a_1}{2} = \frac{2 \left(\frac{2}{1} a_0 \right)}{2} = \frac{2^2}{(2)(1)} a_0$$

$$k = 2: \quad a_3 = \frac{2 a_2}{3} = \frac{2 \left[\frac{2^2}{(2)(1)} a_0 \right]}{3} = \frac{2^3}{(3)(2)(1)} a_0$$

$$k = 3: \quad a_4 = \frac{2 a_3}{4} = \frac{2 \left[\frac{2^3}{(3)(2)(1)} a_0 \right]}{4} = \frac{2^4}{(4)(3)(2)(1)} a$$

The pattern is clear :
$$a_k = \frac{2^k}{k!} a_0.$$

Thus the solution of our differential equation is:

$$y = \sum_{k=0}^{\infty} \frac{2^k}{k!} a_0 x^k = a_0 \sum_{k=0}^{\infty} \frac{2^k}{k!} x^k.$$

This power series has radius of convergence $R = \infty$ and is a solution of the given differential equation for all $x \in \Re.$[2]□

In the first subsection, we use the method of this example to obtain power series solutions of homogeneous first order linear differential equations $y' + r(x)y = 0$. In the second subsection, we extend this method to solve homogeneous second order linear differential equations $y'' + q(x)y' + r(x)y = 0$. The justification in this case is given in the appendix to this section. The methods of this section can be generalized to obtain power series solutions of homogeneous linear differential equations of all orders.

First Order Differential Equations

We show every differential equation $y' + r(x)y = 0$ has a power series solution. Then we list a six step procedure to find this solution. Those who have read Section 4.11 know this linear differential equation has solution $y = C \exp(-\int r(x) \, dx)$ for $C \in \Re$. Nevertheless, the method of this subsection is useful when we can not integrate $r(x)$. Even when we can integrate $r(x)$, a power series representation of the solution is useful in an application which requires explicit approximations. Since we are finding the Taylor series of the solution, it is reasonable to require that the coefficient $r(x)$ of y is *analytic*, i.e. has a Taylor series expansion.

Definition 5.8.2 (a) *The function $f(x)$ is called analytic at x_0 if there is a power series $\sum_{n=0}^{\infty} a_n (x - x_0)^n$ and an open interval $(x_0 - R, x_0 + R)$, with $R > 0$, such that:*

$$f(x) = \sum_{n=0}^{\infty} a_n (x - x_0)^n \quad \text{for } x \in (x_0 - R, x_0 + R). \tag{5.8.1}$$

R is called a radius of convergence of $f(x)$ at $x = x_0$.
(b) *x_0 is called an ordinary point of the differential equation $y'(x) + r(x)y(x) = 0$ if $r(x)$ is analytic at x_0.*

By Corollary 5.7.14, the power series $P(x) = \sum_{n=0}^{\infty} a_n (x - x_0)^n$ in Definition 5.8.2 is the Taylor series of $f(x)$ at $x = x_0$. The radius of convergence of $P(x)$ is at least R, but R is not necessarily the largest positive number for which (5.8.1) is valid. This is convenient in the statements of the theorems below.

Examples 5.8.3 (1) All polynomials are analytic at every $x_0 \in \Re$.

(2) By Proposition 5.5.13: e^x, $\sin x$, $\cos x$ and $(1 + x)^p$ are analytic at $x = 0$.

(3) By Example 5.5.14 (4), $f(x) = \left\{ \begin{array}{ll} e^{-1/x^2} & \text{if } x \neq 0 \\ 0 & \text{if } x = 0 \end{array} \right\}$ is not analytic at $x = 0$. □

---

[2]In this case, we can identify the solution we have found in terms of a familiar function:

$$y = a_0 \sum_{k=0}^{\infty} \frac{(2x)^k}{k!} = a_0 e^{2x}$$

by Proposition 5.5.13(a). In fact, we could have easily solved this differential equation $y' = 2y$ for exponential growth. However, the purpose of this example is to illustrate the method of using power series to solve a differential equation rather than to obtain the solution in the easiest manner.

We use the trick of Example 5.8.1 to add power series by changing the name of dummy variable. Note the dummy variable of a sum is used to describe the summ but does not appear in the sum itself. For example $\sum_{n=0}^{\infty} a_n x^n$ and $\sum_{k=0}^{\infty} a_k x^k$ represent the series $a_0 + a_1 x + a_2 x^2 + \cdots$. Thus we can change the dummy var in an infinite series without changing the sum. For example, to add

$$\sum_{n=0}^{\infty} a_n x^n + \sum_{n=1}^{\infty} n a_n x^{n-1}$$

change the dummy variable in the second series to $k = n - 1$, i.e. $n = k + 1$, so the powers of x in both series have exponent equal to the dummy variable. The two series can be added and written as one series:

$$\sum_{n=0}^{\infty} a_n x^n + \sum_{n=1}^{\infty} n a_n x^{n-1} = \sum_{n=0}^{\infty} a_n x^n + \sum_{k=0}^{\infty} (k+1) a_{k+1} x^k = \sum_{n=0}^{\infty} [a_n + (n+1) a_{n+1}]$$

The following theorem says that the differential equation $y' + r(x)y = 0$ h power series solution in $x - x_0$ at each ordinary point x_0.

Theorem 5.8.4 *Assume x_0 is an ordinary point of the differential equation*

$$y'(x) + r(x)y(x) = 0. \tag{5}$$

Let $R > 0$ be a radius of convergence of $r(x)$. Then there is a function $u(x)$ whi analytic at x_0, with a radius of convergence R, such that the solution of this differe equation on the interval $(x_0 - R, x_0 + R)$ is:

$$y(x) = Cu(x) \text{ for } C \in \Re.$$

Proof For $x \in (x_0 - R, x_0 + R)$, write

$$r(x) = \sum_{m=0}^{\infty} r_m (x - x_0)^m. \tag{5.}$$

We want to find a power series, with a radius of convergence R,

$$y(x) = \sum_{n=0}^{\infty} y_n (x - x_0)^n$$

which is a solution of the given differential equation. That is, we require:

$$0 = \left[\sum_{n=0}^{\infty} y_n (x - x_0)^n \right]' + \left[\sum_{m=0}^{\infty} r_m (x - x_0)^m \right] \left[\sum_{n=0}^{\infty} y_n (x - x_0)^n \right]$$

$$0 = \sum_{n=1}^{\infty} n y_n (x - x_0)^{n-1} + \sum_{k=0}^{\infty} \sum_{n=0}^{k} r_{k-n} y_n (x - x_0)^k.$$

The power series for $r(x)$ and $y(x)$ are absolutely convergent for $x \in (x_0 - R, x_0 +$ By Theorem 5.7.18, their Cauchy product converges absolutely to $r(x)y(x)$ on interval. To add the two power series above, change the dummy variable in the f series to $k = n - 1$. Then

$$0 = \sum_{k=0}^{\infty} (k+1) y_{k+1} (x - x_0)^k + \sum_{k=0}^{\infty} \sum_{n=0}^{k} r_{k-n} y_n (x - x_0)^k$$

$$= \sum_{k=0}^{\infty} \left[(k+1) y_{k+1} + \sum_{n=0}^{k} r_{k-n} y_n \right] (x - x_0)^k.$$

This power series equals zero if and only if the coefficient of each $(x - x_0)^k$ is zero. Thus $\sum_{n=0}^{\infty} y_n (x - x_0)^n$ is a solution of our differential equation if and only if

$$(k + 1)y_{k+1} = -\sum_{n=0}^{k} r_{k-n} y_n \quad \text{for } k \geq 0. \tag{5.8.4}$$

This recursion relation gives the value of y_{k+1} in terms of $y_0, y_1, \ldots, y_k$.

The difficulty is to show the power series $y = \sum_{n=0}^{\infty} y_n (x - x_0)^n$, defined by the recursion relation (5.8.4), is absolutely convergent for $|x - x_0| < R$. Choose ρ with $|x - x_0| < \rho < R$. Since the terms of the convergent power series (5.8.3) for $r(x_0 + \rho)$ have limit zero, the terms of this series are bounded. That is, there is $B > 0$ such that

$$|r_n| \rho^n \leq B \quad \text{for } n \geq 0.$$

Multiply the right side of (5.8.4) by $\frac{\rho^k}{\rho^k}$:

$$(k + 1)|y_{k+1}| \leq \frac{1}{\rho^k} \sum_{n=0}^{k} |r_{k-n}| \rho^{k-n} |y_n| \rho^n \leq \frac{B}{\rho^k} \sum_{n=0}^{k} |y_n| \rho^n. \tag{5.8.5}$$

Define $a_0 = |y_0|$ and define a_{k+1} for $k \geq 0$ by:

$$(k + 1)a_{k+1} = \frac{B}{\rho^k} \sum_{n=0}^{k} a_n \rho^n. \tag{5.8.6}$$

If $|y_n| \leq a_n$ for $0 \leq n \leq k$, then $|y_{k+1}| \leq a_{k+1}$ by (5.8.5) and (5.8.6). Hence

$$0 \leq |y_k| \leq a_k \quad \text{for all } k \geq 0. \tag{5.8.7}$$

We apply the Ratio Test to show the power series $\sum_{n=0}^{\infty} a_n (x - x_0)^n$ is absolutely convergent for $|x - x_0| < \rho$. Take equation (5.8.6) with $k = m$ and multiply it by ρ. Then subtract from it equation (5.8.6) with $k = m - 1$:

$$\rho(m + 1)a_{m+1} = \frac{B}{\rho^{m-1}} \sum_{n=0}^{m} a_n \rho^n = \frac{B}{\rho^{m-1}} \sum_{n=0}^{m-1} a_n \rho^n + \frac{B}{\rho^{m-1}} (a_m \rho^m)$$

$$m a_m = \frac{B}{\rho^{m-1}} \sum_{n=0}^{m-1} a_n \rho^n$$

$$\rho(m + 1)a_{m+1} - m a_m = B \rho a_m.$$

Solve this equation for $\frac{a_{m+1}}{a_m}$:

$$\rho(m + 1)a_{m+1} = (m + B\rho)a_m$$

$$\frac{a_{m+1}}{a_m} = \frac{m + B\rho}{\rho m + \rho}.$$

Then $\qquad \lim_{m \to \infty} \frac{a_{m+1}|x - x_0|^{m+1}}{a_m |x - x_0|^m} = \lim_{m \to \infty} \frac{m + B\rho}{\rho m + \rho} |x - x_0| = \frac{|x - x_0|}{\rho} < 1$

because $|x - x_0| < \rho$. By the Ratio Test, $\sum_{m=0}^{\infty} a_m (x - x_0)^m$ is absolutely convergent. By the Comparison Test and (5.8.7), $\sum_{m=0}^{\infty} y_m (x - x_0)^m$ is absolutely convergent.

The recursion relation (5.8.4) defines all the y_{k+1} in terms of y_0. Relation (5.8.4) implies that if the y_n, for $0 \leq n \leq k$, are multiples of y_0, then y_{k+1} is also a multiple of y_0. Hence all the y_t are multiples of y_0. Write $y_t = u_t y_0$. Then

$$y(x) = \sum_{t=0}^{\infty} y_t (x - x_0)^t = \sum_{t=0}^{\infty} u_t y_0 (x - x_0)^t = y_0 \sum_{t=0}^{\infty} u_t (x - x_0)^t = y_0 u(x)$$

where $u(x) = \sum_{t=0}^{\infty} u_t(x - x_0)^t$.

In an example, the power series solution of the differential equation $y' + r(x)y =$
found by the method used to prove Theorem 5.8.4. There are 6 steps in this proce

Step 1 Substitute the power series $y(x) = \sum_{n=0}^{\infty} y_n(x - x_0)^n$, and
$y'(x) = \sum_{n=1}^{\infty} n a_n(x - x_0)^{n-1}$ into the differential equation.

Step 2 Multiply the power series $r(x)$ and $y(x)$. This will be straightforwa
those examples where $r(x)$ is a polynomial.

Step 3 If necessary, change the dummy variable in the power series $y'(x)$ or $r(x$
so that the powers of $x - x_0$ have the same exponent in each of these series.
the resulting series to obtain one series for $y' + r(x)y$ which equals zero.

Step 4 Set the coefficient of $(x - x_0)^k$ equal to zero. Solve the resulting equatio
y_{k+1} to obtain a recursion relation for y_{k+1} in terms of the y_n with $0 \le n$

Step 5 Solve this recursion relation to obtain a formula for the y_t, with $t \ge$
multiples of y_0: $y_t = y_0 u_t$ for $t \ge 1$. Note $u_0 = 1$.

Step 6 Write the solution of the differential equation as: $y(x) = y_0 \sum_{t=0}^{\infty} u_t(x -$

We illustrate this procedure with three examples.

Examples 5.8.5 (1) Find the power series solution in x of $y' + 5xy = 0$.

Solution The coefficient of y is the power series $r(x) = 0 + 5x + 0x^2 + \cdots$
infinite radius of convergence. Hence our solution will be valid for all x
In Step 1, substitute $y(x) = \sum_{n=0}^{\infty} y_n x^n$ and $y'(x) = \sum_{n=1}^{\infty} n y_n x^{n-1}$ into
differential equation:

$$0 = \sum_{n=1}^{\infty} n y_n x^{n-1} + 5x \sum_{n=0}^{\infty} y_n x^n.$$

In Step 2, multiply the second series by $5x$:

$$0 = \sum_{n=1}^{\infty} n y_n x^{n-1} + \sum_{n=0}^{\infty} 5 y_n x^{n+1}.$$

To add these series in Step 3, change the dummy variable in the first serie
$k = n - 1$ and the dummy variable in the second series to $p = n + 1$:

$$0 = \sum_{k=0}^{\infty} (k+1) y_{k+1} x^k + \sum_{p=1}^{\infty} 5 y_{p-1} x^p = y_1 x^0 + \sum_{k=1}^{\infty} [(k+1) y_{k+1} + 5 y_{k-1}] x^k$$

In Step 4, set the coefficient of each x^k equal to zero:

$$y_1 = 0 \quad \text{and} \quad (k+1) y_{k+1} + 5 y_{k-1} = 0 \text{ for } k \ge 1.$$

In Step 5, solve for y_{k+1} in terms of y_{k-1}:

$$y_{k+1} = -\frac{5 y_{k-1}}{k+1} \text{ for } k \ge 1.$$

Since $y_1 = 0$, it follows that $y_{2h+1} = 0$ for all $h \geq 0$. Also

$$y_2 = -\frac{5y_0}{2},$$

$$y_4 = -\frac{5y_2}{4} = -\frac{5\left(-\frac{5y_0}{2}\right)}{4} = (-1)^2 \frac{5^2 y_0}{2^2(2)(1)},$$

$$y_6 = -\frac{5y_4}{6} = -\frac{5\left[(-1)^2 \frac{5^2 y_0}{2^2(2)(1)}\right]}{6} = (-1)^3 \frac{5^3 y_0}{2^3(3)(2)(1)}.$$

The pattern is clear : $y_{2n} = (-1)^n \frac{5^n y_0}{2^n(n)(n-1)\cdots(2)(1)} = \left(-\frac{5}{2}\right)^n \frac{y_0}{n!}.$

Hence the solution of this differential is:

$$y(x) = \sum_{n=0}^{\infty} \left(-\frac{5}{2}\right)^n \frac{y_0}{n!} x^{2n} = y_0 \sum_{n=0}^{\infty} \left(-\frac{5}{2}\right)^n \frac{x^{2n}}{n!} \quad \text{for all } x \in \Re.$$

By (5.5.14), we can identify this power series as $y(x) = y_0 e^{-5x^2/2}$.

(2) Find the first five terms of the power series solution in $x - 2$ of the initial value problem $y' + xy = 0$, $y(2) = -5$.

Solution The coefficient of y is a power series with infinite radius of convergence: $r(x) = x = 2 + 1(x - 2) + 0(x - 2)^2 + 0(x - 2)^3 + \cdots$ Hence our power series solution will be valid for all $x \in \Re$. In Step 1, substitute $y(x) = \sum_{n=0}^{\infty} y_n(x-2)^n$ and $y'(x) = \sum_{n=1}^{\infty} n y_n(x-2)^{n-1}$ into the differential equation:

$$0 = \sum_{n=1}^{\infty} n y_n(x-2)^{n-1} + [2 + (x-2)] \sum_{n=0}^{\infty} y_n(x-2)^n.$$

In Step 2, carry out the multiplication of $2 + (x-2)$ times $y(x)$:

$$0 = \sum_{n=1}^{\infty} n y_n(x-2)^{n-1} + \sum_{n=0}^{\infty} 2y_n(x-2)^n + \sum_{n=0}^{\infty} y_n(x-2)^{n+1}.$$

To add the series in Step 3, change the dummy variable in the first series to $k = n - 1$ and the dummy variable in the third series to $m = n + 1$:

$$0 = \sum_{k=0}^{\infty} (k+1)y_{k+1}(x-2)^k + \sum_{n=0}^{\infty} 2y_n(x-2)^n + \sum_{m=1}^{\infty} y_{m-1}(x-2)^m$$

$$0 = (y_1 + 2y_0)(x-2)^0 + \sum_{k=1}^{\infty} [(k+1)y_{k+1} + 2y_k + y_{k-1}](x-2)^k.$$

In Step 4, set the coefficient of each $(x-2)^k$ equal to zero:

$$y_1 + 2y_0 = 0 \quad \text{and} \quad (k+1)y_{k+1} + 2y_k + y_{k-1} = 0 \quad \text{for } k \geq 1.$$

Do Step 5 by solving for y_{k+1} in terms of y_k and y_{k-1}:

$$y_1 = -2y_0 \quad \text{and} \quad y_{k+1} = -\frac{2y_k + y_{k-1}}{k+1} \quad \text{for } k \geq 1.$$

Apply this recursion relation for $k = 1$, $k = 2$ and $k = 3$:

$$y_2 = -\frac{2y_1 + y_0}{2} = -\frac{2(-2y_0) + y_0}{2} = \frac{3}{2}y_0,$$

$$y_3 = -\frac{2y_2 + y_1}{3} = -\frac{2\left(\frac{3}{2}y_0\right) - 2y_0}{3} = -\frac{1}{3}y_0,$$

$$y_4 = -\frac{2y_3 + y_2}{4} = -\frac{2\left(-\frac{1}{3}y_0\right) + \frac{3}{2}y_0}{4} = -\frac{5}{24}y_0.$$

Hence the solution of our differential equation is

$$y(x) = y_0 \left[1 - 2(x-2) + \frac{3}{2}(x-2)^2 - \frac{1}{3}(x-2)^3 - \frac{5}{24}(x-2)^4 + \cdots \right]$$

for all $x \in \Re$. Since $y_0 = y(2) = -5$, the solution of our initial value proble

$$y(x) = -5 + 10(x-2) - \frac{15}{2}(x-2)^2 + \frac{5}{3}(x-2)^3 + \frac{25}{24}(x-2)^4 + \cdots .$$

(3) Find the power series solution in x of the differential equation $(1+x)y' + 2y$

Solution To apply Theorem 5.8.4, rewrite this differential equation as:

$$y' + \frac{2}{1+x}y = 0. \qquad (5$$

The Taylor series of $\frac{2}{1+x}$ is a geometric series with ratio $-x$ which conv
absolutely for $|x| < 1$. Thus the power series solution of this differential equa
which we find will be valid for $|x| < 1$. If we did Step 1 by substituting the s
of $y(x)$ and $y'(x)$ into (5.8.8), we would have to compute the Cauchy pro
of the series of $y(x)$ and the geometric series $\frac{2}{1+x} = \sum_{n=0}^{\infty} (-1)^n 2x^n$. To a
this complication, we substitute $y(x) = \sum_{n=0}^{\infty} y_n x^n$ and $y'(x) = \sum_{n=1}^{\infty} n y_n$
into the original differential equation:

$$0 = (1+x) \sum_{n=1}^{\infty} n y_n x^{n-1} + 2 \sum_{n=0}^{\infty} y_n x^n.$$

Step 2 is the multiplication of the first series by $1 + x$:

$$0 = \sum_{n=1}^{\infty} n y_n x^{n-1} + \sum_{n=1}^{\infty} n y_n x^n + \sum_{n=0}^{\infty} 2 y_n x^n.$$

To add these series in Step 3, change the first dummy variable to $k = n - 1$

$$0 = \sum_{k=0}^{\infty} (k+1) y_{k+1} x^k + \sum_{n=1}^{\infty} n y_n x^n + \sum_{n=0}^{\infty} 2 y_n x^n$$

$$0 = (y_1 + 2y_0) x^0 + \sum_{k=1}^{\infty} [(k+1)y_{k+1} + k y_k + 2 y_k] x^k.$$

In Step 4, set the coefficient of each x^k equal to zero:

$$y_1 + 2y_0 = 0 \quad \text{and} \quad (k+1)y_{k+1} + (k+2)y_k = 0 \text{ for } k \geq 1.$$

In Step 5, solve for y_{k+1} in terms of y_k:

$$y_1 = -2y_0 \quad \text{and} \quad y_{k+1} = -\frac{k+2}{k+1} y_k \text{ for } k \geq 1.$$

Then $y_2 = -\frac{3}{2} y_1 = 3y_0$, $y_3 = -\frac{4}{3} y_2 = -4y_0$, $y_4 = -\frac{5}{4} y_3 = 5y_0$ and

$$y_k = (-1)^k (k+1) y_0 \quad \text{for } k \geq 0.$$

Thus $y(x) = \sum_{k=0}^{\infty} (-1)^k (k+1) y_0 x^k = y_0 \sum_{k=0}^{\infty} (-1)^k (k+1) x^k$ for $-1 < x < 1.$

Second Order Differential Equations

The method of the preceding subsection generalizes to find power series solutions of second order differential equations of the form:

$$y'' + q(x)y' + r(x)y = 0. \tag{5.8.9}$$

In Theorem 5.8.7, we state the generalization of Theorem 5.8.4 on the existence of power series solutions of this equation. Then we outline the procedure for finding these power series and apply it to several examples. The proof of Theorem 5.8.7 is analogous to the proof of Theorem 5.8.4 and is deferred to the appendix of this section.

Recall the first order differential equation $y' + r(x)y = 0$ has a solution which is a power series in $x - x_0$ when x_0 is an ordinary point, i.e. $r(x)$ is analytic at $x = x_0$. We define ordinary points of second order equations.

Definition 5.8.6 *If $q(x)$ and $r(x)$ are analytic at $x = x_0$, then x_0 is called an ordinary point of the differential equation $y'' + q(x)y' + r(x)y = 0$.*

The following theorem says the differential equation (5.8.9) has a power series solution in $x - x_0$ at each ordinary point x_0.

Theorem 5.8.7 *Assume x_0 is an ordinary point of the differential equation*

$$y'' + q(x)y' + r(x)y = 0.$$

Let $R > 0$ denote a radius of convergence of both $q(x)$ and $r(x)$. Then there are functions $u(x)$ and $v(x)$ which are analytic at x_0 with a radius of convergence R such that the solution of this differential equation on the interval $(x_0 - R, x_0 + R)$ is:

$$y = Au(x) + Bv(x) \quad \text{for } A, B \in \Re.$$

The six step procedure for finding power series solutions of the differential equation $y'' + q(x)y' + r(x)y = 0$ is analogous to the procedure we used to find power series solutions of a first order differential equation.

Step 1 Substitute the power series $y = \sum_{k=0}^{\infty} y_k(x-x_0)^k$, $y' = \sum_{k=1}^{\infty} ka_k(x-x_0)^{k-1}$ and $y'' = \sum_{k=2}^{\infty} k(k-1)a_k(x-x_0)^{k-2}$ into the differential equation.

Step 2 Multiply the power series $q(x)y'$ and $r(x)y$. This will be straightforward in those examples where $q(x)$ and $r(x)$ are polynomials.

Step 3 If necessary, change the dummy variables in some of these series so that the powers of $x - x_0$ have the same exponent in all the series. Add the resulting series to obtain one power series for $y'' + q(x)y' + r(x)y$ which equals zero.

Step 4 Set the coefficient of $(x - x_0)^k$ equal to zero. Solve the resulting equation for y_{k+2} to obtain a recursion relation for y_{k+2} in terms of the y_n with $0 \le n \le k+1$.

Step 5 Solve this recursion relation for the y_t as linear combinations of y_0 and y_1:

$$y_t = u_t y_0 + v_t y_1 \quad \text{for } t \ge 2.$$

Note $y_0 = 1y_0 + 0y_1$ and $y_1 = 0y_0 + 1y_1$. Hence $u_0 = 1$, $v_0 = 0$ and $u_1 = 0$, $v_1 = 1$.

Step 6 Write the solution of the differential equation as:

$$y = y_0 \sum_{t=0}^{\infty} u_t(x - x_0)^t + y_1 \sum_{t=0}^{\infty} v_t(x - x_0)^t.$$

We illustrate this procedure with three examples.

Examples 5.8.8 (1)   Find the power series solution in $x - 1$ of $y'' + y = 0$.

Solution Since the coefficient 1 of y is the power series $1 + 0(x - 1) + \cdots$, infinite radius of convergence, our solution will be valid for all $x \in \Re$. In S substitute the power series for y and y'' into the differential equation:

$$0 = \sum_{k=2}^{\infty} k(k-1) y_k (x-1)^{k-2} + \sum_{k=0}^{\infty} y_k (x-1)^k.$$

Step 2 is not necessary in this example. To add these series in Step 3, cl the dummy variable in the first series to $n = k - 2$:

$$0 = \sum_{n=0}^{\infty} (n+2)(n+1) y_{n+2} (x-1)^n + \sum_{k=0}^{\infty} y_k (x-1)^k = \sum_{k=0}^{\infty} [(k+2)(k+1) y_{k+2} +$$

In Step 4, set the coefficient of each $(x-1)^k$ equal to zero:

$$0 = (k+2)(k+1) y_{k+2} + y_k \ \text{ for } k \geq 0.$$

In Step 5, solve for y_{k+2} in terms of y_k:

$$y_{k+2} = -\frac{1}{(k+2)(k+1)} y_k \ \text{ for } k \geq 0.$$

Then $y_2 = -\frac{1}{2 \cdot 1} y_0$, $y_4 = -\frac{1}{4 \cdot 3} y_2 = \frac{1}{4 \cdot 3 \cdot 2 \cdot 1} y_0$ and in general

$$y_{2k} = (-1)^k \frac{1}{(2k)!} y_0 \ \text{ for } k \geq 0.$$

Similarly $y_3 = -\frac{1}{3 \cdot 2} y_1$, $y_5 = -\frac{1}{5 \cdot 4} y_3 = \frac{1}{5 \cdot 4 \cdot 3 \cdot 2} y_1$ and in general

$$y_{2k+1} = (-1)^k \frac{1}{(2k+1)!} y_1 \ \text{ for } k \geq 0.$$

Note $u_{2k} = (-1)^k \frac{1}{(2k)!}$, $u_{2k+1} = 0$, $v_{2k} = 0$ and $v_{2k+1} = (-1)^k \frac{1}{(2k+1)!}$ in example. The solution of this differential equation for all $x \in \Re$ is:

$$y = \sum_{n=0}^{\infty} y_n (x-1)^n = \sum_{k=0}^{\infty} y_{2k} (x-1)^{2k} + \sum_{k=0}^{\infty} y_{2k+1} (x-1)^{2k+1}$$

$$y = y_0 \sum_{k=0}^{\infty} (-1)^k \frac{(x-1)^{2k}}{(2k)!} + y_1 \sum_{k=0}^{\infty} (-1)^k \frac{(x-1)^{2k+1}}{(2k+1)!}.$$

Alternatively by (5.5.15) and (5.5.16), $y = y_0 \cos(x-1) + y_1 \sin(x-1)$.

(2) Find the power series solution in x of Airy's equation $y'' = xy$.

Solution Since the power series $x = 0 + 1x + 0x^2 + \cdots$ has infinite radiu convergence, the solution we produce will be valid for all $x \in \Re$. Do Step 1 substituting the power series for y and y'' into Airy's equation:

$$\sum_{k=2}^{\infty} k(k-1) y_k x^{k-2} = x \sum_{k=0}^{\infty} y_k x^k = \sum_{k=0}^{\infty} y_k x^{k+1}.$$

We did Step 2 by multiplying the right power series by x. Do Step 3 by changing the dummy variable in the right series to $n = k + 3$:

$$0 = \sum_{n=3}^{\infty} y_{n-3} x^{n-2} - \sum_{k=2}^{\infty} k(k-1) y_k x^{k-2} = -2y_2 + \sum_{k=3}^{\infty} [y_{k-3} - k(k-1) y_k] x^{k-2}.$$

In Step 4, equate the coefficient of each power of x to zero:

$$2y_2 = 0 \text{ and } y_k = \frac{y_{k-3}}{k(k-1)} \text{ for } k \geq 3.$$

Since $y_2 = 0$, it follows that $y_5 = \frac{y_2}{5 \cdot 4} = 0$, $y_8 = \frac{y_5}{8 \cdot 7} = 0$ and $y_{3m+2} = 0$ for $m \geq 0$. Next, observe that $y_3 = \frac{y_0}{3 \cdot 2}$, $y_6 = \frac{y_3}{6 \cdot 5} = \frac{y_0}{6 \cdot 5 \cdot 3 \cdot 2}$ and in general

$$y_{3k} = \frac{y_0}{[3k][3k-1][3(k-1)][3(k-1)-1] \cdots [3][2]}.$$

Similarly $y_4 = \frac{y_1}{4 \cdot 3}$, $y_7 = \frac{y_4}{7 \cdot 6} = \frac{y_1}{7 \cdot 6 \cdot 4 \cdot 3}$ and in general

$$y_{3k+1} = \frac{y_1}{[3k+1][3k][3(k-1)+1][3(k-1)] \cdots [4][3]}.$$

Thus the solution of Airy's equation for all $x \in \Re$ is:

$$y = \sum_{n=0}^{\infty} y_n x^n = \sum_{k=0}^{\infty} y_{3k} x^{3k} + \sum_{k=0}^{\infty} y_{3k+1} x^{3k+1}$$

$$y = y_0 \sum_{k=0}^{\infty} \frac{x^{3k}}{[3k][3k-1][3(k-1)][3(k-1)-1] \cdots [3][2]}$$

$$+ y_1 \sum_{k=0}^{\infty} \frac{x^{3k+1}}{[3k+1][3k][3(k-1)+1][3(k-1)] \cdots [4][3]}.$$

(3) **(a)** Let $\lambda \in \Re$. Find the power series solution in x of Legendre's equation:

$$(1 - x^2) y'' - 2xy' + \lambda(\lambda+1) y = 0.$$

(b) Find the solution of Legendre's equation with $y(0) = -2$ and $y'(0) = 6$.

Solution **(a)** Divide this differential equation by $1 - x^2$:

$$y'' - \frac{2x}{1 - x^2} y' + \frac{\lambda(\lambda+1)}{1 - x^2} y = 0.$$

The Taylor series of $-\frac{2x}{1-x^2}$ and $\frac{\lambda(\lambda+1)}{1-x^2}$ are geometric series with ratio x^2 and radius of convergence $R = 1$. Thus our solution of Legendre's equation will be valid for $-1 < x < 1$. Substitute the power series for y, y', y'' into the original differential equation and simplify the resulting series:

$$0 = (1 - x^2) \sum_{k=2}^{\infty} k(k-1) y_k x^{k-2} - 2x \sum_{k=1}^{\infty} k y_k x^{k-1} + \lambda(\lambda+1) \sum_{k=0}^{\infty} y_k x^k$$

$$0 = \sum_{k=2}^{\infty} k(k-1) y_k x^{k-2} - \sum_{k=2}^{\infty} k(k-1) y_k x^k - \sum_{k=1}^{\infty} 2k y_k x^k + \sum_{k=0}^{\infty} \lambda(\lambda+1) y_k x^k$$

To sum these series, change the dummy variable in the first series to $n = k - 2$:

$$0 = \sum_{n=0}^{\infty} (n+2)(n+1) y_{n+2} x^n - \sum_{k=0}^{\infty} k(k-1) y_k x^k - \sum_{k=0}^{\infty} 2k y_k x^k + \sum_{k=0}^{\infty} \lambda(\lambda+1) y_k x^k$$

$$0 = \sum_{k=0}^{\infty} [(k+2)(k+1) y_{k+2} - k(k-1) y_k - 2k y_k + \lambda(\lambda+1) y_k] x^k$$

Now set the coefficient of each x^k equal to zero and solve for y_{k+2}:

$$0 = (k+2)(k+1)y_{k+2} - k(k-1)y_k - 2ky_k + \lambda(\lambda+1)y_k$$

$$y_{k+2} = \frac{[k(k+1) - \lambda(\lambda+1)]\,y_k}{(k+2)(k+1)} = -\frac{[\lambda^2 + \lambda - k(k+1)]\,y_k}{(k+2)(k+1)}$$

$$y_{k+2} = -\frac{[\lambda - k][\lambda + k + 1]}{(k+2)(k+1)}y_k.$$

Hence $y_2 = -\dfrac{[\lambda - 0][\lambda + 1]}{2\cdot 1}y_0,\ y_4 = -\dfrac{[\lambda - 2][\lambda + 3]}{4\cdot 3}y_2 = \dfrac{[\lambda - 2][\lambda + 3][\lambda}{4\cdot 3\cdot 2}$

and in general $y_{2k} = (-1)^k \dfrac{\lambda\,[\lambda - 2]\cdots[\lambda - (2k-2)]\,[\lambda+1]\,[\lambda+3]\cdots[\lambda -}{(2k)!}$

Also, $y_3 = -\dfrac{[\lambda - 1][\lambda + 2]}{3\cdot 2}y_1,\ y_5 = -\dfrac{[\lambda - 3][\lambda + 4]}{5\cdot 4}y_3 = \dfrac{[\lambda - 3][\lambda + 4][\lambda -}{5\cdot 4\cdot 3\cdot}$

and in general $y_{2k+1} = (-1)^k \dfrac{[\lambda - 1][\lambda - 3]\cdots[\lambda - (2k-1)]\,[\lambda + 2]\,[\lambda + 4}{(2k+1)!}$

Thus the solution to Legendre's equation for $-1 < x < 1$ is:

$$y = \sum_{n=0}^{\infty} y_n x^n = \sum_{k=0}^{\infty} y_{2k}x^{2k} + \sum_{k=0}^{\infty} y_{2k+1}x^{2k+1}$$

$$y = y_0 \sum_{k=0}^{\infty}(-1)^k \frac{\lambda\,[\lambda - 2]\cdots[\lambda - (2k-2)]\,[\lambda + 1]\,[\lambda + 3]\cdots[\lambda + 2k - 1]}{(2k)!}$$

$$+ y_1 \sum_{k=0}^{\infty}(-1)^k \frac{[\lambda - 1][\lambda - 3]\cdots[\lambda - (2k-1)]\,[\lambda + 2]\,[\lambda + 4]\cdots[\lambda + 2}{(2k+1)!}$$

Observe if $\lambda = 2t$ is a non–negative even integer, then the coefficient of y_0 *even polynomial* of degree $2t$. If $\lambda = 2t + 1$ is a positive odd integer, then coefficient of y_1 is an *odd polynomial* of degree $2t + 1$. These polynomials called Legendre polynomials.

(b) $y_0 = y(0) = -2$ and $y_1 = y'(0) = 6$. Thus the solution of this initial v problem for $-1 < x < 1$ is:

$$y = \sum_{k=0}^{\infty}(-1)^{k+1} \frac{2\lambda\,[\lambda - 2]\cdots[\lambda - (2k-2)]\,[\lambda + 1]\,[\lambda + 3]\cdots[\lambda + 2k - 1]}{(2k)!}x^{2k}$$

$$+ \sum_{k=0}^{\infty}(-1)^k \frac{6\,[\lambda - 1]\,[\lambda - 3]\cdots[\lambda - (2k-1)]\,[\lambda + 2]\,[\lambda + 4]\cdots[\lambda + 2k]}{(2k+1)!}x^{2k+1}.$$

Summary

The reader should know how to find power series representations of the solution first and second order linear differential equations.

Basic Exercises

1. Determine whether each function is analytic at $x = 0$.

   (a) $f(x) = \frac{x}{x+1}$ **(b)** $g(x) = \frac{1}{x^2-1}$

   (c) $h(x) = \sin\frac{1}{x}$ for $x \neq 0$, $h(0) = 0$.   **(d)** $k(x) = x\sin\frac{1}{x}$ for $x \neq 0$, $k(0) =$

   (e) $m(x) = x^2 \sin\frac{1}{x}$ for $x \neq 0$, $m(0) = 0$.   **(f)** $n(x) = \frac{\sin x}{x}$ for $x \neq 0$, $n(0) = 1$

   (g) $p(x) = \frac{1-\cos x}{x}$ for $x \neq 0$, $p(0) = 0$.   **(h)** $q(x) = \frac{e^x-1}{x}$ for $x \neq 0$, $q(0) = 1$

2. Show each function is analytic at $x = x_0$ for every $x_0 \in \Re$.

(a) $f(x) = e^x$ (b) $g(x) = \sin x$ (c) $h(x) = \cos x$

3. Change the dummy variables, where necessary, to add the following series:

(a) $\sum_{n=0}^{\infty}(n-1)^2 x^n + \sum_{n=1}^{\infty} n^2 x^{n-1}$ (b) $\sum_{n=0}^{\infty} nx^{n+1} - \sum_{n=1}^{\infty} 3x^{n-1}$

(c) $\sum_{n=1}^{\infty}(3n+2)x^{n-1} - \sum_{n=1}^{\infty}(5n-1)x^{n-1} + \sum_{n=2}^{\infty}(4n-3)x^{n-2}$

(d) $\sum_{n=2}^{\infty} \frac{n}{n-1}x^n - \sum_{n=2}^{\infty} \frac{n}{n-1}x^{n-2}$ (e) $\sum_{n=0}^{\infty} 5nx^{n+2} + \sum_{n=1}^{\infty} 3nx^{n-1} - \sum_{n=1}^{\infty} 2nx^{n+1}$

(f) $\sum_{n=0}^{\infty}(n^2 - 1)x^{n+1} + \sum_{n=1}^{\infty} n^2 x^{n-1}$

4. Find the ordinary points of each differential equation.

(a) $y' + y\tan x = 0$ (b) $y' + y\ln(1+x^2) = 0$

(c) $xy' + y = 0$ (d) $e^{-x^2}y' + xy = 0$

(e) $y'' + \frac{y'}{1-x^2} + \frac{y}{1+x^2} = 0$ (f) $y'' + \frac{xy'}{e^x - 1} + \frac{y}{e^x + 1} = 0$

(g) $(x^4 - 1)y'' + (x+1)y' + (x-1)y = 0$ (h) $xy'' + y\sin x = 0$

(i) $y''\tan x + xy = 0$ (j) $(1-2x)y'' + y'\cos\pi x = 0$

5. Assume $f(x) = \sum_{k=0}^{\infty} a_k(x-c)^k$ has radius of convergence $R > 0$. If $f(x) = 0$ for $x \in (c - R, c + R)$, show $a_k = 0$ for $k \geq 0$.

6. Find the first five terms of the power series solution in $x - x_0$ of each first order differential equation. Specify the values of x for which your solution is valid.

(a) $y' + (3 + 2x)y = 0$, $x_0 = 0$ (b) $y' + (x^2 - 5)y = 0$, $x_0 = 1$

(c) $(1+x)y' + x^2 y = 0$, $x_0 = 3$ (d) $y' + e^x y = 0$, $x_0 = 0$

(e) $y'\sin x + y\cos x = 0$, $x_0 = \frac{\pi}{2}$ (f) $e^x y' + y\ln(1+x) = 0$, $x_0 = 0$

7. Find the first five terms of a power series solution of each first order initial value problem. Specify the values of x for which your solution is valid.

(a) $y' + (x^2 + 1)y = 0$ with $y(0) = 4$ (b) $y' + (x-3)y = 0$ with $y(2) = -1$

(c) $(x+1)y' + x^2 y = 0$ with $y(0) = 2$ (d) $e^x y' + e^{-x}y = 0$ with $y(0) = -3$

(e) $\frac{y'}{x+1} + \frac{y}{x^2 - 2x} = 0$ with $y(1) = 7$ (f) $y'\ln(1-x) + xye^{x^2} = 0$ with $y(0) = 5$

8. Find the power series solution in $x - x_0$ of each first order differential equation. Specify the values of x for which your solution is valid.

(a) $y' = -7y$, $x_0 = 0$ (b) $y' + 3xy = 0$, $x_0 = 0$ (c) $xy' = 2y$, $x_0 = 1$

(d) $xy' + y = 0$, $x_0 = -1$ (e) $y' + x^2 y = 0$, $x_0 = 0$ (f) $xy' + 3y = 0$, $x_0 = 2$

9. Find a power series solution of each first order initial value problem. Specify the values of x for which your solution is valid.

(a) $y' + 3y = 0$ with $y(2) = 9$ (b) $y' + (x-1)y = 0$ with $y(1) = 4$

(c) $(x - 5)y' + y = 0$ with $y(-5) = 2$ (d) $y' = 12x^3 y$ with $y(0) = 7$

(e) $y' + (x^2 - 2x + 1)y = 0$ with $y(1) = 5$ (f) $(x+2)y' + 2y = 0$ with $y(3) = 4$

10. Find the first five terms of the power series solution in $x - x_0$ of each second order differential equation. Specify the values of x for which your solution is valid.

(a) $y'' + xy' - x^2 y = 0$, $x_0 = 0$ (b) $xy'' - xy' + y = 0$, $x_0 = 2$

(c) $5y'' + xy' + (x-1)^2 y = 0$, $x_0 = 1$ (d) $y'' + y' + e^x y = 0$, $x_0 = 0$

(e) $y'' + y'\sin x + y = 0$, $x_0 = 0$ (f) $e^x y'' + y'\ln(1+x) + y\sin x = 0$, $x_0 = 0$

11. Find the first five terms of a power series solution of each second order initial value problem. Specify the values of x for which your solution is valid.

(a) $(x^2 + 1)y'' - xy = 0$ with $y(0) = -2$ and $y'(0) = 1$

(b) $y'' + 3y' - xy = 0$ with $y(1) = 4$ and $y'(1) = 2$

(c) $x^3 y'' + 4y' - 2y = 0$ with $y(-2) = 3$ and $y'(-2) = 5$

(d) $y'' - e^x y' + 3y = 0$ with $y(0) = 1$ and $y'(0) = -6$

(e) $y'' + y'\ln(1-x) + y\ln(1+x) = 0$ with $y(0) = 5$ and $y'(0) = 3$

12. Find the power series solution in $x - x_0$ of each second order differential equ
Specify the values of x for which your solution is valid.

(a) $y'' + xy' + y = 0$, $x_0 = 0$ (b) $(1 + x)^2 y'' - 2(1 + x)y' + 2y = 0$, $x_0 =$
(c) $y'' + x^2 y' + 2xy = 0$, $x_0 = 0$ (d) $y'' + (3x - 6)y' + 3y = 0$, $x_0 = 2$
(e) $y'' - xy' - y = 0$, $x_0 = 0$ (f) $(1 - x^2)y'' - xy' + \lambda^2 y = 0$, $x_0 = 0$

13. Find a power series solution of each second order initial value problem. Sp
the values of x for which your solution is valid.

(a) $y'' + 4y = 0$ with $y(\pi) = 3$ and $y'(\pi) = -1$
(b) $y'' + (x - 1)^2 y' + (x - 1)y = 0$ with $y(1) = 2$ and $y'(1) = -3$
(c) $(1 + x^2)y'' - xy' + y = 0$ with $y(0) = 4$ and $y'(0) = -2$
(d) $(2 - x^2)y'' + 2y = 0$ with $y(0) = 1$ and $y'(0) = 5$
(e) $y'' + 9xy' = 0$ with $y(0) = 8$ and $y'(0) = 7$

Challenging Problems

1. **(a)** The Legendre polynomial $P_n(x)$ is the polynomial solution of the Lege
equation for $\lambda = n$ an integer and $P_n(1) = 1$. Find $P_n(x)$ for $0 \le n \le 4$.
(b) Show that the Legendre polynomials, for $n \ge 0$, are given by:

$$P_n(x) = \sum_{k=0}^{[n/2]} (-1)^k \frac{(2n - 2k)!}{k!(n - k)!(n - 2k)!} x^{n-2k}.$$

(c) Show the Legendre polynomials satisfy the equation $P_n(x) = \frac{1}{2^n n!} \frac{d^n}{dx^n}(x^2 - $

2. **(a)** Show the Legendre polynomial $P_n(x)$ satisfies $\int_{-1}^{1} P_n(x)^2 \, dx = \frac{2}{2n+1}$ for n
(b) Derive the *orthogonality property*: $\int_{-1}^{1} P_s(x)P_t(x) \, dx = 0$ if $s \ne t$.
(c) Let $P(x)$ be any polynomial of degree at most n. Show
$P(x) = a_0 P_0(x) + \cdots + a_n P_n(x)$ where $a_k = \frac{2k+1}{2} \int_{-1}^{1} P(x)P_k(x) \, dx$ for $0 \le k \le$

3. **(a)** Find the power series solution in x of Hermite's equation $y'' - 2xy' + 2\lambda y$
(b) Show there is a polynomial solution of Hermite's equation of degree λ if $\lambda =$
a non–negative integer. The Hermite polynomial $H_n(x)$ denotes the multiple of
polynomial which has the form $H_n(x) = 2^n x^n + \cdots$.
(c) Find $H_n(x)$ for $0 \le n \le 4$.

4. **(a)** Show the Hermite polynomials satisfy: $H_n(x) = (-1)^n e^{x^2} \frac{d^n}{dx^n} e^{-x^2}$.
(b) Show $H_n(x) = \sum_{k=0}^{[n/2]} (-1)^k \frac{n!}{k!(n-2k)!}(2x)^{n-2k}$.
(c) Let $h_n(x) = e^{-x^2/2} H_n(x)$ for $n \ge 0$. Show the $h_n(x)$ satisfy the orthogona
relation $\int_{-\infty}^{\infty} h_s(x)h_t(x) \, dx = 0$ for $s \ne t$.
(d) Assuming $\int_{-\infty}^{\infty} e^{-x^2} \, dx = \sqrt{\pi}$, show $\int_{-\infty}^{\infty} h_n(x)^2 \, dx = 2^n n! \sqrt{\pi}$.
(c) Let $P(x)$ be any polynomial of degree at most n. Show for $0 \le k \le n$:
$P(x) = a_0 H_0(x) + \cdots + a_n H_n(x)$ where $a_k = \frac{1}{2^n n! \sqrt{\pi}} \int_{-\infty}^{\infty} e^{-x^2} P(x)H_k(x) \, dx$.

Appendix

In this appendix, we prove Theorem 5.8.7 which states the existence of power se
solutions of linear second order differential equations. Its proof generalizes the p
of the corresponding result, Theorem 5.8.4, for first order differential equation.

Theorem 5.8.7 *Assume x_0 is an ordinary point of the differential equation*

$$y'' + q(x)y' + r(x)y = 0.$$

*Let $R > 0$ denote a radius of convergence of both $q(x)$ and $r(x)$. Then there
functions $u(x)$ and $v(x)$ which are analytic at x_0 with a radius of convergence R s*

that the solution of this differential equation on the interval $(x_0 - R, x_0 + R)$ is:

$$y = Au(x) + Bv(x) \quad \text{for } A, B \in \Re.$$

Proof For $x \in (x_0 - R, x_0 + R)$ write

$$q(x) = \sum_{m=0}^{\infty} q_m (x - x_0)^m, \qquad r(x) = \sum_{m=0}^{\infty} r_m (x - x_0)^m. \tag{5.8.10}$$

We want to find a power series $y(x) = \sum_{n=0}^{\infty} y_n (x - x_0)^n$, with a radius of convergence R, which is a solution of the given differential equation. That is, we require:

$$0 = \sum_{n=2}^{\infty} n(n-1) y_n (x - x_0)^{n-2} + \left[\sum_{m=0}^{\infty} q_m (x - x_0)^m \right] \left[\sum_{n=1}^{\infty} n y_n (x - x_0)^{n-1} \right]$$

$$+ \left[\sum_{m=0}^{\infty} r_m (x - x_0)^m \right] \left[\sum_{n=0}^{\infty} y_n (x - x_0)^n \right]$$

$$0 = \sum_{n=2}^{\infty} n(n-1) y_n (x - x_0)^{n-2} + \sum_{k=0}^{\infty} \sum_{n=0}^{k} (n+1) q_{k-n} y_{n+1} (x - x_0)^k$$

$$+ \sum_{k=0}^{\infty} \sum_{n=0}^{k} r_{k-n} y_n (x - x_0)^k.$$

The power series $q(x)$, y', $r(x)$, y are absolutely convergent for $x \in (x_0 - R, x_0 + R)$. By Theorem 5.7.18, the Cauchy products for $q(x)y'$ and $r(x)y$ converge absolutely for $x \in (x_0 - R, x_0 + R)$. To add the three power series above, change the dummy variable in the first series to $k = n - 2$. Then

$$0 = \sum_{k=0}^{\infty} (k+2)(k+1) y_{k+2} (x - x_0)^k + \sum_{k=0}^{\infty} \sum_{n=0}^{k} [(n+1) q_{k-n} y_{n+1} + r_{k-n} y_n] (x - x_0)^k$$

$$0 = \sum_{k=0}^{\infty} \left[(k+2)(k+1) y_{k+2} + \sum_{n=0}^{k} (n+1) q_{k-n} y_{n+1} + \sum_{n=0}^{k} r_{k-n} y_n \right] (x - x_0)^k.$$

This power series equals zero if and only if the coefficient of each $(x - x_0)^k$ is zero. Thus $\sum_{n=0}^{\infty} y_n (x - x_0)^n$ is a solution of our differential equation if and only if

$$(k+2)(k+1) y_{k+2} = - \sum_{n=0}^{k} [(n+1) q_{k-n} y_{n+1} + r_{k-n} y_n]. \tag{5.8.11}$$

This recursion relation gives the value of each y_{k+2}, for $k \geq 0$, in terms of $y_0, y_1, \ldots, y_{k+1}$.

It remains to show the power series $y(x) = \sum_{n=0}^{\infty} y_n (x - x_0)^n$, defined by the recursion relation (5.8.11), is absolutely convergent for $x \in (x_0 - R, x_0 + R)$. Choose ρ with $|x - x_0| < \rho < R$. Since the nth terms of the convergent power series (5.8.10) for $q(x_0 + \rho)$ and $r(x_0 + \rho)$ approach zero, the terms of these two series are bounded. That is, there is $B > 0$ such that:

$$|q_n| \rho^n \leq B \quad \text{and} \quad |r_n| \rho^n \leq B \text{ for } n \geq 0.$$

Multiply the right side of (5.8.11) by $\frac{\rho^k}{\rho^k}$:

$$(k+2)(k+1) |y_{k+2}| \leq \frac{1}{\rho^k} \sum_{n=0}^{k} [(n+1) |q_{k-n}| \rho^{k-n} |y_{n+1}| + |r_{k-n}| \rho^{k-n} |y_n|] \rho^n$$

$$\leq \frac{B}{\rho^k} \sum_{n=0}^{k} [(n+1) |y_{n+1}| + |y_n|] \rho^n + B |y_{k+1}| \rho. \tag{5.8.12}$$

The additional summand on the right side of the preceding inequality will si
the computations below. Define $a_0 = |y_0|$, $a_1 = |y_1|$ and define a_{k+2} for $k \geq 0$ b

$$(k+2)(k+1)a_{k+2} = \frac{B}{\rho^k} \sum_{n=0}^{k} \left[(n+1)a_{n+1} + a_n \right] \rho^n + Ba_{k+1}\rho \quad \text{for } k \geq 0. \quad (5$$

If $|y_n| \leq a_n$ for $0 \leq n \leq k+1$, then $|y_{k+2}| \leq a_{k+2}$ by (5.8.12) and (5.8.13). Her

$$0 \leq |y_k| \leq a_k \quad \text{for all } k \geq 0. \quad (5$$

We apply the Ratio Test to show the power series $\sum_{k=0}^{\infty} a_k(x - x_0)^k$ is abso
convergent for $|x - x_0| < \rho$. Take equation (5.8.13) with $k = m - 1$ and multi
by ρ. Then subtract from it equation (5.8.13) with $k = m - 2$:

$$\rho(m+1)ma_{m+1} = \frac{B}{\rho^{m-2}} \sum_{n=0}^{m-1} \left[(n+1)a_{n+1} + a_n \right] \rho^n + Ba_m\rho^2$$

$$= \frac{B}{\rho^{m-2}} \sum_{n=0}^{m-2} \left[(n+1)a_{n+1} + a_n \right] \rho^n$$

$$+ \frac{B}{\rho^{m-2}} \left[ma_m + a_{m-1} \right] \rho^{m-1} + Ba_n$$

$$m(m-1)a_m = \frac{B}{\rho^{m-2}} \sum_{n=0}^{m-2} \left[(n+1)a_{n+1} + a_n \right] \rho^n + Ba_{m-1}$$

$$\rho(m+1)ma_{m+1} - m(m-1)a_m = \frac{B}{\rho^{m-2}} \left[ma_m + a_{m-1} \right] \rho^{m-1} + Ba_m\rho^2 - Ba_n$$

$$= mB\rho a_m + B\rho^2 a_m.$$

Observe the extra summand $Ba_{k+1}\rho$ in the definition of the a_{k+2} in (5.8.13) prod
the cancellation of the terms $\pm Ba_{m-1}\rho$ above. Solve the preceding equation for $\frac{a}{}$

$$m(m+1)\rho a_{m+1} = \left[m(m-1) + mB\rho + B\rho^2 \right] a_m$$

$$\frac{a_{m+1}}{a_m} = \frac{m(m-1) + mB\rho + B\rho^2}{m(m+1)\rho}.$$

Then $\lim_{m\to\infty} \dfrac{a_{m+1}|x-x_0|^{m+1}}{a_m|x-x_0|^m} = \lim_{m\to\infty} \dfrac{m(m-1) + mB\rho + B\rho^2}{m(m+1)\rho} |x-x_0| = \dfrac{|x-}{}$

because $|x - x_0| < \rho$. By the Ratio Test, $\sum_{m=0}^{\infty} a_m(x-x_0)^m$ is absolutely converg
By the Comparison Test and (5.8.14), $\sum_{m=0}^{\infty} y_m(x-x_0)^m$ is absolutely converge

   Recursion relation (5.8.11) defines the y_k in terms of y_0 and y_1. In addition, i
for $0 \leq n \leq k+1$, are linear combinations of y_0 and y_1, then so is y_{k+2}. Hence all
y_t are linear combinations of y_0 and y_1. Write $y_t = u_t y_0 + v_t y_1$. Then

$$y(x) = \sum_{t=0}^{\infty} y_t(x-x_0)^t = \sum_{t=0}^{\infty} (u_t y_0 + v_t y_1)(x-x_0)^t$$

$$= y_0 \sum_{t=0}^{\infty} u_t(x-x_0)^t + y_1 \sum_{t=0}^{\infty} v_t(x-x_0)^t = y_0 u(x) + y_1 v(x)$$

where $u(x) = \sum_{t=0}^{\infty} u_t(x-x_0)^t$ and $v(x) = \sum_{t=0}^{\infty} v_t(x-x_0)^t$.

Theory

5.9   Numerical Series

Prerequisite: Section 5.7

In this section we present several tests to determine whether a numerical series $\sum_{k=p}^{\infty} a_k$ converges or diverges. When all the a_k are positive, the partial sums $S_n = a_p + \cdots + a_n$ of this series form an increasing sequence. When $S = \lim_{n \to \infty} S_n$ is finite, the series converges with sum S. When this limit is infinite, the series diverges. Two tests are presented to determine whether a series of positive terms converges or diverges. In the first subsection, the Integral Test compares a given series with a corresponding improper integral. In the second subsection, the Limit Comparison Test compares a given complicated series with a simpler series of the same type. In the third subsection, we study series with both positive and negative terms. The Alternating Series Test says that this type of series often converges through cancellation between its positive and negative summands. In the last subsection, we consider a power series, with radius of convergence $0 < R < \infty$. We apply these tests at $x = \pm R$ to determine the interval of convergence.

Integral Test

Assume there is a continuous function f such that the terms of the series $\sum_{k=p}^{\infty} a_k$ are given by $a_k = f(k)$. Then

$$
\begin{aligned}
a_p + a_{p+1} + \cdots + a_k + \cdots &= (a_p)(1) + (a_{p+1})(1) + \cdots + (a_k)(1) + \cdots \\
&= f(p)(1) + f(p+1)(1) + \cdots + f(k)(1) + \cdots
\end{aligned}
$$

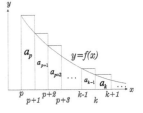

The latter sum is the sum of the areas of the rectangles of width one in Figure 5.9.1. These rectangles approximate the area under the graph of $y = f(x)$ for $x \geq p$ which is given by the improper integral $\int_p^{\infty} f(x)\, dx$. Thus

$$
\sum_{k=p}^{\infty} a_k \approx \int_p^{\infty} f(x)\, dx.
$$

Figure 5.9.1
Integral Test

The Integral Test says: the series and improper integral both converge or both diverge.

Proposition 5.9.2 (Integral Test) *Let f be a continuous function which has domain $[p, \infty)$ with p an integer. Assume:*

   (a) $f(x) \geq 0$ for $x \geq p$;   *(b) f is a decreasing function.*

Define $a_k = f(k)$ for $k \geq p$. Then the series $\sum_{k=p}^{\infty} a_k$ converges if and only if the improper integral $\int_p^{\infty} f(x)\, dx$ converges.

Proof   Assume the improper integral $\int_p^{\infty} f(x)\, dx$ diverges. We show the series $\sum_{k=p}^{\infty} a_k$ also diverges. Let $S_n = \sum_{k=p}^{n} a_k$ denote the nth partial sum of this series. Consider the partition $P_n = \{p,\, p+1,\, p+2,\, \ldots,\, n,\, n+1\}$ of the interval $[p, n+1]$. Note the summands of the upper Riemann sum $U(P_n, f)$ are the areas $a_k \times 1$, for $p \leq k \leq n$, of the rectangles in the left diagram of Figure 5.9.3. Then

$$
\begin{aligned}
\lim_{n \to \infty} S_n &= \lim_{n \to \infty} [a_p \times 1 + \cdots + a_n \times 1] = \lim_{n \to \infty} U(P_n, f) \\
&\geq \lim_{n \to \infty} \int_p^{n+1} f(x)\, dx = \int_p^{\infty} f(x)\, dx = \infty.
\end{aligned}
$$

Thus $\lim_{n\to\infty} S_n = \infty$ and the series $\sum_{k=p}^{\infty} a_k$ diverges.

Now assume the improper integral $\int_p^{\infty} f(x)\ dx$ converges. We show the $\sum_{k=p}^{\infty} a_k$ also converges. Consider the partition $Q_n = \{p,\ p+1,\ \ldots,\ n-1,\ n\}$ interval $[p, n]$. Note the summands of the lower Riemann sum $L(Q_n, f)$ are the $a_{k+1} \times 1$, $p \le k \le n-1$, of the rectangles in the right diagram of Figure 5.9.3.

$$S_n - a_p \quad = \quad a_{p+1} \times 1 + \cdots + a_n \times 1 = L(Q_n, f) \le \int_p^n f(x)\ dx$$

$$S_n \quad \le \quad a_p + \lim_{n\to\infty} \int_p^n f(x)\ dx = a_p + \int_p^{\infty} f(x)\ dx.$$

Thus the sequence of partial sums $\{S_n\}$ is bounded above by the number $a_p + \int_p^{\infty} f(x)\ dx$, and the increasing sequence $\{S_n\}$ converges to its least upper bo Therefore the series $\sum_{k=p}^{\infty} a_k$ converges.

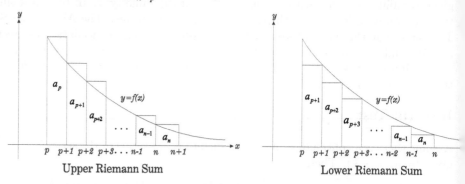

Upper Riemann Sum   Lower Riemann Sum

Figure 5.9.3 Comparing Partial Sums with Riemann Sums

We use improper integrals, computed in Section 4.9, to analyze infinite series

Examples 5.9.4 (1) Let p be a positive number. Determine the values of p which the p-series $\sum_{k=1}^{\infty} \frac{1}{k^p}$ converges.

Solution Let $f(x) = \frac{1}{x^p}$ for $x \ge 1$. Then f is a continuous decreasing func with $f(x) > 0$ for $x \ge 1$. By Example 4.9.7 (1), the improper integral $\int_1^{\infty} \frac{1}{x}$ converges if $p > 1$ and diverges if $p \le 1$. By the Integral Test, the p-s converges if $p > 1$ and diverges if $p \le 1$.

(2) Determine whether the series $\sum_{k=2}^{\infty} \frac{1}{k \ln k}$ converges or diverges.

Solution Let $g(x) = \frac{1}{x \ln x}$ for $x \ge 2$. Then g is a continuous decreasing fun with $g(x) > 0$ for $x \ge 2$. By Example 4.9.7 (5), the improper integral $\int_2^{\infty} \frac{1}{x \ln x}$ diverges. By the Integral Test, the series $\sum_{k=2}^{\infty} \frac{1}{k \ln k}$ also diverges.

The p-series are an interesting family of numerical series. For example, w $p = 1$, we have the harmonic series $\sum_{k=1}^{\infty} \frac{1}{k} = 1 + \frac{1}{2} + \frac{1}{3} + \cdots + \frac{1}{k} + \cdots$ Thus we h a new proof that the harmonic series diverges.

When $p = \frac{1}{2}$ we have the divergent series $\sum_{k=1}^{\infty} \frac{1}{\sqrt{k}} = 1 + \frac{1}{\sqrt{2}} + \frac{1}{\sqrt{3}} + \cdots + \frac{1}{\sqrt{k}} +$

When $p = 2$ we have the convergent series $\sum_{k=1}^{\infty} \frac{1}{k^2} = 1 + \frac{1}{2^2} + \frac{1}{3^2} + \cdots + \frac{1}{k^2} + \cdots$

Comparison Tests

Part of the Comparison Test was given in Proposition 5.7.2 and was used in Section 7 to prove properties of power series. We begin with a full statement of the Comparison Test. Its use is illustrated, and its shortcomings are noted. Then we present the Limit Comparison Test which applies to a wide class of examples.

Proposition 5.9.5 (Comparison Test) *Consider the infinite series $\sum_{k=p}^{\infty} a_k$ and $\sum_{k=q}^{\infty} b_k$. Assume there is a positive integer N such that*

$$0 \le a_k \le b_k \quad \text{for } k \ge N.$$

(a) *If the series $\sum_{k=q}^{\infty} b_k$ converges, then the series $\sum_{k=p}^{\infty} a_k$ also converges.*

(b) *If the series $\sum_{k=p}^{\infty} a_k$ diverges, then the series $\sum_{k=q}^{\infty} b_k$ also diverges.*

Proof Let $S_n = \sum_{k=N}^{n} a_k$ and $T_n = \sum_{k=N}^{n} b_k$ denote the partial sums of the two series. Both sequences of partial sums $\{S_n\}$ and $\{T_n\}$ are increasing, and

$$0 \le S_n \le T_n \text{ for } n \ge N.$$

(a) Since the series $B = \sum_{k=N}^{\infty} b_k = \sum_{k=q}^{\infty} b_k - b_q - \cdots - b_{N-1}$ converges, the sequence $\{T_n\}$ has B as its least upper bound. Then B is an upper bound of the sequence $\{S_n\}$ which therefore converges to its least upper bound. Hence the series $\sum_{k=N}^{\infty} a_k$ converges, and the series $\sum_{k=p}^{\infty} a_k = a_p + \cdots + a_{N-1} + \sum_{k=N}^{\infty} a_k$ also converges.
(b) The series $\sum_{k=N}^{\infty} a_k = \sum_{k=p}^{\infty} a_k - a_p - \cdots - a_{N-1}$ diverges. Hence the sequence $\{S_n\}$ has infinite limit. Therefore the sequence $\{T_n\}$ also has limit infinity, and the series $\sum_{k=N}^{\infty} b_k$ diverges. Thus $\sum_{k=q}^{\infty} b_k = b_q + \cdots + b_{N-1} + \sum_{k=N}^{\infty} b_k$ also diverges. $\square$

The Comparison Test works well on certain examples.

Examples 5.9.6 (1) Determine whether the series $\sum_{k=2}^{\infty} \frac{1}{\ln k}$ converges or diverges.

Solution Observe $0 < \frac{1}{k \ln k} < \frac{1}{\ln k}$ for $k \ge 2$. In Example 5.9.4 (2), we used the Integral Test to show the series $\sum_{k=2}^{\infty} \frac{1}{k \ln k}$ diverges. By the Comparison Test, $\sum_{k=2}^{\infty} \frac{1}{\ln k}$ also diverges.

(2) Determine whether the series $\sum_{k=1}^{\infty} \frac{\sin k}{k^3}$ converges or diverges.

Solution Observe $0 < \frac{|\sin k|}{k^3} \le \frac{1}{k^3}$ for $k \ge 1$. The series $\sum_{k=1}^{\infty} \frac{1}{k^3}$ converges because it is the p–series with $p = 3 > 1$. By the Comparison Test, $\sum_{k=1}^{\infty} \frac{|\sin k|}{k^3}$ converges. Thus $\sum_{k=1}^{\infty} \frac{\sin k}{k^3}$ is absolutely convergent, hence convergent.

(3) Determine whether the series $\sum_{k=0}^{\infty} (-1)^k \frac{1}{e^{k^2}}$ converges or diverges.

Solution Observe $\frac{1}{e^{k^2}} \le \frac{1}{e^k}$ for $k \ge 0$. The series $\sum_{k=0}^{\infty} \frac{1}{e^k}$ converges because it is a geometric series with ratio $\frac{1}{e} < 1$. By the Comparison Test, $\sum_{k=0}^{\infty} \frac{1}{e^{k^2}}$ also converges. Thus $\sum_{k=0}^{\infty} (-1)^k \frac{1}{e^{k^2}}$ is absolutely convergent, hence convergent. $\square$

Note the Comparison Test does not apply to some simple series. For example, we know the geometric series $\sum_{k=1}^{\infty} \frac{1}{5^k}$ converges, yet we can not conclude from the Comparison Test that the similar series $\sum_{k=1}^{\infty} \frac{1}{5^k - 1}$ also converges. Similarly, we know the p–series $\sum_{k=1}^{\infty} \frac{1}{\sqrt{k}}$ diverges, yet we can not conclude from the Comparison Test that the similar series $\sum_{k=1}^{\infty} \frac{1}{1 + \sqrt{k}}$ also diverges. The Limit Comparison Test applies to these examples.

Proposition 5.9.7 (Limit Comparison Test) *Let $\sum_{k=p}^{\infty} a_k$ and $\sum_{k=q}^{\infty} b_k$ be two of positive numbers. Assume*

$$\lim_{k\to\infty} \frac{a_k}{b_k} = L$$

with $L \neq 0$ and $L \neq \infty$. Then both $\sum_{k=p}^{\infty} a_k$ and $\sum_{k=q}^{\infty} b_k$ converge or both dive

Proof Find a positive integer N such that if $k \geq N$, then $\frac{L}{2} < \frac{a_k}{b_k} < \frac{3L}{2}$. Thus

$$\frac{L}{2}b_k < a_k < \frac{3L}{2}b_k \text{ for } k \geq N.$$

By the Comparison Test:

- if $\sum_{k=p}^{\infty} a_k$ converges, then $\sum_{k=q}^{\infty} \frac{L}{2}b_k = \frac{L}{2}\sum_{k=q}^{\infty} b_k$ and hence $\sum_{k=q}^{\infty} b_k$ conver
- if $\sum_{k=q}^{\infty} b_k$ converges, then $\frac{3L}{2}\sum_{k=q}^{\infty} b_k = \sum_{k=q}^{\infty} \frac{3L}{2}b_k$ and hence $\sum_{k=p}^{\infty} a_k$ conv

Observe the hypothesis of the Limit Comparison Test is symmetric:

$$\lim_{k\to\infty} \frac{a_k}{b_k} \text{ is neither 0 nor } \infty \text{ if and only if } \lim_{k\to\infty} \frac{b_k}{a_k} \text{ is neither 0 nor } \infty.$$

We begin with the two examples mentioned above which can not be analyzed b; Comparison Test. Then we apply the Limit Comparison Test to a series $\sum_{k=1}^{\infty}$ defined by a rational function $\frac{f(x)}{g(x)}$. We compare this series to the p–series w: equal to the degree of the denominator $g(x)$ minus the degree of the numerator

Examples 5.9.8 (1) Determine if the series $\sum_{k=1}^{\infty} \frac{1}{5^k-1}$ converges or diverges.

Solution Apply the Limit Comparison Test to $\sum_{k=1}^{\infty} \frac{1}{5^k-1}$ and $\sum_{k=1}^{\infty} \frac{1}{5^k}$:

$$\lim_{k\to\infty} \frac{1/(5^k-1)}{1/5^k} = \lim_{k\to\infty} \frac{5^k}{5^k-1} = \lim_{k\to\infty} \frac{1}{1-5^{-k}} = 1$$

which is neither 0 nor ∞. The geometric series $\sum_{k=1}^{\infty} \frac{1}{5^k}$ with ratio $\frac{1}{5}$ conve because $\frac{1}{5} < 1$. By the Limit Comparison Test, $\sum_{k=1}^{\infty} \frac{1}{5^k-1}$ also converges.

(2) Determine whether the series $\sum_{k=1}^{\infty} \frac{1}{1+\sqrt{k}}$ converges or diverges.

Solution Apply the Limit Comparison Test to $\sum_{k=1}^{\infty} \frac{1}{1+\sqrt{k}}$ and $\sum_{k=1}^{\infty} \frac{1}{\sqrt{k}}$:

$$\lim_{k\to\infty} \frac{1/(1+\sqrt{k})}{1/\sqrt{k}} = \lim_{k\to\infty} \frac{\sqrt{k}}{1+\sqrt{k}} = \lim_{k\to\infty} \frac{1}{\frac{1}{\sqrt{k}}+1} = 1$$

which is neither 0 nor ∞. The p–series $\sum_{k=1}^{\infty} \frac{1}{\sqrt{k}}$ with $p = \frac{1}{2}$ diverges beca $\frac{1}{2} \leq 1$. By the Limit Comparison Test, the series $\sum_{k=1}^{\infty} \frac{1}{1+\sqrt{k}}$ also diverges.

(3) Determine whether the series $\sum_{k=1}^{\infty} \frac{5k^6+2k^4+8}{3k^8+5k^3+2}$ converges or diverges.

Solution Apply the Limit Comparison Test to the series $\sum_{k=1}^{\infty} \frac{5k^6+2k^4+8}{3k^8+5k^3+2}$ the p–series $\sum_{k=1}^{\infty} \frac{k^6}{k^8} = \sum_{k=1}^{\infty} \frac{1}{k^2}$:

$$\lim_{k\to\infty} \frac{(5k^6+2k^4+8)/(3k^8+5k^3+2)}{1/k^2} = \lim_{k\to\infty} \frac{5k^8+2k^6+8k^2}{3k^8+5k^3+2} = \frac{5}{3}$$

which is neither 0 nor ∞. The p–series $\sum_{k=1}^{\infty} \frac{1}{k^2}$ with $p = 2$ converges beca $2 > 1$. By the Limit Comparison Test, $\sum_{k=1}^{\infty} \frac{5k^6+2k^4+8}{3k^8+5k^3+2}$ also converges.

(4) Determine whether the series $\sum_{k=1}^{\infty} \sqrt{\frac{2k^3-5k+7}{8k^5+2k^2+9}}$ converges or diverges.

Solution Apply the Limit Comparison Test to the series $\sum_{k=1}^{\infty} \sqrt{\frac{2k^3-5k+7}{8k^5+2k^2+9}}$

and $\sum_{k=1}^{\infty} \sqrt{\frac{k^3}{k^5}} = \sum_{k=1}^{\infty} \frac{1}{k}$:

$$\lim_{k\to\infty} \frac{\sqrt{(2k^3-5k+7)/(8k^5+2k^2+9)}}{1/k} = \lim_{k\to\infty} \sqrt{\frac{2k^5-5k^3+7k^2}{8k^5+2k^2+9}} = \sqrt{\frac{2}{8}} = \frac{1}{2}$$

which is neither 0 nor ∞. The harmonic series $\sum_{k=1}^{\infty} \frac{1}{k}$ diverges. By the Limit Comparison Test, $\sum_{k=1}^{\infty} \sqrt{\frac{2k^3-5k+7}{8k^5+2k^2+9}}$ also diverges. $\square$

Although the Limit Comparison Test seems to apply to a broader class of series than the Comparison Test, there are examples of series which require the Comparison Test. In Examples 5.9.6 we used the Comparison Test to analyze the series $\sum_{k=2}^{\infty} \frac{1}{\ln k}$ and $\sum_{k=1}^{\infty} \frac{\sin k}{k^3}$. These series can not be analyzed by the Limit Comparison Test.

Alternating Series Test

Consider a series where the signs of the summands alternate. The Alternating Series Test says that this series converges through cancellation between its positive and negative summands as long as the limit of its nth term equals zero. In addition, this test produces a bound on the error of approximating the sum of this series by one of its partial sums.

Proposition 5.9.9 (Alternating Series Test) *Consider the series* $\sum_{k=p}^{\infty}(-1)^k a_k$ *where:*

(a) $a_k > 0$ *for* $k \geq p$;   **(b)** *the sequence* $\{a_k\}$ *is decreasing;*   **(c)** $\lim_{k\to\infty} a_k = 0$.

Then the series $\sum_{k=p}^{\infty}(-1)^k a_k$ *converges. Moreover, the approximation of the sum* $S = \sum_{k=p}^{\infty}(-1)^k a_k$ *by the partial sum* $S_n = \sum_{k=p}^{n}(-1)^k a_k$ *has error less than* a_{n+1}.

Proof Note the partial sums of this series satisfy the following inequalities:

(1) $S_{2n} - S_{2n+2} = a_{2n+1} - a_{2n+2} > 0$;

(2) $S_{2n+1} - S_{2n-1} = a_{2n} - a_{2n+1} > 0$.

(3) For $n \geq k$: $S_{2n} - S_{2k-1} = a_{2n} + (S_{2n-1} - S_{2k-1}) > 0$;

These observations determine the relative positions of the partial sums S_m on the number line as depicted in Figure 5.9.10.
- By (1), the even partial sums are a decreasing sequence.
- By (2), the odd partial sums are an increasing sequence.
- By (3), the odd partial sums are to the left of the even ones.

By (3) with $n = k$: $\lim_{n\to\infty} (S_{2n} - S_{2n-1}) = \lim_{n\to\infty} a_{2n} = 0$. Thus the least upper bound S of the odd partial sums is also the greatest lower bound of the even partial sums.

Then $S = \lim_{n\to\infty} S_{2n-1} = \lim_{n\to\infty} S_{2n}$,   and $S = \lim_{n\to\infty} S_n$.

Hence the series $\sum_{k=p}^{\infty}(-1)^k a_k$ converges with sum S.

Since $S_{2n-1} \leq S \leq S_{2n}$, the error $S - S_{2n-1}$ in the approximation $S \approx S_{2n-1}$ is less than $S_{2n} - S_{2n-1} = a_{2n}$. Similarly, $S_{2n+1} \leq S \leq S_{2n}$. Hence the error $S_{2n} - S$ in the approximation $S \approx S_{2n}$ is less than $S_{2n} - S_{2n+1} = a_{2n+1}$. $\square$

Recall that a series $\sum_{k=p}^{\infty} a_k$ is called absolutely convergent if the series $\sum_{k=p}^{\infty} |a_k|$, which does not depend on cancellation, converges. In this case the original series also

converges. If the series $\sum_{k=p}^{\infty} |a_k|$ diverges but the series $\sum_{k=p}^{\infty} a_k$ converges, th
call the latter series conditionally convergent. A conditionally convergent series
converges because of the cancellation between its positive and negative summar

$$S_1 \quad S_3 \cdots S_{2n-1} \, S_{2n+1} \cdots \longrightarrow S \longleftarrow \cdots S_{2n+2} \quad S_{2n} \cdots S_4 \, S_2$$

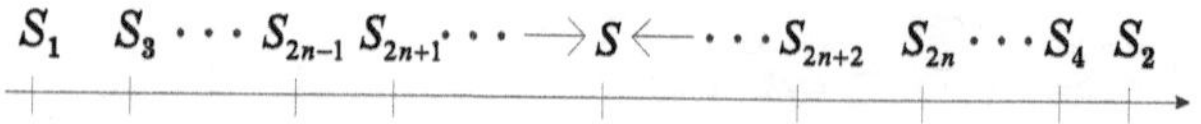

Figure 5.9.10 The Alternating Series Test

Examples 5.9.11 (1) Determine whether the series $\sum_{k=1}^{\infty}(-1)^k \frac{\sqrt{k}}{3k-1}$ is absol
convergent, conditionally convergent or divergent.

Solution Use the Limit Comparison Test to check for absolute converg
That is, compare the series $\sum_{k=1}^{\infty} \frac{\sqrt{k}}{3k-1}$ and $\sum_{k=1}^{\infty} \frac{\sqrt{k}}{k} = \sum_{k=1}^{\infty} \frac{1}{\sqrt{k}}$:

$$\lim_{k\to\infty} \frac{\sqrt{k}/(3k-1)}{1/\sqrt{k}} = \lim_{k\to\infty} \frac{k}{3k-1} = \frac{1}{3}$$

which is neither 0 nor ∞. Note $\sum_{k=1}^{\infty} \frac{1}{\sqrt{k}}$ is the p–series with $p = \frac{1}{2}$ whic
verges because $\frac{1}{2} \leq 1$. By the Limit Comparison Test, $\sum_{k=1}^{\infty} \frac{\sqrt{k}}{3k-1}$ also dive
and $\sum_{k=1}^{\infty}(-1)^k \frac{\sqrt{k}}{3k-1}$ is not absolutely convergent. By the quotient rule:

$$D\left(\frac{\sqrt{k}}{3k-1}\right) = \frac{\frac{1}{2\sqrt{k}}(3k-1) - \sqrt{k}(3)}{(3k-1)^2} = \frac{-3k-1}{2\sqrt{k}(3k-1)^2} < 0 \text{ for } k \geq 1.$$

Hence $\left\{\frac{\sqrt{k}}{3k-1}\right\}$ is a decreasing sequence of positive numbers with limit zero.
$\sum_{k=1}^{\infty}(-1)^k \frac{\sqrt{k}}{3k-1}$ converges by the Alternating Series Test. Therefore this s
is conditionally convergent.

(2)   Determine whether the series $\sum_{k=1}^{\infty}(-1)^k \frac{k!}{(2k)!}$ is absolutely convergent, co
tionally convergent or divergent.

Solution Check for absolute convergence by the Ratio Test:

$$\lim_{k\to\infty} \left| \frac{(-1)^{k+1}(k+1)!/[2(k+1)]!}{(-1)^k k!/(2k)!} \right| = \lim_{k\to\infty} \frac{(k+1)!}{k!} \frac{(2k)!}{(2k+2)!}$$

$$= \lim_{k\to\infty} \frac{(k+1)k!}{k!} \frac{(2k)!}{(2k+2)(2k+1)(2k)!} = \lim_{k\to\infty} \frac{k+1}{2(k+1)(2k+1)}$$

$$= \lim_{k\to\infty} \frac{1}{4k+2} = 0 < 1.$$

By the Ratio Test, the series $\sum_{k=1}^{\infty}(-1)^k \frac{k!}{(2k)!}$ is absolutely convergent.

(3)   Determine whether the series $\sum_{k=1}^{\infty}(-1)^k \frac{k}{2k-1}$ is absolutely convergent, co
tionally convergent or divergent.

Solution Observe $\lim_{k\to\infty} \frac{k}{2k-1} = \frac{1}{2}$. Thus $\lim_{k\to\infty}(-1)^k \frac{k}{2k-1}$ does not e:
and the series $\sum_{k=1}^{\infty}(-1)^k \frac{k}{2k-1}$ diverges by the nth Term Test. The Alterna
Series Test does not apply to this series because the decreasing sequence $\left\{ \frac{k}{2k-}\right.$
of positive numbers does not have limit zero.

The last statement in the Alternating Series Test is useful: the error in appr
mating the sum of an alternating series by summing its first n terms is less than
absolute value of its $(n+1)^{\text{st}}$ term.

Examples 5.9.12 (1) Determine a bound on the error in the approximation of e^{-1}
by adding the first four terms of the series $e^x = \sum_{k=0}^{\infty} \frac{x^k}{k!}$ with $x = -1$.

Solution Since $e^{-1} = \sum_{k=0}^{\infty} (-1)^k \frac{1}{k!}$ is an alternating series, the approximation

$$e^{-1} \approx 1 - 1 + \frac{1}{2} - \frac{1}{6} = \frac{1}{3}$$

has error less than the absolute value of the fifth term $\left| (-1)^4 \frac{1}{4!} \right| = \frac{1}{24}$.

(2) Use a power series to estimate $\int_0^1 \frac{\sin x}{x} \, dx$ with error less than .01.

Solution By Prop. 5.5.13)(b), $\sin x = \sum_{k=0}^{\infty} (-1)^k \frac{x^{2k+1}}{(2k+1)!}$ for $x \in \Re$. Hence

$$\int_0^1 \frac{\sin x}{x} \, dx = \int_0^1 \frac{1}{x} \sum_{k=0}^{\infty} (-1)^k \frac{x^{2k+1}}{(2k+1)!} \, dx = \int_0^1 \sum_{k=0}^{\infty} (-1)^k \frac{x^{2k}}{(2k+1)!} \, dx$$

$$= \sum_{k=0}^{\infty} (-1)^k \frac{x^{2k+1}}{(2k+1)(2k+1)!} \bigg|_0^1 = \sum_{k=0}^{\infty} (-1)^k \frac{1}{(2k+1)(2k+1)!} .$$

The first few terms of this alternating series are: $1, -\frac{1}{18}, \frac{1}{600}$. Since the third
term $\frac{1}{600} < .002$ is less than .01:

$$\int_0^1 \frac{\sin x}{x} \, dx \approx 1 - \frac{1}{18} = \frac{17}{18} \approx 0.944$$

with error less than .002. $\qquad\qquad\qquad\qquad\qquad\qquad\qquad\qquad \Box$

Interval of Convergence of a Power Series

Recall that a power series $P(x) = \sum_{k=0}^{\infty} a_k (x - c)^k$ has a radius of convergence R,
$0 \leq R \leq \infty$, such that $P(x)$ is absolutely convergent for $c - R < x < c + R$. If $R < \infty$,
then $P(x)$ is divergent for $x < c - R$ or $x > c + R$. As the examples below show, we
can have absolute convergence, conditional convergence or divergence when $x = c - R$
or $x = c + R$.

Definition 5.9.13 Let $P(x) = \sum_{k=0}^{\infty} a_k (x - c)^k$ be a power series with radius of
convergence R. The set of numbers for which $P(x)$ converges is called its interval of
convergence. The interval of convergence has the form:

$$(c - R, c + R), \quad [c - R, c + R], \quad (c - R, c + R] \quad or \quad [c - R, c + R).$$

To find the interval of convergence of a power series $P(x)$, we first find its radius
of convergence by the Ratio Test. If $0 < R < \infty$, we use other tests to determine
whether the numerical series $P(c - R)$ and $P(c + R)$ converge or diverge.

Examples 5.9.14 (1) Find the interval of convergence of the power series
$P(x) = \sum_{k=0}^{\infty} \frac{kx^k}{(k^2+1)3^k}$.

Solution Apply the Ratio Test to determine the radius of convergence:

$$\lim_{k \to \infty} \left| \frac{(k+1)x^{k+1} / \left[((k+1)^2 + 1) 3^{k+1} \right]}{kx^k / \left[(k^2 + 1)3^k \right]} \right| = \lim_{k \to \infty} \frac{|x|^{k+1}}{|x|^k} \frac{3^k}{3^{k+1}} \frac{(k+1)(k^2+1)}{k\left[(k+1)^2 + 1 \right]}$$

$$= \lim_{k \to \infty} \frac{|x|}{3} \frac{k^3 + k^2 + k + 1}{k^3 + 2k^2 + 2k} = \frac{|x|}{3} .$$

Hence $P(x)$ is absolutely convergent for $\frac{|x|}{3} < 1$ and divergent for $\frac{|x|}{3} >$
$P(x)$ is absolutely convergent for $-3 < x < 3$ and divergent for $x < -3$ or x

When $x = 3$, $P(3) = \sum_{k=0}^{\infty} \frac{k3^k}{(k^2+1)3^k} = \sum_{k=1}^{\infty} \frac{k}{k^2+1}$. We compare this series
the harmonic series $\sum_{k=1}^{\infty} \frac{1}{k}$ by the Limit Comparison Test:

$$\lim_{k \to \infty} \frac{k/(k^2+1)}{1/k} = \lim_{k \to \infty} \frac{k^2}{k^2+1} = 1$$

which is neither 0 nor ∞. Since the harmonic series diverges, the series
also diverges.

When $x = -3$, $P(-3) = \sum_{k=0}^{\infty} \frac{k(-3)^k}{(k^2+1)3^k} = \sum_{k=1}^{\infty} (-1)^k \frac{k}{k^2+1}$. Since $\left\{ \frac{1}{k} \right.$
is a decreasing sequence of positive numbers with limit zero, the series P
converges by the Alternating Series Test.

Thus the interval of convergence of $P(x)$ is $[-3, 3)$.

(2) Find the interval of convergence of the power series $Q(x) = \sum_{k=1}^{\infty} (-1)^k \frac{(x-}{k^3}$

Solution Apply the Ratio Test to determine the radius of convergence:

$$\lim_{k \to \infty} \left| \frac{(-1)^{k+1}(x-2)^{k+1} / [(k+1)^3 5^{k+1}]}{(-1)^k (x-2)^k / (k^3 5^k)} \right| = \lim_{k \to \infty} \frac{|x-2|^{k+1}}{|x-2|^k} \frac{5^k}{5^{k+1}} \frac{k}{(k+}$$

$$= \lim_{k \to \infty} \frac{|x-2|}{5} \frac{k^3}{k^3 + 3k^2 + 3k + 1} = \frac{|x-2|}{5}.$$

Hence $Q(x)$ is absolutely convergent for $\frac{|x-2|}{5} < 1$, i.e. for $-3 < x < 7$,
$Q(x)$ is divergent for $\frac{|x-2|}{5} > 1$, i.e. for $x < -3$ or $x > 7$.

When $x = 7$, $Q(7) = \sum_{k=1}^{\infty} (-1)^k \frac{5^k}{k^3 5^k} = \sum_{k=1}^{\infty} (-1)^k \frac{1}{k^3}$ which converges by
Alternating Series Test.

When $x = -3$, $P(-3) = \sum_{k=1}^{\infty} (-1)^k \frac{(-5)^k}{k^3 5^k} = \sum_{k=1}^{\infty} \frac{1}{k^3}$ which is the conver
p–series with $p = 3 > 1$.

Thus the interval of convergence of $Q(x)$ is $[-3, 7]$.

(3) Find the interval of convergence of the power series $S(x) = \sum_{k=0}^{\infty} \frac{(k!)^2 x^k}{(2k)!}$.

Solution Apply the Ratio Test to determine the radius of convergence:

$$\lim_{k \to \infty} \left| \frac{[(k+1)!]^2 x^{k+1} / [(2(k+1))!]}{(k!)^2 x^k / (2k)!} \right| = \lim_{k \to \infty} \frac{|x|^{k+1}}{|x|^k} \left[\frac{(k+1)!}{k!} \right]^2 \frac{(2k)!}{(2k+2)!}$$

$$= \lim_{k \to \infty} |x| \left[\frac{(k+1)k!}{k!} \right]^2 \frac{(2k)!}{(2k+2)(2k+1)(2k)!} = \lim_{k \to \infty} |x| \frac{k+1}{4k+2} = \frac{|x|}{4}.$$

Hence $S(x)$ is absolutely convergent for $\frac{|x|}{4} < 1$ and divergent for $\frac{|x|}{4} > 1$,
$S(x)$ is absolutely convergent for $-4 < x < 4$ and divergent for $x < -4$ or x

When $x = 4$, $S(4) = \sum_{k=0}^{\infty} \frac{(k!)^2 4^k}{(2k)!}$. Observe $2^k (k!)$ equals the product of
even factors of $(2k)!$. Thus

$$\lim_{k \to \infty} \frac{(k!)^2 4^k}{(2k)!} = \lim_{k \to \infty} \frac{(k! 2^k)^2}{(2k)!} = \lim_{k \to \infty} \frac{(2k)(2k-2)\cdots(4)(2)}{(2k-1)(2k-3)\cdots(3)(1)}$$

$$= \lim_{k \to \infty} \left(\frac{2k}{2k-1} \right) \left(\frac{2k-2}{2k-3} \right) \cdots \left(\frac{4}{3} \right) \left(\frac{2}{1} \right)$$

$$= \lim_{k \to \infty} \left(1 + \frac{1}{2k-1}\right)\left(1 + \frac{1}{2k-3}\right)\cdots\left(1 + \frac{1}{3}\right)\left(1 + \frac{1}{1}\right)$$

$$\geq \lim_{k \to \infty} \left(\frac{1}{2k-1} + \frac{1}{2k-3} + \cdots + \frac{1}{3} + \frac{1}{1}\right) = \infty$$

because the series $\sum_{k=1}^{\infty} \frac{1}{2k-1}$ diverges. By the nth Term Test, $S(4)$ diverges.

When $x = -4$, $S(-4) = \sum_{k=0}^{\infty} \frac{(k!)^2(-4)^k}{(2k)!} = \sum_{k=0}^{\infty}(-1)^k \frac{(k!)^2 4^k}{(2k)!}$. By the preceding computation, the limit of the nth term of this series does not exist, and $S(-4)$ diverges by the nth Term Test.

Thus the interval of convergence of $S(x)$ is $(-4, 4)$.

(4) Find the interval of convergence of the power series $T(x) = \sum_{k=0}^{\infty} \frac{x^k}{k!}$.

Solution By Proposition 5.5.13(a), $T(x)$ converges with sum e^x for all $x \in \Re$. Hence the interval of convergence of $T(x)$ is $(-\infty, \infty) = \Re$.

(5) Find the interval of convergence of the power series $U(x) = \sum_{k=0}^{\infty} k^k x^k$.

Solution Apply the Ratio Test to find the radius of convergence:

$$\lim_{k \to \infty} \left| \frac{(k+1)^{k+1} x^{k+1}}{k^k x^k} \right| = \lim_{k \to \infty} \frac{(k+1)^{k+1}}{k^k} \frac{|x|^{k+1}}{|x|^k} = \lim_{k \to \infty} (k+1)\left(\frac{k+1}{k}\right)^k |x|$$

$$= \lim_{k \to \infty} (k+1)|x| \lim_{k \to \infty} \left(1 + \frac{1}{k}\right)^k = \lim_{k \to \infty} (k+1)|x|e = \left\{ \begin{array}{ll} \infty & \text{if } x \neq 0 \\ 0 & \text{if } x = 0 \end{array} \right\}.$$

Thus $U(x)$ converges if $x = 0$ and diverges if $x \neq 0$. Hence the interval of convergence of $U(x)$ is $[0, 0] = \{0\}$. $\square$

Historical Remarks

Using infinite series was a basic tool in the development of calculus in the 17th century and in the applications of calculus in the 18th century. However, little attention was given to define convergence or to develop methods that determine whether a given series converges. The rigorous treatment of infinite series was set out by Augustin–Louis Cauchy in 1821. In his text *cours d'analyse*, he defines convergence of an infinite series as in Def. 5.4.2 and presents the basic convergence tests: the Root Test, Comparison Test, Limit Comparison Test and Integral Test. He also gives proofs of the Ratio Test and the Alternating Series Test. The former had been found by Edward Waring in 1776, while the latter had been discovered by Gottfried Wilhelm Leibniz in 1713.

Summary

The reader should be able to apply the nth Term Test, Ratio Test, Integral Test, Comparison Test, Limit Comparison Test and Alternating Series Test to decide whether a given series is absolutely convergent, conditionally convergent or divergent. She should also know how to use these tests to determine the behavior of a power series at the endpoints of its interval of convergence.

Basic Exercises

Determine whether each of these 36 series is absolutely convergent, conditionally convergent or divergent.

1. $\sum_{k=2}^{\infty} \frac{1}{k(\ln k)^2}$ 2. $\sum_{k=1}^{\infty}(-1)^k \left(\sqrt{k+1} - \sqrt{k}\right)$ 3. $\sum_{k=1}^{\infty} \cos(\pi k)\frac{k}{e^{k^2}}$

4. $\sum_{k=1}^{\infty} \frac{\sqrt{k}}{e^{k^3}}$ 5. $\sum_{k=1}^{\infty}(-1)^k \sqrt{\frac{2^k+7}{3^k+8}}$ 6. $\sum_{k=2}^{\infty}(-1)^k \frac{1}{k(\ln k)(\ln \ln k)}$

7. $\sum_{k=2}^{\infty} \sin\left(\frac{\pi k}{2}\right) \frac{1}{\sqrt{\ln k}}$ **8.** $\sum_{k=1}^{\infty} (-1)^k \ln\left(1 + \frac{1}{k}\right)$ **9.** $\sum_{k=1}^{\infty} \frac{1}{k 2^{\sqrt{k}}}$

10. $\sum_{k=1}^{\infty} (-1)^k \frac{k^3 - 4k^2 + k - 7}{10k^4 + 3k^2 + 9}$ **11.** $\sum_{k=0}^{\infty} \frac{3^k}{k!}$ **12.** $\sum_{k=1}^{\infty} \frac{k^2 + 5}{k^k}$

13. $\sum_{k=1}^{\infty} \frac{k^2 + 6k + 11}{k^5 + 7k^3 - 4}$ **14.** $\sum_{k=1}^{\infty} (-1)^k \arcsin(1/k)$ **15.** $\sum_{k=1}^{\infty} \frac{\ln k}{k^2}$

16. $\sum_{k=1}^{\infty} \frac{k!}{k^k}$ **17.** $\sum_{k=1}^{\infty} (-1)^k e^{-\frac{k^4}{k^2 + 1}}$ **18.** $\sum_{k=1}^{\infty} (-1)^k \frac{k}{3 + \sqrt[3]{k^5 -}}$

19. $\sum_{k=1}^{\infty} (-1)^k \left(\frac{k-1}{k}\right)^{k^2}$ **20.** $\sum_{k=1}^{\infty} (-1)^k \sin(1/k)$ **21.** $\sum_{k=1}^{\infty} \frac{\cos k}{k^2}$

22. $\sum_{k=1}^{\infty} (-1)^k \arctan\frac{1}{k}$ **23.** $\sum_{k=1}^{\infty} (-1)^k \frac{k!}{5^k}$ **24.** $\sum_{k=1}^{\infty} (-1)^k \frac{k^2}{50k^2 + 8}$

25. $\sum_{k=1}^{\infty} (-1)^k \frac{\arctan k}{k!}$ **26.** $\sum_{k=1}^{\infty} (-1)^k \left(\frac{1}{\sqrt{k}} - \frac{1}{\sqrt{k+1}}\right)$ **27.** $\sum_{k=1}^{\infty} (-1)^k \frac{2^{2k}(}{(2k}$

28.[3] $\sum_{k=1}^{\infty} (-1)^k \frac{(2k)!}{2^{2k}(k!)^2}$ **29.** $\sum_{k=2}^{\infty} (-1)^k \frac{1}{(\ln k)^5}$ **30.** $\sum_{k=1}^{\infty} (1)^k \sqrt{\frac{k^3 + k}{k^4 + 2k^2}}$

31. $\sum_{k=2}^{\infty} (-1)^k \frac{1}{\sqrt[k]{\ln k}}$ **32.** $\sum_{k=1}^{\infty} (-1)^k \tan e^{-k}$ **33.** $\sum_{k=1}^{\infty} (-1)^k \frac{1}{\sqrt[k]{k}}$

34. $\sum_{k=1}^{\infty} \frac{7^k}{k! + \sqrt{k}}$ **35.** $\sum_{k=2}^{\infty} \frac{7^k}{k! - \sqrt{k}}$ **36.** $\sum_{k=2}^{\infty} (-1)^k \frac{7^k}{\sqrt{k} + \ln k}$

37. Determine the values of the positive number p for which each series conver
(a) $\sum_{k=2}^{\infty} \frac{1}{k(\ln k)^p}$ (b) $\sum_{k=2}^{\infty} \frac{1}{(\ln k)^p}$ (c) $\sum_{k=2}^{\infty} \frac{\ln k}{k^p}$ (d) $\sum_{k=1}^{\infty} \frac{\sin \frac{1}{k}}{k^p}$

38. (a) If $\sum_{k=p}^{\infty} a_k$ is absolutely convergent, show $\sum_{k=p}^{\infty} a_k \sin k$ is also absol
convergent.
(b) If $\sum_{k=p}^{\infty} b_k$ diverges and $b_k \geq 0$ for $k \geq p$, show $\sum_{k=p}^{\infty} \frac{b_k}{\sin^2 k}$ also diverges.

39. Let $f(x)$ be a decreasing function with domain $[1, \infty)$ that has positive va
If $\sum_{k=1}^{\infty} f(k)$ converges, show $\sum_{k=1}^{\infty} f(k^2)$ also converges.

Determine the interval of convergence of each power series.

40. $\sum_{k=0}^{\infty} \frac{x^k}{4^k + 7}$ **41.** $\sum_{k=0}^{\infty} \frac{x^{2k}}{8 + \sqrt{k}}$ **42.** $\sum_{k=0}^{\infty} (-1)^k \frac{k^5 x^k}{k!}$

43. $\sum_{k=0}^{\infty} (-1)^k \frac{4^k (x-3)^k}{k^2 + 6}$ **44.** $\sum_{k=1}^{\infty} \frac{k! x^k}{k^k}$ **45.** $\sum_{k=0}^{\infty} (-1)^k (x+5)^k$

46. $\sum_{k=1}^{\infty} \frac{x^{2k}}{6^k \sqrt{k}}$ **47.** $\sum_{k=1}^{\infty} \frac{x^k}{k \arctan k}$ **48.** $\sum_{k=0}^{\infty} \frac{k!(x+9)^k}{3^k + 5}$

49. $\sum_{k=2}^{\infty} \frac{(x-2)^k}{4^k \ln k}$ **50.** $\sum_{k=1}^{\infty} \left(1 + \frac{1}{k}\right)^{k^2} x^k$ **51.** $\sum_{k=0}^{\infty} \frac{(k!)^3 x^k}{(3k)!}$

52. The Limit Comparison Test is applied to the series with positive terms $\sum_{k=}^{\infty}$
and $\sum_{k=q}^{\infty} b_k$. Assume the resulting limit $L = \lim_{k \to \infty} \frac{a_k}{b_k}$ exists.
(a) If $L = 0$, what conclusions can you make about these two series?
(b) If $L = \infty$, what conclusions can you make about these two series?

53. Let $\sum_{k=p}^{\infty} a_k$ be a series of positive terms.
(a) Show if this series converges, then $\sum_{k=p}^{\infty} a_k^2$ also converges.
(b) Give an example where the series $\sum_{k=p}^{\infty} a_k^2$ converges, but $\sum_{k=p}^{\infty} a_k$ diverges

54. (a) Let f be a continuous function with $f(k) = a_k$ for $k \geq p$. Assume the Inte
Test applies to show the series $\sum_{k=p}^{\infty} a_k$ converges. Let S denote the sum of this se
and let S_n be the nth partial sum. Show the error $S - S_n$ of the approximation
by S_n satisfies $|S - S_n| \leq \int_n^{\infty} f(x)\, dx$.
(b) Find a bound on the error of the approximation S_{100} of $S = \sum_{k=1}^{\infty} \frac{1}{k^2}$.

55. (a) Give an example of a convergent series $\sum_{k=p}^{\infty} a_k$ for which the series
$\sum_{k=p}^{\infty} (-1)^k a_k$ diverges.

---

[3]You may assume the limit of the nth term is zero.

(b) Give an example of a conditionally convergent series $\sum_{k=p}^{\infty} b_k$ such that the series $\sum_{k=p}^{\infty} (-1)^k b_k$ also converges.

56. Estimate the sum of each series by adding the first n terms. Give a bound on the error of your estimate.

(a) $\ln 2 = \sum_{k=1}^{\infty} (-1)^{k-1} \frac{1}{k}$ with $n = 5$ (b) $\pi = \sum_{k=0}^{\infty} (-1)^k \frac{4}{2k+1}$ with $n = 4$

(c) $\frac{1}{e} = \sum_{k=0}^{\infty} (-1)^k \frac{1}{k!}$ with $n = 3$ (d) $\frac{1}{\sqrt{e}} = \sum_{k=0}^{\infty} (-1)^k \frac{1}{2^k k!}$ with $n = 4$

(e) $\sum_{k=0}^{\infty} (-1)^k \frac{3^k}{(2k)!}$ with $n = 3$ (f) $\sum_{k=1}^{\infty} (-1)^k \frac{k}{k^2+1}$ with $n = 2$

57. Sum the first few terms of a power series to estimate each integral with error less than .01. Justify that your estimate has the required accuracy.

(a) $\int_0^1 \cos \sqrt{x} \, dx$ (b) $\int_0^1 \ln(1+x^2) \, dx$ (c) $\int_0^{1/2} \frac{1}{\sqrt{1+x^3}} \, dx$

(d) $\int_0^1 e^{-x^2} \, dx$ (e) $\int_0^{1/2} \frac{\arctan x}{x} \, dx$ (f) $\int_0^{4/5} \frac{1}{1+x^3} \, dx$

58. Let $\{a_k\}$, for $k \geq 1$, be a decreasing sequence of positive numbers with limit zero. Show the series $\sum_{k=1}^{\infty} (-1)^{k(k+1)/2} a_k$ converges.

Challenging Problems

1. (a) Show a power series $\sum_{k=0}^{\infty} a_k x^k$ is continuous on its interval of convergence.
(b) Deduce the series (5.4.11) for $\ln 2$ and the series (5.4.13) for $\arctan \frac{\pi}{4}$ without consideration of the remainder terms.

2. (a) Let $\sum_{k=p}^{\infty} a_k$ be a conditionally convergent series. Let $S \in \Re$. Show there is a rearrangement of this series which converges to S as well as a rearrangement of this series which diverges.
Hint: Use the idea of the proof of the Alternating Series Test.
(b) Let $\sum_{k=q}^{\infty} b_k$ be an absolutely convergent series with sum S. Show every rearrangement of this series is absolutely convergent with sum S.

3. Consider the series $\sum_{k=1}^{\infty} (-1)^{k-1} a_k$ where $a_{2k-1} = \frac{1}{k}$ and $a_{2k} = \int_k^{k+1} \frac{1}{x} \, dx$.
(a) Show this series converges by the Alternating Series Test with sum γ.
(b) Let S_n denote the nth partial sum of the harmonic series. Show the $(2n-1)^{\text{th}}$ partial sum of the alternating series $\sum_{k=1}^{\infty} (-1)^{k-1} a_k$ equals $S_n - \ln n$.

(c) Show $\gamma = \lim_{n \to \infty} \left(1 + \frac{1}{2} + \frac{1}{3} + \cdots + \frac{1}{n} - \ln n \right)$.

5.10   Complex Numbers

The real numbers are not *algebraically complete* in the sense that there are polynomials $P(x)$ with real coefficients which have no real roots. That is, there is no $r \in \Re$ with $P(r) = 0$. The simplest example of such a polynomial is $P(x) = x^2 + 1$. We introduce a set of numbers $\mathcal{C}$, called the *complex numbers*, which contains $\Re$. The Fundamental Theorem of Algebra says that the set of complex numbers is complete.

Theorem 5.10.1 (Fundamental Theorem of Algebra) *Every polynomial with coefficients in $\mathcal{C}$ has a root in $\mathcal{C}$.*

In the first subsection, we give an algebraic definition of complex numbers and depict complex numbers as points in the plane. In the second subsection, we use the latter viewpoint to define the polar representation of complex numbers and prove DeMoivre's formula for finding powers of complex numbers. In the third subsection,

we apply this formula to find all the roots of the polynomial $x^n - z$ for $z \in C$
also use complex numbers, in the method of partial fractions, to integrate a ra
function whose denominator has quadratic factors with complex roots.

Arithmetic of Complex Numbers

We define complex numbers as well as the operations of addition, subtraction,
plication, division, conjugation and absolute value. We interpret these operatic
terms of the geometric description of complex numbers as points in the plane.

Definition 5.10.2 (a) *The set of complex numbers C consists of all symbols
form $a + bi$ for $a, b \in \Re$.*
(b) *Consider the real numbers $\Re$ to be a subset of C by identifying $r \in \Re$
$r + 0i \in C$.*
(c) *Complex numbers of the form $0 + bi$ are called imaginary.*

Notes (1) The plus sign in the symbol $z = a + bi$ has no meaning yet as a
since we have not defined the addition of complex numbers. We could just as
have used a comma instead of a plus sign to write z as a, bi.
(2) Two complex numbers $a + bi$ and $a' + b'i$ are equal if and only if $a = a'$ and b

Examples 5.10.3 Find all real numbers x and y for which

$$(x - 2y + 3) + (2x + y + 1)i = (2x - y - 2) + (x + 2y)i.$$

Solution Equate the real and imaginary parts of this equation:

$$x - 2y + 3 = 2x - y - 2$$
$$2x + y + 1 = x + 2y.$$

Add these equations: $3x - y + 4 = 3x + y - 2$ and $y = 3$. Substitute $y = 3$ into
first equation above: $x - 3 = 2x - 5$ and $x = 2$.

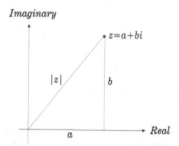

Figure 5.10.4 The Complex Number z as a Point in the Plane

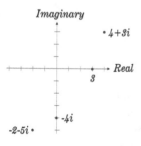

Figure 5.10.5
Example 5.10.6

Identify the set of complex numbers C with the coordinate plane by letting
complex number $a + bi$ correspond to the point (a, b) in the plane. See Figure 5.1
Under this correspondence the real numbers $a + 0i \in \Re$ correspond to the points (
on the x-axis while the imaginary numbers $0 + bi$ correspond to the points $(0, b$
the y-axis. In this context, the x-axis is called the *real axis* and the y-axis is ca
the *imaginary axis*. We also refer to the xy-plane as the complex plane.

Examples 5.10.6 Plot the points $4 + 3i$, $-2 - 5i$, 3 and $-4i$ in the complex pla

Solution These points are depicted in Figure 5.10.5.

The absolute value $|x|$ of a real number $x \in \Re$ can be interpreted as the distance between the point x on the real number line and the origin. Analogously the *absolute value* $|z|$ of a complex number $z = a + bi$ is the distance between the corresponding point (a, b) in the complex plane and the origin, i.e. the length of the line segment joining (a, b) and the origin. Apply the Pythagorean Theorem to the triangle in Figure 5.10.4: $|z| = \sqrt{a^2 + b^2}$. We take this formula as the definition of of $|z|$.

Definition 5.10.7 *The absolute value of the complex number $z = a + bi$ is given by:*

$$|z| = |a + bi| = \sqrt{a^2 + b^2}.$$

$|z|$ *is also called the norm or length of z.*

Observe the absolute value of a real number $a + 0i$ is $\sqrt{a^2 + 0^2} = \sqrt{a^2}$ which agrees with the usual definition of $|a|$.

Examples 5.10.8 (1) Find the absolute value of the complex number $3 - 4i$.

  Solution $|3 - 4i| = \sqrt{3^2 + (-4)^2} = \sqrt{25} = 5$.

(2) Show that $|z| = 0$ if and only if $z = 0$.

  Solution Write $z = a + bi$. Then $0 = |z| = \sqrt{a^2 + b^2}$ if and only if $a^2 + b^2 = 0$. Since the sum of two positive numbers is positive, the only way that $a^2 + b^2$ can have sum zero is if $a = b = 0$.

(3) Given a specific complex number z_0, identify the set S of all complex numbers z which have the same absolute value as z_0.

  Solution Let $|z_0| = R$. The set S consists of all points $z = a + bi$ for which the corresponding point (a, b) in the complex plane has distance R from the origin. Thus S is the circle with center the origin and radius R.   □

The arithmetic operations of addition and subtraction extend from real numbers to complex numbers.

Definition 5.10.9 *Define the sum and difference of complex numbers by:*

$$
\begin{aligned}
(a + bi) + (a' + b'i) &= (a + a') + (b + b')i, \\
(a + bi) - (a' + b'i) &= (a - a') + (b - b')i \quad \text{for } a, b, a', b' \in \Re.
\end{aligned}
$$

Let $a = a + 0i$ and $a' = a' + 0i$ be real numbers. Their sum and difference is the same whether we compute them as real numbers or as complex numbers. That is, $(a + 0i) + (a' + 0i) = (a + a') + 0i = a + a'$ and $(a + 0i) - (a' + 0i) = (a - a') + 0i = a - a'$.

Examples 5.10.10 (1) Compute $(3 + 4i) + (-7 + 2i)$.

  Solution $(3 + 4i) + (-7 + 2i) = (3 - 7) + (4 + 2)i = -4 + 6i$.

(2) Compute $(6 - 5i) - (8 - 9i)$.

  Solution $(6 - 5i) - (8 - 9i) = (6 - 8) + (-5 + 9)i = -2 + 4i$.   □

It is straightforward to verify that the addition of complex numbers has the usual properties of the addition of real numbers. We leave the verification of these properties for the exercises. The operations of addition and subtraction of complex numbers have geometric interpretations in terms of the representative points of these numbers in the

complex plane. These interpretations in terms of vector arithmetic are given
appendix of Section 6.2.

We want the multiplication of complex numbers to satisfy the distributive pro
Thus the product of two complex numbers is determined by choosing a value
Define $i^2 = -1$, i.e. i is a root of the polynomial $x^2 + 1$.

Definition 5.10.11 *Define the product of complex numbers by:*

$$(a + bi) \cdot (a' + b'i) = (aa' - bb') + (ab' + a'b)i \ \ for \ a, b, a', b' \in \mathfrak{R}.$$

Observe the product of two real numbers $a = a + 0$ and $a' = a' + 0$
same whether we compute their product as real numbers or as complex num
i.e. $(a + 0i) \cdot (a' + 0i) = (aa') + 0i = aa'$.

Examples 5.10.12 (1) Compute $(5 + 8i)(3 - 2i)$.

Solution $(5 + 8i)(3 - 2i) = [(5)(3) - (8)(-2)] + [(5)(-2) + (8)(3)]\,i = 31 + 14i$

(2) Compute $(1 - 2i)^3$.

Solution $(1 - 2i)^3 = [(1 - 2i)(1 - 2i)]\,(1 - 2i) = (-3 - 4i)(1 - 2i) = -11 + 2i$.

The straightforward verification that multiplication of complex numbers ha
usual properties of multiplication of real numbers is left for the exercises. The follo
proposition says that the absolute values of the sum and product of complex num
have the same properties as for real numbers.

Proposition 5.10.13 *Let $z, w \in C$. Then*

$\quad$ **(a)** $|z + w| \le |z| + |w|;$ $\qquad$ **(b)** $|zw| = |z| \cdot |w|.$

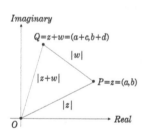

Imaginary

$Q = z + w = (a + c, b + d)$

$|w|$

$|z + w|$ $\quad\bullet\, P = z = (a, b)$

$|z|$

O $\qquad\qquad\qquad\longrightarrow$ *Real*

Figure 5.10.14
Triangle OPQ

Proof Write $z = a + bi$ and $w = c + di$.
(a) Let $P = (a, b)$ and $Q = (a + c, b + d)$ be the representations of $z = a + bi$
$z + w = (a + c) + (b + d)i$ in the complex plane. Consider the lengths of the sid
the triangle OPQ in Figure 5.10.14. Clearly $|OP| = |z|$ and $|OQ| = |z + w|$. $|P$
the distance between $P = (a, b)$ and $Q = (a + c, b + d)$:

$$|PQ| = \sqrt{[(a + c) - a]^2 + [(b + d) - b]^2} = \sqrt{c^2 + d^2} = |w|.$$

Since a line segment is the shortest distance between two points,

$$|z + w| = |OQ| \le |OP| + |PQ| = |z| + |w|.$$

(b) Since $zw = (ac - bd) + (ad + bc)i$,

$$
\begin{aligned}
|zw| &= \sqrt{(ac - bd)^2 + (ad + bc)^2} = \sqrt{a^2c^2 + b^2d^2 + a^2d^2 + b^2c^2} \\
&= \sqrt{a^2 + b^2}\sqrt{c^2 + d^2} = |z|\,|w|.
\end{aligned}
$$

Note The inequality in (a) is called the *triangle inequality* because it is derived f
the geometry of the triangle OPQ in Figure 5.10.14.

The operation of conjugation will be used to define the quotient of complex numb

Definition 5.10.15 *Define the conjugate of the complex number $z = a + bi$ by:*

$$\overline{z} = \overline{a + bi} = a - bi.$$

Observe the geometric interpretation of conjugation is to send the point (a, b) which represents $z = a + bi$ to the point $(a, -b)$ which represents $\overline{z} = a - bi$. This is precisely reflection about the x–axis. See Figure 5.10.17.

Examples 5.10.16 (1) Find the conjugate of $9 - 6i$.

Solution $\overline{9 - 6i} = 9 - (-6)i = 9 + 6i$.

(2) Show $z = \overline{z}$ if and only if z is real.

Solution Write $z = a + bi$. Then $a + bi = z = \overline{z} = a - bi$ if and only if $b = -b$. This occurs if and only if $b = 0$, i.e. when z is real.

(3) Show $|\overline{z}| = |z|$.

Solution Write $z = a + bi$. Then $|\overline{z}| = |a - bi| = \sqrt{a^2 + (-b)^2} = \sqrt{a^2 + b^2} = |z|$. □

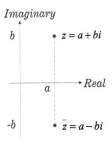

Figure 5.10.17
Conjugation

The following property of conjugation makes this operation very useful.

Proposition 5.10.18 *Let z be a complex number. Then*

$$z\overline{z} = |z|^2.$$

Proof Write $z = a + bi$. Then

$$z\overline{z} = (a + bi)(a - bi) = \left[a^2 - (-b^2)\right] + \left[a(-b) + b(a)\right]i = a^2 + b^2 = |a + bi|^2 = |z|^2. \;\square$$

Note, although z and $\overline{z}$ may be complex numbers, their product $z\overline{z} = |z|^2$ is always a non–negative real number. Thus for any nonzero complex number z:

$$z\left(\frac{1}{|z|^2}\overline{z}\right) = \left(\frac{1}{|z|^2}\overline{z}\right)z = 1. \tag{5.10.1}$$

We call the complex number $\frac{1}{|z|^2}\overline{z}$ the *reciprocal* z^{-1} of z:

$$z^{-1} = \frac{1}{|z|^2}\overline{z}. \tag{5.10.2}$$

Examples 5.10.19 (1) Verify the identity $z\overline{z} = |z|^2$ for $z = 3 - 4i$.

Solution $z\overline{z} = (3 - 4i)(3 + 4i) = (9 + 16) + (12 - 12)i = 25$ and $|z|^2 = 3^2 + (-4)^2 = 25$.

(2) Find a complex number w such that $(5 + 12i)w = 1$.

Solution By (5.10.1), take

$$w = (5 + 12i)^{-1} = \frac{1}{|5 + 12i|^2}\overline{5 + 12i} = \frac{1}{5^2 + 12^2}(5 - 12i) = \frac{5}{169} - \frac{12}{169}i.$$

(3) Solve the equation $(5 + 12i)z = 3 - 2i$ for $z \in \mathcal{C}$.

Solution Multiply the given equation by the number w of the preceding example:

$$\left(\frac{5}{169} - \frac{12}{169}i\right)(5 + 12i)z = \left(\frac{5}{169} - \frac{12}{169}i\right)(3 - 2i)$$

$$z = \frac{15 - 24}{169} + \frac{-36 - 10}{169}i = -\frac{9}{169} - \frac{46}{169}i \qquad \square$$

We define the quotient $\frac{w}{z}$ of complex numbers as the product of w with the
rocal of z.

Definition 5.10.20 *Let* $z, w \in \mathcal{C}$ *with* $z \neq 0$. *Define the quotient* w *divided by*

$$\frac{w}{z} = z^{-1}w = \frac{1}{|z|^2}\overline{z}w.$$

Examples 5.10.21 (1) Evaluate $\frac{5-2i}{4+3i}$.

   Solution By definition of this quotient:

$$\frac{5-2i}{4+3i} = \frac{1}{|4+3i|^2}(\overline{4+3i})(5-2i) = \frac{1}{25}(4-3i)(5-2i) = \frac{1}{25}(14-23i) = \frac{14}{25} -$$

(2) Show $\left|\frac{w}{z}\right| = \frac{|w|}{|z|}$.

   Solution By definition of $\frac{w}{z}$:

$$\left|\frac{w}{z}\right| = \left|\frac{1}{|z|^2}\overline{z}w\right| = \frac{1}{|z|^2}\,|\overline{z}|\,|w| = \frac{1}{|z|^2}|z|\,|w| = \frac{|w|}{|z|}.$$

Polar Representation

A convenient method to describe the position of a complex number in the com
plane is its polar representation. This representation gives geometric interpreta
of the product and quotient of complex numbers. In particular, DeMoivre's form
the geometric interpretation of z^n, is especially useful.

   Each complex number $z = a + bi$ is represented by the point $P = (a,b)$ in
complex plane or, equivalently, by the line segment L joining P with the origin.
Figure 5.10.22. The line segment L is determined by its length $|z|$ and by the an
between the real axis and L. As in the definition of the trigonometric functions, c
terclockwise angles are considered to be positive and clockwise angles are consid
to be negative. By the definition of the trigonometric functions, the point where
line segment L intersects the unit circle with center at the origin is $(\cos\theta, \sin\theta)$. F
the large right triangle in Figure 5.10.22: $a = |z|\cos\theta$ and $b = |z|\sin\theta$. Thus

$$P = (|z|\cos\theta, |z|\sin\theta) \quad \text{and} \quad z = |z|\cos\theta + i|z|\sin\theta = |z|(\cos\theta + i\sin\theta).$$

This description of z, in terms of $|z|$ and θ, is called its *polar representation*.

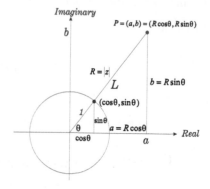

Figure 5.10.22 Polar Representation $P(R, \theta)$ of $z = a + bi$

Definition 5.10.23 *If* $R, \theta \in \mathfrak{R}$ *with* $R \geq 0$, *define the complex number*

$$z = P(R, \theta) = R(\cos \theta + i \sin \theta). \qquad (5.10.3)$$

$P(R, \theta)$ *is called the polar representation of* z.

Observe if n is an integer, then $P(R, \theta + 2n\pi) = P(R, \theta)$. From Figure 5.10.22, we see that if $z = a + bi$ has polar representation $P(R, \theta)$, then

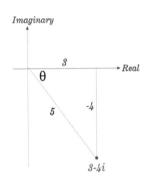

$$R = |z| = \sqrt{a^2 + b^2}, \quad \theta = \arcsin \frac{b}{|z|} = \arcsin \frac{b}{\sqrt{a^2 + b^2}} \qquad (5.10.4)$$

$$a = R \cos \theta, \qquad b = R \sin \theta. \qquad (5.10.5)$$

Examples 5.10.25 (1) Find the polar representation of $z = 3 - 4i$.

Solution Apply (5.10.4) with $a = 3$ and $b = -4$. Then $R = \sqrt{3^2 + (-4)^2} = 5$. By Figure 5.10.24, $\sin \theta = -\frac{4}{5}$. Thus $z = P\left(5, \arcsin -\frac{4}{5}\right)$.

(2) Find the complex number z whose polar representation is $P\left(8, \frac{\pi}{6}\right)$.

Solution By Figure 5.10.26,

$$z = 8 \cos \frac{\pi}{6} + i8 \sin \frac{\pi}{6} = 8 \left(\frac{\sqrt{3}}{2}\right) + i8 \left(\frac{1}{2}\right) = 4\sqrt{3} + 4i. \qquad \square$$

Figure 5.10.24
$3 - 4i$

The following proposition shows that multiplication, division and exponentiation of complex numbers are easily performed using their polar representations.

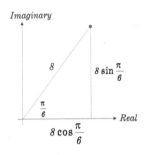

Proposition 5.10.27 (a) $P(R, \theta)P(S, \phi) = P(RS, \theta + \phi)$.

(b) *For* $S > 0$: $\dfrac{P(R, \theta)}{P(S, \phi)} = P\left(\dfrac{R}{S}, \theta - \phi\right)$.

(c) *For* n *a positive integer:* $P(R, \theta)^n = P(R^n, n\theta)$. *Equivalently,*

$$(\cos \theta + i \sin \theta)^n = \cos n\theta + i \sin n\theta. \qquad (5.10.6)$$

This equation is called DeMoivre's formula.

Figure 5.10.26
$P\left(8, \frac{\pi}{6}\right)$

Proof Apply (5.10.3).

(a) $\quad P(R, \theta)P(S, \phi) = (R \cos \theta + iR \sin \theta)(S \cos \phi + iS \sin \phi)$
$\qquad\qquad\qquad\quad = RS \left[(\cos \theta \cos \phi - \sin \theta \sin \phi) + i(\cos \theta \sin \phi + \sin \theta \cos \phi)\right]$
$\qquad\qquad\qquad\quad = RS \left[\cos(\theta + \phi) + i \sin(\theta + \phi)\right] = P(RS, \theta + \phi)$.

(b) $\quad \dfrac{1}{P(S, \phi)} = \dfrac{1}{S(\cos \phi + i \sin \phi)} = \dfrac{1/S}{\cos \phi + i \sin \phi} \dfrac{\cos \phi - i \sin \phi}{\cos \phi - i \sin \phi}$

$\qquad\qquad = \dfrac{P\left(\frac{1}{S}, -\phi\right)}{\cos^2 \phi + \sin^2 \phi} = P\left(\dfrac{1}{S}, -\phi\right)$.

Now apply the formula of (a): $\frac{P(R, \theta)}{P(S, \phi)} = P(R, \theta)P\left(\frac{1}{S}, -\phi\right) = P\left(\frac{R}{S}, \theta - \phi\right)$.

(c) Compute $P(R, \theta)^n$ by applying the formula of (a) $n - 1$ times. $\qquad \square$

Examples 5.10.28 Let $z = 5 - 5i$ and $w = 6\sqrt{3} + 6i$.

(1) Find the polar representation of zw.

> **Solution** Since $|z| = \sqrt{5^2 + (-5)^2} = \sqrt{50} = 5\sqrt{2}$, write
>
> $$z = 5\sqrt{2}\left(\frac{1}{\sqrt{2}} - \frac{i}{\sqrt{2}}\right) = 5\sqrt{2}\left(\cos\frac{7\pi}{4} + i\sin\frac{7\pi}{4}\right) = P\left(5\sqrt{2}, \frac{7\pi}{4}\right).$$
>
> Since $|w| = \sqrt{(6\sqrt{3})^2 + 6^2} = \sqrt{144} = 12$, write
>
> $$w = 12\left(\frac{\sqrt{3}}{2} + \frac{i}{2}\right) = 12\left(\cos\frac{\pi}{6} + i\sin\frac{\pi}{6}\right) = P\left(12, \frac{\pi}{6}\right).$$
>
> By Proposition 5.10.27(a), the polar representation of zw is:
>
> $$zw = P\left(5\sqrt{2}, \frac{7\pi}{4}\right) \cdot P\left(12, \frac{\pi}{6}\right) = P\left(5\sqrt{2} \cdot 12, \frac{7\pi}{4} + \frac{\pi}{6}\right) = P\left(60\sqrt{2}, \frac{23}{1}\right.$$

(2) Find the polar representation of $\frac{z}{w}$.

> **Solution** By Proposition 5.10.27(b), the polar representation of $\frac{z}{w}$ is:
>
> $$\frac{z}{w} = \frac{P\left(5\sqrt{2}, \frac{7\pi}{4}\right)}{P\left(12, \frac{\pi}{6}\right)} = P\left(\frac{5\sqrt{2}}{12}, \frac{7\pi}{4} - \frac{\pi}{6}\right) = P\left(\frac{5\sqrt{2}}{12}, \frac{19\pi}{12}\right).$$

(3) Find the polar representation of z^5.

> **Solution** By Proposition 5.10.27(c), the polar representation of z^5 is:
>
> $$z^5 = P\left(5, \frac{7\pi}{4}\right)^5 = P\left(5^5, 5\frac{7\pi}{4}\right) = P\left(3125, \frac{35\pi}{4}\right) = P\left(3125, \frac{3\pi}{4}\right).$$

Applications to Polynomial Roots

We use conjugation to prove that complex roots of polynomials occur in conju
pairs. Then we use DeMoivre's formula to find all the roots of the polynomials x^n
We conclude by showing how complex numbers simplify the partial fractions con
tation for a rational function whose denominator has complex roots.

Proposition 5.10.29 *If $z = a + bi$ is a root of the polynomial $P(x)$ with real co
cients, then $\bar{z} = a - bi$ is also a root of $P(x)$.*

Proof Write $P(x) = a_n x^n + \cdots + a_1 x + a_0$ with $a_n, \ldots, a_0 \in \Re$. Since the a_k are
$\bar{a}_k = a_k$ for $0 \le k \le n$. Note z is a root of $P(x)$ means that $0 = P(z)$. Conjugate
equation using the formulas of Exercises 7(e) and 8(f):

$$\begin{aligned} 0 &= \bar{0} = \overline{P(z)} = \overline{a_n z^n + \cdots + a_1 z + a_0} = \bar{a}_n \ \bar{z}^n + \cdots + \bar{a}_1 \ \bar{z} + \bar{a}_0 \\ &= a_n \bar{z}^n + \cdots a_1 \bar{z} + a_0 = P(\bar{z}). \end{aligned}$$

Thus $\bar{z}$ is also a root of $P(x)$.

Assuming the Fundamental Theorem of Algebra, we describe the factorizatio
a polynomial as a product of linear and quadratic factors.

Corollary 5.10.30 *Every polynomial $P(x)$ with real coefficients can be written
product of a number, linear polynomials of the form $x - A$ and quadratic polynom
of the form $x^2 + Bx + C$ with $B^2 - 4C < 0$.*

Proof Write $P(x) = a_n x^n + \cdots + a_1 x + a_0 = a_n Q(x)$ where $a_n \neq 0$ and

$$Q(x) = x^n + \frac{a_{n-1}}{a_n} x^{n-1} + \cdots + \frac{a_1}{a_n} x + \frac{a_0}{a_n}.$$

By the Fundamental Theorem of Algebra, $Q(x)$ has n roots in $\mathcal{C}$. Each real root $x = A$ of $Q(x)$ produces a linear factor $x - A$ of $Q(x)$. Every pair of complex roots $x = a + bi$, $x = a - bi$ of $Q(x)$, with $b \neq 0$, produces a quadratic factor $[x - (a + bi)][x - (a - bi)] = x^2 - 2ax + (a^2 + b^2)$ of $Q(x)$ with $(-2a)^2 - 4(a^2 + b^2) = -4b^2 < 0$. □

If $P(x)$ is a polynomial of degree $2n + 1$, then $P(x)$ has at most n pairs of complex roots of the form $a + bi$, $a - bi$ with $b \neq 0$. Hence $P(x)$ must have a real root.

Corollary 5.10.31 *If $P(x)$ is a polynomial of odd degree with real coefficients, then $P(x)$ has at least one real root.*

Examples 5.10.32 Find all the roots of $P(x) = x^5 - 4x^4 + 11x^3 - 4x^2 - 50x + 100$, given that $2 + i$ and $1 - 3i$ are two of its roots.

Solution $P(x)$ must also have $\overline{2 + i} = 2 - i$ and $\overline{1 - 3i} = 1 + 3i$ as roots. Therefore $P(x)$ is divisible by

$$Q(x) = [x - (2 + i)][x - (2 - i)][x - (1 - 3i)][x - (1 + 3i)] = x^4 - 6x^3 + 23x^2 - 50x + 50.$$

Since $\frac{P(x)}{Q(x)} = x + 2$, it follows that $P(x) = Q(x)(x + 2)$ and all the roots of $P(x)$ are -2, $2 - i$, $2 + i$, $1 - 3i$ and $1 + 3i$. □

DeMoivre's formula determines n complex numbers whose n^{th} powers are a given number $z \in \mathcal{C}$, thereby finding all the roots of the polynomial $x^n - z$.

Corollary 5.10.33 *Let $z = P(R, \theta)$. The roots of the polynomial $x^n - z$ are*

$$P\left(\sqrt[n]{R}, \frac{\theta + 2k\pi}{n}\right) \quad \text{for } k \text{ an integer with } 0 \leq k < n.$$

Proof By DeMoivre's formula:

$$P\left(\sqrt[n]{R}, \frac{\theta + 2k\pi}{n}\right)^n = P(R, \theta + 2k\pi) = P(R, \theta) = z$$

for every integer k. If $0 \leq i < j < n$, then $0 \leq \frac{2i\pi}{n} < \frac{2j\pi}{n} < 2\pi$. Therefore $P(\sqrt[n]{R}, \frac{\theta + 2\pi i}{n}) \neq P(\sqrt[n]{R}, \frac{\theta + 2\pi j}{n})$. Thus $P(\sqrt[n]{R}, \frac{\theta + 2k\pi}{n})$, for $0 \leq k < n$, are n distinct roots of the degree n polynomial $x^n - z$. Since a polynomial of degree n has at most n distinct roots, these must be all the roots of $x^n - z$. □

The n roots of $x^n - z$ in the complex plane are the vertices of a regular n–sided polygon as illustrated in the following examples.

Examples 5.10.35 (1) Find all values of $\sqrt[6]{-64}$.

Solution Since $-64 = P(64, \pi) = P(2^6, \pi)$, the sixth roots of 64 are:

$$P\left(2, \tfrac{\pi}{6}\right) = \sqrt{3} + i,$$
$$P\left(2, \tfrac{\pi + 2\pi}{6}\right) = P\left(2, \tfrac{\pi}{2}\right) = 2i,$$
$$P\left(2, \tfrac{\pi + 4\pi}{6}\right) = P\left(2, \tfrac{5\pi}{6}\right) = -\sqrt{3} + i,$$
$$P\left(2, \tfrac{\pi + 6\pi}{6}\right) = P\left(2, \tfrac{7\pi}{6}\right) = -\sqrt{3} - i,$$
$$P\left(2, \tfrac{\pi + 8\pi}{6}\right) = P\left(2, \tfrac{3\pi}{2}\right) = -2i,$$
$$P\left(2, \tfrac{\pi + 10\pi}{6}\right) = P\left(2, \tfrac{11\pi}{6}\right) = \sqrt{3} - i.$$

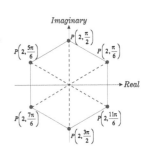

Figure 5.10.34
Sixth Roots of -64

These six roots are the vertices of the regular hexagon depicted in Figure 5.10.34.

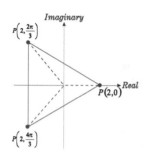

Figure 5.10.36
Roots of $x^3 - 8$

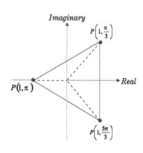

Figure 5.10.37
Roots of $x^3 + 1$

(2) Find all the roots of the polynomial $x^6 - 7x^3 - 8$.

Solution Observe $x^6 - 7x^3 - 8 = (x^3 - 8)(x^3 + 1)$. Since $8 = P(2^3, 0$
polynomial $x^3 - 8$ has roots:

$$P(2,0) = 2, \quad P\left(2, \frac{2\pi}{3}\right) = -1 + \sqrt{3}i, \quad P\left(2, \frac{4\pi}{3}\right) = -1 - \sqrt{3}i.$$

These 3 roots form the vertices of the equilateral triangle depicted in Figure
Since $-1 = P(1^3, \pi)$, the polynomial $x^3 + 1$ has roots:

$$P\left(1, \frac{\pi}{3}\right) = \frac{1}{2} + \frac{\sqrt{3}}{2}i, \qquad P\left(1, \frac{\pi + 2\pi}{3}\right) = P(1, \pi) = -1,$$

$$P\left(1, \frac{\pi + 4\pi}{3}\right) = P\left(1, \frac{5\pi}{3}\right) = \frac{1}{2} - \frac{\sqrt{3}}{2}i.$$

These 3 roots form the vertices of the equilateral triangle depicted in Figure
Thus the polynomial $x^6 - 7x^3 - 8$ has the six roots:

$$2, \quad -1, \quad -1 + \sqrt{3}\,i, \quad -1 - \sqrt{3}\,i, \quad \frac{1}{2} + \frac{\sqrt{3}}{2}i, \quad \frac{1}{2} - \frac{\sqrt{3}}{2}i.$$

Recall, in Section 4.7 we used the method of partial fractions to integrate a rat
function $R(x) = \frac{P(x)}{Q(x)}$ where $P(x)$ and $Q(x)$ are polynomials. Consider the case $\tt w$
the degree of $P(x)$ is less than the degree of $Q(x)$, and $Q(x)$ has no repeated roots
wrote $R(x)$ as a sum so that for each linear factor $x - a_i$ of $Q(x)$ there is a summ
$\frac{A_i}{x - a_i}$ of $R(x)$. For each quadratic factor $x^2 + b_j x + d_j$ of $Q(x)$ with complex roots
is a summand $\frac{B_j x_j + D_j}{x^2 + b_j x + d_j}$ of $R(x)$. Add these summands. We equated the nume
$P(x)$ of $R(x)$ with the numerator of this sum to produce an equation involving
unknown numbers A_i, B_j, D_j. We substituted $x = a_i$ into this equation to find
value of A_i. Similarly, substituting the two complex roots of $x^2 + b_j x + d_j$ proc
the values of B_j and D_j. The following examples illustrate this procedure.

Examples 5.10.38 (1) Evaluate $\int \frac{3x^2 + 2x - 18}{x^3 + 9x}\, dx$.

Solution Since $x^3 + 9x = x(x^2 + 9)$, write

$$\frac{3x^2 + 2x - 18}{x^3 + 9x} = \frac{A}{x} + \frac{Bx + D}{x^2 + 9} = \frac{A(x^2 + 9) + x(Bx + D)}{x^3 + 9x}.$$

Multiply this equation by $x^3 + 9x$:

$$3x^2 + 2x - 18 = A(x^2 + 9) + x(Bx + D). \tag{5.1}$$

Substitute $x = 0$ into this equation: $-18 = 9A$ and $A = -2$. Since $x^2 + 9$
roots $3i$ and $-3i$, substitute $x = 3i$ into equation (5.10.7):

$$3(3i)^2 + 2(3i) - 18 = 0 + 3i(3iB + D)$$
$$-45 + 6i = -9B + 3Di.$$

Equate the real and imaginary parts of this equation: $-9B = -45$ and $3D$
That is, $B = 5$ and $D = 2$. Thus

$$\int \frac{3x^2 + 2x - 18}{x^3 + 9x}\, dx = \int \frac{-2}{x} + \frac{5x + 2}{x^2 + 9}\, dx = -2\ln|x| + \frac{5}{2}\ln\left(x^2 + 9\right) + \frac{2}{3}\,\text{arcta}$$

(2) Evaluate $\int \frac{7x^2+10x+23}{(x+1)(x^2+4x+13)}\,dx$.

Solution Write

$$\frac{7x^2+10x+23}{(x+1)(x^2+4x+13)} = \frac{A}{x+1} + \frac{Bx+D}{x^2+4x+13} = \frac{A(x^2+4x+13)+(Bx+D)(x+1)}{(x+1)(x^2+4x+13)}.$$

Multiply this equation by $(x+1)(x^2+4x+13)$:

$$7x^2+10x+23 = A(x^2+4x+13)+(Bx+D)(x+1). \qquad (5.10.8)$$

Substitute $x=-1$ into this equation: $20 = 10A$ and $A=2$. The polynomial $x^2+4x+13$ has roots $-2+3i$ and $-2-3i$. Substitute $x=-2+3i$ into (5.10.8):

$$7(-2+3i)^2+10(-2+3i)+23 \;=\; 0+[B(-2+3i)+D][-1+3i]$$
$$-32-54i \;=\; (-7B-D)+(-9B+3D)i.$$

Equate the real and imaginary parts of this equation:

$$-32 = -7B-D \quad \text{and} \quad -54 = -9B+3D.$$

Thus $B=5$ and $D=-3$. Now

$$\int \frac{7x^2+10x+23}{(x+1)(x^2+4x+13)}\,dx = \int \frac{2}{x+1} + \frac{5x-3}{x^2+4x+13}\,dx$$

$$= 2\ln|x+1| + \int \frac{\frac{5}{2}(2x+4)}{x^2+4x+13} - \frac{13}{(x+2)^2+9}\,dx$$

$$= 2\ln|x+1| + \frac{5}{2}\ln\left(2x^2+4x+13\right) - \frac{13}{3}\arctan\frac{x+2}{3} + C. \qquad \square$$

Historical Remarks

Complex numbers first appeared in the methods of Italian mathematicians for finding roots of polynomials. In 1545, Gerolamo Cardano showed how to find the real roots of cubic polynomials and pointed out that he was purposely ignoring the complex roots. On the other hand, he did solve a quadratic equation by showing the two numbers whose sum equals 10 and whose product equals 40 are $5+\sqrt{-15}$ and $5-\sqrt{-15}$. In 1560, Rafael Bombelli extended Cardano's methods to find all roots of cubic equations, including the complex ones. In addition, he demonstrated how to perform algebraic operations on complex numbers. Complex roots were included by Albert Girard in his statement of the Fundamental Theorem of Algebra in 1629. However, 17th century mathematicians described these roots as "useless" or "impossible".

Complex numbers were first used in calculus in the 1740s in the work of Leonhard Euler. We will establish several of his interesting formulas in the next section. In 1767, he introduced the term *imaginary numbers* for the square roots of negative numbers and argued that these numbers exist in our minds. The representation of complex numbers as points in the plane originated with the Norwegian surveyor Casper Wessel in 1797 and the Swiss bookkeeper Jean-Robert Argand in 1806. Their work was not generally known, and these ideas were disseminated through Carl Friedrich Gauss's proofs of the Fundamental Theorem of Algebra. Although the geometric interpretation of complex numbers was implicit in his earlier work, Gauss did not published an explicit description until 1831.

A rigorous algebraic description of complex numbers as ordered pairs of real numbers (a,b) was given by William Rowan Hamilton in 1837. He defined the sum, product and length of ordered pairs by

$(a, b) + (c, d) = (a + c, b + d)$, $(a, b) \cdot (c, d) = (ab - cd, ad + bc)$, $|(a, b)| = \sqrt{a^2}$

and showed these operations satisfy the usual properties. He searched unsucces
for a similar algebraic structure on ordered triples. His motivation is explained
Historical Remarks of Section 6.2. However, in 1843 Hamilton defined an alg
structure on ordered quadruplets which he called *quaternions*, and Arthur C
defined an algebraic structure on 8–tuples. The multiplication of quaternions
commutative, while the multiplication of Cayley numbers is neither associati
commutative. The basic properties of quaternions are given as Problem 3. In
Frank Adams, using sophisticated methods of algebraic topology, showed th
numbers, complex numbers, quaternions and Cayley numbers are the only exa
of reasonable algebraic structures on n–tuples of real numbers.

Summary

The reader should understand the algebraic definition of complex numbers, the
scription as points in the complex plane and their polar representations. She s
be able to add, subtract, multiply and divide two complex numbers as well a
the length, powers and roots of a complex number. In addition, the basic prop
of these operations should be known. She should understand that complex ro
polynomials with real coefficients occur in conjugate pairs. She should be able t
the complex roots of the quadratic factors of the denominator of a rational fun
to simplify partial fractions computations.

Basic Exercises

1. Solve each equation for x and y.
(a) $x - 4yi = y + 12i$
(b) $(2x + y + 4) + (3x - 2y + 5)i = (6x - 3y + 3) + (5x - 2y + 2)i$
(c) $(x - 5y + 1) + (2x + 2y - 6)i = (5x - 2y - 5) + (2x + 3y - 4)i$
(d) $(3x + 4y - 2) + (x - 5y + 7)i = (2x + y + 9) + (4x - 2y - 2)i$

2. Find all complex numbers which are both real and imaginary.

3. Plot these complex numbers as points in the complex plane.
(a) -7 **(b)** $5i$ **(c)** $2 + 3i$ **(d)** $3 - 4i$ **(e)** $-1 - 2i$ **(f)** $-4 + i$

4. Find the absolute value of each complex number in Exercise 3.

5. Find the conjugate of each complex number in Exercise 3.

6. Evaluate each expression.
(a) $(2 + 7i) + (5 - 3i)$ **(b)** $(3 - 3i) - (4 + 8i)$ **(c)** $(7 + 2i)(4 - 5i)$
(d) $\frac{2+5i}{1-4i}$ **(e)** $[(3 - 2i) + (5 + 4i)][(6 + i) - (4 - 3i)]$ **(f)** $\frac{(6-i)-(3-2i)}{(5-4i)+(2+7i)}$
(g) $(2 - i)^3$ **(h)** $(3 + 2i)^{-3}$ **(i)** $\left(\frac{4+i}{2-i}\right)^2$ **(j)** $\frac{(6-5i)(1+2i)^2}{(4+3i)(1+i)^3}$

7. Verify these properties of the addition of complex numbers:
(a) $z + w = w + z$ for $z, w \in \mathcal{C}$ (commutative property);
(b) $(z + v) + w = z + (v + w)$ for $z, v, w \in \mathcal{C}$ (associative property);
(c) $0 + z = z$ for $z \in \mathcal{C}$ (identity property);
(d) if $z \in \mathcal{C}$, then there is $-z \in \mathcal{C}$ such that $z + (-z) = 0$ (inverse property);
(e) $\overline{z + w} = \overline{z} + \overline{w}$.

8. Verify these properties of the multiplication of complex numbers:
(a) $zw = wz$ for $z, w \in \mathcal{C}$ (commutative property);
(b) $(zv)w = z(vw)$ for $z, v, w \in \mathcal{C}$ (associative property);
(c) $1z = z$ for $z \in \mathcal{C}$ (identity property);

(d) if $z \in C$, with $z \neq 0$, then there is $z^{-1} \in C$ such that $zz^{-1} = 1$ (inverse property);
(e) $(z + v)w = zw + vw$ for $z, v, w \in C$ (distributive property);
(f) $\overline{zw} = \overline{z}\,\overline{w}$.

9. Find the polar representation of each complex number.
(a) -5 **(b)** $-8i$ **(c)** $5 - 12i$ **(d)** $4 - 4i$ **(e)** $-6 + 6\sqrt{3}i$

10. Express each complex number in the form $a + bi$.
(a) $P\left(7, \frac{\pi}{2}\right)$ **(b)** $P(5, \pi)$ **(c)** $P\left(6, \frac{5\pi}{4}\right)$ **(d)** $P\left(10, \frac{10\pi}{3}\right)$ **(e)** $P\left(8, -\frac{7\pi}{6}\right)$

11. Find the polar representation of each expression.

(a) $P\left(3, \frac{3\pi}{10}\right) P\left(5, \frac{4\pi}{5}\right)$ **(b)** $\dfrac{P\left(10, \frac{5\pi}{12}\right)}{P\left(2, \frac{3\pi}{8}\right)}$ **(c)** $P\left(2, \frac{\pi}{10}\right)^5$

(d) $P\left(3, \frac{3\pi}{4}\right) P\left(2, \frac{5\pi}{6}\right) P\left(5, \frac{\pi}{8}\right)$ **(e)** $\dfrac{P\left(6, \frac{\pi}{12}\right) P\left(8, \frac{3\pi}{4}\right)}{P\left(2, \frac{\pi}{6}\right) P\left(3, \frac{2\pi}{3}\right)}$ **(f)** $P\left(2, \frac{\pi}{6}\right)^3 P\left(3, \frac{\pi}{4}\right)^2 P\left(1, \frac{2\pi}{3}\right)^5$

(g) $\dfrac{P\left(3, \frac{\pi}{4}\right)^4 P\left(2, \frac{\pi}{6}\right)^5}{P\left(1, \frac{5\pi}{6}\right)^6 P\left(6, \frac{\pi}{8}\right)^3}$ **(h)** $(6 - 6i)(4 - 4i\sqrt{3})$ **(i)** $\dfrac{20\sqrt{3} + 20i}{5\sqrt{3} - 5i}$

(j) $(-3 + 3i)^4$ **(k)** $\dfrac{(2+2i)^4(-\sqrt{3}+i)^3}{(-1+i)^5(1-i\sqrt{3})^4}$ **(l)** $\dfrac{(3-3i)^3(2-2i\sqrt{3})^5}{(1-i)^4(\sqrt{3}+i)^3}$

12. Show that DeMoivre's formula extends to rational powers: $P(R, \theta)^q = P(R^q, q\theta)$ for all rational numbers q.

13. Find all values of the indicated roots.
(a) $\sqrt{-9}$ **(b)** $\sqrt[4]{-81}$ **(c)** $\sqrt[3]{3 - 3i}$ **(d)** $\sqrt[3]{4\sqrt{3} + 4i}$ **(e)** $\sqrt[4]{8 - 8i\sqrt{3}}$

14. Find all the roots of each polynomial:
(a) $x^4 - 8x^3 + 27x^2 - 50x + 50$ which has roots $3 + i$ and $1 + 2i$;
(b) $x^3 - 8x^2 + 9x + 58$ which has the root $5 + 2i$;
(c) $x^4 - 9x^3 + 21x^2 - 7x + 100$ which has the root $4 - 3i$;
(d) $x^5 - 9x^4 + 33x^3 - 63x^2 + 64x - 30$ which has roots $1 - i$ and $2 + i$;
(e) $x^6 - 10x^5 + 45x^4 - 120x^3 + 204x^2 - 200x + 100$ which has roots $1 + 2i$ and $3 + i$;
(f) $x^5 + 32$;   **(g)** $x^4 - 2x^2 - 24$;   **(h)** $x^6 + 3x^3 - 28$;   **(i)** $x^8 - x^4 - 12$;
(j) $x^{12} - 3x^8 + 3x^4 - 1$.

15. Evaluate each integral.
(a) $\int \frac{x^2 - 2x + 4}{x^3 + 4x^2}\, dx$ **(b)** $\int \frac{2x^3 + x - 7}{x^4 + x^2}\, dx$ **(c)** $\int \frac{6x^2 - 3x + 2}{x^3 + 2x^2 + 10x}\, dx$

(d) $\int \frac{-4x^3 + 3x^2 - 21x + 58}{x^4 + 4x^3 + 29x^2}\, dx$ **(e)** $\int \frac{5x^2 + 18x + 61}{(x-3)(x^2+6x+13)}\, dx$ **(f)** $\int \frac{x^3 + 4x^2 + 40x + 4}{(x+4)(x^2-4x+20)}\, dx$

(g) $\int \frac{6x^3 + 18x^2 + 26x - 2}{(x^2+2x+2)(x^2+4x+8)}\, dx$ **(h)** $\int \frac{3x^3 + 34x^2 - 14x + 120}{(x^2+6x+10)(x^2+2x+26)}\, dx$

16. Solve the quadratic equation $zx^2 + wx + v = 0$ where $z, w, v \in C$.

17. Let z and w be complex numbers. Show $|z - w|^2 + |z + w|^2 = 2|z|^2 + 2|w|^2$.

18. Let z and w be complex numbers. Show the distance between the representative points of z and w in the complex plane is $|z - w|$.

19. Let $R > 0$ be a fixed real number, and let z_0 be a fixed complex number. Describe the set of complex numbers z in the complex plane which satisfy $|z - z_0| = R$.

20. Let $R > 0$ be a fixed real number.
(a) Describe the set of complex numbers in the complex plane which satisfy $|z| \leq R$.
(b) Describe the set of complex numbers in the complex plane which satisfy $|z| \geq R$.

21. Prove Corollary 5.10.31 as a consequence of the Intermediate Value Theorem.

Challenging Problems

1. Show the representatives of the complex numbers z, v, w in the complex pla
the vertices of an equilateral triangle if and only if $z^2 + v^2 + w^2 = zw + zv + v$

2. Derive the triangle inequality of Proposition 5.10.13(a)

$$\sqrt{(a+c)^2 + (b+d)^2} \leq \sqrt{a^2 + b^2} + \sqrt{c^2 + d^2} \text{ for } a, b, c, d \in \Re$$

algebraicly without reference to the geometric representation of complex numbe

3. Define the quaternions as the set of all quadruples (a, b, c, d) with $a, b, c, d \in$
(a) Addition of quaternions is defined by:

$$(a, b, c, d) + (a', b', c', d') = (a + a', b + b', c + c', d + d').$$

Let $1 = (1, 0, 0, 0)$, $i = (0, 1, 0, 0)$, $j = (0, 0, 1, 0)$ and $k = (0, 0, 0, 1)$.
Find the formula for $(a + bi + cj + dk) + (a' + b'i + c'j + d'k)$.
(b) Show addition of quaternions is associative, commutative, has 0 as an ad
identity and has additive inverses.
(c) Multiplication of quaternions satisfies

$$i^2 = -1, \quad j^2 = -1, \quad k^2 = -1, \quad ij = k, \quad jk = i, \quad ki = j, ji = -k, \quad kj = -i, \quad ik =$$

If multiplication of quaternions is to be distributive, find the formula for
$(a + bi + cj + dk)(a' + b'i + c'j + d'k)$. This is the definition of quaternion multiplica
(d) Show multiplication of quaternions is associative, distributive, has 1 as a
tiplicative identity and nonzero quaternions have reciprocals. Also verify: $r(q$
$(rq)(w) = q(rw)$ for q, w quaternions and $r \in \Re$.
(e) Define the length of a quaternion by

$$|a + bi + cj + dk| = \sqrt{a^2 + b^2 + c^2 + d^2}.$$

Show $|z + w| \leq |z| + |w|$ and $|zw| = |z||w|$ for all quaternions z, w.
(f) Show the inclusion of $\Re$ in the quaternions as the numbers $r + 0i + 0j + 0k$ resp
the operations of addition, multiplication and length.
(g) Show the inclusion of $\mathcal{C}$ in the quaternions as the numbers $a + bi + 0j + 0k$ resp
the operations of addition, multiplication and length.

5.11   Complex Power Series

Prerequisites: Sections 1.10, 5.7 and 5.10

The study of functions of a complex variable is an important area of mathema
Many constructions of calculus generalize directly to this setting. In the first sul
tion, we extend the topological concepts of neighborhood, open set and sequenc
this context. In the second subsection, we define limits and continuity for funct
of a complex variable. In the third subsection, we define the derivative of a func
of a complex variable and establish its basic properties. In the fourth subsection
show a complex valued power series has a radius of convergence, can be differenti
term by term and is the Taylor series of its sum. In the next three subsections
use power series to extend the domains of definition of the exponential, logari
and trigonometric functions to complex numbers and establish the basic prope
of these complex valued functions. In the last subsection, we state of two impor
theorems which illustrate significant differences between differentiable functions
complex variable and differentiable functions of a real variable.

Throughout this section, we identify a complex number $a + bi$ with the corresponding point (a, b) in the complex plane.

Topology of the Complex Plane

The definition of the limit of a function of a complex variable in the next subsection is a direct generalization of the definition of the limit of a function of a real variable. In particular, this definition uses the topological concepts of neighborhood and open set in the complex plane. In this subsection, we use the distance function in the complex plane to define these concepts. We also define convergence of complex valued sequences. Just as with real valued sequences, we identify convergent complex valued sequences with Cauchy sequences.

Let (a, b) and (c, d) be the points in the complex plane which represent the complex numbers $z = a + bi$ and $w = c + di$. By the Pythagorean Theorem, the distance between these two points is $\sqrt{(a - c)^2 + (b - d)^2} = |(a - c) + (b - d)i| = |z - w|$. We define $|z - w|$ to be the distance between the complex numbers z and w.

Definition 5.11.1 *The distance between the complex numbers z and w is:*

$$d(z, w) = |z - w|.$$

Observe this definition extends the definition of distance between real numbers of Definition 1.10.2: $d(a + 0i, c + 0i) = |(a + 0i) - (c + 0i)| = |a - c| = d(a, c)$.

Examples 5.11.2 Find the distance between $8 - 7i$ and $5 - 3i$.

Solution $d(8 - 7i, 5 - 3i) = |(8 - 7i) - (5 - 3i)| = |3 - 4i| = \sqrt{3^2 + (-4)^2} = 5$.   □

The distance between complex numbers satisfies the properties of the distance between real numbers given in Proposition 1.10.3.

Proposition 5.11.3 *Let $z, v, w \in C$. Then*
(a) $d(z, w) = d(w, z)$;
(b) $d(z, w) \geq 0$ while $d(z, w) = 0$ if and only if $z = w$;
(c) $d(z, v) \leq d(z, w) + d(w, v)$ (*Triangle Inequality*).

Proof (a) $d(z, w) = |z - w| = |w - z| = d(w, z)$.
  (b) $0 = d(z, w) = |z - w|$ if and only if $z - w = 0$, i.e. $z = w$.
  (c) $d(z, v) = |z - v| = |(z - w) + (w - v)| \leq |z - w| + |w - v| = d(z, w) + d(w, z)$. □

Note The triangle inequality receives its name from its interpretation in the complex plane: in the triangle with vertices z, v, w the length of the side zv is less than or equal to the sum of the lengths of the sides zw and wv. See Figure 5.11.4.

For $x \in \Re$, the neighborhood $N(x; r)$ of x of radius r is the open interval $(x - r, x + r)$ which consists of the points on the real line whose distance from x is less than r. We use the same idea to define neighborhoods of complex numbers.

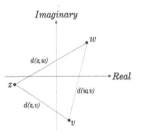

Figure 5.11.4
Triangle Inequality

Definition 5.11.5 *Let $w \in C$ with r a positive real number. The neighborhood $N(w; r)$ of w with radius r denotes the set of all $z \in C$ such that $d(z, w) < r$.*

The picture of a real neighborhood $N(x; r)$ is the open interval with center x and width $2r$. So too, the picture of a complex neighborhood $N(w; r)$ is the open disc with center w and diameter $2r$, i.e. r is the radius of this disc. See Figure 5.11.7. I from Definition 1.10.8, that a subset of $\Re$ is called open if it contains a neighbo of each of its points. We define open subsets of $\mathcal{C}$ in the same way.

Definition 5.11.6 *A subset U of $\mathcal{C}$ is called open if for each $z \in U$ there neighborhood $N(z;r)$ of z with $N(z,r) \subset U$.*

Notes   **(1)** The value of r may be different for each $z \in U$.
(2) An open subset U of the complex plane does not contain any points of its b ary because each neighborhood of a boundary point contains points outside of

Examples 5.11.8 Show the neighborhood $N(w;s)$ is an open set.

Solution Let $z \in N(w;s)$. If r is the smaller of $d(z,w)$ and $s - d(z,w)$, $N(z;r) \subset N(w;s)$. See Figure 5.11.9.

The definitions of convergent sequences and Cauchy sequences generalize to plex valued sequences.

Definition 5.11.10 **(a)** *The sequence $\{z_n\}$ of complex numbers converges to if every neighborhood $N(z;\epsilon)$ contains all but a finite number of the z_n.*
(b) *The sequence $\{z_n\}$ of complex numbers is a Cauchy sequence if for every there is a positive integer N such that if $m, n \geq N$, then $d(z_m, z_n) < \epsilon$.*

Examples 5.11.11 Let $|z| < 1$. Show the sequence $\{z^n\}$ converges to zero.

Solution Since $|z| < 1$, the sequence of real numbers $\{|z^n|\} = \{|z|^n\}$ converg zero. Given $\epsilon > 0$, find a positive integer N such that $|z|^N < \epsilon$. If $n \geq N$, $d(z^n, 0) = |z^n| = |z|^n \leq |z|^N < \epsilon$ and $z^n \in N(0;\epsilon)$. Hence $\{z^n\}$ converges to zerc

Theorem 1.10.38, the identification of convergent sequences and Cauchy seque of real numbers, generalizes to complex valued sequences.

Theorem 5.11.12 *The complex valued sequence $\{z_n\}$ converges if and only if i. Cauchy sequence.*

Proof Assume $\{z_n\}$ is a convergent sequence with limit z. Given $\epsilon > 0$, find a pos integer N such that if $t \geq N$, then $z_t \in N(z;\frac{\epsilon}{2})$, i.e. $d(z_t, z) < \frac{\epsilon}{2}$. If $m, n \geq N$, th

$$d(z_m, z_n) \leq d(z_m, z) + d(z, z_n) < \frac{\epsilon}{2} + \frac{\epsilon}{2} = \epsilon.$$

Thus $\{z_n\}$ is a Cauchy sequence.

Assume $\{z_n\}$ is a Cauchy sequence. Write $z_n = a_n + b_n i$ with $a_n, b_n \in \Re$. Sir

$$d(a_m, a_n) \leq d(z_m, z_n) \quad \text{and} \quad d(b_m, b_n) \leq d(z_m, z_n),$$

$\{a_n\}$ and $\{b_n\}$ are Cauchy sequences of real numbers. By Theorem 1.10.38, the sequences $\{a_n\}$ and $\{b_n\}$ converge with limits A and B. We show the complex sequ $\{z_n\}$ also converges with limit $z = A + Bi$. Let $\epsilon > 0$. Find P such that if $n \geq P$, $a_n \in N(A; \frac{\epsilon}{\sqrt{2}})$. Find Q such that if $n \geq Q$, then $b_n \in N(B; \frac{\epsilon}{\sqrt{2}})$. Let $N = P + Q$ $n \geq N$, then $a_n \in N(A; \frac{\epsilon}{\sqrt{2}})$ and $b_n \in N(B; \frac{\epsilon}{\sqrt{2}})$. Hence

$$d(z_n, z) = \sqrt{(a_n - A)^2 + (b_n - B)^2} < \sqrt{\frac{\epsilon^2}{2} + \frac{\epsilon^2}{2}} = \sqrt{\epsilon^2} = \epsilon,$$

and $z_n \in N(z;\epsilon)$. Thus the sequence $\{z_n\}$ converges to z.

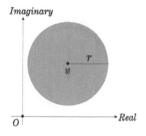

Figure 5.11.7
Neighborhood $N(w;r)$

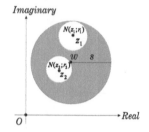

Figure 5.11.9
$N(w;s)$ is Open

Limits

In this subsection, we define the limit of a complex valued function and give the properties of these limits. Then we define continuous complex valued functions and list their properties. In particular, we establish the continuity of complex polynomials and complex rational functions as defined in the following examples.

Examples 5.11.13 (1) Let $c_0, \ldots, c_n \in C$. Define the complex polynomial

$$f(z) = c_n z^n + \cdots + c_1 z + c_0 \text{ for } z \in C.$$

In particular $f_1(z) = 3z^2 - 5z + 2$ and $f_2(z) = (5 - 4i)z^3 - 3iz^2 + 7iz - 10$ are complex polynomials.

(2) Let $p(z)$ and $q(z)$ be complex polynomials. Define the complex rational function $g(z) = \frac{p(z)}{q(z)}$. The domain of g is the set of all complex numbers with the roots of the polynomial $q(z)$ deleted. For example, $g(z) = \frac{5z - 8i}{z^2 + 9}$ is a complex rational function with domain all complex numbers other than $3i$ and $-3i$. □

Let f be a real valued function with domain an open interval I and $c \in I$. Recall from (1.10.4) that $\lim_{x \to c} f(x) = L$ if $f(x)$ get close to L as x approaches c. For ϵ a small positive number, the neighborhood $N(L; \epsilon)$ consists of numbers which are close to L. We require that for each such neighborhood there is a corresponding neighborhood $N(c; \delta)$, of numbers close to c, such that if $x \in N(c; \delta)$, with $x \neq c$, then $f(x) \in N(L; \epsilon)$. We use the same criterion to define the limit of a complex valued function.

Definition 5.11.14 *Let f be a complex valued function with domain the open subset D of C. Let $w \in D$ and $L \in C$. Define*

$$\lim_{z \to w} f(z) = L$$

if for every neighborhood $N(L; \epsilon)$ there is a corresponding neighborhood $N(w; \delta)$ such that $N(w; \delta) \subset D$ and if $z \in N(w; \delta)$, with $z \neq w$, then $f(z) \in N(L; \epsilon)$.

Examples 5.11.15 (1) Let $c \in C$. Show $\lim_{z \to w} c = c$.

Solution Observe $f(z) = c$, $D = C$ and $L = c$ in this example. Given $N(c; \epsilon)$, choose any $\delta > 0$. If $z \in N(w; \delta)$, then $f(z) = c \in N(c; \epsilon)$.

(2) Show $\lim_{z \to w} z = w$.

Solution Note $f(z) = z$, $D = C$ and $L = w$ in this example. Given $N(c; \epsilon)$, choose $\delta = \epsilon$. If $z \in N(w; \delta)$, then $f(z) = z \in N(w; \epsilon)$ because $\epsilon = \delta$. □

Limits of complex valued functions have the same properties as the limits of real valued functions which we established in Section 1.6. The proofs of Section 1.6 generalize directly to complex limits. We leave the details for the exercises.

Proposition 5.11.16 *Let f and g be complex valued functions with domain the same open subset D of C. Assume that $\lim_{z \to w} f(z) = L$ and $\lim_{z \to w} g(z) = M$. Then*

(a) $\lim_{z \to w} f(z)$ *is unique;*

(b) $\lim_{z \to w} [f(z) + g(z)] = L + M$;

(c) $\lim_{z \to w} [f(z)g(z)] = LM$;

(d) $\lim_{z \to w} \dfrac{f(z)}{g(z)} = \dfrac{L}{M}$ *if $M \neq 0$.*

The definition of continuity of a complex valued function is modeled on the nition of a continuous real valued function.

Definition 5.11.17 *Let f be a complex valued function with open domain $D \subset$*
(a) *f is continuous at $w \in D$ if $\lim\limits_{z \to w} f(z) = f(w)$.*
(b) *We say f is continuous if f is continuous at each $w \in D$.*

Continuous complex valued functions have the same basic properties as conti real valued functions. The first three properties listed below follow easily from F sition 5.11.16, while the fourth one is proved in the same manner as the correspo property for real valued functions. The details are left as an exercise.

Corollary 5.11.18 *Let f and g be continuous complex valued functions with de the same open subset D of C. Then*
(a) *$f + g$ is continuous;* **(b)** *fg is continuous;* **(c)** *$\frac{f}{g}$ is continuous when $0 \notin g$*
(d) *Let F, G be complex valued continuous functions with open domains such $F \circ G$ is defined. Then $F \circ G$ is continuous.*

Examples 5.11.19 (1) Show every complex polynomial is continuous.

> **Solution** By Examples 5.11.15 (1), (2), the functions $f(z) = c$ and $g(z) =$ continuous. A polynomial is constructed by taking sums and products of functions. Hence a polynomial is continuous by Corollary 5.11.18 (a), (b).

(2) Show every complex rational function is continuous.

> **Solution** A rational function is the quotient of two polynomials. Theref rational function is continuous by Example 1 and Corollary 5.11.18 (c).

Derivatives

In this subsection, we define the derivative of a complex valued function f as the of the average change of f. The basic rules for computing derivatives of real va functions generalize to derivatives of complex valued functions. A complex va function whose derivative exists is called *holomorphic*.

Definition 5.11.20 *f is a complex valued function with open domain $D \subset C$.*
(a) *f is holomorphic at $w \in D$ if the following limit exists:*

$$\lim_{z \to w} \frac{f(z) - f(w)}{z - w} = f'(w).$$

$f'(w)$ is called the derivative of $f(z)$ at $z = w$.
(b) *f is holomorphic if it is holomorphic at every $w \in D$.*

The proofs of the basic rules for finding derivatives of complex valued funct are analogous to the proofs of the corresponding results for real valued functions. relegate these proofs to the exercises.

Proposition 5.11.21 *Assume f and g are complex valued functions, with dor the same open subset D of C, which are holomorphic at $w \in D$. Then*
(a) *$f + g$ is holomorphic at w and $(f + g)'(w) = f'(w) + g'(w)$;*
(b) *fg is holomorphic at w and $(fg)'(w) = f'(w)g(w) + f(w)g'(w)$.*

(c) *If $g(w) \neq 0$, then $\frac{f}{g}$ is holomorphic at w and $\left(\dfrac{f}{g}\right)'(w) = \dfrac{f'(w)g(w) - f(w)g'}{g(w)^2}$*

Examples 5.11.22 (1) Let $f(z) = c$ be the constant function with $c \in \mathcal{C}$. Show f is holomorphic with $f'(z) = 0$.

Solution
$$f'(w) = \lim_{z \to w} \frac{f(z) - f(w)}{z - w} = \lim_{z \to w} \frac{c - c}{z - w} = 0.$$

(2) Show the function $g(z) = z$ is holomorphic with $g'(z) = 1$.

Solution
$$g'(w) = \lim_{z \to w} \frac{g(z) - g(w)}{z - w} = \lim_{z \to w} \frac{z - w}{z - w} = 1.$$

(3) Show $h(z) = z^n$ is holomorphic with $h'(z) = nz^{n-1}$ for n a positive integer.

Solution When $n = 1$, we verified this formula in the preceding example. If this formula is true for z^n, with $n \geq 1$, then by Proposition 5.11.21(b):

$$D\left(z^{n+1}\right) = D\left(z \cdot z^n\right) = D(z)z^n + zD\left(z^n\right) = 1 \cdot z^n + z \cdot nz^{n-1} = (n+1)z^n.$$

Thus the formula is true for $n + 1$ and hence for all positive integers n.

(4) **(a)** Show every complex polynomial is holomorphic.
(b) Find the derivative of $f(z) = (8i + 3)z^7 - 5iz^4 + 11$.

Solution **(a)** A polynomial is constructed by taking sums and products of the holomorphic functions of Examples 1 and 2. Thus a polynomial is holomorphic by Proposition 5.11.21 (a), (b).
(b) $f'(z) = (8i + 3)(7z^6) - 5i(4z^3) + 0 = (56i + 21)z^6 - 20iz^3$.

(5) **(a)** Show every complex rational function is holomorphic.
(b) Find the derivative of $g(z) = \frac{iz^2 + 3}{(3i-2)z^2 + (6i+5)}$.

Solution **(a)** A rational function is the quotient of two polynomials. Hence a rational function is holomorphic by Proposition 5.11.21 (c).
(b) By the quotient rule,

$$
\begin{aligned}
g'(z) &= \frac{2iz\left[(3i - 2)z^2 + (6i + 5)\right] - \left(iz^2 + 3\right)(3i - 2)2z}{\left[(3i - 2)z^2 + (6i + 5)\right]^2} \\
&= \frac{\left[(-6 - 4i)z^3 + (-12 + 10i)z\right] - \left[(-6 - 4i)z^3 + (-12 + 18i)z\right]}{\left[(3i - 2)z^2 + (6i + 5)\right]^2} \\
&= \frac{-8iz}{\left[(3i - 2)z^2 + (6i + 5)\right]^2} \qquad \square
\end{aligned}
$$

Real valued differentiable functions are continuous, and complex valued functions also have this property.

Proposition 5.11.23 *If $f(z)$ is holomorphic at $z = w$, then $f(z)$ is continuous at $z = w$.*

Proof $\lim_{z \to w} [f(z) - f(w)] = \lim_{z \to w} \frac{f(z) - f(w)}{z - w} \cdot \lim_{z \to w} (z - w) = f'(w) \cdot 0 = 0$. Thus $\lim_{z \to w} f(z) = \lim_{z \to w} f(w) = f(w)$, and f is continuous at $z = w$. $\qquad \square$

The following special case of the chain rule is sufficient for our purposes.

Proposition 5.11.24 *Let f and g be complex valued functions such that $f \circ g$ is defined. If $g(z)$ is holomorphic at $z = w$ with $g'(w) \neq 0$ and $f(u)$ is holomorphic at $u = g(w)$, then $(f \circ g)(z)$ is holomorphic at $z = w$ and*

$$(f \circ g)'(w) = f'(g(w))\, g'(w).$$

Proof Let $r = \left| \frac{g'(w)}{2} \right|$. By Proposition 5.11.23, the holomorphic function g continuous at $z = w$. Since $g'(w) \neq 0$, there is a neighborhood $N(w; \delta)$ such if $z \in N(w; \delta)$, then $\frac{g(z) - g(w)}{z - w} \in N(g'(w); r)$ which does not contain zero. $g(z) - g(w) \neq 0$ for $z \in N(w; \delta)$, and

$$
\begin{aligned}
(f \circ g)'(w) &= \lim_{z \to w} \frac{f(g(z)) - f(g(w))}{z - w} = \lim_{z \to w} \frac{f(g(z)) - f(g(w))}{g(z) - g(w)} \frac{g(z) - g(w)}{z - w} \\
&= \lim_{z \to w} \frac{f(g(z)) - f(g(w))}{g(z) - g(w)} \cdot \lim_{z \to w} \frac{g(z) - g(w)}{z - w}.
\end{aligned}
$$

Since the function $g(z)$ is continuous at $z = w$, $g(z)$ approaches $g(w)$ as z appro w. Hence $(f \circ g)'(w)$ exists, and

$$
(f \circ g)'(w) = \lim_{g(z) \to g(w)} \frac{f(g(z)) - f(g(w))}{g(z) - g(w)} \cdot \lim_{z \to w} \frac{g(z) - g(w)}{z - w} = f'(g(w))g'(w)
$$

Power Series

Convergent and absolutely convergent series of complex numbers are defined i same way as for series of real numbers. As with real series, an absolutely conve complex series is convergent. We prove that every complex valued power serie a radius of convergence R such that the power series is absolutely convergent $ open disc of radius R. We show a complex valued power series is holomorphic its derivative is obtained through term by term differentiation.

Definition 5.11.25 (a) *Let $\sum_{k=p}^{\infty} z_k$ be a series of complex numbers. This converges and has sum $S \in C$ if the sequence of partial sums*

$$
S_n = z_p + \cdots + z_n, \quad \text{for } n \geq p,
$$

has limit S, i.e. $\lim_{n \to \infty} S_n = S$. *A series which is not convergent is called diverge* **(b)** *The series of complex numbers $\sum_{k=p}^{\infty} z_k$ is absolutely convergent if the s of real numbers $\sum_{k=p}^{\infty} |z_k|$ converges. A convergent series which is not absol convergent is called conditionally convergent.*

Examples 5.11.26 Determine whether each series is absolutely convergent, cc tionally convergent or divergent.

(1) $\sum_{k=1}^{\infty} \frac{i^k}{2^k}$

Solution Since $\sum_{k=1}^{\infty} \left| \frac{i^k}{2^k} \right| = \sum_{k=1}^{\infty} \frac{1}{2^k}$ is a geometric series with ratio $\frac{1}{2} <$ converges. Thus the series $\sum_{k=1}^{\infty} \frac{i^k}{2^k}$ is absolutely convergent.

(2) $\sum_{k=1}^{\infty} \frac{1}{i^k}$

Solution The even summands of this series are $\sum_{k=1}^{\infty} (-1)^k$, and the odd s mands are $\sum_{k=1}^{\infty} (-1)^k i = i \sum_{k=1}^{\infty} (-1)^k$. Thus $\sum_{k=1}^{\infty} \frac{1}{i^k} = (1 + i) \sum_{k=1}^{\infty} (- $ which diverges.

(3) $\sum_{k=1}^{\infty} \frac{i^k}{k}$

Solution $\sum_{k=1}^{\infty} \left| \frac{i^k}{k} \right|$ is the harmonic series $\sum_{k=1}^{\infty} \frac{1}{k}$ which diverge Hence given series is not absolutely convergent. The even summands of the g series are $\sum_{k=1}^{\infty}(-1)^k\frac{1}{2k}$ which converges to $-\frac{\ln 2}{2}$ while the odd summands are $\sum_{k=1}^{\infty}(-1)^{k-1}\frac{i}{2k-1} = i\sum_{k=1}^{\infty}(-1)^{k-1}\frac{1}{2k-1}$ which converges to $i\arctan 1 = i\frac{\pi}{4}$. Thus the given series is conditionally convergent with $\sum_{k=1}^{\infty}\frac{i^k}{k} = -\frac{\ln 2}{2} + \frac{\pi}{4}i = \frac{1}{4}(-\ln 4 + \pi i)$. □

The first two parts of the next proposition are the nth Term Test for complex series. The last part states that absolute convergence implies converges for complex series.

Proposition 5.11.27 *Let $\sum_{k=0}^{\infty} z_k$ be a series of complex numbers.*

(a) *If $\sum_{k=0}^{\infty} z_k$ converges, then $\lim_{k\to\infty} z_k = 0$.*

(b) *If $\lim_{k\to\infty} z_k \neq 0$, then $\sum_{k=0}^{\infty} z_k$ diverges.*

(c) *If $\sum_{k=0}^{\infty} z_k$ is absolutely convergent, then it converges.*

Proof **(a)** Let S be the sum of the series and let S_n denote its nth partial sum. Then $z_n = S_n - S_{n-1}$ and

$$\lim_{n\to\infty} z_n = \lim_{n\to\infty} (S_n - S_{n-1}) = \lim_{n\to\infty} S_n - \lim_{n\to\infty} S_{n-1} = S - S = 0.$$

(b) This statement is a reformulation of (a).

(c) We show the sequence of partial sums $\{S_n\}$ of the series $\sum_{k=0}^{\infty} z_k$ is a Cauchy sequence. Let $\overline{S}_n$ denote the nth partial sum of the series of real numbers $\sum_{k=0}^{\infty} |z_k|$. Let $\epsilon > 0$. If $h < q$, then

$$|S_q - S_h| = |z_{h+1} + \cdots + z_q| \leq |z_{h+1}| + \cdots + |z_q| = \overline{S}_q - \overline{S}_h. \tag{5.11.1}$$

Since the series $\sum_{k=0}^{\infty} |z_k|$ converges, its sequence of partial sums converges and is therefore a Cauchy sequence, i.e. there is a positive integer N such that if $q \geq h \geq N$, then $\overline{S}_q - \overline{S}_h < \epsilon$. By (5.11.1), $|S_q - S_h| < \epsilon$, and $\{S_n\}$ is a Cauchy sequence. Hence the sequence of partial sums $\{S_n\}$ converges, and $\sum_{k=0}^{\infty} z_k$ converges. □

A real power series in $x - x_0$ is absolutely convergent on an open interval $N(x_0; R)$ in $\mathcal{R}$. Analogously a complex power series in $z - z_0$ is absolutely convergent on an open disc $N(z_0; R)$ in $\mathcal{C}$.

Proposition 5.11.28 *Let $P(z) = \sum_{n=0}^{\infty} c_n(z - z_0)^n$ be a power series with z_0 and the c_n in $\mathcal{C}$. Then there is $0 \leq R \leq \infty$ such that $P(z)$ is absolutely convergent for $|z - z_0| < R$ and divergent for $|z - z_0| > R$.*

Proof We show if $P(w)$ is convergent and $|v - z_0| < |w - z_0|$, then $P(v)$ is absolutely convergent. By Proposition 5.11.27(a), $\lim_{n\to\infty} c_n(w - z_0)^n = 0$. Hence there is a positive integer N such that if $n \geq N$, then $|c_n(w - z_0)^n| < 1$. Thus

$$\sum_{n=N}^{\infty} |c_n(v - z_0)^n| = \sum_{n=N}^{\infty} |c_n(w - z_0)^n| \left|\frac{(v - z_0)^n}{(w - z_0)^n}\right| \leq \sum_{n=N}^{\infty} \left|\frac{v - z_0}{w - z_0}\right|^n.$$

The latter series is a real valued geometric series with ratio $\frac{|v-z_0|}{|w-z_0|} < 1$ which converges. By the Comparison Test, the real series $\sum_{n=0}^{\infty} |c_n(v - z_0)^n|$ also converges, and the complex series $\sum_{n=0}^{\infty} c_n(v - z_0)^n$ is absolutely convergent.

Let R be the least upper bound of the set S of all the real numbers $|z - z_0|$ such that $P(z)$ converges. If $|v - z_0| < R$, then there is $w \in C$ with $|v - z_0| < |w - z_0| < R$

such that $|w - z_0| \in S$, i.e. $P(w)$ converges. By the preceding paragraph, P absolutely convergent. If $R < |v - z_0|$, then $|v - z_0| \notin S$, i.e. $P(v)$ diverges.

The R of Proposition 5.11.28 is called the *radius of convergence* of the power $P(z) = \sum_{n=0}^{\infty} c_n(z - z_0)^n$. If $R = \infty$, then $P(z)$ is absolutely convergent for complex number z. If $R = 0$, then $P(z)$ only converges for $z = z_0$. If $0 < R$ then $P(z)$ is absolutely convergent for z in the open disc $N(z_0; R)$ which is the in of the circle with center z_0 and radius R, while $P(z)$ is divergent for z outsid circle. (We have no information on $P(z)$ for z on the circle itself.) If z_0 and z coefficients c_n are real numbers, then R is also the radius of convergence of th power series $\sum_{n=0}^{\infty} c_n(z - z_0)^n$. Its interval of convergence I, ignoring endpoints, intersection of the disc $N(z_0; R)$ with the real axis. See Figure 5.11.29. As wit power series, we find the radius of convergence of $P(z)$ by applying the Ratio T

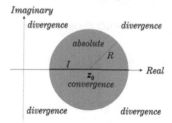

Figure 5.11.29 Disc of Convergence of a Power Series with Real Coefficients

Examples 5.11.30 (1) Find the disc of convergence of $P(z) = \sum_{n=0}^{\infty} \frac{(n!)^2}{(2n+1)!}(z$

Solution Apply the Ratio Test to the real series $\sum_{n=0}^{\infty} \frac{(n!)^2}{(2n+1)!}|z - 3|^n$:

$$\lim_{n\to\infty} \frac{\frac{[(n+1)!]^2}{[2(n+1)+1]!}|z - 3|^{n+1}}{\frac{(n!)^2}{(2n+1)!}|z - 3|^n} = \lim_{n\to\infty} \left[\frac{(n+1)!}{n!}\right]^2 \frac{(2n+1)!}{(2n+3)!}\frac{|z - 3|^{n+1}}{|z - 3|^n}$$

$$= \lim_{n\to\infty} (n+1)^2 \frac{1}{(2n+3)(2n+2)}|z - 3| = \lim_{n\to\infty} \frac{n+1}{4n+6}|z - 3| = \frac{|z -}{4}$$

Hence $P(z)$ is absolutely convergent for $\frac{|z-3|}{4} < 1$. For $\frac{|z-3|}{4} > 1$, the lim the summands of $\sum_{n=0}^{\infty} \frac{(n!)^2}{(2n+1)!}|z - 3|^n$ is nonzero, and $P(z)$ diverges by th Term Test. Thus, $P(z)$ is absolutely convergent for $|z - 3| < 4$ and diverge $|z - 3| > 4$. Hence the disc of convergence of the power series $P(z)$ is $N(3;$

(2) Find the disc of convergence of $Q(z) = \sum_{n=1}^{\infty} \frac{n!(3i+4)^n}{n^n}(z - 2i + 9)^n$.

Solution Apply the Ratio Test to the real series $\sum_{n=1}^{\infty} \frac{n!|3i+4|^n}{n^n}|z - 2i + 9$ $\sum_{n=1}^{\infty} \frac{n!5^n}{n^n}|z - 2i + 9|^n$:

$$\lim_{n\to\infty} \frac{\frac{(n+1)!5^{n+1}}{(n+1)^{n+1}}|z - 2i + 9|^{n+1}}{\frac{n!5^n}{n^n}|z - 2i + 9|^n} = \lim_{n\to\infty} \frac{(n+1)!}{n!}\frac{5^{n+1}}{5^n}\frac{n^n}{(n+1)(n+1)^n}\frac{|z - 2i}{|z - 2}$$

$$= \lim_{n\to\infty} 5\left(\frac{n}{n+1}\right)^n|z - 2i + 9| = \lim_{n\to\infty} 5\frac{1}{\left(1 + \frac{1}{n}\right)^n}|z - 2i + 9| = \frac{5|z -}{}$$

Thus $Q(z)$ is absolutely convergent for $\frac{5|z-2i+9|}{e} < 1$. For $\frac{5|z-2i+9|}{e} > 1$, limit of the summands of $\sum_{n=1}^{\infty} \frac{n!5^n}{n^n}|z - 2i + 9|^n$ is nonzero, and $P(z)$ dive by the nth Term Test. Thus, $Q(z)$ is absolutely convergent for $|z - (2i - 9)|$ and diverges for $|z - (2i - 9)| > \frac{e}{5}$. Therefore the disc of convergence of power series $Q(z)$ is $N\left(2i - 9; \frac{e}{5}\right)$.

Theorem 5.7.12 says that a real power series can be differentiated term by term. Its proof generalizes directly to complex power series and is relegated to the exercises.

Theorem 5.11.31 *Assume the complex power series $P(z) = \sum_{n=0}^{\infty} c_n(z - z_0)^n$ has radius of convergence $R > 0$. Then the power series $\sum_{n=1}^{\infty} nc_n(z - z_0)^{n-1}$ is absolutely convergent for $z \in N(z_0; R)$, the function $P(z)$ is holomorphic and*

$$P'(z) = \sum_{n=1}^{\infty} nc_n(z - z_0)^{n-1}.$$

We apply this theorem to the power series of the preceding example.

Examples 5.11.32 (1) Find the derivative of $P(z) = \sum_{n=0}^{\infty} \frac{(n!)^2}{(2n+1)!}(z - 3)^n$.

Solution For $z \in N(3; 4)$: $P'(z) = \sum_{n=1}^{\infty} \frac{(n!)^2 n}{(2n + 1)!}(z - 3)^{n-1}$.

(2) Find the derivative of $Q(z) = \sum_{n=1}^{\infty} \frac{n!(3i+4)^n}{n^n}(z - 2i + 9)^n$.

Solution For $z \in N\left(2i - 9; \frac{e}{5}\right)$:

$$Q'(z) = \sum_{n=1}^{\infty} \frac{n!(3i + 4)^n n}{n^n}(z - 2i + 9)^{n-1} = \sum_{n=1}^{\infty} \frac{n!(3i + 4)^n}{n^{n-1}}(z - 2i + 9)^{n-1}. \quad \square$$

We identify every power series as a Taylor series.

Corollary 5.11.33 *Assume the power series $P(z) = \sum_{k=0}^{\infty} c_k(z - z_0)^k$ has radius of convergence $R > 0$. Then the nth derivative of $P(z)$ at $z = z_0$ exists, and*

$$c_n = \frac{P^{(n)}(z_0)}{n!} \quad \text{for } n \geq 0.$$

Proof By Theorem 5.11.31 the nth derivative of $P(z)$ is given by:

$$P^{(n)}(z) = \sum_{k=n}^{\infty} k(k - 1) \cdots (k - n + 1)c_k(z - z_0)^{k-n} \quad \text{for } z \in N(z_0; R).$$

$P^{(n)}(z_0)$ is the constant term $n!c_n$ of this power series. Hence $c_n = \frac{P^{(n)}(z_0)}{n!}$. $\quad \square$

We apply this corollary to the power series of the preceding example.

Examples 5.11.34 (1) Let $P(z) = \sum_{n=0}^{\infty} \frac{(n!)^2}{(2n+1)!}(z - 3)^n$. Find $P^{(n)}(3)$ for $n \geq 0$.

Solution $P^{(n)}(3) = \frac{(n!)^2}{(2n + 1)!}n! = \frac{(n!)^3}{(2n + 1)!}$ for $n \geq 0$.

(2) Let $Q(z) = \sum_{n=1}^{\infty} \frac{n!(3i+4)^n}{n^n}(z - 2i + 9)^n$. Find $Q^{(n)}(2i - 9)$ for $n \geq 1$.

Solution $Q^{(n)}(2i - 9) = \frac{n!(3i + 4)^n}{n^n}n! = \frac{(n!)^2(3i + 4)^n}{n^n}$ for $n \geq 1$. $\quad \square$

We introduce the term *analytic* to describe complex valued functions which have a power series representation.

Definition 5.11.35 $f(z)$ *is a complex valued function with open domain* $D \subset$
(a) $f(z)$ *analytic at* $z = z_0 \in D$ *if there is a complex power series* $\sum_{n=0}^{\infty} c_n(z \cdot$
and $R > 0$ *such that* $f(z) = \sum_{n=0}^{\infty} c_n(z - z_0)^n$ *for* $z \in N(z_0; R)$.
(b) $f(z)$ *is analytic if it is analytic at every* $z \in D$.

We show that a complex power series is not just analytic at the center of i
of convergence but is in fact analytic at all points of this open disc.

Corollary 5.11.36 *If* $P(z) = \sum_{n=0}^{\infty} c_n(z - z_0)^n$ *is analytic at* $z = z_0$ *and has*
of convergence $R > 0$, *then* $P(z)$ *is analytic at every* $z_1 \in N(z_0; R)$.

Proof For $z_1 \in N(z_0; R)$, let R_1 denote the smaller of the numbers $|z_1 - z_0|$
$R - |z_1 - z_0|$. Observe from Figure 5.11.9 that $N(z_1; R_1) \subset N(z_0; R)$. Then

$$P^{(k)}(z_1) = \sum_{n=k}^{\infty} c_n n(n-1) \cdots (n-k+1)(z_1 - z_0)^{n-k} = \sum_{n=k}^{\infty} c_n \frac{n!}{(n-k)!}(z_1 - z_0$$

by Theorem 5.11.31. For $z \in N(z_1; R_1)$, we have the absolutely convergent serie

$$\begin{aligned}
P(z) &= \sum_{n=0}^{\infty} c_n(z - z_0)^n = \sum_{n=0}^{\infty} c_n \left[(z - z_1) + (z_1 - z_0)\right]^n \\
&= \sum_{n=0}^{\infty} \sum_{k=0}^{n} c_n \binom{n}{k} (z - z_1)^k (z_1 - z_0)^{n-k} \\
&= \sum_{k=0}^{\infty} \sum_{n=k}^{\infty} c_n \frac{n!}{k!(n-k)!} (z_1 - z_0)^{n-k} (z - z_1)^k \\
&= \sum_{k=0}^{\infty} \frac{1}{k!} \left[\sum_{n=k}^{\infty} c_n \frac{n!}{(n-k)!} (z_1 - z_0)^{n-k} \right] (z - z_1)^k = \sum_{k=0}^{\infty} \frac{P^{(k)}(z_1)}{k!} (z - z_1
\end{aligned}$$

Hence $P(z)$ is analytic at $z = z_1$.

We use power series to extend the definition of a real valued function from
interval of convergence of its Taylor series to an analytic function on the correspor
open disc. The remainder of this section is devoted to applications of this result

Corollary 5.11.37 *Assume the smooth real valued function* $f(x)$ *equals its* T
series $\sum_{n=0}^{\infty} a_n(x - x_0)^n$ *for* $|x - x_0| < R$. *Then there is a unique analytic funct*

$$F(z) = \sum_{n=0}^{\infty} a_n(z - x_0)^n$$

with domain the open disc $N(x_0; R)$ *such that* $F(x) = f(x)$ *for* $x \in (x_0 - R, x_0 +$

Proof Note the real series $\sum_{n=0}^{\infty} |a_n||z - x_0|^n$ is convergent for $|z - x_0| < R$. H
the complex power series $F(z)$ is absolutely convergent for $z \in N(x_0; R)$. Let $G(z$
an analytic function with domain the open disc $N(x_0; R)$ such that $G(x) = f(x$
$x \in (x_0 - R, x_0 + R)$. We must show $G(z) = F(z)$ for $z \in N(x_0; R)$. We verify
case when $G(z)$ is the power series $G(z) = \sum_{n=0}^{\infty} b_n(z - x_0)^n$ for $z \in N(x_0; R)$.
general case is given as a Challenging Problem. For each $n \geq 1$, the complex deriva
$G^{(n)}(x_0)$ exists using limits as z approaches x_0 for $z \in \mathcal{C}$. Hence the real deriva
$G^{(n)}(x_0)$ exists because it uses the same limits as x approaches x_0 for $x \in \Re$. Morec

these two derivatives have the same value. Since $G(x) = f(x)$ for $x \in (x_0 - R, x_0 + R)$, the real derivative $G^{(n)}(x_0)$ is the same as $f^{(n)}(x_0)$ which equals $n! a_n$ by Cor. 5.7.14. The complex derivative $G^{(n)}(x_0)$ equals $n! b_n$ by Cor. 5.11.33. Hence $a_n = b_n$ for $n \geq 0$, and $G(z) = F(z)$. $\qquad \square$

Exponential Function

Recall the real valued function e^x has Taylor series $e^x = \sum_{n=0}^{\infty} \frac{x^n}{n!}$ for $x \in \Re$. By Corollary 5.11.37 there is a unique analytic function which extends the domain of this function from $\Re$ to $\mathcal{C}$ given by:

$$e^z = \sum_{n=0}^{\infty} \frac{z^n}{n!} \quad \text{for } z \in \mathcal{C}. \tag{5.11.2}$$

In this subsection, we establish the basic properties of this complex valued exponential function. We begin with its derivative. Just as the real valued function e^x equals its derivative, so too the complex function e^z equals its derivative.

Proposition 5.11.38 *The function e^z is holomorphic with $D(e^z) = e^z$.*

Proof By Corollary 5.11.36, e^z is an analytic function. By Theorem 5.11.31 e^z is holomorphic and its derivative is obtained using term by term differentiation:

$$D(e^z) = \sum_{n=0}^{\infty} D\left(\frac{z^n}{n!}\right) = \sum_{n=1}^{\infty} \frac{nz^{n-1}}{n!} = \sum_{n=1}^{\infty} \frac{z^{n-1}}{(n-1)!} = \sum_{m=0}^{\infty} \frac{z^m}{m!} = e^z \text{ where } m = n-1. \; \square$$

We establish the property $e^z e^w = e^{z+w}$ by taking the Cauchy product of the power series of e^z and e^w. The Cauchy product of two absolutely convergent complex series with sums A, B is absolutely convergent with sum AB. We leave the proof as an exercise. It is a direct generalization of the proof of Theorem 5.7.18, the corresponding fact for real power series.

Corollary 5.11.39 *Let $w, z \in \mathcal{C}$. Then*

(a) $e^z e^w = e^{z+w}$;

(b) $\frac{e^z}{e^w} = e^{z-w}$.

Proof (a) Since the series $e^z = \sum_{n=0}^{\infty} \frac{z^n}{n!}$ and $e^w = \sum_{n=0}^{\infty} \frac{w^n}{n!}$ are absolutely convergent, their Cauchy product is absolutely convergent with sum $e^z e^w$:

$$e^z e^w = \sum_{n=0}^{\infty} \left(\sum_{m=0}^{n} \frac{z^m}{m!} \frac{w^{n-m}}{(n-m)!} \right) = \sum_{n=0}^{\infty} \left[\sum_{m=0}^{n} \frac{n!}{m!(n-m)!} \frac{z^m w^{n-m}}{n!} \right]$$

$$= \sum_{n=0}^{\infty} \frac{1}{n!} \left[\sum_{m=0}^{n} \binom{n}{m} z^m w^{n-m} \right] = \sum_{n=0}^{\infty} \frac{1}{n!} (z+w)^n = e^{z+w}.$$

(b) By (a), $e^w e^{-w} = e^{w+(-w)} = e^0 = 1$ and $e^{-w} = \frac{1}{e^w}$. Apply (a) again: $\frac{e^z}{e^w} = e^z e^{-w} = e^{z-w}$. $\qquad \square$

We use this corollary to determine the polar representation of e^{x+iy}.

Corollary 5.11.40 *Let x, y, θ be real numbers. Then*

(a) $e^{i\theta} = \cos\theta + i\sin\theta$;

(b) $e^{x+iy} = e^x (\cos y + i\sin y) = P(e^x, y)$;

(c) $|e^{x+iy}| = e^x$;

(d) e^z *is a periodic function with period $2\pi i$, i.e. $e^{z+2\pi i} = e^z$ for all $z \in \mathcal{C}$.*

Proof (a) Separate the even and odd terms in the power series of $e^{i\theta}$:

$$e^{i\theta} = \sum_{n=0}^{\infty} \frac{(i\theta)^n}{n!} = \sum_{n=0}^{\infty} (i)^{2n} \frac{\theta^{2n}}{(2n)!} + \sum_{n=0}^{\infty} (i)^{2n+1} \frac{\theta^{2n+1}}{(2n+1)!}$$

$$= \sum_{n=0}^{\infty} (-1)^n \frac{\theta^{2n}}{(2n)!} + \sum_{n=0}^{\infty} (-1)^n i \frac{\theta^{2n+1}}{(2n+1)!} = \cos\theta + i\sin\theta.$$

(b) $e^{x+iy} = e^x e^{iy} = e^x(\cos y + i\sin y) = P(e^x, y)$.

(c) $\left|e^{x+iy}\right| = |P(e^x, y)| = e^x$.

(d) $e^{z+2\pi i} = e^z e^{2\pi i} = e^z(\cos 2\pi + i\sin 2\pi) = e^z(1 + 0i) = e^z$.

We use Corollary 5.11.40(b) to compute the value of e^z for specific z.

Example 5.11.41 Compute $e^{5 + \frac{2\pi}{3}i}$.

Solution By Corollary 5.11.40(b):

$$e^{5 + \frac{2\pi}{3}i} = e^5 \left(\cos\frac{2\pi}{3} + i\sin\frac{2\pi}{3} \right) = e^5 \left(-\frac{1}{2} + \frac{\sqrt{3}}{2}i \right) = -\frac{e^5}{2} + \frac{\sqrt{3}\, e^5}{2}i.$$

Logarithm Function

We define the logarithm as the inverse function of a restriction of the expon● function. We show the usual identities for $\log(zw)$ and $\log\frac{z}{w}$ are valid up to a mu● of $2\pi i$. We also compute the derivative of $\log z$ and its Taylor series.

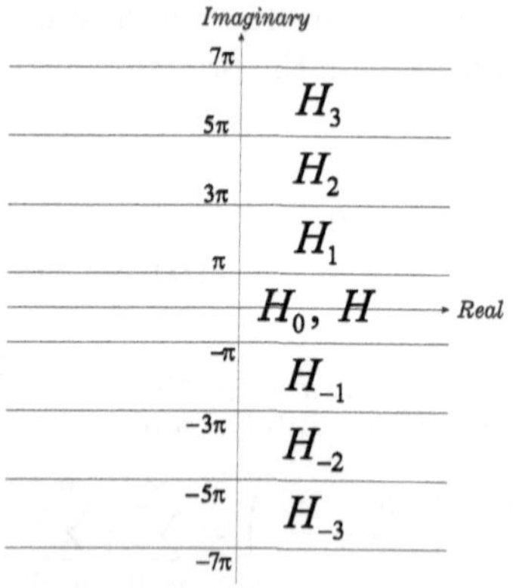

Figure 5.11.42 The Horizontal Strips H_n

For each integer n, let H_n be the horizontal strip of width 2π and infinite leng● the complex plane consisting of those (x, y) with $(2n - 1)\pi < y \leq (2n + 1)\pi$. Figure 5.11.42. By Corollary 5.11.40(b),(d) the value of e^z on the strip H_n is:

$$e^{\ln R + i(\theta + 2n\pi)} = P(R, \theta) \text{ for } R > 0 \text{ and } -\pi < \theta \leq \pi.$$

Thus e^z is one–to–one on each H_n with range $\mathcal{C}'$, the complex plane with the o● deleted. Let H be the interior of the strip H_0. That is, H is the set of all (x, y) $-\pi < y < \pi$. The image S of H under e^z is the complex plane with the negative axis, the set of $(x, 0)$ with $x \leq 0$, removed. Let $\exp z$ denote the restriction of function e^z to H:

$$\exp : H \to S \text{ by } \exp z = e^z.$$

Define the logarithm as the the inverse function of the one–to–one function exp:

$$\log : S \to H \quad \text{by} \quad \log z = w \text{ if } z = \exp w.$$

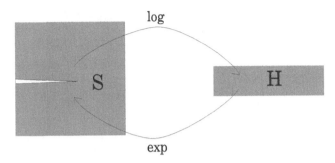

log

S

H

exp

Figure 5.11.43 The Inverse Functions $\exp w$ and $\log z$

By Corollary 5.11.40(b), $\log P(R, \theta) = \log \exp(\ln R + i\theta)$. Hence

$$\log P(R, \theta) = \ln R + i\theta \tag{5.11.3}$$

for $R, \theta \in \Re$ with $R > 0$ and $-\pi < \theta < \pi$. See Figure 5.11.43. The preceding formula is useful for computing the value of $\log z$ for a specific complex number z. Note

$$\begin{aligned}
\text{Domain } \log &= \text{Range } \exp = S, &(5.11.4)\\
\text{Range } \log &= \text{Domain } \exp = H, &(5.11.5)\\
\exp(\log z) &= z \text{ for } z \in S, &(5.11.6)\\
\log(\exp w) &= w \text{ for } w \in H. &(5.11.7)
\end{aligned}$$

As depicted in Figure 5.11.44, H_0 is the union of H and the line $y = \pi i$. Since e^z is one–to–one on H_0, it has an inverse function $\log' : C' \to H_0$, which extends the log function. However, the function $\log'$ is not continuous on the negative x–axis. Points in the second quadrant near the negative x–axis are sent by $\log'$ to points near the line $y = \pi$. On the other hand, points in the third quadrant near the negative x–axis are sent by $\log'$ to points near the line $y = -\pi$. This is why we use H, and not H_0, as the domain of exp.

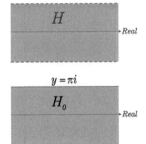

H

• Real

$y = \pi i$

H_0

• Real

Figure 5.11.44
H and H_0

Example 5.11.45 Compute $\log \left(-\sqrt{\frac{e}{2}} - \sqrt{\frac{e}{2}} i \right)$.

Solution By (5.11.3):

$$\begin{aligned}
\log \left(-\sqrt{\frac{e}{2}} - \sqrt{\frac{e}{2}} i \right) &= \log \left[\sqrt{e} \left(-\frac{1}{\sqrt{2}} - \frac{1}{\sqrt{2}} i \right) \right] = \log P \left(\sqrt{e}, -\frac{3\pi}{4} \right)\\
&= \ln \sqrt{e} - \frac{3\pi}{4} i = \frac{1}{2} - \frac{3\pi}{4} i. \qquad \square
\end{aligned}$$

We must be careful in computing $\log zw$ and $\log \frac{z}{w}$, because sums or differences of numbers in H may lie in H_{-1}, H_0 or H_1.

Proposition 5.11.46 *Let* $z, w \in S$.

(a) *If* $zw \in S$, *then* $\log z + \log w - \log(zw)$ *is either* $-2\pi i$, 0 *or* $2\pi i$.

(b) *If* $\frac{1}{w} \in S$, *then* $\log \frac{1}{w} = -\log w$.

(c) *If* $\frac{z}{w} \in S$, *then* $\log z - \log w - \log \frac{z}{w}$ *is either* $-2\pi i$, 0 *or* $2\pi i$.

Proof (a) By (5.11.6):

$$\exp(\log(zw)) = zw = \exp(\log z)\exp(\log w) = \exp(\log z + \log w).$$

Thus $\Delta = \log z + \log w - \log(zw)$ is a multiple of $2\pi i$. Since $\log z, \log w \in H$, $\log z$ is in either H_{-1}, H_0 or H_1 while $\log zw \in H_0$. If $\log z + \log w \in H_{-1}$ then $\Delta =$ if $\log z + \log w \in H_0$ then $\Delta = 0$, and if $\log z + \log w \in H_1$ then $\Delta = 2\pi i$.

(b) $e^{-\log w} = \frac{1}{e^{\log w}} = \frac{1}{w} = e^{\log \frac{1}{w}}$. Since $\log w \in H$, $-\log w \in H$. Also $\log \frac{1}{w}$ Hence $-\log w = \log \frac{1}{w}$.

(c) By (a), (b): $\log z - \log w - \log \frac{z}{w} = \log z + \log \frac{1}{w} - \log \frac{z}{w}$ is either $-2\pi i$, 0 o

Example 5.11.47 Let $z = w = e^{\frac{3\pi}{4}i}$. Then $zw = e^{\frac{3\pi}{2}i}$ and

$$\log zw \;=\; \log e^{\frac{3\pi}{2}i} = \log e^{-\frac{\pi}{2}i} = -\frac{\pi}{2}i.$$

On the other hand, $\log z + \log w \;=\; \dfrac{3\pi}{4}i + \dfrac{3\pi}{4}i = \dfrac{3\pi}{2}i.$

Thus $\log z + \log w - \log zw = 2\pi i$ in this case.

We compute the derivative of the complex logarithm function.

Proposition 5.11.48 *The function* $\log z$ *is holomorphic on* S *with*

$$D(\log z) = \frac{1}{z} \quad \text{for } z \in S.$$

Proof Note $z = \exp w$ is never 0 for $w \in H$. Apply the chain rule:

$$\frac{d}{dw}(w) \;=\; \frac{d}{dw}\log(\exp w) \;=\; D(\log)(\exp w)\cdot D(\exp w)$$
$$1 \;=\; D(\log)(\exp w)\cdot \exp w$$
$$D(\log)(z) \;=\; D(\log)(\exp w) \;=\; \frac{1}{\exp w} = \frac{1}{z}.$$

We find the higher derivatives of the logarithm and its Taylor series $P(z)$ at with disc of convergence $N(c;|c|)$. However, we have not yet studied the met required to determine the values of z for which $\log z$ equals $P(z)$. In particul. $c = a + bi$ with $a < 0$ then $N(c;|c|)$ includes points on the negative x-axis whic not in the domain S of $\log z$. In this case, $\log z$ can not equal $P(z)$ on its entire of convergence. We state when $\log z$ equals $P(z)$ in Corollary 5.11.54.

Corollary 5.11.49 *If* $c \in S$*, the Taylor series of* $\log z$ *is given by:*

$$P(z) = \log c + \sum_{n=1}^{\infty}(-1)^{n-1}\frac{(z-c)^n}{nc^n}$$

with radius of convergence $|c|$.

Proof Let $f(z) = \log z$. Then $f(c) = \log c$ and $f'(z) = \frac{1}{z}$. For $n \geq 2$,

$$f^{(n)}(z) = \frac{d^{n-1}}{dz^{n-1}}\left(z^{-1}\right) = (-1)^{n-1}(n-1)!\,z^{-n}.$$

The coefficient of z^n in the Taylor series of $f(z)$ at $z = c$ is $\frac{f^{(n)}(c)}{n!}$, which equals $\log c$ if $n = 0$ or $(-1)^{n-1}\frac{(n-1)!c^{-n}}{n!} = (-1)^{n-1}\frac{1}{nc^n}$ if $n \geq 1$. To determine the radius of convergence, apply the Ratio Test to the series $|\log c| + \sum_{n=1}^{\infty}\frac{|z-c|^n}{n|c|^n}$:

$$\lim_{n\to\infty}\frac{\frac{|z-c|^{n+1}}{(n+1)|c|^{n+1}}}{\frac{|z-c|^n}{n|c|^n}} = \lim_{n\to\infty}\frac{n}{n+1}\frac{|c|^n}{|c|^{n+1}}\frac{|z-c|^{n+1}}{|z-c|^n} = \lim_{n\to\infty}\frac{n}{n+1}\frac{|z-c|}{|c|} = \frac{|z-c|}{|c|}.$$

Hence this Taylor series is absolutely convergent for $|z - c| < |c|$ and divergent for $|z - c| > |c|$. Thus its radius of convergence is $R = |c|$. □

Trigonometric Functions

We define $\sin z$ and $\cos z$ as power series and establish their basic properties. The other four trigonometric functions are defined in terms of $\sin z$ and $\cos z$. We leave the properties of those functions for the exercises.

By Proposition 5.5.13(b),(c), the Taylor series of the real valued functions $\sin x$ and $\cos x$ are:

$$\sin x = \sum_{n=0}^{\infty}(-1)^n\frac{x^{2n+1}}{(2n+1)!}, \qquad \cos x = \sum_{n=0}^{\infty}(-1)^n\frac{x^{2n}}{(2n)!} \quad \text{for } x \in \Re.$$

By Corollary 5.11.37 there are unique analytic functions which extend the domains of these functions from $\Re$ to $\mathcal{C}$ given by:

$$\sin z = \sum_{n=0}^{\infty}(-1)^n\frac{z^{2n+1}}{(2n+1)!}, \qquad \cos z = \sum_{n=0}^{\infty}(-1)^n\frac{z^{2n}}{(2n)!} \quad \text{for } z \in \mathcal{C}. \qquad (5.11.8)$$

We establish the basic properties of these functions.

Proposition 5.11.50 *Let* $z, w \in \mathcal{C}$. *Then*

(a) $\sin z = \frac{e^{iz}-e^{-iz}}{2i}$;   (b) $\cos z = \frac{e^{iz}+e^{-iz}}{2}$;   (c) $\sin^2 z + \cos^2 z = 1$;

(d) $\sin z$ *is a holomorphic function with* $D(\sin z) = \cos z$;

(e) $\cos z$ *is a holomorphic function with* $D(\cos z) = -\sin z$;

(f) $\sin(z + w) = \sin z \cos w + \cos z \sin w$;

(g) $\cos(z + w) = \cos z \cos w - \sin z \sin w$;

(h) $\sin z$ *and* $\cos z$ *are periodic with period* 2π:

$$\sin(z + 2\pi) = \sin z \quad \text{and} \quad \cos(z + 2\pi) = \cos z.$$

Proof **(a)** When n is odd, write $n = 2k + 1$. Then

$$\frac{1}{2i}\left(e^{iz} - e^{-iz}\right) = \frac{1}{2i}\left(\sum_{n=0}^{\infty}\frac{(iz)^n}{n!} - \sum_{n=0}^{\infty}\frac{(-iz)^n}{n!}\right) = \frac{1}{2i}\sum_{n=0}^{\infty}[1 - (-1)^n]\,i^n\frac{z^n}{n!}$$

$$= \frac{1}{2i}\sum_{k=0}^{\infty}2(-1)^k i\frac{z^{2k+1}}{(2k+1)!} = \sum_{k=0}^{\infty}(-1)^k\frac{z^{2k+1}}{(2k+1)!} = \sin z.$$

(b) When n is even, write $n = 2k$. Then

$$\frac{1}{2}\left(e^{iz} + e^{-iz}\right) = \frac{1}{2}\left(\sum_{n=0}^{\infty}\frac{(iz)^n}{n!} + \sum_{n=0}^{\infty}\frac{(-iz)^n}{n!}\right) = \frac{1}{2}\sum_{n=0}^{\infty}[1 + (-1)^n]\,i^n\frac{z^n}{n!}$$

$$= \frac{1}{2}\sum_{k=0}^{\infty}2(-1)^k\frac{z^{2k}}{(2k)!} = \sum_{k=0}^{\infty}(-1)^k\frac{z^{2k}}{(2k)!} = \cos z.$$

(c) By (a) and (b): $\sin^2 z + \cos^2 z = \frac{e^{2iz}-2+e^{-2iz}}{-4} + \frac{e^{2iz}+2+e^{-2iz}}{4} = \frac{4}{4} = 1.$

(d), (e) Apply (a) and (b):

$$D(\sin z) = D\left(\frac{e^{iz}-e^{-iz}}{2i}\right) = \frac{ie^{iz}+ie^{-iz}}{2i} = \frac{e^{iz}+e^{-iz}}{2} = \cos z$$

$$D(\cos z) = D\left(\frac{e^{iz}+e^{-iz}}{2}\right) = \frac{ie^{iz}-ie^{-iz}}{2} = -\frac{e^{iz}-e^{-iz}}{2i} = -\sin z.$$

(f), (g) These formulas follow from (a) and (b) by straightforward computat We leave the details for the exercises.

(h) Since the exponential function is periodic with period $2\pi i$, the functions e^{iz} e^{-iz} are periodic with period 2π. Therefore the linear combinations $\sin z$ and c e^{iz} and e^{-iz} are also periodic with period 2π.

Note The proof of (h) shows the periodicity of exp with period $2\pi i$ is the sou the periodicity of sine and cosine with period 2π.

We use the formulas of Proposition 5.11.50(a),(b) to compute $\sin z$ and $\cos z$

Examples 5.11.51 Compute $\sin\left(\frac{\pi}{6}+2i\right)$ and $\cos\left(\frac{\pi}{6}+2i\right)$.

Solution By Proposition 5.11.50(a),(b):

$$\sin\left(\frac{\pi}{6}+2i\right) = \frac{1}{2i}\left(e^{\frac{\pi}{6}i-2}-e^{2-\frac{\pi}{6}i}\right) = -\frac{i}{2}\left[e^{-2}\left(\cos\frac{\pi}{6}+i\sin\frac{\pi}{6}\right)-e^{2}\left(\cos\frac{\pi}{6}-i\right.\right.$$

$$= -\frac{i}{2}\left[e^{-2}\left(\frac{\sqrt{3}}{2}+\frac{i}{2}\right)-e^{2}\left(\frac{\sqrt{3}}{2}-\frac{i}{2}\right)\right] = \frac{1}{4}\left(e^{2}+e^{-2}\right)+\frac{\sqrt{3}}{4}\left(e^{2}\right.$$

$$\cos\left(\frac{\pi}{6}+2i\right) = \frac{1}{2}\left(e^{\frac{\pi}{6}i-2}+e^{2-\frac{\pi}{6}i}\right) = \frac{1}{2}\left[e^{-2}\left(\cos\frac{\pi}{6}+i\sin\frac{\pi}{6}\right)+e^{2}\left(\cos\frac{\pi}{6}-i\right.\right.$$

$$= \frac{1}{2}\left[e^{-2}\left(\frac{\sqrt{3}}{2}+\frac{i}{2}\right)+e^{2}\left(\frac{\sqrt{3}}{2}-\frac{i}{2}\right)\right] = \frac{\sqrt{3}}{4}\left(e^{2}+e^{-2}\right)+\frac{1}{4}\left(e^{-2}-\right.$$

The other trigonometric functions are defined in the usual way from $\sin z$ and

$$\tan z = \frac{\sin z}{\cos z} = i\frac{e^{-iz}-e^{iz}}{e^{iz}+e^{-iz}}, \qquad \cot z = \frac{\cos z}{\sin z} = i\frac{e^{iz}+e^{-iz}}{e^{iz}-e^{-iz}}, \qquad (5.$$

$$\sec z = \frac{1}{\cos z} = \frac{2}{e^{iz}+e^{-iz}}, \qquad \csc z = \frac{1}{\sin z} = \frac{2i}{e^{iz}-e^{-iz}}. \qquad (5.1$$

Further Results

The subject of holomorphic functions is extensive, and we have merely consid those results which generalize analogous facts about real valued functions. There however, additional properties of holomorphic functions which are not true for valued functions. The following two theorems give examples of these properties. T are deep theorems which require powerful new methods for their proofs. We pre the proofs in Section 9.12.

If a real valued function is differentiable, then its second derivative may not e For example, if $f(x) = x^{5/3}$ then $f'(x)$ exists for all $x \in \Re$ but $f''(0)$ does not e The following theorem says that a holomorphic function never has such a problem its first derivative exists, then all its derivatives exist.

Theorem 5.11.52 *If $f(z)$ is holomorphic on an open disc, then all the higher de tives of $f(z)$ exist on this disc.*

If all the derivatives of a real valued function exist, the function may only equal its Taylor series in $x - c$ when $x = c$. This is the case with the Maclaurin series of the function $f(x) = e^{-1/x^2}$ for $x \neq 0$ and $f(0) = 0$. However, a complex valued function which is holomorphic on an open disc $N(z_0; R)$ always equals its Taylor series at z_0 on the entire open disc $N(z_0; R)$.

Theorem 5.11.53 *If $f(z)$ is holomorphic on the open disc $N(z_0; R)$, then $f(z)$ is analytic on this open disc. In particular, for all $z \in N(z_0; R)$, its Taylor series $P(z) = \sum_{n=0}^{\infty} \frac{f^{(n)}(z_0)}{n!}(z - z_0)^n$ is absolutely convergent and $f(z) = P(z)$.*

This theorem allows us to conclude that the logarithm function equals its Taylor series which we computed in Corollary 5.11.49.

Corollary 5.11.54 *Let $c = a + bi \in S$. If $a > 0$ define $R = |c|$ while if $a \leq 0$ define $R = |b|$. Then*

$$\log z = \log c + \sum_{n=1}^{\infty} (-1)^{n-1} \frac{(z - c)^n}{nc^n} \quad \text{for } z \in N(c; R).$$

Proof The function $\log z$ is holomorphic for all z in its domain S, the complex plane with negative x–axis N removed. By the preceding theorem, $\log z$ equals its Taylor series at $z = c$ on the largest open disc $N(c; R)$ which is contained in S, i.e. $N(c; R)$ can not contain any points of N. If $c = a + bi$, then the point of N closest to c is the origin if $a \geq 0$ or $a + 0i$ if $a < 0$. Hence $R = d(c, 0) = |c|$ if $a > 0$ while $R = d(c, a + 0i) = |b|$ if $a \leq 0$. $\qquad\square$

The fundamental reason for the remarkably good behavior of holomorphic functions is that these functions are very special, i.e. they are a restricted class of well behaved complex valued functions. On the other hand, differentiable real valued functions, and even smooth real valued functions, are a large class of functions which include bad examples along with the good ones.

Historical Remarks

Complex numbers were first used in calculus in the 1740s in the work of Leonhard Euler. He defined the complex function e^z from the Taylor series $e^x = \sum_{n=0}^{\infty} \frac{x^n}{n!}$ of the real valued function, as we did, by extending its domain from real numbers x to complex numbers z. His method for solving linear differential equations with constant coefficients indicated that the unique solution of the initial value problem $y'' = -y$, $y(0) = 0$, $y'(0) = 1$ is given by both $y = \sin x$ and $y = \frac{e^{ix} - e^{-ix}}{2i}$. He concluded that $\sin x = \frac{e^{ix} - e^{-ix}}{2i}$. He went on to establish all the concrete results of this section, and more, although he did not deal with the theoretical foundations of his calculations. The fundamental identity $e^{i\theta} = \cos\theta + i\sin\theta$ is called *Euler's Formula*.

The study of general functions of a complex variable was begun by Carl Friedrich Gauss in 1811 and independently by Augustin-Louis Cauchy in 1814. Many of the basic properties of these functions were published by Cauchy in the 1820s. Several of these results are presented in Section 9.12. In particular, the proofs of the two theorems of the preceding subsection are direct consequences of Cauchy's work.

Summary

The concepts of distance, neighborhood and open set in the complex plane should be understood. The reader should be able to determine whether a given sequence or

series of complex numbers is convergent or divergent. She should know how to co
limits and derivatives of complex valued functions. She should be able to dete
the radius of convergence of a complex power series and compute its derivativ
should also understand the definitions and basic properties of the complex
exponential, logarithm and trigonometric functions.

Basic Exercises

1. Find the domain of each complex valued function.

   (a) $f(z) = 2z^7 - (8i + 3)z + 4i - 6$ (b) $g(z) = \frac{z-3i}{iz+7}$

   (c) $h(z) = \frac{(2+5i)z+9i-2}{z^2+i}$ (d) $k(z) = \frac{iz-8+9i}{(3-4i)z-7+6i}$

   (e) $m(z) = \frac{z+3}{(12+5i)z^2+7+17i}$ (f) $n(z) = \frac{z}{z^3-4+4i}$

   (g) $p(z) = \frac{1}{z^4+8\sqrt{3}+8i}$ (h) $q(z) = \log(1 + z^2)$

2. Find the distance between each pair of complex numbers:

   (a) $3 - 2i$ and $4 + 3i$;   (b) 5 and $1 - 6i$;   (c) $7 + 3i$ and $2 - 2i$;
   (d) $4 + 8i$ and $5i$;   (e) $2 - 4i$ and $-6 + 3i$;   (f) $5 - 4i$ and $-1 - i$.

3. Let $z \in \mathcal{C}$ with $r, s > 0$.
   (a) Show $N(z; r) \cup N(z; s)$ is a neighborhood of z.
   (b) Show $N(z; r) \cap N(z; s)$ is a neighborhood of z.

4. Show each set is open in the complex plane:
   (a) $\mathcal{C}$;   (b) the set A of all $a + bi$ with $a > 0$;
   (c) the set B of all $a + bi$ with $a > 0$ and $b > 0$;
   (d) the set D of all $a + bi$ with $-3 < a < 5$ and $2 < b < 6$;
   (e) the set E of all $z \in \mathcal{C}$ with $4 < |z| < 7$.

5. Let U and V be open subsets of C.
   (a) Show $U \cup V$ is open.   (b) Show $U \cap V$ is open.

6. Define a subset of C to be *closed* if its complement is open. Show each of
subsets of C is closed:
   (a) the set A of all $z \in \mathcal{C}$ with $|z - 2 + 4i| \leq 7$;
   (b) the set B of all $a + bi$ with $a \geq 0$ and $b \geq 0$;
   (c) the set D of all $a + bi$ with $3 \leq a \leq 8$ and $-5 \leq b \leq -1$;
   (d) the imaginary axis;   (e) the set E of all $a + ai$.

7. (a) F and G are closed subsets of C. Show $F \cup G$ and $F \cap G$ are also closed
   (b) Show a subset F of C is closed if and only if whenever $\{z_n\}$ is a conver
sequence of points of F with limit z then $z \in F$.

8. Determine whether each sequence of complex numbers converges. Find the
of each convergent sequence.

   (e) $\left\{\left(\frac{1-4i}{5+i}\right)^n\right\}$ (a) $\{(2 - 3i)^n\}$ (b) $\left\{\frac{(5+9i)^n}{n!}\right\}$ (c) $\left\{\frac{7ni-2}{9ni+1}\right\}$ (d) $\left\{\frac{(2-5i)n^2+5i}{(1+3i)n^3+5n}\right\}$

9. Prove the properties of limits, Proposition 5.11.16.

10. Prove the properties of continuous functions, Corollary 5.11.18.

11. Verify each limit directly from the definition of the limit:
   (a) $\lim\limits_{z \to 7+3i} 5z = 35 + 15i$;   (b) $\lim\limits_{z \to 2-6i} 2iz = 12 + 4i$;
   (c) $\lim\limits_{z \to 4+i} (3iz - 5) = 12i - 8$;   (d) $\lim\limits_{z \to 3-5i} [(2i + 3)z - 6 + 7i] = 13 - 2i$.

12. Explain why each of these functions is continuous:
   (a) $f(z) = \tan z$;   (b) $g(z) = e^{z^2-i}$;   (c) $h(z) = e^z \sin z$;
   (d) $k(z) = \log(iz - 3)$;   (e) $m(z) = \frac{e^{iz}}{z^2-3iz+8}$;   (f) $n(z) = \sin\log z$.

13. Prove Proposition 5.11.21.

14. Find the derivative of each holomorphic function.

(a) $f(z) = 8iz^9 - (3+5i)z^6 + (6-4i)z^3 - 5z + 9i - 11$ (b) $g(z) = \frac{(6i-3)z^3 - 4i + 7}{z^4 - (8+9i)z + 2}$

(c) $h(z) = e^{\sin z}$ (d) $k(z) = (\cos z)(\log z)$

(e) $m(z) = \frac{e^{2iz}}{z^2 + 6i}$ (f) $n(z) = \sin\left(z^2 + e^z\right)$

(g) $p(z) = \cos^2 \log(iz^3 + 8)$ (h) $q(z) = \frac{e^z}{\cos z + i \sin z}$

(i) $r(z) = \log\left(\sin^2 z + \cos^2 iz\right)$ (j) $s(z) = e^{z^2 + i} \sin(iz^2 + 1)$

15. Let $f(z)$ be a holomorphic function having domain the open set D, and let n be a positive integer. Show $f(z)^n$ is also holomorphic with $D\left[f(z)^n\right] = nf(z)^{n-1}f'(z)$.

16. Let $F(z)$ be a holomorphic complex valued function with domain the open set U. Assume U contains an open interval I of real numbers. Define a real valued function f with domain I by $f(x) = F(x)$ for $x \in I$. Let F' denote the complex derivative of F. Show f is a differentiable real valued function with $f'(x) = F'(x)$ for $x \in I$.

17. Let $A = \sum_{k=p}^{\infty} a_n$ and $B = \sum_{k=p}^{\infty} b_n$ be convergent series of complex numbers.
(a) Show $\sum_{k=p}^{\infty}(a_n + b_n)$ is a convergent series with sum $A + B$.
(b) If $c \in \mathcal{C}$, show $\sum_{k=p}^{\infty} ca_n$ is a convergent series with sum cA.

18. Determine whether each series is absolutely convergent, conditionally convergent or divergent.

(a) $\sum_{k=1}^{\infty} i^k$ (b) $\sum_{k=1}^{\infty} \frac{i^k}{\sqrt{k}}$ (c) $\sum_{k=1}^{\infty} \frac{1}{(k+i)^2}$ (d) $\sum_{k=2}^{\infty} \frac{i^{2k+1}}{\ln k}$

(e) $\sum_{k=1}^{\infty} \frac{\sin(ik)}{k^3}$ (f) $\sum_{k=1}^{\infty} \frac{i^k}{2^k + 1}$ (g) $\sum_{k=1}^{\infty} \frac{1}{e^{ik}}$ (h) $\sum_{k=1}^{\infty} \frac{\sin(k+ik)}{\cos(k-ik)}$

(i) $\sum_{k=1}^{\infty} \frac{1}{e^{ik}}$ (j) $\sum_{k=1}^{\infty} \frac{e^{ik}}{k!}$ (k) $\sum_{k=1}^{\infty} \frac{1}{\log(ki)}$ (l) $\sum_{k=1}^{\infty} \frac{\log(k+ki)}{k^2}$

19. Find the radius of convergence of each power series.

(a) $\sum_{n=0}^{\infty} \frac{(z+i)^n}{\sqrt{2^n + 1}}$ (b) $\sum_{n=0}^{\infty} \frac{n!(z-2i+5)^n}{(2n)!}$ (c) $\sum_{n=0}^{\infty} \frac{(z+6i-7)^{2n}}{(4i+3)^n}$

(d) $\sum_{n=1}^{\infty} \frac{(z+3i-8)^n}{n}$ (e) $\sum_{n=0}^{\infty} \frac{(iz+7)^n}{ni+5}$ (f) $\sum_{n=0}^{\infty} \frac{nz^n}{(i+n)^n}$

(g) $\sum_{n=0}^{\infty} \frac{[(12-5i)z+2+3i]^n}{(3i-4)^n + 9i}$ (h) $\sum_{n=0}^{\infty} \frac{(3n)!(iz+7)^{2n}}{n!(2n)!}$ (i) $\sum_{n=0}^{\infty} \frac{z^n}{(n+i)(n-1+i)\cdots(1+i)i}$

20. Find power series representations of the first and second derivatives of each power series in the preceding example.

21. Prove Theorem 5.11.31.

22. Prove Theorem 5.7.18 for complex valued series: if $A = \sum_{n=0}^{\infty} a_n$ and $B = \sum_{n=0}^{\infty} b_n$ are absolutely convergent series of complex numbers, then the series $\sum_{n=0}^{\infty} \left(\sum_{k=0}^{n} a_k b_{n-k}\right)$ is also absolutely convergent with sum AB.

23. Use a calculator to approximate e^x, $\ln x$, $\sin x$ and $\cos x$ for $x \in \mathcal{R}$ in order to estimate each of these complex numbers.

(a) e^{3+2i} (b) e^{7i-6} (c) $\log(3i-4)$ (d) $\log(-i-2)$
(e) $\sin(5+2i)$ (f) $\sin(8-6i)$ (g) $\cos(8i-9)$ (h) $\cos(-7i)$

24. Compute $\log z + \log w - \log zw$ and $\log z - \log w - \log \frac{z}{w}$ in each case:
(a) $z = 2 - 3i$ and $w = 4 + 5i$;   (b) $z = -4 + 6i$ and $w = 1 + 2i$;
(c) $z = -5 - 2i$ and $w = 1 + 3i$;   (d) $z = -1 - 7i$ and $w = -2 - 4i$.

25. Prove Proposition 5.11.50(f), (g).

26. (a) Find the domains of $\tan z$, $\cot z$, $\sec z$ and $\csc z$.
(b) Show $1 + \tan^2 z = \sec^2 z$ and $1 + \cot^2 z = \csc^2 z$.
(c) Show these four functions are periodic, and determine their periods.
(d) Show these four functions are holomorphic, and find their derivatives.

The following three exercises are for those who have studied Section

27. For all $z \in \mathcal{C}$, define $\sinh z = \sum_{n=0}^{\infty} \frac{z^{2n+1}}{(2n+1)!}$ and $\cosh z = \sum_{n=0}^{\infty} \frac{z^{2n}}{(2n)!}$.

(a) Show $\sinh z = \frac{e^z - e^{-z}}{2}$.   **(b)** Show $\cosh z = \frac{e^z + e^{-z}}{2}$.   **(c)** Show $\cosh^2 z - \sinh^2$
(d) Show $\cosh z$ and $\sinh z$ are periodic with period $2\pi i$.
(e) Show $\sinh z$ is holomorphic with $D(\sinh z) = \cosh z$.
(f) Show $\cosh z$ is holomorphic with $D(\cosh z) = \sinh z$.

28. Verify the following identities for $z \in \mathcal{C}$ and $a, b \in \Re$.

(a) $\cos z = \cosh iz$ **(b)** $i \sin z = \sinh iz$ **(c)** $\cosh z = \cos iz$ **(d)** $i \sinh z$
(e) $\sin(a + bi) = \sin a \cosh b + i \cos a \sinh b$ **(f)** $\cos(a + bi) = \cos a \cosh b - $
(g) $|\sin(a + bi)| = \sqrt{\sin^2 a + \sinh^2 b}$ **(h)** $|\cos(a + bi)| = \sqrt{\cos^2 a + c}$
(i) $\cosh(a + bi) = \cosh a \cos b + i \sinh a \sin b$ **(j)** $|\cosh(a + bi)| = \sqrt{\sinh^2 a + }$

29. Let $f(z) = \sin z$ with $z = x + iy$.

(a) Show $f(z)$ maps each vertical line $x = a$, for a not a multiple of π, ont
branch of the hyperbola $\frac{x^2}{\sin^2 a} - \frac{y^2}{\cos^2 a} = 1$.

(b) Show $f(z)$ maps each horizontal line $y = b$, for $b \neq 0$, onto the ellipse
$\frac{x^2}{\cosh^2 b} + \frac{y^2}{\sinh^2 b} = 1$.

Challenging Problems

1. Verify these limits directly from the definition of a limit.

(a) $\lim\limits_{z \to w} z^2 = w^2$ for $w \in \mathcal{C}$.   **(b)** $\lim\limits_{z \to w} \frac{1}{z} = \frac{1}{w}$ for $0 \neq w \in \mathcal{C}$.

2. **(a)** Show if $\lim\limits_{z \to w} f(z) = L$, then $\lim\limits_{z \to w} \overline{f(z)} = \overline{L}$.

(b) Let $f(z)$ be holomorphic on an open set D which is closed under conjuga
Show $\overline{f(z)}$ is also holomorphic on D.

(c) Let $f(z)$ be analytic on an open set D which is closed under conjugation.
$\overline{f(z)}$ is also analytic on D.

3. Show $\lim\limits_{n \to \infty} \left(1 + \frac{z}{n}\right)^n = e^z$ for all $z \in \mathcal{C}$.

4. **(a)** Show the function $f(z) = |z|$ is not holomorphic for any value of $z \in \mathcal{C}$.
(b) Show the function $f(z) = |z|^2$ is only holomorphic at $z = 0$.

5. **(a)** Show the function $f(z) = z^2$ is one–to–one on the right half plane P consi
of all $a + bi$ with $a > 0$. Show the range $f(P)$ of f equals S, the complex plane
the negative x–axis removed.

(b) Let $\sqrt{z}$ denote the inverse function of f. Show $\sqrt{z}$ is continuous and holomor
Find its derivative.

(c) Show the domain of $\sqrt{z}$ can not be extended from S to define a contin
function with domain the entire complex plane.

(d) Find the Taylor series of $\sqrt{z}$ at $z = a + bi$ with $a > 0$ or $b \neq 0$. What is
radius of convergence of this power series?

6. Let $f(z)$ be holomorphic at $z = z_0$ with the radius of convergence of its Ta
series at $z = z_0$ a positive number R (not infinity). Assuming Theorem 5.11.53, s
there is a point on the circle $|z - z_0| = R$ at which f is not holomorphic.

7. Complete the proof of Corollary 5.11.37. That is, assume the smooth real va
function $f(x)$ equals its Taylor series $\sum_{n=0}^{\infty} a_n(x - x_0)^n$ for $|x - x_0| < R$. Let $F($
$\sum_{n=0}^{\infty} a_n(z - x_0)^n$ for $z \in N(x_0; R)$. Let $G(z)$ be an analytic function with dor
the open disc $N(x_0; R)$ such that $G(x) = f(x)$ for $x \in (x_0 - R, x_0 + R)$. Show $G($
$F(z)$ for $z \in N(x_0; R)$.

5.12   Review Exercises for Chapter 5

These exercises are based on Sections 2–7 and 9 of this chapter.

Decide whether each of these 50 statements is True or False. Justify your answers.

1. A sequence which is both increasing and decreasing is bounded.

2. A sequence of monomials $\{m_n(x)\}$ of positive degree is always bounded for some value of x.

3. Let $f(x)$ be a continuous function with domain $[1, \infty)$. If the sequence $\{f(n)\}$ is increasing, then $f(x)$ is an increasing function.

4. The sequence of partial sums of an infinite series is always increasing.

5. A convergent sequence must be either increasing or decreasing.

6. If the sequence $\{a_n\}$ is bounded, then the sequence $\left\{\frac{a_n}{n}\right\}$ converges.

7. If the sequence $\{|a_n|\}$ converges, then the sequence $\{a_n\}$ converges.

8. If the sequence $\{\exp(a_n)\}$ converges, then the sequence $\{a_n\}$ converges.

9. A sequence which is bounded below and bounded above must converge.

10. If the series $\sum_{k=1}^{\infty} a_k$ converges, then the sequence $\{a_k\}$ converges.

11. If the domain of the continuous function f is an open interval, then the range of f is also an open interval.

12. The series $\sum_{k=1}^{\infty} \frac{1}{k!}$ converges by the nth Term Test.

13. If a geometric series has ratio one, then the Ratio Test does not determine if it converges or diverges.

14. The sum of a geometric power series is always a rational function.

15. If the partial sums S_n of a series are given by $S_n = f(n)$ where $f(x)$ is a bounded rational function, then the series converges.

16. If the partial sums S_n of a series are given by $S_n = \arctan n$, then the series converges with sum $\frac{\pi}{2}$.

17. If the sequence $\{a_n\}$ of nonzero numbers is increasing, then the series $\sum_{n=1}^{\infty} a_n$ diverges.

18. If the series $\sum_{k=1}^{\infty} a_k$ converges, then the series $\sum_{k=1}^{\infty} a_{2k}$ also converges.

19. If the series $\sum_{k=1}^{\infty} a_k$ is absolutely convergent, then the series $\sum_{k=1}^{\infty} a_{2k}$ is also absolutely convergent.

20. If the series $\sum_{k=1}^{\infty} a_k$ is absolutely convergent, then the series $\sum_{k=1}^{\infty} a_k^2$ converges.

21. If the series $\sum_{k=1}^{\infty} a_k$ and $\sum_{k=1}^{\infty} b_k$ are divergent, then the series $\sum_{k=1}^{\infty} (a_k + b_k)$ is divergent.

22. If the series $\sum_{k=1}^{\infty} a_k^2$ converges, then the sereis $\sum_{k=1}^{\infty} a_k$ converges.

23. Let $\{a_k\}$ be a sequence of nonzero numbers. If the series $\sum_{k=1}^{\infty} a_k$ converges, then the series $\sum_{k=1}^{\infty} \frac{1}{a_k}$ diverges.

24. $\sum_{k=1}^{\infty} \frac{1}{k(k+1)} = \sum_{k=1}^{\infty} \frac{1}{2^k}$.

25. The series $\sum_{k=1}^{\infty} \left(1 + \frac{x}{k}\right)^k$ converges to e^x.

26. The sequence $\left\{\frac{x^n}{n!}\right\}$ converges to e^x.

27. An absolutely convergent series is conditionally convergent.

28. A power series is conditionally convergent for at most two values of x.

29. If $\sum_{k=1}^{\infty} a_k$ is absolutely convergent and $\sum_{k=1}^{\infty} b_k$ is convergent, then $\sum_{k=}^{\infty}$ is convergent.

30. The integral of a geometric series with ratio x^n is a power series which con for $-1 < x < 1$.

31. A power series in $x - c$, with radius of convergence $R > 0$, is smooth at x

32. Let $T(x)$ be the Taylor series of the smooth function $f(x)$ at $x = c$. If T absolutely convergent for $x \in \Re$, then $f(x) = T(x)$ for $x \in \Re$.

33. Let $T(x)$ be the Taylor series of the smooth function $f(x)$ at $x = c$. Assum $f^{(n)}(x) \le e^{-x^2}$ for $x \in \Re$ and $n \ge 1$. Then $f(x) = T(x)$ for all x.

34. Only the function $f(x) = 0$ has the Taylor series $T(x) = 0$.

35. For $x \in \Re$, the smooth function $f(x)$ equals its Taylor series $T(x)$ at $x =$ $f^{(k)}(c) = 0$ for $0 \le k < n$, then $\frac{f(x)}{(x-c)^n}$ is a smooth function.

36. The Ratio Test says that the series $\sum_{k=1}^{\infty} \frac{1}{k^2}$ converges.

37. A power series in $x - c$ converges for at least one value of x.

38. If the power series $P(x)$ converges at $x = n$ for every positive integer n, $P(x)$ converges for all $x \in \Re$.

39. Let $P(x)$ and $Q(x)$ be two power series in $x - c$ with radius of convergen Then $P(x) + Q(x)$ also has radius of convergence R.

40. Let $P(x)$ and $Q(x)$ be two nonzero power series in $x - c$ with radius of conver $R > 0$. If $Q^{(n)}(c) = 0$ implies that $P^{(n)}(c) = 0$, then $\lim\limits_{x \to c} \dfrac{P(x)}{Q(x)}$ exists.

41. A power series in $x - c$, with radius of convergence $R > 0$, is continuo $(c - R, c + R)$.

42. If the power series $\sum_{n=0}^{\infty} a_n x^n$ and $\sum_{n=0}^{\infty} b_n x^n$ have radii of convergence R S with $R < S$, then the power series $\sum_{n=0}^{\infty} (a_n + b_n) x^n$ has radius of convergenc

43. The improper integral $\int_0^{\infty} \frac{x^{100}}{2^x}\, dx$ converges.

44. The series $\sum_{k=1}^{\infty} \frac{\arctan k}{k^p}$ converges if and only if $p > 1$.

45. If $\sum_{k=1}^{\infty} a_k = A$ and $\sum_{k=1}^{\infty} b_k = B$, then $\sum_{k=1}^{\infty} a_k b_k = AB$.

46. If the series $\sum_{k=0}^{\infty} a_k$ of positive terms diverges, then the series $\sum_{k=0}^{\infty} (-1)^k$ conditionally convergent.

47. The series $\sum_{k=0}^{\infty} \frac{1}{e^{k^2}}$ converges.

48. If the series $\sum_{k=1}^{\infty} a_k$ converges, then the series $\sum_{k=1}^{\infty} (-1)^k a_k$ also converge

49. If $\sum_{k=1}^{\infty} \frac{1}{a_k}$ is absolutely convergent, then $\sum_{k=1}^{\infty} \frac{\sin k}{a_k}$ converges.

50. If the power series $\sum_{n=0}^{\infty} a_n x^n$ has interval of convergence $[-R, R]$, the derivative $\sum_{n=1}^{\infty} n a_n x^{n-1}$ also has interval of convergence $[-R, R]$.

Determine whether each sequence converges or diverges. If it converges, find its limit.

51. $\left\{ \frac{5^n}{n!} \right\}$ **52.** $\left\{ \left(1 + \frac{5}{n} \right)^n \right\}$ **53.** $\left\{ \left(2 + \frac{5}{n} \right)^n \right\}$

54. Find the annual rate of interest of money compounded continuously at 4%.

55. Find the radius of convergence of the power series $\sum_{n=1}^{\infty} \frac{1}{\sin \frac{1}{n}} x^{2n}$.

56. Let $f(x) = \sqrt{4x - x^2}$. Find $f^{(6)}(2)$.

Find the Taylor series at $x = 0$ of each function.

57. $f(x) = \frac{x}{e^{x^2}}$ **58.** $g(x) = \sqrt[3]{8 - x^2}$ **59.** $h(x) = \sqrt{x} \arctan \sqrt{x}$

Find the interval of convergence of each power series.

60. $\sum_{n=2}^{\infty} \frac{x^n}{\ln n}$ **61.** $\sum_{n=0}^{\infty} (-1)^n \frac{x^{3n}}{27^n - 8^n}$ **62.** $\sum_{n=0}^{\infty} (-1)^n \frac{n! x^n}{2^n + (2n)!}$

Use a partial sum of an infinite series to approximate each integral to the nearest .001.

63. $\displaystyle\int_0^{1/2} \sqrt{x} \sin \sqrt{x} \, dx$ **64.** $\displaystyle\int_0^{1/10} e^{x^2} \, dx$

Determine if each series converges absolutely, converges conditionally or diverges.

65. $\sum_{k=1}^{\infty} \sin \frac{1}{k}$ **66.** $\sum_{k=1}^{\infty} (-1)^k \left(1 + \frac{1}{k} \right)^k$ **67.** $\sum_{k=2}^{\infty} \frac{\ln k}{k^2}$

68. $\sum_{k=2}^{\infty} (-1)^k \frac{1}{k \ln k}$ **69.** $\sum_{k=2}^{\infty} (-1)^k \frac{1}{k(\ln k)^2}$ **70.** $\sum_{k=1}^{\infty} (-1)^k \frac{(3k)!}{(k!)^3}$

Answers to the Basic Exercises

Chapter 1

Section 1.2

1. **(a)** **(i)** pair of pants, black shoe, blue shirt **(ii)** green dress, toaster, red hat
(iii) set of shirts in my closet; set of shoes in my closet
(b) **(i)** Visa Card, driver's license, $20 bill **(ii)** Master Card, photo of my children,
$100 bill **(iii)** set of credit cards in my wallet; set of bills in my wallet
(c) **(i)** Canada, US, England **(ii)** Toronto, Hawaii, Newfoundland
(iii) set of European countries; set of third world countries
(d) **(i)** blue, red, violet **(ii)** pink, chartreuse, beige **(iii)** blue, yellow and green

2. $S \subset E \subset P \subset Q$ and $S \subset R \subset P \subset Q$

3. **(a)** -1.5 **(b)** $0.\overline{6}$ **(c)** $.071$ **(d)** $.\overline{428571}$ **(e)** $-.\overline{18}$ **(f)** 5.873 **(g)** 10 **(h)** -4800

4. **(a)** natural number, integer, rational number **(b)** integer, rational number
(c) natural number, integer, rational number **(d)** rational number
(e) rational number **(f)** irrational number
(g) natural number, integer, rational number **(h)** rational number
(i) rational number **(j)** irrational number

5.

6. **(a)** (i) $2, 9$ (ii) 7 (iii) **(b)** (i) $-3, 2$ (ii) 5 (iii)

(c) (i) $-8, -1$ (ii) 7 (iii) **(d)** (i) $-4, 7$ (ii) 11 (iii)

(e) (i) 1 (ii) ∞ (iii) **(f)** (i) 6 (ii) ∞ (iii)

(g) (i) -5 (ii) ∞ (iii) **(h)** (i) -2 (ii) ∞ (iii)

7. (a), (c) and (f) contain 3. **(a)** **(b)** **(c)**

(d) **(e)** **(f)**

8. **(a)** $A = (-\infty, 0)$ **(b)** $B = (-2, 2)$ **(c)** $C = [0, 9\pi]$ **(d)** $D = (25, 49]$
(e) $E = (-\infty, -4)$

9. (a) $[-3, 6)$ (b) $[-3, 5]$ (c) $(-4, 9]$ (d) $(-6, -4) \cup [-2, -3]$ (e) $(-$
(f) $(-\infty, 6]$ (g) $\Re$ (h) $(-\infty, -6] \cup (-5, \infty]$ (i) $(-3, 5]$

10. (a) $[-7, \infty)$ (b) $(-4, 7)$ (c) $[-3, 2)$ (d) $(-2, 0) \cup$

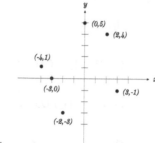

-7   -4   7   -3   2   -2   0

13. (a) True (b) True (c) True (d) True (e) False (f) False (g)
(h) False (i) True (j) False (k) False (l) True

14. (a) ∞ (b) not defined (c) ∞ (d) ∞ (e) not defined
(f) not defined (g) ∞ (h) $-\infty$ (i) not defined

16. (a) f is a function with domain $\Re$ and range $\Re$ because $f(x) = \sqrt[3]{x}$ is un
defined for each real number x.
(b) g is not a function because $g(x) = \sqrt[4]{x}$ is not defined when $x < 0$.
(c) $h(x) = \frac{1}{2}(1 - 3x)$ is a function with domain $(-\infty, 0)$ and range $(\frac{1}{2}, \infty)$.
(d) $j(x) = \sqrt[3]{1 - x^3}$ is a function with domain $(0, \infty)$ and range $(-\infty, 1)$.
(e) k is not a function because $y = k(x) = \pm\sqrt{1 - x^4}$.
(f) m is a function with domain $(-2, 4)$ and range $[1, 273)$.
(g) n is a function because the smallest integer in $(-1, 5]$ is zero.
(h) p is not a function because $p(x)$ is not defined for $x < 0$. Moreover, $p(2)$
assigns two numbers to $x = 2$.

17. $p(t) = 200 + 60t$, domain $[0, 3]$. **18.** $s(t) = -16t^2 - 32t + 240$, domain $[0, 3]$

19. (a) $f : D \to \Re$ by $f(x) = x(20 - x)$ with $D = \{0,\ 1,\ 2,\ 3,\ 4,\ 5,\ 6,\ 7,\ 8,\ 9,$
(b) $g : S \to \Re$ by $g(x) = x(20 - x)$ with $S = \{0,\ 0.1,\ 0.2,\ 0.3,\ \dots, 9.8,\ 9.9, 10$
(c) $h : [0, 10] \to \Re$ by $h(x) = x(20 - x)$.

20. (a) $[1, 8]$ (b) $[7, 11]$ (c) $[0, 25]$ (d) $[3, 5]$ (e) $(0, 1]$ (f) $\{-1,$

21. (a) $(\frac{1}{5}, 1]$ (b) $[\frac{1}{2}, \infty)$ (c) $(0, \frac{1}{3})$ (d) $(0, \infty)$

22. (a) Domain f = Range $f = \Re$. (b) Domain g = Range $g = [0, \infty)$.
(c) Domain $h = [0, 10]$ and Range $h = [0, 40]$.

23. (a) $\Re$ (b) $(-\infty, -5) \cup (-5, \infty)$ (c) $(-\infty, 8]$ (d) $\Re$
(e) $(-\infty, -2) \cup (-2, 3) \cup (3, \infty)$ (f) $(-\infty, -6] \cup [2, \infty)$ (g) $\Re$
(h) $(-\infty, -2] \cup (3, \infty)$ (i) $(-\infty, -3) \cup (-3, -2) \cup (-2, 2) \cup (2, 3) \cup (3, \infty)$
(j) $(-\infty, -2) \cup \{-1\} \cup (4, \infty)$ (k) $(-3, 2] \cup (3, \infty)$ (l) $(-\infty, -7] \cup (-2, 0) \cup ($

24.

(0,5)
(2,4)
(-4,1)
(-3,0)
(3,-1)
(-2,-3)

25.

11
$f(x) = 2x + 3$
5
1   4

(a)

$g(x) = x^2 + 1$
10
1
-3

(b)

5
2
$h(x) = |x|$
-2   5

(c)

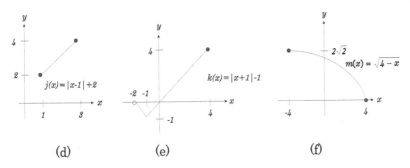

(d) (e) (f)

26. (a) This curve represents a function with domain $[-6, 6]$ and range $[-6, 5]$.
(b) This curve does not represent a function.
(c) This curve represents a function with domain $[-7, 6]$ and range $[-3, 6]$.
(d) This curve represents a function with domain $[-4, 5]$ and range $[-5, 4]$.
(e) This curve does not represent a function.
(f) This curve represents a function with domain $(-\infty, -1) \cup (2, \infty)$ and range $(-\infty, -2) \cup (1, \infty)$.
(g) This curve represents a function with domain $\Re$ and range $(0, 5]$.
(h) This curve represents a function with domain $\Re$ and range $(-5, 6)$.

27. (a) odd (b) even (c) neither (d) odd (e) even (f) odd (g) neither
(h) even (i) odd   28. (a) even (b) neither (c) odd (d) even

29. (a) $f(x) = 3 + (x - 6)^2$ (b) $g(x) = -2 + (x + 9)^3 + 5(x + 9) + 7$
(c) $h(x) = 8 + \frac{x+4}{5(x+4)^2+9}$ (d) $k(x) = -7 + \sqrt{(x-5)^5 - 2(x-5)^2 + 3}$

30.

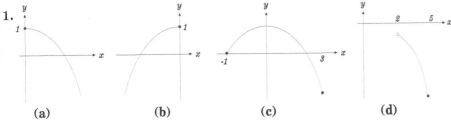

(a) $y = (x+3)^2 - 10$ (b) $y = 4 - (x-2)^2$ (c) $y = 2 + (x-4)^2$ (d) $y = 13 - (x+1)^2$

Section 1.3

1.

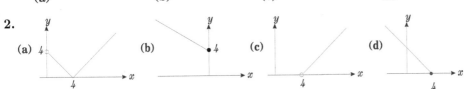

(a) (b) (c) (d)

2.

(a) (b) (c) (d)

3.

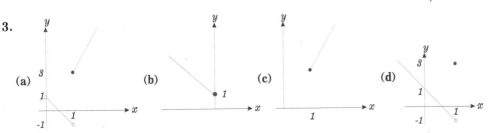

(a) (b) (c) (d)

4.

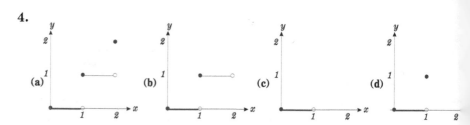

5. **(a)** $[0,6]$ **(b)** $s \mid [0,2]$ **(c)** $s \mid [2,6]$

6. **(a)** $[0,\infty)$ **(b)** $T \mid [0,60]$ **(c)** $T \mid [0,1440]$ **(d)** $T \mid [0,50]$

7. (a) All these functions have domain $[1,\infty)$.
   (i) $(f+g)(x) = x^2 + 5x + 3$ **(ii)** $(fg)(x) = 5x^3 + 3x^2$
   (iii) $(f/g)(x) = \frac{x^2}{5x+3}$ **(iv)** $(3f - 2g)(x) = 3x^2 - 10x - 6$
   (v) $g^3(x) = 125x^3 + 225x^2 + 135x + 27$ **(vi)** $((f+g)/(fg))(x) = \frac{x^2+5x+3}{5x^3+3x^2}$
   (b) All these functions have domain $\Re$.
   (i) $(f+g)(x) = \sqrt{x^2+1} + x^2 + 1$ **(ii)** $(fg)(x) = (x^2+1)^{3/2}$
   (iii) $(f/g)(x) = \frac{1}{\sqrt{x^2+1}}$ **(iv)** $(3f - 2g)(x) = 3\sqrt{x^2+1} - 2x^2 -$
   (v) $g^3(x) = x^6 + 3x^4 + 3x^2 + 1$ **(vi)** $((f+g)/(fg))(x) = \frac{1+\sqrt{x^2+1}}{x^2+1}$
   (c) All these functions have domain $[2,\infty)$.
   (i) $(f+g)(x) = \frac{-2x^3+2x^2+5x}{x^2-1}$ **(ii)** $(fg)(x) = -\frac{6x+4}{x+1}$
   (iii) $(f/g)(x) = \frac{3x+2}{-2x^3+2x^2+2x-2}$ **(iv)** $(3f - 2g)(x) = \frac{4x^3-4x^2+5x+10}{x^2-1}$
   (v) $g^3(x) = -8x^3 + 24x^2 - 24x + 8$ **(vi)** $((f+g)/(fg))(x) = \frac{2x^3-2x^2-5x}{6x^2-2x-4}$

8. **(a)** Domain $g_0 = (-\infty, -5) \cup (-5, 2) \cup (2, \infty)$ and $(f/g_0)(x) = \frac{x^2-x-6}{x^2+3x-10}$.
   (b) Domain $g_0 = (-\infty, -1) \cup (-1, \infty)$ and $(f/g_0)(x) = \frac{x^3-1}{x^3+1}$.
   (c) Domain $g_0 = [2,4) \cup (4, \infty)$ and $(f/g_0)(x) = \sqrt{\frac{x^2-4}{x^2-8x+16}}$.
   (d) Domain $g_0 = (-\infty, 0) \cup (0, \infty)$ and $(f/g_0)(x) = \frac{x}{|x|}$.
   (e) Domain $g_0 = [0, 1)$ and $(f/g_0)(x) = \sqrt{1-x^2}$.

9. **(a)** $F(s) = \begin{cases} 6 & \text{if } 0 \le s \le 3 \\ 2 & \text{if } 3 < s \le 8 \\ 3 & \text{if } 8 < s \le 12 \end{cases}$

   (b) $W(s) = p(s)F(s)$ where $p(s) = \begin{cases} s & \text{if } 0 \le s \le 3 \\ s+6 & \text{if } 3 < s \le 8 \\ s+\frac{4}{3} & \text{if } 8 < s \le 12 \end{cases}$

   (c) $W(s) = p(s)F(s) + q(s)G(s)$ where $G(s) = \begin{cases} 4 & \text{if } 0 \le s \le 6 \\ 6 & \text{if } 6 < s \le 12 \end{cases}$ and

   $q(s) = \begin{cases} s & \text{if } 0 \le s \le 6 \\ s-2 & \text{if } 6 < s \le 12 \end{cases}$.   **10.** $G(x) = \frac{1}{3}F_1(x) + \frac{1}{3}F_2(x) + \frac{1}{3}F_3(x)$ for x

12. **(a)** $f^5(x) = x^{10}$ **(b)** $g^3(x) = 8x^3 - 36x^2 + 54x - 27$
   (c) $h^4(x) = x^4 + 4x^2 + 6 + \frac{4}{x^2} + \frac{1}{x^4}$ **(d)** $k^2(x) = \frac{x^2+4x\sqrt{x}+6x+4\sqrt{x}+1}{x^2-2x-1}$

14. **(a)** Range $g = \Re \subset$ Domain $f = \Re$ and $(f \circ g)(x) = 10 - 15x$.
   (b) Range $g = [-1, \infty) \subset$ Domain $f = \Re$ and $(f \circ g)(x) = x^4 - 2x^2 + 2$.
   (c) Range $g = [-4, \infty) \subset$ Domain $f = \Re$ and $(f \circ g)(x) = \frac{1}{x^4-7x^2+13}$.
   (d) Range $g = (-\infty, 5] \subset$ Domain $f = \Re$ and $(f \circ g)(x) = \sqrt{x^4 - 10x^2 + 26}$.
   (e) Range $g = (0, 1] \subset$ Domain $f = (-3, 2]$ and $(f \circ g)(x) = \sqrt{\frac{2x^2+1}{3x^2+4}}$.
   (f) Range $g = [0, \infty) \subset$ Domain $f = [-1, \infty)$ and $(f \circ g)(x) = \sqrt{1+|x|}$.

15. **(a)** $(f \circ g)(x) = 4x^2 + 2x - 6$ **(b)** Domain $g' = [0, 3/2]$ and $(f \circ g')(x) = \sqrt{3 - }$

(c) $(f \circ g)(x) = \sqrt{x-1}/x$ (d) Domain $g' = [24, \infty)$ and $(f \circ g')(x) = \frac{\sqrt{x+1}+2}{x-15}$.

(e) $(f \circ g)(x) = (x^{3/2} - x^{3/4} - 1)^{2/3}$

16. (a) $B = A \circ f \circ g$ (b) $C = A \circ f \circ f \circ g$ (c) $D = A \circ g \circ g \circ g$

17. (b), (c), (e) are true.
(a) This statement is false. For example, let $f(x) = x^2$ with $g_1(x) = g_2(x) = x$.
(d) This statement is false. For example, let $f(x) = x + 1$ with $g_1(x) = g_2(x) = x$.
(f) This statement is false. For example, let $f(x) = g(x) = x^2$ with $h(x) = 1 + x^2$.

18. (a) $D_0 = \Re$ and $(f \circ f)(x) = x^4 + 2x^2 + 2$
(b) $D_0 = [0, \infty)$ and $(f \circ f)(x) = x^{1/4}$
(c) $D_0 = (-\infty, 0) \cup (0, \infty)$ and $(f \circ f)(x) = x$
(d) $D_0 = (-\infty, -1) \cup (-1, 0) \cup (0, \infty)$ and $(f \circ f)(x) = -\frac{1}{x}$.

19. (a) and (b) are true. (c) and (d) are false. For example, these two identities fail when $f(x) = x + 1$ and $g(x) = x^2$. **20.** (a) $f^{(3)}$ (b) $f^{(12)}$

21. (a), (b), (d) are one–to–one while (c), (e), (f) are not one–to–one.

22. (a) Domain $H = [0, b]$ where the plane is rising during this time interval.
$H^{-1}(y)$ is the time when the plane has altitude y.
(b) Domain $L = [0, b]$ where the ice cube is melting but still exists during this time interval. $L^{-1}(y)$ is the time when the ice cube has side of length y.
(c) Domain $V = [0, b]$ where grain is continuously being added to the silo during this time interval. $V^{-1}(y)$ is the time when the silo contains y cubic meters of grain.
(d) Domain $S = [0, b]$ where the car is accelerating during this time interval. $S^{-1}(y)$ is the time when the car has speed y.
(e) Domain $A = [0, \infty)$. $A^{-1}(y)$ is the radius of a circle with area y.
(f) Domain $W = [0, \infty)$. $W^{-1}(y)$ is the side of a cube with volume y.
(g) Domain P is the set of natural numbers n with $a \leq n \leq b$ where the flu is spreading on days a through b. $P^{-1}(y)$ is the number of people with the flu on day y.
(h) Domain Q is the set of positive integers. $Q^{-1}(y)$ is the number of tosses for which the probability of all tosses being heads equals y.

23, 24. (a) Domain $f^{-1} = \Re$, Range $f^{-1} = \Re$ and $f^{-1}(y) = \frac{y+3}{5}$.
(b) Domain $g^{-1} = (-\infty, 4]$, Range $g^{-1} = [0, \infty)$ and $g^{-1}(y) = +\sqrt{4-y}$.
(c) Dom $h^{-1} = (-\infty, 5) \cup (5, \infty)$, Range $h^{-1} = (-\infty, -4) \cup (-4, \infty)$, $h^{-1}(y) = \frac{3-4y}{y-5}$.

(d) Domain $k^{-1} = [1, \infty)$, Range $k^{-1} = [0, \infty)$ and $k^{-1}(y) = \sqrt{\sqrt{y^2 - \frac{3}{4}} - \frac{1}{2}}$.

(e) Domain $m^{-1} = [0, \infty)$, Range $m^{-1} = [-1, \infty)$ and $m^{-1}(y) = \sqrt[3]{y^2 - 1}$.
(f) Domain $n^{-1} = [4, \infty)$, Range $n^{-1} = [-1, \infty)$ and $n^{-1}(y) = \frac{y-7}{3}$.

25. (a) (i) $f(x) = 7x + 2$ equals $f(x') = 7x' + 2$ implies $7x = 7x'$ and $x = x'$.
Therefore, f is one–to–one. (ii) Domain $f^{-1} = \Re$ (iii) $f^{-1}(x) = (x - 2)/7$
(b) (i) $g(x) = x^3 + 1$ equals $g(x') = (x')^3 + 1$ implies $x^3 = (x')^3$ and $x = x'$.
Therefore, g is one–to–one.
(ii) Domain $g^{-1} = \Re$. (iii) $g^{-1}(x) = \sqrt[3]{x - 1}$
(c) (i) $h(x) = \sqrt[3]{x^2 + 2x + 1}$ equals $h(x') = \sqrt[3]{(x')^2 + 2x' + 1}$ means $(x + 1)^{2/3} = (x' + 1)^{2/3}$. Since x, $x' \geq 0$, it follows that $x + 1 = x' + 1$, $x = x'$ and h is one–to–one.
(ii) Domain $h^{-1} = [1, \infty)$ (iii) $h^{-1}(x) = x^{3/2} - 1$
(d) (i) $j(x) = \frac{1}{x^2 + 1}$ equals $j(x') = \frac{1}{(x')^2 + 1}$ implies $x^2 + 1 = (x')^2 + 1$. Since x, $x' \geq 0$,
$x = x'$ and j is one–to–one. (ii) Domain $j^{-1} = (0, 1]$. (iii) $j^{-1}(x) = \sqrt{\frac{1-x}{x}}$
(e) (i) $k(x) = \frac{2x+5}{3x-1}$ equals $k(x') = \frac{2x'+5}{3x'-1}$ implies $(2x+5)(3x'-1) = (2x'+5)(3x-1)$,
$6xx' - 2x + 15x' - 5 = 6x'x - 2x' + 15x - 5$, $17x' = 17x$ and $x = x'$ and k is one–to–one.
(ii) Domain k^{-1} equals all numbers other than $2/3$. (iii) $k^{-1}(x) = \frac{x+5}{3x-2}$.

26.

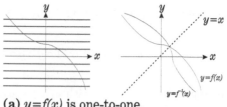

(a) $y=f(x)$ is one-to-one.

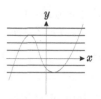

(b) $y=g(x)$ is not one-to-one.

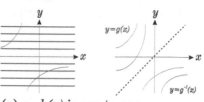

(c) $y=h(x)$ is one-to-one.

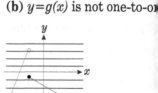

(d) $y=k(x)$ is not one-to-one.

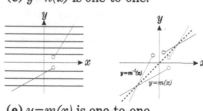

(e) $y=m(x)$ is one-to-one.

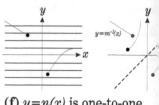

(f) $y=n(x)$ is one-to-one.

27. **(a), (c)** These two restrictions of f are not one-to-one.
(b), (d) Let g denote the restriction of f to $[-5, -3]$ in (b), and let h denote restriction of g to $[-3, 2]$ in (d).

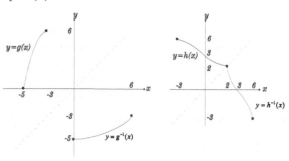

30. **(a)** **(i)** $f(5) = 5 = f(-5)$. **(ii)** The domain of the restriction is $[0, \infty)$. **(iii)** The domain of the inverse function is $[0, \infty)$. **(iv)** $f^{-1}(x) = x$
(b) **(i)** $g(1) = -4 = g(-1)$ **(ii)** The domain of the restriction is $[0, \infty)$.
(iii) The domain of the inverse function is $[-5, \infty)$. **(iv)** $g^{-1}(x) = \sqrt{x + 5}$
(c) **(i)** $h(0) = 3 = h(-2)$ **(ii)** The domain of the restriction is $[-1, \infty)$.
(iii) The domain of the inverse function is $[2, \infty)$. **(iv)** $h^{-1}(x) = \sqrt{x - 2} - 1$
(d) **(i)** $j(-1/2) = 2\sqrt{2} = j(1/2)$ **(ii)** The domain of the restriction is $[0, 2/$
(iii) The domain of the inverse function is $[0, 3]$. **(iv)** $j^{-1}(x) = \sqrt{9 - x^2}/2$
(e) **(i)** $k(2) = 4 = k(-14)$ **(ii)** The domain of the restriction is $[-6, \infty)$.
(iii) The domain of the inverse function is $[0, \infty)$. **(iv)** $k^{-1}(x) = x^{3/2} - 6$.

Section 1.4

1. (a) $\alpha = -4\pi$ rad. $= -720°$ **(b)** $\beta = \frac{\pi}{2}$ rad. $= 90°$ **(c)** $\gamma = -\frac{6\pi}{5}$ rad. $= -216°$
(d) $\delta = 10\pi$ rad. $= 1800°$ **(e)** $\epsilon = -2\pi$ rad. $= -360°$

2. **(a)** π radians **(b)** $\pi/6$ radians **(c)** $-\pi/10$ radians **(d)** 2π rad
(e) $\pi/18$ radians **(f)** $-4\pi/3$ radians **(g)** $2\pi/9$ radians **(h)** $5\pi/4$ r

3. (a) $45°$ (b) $540°$ (c) $-120°$ (d) $180°$
(e) $540/\pi°$ (f) $150°$ (g) $18°$ (h) $-30°$

4. (a) $8\pi/3$ (b) π (c) $55\pi/3$ (d) 7π

5. (a) $\Re$ (b) $\Re$ with all multiples of $\frac{\pi}{2}$ deleted (c) $\Re$ with all multiples of $\frac{\pi}{2}$ deleted (d) the union of $(2N\pi, (2N+1)\pi)$ for all integers N

6. (a) $\sin \pi/3 = \sqrt{3}/2$ $\cos \pi/3 = 1/2$ $\tan \pi/3 = \sqrt{3}$
$\cot \pi/3 = 1/\sqrt{3}$ $\sec \pi/3 = 2$ $\csc \pi/3 = 2/\sqrt{3}$
(b) $\sin \pi = 0$ $\cos \pi - -1$ $\tan \pi = 0$
$\cot \pi$ is not defined $\sec \pi = -1$ $\csc \pi$ is not defined
(c) $\sin 5\pi/4 = -1/\sqrt{2}$ $\cos 5\pi/4 = -1/\sqrt{2}$ $\tan 5\pi/4 = 1$
$\cot 5\pi/4 = 1$ $\sec 5\pi/4 = -\sqrt{2}$ $\csc 5\pi/4 = -\sqrt{2}$
(d) $\sin -\pi/6 = -1/2$ $\cos -\pi/6 = \sqrt{3}/2$ $\tan -\pi/6 = -1/\sqrt{3}$
$\cot -\pi/6 = -\sqrt{3}$ $\sec -\pi/6 = 2/\sqrt{3}$ $\csc -\pi/6 = -2$
(e) $\sin 2\pi/3 = \sqrt{3}/2$ $\cos 2\pi/3 = -1/2$ $\tan 2\pi/3 = -\sqrt{3}$
$\cot 2\pi/3 = -1/\sqrt{3}$ $\sec 2\pi/3 = -2$ $\csc 2\pi/3 = 2/\sqrt{3}$
(f) $\sin 15\pi/2 = -1$ $\cos 15\pi/2 = 0$ $\tan 15\pi/2$ is not defined
$\cot 15\pi/2 = 0$ $\sec 15\pi/2$ is not defined $\csc 15\pi/2 = -1$
(g) $\sin 17\pi/6 = -1/2$ $\cos -17\pi/6 = -\sqrt{3}/2$ $\tan -17\pi/6 = 1/\sqrt{3}$
$\cot -17\pi/6 = \sqrt{3}$ $\sec -17\pi/6 = -2/\sqrt{3}$ $\csc -17\pi/6 = -2$
(h) $\sin 21\pi/4 = -1/\sqrt{2}$ $\cos 21\pi/4 = -1/\sqrt{2}$ $\tan 21\pi/4 = 1$
$\cot 21\pi/4 = 1$ $\sec 21\pi/4 = -\sqrt{2}$ $\csc 21\pi/4 = -\sqrt{2}$

7. (a) $\sin \theta = 5/13$ $\cos \theta = 12/13$ $\tan \theta = 5/12$ **8.** $\theta \approx 37°$
$\cot \theta = 12/5$ $\sec \theta = 13/12$ $\csc \theta = 13/5$
(b) $\sin \phi = 1/\sqrt{5}$ $\cos \phi = 2/\sqrt{5}$ $\tan \phi = 1/2$
$\cot \phi = 2$ $\sec \phi = \sqrt{5}/2$ $\csc \phi = \sqrt{5}$
(c) $\sin \psi = \sqrt{55}/8$ $\cos \psi = 3/8$ $\tan \psi = \sqrt{55}/3$
$\cot \psi = 3/\sqrt{55}$ $\sec \psi = 8/3$ $\csc \psi = 8/\sqrt{55}$

9. approximately 284 meters

10. (a) approximately 1.76 miles ≈ 9310 feet   (b) approximately 10.15 miles

11. (a) approximately 173 meters   (b) 200 meters

12. (a) Let D_1 be the set of all $x \geq 0$ which are not of the form $\frac{\pi}{2} + N\pi$ with N a natural number. $(f + g)(x) = \tan x + \sqrt{x}$ with domain D_1.
$(fg)(x) = \sqrt{x}\tan x$ with domain D_1. $(f/g)(x) = (\tan x)/\sqrt{x}$ domain $x \in D_1$, $x \neq 0$.
$(f \circ g)(x) = \tan \sqrt{x}$ with domain D_1.
$(g \circ f)(x) = \sqrt{\tan x}$ with domain all $x + N\pi$ with $x \in [0, \pi/2)$ and N an integer.
(b) Let D_2 be the set of numbers not of the form $N\pi + \pi/2$ with N an integer.
$(f + g)(x) = x^2 - 1 + \sec x$ with domain D_2. $(fg)(x) = (x^2 - 1)\sec x$ with domain D_2.
$(f/g)(x) = (x^2 - 1)\cos x$ with domain D_2. $(f \circ g)(x) = \tan^2 x$ with domain D_2.
$(g \circ f)(x) = \sec(x^2 - 1)$ with domain all numbers not of the form $\pm\sqrt{1 + N\pi + \pi/2}$ for N an integer.
(c) Let D_3 denote the set of all nonzero real numbers.
$(f + g)(x) = \cos x + 1/x$ with domain D_3. $(fg)(x) = (\cos x)/x$ with domain D_3.
$(f/g)(x) = x \cos x$ with domain D_3. $(f \circ g)(x) = \cos(1/x)$ with domain D_3.
$(g \circ f)(x) = \sec x$ with domain D_2.
(d) Let D_4 be the set of all numbers not of the form $N\pi$ with N an integer.
$(f + g)(x) = \frac{x}{x^2+1} + \cot x$ with domain D_4. $(fg)(x) = \frac{x \cot x}{x^2+1}$ with domain D_4.
$(f/g)(x) = \frac{x \tan x}{x^2+1}$ with domain all numbers not of the form $N\pi/2$ with N an integer.
$(f \circ g)(x) = \cos x \sin x$ with domain D_4. $(g \circ f)(x) = \cot(\frac{x}{x^2+1})$ with domain all $x \neq 0$.

13. (a) odd **(b)** neither **(c)** even **(d)** odd

16. (a) $\tan(\theta+\phi) = (\tan\theta+\tan\phi)/(1-\tan\theta\tan\phi)$. **(b)** $\tan 2\theta = 2\tan\theta/(1-t$

17. (a) $\sin 3\theta = 3\sin\theta - 4\sin^3\theta$. **(b)** $\cos 3\theta = 4\cos^3\theta - 3\cos\theta$.

18. **(a)** $\sin^2\theta = (1-\cos 2\theta)/2$ **(b)** $\cos^2\theta = (1+\cos 2\theta)/2$
  (c) $\sin^4\theta = 3/8 - (\cos 2\theta)/2 + (\cos 4\theta)/8$ **(d)** $\cos^4\theta = 3/8 + (\cos 2\theta)/2$

19. (a) $\alpha \approx .667$, $\beta = \frac{2\pi}{3}$, $\gamma \approx .380$ radians **(b)** $a = 2\sqrt{13+6\sqrt{3}}$, $\phi \approx$
$\psi \approx .208$ **(c)** $\theta = \frac{5\pi}{12}$, $b \approx 5.856$, $c \approx 7.173$ **20.** all three are $\frac{1}{\sqrt{2}}$

21.

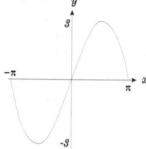

(a) $y = 3\sin x$

(d) $y = 5\sin x$

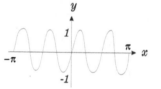

(b) $y = \sin 4x$

(e) $y = \sin 2x$

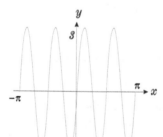

(c) $y = 3\sin 4x$

(f) $y = 5\sin 2x$

22.

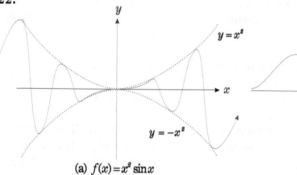

(a) $f(x) = x^2\sin x$

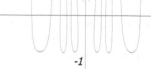

(b) $g(x) = \sin\dfrac{1}{x^2}$

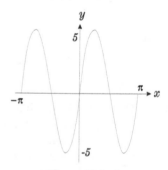

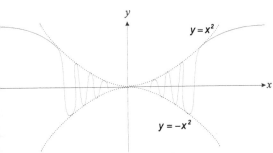

(c) $h(x) = x^2 \sin \dfrac{1}{x^2}$

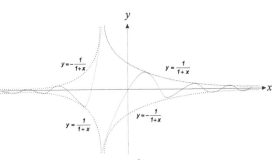

(d) $k(x) = \dfrac{\sin x}{1 + x}$

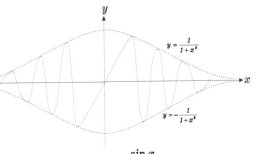

(e) $m(x) = \dfrac{\sin x}{1 + x^2}$

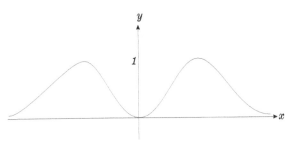

(f) $n(x) = \sin \dfrac{\pi}{1 + x^2}$

23. (a)

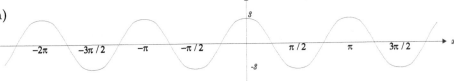

(b)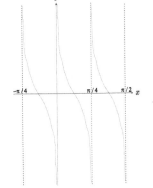

(c)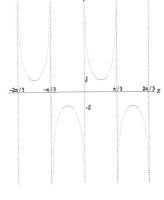

(d)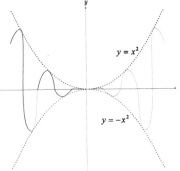

(e)

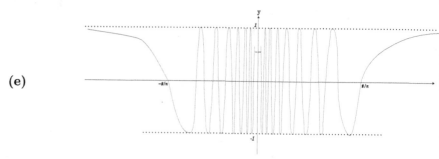

(f)

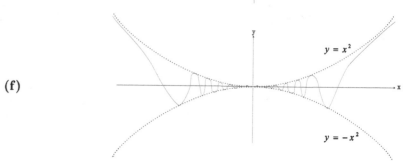

24. (a) $\frac{\pi}{3}$ (b) 0 (c) $\frac{3\pi}{4}$ (d) π (e) $\frac{\pi}{6}$ (f) $\frac{\pi}{2}$

25. (a) $\frac{\pi}{4}$ (b) $-\frac{\pi}{6}$ (c) $\frac{\pi}{2}$ (d) $-\frac{\pi}{3}$ (e) $-\frac{\pi}{4}$ (f) 0

26. (a) $\frac{\pi}{6}$ (b) $\frac{\pi}{4}$ (c) $-\frac{\pi}{3}$ (d) $-\frac{\pi}{6}$ (e) 0

27. (a) $\frac{\pi}{3}$ (b) 0 (c) $\frac{\pi}{4}$ (d) $\frac{2\pi}{3}$ (e) $\frac{3\pi}{4}$ (f) $-\frac{\pi}{3}$

28. (a) 1 (b) $1/10$ (c) $\sqrt{50}$ (d) $-12/5$ (e) $1/3$ (f) $3\pi/4$ (g) $\pi/10$
(h) $\pi/7$ (i) 0

29. (a) $f(x) = \frac{\sqrt{1+x^2}}{|x|}$ (b) $g(x) = \frac{x}{\sqrt{2x+1}}$ (c) $h(x) = \sqrt{\frac{x}{x+1}}$
(d) $k(x) = 2\sqrt{x - x^2}$. (e) $m(x) = |x|\sqrt{x^2 + 2}$. (f) $n(x) = \sqrt{x - 2\sqrt{x} + 2}$

30. (a) $\sin\theta = \frac{2}{x}$ (b) $\theta = \arcsin\frac{2}{x}$ (c) $L = x\cos\left(\arcsin\frac{2}{x}\right) = \sqrt{x^2 - 4}$

31. (a) $\sec\phi = \frac{x}{10}$ (b) $\phi = \operatorname{arcsec}\frac{x}{10}$ (c) $h = 10\tan\left(\operatorname{arcsec}\frac{x}{10}\right) = \sqrt{x^2 - }$

32. (a) $\cos\psi = \frac{x}{15}$ (b) $\psi = \arccos\frac{x}{15}$ (c) $y = 15\sin\left(\arccos\frac{x}{15}\right) = \sqrt{225}$

Section 1.5

1. (a) $f(1.9) = 2.9$, $f(1.99) = 2.99$, $f(2.1) = 3.1$, $f(2.01) = 3.01$, $\lim_{x\to 2} f(x) =$
(b) $f(10) = .12702$, $f(10.9) = .12520$, $f(10.99) = .12502$, $f(12) = .12311$,
$f(11.1) = .12481$, $f(11.01) = .12498$, $\lim_{x\to 11} f(x) = \frac{1}{8}$
(c) $f(1) = .84147$, $f(0.1) = .99833$, $f(-0.5) = .95885$, $f(-0.02) = .9$
$\lim_{x\to 0} f(x) = 1$
(d) $f(0.1) = .04996$, $f(0.01) = .00500$, $f(-0.3) = -.14888$, $f(-0.02) = -.0$
$\lim_{x\to 0} f(x) = 0$
(e) $f(0.1) = .71773$, $f(0.01) = 1.70000$, $f(0.001) = .69339$, $f(-0.1) = .66$
$f(-0.01) = .69075$, $f(-0.001) = .69291$, $\lim_{x\to 0} f(x) \approx 0.693$

2. (a)(i) $\lim_{x\to c} f(x) = 3$. (ii) $\lim_{x\to c} f(x)$ does not exist. (iii) $\lim_{x\to\infty} f(x) = 5$. (iv) $\lim_{x\to c} f($
(b)(i) $\lim_{x\to c^+} f(x) = 3$. (ii) $\lim_{x\to c^+} f(x) = 1$. (iii) $\lim_{x\to c^+} f(x) = 2$. (iv) $\lim_{x\to c^+} f(x) = 2$

c)(i) $\lim\limits_{x\to c^-} f(x) = 3$. **(ii)** $\lim\limits_{x\to c^-} f(x) = 3$. **(iii)** $\lim\limits_{x\to c^-} f(x) = 3$. **(iv)** $\lim\limits_{x\to c^-} f(x) = +\infty$.

3. (a) 15 (b) $\frac{2}{3}$ (c) -1 (d) 0 (e) 2 (f) DNE

4. (a) $\lim\limits_{x\to 3^-} f(x) = -2$, $\lim\limits_{x\to 3^+} f(x) = 10$, $\lim\limits_{x\to 3} f(x)$ DNE.

(b) $\lim\limits_{x\to -3^-} f(x) = \lim\limits_{x\to -3^+} f(x) = \lim\limits_{x\to -3} f(x) = 9$.

(c) $\lim\limits_{x\to 0^-} f(x) = 3$, $\lim\limits_{x\to 0^+} f(x) = 1$, $\lim\limits_{x\to 0} f(x)$ DNE.

(d) $\lim\limits_{x\to 2^-} f(x) = \lim\limits_{x\to 2^+} f(x) = \lim\limits_{x\to 2} f(x) = 3$.

(e) $\lim\limits_{x\to -1^-} f(x) = -1$, $\lim\limits_{x\to -1^+} f(x) = 1$, $\lim\limits_{x\to -1} f(x)$ DNE.

(f) $\lim\limits_{x\to 4^-} f(x) = 2$, $\lim\limits_{x\to 4^+} f(x) = -2$, $\lim\limits_{x\to 4} f(x)$ DNE.

5. (a) never (b) only at $c = \frac{3}{2}$ (c) only at $c = 0$ (d) only at $\frac{\pi}{4} + n\pi$ for $n \in Z$

6. (a) none exist (b) all exist and equal 0 (c) none exist (d) all exist and equal 0

(e) $\lim\limits_{x\to 0^-} \arctan\frac{1}{x} = -\frac{\pi}{2}$, $\lim\limits_{x\to 0^+} \arctan\frac{1}{x} = \frac{\pi}{2}$, $\lim\limits_{x\to 0} \arctan\frac{1}{x}$ DNE.

7. (a) ○——○——○ (b) ○——○——○ (c) ○——○ (d) ○——○
      3   5   7   -7   -6   -5   5   7   -7   -4

(e) ●——○——○ (f) ○——○——● (g) ——————○ (h) ○———→ (i) ○——○——○
      0   2   6   -3   -1   1   -7   6   1   2   5

8.

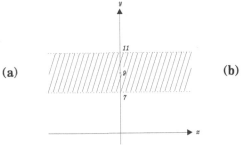

(a)

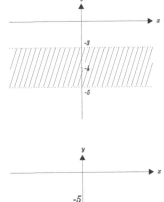

(b)

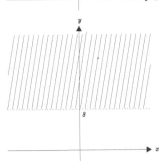

(c)

(d)

9. (a) 0.4 (b) 0.2 (c) 0.06 (d) $\frac{1}{2}$ (e) 1.16 (f) 0.09

11. (a) neither (b) large (c) neither (d) small (e) neither (f) small

12. (a) $\lim\limits_{x\to -\infty} \dfrac{1}{x^2 + 1} = \lim\limits_{x\to \infty} \dfrac{1}{x^2 + 1} = 0$ (b) $\lim\limits_{x\to -\infty} \dfrac{1}{\sqrt[3]{x}} = \lim\limits_{x\to \infty} \dfrac{1}{\sqrt[3]{x}} = 0$

(c) $\lim\limits_{x\to -\infty} \dfrac{|x|}{x + 1} = -1$, $\lim\limits_{x\to \infty} \dfrac{|x|}{x + 1} = 1$ (d) both DNE

(e) both DNE   (f) $\lim\limits_{x\to -\infty} \sin\frac{1}{x} = \lim\limits_{x\to \infty} \sin\frac{1}{x} = 0$

(g) $\lim\limits_{x\to -\infty} \cos\frac{1}{x} = \lim\limits_{x\to \infty} \cos\frac{1}{x} = 1$ (h) $\lim\limits_{x\to -\infty} \text{arccot } x = 0$, $\lim\limits_{x\to \infty} \text{arccot } x = \pi$

(i) $\lim\limits_{x\to -\infty} \text{arcsec } x = \lim\limits_{x\to \infty} \text{arcsec } x = \frac{\pi}{2}$ (j) $\lim\limits_{x\to -\infty} \text{arccsc } x = \lim\limits_{x\to \infty} \text{arccsc } x = 0$

(k) $\lim\limits_{x\to -\infty} \dfrac{\sin x}{x} = \lim\limits_{x\to \infty} \dfrac{\sin x}{x} = 0$ (l) both DNE

13. (a) 15 (b) 6 (c) 25,600 (d) 20 **15.** (a) $\lim_{x \to 0^-} \dfrac{1}{x^4} = \lim_{x \to 0^+} \dfrac{1}{x^4} = \lim_{x \to 0} \dfrac{1}{x^4}$

(b) $\lim_{x \to -1^-} \dfrac{1}{x^2 - 1} = \infty$, $\quad \lim_{x \to -1^+} \dfrac{1}{x^2 - 1} = -\infty$, $\quad \lim_{x \to -1} \dfrac{1}{x^2 - 1}$ DNE.

(c) $\lim_{x \to 1^-} \dfrac{x + 5}{x^3 - x} = -\infty$, $\quad \lim_{x \to 1^+} \dfrac{x + 5}{x^3 - x} = \infty$, $\quad \lim_{x \to 1} \dfrac{x + 5}{x^3 - x}$ DNE.

(d) $\lim_{x \to \pi^-} \cot x = -\infty$, $\quad \lim_{x \to \pi^+} \cot x = \infty$, $\quad \lim_{x \to \pi} \cot x$ DNE.

(e) $\lim_{x \to \frac{\pi}{2}^-} \sec x = \infty$, $\quad \lim_{x \to \frac{\pi}{2}^+} \sec x = -\infty$, $\quad \lim_{x \to \frac{\pi}{2}} \sec x$ DNE.

(f) $\lim_{x \to 0^-} \csc x = -\infty$, $\quad \lim_{x \to 0^+} \csc x = \infty$, $\quad \lim_{x \to 0} \csc x$ DNE.

17.

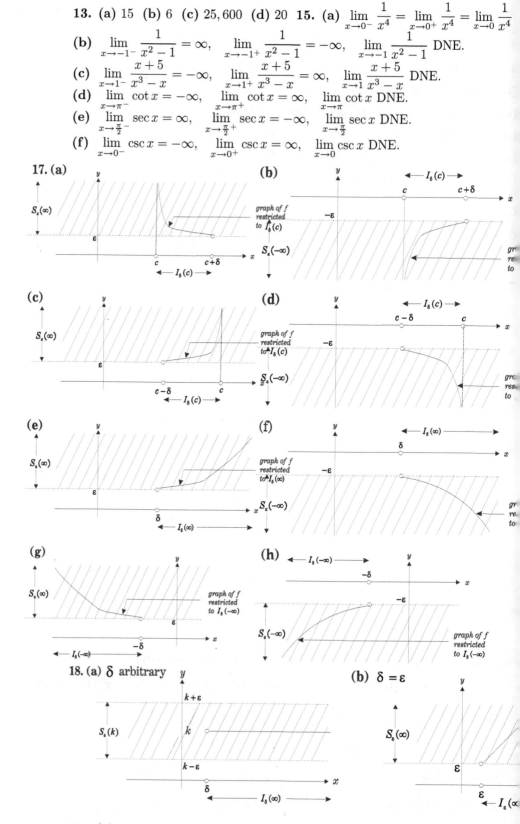

19. (a) For every $\epsilon > 0$, there is a corresponding $\delta > 0$ such that if $2 - \delta < x <$ and $x \neq 2$, then $|x^2 + x - 6| < \epsilon$.

b) For every $\epsilon > 0$, there is a corresponding $\delta > 0$ such that if $-3 - \delta < x < -3 + \delta$ and $x \neq -3$, then $\frac{1}{x^2 + 6x + 9} > \epsilon$.

c) For every $\epsilon > 0$, there is a corresponding $\delta > 0$ such that if $x < -\delta$, then $-\epsilon < \frac{x}{x^2 + 5} < \epsilon$.

d) For every $\epsilon > 0$, there is a corresponding $\delta > 0$ such that if $1 < x < 1 + \delta$, then $\frac{1}{1 - x^2} < -\epsilon$.

e) For every $\epsilon > 0$, there is a corresponding $\delta > 0$ such that if $4 - \delta < x < 4$, then $-1 - \epsilon < \frac{|x-4|}{x-4} < -1 + \epsilon$.

f) For every $\epsilon > 0$, there is a corresponding $\delta > 0$ such that if $x > \delta$, then $\frac{x^2}{1-x} < -\epsilon$.

Section 1.6

1. **(a)** Use Property 1. **(b)** Use Property 1. **(c)** Use Property 1. **(d)** Use Property 2. **(e)** Use Property 2. **(f)** Use Property 1. **(g)** Use Property 2. **(h)** Use Property 2.

2. **(a)** 0 **(b)** 0 **(c)** 0 **(d)** DNE **(e)** DNE **(f)** 0 **(g)** 0 **(h)** DNE

3. **(a)** 4 **(b)** 0 **(c)** 1 **(d)** 0 **(e)** 0 **(f)** 1 **(g)** 8 **(h)** 0 **(i)** 1

4. **(a)** 0 **(b)** 0 **(c)** 0 **(d)** 0 **(e)** 0 **(f)** 1 **(g)** 0

5. -64 using Property 4

6. 28 using Properties 3, 4

7. 3125 using Properties 3, 4

8. 12 using Properties 3, 4

9. 4 using Properties 3, 4, 5

10. $\frac{1}{3}$ using Properties 3, 5

11. $\frac{1}{3}$ using Properties 3, 4, 5

12. $\frac{1}{3}$ using Properties 3, 5

13. 0 using Properties 3, 4, 5

14. ∞ using Properties 3, 4, 5

15. 0 using Properties 3, 4, 5

16. $\frac{3}{4}$ using Properties 3, 4, 5

17. 0 using Properties 3, 4, 5

18. 0 using Property 3

19. $\frac{1}{2}$ using Properties 3, 4, 5

20. $-\infty$ using Properties 3, 4, 5

21. 0 using Properties 3, 4, 5

22. 3 using Properties 3, 4, 5

23. 2 using Properties 3, 4, 6

24. ∞ using Property 5

25. $\sqrt{\frac{5}{2}}$ using Properties 3, 5, 6

26. 2 using Property 3

27. $-\frac{1}{10}$ using Properties 3, 4, 5, 6

28. $\frac{80}{3}$ using Properties 3, 4, 5, 6

29. $-\frac{32}{3}$ using Properties 3, 4, 5, 6

30. $\frac{4}{3}$ using Properties 3, 4, 5, 6

31. 8 using Properties 3, 4, 5, 6

32. $-\frac{1}{12}$ using Properties 3, 4, 5, 6

33. $-\frac{21}{40}$ using Properties 3, 4, 5, 6

34. 0 using Properties 3, 4, 5, 6

35. $\frac{3}{2}$ using Properties 3, 4, 5, 6

36. $-\frac{3}{2}$ using Properties 3, 4, 5, 6

37. $\frac{1}{4}$ using Properties 3, 4, 5, 6

38. 0 using Properties 3, 4, 5, 6

39. 2 using Properties 3, 4, 6, 8

40. $\frac{14}{55}$ using Properties 3, 4, 5, 8

41. $\frac{9}{4}$ using Properties 3, 4, 5, 6, 8

42. 3 using Properties 3, 4

43. 0 using Properties 3, 4, 5, 6

44. 0 using Properties 4, 7

45. 0 using Property 7

46. 0 using Properties 4, 7

47. 1 using Properties 3, 4, 5, 6

48. 0 using Property 7

49. 0 using Properties 5

50. $\frac{1}{5}$ using Property 4

51. 0 using Properties 3, 4, 6

52. $\frac{4}{7}$ using Properties 4, 5

53. 1 using Property 4

54. 1 using Property 5

55. -2 using Properties 3, 4, 5

56. $-\sqrt{2}$ using Properties 3, 4

57. $\frac{1}{2}$ using Properties 4, 5

58. 1 using Property 4

59. 0 using Properties 4, 5

60. 2 using Properties 4, 5

61. $\frac{1}{3}$ using Property 4

62. -1 using Properties 3, 4, 5

63. 0 using Properties 4, 5

64. DNE using Property 2

65. $\frac{1}{2}$ using Properties 3, 4, 5

66. 1 using Property 5

67. 0 using Property 4

68. DNE using Property 2, 3, 4

69. 1 using Property 5   **70.** $\frac{1}{2}$ using Properties 3, 4, 5
71. 0 using Properties 3, 5   **72.** DNE using Property 2

Section 1.7

1. **(a)** f is continuous for $x \neq -3$, 1, has a removable discontinuity at $x = -$
an essential discontinuity at $x = 1$.
(b) g is continuous for $x \neq -2$, 2, 4, has essential discontinuities at $x = -2$,
and a removable discontinuity at $x = 4$.
(c) h is cont. at all $x \neq 3$ in its domain and has an essential discontinuity at
(d) k is continuous for $x \neq 5$ and has an essential discontinuity at $x = 5$. k
defined at $x = -6$, -3, 2, i.e. these points are not in the domain of k.
(e) m is continuous for $x \neq -2$, 3, has an essential discontinuity at $x = 3$
removable discontinuity at $x = -2$.
(f) n is only continuous at $x = -4$, has a removable discontinuity at $x = $
essential discontinuities at all $x \neq -4$, 5.

2. a is continuous on $\Re$.   **3.** b is continuous on $\Re$.

4. c is continuous for $x \neq \pm 1$ and has essential discontinuities at $x = \pm 1$.

5. d is continuous for $x \neq 2$, 3, has an essential discontinuity at $x = 3$
removable discontinuity at $x = 2$.

6. e is continuous for $x \neq -1$ and has a removable discontinuity at $x = -1$.

7. f is continuous for $x \neq \pm 2$nd has removable discontinuities at $x = \pm 2$.

8. g is cont. for $x \neq 3$, has an essential discont. at $x = -5$ and a removable di
at $x = 3$.   **9.** h is continuous on $[0, \infty)$.

10. i is continuous for $x \neq 0$ and has an essential discontinuity at $x = 0$.

11. j is continuous for $x \neq -3$ and has an essential discontinuity at $x = -3$.

12. k is continuous for $x \neq -4$, 5 and has essential discontinuities at $x = -4$,

13. m is continuous for $x \neq -4$ and has a removable discontinuity at $x = -4$.

14. n is continuous on $[0, \infty)$.   **15.** p is continuous on $(0, \infty)$.

16. q is continuous for $x \neq 0$ and has a removable discontinuity at $x = 0$.

17. r is continuous for $x \neq 0$ and has a removable discontinuity at $x = 0$.

18. s is continuous for $x \neq 0$ and has an essential discontinuity at $x = 0$.

19. t is continuous at $x = 0$ and has an essential discontinuity at all $x \neq 0$.

20. u is continuous for $x = -4$, 7. u has an essential discontinuity for $x \neq -4$, 7

21. **(a)** Change $f(1)$ to 1.   **(b)** Change $g(1)$ to 0.   **(c)** Change $h(1)$ to $\frac{\pi}{2}$
(d) Change $k(1)$ to 1.   **(e)** Change $m(1)$ to 1.

24. Use properties 1, 2, 4.   **25.** Use properties 1, 2.   **26.** Use property 2.

27. Use properties 1, 2, 4.   **28.** Use properties 1, 3.   **29.** Use property 4.

30. Use properties 1,2,3,4. **31.** Use properties 1,2,3,4,5. **32.** Use properties 2,4,

33. Use properties 1, 2, 3, 4.   **37.** **(a)** $[5, 11]$ **(b)** $[-4, 4]$ **(c)** $[-1, 4]$ **(d)** $[-$

39. **(a)** Since $1 \notin$ Domain f, the domain of f does not contain an interval w
contains 0 and 2.   **(b)** g is not continuous at $x = 4$.

c) Since $\frac{\pi}{2} \notin$ Domain h, the domain of h does not contain an interval which contains 0 and π. **(d)** k is not continuous. **(e)** m is not continuous at $x = 0$. **(f)** Since $0 \notin$ Domain n, the domain of n soes not contain an interval which contains -1 and 1.

40. **(a)** 1.7 **(b)** 2.8 **(c)** 3.3 **(d)** 1.3 **(e)** 1.7 **(f)** 2.1

41. **(a)** .7 **(b)** .6 **(c)** .8 **(d)** .6 **(e)** .3

42.

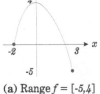

(a) Range $f = [-5,4]$

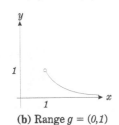

(b) Range $g = (0,1)$

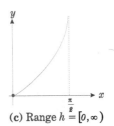

(c) Range $h = [0, \infty)$

(d) Range $k = [0, \frac{1}{2}]$

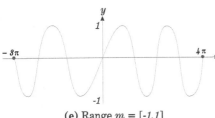

(e) Range $m = [-1,1]$

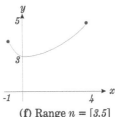

(f) Range $n = [3,5]$

43. $n \geq 2$ **44.** The only cont. functions with a finite range are constant functions.

46. **(a)** $m = 1$, $f(1) = 4$ and $M = -3$, $f(-3) = 20$
(b) $m = -5$, $g(-5) = 0$ and $M_1 = -14$, $M_2 = 4$, $g(-14) = g(4) = 9$
(c) $m = \frac{\pi}{3}$, $h\left(\frac{\pi}{3}\right) = -\frac{3}{2}$ and $M = 0$, $h(0) = 3$ **(d)** $m = -4$, $k(-4) = 3$ and $M = 0$, $k(0) = 5$
(e) $m_1 = -1$, $m_2 = 5$, $p(-1) = p(5) = -2$; $M = 2$, $p(2) = 7$
(f) $m = 6$, $q(6) = \frac{1}{37}$; $M = 0$, $q(0) = 1$

47. **(a)** Domain f is not a closed interval. **(b)** Domain g is not a closed interval.
(c) h is not continuous at $x = 0$. **(d)** Domain k is not a closed interval.
(e) Domain m is not a closed interval. **(f)** n is not continuous at $x = 0$.

Section 1.9

1. **(a)** $A \cup B = (-1, 11)$, $A \cap B = [2, 7]$, $A^C = (-\infty, -1] \cup (7, \infty)$,
$B^C = (-\infty, 2) \cup [11, \infty)$ **(b)** $\mathbb{R}$, $A \cap B = [-3, 8)$, $A^C = [8, \infty)$, $B^C = (-\infty, -3)$
(c) $A \cup B = (-1, 9]$, $A \cap B = \emptyset$, $A^C = (-\infty, -1] \cup [5, \infty)$, $B^C = (-\infty, 5) \cup (9, \infty)$
(d) $A \cup B = \mathbb{R}$, $A \cap B = (-5, -3) \cup [3, 8]$, $A^C = [-3, 3)$, $B^C = (-\infty, -5] \cup (8, \infty)$
(e) $A \cup B = [-4, 5) \cup (7, 15)$, $A \cap B = (0, 3] \cup [10, 11)$, $A^C = (-\infty, -4) \cup (3, 7] \cup [11, \infty)$,
$B^C = (-\infty, 0] \cup [5, 10) \cup [15, \infty)$

2. **(a)**

$A \cup B$

$A \cap B$

A^C

B^C

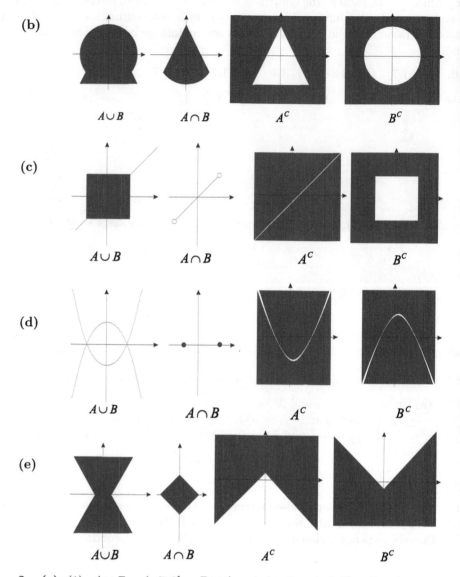

(b)

$A \cup B$ $A \cap B$ A^c B^c

(c)

$A \cup B$ $A \cap B$ A^C B^c

(d)

$A \cup B$ $A \cap B$ A^c B^c

(e)

$A \cup B$ $A \cap B$ A^c B^c

3. **(a)** (1) $A \cup B = (-7, 8] = B \cup A$ and $A \cap B = (-2, 5] = B \cap A$

(2) $(A \cup B) \cup C = (-7, 11) = A \cup (B \cup C)$ and $(A \cap B) \cap C = [0, 5] = A \cap (B \cap C$

(3) $A \cup \phi = (-7, 5] = A$ and $A \cap U = (-7, 5] = A$

(4) $A \cup U = \Re = U$ and $A \cap \emptyset = \emptyset$

(5) $A \cup A = (-7, 5] = A$ and $A \cap A = (-7, 5] = A$

(6) $A \cap (B \cup C) = (-2, 5] = (A \cap B) \cup (A \cap C)$ and $A \cup (B \cap C) = (-7, 8] = (A \cup B) \cap ($

(b) (1) $A \cup B = \Re = B \cup A$ and $A \cap B = \emptyset = B \cap A$

(2) $(A \cup B) \cup C = \Re = A \cup (B \cup C)$ and $(A \cap B) \cap C = \emptyset = A \cap (B \cap C)$

(3) $A \cup \phi = Q = A$ and $A \cap U = Q = A$

(4) $A \cup U = \Re = U$ and $A \cap \emptyset = \emptyset$

(5) $A \cup A = Q = A$ and $A \cap A = Q = A$

(6) $A \cap (B \cup C) = Z = (A \cap B) \cup (A \cap C)$ and $A \cup (B \cap C) = Q = (A \cup B) \cap (A$

c) (1)

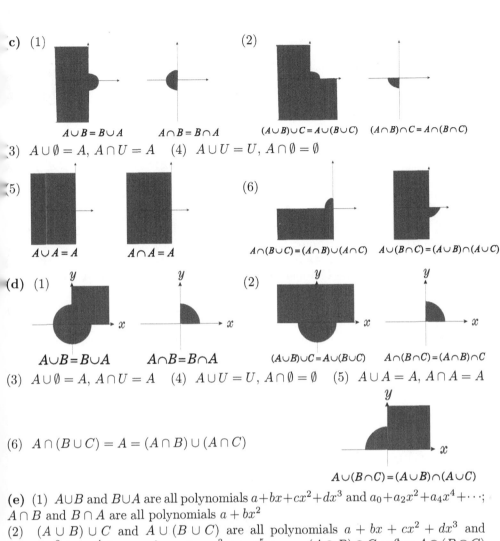

$$A \cup B = B \cup A \qquad A \cap B = B \cap A \qquad (A \cup B) \cup C = A \cup (B \cup C) \quad (A \cap B) \cap C = A \cap (B \cap C)$$

(3) $A \cup \emptyset = A, \, A \cap U = A$ (4) $A \cup U = U, \, A \cap \emptyset = \emptyset$

(5) (6)

$$A \cup A = A \qquad A \cap A = A \qquad A \cap (B \cup C) = (A \cap B) \cup (A \cap C) \quad A \cup (B \cap C) = (A \cup B) \cap (A \cup C)$$

(d) (1) (2)

$$A \cup B = B \cup A \qquad A \cap B = B \cap A \qquad (A \cup B) \cup C = A \cup (B \cup C) \quad A \cap (B \cap C) = (A \cap B) \cap C$$

(3) $A \cup \emptyset = A, \, A \cap U = A$ (4) $A \cup U = U, \, A \cap \emptyset = \emptyset$ (5) $A \cup A = A, \, A \cap A = A$

(6) $A \cap (B \cup C) = A = (A \cap B) \cup (A \cap C)$

$$A \cup (B \cap C) = (A \cup B) \cap (A \cup C)$$

(e) (1) $A \cup B$ and $B \cup A$ are all polynomials $a + bx + cx^2 + dx^3$ and $a_0 + a_2 x^2 + a_4 x^4 + \cdots$; $A \cap B$ and $B \cap A$ are all polynomials $a + bx^2$

(2) $(A \cup B) \cup C$ and $A \cup (B \cup C)$ are all polynomials $a + bx + cx^2 + dx^3$ and $a_0 + a_2 x^2 + a_4 x^4 + \cdots$ and $a_1 x + a_3 x^3 + a_5 x^5 + \cdots$; $(A \cap B) \cap C = \emptyset = A \cap (B \cap C)$

(3) $A \cup \phi$, A and $A \cap U$ are all polynomials $a + bx + cx^2 + dx^3$;

(4) $A \cup U$ and U equal the set of all polynomials while $A \cap \emptyset = \emptyset$

(5) $A \cup A$, A and $A \cap A$ are all polynomials $a + bx + cx^2 + dx^3$;

(6) $A \cap (B \cup C)$ and $(A \cap B) \cup (A \cap C)$ are all polynomials $a + bx^2$ and $cx + dx^3$; $A \cup (B \cap C)$ and $(A \cup B) \cap (A \cup C)$ are all polynomials $a + bx + cx^2 + dx^3$

5. (a) $(0, \infty)$ **(b)** $[0, 1]$ **(c)** Q

(d) the disc of radius 1 and center the origin

(e) the disc of radius 1 and center the origin with the origin removed

(f) the lower half–plane of points (x, y) with $y \le 0$

6. (a) (1) $A \cup A^C = (-7, 5] \cup (-\infty, -7] \cup (5, \infty) = \Re = U$ and
$A \cap A^C = (-7, 5] \cap ((-\infty, -7] \cup (5, \infty)) = \emptyset$

(2) $(A \cup B)^C = (-\infty, 11)^C = [11, \infty) = A^C \cap B^C$ and
$(A \cap B)^C = [1, 5]^C = (-\infty, 1) \cup (5, \infty) = A^C \cup B^C$

(3) $(A^C)^C = (-\infty, 5] = A$ (4) $U^C = \Re^C = \emptyset$ and $\emptyset^C = \Re = U$

(b) (1) $A \cup A^C = Q \cup I = \Re = U$ and $A \cap A^C = Q \cap I = \emptyset$

(2) $(A \cup B)^C = \Re^C = \emptyset = I \cap Q = A^C \cap B^C$ and $(A \cap B)^C = \emptyset^C = \Re = I \cup Q = A^C \cup B^C$

(3) $(A^C)^C = Q = A$ (4) $U^C = \Re^C = \emptyset$ and $\emptyset^C = \Re = U$

(c) (1) $A \cup A^C = \Re^2 = U$ and $A \cap A^C = \emptyset$

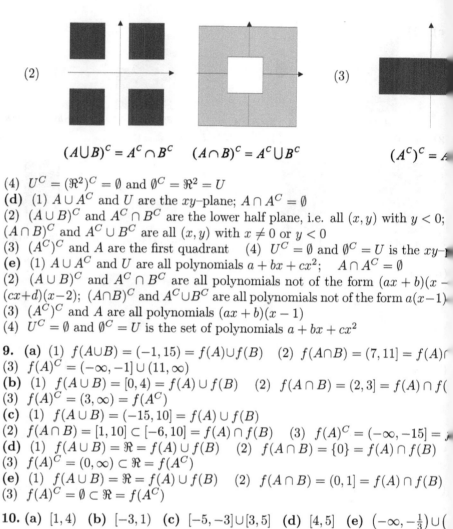

(2)

(3)

$$(A \cup B)^C = A^C \cap B^C \qquad (A \cap B)^C = A^C \cup B^C \qquad (A^C)^C = A$$

(4) $U^C = (\Re^2)^C = \emptyset$ and $\emptyset^C = \Re^2 = U$

(d) (1) $A \cup A^C$ and U are the xy–plane; $A \cap A^C = \emptyset$

(2) $(A \cup B)^C$ and $A^C \cap B^C$ are the lower half plane, i.e. all (x, y) with $y < 0$; $(A \cap B)^C$ and $A^C \cup B^C$ are all (x, y) with $x \neq 0$ or $y < 0$

(3) $(A^C)^C$ and A are the first quadrant   (4) $U^C = \emptyset$ and $\emptyset^C = U$ is the xy–plane

(e) (1) $A \cup A^C$ and U are all polynomials $a + bx + cx^2$; $A \cap A^C = \emptyset$

(2) $(A \cup B)^C$ and $A^C \cap B^C$ are all polynomials not of the form $(ax + b)(x - $ $(cx+d)(x-2)$; $(A \cap B)^C$ and $A^C \cup B^C$ are all polynomials not of the form $a(x-1)$

(3) $(A^C)^C$ and A are all polynomials $(ax + b)(x - 1)$

(4) $U^C = \emptyset$ and $\emptyset^C = U$ is the set of polynomials $a + bx + cx^2$

9. (a) (1) $f(A \cup B) = (-1, 15] = f(A) \cup f(B)$ (2) $f(A \cap B) = (7, 11] = f(A) \cap$

(3) $f(A)^C = (-\infty, -1] \cup (11, \infty)$

(b) (1) $f(A \cup B) = [0, 4) = f(A) \cup f(B)$ (2) $f(A \cap B) = (2, 3] = f(A) \cap f($

(3) $f(A)^C = (3, \infty) = f(A^C)$

(c) (1) $f(A \cup B) = (-15, 10] = f(A) \cup f(B)$

(2) $f(A \cap B) = [1, 10] \subset [-6, 10] = f(A) \cap f(B)$ (3) $f(A)^C = (-\infty, -15] = f$

(d) (1) $f(A \cup B) = \Re = f(A) \cup f(B)$ (2) $f(A \cap B) = \{0\} = f(A) \cap f(B)$

(3) $f(A)^C = (0, \infty) \subset \Re = f(A^C)$

(e) (1) $f(A \cup B) = \Re = f(A) \cup f(B)$ (2) $f(A \cap B) = (0, 1] = f(A) \cap f(B)$

(3) $f(A)^C = \emptyset \subset \Re = f(A^C)$

10. (a) $[1, 4)$ **(b)** $[-3, 1)$ **(c)** $[-5, -3] \cup [3, 5]$ **(d)** $[4, 5]$ **(e)** $(-\infty, -\frac{1}{3}) \cup ($

11. (a) (1) $f^{-1}(A \cup B) = (-2, 2) = f^{-1}(A) \cup f^{-1}(B)$

(2) $f^{-1}(A \cap B) = [-1, 1] = f^{-1}(A) \cap f^{-1}(B)$

(3) $f^{-1}(A^C) = (-\infty, -1) \cup (1, \infty)$ (4) $f(f^{-1}(A)) = [0, 1] \subset A$

(5) $f^{-1}(f(S)) = (-3, 3) \supset [0, 3) = S$

(b) (1) $f^{-1}(A \cup B) = (\frac{1}{9}, \infty) = f^{-1}(A) \cup f^{-1}(B)$

(2) $f^{-1}(A \cap B) = [\frac{1}{7}, 1] = f^{-1}(A) \cap f^{-1}(B)$ (3) $f^{-1}(A^C) = (0, \frac{1}{9}] \cup (1, \infty) = f^{-1}$

(4) $f(f^{-1}(A)) = [1, 9) = A$ (5) $f^{-1}(f(S)) = (1, 5] = S$

(c) (1) $f^{-1}(A \cup B) = [0, 3\pi] = f^{-1}(A) \cup f^{-1}(B)$

(2) $f^{-1}(A \cap B) = \{\frac{\pi}{2}, \frac{3\pi}{2}, \frac{5\pi}{2}\} = f^{-1}(A) \cap f^{-1}(B)$

(3) $f^{-1}(A^C) = (\frac{\pi}{2}, \frac{3\pi}{2}) \cup (\frac{5\pi}{2}, 3\pi] = f^{-1}(A)^C$ (4) $f(f^{-1}(A)) = [0, 1] \subset [0, \infty)$

(5) $S = (\frac{\pi}{2}, \pi] \subset (\frac{\pi}{2}, \frac{3\pi}{2}] \cup (\frac{5\pi}{2}, 3\pi] = f^{-1}(f(S))$

(d) (1) $f^{-1}(A \cup B)$ and $f^{-1}(A) \cup f^{-1}(B)$ are the xy–plane

(2)

(3)

(4) $f(f^{-1}(A)) = [0, \infty$

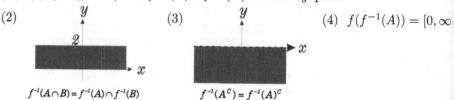

$f^{-1}(A \cap B) = f^{-1}(A) \cap f^{-1}(B)$ $f^{-1}(A^C) = f^{-1}(A)^C$

5)

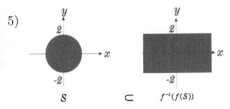

$S \quad \subset \quad f^{-1}(f(S))$

(e) (1)

$f^{-1}(A \cup B) = f^{-1}(A) \cup f^{-1}(B)$

(2)

$y=-x$

$x+y=2$

$f^{-1}(A \cap B) = f^{-1}(A) \cap f^{-1}(B)$

(3)

$y=-x$

$x+y=2$

$f^{-1}(A^C) = f^{-1}(A)^C$

(4) $f(f^{-1}(A)) = [0,2] = A$

(5)

$y=-x$

$S \quad \subset \quad f^{-1}(f(S))$

13. $f(A \cap B) = [1,4] = f(A) \cap f(B)$ **14.** $f(A)^C = (4,\infty) = f(A^C)$

Section 1.10

1.

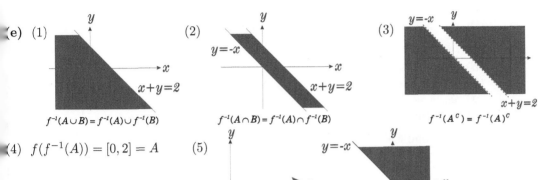

(a) at -2, 0, 2 with $N(0;2)$

(b) at 4.5, 5, 5.5 with $N(5;1/2)$

(c) at -4, -3, -2 with $N(-3;1)$

2. (a) $d(2,7) = 5$ **(b)** $d(-2,-7) = 5$ **(c)** $d(-4,3) = 7$

5. (a) A is open but not closed.   **(b)** B is closed but not open.
(c) C is neither open nor closed.   **(d)** D is closed but not open.
(e) E is open but not closed.   **(f)** F is neither open nor closed.
(g) G is both open and closed.   **(h)** H is closed but not open.
(i) I is open but not closed.   **(j)** J is closed but not open.
(k) K is open but not closed.   **(l)** L is closed but not open.
(m) M is closed but not open.   **(n)** N is neither open nor closed.
(o) V is closed but not open.   **(p)** P is closed but not open.
(q) Q is neither open nor closed.   **(r)** Z is closed but not open.
(s) S is closed but not open.   **(t)** T is closed but not open.
(u) U is neither open nor closed.

6. (a) a_n is convergent with limit 0.   **(b)** b_n is convergent with limit 2/5.
(c) c_n is divergent.   **(d)** d_n is convergent with limit 1/2.
(e) e_n is convergent with limit -1.   **(f)** f_n is convergent with limit 1/5.
(g) g_n is divergent.   **(h)** h_n is convergent with limit 0.
(i) i_n is convergent with limit 0.   **(j)** j_n is convergent with limit 1.
(k) k_n is convergent with limit $\frac{\pi}{2}$.   **(l)** s_n is divergent.

7. (a) $s_n = 5 - \frac{1}{n}$.   **(c)** $s_n = -1 + \frac{1}{n}$.   **(e)** $s_n = -3 + \frac{1}{n}$.   **(f)** $s_n = -3 + \frac{1}{n}$.
(i) $s_n = -\frac{1}{n}$.   **(k)** $s_n = \frac{n}{n+1}$.   **(n)** $s_n = \frac{1}{n} - \frac{1}{2}$.
(q) $s_n \in Q$ such that $s_n^2 > 2$ and $s_n^2 - 2 < 1/n$.   **(u)** $s_n = \frac{1}{3} - \frac{1}{6n}$.

10. (a) A is bounded below and above with $\mathrm{glb}(A) = -3$ and $\mathrm{lub}(A) = 5$.
(b) B is bounded below and above with $\mathrm{glb}(B) = -5$ and $\mathrm{lub}(B) = 11$.
(c) C is bounded below and above with $\mathrm{glb}(C) = 0$ and $\mathrm{lub}(C) = 1$.
(d) D is not bounded below nor above.
(e) E is bounded below and above with $\mathrm{glb}(E) = 1/2$ and $\mathrm{lub}(E) = 1$.
(f) F is bounded below and above with $\mathrm{glb}(F) = -1$ and $\mathrm{lub}(F) = 1$.
(g) G is bounded below but not bounded above.
(h) H is bounded below and above.
(i) I is bounded below and above.

14 (a) a_n is a Cauchy sequence because it converges to 0.
(b) b_n is a Cauchy sequence because it converges to 1.
(c) c_n is not a Cauchy sequence because it diverges.
(d) d_n is a Cauchy sequence because it converges to 1.
(e) e_n is not a Cauchy sequence because it diverges.

15 (a) complete **(b)** complete **(c)** not complete **(d)** complete
(e) not complete **(f)** complete **(g)** complete **(h)** complete

Section 1.11

1. (a) connected **(b)** connected **(c)** connected **(d)** disconnected
(e) connected **(f)** connected **(g)** disconnected **(h)** disconnected
(i) disconnected **(j)** disconnected **(k)** disconnected

2. A finite connected subset of $\Re$ must be the empty set or a set which contains exactly one point.

6. Let $f(x) = 3$ with domain $(0, 1) \cup (3, 4)$. The range of f is the connected set $\{3\}$. Alternatively, let $g(x) = x^2$ with domain $[-2, -1] \cup [1, 2]$. The range of g is the connected set $[1, 4]$.

Section 1.12

1. (a) $\left\{(2n)^3\right\} = \left\{8n^3\right\}$ and $\left\{(3n)^3\right\} = \left\{27n^3\right\}$

(b) $\left\{\frac{1}{\sqrt{(2n)^2+1}}\right\} = \left\{\frac{1}{\sqrt{4n^2+1}}\right\}$ and $\left\{\frac{1}{\sqrt{(3n)^2+1}}\right\} = \left\{\frac{1}{\sqrt{9n^2+1}}\right\}$ **(c)** $\left\{\sin\frac{\pi}{2n}\right\}$ and $\left\{\frac{\sin\pi}{3n}\right\}$

2. $s_n = \left\{(-1)^n \frac{n}{n+1}\right\}$ diverges, and the subsequence $\left\{\frac{2n}{2n+1}\right\}$ converges to 1.

4. (a) (i) $s_n = \{n\}$ has no convergent subsequence.
(ii) $s_n = \left\{-3 + \frac{1}{n}\right\}$ has no convergent subsequence.
(iii) $s_n = \left\{2 + \frac{1}{n}\right\}$ has no convergent subsequence.
(iv) $s_n = \{-n\}$ has no convergent subsequence.
(v) $s_n = \{n\}$ has no convergent subsequence.
(vi) Let s_n be a sequence of elements of D with limit $\frac{\sqrt{2}}{2}$. Then s_n has no convergent subsequence.
(vii) $s_n = \left\{\frac{1}{n}\right\}$ has no convergent subsequence.
(viii) $s_n = \{n\pi\}$ has no convergent subsequence.
(ix) $s_n = \left\{\frac{1}{n\pi}\right\}$ has no convergent subsequence.
(x) $s_n = \left\{n\sin\frac{2}{n\pi}\right\}$ has no convergent subsequence.
(b) (i) Z is not bounded hence not compact. **(ii)** A is not closed hence not com
(iii) B is not closed hence not compact. **(iv)** C is not bounded hence not c
(v) Q is not bounded hence not compact. **(vi)** D is not closed hence not com
(vii) E is not closed hence not compact. **(viii)** F is not bounded hence not
(ix) G is not closed hence not compact. **(x)** H is not closed hence not com

5. (a) compact (b) not compact (c) not compact (d) compact (e) not compact
(f) compact (g) not compact (h) not compact (i) compact (j) compact

Review Exercises

1. True **2.** True **3.** False **4.** True **5.** True **6.** False **7.** False **8.** True
9. True **10.** True **11.** False **12.** True **13.** False **14.** False **15.** False
16. False **17.** False **18.** False **19.** True **20.** False **21.** False **22.** True
23. True **24.** True **25.** False **26.** False **27.** True **28.** False **29.** False
30. True **31.** False **32.** True **33.** True **34.** True **35.** False **36.** True
37. False **38.** True **39.** False **40.** True **41.** True **42.** False **43.** True
44. False **45.** True **46.** False **47.** True **48.** False **49.** False **50.** True

51. $-\frac{4}{3}$ **52.** (a) essential (b) removable

53. (a) Yes (b) Domain $f^{-1} = (0,1]$, Range $f^{-1} = [0,\infty)$ **54.** $\left[0, \frac{2}{\pi}\right]$

55. (b) $f^{-1}(x) = \sin x$ with domain $\left[-\frac{\pi}{2}, \frac{\pi}{2}\right]$ **56.** (a) yes (b) $\pm\sqrt{3}, \pm\sqrt{5}$

57. no **58.** $\frac{11\pi}{6}$ **59.** 1 **60.** $(\tan \circ \arcsin)(x) = \frac{x}{\sqrt{1-x^2}}$

61. (a) $\delta = \frac{\epsilon}{1+\epsilon}$ (b) $\delta = \sqrt[4]{\epsilon}$ **62.** $(-\infty, \infty)$ **63.** $\frac{\sqrt{7}}{3}$ **64.** $\frac{1}{4}$ **65.** 0 **66.** 0

68. removable discontinuity

69. roots between $-\frac{11}{4}, -\frac{5}{2}$, between $\frac{1}{4}, \frac{1}{2}$ and between $\frac{5}{4}, \frac{3}{2}$

70. nothing **71.** (a) no (b) yes

72. $f(x) = x^2 + x$, $g(x) = x+1$ or $f(x) = -x^2 - x$, $g(x) = -x-1$; $-1 \notin$ Domain g

73. $\frac{1}{3}$ **74.** Domain $f = (-\infty, 0) \cup (0, \infty)$, Range $f = \left(-\infty, -\frac{2}{\pi}\right) \cup \left(\frac{2}{\pi}, \infty\right)$

75. Range $f = \left[\frac{1}{17}, \frac{1}{5}\right]$

Chapter 2

Section 2.2

1. (a) 12.3 (b) 25.25 (c) 0.8 (d) 2.3 (e) $-.05$

2. (a) $(5, 125)$ (b) $(5.5, 41.75)$ (c) $(-1.3, -13/7) \approx (-1.3, -1.9)$
(d) $(1.6, \sqrt{8.2}) \approx (1.6, 2.9)$ (e) $(3.2, \sqrt{97/11}) \approx (3.2, 3.0)$
(f) $(.81, 9/19) \approx (.81, .47)$

3. (a) 5, 4.5, 3.80, 3.90 (b) $-65, -77, -138, -96$
(c) $-.037, -.038, -.044, -.041$ (d) .163, .165, .170, .168 (e) 3.3, 2.0, .7, .9

4. (a) 4 (b) -81 (c) -0.04 (d) $1/6$ (e) 1

5. (a) T is $y = 12x - 13$ (b) T is $y = 27x - 54$ (c) T is $y = \frac{7x}{8} - \frac{15}{8}$
(d) T is $y = \frac{7x}{3} + \frac{4}{3}$ (e) T is $y = -\frac{5x}{128} + \frac{11}{64}$

6. (a) N is $y = -\frac{x}{12} + \frac{67}{6}$. (b) N is $y = -\frac{x}{27} + \frac{244}{9}$. (c) N is $y = -\frac{8x}{7} + \frac{1}{7}$.
(d) N is $y = -\frac{3x}{7} + \frac{48}{7}$. (e) N is $y = \frac{128x}{5} + \frac{1029}{20}$.

7. (a) $-\frac{1}{6}$ (b) $\frac{1}{3}$ (c) $\frac{5}{2}$ (d) $-\frac{7}{3}$ (e) $y = -\frac{1}{5}x + \frac{37}{5}$ (f) $y = -4x + 19$

8. (a) $f'(x) = 7$ (b) $g'(x) = 10x$ (c) $h'(x) = 3 - 2x$ (d) $j'(x) = 12x^2 - 1$
(e) $k'(x) = 6x^2 + 6x$ (f) $m'(x) = 4x^3 + 6$

9. (a) $f'(x) = -\frac{1}{x^2}$ (b) $g'(x) = -\frac{8}{(7x-2)^2}$ (c) $h'(x) = -\frac{47}{(3x+4)^2}$

(d) $j'(x) = \frac{2x^2+2x}{(2x+1)^2}$ (e) $k'(x) = \frac{54x}{(5x^2+6)^2}$ (f) $m'(x) = \frac{x^4+3x^2}{(x^2+1)^2}$

10. (a) $f'(x) = \frac{1}{2\sqrt{x}}$ (b) $g'(x) = \frac{3}{2\sqrt{3x-4}}$ (c) $h'(x) = \frac{x}{\sqrt{x^2-8}}$

(d) $j'(x) = -\frac{1}{(2x+1)^{3/2}}$ (e) $k'(x) = -\frac{23}{2\sqrt{4x+5}(3x-2)^{3/2}}$ (f) $m'(x) = -\frac{}{2\sqrt{x}(}$

12. FALSE. For example, $f(x) = |x|$ is continuous at all x but $f'(0)$ does not

13. (a) $f'(x)$ exists if $x \neq 2/3$. In particular, $f'(x) = \begin{cases} 3 & \text{if } x > 2/3 \\ -3 & \text{if } x < 2/3 \end{cases}$.
$f'(2/3)$ does not exist because there is a cusp there: the left derivative equals $-$
the right derivative equals $+3$.

(b) $g'(x) = \begin{cases} -5 & \text{for } x < -\frac{7}{5} \\ 5 & \text{for } x > -\frac{7}{5} \end{cases}$. Hence $g'(-7/5)$ does not exist because t
a cusp there: the left derivative is -5 and the right derivative is $+5$.

(c) $h'(x) = 2x$ for $x < -1$ or $x > 1$ while $h'(x) = -2x$ for $-1 < x < 1$. $h'(-1$
not exist because the left derivative is -2 while the right derivative is $+2$. $h'(1$
not exist because the left derivative is -2 while the right derivative is $+2$.

(d) $i'(x)$ exists and equals $-\frac{7}{2\sqrt{4-7x}}$ for $x < 4/7$. $i'(4/7)$ does not exist becau
is not an interior point of the domain of i.

(e) $j'(x)$ exists for all x. In particular, $j'(x) = \begin{cases} 8 & \text{if } x \geq 1 \\ 8x & \text{if } x < 1 \end{cases}$.

(f) $k'(x)$ exists if $x \neq 4$. In particular, $k'(x) = \begin{cases} 5 & \text{if } x > 4 \\ -4 & \text{if } x < 4 \end{cases}$. $k'(4)$
there is a cusp there: the left derivative is -4 while the right derivative is 5.

(g) $m'(x) = 3$ if $x \neq 2$. However, $m'(2)$ DNE because m is not continuous at

(h) $n'(x) = \frac{1}{\sqrt{3+2x}}$ for $x > -\frac{3}{2}$. $n'(-3/2)$ does not exist because $-\frac{3}{2}$ is not an in
point of the domain of n.

(i) $p'(x) = \begin{cases} 2x - 2 & \text{for } x \leq 1 \\ 2 - 2x & \text{for } x > 1 \end{cases}$.

(j) $q'(x) = \begin{cases} 3x^2 - 10x & \text{for } x < 2 \\ 4x^3 - 40 & \text{for } x > 2 \end{cases}$. $q'(2)$ DNE as q is not continuous at x

(k) $r'(x)$ does not exist when $x \neq 0$ because some fractions $\frac{r(x+h)-r(x)}{h}$ are cl
zero while others are close to $2x$ which is nonzero. However, $r'(0) = 0$.

(l) s is differentiable nowhere because it is discontinuous when $x \neq -2, 0, 1$. At
points neither the left nor right derivatives exist.

15. $f'(x) = x \sin \frac{1}{x}$ with $c = 0$

16. (a) $f'_+(1) = 3$ (b) $g'_-(9) = 248$ (c) $h'_+(6) = +\infty$ (d) $i'_-(3) = -$
(e) $j'_+(-2) = 1$ and $j'_-(-2) = -1$ (f) $k'_+(1/5) = 5$ and $k'_-(1/5) = -5$
(g) $m'_+(2) = 4$ and $m'_-(2) = -4$ (h) $n'_+(-3) = 8$ and $n'_-(5) = -8$

17. (a) left half–tangent line $y = 17 - 16x$ for $x \leq 2$
(b) left half–tangent line $(9, y)$ for $y \geq 0$
(c) right half–tangent line $y = \frac{5}{64}x + \frac{3}{64}$ for $x \geq 1$
(d) left half–tangent $y = 63 - 9x$ for $x \leq 7$, right half–tangent $y = 9x - 63$ for
(e) right half–tangent line $(-4, y)$ for $y \leq 0$
(f) left half–tangent $y = -7x - 35$, $x \leq -5$, right half–tangent $y = 7x + 35$, $x \geq$

18. (a) differentiable (b) differentiable (c) not differentiable (d) different
(e) not differentiable (f) differentiable (g) differentiable (h) differentia

19. **(a)** $x = -7/3$ is a cusp. **(b)** $x = 4/5$ is a cusp.
(c) $x = -6$ and $x = 5$ are cusps. **(d)** $x = -3$ and $x = 5$ are cusps.
(e) $x = 6$ is a vertical tangent. **(f)** $x = -7$ is a one sided vertical tangent.
(g) $x = 8$ is a one sided vertical tangent. **(h)** $x = 3$ is a vertical cusp.
(i) $x = 1$ is a cusp. **(j)** $r'(x)$ exists for all x.

20. **(a)** TRUE. $f'(c)$ must be a number not $-\infty$ or ∞.
(b) TRUE. $f'(c)$ DNE because the left and right derivatives have different values.
(c) TRUE. $f'(c)$ DNE because the left and right derivatives have different values.
(d) FALSE. only the one–sided derivatives $f'_+(a)$ and $f'_-(b)$ are required to exist.
(e) FALSE. only the one–sided derivatives $f'_+(a)$ and $f'_-(b)$ are required to exist and they may be infinite.

21. **(a)** $f'_-(1)$, $f'(3)$, $f'(4)$ do not exist.
(b) $g'_-(1)$, $g'_+(2)$, $g'(2)$, $g'_-(3)$, $g'_+(3)$, $g'(3)$, $g'_+(5)$ DNE. **(c)** $h'(6)$, $h'_+(8)$ DNE.

Section 2.3

1. **(a)** $f'(x) = 28x^3 - 18x$ **(b)** $g'(x) = 10x - \frac{4}{\sqrt{x}}$ **(c)** $h'(x) = -\frac{10}{x^2} - 3x^2$
(d) $k'(x) = 30x^5 - 12x^2$ **(e)** $m'(x) = \frac{3}{\sqrt{x}} + \frac{6}{x^4}$ **(f)** $p'(x) = 15x^4 + \frac{2}{\sqrt{x}} - \frac{10}{x^3}$

2. **(a)** $f'(x) = (3x^2 + 1)(x^2 - x - 1) + (x^3 + x - 1)(2x - 1)$ **(b)** $g'(x) = \frac{1}{2\sqrt{x}}(3x^5 + 5x^3 + 7) + \sqrt{x}(15x^4 + 15x^2)$ **(c)** $h'(x) = -\frac{1}{x^2}(2x^6 - 9x^2 + 4) + \frac{1}{x}(12x^5 - 18x)$
(d) $k'(x) = \left(\frac{1}{2\sqrt{x}} + 1\right)(\sqrt{x} - 3x + 7) + (\sqrt{x} + x - 4)\left(\frac{1}{2\sqrt{x}} - 3\right)$
(e) $m'(x) = \left(-\frac{5}{x^2} - 24x^2\right)\left(\frac{9}{x} + 4x^2 - 3\right) + \left(\frac{5}{x} - 8x^3 + 6\right)\left(-\frac{9}{x^2} + 8x\right)$
(f) $p'(x) = \left(\frac{3}{\sqrt{x}} + \frac{4}{x^2}\right)\left(\frac{2}{x^2} + \frac{6}{x^3}\right) - \left(6\sqrt{x} - \frac{4}{x}\right)\left(\frac{4}{x^3} + \frac{18}{x^4}\right)$

3. **(a)** $f'(x) = -\frac{13}{(3x-2)^2}$ **(b)** $g'(x) = \frac{(20x^4 - 3)(x + 8\sqrt{x} + 1) - (4x^5 - 3x + 1)\left(1 + \frac{4}{\sqrt{x}}\right)}{(x + 8\sqrt{x} + 1)^2}$
(c) $h'(x) = \frac{1 - x^2}{(x^2 + 1)^2}$ **(d)** $k'(x) = -\frac{4x^3 + 2x}{(x^4 + x^2 + 1)^2}$ **(e)** $m'(x) = \frac{12x^{-1/2} - 15x^{5/2}}{(x^3 + 4)^2}$
(f) $p'(x) = \frac{13}{\sqrt{x}(5 - 2\sqrt{x})^2}$

4. **(a)** $f'(x) = (4x + 3)(x^3 - 5x + 2)(x^4 - 3x^2 + 6) + (2x^2 + 3x + 4)(3x^2 - 5)(x^4 - 3x^2 + 6)$
$+ (2x^3 + 3x + 4)(x^3 - 5x + 2)(4x^3 - 6x)$
(b) $g'(x) = 2(7x^4 - 2x^{-3} + 8x^2 - 5\sqrt{x} - 1)\left(28x^3 + 6x^{-4} + 16x - \frac{5}{2\sqrt{x}}\right)$
(c) $h'(x) = 3(5x^2 - 6\sqrt{x} + 7)^2 \left(10x - \frac{3}{\sqrt{x}}\right)$
(d) $k'(x) = \frac{2x(6\sqrt{x} + 3)(8\sqrt{x} + 9) + (x^2 + 1)\left(\frac{3}{\sqrt{x}}\right)(8x\sqrt{x} + 9) - (x^2 + 1)(6\sqrt{x} + 3)\frac{4}{\sqrt{x}}}{(8\sqrt{x} + 9)^2}$
(e) $m'(x) = \frac{35x^4\sqrt{x}(3x^2 - 2) - (7x^5 + 1)\left(\frac{1}{2\sqrt{x}}\right)(3x^2 - 2) - (7x^5 + 1)\sqrt{x}(6x)}{x(3x^2 - 2)^2}$
(f) $p'(x) = \frac{\left(2x + \frac{1}{2\sqrt{x}}\right)(6x - \sqrt{x} + 5)(5x^3 - 1)(4\sqrt{x} - 2) + (x^2 + \sqrt{x} + 1)\left(6 - \frac{1}{2\sqrt{x}}\right)(5x^3 - 1)(4\sqrt{x} - 2)}{(5x^3 - 1)^2(4\sqrt{x} - 2)^2}$
$\frac{-(x^2 + \sqrt{x} + 1)(6x - \sqrt{x} + 5)(15x^2)(4\sqrt{x} - 2) - (x^2 + \sqrt{x} + 1)(6x - \sqrt{x} + 5)(5x^3 - 1)\frac{2}{\sqrt{x}}}{(5x^3 - 1)^2(4\sqrt{x} - 2)^2}$

(g) $q'(x) = -\frac{5x^6 + 12x^{5/2} + 7}{2x^{9/2}(x^{5/2} + 1)^2}$

5. **(a)** $2f(x)f'(x)$ **(b)** $3f(x)^2 f'(x)$
(c) $f'(x)g(x)h(x) + f(x)g'(x)h(x) + f(x)g(x)h'(x)$ **(d)** $-2\frac{f'(x)}{f(x)^3}$ **(e)** $-3\frac{f'(x)}{f(x)^4}$
(f) $\frac{f'(x)g(x)h(x) + f(x)g'(x)h(x) - f(x)g(x)h'(x)}{h(x)^2}$ **(g)** $\frac{f'(x)g(x)h(x) - f(x)g'(x)h(x) - f(x)g(x)h'(x)}{g(x)^2 h(x)^2}$
(h) $\frac{f'(x)g(x)h(x)k(x) + f(x)g'(x)h(x)k(x) - f(x)g(x)h'(x)k(x) - f(x)g(x)h(x)k'(x)}{h(x)^2 k(x)^2}$

6, 7. **(a)** $f'(x) = \frac{5}{2}x^{3/2}$ **(b)** $g'(x) = \frac{7}{2}x^{5/2}$ **(c)** $h'(x) = -\frac{9}{2}x^{-11/2}$ **(d)** $k'(x) = -\frac{3}{2}x^{-5/2}$

9. (a) $f'(x) = -14x^{-9/2} - 15x^{3/2}$

(b) $g'(x) = (11x^{9/2} + 6x^{-5/2})(8x^{-9/2} + 12x^{7/2}) + (2x^{11/2} - 4x^{-3/2})(-36x^{-11/2} + 4$

(c) $h'(x) = \dfrac{25x^{3/2}(4x^{-13/2} + 5x^2 - 1) - (10x^{5/2} + 7)(-26x^{-15/2} + 10x)}{(4x^{-13/2} + 5x^2 - 1)^2}$

(d) $k'(x) = (3x^2 - 45x^{13/2})(4x^{-17/2} + 7x^3 + 8) + (x^3 - 6x^{15/2} + 9)(-34x^{-19/2}$

(e) $m'(x) = \dfrac{(2x - 9x^{-5/2})(x^{7/2} - x^3 + 1) - (x^2 + 6x^{-3/2})(\frac{7}{2}x^{5/2} - 3x^2)}{(x^{7/2} - x^3 + 1)^2}$

(f) $p'(x) = (2x^{7/2} - 3x^{-9/2})(14x^{5/2} + 27x^{-11/2})$

10. (a) $y = x$ **(b)** $y = 7x - 13$ **(c)** $y = 20x - 48$ **(d)** $y = x + 2$
(e) $y = -\frac{2}{147}x + \frac{13}{49}$ **(f)** $y = -\frac{17}{2}x + \frac{29}{2}$

11. (a) $y = -\frac{x}{16} - \frac{255}{8}$ **(b)** $y = \frac{x}{12} + \frac{11}{12}$ **(c)** $y = \frac{1}{27} + 162(x - 9)$
(d) $y = -11 + \frac{1}{116}(x + 2)$ **(e)** $y = \frac{17}{6} + 18(x - 16)$ **(f)** $y = -\frac{3}{20}x + \frac{43}{20}$

12. (a) $y = 0, \ x \geq 0$ **(b)** $y = 5x, \ x \geq 0$ **(c)** $x = 0, \ y \leq \frac{1}{2}$ **(d)** $y = 3 - 6x,$

13. (a) $f''(x) = 84x^5 - 60x^3$. **(b)** $g''(x) = 36x + 264x^{-13} + 168x^{-9}$.
(c) $h''(x) = \frac{315}{4}x^{5/2} - 2x^{-3/2}$. **(d)** $i''(x) = \frac{12x^4 - 12x}{(x^3 + 2)^3}$. **(e)** $j''(x) = \frac{294x^3 +}{(7x^2 -}$
(f) $k''(x) = \frac{\sqrt{x} - 1}{4x\sqrt{x}(\sqrt{x} + 1)^3}$.

14. (a) $f'(x) = 6x, \quad f''(x) = 6, \quad f^{(n)}(x) = 0$ for $n \geq 3$.
(b) $g'(x) = 12x^2 - 9, \quad g''(x) = 24x, \quad g^{(3)}(x) = 24 \quad g^{(n)}(x) = 0$ for $n \geq 4$.
(c) $h'(x) = 8x^3 - 16x + 4, \quad h''(x) = 24x^2 - 16, \quad h^{(3)}(x) = 48x, \quad h^{(4)}(x)$
$h^{(n)}(x) = 0$ for $n \geq 5$. **(d)** $i^{(n)}(x) = (-1)^n \frac{n(n-1)\cdots(3)(2)(1)}{x^{n+1}}$ for $n \geq 1$.
(e) $j'(x) = \frac{1}{2}x^{-1/2}, \quad j^{(n)}(x) = (-1)^{n-1}\frac{(2n-3)(2n-5)\cdots(5)(3)(1)}{2^n}x^{-\left(\frac{2n-1}{2}\right)}$ for $n \geq$
(f) $k^{(n)}(x) = (-1)^{n-1}\frac{n(n-1)\cdots(2)(1)}{(x+1)^{n+1}}$ for $n \geq 1$.
(g) $m^{(n)}(x) = (-1)^{n-1}2\frac{n(n-1)\cdots(1)(1)}{(x+1)^{n+1}}$ for $n \geq 1$. **(h)** $p^{(n)}(x) = (-1)^n 2^n \frac{n(n-}{(2x}$
(i) $q'(x) = \frac{3}{2}\sqrt{x}, \ q^{(2)}(x) = \frac{3}{4}x^{-1/2}, \ q^{(n)}(x) = (-1)^n \frac{(3)(1)(3)(5)(7)\cdots(2n-5)}{2^n}$ for $n \geq$

Section 2.4

1. (a) $\Delta x = \frac{1}{2}, \Delta y = \frac{13}{4}$ **(b)** $\Delta x = -1, \Delta u = \frac{1}{12}$ **(c)** $\Delta y = -\frac{1}{24}$ **(d)** Δ
(e) $\Delta x = \frac{1}{15}$

2. (a) $\frac{\Delta y}{\Delta x} = -\frac{19}{3}$ **(b)** $\frac{\Delta u}{\Delta x} = -\frac{3}{10}$ **(c)** $\frac{\Delta y}{\Delta u} = \frac{76}{9}$ **(d)** $\frac{\Delta w}{\Delta u} = -\frac{1}{36}$ **(e)** $\frac{\Delta}{\Delta}$

3. (a) $\frac{dy}{dx} = 18(5x^7 - 2x^5 + 7)^{17}(35x^6 - 10x^4)$ **(b)** $\frac{dy}{dx} = \frac{12x^7 + 14x^3}{\sqrt{3x^8 + 7x^4 + 3}}$
(c) $\frac{dy}{dx} = -(9x^6 + 5x^2 + 4)^{-3/2}(27x^5 + 5x)$ **(d)** $\frac{dy}{dx} = \frac{4(63x^6 + 135x^4 + 28)}{9(3x^5 - 2)^{5/9}(7x^2 + 9)}$
(e) $\frac{dy}{dx} = -16x(2x^2 - 8)^{-7/3}$ **(f)** $\frac{dy}{dx} = 168x - \frac{9x}{\sqrt{3x^2 + 1}}$ **(g)** $\frac{dy}{dx} = -\frac{160x^3}{(5x^4 - 2)^9}$
(h) $\frac{dy}{dx} = \frac{45}{14\sqrt{x}}(5\sqrt{x} - 1)^{2/7}(8\sqrt{x} + 6)^{7/9} + \frac{28}{9\sqrt{x}}(5\sqrt{x} - 1)^{9/7}(8\sqrt{x} + 6)^{-2/9}$
(i) $\frac{dy}{dx} = \frac{(6x^{-1/2} + 2x^{-5/6})(x^{1/4} + x^{1/3}) - (x^{1/2} + x^{1/6})(3x^{-3/4} + 4x^{-2/3})}{12(x^{1/4} + x^{1/3})^2}$
(j) $\frac{dy}{dx} = \frac{(4x^3 - 12)^{1/4}(57x^8 + 144x^5 + 120x^2)}{(7x^6 + 8)^{9/7}}$ **(k)** $\frac{dy}{dx} = \frac{27x^2(x^3 + 2)^8}{2\sqrt{1 + (x^3 + 2)^9}}$
(l) $\frac{dy}{dx} = -40x^4\frac{(5 - \sqrt{2x^5 + 7})^7}{\sqrt{2x^5 + 7}}$ **(m)** $\frac{dy}{dx} = -\frac{36x^2}{5(x^3 + 1)^{2/3}(\sqrt[3]{x^3 + 1} + 5)^{3/5}(4\sqrt[3]{}}$
(n) $\frac{dy}{dx} = \frac{1}{8\sqrt{x}(5\sqrt{x} + 7)^{3/4}(2\sqrt{x} + 3)^{5/4}}$ **(o)** $\frac{dy}{dx} = \frac{1}{8\sqrt{1 + \sqrt{2 + \sqrt{3 + x}}}}\frac{1}{\sqrt{2 + \sqrt{3 + x}}}\frac{1}{\sqrt{3}}$
(p) $\frac{dy}{dx} = -1260x^3(7 + (8 - (x^4 + 6)^7)^5)^8(8 - (x^4 + 6)^7)^4(x^4 + 6)^6$ **(q)** $\frac{dy}{dx} = \frac{224x^7(1+}{[2 + (1 + x^8}$
(r) $\frac{dy}{dx} = 56\left[5 - \sqrt{4 + (x^2 + 5)^8}\right]^6 \frac{(x^2 + 5)^7}{\sqrt{4 + (x^2 + 5)^8}}$ **(s)** $\frac{dy}{dx} = \frac{4x(x^2 + 1)^{-2/3} + 3x(x^2 + 1)^{-}}{12\sqrt{\sqrt[3]{x^2 + 1} + \sqrt[4]{x^2 + 1}}}$
(t) $\frac{dy}{dx} = -\frac{21x^6\left[3(x^7 + 5)^{-1/4} + 2(x^7 + 5)^{-1/3} + 3(x^7 + 5)^{5/12}\right]}{\sqrt{1 - 12(x^7 + 5)^{3/4}}\left[2 + 18(x^7 + 5)^{2/3}\right]^{3/2}}$

4. (a) $y = 10x - 39$ **(b)** $y = \frac{x}{2} + 2$ **(c)** $y = \frac{x}{216} + \frac{28}{27}$ **(d)** $y = -2016x -$

e) $y = -328x + 2944$

5. **(a)** $y = -\frac{16}{3}x + \frac{35}{6}$ **(b)** $y = \frac{9}{16}x + \frac{45}{4}$ **(c)** $y = \frac{x}{8} + \frac{11}{4}$ **(d)** $y = \frac{4}{33}x + \frac{150}{11}$
e) $y = 429 - 12x$

6. **(a)** f has a 1 sided vertical tangent at $x = 5$. **(b)** g has a vertical cusp at $x = -4$.
c) h has a vertical cusp at $x = 3$. **(d)** j has vertical tangents at $x = 2$ and $x = 4$.
e) k has a cusp at $x = 1/2$. **(f)** m has a vertical cusp at $x = 1/2$.

7. **(a)** $\frac{dy}{dx} = 3f(x)^2 f'(x)$ **(b)** $\frac{dy}{dx} = 4x^3 f'(x^4)$ **(c)** $\frac{dy}{dx} = \frac{5f(\sqrt{x})^4 f'(\sqrt{x})}{2\sqrt{x}}$
(d) $\frac{dy}{dx} = \frac{f'(x)}{2\sqrt{f(x)}}$ **(e)** $\frac{dy}{dx} = f'(f(x))f'(x)$ **(f)** $\frac{dy}{dx} = -\frac{6xf'(3x^2+2)}{[1+f(3x^2+2)]^2}$

(g) $\frac{dy}{dx} = \frac{\frac{f'(x)f(\sqrt{x})}{\sqrt{f(x)}} - f'(x)\sqrt{f(x)} - \frac{\sqrt{f(x)}f'(\sqrt{x})}{\sqrt{x}}}{2[f(x)+f(\sqrt{x})]^2}$ **(h)** $\frac{dy}{dx} = \frac{f'(\sqrt{x})}{4\sqrt{xf(\sqrt{x})}}$

8. $D(f \circ g \circ h \circ k)(x) = Df(g(h(k(x)))) \cdot Dg(h(k(x))) \cdot Dh(k(x)) \cdot Dk(x)$

9. **(a)** $D(f \circ g)(1) = 15$ **(b)** $D(f \circ g)(4) = -56$ **(c)** $D(g \circ f)(3) = -56$
(d) $D(g \circ f)(2) = 12$

10. **(a)** -90 **(b)** -40 **(c)** -40 **(d)** -90

11. **(a)** $y' = -\frac{x^3}{y^3}$ **(b)** $y' = \frac{14y - 15xy^3}{30x^2 y^2 - 28x}$ **(c)** $y' = \frac{35x^4 y - 6y^4}{24xy^3 - 7x^5}$ **(d)** $y' = \frac{10xy^2 \sqrt{3x^2+4y^2} - 3x}{4y - 10x^2 y\sqrt{3x^2+4y^2}}$

(e) $y' = \frac{7y - 30x^4(4y^5 - 7)}{20y^4(6x^5+2) - 7x}$ **(f)** $y' = \frac{(8y+9)[15x^2 - 14xy(8y+9)]}{7x^2(8y+9)^2 + 8(5x^3 - 6)}$
(g) $y' = \frac{\frac{5}{3}(30x^5 - 8x^3 y^7)(5x^6 - 2x^4 y^7)^{2/3} - 16xy}{8x^2 + 14\frac{5}{3}x^4 y^6(5x^6 - 2x^4 y^7)^{2/3}}$ **(h)** $y' = \frac{18y^2}{5(5y-3)^{-5/6} - 36xy}$
(i) $y' = \frac{3x^{1/5}y^{7/5} - 4x^{3/5}y^{3/5}}{1 - 2x^{6/5}y^{2/5}}$

12. **(a)** $y = -\frac{5}{2}x + \frac{21}{2}$ **(b)** $y = \frac{7}{11}x - \frac{25}{11}$ **(c)** $y = -\frac{14}{33}x + \frac{25}{11}$ **(d)** $y = \frac{200}{33}x + \frac{332}{33}$
(e) $y = -\frac{174}{13}x + \frac{387}{13}$

13. **(a)** $y = -27 + \frac{25}{9}(x - 125)$ **(b)** $y = 1 + \frac{51}{5}(x - 9)$ **(c)** $y = -4 - \frac{84}{53}(x - 3)$
(d) $y = x + 2$ **(e)** $y = 16 - \frac{59}{16}(x - 81)$

14. **(a)** $y'' = -\frac{2xy^3 + 2x^4}{y^5}$ **(b)** $y'' = \frac{2(y-2x)(2y-x) - 2(y-2x)^2 - 2(2y-x)^2}{(2y-x)^3}$
(c) $y'' = -\frac{2(3y+1)}{(3y^2 + 2y + 1)^3}$ **(d)** $y'' = 50\frac{(y+2)^{-3/2} + 24y}{[(y+2)^{-1/2} - 6y^2]^3}$
(e) $y'' = 5x^{1/4}y^{1/4} + x^{-2}y + (x^{5/4}y^{-3/4} - x^{-1})(4x^{5/4}y^{1/4} - x^{-1}y)$
(f) $y'' = 2 - \frac{3}{2}x^{-1/2} + \frac{1}{2}x^{-3/2}$

15. **(a)** $\frac{dy}{dx} = 3f(x)^2 f'(x)$ **(b)** $\frac{dy}{dx} = 4x^3 f'(x^4)$ **(c)** $\frac{dy}{dx} = \frac{5f(\sqrt{x})^4 f'(\sqrt{x})}{2\sqrt{x}}$
(d) $\frac{dy}{dx} = \frac{y - 2xf'(x^2+y^2)}{2yf'(x^2+y^2) - x}$ **(e)** $\frac{dy}{dx} = \frac{80x^4\sqrt{x+y}}{3f(\sqrt{x+y})^2 f'(\sqrt{x+y})} - 1$ **(f)** $\frac{dy}{dx} = \frac{yf'(xy) - 2xf'(x^2)}{2yf'(y^2) - xf'(xy)}$

Section 2.5

1. **(a)** $\frac{dy}{dx} = 5\sec^2(5x+3)$ **(b)** $\frac{dy}{dx} = 2\sec^2 x \tan x$
(c) $\frac{dy}{dx} = -54x^2 \cos^2(6x^3 - 7)\sin(6x^3 - 7)$ **(d)** $\frac{dy}{dx} = -128x^3 \csc^4(8x^4 - 5)\cot(8x^4 - 5)$
(e) $\frac{dy}{dx} = 2x\cos(x^2+1)\sec(x^4-1) + 4x^3 \sin(x^2+1)\sec(x^4-1)\tan(x^4-1)$
(f) $\frac{dy}{dx} = -20\cot^3(5x-2)\csc^5(5x-2) - 15\cot^5(5x-2)\csc^3(5x-2)$
(g) $\frac{dy}{dx} = \frac{9x^2\cos(3x^3+4)\tan(6x^2+3) - 12x\sec^2(6x^2+3)\sin(3x^3+4)}{\tan^2(6x^2+3)}$
(h) $\frac{dy}{dx} = -\frac{[12x^5 \csc^2(2x^6-1)](3x^5+1) + 15x^4\cot(2x^6-1)}{(3x^5+1)^2}$ **(i)** $\frac{dy}{dx} = \frac{4x\sec^2\sqrt{4x^2-3}}{\sqrt{4x^2-3}}$
(j) $\frac{dy}{dx} = \frac{21x^6\cos(6x^7+5)}{\sqrt{\sin(6x^7+5)}}$ **(k)** $\frac{dy}{dx} = -\frac{x\csc^2\sqrt{x^2+1}}{2\sqrt{(x^2+1)}\cot\sqrt{x^2+1}}$

2. **(a)** $\frac{dy}{dx} = \frac{y - 2x\cos(x^2+y^2)}{2y\cos(x^2+y^2) - x}$ **(b)** $\frac{dy}{dx} = \frac{xy^2 - \tan(x+y)\sec^2(x+y)}{\tan(x+y)\sec^2(x+y) - x^2 y}$

(c) $\dfrac{dy}{dx} = -\dfrac{y^2 + 2xy + 2y\sqrt{xy^2 + x^2 y}\,\csc(xy)\cot(xy)}{2xy + x^2 + 2x\sqrt{xy^2 + x^2 y}\,\csc(xy)\cot(xy)}$ **(d)** $\dfrac{dy}{dx} = -\dfrac{4x^{3/2}y^{5/2} + y\sin\sqrt{xy}}{4x^{5/2}y^{3/2} + x\sin\sqrt{xy}}$

(e) $\dfrac{dy}{dx} = \dfrac{x^2 y\csc^2\frac{x}{y} - y^3\sec\frac{y}{x}\tan\frac{y}{x}}{x^3\csc^2\frac{x}{y} - xy^2\sec\frac{y}{x}\tan\frac{y}{x}}$ **(f)** $\dfrac{dy}{dx} = \dfrac{\sec^2(x+y) - y\cos(xy)}{x\cos(xy) - \sec^2(x+y)}$

3. (a) $y = \frac{1}{2} + \frac{\sqrt{3}}{2}\left(x - \frac{\pi}{6}\right)$ **(b)** $y = -1 + 6\left(x - \frac{\pi}{4}\right)$ **(c)** $y = -8 - 48\sqrt{3}\left(x - \right.$
(d) $y = \sqrt[4]{2} - \frac{\pi}{2^{3/4}}\left(x - \frac{1}{2}\right)$ **(e)** $y = \pi - x$ **(f)** $y = 1 - \frac{1+\pi}{2+\pi}(x-2)$

4. (a) $y = \frac{\sqrt{2}}{2} - \sqrt{2}\left(x - \frac{\pi}{4}\right)$ **(b)** $y = \frac{2}{3} + \frac{3\sqrt{3}}{10}\left(x - \frac{\pi}{3}\right)$ **(c)** $y = 1 + \frac{\pi}{4} - x$
(d) $x + 3y = 2$ **(e)** $x = 0$

5. (a) $f^{(4k)}(x) = \sin x,\ f^{(4k+1)}(x) = \cos x,\ f^{(4k+2)}(x) = -\sin x,\ f^{(4k+3)}(x) = -$
(b) $g^{(4k)}(x) = \cos x,\ g^{(4k+1)}(x) = -\sin x,\ g^{(4k+2)}(x) = -\cos x,\ g^{(4k+3)}(x) =$

6. (a) $Df^{-1}(4) = \frac{1}{192}$ **(b)** $Dg^{-1}(0) = -1$ **(c)** $Dh^{-1}\left(\frac{1}{3}\right) = 9$ **(d)** $Dk^{-1}(69) =$
(e) $Dm^{-1}(1) = 2 + 2\tan^2 1$

7. (a) $\frac{dy}{dx} = \frac{1}{3(\sqrt[3]{x})^2}$ for $x \neq 0$ **(b)** $\frac{dy}{dx} = \frac{x}{\sqrt{x^2+1}}$ **(c)** $\frac{dy}{dx} = \frac{1}{(x+1)^2}$
(d) $\frac{dy}{dx} = \frac{1}{\sqrt{(1-x^2)}\arcsin x}$ **(e)** $\frac{dy}{dx} = \frac{1}{2\sqrt{x}+2x^{3/2}}$

8. (a) $y = \frac{x}{7} - \frac{2}{7}$ **(b)** $y = \frac{x}{5} + \frac{8}{5}$ **(c)** $y = 3x - 1$ **(d)** $y = \frac{x}{4} + \frac{3}{2}$ **(e)** y

9. (a) $D(f^{-1}g^{-1})(x) = \frac{g^{-1}(x)}{Df(f^{-1}(x))} + \frac{f^{-1}(x)}{Dg(g^{-1}(x))}$
(b) $D\left(\frac{f^{-1}}{g^{-1}}\right)(x) = \frac{g^{-1}(x)Dg(g^{-1}(x)) - f^{-1}(x)Df(f^{-1}(x))}{[g^{-1}(x)]^2 Df(f^{-1}(x))Dg(g^{-1}(x))}$
(c) $D(f^{-1}\circ g^{-1})(x) = \frac{1}{Df(f^{-1}(g^{-1}(x)))Dg(g^{-1}(x))}$
(d) $D(ff^{-1})(x) = Df(x)f^{-1}(x) + \frac{f(x)}{Df(f^{-1}(x))}$
(e) $D\left(\frac{f^{-1}}{fg^{-1}}\right)(x) = \frac{f(x)g^{-1}(x)Dg(g^{-1}(x)) - f^{-1}(x)g^{-1}(x)Df(x)Df(f^{-1}(x))Dg(g^{-1}(x)) - f(x)}{f(x)^2[g^{-1}(x)]^2 Df(f^{-1}(x))Dg(g^{-1}(x))}$

10. (a) $\frac{dy}{dx} = \frac{10x}{25x^4 - 10x^2 + 2}$ **(b)** $\frac{dy}{dx} = \frac{(\arcsin x)^{-4/5}}{5\sqrt{1-x^2}}$ **(c)** $\frac{dy}{dx} = \frac{12[\text{arcsec } (6x-5)]}{|6x-5|\sqrt{9x^2-15x+}}$
(d) $\frac{dy}{dx} = \frac{3\arctan(3x+2)}{\sqrt{-12x-9x^2-3}} + \frac{3\arcsin(3x+2)}{9x^2+12x+5}$ **(e)** $\frac{dy}{dx} = \frac{4x[1-2\sqrt{4x^4+14x^2+12}\text{arcsec }(4x^2}{(4x^2+7)^2\sqrt{4x^4+14x^2+12}}$
(f) $\frac{dy}{dx} = \frac{\sqrt{1-x^2}\arcsin x - (1+x^2)\arctan x}{(1+x^2)\sqrt{1-x^2}(\arcsin x)^2}$ **(g)** $\frac{dy}{dx} = \frac{\arcsin 2x}{|x|\sqrt{x^2-1}} + \frac{2\text{arcsec }x}{\sqrt{1-4x^2}}$
(h) $\frac{dy}{dx} = \frac{1}{1+(\arctan x)^2} \cdot \frac{1}{1+x^2}$ **(i)** $\frac{dy}{dx} = \frac{\sec^2}{\sqrt{1-\tan^2 x}}$ **(j)** $\frac{dy}{dx} = \frac{1}{2x\sqrt{x-1}}$
(k) $\frac{dy}{dx} = \frac{1}{2(1+\arcsin x)\sqrt{(1-x^2)}\arcsin x}$

11. (a) $\frac{dy}{dx} = -\sqrt{\frac{1-y^2}{1-x^2}}\,\frac{y+\sqrt{1-x^2}\arcsin y}{x+\sqrt{1-y^2}\arcsin x}$ **(b)** $\frac{dy}{dx} = \frac{|xy^3|\sqrt{x^2y^2-1}-y}{x-2xy|xy|\sqrt{x^2y^2-1}}$
(c) $\frac{dy}{dx} = \frac{y[1+(x+y)^2] - \sqrt{1-x^2y^2}}{\sqrt{1-x^2y^2}-x[1+(x+y)^2]}$ **(d)** $\frac{dy}{dx} = \frac{2-4x^3y\sqrt{1-(2x-3y)^2}}{x^4\sqrt{1-(2x-3y)^2}+3}$
(e) $\frac{dy}{dx} = \frac{x^{16}y^{12} + 4x^8 y^6 + 5 - 12x^7 y^6 (x+y)^{1/3}}{9x^8 y^5 (x+y)^{1/3} - x^{16}y^{12} - 4x^8 y^6 - 5}$ **(f)** $\frac{dy}{dx} = \frac{2x(1+x^2y^2) - y\sqrt{1-(x^2+y^2)^2}}{x\sqrt{1-(x^2+y^2)^2} - 2y(1+x^2y^2)}$

12. (a) $y = \frac{\pi}{12} + \left(\frac{\pi}{6} + \frac{1}{\sqrt{3}}\right)\left(x - \frac{1}{2}\right)$ **(b)** $y = \frac{\pi^2}{16} - \frac{\pi}{2} + \frac{\pi}{4}\sqrt{2}x$
(c) $y = \frac{\pi}{4} + \left(2 - \frac{\pi^2}{2}\right)\left(x - \frac{1}{4}\right)$ **(d)** $y = 2x - 2$ **(e)** $y = \frac{\pi}{6} + 2\sqrt{3}\left(x + \frac{13}{6}\right)$ **(f)** $x =$

13. (a) $y = \frac{3\sqrt{3}}{\pi} + \frac{4\pi^2}{9\sqrt{3}-12\pi}(x - \sqrt{3})$ **(b)** $y = \frac{\pi}{6} + \frac{\sqrt{3}}{4} - \frac{1}{2}\sqrt{\frac{3}{2}}x$
(c) $y = \frac{\pi^2}{24} - \frac{3}{\pi}(\sqrt{3} - 1)\left(x - \frac{1}{2}\right)$ **(d)** $y = \frac{4}{\pi} + \frac{\pi^2\sqrt{2}}{16}(x - \sqrt{2})$ **(e)** $y = \frac{\pi}{4} - \frac{2}{\pi^2+4}\left(x - \right.$
(f) $y = \frac{2}{3} - \frac{3\pi}{16(\sqrt{3}-1)}(2x - 1)$

Section 2.6

1. $f'(0) = 0$ **2.** no

764

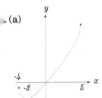

(a)

$f(x) = x^2 + 6x$

ocal min. at $x = -3$

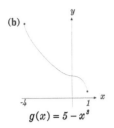

(b)

$g(x) = 5 - x^3$

no local extrema

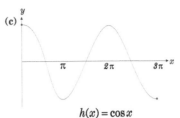

(c)

$h(x) = \cos x$

local min. at $x = \pi$, local max. at $x = 2\pi$

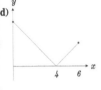

d)

$j(x) = |x - 4|$

local min. at $x = 4$

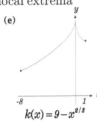

(e)

$k(x) = 9 - x^{2/3}$

local max. at $x = 0$

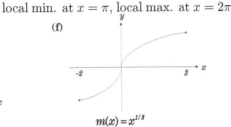

(f)

$m(x) = x^{1/3}$

no local extrema

5. $f'(-3) = 0$ **(b)** – **(c)** $h'(\pi) = 0$, $h'(2\pi) = 0$ **(d)** $j'(4)$ DNE **(e)** $k'(0)$ DNE **(f)** –

6. (a) f has its minimum value of 0 at $x = 0$ and its maximum value of 9 at $x = 3$. $f'(0) = 0$ with $x = 0$ an interior point of $[-1, 3]$ while $f'(3) = 6 \neq 0$ because $x = 3$ is an endpoint of $[-1, 3]$.
(b) g has its min. value of -1 at $x = 3\pi/2$ and its max. value of 1 at $x = \pi/2$. $g'(-3\pi/2) = g'(\pi/2) = 0$ with $x = -3\pi/2$ and $x = \pi/2$ interior points of $[0, 2\pi]$.
(c) h has its minimum value of -1 at $x = -1$ and its maximum value of 8 at $x = 2$. $h'(-1) = 3 \neq 0$ and $h'(2) = 12 \neq 0$ because $x = -1$ and $x = 2$ are endpoints of $[-1, 2]$.
(d) j has its minimum value of 1 at $x = 0$ and its maximum value of 2 at $x = \pi/3$. $j'(0) = 0$ with $x = 0$ an interior point of $[-\pi/4, \pi/3]$. $j'(\pi/3) = 2\sqrt{3} \neq 0$ because $x = \pi/3$ is an endpoint of $[-\pi/4, \pi/3]$.
(e) k has its minimum value of 0 at $x = 2$ and its maximum value of 16 at $x = -2$. $k'(0) = 0$ with $x = 0$ an interior point of $[-2, 4]$. $k'(-2) = -8 \neq 0$ because $x = -2$ is an endpoint of $[-2, 4]$.
(f) m has its minimum value of zero at $x = 2$ and its maximum value of 3 at $x = 5$. $m'(2)$ DNE with $x = 2$ an interior point of $[1, 5]$ while $m'(5) = 1 \neq 0$ because $x = 5$ is an endpoint of $[1, 5]$.

7. (a) Rolle's Theorem applies and $f'(0) = 0$.
(b) Rolle's Theorem does not apply because $\pi/2 \notin Domain\ g$.
(c) Rolle's Theorem applies and $h'(4/3) = 0$. [Note that $h'(0) = 0$ but $0 \notin (0, 2)$.]
(d) Rolle's Theorem applies and $j'(0) = 0$.
(e) Rolle's Theorem does not apply because $k'(0)$ does not exist.
(f) Rolle's Theorem does not apply because $m'(5)$ does not exist.

9. (a) $c = 3$ **(b)** $c = +\sqrt{13/3}$ [Note that $-\sqrt{13/3} \notin (-1, 4)$.]
(c) $c = 0$, $c = \pi$, $c = 2\pi$, $c = 3\pi$ and $c = 4\pi$. **(d)** $c = 1 + \frac{\sqrt{3}}{3}$ [Note that $1 - \frac{\sqrt{3}}{3} \notin (1, 2)$.] **(e)** $c = \sqrt{\frac{4}{\pi} - 1}$ [Note that $-\sqrt{\frac{4}{\pi} - 1} \notin (0, 1)$.] **(f)** $c = 2$

10. (a) $f(x) = \frac{3}{5}x^5 - \frac{5}{3}x^3 + 6x + C$ **(b)** $g(x) = -\frac{2}{x^2} + C$
(c) $h(x) = 8x^{3/2} + C$ **(d)** $j(x) = \frac{5}{6}\sin 6x + C$
(e) $k(x) = 4\arcsin x + C$ **(f)** $m(x) = 2\tan\sqrt{x} + C$

11. (a) $f(x) = x^4 - 3x^2 + 4x + 5$ **(b)** $g(x) = \frac{18}{5}x^{5/3} - \frac{18}{5}x^{5/2} + 2$
(c) $h(x) = 8\arctan x - \pi$ **(d)** $j(x) = 2\tan x + 3$
(e) $k(x) = 11 - 4\sqrt{1 - x^2}$ **(f)** $m(x) = 8\arctan x - 2\pi$

13. (a) $f(x) = -16x^2 + Ax + B$ (b) $g(x) = 2x^4 - 3x^2 + Ax + B$
(c) $h(x) = -\sin x + Ax + B$ (d) $j(x) = \frac{4}{15}x^{5/2} + Ax + B$ (e) $k(x) = \frac{1}{6x^2} +$
(f) $m(x) = -\frac{1}{2}\arctan x + Ax + B$

14. (a) $f(x) = -16x^2 + 3x - 2$ (b) $g(x) = 3x^6 - x + 5$ (c) $h(x) = \frac{4}{x^5} + 24x$
(d) $j(x) = 36x^{7/3} - 6x - 4$ (e) $k(x) = x - \cos x$ (f) $m(x) = 4\arcsin x + 6x +$

15. (a)

(b)

(c)

(d)

(e)

(f)

(g)

(h)

(i)

(j)

16. (a) $x = -4$ is a local minimum.   (b) $x = 0$, $x = 2\pi$, $x = 4\pi$ are local ma
while $x = \pi$, $x = 3\pi$ are local minima.   (c) h has no critical points.
(d) $x = 6$ is a local min. (e) $x = 0$ is not a local extremum. (f) $x = 0$ is a local

17. (a) $x = 4$ is a local min. (b) $x = -3$ is a local max. while $x = -2$ is a local
(c) There are local minima at $x = -1$, $x = 2$ and local maxima at $x = -2$, $x =$
(d) There is a local minimum at $x = -1/2$ and a local maximum at $x = 1/2$.
(e) $x = 0$ is a local max.   (f) The only critical point $x = 0$ is not a local extrem
(g) The only critical point $x = -3$ is not a local extremum.
(h) There is a local min. at $x = \sqrt{3\pi/2}$ and local max. at $x = \sqrt{\pi/2}$, $x = \sqrt{5\pi}$

766

i) $x = 1/2$ is a local minimum.

j) There are local minima at $x = 1$, $x = 4$ and a local maximum at $x = 5/2$.

k) There are local minima at $x = -2$, $x = 8$ and a local maximum at $x = 3$.

l) There are local min. at $x = -2$, -1, $x = 1$, $x = 2$ and local max. at $x = 0$, $\pm\sqrt{\frac{5}{2}}$.

Section 2.7

1. **(a)** The x–intercepts are $x = -2$, $x = 0$ and $x = 4$. The y–intercept is $y = 0$.

(b) The x–intercepts are $x = -2$ and $x = -1$. The y–intercept is $y = 2/5$.

(c) There are no x–intercepts. The y–intercept is $y = -1$.

(d) The x–intercept is $x = -1$. There are no y–intercepts.

(e) There are no x–intercepts. The y–intercept is $y = \sqrt{6}$.

(f) The x–intercepts are $x = \pm 1$. The y–intercept is $y = -\frac{\pi}{2}$.

2. A circle is not a function.

3. **(a)** x^3 **(b)** $x^3 - x^2$ **(c)** $x^3 - x$ **(d)** Complex roots occur in conjugate pairs.

4. All these functions have the same asymptote on the left and the right.

(a) $y = 3$ **(b)** $y = 0$ **(c)** None **(d)** $y = 0$

(e) $y = \sqrt{3}$ **(f)** $y = 1$ **(g)** $y = \frac{\pi}{2} + 1$ **(h)** $y = \frac{4}{3}$ **(i)** $y = \frac{\pi}{4}$

6. **(a)** $x = -3$ and $x = 4$ **(b)** $x = 1$, $x = 2$ and $x = 3$ **(c)** $x = 1$

(d) $x = n$ for n an integer. **(e)** $x\frac{(2n+1)\pi}{4} - \frac{1}{2}$ for n an integer. **(f)** $x = 0$

7. **(a)** $y = 5x - 2$ **(b)** $y = 3x - \frac{1}{2}$ **(c)** None **(d)** None **(e)** $y = 2x + 2$

(f) $y = 2x$

8. Except for (h), all horizontal asymptotes below are on the left and the right.

(a) Horizontal asymptote $y = 0$; vertical asymptotes $x = \pm 3$; no oblique asymptotes.

(b) Horiz asymp $y = 2$; vert asymp $x = -2$, $x = 0$, $x = 3$; no oblique asymptotes.

(c) No horizontal asymptotes; vertical asymptote $x = 1$; oblique asymptote $y = 8x$.

(d) Horiz. asymptote $y = 3$; vert. asymptotes $x = -2$, $x = 5$; no oblique asymptotes.

(e) Horizontal asymptote $y = \pi/2$; no vertical or oblique asymptotes.

(f) Horizontal asymptote $y = 0$; no vertical or oblique asymptotes.

(g) Horizontal asymptote $y = 1$; vertical asymptote $x = \pm 1$; no oblique asymptotes.

(h) Horiz. asymptote $y = -\pi/2$ on the left, $y = \pi/2$ on the right; no vert. asymptotes.

(i) Horizontal asymptote $y = 0$; no vertical asymptote.

(j) Horizontal asymptote $y = \sqrt{3}$; no vertical asymptotes.

(k) Horizontal asymptote $y = 1$; vertical asymptote $x = 0$.

(l) Horizontal asymptote $y = -\frac{\pi}{6}$ on left, $y = \frac{\pi}{6}$ on right; no vertical asymptotes.

9. **(a)** f is concave down everywhere and has no inflection points.

(b) g is concave up for $x < -4$, concave down for $x > -4$ and has infl. point $x = -4$.

(c) h is concave up everywhere and has no inflection points.

(d) k is concave up for $x < 0$, concave down for $x > 0$ and has inflection point $x = 0$.

(e) m is concave down for $x < 1$, concave up for $x > 1$ and has inflection point $x = 1$.

10. All horizontal asymptotes below are on the left and the right.

(a) Horizontal asymptote $y = -1$; vertical asymptote $x = -1$.

(b) Horizontal asymptote $y = 0$; vertical asymptote $x = 0$.

(c) Horizontal asymptote $y = 0$; no vertical asymptotes.

(d) Oblique asymptote $y = -x$; vertical asymptote $x = -1$.

(e) Horizontal asymptote $y = 0$; vertical asymptotes $x = \pm 1$.

(f) Horizontal asymptote $y = -1$; vertical asymptote $x = -1$.

11. **(a)** f is concave down for $x < 2/3$, concave up for $x > 2/3$ and $x = 2/3$ is an inflection point.

(b) g is concave up for $x < 1$ or $x > 2$ while g is concave down for $1 < x < 2$ inflection points at $x = 1$ and $x = 2$.

(c) h is concave up everywhere and has no inflection points.

(d) j is concave up for $x < -3/2$ or $x > 3/2$ while j is concave down for $-3/2$ $3/2$. j has no inflection points.

(e) k is concave up if $-1 < x < 0$ or $x > 1$ while k is concave down if $0 < x$ $x < -1$. There are inflection points at $x = -1$, $x = 0$ and $x = 1$.

(f) m is concave up for $x < -1$ or $x > 1$ while m is concave down if $-1 <$ There are no inflection points.

(g) n is concave up for $\pi/2 < x < 3\pi/2$ while n is concave down for $0 < x < \pi$ $3\pi/2 < x < 2\pi$. There are inflection points at $x = \pi/2$ and $x = 3\pi/2$.

(h) p is concave up for $x < 0$, concave down for $x > 0$ and $x = 0$ is an inflection

(i) q is concave up for $x > 6$ or $x < -2$ while q is concave down for $-2 <$ There are inflection points at $x = -2$ and $x = 6$.

(j) r is concave down if $x < 1$, r is concave up if $x > 1$. $x = 1$ is an inflection

12. $f(x) = x^4$ **13.** Use the fact: $\frac{d^2}{dx^2}[-f(x)] = -f''(x)$.

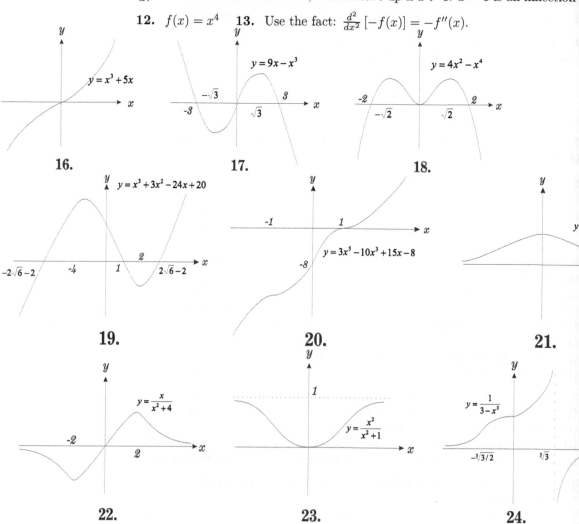

768

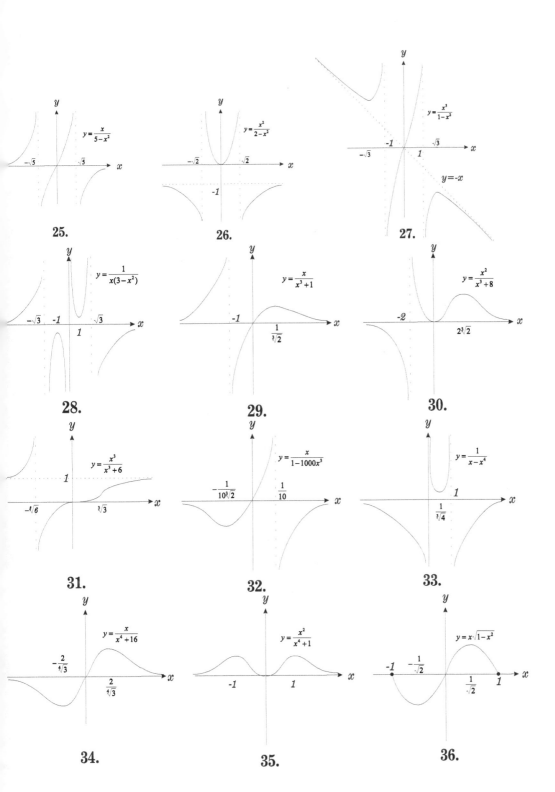

25.

26.

27.

28.

29.

30.

31.

32.

33.

34.

35.

36.

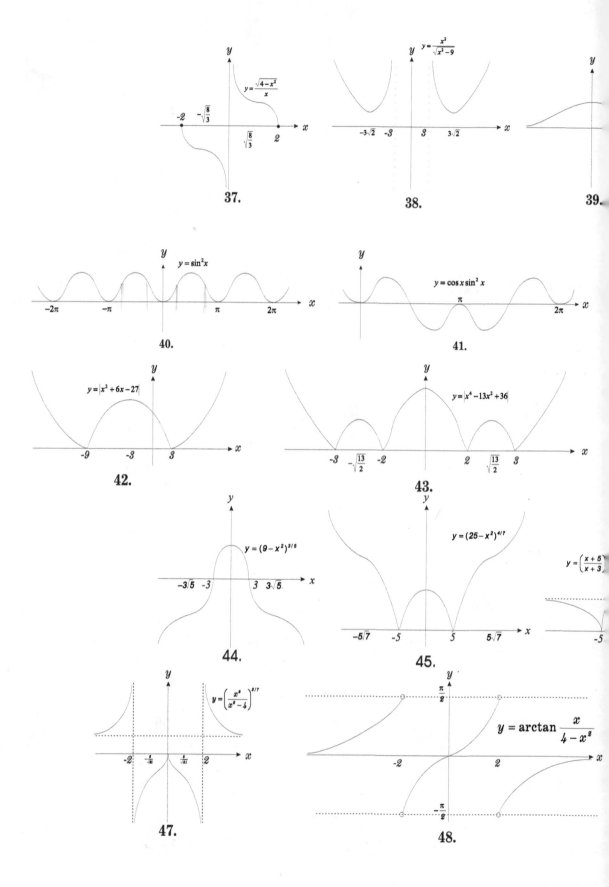

37.

$y = \dfrac{\sqrt{4-x^2}}{x}$

-2 $-\sqrt{\dfrac{8}{3}}$ $\sqrt{\dfrac{8}{3}}$ 2

38.

$y = \dfrac{x^2}{\sqrt{x^2-9}}$

$-3\sqrt{2}$ -3 3 $3\sqrt{2}$

39.

40.

$y = \sin^2 x$

-2π $-\pi$ π 2π

41.

$y = \cos x \sin^2 x$

π 2π

42.

$y = |x^2 + 6x - 27|$

-9 -3 3

43.

$y = |x^4 - 13x^2 + 36|$

-3 $-\sqrt{\dfrac{13}{2}}$ -2 2 $\sqrt{\dfrac{13}{2}}$ 3

44.

$y = (9 - x^2)^{3/5}$

$-3\sqrt{5}$ -3 3 $3\sqrt{5}$

45.

$y = (25 - x^2)^{4/7}$

$-5\sqrt{7}$ -5 5 $5\sqrt{7}$

$y = \left(\dfrac{x+5}{x+3}\right)$

-5

47.

$y = \left(\dfrac{x^3}{x^3 - 4}\right)^{3/7}$

-2 $-\dfrac{3}{\sqrt{81}}$ $\dfrac{3}{\sqrt{81}}$ 2

48.

$y = \arctan \dfrac{x}{4 - x^2}$

$\dfrac{\pi}{2}$ $-\dfrac{\pi}{2}$ -2 2

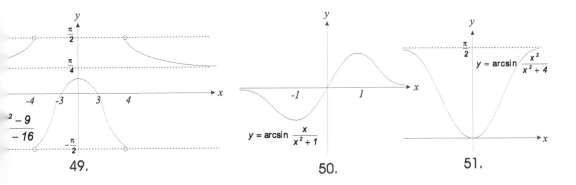

49.

50.

51.

Section 2.8

1. $-2/71$ **2.** $103/15$ **3.** $-2/3$ **4.** 1 **5.** $36/49$ **6.** 1 **7.** 0
8. $7/3$ **9.** $9/5$ **10.** 0 **11.** $7/8$ **12.** $1/2$ **13.** $3/5$ **14.** $10/\pi$
15. DNE **16.** $-\infty$ **17.** DNE **18.** 0 **19.** $-\infty$ **20.** 1 **21.** 0
22. $\sqrt{3/5}$ **23.** 0 **24.** 0 **25.** $-1/3$ **26.** -2 **27.** -2 **28.** ∞
29. π **30.** 2 **31.** $\frac{1}{2}$ **32.** $1/2$ **36.** 1 **37.** 0 **38.** $-\infty$
39. $-1/e$ **40.** 0 **41.** $-\infty$ **42.** 0 **43.** 0 **44.** 0 **45.** $\frac{7}{2}\ln 7$
46. 0 **47.** ∞

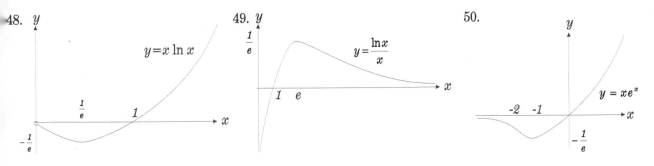

48.

49.

50.

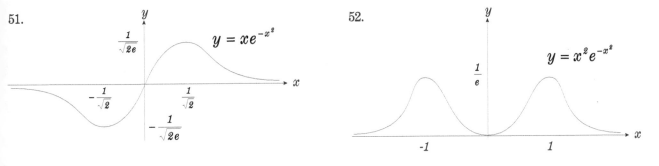

51.

52.

Section 2.9

	Time Interval	Average Rate of Change of Temp	Time Interval	Average Rate of Change of Temp
	$3:00-3:10$	$+.01\,°F/min$	$4:00-4:10$	$-.02\,°F/min$
	$3:10-3:20$	$+.02\,°F/min$	$4:10-4:20$	$-.03\,°F/min$
1. (a)	$3:20-3:30$	$+.03\,°F/min$	$4:20-4:30$	$-.04\,°F/min$
	$3:30-3:40$	$+.05\,°F/min$	$4:30-4:40$	$-.01\,°F/min$
	$3:40-3:50$	$+.02\,°F/min$	$4:40-4:50$	$+.02\,°F/min$
	$3:50-4:00$	$+.01\,°F/min$	$4:50-5:00$	$-.01\,°F/min$

(b)

Time Interval	Average Rate of Change of Temp	Time Interval	Average Rate of Change of Temp
$3:00 - 3:20$	$+.015 \, °F/min$	$4:00 - 4:20$	$-.025 \, °F/min$
$3:20 - 3:40$	$+.040 \, °F/min$	$4:20 - 4:40$	$-.025 \, °F/min$
$3:40 - 4:00$	$+.015 \, °F/min$	$4:40 - 5:00$	$.005 \, °F/min$

2. (a)

Time Interval	Average Rate of Change of Alt	Time Interval	Average Rate of Change of Alt
$0 - 15 \ sec$	$2.7 \, meters/sec$	$120 - 135 \ sec$	$40.7 \, meters/sec$
$15 - 30 \ sec$	$10.7 \, meters/sec$	$135 - 150 \ sec$	$35.3 \, meters/sec$
$30 - 45 \ sec$	$16.7 \, meters/sec$	$150 - 165 \ sec$	$31.3 \, meters/sec$
$45 - 60 \ sec$	$20.0 \, meters/sec$	$165 - 180 \ sec$	$28.0 \, meters/sec$
$60 - 75 \ sec$	$28.7 \, meters/sec$	$180 - 195 \ sec$	$25.3 \, meters/sec$
$75 - 90 \ sec$	$36.0 \, meters/sec$	$195 - 210 \ sec$	$23.3 \, meters/sec$
$90 - 105 \ sec$	$42.0 \, meters/sec$	$210 - 225 \ sec$	$21.3 \, meters/sec$
$105 - 120 \ sec$	$36.0 \, meters/sec$		

(b)

Time Interval	Average Rate of Change of Alt	Time Interval	Average Rate of Change of Alt
$0 - 30 \ sec$	$6.7 \, meters/sec$	$120 - 150 \ sec$	$38.0 \, meters/sec$
$30 - 60 \ sec$	$18.3 \, meters/sec$	$150 - 180 \ sec$	$29.7 \, meters/sec$
$60 - 90 \ sec$	$32.3 \, meters/sec$	$180 - 210 \ sec$	$24.3 \, meters/sec$
$90 - 120 \ sec$	$39.0 \, meters/sec$		

(c)

Time Interval	Average Rate of Change of Alt	Time Interval	Average Rate of Change of Alt
$0 - 45 \ sec$	$10.0 \, meters/sec$	$135 - 180 \ sec$	$31.6 \, meters/sec$
$45 - 90 \ sec$	$28.2 \, meters/sec$	$180 - 225 \ sec$	$23.3 \, meters/sec$
$90 - 135 \ sec$	$39.6 \, meters/sec$		

3. (a)

Year Interval	Average Rate of Change of Population	Year Interval	Average Rate of Change of Pop
$1910 - 1920$	$+27.1 \, people/year$	$1960 - 1970$	$+181.6 \, people/$
$1920 - 1930$	$-62.2 \, people/year$	$1970 - 1980$	$-18.2 \, people/$
$1930 - 1940$	$-11.3 \, people/year$	$1980 - 1990$	$-92.2 \, people/$
$1940 - 1950$	$+71.9 \, people/year$	$1990 - 2000$	$-16.2 \, people/$
$1950 - 1960$	$+48.3 \, people/year$		

(b)

Year Interval	Average Rate of Change of Population	Year Interval	Average Rate of of Populat
$1920 - 1940$	$-36.75 \, people/year$	$1960 - 1980$	$+81.7 \, people/$
$1940 - 1960$	$+60.1 \, people/year$	$1980 - 2000$	$-54.2 \, people/$

4. (a)

Week Interval	Average Rate of Change of Batting Average	Week Interval	Average Rate of Ch of Batting Avera
$0 - 2$	$+.1175 \, points/week$	$14 - 16$	$+.0025 \, points/we$
$2 - 4$	$+.0135 \, points/week$	$16 - 18$	$-.0005 \, points/we$
$4 - 6$	$+.0015 \, points/week$	$18 - 20$	$-.0060 \, points/we$
$6 - 8$	$+.0030 \, points/week$	$20 - 22$	$+.0015 \, points/we$
$8 - 10$	$+.0005 \, points/week$	$22 - 24$	$+.0050 \, points/we$
$10 - 12$	$+.0035 \, points/week$	$24 - 26$	$+.0025 \, points/we$
$12 - 14$	$-.0030 \, points/week$	$26 - 28$	$-.0015 \, points/we$

b)

Week Interval	Average Rate of Change of Batting Average	Week Interval	Average Rate of Change of Batting Average
0 − 4	+.65500 points/week	16 − 20	−.00325 points/week
4 − 8	+.00225 points/week	20 − 24	+.00325 points/week
8 − 12	+.00200 points/week	24 − 28	+.00050 points/week
12 − 16	−.00025 points/week		

5. (a)

Pencils	Average Rate of Change of Cost	Pencils	Average Rate of Change of Cost
0 − 1000	1.979 cents/pencil	5000 − 6000	2.236 cents/pencil
1000 − 2000	2.228 cents/10^3 pencils	6000 − 7000	2.236 cents/10^3 pencils
2000 − 3000	2.233 cents/10^3 pencils	7000 − 8000	2.236 cents/10^3 pencils
3000 − 4000	2.235 cents/10^3 pencils	8000 − 9000	2.236 cents/10^3 pencils
4000 − 5000	2.235 cents/10^3 pencils	9000 − 10000	2.236 cents/10^3 pencils

(b)

Pencils	Average Rate of Change of Cost	Pencils	Average Rate of Change of Cost
0 − 2000	2.104 cents/10^3 pencils	6000 − 8000	2.236 cents/10^3 pencils
2000 − 4000	2.234 cents/10^3 pencils	8000 − 10000	2.236 cents/10^3 pencils
4000 − 6000	2.235 cents/10^3 pencils		

(c) $MC(0) = .009$ $MC(1000) = 2.220$ $MC(2000) = 2.232$ $MC(3000) = 2.234$
$MC(4000) = 2.235$ $MC(5000) = 2.235$ $MC(6000) = 2.236$ $MC(7000) = 2.236$
$MC(8000) = 2.236$ $MC(9000) = 2.236$ $MC(10,000) = 2.236$

6. (a) $\frac{dA}{ds} = 2s$ **(b)** $\frac{dP}{ds} = 4$ **7. (a)** $\frac{dA}{ds} = \frac{s\sqrt{3}}{2}$ **(b)** $\frac{dP}{ds} = 3$

8. (a) $\frac{dV}{ds} = 3s^2$ **(b)** $\frac{dA}{ds} = 12s$ **9. (a)** $\frac{dV}{dr} = 4\pi r^2$ **(b)** $\frac{dA}{dr} = 8\pi r$

10. (a) $\frac{dV}{dr} = 2\pi rh$ **(b)** $\frac{dA}{dr} = 4\pi r + 2\pi h$ **(c)** $\frac{dV}{dh} = \pi r^2$ **(d)** $\frac{dA}{dh} = 2\pi r$

11. (a) $\frac{dV}{dr} = \frac{2}{3}\pi rh$ **(b)** $\frac{dV}{dh} = \frac{1}{3}\pi r^2$ **12.** $W = Fs$ so $\frac{dW}{dt} = \frac{d}{dt}(Fs) = F\frac{ds}{dt} = Fv.$

13. $c'(t) = \frac{4}{5}c(t)^2$

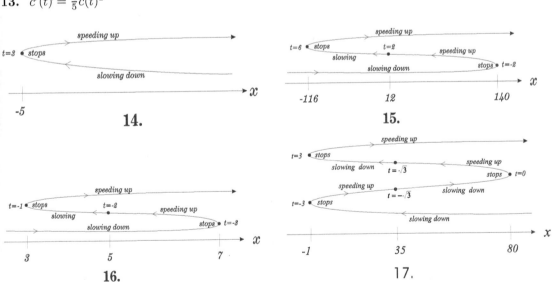

14. **15.** **16.** **17.**

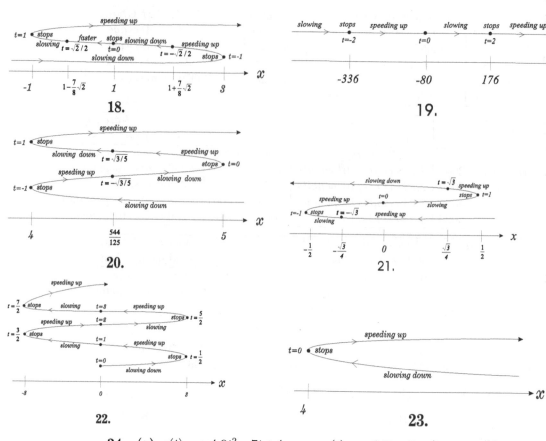

18. **19.**

20. **21.**

22. **23.**

24. **(a)** $s(t) = -4.9t^2 - 7t + 4$ m $\quad v(t) = -9.8t - 7$ m/sec $\quad a(t) = -9.8$ m
(b) $s(t) = -16t^2 + 5t + 19$ ft $\quad v(t) = -32t + 5$ ft/sec $\quad a(t) = -32$ ft/sec²
(c) $s(t) = -4.9t^2 + 19.8t - 5.8$ m $\quad v(t) = -9.8t + 19.8$ m/sec $\quad a(t) = -9.8$ m
(d) $s(t) = -16t^2 + \frac{158}{3}t + 2$ ft $\quad v(t) = -32t + \frac{158}{3}$ ft/sec $\quad a(t) = -32i$ ft,
(e) $s(t) = -4.9t^2 + 32.9t - 47.2$ m $\quad v(t) = -9.8t + 32.9$ m/sec $\quad a(t) = -9.8$ m
(f) $s(t) = -16t^2 + 224t - 632$ ft $\quad v(t) = -32t + 224$ ft/sec $\quad a(t) = -32$ f

25. The watch hits the ground in 4.9 seconds with velocity 48 meters per seco
26. The bird must fly at least 1.5 meters per second to avoid being shot.
27. The depth of the pit is 256 feet or 78.4 meters.
28. The muffin was thrown upwards at 9.9 meters per second and hits the g
with a velocity of 26.2 meters per second.
29. The ball will hit the ground in 5.5 seconds at 89 feet per second.
30. The nest is 144 feet above the ground.
31. The ball was kicked upwards at 64 feet per second and will rise 64 feet.

33. **(i)** $P(q) = q^2 - 1800q$ **(ii)** $MC(q) = 800$, $MR(q) = 2q - 1000$, $MP(q) = 2q -$
(iii) $q = 0$ tons

34. **(i)** $P(q) = 6q^2 + 15q - \frac{q^3}{1000}$ **(ii)** $MC(q) = \frac{3q^2}{1000} + 85$, $MR(q) = 12q + 10$
$MP(q) = 12q + 15 - \frac{3q^2}{1000}$ **(iii)** $q = 5000$ barrels

35. **(i)** $P(q) = \frac{3q}{20} - \frac{q^2}{100,000} - 5,000$
(ii) $MC(q) = \frac{1}{10}$, $MR(q) = \frac{1}{4} - \frac{q}{50,000}$, $MP(q) = \frac{3}{20} - \frac{q}{50,000}$ **(iii)** 7500 litres.

36. **(i)** $P(q) = \frac{4q}{1 + q^2/70,000} - 2q - 40$ **(ii)** $MC(q) = 2$, $MR(q) = \frac{4 - q^2/17}{(1 + q^2/70,}$
$MP(q) = \frac{4 - q^2/17,500}{(1 + q^2/70,000)^2} - 2$ **(iii)** $q = 128.5$ gallons

7. (i) $P(q) = 8,000,000 \arctan(q/10,000) - 200q - 1,000$ (ii) $MC(q) = 300$, $MR(q) = \frac{800}{1+q^2/10^8} + 100$, $MP(q) = \frac{800}{1+q^2/10^8} - 200$ (iii) $q = 17,000$ tons

8. (i) $P(n) = 5n - \frac{n^2}{10,000} - 1,000$ (ii) $MC(n) = 15$, $MR(n) = 20 - \frac{2n+1}{10,000}$, $MP(n) = 5 - \frac{2n+1}{10,000}$ (iii) $n = 25,000$ cars

9. (i) $P(n) = 40n - \frac{3n^2}{1400} - 800$
(ii) $MC(n) = 50$, $MR(n) = 90 - \frac{6n+3}{1400}$, $MP(n) = 40 - \frac{6n+3}{1400}$ (iii) $n = 9333$ coats

10. (i) $P(n) = n^3 + 555n - 63n^2$ (ii) $MC(n) = 126n + 263$, $MR(n) = 3n^2 + 3n + 756$, $MP(n) = 3n^2 - 123n + 493$ (iii) $n = 5$ rings

11. (i) $P(n) = 230n^{3/8} - 90\sqrt{n} - 4,000$ (ii) $MC(n) = 90\sqrt{n+1} - 90\sqrt{n}$, $MR(n) = 230(n+1)^{3/8} - 230n^{3/8}$, $MP(n) = 230(n+1)^{3/8} - 230n^{3/8} - 90\sqrt{n+1} + 90\sqrt{n}$
(iii) $n = 182$ plates

12. (i) $P(n) = \frac{1500n}{1+n/10,000} - 700n - 2,000$ (ii) $MC(n) = 700$, $MR(n) = \frac{1500n+1500}{1+(n+1)/10,000} - \frac{1500n}{1+n/10,000}$, $MP(n) = \frac{1500n+1500}{1+(n+1)/10,000} - \frac{1500n}{1+n/10,000} - 700$
(iii) $n = 4,639$ radios

Section 2.10

1. (a) 24.0 m/sec   (b) 199 m/sec   (c) .019 m/sec   (d) -8.7 m/sec

2. 1131 cm^3/min   **3.** .0071 cm/sec   **4.** 3/8 mile/year   **5.** 0.048 cm/min
6. 180 cm^2/hr   **7.** 37.6 cm^3/min   **8.** 2,890,000 cm^3/hr   **9.** -0.22 cm/sec
10. 0.01 in/sec   **11.** 95.6 mph   **12.** 492.6 km/hr   **13.** 82.4 km/hr
14. 355 km/hr   **15.** 71 cm/sec, .003 rad/sec   **16.** 76 ft/sec   **17.** 11.6 m/min

18. (a) 1.1 cm/min   (b) 1.9 cm/min   (c) 0.64 cm/min

19. 77.7 mph   **20.** 18.4 cm/hr   **21.** 53 m/min   **22.** 2.5 in/min
23. -1.92 radians/min   **24.** -83.8 km/min

Section 2.11

1. (a) $dx = +1$, $dy = 54$, $\Delta y = 74$ (b) $dx = -2$, $dy = -2.75$, $\Delta y = -2.71$
(c) $dx = -1$, $dy = .036$, $\Delta y = .043$ (d) $dx = \pi/24$, $dy = .113$, $\Delta y = .109$
(e) $dx = 2$, $dy = 1$, $\Delta y = .464$ (f) $dx = \pi/16$, $dy = -.196$, $\Delta y = -.191$

2. (a) $d(f+g) = 12x^2\, dx$ (b) $d(7f) = (42x^2 + 42)\, dx$
(c) $f(x) = 2x^3 + 6x - 70$ and $g(x) = 2x^3 - 6x - 389$
(d) $d(fg) = \left[(6x^2 + 6)(2x^3 - 6x - 389) + (6x^2 - 6)(2x^3 + 6x - 70)\right]\, dx$
(e) $d(f/g) = \left[\frac{(6x^2+6)(2x^3-6x-389)-(2x^3+6x-70)(6x^2-6)}{(2x^3-6x-389)^2}\right]\, dx$
(f) $d(f \circ g) = \left[6(2x^3 - 6x - 389)^2 + 6\right]\left[6x^2 - 6\right]\, dx$

4. (a) $\sqrt{26} \approx 5.100$ with error $< .001$.   (b) $\sqrt{48} \approx 6.92857$ with error $< .00039$.
(c) $\sqrt[3]{128} \approx 5.04000$ with error $< .00032$.   (d) $\sqrt[4]{80} \approx 2.990741$ with error $< .000045$.
(e) $\sqrt[5]{33} \approx 2.01250$ with error $< .00016$.   (f) $65^{3/2} \approx 524.000$ with error $< .047$.
(g) $213^{2/3} \approx 35.6667$ with error $< .0008$.   (h) $245^{-3/4} \approx .016129$ with error $< .000024$.
(i) $30^{-7/5} \approx .008496$ with error $< .000032$.   (j) $62^{5/6} \approx 31.1667$ with error $< .003$.

5. (a) $\sin 44° \approx .69477$ with error $< .00011$.   (b) $\cos 31° \approx .85730$ with error $< .00013$.
(c) $\tan 46° \approx 1.03491$ with error $< .00076$.   (d) $\sec 59° \approx 1.9395$ with error $< .0022$.
(e) $\arctan 1.1 \approx .83540$ with error $< .00275$.   (f) $\arcsin 0.4 \approx .4081$ with error $< .0039$.

6. (a) 1.75 **(b)** 1.094 **(c)** 3 **(d)** .247 **(e)** 3.000625 **(f)** .686 **(g)** 1.0443 **(h**

7. 0.2 cm **8.** The space is 1.9 inches exactly (no error).
9. The volume of tin is approximately 2356 cm^3 with an error less than 15 c
10. The extra paint needed is approximately 1 cm^3 with an error less than .000
11. The volume of ice cream is 142 cm^3 with an error less than 2.72 cm^3.

12. (a) 3.75 cm^2 **(b)** 4.77 cm^2 **(c)** 2.89 cm^2 **13.** $18.49

14. .02500 radians, error $<$ .00026 radians. **15.** .0128 radians, error $<$.
radians.

16. (a) $E(y) = 5E(x)$ **(b)** $E(y) = \frac{18x^6 - 12x^3}{3x^6 - 4x^3 + 8}E(x)$ **(c)** $E(y) = -\frac{59x}{(7x-2)(5x-}$
(d) $E(y) = \frac{2x^4}{x^4+3}E(x)$ **(e)** $E(y) = |\pi x \cot(\pi x)|\ E(x)$

17. The error in computing the volume is less than 9%.
18. The error in computing the height of the building is less than 7%.
19. The percent error in computing force is twice the percent of error in measu
20. The error in computing the revenue is less than 1.67%.
21. The error in computing the area is less than 2.4%.
22. The error in saying the bag has volume 15.7 cm^3 is $< 15\%$. **23.** $E(P) = -$

24. (a) $x_{n+1} = x_n - \frac{x_n^2 + x_n - 5}{2x_n + 1}$, $x_1 = 1.80 \pm .023$ and $x_2 = 1.79130 \pm .00019$.
(b) $x_{n+1} = x_n - \frac{2x_n^3 - x_n + 2}{6x_n^2 - 1}$, $x_1 = -1.200 \pm .058$ and $x_2 = -1.1668 \pm .0048$.
(c) $x_{n+1} = x_n - \frac{x_n^3 - 3x_n - 1}{3x_n^2 - 3}$, $x_1 = 1.889$, error $< .033$ and $x_2 = 1.87945$, error $< .$
(d) $x_{n+1} = x_n - \frac{x_n^4 - 8x_n^2 + 11}{4x_n^3 - 16x_n}$, $x_1 = 2.67$, error $< .57$ and $x_2 = 2.53$, error $< .51$.
(e) $x_{n+1} = x_n - \frac{3x_n^{5/3} + 3x^{4/3} - 147}{5x^{2/3} + 4x^{1/3}}$, $x_1 = 8.10714$ with error $< .00016$ and
$x_2 = 8.10673668896$ with error $< 1.1 \times 10^{-10}$.
(f) $x_{n+1} = x_n - \frac{4x^{7/4} - 16x^{5/4} - 4}{7x^{3/4} - 20x^{1/4}}$, $x_1 = 16.2500 \pm .0056$ and $x_2 = 16.24621 \pm .00$
(g) $x_{n+1} = x_n - \frac{4x - 3\tan x}{4 - 3\sec^2 x}$, $x_1 = .856$, error $< .013$ and $x_2 = .84526$, error $< .0$
(h) $x_{n+1} = x_n - \frac{3x - 4\arctan x}{3 - 4/(1+x^2)}$, $x_1 = 1.142 \pm .023$ and $x_2 = 1.12647 \pm .00051$.

25. (a) $\sqrt{24} \approx 4.8989796$ with error less than .0000055.
(b) $\sqrt{38} \approx 6.16441408$ with error less than .00000044.
(c) $\sqrt[3]{62} \approx 3.9578916$ with error less than .0000017.
(d) $\sqrt[3]{-26} \approx -2.9624961$ with error less than .0000025.
(e) $\sqrt[4]{17} \approx 2.0305436$ with an error less than .0000011.
(f) $\sqrt[5]{33} \approx 2.0123466$ with an error less than .0000016.

26. (a) The fixed point of f is approximately 3.6190 with error less than .00
(b) The fixed point of g is approximately -0.65 with error less than .22.
(c) The fixed point of h is approximately 30.56417168 with error less than .0000
(d) The fixed point of j is approximately 8.98849 with error less than .00011.
(e) The fixed point of k is approximately 1.065382 with error less than .00003
(f) The fixed point of m is approximately 0.739085 with error less than .00003

Appendix

1. (a) $h = 5/3$, k any number in $(1, 3)$. **(b)** $h = -\sqrt{2}$, $k = -3/2$.
(c) $h = \sqrt[3]{2}$, $k = \sqrt[4]{2}$. **(d)** $h = .346$ radians, $k = .405$ rad
(e) $h = .553$ radians, $k = .582$ radians. **(f)** $h = \left(\frac{75}{4}\right)^{2/3} - 6$, $k = \left(\frac{675}{886}\right)^{2/5}$

3. (a) $f(x) = 11 + 8(x-2) + 3(x-2)^3$ **(b)** $g(x) = 74 + 93(x-2) + 36(x-2)^2 + 5(x$
(c) $h(x) = -7 + 13(x - 2) + 22(x - 2)^2 + 11(x - 2)^3 + 2(x - 2)^4$
(d) $k(x) = 32 + 80(x - 2) + 160(x - 2)^2 + 240(x - 2)^3 + 240(x - 2)^4 + 120(x -$

. $x = -1$ is a local maximum and $x = 2$ is a local minimum.
. $x = -1$ is a local maximum and $x = 1$ is a local minimum.
. $x = 0$ is a local maximum.   4. $x = 0$ is a local minimum.
. $x = \frac{(2n+1)\pi}{2}$ is a local min. for n even and a local max. for n odd where $n \in Z$.
. $x = 0$ is a local minimum.
. (a) $x = 0$ is a local minimum.   (c) $x = -5$ is a local maximum.
b), (d), (e), (f) There is no conclusion from the Second Derivative Test.

8. min at $x = 6$, max at $x = 4$
9. no min, no max
10. min at $x = -2$, max at $x = 1$
11. no min, max at $x = 2$
12. min at $x = 4$, no max
13. no min, no max
14. no min, no max
15. no min, max at $x = 4$
16. no min, max at $x = 3$
17. no min, max at $x = 1$
18. no min, no max
19. no min, max at $x = 3$
20. min at $x = 5$ and max at $x = -1$.
21. min at $x = 0$ and max at $x = 3$.
22. no min and max at $x = 3 + \sqrt{5}$.
23. no min and no max
24. min at $x = -1$ and max at $x = 1$.
25. min at $x = \frac{1}{3}$ and max at $x = -5$.
26. min at $x = 0$ and no max
27. no min and max at $x = 0$.
28. min at $x = 0$ and no maximum.
29. min at $x = -1$ and max at $x = 6$.

30. (a) 30, 30 (b) $\sqrt{10}$, $\sqrt{10}$ or $-\sqrt{10}$, $-\sqrt{10}$ (c) 225, 225 (d) 5, 5 or -5, -5
(e) There is no solution: there are arbitrarily large products.
31. (a) This fence is a square with side 10 meters.
(b) This fence consists of two sides of length $5\sqrt{3}$ ft and one side of length $10\sqrt{3}$ ft.
(c) Each of the two sides of this fence has length approximately 21.9 meters.

32. (a) This box has length and width $\sqrt[3]{\frac{12}{7}}$ meters and height $\frac{3}{2}\sqrt[3]{\frac{49}{18}}$ meters.

(b) This box has radius $10/\sqrt[3]{60\pi}$ cm and height $\sqrt[3]{450/\pi}$ cm.
(c) This box has radius $\sqrt[6]{450/\pi^2}$ in and height $\sqrt[3]{60/\pi}$ in.

33. To maximize the area bend the entire wire into a circle. To minimize the area bend a piece of wire of length $\frac{\pi}{4+\pi}$ into a circle and a piece of length $\frac{4}{4+\pi}$ into a square.

34. (a) $(\frac{29}{10}, \frac{97}{10})$.   (b) $(\sqrt{\frac{7}{2}}, \frac{7}{2})$ and $(-\sqrt{\frac{7}{2}}, \frac{7}{2})$.   (c) $(4, 2)$ (d) $(2, 1)$

35. The rectangle has length $1/6$ and width $1/10$.
36. The triangle of smallest area has legs of length 14 and 8. There is no such triangle of largest area.

37. (a) The volume of this cylinder is $4\pi\sqrt{3}R^3/9$.
(b) The volume of this cone is $32\pi R^3/81$.
(c) The volume of this cone is $8\pi R^3/3$.
(d) This cylinder has radius and height equal to $16/3$.

38. The width of the window is $\frac{20}{4+\pi}$ meters, and its height is $\frac{10}{4+\pi}$ meters.

39. $(2 + \sqrt[3]{18})\sqrt{1 + \frac{1}{2}\sqrt[3]{18}}$ **40.** The depth of this trough is $5\sqrt{3}$ inches. **41.** $M = Q$

Section 2.13

1. (a) (t, t^2) for $t \in [-1, 4]$ (b) $(\theta, \cos\theta)$ for $\theta \in [0, \pi]$ (c) $(11t - 5, t + 3)$ for $t \in [0, 1]$
(d) $(2 + 3\cos\theta, 3\sin\theta)$ for $\theta \in [\frac{\pi}{2}, \frac{3\pi}{2}]$ (e) $(-3 + 5\cos\theta, 4 + 5\sin\theta)$ for $\theta \in [0, 2\pi]$
(f) $(2\cos\theta, 2\sin\theta)$ for $\theta \in (\frac{\pi}{2}, 2\pi)$

2.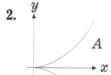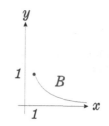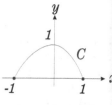

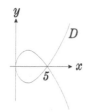 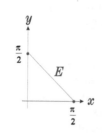 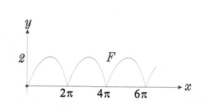

3.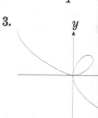

4. (a) $y = \frac{3}{20}x + \frac{46}{5}$ (b) $y = -\frac{25}{252}x - \frac{8}{63}$ (c) $y = 51 - 4x$ (d) $x = 3$
(e) $y = x\sqrt{2} - 4$ (f) $y = x$

5. (a) Slope $T_x = \frac{2}{3x}$, and S is parallel to T_x when $x = 2$, i.e. at $(7,5)$ on C.
(b) Slope $T_x = \frac{10x-7}{6x-1}$, and S is parllel to T_x when $x = -3/2$, i.e. at $(13,/4, 9$

(c) Slope $T_x = \frac{5}{3}\sqrt{\frac{3x^2+1}{5x^2+4}}$ and $S \parallel T_x$ when $x = \sqrt{11/30}$, i.e. at $(\sqrt{21/10}, \sqrt{3}$

(d) Slope $T_x = -\tan \pi x$, and $S \parallel T_x$ when $x = \frac{1}{\pi}\arctan(\sqrt{2}-1)$, i.e. at $\left(\frac{\sqrt{2}-1}{\sqrt{4-2\sqrt{}}}\right.$

(e) Slope $T_x = \frac{1}{x}\sqrt{\frac{4-x^2}{x^2-1}}$, and S is parallel to T_x when $x = \sqrt{2}$, i.e. at $(\pi/4, \pi/$

6. (a) $k = \frac{3}{2}$ (b) $k = \frac{2+\sqrt{7}}{3}$ (c) $k = \frac{3}{2}$ (d) $k = \frac{\pi}{2}, \frac{3\pi}{2}, \frac{5\pi}{2}$ (e) $k = \sqrt{2\sqrt{3}-}$

8. The Mean Value Theorem may give different values of k for f and g.

Review Exercises

1. T	**2.** T	**3.** F	**4.** F	**5.** T	**6.** F	**7.** T	**8.** F
9. F, F	**10.** T	**11.** T	**12.** F	**13.** T	**14.** T	**15.** F	**16.** T
17. T	**18.** T	**19.** F	**20.** F	**21.** T	**22.** T	**23.** F	**24.** T
25. F	**26.** F	**27.** F	**28.** T	**29.** F	**30.** F	**31.** F	**32.** F
33. T	**34.** T	**35.** F	**36.** F	**37.** T	**38.** F	**39.** T	**40.** F

42. vertical cusps at $x = \pm 1$ **43.** $\frac{39}{2}$ **44.** $y = \frac{431}{28} - \frac{64}{35}x$ **45.** $-\frac{1}{2\sqrt{x}}$

46. $\frac{1}{f'(f^{-1}(x))f'((f\circ f)^{-1}(x))}$ **47.** $y = \frac{11}{5} - \frac{3}{5}x$ **48.** $99x + 2^{100}y = 100$

49. $y' = \cos(\sin(\sin x)) \cdot \cos(\sin x) \cdot \cos x$ **50.** $y' = \dfrac{x^3}{\sqrt{(1+\sqrt{1+x^4})(1+x^4)}}$

51. $y' = \dfrac{(\sec^2 x)(\arctan x)(1+x^2)+\tan x-(2x\tan x)(\arctan x)}{(1+x^2)^2}$

52. $x = 0$ is not a local extremum.

54. There is $0 < c < 1$ with $\dfrac{(1+c^2)(1+\arctan c)-\sqrt{1-c^2}\arcsin c}{(1+c^2)(1+\arctan c)^2\sqrt{1-c^2}} = \dfrac{2\pi}{\pi+4}$.

55. $x = 0$ is a local minimum. **56.** $y = 1 + \sqrt{2} - \cot x \csc x$

57. vertical asymptotes at $x = -2$ and $x = 4$; oblique asymptote $y = x + 2$ o
left and the right.

58. (a) $x = 1$ (b) concave up for $x > 0$ and concave down for $x < 0$ (c) $x =$

59.

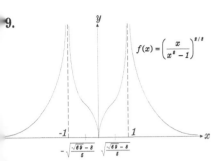

$f(x) = \left(\dfrac{x}{x^6 - 1}\right)^{3/8}$

60.

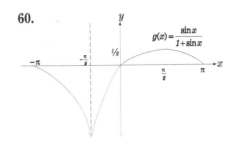

$g(x) = \dfrac{\sin x}{1 + \sin x}$

61. $+\infty$ **62.** 1 **63.** $\frac{1}{4}$ **64.** $\frac{1}{5}$ **65.**

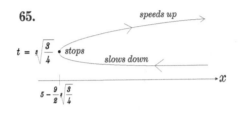

$t = 8\sqrt{\dfrac{3}{4}} \bullet$ stops

speeds up

slows down

$5 - \frac{9}{2}\sqrt{\frac{3}{4}}$

66. $-\frac{12}{25}$ radian/sec **67.** $\frac{4896}{13}$ meters/hour

68. f has minimum value $\frac{\pi}{4}$ at $x = 0$ but no maximum value.

69.

$y = f(x)$

c

70. $6\sqrt{7}$ feet

Chapter 3

Section 3.2

1. (a) $A = \sum_{k=1}^{20}(2k-1)$ (b) $B = \sum_{k=-49}^{49} k^3$ (c) $C = \sum_{k=1}^{30} x^{3k}$

(d) $D = \sum_{k=1}^{40} k \sin^{k-1} x \cos x$ (e) $E = \sum_{k=1}^{90} k^2$ (f) $F = \sum_{k=1}^{90} \pi \frac{k^2}{16}$

2. (a) 52 (b) 15 (c) $\sqrt{6} + \sqrt{7} + 2\sqrt{2}$ (d) $-\frac{7}{12}$ (e) -1

(f) $1 + \frac{x}{2} + \frac{x^2}{3} + \frac{x^3}{4} + \frac{x^4}{5}$ (g) $x^3 + \frac{x^5}{2} + \frac{x^7}{5} + \frac{x^9}{10}$ (h) $-1 + 2^x - 3^x + 4^x$ (i) $2x^2$

3. (a) $\sum_{n=0}^{9}(5n^2 - 12n + 14)$ (b) (b) $\sum_{n=1}^{102}\left[2^{n+3} + 3^{n+2} - (-1)^n\right]$

(c) $\sum_{n=0}^{6}(3x^{4n} + 8x^{2n} - 8)$ (d) $\sum_{n=1}^{42}\left(\frac{x^{2n+1}}{n} - \frac{x^{3n+2}}{n+1} + \frac{x^{4n+3}}{n+2}\right)$

4. (a) $\sum_{n=1}^{4} n^3$ (b) $\sum_{n=2}^{6}(-1)^n 2n^2$ (c) $\sum_{n=1}^{5} \arctan(2n+1)$

(d) $\sum_{n=1}^{4}(-1)^{n-1}\sin\frac{2\pi n}{3}$ (e) $\sum_{n=0}^{2}(n+1)x^{3n+2}$ (f) $\sum_{n=1}^{4}(-1)^{n-1}\frac{x^{3n}}{2(n!)}$

5. (a) 190 (b) 9455 (c) 285 (d) 59,016 (e) 7779 (f) -2079

6. (a) $(x+1)\sum_{k=1}^{2n}(-1)^k x^k$ (b) $(x-1)(x+1)(x^2+1)\sum_{k=0}^{n-1} x^{4k}$ (c) $(x+1)\sum_{k=0}^{n} x^{2k}$

(d) $(1-x)\sum_{k=0}^{n} x^{2k}$ **7.** (a) $I = \sum_{k=0}^{99}(500 - 5k)$ (b) I=\$25,250

8. (a) $W = \frac{1}{1000}\sum_{k=1}^{10}(1100k - 100k^2)$ (b) W=\$22.00 **9.** $\sum_{k=1}^{10} 2(6^{k-1})$

10. $\sum_{k=1}^{24}(25-k)\left[(101-k)^3 - (100-k)^3\right]$ cm^3

11.(a) $\{0,1,2,3,6\}$, $\{0,2,3,4,6\}$, $\{0,3,4,5,6\}$

(b) $\{-3,-2,-1,0,1,2,3,4,4.1,4.2,5\}$, $\{-3,-2.5,-2,-1.5,-1,0,1,2,3,4,5\}$,
$\{-3,-2,-1,-0.5,0,0.5,1,2,3,4,5\}$

(c) $\{-4,-3,-2,-1,0,4\}$, $\{-4,0,1,2,3,4\}$, $\{-4,-2,-1,1,2,4\}$

(d) $\{-6,-5.5,-5,-4.5,-4,-3.5,-3\}$, $\{-6,-5,-4.7,-4.2,-4.1,-4,-3\}$,
$\{-6,-5.8,-5.2,-4.3,-3.4,-3.1,-3\}$

12. (a) 3, 2, 3 (b) 1, 1, 1 (c) 4, 4, 2 (d) 0.5, 1, 0.9

13. (a) $\{0, 0.75, 1.5, 2.25, 3\}$ (b) $\{2, 2\frac{1}{3}, 2\frac{2}{3}, 3, 3\frac{1}{3}, 3\frac{2}{3}, 4, 4\frac{1}{3}, 4\frac{2}{3}, 5\}$

(c) $\{-10, -8.5, -7, -5.5, -4, -2.5, -1\}$ (d) $\{-4, -2, 0, 2, 4, 6, 8, 10\}$

14. (a) $\frac{3}{4}$ (b) $\frac{1}{3}$ (c) $\frac{3}{2}$ (d) 2

15. (a)

$R(P,T,f) = 0.274$

(b)

$R(P,T,f) = 37.5$

(c)

$R(P,T,f) = 0.256$

(d)

$R(P,T,f) = -1.19$

(e)

$R(P,T,f) = 19.5$

(f)

$R(P,T,f) = 17.$

(g)

$R(P,T,f) = 1.37$

(h)

$R(P,T,f) = -0.645$

(i)

$R(P,T,f) = 0.7$

(j)

$R(P,T,f) = 7.76$

16. (a) $L(P,f) = 36$, $U(P,f) = 99$ (b) $L(P,f) = 1.075$, $U(P,f) = 1.85$

(c) $L(P,f) = -0.796$, $U(P,f) = 0.715$ (d) $L(P,f) = -1.70$, $U(P,f) = 0.72$

(e) $L(P,f) = 28$, $U(P,f) = 109$ (f) $L(P,f) = -925$, $U(P,f) = 2$

(g) $L(P,f) = 1.2$, $U(P,f) = 3.7$ (h) $L(P,f) = 1.98$, $U(P,f) = 2.61$

17. (a) $T = \left\{\frac{1}{2}, 2, 3\frac{1}{2}, 4\frac{1}{2}\right\}$, $R(P,T,f) = 150$

(b) $T = \left\{-3\frac{1}{2}, -2, -\frac{1}{2}, 1, 3\frac{1}{2}, 5\frac{1}{2}\right\}$, $R(P,T,f) = 1549$

(c) $T = \left\{\frac{9}{8}, \frac{3}{2}, 1\frac{7}{8}, 2\frac{1}{4}, 2\frac{3}{4}, 3\frac{1}{6}, 3\frac{2}{3}\right\}$, $R(P,T,f) = 17.9$

(d) $T = \left\{-1\frac{2}{3}, -\frac{11}{12}, -\frac{1}{4}, \frac{3}{8}, 1\frac{3}{8}, 2\frac{1}{2}\right\}$ $R(P,T,f) = -6.20$

(e) $T = \{3, 5, 7\}$, $R(P,T,f) = 1.35$ (f) $T = \{-9, -7, -5, -3, -1\}$, $R(P,T,f) = 51.6$

(g) $T = \left\{-1\frac{3}{4}, -1\frac{1}{4}, -\frac{3}{4}, -\frac{1}{4}, \frac{1}{4}, \frac{3}{4}\right\}$, $R(P,T,f) = -.658$

(h) $T = \left\{\frac{\pi}{6}, \frac{\pi}{2}, \frac{5\pi}{6}\right\}$, $R(P,T,f) = 2.09$ (i) $T = \left\{\frac{\pi}{32}, \frac{3\pi}{32}, \frac{5\pi}{32}, \frac{7\pi}{32}\right\}$, $R(P,T,f) = .345$

780

j) $T = \{-\frac{4\pi}{15}, -\frac{2\pi}{15}, 0, \frac{2\pi}{15}, \frac{4\pi}{15}\}$, $R(P, T, f) = 2.59$

8, 19. (a) $L(P, f) = 93$, $U(P, f) = 244$ and $93 \leq 150 \leq 244$.
b) $L(P, f) = 756$, $U(P, f) = 3622$ and $756 \leq 1549 \leq 3622$.
c) $L(P, f) = 14.3$, $U(P, f) = 22.0$ and $14.3 \leq 17.9 \leq 22.0$.
d) $L(P, f) = -10.7$, $U(P, f) = .468$ and $-10.7 \leq -6.20 \leq .468$.
e) $L(P, f) = 1.08$, $U(P, f) = 1.83$ and $1.08 \leq 1.35 \leq 1.83$.
f) $L(P, f) = 43.0$, $U(P, f) = 61.1$ and $43.0 \leq 51.6 \leq 61.1$.
g) $L(P, f) = -.983$, $U(P, f) = -.383$ and $-.983 \leq -.658 \leq -.383$.
h) $L(P, f) = .907$, $U(P, f) = 2.86$ and $.907 \leq 2.09 \leq 2.86$.
i) $L(P, f) = .251$, $U(P, f) = .447$ and $.251 \leq .345 \leq .447$.
j) $L(P, f) = 2.31$, $U(P, f) = 3.14$ and $2.31 \leq 2.59 \leq 3.14$.

20. **(a)** area 169, error less than 75.5 **(b)** area 2189, error less than 1433
(e) area 1.46, error less than .38 **(f)** area 52.1, error less than 9.1
(h) area 1.88, error less than .98 **(i)** area .349, error less than .098
(j) area 2.73, error less than .42

21. **(a)** -6 **(b)** 2.45 **(c)** 440 **(d)** 1.5 **(e)** 24

22. **(a)** $\text{Area}(S_1) \approx \frac{1}{3} + \frac{1}{6n^2}$ with $n = 51$: $\text{Area}(S_1) \approx .33$
(b) $\text{Area}(S_2) \approx 231$ with $n = 147,001$: $\text{Area}(S_2) \approx 231.000$
(c) $\text{Area}(S_3) \approx \frac{20}{3} + \frac{4}{3n^2}$ with $n = 2,001$: $\text{Area}(S_3) \approx 6.667$
(d) $\text{Area}(S_4) \approx 24$ with $n = 1,801$: $\text{Area}(S_4) \approx 24.000$
(e) $\text{Area}(S_5) \approx 132 + \frac{8}{n^2}$ with $n = 2,001$: $\text{Area}(S_5) \approx 132.00$
(f) $\text{Area}(S_6) \approx 168 + \frac{64}{n^2}$ with $n = 9,601$: $\text{Area}(S_6) \approx 168.00$
(g) $\text{Area}(S_7) \approx 19\frac{1}{3} - \frac{4}{3n^2}$ with $n = 134$: $\text{Area}(S_7) \approx 19.33$
(h) $\text{Area}(S_8) \approx 42 - \frac{9}{n^2}$ with $n = 13,501$: $\text{Area}(S_8) \approx 42.000$

23. (b) No. $R(P_{2n}, T_{2n}, f)$ calculates area in the third quadrant as negative and area in the first quadrant as positive.

Section 3.3

1. (a) 48 **(b)** 72 **(c)** 0 **(d)** 88 **(e)** -30 **(f)** 0 **(g)** $\frac{9}{2}\pi$ **(h)** -14 **(i)** $\frac{75}{2}$ **(j)** 0
(k) 0 **(l)** 90 **(m)** 0 **(n)** 9 **(o)** -75 **(p)** 4π **(q)** 0 **(r)** -7 **(s)** -2π **(t)** 90 **(u)** 0

2. (a) $L(P_n, f) = 28$, $U(P_n, f) = 28$, $\int_2^6 7\,dx = 28$
(b) $L(P_n, f) = -25$, $U(P_n, f) = -25$, $\int_{-3}^2 -5\,dx = -25$
(c) $L(P_n, f) = 24 - \frac{24}{n}$, $U(P_n, f) = 24 + \frac{24}{n}$, $\int_0^4 3x\,dx = 24$
(d) $L(P_n, f) = 70 - \frac{25}{n}$, $U(P_n, f) = 70 + \frac{25}{n}$, $\int_0^5 2x + 9\,dx = 70$
(e) $L(P_n, f) = 144 - \frac{108}{n}$, $U(P_n, f) = 144 + \frac{108}{n}$, $\int_1^7 6x\,dx = 144$
(f) $L(P_n, f) = 126 - \frac{90}{n}$, $U(P_n, f) = 126 + \frac{90}{n}$, $\int_2^8 5x - 4\,dx = 126$
(g) $L(P_n, f) = \frac{64}{3} - \frac{32(3n-1)}{3n^2}$, $U(P_n, f) = \frac{64}{3} + \frac{32(3n+1)}{3n^2}$, $\int_0^4 x^2\,dx = \frac{64}{3}$
(h) $L(P_n, f) = 54 - \frac{27(3n-1)}{n^2}$, $U(P_n, f) = 54 + \frac{27(3n+1)}{n^2}$, $\int_{-3}^0 6x^2\,dx = 54$
(i) $L(P_n, f) = 20 - \frac{4(6n-1)}{n^2}$, $U(P_n, f) = 20 + \frac{4(6n+1)}{n^2}$, $\int_0^2 3x^2 + 6x\,dx = 20$
(j) $L(P_n, f) = \frac{188}{3} - \frac{64(3n-1)}{3n^2}$, $U(P_n, f) = \frac{188}{3} + \frac{64(3n+1)}{3n^2}$, $\int_0^4 5 + 2x^2\,dx = \frac{188}{3}$
(k) $L(P_n, f) = \frac{11}{6} - \frac{6n-1}{6n^2}$, $U(P_n, f) = \frac{11}{6} + \frac{6n+1}{6n^2}$, $\int_0^1 x^2 + x + 1\,dx = \frac{11}{6}$
(l) $L(P_n, f) = 300 - \frac{36(11n-3)}{n^2}$, $U(P_n, f) = 300 + \frac{36(11n+3)}{n^2}$, $\int_{-6}^0 2 - 4x + 3x^2\,dx = 300$

3. (a) f is continuous at $x = 0$, hence on $[0, 1]$, because $\lim_{x \to 0} x \sin \frac{1}{x} = 0 = f(0)$. Apply Theorem 3.3.16.

(b) g is cont. at $x = 0$, hence on $[-\pi, \pi]$, as $\lim_{x \to 0} \frac{\sin x}{x} = 1 = g(0)$. Apply Thm. 3.3.16.

(c) h is continuous at $x = 1$, hence on $[0, 2]$, because $\lim\limits_{x \to 1} \dfrac{\sqrt{x+3}-2}{x-1} = \dfrac{1}{4}$.
Apply Theorem 3.3.16.

5. **(a)** 64 **(b)** 72 **(c)** $\frac{125}{2}$ **(d)** $8 + \frac{\pi}{4}$ **(e)** 11

6. **(a)** 3 **(b)** 0 **(c)** 9 **(d)** -5 **(e)** $\frac{201}{14}$ **(f)** 0

7. **(a)** $c = 3$ **(b)** $c = \frac{5}{2}$ **(c)** $c = 3$ **(d)** $c = -1, c = 0, c = 1$ **(e)** $c = -2, c = 0,$
(f) $c = -\frac{5\pi}{2}, c = -\frac{3\pi}{2}, c = -\frac{\pi}{2}, c = \frac{\pi}{2}, c = \frac{3\pi}{2}, c = \frac{5\pi}{2}, c = \frac{7\pi}{2}$

8. **(a)** $R(P_{2n}, T_{2n}, G) = 0$
(b) By (a), 0 and $\int_{-a}^{a} G(x)\,dx$ are between every $L(P_{2n}, G)$ and $U(P_{2n}, G)$. If M
maximum value of G on $[-a, a]$, then $U(P_{2n}, G) - L(P_{2n}, G) \le 2M\frac{a}{n}$. Hence the
unique number between every $L(P_{2n}, G)$ and $U(P_{2n}, G)$. Therefore $\int_{-a}^{a} G(x)\,dx$

Section 3.4

1. **(a)** 6 **(b)** 54 **(c)** 24 **(d)** 0

2. **(a)** $F(x) = \frac{3}{2}x^2$ **(b)** $F(x) = -\frac{5}{2}x^2$ **(c)** $F(x) = 8x$ **(d)** $F(x) = -4x$
(e) $F(x) = 3x^2 + 12x$ **(f)** $F(x) = x^2 - 8x$ **(g)** $F(x) = 24 - \frac{3}{2}x^2$ **(h)** $F(x) = -2x$

3. **(a)** $F'(x) = \sqrt{x^3+8}$ **(b)** $G'(x) = \tan(2x+1)$ **(c)** $H'(x) = (x^2+4)^5$
(d) $I'(x) = -1/x$ **(e)** $J'(x) = -|\sin x|$ **(f)** $K'(x) = 1/(2x^{1/2} + 2x^{3/2})$
(g) $L'(x) = \sin x \arctan(\cos x)$ **(h)** $M'(x) = 1/(1 - x^2)$ **(i)** $N'(x) = 2x \cos x$
$\cot \sqrt{x}$
(j) $P'(x) = 5\sin^4(5x - 4) - 2\sin^4(2x + 3)$
(k) $Q'(x) = \frac{1}{3}x^{-2/3}(x^{2/3} + 1)^{100} - \frac{1}{4}x^{-3/4}(\sqrt{x} + 1)^{100}$
(l) $R'(x) = \sqrt[5]{3\tan x + 4}\ \sec^2 x + \sqrt[5]{3\cot x + 4}\ \csc^2 x$

4. **(a)** 480 **(b)** -33 **(c)** $24,208/15$ **(d)** $52/3$ **(e)** $57/2$ **(f)** 1
(g) $\frac{2}{3}\sqrt{3} - \sqrt{2}$ **(h)** $\pi/2$ **(i)** $2 - \frac{\sqrt{2}}{2}$ **(j)** $4\pi/3$ **(k)** 512 **(l)** 2

5. **(a)** $40/3$ **(b)** $53/2$ **(c)** $266/5$ **(d)** 5 **(e)** 5
(f) $678/5$ **(g)** $19/4$ **(h)** $2 + \frac{\sqrt{2}}{2} + \frac{\sqrt{3}}{2}$ **(i)** $\sqrt{2}$ **(j)** $6 + \frac{\sqrt{2}}{2} - \frac{\sqrt{3}}{2}$

6. **(a)** $343/6$ **(b)** 72 **(c)** $1/6$ **(d)** 8 **(e)** $407/4$
(f) $37/12$ **(g)** $2 - \frac{\pi}{2}$ **(h)** $\frac{3\pi}{8} - 1$ **(i)** $6\sqrt{2}$ **(j)** $\frac{16\pi}{3} - 4\sqrt{3}$

7. **(a)** 12 **(b)** $\sqrt{3}$ **(c)** $4/3$ **(d)** 36 **(e)** $64/3$ **(f)** $253/12$
(g) $1 - \frac{\pi}{4}$ **(h)** $\frac{\pi}{2} - \frac{1}{2}$

8. **(a)** 6 **(b)** 12 **(c)** $38/15$ **(d)** $\frac{2}{\pi}$ **(e)** $\frac{\pi}{4}$ **(f)** $\frac{12(2-\sqrt{2})}{\pi}$

9. No – the Second Fundamental Theorem only applies to continuous integr
The integrand $\frac{2}{x^3}$ has a vertical asymptote at $x = 0$ and is not continuous there

Section 3.5

1. **(a)** $\int \frac{61}{(8x+7)^2}\,dx = \frac{3x-5}{8x+7} + C$ **(b)** $\int \frac{x+2}{\sqrt{x^2+4x+1}}\,dx = \sqrt{x^2+4x}$
(c) $\int 60x^2(x^3+7)^{19}\,dx = (x^3+7)^{20} + C$ **(d)** $\int 5\cos^4 x \sin x\,dx = -\cos^5 x$
(e) $\int 6\cot^5 x \csc^2 x\,dx = -\cot^6 x + C$ **(f)** $\int \frac{21}{9x^2-12x+53}\,dx = \arctan\frac{3x-}{7}$

2. $\frac{5}{8}x^8 + C$ **3.** $\frac{3}{10}x^{10} - \frac{4}{7}x^7 - 5x + C$

4. $-\frac{3}{x^2} - \frac{12}{x} + C$ **5.** $2x^{3/2} - 10x^{1/2} + C$ $x^{2/3 + \wedge i3 + 1}$

6. $10x^{2/5} + \frac{12}{x^{2/3}} + C$ **7.** $\frac{3}{1-x} + C$

8. $4\tan x - x + C$ **9.** $-6\csc x + C$

10. $-6\cos x - 4\sin x + C$ **11.** $5\arctan x + C$

12. $\frac{x^5}{5} - 2x^3 + 9x + C$ **13.** $-\frac{5}{2x^2} + \frac{1}{x^3} + \frac{1}{4x^4} - \frac{1}{x^5} + C$

14. $2x^4 + 20x^3 + 75x^2 + 125x + C$ **15.** $\frac{1}{2}\arctan(2x) + C$

16. $\frac{2}{5}x^{5/2} - \frac{14}{3}x^{3/2} + 6x^{1/2} + C$ **17.** $\frac{2}{5}x^{5/2} + 4x^{3/2} + 18x^{1/2} + C$

18. $5x^3 - \frac{7}{2}x^2 - 2x + C$ **19.** $\frac{1}{15}\arctan\frac{3x}{5} + C$

20. $2x^{5/2} - \frac{16}{3}x^{3/2} + 18x^{1/2} + C$ **21.** $\arcsin\frac{x}{2} + C$

22. $\frac{3}{5}x^{5/3} - \frac{6}{7}x^{7/6} - 9x^{2/3} + C$ **23.** $\frac{1}{12}\arctan\frac{3x}{4} + C$

24. $-\frac{1}{x} + \frac{8}{3}x^{-3/2} - \frac{2}{x^2} + C$ **25.** $\frac{8}{9}\arctan\frac{x}{9} + C$

26. $\frac{4}{27}x^{27/4} - \frac{12}{19}x^{19/4} + \frac{12}{11}x^{11/4} - \frac{4}{3}x^{3/4} + C$ **27.** $\frac{12}{5}\arctan 5x + C$

28. $-\frac{1}{x} - \frac{1}{3x^3} + \frac{12}{5x^5} + C$ **29.** $\frac{1}{3}\arcsin(3x) + C$

30. $-\frac{8}{x} - \frac{6}{x^2} - \frac{2}{x^3} - \frac{1}{4x^4} + C$ **31.** $\arcsin\frac{x}{6} + C$

32. $\frac{6}{5}x^{5/2} - \frac{4}{3}x^{3/2} + 24\sqrt{x} + \frac{16}{\sqrt{x}} + C$ **33.** $\frac{1}{4}\arcsin(2x) + C$

34. $\frac{1}{7}\arcsin\frac{7x}{3} + C$ **35.** $\frac{1}{6}x^6 - \frac{2}{5}x^5 - 2x^4 + \frac{16}{3}x^3 + 8x^2 - 32x + C$

36. $\frac{2}{5}x^{5/2} + \frac{1}{2}x^2 - \frac{2}{3}x^{3/2} - x + C$ **37.** $\frac{3}{5}x^{5/3} + \frac{3}{4}x^{4/3} + x + C$

38. 198 **39.** -72 **40.** $12 - \frac{8}{3}\sqrt{3} - 2\sqrt{2}$ **41.** $\frac{472}{15}$ **42.** $\frac{167}{5}$ **43.** $\frac{18}{5}$ **44.** π **45.** $\frac{\pi}{120}$

46. $\frac{1}{a}\text{arcsec}\frac{x}{a} + C$ **47.** (a) $\frac{175}{4}$ (b) $\frac{1126}{45}$ (c) $\frac{\pi}{18}$ (d) $\frac{\pi}{64}$ (e) $\frac{241}{81}$ (f) $\frac{\pi}{36}$

48. $\frac{1}{2}$ **49.** $\frac{8}{9}(7 - 4\sqrt{2})$ **50.** $4\sqrt{2} - 2\arcsin\frac{2\sqrt{2}}{3}$ **51.** $\frac{13}{5}\arctan\frac{1}{5} - \frac{1}{6}$

Section 3.6

1. $\frac{1}{5}\sin(5x + 3) + C$ **2.** $\frac{1}{40}(4x + 1)^{5/2} - \frac{1}{24}(4x + 1)^{3/2} + C$ **3.** $\frac{(8x^2 + 3)^{11}}{176} + C$

4. $\frac{5}{81}(9x + 7)^{9/5} + C$ **5.** $-\frac{1}{30(5x^3 + 7)^2} + C$ **6.** $\frac{1}{30}(5x^4 + 6)^{3/2} + C$

7. $\frac{2}{135}(3x + 5)^{5/2} - \frac{20}{81}(3x + 5)^{3/2} + \frac{50}{27}(3x + 5)^{1/2} + C$ **8.** $\frac{x}{2} + \frac{1}{4}\sin 2x + C$

9. $\frac{1}{5}\sec^5 x + C$ **10.** $\frac{1}{5}\sin^5 x - \frac{1}{7}\sin^7 x + C$

11. $\frac{1}{9}\sin^9 x - \frac{2}{11}\sin^{11} x + \frac{1}{13}\sin^{13} x + C$ **12.** $\frac{x}{16} - \frac{1}{64}\sin 4x - \frac{1}{48}\sin^3 2x + C$

13. $\frac{1}{4}\sin 2x + \frac{1}{28}\sin 14x + C$ **14.** $\frac{1}{5}\sin^5 x - \frac{3}{7}\sin^7 x + \frac{1}{3}\sin^9 x - \frac{1}{11}\sin^{11} x + C$

15. $-\cos x + \frac{1}{3}\cos^3 x + C$ **16.** $\frac{1}{4}\arctan(\frac{x+3}{4}) + C$

17. $\frac{3x}{8} + \frac{1}{4}\sin 2x + \frac{1}{32}\sin 4x + C$ **18.** $-\frac{1}{7}\cos^7 x + \frac{2}{9}\cos^9 x - \frac{1}{11}\cos^{11} x + C$

19. $\frac{1}{512}\left[-\frac{1}{(8x+2)^3} + \frac{3}{(8x+2)^4} + \frac{308}{5(8x+2)^5}\right] + C$ **20.** $\sin x - \frac{2}{3}\sin^3 x + \frac{1}{5}\sin^5 x + C$

21. $\frac{1}{2}\cos x - \frac{1}{14}\cos 7x + C$ **22.** $\frac{1}{8}(4x + 9)^{3/2} + \frac{1}{8}(4x + 9)^{1/2} + C$

23. $\frac{2}{5}\sqrt{5\sin x + 3} + C$ **24.** $\frac{3}{20}(4x^2 - 3)^{5/3} + \frac{27}{16}(4x^2 - 3)^{2/3} + C$

25. $\frac{1}{20}(5\tan x + 8)^4 + C$ **26.** $\arcsin(\frac{x+1}{3}) + C$

27. $\frac{5x}{16} - \frac{1}{24}\sin 12x + \frac{1}{128}\sin 24x + \frac{1}{288}\sin^3 12x + C$ **28.** $\frac{1}{3}\arctan(\frac{x+2}{3}) + C$

29. $-\frac{1}{4}\cos 2x - \frac{1}{24}\cos 12x + C$ **30.** $\frac{3x}{128} - \frac{1}{128}\sin 4x + \frac{1}{1024}\sin 8x + C$

31. $\frac{1}{2}(\arctan x)^2 + C$ **32.** $-\frac{1}{12}\cos^6 2x + \frac{3}{16}\cos^8 2x - \frac{3}{20}\cos^{10} 2x + \frac{1}{24}\cos^{12} 2x + C$

33. $\frac{1}{8}\arctan(\frac{2x+5}{4}) + C$ **34.** $\frac{1}{1+\csc x} + C$ **35.** $\frac{1}{2}\arcsin(\frac{x^2+3}{\sqrt{11}}) + C$

36. $-\frac{1}{15}\csc^3 5x + C$ **37.** $-\cos x + \cos^3 x - \frac{3}{5}\cos^5 x + \frac{1}{7}\cos^7 x + C$

38. $\frac{5x}{128} + \frac{1}{144}\sin^3 6x - \frac{1}{384}\sin 12x - \frac{1}{3072}\sin 24x + C$ **39.** $-\frac{2}{15}(5\cot x + 2)^{3/2} + C$

40. $-\cot x + C$ **41.** $\frac{3x^2+3}{16} - \frac{1}{8}\sin(2x^2 + 2) + \frac{1}{64}\sin(4x^2 + 4) + C$

42. $\frac{1}{10}(3x^5 + 5x^3 + 1)^{2/3} + C$ **43.** $\frac{1}{4}\arctan(x^2/2) + C$

44. $\sin\sqrt{x} - \frac{1}{9}\sin 9\sqrt{x} + C$ **45.** $\frac{1}{\sin x + \cos x} + C$

46. $-\frac{1}{7}\cot^7 x + C$ **47.** $\frac{1}{2}(\arcsin x)^2 + C$ **48.** $\frac{1}{3}x^3 + C$

49. $|\sec x| + C$ **50.** $\frac{1}{3}\arcsin(\frac{3x+2}{3}) + C$ **51.** $\frac{1}{2}\arctan(\frac{1+\sin x}{2}) + C$

52. $\sqrt{10}-\sqrt{2}$ **53.** $\frac{16}{3}-2\sqrt{3}$ **54.** $\frac{233}{4480}$ **55.** $\frac{1}{202}(9^{101}-3^{101})$

56. $\frac{1}{4}$ **57.** $\frac{11728}{315}$ **58.** $\frac{1}{5}(54^{5/3}-17^{5/3})$ **59.** $\frac{1}{\sqrt{6}}\arctan\sqrt{6}-\frac{1}{\sqrt{6}}\arctan$

60. $\frac{31}{5}$ **61.** $\frac{\pi}{4}-\arctan\frac{1}{2}$ **62.** $\frac{\pi}{6}$ **63.** $-\frac{1}{2}$

64. $\frac{802}{125}$ **65.** $\frac{8}{3\pi}$ **66.** $\frac{\pi^2}{16}$ **67.** $\frac{5}{3\pi}$ **68.** $\frac{1}{128}\left(35+\frac{507\sqrt{3}}{16\pi}\right)$ **69.** $\frac{1}{55\pi}(19+14\sqrt{2})$

70. $3/22$ **71.** 18 **72.** $\frac{64\pi}{3}-8\sqrt{3}$ **73.** 3

74. $\frac{8}{51}5^{51/50}-\frac{16}{51}3^{51/50}-\frac{1}{101}5^{101/50}+\frac{2}{101}3^{101/50}-15(5^{1/50})+30(3^{1/50})$

75. $\frac{4}{15}$ **76.** $\frac{2}{3}\sqrt{6}-\frac{3}{2}$ **77.** $\frac{10}{3}-\pi$ **78.** $17\arctan 4-4$

Section 3.7

1. (a) $\frac{9}{2}$ (b) $\frac{9\pi}{16}$ (c) $\frac{9\sqrt{3}}{8}$ (d) $\frac{24\sqrt{3}}{5}$ (e) $\frac{6\sqrt{3}}{5}$

2. (a) 120 (b) 8π (c) 8 (d) 48 (e) 6π

3. (a) $\frac{512}{15}$ (b) $\frac{256\pi}{15}$ (c) $\frac{256\sqrt{3}}{15}$ (d) 64 (e) $16\sqrt{3}$

4. (a) $\frac{16ab^2}{3}$ (b) $\frac{2\pi ab^2}{3}$ (c) $\frac{4ab^2\sqrt{3}}{3}$ (d) $\frac{16a^2b}{3}$ (e) $\frac{4a^2b}{3}$ **5.** (a) $\frac{S^2H}{3}$ (b) $\frac{\ }{1}$

6. (a) $\frac{128\pi}{7}$ (b) 2π (c) $\frac{\pi^2}{2}$ (d) $\pi\sqrt{3}$ (e) $\frac{256\pi}{5}$ (f) $\frac{369\pi}{10}$
(g) $\frac{8\pi^2}{3}-2\pi\sqrt{3}$ (h) 160π (i) $4\pi\sqrt{2}$ (j) $\frac{76\pi}{3}$

7. (a) $\frac{40\pi}{3}$ (b) 625π (c) 45π (d) 512π (e) $\frac{324\pi}{5}$

8. (a) $\frac{\pi}{6}$ (b) $\frac{2\pi}{3}(10\sqrt{10}-1)$ (c) $\frac{17,408\pi}{15}$ (d) 216π (e) $\frac{913\pi}{15}$ (f) 8π (g)
(h) $2\pi\sqrt{2}$ (i) $\frac{1944\pi}{5}$ (j) $\frac{72\pi}{5}$

9. (a) $\frac{162\pi}{5}$ (b) $\frac{640\pi}{3}$ (c) $\frac{256\pi}{3}$ (d) 80π (e) $\frac{4\pi^2}{3}-\pi\sqrt{3}$

Section 3.8

1. (a) 1.218 (b) $.996$ (c) 1.095 (d) 1.107 (e) $.996$ (f) 1.218
(g) 1.0989 (h) $.019$ (i) $.037$ (j) $.0033$

2. (a) $.717$ (b) 1.05 (c) $.910$ (d) $.883$ (e) $.717$ (f) 1.05 (g)
(h) $.042$ (i) $.084$ (j) $.017$

3. (a) 3.0896 (b) 3.2056 (c) 3.1386 (d) 3.1477 (e) 3.0896
(f) 3.1476 (g) 3.141698 (h) $.0061$ (i) $.012$ (j) $.00024$

4. (a) $.845$ (b) $.927$ (c) $.879$ (d) $.886$ (e) $.845$ (f) $.927$ (g)
(h) $.0054$ (i) $.011$ (j) $.0005$

5. (a) $.32074$ (b) $.37074$ (c) $.34699$ (d) $.34574$ (e) $.32074$
(f) $.37074$ (g) $.34658$ (h) $.00061$ (i) $.0012$ (j) $.00024$

6. (a) 2.8080 (b) 3.1171 (c) 2.9556 (d) 2.9782 (e) 2.8080
(f) 3.1171 (g) 2.95788 (h) $.0052$ (i) $.010$ (j) $.00013$

7. (a) $.39914$ (b) $.47768$ (c) $.43903$ (d) $.43841$ (e) $.39914$
(f) $.47768$ (g) $.4388260$ (h) $.0011$ (i) $.0022$ (j) $.0000026$

8. (a) 1.083 (b) 1.186 (c) 1.109 (d) 1.117 (e) 1.083 (f) 1.1
(g) 1.1114 (h) $.0039$ (i) $.0078$ (j) $.0017$

9. (a) $.251$ (b) $.251$ (c) $.260$ (d) $.251$ (e) $.184$ (f) $.317$
(g) $.2567$ (h) $.016$ (i) $.032$ (j) $.0017$

10. (a) $.47$ (b) $.47$ (c) $.52$ (d) $.47$ (e) $.23$ (f) $.72$ (g) $.5$
(h) $.12$ (i) $.24$ (j) $.064$

11. (a) 10.9 (b) 10.7 **12.** (a) 1.8 (b) 1.8 **13.** (a) -4.8 (b)

14. (a) 5.9 (b) 5.9 **15.** (a) 2.3 (b) 2.4

. (a) 48 miles (b) 26 km (c) 19.5 km (d) $14\frac{2}{3}$ miles (e) 26.5 miles
(f) 36 km (g) 3.5 km (h) $\frac{9658}{35} \approx 276$ miles (i) $\frac{8}{\pi}$ miles (j) $\frac{1}{240}$ km

2. $\frac{70}{\pi}$ cm 3. 288 feet 4. $453\frac{1}{8}$ feet 5. $40\sqrt{6} \approx 98$ feet/sec 6. $64\sqrt{\frac{2}{3}} + 48 \approx 100$ feet

7. 128 pounds 8. $\frac{4}{\pi}$ kg 9. $\frac{9}{4}$ kg 10. $\frac{2\pi}{3}$ pounds 11. $\frac{\pi}{4}$ pounds

12. 12 Joules 13. 30 foot–pounds 14. $\frac{15}{2}$ foot–pounds 15. $\frac{\pi}{8}$ joules

16. $\frac{5}{8}$ foot–pounds 17. 90 foot–pounds 18. $1,470,000$ joules 19. $196,000,000$ joules

20. $1.99 \times 10^{15} \int_{6.37 \times 10^6}^{7.87 \times 10^6} \frac{6.372 \times 10^6 - x}{x^2}\, dx$ 21. (a) $117,600$ Ntns (b) $156,800$ Ntns

22. (a) $1,764,000$ Newtons (b) $3,034,080$ Newtons (c) $3,669,120$ Newtons

23. (a) $156,800/3 \approx 52,267$ Newtons (b) $246,300$ Newtons

24. (a) 211 pounds (b) 105.5 pounds

25. .94 Newtons 26. 325 pounds 27. $26,761,000$ Newtons 28. .64 inches

29. The force is $10,000$ pounds on the wall at the shallow end, $62,500$ pounds on the wall at the deep end and approximately $81,250$ pounds on each of the sides.

30. 2.45 meters

Section 3.10

1. (a) $X_5 = -9$ kg–mtrs (b) $X_{-4} = 30$ ft–lbs (c) $X_{-8} = 8$ gm–cm (d) $X_7 = 12$ in–oz
(e) $X_0 = -15$ kg-cm (f) $X_{-5} = 15$ ft–lbs

2. (a) $X_0 = 20$ ft–lbs (b) $X_5 = -43$ ft–lbs (c) $X_{-3} = -15$ gram–cm
(d) $X_2 = -28$ gram–cm (e) $X_{-4} = 33$ kg–meters

3. (a) $\frac{20}{9}$ ft (b) $\frac{27}{14}$ ft (c) $-\frac{57}{14}$ cm (d) $\frac{3}{5}$ cm (e) $-\frac{59}{23}$ meters

5. (a) $\frac{46}{3}$ kg–meters (b) $\frac{11-3^{23}}{48}$ gm–cm (c) $\frac{10,064}{135}$ ft–lbs
(d) $2\sqrt{6} - 3$ ft–lbs (e) $\frac{5216}{35}\sqrt{3} - \frac{20192}{105}$ gm–cm (f) $-\frac{68}{15}$ ft–lbs
(g) $\frac{6\sqrt{3}}{\pi}$ inch–ounces (h) $\frac{1}{\pi}$ kg–meters

6. (a) $\frac{13}{10}$ meters (b) $\frac{7 \cdot 3^{24} + 25}{24(3^{22} - 1)}$ cm (c) $\frac{2188}{465}$ ft (d) $\frac{2\sqrt{6} - 3}{\arcsin(4/5) - \arcsin(1/5)}$ ft
(e) $\frac{10(93\sqrt{3} - 103)}{7(19 - 21\sqrt{3})}$ cm (f) $-\frac{7}{5}$ ft

8. (a) $X_{-2}(P) = 390$, $Y_4(P) = -150$. (b) $X_{-1}(P) = \frac{7\pi r^2}{2}$, $Y_{-1}(P) = \frac{14r^3}{3} + \frac{7\pi r^2}{2}$.
(c) $X_2(P) = -\frac{475}{6}$ and $Y_1(P) = \frac{95}{3}$. (d) $X_1(P) = \frac{11,232}{25}$ and $Y_{-2}(P) = 612$.
(e) $X_4(P) = -20$ and $Y_{-3}(P) = \frac{1720}{81}$.
(f) $X_{-5}(P) = \frac{352}{5}\sqrt[3]{2}$ and $Y_4(P) = \frac{64}{7}\sqrt[3]{4} - \frac{256}{5}\sqrt[3]{2}$.
(g) $X_2(P) = 950$ and $Y_3(P) = 300$. (h) $X_{-3}(P) = \frac{247}{2}$ and $Y_2(P) = -\frac{57}{2}$.

9. (a) $\left(\frac{9}{2}, \frac{3}{2}\right)$ (b) $\left(0, \frac{4r}{3\pi}\right)$ (c) $\left(-\frac{39}{28}, \frac{33}{14}\right)$ (d) $\left(\frac{103}{25}, \frac{9}{4}\right)$ (e) $\left(1, \frac{5}{27}\right)$
(f) $\left(\frac{1}{2}, \frac{5}{7}\sqrt[3]{2}\right)$ (g) $\left(\frac{25}{3}, 5\right)$ (h) $\left(-\frac{122}{123}, \frac{63}{41}\right)$

11. (a) $X_2(P) = -\frac{243}{4}$, $Y_1(P) = \frac{243}{10}$ (b) $X_0(P) = 256\pi$, $Y_0(P) = 160\pi$
(c) $X_2(P) = -\frac{100}{3}$, $Y_3(P) = \frac{892}{3}$ (d) $X_{-1}(P) = 864$, $Y_5(P) = 864$
(e) $X_2(P) = \frac{96}{5}\sqrt{6} - 48\sqrt{3}$, $Y_1(P) = 45 + 24\sqrt{3} - 48\sqrt{6}$ (f) $X_{-2}(P) = 450$, $Y_{-4}(P) = \frac{3168}{5}$
(g) $X_3(P) = -1440$, $Y_5(P) = 5040$ (h) $X_0(P) = 60\pi - 80$, $Y_0(P) = 64$

12. (a) $\left(\frac{1}{2}, \frac{8}{5}\right)$ (b) $(8, 5)$ (c) $\left(\frac{11}{8}, \frac{343}{40}\right)$ (d) $(3, 9)$ (e) $\left(\frac{3(4\sqrt{6} - 5\sqrt{3})}{5(\sqrt{6} - \sqrt{3})}, \frac{9}{8(\sqrt{6} - \sqrt{3})}\right)$
(f) $\left(\frac{17}{4}, \frac{24}{5}\right)$ (g) $(1, 12)$ (h) $\left(2, \frac{32}{15\pi - 20}\right)$

15. (a) $\left(-\frac{1}{2}, -\frac{1}{3}\right)$ (b) $(4, -8)$ (c) $\left(\frac{13}{3}, 2\right)$ (d) $\left(\frac{8}{5}, \frac{1}{2}\right)$ (e) $(0, 15)$
(f) $\left(-\frac{163}{185}, -\frac{103}{1295}\right)$ (g) $\left(\frac{3\pi}{4}, 0\right)$ (h) $\left(\frac{7}{2}, \frac{21}{2}\right)$

18. (a) $\frac{4\pi R^3}{3}$ (b) $\pi R^2 H$ (c) $\frac{\pi R^2 H}{3}$ (d) $\frac{\pi H}{3}(R^2 + rR + r^2)$ (e) $\frac{\pi R}{3}(HR + 2R^2)$

19. (a) $\frac{72\pi}{5}$ (b) $160\pi^2$ (c) $\frac{2744\pi}{15}$ (d) 1296π (e) 9π (f) $\frac{432\pi}{5}$ (g) 1728π (h) $\frac{128\pi}{15}$

20. (a) $\frac{9\pi}{2}$ (b) $256\pi^2$ (c) $\frac{88\pi}{3}$ (d) 432π (e) $\frac{24\pi}{5}\sqrt{3}(4\sqrt{2} - 5)$ (f) $\frac{153\pi}{2}$
(g) 144π (h) $\frac{8\pi}{3}(3\pi - 4)$

Section 3.11

1. **(a)** Taking $\delta = \epsilon/5$ shows that f is uniformly continuous.
(b) g is uniformly continuous on $[0, \pi/4]$ by Thoerem 3.11.4.
(c) h is not uniformly continuous on $(0, \pi/2)$ because h has a vertical asy
$x = \pi/2$. Therefore, any fixed number $\delta > 0$ will not work at $\pi/2 - \delta/2$.
(d) k is uniformly continuous on $[1, 4]$ by Theorem 3.11.4.
(e) m is not uniformly continuous on $[0, 2]$ because m is not continuous at x
(f) n is uniformly continuous on $[0, \pi]$ by Theorem 3.11.4.
(g) Taking $\delta = \epsilon/9$ shows that p is uniformly continuous on $\mathfrak{R}$.
(h) q is not uniformly continuous on $[0, \infty)$. Given $\epsilon > 0$, any fixed number $\delta >$
not work for $x > \epsilon/\delta$. The reason is that $(x + \frac{\delta}{2})^2 < x^2 + \epsilon$ implies that $\delta x < \epsilon -$
and $x < \epsilon/\delta$.

7. **(a)** 2 **(b)** 4 **(c)** $\frac{3}{2} + 3\sqrt[3]{2} - 3\sqrt[3]{4}$ **(d)** 144 **(e)** $43 + 20\sqrt{10}$

Section 3.12

4. **(a)** $\frac{40}{119}$ **(b)** $\tan 14$ **(c)** $\frac{\sqrt{17}}{3}$ **(d)** $-\frac{\sqrt{2}}{4} \sec \frac{\pi^2}{24}$ **(e)** 1 **(f)** $-\frac{\pi}{4}$
6. **(a)** $g(x_1, x_2) = x_1^2 f(x_2)$ **(b)** $g(x_1) = f(x_1)^2$ **(c)** $g(x_1, x_2) = \frac{f(x_1)}{1 + f(x_2)^2}$
(d) $g(x_1, x_2, x_3) = \frac{x_1 f(x_2)}{1 + f(x_3)^2}$ **(e)** $g(x_1, x_2) = x_1 \arctan f(x_2)$ **(f)** $g(x_1, x_2) = \sqrt{x_1^2}$
8. **(a)** 3 **(b)** 4 **(c)** 5 **(d)** 2
10. **(a)** square with center $\mathbf{c}$ and vertices at $(c_1 \pm k, c_2 \pm k)$
(b) cube with center $\mathbf{c}$ and vertices at $(c_1 \pm k, c_2 \pm k, c_3 \pm k)$
11. $d((0, 0), (1, 1)) = 1$ while the usual distance is $\sqrt{2}$.
12. **(a), (c), (d), (e)** uniformly continuous **(b), (f)** not uniformly contin
13. **(a)** 234 **(b)** 445 **(c)** 113 **(d)** $\frac{67}{8}$ **(e)** 136

Review Exercises

1. T **2.** T **3.** F **4.** F **5.** T **6.** T **7.** F **8.** T
9. F **10.** F **11.** F **12.** T **13.** T **14.** T **15.** F **16.** T
17. T **18.** F **19.** F **20.** F **21.** T **22.** T **23.** T **24.** F
25. T **26.** T **27.** T **28.** T **29.** T **30.** F **31.** T **32.** T
33. F **34.** T **35.** F **36.** F **37.** F **38.** T **39.** F **40.** T

41. $n = 1$ **42.** $\frac{2}{3}$ **43.** $\frac{1}{\pi}$ **44.** 4.67 **45.** $\int_{-\pi/3}^{\pi/4} \sqrt{\tan^6 x}\, dx$ **46.** $\frac{2x^3}{1 - x^4}$

47. 4 **48.** $64\pi - 16\sqrt{3}$ **49.** 1 **50.** $\int_{\pi/4}^{\pi/3} \tan^2 \theta \sec \theta\, d\theta$

51. For $x \le 0$, $|x| = -x$ and $\int |x|\, dx = \int -x\, dx = -\frac{x^2}{2} + C$ is negative.
52. (a) The diagonals divide the quadrilateral into 4 right triangles whose legs
lengths: (1) x, y; (2) x, $t - y$; (3) $s - x$, y; (4) $s - x$, $t - y$. The sum of the ar
these triangles is: $\frac{1}{2}xy + \frac{1}{2}x(t - y) + \frac{1}{2}(s - x)y + \frac{1}{2}(s - x)(t - y) = \frac{1}{2}st$. **(b)**
53. $\frac{496}{15}\pi$ **54.** 6π cm^3/hr **55.** $\frac{256}{3}\pi$ cm^3 **56.** $x - \frac{2}{3}x^{3/2} + \frac{1}{2}x^2 + C$ **57**
58. $-\frac{2}{\sqrt{\tan x}} + 6\sqrt{\tan x} + 2\tan^{3/2} x + \frac{2}{5}\tan^{5/2} x + C$
59. $\frac{2}{135}(3x + 2)^{5/2} - \frac{8}{81}(3x + 2)^{3/2} + \frac{8}{27}\sqrt{3x + 2} + C$
60. $\frac{1}{\sin x + \cos x} + C$ **61.** $\frac{1}{3}\arctan \frac{\sin x}{3} + C$ **62.** $\frac{1}{2}\arcsin\left(\frac{1}{2}\sin^2 x\right) + C$
63. $\frac{2}{3}\sin^3 \theta - \frac{2}{3}\cos^3 \theta - \cos \theta + \sin \theta + C$ **64.** $\frac{6}{7}x^{7/6} + \frac{24}{13}x^{13/12} + x + C$ **6**
66. $2\sqrt{2}$ **67.** 0 **68.** **(a)** 1.0976 **(b)** 1.0983 **69.** $\frac{892}{27}$ cm **70.** 1280 ft–lbs

Chapter 4

Section 4.2

1. (a) $\frac{1}{8}$ (b) 9 (c) $\frac{1}{8}$ (d) $\frac{1}{625}$ (e) $\frac{1}{32}$

2. (a) 11.664 (b) 8.827 (c) .0906 (d) 114.5 (e) .103

3. (a) 3 (b) $\frac{5}{3}$ (c) $\frac{3}{4}$ (d) $-\frac{2}{3}$ (e) $-\frac{2}{3}$

4. (a) 1 (b) 0 (c) 3 (d) $\frac{1}{2}$ (e) -2

5. (a) π (b) 3 (c) 1 (d) $\log_{10} 2$ (e) 8^x

6. (a) $\exp(\sqrt{2}\ln 5)$ (b) $\exp(e\ln\pi)$ (c) $\frac{1}{\ln 7}$ (d) $\frac{1}{2}\ln\pi$

(e) $\frac{1}{\ln 2}(9\ln 3 + 4\ln 7 - 8\ln 5 - 12\ln 2)$ (f) $\frac{1}{\ln 5}\left(4\ln 2 - 3\ln 3 - \frac{3}{2}\ln 7\right)$

(g) $\frac{1}{2\ln 3}(7\ln 5 - 3\ln 2)$ (h) $\frac{3}{2\ln 2}(\ln 2 + 2\ln 3)$

7. (a) e (b) $\exp(4)$ (c) $\ln 12$ (d) $\frac{7}{2}\ln 2$ (e) $\frac{9}{2}$ (f) $-\frac{55}{6}$ (g) $\frac{64}{25}$ (h) $\frac{3}{2}\sqrt{2}$ (i) $2\sqrt{2}$

8. (a) $\ln 4 = \int_1^4 \frac{1}{t}\,dt$ (b) $\exp 2$ is the unique number x such that $\ln x = 2$.

(c) $3^\pi = \exp(\pi\ln 3)$ (d) $\log_{10} 3$ is the unique number y such that $10^y = 3$.

9. (a) 2.71 (b) $-.25$ (c) 2.05 (d) 1.57 (e) -2.22

10. (a) $1.3 < \ln 5 < 2.1$ (b) $1.4 < \ln 5 < 1.8$ (c) $1.2 < \ln 4 < 1.6$

(d) $-.76 < \ln\frac{1}{2} < -.63$ (e) $-1.2 < \ln\frac{1}{3} < -1.0$ (f) $-0.30 < \ln\frac{3}{4} < -0.27$

11. (a) $\Re$ (b) $\left(-\infty, \frac{1}{3}\right) \cup \left(\frac{1}{3}, \infty\right)$ (c) $(-\infty, 0) \cup (0, \infty)$ (d) $\left(-\infty, \frac{34}{9}\right) \cup \left(\frac{34}{9}, \infty\right)$

(e) $(-\infty, 0]$ (f) $[2, \infty)$

12. (a) $(\ln\pi)\pi^x$ (b) $\pi x^{\pi-1}$ (c) $\frac{1}{2}(\ln 3)\sqrt{3^x}$ (d) $\sqrt{3}x^{\sqrt{3}-1}$

13. (a) $(\sqrt{5}-3)x^{\sqrt{5}-4}$ (b) $\cot x$ (c) $\ln x + 1$ (d) 0 (e) $(\ln 10)10^x$

(f) $2xe^{x^2}$ (g) $-\frac{4}{(e^x - e^{-x})^2}$ (h) $(\ln 5)5^x x^5 + 5^{x+1}x^4$ (i) $\frac{1}{\ln 2}\frac{(\ln 3)3^x + (\ln 7)7^x}{3^x + 7^x}$

(j) $\frac{1}{x\ln x}$ (k) e^{x+e^x} (l) $-(\ln 8)8^{\cos e^x}\sin(e^x)e^x$ (m) $-\frac{1}{x(\ln x)^2}$

(n) $\frac{\arctan\ln x}{x\sqrt{1-(\ln x)^2}} + \frac{\arcsin x}{x[1+(\ln x)^2]}$ (o) $\frac{\ln 2 + \ln 3}{2x(\ln 2)(\ln 3)\sqrt{\log_2 x + \log_3 x}}$

14. (a) $e^{-x^2}(12x - 8x^2)$ (b) $(\sin x)(\ln 2)2^{\cos x}\left[1 + 3(\ln 2)\cos x - (\ln 2)^2\sin^2 x\right]$

(c) $2\sec^2 x\tan x + 2\csc^2 x\cot x$ (d) $\frac{1}{x^3(\ln 4)^3}\left[(2(\ln 4)^2 - 1)\cos\log_4 x + 3(\ln 4)\sin\log_4 x\right]$

(e) $\frac{8(\ln x)^2 + 6\ln x + 3}{8x^3(\ln x)^{5/2}}$

15. (a) $f^{(n)}(x) = \exp(x)$ (b) $g^{(n)}(x) = (\ln 5)^n 5^x$ (c) $h^{(n)}(x) = (-1)^{n-1}\frac{(1)(2)(3)\cdots(n-1)}{x^n}$

(d) $k^{(n)}(x) = (-1)^{n-1}\frac{(1)(2)(3)\cdots(n-1)}{(\ln 3)x^n}$ (e) $m^{(n)}(x) = \pi(\pi-1)(\pi-2)\cdots(\pi-n+1)x^{\pi-n}$

16. (a) $\frac{dy}{dx} = \frac{y}{x}$ (b) $\frac{dy}{dx} = \frac{y\cos\ln(xy) - xy^2\cot(xy)}{x^2 y\cot(xy) - x\cos\ln(xy)}$ (c) $\frac{dy}{dx} = \frac{y^2(x^2+y^2)[1+(\ln(x^2+y^2))^2] - 2x}{2y - 2xy(x^2+y^2)[1+(\ln(x^2+y^2))^2]}$

(d) $\frac{dy}{dx} = \frac{2(\ln 6)^2 6^{x+y}(xy+\sqrt{xy}) - y}{x - 2(\ln 6)^2 6^{x+y}(xy+\sqrt{xy})}$ (e) $\frac{dy}{dx} = \frac{xy^2\ln 4 - 2y\exp(\log_2 x + \log_4 y)}{x\exp(\log_2 x + \log_4 y) - x^2 y\ln 4}$

(f) $\frac{dy}{dx} = -\frac{(\ln 10)\csc^2(x+y) + 2xy^2}{(\ln 10)\csc^2(x+y) + 2x^2 y}$

17. (a) $\frac{d^2 y}{dx^2} = -y\frac{(x-1)^2 + (y-1)^2}{x^2(y-1)^3}$ (b) $\frac{d^2 y}{dx^2} = -\frac{(\ln 2)^2(x+y)}{[(\ln 2)(x+y) - 1]^3}$

(c) $\frac{d^2 y}{dx^2} = -\frac{2}{2y - x^2 - y^2} - \frac{4x^2}{(2y - x^2 - y^2)^2} + \frac{8x^2(y-1)}{(2y - x^2 - y^2)^3}$

18. (a) vertical cusp (b) vertical tangent (c) cusp (d) vertical tangent

(e) cusp (f) cusp

19. (a) Both answers are correct because $\ln e = 1$.

(b) Both answers are correct because $(\ln e)e^x = (1)\exp(x) = \exp(x)$.

20. (a) Both have derivative $\frac{1}{x\ln x}$.

(b) $\ln(\log_2 x) = \ln\left(\frac{\ln x}{\ln 2}\right) = \ln\ln x - \ln\ln 2$ and $\ln\ln x$ are equal up to a constant.

21. (a) $\frac{x^{\pi-\sqrt{2}+1}}{\pi - \sqrt{2} + 1} + C$ (b) $\frac{1}{5}e^{5x+1} + C$ (c) $-\frac{1}{2}e^{-x^2} + C$ (d) $\frac{1}{\ln\pi}\pi^x + C$ (e) $\ln 3$

787

(f) $\frac{1}{11}\ln(11x-5)+C$ **(g)** $\frac{1}{2}(\ln x)^2+C$ **(h)** $-2\sqrt{1-e^x}+C$ **(i)** $\frac{1}{2}e^{2x}+2e^x$
(j) $\frac{1}{2}\ln(x^2+1)+C$ **(k)** $\frac{1}{12\ln 5}\left[\ln(3x+1)\right]^2+C$ **(l)** $\ln(e^x+1)+C$

22.

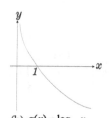

(a) $f(x)=\log_3 x$ (b) $g(x)=\log_{1/5}x$ (c) $h(x)=\ln(1+x)$ (d) $i(x)=4^x$

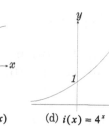

(e) $j(x)=3^{-x}$ (f) $k(x)=e^{1+x}$ (g) $m(x)=e^{-5x}$ (h) $n(x)=\ln\sqrt{5^{8x}}=$

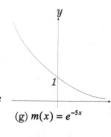

24. **(a)** $D(\ln x)$ at $x=1$ equals 1.   **(b)** $D(\ln x)$ at $x=e$ equals $\frac{1}{e}$.
(c) $D(\exp x)$ at $x=0$ equals 1.   **(d)** $D(\exp x)$ at $x=1$ equals e.
(e) $D(5^x)$ at $x=0$ equals $\ln 5$.   **(f)** $D(7^x)$ at $x=1$ equals $7\ln 7$.
(g) $D(\log_3 x)$ at $x=1$ equals $\frac{1}{\ln 3}$.   **(h)** $D(\log_2 x)$ at $x=2$ equals $\frac{1}{\ln 4}$.

25. $x^x(1+\ln x)$

Section 4.3

1. **(a)** $y'=(x^2+1)(5x^3-2)(4x^5+7)\left[\frac{2x}{x^2+1}+\frac{15x^2}{5x^3-2}+\frac{20x^4}{4x^5+7}\right]$

(b) $y'=\frac{(3x^4+5)(6x^8-3)}{(2x^2+5)(7x^6-8)}\left[\frac{12x^3}{3x^4+5}+\frac{48x^7}{6x^8-3}-\frac{4x}{2x^2+5}-\frac{42x^5}{7x^6-8}\right]$

(c) $y'=(5x+4)^9(2x^2+1)^6(8x^3-7)^3\left[\frac{45}{5x+4}+\frac{24x}{2x^2+1}-\frac{72x^2}{8x^3-7}\right]$

(d) $y'=\sqrt{(3x^2-1)^9(5x^3+2)^7}\left[\frac{27x}{3x^2-1}+\frac{105}{10x^3+4}\right]$

(e) $y'=\sqrt[3]{\frac{(x^3+3)^4(5x^4-6)^8}{(2x^2-7)^2(6x^6-9)^5}}\left[\frac{4x^2}{x^3+3}+\frac{160x^3}{15x^4-18}-\frac{8x}{6x^2-21}-\frac{20x^5}{2x^6-3}\right]$

(f) $y'=\sin^3 x\cos^4 x\tan^2 x\left[3\cot x-4\tan x+\frac{2\sec^2 x}{\tan x}\right]$

(g) $y'=\frac{(3\sin x+8)^5(5\tan x-9)^3}{(6\cos x-1)^2(4\cot x-3)^6}\left[\frac{15\cos x}{3\sin x+8}+\frac{15\sec^2 x}{5\tan x-9}+\frac{12\sin x}{6\cos x-1}+\frac{24\csc^2 x}{4\cot x-3}\right]$

(h) $y'=(3e^x+7)^5(4e^{-x}+1)^{10}(e^{3x}-8)^4\left[\frac{15e^x}{3e^x+7}-\frac{40e^{-x}}{4e^{-x}+1}+\frac{12e^{3x}}{e^{3x}-8}\right]$

(i) $y'=\frac{(2^x-1)^5(5^{-x}+7)^8}{(3^x+2)^4(7^{-x}-8)^3}\left[\frac{5(\ln 2)2^x}{2^x-1}-\frac{8(\ln 5)5^{-x}}{5^{-x}+7}-\frac{4(\ln 3)3^x}{3^x+2}+\frac{3(\ln 7)7^{-x}}{7^{-x}-8}\right]$

(j) $y'=\sqrt{\frac{[3+\ln(x^5)]^7[\ln(5x+2)]^3}{(\ln\sin x)^5}}\left[\frac{35}{2x(3+5\ln x)}+\frac{15}{2(5x+2)\ln(5x+2)}-\frac{5\cos x}{2\sin x\ln(\sin x)}\right]$

(k) $y'=-x^{-x}(1+\ln x)$ **(l)** $y'=(\sin x)^x(\ln\sin x+x\cot x)$ **(m)** $y'=x^{\sin x}\left[(c\right.$

(n) $y'=(\arcsin x)^x\left(\ln\arcsin x+\frac{x}{(\arcsin x)\sqrt{1-x^2}}\right)$ **(o)** $y'=x^{\arctan x}\left(\frac{\ln x}{1+x^2}+a\right.$

(p) $y'=(\arctan x)^{\tan x}\left[(\sec^2 x)\ln\arctan x+\frac{\tan x}{(\arctan x)(1+x^2)}\right]$ **(q)** $y'=(\ln x)^x$

(r) $y'=(\ln x)^{\ln x}\left(\frac{1+\ln\ln x}{x}\right)$

2. **(a)** $\frac{1}{5}\ln|x^5+2|+C$ **(b)** $\frac{1}{6}\ln|2x^3+6x-5|+C$ **(c)** $-\frac{1}{3}\ln(3\cos x+$
(d) $\frac{1}{4}\ln|4\tan x+3|+C$ **(e)** $\frac{1}{\ln 2}\ln(2^x+5)+C$ **(f)** $(\ln 3)\ln|\ln x|+$
(g) $\frac{2}{3}\ln(3\sqrt{x}+2)+C$ **(h)** $\frac{2}{7}\ln(x^{7/2}+1)+C$ **(i)** $\ln|\ln\ln x|+C$
(j) $\ln|\arcsin x|+C$ **(k)** $\frac{1}{\ln 2}\ln(2^x+2^{-x})+C$ **(l)** $-\ln|\ln\cos x|+$

3. (a) $2\ln(x^2 - 8x + 17) + 23\arctan(x - 4) + C$ (b) $\ln(x^2 - 4x + 13) + \frac{5}{3}\arctan\left(\frac{x-2}{3}\right) + C$
(c) $\frac{3}{2}\ln(x^2 + 2x + 5) - \frac{1}{2}\arctan\left(\frac{x+1}{2}\right) + C$ (d) $\frac{5}{2}\ln(x^2 + 6x + 25) - \frac{9}{2}\arctan\left(\frac{x+3}{4}\right) + C$
(e) $\frac{1}{4}\ln(2x^2 - 6x + 27) - \frac{\sqrt{5}}{6}\arctan\left(\frac{2x-3}{3\sqrt{5}}\right) + C$ (f) $\frac{1}{6}\ln(3x^2 - 18x + 45) + \frac{1}{\sqrt{6}}\arctan\left(\frac{x-3}{\sqrt{6}}\right) + C$

. (a) 0 (b) $\ln\frac{3}{2}$ (c) 0 (d) 1 (e) 0 (f) e (g) e^3 (h) 1 (i) $-\infty$
j) 1 (k) e^{e+1} (l) e (m) e^{-1} (n) k (o) ∞ (p) e^{-1} (q) e (r) 1

.

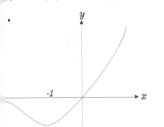

(a) $f(x) = xe^x$

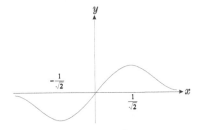

(b) $g(x) = xe^{-x^2}$

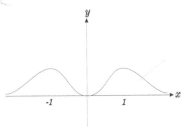

(c) $h(x) = x^2 e^{-x^2}$

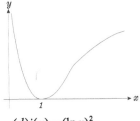

(d) $i(x) = (\ln x)^2$

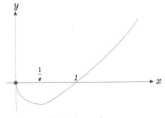

(e) $j(x) = x\ln x$

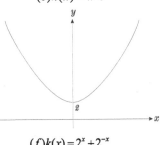

(f) $k(x) = 2^x + 2^{-x}$

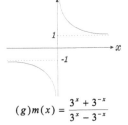

(g) $m(x) = \dfrac{3^x + 3^{-x}}{3^x - 3^{-x}}$

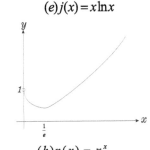

(h) $n(x) = x^x$

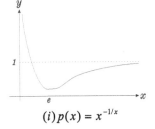

(i) $p(x) = x^{-1/x}$

6. $e - 1$ **7.** $2 - \frac{4}{e}$ **8.** $\frac{1}{2} - \frac{1}{e}$ **9.** $\frac{(e^2 - 1)^2}{2e^2}$ **10.** $\frac{8}{3}\pi$ **11.** $\frac{\pi}{2}e^2 - \frac{\pi}{2}$ **12.** $\pi - \frac{\pi}{e}$

13. (a) $2\ln 7 + 4\ln 4 - 2\ln 3$ (b) $\frac{1}{3}\ln\frac{891}{29}$ (c) $2\ln 3$ (d) $\ln\ln 5 - \ln\ln 2$

14. $\frac{1}{2}\ln(e^4 + 1) - \frac{1}{2}\ln 2$ kg.   **15.** $\frac{1}{4}\ln 2$ joules

16. (a) $\bar{x} = \frac{6}{\pi}\ln 2$ m.   (b) $\bar{x} = \frac{4 - 2\arctan 2}{\ln 5}$ ft.   (c) $\bar{x} = \frac{9 - 5\ln 17 + 5\ln 8}{3\ln 17 - 3\ln 8}$ ft.

17. $\bar{x} = \frac{6 - \ln 7}{2\ln 7}$, $\bar{y} = \frac{3}{7\ln 7}$

Section 4.4

1. $-x\cos x + \sin x + C$

2. $\frac{x}{\ln 5}5^x - \frac{1}{(\ln 5)^2}5^x + C$

3. $x^2\sin x + 2x\cos x - 2\sin x + C$

4. $\frac{x^2}{2}\log_{10} x - \frac{x^2}{4\ln 10} + C$

5. $x^2 e^x - 2xe^x + 2e^x + C$

6. $\frac{x^8}{8}\ln x - \frac{x^8}{64} + C$

7. $x\arcsin x + \sqrt{1 - x^2} + C$

8. $\frac{x^3}{3}e^{x^3} - \frac{1}{3}e^{x^3} + C$

9. $x(\ln x)^2 - 2x\ln x + 2x + C$

10. $\frac{1}{2}e^x\cos x + \frac{1}{2}e^x\sin x + C$

11. $\frac{2}{29}e^{2x}\sin 5x - \frac{5}{29}e^{2x}\cos 5x + C$

12. $x\ln(1 + x^2) - 2x + 2\arctan x + C$

13. $\frac{x}{2}\sin\ln x - \frac{x}{2}\cos\ln x + C$

14. $\tan x + \ln|\cos x| + C$

15. $\frac{1}{3}x^3\arcsin x + \frac{1}{3}\sqrt{1 - x^2} - \frac{1}{9}(1 - x^2)^{3/2} + C$ **16.** $x\log_{10} x - \frac{x}{\ln 10} + C$

17. $\frac{\ln 3}{\pi^2+(\ln 3)^2}3^x\cos\pi x+\frac{\pi}{\pi^2+(\ln 3)^2}3^x\sin\pi x+C$ **18.** $\frac{\ln 6}{1+(\ln 6)^2}6^x\sin x-\frac{1}{1+(\ln 6)^2}6^x$

19. $\ln x\ln\ln x-\ln x+C$ **20.** $\frac{1}{3}x^3+\frac{1}{2}x+-\frac{1}{2}\sin x\cos x-2x\cos x+2\sin$

21. $-2x\cos\sqrt{x}+4\sqrt{x}\sin\sqrt{x}+4\cos\sqrt{x}+C$ **22.** $\frac{1}{2}xe^x\sin x-\frac{1}{2}xe^x\cos x-\frac{1}{2}e^x$ c

23. $xe^x\ln x-e^x+C$ **24.** $-\frac{1}{2}\cot x\csc x-\frac{1}{2}\ln|\cot x+\csc x|+C$

25. $9\ln 3-4\ln 2-\frac{5}{2}$ **26.** $\frac{\pi^2}{72}+\frac{\pi\sqrt{3}}{6}-1$ **27.** $\frac{\pi}{4}-\frac{1}{2}$

28. $\frac{e^2}{2}+\frac{2}{3}+\frac{4}{e}-\frac{1}{2e^2}$ **29.** $\frac{\sqrt{3}}{\pi}+\frac{1}{2\pi}\ln(2+\sqrt{3})$ **30.** 1

31. **(a)** $\int\cos^N x\,dx=\frac{1}{N}\cos^{N-1}x\sin x+\frac{N-1}{N}\int\cos^{N-2}x\,dx$
(b) $\int(\ln x)^N\,dx=x(\ln x)^N-N\int(\ln x)^{N-1}\,dx$
(c) $\int x^N\sin x\,dx=-x^N\cos x+Nx^{N-1}\sin x-N(N-1)\int x^{N-2}\sin x\,dx$
(d) $\int x^N\cos x\,dx=x^N\sin x+Nx^{N-1}\cos x-N(N-1)\int x^{N-2}\cos x\,dx$
(e) $\int\sec^N x\,dx=\frac{1}{N-1}\tan x\sec^{N-2}x+\frac{N-2}{N-1}\int\sec^{N-2}x\,dx$
(f) $\int x^M(\ln x)^N\,dx=\frac{x^{M+1}}{M+1}(\ln x)^N-\frac{N}{M+1}\int x^M(\ln x)^{N-1}\,dx$

32. $\frac{e^2}{2}-\frac{3}{2}$ **33.** $\frac{\sqrt{e}}{1+\pi^2}+\sqrt{e}+\frac{\pi}{1+\pi^2}-2$ **34.** $\frac{1}{2}-\frac{6}{5e}$ **35.** $\frac{\pi}{32}-\frac{\ln 2}{8}$ **36.**

37. 2π **38.** **(a)** 2π **(b)** $\frac{\pi}{2}(e^2-1)$ **39.** $2\pi^2$ **40.** $\frac{\pi^2}{2}-\pi\ln 2$

41. **(a)** $\frac{431}{2\ln 2}-\frac{435}{2(\ln 2)^2}+\frac{141}{(\ln 2)^3}-\frac{39}{(\ln 2)^4}$ miles **(b)** $\frac{25\pi}{144}+\frac{\sqrt{3}}{8}-\frac{4}{9}$ km **(c)** $\frac{7\pi}{2}$
(d) $4\sqrt{2}+3\ln 3-6.2+2\ln(\sqrt{2}-1)-2\ln(\sqrt{2}+1)+1.1\ln.21-\ln.1+\ln 2.1$

42. $\frac{2}{(\ln 5)^2+4}\left(5^{\pi/2}+1\right)$ pounds   **43.** $\frac{\sqrt{2}}{2}+\frac{1}{2}\ln(\sqrt{2}+1)+\frac{1}{3}+\frac{\ln 3}{4}$ joules

44. **(a)** $\frac{40,000\ln 10-9,999}{800\ln 10-396}$ ft **(b)** $\frac{192\ln 4-63}{63\ln 4}$ ft **(c)** $\frac{\pi-2}{\pi-2\ln 2}$ m **(d)** $\frac{\pi^2-8}{2\pi-4}$ m

45. **(a)** $\left(\frac{e^2}{4}+\frac{1}{4},\frac{e}{2}-1\right)$ **(b)** $\left(\frac{24\ln 2-7}{7\ln 2},\frac{9}{4}\right)$ **(c)** $\left(\frac{1}{2},\frac{\pi}{8}\right)$ **(d)** $\left(e-2,\frac{e^2}{8}-\right.$

46. **(a)** $x^2\cosh x-2x\sinh x+2\cosh x+C$ **(b)** $\frac{1}{4}e^{2x}+\frac{x}{2}+C$
(c) $x\sinh^{-1}x-\sqrt{x^2+1}+C$ **(d)** $x\tanh^{-1}x+\frac{1}{2}\ln(1-x^2)+C$

47. $\frac{\pi}{2}-\frac{2}{5}-\arcsin\frac{3}{5}$ **48.** $\frac{6}{5}(\ln 5)^2-\frac{13}{5}\ln 5+\frac{12}{5}$

49. **(a)** $\pi\ln(2+\sqrt{3})+2\pi\sqrt{3}$ **(b)** $2\pi+2\pi\left[\ln(2+\sqrt{3})\right]^2-2\pi\sqrt{3}\ln(2+\sqrt{}$

Section 4.5

1. $\tan x-x+C$ **2.** $\frac{1}{2}\sec x\tan x-\frac{1}{2}\ln|\sec x+\tan x|+C$

3. $\frac{1}{2}\tan^2 x+\ln|\cos x|+C$ **4.** $-\frac{1}{6}\csc^6 x+\frac{1}{4}\csc^4 x+C$

5. $-\frac{1}{9}\csc^9 x+\frac{2}{7}\csc^7 x-\frac{1}{5}\csc^5 x+C$ **6.** $\tan x+\frac{1}{3}\tan^3 x+C$

7. $-\frac{1}{4}\cot x\csc^3 x-\frac{3}{8}\cot x\csc x-\frac{3}{8}\ln|\cot x+\csc x|+C$

8. $\frac{1}{4}\tan x\sec^3 x-\frac{1}{8}\tan x\sec x-\frac{1}{8}\ln|\tan x+\sec x|+C$

9. $\frac{1}{9}\sec^9 x-\frac{2}{7}\sec^7 x+\frac{1}{5}\sec^5 x+C$ **10.** $\frac{1}{7}\tan^7 x+\frac{1}{5}\tan^5 x+C$

11. $\frac{1}{6\pi}\tan\pi x\sec^5\pi x+\frac{5}{24\pi}\tan\pi x\sec^3\pi x+\frac{5}{16\pi}\tan\pi x\sec\pi x+\frac{5}{16\pi}\ln|\tan\pi x+$

12. $-\frac{1}{12}\cot(3x-7)\csc^3(3x-7)+\frac{1}{24}\cot(3x-7)\csc(3x-7)$
$\qquad\qquad\qquad +\frac{1}{24}\ln|\csc(3x-7)+\cot(3x-7)|+C$

13. $\frac{1}{8}\tan^8(e^x)+\frac{3}{10}\tan^{10}(e^x)+\frac{1}{4}\tan^{12}(e^x)+\frac{1}{14}\tan^{14}(e^x)+C$

14. $\frac{1}{8}\tan(x^2)\sec^3(x^2)-\frac{5}{16}\tan(x^2)\sec(x^2)+\frac{3}{16}\ln|\tan(x^2)+\sec(x^2)|+C$

15. $-\frac{1}{7}\cot^7(\ln x)-\frac{2}{9}\cot^9(\ln x)-\frac{1}{11}\cot^{11}(\ln x)+C$

16. $\frac{1}{\ln 2}\left[\frac{1}{9}\sec^9(2^x)-\frac{4}{7}\sec^7(2^x)+\frac{6}{5}\sec^5(2^x)-\frac{4}{3}\sec^3(2^x)+\sec(2^x)\right]+C$

7. $-\frac{1}{8\pi}\cot^8 \pi x + \frac{1}{6\pi}\cot^6 \pi x - \frac{1}{4\pi}\cot^4 \pi x + \frac{1}{2\pi}\cot^2 \pi x + \frac{1}{\pi}\ln|\sin \pi x| + C$

8. $\frac{1}{6}\tan^4(x^{3/2}) + \frac{2}{9}\tan^6(x^{3/2}) + \frac{1}{12}\tan^8(x^{3/2}) + C$

9. $\frac{1}{24}\tan(x^4)\sec^5(x^4) - \frac{13}{96}\tan(x^4)\sec^3(x^4) + \frac{11}{64}\tan(x^4)\sec(x^4)$
$\qquad\qquad\qquad\qquad\qquad - \frac{5}{64}\ln|\sec(x^4) + \tan(x^4)| + C$

0. $\frac{1}{4}\tan^8 \sqrt{x} - \frac{1}{3}\tan^6 \sqrt{x} + \frac{1}{2}\tan^4 \sqrt{x} - \tan^2 \sqrt{x} - 2\ln|\cos \sqrt{x}| + C$

1. $\frac{1}{\ln 5}\left[\frac{1}{7}\tan^7(5^x) + \frac{1}{3}\tan^9(5^x) + \frac{3}{11}\tan^{11}(5^x) + \frac{1}{13}\tan^{13}(5^x)\right] + C$

2. $\frac{1}{5}\tan^5 x + \frac{3}{7}\tan^7 x + \frac{1}{3}\tan^9 x + \frac{1}{11}\tan^{11} x + C$

3. $\frac{8}{3}$ **24.** $\frac{26}{15} - \frac{\pi}{2}$ **25.** $2\pi^2 - \frac{8\pi}{3}$ **26.** $\frac{\pi^2}{2} - \frac{13\pi}{15}$

8. (a) $2 - \frac{3}{2}\ln 2$ miles (b) $4 - \frac{8\sqrt{3}}{135}$ miles **29.** $\frac{1}{2\pi}(1 - \ln 2)$ **30.** $\frac{28}{15}$ joules

1. (a) $\frac{9\pi^2 - 16\pi - 16 - 64\ln 2}{24\pi - 64}$ meters (b) $\frac{4\pi\sqrt{3} - 3 - 4\ln 2}{12\sqrt{3}}$ meters

2. (a) $\left(\frac{8\pi - \pi^2 - 16\ln 2}{32 - 8\pi}, \frac{3\pi - 8}{24 - 6\pi}\right)$ (b) $(0, 1)$

3. (a) $-\frac{1}{3}\text{sech}^3 x + \frac{1}{5}\text{sech}^5 x + C$ (b) $-\frac{1}{7}\coth^7 x + \frac{1}{5}\coth^5 x + C$
(c) $\tanh x - \frac{2}{3}\tanh^3 x + \frac{1}{5}\tanh^5 x + C$ (d) $-\ln|\coth x + \text{csch}\, x| + C$
(e) $2\arctan(e^x) + C$ (f) $-\frac{1}{2}\coth x \,\text{csch}\, x + \frac{1}{2}\ln|\coth x + \text{csch}\, x| + C$

4. $24\arctan(3 + 2\sqrt{2}) - 4\sqrt{2} + 2\ln(3 + 2\sqrt{2}) - 6\pi$

5. (a) $2e^2 - 6$ (b) $\frac{8\pi}{3(e^2+1)^2}(e^8 - 5e^6 - 7e^2 - 1)$

Section 4.6

1. $\frac{x}{2}\sqrt{x^2 + 9} + \frac{9}{2}\ln|x + \sqrt{x^2 + 9}| + C$ **2.** $\frac{25}{2}\arcsin\left(\frac{x}{5}\right) + \frac{x}{2}\sqrt{25 - x^2} + C$

3. $\frac{3}{25,000}\arctan\left(\frac{x}{5}\right) + \frac{3x}{5,000(x^2+25)} + \frac{x}{100(x^2+25)^2} + C$

4. $\frac{x^3}{4}\sqrt{x^2 - 1} - \frac{x}{8}\sqrt{x^2 - 1} - \frac{1}{8}\ln|x + \sqrt{x^2 - 1}| + C$ **5.** $8\arcsin\left(\frac{x}{4}\right) - \frac{x}{2}\sqrt{16 - x^2} + C$

6. $\frac{1}{4}(x + 1)(x^2 + 2x + 5)^{3/2} + \frac{3}{2}(x + 1)\sqrt{x^2 + 2x + 5} + 6\ln|x + 1 + \sqrt{x^2 + 2x + 5}| + C$

7. $\frac{1}{4}(x + 3)^3\sqrt{x^2 + 6x} - \frac{45}{8}(x + 3)\sqrt{x^2 + 6x} + \frac{243}{8}\ln|x + 3 + \sqrt{x^2 + 6x}| + C$

8. $\frac{1}{54}\arctan\left(\frac{x-2}{3}\right) + \frac{x-2}{18(x^2-4x+13)} + C$

9. $\frac{1}{6}(x - 4)^5\sqrt{x^2 - 8x + 9} - \frac{91}{24}(x - 4)^3\sqrt{x^2 - 8x + 9} + \frac{539}{16}(x - 4)\sqrt{x^2 - 8x + 9}$
$\qquad\qquad\qquad\qquad\qquad - \frac{1715}{16}\ln|x - 4 + \sqrt{x^2 - 8x + 9}| + C$

10. $\frac{x}{4(4-x^2)^2} - \frac{x}{32(4-x^2)} - \frac{1}{64}\ln\frac{|x+2|}{\sqrt{4-x^2}} + C$

11. $\sqrt{x^2 + 6x + 15} - 3\ln|x + 3 + \sqrt{x^2 + 6x + 15}| + C$ **12.** $\frac{3-x}{9\sqrt{x^2-6x}} + C$

13. $\frac{x+1}{10\sqrt{9-x^2-2x}} + C$ **14.** $\frac{x^2}{4}\sqrt{x^4 - 9} + \frac{9}{4}\ln|x^2 + \sqrt{x^4 - 9}| + C$

15. $\frac{\tan x}{25\sqrt{5-\tan^2 x}} + \frac{\tan^3 x}{75(5-\tan^2 x)^{3/2}} + C$ **16.** $\frac{1}{16}\arctan\left(\frac{e^x+2}{2}\right) + \frac{e^x+2}{8(e^{2x}+4e^x+8)} + C$

17. $(\sqrt{x} + 2)\sqrt{x + 4\sqrt{x} + 1} - 3\ln(\sqrt{x + 4\sqrt{x} + 1} + \sqrt{x} + 2) + C$

18. $-\frac{1}{7\ln 2}(5 - 2^{x+1} - 4^x)^{7/2} + C$ **19.** $\frac{1}{4}\arctan(x^2) + \frac{x^2}{4(x^4+1)} + C$

20. $\frac{\sqrt{2}}{8}\ln\left(e^x + e^{-x} + \sqrt{2}\right) - \frac{\sqrt{2}}{16}\ln\left(e^{2x} + e^{-2x}\right) - \frac{e^x+e^{-x}}{4(e^{2x}+e^{-2x})} + C$

21. $\ln\left|\ln x + 3 + \sqrt{(\ln x)^2 + 6\ln x}\right| + C$ **22.** $4\sqrt{5} + 2\ln(2 + \sqrt{5})$ **23.** 1

24. $\frac{e}{\sqrt{2}}\left[\frac{1}{2}\sqrt{4e^2-2e}-\frac{1}{\sqrt{2}}-\frac{3}{2}\ln(\sqrt{2e-1}+\sqrt{2e})+\frac{3}{2}\ln(1+\sqrt{2})\right]$

25. $\frac{51\pi}{1010}-\frac{\pi}{2}\arctan\frac{1}{10}$ **26.** $50\pi\sqrt{39}-\frac{\pi}{2}\ln(25+4\sqrt{39})$ **27.** $8\pi-6\pi$ arcs

28. (a) $\frac{4}{15}-\frac{11\sqrt{3}}{160}$ km   (b) $\frac{21\sqrt{10}}{100}+\frac{22\sqrt{5}}{75}$ km   **29.** $\frac{9\sqrt{5}}{4}-\frac{1}{8}\ln(2+\sqrt{5})$

30. $\frac{4\pi}{3}-2\sqrt{3}$ Joules   **31.** (a) $\frac{6}{\pi+2}$ ft   (b) $\dfrac{113,664\sqrt{3}}{32,760\sqrt{3}-420\ln(\sqrt{3}+2)}$ ft

32. (a) $\left(\frac{1}{2}\ln 10,\ \frac{1}{4}\arctan 3+\frac{3}{40}\right)$

   (b) $\left(\dfrac{391\sqrt{13}-120\ln\frac{3+\sqrt{3}}{1+\sqrt{5}}-45\sqrt{5}}{285\sqrt{13}+120\ln\frac{3+\sqrt{3}}{1+\sqrt{5}}-55\sqrt{5}},\ \dfrac{100,596}{35\left(57\sqrt{13}+24\ln\frac{3+\sqrt{3}}{1+\sqrt{5}}-11\sqrt{5}\right)}\right)$

33. (a) $\frac{1}{2}\operatorname{arcsinh}x+\frac{x}{2}\sqrt{1+x^2}+C$ (b) $\frac{1}{2}\ln|x-1|-\frac{1}{2}\ln|x+1|+C$

   (c) $\frac{1}{2}(\arcsin x-x\sqrt{1-x^2})+C$

Section 4.7

1. (a) $1+\frac{5x-8}{x^2-5x+8}$ (b) $\frac{x^3+5x-2}{x^5-4x^2+7}$ (c) $x-3+\frac{7x+1}{x^4+1}$ (d) $x-3-\frac{3}{x-2}$ (e) $x+$
(f) $1-\frac{6}{x+6}$ (g) $x^{10}-7x^8-4x^7+7x^5+5x-3+\frac{9}{x}$ (h) $x^3-4x-1+\frac{4x}{x}$
(i) $x^5-3x^4+11x^3-39x^2+139x-495+\frac{1762x-982}{x^2+3x-2}$ (j) $x^2+3x+5-\frac{24}{x^2-3}$
(k) $\frac{x^8+1}{x^9+x^7+x^5+x^3+x+1}$ (l) $x^7+x^6+x^3+x^2+\frac{1}{x-1}$

2. (a) $(x-5)(x+3)$ (b) $(x-1)(x+2)(x-3)$ (c) $(x+1)^3$ (d) $(x-1)^3($
(e) $(x-1)(x^2+x+1)$ (f) $(x+1)^2(x-2)^3$ (g) $(x-1)(x-3)(x+2)$
(h) $(x-3)(x+3)(x-2)(x+2)$ (i) $(x+2)(x-3)(x-1)^2$ (j) $(x^2+3)(x^2$
(k) $(x+2)(x^2-2x+2)$ (l) $(x^2+x+1)^2$

3. (a) $\frac{1/2}{x-3}+\frac{1/2}{x+3}$ (b) $\frac{1/5}{x-3}-\frac{1/5}{x+2}$ (c) $\frac{3}{x+6}-\frac{2}{x-3}$ (d) $\frac{3}{x}-\frac{4}{x^2}+\frac{5}{x^3}+\frac{6}{x-4}$
(e) $-\frac{3}{x-2}+\frac{3}{(x-2)^2}+\frac{4}{x+1}-\frac{1}{(x+1)^2}-\frac{2}{(x+1)^3}$ (f) $\frac{7}{x}-\frac{8}{x^2}+\frac{4}{x^3}+\frac{2}{x^4}+\frac{18-x}{x^2+2x+4}$
(g) $\frac{3}{x}+\frac{2x-1}{x^2+x+1}$ (h) $\frac{2}{x-3}+\frac{4x-8}{x^2-3x+4}$ (i) $\frac{x-1}{x^2+9}+\frac{3x-2}{(x^2+9)^2}+\frac{5}{x^2+25}$ (j) $\frac{4x-3}{x^2+2x+3}+\frac{1}{x}$
(k) $\frac{3}{x+2}-\frac{2}{(x+2)^2}+\frac{1-x}{x^2-x-5}$ (l) $\frac{x-1}{x^2+1}+\frac{2x-3}{(x^2+1)^2}+\frac{5-4x}{(x^2+1)^3}+\frac{x}{x^2+4}$

4. (a) $3\ln|x-4|+C$ (b) $-\frac{5}{2(x+2)^2}+C$ (c) $-\frac{7}{5(x-4)^5}+C$ (d) $4\ln|x+$
(e) $3\ln(x^2+9)+\frac{7}{3}\arctan\frac{x}{3}+C$ (f) $2\ln(x^2+25)-\frac{3}{5}\arctan\frac{x}{5}+C$
(g) $\ln(x^2+2x+10)+\arctan\frac{x+1}{3}+C$ (h) $4\ln(x^2+6x+25)-\frac{31}{4}\arctan\frac{x+}{4}$
(i) $\frac{1}{16}\arctan\frac{x}{2}+\frac{x-8}{8(x^2+4)}+C$ (j) $-\frac{27x^3+45x+8}{8(1+x^2)^2}-\frac{27}{8}\arctan x+C$
(k) $-\frac{x+8}{2(x^2+4x+13)}-\frac{1}{6}\arctan\frac{x+2}{3}+C$
(l) $-\frac{15}{2048}\arctan\frac{x+3}{4}-\frac{5x+31}{16(x^2+6x+25)^2}-\frac{15(x+3)}{512(x^2+6x+25)}+C$

5. $\frac{1}{2}\ln|x^2-1|+C$ **6.** $\frac{10}{7}\ln|x-5|+\frac{11}{7}\ln|x+2|+C$

7. $\frac{x^3}{3}+16x+32\ln|x-4|-32\ln|x+4|+C$ **8.** $x+\frac{12}{5}\ln|x-3|-\frac{7}{5}\ln|x+$

9. $\frac{5}{4}\ln|x-1|-\frac{5}{4}\ln|x+1|-\frac{7}{2x-2}+C$ **10.** $\frac{1}{3}\ln|x-1|-\ln|x+1|+\frac{8}{3}\ln|x+2|+\frac{}{x-}$

11. $\frac{38}{27}\ln|x-2|-\frac{5}{9}\ln|x+1|-\frac{2}{27}\ln(x^2+x+3)+\frac{254}{27\sqrt{11}}\arctan\frac{2x+1}{\sqrt{11}}+C$

12. $\frac{13}{8}\ln|x-1|-\frac{7}{4x-4}-\frac{7}{4(x-1)^2}-\frac{5}{8}\ln|x+1|+C$

13. $2\ln|x-2|-\ln|x+1|-\frac{3}{x+1}-\frac{1}{2(x+1)^2}+\frac{2}{3(x+1)^3}+C$

14. $\frac{1}{8}\ln|x-2|-\frac{21}{8}\ln|x+2|+\frac{10-9\sqrt{2}}{8}\ln|x-\sqrt{2}|+\frac{10+9\sqrt{2}}{8}\ln|x+\sqrt{2}|+C$

15. $-\frac{4}{9}\ln|x|+\frac{13}{18}\ln(x^2+9)+C$ **16.** $x+\frac{1}{4}\ln|x-1|-\frac{5}{4}\ln|x+1|-\frac{1}{2x+2}$

17. $x+\ln|x-2|-\ln|x+2|-2\arctan\left(\frac{x}{2}\right)+C$

8. $\frac{2}{13}\ln|x| + \frac{11}{26}\ln(x^2 + 4x + 13) - \frac{43}{39}\arctan\left(\frac{x+2}{3}\right) + C$

9. $x - \frac{3}{10}\ln|x + 1| - \frac{67}{20}\ln(x^2 + 6x + 25) - \frac{1}{10}\arctan\left(\frac{x+3}{4}\right) + C$

0. $-\frac{3}{10}\ln(x^2 + 9) + \frac{13}{15}\arctan\left(\frac{x}{3}\right) + \frac{3}{10}\ln(x^2 + 4) - \frac{4}{5}\arctan\left(\frac{x}{2}\right) + C$

1. $\frac{1}{4}\ln|x| - \frac{1}{8}\ln(x^2 + 4) + \frac{1}{4}\arctan\left(\frac{x}{2}\right) + \frac{3-x}{2x^2+8} + C$

2. $\ln|x + 2| + \ln(x^2 + 4x + 29) + \frac{x+12}{10(x^2+4x+29)} - \frac{49}{50}\arctan\left(\frac{x+2}{5}\right) + C$

3. $\ln|x - 4| - \frac{3}{2}\ln(x^2 + 2x + 17) + \frac{x+5}{8(x^2+2x+17)} + \frac{41}{32}\arctan\left(\frac{x+1}{4}\right) + C$

4. $\ln|x| - \frac{27}{16}\arctan x + \frac{5x+8}{16(x^2+1)} - \frac{13x+4}{8(x^2+1)^2} + \frac{5x^3-4}{6(x^2+1)^3} + C$

5. $39 - \frac{40}{3}(\ln 2 + \ln 5)$ **26.** $\frac{15}{2} + 50\arctan 2 - 50\arctan 3$ **27.** $\frac{\pi}{5} - \frac{\pi}{12}\ln 5$

8. (a) $\frac{\pi^2}{8} - \frac{\pi}{4}$ (b) $2\pi - \frac{\pi^2}{2}$

9. (a) $\frac{4}{27}\ln 5 + \frac{20}{27}\arctan\frac{1}{3} - \frac{8}{27}\ln 2 - \frac{10}{27}\arctan\frac{2}{3}$ miles

(b) $-\frac{7}{2} + \frac{203}{50}\ln\frac{(9)(64)(1961)}{(7^4)(169)} - \frac{641\sqrt{3}}{75}\left(3\arctan\frac{\sqrt{3}}{15} - 2\arctan\frac{3\sqrt{3}}{5} + \arctan\frac{13\sqrt{3}}{15}\right)$ cm

0. $2 + \frac{\pi}{2} - 2\arctan 3$ pounds   **31.** $\ln\frac{845\sqrt{5}}{81} + 2\arctan 2$ Joules

2. (a) $2 + \frac{9}{7}\ln 3 - \frac{9}{7}\ln 5 - \frac{16}{7}\ln 2$ feet   (b) $\frac{3}{4} + \ln 5 - 3\ln 2$ feet

3. (a) $\overline{x} = \ln 3 - 1$ and $\overline{y} = \frac{1}{12} - \frac{1}{16}\ln 3$

(b) $\overline{x} = \frac{1}{3}\arctan(2/3) - \frac{1}{3}\arctan(1/3)$ and $\overline{y} = \frac{29}{10,530} + \frac{1}{162}\arctan(1/3) - \frac{1}{162}\arctan(2/3)$

Section 4.8

1. $\ln|1 + \tan(x/2)| + C$ **2.** $2\ln|\tan(x/2)| - \cot(x/2) - 2\ln|\sec(x/2)| + C$

3. $\frac{7}{25}\ln|2 + \tan(x/2)| + \frac{7}{25}\ln|2\tan(x/2) - 1| - \frac{14}{25}\ln|\sec(x/2)| - \frac{x}{25} + C$

4. $\frac{\sqrt{2}}{4}\ln\left[\frac{\sec^2(x/2) - \sqrt{2}\tan(x/2)}{\sec^2(x/2) + \sqrt{2}\tan(x/2)}\right] + \frac{\sqrt{2}}{2}\arctan(1 + \sqrt{2}\tan(x/2)) - \frac{\sqrt{2}}{2}\arctan(1 - \sqrt{2}\tan(x/2)) + C$

5. $-\frac{2}{1+\tan(x/2)} - x - 2\cos^2(x/2) + C$ **6.** $\frac{1}{\sqrt{10}}\arctan\left[\sqrt{\frac{2}{5}}\tan(x/2)\right] + C$

7. $\frac{4}{15}\ln|2 + \tan(x/2)| - \frac{4}{15}\ln|1 + 2\tan(x/2)| + \frac{x}{5} + C$

8. $-\left(1 + \frac{\sqrt{3}}{3}\right)\ln|\tan(x/2) - \sqrt{3}| + \left(\frac{\sqrt{3}}{3} - 1\right)\ln|\sqrt{3} + \tan(x/2)| + 2\ln|\sec(x/2)| + C$

9. $2\ln|1 + \tan(x/2)| - \frac{2}{1+\tan(x/2)} - 2\ln|\sec(x/2)| - x - 2\cos^2(x/2) + C$

10. $\frac{19}{25}\ln|1 + 2\tan(x/2)| - \frac{1}{25}\ln|2 - \tan(x/2)| - \frac{18}{25}\ln|\sec(x/2)| + \frac{12x}{25} + C$

11. $x - \frac{1}{2}\ln(e^{2x} + 1) + \arctan(e^x) + C$ **12.** $-x + \frac{1}{2}\ln|e^{2x} - 1| + C$

13. $x - 2\ln|e^x - 1| + C$ **14.** $2x + e^{-x} - 2\ln(e^x + 1) + C$

15. $-x + 2\ln(e^x - 1) - \frac{1}{2}\ln(e^{2x} + e^x + 1) - \sqrt{3}\arctan\left(\frac{2e^x+1}{\sqrt{3}}\right) + C$

16. $x - \log_2(2^x + 1) + \frac{1}{(\ln 2)(2^x+1)} + C$

17. $\frac{3^x}{\ln 3} - x + 2\log_3|3^x - 1| + C$

18. $x - \frac{1}{2}\log_5(1 + 5^x + 25^x) - \frac{1}{\sqrt{3}\ln 5}\arctan\frac{1+2(5^x)}{\sqrt{3}} + C.$

19. $\frac{2}{3}x^{3/2} - x + 2\sqrt{x} - 2\ln(1 + \sqrt{x}) + C$ **20.** $x - \frac{8}{3}x^{3/4} + 4\sqrt{x} - 8\sqrt[4]{x} + 8\ln(1 + \sqrt[4]{x}) + C$

21. $2\sqrt{x} + \frac{3}{2}\ln|1 - \sqrt[6]{x}| - \frac{3}{2}\ln(1 + \sqrt[6]{x}) + 3\arctan\sqrt[6]{x} + C$

22. $\frac{4}{5}x^{5/4} - x + \frac{4}{3}x^{3/4} - 2\sqrt{x} + 4\sqrt[4]{x} + \frac{12}{13}x^{13/12} - \frac{6}{5}x^{5/6} + \frac{12}{7}x^{7/12} - 3\sqrt[3]{x}$
$+ 12\sqrt[12]{x} - 8\ln(1 + \sqrt[12]{x}) - 2\ln(\sqrt[6]{x} - \sqrt[12]{x} + 1) + 4\sqrt{3}\arctan\left(\frac{1 - 2\sqrt[12]{x}}{\sqrt{3}}\right) + C$

23. $\frac{2}{3}(x+2)^{3/2} - 5x + 46\sqrt{x+2} - 230\ln(5 + \sqrt{x+2}) + C$

24. $-x + \frac{27}{2}(x+5)^{2/3} - 162\sqrt[3]{x+5} + 972\ln|6 + \sqrt[3]{x+5}| + C$

25. $\frac{3}{4}x^{4/3} + \frac{6}{7}x^{7/6} - \frac{6}{5}x^{5/6} - \frac{3}{2}x^{2/3} + 3\sqrt[3]{x} + 6\sqrt[6]{x} - 4\sqrt{3}\arctan\left(\frac{2\sqrt[6]{x}-1}{\sqrt{3}}\right) + C$

26. $x + 2\ln|x-1| + C$ **27.** $-x - \frac{12}{5}x^{5/6} - 3x^{2/3} - 4\sqrt{x} - 6\sqrt[3]{x} - 12\sqrt[6]{x} - 12\ln|1-$

28. $2\pi - 3\sqrt{3}$ **29.** $\frac{3}{8} + \frac{1}{2}\ln 2 - \frac{1}{2}\ln 3$ **30.** $18\ln 3 - 12$ **31.** $\frac{2\pi^2}{2}$
32. $\frac{17\pi}{3} - 8\pi\ln 2$ **33.** $\frac{\pi}{2} + \frac{\pi}{e+1} + \pi\ln 2 - \pi\ln(e+1)$

35. (a) $\frac{13}{10}\ln(4 - \sqrt{3}) - \frac{4}{5}\ln 2 - \frac{13}{20}\ln 3 - \frac{4}{5}\ln(2 - \sqrt{3})$ miles
(b) $16\ln 2 - 10\ln 3 - 6 + \ln(e^2 - e + 1) + 4\ln(e+1) - \frac{2\pi}{\sqrt{3}} + \frac{4}{\sqrt{3}}\arctan\left(\frac{2e-1}{\sqrt{3}}\right)$
(c) $6 - 6\sqrt[6]{2} - 13\ln 2 - 18\ln 3 + 12\ln(1 + \sqrt[6]{2}) + 18\ln(2 + \sqrt[6]{2})$ km

36. $\frac{1}{\pi}\ln\frac{3}{2}$ lbs **37.** $2 + \frac{1}{3}\ln 2 - \frac{1}{3}\ln(e^6 + 1)$ joules **38.** $\frac{\frac{2\sqrt{3}\pi}{3} - \frac{441}{220} - 2\ln 2}{\frac{2\pi}{\sqrt{3}} - \frac{9}{5} - 2\ln 2}$ me

39. $\bar{x} = \frac{\frac{856,052\sqrt{2}}{15,015} + 12\arctan\sqrt{2} + \frac{135,904}{15,015} - 3\pi}{8 - 4\sqrt{2} - 3\pi + 12\arctan\sqrt{2}}$ and $\bar{y} = \frac{2 + 6\ln 3 - 12\ln 2}{8 - 4\sqrt{2} - 3\pi + 12\arctan\sqrt{2}}$

Section 4.9

1. $\frac{1}{8}$ 2. 36 3. diverges 4. 9 5. π
6. -1 7. 1 8. diverges 9. diverges 10.
11. $\frac{\pi}{2}$ 12. diverges 13. $\ln\frac{3}{2}$ 14. diverges 15.
16. diverges 17. diverges 18. diverges 19. $\sqrt{5}$ 20.
21. diverges 22. $\pi + \frac{\ln 15}{4} - 2\arctan 2$ 23. $\frac{\pi}{\sqrt{3}}$ 24. $\frac{2}{3}\sqrt[4]{27}$ 25.
26. 0 27. diverges 28. diverges 29. diverges 30.
31. diverges 32. diverges 33. diverges 34. diverges 35.
36. converges 37. diverges 38. diverges 39. converges 40.
41. diverges 42. converges 43. diverges 44. converges 45.
46. converges 47. converges 48. converges

49. converges for all p **50.** converges for all p **51.** converges for $p > 1$

52. converges for $p > 1$ **53.** converges for all p **54.** converges for $p < 2$

55. (a) $p > \frac{1}{2}$ **(b)** $\frac{1}{2} < p \le 1$ **(c)** $p > 2$

56. (a) $p < \frac{1}{2}$ **(b)** $\frac{1}{2} \le p < 1$ **(c)** $p < 2$ **57.** $\frac{\pi}{\ln 2}$ **58.** π^2 **59.** $\frac{\pi}{e}$

63. (a) $\frac{1}{3}$ mile **(b)** infinite distance **(c)** 1 km **(d)** $\frac{1}{2} + \frac{e^{\pi/2}}{e^\pi - 1}$ km **64.** 2 kg **65.** 4

66. (a) $\frac{\frac{\pi}{3\sqrt{3}} + \frac{1}{3}\ln 2}{\frac{\pi}{3\sqrt{3}} - \frac{1}{3}\ln 2}$ meters **(b)** $\frac{1}{\ln 2}$ meters **(c)** $\frac{23,672/105 - 6\ln 3}{16 - 6\ln 3}$ feet **(d)**

67. (a) $\bar{x} = \frac{3}{2}$ and $\bar{y} = \frac{3}{14}$ **(b)** $\bar{x} = 2$ and $\bar{y} = \frac{1}{8}$ **(c)** $\bar{x} = \frac{1}{4}$ and $\bar{y} = -1$

68. (a) diverges **(b)** π **(c)** diverges **(d)** $\ln 2$ **(e)** $\ln\frac{3\sqrt{3}}{2}$ **(f)** diverge
(g) $\frac{\pi}{2}$ **(h)** $\ln(3 + 2\sqrt{2})$

69. (a) converges **(b)** converges **(c)** converges **(d)** diverges

Section 4.10

1. **(a)** $\sinh 0 = 0$ $\cosh 0 = 1$ $\tanh 0 = 0$
$\coth 0$ not defined $\text{sech } 0 = 1$ $\text{csch } 0$ not defined

(b) $\sinh 1 = 1.18$ $\cosh 1 = 1.54$ $\tanh 1 = 0.766$
$\coth 1 = 1.31$ $\operatorname{sech} 1 = .648$ $\operatorname{csch} 1 = .851$
(c) $\sinh -1 = -1.18$ $\cosh -1 = 1.54$ $\tanh -1 = -0.766$
$\coth -1 = -1.31$ $\operatorname{sech} -1 = .648$ $\operatorname{csch} -1 = -.851$
(d) $\sinh 2 = 3.63$ $\cosh 2 = 3.76$ $\tanh 2 = 0.965$
$\coth 2 = 1.04$ $\operatorname{sech} 2 = .266$ $\operatorname{csch} 2 = .275$
(e) $\sinh \ln 2 = \frac{3}{4}$ $\cosh \ln 2 = \frac{5}{4}$ $\tanh \ln 2 = \frac{3}{5}$
$\coth \ln 2 = \frac{5}{3}$ $\operatorname{sech} \ln 2 = \frac{4}{5}$ $\operatorname{csch} \ln 2 = \frac{4}{3}$

2.(a) $\cosh x = \sqrt{5}$ $\tanh x = -\frac{2\sqrt{5}}{5}$ $\coth x = -\frac{\sqrt{5}}{2}$ $\operatorname{sech} x = \frac{\sqrt{5}}{5}$ $\operatorname{csch} x = -\frac{1}{2}$
(b) $\sinh y = \sqrt{15}$ $\tanh y = \frac{\sqrt{15}}{4}$ $\coth y = \frac{4\sqrt{15}}{15}$ $\operatorname{sech} y = \frac{1}{4}$ $\operatorname{csch} y = \frac{\sqrt{15}}{15}$
(c) $\sinh z = \frac{\sqrt{2}}{4}$ $\cosh z = \frac{3\sqrt{2}}{4}$ $\coth z = 3$ $\operatorname{sech} z = \frac{2\sqrt{2}}{3}$ $\operatorname{csch} z = 2\sqrt{2}$
(d) $\sinh t = -\frac{\sqrt{6}}{12}$ $\cosh t = \frac{5\sqrt{6}}{12}$ $\tanh t = -\frac{1}{5}$ $\operatorname{sech} t = \frac{2\sqrt{6}}{5}$ $\operatorname{csch} t = -2\sqrt{6}$
(e) $\sinh u = -\frac{\sqrt{21}}{2}$ $\cosh u = \frac{5}{2}$ $\tanh u = -\frac{\sqrt{21}}{5}$ $\coth u = -\frac{5\sqrt{21}}{21}$ $\operatorname{csch} u = -\frac{2\sqrt{21}}{21}$
(f) $\sinh v = \frac{1}{3}$ $\cosh v = \frac{\sqrt{10}}{3}$ $\tanh v = \frac{\sqrt{10}}{10}$ $\coth v = \sqrt{10}$ $\operatorname{sech} v = \frac{3\sqrt{10}}{10}$

7. (a) $\tanh(x+y) = \frac{\tanh x + \tanh y}{1 + \tanh x \tanh y}$ **(b)** $\tanh(x-y) = \frac{\tanh x - \tanh y}{1 - \tanh x \tanh y}$
(c) $\tanh 2x = \frac{2\tanh x}{1 + \tanh^2 x}$

8. (a) $\sinh x \cosh y = \frac{1}{2}\sinh(x+y) + \frac{1}{2}\sinh(x-y)$
(b) $\sinh x \sinh y = \frac{1}{2}\cosh(x+y) - \frac{1}{2}\cosh(x-y)$
(c) $\cosh x \cosh y = \frac{1}{2}\cosh(x+y) + \frac{1}{2}\cosh(x-y)$

11. (a) $5\operatorname{sech}^2(5x+1)$ **(b)** $\frac{\sinh x}{2\sqrt{1+\cosh x}}$
(c) $\frac{(2x+\operatorname{sech}^2 x)(x^2-\coth x) - (x^2+\tanh x)(2x+\operatorname{csch}^2 x)}{(x^2-\coth x)^2}$ **(d)** $\operatorname{sech}^3 x - \operatorname{sech} x \tanh^2 x$
(e) $-\cos(\operatorname{csch} x)\operatorname{csch} x \coth x$ **(f)** $\coth x$
(g) $(1+\sec^2 x)\sinh(x+\tan x)$ **(h)** $\frac{\cos x \operatorname{csch}^2 \sqrt{5-\sin x}}{2\sqrt{5-\sin x}}$ **(i)** $\frac{4x}{(x^2+1)^2}$

12. (a) $\ln \cosh x + C$ **(b)** $\ln|\sinh x| + C$ **(c)** $2\arctan(e^x) + C$
(d) $-\ln|\coth x + \operatorname{csch} x| + C$ **(e)** $-\frac{1}{2}\operatorname{csch}^2 x + C$ **(f)** $\frac{1}{4}\sinh^4 x + \frac{1}{6}\sinh^6 x + C$
(g) $\frac{1}{7}\cosh^7 x + C$ **(h)** $-\frac{1}{6}\coth^6 x + C$ **(i)** $-\frac{1}{3}\operatorname{sech}^3 x + \frac{1}{5}\operatorname{sech}^5 x + C$
(j) $\frac{1}{4}\sinh 2x - \frac{x}{2} + C$ **(k)** $-\frac{1}{4}\coth x + C$ **(l)** $\frac{1}{16}\tanh x - \frac{1}{48}\tanh^3 x + C$

17. (a) 24   **(b)** 8   **(c)** 3   **(d)** 11   **(e)** $\frac{7}{2}$ **(f)** $\frac{\sqrt{6}}{12}$ **(g)** $-\frac{\sqrt{3}}{3}$ **(h)** $-\frac{9\sqrt{82}}{82}$ **(i)** $\ln 3$

18. (a) 2.31   **(b)** 3.18   **(c)** .458   **(d)** $-.255$ **(e)** 1.45   **(f)** $-.0998$

19. (a) $\frac{x}{|x|\sqrt{x^2-1}}$ **(b)** $\frac{\frac{\cosh^{-1} x}{\sqrt{x^2+1}} - \frac{\sinh^{-1} x}{\sqrt{x^2-1}}}{(\cosh^{-1} x)^2}$
(c) $-\frac{\operatorname{sech}^{-1} x}{x\sqrt{1-x^2}\sqrt{1+(\operatorname{sech}^{-1} x)^2}}$ **(d)** $\frac{x^2-2}{(1-x^2)(x+\tanh^{-1} x)^2}$ **(e)** $-\csc x$
(f) $-\frac{3(\operatorname{sech}^{-1} x)^2}{x\sqrt{1-x^2}}$ **(g)** $\frac{\sec^2 x}{\sqrt{\tan^2 x-1}}$
(h) $\frac{\operatorname{csch}^2 x - \cosh x}{|\sinh x + \coth x|\sqrt{1+(\sinh x+\coth x)^2}}$ **(i)** $\frac{1}{\cosh^{-1} x + \sinh^{-1} x}\left(\frac{1}{\sqrt{x^2-1}} + \frac{1}{\sqrt{x^2+1}}\right)$

20. (a) $\frac{1}{6}\tanh^{-1}\left(\frac{3x}{2}\right) + C$ **(b)** $-\frac{1}{5}\operatorname{sech}^{-1}\left(\frac{2x}{5}\right) + C$ **(c)** $-\frac{1}{3}\operatorname{csch}^{-1}\left(\frac{4x}{3}\right) + C$
(d) $\frac{1}{2}\cosh^{-1}(2\sin x) + C$ **(e)** $\cosh^{-1}\left(\frac{e^x}{3}\right) + C$ **(f)** $-\frac{1}{\ln 2}\operatorname{csch}^{-1}(2^x) + C$
(g) $\cosh^{-1}\tan x + C$ **(h)** $\cosh^{-1}(2x-1) + C$ **(i)** $\cosh^{-1}\frac{x+3}{3} + C$

21. 12.6 meters   **22.** 269 Newtons

23. (a) .807 sec   **(b)** 8.05 ft   **(c)** 1.614 sec   **(d)** -17.9 ft per sec

24. (a) .311 seconds   **(b)** 4.20 seconds   **(c)** -14.0 meters per second

25. (a) $\ln(1+\cosh 1) + \ln(1+\cosh 2) - 2\ln 2$ miles   **(b)** $\frac{1}{2}(\sinh 4 + \sinh 9)$ miles
(c) $\frac{1}{10\ln 2}\left(\operatorname{sech}^{-1}\frac{1}{10} - \operatorname{sech}^{-1}\frac{4}{5}\right)$ km   **26.** $\frac{1}{48}\sinh^3 4 + \frac{1}{64}\sinh 8 - \frac{1}{8}$ lbs   **27.** $\frac{1}{3}\sinh^{-1} 8$ jls

Section 4.11

1. (a) $y = \frac{1}{2}x^2 \ln x - \frac{1}{4}x^2 + C$ (b) $y = \frac{1}{C-5x}$ (c) $y = \frac{1}{2}(\ln x)^2 + C$
(d) $y = Ae^{2x^3}$ (e) $y = \pm\left(\frac{1}{2}x\sqrt{x^2+1} + \frac{1}{2}\ln|x + \sqrt{x^2+1}|\right) + C$
(f) $y = -\ln(x + C)$ (g) $y = \frac{2}{3}x + \frac{1}{9} + Ce^{-3x}$ (h) $\sec y + \tan y = Ae^{-\cos x}$
(i) $y = \frac{e^x + C}{|x|}$ (j) $y = Ce^{\frac{1}{2}x^2+x} - 1$ (k) $y = 3x^4 e^{\cos x} + Ce^{\cos x}$ (l) $y = \frac{Ax}{Ax}$

2. (a) $y = 3e^{-6x}$ (b) $y = \sin x - x\cos x + 3 - \pi$
(c) $y = -2e^{\pi(x-1)}$ (d) $y = \frac{5\sqrt{2}}{2}\sec x$
(e) $y = \frac{1}{4}x^4 + x^2 + \ln|x| + \frac{11}{4}$ (f) $y = \frac{1}{5}x - \frac{1}{25} + \frac{101}{25}e^{-5x}$
(g) $y = -\ln(e^5 + e^{-2} - e^x)$ (h) $y = \frac{\ln x}{7x} - \frac{1}{49x} + \frac{1}{49x^8}$
(i) $y = -\frac{x}{3} - \frac{1}{9} - \frac{2}{9}e^{3x-6}$ (j) $y = \sqrt[3]{\frac{x}{2} + \frac{52}{x}}$
(k) $y = \sqrt{\ln(x^2+1) + 16} - \ln 2$ (l) $y = \frac{1}{2}x^2 e^{-x} + 3e^{-x}$

3. (a) $625,000$ (b) $24\log_5 2 \approx 10.3$ hours

4. (a) $8000\left(\frac{8}{15}\right)^{5/2} \approx 1662$ bacteria   (b) $8000\left(\frac{15}{8}\right)^{7/2} \approx 72,210$ bacteria

5. (a) $15,000(2^{10/3}) \approx 151,191$ bacteria   (b) $3\log_2(5/3) \approx 2.21$ hours

6. (a) $10(4/5)^{5/4} \approx 7.57$ ounces   (b) $\frac{12}{\log_{10}(5/4)} \approx 124$ hours

7. (a) $12(6/5)^{2/3} \approx 13.6$ grms   (b) $8\frac{1}{3}$ grms   (c) $6 - 9\log_{6/5}(5/12) \approx 49$

9. (a) $5700(2 - \log_2 3) \approx 2366$ years (b) $\frac{100}{2^{204/475}} \approx 74.3\%$ 10. $\$6749$ 11. 9.16%

12. 13.9 yrs 13. 2.996% 14. 5.03 ppm 15. 4% 16. 3.35% 17. 18.3%

18. (a) The concentration of salt is $5e^{-t/25}$ percent in the first tank and $\frac{14}{3}e^{-t/100} - \frac{5}{3}e^{-t/25}$ percent in the second tank.   (b) 3.11 percent

19. 15,000 bacteria 20. 500 bees 21. 4 fish 22. 20,000 bacteria 23. 140 pe

24. $258.71 25. $18.87 26. $P \approx 100 + 112.5\sin\frac{\pi t}{180} + 15.5\cos\frac{\pi t}{180}$ dollars/bus

27. 19 hrs, 15 min, 15 sec   28. 16 min, 40 sec   29. 1.21×10^{-7} sec

31. $h = -\alpha_1 k/\beta$ 32. $h = -\alpha_1 k/\beta$ 33. (a) $T = T_0 + [T(0) - T_0]e^{-kt}$ (b) 3

34. (a) $v = 6528e^{-t/200} - 6400$ ft/sec   (b) 253 ft   (c) 7.92 sec

35. (a) $y = \frac{Ny(0)e^{Nkt}}{N - y(0) + y(0)e^{Nkt}}$ (b) In the long run $y \approx N$, i.e. everyone becomes

36. 31.5% 37. (a) $y = \sinh^{-1}\left(\frac{1}{2}x^2 + C\right)$ (b) $y = Ce^{2x} - \frac{2}{3}\sinh x - \frac{1}{3}\cosh$
(c) $\ln|\sinh y| - \frac{1}{2}\coth^2 y = e^x + C$ (d) $y = 12\sinh x - \frac{12\cosh x}{x} + \frac{C}{x}$

38. (a) $y = \cosh^{-1}\left(\frac{13}{5}e^{2x^4}\right)$ (b) $y = \cosh x - 1 + 2e^{1-\cosh x}$
(c) $\coth y - \operatorname{csch} y = \left(\frac{e-1}{e+1}\right)e^{(e^x-1)}$

Section 4.12

1. (a) $y = -\frac{1}{\pi^2}\sin\pi x + Cx + B$ (b) $y = \arcsin\frac{x}{C} + B$
(c) $y = Cx^2 - 12x + B$ (d) $y = 2A\tan(Ax + B)$ or $y = \frac{A(1 + Be^{Ax})}{1 - Be^{Ax}}$ or $y = \frac{1}{2}$
(e) $y = 0$ or $y = -3(x+A)^2$, or $y = 3/A^2 \cosh^2(Ax+B)$ or $y = -3/A^2 \sinh^2(A$
or $y = -3/A^2 \cos^2(Ax + B)$ (f) $y = B + \frac{C - \ln|x|}{x}$ (g) $y = \frac{x^3}{6} + x\ln|x| + Cx$
(h) $y = -\ln|B - Ax|$

(a) $y = xe^x - 2e^x + 4$ **(b)** $y = 4e^{2x} - 7$ **(c)** $y = \frac{1}{18}(12x+4)^{3/2} - \frac{41}{9}$
(d) $y = \frac{1}{\sqrt{1-\frac{3}{4}e^{2x}}}$ **(e)** $y = \frac{x^2}{6} - \frac{x}{9} - \frac{19}{27}e^{-3x} - \frac{35}{27}$ **(f)** $y = 5x + x\arctan x - \frac{1}{2}\ln(x^2+1) + 3$

(a) $y = Ae^{-4x} + Be^{3x}$ **(b)** $y = Ae^{7x} + Be^{-3x}$ **(c)** $y = A\sin 3x + B\cos 3x$
(d) $y = Ae^{-2x}\sin 4x + Be^{-2x}\cos 4x$ **(e)** $y = Ae^{-3x} + Bxe^{-3x}$ **(f)** $y = Ae^{5x} + Be^{-3x}$
(g) $y = Ae^{4x} + Bxe^{4x}$ **(h)** $y = Ae^{3x}\sin 6x + Be^{3x}\cos 6x$
(i) $y = Ae^{-5x}\cos 3x + Be^{-5x}\sin 3x$

(a) $y = 16e^{3x} - 13e^{4x}$ **(b)** $y = e^{2x}(10x - 2)$ **(c)** $y = 13e^x - 9xe^x$
(d) $y = 6\cos 7x + \frac{8}{7}\sin 7x$ **(e)** $y = e^{4\pi - 4x}(\sin 8x + 2\cos 8x)$
(f) $y = e^{6x - 3\pi}\left(\sin 7x + \frac{9}{7}\cos 7x\right)$

(a) $2x^5 + 16x^3 + 10x$ **(b)** $-\sec x$ **(c)** $\frac{e^x}{x} - e^x \ln x$ **(d)** $1 - 2x\arctan x$

(a) $y = -2x - 1 + Ae^{7x} + Be^{-4x}$ **(b)** $y = \frac{1}{7}e^{4x} + Ae^{3x} + Be^{-3x}$
(c) $y = \frac{9-\pi^2}{(9+\pi^2)^2}\cos \pi x + \frac{6\pi}{(9+\pi^2)^2}\sin \pi x + Ae^{-3x} + Bxe^{-3x}$
(d) $y = 2xe^{3x} - \frac{4}{17}e^{3x} + Ae^{2x}\sin 4x + Be^{2x}\cos 4x$
(e) $y = Ae^{-6x} + Be^{4x} - (7x + 1/5)\sin 2x - (x + 11/10)\cos 2x$
(f) $y = 8e^{-x}\cos 3x + 9e^{-x}\sin 3x + Ae^{-4x}\sin 4x + Be^{-4x}\cos 4x$
(g) $y = \left(x + \frac{181}{85}\right)e^x \sin x - \left(13x + \frac{58}{85}\right)e^x \cos x + Ae^{4x} + Be^{-3x}$
(h) $y = x^2 + \frac{x}{10} + \frac{8}{25} + Ae^x \cos 3x + Be^x \sin 3x$
(i) $y = x^2 e^{2x}\sin x - \frac{4}{5}xe^{2x}\sin x - \frac{1}{25}e^{2x}\sin x - x^2 e^{2x}\cos x + 2xe^{2x}\cos x - \frac{31}{25}e^{2x}\cos x$
$\quad + A + Be^{-x}$ **(j)** $y = \left(2x + \frac{7}{5}\right)e^{-x}\cos x - \left(x + \frac{1}{5}\right)e^{-x}\sin x + A\cos 2x + B\sin 2x$

7. (a) $y = \frac{1}{30}e^{5x} + Ae^{-5x} + Be^{2x}$
(b) $y = \left(-\frac{3x}{8} + \frac{1}{4}\sin 2x - \frac{1}{32}\sin 4x\right)\cos x + \frac{1}{4}\sin^5 x + A\cos x + B\sin x$
(c) $y = e^{5x}(A + Bx - \ln|\cos x|)$ **(d)** $y = \frac{1}{2}\sin 2x\left(\frac{1}{3}\cos^5 x \sin x + \frac{1}{6}\cos^3 x \sin x + \frac{x}{4} + \frac{1}{8}\sin 2x\right)$
$\quad + \frac{1}{6}\cos^6 x \cos 2x + A\sin 2x + B\cos 2x$
(e) $y = \frac{1}{16}e^{-2x} + Ae^{2x} + Bxe^{2x}$ **(f)** $y = \frac{3}{10}e^x \sin x - \frac{13}{30}e^x \cos x + Ae^{-2x} + Be^x$

9. $s(t) = s(0)\cos \frac{k}{\sqrt{m}}t + \frac{\sqrt{m}v(0)}{k}\sin \frac{k}{\sqrt{m}}t$

10. (a) overdamped with $s(t) = \frac{s(0)}{2}\left[(1 + \sqrt{2})e^{-3 + \frac{3}{2}\sqrt{2}} - (\sqrt{2} - 1)e^{-3 - \frac{3}{2}\sqrt{2}}\right]$
(b) underdamped with $s(t) = s(0)\sqrt{\frac{1536}{1511}}e^{-\frac{5t}{12}}\cos \left(\frac{\sqrt{1511}}{12}t - \arctan \frac{5}{\sqrt{1511}}\right)$
(c) critically damped with $s(t) = s(0)\left(1 + \frac{4t}{3}\right)e^{-\frac{4t}{3}}$
(d) underdamped with $s(t) = s(0)\frac{\sqrt{6}}{2}e^{-\frac{2t}{3}}\cos \left(\frac{2}{3}\sqrt{2}t - \arctan \frac{1}{\sqrt{2}}\right)$
(e) critically damped with $s(t) = s(0)\left(1 + \frac{5t}{3}\right)e^{-\frac{5t}{3}}$
(f) overdamped with $s(t) = \frac{s(0)}{3}\left(4e^{-\frac{5t}{4}} - e^{-5t}\right)$

11. (a) $s(t) = \frac{8}{18-\pi^2}\sin \frac{1}{2}(3\sqrt{2} - \pi)t \sin \frac{1}{2}(\pi + 3\sqrt{2})t$
(b) $s(t) = \frac{4}{59}\sin \frac{1}{6}(15 - 4\sqrt{3})t \sin \frac{1}{6}(15 + 4\sqrt{3})t$ **(c)** $s(t) = \frac{3t}{16}\sin 4t$
(d) $s(t) = \frac{5t}{6}\sin 3t$

12. When $\lambda \neq \omega$, $s(t) = \frac{F_0}{m(\omega^2 - \lambda^2)}\cos \lambda t + \frac{v(0)}{\omega}\sin \omega t + \left[s(0) - \frac{F_0}{m(\omega^2 - \lambda^2)}\right]\cos \omega t$.
When $\lambda = \omega$, $s(t) = \frac{F_0}{2m\omega}t \sin \omega t + s(0)\cos \omega t + \frac{v(0)}{\omega}\sin \omega t$

13. When $\lambda \neq \omega$, $s(t) = \frac{F_0}{m\omega(\omega^2 - \lambda^2)}(\omega \sin \lambda t - \lambda \sin \omega t)$.
When $\lambda = \omega$, $s(t) = \frac{F_0}{2m\omega^2}(\sin \omega t - \omega t \cos \omega t)$

14. (a) $s(t) = -\frac{8}{289}e^{-5t} - \frac{5}{17}te^{-5t} + \frac{1}{17}\cos \left(3t - \arctan \frac{15}{8}\right)$ and the steady state
solution is $\frac{1}{17}\cos \left(3t - \arctan \frac{15}{8}\right)$.
(b) $s(t) = \frac{6}{13}e^{-3t} - \frac{12}{25}e^{-\frac{3t}{2}} + \frac{6\sqrt{13}}{65}\cos(2t - \arctan 18)$ and the steady state solution
is $\frac{6\sqrt{13}}{65}\cos(2t - \arctan 18)$.

797

(c) $s(t) = -\frac{8}{17\sqrt{47}}e^{-t}\sin\sqrt{47}t - \frac{4}{17}e^{-t}\cos\sqrt{47}t + \frac{1}{\sqrt{17}}\cos\left(4t - \arctan\frac{1}{4}\right)$
steady state solution is $\frac{1}{\sqrt{17}}\cos\left(4t - \arctan\frac{1}{4}\right)$.

(d) $s(t) = -\frac{13}{80}e^{-4t}\sin 3t - \frac{9}{80}e^{-4t}\cos 3t + \frac{3}{8\sqrt{10}}\cos\left(t - \arctan\frac{1}{3}\right)$ and the stead
solution is $\frac{3}{8\sqrt{10}}\cos\left(t - \arctan\frac{1}{3}\right)$.

15. (a) $y = 4\sinh 5x + Cx + B$ **(b)** $y = \sinh^{-1}(C - x)$
(c) $y = \frac{3x}{8} + \frac{1}{4}\sinh 2x + \frac{1}{32}\sinh 4x + \frac{2}{3}\cosh^3 x + C$ or $y = \frac{35}{8}x + \frac{1}{4}\sinh 2x + \frac{1}{32}\sin$
$\frac{2}{3}\cosh^3 x \pm (4\sinh x + 2\cosh^2 x) + C$

16. (a) $-k$ **(b)** $-ke^{2rx}$ **(c)** $-2k\mathrm{sech}\,kx\,\mathrm{csch}\,kx$

18. (a) $y = \frac{11}{840}e^x\sinh x - \frac{319}{840}e^x\cosh x + Ae^{-4x} + Be^{7x}$
(b) $y = 3xe^{4x} + \frac{15}{8}e^{-4x} + Ae^{-6x} + Be^{4x}$
(c) $y = -\frac{1}{16}\sinh 4x\cosh 2x + \frac{x}{4}\cosh 2x + \frac{1}{16}\cosh 4x\sinh 2x + A\cosh 2x + B\sin$
(d) $y = \frac{25}{4}\ln|\mathrm{csch}\,4x + \coth 4x|\cosh 4x - \frac{25}{4} + A\cosh 4x + B\sinh 4x$

Section 4.13

1. (a) $x = 0$ **(b)** $x = 5, -2$ **(c)** $x = -1, 0, 1$ **(d)** $x = 1$ **(e)** $x = 0$ **(f)** $x =$

2. (a) Range $f = [-1, 0] \subset [-1, 1] = $ Domain f
(b) Range $g = \left[-\frac{1}{3}, 1\right] \subset [-1, 1] = $ Domain g
(c) Range $h = \left[\frac{1}{21}, \frac{194{,}481}{194{,}923}\right] \subset \left[\frac{1}{21}, 2\right] = $ Domain h
(d) Range $k = [0, 1] \subset \left[-\frac{\pi}{2}, \frac{\pi}{2}\right] = $ Domain k
(e) Range $m = \left[\frac{1}{e}, 1\right] \subset [0, 1] = $ Domain m
(f) Range $n \subset \left(-\frac{\pi}{2}, \frac{\pi}{2}\right) \subset [-\pi, \pi] = $ Domain n

3. (a) $d(f(x), f(y)) \leq \frac{3}{4}d(x, y) < \frac{7}{8}d(x, y)$, and the fixed point of f is approx
(b) $d(g(x), g(y)) \leq \frac{1}{2}d(x, y) < \frac{3}{4}d(x, y)$, and the fixed point of g is approx. .52
(c) $d(h(x), h(y)) < .7d(x, y)$, and the fixed point of h is approximately .682.
(d) $d(j(x), j(y)) < \frac{3}{4}d(x, y)$, and the fixed point of j is approximately 1.732.
(e) $d(k(x), k(y)) \leq \frac{1}{2}d(x, y) < \frac{3}{4}d(x, y)$, and the fixed point of k is approx. .88
(f) $d(m(x), m(y)) \leq \frac{1}{3}d(x, y) < \frac{1}{2}d(x, y)$, and the fixed point of m is approx. .
(g) $d(n(x), n(y)) \leq \frac{1}{2}d(x, y) < \frac{3}{4}d(x, y)$, and the fixed point of n is approx. 1.1
(h) $d(p(x), p(y)) \leq \frac{4}{5}d(x, y) < \frac{9}{10}d(x, y)$, and the fixed point of p is approx. .1

6. $x = f(x) = x + \frac{1}{x}$ has no solution, so f has no fixed points. Theorem 4.13.1
not apply because f is not a contraction. In fact, $d(f(x), f(y)) = |1 - \frac{1}{xy}|d(x,$
the numbers $|1 - \frac{1}{xy}|$ can be chosen arbitrarily close to one.

7. (a) $f'(x) = \sin\left(x + \frac{\pi}{10}\right)$ has a maximum value of one, so f is not a contra
However, $D(f^2)(x) = \sin\left(\cos\left(x + \frac{\pi}{10}\right) + \frac{\pi}{10}\right)\sin\left(x + \frac{\pi}{10}\right)$ has a maximum valu
than $\sin\left(1 + \frac{\pi}{10}\right)(1) < .97$. Thus f^2 is a contraction. The fixed point of f is ap
mately .606.
(b) $g'(x) = \cos x$ has a maximum value of one, so g is not a contraction.
$d(g^2(x), g^2(y)) \leq (\sin 1)d(x, y) < .9d(x, y)$. The fixed point of g is approx. 2.31(
(c) Since $|h'(x)| = e^{-x}$ has maximum value one, h is not a contraction. The
mum value of $|D(h^2)|$ for $x \geq 0$ is $1/e$ (at $x = 0$). Since $1/e < 1$, h^2 is a contra
The fixed point of h is approximately .567.
(d) $d(j(2), j(2 - \frac{1}{2^n})) = \frac{1}{2^n}(1 - \frac{1}{2^{n+2}})$. Thus j is not a contraction. $d(j^2(y), j^2($
$\frac{3}{8}d(y, x)$, so j^2 is a contraction. The fixed point of j is approximately .236.
(e) $d(k(y), k(x)) = d(x, y)\left|\frac{2}{xy} + \frac{1}{10}\right|$. When $x = 1$, $y = 1.1$, $\left|\frac{2}{xy} + \frac{1}{10}\right| = \frac{211}{110}$.
k is not a contraction. $d(k^2(y), k^2(x)) < \frac{2021}{2100}d(y, x)$. The fixed point of k is ap
mately 1.491.

f) $m'(0) = 1$, so m is not a contraction. $D(m^2(x)) \leq \frac{1}{5}$ for $x \in [-1,1]$, so m^2 is a contraction. The fixed point of m is approximately 3.274.

8. (a) $f^2(x) = x^{1/4}$. Hence $|f'(x)| = 1/2\, x^{-1/2}$ and $D(f^2)(x) = 1/4\, x^{-3/4}$ have values $\sqrt{10}/2$ and $10^{3/4}/4$, at $x = 0.1$, which are greater than one. Thus f and f^2 are not contractions.
b) $f^3(x) = x^{1/8}$. Hence $D(f^3)(x) = 1/8\, x^{-7/8}$ has maximum value (for $x \geq 0.1$) of $.1^{7/8}/8 < 0.94 < 1$. Thus f^3 is a contraction. **(c)** The fixed point of f is 1.

9. (a) $d(f,g) = 6$ **(b)** $d(h,j) = 84$ **(c)** $d(k,m) = e - \frac{1}{e}$ **(d)** $d(n,p) = \ln 3$
e) $d(q,r) = \sqrt{2}$ **10.** $d(f,g) = 5/4$, $d(f,h) = 1$, $d(h,g) = 2$

1. (a) diverges **(b)** converges to $G(x) = 0$ **(c)** converges to $H(x) = 0$
d) diverges **(e)** diverges **(f)** diverges **(g)** diverges
h) converges to $R(x) = e^x$ **(i)** diverges **(j)** converges to $T(x) = 0$

2. (b), (c), (h), (j) are Cauchy sequences.

5. (a) $d(S(f), S(g)) \leq \frac{1}{3}d(f,g) < .4d(f,g)$ **(b)** $d(T(f), T(g)) \leq \frac{\pi}{4}d(f,g) < .8d(f,g)$
c) $d(U(f), U(g)) \leq \frac{2}{3}d(f,g) < .7d(f,g)$ **(d)** $d(V(f), V(g)) \leq \frac{1}{2}d(f,g) < .6d(f,g)$
e) $d(W(f), W(g)) \leq \frac{4}{5}d(f,g) < .9d(f,g)$

6. (a), (b), (c), (e) The zero function is the unique fixed point of each operator.
d) The unique fixed point of this operator is $f(x) = c$ where c is the unique solution of $e^{-c^2} = 8cM^2$.

19. (a) $d(f(x), f(y)) \leq \frac{1}{2}d(x,y) < \frac{3}{4}d(x,y)$. The unique fixed point of f is approx. .844.
b) $d(g(x), g(y)) \leq \frac{2}{3}d(x,y) < \frac{3}{4}d(x,y)$. The unique fixed point of g is approx. $-.453$.

20. (a) $d(S(f), S(g)) \leq \frac{3}{4}d(f,g) < \frac{7}{8}d(f,g)$. **(b)** $d(T(f), T(g)) \leq \frac{1}{2}d(f,g) < \frac{3}{4}d(f,g)$.
c) $d(U(f), U(g)) \leq \frac{\sinh 1}{9}d(f,g) < 0.14d(f,g)$.

Section 4.14

1. (a) $\bar{y}' = (\bar{x}+1)(\bar{y}+2)$, $\bar{y}(0) = 0$ **(b)** $\bar{y}' = e^{\bar{x}+\bar{y}+3}$, $\bar{y}(0) = 0$
c) $\bar{y}' = (\bar{x}^2 + 4\bar{x} + 3)(\bar{y}^2 + 6\bar{y} + 10)$, $\bar{y}(0) = 0$
d) $\bar{y}' = (\bar{x}+\bar{y}+\frac{\pi}{2}-1)(\bar{x}-\bar{y}+\frac{\pi}{2}+1)$, $\bar{y}(0) = 0$

2. (a) $y(x) = \frac{x^3}{3} + \int_0^x y(t^2)\, dt$ **(b)** $y(x) = \int_0^x \ln[t + y(t) + 1]\, dt$
c) $y(x) = \int_0^x \tan[ty(t)]\, dt$ **(d)** $y(x) = \int_0^x \sqrt{4 - t^2 - y(t)^2}\, dt$

3. (a) $F(g(x)) = \frac{x^3}{3} + \int_0^x g(t^2)\, dt$ **(b)** $F(g(x)) = \int_0^x \ln[t + g(t) + 1]\, dt$
c) $F(g(x)) = \int_0^x \tan[tg(t)]\, dt$ **(d)** $F(g(x)) = \int_0^x \sqrt{4 - t^2 - g(t)^2}\, dt$

4. (a) $F^n(g(x)) = \sum_{k=1}^n \frac{2}{k!5^k}x^k$ **(b)** $F^n(g(x)) = \sum_{k=2}^{n+1} \frac{1}{k!3^{k-1}}x^k$
c) $F^n(g(x)) = \sum_{k=3}^{n+2} \frac{2}{k!7^{k-2}}x^k$ **(d)** $F^3(g(x)) = \frac{5}{2^{10}}x^{10} + \frac{15}{(7)2^{11}}x^7 + \frac{5}{256}x^4 + \frac{1}{4}x$
e) $F(g(x)) = \frac{2}{3}(x+11)^{3/2}$ **(f)** $F(g(x)) = \ln(x+7)$

5. (a) yes, $K = 1$ **(b)** yes, $K = 1$ **(c)** yes, $K = 2$ **(d)** yes, $K = 3$
(e) yes, $K = 3/2$ **(f)** no

6. (a) $n = 4$ **(b)** $n = 15$ **(c)** $n = 9$ **(d)** $n = 7$ **(e)** $n = 22$

7. (a) $F(G)(x) = \int_0^x G(t)^2\, dt - x$, $F^3(g)(x) = \frac{x^7}{63} - \frac{2x^5}{15} + \frac{x^3}{3} - x$
(b) $F(G)(x) = \int_0^x tG(t)^2\, dt + x$, $F^3(g)(x) = \frac{x^{10}}{160} + \frac{x^7}{14} + \frac{x^4}{4} + x$
(c) $F(G)(x) = \int_0^x G(t)^3\, dt + 5x$, $F^3(g)(x) = \frac{5^9}{832}x^{13} + 57\frac{3}{32}x^{10} + 55\frac{3}{28}x^7 + \frac{125}{4}x^4 + 5x$
(d) $F(G)(x) = \int_0^x G(t)e^t\, dt - x$, $F^3(g)(x) = -\frac{1}{2}xe^{2x} - xe^x + \frac{3}{4}e^{2x} - x - \frac{3}{4}$
(e) $F(G)(x) = \int_0^x G(t)\sin t\, dt + x$, $F^3(g)(x) = \frac{x}{4}\cos 2x - \frac{3}{8}\sin 2x - x\cos x + \sin x + \frac{3x}{2}$
(f) $F(G)(x) = \int_0^x G(t)^2\, dt + \frac{x^3}{3}$, $F^3(g)(x) = \frac{x^{15}}{59,535} + \frac{2x^{11}}{2079} + \frac{x^7}{63} + \frac{x^3}{3}$

8. **(a)** If $k \neq 0$: $\alpha = -\frac{3}{4|k|}$, $\beta = \frac{3}{4|k|}$. If $k = 0$, take any $\alpha <$

$$y(x) = \lim_{n \to \infty} \left(\frac{k^n x^n}{n!} + \cdots + kx + 1 \right)$$

(b) $\alpha = 1 - \sqrt{3}$, $\beta = \sqrt{3} - 1$, $y(x) = \lim_{n \to \infty} \left(\frac{x^{n+1}}{(n+1)!} + \cdots + \frac{x^2}{2} \right)$

(c) Take any $\alpha < 0 < \beta$.

$$y(x) = \lim_{n \to \infty} \left(\frac{x^{2n+1}}{(2n+1)(2n-1)\cdots(5)(3)} + \frac{x^{2n}}{(2n)(2n-2)\cdots(6)(4)} + \cdots + \frac{x^4}{4} \right.$$

(d) $\alpha = -\frac{\sqrt{3}}{2\sqrt{2}}$, $\beta = \frac{\sqrt{3}}{2\sqrt{2}}$, $y(x) = \lim_{n \to \infty} \left(\frac{x^{2n}}{n!} + \cdots + \frac{x^2}{1!} \right)$

(e) $\alpha = -\frac{\sqrt{5}}{6}$, $\beta = \frac{\sqrt{5}}{6}$, $y(x) = \lim_{n \to \infty} \left(\frac{2^{n+2} x^{2n}}{n!} + \cdots + \frac{2^{1+2} x^2}{1!} + 4 \right)$

(f) $\alpha = -\frac{1}{\sqrt[3]{36}}$, $\beta = \frac{1}{\sqrt[3]{36}}$, $y(x) = \lim_{n \to \infty} \left(\frac{5 \cdot 2^n x^{3n}}{n!} + \cdots + \frac{5 \cdot 2 x^3}{1!} + 5 \right)$

9. **(a)** Theorem 4.14.14 does not apply. When $k = 0$ there is a unique sc
When $k \neq 0$ there is no solution.
(b) Theorem 4.14.14 applies, and there is a unique solution.
(c) Theorem 4.14.14 does not apply. There is a unique solution for all k.
(d) Theorem 4.14.14 does not apply. When $k = 0$ there is no solution. Wher
there is a unique solution.
(e) Theorem 4.14.14 does not apply. When $k = 0$ there are an infinite num
solutions. When $k \neq 0$ there is no solution.
(f) Theorem 4.14.14 applies, and there is a unique solution.

10. Theorem 4.14.14 applies. Note $y = \frac{1}{4}(x - 2)^2 - 1$ is not a solution.

Review Exercises

1. F	2. F	3. T	4. T	5. T	6. F	7. T	8. F
9. T	10. F	11. F	12. T	13. F	14. T	15. F	16. F
17. T	18. F	19. T	20. F	21. T	22. T	23. F	24. F
25. T	26. F	27. T	28. T	29. F	30. T		

31. **(a)** $\frac{1}{x(\ln x)(\ln \ln x)(\ln \ln \ln x)}$ **(b)** $f(x) = \ln \ln \ln \ln x - \ln \ln 2$

32. ∞ **33.** $f'(x) = \frac{e^x}{2 + 2e^x} > 0$ **34.** $\frac{1}{\ln 10} = \log_{10} e$ **35.** $\sec^2(x^x) \cdot x^x (1 +$

36. $x = 0$ or $x = 1$ **37.**

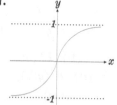

38. $\int \tan^N x \, dx = \frac{1}{N-1} \tan^{N-1} x - \int \tan^{N-2} x \, dx$

39. π **40.** $\frac{\pi^2}{2}$ **41.** ∞ **42.** $\frac{1}{12}(8\pi - 3\pi^2 + 12)$

43. $\lim_{x \to 0} \frac{\sin^2 x}{x^2} = 1$ and $\int_0^1 \pi \frac{\sin^2 x}{x^2} \, dx$ is a proper integral. Since $\frac{\sin^2 x}{x^2} \leq \frac{1}{x^2}$ for
$\pi \int_1^\infty \frac{\sin^2 x}{x^2} \, dx$ converges by the comparison test. Thus $V = \pi \int_0^\infty \frac{\sin^2 x}{x^2} \, dx$ is fin

44. $f(x) = \frac{1}{1 + \ln x}$ **45.** L is $D(\ln(1 + x))$ at $x = 0$ which is 1.

46. 8,000 bacteria **47.** 3.7% in each vat **48.** $\ln|\tan x| + C$ **49.** $C - \frac{1}{2}$

50. $x^2 \sin x + 2x \cos x - 2 \sin x + C$ **51.** $\frac{1}{5} \tan^5 x - \frac{1}{3} \tan^3 x + \tan x - x + C$

52. $\tan x + \frac{2}{3}\tan^3 x + \frac{1}{5}\tan^5 x + C$ **53.** $\frac{43}{2}\arcsin\frac{x-3}{5} - \frac{1}{2}(x+9)\sqrt{6x - x^2 + 16} + C$

54. $\ln|\tan x + \sqrt{4 + \tan^2 x}| + C$ **55.** $\frac{1}{2}(\ln x)\sqrt{(\ln x)^2 - 1} - \frac{1}{2}\ln|\ln x + \sqrt{(\ln x)^2 - 1}| + C$

56. $x + \frac{1}{6}\ln\frac{(x-1)^2}{x^2+x+1} - \frac{1}{\sqrt{3}}\arctan\frac{2x+1}{\sqrt{3}} + C$ **57.** $-\frac{3}{2}\ln\frac{x^2}{|x^2-1|} + \frac{1}{2x^2} + \frac{9}{8(x^2-1)} - \frac{x^2+1}{8(x^2-1)^2} + C$

58. $-\frac{1}{169x} - \frac{1}{54(13^3)}\left(216\ln\frac{x^2}{x^2+4x+13} + 11\arctan\frac{x+2}{3} + 13\frac{15x+138}{x^2+4x+13}\right) + C$

59. $\frac{1}{2}\left(\ln|\sec\theta| + \theta + \ln|\tan\theta - 1|\right) + C$

$4\ln\left|1 - \tan^2\frac{\theta}{2}\right| - (8\sqrt{2} + 12)\ln\left|1 + \sqrt{2} + \tan\frac{\theta}{2}\right| + (8\sqrt{2} - 12)\ln\left|1 - \sqrt{2} + \tan\frac{\theta}{2}\right| + C$

60. $2\arctan\sqrt{e^x} + C$ **61.** $\frac{6}{5}x^{5/6} - 2x^{1/2} + 6x^{1/6} - 6\arctan x^{1/6} + C$

62. $\frac{1}{2}(\ln\tan x)^2 + C$ **63.** $x\left[(\ln x)^3 - 3(\ln x)^2 + 6\ln x - 6\right] + C$

64. $\ln|1 + \arctan x| + C$ **65.** $2\sqrt{x} - 2\arctan\sqrt{x} + C$ **66.** converges **67.** diverges

68. diverges **69.** $y = -\ln(C - e^x)$ **70.** $y = -\frac{1}{2}(x-1) + \frac{3}{2} - \frac{3}{x-1} + \frac{3}{(x-1)^2} + \frac{C}{(x-1)^2 e^x}$

Chapter 5

Section 5.2

1. (a) $\{n^3\}$ for $n \geq 1$ (b) $\{(-1)^n\sqrt{n}\}$ for $n \geq 1$ (c) $\{\frac{1}{n!}\}$ for $n \geq 0$

(d) $\{(-1)^n(2n)!\}$ for $n \geq 0$ (e) $\{\frac{x^n}{n+1}\}$ for $n \geq 1$ (f) $\{(-1)^n\frac{n!}{x^n}\}$ for $n \geq 0$

2. (a) $1, \frac{1}{4}, \frac{1}{9}, \frac{1}{16}, \frac{1}{25}$ (b) $-2, 5, -8, 11, -14$ (c) $1, \frac{1}{2}, \frac{1}{6}, \frac{1}{24}, \frac{1}{120}$

(d) $\frac{1}{2}, \frac{1}{12}, \frac{1}{120}, \frac{1}{1680}, \frac{1}{30,240}$ (e) $-1, \frac{1}{2}, -\frac{1}{3}, \frac{1}{4}, -\frac{1}{5}$ (f) $\frac{x}{3}, \frac{x^2}{6}, \frac{x^3}{9}, \frac{x^4}{12}, \frac{x^5}{15}$

(g) $\sin(x^2), \sin\frac{x^4}{2}, \sin\frac{x^6}{3}, \frac{x^8}{4}, \frac{x^{10}}{5}$ (h) $\ln x, \ln x, \frac{1}{2}\ln x, \frac{1}{6}\ln x, \frac{1}{24}\ln x$

3. (a) $f(x) = x^3 - 2x + 6$ with domain $[1, \infty)$ (b) $f(x) = \frac{e^x}{\ln x}$ with domain $[2, \infty)$

(c) $f(x) = \frac{\sin x}{\arctan x}$ with domain $[1, \infty)$ (d) $f(x) = \sqrt{x^2 - x - 6}$ with domain $[3, \infty)$

(e) $f(x) = x^{1/x}$ with domain $[1, \infty)$ (f) $f(x) = (1 + \frac{1}{x})^x$ with domain $[1, \infty)$

4. (a) $1, 1, 1, 1, 1$ (b) $1, 3, \frac{9}{2}, \frac{9}{2}, \frac{27}{8}$ (c) $0, 1, 4, 9, 16$

(d) $0, -\frac{1}{2}\ln 2, \frac{1}{4}\ln 3, -\frac{1}{8}\ln 4, \frac{1}{16}\ln 5$ (e) $1, \frac{1}{2}, \frac{1}{24}, \frac{1}{720}, \frac{1}{40,320}$ (f) $8, 4, \frac{8}{3}, 2, \frac{8}{5}$

6.

	increasing	decreasing	bounded below	bounded above
(a)	yes	no	yes	no
(b)	no	no	no	no
(c)	no	no	yes	yes
(d)	yes	no	yes	yes
(e)	yes	no	yes	yes
(f)	no	no	yes	yes
(g)	yes	no	yes	yes
(h)	no	yes	yes	yes
(i)	yes	no	yes	yes
(j)	no	no	yes	yes
(k)	no	no	yes	yes
(l)	yes	no	yes	no
(m)	no	no	yes	yes
(n)	yes	yes	yes	yes
(o)	no	no	yes	yes

7. (a) increasing for $x = 0$ or $x \geq 2$; decreasing for $0 \leq x \leq 1$; bounded for $-1 \leq x \leq 1$, bounded below for $x > 1$.

(b) increasing for $-\sqrt{2} \le x \le 0$; decreasing for $0 \le x \le \sqrt{2}$; bounded for all
(c) decreasing and bounded for all x. **(d)** increasing and bounded for all x
(e) decreasing, bounded above for $x \le -1$ or $x \ge 1$; increasing, bounded for $-$
(f) decreasing and bounded for all x.
(g) increasing for $-1 \le x \le 1$; decreasing for $x \le -1$ or $x \ge 1$; bounded for ε
(h) increasing for $1 \le x$; decreasing for $0 \le x \le 1$; bounded for all x.
(i) increasing for $-\frac{2\pi}{3} \le x \le 0$; decreasing for $0 \le x \le \frac{2\pi}{3}$; bounded for all x.
(j) increasing and bounded below for $x \ge 0$; decreasing and bounded for $x \le$
(k) increasing for $x = 0$; decreasing for $-\sqrt{12} \le x \le \sqrt{12}$; bounded for all x.
(l) increasing for $x = 0$; decreasing for $0 \le x \le 12$; bounded for all x.

Section 5.3

1. (a) $\left\{\frac{1}{n^2}\right\}$ converges with limit 0. Take $N_\epsilon > \frac{1}{\sqrt{\epsilon}}$.

(b) $\left\{\frac{2n}{3n+1}\right\}$ converge with limit $\frac{2}{3}$. Take $N_\epsilon \ge \frac{2}{9\epsilon}$.

(c) $\{\sqrt{n}\}$ is unbounded, hence diverges.

(d) $\{2^{-n}\}$ converges with limit 0. Take $N_\epsilon > -\log_2 \epsilon$.

(e) $\left\{(-1)^{n^2}\right\}$ alternates between -1 and $+1$, hence diverges.

(f) $\left\{(-1)^{n(n+1)}\right\} = \{1\}$ converges with limit 1. Take all $N_\epsilon = 1$.

(g) $\left\{(-1)^n \frac{1}{n}\right\}$ converges with limit 0. Take $N_\epsilon > \frac{1}{\epsilon}$.

(h) $\left\{(-1)^n \frac{n}{n+1}\right\}$ diverges. Odd terms are close to -1 and even terms are clos

2. (a) $f(x) = \frac{x}{e^x}$ with $\lim\limits_{x \to \infty} f(x) = 0$. $\left\{\frac{n}{e^n}\right\}$ converges with limit 0.

(b) $f(x) = \frac{\ln x}{e^x}$ with $\lim\limits_{x \to \infty} f(x) = 0$. $\left\{\frac{\ln n}{e^n}\right\}$ converges with limit 0.

(c) $f(x) = \frac{\arctan(2x+1)}{\arctan(5x+3)}$ with $\lim\limits_{x \to \infty} f(x) = 1$. $\left\{\frac{\ln n}{e^n}\right\}$ converges with limit 1.

(d) $f(x) = \frac{x}{\ln x}$ with $\lim\limits_{x \to \infty} f(x) = \infty$. $\left\{\frac{n}{\ln n}\right\}$ diverges.

(e) $f(x) = \cos \pi x$ and $\lim\limits_{x \to \infty} f(x)$ DNE. $\{\cos \pi n\} = \{(-1)^n\}$ diverges.

(f) $f(x) = \tan \pi x$ and $\lim\limits_{x \to \infty} f(x)$ DNE. $\{\tan \pi n\} = \{0\}$ converges with limit 0

(g) $f(x) = \frac{x^2}{x^2-1} - \frac{x}{2x+1}$ with $\lim\limits_{x \to \infty} f(x) = \frac{1}{2}$. $\left\{\frac{n^2}{n^2-1} - \frac{n}{2n+1}\right\}$ converges with li

(h) $f(x) = x \sin \frac{1}{x}$ with $\lim\limits_{x \to \infty} f(x) = 1$. $\left\{n \sin \frac{1}{n}\right\}$ converges with limit 1.

3. (a) converges with limit 0 **(b)** converges with limit 0 **(c)** converges with
(d) converges with limit 0 for all x **(e)** converges with limit 0 for all x
4. (a) -25 **(b)** $12,500$ **(c)** $.005$ **(d)** $5\sqrt{2}$ **(e)** $\ln 5$

5. diverges	**6.** diverges	**7.** converges wi
8. converges with limit 0	**9.** converges with limit 1	**10.** diverges
11. converges with limit $\frac{\pi}{2}$	**12.** converges with limit 0	**13.** converges w
14. diverges	**15.** converges with limit 0	**16.** diverges
17. diverges	**18.** converges with limit 0	**19.** converges w
20. converges with limit 0	**21.** converges with limit 0	**22.** converges w
23. converges with limit 0	**24.** converges with limit $-\frac{2}{15}$	**25.** diverges
26. converges with limit $\frac{\pi}{4}$	**27.** converges with limit $\frac{\pi}{2}$	**28.** converges w
29. diverges	**30.** converges with limit 0	**31.** converges w

32. converges to 0 for $-1 \le x \le 1$ and diverges for $x \le -1$ or $x \ge 1$
33. converges to 0 for all x **34.** converges to 1 for all x **35.** converges to 1 for
36. diverges for $x < -1$ or $x > 1$, converges to $\ln(1 + x^2)$ for $-1 < x <$
converges to 0 for $x = \pm 1$.
37. converges to 0 for all x **38.** converges to 1 for $x \ne 0$ and converges to 0 for
39. converges to 0 for $-1 < x < 1$, converges to $\frac{\pi}{4}$ for $x = 1$, converges to $-$
$x = -1$, converges to $\frac{\pi}{2}$ for $x > 1$ or $x < -1$.

40. converges to 0 for all x

41. converges to 0 for $x < 0$, converges to 1 for $x = 0$ and diverges for $x > 0$

42. converges to 0 for $-1 \le x \le 1$ and diverges for $x < -1$ or $x > 1$.

43. converges to 1 for all x **44.** converges to 1 for all x

45. converges to 0 for $x < -1$ or $x > 1$, converges to 1 for $x = 1$ and diverges for $-1 \le x < 0$ or $0 < x < 1$

46. converges to $-\frac{\pi}{2}$ for $x < 0$, converges to 0 for $x = 0$ and converges to $\frac{\pi}{2}$ for $x > 0$

47. converges to 0 for all x

48. converges to 0 for $x = \pm\sqrt{2}$ and diverges for all other x

49. converges to 0 for $x < 0$, converges to 1 for $x = 0$ and diverges for $x > 0$

50. converges to 0 for $x \ne \pm 1$, converges to $\frac{1}{2}$ for $x = 1$ and diverges for $x = -1$

51. diverges for $x \ne 0$ and converges to 0 for $x = 0$

52. converges to $e^{x/2}$ for all x **53.** diverges for all x

54. converges to 0 for all x **55.** converges to 0 for all x

56. (a) $\lim\limits_{n \to \infty} a_n = \ln 3$ (b) nothing (c) $\lim\limits_{n \to \infty} a_n = 16$ (d) $\lim\limits_{n \to \infty} a_n = e^2$
(e) $\lim\limits_{n \to \infty} a_n = 4$ (f) nothing

57. (a) GLB is 3/4 and LUB is 1. (b) GLB is -1 and LUB is $-1/3$. (c) GLB is 1/3 and there is no LUB. (d) GLB is -1 and LUB is 2/3. (e) GLB is 0 and LUB is $\ln 3)/3$. (f) GLB is 0 and LUB is $3\sqrt{2}/4$. (g) GLB is -1 and LUB is 1. (h) GLB is $-\pi/2$ and LUB is $\pi/2$.

58.

	a_1	a_2	a_3	a_4	a_5	a_6		a_1	a_2	a_3	a_4	a_5	a_6
(a)	2	5	11	23	47	95	(b)	-1	$\frac{1}{2}$	$-\frac{1}{6}$	$\frac{1}{24}$	$-\frac{1}{120}$	$\frac{1}{720}$
(c)	0	1	2	5	26	677	(d)	1	1	2	3	5	8
(e)	1	2	2	4	8	32	(f)	0	1	5	12	22	35
(g)	1	2	3	4	5	6							

59. (a) converges to 0 (b) converges to 1 (c) converges to 0
(d) diverges (e) converges to 0 (f) diverges

63. S is finite.   **64.** $a_n = (-1)^n \frac{1}{n}$

65. $f(x) = \frac{1}{x}$ with domain and range $(-\infty, 0) \cup (0, \infty)$.

66. $g(x) = \frac{1}{x^2}$ with domain $(-\infty, 0) \cup (0, \infty)$ and range $(0, \infty)$.

67. (a) $a_n = 1.d_1 d_2 d_3 \ldots d_n$ where $\sqrt{2} = 1.d_1 d_2 d_3 \ldots$. (b) $a_n = \frac{1}{n}$.

Section 5.4

1. (a) 225 (b) -3 (c) $\frac{41}{32}$ (d) $-\frac{13}{6}$ (e) $\frac{1}{20}$

2. (a), (c) These series diverge. (b), (d), (e), (f) nothing

3.

	convergent	sum		convergent	sum		convergent	sum
(a)	yes	$\frac{1}{25}$	(b)	no	-	(c)	no	-
(d)	no	-	(e)	yes	$\frac{1}{5}$	(f)	no	-
(g)	no	-	(h)	yes	$\frac{3}{5}$	(i)	no	-
(j)	yes	$\frac{3}{4}$	(k)	no	-	(l)	yes	$\frac{4}{7}$
(m)	yes	$\frac{15}{4}$	(n)	no	-	(o)	yes	$\frac{392}{5}$
(p)	yes	$\frac{11}{18}$	(q)	yes	$\ln 4 - 1$	(r)	no	-
(s)	no	-	(t)	yes	$\frac{3}{2} + \ln 4$	(u)	yes	$\frac{21}{4}$
(v)	no	-	(w)	no	-	(x)	no	-

4. (a) $.\overline{153846}$ (b) $4.\overline{72}$ (c) $-.2\overline{6}$ (d) $2.\overline{714285}$ (e) $-7.41\overline{6}$

5. (a) $\frac{43}{99}$ (b) $\frac{16,273}{4,950}$ (c) $\frac{10,417}{1,000}$ (d) $-\frac{936,878}{99,900}$ (e) $\frac{4,082,179}{199,980}$

6. (a) converges for $-7 < x < 7$ to $\frac{7}{7-x}$.

(b) converges for $-\sqrt{3} < x < \sqrt{3}$ to $\frac{3x^3}{9+x^4}$.

(c) converges for $-1 \le x < 1$ to $-\ln(1-x)$

(d) converges for $-1 \le x \le 1$ to $-\ln(1+x^2)$

(e) converges for $-1 < x < 1$ to $\frac{1}{2}\ln\left(\frac{1+x}{1-x}\right)$

(f) converges for $-1 < x < 1$ to $\frac{x}{2}\ln\left(\frac{1+x^2}{1-x^2}\right)$

(g) converges for $-1 \le x \le 1$ to $-x^2\arctan(x^2)$

(h) converges for $-1 < x \le 1$ to $\frac{x}{2+x} + \ln(1+x)$

(i) converges for $-1 \le x \le 1$ with sum 0 when $x = 0$ and sum $-\arctan x - \frac{1}{x}\ln$ when $x \in [-1,0) \cup (0,1]$

7. (a) $\sum_{k=0}^{\infty} x^{4k+1}$ for $-1 < x < 1$ (b) $\sum_{k=0}^{\infty}(-1)^k x^{5k+6}$ for $-1 < x < 1$

(c) $\sum_{k=0}^{\infty}(-1)^k x^{2k+3}$ for $-1 < x < 1$ (d) $\sum_{k=0}^{\infty} x^{3k+4}$ for $-1 < x < 1$

(e) $\sum_{k=0}^{\infty} 7^k x^{3k+1}$ for $-\frac{1}{\sqrt[3]{7}} < x < \frac{1}{\sqrt[3]{7}}$ (f) $\sum_{k=0}^{\infty}(-4)^k x^{5k+2}$ for $-\frac{1}{\sqrt[5]{4}} < x <$

(g) $7\sum_{k=0}^{\infty} \frac{9^k}{16^{k+1}} x^{4k+4}$ for $|x| < \frac{2}{\sqrt{3}}$ (h) $\sum_{k=0}^{\infty} \frac{3}{4}\left(-\frac{5}{8}\right)^k x^{3k+5}$ for $|x| < \frac{2}{\sqrt[3]{5}}$

8. (a) $\sum_{k=0}^{\infty} \frac{(x+5)^k}{7^k} = \sum_{k=0}^{\infty} \frac{y^k}{7^k}$ for $y = x + 5$ and converges with sum $-12 < x < 2$.

(b) $\sum_{k=1}^{\infty} \frac{(x-3)^{2k}}{4^{k+1}} = \sum_{k=1}^{\infty} \frac{y^{2k}}{4^{k+1}}$ for $y = x - 3$ and converges with sum $-\frac{x^2}{4x^2 -}$ for $1 < x < 5$.

(c) $\sum_{k=0}^{\infty} \frac{(x+1)^{3k+2}}{5^{2k+1}} = \sum_{k=0}^{\infty} \frac{y^{3k+2}}{5^{2k+1}}$ for $y = x+1$ and converges with sum $\frac{5x^2+1}{24-x^3-}$ for $-\sqrt[3]{25}-1 < x < \sqrt[3]{25}-1$.

(d) $\sum_{k=0}^{\infty}(-1)^k \frac{(x-5)^{4k+3}}{16^{3k+1}} = \sum_{k=0}^{\infty}(-1)^k \frac{y^{4k+3}}{16^{3k+1}}$ for $y = x - 5$ and converges wi $\frac{256(x-5)^3}{4096-(x-5)^4}$ for $-3 < x < 13$.

(e) $\sum_{k=2}^{\infty} \frac{(2x+5)^k}{8^k} = \sum_{k=2}^{\infty} \frac{2^k y^k}{8^k}$ for $y = x + \frac{5}{2}$ and converges with sum $\frac{4x^2+20x-}{24-16:}$ $-\frac{13}{2} < x < \frac{3}{2}$.

(f) $\sum_{k=1}^{\infty} \frac{(4x-3)^{3k}}{27^{5k}} = \sum_{k=1}^{\infty} \frac{64^k y^{3k}}{27^{5k}}$ for $y = x - \frac{3}{4}$ and converges with sum $\frac{(4x)}{3^{15}-($ for $-60 < x < \frac{123}{2}$.

(g) $\sum_{k=0}^{\infty} \frac{(2x+7)^{5k+3}}{3^{4k+1}} = \sum_{k=0}^{\infty} \frac{2^{5k+3}y^{5k+3}}{3^{4k+1}}$ for $y = x + \frac{7}{2}$ and converges wit $\frac{(6x+21)^3}{81-(2x+7)^5}$ for $-\frac{\sqrt[5]{81}+7}{2} < x < \frac{\sqrt[5]{81}-7}{2}$.

(h) $\sum_{k=0}^{\infty}(-1)^k \frac{(3x+5)^{2k+5}}{4^{3k+7}} = \sum_{k=0}^{\infty}(-1)^k \frac{3^{2k+5}y^{2k+5}}{4^{3k+7}}$ for $y = x + \frac{5}{3}$ and converge sum $\frac{(3x+5)^5}{768(13-3x^2-10x)}$ for $-\frac{13}{3} < x < 1$.

9. (a) .89 with $N = 5$ (b) .20 with $N = 4$ (c) .05026 with $N = 2$
(d) .029 with $N = 3$ (e) .953 with $N = 3$

10. (a) $\sum_{k=0}^{\infty}(-1)^k \frac{x^{4k+2}}{2k+1}$ for $|x| \le 1$ (b) $\sum_{k=1}^{\infty} -\frac{x^k}{k}$ for $-1 \le x < 1$

(c) $\sum_{k=0}^{\infty} \frac{x^{2k+1}}{2k+1}$ for $-1 < x < 1$ (d) $\sum_{k=0}^{\infty}(-1)^k \frac{x^{3k+1}}{3k+1}$ for $-1 < x \le$

(e) $\sum_{k=0}^{\infty}(-1)^k \frac{x^{4k+1}}{4k+1}$ for $-1 \le x \le 1$ (f) $\sum_{k=0}^{\infty} \frac{x^{5k+2}}{5k+2}$ for $-1 \le x < 1$

(g) $\sum_{k=0}^{\infty} \frac{x^{4k+6}}{16^{k+1}(4k+6)}$ for $-2 < x < 2$ (h) $\sum_{k=0}^{\infty} \frac{x^{6k+4}}{27^{k+1}(3k+2)}$ for $-\sqrt{3} < x$

(i) $\sum_{k=0}^{\infty} \frac{2x^{4k+2}}{2k+1}$ for $-1 < x < 1$ (j) $\ln 2 + \sum_{k=1}^{\infty}(-1)^{k-1} \frac{x^{2k}}{k2^k}$ for $|x| \le$

11. (a) absolutely convergent (b) divergent (c) conditionally convergent
(d) divergent (e) absolutely convergent (f) absolutely convergent
(g) divergent (h) conditionally convergent (i) divergent
(j) conditoinally convergent (k) absolutely convergent (l) absolutely conve

12. (a) absolutely convergent for $-2 < x < 2$ and divergent for $x \le -2$ or $x \ge$

b) absolutely convergent for $-1 < x < 1$, conditionally convergent for $x = \pm 1$ and divergent for $x < -1$ or $x > 1$

c) absolutely convergent for $-3 < x < 3$ and divergent for $x \le -3$ or $x \ge 3$

d) absolutely convergent for $-1 < x < 1$ and divergent for $x \le -1$ or $x \ge 1$

e) absolutely convergent for $-125 < x < 125$ and divergent for $x \le -125$ or $x \ge 125$

f) absolutely convergent for $-1 \le x \le 1$ and divergent for $x < -1$ or $x > 1$

3. 20 feet   **14.** 46.8 feet in 10 jumps   **15.** $1,569,293$ dollars

6. (b) $\sum_{k=1}^{\infty} a_k = \sum_{k=1}^{\infty} \frac{1}{k}$ and $\sum_{k=1}^{\infty} b_k = \sum_{k=1}^{\infty} -\frac{1}{k}$ **17. (b)** Let $b_k = 1$ for $k \ge 1$.

0. (c) If $a_k = (-1)^k/k$ and $b_k = (-1)^{k-1}/k$, then $a_k + b_k = 0$.

Section 5.5

. **(a)** $P_0(x) = 1$, $P_1(x) = 1 - x$, $P_2(x) = 1 - x + x^2$, $P_3(x) = 1 - x + x^2 - x^3$, $P_4(x) = 1 - x + x^2 - x^3 + x^4$.

b) $P_0(x) = P_1(x) = 1$, $P_2(x) = P_3(x) = 1 - x^2$, $P_4(x) = 1 - x^2 + \frac{1}{2}x^4$.

c) $P_0(x) = P_1(x) = 0$, $P_2(x) = P_3(x) = x^2$, $P_4(x) = x^2 - \frac{1}{2}x^4$.

d) $P_0(x) = P_1(x) = P_2(x) = P_3(x) = 1$, $P_4(x) = 1 + \frac{1}{2}x^4$.

e) $P_0(x) = 0$, $P_1(x) = P_2(x) = x$, $P_3(x) = P_4(x) = x + \frac{1}{6}x^3$.

f) $P_0(x) = 0$, $P_1(x) = P_2(x) = x$, $P_3(x) = P_4(x) = x - \frac{1}{3}x^3$.

g) $P_0(x) = 0$, $P_1(x) = P_2(x) = x$, $P_3(x) = P_4(x) = x + \frac{1}{3}x^3$.

h) $P_0(x) = P_1(x) = 1$, $P_2(x) = P_3(x) = 1 + \frac{1}{2}x^2$, $P_4(x) = 1 + \frac{1}{2}x^2 + \frac{5}{24}x^4$.

i) $P_0(x) = 1$, $P_1(x) = 1 - 3x$, $P_2(x) = 1 - 3x + 6x^2$, $P_3(x) = 1 - 3x + 6x^2 - 10x^3$, $P_4(x) = 1 - 3x + 6x^2 - 10x^3 + 15x^4$.

j) $P_0(x) = P_1(x) = 1$, $P_2(x) = P_3(x) = 1 - \frac{1}{2}x^2$, $P_4(x) = 1 - \frac{1}{2}x^2 + \frac{1}{24}x^4$.

k) $P_0(x) = 1$, $P_1(x) = 1 - \frac{1}{3}x$, $P_2(x) = 1 - \frac{1}{3}x - \frac{1}{9}x^2$, $P_3(x) = 1 - \frac{1}{3}x - \frac{1}{9}x^2 - \frac{5}{81}x^3$, $P_4(x) = 1 - \frac{1}{3}x - \frac{1}{9}x^2 - \frac{5}{81}x^3 - \frac{10}{243}x^4$.

l) $P_0(x) = 1$, $P_1(x) = 1 + ex$, $P_2(x) = 1 + ex + \frac{e(e-1)}{2}x^2$, $P_3(x) = 1 + ex + \frac{e(e-1)}{2}x^2 + \frac{e(e-1)(e-2)}{6}x^3$, $P_4(x) = 1 + ex + \frac{e(e-1)}{2}x^2 + \frac{e(e-1)(e-2)}{6}x^3 + \frac{e(e-1)(e-2)(e-3)}{24}x^4$.

2. Let $n = \deg f(x)$. For $0 \le k < n$, $P_k(x)$ is the truncation of $f(x)$ obtained by deleting the monomials of degree greater than k. For $k \ge n$, $P_k(x) = f(x)$.

4. (a) 10   **(b)** $-\frac{1}{9}$ **(c)** -20 **(d)** $\frac{\pi(\pi-1)(\pi-2)(\pi-3)}{24}$ **(e)** $-\frac{5}{16}$ **(f)** 1

7. (a) $f^{(4k)}(x) = \cos x$, $f^{(4k+1)}(x) = -\sin x$, $f^{(4k+2)}(x) = -\cos x$, $f^{(4k+3)}(x) = \sin x$, $P_{2n}(x) = P_{2n+1}(x) = \sum_{k=0}^{n}(-1)^k \frac{x^{2k}}{(2k)!}$, $|R_{2n}(x)| = |R_{2n+1}(x)| \le \frac{|x|^{2n+2}}{(2n+2)!}$, $\cos x = \sum_{k=0}^{\infty}(-1)^k \frac{x^{2k}}{(2k)!}$.

(b) $f^{(k)}(x) = (-3)^k e^{-3x}$, $P_n(x) = \sum_{k=0}^{n}(-1)^k \frac{3^k}{k!}x^k$, $|R_n(x)| \le \frac{3^{n+1}|x|^{n+1}}{(n+1)!}$ for $x \ge 0$, $|R_n(x)| \le \frac{3^{n+1}e^{-3x}|x|^{n+1}}{(n+1)!}$ for $x < 0$, $e^{-3x} = \sum_{k=0}^{\infty}(-1)^k \frac{3^k}{k!}x^k$.

(c) $f^{(k)}(x) = \frac{1}{2}(e^x - (-1)^k e^{-x})$, $P_0(x) = 0$, $P_{2n-1}(x) = P_{2n}(x) = \sum_{k=1}^{n} \frac{x^{2k-1}}{(2k-1)!}$, $|R_{2n-1}(x)| = |R_{2n}(x)| \le \frac{(e^{|x|}+1)|x|^{2n+1}}{2(2n+1)!}$, $\frac{1}{2}(e^x - e^{-x}) = \sum_{k=1}^{\infty} \frac{x^{2k-1}}{(2k-1)!}$.

(d) $f^{(k)}(x) = \frac{1}{2}(e^x + (-1)^k e^{-x})$, $P_{2n}(x) = P_{2n+1}(x) = \sum_{k=0}^{n} \frac{x^{2k}}{(2k)!}$, $|R_{2n}(x)| = |R_{2n+1}(x)| \le \frac{(e^{|x|}+1)|x|^{2n+2}}{2(2n+2)!}$, $\frac{1}{2}(e^x + e^{-x}) = \sum_{k=0}^{\infty} \frac{x^{2k}}{(2k)!}$.

(e) $f^{(k)}(x) = (\ln 5)^k 5^x$, $P_n(x) = \sum_{k=0}^{n} \frac{(\ln 5)^k}{k!}x^k$, $|R_n(x)| \le \frac{(\ln 5)^{n+1}5^x|x|^{n+1}}{(n+1)!}$ for $x \ge 0$, $|R_n(x)| \le \frac{(\ln 5)^{n+1}|x|^{n+1}}{(n+1)!}$ for $x < 0$, $5^x = \sum_{k=0}^{\infty} \frac{(\ln 5)^k}{k!}x^k$.

(f) $f^{(4k)}(x) = \pi^{4k}\sin \pi x$, $f^{(4k+1)}(x) = \pi^{4k+1}\cos \pi x$, $f^{(4k+2)}(x) = -\pi^{4k+2}\sin \pi x$, $f^{(4k+3)}(x) = -\pi^{4k+3}\cos \pi x$, $P_0(x) = 0$, $P_{2n-1}(x) = P_{2n}(x) = \sum_{k=0}^{n-1}(-1)^k \frac{\pi^{2k+1}}{(2k+1)!}x^{2k+1}$, $|R_{2n-1}(x)| = |R_{2n}(x)| \le \frac{\pi^{2n+1}|x|^{2n+1}}{(2n+1)!}$, $\sin \pi x = \sum_{k=0}^{\infty}(-1)^k \frac{\pi^{2k+1}}{(2k+1)!}x^{2k+1}$.

(g) $f^{(4k)} = 4^{4k}\cos(4x-\pi)$, $f^{(4k+1)} = -4^{4k+1}\sin(4x-\pi)$, $f^{(4k+2)} = -4^{4k+2}$ c
$f^{(4k+3)} = 4^{4k+3}\sin(4x-\pi)$, $P_{2n}(x) = P_{2n+1}(x) = \sum_{k=0}^{n}(-1)^{k+1}\frac{16^k}{(2k)!}x^{2k}$,
$|R_{2n}(x)| = |R_{2n+1}(x)| \leq \frac{4^{2n+2}|x|^{2n+2}}{(2n+2)!}$, $\cos(4x-\pi) = \sum_{k=0}^{\infty}(-1)^{k+1}\frac{16^k}{(2k)!}x^{2k}$.

(h) $f^{(k)}(x) = (-1)^k\frac{1}{2}(1-\frac{1}{2})\cdots(\frac{1}{2}-k+1)(9-x)^{\frac{1}{2}-k}$, $P_n(x) = \sum_{k=0}^{n}(-1)^k\begin{pmatrix}1/\\k\end{pmatrix}$
$|R_n(x)| \leq \left|\begin{pmatrix}1/2\\n+1\end{pmatrix}\right|\frac{|x|^{n+1}}{(9-x)^{(2n+1)/2}}$, $\sqrt{9-x} = \sum_{k=0}^{\infty}(-1)^k\begin{pmatrix}1/2\\k\end{pmatrix}\frac{x^k}{3^{2k-1}}$.

(i) $f^{(n)}(x) = \left(-\frac{1}{4}\right)\left(-\frac{5}{4}\right)\cdots\left(-\frac{1}{4}-k+1\right)2^k(1+2x)^{-\frac{1}{4}-k}$, $P_n(x) = \sum_{k=0}^{n}2^k\begin{pmatrix} \end{pmatrix}$
$|R_n(x)| \leq \left|\begin{pmatrix}-1/4\\n+1\end{pmatrix}\right|2^{n+1}x^{n+1}$, $\frac{1}{\sqrt[4]{1+2x}} = \sum_{k=0}^{\infty}2^k\begin{pmatrix}-1/4\\k\end{pmatrix}x^k$.

(j) $f^{(k)} = \pi(\pi-1)\cdots(\pi-k+1)5^k(3+5x)^{\pi-k}$, $P_n(x) = \sum_{k=0}^{n}\begin{pmatrix}\pi\\k\end{pmatrix}$
$|R_n(x)| \leq \left|\begin{pmatrix}\pi\\n+1\end{pmatrix}\right|3^\pi\frac{5^{n+1}x^{n+1}}{3^{n+1}}$, $(3+5x)^\pi = \sum_{k=0}^{\infty}\begin{pmatrix}\pi\\k\end{pmatrix}\frac{5^k}{3^{k-\pi}}x^k$.

9. **(a)** $n=3$, $\sqrt{e} \approx 1.65$ **(b)** $n=4$, $\frac{1}{e} \approx .375$ **(c)** $n=3$, $\sin.5 \approx .48$
(d) $n=5$, $\sin\frac{3\pi}{7} \approx .976$ **(e)** $n=2$, $\cos.2 \approx .98$ **(f)** $n=2$, $\sqrt[3]{9} \approx 2.08$
(g) $n=2$, $\frac{1}{\sqrt[4]{1.4}} \approx .92$ **(h)** $n=2$, $42^{.2} \approx 2.11$

11. **(a)** -2 **(b)** 0 **(c)** $-\frac{1}{2}$ **(d)** -4 **(e)** $-\infty$ **(f)** -1 **(g)** 2
(i) $-\frac{1}{6}$ **(j)** $\frac{25}{27}$

12. **(a)** $n=8$, 1.46 **(b)** $n=12$, $.84$ **(c)** $n=2$, $.35$ **(d)** $n=2$, $.49$
(e) $n=6$, $.785$ **(f)** $n=1$, $.609$ **(g)** $n=2$, $.44$ **(h)** $n=4$, $.97$ **(i)** $n=$
(j) $n=9$, $.28$ **(k)** $n=1$, $.25$ **(l)** $n=3$, $.679$

13. **(a)** $c = \ln\frac{3}{4}(e^2-1)$ **(b)** $c = \arcsin\frac{2}{\pi}$ or $c = \pi - \arcsin\frac{2}{\pi}$
(c) $c = \exp\left(\frac{e^2+1}{2(e^2-1)}\right)$ **(d)** $c = \frac{\ln 4}{\pi}$

17. $\sinh x = \sum_{k=0}^{\infty}\frac{x^{2k+1}}{(2k+1)!}$ **18.** $\cosh x = \sum_{k=0}^{\infty}\frac{x^{2k}}{(2k)!}$

19. (a) -1 **(b)** -1 **(c)** 2 **20. (a)** $n=2$, 1.264 **(b)** $n=2$, 1.056 **(c)** $n=2$, $.5$

Appendix (a) $R_n(x) = \frac{1}{n!}\int_0^x e^t(x-t)^n\,dt$ **(b)** $R_{2n}(x) = \frac{(-1)^n}{(2n)!}\int_0^x(x-t)^{2n}$
(c) $R_{2n+1}(x) = \frac{(-1)^{n+1}}{(2n+1)!}\int_0^x(x-t)^{2n+1}\cos t\,dt$ **(d)** $R_n(x) = (-1)^n\int_0^x\frac{(x-t)^n}{(1+t)^{n+1}}$
(e) $R_n(x) = (n+1)\begin{pmatrix}1/2\\n+1\end{pmatrix}\int_0^x\frac{(x-t)^n}{(1+t)^{n+1/2}}\,dt$ **(f)** $R_n(x) = (n+1)\begin{pmatrix}-3\\n+1\end{pmatrix}\int_0$

Section 5.6

1. **(a)** $P_3(x,1) = 2 + 2(\ln 2)(x-1) + (\ln 2)^2(x-1)^2 + \frac{(\ln 2)^3}{3}(x-1)^3$
(b) $P_4(x,1) = (x-1) - \frac{1}{2}(x-1)^2 + \frac{1}{3}(x-1)^3 - \frac{1}{4}(x-1)^4$
(c) $P_5(x,2) = \frac{1}{2} - \frac{1}{4}(x-2) + \frac{1}{8}(x-2)^2 - \frac{1}{16}(x-2)^3 + \frac{1}{32}(x-2)^4 - \frac{1}{64}(x-2)$
(d) $P_4(x,4) = 2 + \frac{1}{4}(x-4) - \frac{1}{64}(x-4)^2 + \frac{1}{512}(x-4)^3 - \frac{5}{16,384}(x-4)^4$
(e) $P_3(x,\frac{\pi}{4}) = 1 + 2(x-\frac{\pi}{4}) + 2(x-\frac{\pi}{4})^2 + \frac{8}{3}(x-\frac{\pi}{4})^3$
(f) $P_3(x,\frac{\pi}{2}) = 1 + \frac{1}{2}(x-\frac{\pi}{2})^2$
(g) $P_3(x,\frac{1}{2}) = \frac{\pi}{6} + \frac{2}{\sqrt{3}}(x-\frac{1}{2}) + \frac{2}{3\sqrt{3}}(x-\frac{1}{2})^2 + \frac{8}{9\sqrt{3}}(x-\frac{1}{2})^3$
(h) $P_3(x,1) = \frac{\pi}{4} + \frac{1}{2}(x-1) - \frac{1}{4}(x-1)^2 + \frac{1}{12}(x-1)^3$
(i) $P_5(x,\pi) = (x-\pi)^2 - \frac{1}{3}(x-\pi)^4$
(j) $P_3(x,8) = 4 + \frac{1}{3}(x-8) - \frac{1}{144}(x-8)^2 + \frac{1}{2592}(x-8)^3$

2. (a) $f(x) = 28 + 30(x-2) + 11(x-2)^2 + (x-2)^3$
(b) $g(x) = 3 - 6(x+1) + 7(x+1)^2 - 4(x+1)^3 + (x+1)^4$

(c) $h(x) = -11 - 6(x-3) + 3(x-3)^2 + (x-3)^4$

(d) $k(x) = 37 - 40(x+2) + 24(x+2)^2 - 8(x+2)^3 + (x+2)^4$

3. (a) $3^x = \sum_{k=0}^{\infty} \frac{9(\ln 3)^k}{k!}(x-2)^k$ (b) $e^{2x-7} = \sum_{k=0}^{\infty} \frac{2^k e}{k!}(x-4)^k$

(c) $\sin x = \sum_{k=0}^{\infty}(-1)^{k+1}\frac{(x-\pi)^{2k+1}}{(2k+1)!}$ (d) $\cos \pi x = \sum_{k=0}^{\infty}(-1)^{k+1}\frac{\pi^{2k}}{(2k)!}(x-3)^{2k}$

(e) $\sqrt[3]{x} = \sum_{k=0}^{\infty}\binom{1/3}{k}(x-1)^k$ (f) $\ln x = \ln c + \sum_{k=1}^{\infty}(-1)^{k-1}\frac{(x-c)^k}{kc^k}$

(g) $p(x) = \sum_{k=0}^{\infty}\binom{-1/2}{k}\frac{(x-2)^{2k}}{3^{2k+1}}$ (h) $\frac{1}{7+x} = \sum_{k=0}^{\infty}(-1)^k\frac{(x+2)^k}{5^{k+1}}$

(i) $\frac{1}{\sqrt{x}} = \sum_{k=0}^{\infty}\binom{-1/2}{k}\frac{(x-9)^k}{3^{2k+1}}$ (j) $\arctan(1+x) = \sum_{k=0}^{\infty}(-1)^k\frac{(x+1)^{2k+1}}{2k+1}$

4. (a) $n=2$, $3^{2.1} \approx 10.04$ (b) $n=10$, $e^{3.2} \approx 24.53$ (c) $n=3$, $\sin\frac{9\pi}{10} \approx .32$
(d) $n=4$, $\cos\frac{17\pi}{5} \approx -.31$ (e) $n=1$, $\sqrt[3]{1.2} \approx 1.07$ (f) $n=1$, $\ln 6.3 \approx 1.84$
(g) $n=1$, $p(2.3) \approx .33$ (h) $n=2$, $q(-1.4) \approx .18$ (i) $n=1$, $\sqrt{15.5} \approx 3.94$

5. (a) e (b) $\frac{1}{2}$ (c) $-\frac{2}{e}$ (d) $\frac{1}{3}$ (e) 0 (f) e (g) $\frac{\pi}{e-e^2}$ (h) $-\frac{5}{6e}$

6. (a) $n=2$, 1.147 (b) $n=6$, $.288$ (c) $n=0$, $.45$ (d) $n=5$, $-.897$ (e) $n=8$, 2.74

7. (a) $\sinh x = \sum_{k=0}^{\infty}\frac{e^2-1}{(2k)!(2e)}(x-1)^{2k} + \sum_{k=0}^{\infty}\frac{e^2+1}{(2k+1)!(2e)}(x-1)^{2k+1}$

(b) $\cosh x = \sum_{k=0}^{\infty}\frac{e^2+1}{(2k)!(2e)}(x-1)^{2k} + \sum_{k=0}^{\infty}\frac{e^2-1}{(2k+1)!(2e)}(x-1)^{2k+1}$

(c) $\sinh x + \cosh x = \sum_{k=0}^{\infty}\frac{e^2}{k!}(x-2)^k$ **8.** 3.549

Section 5.7

1. (a) converges   (b) converges   (c) converges   (d) diverges   (e) converges
(f) converges   (g) converges   (h) converges   (i) converges   (j) diverges
(k) converges   (l) converges

2. (b) (i) diverges   (ii) converges   (iii) diverges   (iv) converges   (v) diverges
(vi) diverges

4. This derivation is not valid. It uses circular reasoning since we used the geometric series to prove the Ratio Test.

5. (a) $R = \sqrt{3}$ (b) $R=1$ (c) $R=\infty$ (d) $R=\frac{1}{8}$ (e) $R=\frac{1}{2}$ (f) $R=0$
(g) $R=\infty$ (h) $R=10$ (i) $R=5$ (j) $R=\sqrt[3]{e/4}$ (k) $R=2$ (l) $R=\frac{1}{3}$

6. (a) $5 \le R \le 10$ (b) $7 \le R \le 8$ (c) $5 \le R \le 6$ (d) $3 \le R \le 4$
(e) $R=\infty$ (f) $0 \le R \le 1$ (g) $R=0$ (h) $R=\infty$

7. (a) $\arctan x = \sum_{k=0}^{\infty}(-1)^k\frac{x^{2k+1}}{2k+1}$, $R=1$ (b) $\ln(1-x^2) = \sum_{k=1}^{\infty}-\frac{x^{2k}}{k}$, $R=1$
(c) $\ln(1-x) = \sum_{k=1}^{\infty}-\frac{x^k}{k}$, $R=1$ (d) $\ln(1+x) = \sum_{k=1}^{\infty}(-1)^{k-1}\frac{x^k}{k}$ $R=1$

9. $R = \frac{1}{L}$ if $L > 0$ and $R = \infty$ if $L = 0$

10. $\cos\sqrt{x} = T(x)$ for $x \ge 0$. When $x < 0$, $\cos\sqrt{x}$ is not defined and $T(x)$ converges.

11. (a) 0 (b) 0

12. (a) $\sum_{k=1}^{\infty}(-1)^k\frac{x^{k-1}}{(k-1)!}$ for $x \in \Re$;   (b) $\sum_{k=1}^{\infty}\frac{x^{2k-1}}{(2k-1)!}$ for $x \in \Re$;
(c) $\sum_{k=1}^{\infty}\frac{x^{k-1}}{k^{k-1}}$ for $x \in \Re$;   (d) $\sum_{k=1}^{\infty}(-1)^k\frac{x^{k-1}}{k!(k-1)!}$ for $x \in \Re$;
(e) $\sum_{k=1}^{\infty}\frac{k^2}{k^2+1}(x+2)^{k-1}$, $-3 < x < -1$;   (f) $\sum_{k=1}^{\infty}(-3)^k\sqrt{k}(x-5)^{k-1}$, $\frac{14}{3} < x < \frac{16}{3}$;
(g) $\sum_{k=1}^{\infty}(\ln k)(x+6)^{k-1}$, $-7 < x < -5$;   (h) $\sum_{k=1}^{\infty}\frac{e^k}{[(k-1)!]^2}(x-3)^{k^2-1}$, $2 < x < 4$;
(i) $3\sum_{k=1}^{\infty}\binom{-2}{k}\frac{kx^{3k-1}}{8^k}$, $-2 < x < 2$.

13. **(a)** $f^{(2n+1)}(0) = 0$ and $f^{(2n)}(0) = (-1)^n \frac{(2n)!}{n!}$ for $n \geq 0$;

(b) $f(0) = 0$ and $f^{(n)}(0) = (-1)^{n-1} \frac{n!}{(2n-1)!}$ for $n \geq 1$;

(c) $f^{(2n+1)}(3) = 0$ and $f^{(2n)}(3) = \frac{(2n)!}{e^9 n!}$ for $n \geq 0$;

(d) $f^{(2n)}(0) = 0$ and $f^{(2n+1)}(0) = (-1)^n \begin{pmatrix} -1/2 \\ n \end{pmatrix} (2n)!$ for $n \geq 0$;

(e) $f^{(n)}(0) = 0$ if n is not divisible by 4 and $f^{(4n)}(0) = \begin{pmatrix} 1/2 \\ n \end{pmatrix} (4n)!$ for n

(f) $f^{(2n+1)}(-4) = (-1)^n \frac{(2n+1)!}{9^{n+1}}$ and $f^{(2n)}(-4) = (-1)^{n+1} \frac{4(2n)!}{9^{n+1}}$ for $n \geq 0$.

14. **(a)** $\sum_{n=0}^{\infty} (-1)^n \frac{1}{n!(3n+1)2^{3n+1}}$ **(b)** $\sum_{n=0}^{\infty} (-1)^n \frac{1}{(2n+1)!(4n+3)3^{4n+3}}$

(c) $\sum_{n=0}^{\infty} (-1)^n \begin{pmatrix} -1/2 \\ n \end{pmatrix} \frac{1}{(5n+1)2^{5n+1}}$ **(d)** $\sum_{n=0}^{\infty} (-1)^n \left(\frac{2}{3}\right)^{n+3} \frac{1}{(2n)!(n+3)}$

(e) $\sum_{n=0}^{\infty} (-1)^n \frac{1}{(2n+1)(n+2)4^{n+2}}$ **(f)** $\sum_{n=1}^{\infty} (-1)^{n-1} \frac{1}{n(2n+1)2^{2n+1}}$

18. **(a)** $1 + x - \frac{x^3}{3} - \frac{x^4}{6}$ **(b)** $x + x^2 + \frac{x^3}{6} - \frac{x^4}{6}$ **(c)** $x + \frac{x^3}{3}$ **(d)** $1 - \frac{x^2}{2}$

(e) $e(x-1) - \frac{e}{2}(x-1)^2 - \frac{2e}{3}(x-1)^3 + \frac{e}{4}(x-1)^4$

(f) $-2(\ln \pi)(x - \pi) + \frac{\pi^2 \ln \pi + 3}{3\pi^2}(x - \pi)^3$

19. **(a)** $f^{(n)}(0) = \frac{n!}{(2n)!}$ **(b)** $f^{(n)}(0) = 0$ if $n - 4$ is not divisible by 8

$f^{(8n+4)}(0) = \frac{(8n+4)!}{(2n+1)!}$ **(c)** $f^{(n)}(0) = 0$ if n is not divisible by 6 while $f^{(6n)}(0) =$

20. **(a)** $\int_0^1 \cosh \sqrt{x}\, dx = \sum_{k=0}^{\infty} \frac{1}{(k+1)(2k)!}$ **(b)** $\int_0^{1/2} \frac{\sinh x}{x}\, dx = \sum_{k=0}^{\infty} \frac{1}{(2k+1)2^{2}}$

(c) $\int_0^{1/3} \sqrt{x} \sinh \sqrt{x}\, dx = \sum_{k=0}^{\infty} \frac{1}{(k+2)3^{k+2}(2k+1)!}$

22. **(a)** $e^x \sinh x = \sum_{n=0}^{\infty} \sum_{k=0}^{[(n-1)/2]} \frac{1}{(2k+1)!(n-2k-1)!} x^n$

(b) $(\sinh x)(\cosh x) = \sum_{n=0}^{\infty} \sum_{k=0}^{n} \frac{1}{(2k)!(2n-2k+1)!} x^{2n+1}$

(c) $(\cosh x)\ln(1+x) = \sum_{n=0}^{\infty} \sum_{k=0}^{[(n-1)/2]} \frac{(-1)^{n-1}}{(2k)!(n-2k)} x^n$

Appendix

1. **(a)** absolutely convergent   **(b)** divergent   **(c)** nothing
(d) absolutely convergent   **(e)** divergent   **(f)** nothing
(g) divergent   **(h)** nothing   **(i)** absolutely convergent

2. **(a)** $R = 2$ **(b)** $R = 1$ **(c)** $R = \infty$ **(d)** $R = 3$ **(e)** $R = 2$ **(f)**

3. $\sum_{k=1}^{\infty} (-1)^k a_k$ diverges.   **4.** $R = \frac{1}{L}$ if $L > 0$ while $R = \infty$ if $L = 0$.

Section 5.8

1. **(a)** analytic   **(b)** analytic   **(c)** not analytic   **(d)** not analytic
(e) not analytic   **(f)** analytic   **(g)** analytic   **(h)** analytic

3. **(a)** $\sum_{n=0}^{\infty} (2n^2 + 2)x^n$ **(b)** $-3 + \sum_{n=1}^{\infty} (n-4)x^n$ **(c)** $\sum_{n=0}^{\infty} (2n +$
(d) $2 + \frac{3x}{2} + \sum_{n=2}^{\infty} \frac{2}{n^2-1} x^n$ **(e)** $3 + 6x + \sum_{n=2}^{\infty} (6n-5)x^n$ **(f)** $\sum_{n=0}^{\infty} (2n^2$

4. **(a)** $x \neq (2n+1)\frac{\pi}{2}$, n an integer   **(b)** all x **(c)** $x \neq 0$ **(d)** all x
(e) $x \neq \pm 1$ **(f)** $x \neq 0$ **(g)** $x \neq \pm 1$ **(h)** all x
(i) $x \neq n\pi$, n a nonzero integer   **(j)** all x

6. **(a)** $y = y_0 \left(1 - 3x + 7x^2 - \frac{15}{2}x^3 + \frac{17}{6}x^4 + \cdots\right)$ for all x.

(b) $y = y_0 \left[1 + 4(x-1) + 7(x-1) - \frac{19}{3}(x-1)^3 + \frac{65}{6}(x-1)^4 + \cdots\right]$ for all x.

(c) $y = y_0 \left[1 - \frac{9}{4}(x-3) + \frac{33}{16}(x-3)^2 - \frac{163}{192}(x-3)^3 + \frac{1}{256}(x-3)^4 + \cdots\right]$, $-1 <$

(d) $y = y_0 \left(1 - x + \frac{1}{6}x^3 + \frac{1}{24}x^4 + \cdots\right)$ for all x.

(e) $y = y_0 \left[1 - \frac{1}{2}\left(x - \frac{\pi}{2}\right)^2 + \frac{1}{24}\left(x - \frac{\pi}{2}\right)^4 + \cdots\right]$ for $0 < x < \pi$.

) $y = y_0 \left(1 - \frac{1}{2}x^2 + \frac{1}{2}x^3 - \frac{5}{24}x^4 + \cdots\right)$ for $-1 < x < 1$.

  (a) $y = 4 - 4x + 2x^2 - 2x^3 + \frac{3}{2}x^4 + \cdots$ for all x.
) $y = -1 - (x-2) + \frac{1}{3}(x-2)^3 + \frac{1}{12}(x-2)^4 + \cdots$ for all x.
) $y = 2 - \frac{2}{3}x^3 + \frac{1}{2}x^4 + \cdots$ for $-1 < x < 1$
) $y = -3 + 3x - \frac{9}{2}x^2 + \frac{11}{2}x^3 - \frac{49}{8}x^4 + \cdots$ for all x.
) $y = 7 + 14(x-1) + \frac{35}{2}(x-1)^2 - \frac{7}{3}(x-1)^3 + \frac{77}{24}(x-1)^4 + \cdots$ for $0 < x < 2$.
) $y = 5 + 5x + \frac{5}{4}x^2 + \frac{10}{9}x^3 + \frac{85}{144}x^4 + \cdots$ for $-1 < x < 1$.

  (a) $y = y_0 \sum_{n=0}^{\infty}(-1)^n \frac{7^n}{n!}x^n$ for all x. (b) $y = y_0 \sum_{n=0}^{\infty}(-1)^n \frac{3^n}{2^n n!}x^{2n}$ for all x.
) $y = y_0 \left[1 + 2(x-1) + (x-1)^2\right]$ for all x. (d) $y = y_0 \sum_{n=0}^{\infty}(x+1)^n$, $-2 < x < 0$.
) $y = y_0 \sum_{n=0}^{\infty}(-1)^n \frac{1}{3^n n!}x^{3n}$ for all x.
) $y = y_0 \sum_{n=0}^{\infty}(-1)^n \frac{(n+2)(n+1)}{2^{n+1}}(x-2)^n$ for $0 < x < 4$.

  (a) $y = \sum_{n=0}^{\infty}(-1)^n \frac{3^{n+2}}{n!}(x-2)^n$ for all x.
) $y = \sum_{n=0}^{\infty}(-1)^n \frac{1}{2^{n-2}n!}(x-1)^{2n}$ for all x.
) $y = \sum_{n=0}^{\infty}\frac{2}{10^n}(x+5)^n$ for $-15 < x < 5$. (d) $y = \sum_{n=0}^{\infty}\frac{3^n 7}{n!}x^{4n}$ for all x.
) $y = \sum_{n=0}^{\infty}(-1)^n \frac{5}{3^n n!}(x-1)^{3n}$ for all x.
) $y = \sum_{n=0}^{\infty}(-1)^n \frac{4n+4}{5^n}(x-3)^n$ for $-2 < x < 8$.

0. (a) $y = y_0 \left(1 + \frac{x^4}{12} + \cdots\right) + y_1 \left(x - \frac{x^3}{6} + \cdots\right)$ for all x.
) $y = y_0 \left[1 - \frac{1}{4}(x-2)^2 - \frac{1}{24}(x-2)^3 - \frac{1}{96}(x-2)^4 + \cdots\right]$
$-y_1 \left[(x-2) + \frac{1}{2}(x-2)^2 + \frac{1}{12}(x-2)^3 + \frac{1}{48}(x-2)^4 + \cdots\right]$ for $0 < x < 4$.
) $y = y_0 \left[1 - \frac{1}{60}(x-1)^4 + \cdots\right] + y_1 \left[(x-1) - \frac{1}{10}(x-1)^2 - \frac{2}{75}(x-1)^3\right]$
$+\frac{7}{1500}(x-1)^4 + \cdots\right]$ for all x.
) $y = y_0 \left(1 - \frac{1}{2}x^2 - \frac{1}{24}x^4 + \cdots\right) + y_1 \left(x - \frac{1}{2}x^2 + \frac{1}{3}x^3 - \frac{1}{6}x^4 + \cdots\right)$ for all x.
) $y = y_0 \left(1 - \frac{1}{2}x^2 + \frac{1}{8}x^4 + \cdots\right) + y_1 \left(x - \frac{1}{3}x^3 + \cdots\right)$ for all x.
) $y = y_0 \left(1 - \frac{1}{6}x^3 + \frac{1}{12}x^4 + \cdots\right) + y_1 \left(x - \frac{1}{6}x^3 + \frac{1}{24}x^4 + \cdots\right)$ for $-1 < x < 1$.

1. (a) $y = -2 + x - \frac{1}{3}x^3 + \frac{1}{12}x^4 + \cdots$ for all x.
) $y = 4 + 2(x-1) - (x-1)^2 + 2(x-1)^3 - \frac{17}{12}(x-1)^4 + \cdots$ for all x.
) $y = 3 + 5(x+2) + \frac{7}{8}(x+2)^2 + \frac{3}{8}(x+2)^3 + \frac{77}{384}(x+2)^4 + \cdots$ for $-4 < x < 0$.
) $y = 1 - 6x - \frac{9}{2}x^2 + \frac{1}{2}x^3 + \frac{1}{4}x^4 + \cdots$ for all x.
) $y = 5 + 3x - \frac{1}{3}x^3 + \frac{1}{12}x^4 + \cdots$ for $-1 < x < 1$.

2. (a) $y = y_0 \sum_{n=0}^{\infty}(-1)^n \frac{x^{2n}}{2^n n!} + y_1 \sum_{n=0}^{\infty}(-1)^n \frac{x^{2n+1}}{(2n+1)(2n-1)\cdots(3)(1)}$ for all x.
(b) $y = y_0 \left(1 - x^2\right) + y_1 \left(x + x^2\right)$ for all x.
(c) $y = y_0 \sum_{n=0}^{\infty}(-1)^n \frac{x^{3n}}{3^n n!} + y_1 \sum_{n=0}^{\infty}(-1)^n \frac{x^{3n+1}}{[3n+1][3(n-1)+1]\cdots(4)(1)}$ for all x.
(d) $y = y_0 \sum_{n=0}^{\infty}(-1)^n \frac{3^n}{2^n n!}(x-2)^{2n} + y_1 \sum_{n=0}^{\infty}(-1)^n \frac{3^n(x-2)^{2n+1}}{(2n+1)(2n-1)\cdots(3)(1)}$ for all x.
(e) $y = y_0 \sum_{n=0}^{\infty}\frac{x^{2n}}{2^n n!} + y_1 \sum_{n=0}^{\infty}\frac{x^{2n+1}}{(2n+1)(2n-1)\cdots(3)(1)}$ for all x.
(f) $y = y_0 \sum_{n=0}^{\infty}\frac{[(2n)^2-\lambda^2]\cdots[2^2-\lambda^2][0^2-\lambda^2]}{(2n)!}x^{2n} + y_1 \sum_{n=0}^{\infty}\frac{[(2n-1)^2-\lambda^2]\cdots[3^2-\lambda^2][1^2-\lambda^2]}{(2n+1)!}x^{2n+1}$
for $-1 < x < 1$.
13. (a) $y = \sum_{n=0}^{\infty}(-1)^n 4^n \left[\frac{3}{(2n)!}(x-\pi)^{2n} - \frac{1}{(2n+1)!}(x-\pi)^{2n+1}\right]$ for all x.
(b) $y = \sum_{n=0}^{\infty}(-1)^n \frac{2[3(n-1)+1]^2\cdots 4^2 1^2}{(3n)!}(x-1)^{3n} - \sum_{n=0}^{\infty}(-1)^n \frac{3[3n-1]^2\cdots 5^2 2^2}{(3n+1)!}(x-1)^{3n+1}$
for all x.
(c) $y = -2x + \sum_{n=0}^{\infty}(-1)^n \frac{(2n)!}{4^{n-1}(n!)^2}x^{2n}$ for all x.
(d) $y = 1 - \frac{1}{2}x^2 + \sum_{n=0}^{\infty} -\frac{5}{2^n(2n+1)(2n-1)}x^{2n+1}$ for $-\sqrt{2} < x < \sqrt{2}$.
(e) $y = 8 + \sum_{n=0}^{\infty}(-1)^n \frac{9^n 7}{(2n+1)2^n n!}x^{2n+1}$ for all x.

Section 5.9

1. absolutely convergent
2. conditionally convergent
3. absolutely convergent
4. absolutely convergent
5. absolutely convergent
6. conditionally convergent
7. conditionally convergent
8. conditionally convergent
9. absolutely convergent
10. conditionally convergent
11. absolutely convergent
12. absolutely convergent
13. absolutely convergent
14. conditionally convergent
15. absolutely convergent
16. absolutely convergent
17. absolutely convergent
18. conditionally convergent
19. absolutely convergent
20. conditionally convergent
21. absolutely convergent
22. conditionally convergent
23. divergent
24. divergent
25. absolutely convergent
26. absolutely convergent
27. divergent
28. conditinally convergent
29. conditionally convergent
30. conditionally convergent
31. divergent
32. absolutely convergent
33. divergent
34. absolutely convergent
35. absolutely convergent
36. divergent

37. (a) converges for $p > 1$ and diverges for $0 < p \le 1$ (b) diverges for p (c) converges for $p > 1$ and diverges for $0 < p \le 1$ (d) converges for p

38. (a) $|a_k \sin k| \le |a_k|$ and $\sum_{k=p}^{\infty} a_k \sin k$ converges absolutely by the Compari (b) $\frac{b_k}{\sin^2 k} \ge b_k$, and $\sum_{k=p}^{\infty} \frac{b_k}{\sin^2 k}$ diverges by the Comparison Test.

39. Since $f(k) \ge f(k^2)$, $\sum_{k=1}^{\infty} f(k^2)$ converges by the Comparison Test. 40.

41. $(-1, 1)$ 42. $\Re$ 43. $\left[\frac{11}{4}, \frac{13}{4}\right]$ 44. $(-e, e)$ 45. $(-6, -4)$ 46. $(-$ 47. $[-1, 1)$ 48. $[-9, -9]$ 49. $[-2, 6)$ 50. $\left(-\frac{1}{e}, \frac{1}{e}\right)$ 51. $(-$

52. (a) If $\sum_{k=q}^{\infty} b_k$ converges then $\sum_{k=p}^{\infty} a_k$ also converges. If $\sum_{k=p}^{\infty} a_k$ d then $\sum_{k=q}^{\infty} b_k$ also diverges. (b) If $\sum_{k=p}^{\infty} a_k$ converges then $\sum_{k=q}^{\infty} b_k$ also converges. If $\sum_{k=q}^{\infty} b_k$ diverge $\sum_{k=p}^{\infty} a_k$ also diverges.

53. (b) $a_k = \frac{1}{k}$ 54. (b) $\frac{1}{100}$

55. (a) $a_k = (-1)^k \frac{1}{k}$ (b) $b_{2k} = (-1)^k \frac{1}{k}$, $b_{2k+1} = 0$ for $k \ge 1$.

56. (a) .78 with error less than .17 (b) 2.90 with error less than .44 (c) 0.50 with error less than .17 (d) 0.6042 with error less than .0026 (e) -0.125 with error less than .038 (f) -0.1 with errror less than .3

57. (a) $S_3 = 0.764$ with error less than the fourth term .0003. (b) $S_6 = 0.2585$ with error less than the the seventh term .0095. (c) $S_1 = 0.5000$ with error less than the second term .0078. (d) $S_4 = 0.7429$ with error less than the fifth term .0046. (e) $S_2 = 0.4861$ with error less than the third term .0013. (f) $S_4 = 0.7168$ with error less than the the fifth term .0042.

58. This series $\sum_{k=1}^{\infty} (-1)^k (a_{2k-1} + a_{2k})$ converges by the Alternating Series

Section 5.10

1. (a) $x = y = -3$ (b) $x = \frac{3}{2}$, $y = \frac{5}{4}$ (c) $x = 3$, $y = -2$ (d) $x = -1$, $y = 4$

Imaginary · 5i · 2+3i · -4+i -7 → Real · -1-2i

(a) 7 **(b)** 5 **(c)** $\sqrt{13}_{3-4i}$ **(d)** 5 **(e)** $\sqrt{5}$ **(f)** $\sqrt{17}$

(a) -7 **(b)** $-5i$ **(c)** $2-3i$ **(d)** $3+4i$ **(e)** $-1+2i$ **(f)** $-4-i$

(a) $7+4i$ **(b)** $-1-11i$ **(c)** $38-27i$ **(d)** $-\frac{19}{17}+\frac{13}{17}i$ **(e)** $8+36i$
f) $\frac{12}{29}-\frac{i}{29}$ **(g)** $2-11i$ **(h)** $-\frac{9}{2116}-\frac{i}{46}$ **(i)** $\frac{13}{25}+\frac{84}{25}i$ **(j)** $\frac{1}{4}-\frac{11}{4}i$

(a) $P(5,\pi)$ **(b)** $P(8,\frac{3\pi}{2})$ **(c)** $P(13,\arcsin-\frac{12}{13})$ **(d)** $P(4\sqrt{2},\frac{7\pi}{4})$ **(e)** $P(12,\frac{2\pi}{3})$

0. **(a)** $7i$ **(b)** -5 **(c)** $-3\sqrt{2}-3\sqrt{2}i$ **(d)** $-5-5\sqrt{3}i$ **(e)** $-4\sqrt{3}+4i$

1. **(a)** $P(15,\frac{11\pi}{10})$ **(b)** $P(5,\frac{\pi}{24})$ **(c)** $P(32,\frac{\pi}{2})$ **(d)** $P(30,\frac{41\pi}{24})$ **(e)** $P(8,0)$
f) $P(72,\frac{\pi}{3})$ **(g)** $P(12,\frac{11\pi}{24})$ **(h)** $P(48\sqrt{2},\frac{17\pi}{12})$ **(i)** $P(4,\frac{\pi}{3})$ **(j)** $P(324,\pi)$
i) $P(4\sqrt{2},\frac{13}{12}\pi)$ **(l)** $P(1728\sqrt{2},\frac{\pi}{12})$

3. **(a)** $3i,\ -3i$ **(b)** $\frac{3\sqrt{2}}{2}+\frac{3\sqrt{2}}{2}i,\ \frac{-3\sqrt{2}}{2}+\frac{3\sqrt{2}}{2}i,\ \frac{-3\sqrt{2}}{2}-\frac{3\sqrt{2}}{2}i,\ \frac{3\sqrt{2}}{2}-\frac{3\sqrt{2}}{2}i$
c) $P(\sqrt{3\sqrt{2}},-\frac{\pi}{8}),\ P(\sqrt{3\sqrt{2}},\frac{7\pi}{8})$, **(d)** $P(2,\frac{\pi}{18}),\ P(2,\frac{13\pi}{18}),\ P(2,\frac{25\pi}{18})$
e) $P(2,\frac{5\pi}{12}),\ P(2,\frac{11\pi}{12}),\ P(2,\frac{17\pi}{12}),\ P(2,\frac{23\pi}{12})$

4. **(a)** $3+i,\ 3-i,\ 1+2i,\ 1-2i$ **(b)** $-2,\ 5+2i,\ 5-2i$
c) $4-3i,\ 4+3i,\ -\frac{1}{2}-\frac{\sqrt{15}}{2}i,\ -\frac{1}{2}+\frac{\sqrt{15}}{2}i$ **(d)** $3,\ 1-i,\ 1+i,\ 2+i,\ 2-i$
e) $1+i,\ 1-i,\ 1+2i,\ 1-2i,\ 3+i,\ 3-i$ **(f)** $-2,\ P(2,\frac{7\pi}{5}),\ P(2,\frac{9\pi}{5}),\ P(2,\frac{11\pi}{5}),\ P(2,\frac{13\pi}{5})$
g) $\sqrt{6},\ -\sqrt{6},\ 2i,\ -2i$ **(h)** $-\sqrt[3]{7},\ P(\sqrt[3]{7},\frac{5\pi}{3}),\ P(\sqrt[3]{7},\frac{7\pi}{3}),\ \sqrt[3]{4},\ P(\sqrt[3]{4},\frac{2\pi}{3}),\ P(\sqrt[3]{4},\frac{4\pi}{3})$
i) $\sqrt{2},\ \sqrt{2}i,\ -\sqrt{2},\ -\sqrt{2}i,\ \sqrt[4]{3}(\frac{\sqrt{2}}{2}+i\frac{\sqrt{2}}{2}),\ \sqrt[4]{3}(\frac{-\sqrt{2}}{2}+i\frac{\sqrt{2}}{2}),\ \sqrt[4]{3}(\frac{-\sqrt{2}}{2}-i\frac{\sqrt{2}}{2}),$
$\sqrt[4]{3}(\frac{\sqrt{2}}{2}-i\frac{\sqrt{2}}{2})$ **(j)** $1,\ -1,i,\ -i$

15. **(a)** $\ln|x|-\arctan\frac{x}{2}+C$ **(b)** $\ln|x|+\frac{7}{x}+\frac{1}{2}\ln(x^2+1)+7\arctan x+C$
c) $\frac{1}{5}\ln|x|+\frac{29}{10}\ln(x^2+2x+10)-\frac{46}{15}\arctan\frac{x+1}{3}+C$
d) $-\ln|x|-\frac{2}{x}-\frac{3}{2}\ln(x^2+4x+29)+\frac{11}{5}\arctan\frac{x+2}{5}+C$
e) $4\ln|x-3|+\frac{1}{2}\ln(x^2+6x+13)-3\arctan\frac{x+3}{2}+C$
f) $-3\ln|x+4|+\frac{7}{2}\ln(x^2-4x+20)+\frac{5}{2}\arctan\frac{x-2}{4}+C$
g) $\ln(2x^2+2x+2)-3\arctan(x+1)+2\ln(x^2+4x+8)-\frac{5}{2}\arctan\frac{x+2}{2}+C$
h) $-\frac{3}{2}\ln(x^2+6x+10)+14\arctan(x+3)+3\ln(x^2+2x+26)-\frac{7}{5}\arctan\frac{x+1}{5}+C$

16. $x=\frac{-w\pm\sqrt{w^2-4zv}}{2z}$ **19.** circle with center z_0 and radius R

20. **(a)** This region is the closed disc with center the origin and radius R.
(b) This region is obtained from the complex plane by deleting the open disc with center the origin and radius R.

Section 5.11

1. **(a)** $\mathcal{C}$ **(b)** $z\neq 7i$ **(c)** $z\neq -\frac{\sqrt{2}}{2}+i\frac{\sqrt{2}}{2},\ \frac{\sqrt{2}}{2}-i\frac{\sqrt{2}}{2}$ **(d)** $z\neq\frac{9}{5}+i\frac{2}{5}$
(e) $z\neq P(\sqrt[4]{2},\frac{5\pi}{8}),\ P(\sqrt[4]{2},\frac{13\pi}{8})$ **(f)** $z\neq P(2,\frac{7\pi}{12}),\ P(2,\frac{15\pi}{12}),\ P(2,\frac{23\pi}{12})$,
(g) $z\neq P(2,\frac{7\pi}{24}),\ P(2,\frac{19\pi}{24}),\ P(2,\frac{31\pi}{24}),\ P(2,\frac{43\pi}{24})$ **(h)** $z\neq ki$ with $k\leq -1$ or $k\geq 1$

2. **(a)** $\sqrt{26}$ **(b)** $2\sqrt{13}$ **(c)** $5\sqrt{2}$ **(d)** 5 **(e)** $\sqrt{113}$ **(f)** $3\sqrt{5}$

8. **(a)** convergent with limit 0 **(b)** divergent **(c)** convergent with limit 0
(d) convergent with limit $\frac{7}{9}$ **(e)** convergent with limit 0

811

14. (a) $72iz^8 - (18 + 30i)z^5 + (18 - 12i)z^2 - 5$

(b) $\dfrac{(3-6i)z^6+(128-26i)z^3+(-18+36i)z^2+(92+31i)}{[z^4-(8+9i)z+2]^2}$ (c) $(\cos z)e^{\sin z}$

(d) $\dfrac{\cos z}{z} - (\sin z)(\log z)$ (e) $\dfrac{(2iz^2-2z-12)e^{2iz}}{(z^2+6i)^2}$ (f) $(2z + e^z)\cos(z^2 + e^z)$

(g) $-2\left[\cos\log(iz^3 + 8)\right]\left[\sin\log(iz^3 + 8\right]\left[\dfrac{3iz^2}{iz^3+8}\right]$ (h) $\dfrac{e^z[\cos z+\sin z+i(\sin z-\text{co}}{\cos 2z+i\sin 2z}$

(i) $\dfrac{\sin 2z-i\sin 2iz}{\sin^2 z+\cos^2 iz}$ (j) $2ze^{z^2+1}\sin(iz^2 + 1) + 2ize^{z^2+i}\cos(iz^2 + 1)$

18. (a) divergent (b) conditionally convergent (c) absolutely converg
(d) conditionally convergent (e) divergent (f) absolutely convergent
(g) divergent (h) divergent (i) divergent (j) absolutely convergent
(k) divergent (l) absolutely convergent

19. (a) 2 (b) ∞ (c) $\sqrt{5}$ (d) 1 (e) 1 (f) ∞ (g) $\frac{5}{13}$ (h) $\frac{2\sqrt{3}}{9}$ (i) ∞

20.

First Derivative	Second Derivative
(a) $\sum_{n=1}^{\infty} \dfrac{n(z+i)^{n-1}}{\sqrt{2^n+1}}$	$\sum_{n=2}^{\infty} \dfrac{n(n-1)(z+i)^{n-2}}{\sqrt{2^n+1}}$
(b) $\sum_{n=1}^{\infty} \dfrac{n!(z-2i+5)^{n-1}}{2(2n-1)!}$	$\sum_{n=2}^{\infty} \dfrac{n!(z-2i+5)^{n-2}}{(8n-4)(2n-3)!}$
(c) $\sum_{n=1}^{\infty} \dfrac{2n(z+6i-7)^{2n-1}}{(4i+3)^n}$	$\sum_{n=2}^{\infty} \dfrac{2n(2n-1)(z+6i-7)^{2n-2}}{(4i+3)^n}$
(d) $\sum_{n=1}^{\infty}(z + 3i - 8)^{n-1}$	$\sum_{n=2}^{\infty}(n - 1)(z + 3i - 8)^{n-2}$
(e) $\sum_{n=1}^{\infty} \dfrac{in(iz+7)^{n-1}}{ni+5}$	$\sum_{n=2}^{\infty} \dfrac{-n(n-1)(iz+7)^{n-2}}{ni+5}$
(f) $\sum_{n=1}^{\infty} \dfrac{n^2 z^{n-1}}{(i+n)^n}$	$\sum_{n=2}^{\infty} \dfrac{n^2(n-1)z^{n-2}}{(i+n)^n}$
(g) $\sum_{n=1}^{\infty} \dfrac{(12-5i)n[(12-5i)z+2+3i]^{n-1}}{(3i-4)^n+9i}$	$\sum_{n=2}^{\infty} \dfrac{(119-120i)n(n-1)[(12-5i)z+2+3i]^{n-2}}{(3i-4)^n+9i}$
(h) $\sum_{n=1}^{\infty} \dfrac{i(3n)!(iz+7)^{2n-1}}{n!(2n-1)!}$	$\sum_{n=2}^{\infty} -\dfrac{(3n)!(iz+7)^{2n-2}}{n!(2n-2)!}$

23. (a) $-8.36+18.27i$ (b) $.00187+.00163i$ (c) $-1.61+2.50i$ (d) $-.805$
(e) $3.608 + 1.03i$ (f) $199.6 + 29.4i$ (g) $-1358 + 614i$ (h) 548

24. (a) 0, 0 (b) $2\pi i$, 0 (c) 0, $-2\pi i$ (d) $-2\pi i$, 0

26. (a) The domain of $\tan z$ and $\sec z$ is all of $\mathcal{C}$ except for odd multiple of
The domain of $\cot z$ and $\csc z$ is all of $\mathcal{C}$ except for integral multiples of π.
(c) $\tan z$ and $\cot z$ have period π while $\sec z$ and $\csc z$ have period 2π.
(d) $D(\tan z) = \sec^2 z$, $D(\cot z) = -\csc^2 z$, $D(\sec z) = \sec z \tan z$,
$D(\csc z) = -\csc z \cot z$

Review Exercises

1. T	2. T	3. F	4. F	5. F	6. T	7. F	8. T
9. F	10. T	11. F	12. F	13. T	14. T	15. T	16. T
17. F	18. F	19. T	20. T	21. F	22. F	23. T	24. T
25. F	26. F	27. F	28. T	29. T	30. T	31. T	32. F
33. T	34. F	35. T	36. F	37. T	38. T	39. F	40. T
41. T	42. T	43. T	44. T	45. F	46. F	47. T	48. F
49. T	50. F						

51. converges with limit 0 **52.** converges with limit e^5 **53.** dinverges

54. 4.08% **55.** 1 **56.** $\frac{45}{32}$ **57.** $\sum_{k=0}^{\infty}(-1)^k \frac{1}{k!}x^{2k+1}$ **58.** $\sum_{k=0}^{\infty}(-1)^k \frac{1}{2^{3k-1}}$

59. $\sum_{k=1}^{\infty}(-1)^{k-1}\frac{1}{2k-1}x^k$ **60.** $[-1,1)$ **61.** $(-3,3)$ **62.** $(-\infty,\infty)$

63. .1181 **64.** .1003 **65.** diverges **66.** diverges **67.** absolutely conv

68. conditionally convergent **69.** absolutely convergent **70.** diverges

Index

IVP, 578

Jordan, Camille, 488
Jyesthadeva, 670

Kepler, Johann, 316

L'Hôpital's Rule
    history of, 227
    indeterminate form $\frac{0}{0}$, 221
    indeterminate form $\frac{\infty}{\infty}$, 224
    proof of, 284
    when c approaches $\pm\infty$, 225
Lagrange, Joseph Louis, 200, 560, 658, 670
Lambert, Johann Heinrich, 520
Legendre polynomials, 688
Leibniz notation
    definition of, 149
    inverse functions, 183
Leibniz, Gottfried Wilhelm, 52, 68, 149, 176, 258, 274, 332, 461, 520, 537, 670, 699
limits
    ϵ–δ definition, 68
    as x approaches ∞, 62
    history of, 69
    infinite, definition of, 65
    of exponential functions, 451
    ordinary, definition of, 58
    Pinching Theorem, 74
    properties
        proofs of, 84–87, 103
        statements of, 72–74
    trigonometric, 78–82
Liouville, Joseph, 587
Lipschitz condition, 581
Lipschitz, Rudolf, 586
local
    extremum, 188
    maximum, 188
    minimum, 188
logarithm
    complex
        definition of, 727
        properties of, 727
        Taylor series of, 728, 731
    definition of, 441
    derivative of, 442
    graph of, 443
    properties of, 441, 442
logarithmic differentiation, 448
loss, 235

Müller, Johannes, 52

Maclaurin polynomials, 634
Maclaurin, Colin, 274, 658
Madhava, 670
marginal
    cost, 235
    profit, 235
    revenue, 235
Marquis de l'Hôpital, 227
maximum value of a function, 98
Maximum Value Theorem
    proof of, 137
    statement of, 98
Mean Value Theorem
    geometric interpretation, 192
    history of, 200
    proof of, 278
    statement, 192
Mean Value Theorem for Integrals
    generalized, 636
    ordinary, 314
Meray, Charles, 137
Mercator, Nicolas, 670
minimum value of a function, 98
mole, 531
moment of
    plate, 390, 392
    point mass, 385
    region, 391
    rod, 387, 388, 419
    set of point masses, 386, 387
moment, additive property, 392
Muhammad ibn Musa al-Khwarizmi

Napier, John, 52, 444
natural logarithm
    definition of, 433
    derivative of, 433
    graph of, 434
    properties of, 433, 435
neighborhood, 120
Newton, Isaac, 68, 158, 176, 237, 258, 332, 374, 461, 537, 643, 658
    second law of motion, 105, 385
Newton-Cotes three-eighths rule, 3
Newton-Raphson method, 255
Nilakantha, Kerala Gargya, 52
nonstandard analysis, 259
normal line, 148
number line, 5
numbers
    integer, 5
    irrational, 5
    natural, 5
    rational, 5

821